Water Resources/Environmental

CIVIL ENGINEERING PRACTICE

5/Water Resources/Environmental

EDITED BY

PAUL N. CHEREMISINOFF
NICHOLAS P. CHEREMISINOFF
SU LING CHENG

IN COLLABORATION WITH

J. R. Amend	S. L. Cheng	D. M. Hartley	K. L. Logani	A. M. A. Salih
A. Anderson	P. N. Cheremisinoff	P. Helliker	G. B. McBride	Z. A. Samani
K. Banerjee	A. B. Cunningham	G. E. Ho	J. A. McCorquodale	P. Sereico
N. L. Benecke	F. de Souza	M. H. Houck	L. Ortolano	N. K. Tyagi
J. K. Bewtra	L. Duckstein	A. A. Jennings	C. N. Patel	H. M. Weagraff
P. R. Bhave	A. A. Dzurik	V. C. Kulandaiswamy	R. D. Pridmore	R. A. Wurbs
A. G. Bhole	D. D. Fangmeier	J. Kuwabara	D. V. Rao	C. T. Yang
I. Bogardi	Y.-S. Fok	C. Larneo	J. C. Rutherford	M. Yitayew
E. K. Can	G. H. Hargreaves			

TECHNOMIC
PUBLISHING CO., INC.

LANCASTER · BASEL

Published in the Western Hemisphere by
Technomic Publishing Company, Inc.
851 New Holland Avenue
Box 3535
Lancaster, Pennsylvania 17604 U.S.A.

Distributed in the Rest of the World by
Technomic Publishing AG

Printed in the United States of America
10 9 8 7 6 5 4 3 2 1

Main entry under title:
   Civil Engineering Practice 5—Water Resources/Environmental

A Technomic Publishing Company book
Bibliography: p.
Includes index p. 809

Library of Congress Card No. 87-50629
ISBN No. 87762-540-9

TABLE OF CONTENTS

PREFACE

While the designation civil engineering dates back only two centuries, the profession of civil engineering is as old as civilized life. Through ancient times it formed a broader profession, best described as master builder, which included what is now known as architecture and both civil and military engineering. The field of civil engineering was once defined as including all branches of engineering and has come to include established aspects of construction, structures and emerging and newer sub-disciplines (e.g., environmental, water resources, etc.). The civil engineer is engaged in planning, design of works connected with transportation, water and air pollution as well as canals, rivers, piers, harbors, etc. The hydraulic field covers water supply/power, flood control, drainage and irrigation, as well as sewerage and waste disposal.

The civil engineer may also specialize in various stages of projects such as investigation, design, construction, operation, etc. Civil engineers today as well as engineers in all branches have become highly specialized, as well as requiring a multiplicity of skills in methods and procedures. Various civil engineering specialties have led to the requirement of a wide array of knowledge.

Civil engineers today find themselves in a broad range of applications and it was to this end that the concept of putting this series of volumes together was made. The tremendous increase of information and knowledge all over the world has resulted in proliferation of new ideas and concepts as well as a large increase in available information and data in civil engineering. The treatises presented are divided into five volumes for the convenience of reference and the reader:

VOLUME 1 Structures
VOLUME 2 Hydraulics/Mechanics
VOLUME 3 Geotechnical/Ocean Engineering
VOLUME 4 Surveying/Transportation/Energy/Economics & Government/Computers
VOLUME 5 Water Resources/Environmental

A serious effort has been made by each of the contributing specialists to this series to present information that will have enduring value. The intent is to supply the practitioner with an authoritative reference work in the field of civil engineering. References and citations are given to the extensive literature as well as comprehensive, detailed, up-to-date coverage.

To insure the highest degree of reliability in the selected subject matter presented, the collaboration of a large number of specialists was enlisted, and this book presents their efforts. Heartfelt thanks go to these contributors, each of whom has endeavored to present an up-to-date section in their area of expertise and has given willingly of valuable time and knowledge.

PAUL N. CHEREMISINOFF
NICHOLAS P. CHEREMISINOFF
SU LING CHENG

CONTRIBUTORS TO VOLUME 5

J. R. AMEND, Montana State University, Bozeman, MT

A. ANDERSON, University of Newcastle upon Tyne, United Kingdom

K. BANERJEE, New Jersey Institute of Technology, Newark, NJ

N. L. BENECKE, Lummus Crest Inc., Bloomfield, NJ

J. K. BEWTRA, University of Windsor, Windsor, Ontario, Canada

P. R. BHAVE, Visvesvaraya Regional College of Engineering, Nagpur, India

A. G. BHOLE, Visvesvaraya Regional College of Engineering, Nagpur, India

I. BOGARDI, University of Nebraska, Lincoln NE

E. K. CAN, Lakehead University, Thunder Bay, Ontario, Canada

S. L. CHENG, New Jersey Institute of Technology, Newark, NJ

P. N. CHEREMISINOFF, New Jersey Institute of Technology, Newark, NJ

A. B. CUNNINGHAM, Montana State University, Bozeman, MT

F. DE SOUZA, University of Ceara, Fortaleza, Brazil

L. DUCKSTEIN, University of Arizona, Tucson, AZ

A. A. DZURIK, Florida A&M University/Florida State University, Tallahassee, FL

D. D. FANGMEIER, University of Arizona, Tucson, AZ

Y.-S. FOK, University of Hawaii at Manoa, Honolulu, HI

G. H. HARGREAVES, Utah State University, Logan, UT

D. M. HARTLEY, U.S. Department of Agriculture, Fort Collins, CO

P. HELLIKER, U.S. Environmental Protection Agency, San Francisco, CA

G. E. HO, Murdoch University, Murdoch, Australia

M. H. HOUCK, Purdue University, W. Lafayette, IN

A. A. JENNINGS, University of Toledo, Toledo, OH

V. C. KULANDAISWAMY, Anna University, Madras, India

J. S. KUWABARA, U.S. Geological Survey, Menlo Park, CA

C. LARNEO, Ecol-Sciences, Rahway, NJ

K. L. LOGANI, Civil and Geotechnical Engineering Consultant, Glenview, IL

G. B. MCBRIDE, Ministry of Works and Development, Hamilton, New Zealand

J. A. MCCORQUODALE, University of Windsor, Windsor, Ontario, Canada

L. ORTOLANO, Stanford University, Stanford, CA

C. N. PATEL, New Jersey Institute of Technology, Newark, NJ

R. D. PRIDMORE, Ministry of Works and Development, Hamlilton, New Zealand

D. V. RAO, St. Johns River Water Management District, Palatka, FL

J. C. RUTHERFORD, Ministry of Works and Development, Hamlilton, New Zealand

A. M. A. SALIH, Khartoum University, Khartoum, Sudan

Z. A. SAMANI, New Mexico State University, Las Cruces, NM

P. SEREICO, University of New England, Biddeford, ME

N. K. TYAGI, Central Soil Salinity Research Institute, Karnal, India

H. M. WEAGRAFF, Consultant in Environmental Hydrology, Fort Collins, CO

R. A. WURBS, Texas A&M University, College Station, TX

C. T. YANG, U.S. Department of the Interior, Denver, CO

M. YITAYEW, Univerity of Arizona, Tucson, AZ

CIVIL ENGINEERING PRACTICE

SECTION ONE

Water Supply and Management

CHAPTER 1

Water Supply

PRAMOD R. BHAVE* AND ANAND G. BHOLE**

INTRODUCTION

Next to air, water is required to maintain life. Of the several words for water in Sanskrit, one is "Jeewanam" which also means life. All the ancient civilizations of the world developed along the rivers. Later, due to the development in techniques for the storage and transportation of water, and extraction of ground water, human settlements started spreading away from the rivers.

Initially, the use of water was limited to drinking and domestic purposes such as bathing, cooking and washing. However, the present day uses of water are varied and include uses for domestic and industrial purposes and also for irrigation, hydropower, navigation and recreation. The term "water supply" now-a-days denotes the urban and rural supply of water and envisages the supply of sufficient volume of the required quality water at adequate pressure.

A modern water supply system may include facilities for collection and storage, transportation, pumping, treatment and distribution. Works for collection and storage include development of the watershed, dams, impounding reservoirs, intakes, wells, infiltration galleries and springs. Transportation works include canals, aqueducts and pumping mains for transporting water from the storage to the consumer centres. Pumping works include pumps and other ancillary units for pumping water. Treatment includes aeration, screening, sedimentation, flocculation, filtration, disinfection, softening, and removal of harmful elements. Distribution works include distribution and equalizing reservoirs, pipes, valves, fire hydrants, meters and other appurtenances.

*Department of Civil Engineering, Visvesvaraya Regional College of Engineering, Nagpur 440 011, India
**Department of Environmental Engineering, Visvesvaraya Regional College of Engineering, Nagpur 440 011, India

Two typical water supply systems are shown in Figure 1. A small-community or a rural water supply system [Figure 1(a)] includes a well, pump, storage and distribution facilities, while a large-community or an urban water supply system [Figure 1(b)] includes many of the facilities cited in the previous paragraph.

WATER CONSUMPTION

Water Uses

Water uses are usually classified as domestic, public, commercial and industrial. Domestic use includes water utilized in residential units such as houses and apartments for drinking, cooking, bathing, washing, heating, cooling, air-conditioning, sanitary purposes and watering the lawns and gardens. It is proportional to the population and accounts for 30–50% of the total water consumption.

Public use of water is for public places and buildings such as public gardens, parks and fountains, public swimming pools, hospitals, schools and other educational institutions, hostels, prisons, public sanitary places, street and sewer flushing, and fire fighting. It amounts to about 5–10% of the total consumption.

Commercial use includes water used in office buildings for sanitary and air-conditioning purposes, in hotels and restaurants, for car washing, in laundries, and at golf courses, shopping centres, bus, railway and air terminals, etc. It is sometimes expressed in terms of the floor area of the buildings, and amounts to 10–30% of the total consumption.

Industrial use is for manufacturing and processing purposes and mainly consists of heat exchange, cooling and manufacturing. It bears no relation to the population of an industrial district and is expressed in terms of the unit of production. It accounts for 20–50% of the total consump-

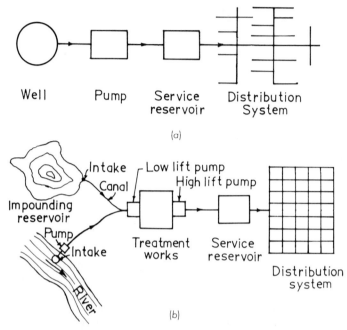

FIGURE 1. Components of typical (a) rural; and (b) urban water supply systems.

tion. As larger cities are generally more industrialized, they show a higher trend of industrial consumption.

In addition to the above direct uses, 10–20% of the total consumption may be unaccounted due to leakage through pipes and valves, evaporation from reservoirs, underregistration of meters, unauthorized use, pilferage of water during transportation and wastage at public stand posts.

Water Demand

FACTORS INFLUENCING WATER DEMAND

Water demand for a community is usually expressed as the average consumption per capita per day and is obtained by dividing the total consumption in a year by the population and the number of days in the year. The average consumption depends upon several factors. It increases with (a) the increase in the size of the community, due to increase in the public, commercial and industrial demands; (b) improvement of the economic status of the community; (c) improvement in the quality of water; (d) increase in the pressure in the distribution system; and (e) provision of sewer facilities. It decreases with (a) increase in the cost of water; and (b) the use of meters. The average consumption also depends upon the climate. In hot and arid climate, the average consumption is more due to frequent bathing, air cooling or air conditioning and heavy lawn sprinkling. In extreme cold climate also, the average consumption increases due to bleeding of water to avoid freezing in the service pipes and the in-

ternal piping systems. Further, the type of supply, i.e., continuous for twenty-four hours or intermittent, also affects the average consumption. Unless the intermittent water supply is for extremely short duration, the average consumption is more for the intermittent supply than for continuous supply. In the intermittent supply, water that is stored but remains unused during no-supply period is thrown away when a fresh supply begins. Further, there is a tendency to keep faucets open and leave them unattended during the no-supply periods. This results in wastage of water when the supply starts.

Several studies have been made to study the effect of some of these factors on water consumption [1–5]. However, no quantitative universally applicable relationship can be established.

WATER DEMAND FOR VARIOUS PURPOSES

The quantity of water required for domestic use is as low as 70 litres per capita per day (Lpcd) in water-scarce areas in India. The average urban domestic demand in India is about 135 Lpcd as shown in Table 1 [6]; while it is about 230–380 Lpcd in U.S.A. as shown in Table 2 [7]. Some typical public and commercial demands as recommended by the Indian Standards Institution are given in Table 3 [8]. Industrial demand of water by some selected industries is shown in Table 4 [9,10,11]. Further information on a wide range of industries is also available [6,10].

Fire demand for fire fighting purposes depends upon the

TABLE 1. Average Urban Domestic Water Use in India.

Purpose	Use (Lpcd)
Drinking	5
Cooking	5
Bathing	55
Washing of Clothes	20
Washing of Utensils	10
Cleaning of houses	10
Toilet flushing	30
Total	135

TABLE 2. Average Urban Domestic Water Use in U.S.A.

Purpose	Use (Lpcd)
Interior	
Drinking and cooking	12
Bathing	68
Laundry	35
Dishwashing	15
Toilet flushing	92
Miscellaneous	8
Total	230
Exterior	
Irrigating, car washing)	0–150
)	
Swimming pools, cleaning)	
Total domestic use	230–380

population of the community, types of industries and duration of fire. The fire demand as established by the American Insurance Association can be expressed as

$$Q = 3{,}865\sqrt{P}\,(1 - 0.01\,\sqrt{P}) \qquad (1)$$

in which Q = rate of fire demand in L/min; and P = population in thousands. The fire-demand rate, fire duration, fire reserve storage and hydrant spacing based on American Insurance Association recommendations are given in Table 5. For population more than 200,000, an additional rate of 7,600–30,400 L/min is required for a second fire.

The average per capita consumption in an Indian town is 270 Lpcd (domestic 50%, public 4%, commercial 8%, industrial 18%, and unaccounted 20%) [6]. Based on the survey of cities with population from 10,000 upwards, the average per capita consumption in U.S.A. in 1970 was 633 Lpcd [12]. It increased at an average rate of 5.08 Lpcd per year from 1945–1970 [13]. According to one estimate [14], the projected per capita consumption of water in U.S.A. for the year 2,000 A.D. will be about 670 Lpcd (domestic 44%, public 9%, commercial and industrial 39% and unaccounted 8%).

FLUCTUATIONS IN WATER DEMAND

Water demand fluctuates with the seasons of the year, the days of the week and the hours of the day. Fluctuations are greater as the size of the community and the duration of time decrease. Fluctuations are expressed as ratios—maximum:average. The fluctuations in demand on a particular day for different users as obtained in a study [15] are shown in Figure 2. Estimates of the fluctuations in U.S.A. are as follows [11]:

Ratio of Rates	Normal Range	Average
Maximum day:Average day	1.5–3.5:1	2.0:1
Maximum hour:Average day	2.0–7.0:1	4.5:1

TABLE 3. Water Requirements for Public and Commercial Uses.

Type of Building	Requirement (L/d)
Factories where bathrooms must be provided	45 per head
Factories where bathrooms need not be provided	30 per head
Hospitals (including laundry):	
a) Number of beds not exceeding 100	340 per bed
b) Number of beds exceeding 100	455 per bed
Nurses' homes and medical quarters	135 per head
Hostels	135 per head
Hotels	180 per bed
Offices	45 per head
Restaurants	70 per seat
Cinemas, concert halls and theatres	15 per seat
Schools:	
a) Day schools	45 per head
b) Boarding schools	135 per head

TABLE 4. Water Demand of Selected Industries.

Industry	Unit of Production	L/unit
Dairy products	kg of finished product	5–20
Cane sugar	kg of cane sugar	80
Beet sugar	kg of beet sugar	40–80
Canned fruits and vegetables	kg of canned product	20–50
Poultry	kg of ready to cook product	20–40
Meat	kg of carcass	180
Beverage alcohol	litre (proof alcohol)	120–170
Cotton goods	kg	160–800
Cellulosic manmade fibres	kg of fibre	600
Organic noncellulosic fibres	kg of fibre	300
Wool	kg of yarn	150–600
Pulp and paper mills	kg	150–500
Leather	kg, tanned	40–80
Fertilizer	kg	80–200
Petroleum	kg of crude oil	5–10
Paints and pigments	kg	5–10
Aluminum	kg	100–150
Steel	kg	200–400
Copper	kg	150–200
Coal	kg	5–20
Hydraulic cement	kg	4
Tyres and tubes	tyre and tube	600
Glass	kg	70
Explosives	kg	800
Automobile	vehicle	40,000

TABLE 5. Fire Flow, Duration, Fire Reserve Storage and Hydrant Spacing Requirements Based on American Insurance Association Recommendations.

Population	Fire Flow		Duration, (h)	Reserve Storage, (ML)	Average Area Served per Hydrant in High-Value Districts, (m^2)	
	L/min	ML/d			Direct Streams	Engine Streams
1,000	3,800	5.47	4	1.14	9,300	11,160
2,000	5,700	8.21	6	2.28	8,370	10,700
4,000	7,600	10.94	8	3.80	7,900	10,230
10,000	11,400	16.42	10	6.84	6,510	9,300
17,000	15,200	21.89	10	9.12	5,110	8,370
28,000	19,000	27.36	10	11.40	3,720	7,900
40,000	22,800	32.83	10	13.68	3,720	7,440
80,000	30,400	43.78	10	18.24	3,720	5,580
125,000	38,000	54.72	10	22.80	3,720	4,460
200,000	45,600	65.66	10	27.36	3,720	3,720

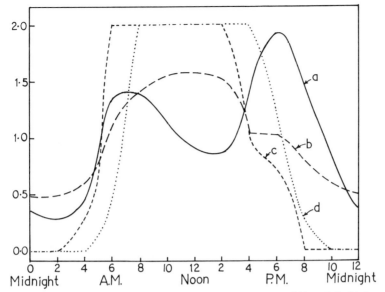

FIGURE 2. Fluctuations in water demand for (a) apartments; (b) hospitals; (c) doctors' office buildings; and (d) commercial purposes.

FORECASTING DEMAND

Estimation of future demands is necessary to design a new or extend an existing water works. Thus, selection of suitable design periods and population forecast is necessary.

Design Period

The future period, expressed in number of years, for which a provision is made in designing the capacity of a structure is its design period. The choice of the design period for a structure depends upon the length of its useful life and possibility of its being obsolete, initial cost and the funds available, ease and cost of expansion, interest and inflation rates, anticipated rate of growth, and possibility in shift of the demand centre.

Water works are normally designed for a 30-yr period after their completion. However, the design period of individual structures may be different because of the several factors cited earlier. For example, tunnels and large aqueducts are difficult and costly to expand and therefore they may be designed for their ultimate requirements. On the other hand, additions can be easily made to water treatment units and therefore their design period may be shorter. Similarly, capacity expansion of pumping mains and distribution networks may be economical and therefore desirable [16,17]. The design periods often employed in practice for different structures of water works are given in Table 6.

Population Forecast

Several methods are available for forecasting population over the design period [6,9,11,18]. These methods are: (1)

Arithmetic increase method in which the population increases at a constant rate; (2) Geometric increase method in which the percentage increase in population remains constant; (3) Incremental increase method in which the average increment in the increase is added to the increase; (4) Decreased rate of increase method in which percentage increase decreases with time; (5) Graphical extension method in which a curve depicting past population versus time is extended; (6) Graphical comparison method in which the population growth of 3 to 4 similar but large cities, having undergone similar phases of development, is compared with the city in question; (7) logistic or S-shaped curve method

TABLE 6. Design Periods for Different Structures of Water Works.

Structure	Design Period (y)
Dams and impounding reservoirs	50–100
Infiltration works, wells	20–30
Pumping machinery:	
a) all prime movers except electric motors	25–30
b) electric motors and pumps	12–15
Water treatment units	15–20
Valves, meters and other appurtenances	20–30
Transportation mains	20–30
Storage and balancing reservoirs	15–30
Distribution system	30–50

in which population growth rate increases initially, reaches a maximum value, decreases thereafter and reaches a saturation limit; (8) demographic method in which population gain due to births, population loss due to deaths and population gain or loss due to migration are considered; and (9) stochastic projection method [19] based on probabilistic rather than deterministic prediction.

Taking into consideration the population forecast over the design period, the per capita consumption, together with public, commercial, industrial and fire demands; the total water demand of a community can be estimated.

STANDARDS FOR WATER QUALITY

The water supplied to the consumers should be free from pathogenic organisms, clear, palatable, free from undesirable taste and odour, of reasonable temperature, neither corrosive nor scale forming, and free from minerals which could produce undesirable physiological effects. Therefore, it is desirable to protect the water sources from being contaminated by pollution. Wherever such protection is difficult or impossible, water should be treated and brought to the level of acceptable standard. Proper care should also be taken, during the transportation and distribution of treated water, to ensure that it does not get polluted through joints or backflow of polluted water from customer piping.

To ensure the quality of water, certain standards have been evolved by the U.S. Public Health Service, World Health Organization, etc. Table 7 gives the physical, chemical and bacteriological standards for maximum acceptable and maximum allowable concentrations [9,20–22]. Methods for collection and examination of samples and further information on the quality standards can be obtained from these references.

Hardness in water is predominantly caused by carbonates and bicarbonates of calcium and magnesium, and sulphates and chlorides of divalent cations. Iron in water causes taste and hardness, discoloration of clothes and plumbing fixtures, and incrustation in water mains. Manganese has effect similar to iron. Manganese is also oxidized into a sediment which clogs pipes, discolors fabrics and stimulates organic growths. Sodium chloride adds salinity to water. Fluorides, when more than 3.0 mg/L (even more than 1.5 mg/L), result in mottled and discolored teeth in children of formative age. Fluorides also reduce the strength of bones.

SOURCES OF WATER

The major sources of water for water supply purposes are surface and ground water. Surface water includes fresh water available in natural lakes, rivers and streams. Ground water includes water obtained from wells, springs and in-filtration galleries. Where these sources are scarce in comparison with the demand, attempts are made to augment water supply through desalination, waste water reclamation, artificial precipitation and also through reduction of evaporation losses in reservoirs.

Factors Influencing Selection of Source

The factors influencing the selection of source for water supply are as follows.

QUANTITY

The quantity of water available should be adequate to meet the requirements during the entire design period. When nearby sources are inadequate, the possibility of augmentation during the latter part of the design period by transporting water from distant sources may also be considered.

QUALITY

The water should be of good quality, i.e., free from toxic, poisonous and other ingredients injurious to health. The impurities should be as few as possible so that they can be easily removed.

RELIABILITY

The reliability of a source [23] depends upon the discharge rate and the elevation. The decreasing order of reliability is: (a) discharge rate adequate and also at sufficient elevation so that supply is feasible without storage and pumping; (b) enough quantity feasible with storage at sufficient elevation so that supply is possible through storage and under gravity only; (c) rate of discharge sufficient but at inadequate elevation so that storage is not necessary but pumping is required; and (d) source that requires both storage and pumping.

DISTANCE OF SOURCE AND GENERAL TOPOGRAPHY OF INTERVENING GROUND

The source should be close to the demand centre. Further, the intervening ground should not be highly uneven. Both these factors will reduce the transportation cost.

LEGAL AND POLITICAL ASPECTS

Legal aspects are important in ground water sources and also in transfer of water from one watershed to another. Political aspects creep in through interstate and international river water disputes and agreements for surface waters.

COST

The cost is the most important factor and should be as low as possible.

TABLE 7. Water Quality Standards.

Substance or Characteristic	Maximum Acceptable Concentration	Maximum Allowable Concentration
Colour on platinum cobalt scale	5	50
Taste	Unobjectionable	—
Odour	Unobjectionable	—
pH range	7.0 to 8.5	6.5 to 9.2
Total dissolved solids	500 mg/L	1500 mg/L
Turbidity in NTU (Nephalometer Turbidity Units)	5	25
Total hardness (as $CaCO_3$)	200 mg/L	600 mg/L
Calcium (as Ca)	75 mg/L	200 mg/L
Chloride (as Cl)	200 mg/L	600 mg/L
Copper (as Cu)	0.05 mg/L	1.5 mg/L
Fluorides (as F)	1.0 mg/L	1.5 mg/L
Iron (as Fe)	0.1 mg/L	1.0 mg/L
Magnesium (as Mg)	30 mg/L	150 mg/L
Manganese (as Mn)	0.05 mg/L	0.5 mg/L
Nitrate (as NO_3)	45 mg/L	45 mg/L
Sulphate (as SO_4)	200 mg/L	400 mg/L
Zinc (as Zn)	5.0 mg/L	15 mg/L
Anionic detergents (as MBAS)	0.2 mg/L	1.0 mg/L
Mineral oil	0.01 mg/L	0.30 mg/L
Phenolic compounds (as phenol)	0.001 mg/L	0.002 mg/L
Toxic substances		
Arsenic (as As)	—	0.05 mg/L
Barium (as Ba)	—	1.0 mg/L
Cadmium (as Cd)	—	0.01 mg/L
Chromium (as hexavalent Cr)	—	0.05mg/L
Cyanide (as CN)	—	0.05 mg/L
Lead (as Pb)	—	0.1 mg/L
Mercury (total as Hg)	—	0.001 mg/L
Selenium (as Se)	—	0.01 mg/L
Polynuclear aromatic hydrocarbons (PAH)	—	0.2 μg/L
Radioactivity levels		
Gross alpha activity	—	3 pCi/L*
Gross beta activity	—	30 pCi/L
Bacteriological levels		
E Coli count in any 100 mL sample	—	0
Coliform organisms in 95% of 100 mL samples in any year	—	0
Coliform organisms in any 100 mL sample	—	10
Coliform organisms in second 100 mL sample, after it is found in first sample	—	0

*pCi = pico (10^{-12}) curie

Effect of Source on Water Quality

Surface waters may contain silts and suspended sediments, dissolved organic matter from top soil, bacteria and disease germs from sewage, chemical impurities from industrial wastes and fertilizers and toxic materials from pesticides. Waters from rivers and streams carry considerable amount of silt and suspended sediment and are coloured during floods, but are comparatively free from turbidity and colour during other seasons. River waters are usually polluted from sewage and industrial wastes. Lake waters, even though free from turbidity, are subject to pollution. Surface waters when stored in impounding reservoirs become clearer due to settling of the suspended sediment and also become colourless due to coagulation and settlement of the suspended matter and bleaching action of the sun rays. Storage also reduces bacteria and pathogenic organisms, however, quality of water in impounding reservoirs might deteriorate due to bacterial decomposition of reservoir floor and solution of iron, manganese and hardness producing calcium and magnesium minerals. Further, water from lower strata of deep lakes and reservoirs contain less oxygen and more carbon dioxide as compared to water from surface layers.

Ground waters are free from silt and suspended load and are also free from sewage and industrial wastes if these pollution sources are not nearby. However, they contain dissolved impurities like salts of calcium, magnesium, iron, manganese, sodium, fluorine, etc. collected from soils and rocks through which they percolate. Deep-well waters are not affected by rainfall and therefore their quality does not change from season to season. They are also relatively free from harmful organisms, but are hard. Shallow-well waters may be relatively soft but are more affected by rainfall and pollution than deep-well waters.

Impounding Reservoirs

Impounding reservoirs store water during high periods of runoff and release it for use during low periods of runoff. The site for an impounding reservoir depends upon the quality and the quantity of water available, geological considerations of the reservoir and dam site; and distance from and elevation relative to the demand centre and the economic considerations.

The amount of storage provided in an impounding reservoir depends upon the height to which a dam can be built, the storage required to satisfy the demand and the net supply, i.e., the safe yield available from the watershed (safe yield is the maximum net volume available from the watershed after deducting the losses due to evaporation from reservoir surface; infiltration, percolation and leakage through the reservoir floor and dam; and withdrawal for downstream riparian owners).

The required capacity of the impounding reservoir can be obtained from the supply and demand mass curves. Typical supply and demand mass curves are shown in Figure 3. If the demand rate is constant, the demand mass curve is a straight line. Salient features of these mass curves are as follows:

1. Curve 1 shows the demand mass curve with a uniform monthly demand of 500,000 m^3 of water.
2. Curve 2 shows the supply mass curve.
3. Slope of the tangent to the mass curve at any point denotes the rate of supply at that instant, e.g., the rate of supply at point A is 1,400,000 m^3/month.
4. When the tangent to the supply mass curve is parallel to the demand mass curve, the supply and demand rates are the same; the reservoir neither fills nor empties (point B). When the tangent is steeper, the reservoir fills (point C) and when it is flatter the reservoir empties (point D).
5. The maximum vertical distance between tangents to the mass curve, drawn parallel to the demand mass curve represents the required capacity of the reservoir. In Figure 3, BEF, GH, JKL and MN are such tangents. EG and KM are the vertical distances, representing deficits. KM is the maximum of such vertical distances and therefore represents the required reservoir capacity which is 700,000 m^3.
6. If the reservoir is full at J, it will be empty at M, again full at L, and overflow between L and P.
7. The slope of the steepest line above the supply mass curve, and tangent to it at two or more points but not intersecting it, represents the maximum rate of demand which can be supplied by the stream. This, however, would require a larger capacity of the impounding reservoir. In Figure 3, RS is such a line, representing a demand of 700,000 m^3/month which would be met if a reservoir with a capacity of 1,300,000 m^3 is provided.

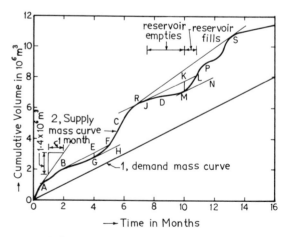

FIGURE 3. Supply and demand mass curves.

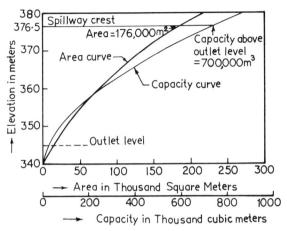

FIGURE 4. Area and capacity curves.

When the demand is variable, the demand mass curve will also be a curve similar to the supply mass curve. In this case, the vertical distances between the two mass curves are plotted on the Y axis with the corresponding times plotted on the X axis. This curve is then considered as the supply mass curve and the horizontal straight line as the demand mass curve. The required reservoir capacity is then obtained from these two curves.

From the contour map of the reservoir site, an elevation-area curve is obtained. The capacity of the reservoir for different elevations may be determined by the prismoidal formula

$$V = \frac{h}{6} (A_1 + 4A_2 + A_3) \qquad (2)$$

in which V = reservoir capacity; h = contour interval; and A_1, A_2, A_3 = areas enclosed within lower, median and upper contours, respectively. Figure 4 shows typical

elevation-area and elevation-capacity curves, or simply area and capacity curves.

Intakes

Water is withdrawn from lakes, reservoirs and rivers through intakes. They may be simple submerged intakes or in the form of towers having intake gates, with openings controlled by racks; and screens, pumps and compressors, chlorinators and other chemical feeders, and discharge measuring devices such as venturimeters.

SIMPLE SUBMERGED INTAKES
Simple submerged intakes (Figure 5) are constructed entirely under water and are used for small water works when the level and quality of water do not fluctuate much. They are lower in cost and pose no obstruction to navigation, ice or floating matter.

INTAKE TOWERS
Intake towers (Figure 6) are used for large water works to withdraw water from lakes, reservoirs or rivers having wide fluctuations in level and quality of water. They may be (a) wet intake towers in which water enters from the entry ports into the tower and then into the conduit pipe through separate gate-controlled openings; or (b) dry intake towers in which water is withdrawn directly into the withdrawal conduit, thus keeping the tower dry.

Intakes should be designed to resist uplift, wave and dynamic forces due to water; ice pressure; impact forces due to floating and submerged objects; and silt pressure due to shifting shoals. Wet intakes are not subjected to uplift forces and are less costly compared to dry intakes. Ports are provided at different levels in high intake towers to facilitate withdrawal of best quality (acceptable sediment load and temperature) water. The size of the pore opening is provided to have the entrance velocity of water around 1.5 m/s.

Location of the intakes should consider the availability of

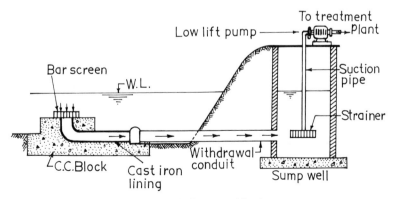

FIGURE 5. Submerged intake.

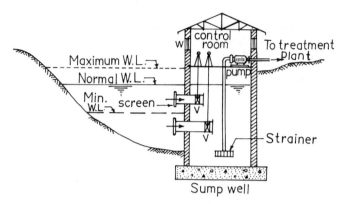

FIGURE 6. Intake tower.

the best quality water, water currents affecting the quality of water and safety of structure, navigation channels, obstruction due to floating ice and debris, formation of bars and shoals, fetch of the wind affecting the height of the waves, and low and high water levels. River intakes are located well upstream of sewage and industrial waste disposal sites, in deep waters, preferably on the outer bank on curves; with stable bottom and banks and having protection against floods, debris and river traffic. Lake and reservoir intakes are provided taking into consideration the sources of pollution, prevailing winds, surface and subsurface currents, depth of water at low and high levels, facilities for selective withdrawal, and shipping lanes.

The conduit leaving from intake to the suction well for the pumps should be laid on a continuous rising or a falling slope to avoid air entrainment. The velocity of water in this conduit may be about 0.6–1.0 m/s.

Ground Water

Ground water is available from: (1) Springs; (2) Wells; and (3) Infiltration Gallery.

SPRINGS

The springs are formed when ground water starts oozing out from the ground surface. Springs may be formed [24] when: (a) An impermeable bed, overlain by a permeable

bed, intercepts the sloping surface of the natural ground [Figure 7(a)]; (b) a sloping permeable bed is intercepted by a dyke [Figure 7(b)]; (c) a sloping permeable bed is intercepted by an impermeable bed due to the presence of a fault [Figure 7(c)]; (d) the water moving along the interconnected joints present in the rock is intercepted by the natural slope of the ground surface [Figure 7(d)]; and (e) the water permeating along the joints in limestone formations keeps on dissolving the rock and widening the joints and is eventually intercepted by the natural slope of the ground surface [Figure 7(e)].

Springs usually supply a very small quantity of water and therefore they may serve as a source of water for small towns in hilly areas.

WELLS

When water is pumped from a well, a cone of depression surrounding the well is formed due to the drawdown of water table. The flow surrounding the well is assumed to be according to the Darcy's law

$$V = ks \qquad (3)$$

in which V = velocity of ground water flow (m/d); k = coefficient of permeability or proportionality constant (m/d); and s = slope of the ground water table or hydraulic gradient.

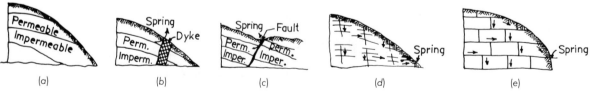

FIGURE 7. Geological formations for springs.

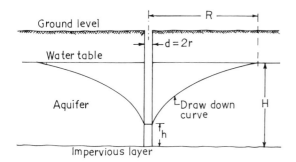

FIGURE 8. *Gravity well under equilibrium conditions.*

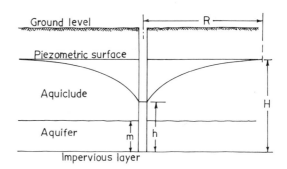

FIGURE 9. *Pressure well under equilibrium conditions.*

Flow into Wells

Aquifer flow into the wells is under two conditions: (1) Equilibrium conditions; and (2) nonequilibrium conditions.

FLOW UNDER EQUILIBRIUM CONDITIONS

(a) Gravity Well Flow Conditions—The flow into the well is under gravity when the surface of the water outside of and surrounding the well is at atmospheric pressure. The rate of flow into a gravity well under equilibrium conditions (Figure 8) is given by

$$Q = \frac{1.36 \ k \ (H^2 - h^2)}{\log \ (R/r)} \qquad (4)$$

in which Q = flowrate into the well (m³/d); k = coefficient of permeability (m/d); H = depth of water in the well before pumping (m); h = depth of water in the well after pumping (m); $H-h$ = draw down (m); R = radius of influence (m); and r = radius of well (m).

(b) Pressure Well Flow Conditions—A well penetrating through an impervious stratum and extending into a confined aquifer is a pressure or an artesian well (Figure 9). The rate of flow into a pressure well under equilibrium conditions is given by

$$Q = \frac{2.72 \ km \ (H-h)}{\log \ (R/r)} \qquad (5)$$

in which m = depth of confined aquifer, and Q, k, H, h, $H-h$, R and r are as described previously.

FLOW UNDER NONEQUILIBRIUM CONDITIONS

The rate of flow to a well under nonequilibrium or unsteady conditions is given by [9,24]

$$p = \frac{114.60 \ Q \ W(u)}{T} \qquad (6)$$

$$\text{and } u = \frac{250S}{T} \ \frac{x^2}{t} \qquad (7)$$

in which p = drawdown (m); Q = uniform rate of pumping (L/min); $W(u)$ = well function of u, values of which can be found from the type curve; S = storage constant; T = coefficient of transmissibility (L/d per meter width); x = distance from well (m); and t = time during which the well has been pumped (d).

As the soil surrounding the wells is generally heterogeneous, it is always advisable to determine the soil properties and the safe yield of the well through pumping tests. When wells are located quite close, the drawdown of one well may affect that of the other and thus influence its discharge. When a well is located on an island surrounded by salt water, the pumping rate should not be too excessive to allow salt water intrusion into it. Wells up to 30 m are usually classed as shallow wells and are dug or bored. Wells more than 30 m deep (deep wells) are bored or drilled.

A well may contain: (1) A screen which allows the flow of water but prevents the entry of larger sand particles into the well and also prevents the collapse of the walls; (2) A casing, extending from above the screen to the ground surface, to contain the walls of the aquifer, to prevent the entry of the polluted water at surface or at shallow depths into the well; and to prevent the loss of pumped water into the higher layers of the surrounding pervious aquifer; and (3) pump and motor to raise the water to the ground surface.

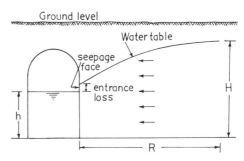

FIGURE 10. *Infiltration gallery under equilibrium conditions.*

INFILTRATION GALLERY

An "infiltration gallery" is a horizontal tunnel or open ditch constructed through the aquifer in a direction nearly normal to the direction of ground water flow. The funnel type gallery is sometimes called a "horizontal well."

The rate of flow into an infiltration gallery under equilibrium conditions [9,25] is given by

$$Q = k L \frac{H^2-h^2}{2R} \qquad (8)$$

in which Q = rate of flow (m³/d); k = coefficient of permeability (m/d); L = length of gallery (m); H = initial depth of water level (m); h = depth of water in gallery (m); and R = radius of influence (m).

PUMPS AND PUMPING MACHINERY

Pumps and pumping machinery are provided in water supply systems for (1) lifting water from surface or ground sources to treatment plants, to service reservoirs, or directly to the distribution system; (2) boosting up pressures to serve high elevation areas, upper floors of multi-storied buildings, and fire fighting engines; and (3) pumping chemicals in treatment units, transporting water through several units of treatment works; backwashing filters, draining settling tanks and removing deposited solids.

Prime Movers

Electric motors and internal combustion engines are the commonly used prime movers. Electric motors are directly coupled to the pumps while internal combustion engines are coupled through belts or gears. Monoblock pumps with motor and pump in the same units are also available for small discharges and heads.

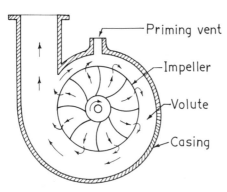

FIGURE 11. *Volute-type centrifugal pump.*

Several types of D.C. and A.C. electric motors such as shunt-wound D.C., series-wound D.C., synchronous A.C., etc. are available. The choice of the type of motor depends upon several factors such as the type of the current and the available circuit; capital and maintenance costs; speed, and its regulation and control; ease in starting; starting current and torque; power factor; rated power and partial load characteristics. Electric motors with any desired horse power with common voltages of 220 V and efficiencies up to 90–95% are available. Proper precaution should be taken (1) to protect motor against overloads by providing appropriate fuses and circuit breakers; and (2) to prevent hazards by grounding and providing lightning conductors. As grounding through piping system deteriorates the quality of water, enhances corrosion and introduces hazards to employees through electric shocks, it is preferable if grounding is done through separate electrodes and not through the piping system [26].

Internal combustion engines, such as diesel engines, are used as prime movers in remote areas where electric supply is not available, or as a standby during electric supply failure.

Pumps

CENTRIFUGAL PUMPS

Centrifugal pumps are commonly used in waterworks pumping. Water enters at the centre of the impeller (Figure 11) and is forced outward towards the casing by centrifugal force. Thus, the mechanical energy or the electrical energy is converted to kinetic energy, which in turn is converted to pressure energy by a gradually diverging discharge tube, called a volute.

Selection of a pump should be such that its capital and operational costs are minimum, there is uniformity of inventory of replacement parts, space requirements are minimum, there is no need for variable-speed control, single pump operation to meet most summer demands is feasible; and the control and operating procedures of the pump are simple [27].

Characteristic Curves

SINGLE PUMP

The relationship between the discharge Q, head H, rotational speed N, the power P supplied by the pump, and the number of impellers can be stated as: (a) At any given stage, Q varies directly as the number of impellers in parallel; (b) at any given speed, H varies directly as the number of impellers in series, i.e., the number of stages; (c) for variable speed, $Q \propto N$, $H \propto N^2$, and $P \propto H^{1.5}$.

In practice, however, these relationships do not strictly hold good.

The performance of centrifugal pumps is described

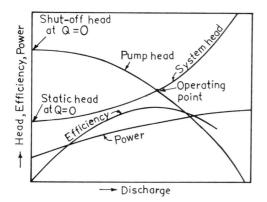

FIGURE 12. Characteristic curves of centrifugal pump.

through characteristic curves (Figure 12) in which the head *H,* power *P* and the efficiency of the pump are plotted against the discharge *Q.* The system head curve, a sum of static head and headlosses, is also plotted. The point of intersection of the head-discharge curve and the system head curve determines the operating point. The pump should be so selected that its efficiency is as large as possible over a wider range of operating conditions.

Pump discharge can be regulated (a) by a valve on the delivery pipe; (b) by varying the speed of the pump; or (c) by having different pump combinations.

PUMP COMBINATIONS

Pumps may be combined so that they run in series or in parallel. In series combination the heads are added, while in parallel the discharges are added.

Series combination of pumps is used in high buildings where water is raised in steps to higher floors through booster pumps at intermediate floors. This reduces pressures on the pipes and pump casings of the lower floors.

Parallel combination of pumps is more common when the pumps have to lift water, varying in quantity (from minimum demand to maximum demand) and head (source at high level and service reservoir empty to source at low level and service reservoir nearly full). Figure 13 shows the head-discharge curves for a three-unit parallel combination [11,28]. Pumps of different capacities should be installed so that they can be run in different combinations to obtain maximum efficiency.

Suction Lift and Cavitation

As pumps lift water from the source to their centre line by partial vacuum, the height to which water can be raised by suction, i.e., suction lift, is limited. Such a height is obtained by subtracting the sum of vapour pressure of water, total losses, and velocity head in suction pipe from the local atmospheric pressure. If this limit is exceeded, cavitation may occur along with vibration, noise and pitting of the pump impellers. To avoid cavitation, it is advisable to limit the suction lift to about 6 m.

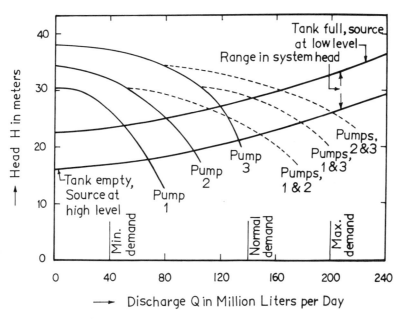

FIGURE 13. Head curves for pumps in parallel.

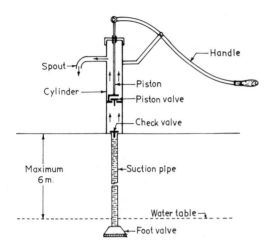

FIGURE 14. Hand-operated reciprocating pump.

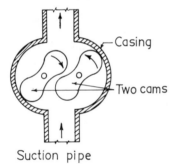

FIGURE 15. Rotary pump with cams.

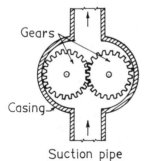

FIGURE 16. Rotary pump with gears.

DISPLACEMENT PUMPS

In displacement pumps, partial vacuum is induced in a chamber so that water is lifted up into this chamber through a suction pipe. This water is then forced out of the chamber into a delivery pipe.

Reciprocating Pumps

Partial vacuum is created in reciprocating pumps through the reciprocating motion of the piston. The simplest type of a reciprocating pump is the hand-operated pump, commonly known as hand pump (Figure 14), still widely used in the rural areas of the developing countries. Where the water table is within 6 m or so, the pump is placed above the ground as shown in Figure 14. When the water table is more than 6 m below the ground surface, the pump cylinder is attached to a drop pipe and lowered below the ground surface.

Reciprocating pumps can also be used for very high but variable heads. However, their initial and maintenance costs are high and are unsuitable for waters containing sediments.

Rotary Pumps

Rotary motion is used in rotary pumps to create partial vacuum. The rotary motion is obtained by using two cams (Figure 15) or two gears (Figure 16).

MISCELLANEOUS TYPES

Air Lift Pumps

In air lift pumps, compressed air is forced into an eduction pipe placed in the well, through a small air pipe, so that air-water mixture rises above in the eduction pipe and is finally discharged. Air lift pumps can be used for lifts up to 150 m but the efficiency is low, lying between 25–50 percent.

Jet Pumps and Hydraulic Rams

Jet pumps and hydraulic rams are sometimes used. However, as their use is not common in water supply practice, they are not described.

FLOW MEASUREMENT

Flow measurements are made in water supply systems to estimate the quantity of water available, the rate or the quantity of water supplied, to fix water charges and to restrain, to some extent, the waste of water as in house plumbing. Several types of devices are used for flow measurement. The selection of the type depends upon the accuracy and range of measurement; simplicity; amount of headloss; cost, durability and ease of repairs.

Flow is measured at several points in a water supply system. Such points and the commonly used flow measuring devices are described below:

1. Flow in a stream or a river is obtained, at a convenient point upstream of the intake, by area-velocity method (velocity measured by floats or current meters), notches and weirs, flumes or dilution method.
2. Yields from wells are estimated from yield tests in which the flow is measured by venturimeters, orifice meters, notches or weirs.
3. Raw water input at intake structures is measured by weirs or flumes for open channel flows; or by venturimeters or orificemeters for pipe flows.
4. Flows in pumping stations are measured by venturimeters.
5. Inflow to a treatment work (usually after aeration if it is practised) is measured by weirs or flumes.
6. Filtrate flow at the outlets of each filter is measured by notches, weirs, orifices or venturimeters.
7. Bulk supplies, from treatment plants or clear water reservoirs, to distribution zones, subzones or industries are measured by venturimeters to obtain flow rates and by positive or inferential type meters to obtain total quantity.
8. Domestic supplies are measured by positive or inferential type meters such as housemeters.
9. Leakage through pipes and plumbing systems are estimated with waste flow measuring meters, or recording meters.

Magnetic flow meters can also be used in place of venturimeters. Water flowing in a pipe also passes through electromagnets of the magnetic flow meter. Voltage that is generated is directly proportional to the velocity of flow, and therefore, magnetic flow meters directly give the rate of flow. Magnetic flow meters have no constrictions as in venturimeters or orificemeters, and therefore the loss of head is less. Furthermore, a magnetic flow meter is very sensitive, can accurately measure fluctuating or even reversed flows.

TRANSPORTATION OF WATER

Water is transported from the source to the areas of consumption under gravity through open channels or closed conduits, or under pressure, through pumping mains. Depending on the topography, local conditions and economic considerations, the mode of transportation of water should be fixed.

General Considerations

OPEN CHANNELS

The flow in open channels is through gravity. The economical sections are trapezoidal with side slopes consistent with the angle of repose of the soil. In rock cutting, rectangular sections with vertical sides can be used.

Open channels suffer from losses due to percolation, evaporation, and unauthorized use. There is also a possibility of pollution. Open channels should never be used for transporting treated water. However, they may be used for transporting raw water or diverting water from one catchment to another to augment impounding-reservoir storage.

GRAVITY AQUEDUCTS AND TUNNELS

Gravity aqueducts and tunnels are used in uneven and hilly areas to shorten the route, conserve the head through reduction and thereby reduce the cost of water transportation. They are usually designed to flow three-quarters full at design capacity, and are usually lined to reduce seepage and headlosses. Gravity tunnels may be horse-shoe shaped for structural reasons. Mean velocities range from 0.30 m/s to 0.60 m/s when unlined and from 1 m/s to 2 m/s when lined.

PRESSURE AQUEDUCTS AND TUNNELS

Pressure aqueducts and tunnels are generally circular in section. Sufficient overburden or reinforcing measures should be provided to resist internal pressure.

PIPELINES

Pipelines are the most common mode of transportation of water in water works. They normally follow the profile of the ground. Several types of pipes as described later are used in practice.

Hydraulic Considerations

When water is transported through channels and pipes, loss of head occurs due to friction. Various formulas, theoretical as well as empirical, are available for estimating frictional headlosses for the given discharge or estimating the velocity or discharge for the given slope of the hydraulic gradient [29]. The most commonly used formulas are described below.

MANNING'S FORMULA

Manning's formula is widely used for open channels, and the gravity aqueducts and tunnels, even though it may be used for pipe flow also. Manning's formula is an empirical formula and is expressed in S.I. units as

$$V = \frac{1}{n} R^{2/3} S^{1/2} \qquad (9)$$

in which V = average velocity of flow (m/s); R = hydraulic radius (m); S = slope of hydraulic gradient; and

TABLE 8. Hazen-Williams Coefficients for Different Pipe Materials.

Pipe Material	Hazen-Williams Coefficient	
	Usual Value	Value Recommended for Design
Cast iron, new	130	100
5 years old	120	
10 years old	110	
15 years old	105	
20 years old	100	
30 years old	90	
40 years old	80	
Galvanized iron, D > 50 mm	120	100
D ≤ 50 mm, used for house service connections	120	55
Steel, rivited joints	110	95
welded joints	140	100
Concrete	140	110
Asbestos cement	150	120
PVC	150	120

n = Manning's coefficient of roughness. The use of Manning's formula and the common values of n [30] are discussed in detail in another section.

DARCY-WEISBACH FORMULA

Darcy-Weisbach formula is commonly used for pressure aqueducts, tunnels and pipelines. It is dimensionally homogeneous and is expressed as

$$h_f = \frac{fL}{D}\frac{V^2}{2g} \qquad (10)$$

$$\text{or } h_f = \frac{16\,fL\,Q^2}{\pi^2\,2g\,D^5} \qquad (11)$$

in which h_f = loss of head due to friction (m); L = length of pipeline (m); D = diameter of pipeline (m); g = gravitational acceleration (m/s²); and f = coefficient of friction, the value of which depends upon the Reynolds number Re and relative roughness e/D (e = average roughness height). The value of f is expressed by Colebrook-White implicit equation, commonly represented by the Moody's chart, as explained in detail in another section. Of the several attempts that have been made to express f explicitly [31–33], the best one is the Swamee-Jain relationship [33] given by

$$f = \frac{0.25}{\left[\log\left(\dfrac{e}{3.7D} + \dfrac{5.74}{Re^{0.9}}\right)\right]^2} \qquad (12)$$

HAZEN-WILLIAMS FORMULA

Hazen-Williams formula is widely used in water supply engineering for flow through pressure conduits and pipe-lines. It is an empirical one and is expressed in SI units as

$$V = 0.849\ C_{HW}R^{0.63}S^{0.54} \qquad (13)$$

$$\text{or } h_f = \frac{10.68\ L\ Q^{1.852}}{C_{HW}^{1.852}\ D^{4.87}} \qquad (14)$$

in which V = average velocity of flow (m/s); R = hydraulic radius (m); S = slope of hydraulic gradient, C_{HW} = Hazen-William's coefficient; h_f = head loss (m); L = length of pipeline (m); Q = discharge (m³/s); and D = pipe diameter (m). The usual values of the Hazen-Williams coefficient C_{HW}, and the values recommended for design purposes are given in Table 8.

Pipe Materials

The materials commonly used in the manufacture of pipes are: (1) cast iron; (2) steel; (3) asbestos-cement; (4) concrete; (5) prestressed concrete; and (6) plastic.

CAST IRON PIPE

The cast iron pipe, commonly known as C.I. pipe, is the most common pipe in water supply. The pipe is manufactured either by pit casting or centrifugal casting. Pit casting consists of pouring the molten metal into a vertical sand mold. In centrifugal casting, the molten metal is poured into a metal or a sand-lined mold, rotating at high speed on rollers. Generally, the centrifugally cast pipes are thinner than the pit cast pipes. Larger diameter C.I. pipes are coated with bituminous or cement mortar lining; while smaller diameter pipes are coated with zinc and are known as galvanized iron pipes. C.I. pipes are commonly available in sizes from 80 mm to 1200 mm.

The types of joints usually employed in C.I. pipes are: (a) standard lead joint in bell and spigot pipe; (b) push on joint; (c) flanged joint; (d) bell joint; (e) threaded I.P.S. joint; (f) victaulic joint; (g) dresser coupling; (h) flexible joint; (i) mechanical joint; (j) precalked joint; and (k) split sleeves coupling.

The C.I. pipes are strong, durable, corrosion resistant and can generally withstand internal pressures up to 240 m of water. However, they are costly, heavy, difficult to handle and likely to break during transportation or while making connections.

The allowable leakage during the maintenance stage should not exceed

$$Q_1 = \frac{ND\sqrt{H_p}}{36} \qquad (15)$$

in which Q_1 = allowable leakage (L/d); N = number of joints in the length of pipeline; D = pipe diameter (mm); and H_p = average pressure head during test (m).

STEEL PIPE

Steel pipes are frequently used for pipe lines, trunk mains and inverted siphons where pressures are high and sizes large. Riveted-steel pipes are generally made with longitudinal lap seams, single riveted for pipes under 1200 mm in diameter and double riveted for pipes above that size. If the steel plates are over 15 mm thick, butt joints are used. Spiral riveted pipe is made in diameters from 75 mm to 1000 mm and of various thicknesses.

Steel pipes are cheaper and easier to construct and transport than C.I. pipes. They are also flexible and adapt to small relative changes in ground level without failure. However, they cannot withstand large external loads and are susceptible to collapse or distortion if a partial vacuum is caused while emptying or sudden release of water through pipe breaks. Steel pipes are also susceptible to corrosion but can be protected against corrosion through bituminous, asphalt or cement lining.

ASBESTOS-CEMENT PIPE

Asbestos-cement pipe, commonly known as A.C. pipe, is composed of a mixture of asbestos fibre and cement, in the approximate proportions of 15 percent and 85 percent, respectively, by weight; and compressed by steel rollers to form a dense, homogeneous laminated material of great strength and density. These pipes came in use since 1928 and expected life is more than 30 years, and can be as high as 75 years.

Asbestos-cement pipes are immune to the action of ordinary soil and also the acids and salts. The pipe, being bad conductor of electricity, is not affected by electrolysis. Tuberculation does not occur and pipe remains smooth even when carrying corrosive water. Joints with two rubber rings compressed between the sleeve and pipe barrels are highly flexible, capable of being deflected to an angle of 12° and

are claimed to allow high pressure without leakage. Laying and jointing of smaller pipes cost less compared to C.I. or steel pipes of similar sizes. A.C. pipe can be drilled and tapped for connections but has not the same strength or suitability for threading as C.I. or steel pipes. Further, any leakage at the threads will become worse with time. However, this difficulty can be overcome by screwing the ferrules through malleable iron saddles fixed at the point of service connections. A.C. pipes are not suitable for use in sulphate soils. The available safety against bursting under pressure, though less than that for the spun iron pipes, is nevertheless adequate for most practical purposes, and also increases as the pipe ages.

The types of joints commonly adopted with A.C. pipes are: (a) C.I. detachable joint; and (b) A.C. coupling joint.

CONCRETE PIPE

Plain cement-concrete pipes are manufactured in small sizes, generally up to 600 mm in diameter; while they are reinforced for larger sizes. Concrete pipes are used on long conduits and aqueducts but not commonly used in distribution systems.

The advantages of a concrete pipe are: (a) It resists loads due to back filling and due to collapsing external pressures or internal partial vacuum; (b) Maintenance cost is low; (c) Tendency to float when empty is less because of its weight; (d) It is not subject to corrosion when buried in ordinary soil; (e) It is not subject to tuberculation and therefore its carrying capacity remains high; (f) Expansion joints are not normally required; and (g) Specially skilled labour is not required for its construction.

The disadvantages of a concrete pipe are: (a) It is difficult to repair; (b) It is prone to corrosion in the presence of acids, alkalies or sulphur compounds; (c) It is heavy and bulky, and therefore difficult to transport; and (d) It has a tendency to leak due to porosity and shrinkage cracks.

Concrete pipes are joined by any one of the following four types of joints: (a) bandage joint; (b) spigot and socket joint (rigid and semiflexible); (c) collar joint (rigid and semiflexible); and (d) flush joint (internal and external).

PRESTRESSED CONCRETE PIPE

In a prestressed concrete pipe, permanent internal stresses are deliberately introduced by tensional steel to counteract, to the desired degree, the stresses caused in the pipe under service. These stresses are entirely independent of the stresses caused by the external loads or the internal pressures. The amount of steel is such that its maximum stress is below the elastic limit. Pipes may be constructed with a steel cylinder surrounding a concrete shell or the steel cylinder may be spirally bound with a spirally wound high tensile wire. The wire is covered, in turn, with a protective but unstressed layer of concrete which forms the outer skin of the pipe. An economical design results by prestressing to such an extent that the elastic limits of the steel cylinder and the wire are reached simultaneously.

The prestressed concrete pipes are available in sizes from 80 mm to 1800 mm. They can withstand more internal pressures and are also thinner and therefore lighter than the concrete or reinforced concrete pipes. The prestressed concrete pipes are provided with flexible joints, the joints being made with rubber gaskets. The pipes have socket spigot ends to suit the rubber ring joints.

PLASTIC PIPE

The plastic pipe is usually the polyvinyl chloride (commonly known as PVC) and high density polyethylene pipes. The pipe usually ranges from 15 mm to 150 mm and occasionally even up to 350 mm in diameter and is used for water distribution systems.

The PVC pipes are much lighter than the C.I. or the A.C. pipes and therefore are easy to handle, transport and install. They can resist pitting, tuberculation and a wide range of chemicals, and also are unaffected by fungi or bacteria. They are immune to galvanic or electrolytic attack, when buried in corrosive soils or near brackish waters. PVC pipes are elastic and comparatively more resistant to deformations resulting from earth movements as compared to A.C. pipes. Solvent cementing technique for jointing PVC pipes is cheap, simple and quite efficient. PVC pipes are not suitable in soils containing aromatic compounds.

Corrosion

Corrosion is the phenomenon of the interaction of a metal with its environment, resulting in its deterioration or destruction. The environment may be water, soil or air. Metals tend to revert, through corrosion, to their more stable compounds.

CAUSES OF CORROSION

The important causes of corrosion are: (a) Galvanic corrosion; (b) hydrogenation; (c) electrolysis; (d) chemical reaction; (e) direct oxidation; (f) stress corrosion; (g) stray current electrolysis; and (h) bacterial corrosion.

As corrosion tends to reduce the strength and life of metals and therefore of structures and thus results in enormous monetary loss every year, methods to protect the pipes from corrosion and to control corrosion should be used.

PROTECTION FROM CORROSION

Protection from corrosion can be obtained through protective coating and lining. Materials used for such coating and lining are the asphaltic materials, enamels, resins, lacquers, zinc, galvanizing, plastics, paints, coal-tar enamel, vitreous coatings, cement linings, etc. Protective linings and coatings are preventive measures and must be adopted at the time of manufacturing of pipes and construction of steel structures.

CONTROL OF CORROSION

Cathodic Protection

Cathodic protection [34,35] is the application of electricity through an external power supply or by use of galvanic methods for combating electrochemical corrosion. Cathodic protection should be used as a supplement and not as an alternative to other methods. The D.C. power requirements vary from 0.4 to 10 kW in most cases. The main power loss occurs in the anode earthing. The earthing can be carried out by any metal (pure or scrap); carbon, coke or graphite.

Protection by Galvanic Anodes

Galvanic anodes serve the same function as the cathodic protection system, but do not require continuous electric supply as in the case of cathodic protection. The required current is supplied by an artificial galvanic couple in which the part to be protected (usually steel) is made the cathode by choosing the other metal, having the higher galvanic potential, as the anode. Zinc, aluminum and magnesium (of sufficient purity) or their alloys which are higher up in the galvanic series must be used for this purpose. Such anodes are generally spaced at 4–6 m along the pipe line.

Deposition of Protective Coatings

A thin protective layer of $CaCO_3$ is deposited by the water inside the surface of pipes. This is accomplished by adjusting the pH value and alkalinity of water, to keep the Langelier Saturation Index to a slightly positive value. Lime or soda ash or both can be used to raise the pH and alkalinity.

Small amounts of sodium silicate (Na_2SiO_3) will deposit dense, adherent but slightly permeable film [9]. A dosage of 12–16 mg/L is maintained in the beginning, and is gradually reduced to 3–4 mg/L.

Treatment of Water

Treatment of water like adjustment of pH, removal of CO_2 and excess O_2, increase in calcium ion and carbonate ion concentrations and addition of inhibitors can overcome to a large extent the corrosive tendency of water. Chemical treatment can be effective as only a supplement to the other methods like protective coatings.

Freezing and Thawing

When the temperature falls below 0°C, the water in the pipelines may freeze. The factors which influence the rate of freezing are: original and final differences of temperatures between water and air, time for change of temperature, volume of water per unit length in the pipe line, velocity of flow, type of insulation and the rate of loss of heat.

Melting of the frozen water, i.e., ice is termed thawing.

Methods that can be adopted for the prevention of freezing and the thawing are [36]:

1. Keeping the water in the pipelines flowing at a minimum velocity of 1.3 m/s;
2. Laying the pipelines below the frost line;
3. Wrapping the pipelines in rags and thus provide some insulation;
4. Digging down to the pipe and building a fire in the trench over it;
5. Pouring hot water on pipes;
6. Blowing steam on or into the pipe; and
7. Electricity.

Keeping the water in the pipeline flowing is one of the simplest methods of preventing freezing even in an unprotected welded steel pipe at 0°C. With lower outside temperatures, water is likely to freeze irrespective of the velocity in long mains. Such pipes should be laid below the frost line.

Thawing of the pipelines by fire, hot water or steam may be hazardous. Thawing of metal pipes by electricity is quick and more effective. Cast iron pipes heat most readily and copper pipes most slowly. Care must be taken to use the lowest amount of current suitable for the pipe material to avoid melting lead in the pipe joints. Sulphur joints do not prevent the electrical heating of pipes, although a higher voltage may be required.

In thawing service pipe lines, the connection is made on the street side, the pipe being separated from the house piping to insure against damage to the electrical appliances, grounded on the water piping, when the thawing current is turned on. The amperages needed for thawing pipes, as suggested by Sheppard [37,38] are given in Table 9.

Forces on Pipes

Pipes are subjected to several forces which induce stresses in them. The pipes should be strong enough to resist these stresses.

INTERNAL WATER PRESSURE

When water is carried under pressure through a pipe, the pipe is subjected to water pressure which tries to produce tensile stresses and burst the pipe. The maximum internal pressure likely to develop is the sum of full static pressure and the water hammer pressure. The full static pressure is due to the water column between the centre line of the pipe and the horizontal static head line for no flow in pipe. The water hammer pressure is caused due to the closure of the valves. The phenomenon of water hammer and the resulting water hammer pressure for sudden as well as gradual valve closures are described in detail in another section. In water supply systems, however, water hammer pressures are seldom important except for very long pipelines. Further, water hammer pressures can be kept under control by pro-

TABLE 9. Amperages for Thawing Pipes.

Street Mains		Service Pipes		
			Amperage	
Size of C.I. Pipe (mm)	Amperage	Size of Pipe (mm)	Wrought iron, Steel, Lead Pipe	Copper Pipe
100	350–400	12	200	500
150	500–600	19	250	625
200	700–900	25	300	750
250	1,000–1,300	31	450	1,000
—	—	38	600	1,500
—	—	50	800	2,000

viding slow-closing valves, pressure relief valves, surge towers and surge tanks.

PRESSURE DUE TO EXTERNAL LOADS

Pipes are often placed in trenches and backfilled. Thus, the pipe is subjected to the weight of the back fill and the superimposed traffic load, if any. This produces compressive stresses in the pipe, its value being maximum when the pipe is empty and the maximum backfill and traffic loads are acting. The magnitude of the induced stress depends upon the type of foundation provided for the pipe in the backfill, the specific weight of soil, nature of the soil—whether cohesive or noncohesive, the diameter of the pipe, the depth and width of the trench, etc.

TEMPERATURE STRESSES

Longitudinal stresses are produced in pipelines, especially when laid on the ground, due to the variation in temperature. Pipes expand during day and contract at night. Expansion joints must be provided to counteract these stresses.

UNBALANCED FORCES AT SECTIONS

When the velocity of water changes at a section, either in magnitude as at a reducer or in direction as at a bend, a change in momentum takes place. This causes unbalanced forces. These forces are usually resisted by the pipes and the joints. However, for large diameter pipes, as in water transportation mains, it might be necessary to hold the pipe firmly by anchoring it to the foundation by iron spikes or rings or encasing the pipe in massive concrete or masonry blocks.

BENDING STRESSES

When pipe lines are carried on supports such as trestles or piers, bending stresses are developed in the pipelines.

Such stresses are also developed when soil under some portion of the pipe line is washed away so that this portion behaves as a beam. The pipe should be strong enough to resist bending stresses.

Stresses may also develop in the pipelines due to unequal settlements of the subgrade [39]. The unequal settlements may be caused due to nonuniformity in the depth of cover, soil properties and construction conditions. In earthquake-prone areas, the earthquake effect should be considered in determining the stresses, strains and the expected joint movements [40]. The pipe alignment should be examined carefully for faults, geologic indications of slides, slips or liquefaction.

Design Considerations

The main design considerations for a transportation system are safety and economy. The transportation system should be safe hydraulically and structurally and also be as economical as feasible.

The economical diameter of a pipe, transporting water under gravity, is the one that shall consume all the available head for friction and other losses during transportation. For a long transportation main during an uneven terrain, optimal diameter can be obtained taking into consideration the slope of the terrain, the slope of the hydraulic gradient, permissible pressure in the pipe and the cost of the pipe [41].

In the design of a pumping system, the capital costs of the pipes and pumps and the recurring expenditure on operation, maintenance and energy must be considered. Deb [42] has shown how the optimal diameter of a rising main can be obtained by considering these costs; and also the salvage value of the pipes and pumps at the end of the design period, inflation rate and the uniform decrease in the Hazen-Williams coefficient over the design period (e.g., from 130 to 100). Assuming that the water demand increases exponentially over the design period, Deb has shown that it is cheaper to install the water transportation system in two phases, and has suggested the method of determining the optimal year for second-phase installation.

The optimal choice of the route of a transportation system can be made by dynamic programming [43,44].

VALVES, HYDRANTS AND FITTINGS

Valves

Valves provided in water supply works can be classified, according to their function, in two categories: (1) Isolating valves which are used to separate or shut off sections of pipes, pumps and control devices from the remaining distribution system for inspection or repair purposes; and (2) control valves which are used to control the flow rate or the pressure; or to facilitate the entry of air; or the disposal of entrapped air, or sediment in the system.

ISOLATING VALVES

Gate Valves

A gate valve is widely used because of its low cost, availability and low frictional headloss when fully open. It may be inside-screw, i.e., non-rising stem type; or outside-screw, i.e., rising-stem type, commonly available in sizes up to 1600 mm with operating pressure heads up to about 100 m. Small valves are operated manually. Medium-size valves have gear reduction devices to facilitate manual operation. Large valves are operated by electric power. Gate valves are not efficient as control or throttling devices because of soft wear, downstream deflection and vibration of the gate disk.

Plug Valves

A plug valve consists of a cylindrically shaped plug placed in a close-fitting cylindrical seat perpendicular to the direction of flow. Cone and spherical valves are special types of plug valves. A plug valve may be used for both isolation and control purposes.

Sluice Valves

Sluice valves usually range in size from 50 mm (pressure head of 160 m) to 1200 mm (pressure head of 60 m). Sluice valves smaller than the conduit size are usually installed for the sake of economy. They are easy to operate and have low cost; however problems arise due to leakage and corrosion of gate frames. Sluice valves have circular, square or rectangular opening and are used for intakes, in open channels or closed conduits.

Butterfly Valves

A butterfly valve has no sliding parts. It consists of a circular disc mounted on a horizontal shaft. The disc is rotated by hand or electric power applied through a gear reduction device. Butterfly valves are easy to operate, low in cost, compact in size due to reduced size of chamber or valve house and have improved closing and retarding characteristics. The maximum permissible velocity in fully open condition is 5 m/s for rubber seated valves and 17 m/s for metal seated ones. Butterfly valves are usually available in size ranging from 350 mm to 1600 mm with a working pressure head of about 100 m.

CONTROL VALVES

Globe Valves

A globe valve consists of a screw-operated disk that is forced down on a circular seat. Globe valves have relatively high head losses, and therefore they are commonly used for pressure regulation rather than isolation purposes. They are generally used on small size pipes.

Needle Valves

A needle valve consists of a streamlined needle that fits into a small orifice. Needle valves have relatively high head losses. They can accurately control the flow of water and therefore they are better suited as control valves.

Check Valves

Check valves, also called "Non-return" or "reflux" valves allow the flow of water in one direction only. They are mainly used on suction and delivery pipes of pumps. A "foot valve" is a check valve fitted on the inlet of the suction pipe of a centrifugal pump. It holds water in the suction pipe and thus obviates or reduces the priming effort. Swing type check valve is the commonest type. Typical range for single door type is 50 mm (pressure head 150 m) to 600 mm (pressure head 80 m); and for multidoor type is 600 mm to 1200 mm (pressure head 50 m).

Flow Dividing Valves

Flow dividing valves, specially designed and constructed, may be used at junctions in mains to ensure that an assigned flow is always maintained in a subsidiary. They contain diaphragms having openings which increase or decrease depending upon the upstream pressure in the main, but independent of the downstream pressure in the subsidiary.

Flowlimiters

A flowlimiter, also called a "flow restricter" or a "maximum demand controller" permits flows up to a predetermined value, but automatically assumes control when the flow exceeds this value and prevents excess withdrawals. A flowlimiter can be used where two or more users, taking water from a common source, are to be prevented from consuming water in excess of a set figure. The performance characteristics of a typical flowlimiter [45] is shown in Figure 17.

Pressure Reducing Valves

A pressure reducing valve [46] can maintain a constant set pressure downstream of it, irrespective of how large the upstream pressure becomes. Thus, several pressure reducing valves together can reduce pressure in the excessive pressure zones of a water distribution system. There are two exceptions to this normal, pressure reducing behaviour of a pressure reducing valve. These exceptions are: (1) If the pressure becomes less than the pressure setting of the valve, it becomes inoperative and has no effect on the flow (except the loss of head through it); and (2) if the pressure reducing valve is byepassed and the downstream pressure exceeds its pressure setting, the pressure reducing valve acts as a check valve and prevents flow in the reverse direction. By preventing reverse flow, the pressure reducing valve allows the pressure immediately downstream from it to exceed its pressure setting. In this case, several pressure reducing valves can effectively control the sources from which supply

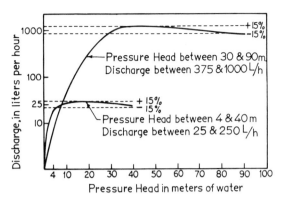

FIGURE 17. *Performance characteristics of flowlimiter.*

is withdrawn under different demand conditions. As the network demand increases, additional sources of supply become operative.

Pressure Relief Valves

Pressure relief valves are used to release excess pressures, usually developed by the sudden closure of valves. They are provided in one or more summits of the transportation mains and keep the pressure within predetermined value by allowing the outflow of water. They are usually spring-loaded or weight-loaded, but are not sufficiently responsive to rapid fluctuations of pressure.

Air Cushion Valves

Air cushion valves [47] are placed on transportation mains where separation of water column is likely to occur. It admits air into the pipeline when separation takes place, allows it to be compressed, and expels air at a predetermined pressure. Thus, air cushion valves help in dissipating energy of returning water column during a water hammer phenomenon.

Zero Velocity Valves

A zero velocity valve has a spring-controlled closing disk. It is fitted on a pumping main. When the pump trips and the forward velocity of water in the main begins to decrease, the disk starts closing and comes to the closed final position when the forward velocity is nearly zero. The valve effectively prevents the return of the water column during a water hammer phenomenon.

Opposed Poppet Valves

An opposed poppet valve, having two poppets of slightly differing diameter, is kept in closed position by a light spring. The valve opens when pressure in the main exceeds a predetermined value.

This valve is used in combination with the air cushion and the zero velocity valves to effectively control water hammer in pumping mains. In a typical 15,000 m long, 1,200 mm diameter pumping main [47] transporting about 140 MLd, through a hilly terrain, 5 zero velocity valves, 9 air cushion valves and 4 opposed poppet valves were installed to limit the water hammer pressure within twice the working pressure. Actually, the highest surge pressure recorded was only 1.19 times the working pressure.

Automatic Shutoff Valves

Automatic shutoff valves may also be used on the pumping mains to close automatically when the velocity in the main exceeds a predetermined value.

Air Valves

Air valves are fitted on pipelines to automatically release the air when the pipeline is being filled; and permit the entry of air when the pipeline is being emptied. Air valves, when fitted on summit, also release entrained air at these points during normal operating conditions. Automatic air valves, fitted in urban areas, may present a serious contamination risk. Due precautions, therefore, should be taken to prevent the entry of polluted surface water into the mains. The ratio of the required air valve diameter to conduit diameter is given as [9,11]:

- For release of air only 1:12
- For admission as well as entry of air 1:8

(1) Air Release Valves—Air release valves are provided in pipelines at summits to remove air that is mechanically entrained as water is drawn into the pipeline. The accumulated air tends to collect at high points, reduces the available cross-sectional area of flow and thus obstructs and reduces the flow carrying capacity.

(2) Air Inlet Valves—When a pipeline is drained, or when a break occurs in a line at a lower end of a slope, water rushes from the upstream to the downstream causing vacuum at higher elevations on the upstream. This may cause a collapse of the pipeline. Air inlet valves, which are normally held shut by water pressure, automatically open when vacuum is developed in the pipeline. They allow the entry of air, reduce vacuum and thereby prevent the collapse of the pipeline.

Scour Valves

Scour valves, also known as "blow-off" valves are small gate valves provided at low points in the pipelines such that the pipeline can be drained or emptied completely. They discharge into natural drainage channels or sumps from which water can be pumped to waste. Care should be taken to prevent entry of outside polluted water.

Altitude Valves

Altitude valves, also called "ball" valves control water levels in service reservoirs, stand pipes and elevated tanks. An altitude valve contains a float which follows the water

level in the reservoir and permits the entry of water when the reservoir is not completely full, but closes and stops the inflow of water when the reservoir becomes full.

Fire Hydrants

Fire hydrants are provided on water distribution systems to obtain water for fire fighting purposes. They are of three general types: (1) The post hydrant with a vertical barrel extending about 600 mm to 900 mm above the ground surface; (2) the flush hydrant which is underground with its cover being flush with the ground surface; and (3) the wall hydrant placed on the wall of a building. The post hydrant is more common in practice. A fire hydrant is made of cast iron with brass fittings that come in contact with water. A gate valve should preferably be placed on the connection to the distribution system to allow for the shut off and repair of the hydrant.

Hydrants have two, three or four hose outlets of 63.5 mm diameter and are correspondingly termed as two-way, three-way and four-way hydrants, having a minimum barrel diameter of 100 mm, 125 mm and 150 mm, respectively. For a flow of 950 L/min, the loss of head in the two-way, three-way and four-way hydrants should not exceed 1.23 m, 1.51 m and 2.11 m, respectively. Standards have been established by American Water Works Association and Insurance Service Office [48]. Some requirements such as fire flow, duration, fire reserve storage and hydrant spacing have already been given in Table 5.

Fittings

Several types of fittings are employed in water distribution systems to change the direction of flow, to reduce the pipe size, to provide junctions for three or more pipes, etc. Such fittings include several types of bends, elbows, tees, cross connections, lateral connections, wyes, etc.

Head Losses

The head losses incurred in the valves and fittings are minor as compared to the frictional head loss in pipes, especially when the pipe lines are long. However, they may be considered when they form a considerable percentage (e.g., more than 5 per cent) of the total head loss. The head loss in such fittings, which is usually the minor head loss, is expressed as

$$h_m = K_m \frac{V^2}{2g} \tag{16}$$

in which h_m = minor head loss (due to fitting); K_m = minor head-loss coefficient; and V = average velocity of flow in the pipeline. The head-loss coefficients for valves and fittings are given in Table 10.

TABLE 10. Head-Loss Coefficients for Valves and Fittings.

Device	Head-Loss Coefficient
Globe valve, fully open	10
Angle valve, fully open	5
Gate valve, fully open	0.3–0.4[a]
Gate valve, three quarters open	1.5
Gate valve, half open	5.6
Gate valve, quarter open	24
Ball check valve, fully open	70
Foot valve, fully open	15
Swing check valve, fully open	2.5
Venturimeter	0.3
Orifice meter	1.0
Close return bend	2.2
Tee, 90° take off	1.5
Tee, straight run	0.3
Elbow, 90°	0.5–1.0[b]
Elbow, 45°	0.4–0.75[b]
Elbow, 22.5°	0.25–0.50[b]
Sudden contraction	0.3–0.5[c]
Gradual contraction	0.05–0.2[c]
Sudden enlargement	0.5–1.0[c]
Gradual enlargement	0.2–0.7[c]
Pipe entrance, sharp cornered	0.50
Pipe entrance, bell mouth	0.05

[a]varying with radius ratios.
[b]varying with radius of bend to pipe diameter ratios.
[c]varying with smaller to larger pipe diameter ratios.

WATER TREATMENT

The process through which the quality of water is improved so that it is brought to confirm or nearly confirm the quality standards (Table 7) is termed water treatment. Surface water contains impurities which are mostly colloidal in nature and therefore the conventional treatment of surface water may include: (1) Aeration; (2) sedimentation; (3) flocculation; (4) filtration; and (5) disinfection. The ground water, on the other hand, contains only dissolved impurities and therefore its treatment is somewhat different as discussed later.

Surface Water

AERATION
Aeration is a process of treatment whereby water is brought into intimate contact with air for (a) increasing the oxygen content; (b) reducing the carbon dioxide content; and (c) removing hydrogen sulphide, methane and various volatile organic compounds responsible for the taste and odours.

Aerators used for aeration purposes are normally classified as [49]: (a) waterfall aerators and (b) bubble aerators.

Waterfall Aerators
Waterfall aerators of significant types are as follows:

(*1*) *Multiple Tray Aerator*—A multiple tray aerator consists of 4–8 trays, with perforated bottoms, at intervals of 300–500 mm. The rate of fall is 0.02 m^3/m^2 of tray surface. For finer dispersion of water, the aerator trays are filled with coarse gravel, about 100 mm deep.

(*2*) *Cascade Aerator*—A cascade aerator consists of a flight of 4–6 steps, each about 300 mm high with a capacity of about 0.01 m^3/s per meter width. The cascade aerator is either circular or rectangular in area. A multiple-platform aerator uses the same principles. Sheets of falling water are formed for full exposure of the water to the air.

(*3*) *Spray Aerator*—A spray aerator consists of stationary nozzles connected to a distribution grid through which the water is sprayed into the surrounding air at a velocity of 5–7 m/s.

Bubble Aerators
Bubble aerators include diffused plate and diffused tube aerators which are provided along one side of the length of the tank and suspended at mid-liquid depth. A venturi aerator, which is also a bubble aerator, requires a throat in the water pipe and a perforated air supply pipe. The amount of air required for bubble aeration is 0.3–0.5 m^3 per m^3 of water.

The efficiency of an aerator is measured in terms of the percentage of CO_2 removed from water. The spray aerators are the most efficient.

SEDIMENTATION
Sedimentation is a process in which the suspended particles that are heavier than water are separated from water by gravitational settling. The terms sedimentation and settling are used interchangeably. A sedimentation basin may also be referred to as a sedimentation tank, settling basin or a settling tank.

Depending on the concentration and the tendency of particles to interact, four types of settling can occur [50]: (1) Discrete particle settling; (2) flocculant settling; (3) hindered (also called zone) settling; and (4) compression settling. During a sedimentation operation it is common to have more than one type of settling occuring at a given time; and it is possible to have all the four types occuring simultaneously.

Discrete Particle Settling (Type I)
Discrete particle settling refers to the sedimentation of discrete particles in a low-solids-concentration suspension. Particles settle as individual entities without any significant interaction with neighbouring particles.

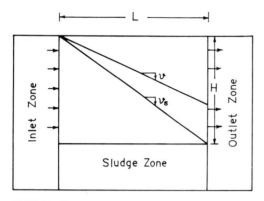

FIGURE 18. Discrete particle settling in an ideal basin.

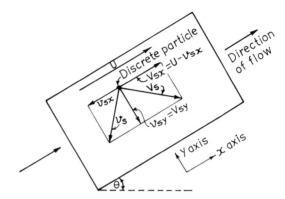

FIGURE 19. Typical tube settler.

The terminal settling velocity of an individual particle v_s is given by

$$v_s = \left[\frac{4}{3} g \frac{(\varrho_s - \varrho) d}{C_D \varrho} \right]^{1/2} \quad (17)$$

in which g = gravitational acceleration; ϱ_s = density of particle, ϱ = density of fluid; d = average diameter of particle; and C_D = drag coefficient that depends on viscous and eddy resistances.

For Reynolds number, $Re \leq 1$, the drag is entirely viscous and the drag coefficient, $C_D = 24/Re$. Substituting this value of C_D in Equation (17), Stokes law is obtained which is

$$v_s = \frac{g (\varrho_s - \varrho) d^2}{18\mu} \quad (18)$$

In the design of sedimentation basins, the usual procedure is to select a particle with a terminal velocity v_s, and to design the basin so that all particles that have a terminal velocity equal to or greater than v_s will be removed. The rate at which clarified water is produced is given by

$$Q = A \cdot v_s \quad (19)$$

in which Q = rate of flow (m³/s); A = surface area of the sedimentation tank (m²); and v_s = terminal settling velocity (m/s).

Equation (19) yields

$$v_s = \frac{Q}{A} \quad (20)$$

in which v_s = overflow rate (m³/m²/s).

Equation (20) shows that the overflow rate or the surface loading rate is equivalent to the settling velocity. Equation (19) also indicates that the flow capacity is independent of the depth for Type I settling.

In an ideal settling basin (Figure 18), the particles with a settling velocity v less than v_s will be removed in the ratio

$$y = \frac{v}{v_s} \quad (21)$$

in which y = fraction of particles removed with settling velocity v.

For a given clarification rate Q, where $Q = v_s A$, only those particles with a velocity equal to or greater than v_s will be completely removed. The remaining particles will be removed in the ratio v/v_s. The total fraction of particles removed is given by

$$\text{Fraction removed} = (1 - y_s) + \int_{o}^{y_s} \frac{v}{v_s} \, dy \quad (22)$$

in which y_s = fraction of particles with velocity v_s. Herein the two terms (the first within the parenthesis, and the second under the integral) represent fraction of particles with velocity v greater than and less than v_s, respectively.

Better utilization of tank volume could be obtained by subdividing it vertically through the addition of horizontal trays. Although basins containing trays are not normally built, a number of proprietary devices take advantage of this principle. Among these devices, the tube settlers are the most common.

A typical tube settler is shown in Figure 19. The effect of the velocity of flow and angle of inclination of the tube on the settling velocity of the particle is shown. The velocity components of the particle are:

$$V_{sx} = U - v_s \sin \theta \quad (23)$$

and

$$V_{sy} = -v_s \cos \theta \quad (24)$$

Flocculant Settling (Type II)

Flocculant settling refers to the sedimentation that takes place in relatively dilute suspension in which the particles do not act as discrete particles but flocculate during sedimentation. By flocculation the particles increase in weight and mass and settle at a faster rate. To determine the settling characteristics of a suspension of flocculant particles, a settling column may be used. A settling column and the typical settling results are shown in Figure 20.

Percent removal of flocculant particles in detention time t_3

$$
= \frac{\Delta d_1}{d_7}\left(\frac{P_1 + P_2}{2}\right) + \frac{\Delta d_2}{d_7}\left(\frac{P_2 + P_3}{2}\right)
$$

$$
+ \frac{\Delta d_3}{d_7}\left(\frac{P_3 + P_4}{2}\right) + \frac{\Delta d_4}{d_7}\left(\frac{P_4 + P_5}{2}\right)
$$

$$
+ \frac{\Delta d_5}{d_7}\left(\frac{P_5 + P_6}{2}\right)
$$

in which t_3 = detention time in a column with liquid depth d_7; and P = percentage of particles removed.

Hindered Settling (Type III)

Hindered settling refers to the sedimentation that takes place in suspensions of intermediate concentration in which the settling of neighbouring particles is hindered because of interparticle forces. The particles tend to remain in fixed positions with respect to one another. The mass of flocs settles as a unit. At the top of the settling mass, a solid-liquid interface develops.

Compression Settling (Type IV)

Compression settling refers to the sedimentation in which the flocs are of high concentration and therefore a mass of flocs is formed. Settling can occur only by compression of the floc volume. Compression takes place because of the constant addition of the flocs and their weight.

When the system contains a high concentration of suspended solids, both hindered or zone settling and compression settling normally occur in addition to discrete and flocculant settlings. The settling phenomenon that occurs with high concentration of suspension is shown in Figure 21.

The clarifier area requirement for sludge thickening could be determined by adopting Talmadge and Fitch method [51], in which

$$
A = \frac{Q \cdot t_u}{D_o} \tag{25}
$$

in which A = area required for sludge thickening (m²); Q = rate of flow into the tank (m³/s); t_u = time to reach desired underflow concentration, C_u (s); and D_o = initial height of interface in column (m).

The value of t_u is found out with the help of curve (Figure 22) in which D_u corresponds to the depth at which the solids

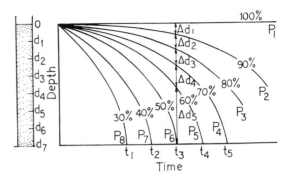

FIGURE 20. Flocculant particle settling.

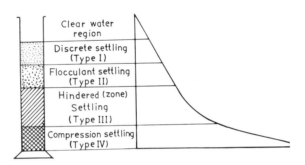

FIGURE 21. Four types of settling.

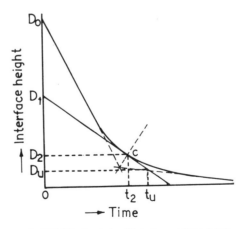

FIGURE 22. Analysis of interface settling curve.

are at the desired underflow concentration C_u. D_u is obtained from

$$D_u = \frac{C_o D_o}{C_u} \qquad (26)$$

in which C_o = initial concentration. The process of determining t_u in brief is: (1) Draw tangents to the flocculant settling and compression settling zones; (2) allow them to intersect; (3) bisect the angle thus formed; (4) let the bisector cut the curve at C; (5) draw a tangent at C; (6) draw a horizontal line at depth d_u; and (7) draw a vertical line from the point of intersection of the tangent at C and the horizontal line through D_u, to obtain the value of t_u.

The volume required for the sludge in the compression region can also be determined by settling tests represented by

$$D_t - D_\infty = (D_2 - D_\infty)\, e^{i(t - t_2)} \qquad (27)$$

in which D_t = sludge depth at time t; D_∞ = sludge depth after a long period (say, 24 h); D_2 = sludge depth at time t_2; and i = constant for a given suspension.

Flotation

Sometimes, solid particles are separated from liquids through flotation in which fine gas (usually air) bubbles are introduced in the liquids. These bubbles attach to the solid particles and cause them to rise to the surface because of buoyant force.

(1) Dissolved Air Flotation—In this system, air is dissolved in water under a pressure of several atmospheres, followed by the release of the pressure to the atmospheric pressure. In small systems, the entire flow can be pressurized by a pump with compressed air added at the pump suction. The entire flow is held in a retention tank under pressure for several minutes to allow time to dissolve the air. The solution is then admitted to a flotation tank through a pressure reducing valve. The air comes out of the solution in the form of minute bubbles throughout the entire volume of the liquid. For large volumes, a portion of the water is recycled, pressurized, and partially saturated with air. The recycled flow is mixed with the unpressurized main stream prior to admission to the flotation tank.

(2) Air Flotation—In this system, air bubbles are formed by introducing the gas phase into the liquid phase through a suitable device like revolving impeller or through diffusers. Short period aeration is not effective.

(3) Vacuum Flotation—Initially water is saturated with air and later partial vacuum is formed. The dissolved air, in the form of minute bubbles along with attached solid particles, rises to the water surface.

(4) Chemical Additives—Additives such as alluminium, ferric salts and activated silica can be used to bind the particulate matter together which can easily entrap the air bubbles.

Design Criteria for Sedimentation Tanks

The following criteria may be adopted for the design of sedimentation tanks [52]:

1. Depth of water in tank : 3–4.5 m
2. Detention time : 1.5–4 h (normally 3 h)
3. Number of units : 2 or more
4. Velocity of flow : 0.3 m/min
5. Surface loading : 1.5 m/h or 36 m³/m²/d (average); Maximum value 50 m³/m²/d
6. Weir loading : 100–200 m³/m/d
7. Extra capacity for storage of sludge : 25% (manually cleaned)
8. Floor slope : 1 in 12 or 8%, for mechanically scraped circular or square tank; 1% for mechanically scraped rectangular tanks (length/width ≥ 2); 1.2:1–2:1 (vertical:horizontal) for hoppers without mechanical scraper
9. Collecting weir should have : 90°, 50 mm deep V-notches at 150–300 mm apart; or circular orifices at 150–300 mm apart
10. Velocity of water at outlet conduit : not more than 0.4 m/s
11. Scraper velocity : One revolution in 45–80 min

FLOCCULATION

Flocculation is a slow mixing process in which destabilized colloidal particles are brought into intimate contact to promote their agglomeration. The flocculation process depends upon physical factors [53–55] such as time of flocculation, velocity gradient, concentration of particles per unit volume, and size and nature of particles; and also chemical factors such as alkalinity, pH, type of coagulant etc. The time of flocculation is normally between 30 to 40 min, 30 min being the most common. For rapid mix, the time varies between 30 to 60 s, 60 s being the most common. The range of velocity gradient for rapid mix is 700–1200 s⁻¹. For flocculation which follows rapid mix, the range of velocity gradient is 10–80 s⁻¹. The dimensionless product of time of flocculation t, and velocity gradient G should be in the range of 10^4 to 10^5. Concentration of suspension is the turbidity present in water. Higher the turbidity, more efficient is the process of flocculation.

Rapid Mix Units

Rapid mixing is the first step in water flocculation process. It is carried out in rapid mix units, in which the destabilization of the colloidal particles is achieved. For optimizing the rapid mix operation, Letterman et al. [56] suggested an empirical relationship

$$G\, T_{opt}\, C^{1.46} = 5.9 \times 10^6 \qquad (28)$$

in which G = velocity gradient (s⁻¹); T_{opt} = optimum rapid mix period (s); and C = concentration of alum dose (mg/L).

TABLE 11. Contact Time and Velocity Gradient in Rapid Mixing.

Contact time (s)	20	30	40	>40
Velocity Gradient (s⁻¹)	1,000	900	790	700

The following types of rapid mix units are used.

(1) *Mechanical Mixer*—This is the most commonly used type with a propeller impeller. The contact time in such a device reduces with increase in the velocity gradient. The approximate relationship is given in Table 11 [57].

(2) *Diffusers and Injection Devices.*—A rectangular grid of diffusers, consisting of a series of tubes with orifices to develop the turbulence for mixing in downstream wakes, is used [58]. The coagulant is fed through the orifices. A study indicated that the diffuser system was superior to the mechanically stirred flash mixers. Design criteria suggested by Kawamura [59] are: $G = 750-1000$ s⁻¹; dilution ratio at maximum alum dosage = 100:1; velocity at injection nozzles = 6–7.6 m/s; and mixing time = 1 s.

(3) *In-Line Blenders.*—This type of rapid mix can achieve instantaneous mixing; requires no head-loss computations and is low in cost since conventional rapid mix facility can be omitted. It can generate velocity gradient in the range 3,000–5,000 s⁻¹. A residence time, as small as 0.5 s is sufficient.

(4) *Hydraulic Mixing.*—In this type of mixing, the flow is measured at a Parshall flume, followed by a hydraulic jump. Velocity gradient is about 800 s⁻¹. The minimum residence time required is 2 s. It requires no mechanical equipment.

The mixing can also be achieved through a recirculating pump. The delivery side has a nozzle operating under 8 m of head.

Flocculators

A flocculator follows a rapid mix device. It helps in building up dense and large size flocs which in turn settle down quickly in the clarifier.

Von Smoluchowski showed in 1917 [60] that orthokinetic flocculation is characterized by the equation.

$$\frac{dN}{dt} = \frac{G}{6} n_i n_j (d_i + d_j)^3 \qquad (29)$$

in which dN/dt = rate of collision between *i*-fold and *j*-fold particles per unit time and volume; n_i, n_j = number of *i*-fold and *j*-fold particles, respectively, per unit volume; d_i, d_j = diameter of *i*-fold and *j*-fold particles, respectively; and G = uniform velocity gradient.

For estimating the mean temporal velocity gradient G,

Camp and Stein [60] suggested

$$G = \left(\frac{P}{\mu \cdot Vol}\right)^{1/2} \qquad (30)$$

in which P = total power input; μ = dynamic viscosity of the liquid; and Vol = volume of water in the flocculator.

Mixing Devices

Several types of flocculators, depending upon their mixing devices are available. Such flocculators and their mixing devices are now briefly described.

HYDRAULIC FLOCCULATORS

Several types of hydraulic flocculators are used in practice. They are:

(a) *Horizontal Flow Baffled Flocculator*—Figure 23 shows the plan of a typical horizontal flow baffled flocculator. This flocculator consists of several around-the-end baffles with in-between spacing not less than 0.45 m to permit cleaning. Clear distance between the end of each baffle and the wall is about 1.5 times the distance between the baffles, but never less than 0.6 m. Water depth is not less than 1.0 m and the water velocity is in the range 0.10–0.30 m/s. The detention time is between 15–20 min. This flocculator is well suited for very small treatment plants. It is easier to drain and clean. The head loss can be changed as per requirement by altering the number of baffles.

The velocity gradient for this flocculator can be calculated from [61]

$$G = \left[\frac{n^2 \varrho (1.44 + f)}{2\mu t} \cdot \frac{Q^2}{H^2 L^2}\right]^{1/2} \qquad (31)$$

in which G = velocity gradient (s⁻¹); H = depth of water in the channel (m); L = length of channel (m); Q = flow rate (m³/s); t = time of flocculation (s); μ = dynamic vis-

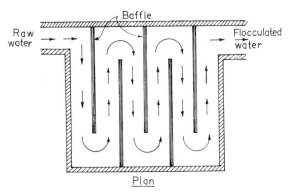

FIGURE 23. Horizontal flow baffled flocculator (plan).

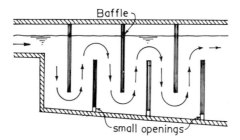

FIGURE 24. Vertical flow baffled flocculator (cross section).

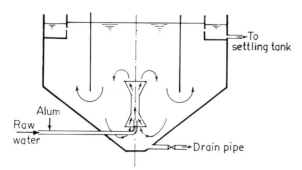

FIGURE 25. Jet flocculator.

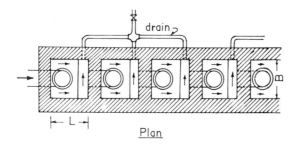

Plan

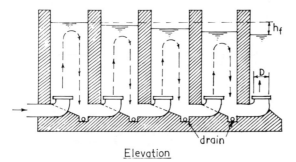

Elevation

FIGURE 26. Alabama-type flocculator.

cosity of water (kg/m·s); ϱ = density of water (kg/m³); f = coefficient of friction of baffles; and n = number of baffles.

The velocity gradient can be achieved in the range 10–100 s⁻¹.

(b) Vertical Flow Baffled Flocculator—Figure 24 shows the cross section of a typical vertical flow baffled flocculator. The distance between the baffles is not less than 0.45 m. Clear space between the upper edge of the baffles and the water surface or the lower edge of the baffles and the basin bottom is about 1.5 times the distance between the baffles. Water depth varies between 2 to 3 times the distance between the baffles and the water velocity is in the range 0.1–0.2 m/s. The detention time is between 10–20 min. This flocculator is mostly used for medium and large size treatment plants.

The velocity gradient can be calculated from

$$G = \left[\frac{n^2 \cdot \varrho \, (1.44 + f)}{2\mu t} \cdot \frac{Q^2}{W^2 L^2} \right]^{1/2} \qquad (32)$$

in which W = width of basin (m); and other notation being the same as that for Equation (31).

(c) Hydraulic Jet Action Flocculator—This is a less known type of hydraulic flocculators and is suitable for small treatment plants.

In one type of jet flocculators, shown in Figure 25, the co-agulant (alum) is injected in the raw water using a special orifice device at the inlet bottom of the tank. Water is then jetted into this hoppered tank. Heliocoidal-flow (also called tangential-flow or spiral-flow) type as well as staircase type flocculators can also be used [61].

(d) Alabama-Type Flocculator—An Alabama-type flocculator [49], shown in Figure 26 is a hydraulic flocculator having separate chambers in series, through which the water flows in two directions. Water flows from one chamber to another, entering each adjacent partition at the bottom end through outlets facing upwards. This flocculator was initially developed in Alabama State, U.S.A. (hence the name) and was later introduced in Latin America.

For effective flocculation in each chamber, the outlets are placed at a depth of about 2.50 m below the water level. The loss of head h_f is normally about 0.35–0.50 m for the entire unit. The range for the velocity gradient is 40–50 s⁻¹. The common design criteria are: Rated capacity per unit chamber: 25–50 L/s/m²; velocity at turns: 0.40–0.60 m/s; length of unit chamber, L: 0.75–1.50 m; width, B: 0.50–1.25 m; depth, h: 2.50–3.50 m; and detention time, t: 15–25 min. Practical guidelines for the design of an Alabama-type flocculator are given in Table 12.

PIPE FLOCCULATORS

The turbulence during the flow through a pipe can create velocity gradients leading to flocculation. The mean veloc-

TABLE 12. Practical Guidelines for Design of Alabama-Type Flocculator.

Flow Rate, Q (L/s)	Width, B (m)	Length, L (m)	Diameter, D (mm)	Unit Chamber Area (m²)	Unit Chamber Volume (m³)
10	0.60	0.60	150	0.35	1.1
20	0.60	0.75	250	0.45	1.3
30	0.70	0.85	300	0.6	1.8
40	0.80	1.00	350	0.8	2.4
50	0.90	1.10	350	1.0	3.0
60	1.00	1.20	400	1.2	3.6
70	1.05	1.35	450	1.4	4.2
80	1.15	1.40	450	1.6	4.8
90	1.20	1.50	500	1.8	5.4
100	1.25	1.60	500	2.0	6.0

ity gradient is calculated from

$$G = \left[\frac{\varrho g\, Q\, h_f}{\mu\, Vol} \right]^{1/2} \qquad (33)$$

in which Q = flow rate (m³/s); Vol = volume of pipe of length L (m³); and h_f = headloss in pipe of length L (= $fL\, V^2/2gD$)

PADDLE FLOCCULATORS

Paddle flocculators are widely used in practice. The design criteria [52] are: —depth of tank: 3–4.5 m; detention time, t: 15–30 min, normally 30 min; velocity of flow: 0.2–0.8 m/s, normally 0.4 m/s; total area of paddles: 10–25% of the cross-sectional area of the tank; range of peripherial velocity of blades: 0.2–0.6 m/s, 0.3–0.4 m/s is recommended; range of velocity gradient, G: 10–75 s^{-1}; range of dimensionless factor Gt: 10^4–10^5; and power consumption: 0.06 to 0.08 kW/m³/min (0.09–0.12 kW/MLd).

For paddle flocculator, the velocity gradient is given by

$$G = \left[\frac{C_D \cdot A_p \varrho\, (V-v)^3}{2\mu\, Vol} \right]^{1/2} \qquad (34)$$

in which C_D = coefficient of drag; A_p = area of paddle (m²); Vol = volume of water in the flocculator (m³); V = velocity of the tip of paddle (m/s); v = velocity of the water adjacent to the tip of paddle (m/s).

The value of C_D was assumed to be constant, but recent experiments [62] have proved that it varies inversely with the value of G.

The optimum value of G can be calculated [63] from

$$G_{opt}^{2.8}\, t\, C = 44 \times 10^5 \qquad (35)$$

in which G_{opt} = optimum velocity gradient s^{-1}; t = time of flocculation (min); and C = alum concentration (mg/L).

The shape of the container also affects the process of flocculation [64]. For the same volume and height of water in the containers of several shapes such as circular, triangular, square, pentagonal and hexagonal, it was observed that the pentagonal shape gave the best performance [64].

Introduction of stators in the flocculator helps to improve the performance of flocculation [65].

Polasek [66] introduced 'pendulum type agitator' to obtain pendulum flocculator, Figure 27, as a variety of paddle flocculator. In this flocculator, the motion of the blades is through small arcs, increasing with distance down the pendulum frame from the pendulum support. The blades are distributed logarithmically along the pendulum to maintain a uniform intensity of agitation over the entire depth.

PEBBLE BED FLOCCULATOR

The pebble bed flocculator [67–69] contains pebbles of size ranging from 1 mm to 50 mm. Smaller the size of the pebbles, better is the efficiency, but faster is the build up of the headloss; and vice versa. The depth of the flocculator is between 0.3 to 1.0 m. The velocity gradient is given by

$$G = \left[\frac{\varrho\, g\, Q\, h_f}{\alpha\, \mu\, A\, L} \right]^{1/2} \qquad (36)$$

in which h_f = head loss across the bed (m); α = porosity of bed; A = area of flocculator (m²); and L = length of the bed (m).

The main advantage of the pebble bed flocculator is that it requires no mechanical devices and electrical power. The operation and maintenance cost is also low and therefore it is useful for rural areas in developing countries. However,

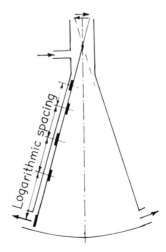

FIGURE 27. Pendulum flocculator.

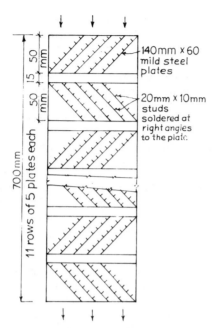

FIGURE 28. Surface contact flocculator.

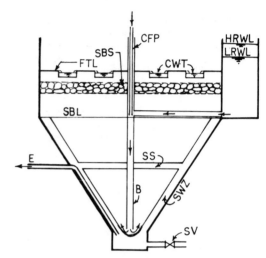

FIGURE 30. Modified sludge blanket clarifier: HRWL—high raw water level; LWRL—low raw water level; FTL—full tank level; SBL—sludge blanket level; CFP—chemical feed pipe; CWT—clarified water troughs; SBS—sludge blanket stabilizer; SS—slurry slit; SWZ—sludge withdrawal zone; B—raw water; E—sludge withdrawal pipe; SV—scour valve.

the drawback of this flocculator is that there is gradual build up of the head loss across the pebble bed.

FLUIDIZED BED FLOCCULATOR

In a fluidized bed flocculator [70,71], the sand bed is in the fluidized form. Even a 10% expansion of the sand bed is enough to create the required turbulence without choking the media. The sand size is between 0.2 to 0.6 mm. The

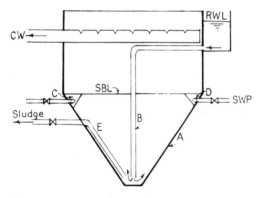

FIGURE 29. Candy type upflow sludge blanket clarifier: RWL—raw water level; SBL—sludge blanket level; SWP—slurry withdrawal pipe; CW—clarified water; A—upflow basin; B—raw water inlet pipe; C—slurry pocket; D—perforated slurry collection pipe; E—sediment withdrawal pipe.

flow of water is naturally upwards. This flocculator also does not require any mechanical equipment or electrical power. Further, there is no build up of the head loss across the bed.

PNEUMATIC FLOCCULATOR

In a pneumatic flocculator [66], air bubbles are allowed to rise through a suspension. This creates velocity gradient useful for flocculation. The velocity gradient can be calculated from

$$G = 0.236 \frac{g \varrho D}{\mu} \left(\frac{Vol_A}{Vol}\right)^{1/2} \qquad (37)$$

in which D = diameter of air bubbles (m); and Vol_A/Vol = volume of air supplied per unit water volume.

The flocculator needs air compressor and the problem of clogging of diffusor is quite common. It is less efficient than the paddle flocculator and therefore less used in practice.

SURFACE CONTACT FLOCCULATOR

The surface contact flocculator was studied experimentally in India [72], to overcome the inherent problem of sludge choking, which increases the head loss over time, in pebble bed flocculators.

The surface contact flocculators consist of studded plates, placed in a zigzag form along the direction of flow. An experimental flocculator [71], shown in Figure 28, comprised of 55 mild steel plates, 140 mm × 60 mm in size, arranged

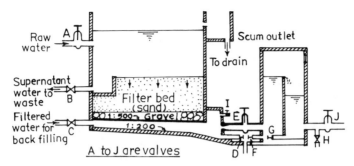

FIGURE 31. Diagramatic section of a slow sand filter.

in 11 rows of 5 plates each. These plates were fixed at 45° to a base plate in zigzag fashion. The flocculator was tested in a continuous downflow system with velocity of flow ranging from 5 m/h to 25 m/h, and turbidity ranging from 50 to 1600 NTU (Nephalometer turbidity unit, a measure of turbidity).

INLINE FLOCCULATOR

An inline static flocculator, or an inline static mixer [73] is a relatively recent device. It is housed in a gravity main and is static. The head loss in an inline flocculator is comparatively less and the maintenance cost is also almost negligible. Only occasional flushing is necessary since deposition of some flocs takes place. The capitalized cost of a typical inline flocculator is one-third of the capitalized cost of the conventional mixing impellers. Laboratory experiments [73] showed that twisted aluminium plates as static mixer in the pipeline gave better performance compared to the semicircular plates or the inclined plane plates.

SLUDGE BLANKET CLARIFIER

A sludge blanket clarifier includes both—flocculation and clarification. Most of these plants are of Candy type as shown in Figure 29. The different problems involved in the conventional Candy type clarifier are in connection with the dosing and mixing; desludging; and the stability of the blanket. An attempt was made in India [74] to overcome these inherent defects, through a modified sludge blanket clarifier, shown in Figure 30.

The velocity gradient of the sludge blanket can be calculated from [75]

$$G = \left[\frac{g \varrho}{\mu} (S_s - 1) (1 - \alpha) h \middle/ \left(\frac{C}{Q} \right) \right]^{1/2} \quad (38)$$

in which S_s = specific gravity of flocs; α = porosity of blanket, h = depth of blanket (m); C = capacity of clarifier (m³); and Q = rate of flow (m³/s).

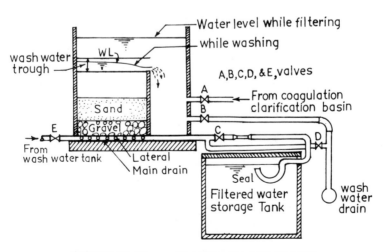

FIGURE 32. Diagramatic section of a rapid sand filter.

TAPERED VELOCITY GRADIENT FLOCCULATOR

In a tapered velocity gradient flocculator, the water is initially subjected to a high velocity gradient and finally to a low velocity gradient, thus generating dense, large size and tough flocs which in turn settle more quickly.

Recent studies [76,77] indicated that the efficiency of a tapered velocity gradient flocculator increases when (a) there is increase in the range of the velocity gradient, the mean value of G remaining the same; (b) there is gradual decrease in the velocity gradient and no sudden decrease along the direction of flow; and (c) dual tapering—tapering

in velocity gradient as well as time of flocculation—is achieved, i.e., highest velocity gradient for the shortest time, followed by little lower value of G, velocity gradient, for a little more time and so on, so that in the end the values of the velocity gradient is the least with the maximum time of flocculation.

FILTRATION

Filtration is a physical, chemical and (sometimes) biological process that removes suspended impurities from water when it passes through porous media. Two general types of

TABLE 13. Comparative Study of Conventional Slow and Rapid Sand Filters.

Characteristic	Slow Sand Filters	Rapid Sand Filters
Rate of filtration	0.1–0.4 m/h (0.2 m/h)[a]	4–20 m/h (5 m/h)
Size of bed	Large, 2,000 m²	Small, 40–400 m²
Depth of bed	0.3 m gravel, 0.9–1.1 m sand, usually reduced to not less than 0.5–0.8 m by scraping	0.3–0.45 m of gravel, 0.6–0.9 m of sand, not reduced by washing
Size of sand	Effective size: 0.20–0.30 mm Uniformity coefficient: 2–3.5 (2.5)	Effective size: ≥ 0.45 mm Uniformity coefficient: ≤ 1.5, depending upon underdrainage system
Grain size distribution of sand in filter	Unstratified	Stratified with smallest or lightest grains at top and coarsest or heaviest at bottom.
Underdrainage system	1. Split tile laterals laid in coarse stone and discharging into tile or concrete main drains 2. Perforated pipe laterals discharging into pipe mains	1. Perforated pipe laterals discharging into pipe mains 2. False floor type, with orifices 3. Many others, generally proprietary
Loss of head	60 mm initial to 1.2 m final	0.3 m initial to 2.4–2.75 m final
Length of run between cleanings	20–60 d (30 d)	12–72 h (24 h)
Penetration of suspended matter	Superficial	Deep, particularly with dual or mixed media
Method of cleaning	Scraping of surface layer of sand and washing and storing cleaned sand for periodic resanding of bed	Dislodging and removing suspended matter by upward flow or backwashing which fluidizes the bed. Possible use of auxiliary scour system.
Amount of water used in cleaning sand	0.2–0.6% of water filtered	1–6% (4%) of water filtered
Preparatory treatment of water	Generally none when raw water turbidity < 50 NTU	Coagulation, flocculation and sedimentation
Supplementary treatment of water	Chlorination	Chlorination
Cost of construction	Relatively high	Relatively low
Cost of operation	Relatively low where sand is cleaned in place	Relatively high
Depreciation cost	Relatively low	Relatively high

[a]Common value shown within parentheses.

filters—slow sand filter and rapid sand filter—are commonly used for water treatment.

A slow sand filter (Figure 31) consists of a layer of ungraded fine sand. Water is filtered through it at a low rate. Due to the low rate of filtration, an active layer of microorganisms, called schmutzdecke, is formed on the top of the sand bed. This layer provides biological treatment and is particularly effective in the removal of microorganisms, including pathogens, from water. The filter is periodically cleaned, when predetermined head loss is built up, by scraping a thin layer of dirty sand from the surface. After several scrapings, the sand is washed and returned to the filter.

A rapid sand filter (Figure 32) on the other hand, consists of a layer of graded sand. Water is filtered through it at a much higher rate. Due to this, the space requirement of a rapid sand filter is about 2–4% of that of a slow sand filter of the same capacity. However, because of fluctuating and high raw-water turbidity a rapid sand filter requires pretreatment of water, while a slow sand filter generally does not require it because of constant, low raw-water turbidity. The rapid sand filter is cleaned by backwashing with water.

Rapid sand filters are extensively used in developed countries, although some of them, (e.g., England and Switzerland) still use slow sand filters. Simple rapid sand filter designs employ manual controls. Modern rapid sand filters are usually costly and complex to construct, operate and maintain. They are often fully automated, especially in developed countries, to reduce labour costs. The automatic operation can shut off a filter at a predetermined head-loss, backwash the filter, and put it back into service; although provision is made to manually override it.

A comparative study regarding the general features of construction and operation of slow and rapid sand filters is given in Table 13. As rapid sand filters are more popular, they are described here first.

Rapid Sand Filters

BASIC COMPONENTS AND OPERATIONS [78]

(a) Filter Media—Sand has been used traditionally as filter medium because it is easily available, inert, hard, durable, almost spheroid in shape and much heavier (sp. gr. 2.65) than water. The effective size D_{10} (the size corresponding to 10% passing through) of sand is in the range of 0.45–0.55 mm, 0.45 mm being the most common. The uniformity coefficient ($D_{60}:D_{10}$) is 1.5 or less. The depth of sand should not be less than 0.6 m and is generally not more than 0.9 m. Average depth of sand is 0.75 m.

(b) Gravel Bed—The gravel surrounds the underdrain system, supports the sand bed above it and also helps for uniform distribution of backwash water. The gravel should preferably be rounded pebbles and not crushed stone. The ranges in sizes and corresponding depths of gravel layers, as suggested by Cox [22] are given in Table 14. Gravel bed is

TABLE 14. Sizes and Depths of Gravel Layers for Rapid Sand Filters.

Range in Size (mm)	Range in Depth (mm)
65–38	130–200
38–20	80–130
20–12	80–130
12–5	50–80
5–2	50–80

not needed when the filter is equipped with porous filter plates which directly support the sand bed.

(c) Underdrain System—The underdrain system collects the filtrate and carries it to the clear water storage; and also distributes the wash water uniformly to the filter bed during the back-wash operations. It commonly comprises of perforated pipe laterals and a centrally located manifold and is therefore known as "lateral and manifold" system.

For uniform backwashing of the filter, the discharge through the orifices should almost be the same, whatever their distance from the manifold. For this purpose, (1) the allowable friction loss h_f in a lateral with n orifices must equal $(1 - m^2)$ times the driving head h_d on the first orifice reached by the washwater if the head on the last or the nth orifice is not to exceed m^2 times the driving head, where m is almost equal to unity, say 0.9; and (2) in terms of the lost head and neglecting the side effects, the friction in the perforated lateral of unvarying size must be numerically equal to the friction exerted by the full incoming flow in passing through one-third the length of the lateral. For optimal hydraulic system, 25 per cent of the overall head-loss should occur in delivering the required flows to the orifices, and the remaining 75 percent should be expended in driving the wash water, in succession, through the orifices, gravel and fluidized bed.

The thumb rule for the design of the underdrain system for filters washed at the rate of 0.15–0.90 m/min is [22] : (1) ratio of the area of orifice to area of bed served—(1.5–5.0) $\times$ 10^{-3}:1; (2) ratio of the area of laterals to area of orifices served—(2–4):1; (3) ratio of area of manifold to area of laterals served:—(1.5–3):1; (4) diameter of orifices: 6–18 mm; (5) spacing of orifices: 75 mm for 6 mm diameter—300 mm for 18 mm diameter; (6) spacing of laterals: closely approximating spacing of orifices; (7) length of laterals on each side of manifold—not more than 60 times their diameter; (8) the orifices are oriented so that wash water is directed downward at an angle of 30°–60° with vertical. Erosion or corrosion of the metal around these holes may be minimized by lining the holes with a brass or bronze bushing.

(d) Filter Floors—Filter floors, also known as "false bottoms" or "false floors" may be used to replace manifold and

lateral system. They (1) support the filter sand bed possibly without stone and gravel in transitional layer below the filter bed proper; and (2) create a single box-like waterway beneath the filter to collect the filtered water or to provide the wash water. The floor is perforated by short tubes or orifices for even distribution of wash water.

(e) Filter Back Wash—There are two methods of filter back wash: (1) Back wash with water alone. This method is also known as hard wash; (2) Air-water wash system in which air agitation followed by water wash is adopted.

In hard wash, the rate of back wash for 50% expansion is about 72 m/h. This method is inefficient and more costly compared to the second system and is therefore becoming obsolete. In the second system, free air at the rate of 0.9–1.5 m³/m² of filter area/min is forced through the underdrains for a period of 5 min for agitation of the sand. Wash water is then introduced through the same underdrains at the rate of 19–29 m³/m² of filter area/h.

(f) Auxiliary Scour—Three methods for auxiliary scour are available. (1) "Surface wash system" which is used in addition to the conventional system of back wash. The primary pipe grid system consists of horizontal pipes suspended on wash water troughs. The secondary vertical pipes are connected to horizontal headers. The vertical pipes are extended to within about 100 mm of the surface of the unfluidized bed. The bottom end of the vertical pipes are 0.6–0.75 m c/c and work under a head of 7–20 m of water. (2) Use of "filter bed agitator" which consists of a horizontal revolving pipe supported at its centre. Several nozzles are attached to this pipe. Such units, either one or more, are located at a height of 25 mm above the unexpanded sand surface. The nozzles face downward at 30° to the horizontal. The resulting jet action causes the arm to revolve up to 20 rpm depending upon the length of the pipe. (3) Use of "mechanical rakes" which are of revolving type and are forced through the sand bed. The normal speed of the rake is 10–12 rpm.

(g) Wash Water Troughs—Wash water troughs are located above the 50 per cent expanded sand level to collect the wash water and discharge it into a drain pipe. The maximum horizontal travel of wash water should not exceed 0.9 m and therefore the spacing between the troughs should not exceed

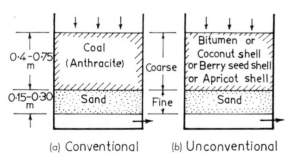

(a) Conventional   (b) Unconventional

FIGURE 33. Dual media filters.

1.8 m. The free fall discharge through the gutter is given by $Q = 1.375 \, b \, h_0^{3/2}$ in which Q = rate of discharge (m³/s); b = width of the gutter (m); and h_0 = initial depth of water (m).

(h) Wash Water Tanks—The capacity of a wash water tank should be 1–6 percent of the filtered water and should be sufficient for at least 10 min wash of one filter or 5–6 min wash of two filters without refilling. The bottom of the tank is normally located about 12 m above the underdrain system. The wash water head should be around 4.5 m just outside the underdrain system.

(i) Filter Appurtenances [75]—Filter appurtenances include manually, hydraulically, pneumatically, or electrically operated gates on the influent, effluent, drain and wash water lines; measuring and rate-control devices; head-loss gages; and wash water controllers and indicators.

Among these appurtenances, the automatic effluent-flow regulator is quite important. The valve mechanism automatically opens or closes to keep the discharge rate constant. Pressure differentials between the venturi throat and outlet are translated into valve movements by a balancing diaphragm or a piston.

Problems associated with rapid sand filters are: (1) air binding; (2) mud accumulation; (3) sand incrustation; and (4) negative head. Remedial measures are described in any standard book, e.g., [75].

CLASSIFICATION AND TYPES

Rapid sand filters can be classified in several ways. They may be classified according to (a) the type of filter media employed; (b) the type of filter rate control system employed; (c) the direction of flow through the bed; and (d) whether they operate under gravity (free-surface) or pressure, i.e., whether they operate under atmospheric or more than atmospheric pressure.

(a) According to Filter Media—In the conventional rapid sand filters, the pore size of the filter bed is the least in the topmost layers. Therefore, the floc particles removed during the filtration process concentrate in these layers and most of the lower layers remain unused. This led to the introduction of dual and multimedia filters, in which lighter media of larger size occupy the upper layers. Thus, flocs can penetrate deeper and lower layers of the filter become effective.

(1) *Dual Media Filters*—A dual media filter (Figure 33(a)) is generally composed of a coarse coal upper layer (sp.gr.: 1.45–1.55; effective size: 1.0–1.6 mm; uniformity coefficient: 1.3–1.7; and depth: 0.40–0.75 m) on top of a lower sand layer (sp.gr.: 2.65; effective size: 0.45–0.80 mm; uniformity coefficient: 1.3–1.7; and depth: 0.15–0.30 m). As the specific gravities of the two materials are different, the two layers retain their respective positions after backwashing, although some intermixing of the layers occur at the interface.

The advantages of the dual-media filters over the conventional filters are: (a) higher filtration rates (10–15 m/h) and therefore the reduction in the total filter area and cost for a given design capacity; (b) longer filter run at any given loading; and (c) feasibility of increasing, at low cost, the capacity of the existing filters by their conversion to dual media filters.

Anthracite is the most-accepted coarse medium in U.S.A., U.K., and the other developed countries. Other countries have used indigenous coals and other unconventional filter media (Figure 33(b)). Brazil has used bituminous coal [61]. In India, high-grade bituminous coal [79], crushed coconut shell [80], crushed berry seeds [81], and kernels of apricots [82] were found suitable as coarse filter media. A dual-cum-mixed media filter (Figure 34), composed of a coarse upper layer of crushed coconut shell (size: 1–2 mm) and mixed bottom layer of 30 percent sand (size: 0.5–0.85 mm) and 70 percent boiler clinker (size: 1–2 mm, sp.gr.: 1.9) was found to develop half the head loss compared to the dual media filter [83].

(2) Multi Media Filters—In a multimedia filter (Figure 35) five media were used [84]. They were: polystyrene (sp.gr.: 1.04); anthracite (sp.gr.: 1.4); crushed flint sand (sp.gr.: 2.65); garnet (sp.gr.: 3.8); and fused alumina (sp.gr.: 3.96). This filter is still not used in the field since it is difficult to procure the media in the required quantity.

(b) According to Rate of Filtration—(1) Constant Rate Filters—A constant rate filter gives filtered water at a constant rate. Control measures in the form of throttling valves tied to flow-measuring, differential pressure, or water level devices are introduced to control the rate at the inlet or the outlet. The simplest method of flow control is to distribute the raw water ("flow splitting") equally over the filter units. The water level rises during filtration to compensate for the head-loss buildup in the filter bed. When water level reaches the maximum permissible level above the filter bed, the filter unit is taken out for backwashing. When it is desirable to maintain a constant water surface in the filters, a rate controller must be provided at the outlet. A simple control device consists of a float connected by a small cable running over sheaves to a butterfly valve on the filter outlet pipe.

(2) Declining Rate Filters—When no rate controllers are used and the filter inlet is placed below the minimum water level, during filtration as the headloss increases the filtration rate decreases. However, the breakthrough of particles is less likely at the end of the run compared to the former type.

(c) According to Direction of Flow—(1) Downflow Filters—The conventional rapid sand filter is a classical example of a downflow filter.

(2) Upflow Filters—In an upflow filter (Figure 36) water flows upwards, under pressure, through coarse-to-fine media. As the entire filter depth (1.8–3.0 m) is utilized, the head loss is low and the filter runs are longer. Further, the rate controller and negative head are absent. As the filter acts as a flocculator at the inlet zone, separate flocculator can be eliminated when raw-water turbidity is less than 200 NTU. However, due to fluidization of the top fine layer, loss

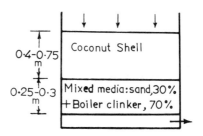

FIGURE 34. Dual-cum-mixed media filter.

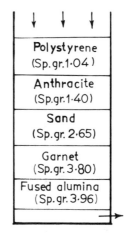

FIGURE 35. Multimedia filter.

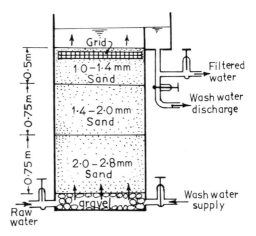

FIGURE 36. Single medium upflow filter.

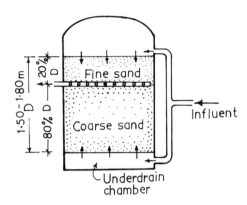

FIGURE 37. Single medium biflow filter.

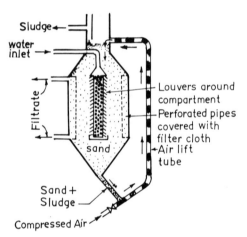

FIGURE 38. Radial flow filter.

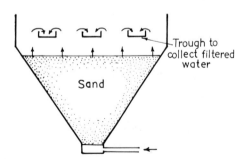

FIGURE 39. Decline velocity filter.

of sand and the consequent deterioration of the filtrate quality may occur. To prevent this, a grid consisting of parallel vertical plates is placed within the bed and slightly below the top of the media. The grid spacing is 100–150 times the size of the fine sand. Formation of the inverted arches of sand minimizes the fluidization of sand. The rate of filtration can be 15–30 m³/m²/h.

(3) Biflow Filters—The problem of the fluidization of the upper layer in an upflow filter is solved in a biflow filter (Figure 37). Here the effluent-collecting pipe is placed in the upper layers of the sand bed and the filtration takes place simultaneously—from above, through the top layer of fine sand; and from below, through the bulk of the media. The effluent pipe is located at a depth of about 20 percent of the total depth of the filter. The depth of sand is 1.50–1.80 m. The total filtration rate can be as high as 18 m³/m²/h, the downward rate being half to two-thirds the upward rate. The back-wash rate is 54 m³/m²/h for a period of 6 min.

(4) Radial Flow Filters—In a radial flow filter, also known as "Simater" continuous sand filter [85], water passes radially through a gradually descending column of sand (Figure 38). The impurities are removed continuously and there are no stoppages for backwashing. Chemically treated water enters into a central hollow column, permeates radially through sand, reaches the peripheral ducts, and flows out of the shell. The dirty sand is continuously drawn from the bottom, washed and lifted with air, and placed again from the top. The sludge separated from the sand is withdrawn from the top.

(5) Decline Velocity Filters—The decline velocity filter (Figure 39) is placed in a trapezoidal container and the direction of flow is from the bottom to the top. As the water travels upwards, the velocity of water decreases giving better opportunity for the suspended particles to get removed in the filter bed. Thus, this filter combines the features of upflow and radial flow filters [86].

(6) Multi Inlet Multi Outlet Filters—Several inlets and outlets are provided in a multi inlet multi outlet filter (Figure 40) to give parallel filter units, one over the other. Uniform sand with uniformity coefficient 1.2 and 0.45–0.60 mm size is used. The effective depth of each filter layer is about 0.2–0.3 m. The location of the outlet pipe corresponding to the inlet pipe is above the inlet pipe and therefore negative head is not developed. Mud balls are also absent. The effective filtration rate is only 2–3 m³/m²/h but the yield is higher per unit occupied area [87].

(d) According to Pressure—(1) Gravity Filters—All downflow filters described earlier act under gravity only and therefore they are gravity filters.

(2) Pressure Filters—Pressure filters work on the same principle as the gravity type rapid sand filters, but the water is passed through the filter under pressure. The pressure filter (Figure 41) is housed in a cylindrical tank, usually made of steel or cast iron. The tank axis may be either vertical or horizontal. The pressure filters are compact, and can

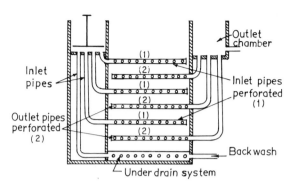

FIGURE 40. Multi inlet multi outlet filter.

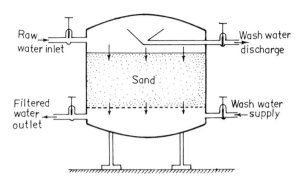

FIGURE 41. Pressure filter (horizontal type).

be prefabricated and moved to site. Coagulation and floc-culation is normally achieved in the top portion of the filter. Economy is further possible by avoiding double pumping. These filters are most suitable for small townships and swimming pools. However, the effectiveness of the back wash cannot be directly observed; and it is also difficult to inspect, clean and replace the sand, gravel and the under-drains.

OTHER TYPES

(a) Direct Filters—Direct filters work on the principle of direct filtration, which involves a series of unit processes in-corporating coagulation, flocculation and filtration without the use of sedimentation. Direct filtration can be used in a variety of applications [88]. Direct filtration reduces alum usage and sludge production. In a direct filtration at a rate upto 4 mm/s on reverse graded, i.e., coarse-to-fine mixed-media filter, satisfactory results were obtained with only 5–6 mg/L of alum and 0.5 mg/L of a polymer [89]. To obtain the same results, conventional treatment required 17 mg/L of alum, and also sedimentation.

Direct filtration process can consistently produce high quality filtered water with significantly low annual operating cost. At a filtration rate of 1.4–4 mm/s, direct filtration can reduce turbidity of 1-100NTU to 0.1–0.3 NTU, comparable

to that produced by conventional treatment. Process control is also somewhat easier than that with conventional treat-ment.

Direct filtration (DF) and conventional filtration (CF) were compared for the treatment of lake Erie water [90]. The results were: (1) average chemical usage: CF − alum: 15 mg/L; DF − alum: 5 mg/L, polymer: 0.15 mg/L; (2) wash water usage: CF − 1%, DF − 2.5%.

(b) Diatomaceous Filters—A diatomaceous filter (not strictly a rapid sand filter but comparable to it in perform-ance) has a medium consisting of diatomaceous earths which are skeletons of diatoms mined from deposits laid down in seas. The filtering medium is a layer of diatoma-ceous earth built up on a porous septum by recirculating a slurry of diatomaceous earth until a firm layer is formed on the septum. The precoat thus formed is used for straining the turbidity in water. For this, the diatomaceous earth is ap-plied at 0.5–2.5 kg/m² of the septum. Sometimes, when the turbidity is very high, diatomaceous earth is added to the in-coming water as body feed. Filtration rate is 7.2–18 m³/m²/h. Diatomaceious filters are of two types: (1) pressure type; and (2) vacuum type (Figure 42).

(c) Valveless, Self-Washing Filters with Automatic Siphon [91]—Valveless, self-washing filters with automatic siphon (Figure 43) work independently and automatically, without

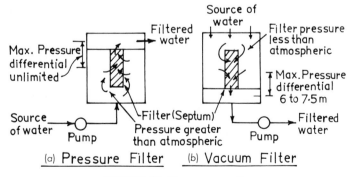

FIGURE 42. Diatomaceous filters.

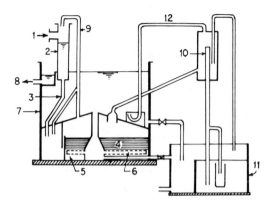

FIGURE 43. Valveless self washing filter: (1) raw water inlet; (2) header tank; (3) raw water inlet pipe; (4) filter layer; (5) supporting layer; (6) filtrate pipe; (7) filtrate storage tank; (8) filtrate outlet; (9) vent pipe; (10) wash water siphon; (11) wash water return; (12) unpriming pipe.

the use of electricity or any auxillary fluid systems, during both the filtration and washing cycles. They have been supplied by DEGREMONT for more than 50 years.

The water in the buffer tank, after filtration, rises from the bottom of the filter to the filtrate storage tank at the top. When the storage tank is full, the filtrate is supplied through the outlet. When the filter bed is clogged, water level rises in the buffer tank and also in the upstream arm of the siphon. When the maximum design head-loss is reached the siphon is primed, and the water in the filtered water storage tank flows back, thus back-washing the filter. The capacity of the storage tank is designed to ensure adequate washing of the filter.

The back-washing takes place automatically at a predetermined fixed head loss and therefore the filter bed is never excessively clogged. Further, compressed air and electrical power are not required. However, the filtrate supply is not available through the period during which the storage tank gets refilled. It is suitable for waters with low or medium suspended solids content. The maximum filtration rate is about 10 m³/m²/h.

Slow Sand Filters

The main features and some design criteria of a slow sand filter are already given in Table 13. Additional design criteria are [92]: (1) Number of filter units, $N = 0.25\sqrt{Q}$, in which N = number of filter units (never less than 2); and Q = flow rate (m³/h); (2) gravel bed: four layers, each 60–100 mm thick with total bed thickness not less than 300 mm; normal size of gravel in layers (top to bottom) are 1–2 mm, 3–6 mm, 9–18 mm and 27–54 mm, respectively; (3) Water depth: 1–1.5 m; (4) effluent control: through a valve that is adjusted as the resistance increases.

The maximum turbidity that a slow sand filter can handle is 100–200 NTU for a few days, 50 NTU is the maximum that should be permitted for longer periods. Therefore, surface waters can be treated by slow sand filters when the turbidity is brought down to this level through flocculation and sedimentation or in addition to this, through "roughing filters." Turbidity can also be lowered by a presedimentation tank.

Roughing Filters

Roughing filters [49] are used for pretreatment rather than for treatment itself. In the case of high turbidity, they precede slow sand filters in general, or sometimes even rapid sand filters.

In an upflow filter (Figure 36), three layers having grain sizes of 15–10 mm, 10–7 mm and 7–4 mm, from bottom upward and with simple underdrain system would serve as roughing filter. Such a filter would have large pores and hence gradual clogging. Filtration rate may be upto 20 m/h. However, the time for back wash is relatively longer, i.e., about 20–30 min.

A roughing filter can also be horizontal (Figure 44), having a depth of 1–2 m. The filter is subdivided into three zones, each about 5 m long and composed of gravel with sizes 30–20 mm, 20–15 mm, and 15–10 mm, respectively. The horizontal water flow rate, computed over the full depth, would be 0.5–1.0 m/h, giving a surface loading of only 0.03–0.1 m/h. The clogging is very gradual.

Coconut fibres were used as filter material in an experimental filter unit, similar to a sand filter. The filter depth was 0.3–0.5 m with supernatant water depth of 1 m. The filter can run for weeks when operated at 0.5–1 m/h.

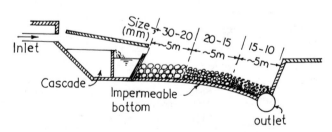

FIGURE 44. Horizontal roughing filter.

DISINFECTION

Unit processes like flocculation, sedimentation and filtration render the water chemically and aesthetically acceptable with some reduction in the bacterial content also. However, they do not provide a 100 percent safe water and therefore it is necessary to disinfect the water to destroy, or at least to inactivate, all the disease-producing organisms.

Factors Influencing Disinfection

TIME OF CONTACT

According to Chick's law, the number of organisms destroyed in unit time is proportional to the number of organisms remaining. Thus,

$$\ln \frac{N_t}{N_o} = -k^{+1/2} t^m \qquad (39)$$

in which N_o = initial number of organisms; N_t = number of organisms remaining after time t; k = rate constant; t = time; and m = constant ($>1, <1, = 1$ when rate of kill rises in time, falls in time, remains constant, respectively.)

CONCENTRATION OF DISINFECTANT

The concentration of disinfectant is given by

$$C^n t_p = \text{constant} \qquad (40)$$

in which C = concentration of disinfectant; t_p = time required to effect a constant percentage (usually 99%) kill of organisms; and n = a coefficient of dilution, or a measure of order of reaction ($= 0.86$ for most of the organisms).

CONCENTRATION OF ORGANISMS

The concentration of organisms can be expressed as

$$C^q N_p = \text{constant} \qquad (41)$$

in which C = concentration of disinfectant; N_p = concentration of organisms reduced by a given percentage (usually 99%) in a given time; and q = coefficient of disinfectant strength.

TEMPERATURE

The effect of temperature is given by

$$\log \frac{t_1}{t_2} = \frac{E(T_2 - T_1)}{4.56 \, T_1 T_2} \qquad (42)$$

in which t_1, t_2 = times required for equal percentage of kill at fixed concentrations of disinfectant; T_1, T_2 = two absolute temperatures for which the rates are to be compared; and E = activation energy (calories).

Types of Disinfection

PHYSICAL DISINFECTION

Physical disinfection can be obtained by the following methods.

(a) By Heat — Boiling is a safe and time-honoured practice that destroys pathogenic micro-organisms. It is effective as a household treatment but impractical for community water supplies.

(b) By Light — Sunlight is a natural disinfectant, principally as a desiccant. Irradiation by ultraviolet light intensifies disinfection. It involves the exposition of a film of water, about 120 mm thick, to one or several quartz mercury vapour arc lamps emitting invisible light. In this method, the exposure is for short periods, no foreign matter is introduced, no taste and odour are produced; and overexposure does not result in any harmful effects. However, its effectiveness is significantly reduced when the water is turbid or contains nitrates, sulphates and ferrous ions. Further, it produces no residual that would protect the water against new contamination. This method is used in several developed countries.

(c) By Ultrasonic Waves — Ultrasonic waves [38] at 20–400 kHz frequency provide complete sterilization of water at retention time of 60 min, and a very high percent reduction at retention time of even 2 s. However, it is expensive but combination of short time sonation and ultraviolet light may be economical.

CHEMICAL DISINFECTION

Chemical disinfection can be attained by using several chemicals.

(a) Oxidizing Chemicals — Oxidizing chemicals are: (1) the halogens — chlorine, bromine and iodine; (2) ozone; and (3) other oxidants such as potassium permanganate and hydrogen peroxide.

Among the halogens, gaseous chlorine and several chlorine compounds are economically the most useful. Chlorine dioxide (ClO_2) produces no taste and odour, but has to be used instantly as it is unstable. ClO_2 is superior to an equivalent concentration of free Cl_2 when dealing with alkaline waters (pH $<$ 7.5). Bromine (Br_2) and iodine (I_2) have been used on a very limited scale for the disinfection of swimming pool waters. Iodine tablets are used at individual level. Ozone is good but relatively expensive, and requires instant use but has no taste and odour problems. Potassium permanganate is relatively less efficient, more expensive and imparts taste, odour and colour to water. Hydrogen peroxide is a strong oxidant but a poor disinfectant.

(b) Metal Ions — Silver ions are bactericidal but silver is very costly. Copper ions are strongly algicidal but only weakly bactericidal. However, copper is relatively cheaper.

(c) Alkalis and Acids — Pathogenic bacteria do not survive for long in highly alkaline (pH $>$ 11) or highly acidic

(pH < 3) waters. Incidental destruction of bacteria in lime softening process is an example.

(d) Surface Active Chemicals — Detergents are surface active chemicals. The cationic detergents are strongly destructive, neutral detergents are intermediate, while the anionic detergents are weak in their action. Detergents are used only in the wash waters and the rinse waters of eating establishments, dairies, soft drink establishments, etc.

Among the several disinfection methods described earlier, chlorination is the universally accepted method of disinfection. It is therefore described in detail.

Chlorination

Water disinfection by chlorination was first used in 1908 for Jersey City in U.S.A. [75]. Chlorine is a greenish yellow toxic gas, found in nature only in the combined state, chiefly with sodium as common salt. It has penetrating and irritating odour, is heavier than air and can be compressed to form a clear, amber-coloured liquid. As a dry gas chlorine is noncorrosive but in the presence of moisture, it becomes highly corrosive. Chlorine is slightly soluble in water.

CHLORINE-WATER REACTIONS

(a) Free Residual Chlorine — Chlorine reacts with water to form hypochlorus acid (HOCl) and hydrochloric acid (HCl) according to

$$Cl_2 + H_2O \rightleftharpoons HOCl + HCl$$
$$HCl \rightleftharpoons H^+ + Cl^- \tag{43a}$$

The hypochlorous acid dissociates into hydrogen ions (H^+)

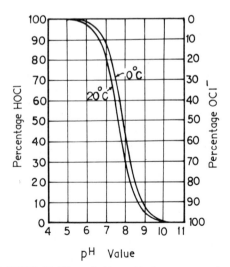

FIGURE 45. *Effect of pH on chlorine-water reactions.*

and hypochlorite ions (OCl^-) according to

$$HOCl \rightleftharpoons H^+ + OCl^- \tag{43b}$$

The undissociated HOCl is about 80–100 times effective as a disinfectant than OCl ions. At pH value of 5.5 and below, it is practically 100% unionised HOCl while above pH 9.5, it is all OCl ions. Between pH 6.0 to 8.5, a very sharp change occurs as shown in Figure 45. The addition of chlorine does not produce any significant change in the pH of the natural waters because of their buffering capacity. Free available chlorine may be defined as the chlorine existing in water as hypochlorous acid and hypochlorite ions.

(b) Combined Residual Chlorine — Free chlorine reacts with ammonia, proteins, amino acids and phenol, if present in water, to form chloramines and chloroderivatives to provide combined available chlorine. The pertinent equations with ammonia are

$$Cl_2 + H_2O \rightleftharpoons HOCl + HCL$$
$$NH_3 + HOCl \rightleftharpoons NH_2Cl + H_2O$$
$$NH_2Cl + HOCl \rightleftharpoons NHCl_2 + H_2O \tag{44}$$
$$NHCl_2 + HOCl \rightleftharpoons NCl_3 + H_2O$$

Monochloramine (NH_2Cl) and dichloramine ($NHCl_2$) have disinfectant properties. Their effect is 25 times less but lasts longer than that of free chlorine. Trichloramine (NCl_3) has no disinfectant properties at all. Trichloramine is found for pH below 4.4, dichloramine predominates for pH between 5 and 6.5, and monochloramine predominates for pH over 7.5. The ratio of ammonia to chlorine is 1:4 for optimum results. The minimum concentrations of free and combined residual chlorine for effective disinfection are given in Table 15 [22].

The reactions of chlorine with water are shown in Figure 46. Stage I represents the destruction of chlorine by reducing compounds. Here the chlorine residual is practically nil. In stage II, the chloro-organic compounds and the chloramines are formed and the combined residual chlorine increases. In stage III, due to the destruction of the chloramines and chloro-organic compounds, the combined residual chlorine decreases. When complete destruction takes place, the drop in the combined residual chlorine stops and break point B is reached. Afterwards, in stage IV, free residual chlorine is available and it increases with the addition of chlorine.

The difference between the amount of chlorine added to water and the amount of residual chlorine after a specified contact period is termed "chlorine demand." In Figure 46, the chlorine demand is about 0.55 ppm.

Apart from chlorine, either gas or liquid, some compounds give out chlorine. They are: (1) Bleaching powder, a

TABLE 15. Minimum Concentrations of Free and Combined Residual Chlorine for Effective Disinfection.

pH Value	Minimum Concentration of Residual Chlorine (ppm)	
	Free[a]	Combined[b]
6–7	0.2	1.0
7–8	0.2	1.5
8–9	0.4	1.8
9–10	0.8	not recommended
>10	>0.8 with longer contact time	not recommended

[a]minimum contact time of 10 min.
[b]minimum contact time of 60 min.

loose combination of lime and chlorine, containing 25–35% chlorine; and (2) high strength calcium hypochlorite, a relatively stable compound, containing 60–70% chlorine by weight.

CHLORINATION PRACTICES

(a) Plain Chlorination—In plain chlorination, chlorine is applied to water as the only treatment for public health protection. It can be practised when (1) turbidity and colour of the raw water are low, (turbidity:5–10 NTU); (2) raw-water source is relatively unpolluted; (3) organic matter is less; and iron and manganese do not exceed 0.3 mg/L.

(b) Prechlorination—In prechlorination, chlorine is applied to water prior to any unit treatment process. The chlorine may be added in the suction pipes of the raw-water pumps or to the water as it enters the mixing chamber. Prechlorination improves coagulation and reduces taste, odour, algae and other organisms in all the treatment units.

(c) Postchlorination—Postchlorination usually refers to the addition of chlorine to water after all other treatments. It is a standard practice at the rapid sand filter plants.

(d) Superchlorination—Heavy chlorination required for quick disinfectant action or destruction of tastes and odours may produce high residuals. This process is termed super-chlorination. It is practised in emergency like a break-down or when the water is heavily polluted. The dose of chlorine may be as high as 10–15 mg/L with a contact period of 10–30 min. The undesirable excess chlorine in water has to be removed by dechlorination.

(e) Dechlorination—The excess chlorine resulting from superchlorination is removed through dechlorination, by some reducing agents like sulphur dioxide, sodium bisulphite ($NaHSO_3$), sodium sulphite (Na_2SO_3), sodium thiosulphate ($Na_2S_2O_3$); oxidising agent like activated carbon; and aeration.

(f) Rechlorination—When the distribution system is large, it may be difficult to maintain a residual chlorine of 0.2 mg/L at the farthest end, unless a very high dosage is applied at the post chlorination stage. This is costly and makes the water unpalatable, at least in the earlier reaches of the distribution system. Therefore, rechlorination is carried out in the distribution system at service reservoirs, booster pumping stations or any other convenient points.

APPLICATION OF CHLORINE

The most common practice in public water supplies is the addition of chlorine either in gaseous form or in the form of a solution made by dissolving gaseous chlorine in a small auxiliary flow of water, the chlorine being obtained from cylinders containing gas under pressure. The normal chlorine dosage required to disinfect water not subject to significant pollution, would not exceed 2 mg/L. The actual chlorine dose requirement is obtained from the chlorine demand tests.

(a) Chlorinators—A chlorinator is a device designed for feeding chlorine to water. Its function is to regulate the flow of gas from the chlorine container. The usual fittings and parts of a chlorinator are: (1) chlorine cylinders with 45–1,000 kg capacity or tank cars with 15,000; 30,000 or 50,000 kg capacity; (2) fusible plug, a safety device, provided to the cylinders designed to melt or soften between 70°C to 75°C to avoid the pressure build-up; (3) reducing valve to bring down the gas pressure to 700–300 N/mm²; (4)

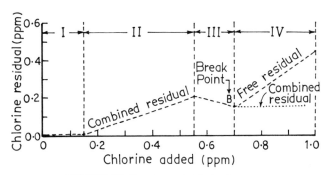

FIGURE 46. Reactions of chlorine in water.

pressure gauges indicating the cylinder and delivery pressures; (5) orifice meter or rotameter to measure the flow rate; and (6) a "desiccator valve" or a "non return valve" containing concentrated sulphuric acid or calcium chloride through which the chlorine must pass to free it from moisture to prevent corrosive action on the fittings.

(b) Piping system—As the chlorine gas is very corrosive in the presence of moisture, the piping is of a suitable PVC material.

(c) Chlorine vaporisers—When the chlorine requirement exceeds 900 kg/d, or when the rate of gasification is less than required because of low room temperature, chlorine vaporisers such as thermostatically controlled hot water baths are provided.

(d) Chlorine feeders—Chlorine feeders are used to control or measure the flow of chlorine in the gaseous state. Direct feed equipment and solution feed equipment are the common types.

(e) Chlorine housing—The chlorine cylinders and feeders should be housed in a fire resistant building that is easily accessible, close to the point of application, and convenient for truck loading and unloading. It should also have proper ventilation.

(f) Safety considerations—The necessary safety precautions [9] should be observed during the chlorination process.

PACKAGE WATER TREATMENT PLANTS

Using technological advances, many of the unit operations in water treatment are combined recently to obtain compact package water treatment plants, especially useful for isolated small communities. Such plants can be fully automatic for use in developed countries or could be devoid of mechanical equipments and electrical energy (except for pumps) for use in developing countries. They are preassembled or prefabricated for direct use on site.

Figure 47 shows a package water treatment plant suggested by Bhole [61] for use in rural areas of developing countries. The plant consists of (1) an alum dosing unit having a large size plastic bucket B_1 for storing alum solution; (2) a pebble bed flocculator having sections of increasing cross-sectional areas to produce tapered velocity gradients; (3) an inclined plate settling tank having 26 plates located below a V-notched weir that conveys the settled water to the filter unit; (4) a filter unit of sand and supporting gravel media; and (5) a chlorination unit, similar to the alum unit, but containing a solution of bleaching powder in bucket B_2.

Raw water is pumped by a pump P_1 to the elevated tank (ET) which provides necessary head for gravity flow through the treatment unit. Another pump P_2 is used for cleaning the filter and the settling tank. During filtration, valves V_1, V_2 and V_5 are opened and the remaining valves are closed. During backwashing, valves V_4 and V_6 are opened, the remaining valves are closed and the pump P_2 is started. The washwater flows upward through the filter, collects in the troughs and flows into the settling tank where it is drained via a floor drain. To drain the flocculator unit, valves V_1 and V_4 are closed and valves V_2 and V_3 are opened.

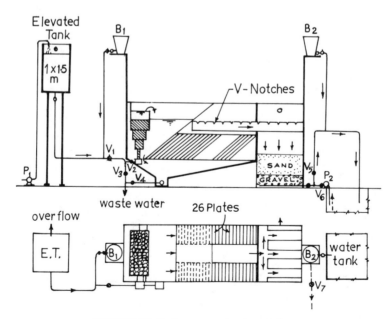

FIGURE 47. Package water treatment plant.

This package plant is 5.3 m × 1.25 m × 1.25 m and weighs about 13 kN. It has a capacity of 270 m^3/d, sufficient for a population of 2,000 people with an average consumption of 135 Lpcd. Several package plants can be operated in parallel to meet larger demands. The cost of the plant is about 20,000 Rupees ($\cong$ US $2,000).

Similar package water treatment plants have also been suggested. They are: (1) The APS package water treatment plant, designed by APS Technical Services Ltd., England [61]; (2) the MIMO water treatment plant, designed by Patwardhan using the multi-inlet multi-outlet approach [87]; (3) WATER BOY package plant [63]; and (4) Waterman package plant developed by M/s Candy Filters (India) Ltd., [93]. A plant suggested by Manevaskan [94] uses aluminium electrodes, which provide aluminium hydroxide when an AC current is passed through them. The flocs are formed in 30–45 min and settle at the bottom.

Ground Water

The ground water mainly contains soluble impurities and not suspended impurities. Therefore, the treatment of ground water differs from that of surface water. The major impurities in ground water are the salts of calcium, magnesium, iron, manganese, sodium and fluorine. Removal of these salts is achieved by adopting one or more of the following processes: (1) Chemical precipitation followed by sedimentation; (2) chemical oxidation followed by sedimentation; (3) adsorption; (4) ion exchange; (5) electrodialysis; (6) evaporation; (7) reverse osmosis; (8) freezing; and (9) disinfection.

CHEMICAL PRECIPITATION

Hardness Removal

Hardness is of two types: (1) Carbonate hardness which is caused by calcium bicarbonate $Ca(HCO_3)_2$; calcium carbonate $CaCO_3$; magnesium bicarbonate $Mg(HCO_3)_2$; and magnesium carbonate $MgCO_3$; and (2) noncarbonate hardness which is caused by calcium sulphate $CaSO_4$; calcium chloride $CaCl_2$; magnesium sulphate $MgSO_4$; and magnesium chloride $MgCl_2$. Carbonate hardness, also known as temporary hardness, can be removed even by boiling; while noncarbonate hardness, also known as permanent hardness, cannot be removed by boiling.

The terms "soft" and "hard" are used in USA [22] as follows: (1) Soft: hardness less than 50 ppm; (2) moderately hard: hardness 50–150 ppm; (3) hard: hardness 150–300 ppm; and (4) very hard: hardness more than 300 ppm.

CHEMICAL REACTIONS

Lime and soda ash process is used to change the soluble calcium and magnesium compounds into nearly insoluble compounds (calcium carbonate is theoretically soluble only

to the extent of 17 ppm at pH 9.4). The chemical reactions are:

Carbonate hardness

$$CO_2 + Ca(OH)_2 = CaCO_3\downarrow + H_2O$$

$$Ca(HCO_3)_2 + Ca(OH)_2 = 2CaCO_3\downarrow + 2H_2O$$

$$Mg(HCO_3)_2 + Ca(OH)_2 = CaCO_3\downarrow + MgCO_3 + 2H_2O$$

$$MgCO_3 + Ca(OH)_2 = Mg(OH)_2\downarrow + CaCO_3\downarrow \quad (45)$$

Noncarbonate hardness

$$MgSO_4 + Ca(OH)_2 = Mg(OH)_2\downarrow + CaSO_4$$
$$\quad (46)$$
$$CaSO_4 + Na_2CO_3 = CaCO_3\downarrow + Na_2SO_4$$

The nonsoluble compounds are then flocculated, settled and filtered as with the conventional coagulation for surface water.

TREATMENT METHODS

(a) Excess Lime Treatment—An excess of lime (10–50 ppm) is added to produce a pH of 10.6 for the precipitation of calcium and magnesium. Soda ash is then added to convert the excess of lime to sodium hydroxide and calcium carbonate. Sodium hydroxide (caustic alkalinity) is also undesirable and therefore split treatment (described later) is preferred unless water is recarbonated to remove caustic alkalinity and produce a pH of 8.

(b) Split Treatment—A major portion of the raw water is given the excess lime treatment and the remaining portion of raw water is added after the softening reactions have occurred. The CO_2 and bicarbonates in this portion of raw water react with the excess of lime and precipitate the calcium at pH 9.4 in the secondary basin.

(c) Excess Lime Treatment Followed by Recarbonation—A more elaborate process consists of primary flocculation and sedimentation, with lime treatment to produce a pH of 10.6 to precipitate the magnesium. The water is then recarbonated with CO_2 to lower the pH to 9.4 and precipitate $CaCO_3$ in the secondary basin; and soda ash is added as needed to precipitate noncarbonate hardness. Finally CO_2 is added to the settled water to produce a pH of about 8.7 so as to convert the residual $CaCO_3$ into soluble bicarbonate and thus prevent the after-precipitation on the filter sand.

Factors for computing doses of hydrated lime or quick lime needed to remove 1 ppm of CO_2, bicarbonate alkalinity or magnesium are given in Table 16. Normally, the purity of hydrated lime is 95% and that of quick lime is 88%. Soda ash is commercially available as 100% sodium carbonate. The factor of 1.06 is used in computing the dose of soda ash required to remove the noncarbonate hardness.

TABLE 16. Multiplying Factors for Computing Doses of Hydrated Lime and Quick Lime.

Sr. No.	Constituent	Hydrated Lime (100% pure)	Quick Lime (100% pure)
1.	Free carbon dioxide as CO_2	1.68	1.27
2.	Bicarbonate alkalinity as $CaCO_3$	0.74	0.56
3.	Magnesium as Mg	3.04	2.30

The chemical dosages required to precipitate can also be calculated by considering the ionic balance of a water with the help of a bar diagram [95]. The Caldwell-Lawrence (C-L) diagrams [96] provide the diagrammatic solutions to the problems involving hardness and $CaCO_3$ equilibria.

Fluoride Removal

NALGONDA TECHNIQUE

Nawlakhe et al. [97] have found that alum and lime remove fluorides from ground water. At Kadiri, India, the dose of alum required was around 500 mg/L to reduce the fluorides from 4.5 mg/L to 1.0 mg/L [98].

COMBINATION OF ALUMINIUM CHLORIDE AND ALUMINIUM SULPHATE

Defluoridation with aluminium chloride alone or in combination with aluminium sulphate was recently found to be an effective method. The data [99] deal with the doses of aluminium chloride, aluminium sulphate or their combination necessary to achieve the required reduction of fluorides for various values of alkalinity.

CHEMICAL OXIDATION

The lower oxides of iron and manganese which are in soluble form are converted to higher oxides of iron and manganese. Aeration is one of the simple methods to achieve this. Another method is by increasing the carbonate content and pH of water at the same time, e.g., by addition of soda ash. For economic reasons, the use of lime and soda ash for deferrization and demanganization is common. This is followed by filtration.

When iron and manganese are loosely bound to organic matter, one of the following methods is adopted [22]: (1) Oxidation by contact aerator using coke, gravel or crushed pyrolusite, followed by settling basin and sand filter; (2) oxidation by aerator followed by filter bed of manganese-coated sand, birm or crushed pyrolusite ore; (3) oxidation by aerator followed by chlorination or chlorination alone; and (4) oxidation with effective aerator followed by lime-mixing basin, settling basin and sand filter.

Organic matter present in water is normally responsible for taste and odour. Oxidation of such matter is possible with chlorine, ozone, potassium permanganate, or any other oxidizing agent such as chlorine dioxide.

ADSORPTION

Odours in water are caused by dissolved gases like hydrogen sulphide, organic matter derived from algae and other micro-organisms, decomposing organic matter in general, industrial wastes, phenol, chlorine, etc. Taste is also added to water.

Activated carbon is normally used to remove taste and odour. Activated carbon is made from raw materials such as coal, wood, nut shells, peat, lignite, saw dust, and the residue from paper, pulp, and petroleum processes. The raw material is initially heated in a closed retort and later oxidized or "activated" with hot air or steam to remove the hydrocarbons which would interfere with adsorption of organic matter. Activated carbon has surface area in excess of 2,000 m^2/g. It is very porous and has many carbon atoms with free valencies. At low pH, activated carbon is positively charged due to adsorption of hydrogen ion and therefore adsorption is greatest at low pH. Other factors which increase the adsorption are temperature, high concentration of organic matter and coiled structure of molecules.

The rate of removal decreases with increasing molecular size, molecular weight and complexity of molecular structure. In mixtures of more than one adsorbate, a degree of preferential adsorption may occur with smaller lighter molecules being more completely removed.

Activated carbon is either powdered or granular. Powdered activated carbon is generally less than 0.075 mm in size and therefore has an extremely high ratio of area to volume. The points of application of powdered activated carbon include: (1) the raw water, as early as possible in the plant; (2) the mixing basin; and (3) split feed, with a portion in the mixing basin and the balance just ahead of the filter. Although activated carbon can be regenerated, this is not commonly done in water treatment plants adding activated carbon on a seasonal basis. Polyelectrolytes should be used as the sole coagulant, if the activated carbon is to be recovered from the sludge. Powdered activated carbon filters are backwashed with high quality water.

Granular activated carbon may be installed in conventional sand filters or post-filtration columns. These carbon filters may be operated in either upflow or downflow direction.

Other materials which are used for adsorption include various types of clays, fly ash, baggas, barks of certain trees, rice husk, etc. [100]. National Environmental Engineering Research Institute (India) has studied Defluoron-1, Defluoron-2, serpentine, activated magnesia, activated alumina as materials for adsorption of flourides [101].

The adsorption process is described by Freundlich isotherm, Langmuir isotherm or BET isotherm.

ION EXCHANGE

The reactions involved in the ion-exchange process are as follows:

$$Ca^{++} + 2\ (Na^+R^-) \rightarrow (Ca^{++}\ R_2^-) + 2\ Na^+$$

$$Mg^{++} + 2\ (Na^+R^-) \rightarrow (Mg^{++}\ R_2^-) + 2\ Na^+$$

$$(47)$$

$$OH^- + (R^+Cl^-) \rightarrow (R^+OH^-) + Cl^-$$

$$SO_4^- + 2\ (R^+OH^-) \rightarrow (R_2^+\ SO_4^-) + 2\ OH^-$$

In these reactions, the letter R is used to indicate the anionic or cationic portion of the material which does not enter into the reactions. Thus, NaR means the base exchange material containing sodium, whereas CaR_2 and MgR_2 represent the same material after the sodium has been exchanged for calcium and magnesium by the removal of these minerals from water and release of an equivalent amount of sodium to water. To restore the sodium, regeneration is practised in which solutions of common salts are used.

The advantages of the ion-exchange process are simplicity of operation and control, compactness in size, and ability to demineralize waters to the desired extent. Non-carbonate hardness is more cheaply removed by the ion-exchange process than with soda ash. However, the process is uneconomical when total hardness exceeds 850–1000 ppm. The total solids content should be below 3,000–5,000 ppm. The applied water should not contain turbidity more than 5 NTU. Green sand and zeolites should not be used to soften waters of high sodium alkalinity or those with a pH below 6.0 or even 8.0. Residual chlorine reacts with some of the organic exchange materials and therefore chlorine should not be applied to raw waters when treated by such materials.

Alumina, SiO_2, MnO_2, metal phosphates and sulphides, lignin, proteins, cellulose, wool, living cells, carbon and resins have ion exchange capacity. Synthetic zeolites are reasonably stable and have high capacities, hence they are commercially used.

Exchangers are of two types: (1) Cation exchangers; and (2) anion exchangers. Cation exchangers contain negatively charged functional groups such as SO_3^- from H_2SO_4 or carboxylic group derived from weak acid. Anion exchangers contain positively charged functional groups, such as quaternary ammonium (NR_3^+), imino (NRH_2^+), phosphonium (PR_3^+), and sulphonium (SR_3^+).

Demineralization

Demineralization is effected in two-step process in which water is passed successively through a cation exchanger in the H^+ form, (H^+R^-); and an anion exchanger in the OH^- form, (R^+OH^-).

Mixed-bed exchanger is a more recent development where a single column contains a mixture of equivalent quantities of cation and anion exchangers. The effluent is generally superior in quality. To regenerate a mixed bed, the resins must be separated. This can be done by differential backwashing, because cation-exchange and anion-exchange resins have different densities. The cation-exchange resins are regenerated with a strong acid, e.g., H_2SO_4. If weak acids such as CO_2 and silicic acid $Si(OH)_4$ are to be removed, strongly basic anion exchangers must be employed. These must be regenerated with sodium hydroxide. Some weakly basic anion exchangers can be regenerated with soda ash.

Ion-exchange process has been used for removal of fluorides [101]. Some of the materials used are polystyrene anion exchange resins, saw dust carbon, carbion, waso-resin-14, and granular tricalcium phosphate.

ELECTRODIALYSIS

Electrodialysis is an electrochemical process of desalination of saline water. The membranes used for the process are selectively permeable to cations or anions. Highly selective membranes have been prepared by casting ion-exchange resins as thin films. Dialysis is the fractionation of solutes made possible by differences in the rate of diffusion of the specific solutes through porous membranes. Semipermeable membranes which are thin barriers offer easy passage to some constituents of a solution but are highly resistant to the passage of other constituents. Membranes made from cation-exchange resins are cation permeable, those made from anion-exchange are anion permeable. Diffusion of ions through membranes is accelerated by applying a voltage across the membrane and hence is termed as electrodialysis. The unit consists of a series of chambers including alternating anion-permeable and cation-permeable membranes together with inert electrodes attached to the outermost compartments. Water introduced in alternate chambers is demineralized by passage of a direct current through the battery of compartments.

An electrodialysis process, known as 'neutral membrane electrodialysis' uses a nonselective membrane or neutral membrane in place of either the anion-permeable or cation-permeable membrane. Another type, known as 'electro-gravitation process' or 'electrodecantation process' uses only one kind of membrane, either the cation-permeable or the anion-permeable.

In electrodialysis, the cost of electrical energy rises with increasing salinity. Therefore, only waters containing less than 5,000 to 10,000 mg/L of dissolved solids are desalted electrochemically.

EVAPORATION [102]

Evaporation is normally adopted for desalination. Evaporation can be achieved either by distillation or by solar energy.

Evaporation by distillation can be achieved through (1) single effect distillation; (2) multi effect distillation, (3) multi stage flash distillation; (4) vapour compression process; or (5) vapour reheat distillation. In all processes, except the first one, the latent heat of vapourization is recaptured and used for heating purposes.

Evaporation by solar energy is achieved through (1) glass-covered solar still; (2) large plastic-covered still; (3) tilted-wick still; or (4) multiple ledge-tilted still. The efficiency of a solar still depends on (1) depth of saline water; (2) temperature of saline water; (3) saline-water-to-cover temperature difference; (4) solar radiation; and (5) latitude, altitude and sky clearance factor of the place.

REVERSE OSMOSIS

Osmotic pressure drives water molecules through permeable membrane from a dilute to a concentrated solution in search of equilibrium. This natural process can be reversed by placing the salt water under hydrostatic pressures higher than the osmotic pressure. Therefore, such process is termed 'reverse osmosis'. These hydrostatic pressures range from 5 to 50 times the osmotic pressure of water [103].

Reverse osmosis systems normally include the membrane, a support structure, a pressure vessel and a pump. The optimum membrane configuration is the hollow fiber, which has an area-to-volume ratio of upto 30,000 m^2/m^3, and requires no separate support.

The commercial reverse-osmosis membranes are made of cellulose triacetate having an acetyl content of 43.2 per cent, or of polyamide polymer. The cellulose triacetate membranes suffer excessive compaction at the driving pressures required for sea water. Therefore, a new generation of membranes has come in market. These membranes are the thin-film composite membranes in which a very thin film (about 200 Å) of a polymer is formed on a porous substrate of a relatively incompressible substance. They have been made of several new polyamide type materials as well as cellulose triacetate. They have produced potable water from sea water, in a single pass through the membrane at pressures of 5,500–6,900 kN/m^2.

Four distinctly different modular designs have been developed. They are: (1) plate and frame type; (2) tubular; (3) spiral wound; and (4) hollow fine fiber.

The plate and frame type resembles a filter press. The tubular concept consists of small diameter porous or perforated tubes with the membrane placed inside the tubes, which also serve as membrane support and pressure vessel. This design has high initial and operating cost.

The spiral-wound membrane unit was developed by Fluid Systems Division, UOP Inc., U.S.A. [63]. The spiral-wound unit consists of one or more membrane envelopes each formed by enclosing a channelized, product-water-carrying material between two large flat membrane sheets. The membrane envelope is sealed on three edges with a special adhesive and is attached to a small diameter pipe having openings to collect the permeate. The envelops are wound round the pipe to form a cylinder of 50, 100, 200 or 300 mm in diameter and upto 1.02 m in length. A polypropylene screen is used to form the feed-water channel between the membrane envelops. The centre tube or pipe is also the permeate collecting channel.

The hollow fine fibre membrane unit was developed by the Du Pont Company [63] using a polyamide polymer. The polyamide fibres were made with diameters of 50–80 μm with the inner diameter about half the outer diameter. Triacetate fibres are estimated to be about 200–300 μm outer diameter. The membrane and pressure vessel are an integrated unit. The fibres are formed into a U-shaped bundle with open ends of fibres potted in an epoxy tube sheet. The bundle attached to the tube sheet is arranged in a cylindrical pressure vessel. Feed water enters the center of the vessel through a porous or a perforated pipe and is distributed radially through the fiber bundle. Under pressure, the water flows into the hollow fine fibers and flows out through the capillaries.

The spiral wound and hollow fine fiber equipments are presently in extensive use throughout the world. Each type can treat about 1.3 m^3/s of water (112,000,000 L/d).

FREEZING

Freezing process [104] is used for desalination and is effected by the application of vacuum processes in which evaporation of a portion of water or of a miscible secondary refrigerant cools the flow. The influent to the process is cooled in a heat exchanger by product stream and waste stream close to the freezing point. The evaporator withdraws the refrigerant, dropping the temperature and producing ice crystals which are removed and washed with a portion of the product water. The ice crystals are used to condense the vapour stream from the evaporator, or cool it following its compression and this in turn melts the ice. A portion of the product is used to rinse the crystals; and the rinse-water, product and brine are used to chill the incoming flow. Freezing, due to the low temperature and lack of direct surface heat exchange, minimizes corrosion and scaling, and is generally more economical than evaporation.

Common methods of freezing are: (1) indirect refrigeration method; (2) vacuum freez-vapour compression method; and (3) vapour absorption or direct refrigeration method.

DISINFECTION

Methods described earlier for the disinfection of surface water can be used for the disinfection of ground water also. However, some simple methods that can be used for the disinfection of well water in rural areas of developing countries are described in the following [105].

Single Pot Method

An earthen pot with two holes of 6 mm diameter in the middle or seven holes at the bottom is filled with 1.5 kg of bleaching powder mixed with 3 kg of coarse sand (2 mm and above), and covered with polythene. The pot is suspended 1

TABLE 17. Common Treatment Processes[a] for Removal of Major Dissolved Impurities.

Parameter	Chemical Precipitation	Chemical Oxidation	Adsorption	Ion Exchange	Electro-dialysis	Evaporation	Reverse Osmosis	Freezing	Disinfection
Colour	x	x	x			x			
Taste and odour	x	x	x	x					
Dissolved solids	x	x	x	x	x	x	x	x	
Total hardness (as CaCO₃)	x		x	x	x		x		
Calcium (as Ca)	x		x	x	x		x		
Chlorides (as Cl)				x	x	x	x	x	
Copper (as Cu)	x		x	x	x	x	x		
Fluorides (as F)	x		x	x	x		x		
Iron (as Fe)	x	x		x	x				
Magnesium (as Mg)	x			x	x		x		
Manganese (as Mn)	x	x		x	x		x		
Nitrate				x	x		x		
Sulphate	x			x	x		x		
Micro-organisms		x							x

[a]Common treatment process is shown by x.

m below the water level. One unit is found to be sufficient to chlorinate a well of 9,000–13,000 L capacity, and having a withdrawal rate of 800–1,200 L/d. The unit needs replenishing once in a week.

Double Pot Method

The unit consists of two pots, with the inner pot filled with 1 kg of bleaching powder and 2 kg of coarse sand and placed inside an outer pot. The inner pot has a 10 mm diameter hole at its upper portion while the outer pot has a 10 mm diameter hole at the bottom. The mouth is loosely covered with polythene and is suspended 1 m below the water level. One unit can chlorinate a well of 3,500–5,000 L capacity and having a withdrawal rate of 400–500 L/d. The unit needs replenishing once in three weeks.

Drip Chlorinator

A drip chlorinator consists of a chamber with a solution of bleaching powder, and fitted with a float, connected to a delivery tube to facilitate the discharge of the solution at a constant rate. It can be used to disinfect either an open dug well or a covered well of 20,000–60,000 L capacity, and having a withdrawal rate of 2,000–6,000 L/d.

Before starting the chlorinator, sufficient quantity of bleaching powder must be added to the well water to satisfy its chlorine demand.

Differential Pressure Type Chlorinator

Differential pressure type chlorinator is commercially available for use with piped water supplies. A solution containing bleaching powder and soda ash (5:1 proportion) filled in a rubber bag (housed in a metal container) is grad-

ually squeezed out in proportion to the differential pressure across an orifice plate. The rate is adjusted by a needle valve. This type of chlorinator is quite satisfactory but is expensive and needs replacement of the rubber bag every 4–6 months.

The impurities found in water and the methods of their removal are given in Table 17.

WATER DISTRIBUTION

Water of the desired quality is distributed to the consumers, in sufficient quantity and at adequate pressure, through a water distribution system. It includes pipes, valves, hydrants, booster pumps, reservoirs for distribution and equalizing purposes, service pipes to the consumers and meters, and all other appurtenances after the water leaves the main pumping station or the main reservoir.

Layout of Distribution Systems

The layout of a distribution system should follow the street pattern with interconnected mains providing loops. The loops provide alternate routes in the event of pipe failure or pipe closure and therefore they increase the reliability of water distribution. The dead ends, when provided, should be as few as possible as they cause stagnation of water and water quality problems. They should be periodically flushed. A branching or a tree system is cheaper (even though less reliable) as compared to a looped system and therefore is used for water supply to small communities.

For equalization of water supply in a large water distribution system, zoning in the distribution system must be provided. The zoning depends upon the density of population, type of locality, topography and the facility of isolation for waste assessment and leak detection. A separate system should be provided for each zone having an average elevation difference of 40–50 m. The neighbouring zones are interconnected to provide emergency supplies; but the valves in the interconnecting pipes should normally be kept closed.

The minimum pipe sizes of 150 mm in residential areas and 200 mm in high-value districts should be provided. However, 100 mm pipe sizes can be used for short lengths in residential areas and for distribution systems of small communities. The maximum length between cross connections should be limited to 180 m. Valves should be so located that in case of pipe breakage, the pipe length required to be shut off should not exceed 150 m in high-value districts and 240 m in other areas.

Distribution and Equalizing Reservoirs

Reservoirs are provided in distribution systems to ensure sufficient quantity and pressure. They are connected to the distribution system by separate inlet and outlet pipes, or by a single pipe such that water enters and leaves by the same pipe. In the latter case, the reservoir is said to be "floating on the system." When the supply rate exceeds the demand rate water enters into the reservoir, and when the demand rate exceeds the supply rate water flows out of the reservoir.

The reservoirs can be surface reservoirs or elevated reservoirs. Surface reservoirs are used when topography permits. They are mostly square, rectangular or circular in shape, and are built in earth, masonry, concrete or reinforced concrete. The depths vary from 3 m to 10 m; usually increasing with the capacity.

When the height of a surface reservoir exceeds its diameter, it is called a "standpipe". It is built in steel or reinforced concrete, however, steel is more preferable because of its watertightness.

Elevated reservoirs or tanks of different sizes and shapes in steel, reinforced concrete and prestressed concrete are used.

Altitude valves which automatically shut when water reaches a predetermined level are provided. An overflow with drainage capacity equal to the maximum rate of inflow is provided for safety against the failure of the altitude valve. Water-level indicators, preferably with remote-recording equipment, should be provided on the reservoirs to know their performance.

Reservoirs which mainly provide sufficient quantity of water are termed "service", "storage", or "distribution" reservoirs; while the reservoirs which mainly ensure adequate pressure are termed "balancing" or "equalizing" reservoirs. Incidentally, distribution reservoirs serve as balancing reservoirs and balancing reservoirs as distribution reservoirs.

DISTRIBUTION RESERVOIRS

Purpose

Distribution reservoirs provide:

(1) Storage for equalizing supply and demand over periods of fluctuating consumption in a day. The reservoir fills when the rate of filtration or pumping exceeds the demand rate and empties in reverse conditions. This allows the pumps and the treatment plants to operate at a constant rate and thereby to reduce their capacity. The required storage is obtained for hourly demand fluctuations on a maximum day and for the proposed period of filtration or pumping; and is computed in a tabular form or determined graphically from either the mass curves as explained earlier, or from the maximum area between the supply-rate and demand-rate curves.

(2) Fire storage for immediate use in large quantities for fire-fighting purposes. The required fire storage depends upon the population served as indicated earlier in Table 5.

(3) Emergency storage for provision during failure of intake, supply conduit or power. The required emergency storage depends upon the extent of damage and the corresponding time of repair, and also the capacity of the supply conduit.

Location and Capacity

In a distribution system with a single distribution reservoir, its ideal location is a central place to have economical pipe sizes and equitable pressures [106]. When the system is fed through direct pumping and also through a distribution reservoir, its location should be at the tail end of the system. The capacity of the distribution storage is obtained by reasonably combining the above three purposes. A major fire may occur on a day of maximum demand and therefore combined capacity for these two purposes should be provided. The simultaneous requirement of emergency storage is less likely and it would also require much larger capacity and therefore is not considered. However, emergency storage is considered separately and a reasonable choice of the capacity for the distribution reservoir is made.

EQUALIZING RESERVOIRS

Purpose

Equalizing reservoirs are used for:

(1) Equalizing pressures in the distribution system and thereby reducing fluctuations in pressure caused by fluctuating demand. This provides better service to consumers and pressures for fire hydrants.

(2) Raising pressures at points remote from distribution reservoirs and pumping stations and improved service during periods of peak demand.

(3) Equalizing head on pumps. When equalizing reservoirs are located near pumping stations, the pumping heads

are more uniform. This results in better selection of pumps and their operation at higher efficiency.

Location and Capacity
Equalizing reservoirs are usually built at the opposite end of the distribution system from the source of supply. They should have sufficient elevation to provide adequate pressure in the system. The required capacity is obtained from the mass curves or the supply-rate and demand-rate curves.

Analysis of Distribution Systems

The analysis of a water distribution system involves the determination of the unknown parameters with the help of known parameters. The length, diameter, and roughness characteristics of a pipe are usually known and therefore are grouped into one parameter—the pipe resistance constant. Therefore, the head loss h_p in a pipe p can be expressed by a general relationship

$$h_p = H_i - H_j = R_p Q_p^n \qquad (48)$$

in which H_i, H_j = hydraulic gradient level (HGL) values at the end nodes i and j, respectively, of the pipe; R_p = resistance constant of pipe p; Q_p = discharge in pipe p; and n = an exponent which is 1.85 in the Hazen-Williams relationship and 2.0 in the Darcy-Weisbach relationship.

Water is withdrawn along the lengths of the pipes through house connections. However, for simplicity, it is common to assume in the analysis (as well as in the design) that the withdrawals are concentrated at their ends. Thus, all the withdrawals are assumed to be concentrated at pipe junctions or nodes. These nodal withdrawals or nodal demands are determined by considering the area served by each node, density of population, per capita consumption, peak demand factor; and specific commercial, public, industrial and fire demands. Even though the nodal demands fluctuate with time, it is common to assume the demand pattern and the flow in the distribution system to be steady.

For distribution systems, the involved parameters are the pipe resistance constants, the nodal flows (inflows at supply nodes, i.e., reservoirs and pumps; and outflows at demand nodes), the pipe discharges and the HGL values at the nodes. Some of these parameters are known and the remaining parameters are determined in the analysis. Any combination of known and unknown parameters can be tackled, provided the combination satisfies certain conditions [107–109]. The "analysis of distribution system" usually involves the determination of the inflows at the supply nodes and the HGL values (or the pressures) at the demand nodes.

PARAMETER RELATIONSHIPS
The different parameters involved in the analysis are interrelated such that they satisfy, in addition to the general

headloss relationship, Equation (48), the following two basic relationships:

a. Node Flow Continuity Relationships
For steady flow in the network, the algebraic sum of the flows into any node must be zero, i.e.,

$$\sum_{\substack{p \text{ incident} \\ \text{on } j}} Q_p + q_j = 0 \text{ for all nodes } j \qquad (49a)$$

or

$$\sum_{\substack{p \text{ incident} \\ \text{on } j}} \left(\frac{H_i - H_j}{R_p} \right)^{1/n} + q_j = 0 \text{ for all nodes j} \qquad (49b)$$

b. The algebraic sum of the headlosses in pipes forming any circuit or loop must be zero, i.e.,

$$\sum_{p \in c} h_p = \sum_{p \in c} R_p Q_p^n = 0 \text{ for all loops} \qquad (50)$$

The total number of equations provided by Equation (49a) or Equation (49b) and Equation (50) is sufficient for the analysis. However, as some of them are nonlinear, their solution is difficult. In the analysis of the distribution systems, some unknown parameters are treated as "basic" unknown parameters. Once these basic unknown parameters are determined, the remaining parameters are easily obtained. Different parameters are used as basic unknown parameters and therefore different basic equations are used in the analysis. These basic equations are the H-equations, Q-equations and the ΔQ-equations.

FORMULATION OF BASIC EQUATIONS [110–112]
The method for formulating the different basic equations is illustrated through a two-looped, two-source, four-demand node distribution system of Figure 48. Nodes 1 and 3 are source nodes with known HGL values of 100.00 m and 95.00 m, respectively. Nodes 2, 4, 5 and 6 are demand nodes with demands of 0.3 m³/s, 0.2 m³/s, 0.4 m³/s and 0.1 m³/s, respectively. The resistance constants and the assumed flow directions for the seven pipes (labeled 1–7) are as shown in the figure. The Hazen-Williams headloss relationship is used so that $n = 1.85$.

H-Equations
The unknown nodal HGL values are taken as the basic unknown parameters in formulating the H-Equations. For the illustrative network, H_2, H_4, H_5 and H_6 are such parameters. Using the node flow continuity relationships, Equation

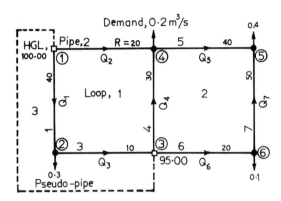

FIGURE 48. *Distribution system for illustrative example.*

(49b) for nodes 2, 4, 5 and 6, the H-equations, respectively, are:

$$\left(\frac{100 - H_2}{40}\right)^{0.54} - \left(\frac{H_2 - 95}{10}\right)^{0.54} - 0.3 = 0 \quad (51a)$$

$$\left(\frac{100 - H_4}{20}\right)^{0.54} + \left(\frac{95 - H_4}{30}\right)^{0.54}$$
$$- \left(\frac{H_4 - H_5}{40}\right)^{0.54} - 0.2 = 0 \quad (51b)$$

$$\left(\frac{H_4 - H_5}{40}\right)^{0.54} + \left(\frac{H_6 - H_5}{50}\right)^{0.54} - 0.4 = 0 \quad (51c)$$

$$\left(\frac{95 - H_6}{20}\right)^{0.54} - \left(\frac{H_6 - H_5}{50}\right)^{0.54} - 0.1 = 0 \quad (51d)$$

In these equations, the flow to a node is positive, while the flow out of it is negative. The exponent 0.54 is the reciprocal of 1.85, the n value. Here four independent nonlinear equations are available for determining the four basic variables. Once these basic variables are determined, the pipe flows $Q_1, \ldots, Q_7$, and the inflows at the source nodes, q_1 and q_3 can be easily determined.

Q-Equations

In formulating the Q-equations, the basic variables are the seven pipe flows, $Q_1, \ldots, Q_7$. The Q-equations are obtained as follows.

1. The node flow continuity relationships (Equation 49a) for nodes 2, 4, 5 and 6 yield, respectively,

$$Q_1 - Q_3 - 0.3 = 0 \quad (52a)$$

$$Q_2 + Q_4 - Q_5 - 0.2 = 0 \quad (52b)$$

$$Q_5 + Q_7 - 0.4 = 0 \quad (52c)$$

$$Q_6 - Q_7 - 0.1 = 0 \quad (52d)$$

2. Considering clockwise direction positive, the loop headloss nonlinear relationships (Equation 50) for the two basic loops 1 and 2 are:

$$20 Q_2^{1.85} - 30 Q_4^{1.85} - 10 Q_3^{1.85} - 40 Q_1^{1.85} = 0 \quad (52e)$$

$$40 Q_5^{1.85} - 50 Q_7^{1.85} - 20 Q_6^{1.85} + 30 Q_4^{1.85} = 0 \quad (52f)$$

3. To obtain seventh independent equation to solve for the seven basic variables, a pseudopipe (shown by dashed line in Figure 48) is introduced to obtain a pseudoloop, loop 3. For this pseudoloop,

$$40 Q_1^{1.85} + 10 Q_3^{1.85} + (95 - 100) = 0 \quad (52g)$$

Thus, seven independent equations—four linear and three nonlinear—are obtained to determine the seven basic variables.

ΔQ-Equations

In formulating the ΔQ-equations, the pipe flows are initially assumed such that they satisfy the node-flow continuity relationships, Equation (49a). Therefore, assuming the flow directions as shown in Figure 48, let $Q_1 = 0.35$ m³/s, $Q_2 = 0.45$ m³/s, $Q_3 = 0.05$ m³/s, $Q_4 = 0.05$ m³/s, $Q_5 = 0.30$ m³/s, $Q_6 = 0.20$ m³/s, and $Q_7 = 0.10$ m³/s. Although these values satisfy the node flow continuity relationships, Equation (49a), they need not necessarily satisfy the loop headloss relationships, Equation (50). Therefore, let ΔQ_1, ΔQ_2, and ΔQ_3 be the flow corrections, additive in clockwise direction, for loops 1, 2 and 3, respectively. Therefore, the ΔQ-equations are:

$$20(0.45 + \Delta Q_1)^{1.85} - 30(0.05 - \Delta Q_1 + \Delta Q_2)^{1.85}$$
$$- 10(0.05 - \Delta Q_1 + \Delta Q_3)^{1.85} \quad (53a)$$
$$- 40(0.35 - \Delta Q_1 + \Delta Q_3)^{1.85} = 0$$

$$40(0.30 + \Delta Q_2)^{1.85} - 50(0.10 - \Delta Q_2)^{1.85}$$
$$- 20(0.20 - \Delta Q_2)^{1.85} \quad (53b)$$
$$+ 30(0.05 - \Delta Q_1 + \Delta Q_2)^{1.85} = 0$$

$$40(0.35 - \Delta Q_1 + \Delta Q_3)^{1.85}$$
$$+ 10(0.05 - \Delta Q_1 + \Delta Q_3)^{1.85} + (95 - 100) = 0 \quad (53c)$$

Thus, three nonlinear equations for three basic unknowns are obtained. Note that pipe 4 is common to both the loops

1 and 2, and therefore the flow in it is affected by both ΔQ_1 and ΔQ_2. Similarly, pipes 1 and 3 are common to loops 1 and 3.

In all the H-, Q-, and ΔQ-equations, as nonlinear equations are encountered, no direct simultaneous solution is feasible. However, note that all the nonlinear terms are similar in nonlinearity, and do not involve nonlinearity such as product of different parameters, logarithmic or trigonometric terms, etc. This is a helpful feature in the solution of these equations.

ANALYSIS METHODS

The different analytical methods for the analysis of water distribution systems can be grouped under three categories: (a) Hardy Cross method; (b) Newton-Raphson method; and (c) linear theory method.

Hardy Cross Method [110, 112.113]

Hardy Cross was perhaps the first to suggest a systematic iterative procedure for the solution of the network analysis problem [113]. His method attempts to solve the ΔQ-equations on the following principles:

1. Only one ΔQ-equation from the available set of ΔQ-equations is considered at a time.

2. The effect of adjacent loops is neglected and therefore each ΔQ-equation contains only one unknown. For example, Equations (53a), (53b), and (53c) will contain only ΔQ_1, ΔQ_2 and ΔQ_3 as unknowns, respectively, i.e., the other ΔQ's are assumed to be zero.

3. Each term of the now-modified ΔQ-equation is expanded in Taylor's series, the linear ΔQ terms are considered and the nonlinear terms are neglected. This gives the correction term as

$$\Delta Q_c = -\frac{\Sigma R\,Q^n}{\Sigma\,|RnQ^{n-1}|} \quad \text{for all } c \qquad (54)$$

in which ΔQ_c = pipe flow correction for loop c.

In the method, originally suggested, the correction ΔQ_c is calculated for a loop, the pipe flows are adjusted and then a second loop is considered. Thus, the loop corrections are calculated and applied sequentially, until all the loops are covered, to complete one iteration. However, in the current practice, the correction terms for all the loops are initially calculated and later applied simultaneously to complete one iteration. As the method balances the headlosses in a loop, the method is known as the "method of balancing heads."

Cornish [114] proposed similar approach for H-equations. The unknown nodal heads are initially assumed and successively corrected. The corrections are given by

$$\Delta H_j = -\frac{n\,\Sigma\,r_j}{\Sigma\,(Q/h)} \quad \text{for all } j \qquad (55)$$

in which Σr_j = residue giving the imbalance of the node

flow continuity relationship, Equation (49b) for node j; Q and h = discharge and head loss in a pipe connected to node j, respectively. As the nodal flows are being balanced in this version, it is known as the "method of balancing flows." Both these versions are now grouped under the same name of Hardy Cross method.

In the Hardy Cross method, as the effect of adjacent loops (or nodes) and the nonlinear terms in the Taylor's series are ignored, the convergence is slow. However, generally, the correction terms successively become smaller and smaller and the convergence is achieved.

The convergence in Hardy Cross method depends upon the initial guess of the pipe discharges. For better initial guess, Conklin [115] suggested the use of distributional factor, for each pipe, as a function of its diameter and length. Williams [116] suggested the use of "convergence acceleration factor" which enhances the convergence of the solution. Voyles and Wilke [117] have shown that the convergence is improved when the loops are selected such that the sum of the common flow resistance factors is minimum. For example, in Figure 48, if R_4 is large (300 instead of 30), it is better to select loop 2 comprising pipes 1, 2, 5, 7, 6 and 3 instead of pipes 4–7. This necessitates selection of overlapping loops instead of the nonoverlapping loops as is usually done.

In spite of the drawbacks of the Hardy Cross method (necessity of initial guess for pipe flows, and slow convergence as far as the number of iterations is considered), it is popular. It is suitable for hand calculation and computer programs are also available. For hand calculation, the calculations are carried out in a tabular form. One iteration of the Hardy Cross method for the network of Figure 48 is illustrated in Table 17. The final solution is shown in Figure 49.

Newton-Raphson Method [110,112]

Consider a function of x, i.e., $F(x)$ as shown in Figure 50. Let "a" be one of the roots so that $F(a) = 0$. To find "a" by trial and error procedure, assume that x is taken as x_1 for the

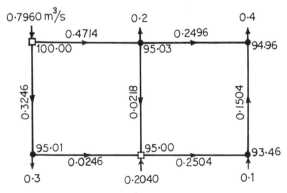

FIGURE 49. Solution of illustrative example.

first trial. Naturally, $F(x_1) \neq 0$. Let Δx be the correction so that $F(x_1 + \Delta x) = 0$.

Expanding by Taylor's theorem

$$F(x_1) + F'(x_1) \, \Delta x + F''(x_1) \frac{(\Delta x)^2}{2!} + \ldots = 0 \quad (56)$$

in which $F'(x_1)$, $F''(x_1) \ldots$ are the first, second . . . derivatives of $F(x)$ with value of x_1 substituted for x. If Δx is small compared to x_1 (this happens as x_1 approaches "a"), the third and subsequent terms of the expansion can be neglected so that

$$\Delta x = - \frac{F(x_1)}{F'(x_1)} \quad (57)$$

Raphson extended this to deal with sets of simultaneous equations involving more than one variable. Thus, for two variables, x and y, let $F_1(x,y) = 0$; and $F_2(x,y) = 0$. If x_1, y_1 are the first trial values and if Δx, Δy are the corrections then $F_1 (x_1 + \Delta x, Y_1 + \Delta y) = 0$; and $F_2 (x_1 + \Delta x, y_1 + \Delta y) = 0$.

Expanding, neglecting higher order terms, and writing in a matrix form.

$$\begin{bmatrix} \dfrac{\partial F_1}{\partial x_1} & \dfrac{\partial F_1}{\partial y_1} \\[2ex] \dfrac{\partial F_2}{\partial x_1} & \dfrac{\partial F_2}{\partial y_1} \end{bmatrix} \begin{bmatrix} \Delta x \\[2ex] \Delta y \end{bmatrix} = - \begin{bmatrix} F_1 \\[2ex] F_2 \end{bmatrix} \quad (58)$$

in which $F_1 = F_1 (x_1, y_1)$; and $F_2 = F_2 (x_1, y_1)$.

Applying this technique to ΔQ-equations for a network having C loops, the variables are ΔQ_1, ΔQ_2, . . . ΔQ_c which are the flow corrections to loops 1,2, . . . C, respectively. If F_1, F_2, . . . F_c are the loop headloss equation values, then

$$\begin{bmatrix} \dfrac{\partial F_1}{\partial \Delta Q_1} & \dfrac{\partial F_1}{\partial \Delta Q_2} & \cdots & \dfrac{\partial F_1}{\partial \Delta Q_c} \\[2ex] \dfrac{\partial F_2}{\partial \Delta Q_1} & \dfrac{\partial F_2}{\partial \Delta Q_2} & \cdots & \dfrac{\partial F_2}{\partial \Delta Q_c} \\[2ex] \cdot & \cdot & \cdot & \cdot \\ \cdot & \cdot & \cdot & \cdot \\ \cdot & \cdot & \cdot & \cdot \\ \cdot & \cdot & \cdot & \cdot \\ \cdot & \cdot & \cdot & \cdot \\ \cdot & \cdot & \cdot & \cdot \\[2ex] \dfrac{\partial F_c}{\partial \Delta Q_1} & \dfrac{\partial F_c}{\partial \Delta Q_2} & \cdots & \dfrac{\partial F_c}{\partial \Delta Q_c} \end{bmatrix} \begin{bmatrix} z_1 \\[2ex] z_2 \\[2ex] \cdot \\ \cdot \\ \cdot \\ \cdot \\ \cdot \\ \cdot \\[2ex] z_c \end{bmatrix} = - \begin{bmatrix} F_1 \\[2ex] F_2 \\[2ex] \cdot \\ \cdot \\ \cdot \\ \cdot \\ \cdot \\ \cdot \\[2ex] F_c \end{bmatrix} \quad (59)$$

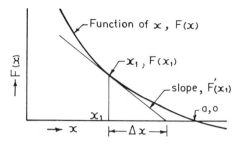

FIGURE 50. Newton's method of finding roots of an equation.

in which z_1, z_2, . . . z_c are the corrections to the assumed values of ΔQ_1, ΔQ_2, . . . ΔQ_c, respectively. The first matrix is the coefficient matrix and is symmetric, i.e.,

$$\frac{\partial F_i}{\partial \Delta Q_j} = \frac{\partial F_j}{\partial \Delta Q_i} , i \neq j$$

For the network of Figure 48, $F_1 = $ left hand side of Equation (53a), $F_2 = $ left hand side of Equation (53b), and $F_3 = $ left hand side of Equation (53c). Therefore,

$$\frac{\partial F_1}{\partial \Delta Q_1} = 37 (0.45 + \Delta Q_1)^{0.85} + 55.5 (0.05 - \Delta Q_1 + \Delta Q_2)^{0.85}$$
$$+ 18.5 (0.05 - \Delta Q_1 + \Delta Q_3)^{0.85} +$$
$$+ 74 (0.35 - \Delta Q_1 + \Delta Q_3)^{0.85}$$

$\cdot$

$\cdot$

$\cdot$

$\cdot$

$$\frac{\partial F_3}{\partial \Delta Q_3} = 74(0.35 - \Delta Q_1 + \Delta Q_3)^{0.85} +$$
$$+ 18.5(0.05 - \Delta Q_1 + \Delta QW_3)^{0.85}$$

Taking $\Delta Q_1 = \Delta Q_2 = \Delta Q_3 = 0$, for the first trial, Equation (59) becomes Equation (60) below

$$\begin{bmatrix} 54.885 & -4.349 & -31.767 \\ -4.349 & 53.43 & 0 \\ -31.767 & 0 & 31.767 \end{bmatrix} \begin{bmatrix} z_1 \\ z_2 \\ z_3 \end{bmatrix} = - \begin{bmatrix} -1.328 \\ 2.707 \\ 0.775 \end{bmatrix}$$

$$(60)$$

solution of which yields $z_1 = 0.0146$, $z_2 = -0.0495$ and $z_3 = -0.0098$. These values are added to the assumed values ($= 0$) of ΔQ_1, ΔQ_2 and ΔQ_3. The procedure is repeated and the solution given in Figure 49 is obtained at the end of third trial.

Newton-Raphson method was first introduced by Martin and Peters [118] for the analysis of water distribution systems. Epp and Fowler [119] suggested the procedure for numbering the loops so that computational work reduces.

The method can also be used for solving the H-equations. The method using the ΔQ- or H-equations is widely used in practice and computer programs are available. As this method considers all the loops simultaneously, the convergence, as far as the number of iterations is considered, is faster compared to the Hardy Cross method. However, the computational work and time per iteration are more. Like the Hardy Cross method, Newton-Raphson method also requires the initial guess of the Q or H values to start the iterative procedure.

Linear Theory Method [110,112]

Wood and Charles [120] proposed a linear theory method in which the Q-equations (Equation 52 a–g) are used. The node flow continuity relationships are already linear, but the loop headloss relationships are nonlinear. They are linearized by merging the nonlinear part of Q in the R values, so that the pipe headloss equation, Equation (48) becomes

$$h_p = R_p Q_p^n = (R_p Q_p^{n-1})\, Q_p = R_p' Q_p \qquad (61)$$

Thus, the pipe resistance constant R_p is replaced by R_p' ($= R_p Q_p^{n-1}$) for every iteration. This linearizes the nonlinear loop headloss equations. All the Q-equations, which are now linear are solved simultaneously.

In the first iteration all Q_p values for estimating R_p' values are taken as 1, giving $R_p' = R_p$. To obviate oscillations in the Q_p values, for all the iterations after the first two iterations, initial Q_p values are taken as the average of Q_p values

obtained from the past two iterations, i.e., for the tth iteration,

$$_{(t)}Q_p = \frac{_{(t-1)}Q_p + _{(t-2)}Q_p}{2} \qquad (62)$$

in which the prefixing subscripts t, $t-1$ and $t-2$ denote the tth, $(t-1)^{th}$ and $(t-2)^{th}$ iterations, respectively.

For the network of Figure 48, the Q-equations are given by Equations (52 a–g). Table 19 illustrates, for the first few iterations, how the R_p' and Q_p values are obtained for this network.

The linear theory method does not require initialization of the pipe flows and also it converges in a few iterations. However, the number of equations to be solved is large as compared to the other two methods.

The finite element method proposed by Collins and Johnson [121] is similar to the linear theory method. However, it linearizes the H-equations.

Analysis through optimization

Analysis of a distribution network can be carried out through optimization. This concept is based on the conservation of energy principle: "Flows in closed distribution networks will adjust so that the expenditure of the system energy is the minimum". Arora [122] used this concept for a 2-pipe looped network and showed that the result obtained is the same as that obtained by the Hardy Cross method based on balancing the headlosses in a loop.

Collins, et al. [123,124] have developed optimization models for large networks and have proved that the optimization of these models yields the solution of the pipe network analysis problem. They developed two optimiza-

TABLE 18. One Iteration of Hardy Cross Method for Network of Figure 48.

Loop	Pipe	R	Assumed Discharge, (m³/s)	Headloss[a], h = RQ^n (m)	ΣRQ^n	|nRQ^{n-1}|	Σ|nRQ^{n-1}|	ΔQ, (m³/s)	Corrected Discharge, (m³/s)
1	1	40	0.35	−5.736	−1.328	30.317	54.885	0.024	0.350
	2	20	0.45	4.565		18.769			0.474
	3	10	0.05	−0.039		1.450			0.050
	4	30	0.05	−0.118		4.349			0.023
2	4	30	0.05	0.118	2.707	4.349	53.43	−0.051	0.023
	5	40	0.30	4.313		26.594			0.249
	6	20	0.20	−1.018		9.421			0.251
	7	50	0.10	−0.706		13.066			0.151
3	1	40	0.35	5.736	0.775	30.317	31.767	−0.024	0.350
	3	10	0.05	0.039		1.450			0.050
	pseudo pipe, 8	—	—	95 − 100 = −5	—				

[a]clockwise headloss is positive, and anticlockwise negative.

TABLE 19. Some Iterations of Linear Theory Method for Network of Figure 48.

Pipe, i	R_p	Iteration I				Iteration II				Iteration III	
		Initial Q_p	R_p'	Obtained Q_p		Initial Q_p	R_p'	Obtained Q_p		Initial Q_p	R_p'
1	40	1	40	0.16		0.16	8.42	0.67		0.415	18.94
2	20	1	20	0.37		0.37	8.59	0.535		0.452	10.18
3	10	1	10	−0.14		−0.14	−1.88	0.37		0.115	1.59
4	30	1	30	0.08		0.08	3.51	−0.075		0.003	0.22
5	40	1	40	0.25		0.25	12.31	0.26		0.255	12.52
6	20	1	20	0.25		0.25	6.16	0.24		0.245	6.05
7	50	1	50	0.15		0.15	9.97	0.14		0.145	9.69

tion models: (1) A model in which the nodal HGL values are the decision variables and the objective function is the minimization of the sum of the headlosses along various elements of the network; and (2) a model in which the pipe flows are taken as the decision variables and the objective function is the minimization of the sum of these flows. They have suggested several algorithms for the solution of these models.

Electrical Analyzers

Similarities exist between the water distribution networks and the electrical circuits. These similarities are: (1) Discharge Q and the current I; (2) headloss h and voltage drop V; (3) pipe resistance R and electrical resistance R; (4) the algebraic sum of the flows at a junction is zero (node flow continuity relationship) and the algebraic sum of the currents at a junction is zero (Kirchoff's first law); and (5) the algebraic sum of the head losses along a loop is zero (loop head-loss relationship) and the algebraic sum of the voltage drops along an electrical circuit is zero (Kirchoff's second law). However, for water distribution networks, the headloss is given by a nonlinear relationship, Equation (48), while for electrical circuits $V = RI$.

Camp and Hazen [125] were the first to introduce electrical analyzers for the analysis of water distribution networks. They used linear resistors ($V = RI$) and therefore adjustments of the resistors by trial and error was necessary. McIlroy [126] succeeded in obtaining nonlinear resistors, termed fluistors, in which the voltage drop changed nonlinearly with the current.

Electrical analyzers require special equipment, services of trained personnel and have high initial cost. The accuracy is within ± 3%. However, they provide physical feel of the network to the designer and are well suited for sensitivity analysis. Several water departments such as at Baltimore and Philadelphia and institutions like Cornell University in U.S.A. installed electrical analyzers as soon as they were commercially available. However, electrical analyzers are not in vogue these days, because computers give faster, cheaper and more accurate results.

Of the five analysis methods described here, the Hardy Cross, Newton-Raphson, and linear theory methods are widely used in practice. Wood and Rayes [127] have studied these methods in detail and have compared their reliability. Procedures to include pumps, check valves and pressure reducing valves in the analysis are also available [46,110]. Collins [128] has described the pitfalls in the analysis of distribution networks. Rao, et al. [129,15] have described the procedure for simulating the behaviour of a water distribution system over time by considering the variation in demand pattern and reservoir water levels.

NODE FLOW ANALYSIS [130–133]

A distribution system, although initially adequately designed, may subsequently be found to be inadequate due to some reasons [130] such as: (1) the demand being more than the anticipated one due to accelerated rate of growth; (2) imbalance in the demand growth pattern; (3) continued operation of the distribution system even after its design life; (4) unauthorized large diameter withdrawal connections or installation of pumps; and (5) closure of some pipes for cleaning repairs or replacement. When such a distribution network is analyzed by any one of the analysis techniques described earlier, it is assumed that the consumptions at all the demand nodes are satisfied and accordingly the flows in the various elements and the corresponding nodal heads are estimated. This type of analysis usually deals with the estimation of the nodal heads (HGL values) at the demand nodes and therefore was termed by Bhave [130–133] as "Node Head Analysis, (NHA)". In the NHA, it is assumed that additional heads or booster pumps will be provided, wherever necessary, to overcome the head deficiencies. Thus, in the NHA, the nodal heads are estimated and then compared with the required ones. The NHA helps in locating the nodes that are deficient in head requirements, in estimating the nodal head deficiency, and also in estimating the required additional heads at the source nodes or boosting pressures in the system to ensure the adequate nodal heads.

When additional heads or booster pumps are not

available, the distribution system fails to simultaneously satisfy the consumptions at all the demand nodes. Although the demands may be completely satisfied at some nodes, they may be only partially satisfied at some other nodes, while there may not be any outflow at the remaining demand nodes. To analyze such a situation, it is necessary to consider the system as it is and then estimate the actual outflows available at the demand nodes. In contrast to the NHA, this type of analysis is termed by Bhave as "Node Flow Analysis, (NFA)." In the NFA, it is assumed that: (1) neither any additional heads at source nodes nor any additional boosting pressures are provided; and (2) flow at a node occurs only if the available nodal head is at least equal to the minimum required one.

Bhave has developed the NFA techniques for serial [130], branching [131], and looped [132] networks. The general NFA problem is treated as an optimization problem [132] and is based on the converse of the conservation of energy principle, stated earlier. Thus, "for a given system energy, the flows in the distribution system will adjust so that the system outflow is maximum." The nodal outflows are taken as the decision variables and the objective function involves maximization of the sum of the nodal outflows. The constraints include the head-loss relationships for the different elements, the node-flow-continuity relationships, and the loop head-loss relationships. In addition, there are alternate constraints which state that (1) the outflow at a demand node is equal to the required one if the available HGL at the demand node is equal to or more than the minimum required one; (2) the nodal outflow lies between the required value and zero if the nodal HGL is equal to the required one; and (3) the nodal outflow is zero if the nodal HGL is less than the minimum required one.

Collins, et al. [123] have shown that the network flow optimization model yields the unique solution of the distribution system analysis. Conversely, therefore, the distribution system analysis should yield the solution of the optimization

TABLE 20. Distribution System Details for Node Flow Analysis.

		Design Solution		NFA Solution	
Node	Required HGL (m)	Consumption (L/min)	Available HGL (m)	Outflow (L/min)	HGL (m)
1	100.00	−5,000	100.00	−3,108	100.00
2	87.00	700	89.92	0	86.70
3	85.00	800	87.30	800	85.48
4	90.00	400	93.20	400	92.27
5	88.00	800	89.69	0	87.93
6	85.50	1,400	87.01	1,381	85.50
7	86.00	900	86.98	527	86.00

problem. Therefore, instead of directly obtaining the solution of the optimization problem and carry out the NFA, the NHA methods, described earlier, can be used. However, because of the alternate constraints, the NHA must be carried out several times, to obtain in the end, a unique solution which satisfies the proper alternate constraints. A systematic procedure to carry out the NHA is available [132].

A distribution system is shown in Figure 51. The node details are given in Table 20. As the available nodal HGL values for the consumptions are more than the minimum required values, the design solution as obtained by the usual NHA is satisfactory. When pipe 1-2 is closed for repairs, the NFA solution (obtained after three NHA iterations) is obtained as shown in Figure 52 and Table 20. Nodes 3 and 4 completely satisfy the demand, nodes 6 and 7 partially satisfy the demand, while nodes 2 and 5 have no outflow at all.

Calibration of Distribution Systems

Calibration of an existing water distribution system [134–136] is necessary to predict its behaviour under

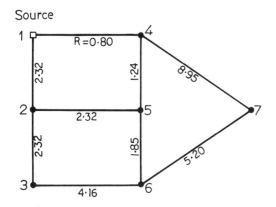

FIGURE 51. Distribution system for node flow analysis.

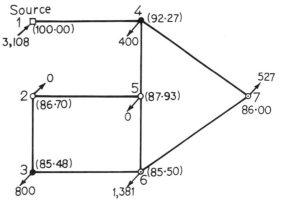

FIGURE 52. Node flow analysis solution.

different field conditions. The physical and operating characteristics of an existing system should be simulated, as accurately as possible, in the mathematical model so that the input to the model would predict realistic results. For calibration of an existing water distribution system, it is necessary to accurately estimate (1) the nodal outflows, (2) the head-loss characteristics of pipes and other elements; and (3) the heads supplied by reservoirs and pumps. To reduce the computational work and make the analysis tractable, it is common to skeletonize the distribution system by removing small size pipes in mathematical modeling.

A distribution system is considered to be calibrated when the predicted nodal heads are satisfactorily close to the heads obtained from field gauge readings for several operating conditions and water consumptions. If a good data set is available, a head difference in the range of 1.5–5.0 m is satisfactory; however, if the data set is poor an accuracy in the range of 3–10 m is a reasonable target. In fixing the permissible accuracy, the head loss from the source to a node should be considered. For a satisfactory calibration, the predicted head at a node should be such that the head loss from the source to the node should not be more than 10% away from the headloss obtained from the field results.

In calibrating a water distribution system model, it is customary to adjust the resistance constant R for the pipes and the consumption q at the demand nodes so that the heads and flows predicted from the model agree with the values observed in the field. Walski [135] gives several formulas which are useful for the adjustment of the R and q values. Qualitatively, (1) adjust R if there is similar error at high and low flows; (2) adjust R and q by similar amounts if the model is accurate at low flow but inaccurate at high flow; and (3) adjust q if the model is accurate at high flow but inaccurate at low flow.

Design of Distribution Systems

INTRODUCTION

A water distribution system must supply the desired quantity of water to the consumers at heads, at least equal to the minimum required ones. Numerous solutions satisfying these requirements are feasible. However, the cost incurred on the distribution system may be 50% or even more of the total cost incurred on the entire water supply system. Therefore, the design of a water distribution system should be as economical as possible.

Several techniques are available for the optimal design of water distribution systems. These techniques can be classified into two categories: (a) Check design and optimization techniques; and (b) direct design and optimization techniques.

Check Design and Optimization Techniques

The check design and optimization techniques have been used traditionally. The designer, depending upon his experience and judgement, assumes the pipe sizes, the heights of the reservoirs and the pumping pressures. He then analyzes the assumed distribution system by one of the methods described earlier, and checks whether the design is hydraulically satisfactory. If the design is not satisfactory, he changes the assumed values until a hydraulically satisfactory distribution system is obtained. Thus, this method is based on a trial-and-error approach, and the designer has to carry out several analyses (unless he initially selects large pipe sizes, heads and pumping pressures) to obtain a single hydraulically satisfactory system. Repeating this procedure, the designer obtains several hydraulically satisfactory designs and selects the least costly one. Naturally, the solution that he selects is one of the few that he has obtained, and therefore the selected solution may be far from the optimal one.

Direct Design and Optimization Techniques

In the direct design and optimization techniques, the designer starts from a hydraulically satisfactory network. Using one of the several optimization techniques available, he successively improves his solution. Each successive solution is less costly than the previous one. In the end he reaches a solution which cannot be improved further. This solution is the optimum solution. The optimum solution he obtains, sometimes depends upon the initial solution. A different initial solution may lead to a different optimum solution. All such optimum solutions are the local optimum solutions and usually only one of them is the global optimum one.

Some of the direct design and optimization techniques are described here. As it is easier to optimize branching distribution systems, they are initially considered, and the looped distribution systems are considered later.

OPTIMIZATION OF BRANCHING DISTRIBUTION SYSTEMS

In branching distribution systems, for the design nodal demands, the discharge in all the pipes can be uniquely fixed. It is therefore possible to obtain the global optimum solution for branching systems. Some optimization methods are described and also partly illustrated through a simple branching network.

Differential Calculus Approach [137]

In the differential calculus approach, the pipe diameter is taken as a continuous variable and the pipe cost is expressed as

$$C = K L D^m \qquad (63)$$

in which C = capital cost of pipe of length L and diameter D; K = pipe cost constant; and m = an exponent usually lying between 1 and 2. The values of K and m are obtained by regression analysis of the available pipe sizes. Usually, only one relationship for the entire set of the available pipe sizes is sufficient. However, for more accuracy, different

TABLE 21. Pipe Diameter-Cost Relationship.

Pipe Diameter (mm)	Unit Cost (rupees)	Range of Pipe Diameter (mm)	Cost Function Relating C (rupees), L (m) and D (mm)
80	87		
100	106		
125	134	80–200	$C = 0.703\ L\ D^{1.093}$
150	167		
200	236		
250	318		
300	415	200–400	$C = 0.122\ L\ D^{1.426}$
350	516		
400	636		

relationships, each applicable to a small range of pipe sizes, can be obtained and used.

The head-loss relationship for a pipe is expressed as

$$H_L = \frac{K_1\ L\ Q^p}{D^r} \qquad (64)$$

in which H_L = head loss through a pipe of length L and diameter D while carrying a discharge Q; K_1 = a coefficient that depends upon the pipe material and diameter, type of flow and the units of other terms; and p, r = exponents.

Using Equations (63) and (64) for a distribution system, it has been shown by Bhave [137–138] that for optimality,

$$\sum \left(\frac{mC}{H_L} \right)_{ij} = \sum \left(\frac{mC}{H_L} \right)_{jk} \quad \text{for all intermediate nodes } j$$

$$(65)$$

in which suffixes ij and jk denote upstream pipe ij and downstream pipe jk, respectively, for intermediate node j.

In the optimization procedure, the nodal HGL values H_j

are taken as the decision variables. They are initially assumed, preferably by using the critical path concept [137,139,140], illustrated later; and are successively corrected applying an additive correction $\triangle H_j$ given by

$$\Delta H_j = \frac{- \sum \left(\dfrac{mC}{H_L} \right)_{ij} + \sum \left(\dfrac{mC}{H_L} \right)_{jk}}{\sum \left[\dfrac{mC}{(H_L)^2} \right]_{ij} + \sum \left[\dfrac{mC}{(H_L)^2} \right]_{jk}} \qquad (66)$$

for all intermediate nodes j.

The correction ΔH_j is restricted such that the H_j value obtained after applying the correction satisfies the node HGL constraint

$$H_j^{max} \geq H_j \geq H_j^{min} \qquad (67)$$

in which H_j^{max}, H_j^{min} = maximum and minimum permissible HGL values at node j, respectively. The iterative procedure is continued until all the ΔH_j values become insignificantly small or the saving in system cost is negligible. The method is iterative and suitable for hand calculation.

ILLUSTRATIVE EXAMPLE

The branching distribution system shown in Figure 53 is optimized using the Hazen-Williams formula with coefficient = 100. The available pipe sizes and their unit costs are given in Table 21.

In the distribution system (Figure 53), demand nodes 2, 4 and 5 are the end nodes. Therefore, for optimality $H_j = H_j^{min}$ at these nodes. The optimal H_j values at nodes 1 and 3 should be determined. The first trial H_1 and H_3 values are determined by critical path concept as in Table 22. Path 0-1-2 (Table 22) has the minimum available friction slope of 0.00500 (shown within parentheses) and therefore path 0-1-2 is the critical path. The slope of the hydraulic grade line for path 0-1-2 is 0.00500 and therefore for intermediate node 1, H_1 = 100.00-0.00500 × 400, i.e., 98.00 m for the first

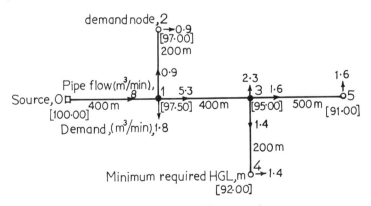

FIGURE 53. Branching network.

TABLE 22. Determination of Critical Paths and Critical Subpaths.

Distribution System	Path	Length of Path (m)	HGL at Source of Path (m)	HGL at End of Path (m)	Maximum Permissible Friction Head (m)	Available Friction Slope for Path
			Distribution system			
1	0-1	400	100.00	97.50	4.00	0.00625
	0-1-2	600		97.00	3.00	(0.00500)[a]
	0-1-3	800		95.00	5.00	0.00625
	0-1-3-4	1000		92.00	8.00	0.00800
	0-1-3-5	1300		91.00	9.00	0.00692
			Distribution subsystem of first order			
1.1	1-3	400	98.00	95.00	3.00	(0.00750)
	1-3-4	600		92.00	6.00	0.01000
	1-3-5	900		91.00	7.00	0.00777
			Distribution subsystems of second order			
1.1.1	3-4	200	95.00	92.00	3.00	(0.01500)
1.1.2	3-5	500		91.00	4.00	(0.00800)

[a]Critical slope is shown within parentheses.

TABLE 23. Estimation of HGL Corrections and System Cost.

Demand node, j	HGL Minimum Required H_j^{min} (m)	HGL Assumed H_j (m)	Pipe	Nature of Pipe	Discharge Q (m³/min)	Length L (m)	Head Loss H_L (m)	Diameter D (mm)	Index m	Pipe Cost C (rupees)	$\frac{mC}{H_L}$	$\frac{mC}{(H_L)^2}$
1	97.50	98.00	0-1	Supply	8	400	2.00	388.9	1.426	240,738	(−)171,646	85,823
			1-2	Distri-bution	0.9	200	1.00	169.6	1.093	38,435	42,009	42,009
			1-3	Distri-bution	5.3	400	3.00	306.0	1.426	171,048	81,305	27,102
											−48,332	154,934

$\Delta H_1 = -48,332/154,934 = -0.31$ m; therefore, $H_1 = 98.00 - 0.31 = 97.69$ m > 97.50 m

Demand node, j	HGL Minimum Required H_j^{min} (m)	HGL Assumed H_j (m)	Pipe	Nature of Pipe	Discharge Q (m³/min)	Length L (m)	Head Loss H_L (m)	Diameter D (mm)	Index m	Pipe Cost C (rupees)	$\frac{mC}{H_L}$	$\frac{mC}{(H_L)^2}$
3	95.00	95.00	1-3	Supply	—	—	—	—	—	—	(−)81,305	27,102
			3-4	Distri-bution	1.4	200	3.00	160.1	1.093	36,084	13,147	4,382
			3-5	Distri-bution	1.6	500	4.00	191.6	1.093	109,800	30,003	7,501
											−38,155	38,985

$\Delta H_3 = -38,155/38,985 = -0.98$ m; therefore, $H_3 = 95.00 - 0.98 = 94.02$ m

But $H_3^{min} = 95.00$; therefore $H_3 = 95.00$ m

Total pipe cost = 596,105 rupees

TABLE 24. Selection of Pipe Sizes in Optimization by Linear Programming.

Link	Discharge, (m^3/min)	Length (m)	Friction Slope as Obtained from Critical Path Concept	Pipe Size	Adjacent Pipe Sizes		Friction Slopes of Pipe Sizes	
					Larger	Smaller	Larger	Smaller
1	8.0	400	0.00500	388.9	400	350	0.00436	0.00835
2	0.9	200	0.00500	169.6	200	150	0.00224	0.00909
3	5.3	400	0.00750	306.0	350	300	0.00390	0.00826
4	1.4	200	0.01500	160.1	200	150	0.00507	0.02059
5	1.6	500	0.00800	191.6	200	150	0.00649	0.02636

iteration. The critical path 0-1-2 is removed from the system leaving behind distribution subsystems of the first order. The path concept is continued further (Table 22) and the critical subpaths for all the distribution subsystems of the first and higher orders are obtained and the first-trial HGL values for the intermediate nodes are determined. The use of the critical path concept in determining the first-trial HGL values is found to reduce the optimization iteractions.

For the Hazen-Williams formula with coefficient = 100, H_L and L in meters, Q in cubic meters per minute and D in millimeters, in Equation (64), $K_1 = 4.3734 \times 10^8$, $p = 1.85$ and $r = 4.87$. The first optimization iteration is carried out as shown in Table 23. The system cost is Rs. 596,105. The next iteration is carried out with $H_1 = 97.69$ m; and $H_3 = 95.00$ m. The iterative procedure is continued until the decrease in system cost is negligible.

The optimal solution has pipes of odd sizes, which are replaced by an equivalent series combination of nearer larger and smaller pipe sizes. Care should be taken to ensure that the node HGL constraints (Equation 67) are not violated.

Linear Programming Approach

Karmeli, Gadish and Meyers [141], Gupta [142], Calhoun [143], etc., suggested linear programming (LP) approach for the optimization of the branching distribution systems. As the head loss in, and the cost of a pipe are the linear functions of its length, the different pipe lengths constituting a link connecting two nodes are taken as the decision variables in the LP model. The objective function is to minimize the sum of the capital costs of the various links. The constraints are: (1) For all links, the sum of the lengths of the pipes selected must equal the length of the link; (2) at each demand node, the available HGL value must be equal to or greater than the minimum required HGL value at the node; and (3) all pipe-lengths must be non-negative.

It can be proved that in the LP solution, each link must consist of either one or two pipe sizes only. However, for global optimality of the solution, the entire set of the available pipe sizes must be used in the LP formulation. This considerably increases the size of the LP model. Therefore, in practice, two or three pipe sizes are selected for each link and the LP model is formulated and solved. Bhave [144] has shown that the critical path concept,

described earlier, can also be used to select such pipe sizes. The selection of pipe sizes and the formulation of the LP model are described for the illustrative example considered earlier.

ILLUSTRATIVE EXAMPLE

The branching distribution system (Figure 53) is optimized by LP. Applying the critical path concept (Table 22), the critical slopes for all the links (shown in parentheses in the last column of Table 22) are obtained. Using these critical slopes, the pipe sizes are obtained and the nearest available larger and smaller commercial size pipes (Table 21) are selected for each link. Friction slopes of these commercial pipe sizes are also calculated (Table 24). Using these data, the LP model is formulated as follows:

$$\text{Minimize:} \quad 636 L_{1-400} + 516 L_{1-350} + \ldots$$
$$+ 236 L_{5-200} + 167 L_{5-150} \quad (68)$$

$$\text{Subject to:} \quad L_{1-400} + L_{1-350} = 400$$
$$\vdots$$
$$\vdots \quad (69)$$
$$\vdots$$
$$L_{5-200} + L_{5-150} = 500$$

$$0.00436 L_{1-400} + 0.00835 L_{1-350} \leq 100.00 - 97.50$$
$$\vdots$$

$$0.00436 L_{1-400} + 0.00835 L_{1-350} + 0.00390 L_{3-350}$$
$$+ 0.00826 L_{3-300} + 0.00649 L_{5-200} \quad (70)$$
$$+ 0.02636 L_{5-150} \leq 100.00 - 91.00$$

$$L_{1-400}, L_{1-350}, \ldots, L_{5-200} \ L_{5-150} \geq 0 \quad (71)$$

The objective function is given by Equation (68) in which L_{1-400} represents the length of pipe of 400 mm diameter in link 1 and 636 is the unit cost of this pipe (Table 21). Similarly, other terms in Equation (68) are obtained. Equation (69) represents a set of 5 link-length constraints, one for each link 1, . . . 5. Here, the first constraint states that the sum of the pipe lengths of 400 mm and 350 mm diameter in link 1 should be 400 m, the length of link 1. Similarly, the link-length constraints for links 2, . . . , 5 are obtained. Equation (70) represents a set of 5 path head-loss constraints, one for each path starting from source 0 and ending at nodes 1, . . . 5, respectively. For example, the first inequality in Equation (70) states that the sum of the head losses in the two pipes of 400 mm and 350 mm diameter in link 1, having 0.00436 and 0.00835 friction slopes, respectively (Table 24), should be less than or equal to $(100.00 - 97.50)$m, the maximum permissible head loss on this path, without violating the HGL constraint for node 1. Equation (71) represents the usual non-negativity constraints. The LP model is solved, usually by the Simplex algorithm, to obtain the optimal solution.

For the illustrative example, the critical path concept gave two pipe sizes for each link and therefore, the number of decision variables is 10. Thus, the critical path concept has helped in reducing the number of decision variables from $9 \times 5 = 45$ (the product of the number of available pipe sizes in Table 21 and the number of the links in the distribution system).

The linear programming approach can be extended to multiple source nodes [145], multiple demands [146] and to a system having variable pumping head or reservoir height, pressure reducing valves and pipes having different pressure ratings [147].

Other Approaches

Based upon Cowan's criterion [a particular case of the cost-headloss criterion, Equation (65)] of checking trunk main design [148], Deb [149] formulated a procedure for designing a water main system in series. The pipe-diameter is treated as a continuous variable and is replaced in the end by commercially available pipes. Later, Deb [150] extended these techniques for the optimal design of branching systems.

Perold [151] suggested a method in which a trial design is checked by inspecting the economy of changing a pipe-size to the next size in one part of the system accompanied by a similar compensating change in size in another part. For a unit change in head loss, the corresponding changes in the pipe lengths and hence the costs are compared. Based on this information, appropriate modifications in the pipe diameters are made. Perold [152] has also considered pumped systems and presented charts for their economical design. Lekale, et al. [153] provided algorithm for long-term optimization of branching distribution systems.

Liang [154] suggested the use of dynamic programming for optimization of a serial system. Each link of the system is treated as a dynamic programming stage. The HGL values at the inlet and outlet of each link are the input and output state variables and the link diameter is the decision variable of each stage. The sum of link-capital cost, power cost and wasted-water cost of all stages was used as a criterion for measuring the overall system effectiveness. This procedure was later extended by Yang, Liang and Wu [155] for optimization of branching systems.

OPTIMIZATION OF LOOPED DISTRIBUTION SYSTEMS

The discharge in the links of looped distribution systems cannot be fixed uniquely; therefore their optimal design is rather difficult. However, when a looped distribution system is optimized, two points are observed: (1) If no constraints are imposed, some links are eliminated and the looped system is converted to a branching one; and (2) the obtained branching system is not unique but depends upon the flow distribution or the nodal heads, assumed as the initial solution in the optimization procedure.

To maintain the looped nature of the distribution systems, one of the two constraints are imposed: (1) "Specified-diameter constraint" in which each link of the distribution system must be at least equal to a minimum or specified diameter [137,156]; or (2) "specified-discharge constraint" in which each link must carry a minimum or specified discharge [137,157]. The links so retained are the loop-forming "secondary" links and the links on the branching configuration are the "primary" links.

Each branching configuration and the secondary links give a local optimum solution. Usually, only one of them is the global optimum solution. As of 1985, no method is available that can identify the branching configuration and the secondary links which would yield the global optimum solution. Even though it is theoretically possible to obtain optimal solution corresponding to each of all the possible branching configurations, it is impractical to use this approach. Thus, obtaining the global optimum solution for looped distribution systems has remained elusive in practice. The different techniques employed for optimizing looped systems use, rather directly or indirectly, a branching configuration that may lead to a fairly good but still a local optimum solution.

Indirect Techniques

In the indirect techniques of optimization of looped systems, the branching configuration is not obtained separately but is used, rather indirectly, in the optimization algorithm. The conversion of the looped system to a branching system is carried out simultaneously with the optimization of the system.

EQUIVALENT PIPE APPROACHES

Tong, et al. [158] presented an equivalent length approach based on the contention that the total amount of pipe in a network is minimum for the given head, inflows and outflows, if the algebraic sum of equivalent pipe lengths of the same size and roughness in a loop is zero, i.e., $\Sigma L_e = 0$. Raman and Raman [159] demonstrated that for equivalent pipe sizes, instead of ΣL_e, $\Sigma(L_e/Q)$ should be zero for each loop.

In contrast to the equivalent length approach, Deb and Sarkar [160] suggested an equivalent diameter approach in which all pipes of the system are replaced by equal lengths of equivalent diameter pipes. A parabolic pressure surface, from source to the farthest demand node is generated over the network, the nodal HGL values and corresponding pipe sizes are calculated, and the system cost is obtained. The procedure is repeated for several pressure surfaces to obtain the least costly system.

It has been shown that the equivalent pipe approaches do not yield even local optimum solutions [161–164].

LINEAR PROGRAMMING APPROACH

Kally [165,166] suggested the use of LP for looped systems. The pipe lengths in a link are the decision variables. The system is initially balanced, the decision variables are changed by 1 unit, the new system is solved and the changes in all the HGL values are noted. These are then used as the coefficients in the linear inequalities, the LP model is formed and then solved. The process is repeated until it converges, i.e., the solution of LP model does not change on successive iterations.

LINEAR PROGRAMMING GRADIENT APPROACH

Alperovits and Shamir [156] suggested an approach in which the link flows (the decision variables) are initially assumed and optimal solution is obtained by LP. Information available from this solution is used to calculate a linear programming gradient to modify the link flows and obtain a reduction in the system cost. A new LP model is then formulated and solved. The process is repeated until it converges to an optimal solution. Quindry, et al. [167] suggested a correction to the gradient-search expression, but it increased the computational work.

Following the Lai and Schaake formulation [168], Quindry, et al [169,170] suggested another gradient approach in which the nodal HGL values are taken as the decision variables, and are corrected successively using the linear programming gradient approach to obtain the optimal solution.

NONLINEAR PROGRAMMING AND OTHER APPROACHES

The looped distribution system optimization model is formulated as a nonlinear programming problem. It is then solved by different iterative techniques, in which starting from one feasible solution, the objective function is successively minimized.

Pitchai [171] used random sampling technique to search for the optimal solution. Jacoby [172] used numerical gradient approach, Lam [173] used discrete gradient approach,

and Watanatada [174] used gradient balancing approach to obtain the optimal solution. Rasmusen [175] suggested an iterative procedure in which diameter of a pipe is sequentially adjusted according to the ratio between marginal pipeline and energy costs. Cembrowicz and Harrington [176], Shamir [177] and several other investigators have also suggested different approaches.

Direct Techniques

In direct techniques, the optimization of a distribution system is carried out in two stages. In the first stage the branching configuration and the loop forming links, leading to a fairly good optimal solution, are identified. In the second stage the distribution system is optimized. Some methods using this approach are described.

PATH CONCEPT APPROACH

Bhave [137] suggested an approach, based on path concept, for identifying the branching configuration for single-source systems. This approach is based on the assumption that it is cheaper to transport water from the source to a demand node by the shortest route, i.e., path. When all the paths are identified, a branching configuration is obtained. The loop-forming links are provided such that they satisfy either the specified-diameter constraint or the specified-discharge constraint. Taking the link-diameter as a continuous variable and the nodal HGL values as the decision variables, the distribution system is optimized by the differential calculus approach, described earlier for branching systems.

MINIMUM SPANNING TREE APPROACH

Ridgik and Lauria [178] suggested the use of minimum spanning tree approach. Here all the nodes are connected by a branching configuration such that the sum of the lengths of the links on this configuration is the minimum.

LINEAR PROGRAMMING APPROACH

To select branching configuration for multisource systems, Bhave [179,180] suggested an approach that combines the path concept and linear programming principles. Path concept is used to obtain branching configuration from each source to the demand nodes. Treating the pipe diameter as a continuous variable an LP model, similar to the classical transportation model, is then formulated and solved to select the branching configuration. Using the commercial pipe sizes, another LP model is formed and solved to obtain the optimum solution.

NONLINEAR PROGRAMMING APPROACH

Treating the link diameter as a continuous variable, Rowell and Barnes [181] suggested an approach in which the branching configuration is obtained by formulating and solving a nonlinear programming model. Later, using the commercially available pipe sizes, an LP model is formed and solved to obtain the optimal solution. A 0–1 integer programming model is used to determine the diameters of the loop-forming links to provide a specified level of reliability in case of failure of the links of the branching configuration.

OPTIMAL EXPANSION OF WATER DISTRIBUTION SYSTEMS

A water distribution system may become inadequate, even during its design life, if accelerated or uneven growth occurs. Further, since it is extremely costly to replace an existing water distribution system after its design life with an entirely new system, attempts are made to improve the delivery capability of the existing system. Thus, optimal expansion of existing water distribution systems is important. Many of the approaches described earlier can be used for this purpose. However, Bhave [182,183] has exclusively considered this aspect. The approach described earlier using the path concept and differential calculus [137] is extended so that an existing distribution system can be expanded to include (1) additional gravity or pumped source nodes; (2) layout modifying links; (3) new links parallel to some existing links; (4) replacement of existing links; and (5) additional demand nodes and corresponding links to cover additional localities.

In the suggested algorithm the designer initially specifies (1) location of the additional gravity or pumped source nodes; (2) layout modifying links connecting existing and new nodes; and (3) links requiring replacement. The algorithm optimally selects (1) the HGL values at the pumped source nodes; (2) the existing links that need strengthening; and (3) the commercially available sizes of the new pipes.

The method is iterative, rapidly locates the links which need strengthening, and is even suitable for hand calculation for small distribution systems. Optimal design of a new distribution system is a particular case of this method. It can also be extended to consider time-varying demand pattern.

GENERAL REMARKS

It should be noted that when a distribution system is optimized for a particular demand pattern, as the solution approaches the global optimum one, the redundancy present in the system becomes less and less. Therefore, for a particular demand pattern nearer an optimal solution is to the global optimum one, poorer is its performance under emergency conditions. Thus, optimality and reliability are two conflicting aspects in the design. The optimality reduces redundancy and hence reliability of the system, while reliability requires more redundancy and thereby increases the cost of the system. It is therefore necessary for the designer to initially identify the conditions under which he wants to study the performance of the system; and include these conditions in the system optimization so that he achieves a satisfactory combination of reliability and optimality.

Maintenance of Distribution Systems

RECORDS

Proper records, usually in the form of comprehensive maps, should be prepared at the time of laying of a distribution system. These records should include (1) type and size of the various pipes with provided covers; (2) location of the fire hydrants; (3) type and location of the various valves; and (4) location of other appurtenances. Standard symbols and suitable colour combinations can be used. The records should be properly kept and should be readily available throughout the life of the system. Replacements and additions should be properly recorded with dates.

DISINFECTION AND CLEANING

New and Repaired Pipes

The disinfection of new and repaired pipes includes: (1) Cleaning and flushing the pipe before disinfection; (2) disinfecting jute, yarn or other jointing material by soaking the material in a 50 ppm strong chlorine solution for at least 30 min; and (3) subjecting the pipe to a very high dose of chlorine (10–200 ppm) for a period of 12–24 h.

When dry calcium hypochlorite compounds are used, weighed quantities of the compounds may be placed in the new pipes. When water is introduced later on, the pipe is subjected to high concentration of chlorine solution. When gaseous chlorine is used, special diffusers and silver tubing are needed for introducing the gas into the pipes. Cox [22] has provided tables giving the quantity of dry disinfectant or chlorine gas required to provide a dose of about 50 ppm.

Pipes should be periodically cleaned by passing a scraping machine or using chemicals. Care should be taken to see that pipe coating is not impaired. Water pipes can be inspected from inside by "fibrescope", an equipment based on the fibre optics technology used in the medical field [184]. Dead ends in the distribution system should be periodically flushed.

Elevated Tanks and Other Structures

Elevated tanks and other structures can be disinfected (1) by direct application of a strong chlorine solution (120 g chloride of lime or 45 g calcium hypochlorite per 150 L of water) to the surfaces of the structures; or (2) by direct application of chloride of lime or calcium hypochlorite (dose 50 ppm) to the water used to fill the structure.

LOCATING UNDERGROUND PIPES

Underground pipes can be easily located if the records are properly maintained. In the absence of well-maintained records, underground pipes can be located by using the pipe as a part of a closed electric circuit and inducing a magnetic field surrounding the pipe. When an observer approaches the pipe, the sound in the earphones increases, and when he is directly above the pipe, the sound stops. Radio transmitter to pick up the magnetic field induced in the pipe can also be used.

LEAK DETECTION AND PREVENTION

Loss of water through leakage may be as high as 50–70% of the total water, as observed in a recent Norwegian survey [185]. Therefore attempts should be made to locate and reduce leakage.

Some of the methods used for leak detection are: (1) direct observation; (2) driving sounding rod into the ground, and withdrawing it to observe whether it is wet or muddy; (3) employing sound amplifiers to observe the sound of leaking water; (4) observing the hydraulic gradient; (5) thermal photography; (6) induced vibrations; etc.

Leaking pipes and appurtenances should be repaired, or should be replaced if leakage occurs frequently. Techniques have been developed [186] to repair CI, steel, concrete or plastic pipes of 50–600 mm diameter without extensive excavation and exposing the pipe, but by internally injecting a plastic resin into the area surrounding the damaged portion, through a travelling injection head.

REFERENCES

1. Dunn, D. F. and T. E. Larson, "Relationship of Domestic Water Use to Assessed Valuation with Selected Demographic and Socio-economic Variables," *Jour. Am. Wat. Wks. Assoc.,* *53,* 441 (1963).

2. Linaweaver, F. P. Jr., et al., "Summary Report on the Residential Water Use Project," *Jour. Am. Wat. Wks. Assoc., 59,* 267 (1967).

3. Gupta, V. P. and P. R. Bhave, "A Study of Domestic Water Consumption in Nagpur," *Jour., Inst. of Engrs. (India), 51* (PH1): 5 (1970).

4. Agthe, D. E. and R. B. Billings, "Dynamic Models of Residential Water Demand," *Wat. Res. Research, 16:* 476 (1980).

5. Carver, P. H. and J. J. Boland, "Short and Long-run Effects of Price on Municipal Water Use," *Wat. Res. Research, 16:* 609 (1980).

6. Garg, S. K., *Water Supply Engineering,* Khanna Publishers, New Delhi, India (1977).

7. Deb, A. K. and M. P. Evans, "Dual Distribution System Analysis," *Jour. Am. Wat. Wks. Assoc., 72:* 103 (1980).

8. "Indian Standard Code of Basic Requirements for Water Supply, Drainage and Sanitation," *Indian Standards Institution,* IS:1172 (1971).

9. "Manual on Water Supply and Treatment," *Ministry of Works and Housing,* Govt. of India, New Delhi (1976).

10. Kollar, K. L. and P. Mac Auley, "Water Requirements for Industrial Development", *Jour. Am. Wat. Wks. Assoc.,* 76: 2 (1980).

11. Fair, G. M., J. C. Geyer, and D. A. Okun, *Elements of Water Supply and Wastewater Disposal,* Second ed., John Wiley and Sons, Inc. (1971).

12. Seidel, H. F., "A Statistical Analysis of Water Utility Operating Data for 1965 and 1970", *Jour. Am. Wat. Wks. Assoc., 70:* 315 (1978).

13. Maidment, D. R. and E. Parzen, "Time Patterns of Water Use in Six Texas Cities," *Jour. of Wat. Res. Planning and Management,* ASCE, *110:* 90 (1984).

14. "The Nation's Water Resources," *U.S. Water Resources Council,* (1968).

15. Rao, H. S., L. C. Merkel and D. W. Bree, Jr., "Extended Period Simulation of Water Systems—Part B," *Jour. Hyd. Div.,* ASCE, *103:* 281 (1977).

16. "Research Needs for Capacity Planning," Subcommittee Report, *Jour. Am. Wat. Wks. Assoc., 75:* 15 (1983).

17. Scarato, R. F., "Time-Capacity Expansion of Urban Water Systems," *Wat. Res. Research, 5:* 929 (1969).

18. McJunkin, F. E., "Population Forecasting by Sanitary Engineers," *Jour. San. Engg. Div.,* ASCE, *90:* 31 (Aug. 1964).

19. Meier, P. M., "Stochastic Population Projection at Design Level," *Jour. San. Engrg. Div.,* ASCE, *98:* 883 (1972).

20. "International Standards for Drinking-Water," *World Health Organization,* (1971).

21. "Drinking Water Standards," U.S. Public Health Service, No. 956 (1962).

22. Cox, C. R., "Operation and Control of Water Treatment Processes," *World Health Organization,* (1969).

23. Babbit, H. E., J. J. Doland and J. L. Cleasby, "Water Supply Engineering," Mc-Graw Hill Book Company, New York (1962).

24. Raghunath, H. M., *Ground Water,* Wiley Eastern Limited (1982).

25. Bhole, A. G., "Design of Infiltration Gallery," *Jour. Indian Wat. Wks. Assoc., 6:* 207 (1974).

26. Guerrera, A. A., "Grounding of Electric Circuits to Water Services: One Utility's Experience," *Jour. Am. Wat. Wks. Assoc., 72:* 82 (1980).

27. Robinson, M. P. and R. E. Blair, Jr., "Pump Station Design: The Benefits of Computer Modelling," *Jour. Am. Wat. Wks. Assoc., 76:* 70 (1984).

28. Hazen, R., "Pumps and Pumping Stations," *Jour. New England Wat. Wks. Assoc., 67:* 121 (1953).

29. Lamont, P. A., "Common Pipe Flow Formulas Compared with the Theory of Roughness," *Jour. Am. Wat. Wks. Assoc., 73:* 274 (1981).

30. Chow, V. T., *Open-Channel Hydraulics,* Mc Graw-Hill Book Company, Inc., International Student Edition (1959).

31. Moody, L. F., "Friction Factors for Pipe Flow," *Trans. Am. Soc. of Mech. Engrs.,* 671 (1944).

32. Wood, D. J., "An Explicit Friction Factor Relationship," *Civ. Engineering,* ASCE, *36:* 60 (1966).

33. Swamee, P. K. and A. K. Jain, "Explicit Equations for Pipe Flow Problems," *Jour. Hyd. Div.*, ASCE, *102:* 657 (1976).

34. Shidhaye, V. M., "Recent Developments in the Field of Cathodic Protection," *Jour. Inst. of Engrs. (India), 55* (CH2): 44 (1975).

35. Shidhaye, V. M., "Protection of Underground Water Mains Against Corrosion", *Jour. Inst. of Engrs. (India), 57* (CH3): 127 (1977).

36. Bhole, A. G., "Remedial Measures to Prevent Freezing in Water Pipe Lines," *Proc. Symp. on Water Supply and Waste Disposal at High Altitudes;* Central Pub. Hlth. Engrg. Res. Inst., 23 (1964).

37. Sheppard, F., "Thawing Water Pipes Electrically", *Jour. Wat. Wks. Engrg., 87:* 11 (1934).

38. Steel, E. W., T. J. McGhee, *Water Supply and Sewerage,* 5th ed., Mc Graw Hill Kogakusha Ltd., (1979).

39. Scarino, J. H., "Buried Pipelines: Settlement Modification and Load Transfer," *Transp. Engrg. Jour.*, ASCE, *107:* 469 (1981).

40. Wang, L. R. L. and H. A. Cornell, "Evaluating the Effects of Earthquakes on Buried Pipelines," *Jour. Am. Wat. Wks. Assoc., 72:* 201 (1980).

41. Canales-Ruiz, R., "Optimal Design of Gravity Flow Water Conduits," *Jour. Hyd. Div.*, ASCE, *106:* 1489 (1980).

42. Deb, A. K., "Optimization in Design of Pumping Systems," *Jour. Env. Engrg. Div.*, ASCE, *104:* 127 (1978).

43. Hall, W. A. and J. S. Hammond, "Preliminary Optimization of an Aqueduct Route," *Jour. Irrig. and Drainage Div.*, ASCE, *91:* 45 (1965).

44. Buras N. and Z. Schweig, "Aqueduct Route Optimization by Dynamic Programming," *Jour. Hyd. Div.*, ASCE, *95:* 1615 (1969).

45. "Flowlimiters," WISA, Arnhem, Holland.

46. Jeppson, R. W. and A. L. Davis, "Pressure Reducing Valves in Pipe Network Analysis," *Jour. Hyd. Div.*, ASCE, *102:* 987 (1976).

47. Kulkarni, V. P., "Water Hammer in Pumping Mains: Cause, Effect and Control," *Jour. Ind. Wat. Wks. Assoc., 7:* 115 (1975).

48. "Fire Suppressing Rating Schedule," Insurance Service Office, Philadelphia, U.S.A.

49. "Small Community Water Supplies," *WHO Publication*, Technical Paper 18 (1981).

50. Metcalf and Eddy, *Wastewater Engineering: Treatment, Disposal, Reuse,* Tata-Mc Graw Hill Publishing Co., Ltd., New Delhi (1979).

51. Talmadge W. P. and E. B. Fitch, "Determining Thickener Unit Areas," *Industrial Engrg. Chem., 47:* 1 (1955).

52. Bhole, A. G., "Design of Water Treatment Plants: Part I," *Jour. Ind. Wat. Wks. Assoc., 7:* 189 (1975).

53. Bhole, A. G. and V. S. Prasad, "The Physical Factors Affecting Flocculation in Water Treatment: Part I," *Jour. Ind. Wat. Wks. Assoc., 5:* 185 (1973).

54. Bhole, A. G. and V. S. Prasad, "The Physical Factors Affect-

ing Flocculation in Water Treatment: Part II," *Jour. Ind. Wat. Wks. Assoc., 5:* 231 (1973).

55. Bhole, A. G. and V. S. Prasad, "A Study of the Effect of Time of Flocculation in a Jar Test," *Ind. Jour. of Env. Hlth., 16:* 35 (1974).

56. Letterman, R. D., J. E. Quon and R. S. Gemmell, "Influence of Rapid Mix Parameters on Flocculation," *Jour. Am. Wat. Wks. Assoc., 65:* 716 (1973).

57. ASCE, AWWA and CSSE, "Water Treatment Plant Design," *Am. Wat. Wks. Assoc., Inc.* (1969).

58. Stenquist, R. and W. Kaufman, "Initial Mixing in Coagulation Processes," *San. Engrg. Research Lab. Rep. No. 72-2,* Univ. of Calif, Berkeley (1972).

59. Kawamura, S., "Consideration on Improving Flocculation," *Jour. Am. Wat. Wks. Assoc., 68:* 328 (1976).

60. Camp, T. R. and P. C. Stein, "Velocity Gradient and Internal Work in Fluid Motion," *Jour. Boston Soc. Civil Engrs., 30:* 219 (1943).

61. Shulz, R. C. and D. A. Okun, *Surface Water Treatment for Communities in Developing Countries,* John Wiley and Sons (1984).

62. Bhole, A. G., "Measuring the Water Velocity in a Paddled Flocculator," *Jour. Am. Wat. Wks. Assoc., 72:* 109 (1980).

63. Sanks, R. L. (Ed.), *Water Treatment Plant Design,* Ann Arbor Sc. Pub. Inc. (1978).

64. Bhole, A. G. and P. Limaye, "Effect of Shape of Paddle and Container on Flocculation Process," *Jour. Inst. of Engrs. (India), 57* (EE2): 52 (1977).

65. Bhole, A. G. and G. N. Warade, "Effect of Stators on the Process of Flocculation," *presented at Annual Paper Meeting of Env. Engrg. Div. of Inst. of Engrs. (India)* (1980).

66. Ives, K. J., "Coagulation and Flocculation II" in *Orthokinetic Flocculation in Solid-liquid Separation,* L. Svarovsky (Ed.), 2nd Ed., Butterworths, London (1981).

67. Bhole, A. G. and S. V. Dahasahasra, "Hydrodynamics of Pebble Bed Flocculator—An Energy Saving Device," presented at All India Seminar on Conservation of Energy and Resources, *Inst. of Engrs. (India), Nagpur Centre* (1982).

68. Bhole, A. G. and D. N. Potdukhe, "Performance Study of Upflow Pebble Bed Flocculator," *Ind. Jour. of Env. Hlth, 25:* 41 (1983).

69. Bhole, A. G. and S. V. Dahasahasra, "A Theoretical Study of Particle Removal Mechanism in Pebble Bed Flocculator," *Asian Environment, 30* (Apr. 1983).

70. Bhole, A. G. and V. A. Mhaisalkar, "Study of a Low Cost Sand-bed Flocculator for Rural Areas," *Ind. Jour. of Env. Hlth, 19:* 33 (1977).

71. Bhole, A. G., et al., "Expanded Sand Bed Flocculator," *Jour. Inst. of Engrs. (India), 56* (PHI): 3 (1975).

72. Bhole, A. G. and V. A. Ughade, "Study of Surface Contact Flocculator," *Jour. Ind. Wat. Wks. Assoc., 13:* 109 (1981).

73. Bhole, A. G., et al., "Some Aspects of Inline-Static Mixers," *Jour. Ind. Wat. Wks. Assoc., 16:* 65 (1984).

74. Dhabadgaonkar, S. M. and A. G. Bhole, "Modified Sludge Blanket Clarifier," *Jour. Ind. Wat. Wks. Assoc., 6:* 149 (1974).

75. Fair, G. M., J. C. Geyer and D. A. Okun, *Water and Waste Water Engineering, Vol. 2,* John Wiley and Sons., Inc. (1968).

76. Bhole, A. G. and V. Mohankrishna, "Significance of Tapered Flocculation in Water Treatment," *Jour., Inst. of Engr. (India), 58* (EE2): 33 (1978).

77. Bhole, A. G. and S. D. Harne, "Significance of Tapering of Time and Velocity Gradient on the Process of Flocculation," *Jour. Ind. Wat. Wks. Assoc., 12:* 57 (1980).

78. Bhole, A. G., "Design of Water Treatment Plants–Part II," *Jour. Ind. Wat. Wks. Assoc., 17:* 189 (1975).

79. Paramsivam, R., "Bitumen Coal–A Substitute for Anthracite in Two Layer Filtration of Water," *Ind. Jour. of Env. Hlth., 15:* 178 (1973).

80. Kardile, J. N., "Crushed Coconut Shell as a New Filter Media for Dual and Multilayer Filters," *Jour. Ind. Wat. Wks. Assoc., 4:* 28 (1972).

81. Bhole, A. G. and J. T. Nashikkar, "Berry-seed Shell as Filter Media," *Jour. Inst. of Engrs. (India), 54* (PH2): 45 (1974).

82. Ranade, S. V. and G. D. Agrawal, "Use of Vegetable Wastes as a Filter Media", Paper presented at Conference on Engineering Materials and Equipment, *Assoc. of Engrs.,* Calcutta, India (1974).

83. Bhole, A. G. and S. F. Rahate, "Performance Study of a Dual-cum-mixed Media Filter," *Jour. Ind. Wat. Wks. Assoc., 9:* 31 (1977).

84. Mohanka, S. S., "Multilayer Filtration," *Jour. Am. Wat. Wks. Assoc., 61:* 501 (1960).

85. Anonymous, "The Simater Continuous Sand Filter," *Jour. Ind. Wat. Wks. Assoc., 2:* 283 (1970).

86. Dhabadgaonkar, S. M., "Declining Rate Upflow Filtration–A New Concept," *Jour. Ind. Wat. Wks. Assoc., 8:* 17 (1976).

87. Patwardhan, S. V., "MIMO Water Treatment Plants," *Jour. Ind. Wat. Wks. Assoc., 10:* 73 (1978).

88. Tate, C. H. and R. R. Trussell, "Recent Development in Direct Filtration," *Jour. Am. Wat. Wks. Assoc., 72:* 165 (1980).

89. Peterson, D. L., D. L. Schleppenbach and Zaudtke, "Studies of Asbestos Removal by Direct Filtration of a Lake Superior Water," *Jour. Am. Wat. Wks. Assoc., 72:* 155 (1980).

90. Westerhoff, G. P., A. F. Hess and M. J. Barnes, "Plant Scale Comparison of Direct Filtration Vs. Conventional Treatment of a Lake Erie Water," *Jour. Am. Wat. Wks. Assoc., 72:* 148 (1980).

91. Degremont, *Water Treatment Hand Book,* 5th Ed., John Wiley and Sons (1979).

92. Bhole, A. G., "Design of Water Treatment Plants: Part III," *Jour. Ind. Wat. Wks. Assoc., 7:* 249 (1975).

93. "Package Water Treatment Plant," M/s. Candy Filters, India.

94. Newsletter, *Ind. Assoc. for Wat. Pollution Control, 21:* 2 (1984).

95. McGhee, J. Terence, "Heuristic Analysis of Lime Soda Softening Processes," *Jour. Am. Wat. Wks. Assoc., 67:* 626 (1975).

96. Caldwell, D. H. and Lawrence, "Water Softening and Conditioning Problems," *Industrial Engrg. Chem., 45:* 535 (1953).

97. Nawalakhe, W. G., et al., "Settling Characteristics of Flocculant Suspension in Defluoridation of Water by Nalgonda Technique," *Jour. Inst. of Engrs (India), 61* (EN2): 85 (1981).

98. Bulusu, K. R., et al., "Performance of Defluoridation Plant at Kadiri," *Jour. Inst. of Engrs. (India), 64* (EN2): 35 (1983).

99. Bulusu, K. R., "Defluoridation of Waters Using Combination of Aluminium Chloride and Aluminium Sulphate," *Jour. Inst. of Engrs. (India), 65* (EN1): 22 (1984).

100. Kulkarni, P. B., A. G. Bhole and A. S. Bal, "Adsorption in Water Treatment–A State of Art," *Jour. Ind. Wat. Wks. Assoc., 17:* 29 (1985).

101. Bulusu, K. R., et al., "Fluorides, Its Incidence in Natural Water and Defluoridation Methods–Comparative Evaluation and Cost Analysis," *NEERI Silver Jubilee Commemoration Volume,* 93 (1985).

102. Porteons, A., "Saline Water Distillation Processes," *Longman* (1975).

103. Lynch, M. A. Jr., and M. S. Mintz, "Membrane and Ion Exchange Processes–A Review," *Jour. Am. Wat. Wks. Assoc., 64:* 711 (1972).

104. Spiegler, K. S., "Principles of Desalination," *Academic Press,* 3rd Printing (1969).

105. "Disinfection for Small Community Water Supplies," *National Env. Engrg. Research Inst.,* Nagpur, India (1974).

106. Deb, A. K., "Optimization of Water Distribution Network Systems," *Jour. Env. Engrg. Div.,* ASCE, *102:* 837 (1976).

107. Shamir, U. and C. D. D. Howard, "Water Distribution System Analysis," *Jour. Hyd. Div.,* ASCE, *94,* 219 (1968).

108. Shamir, U. and C. D. D. Howard, closure to "Water Distribution System Analysis," *Jour. Hyd. Div.,* ASCE, *96:* 577 (1970).

109. Shamir U. and C. D. D. Howard, "Engineering Analysis of Water Distribution Systems," *Jour. Am. Wat. Wks. Assoc., 69:* 510 (1977).

110. Jeppson, R. W., *Analysis of Flow in Pipe Networks,* Ann Arbor Science, Ann Arbor, Mich., U.S.A. (1975).

111. Bhave, P. R., "Analysis of Water Distribution Networks: Part I," *Jour. Ind. Wat. Wks. Assoc., 13:* 149 (1981).

112. Bhave, P. R., "Analysis of Water Distribution Networks: Part II," *Jour. Ind. Wat. Wks. Assoc., 13:* 245 (1981).

113. Cross, H., "Analysis of Flow in Networks of Conduits or Conductors" *Experimental Station Bulletin* 286, Univ. of Illinois, Urbana, Ill., U.S.A. (1936).

114. Cornish, R. J., "The Analysis of Flow in Networks of Pipes," *Jour. Inst. of Civ. Engrs.,* England, *13:* 147 (1939).

115. Conklin, J., "Flow in Pipe Networks by Direct Determination," *Eng. News Rec., 127:* 370 (1941).

116. Williams, G. N., "Enhancement of Convergence in Pipe Network Solutions," *Jour. Hyd. Div.* ASCE, *99:* 1057 (1973).

117. Voyles, C. F. and H. R. Wilke, "Selection of Circuit Arrangements for Distribution Network Analysis by the Hardy Cross Method," *Jour. Am. Wat. Wks. Assoc., 54:* 285 (1962).

118. Martin, D. W. and G. Peters, "The Application of Newton's Method to Network Analysis by Digital Computer," *Jour. Inst. of Wat. Engrs., 17:* 115 (1963).

119. Epp, R. and A. G. Fowler, "Efficient Code for Steady State Flows in Networks," *Jour. Hyd. Div.*, ASCE, *96*:43 (1970).

120. Wood, D. J. and O. A. Charles, "Hydraulic Network Analysis Using Linear Theory," *Jour. Hyd. Div.*, ASCE, *98*: 1157 (1972).

121. Collins, A. G. and R. L. Johnson, "Finite Element Method for Water Distribution Networks," *Jour. Am. Wat. Wks. Assoc.*, *67*: 385 (1975).

122. Arora, M. L., "Flows Split in Closed Loops Expending Least Energy," *Jour. Hyd. Div.*, ASCE, *102*: 455 (1976).

123. Collins, M. A., et al., "Solution of Large Scale Pipe Networks by Improved Mathematical Approaches," *Tech. Rep. IEOR 77016-WR 77001*, School of Engg. and Appl Sc., Southern Methodist Univ., Dallas, Tex, U.S.A. (1978).

124. Collins, M. A., L. Cooper and J. L. Kennington, "Multiple Operating Points in Complex Pump Networks," *Jour. Hyd. Div.*, ASCE, *105*: 229 (1979).

125. Camp, T. R. and H. L. Hazen, "Hydraulic Analysis of Water Distribution Systems by Means of an Electric Network Analyzer," *Jour. New Eng. Wat. Wks. Assoc.*, *48*: 383 (1934).

126. McIlroy, M. S., "Direct Reading Electrical Analyzer for Pipeline Networks," *Jour. Am. Wat. Wks. Assoc.*, *42*: 347 (1950).

127. Wood, D. J. and A. G. Rayes, "Reliability of Algorithms for Pipe Network Analysis," *Jour. Hyd. Div.*, ASCE, *107*: 1145 (1981).

128. Collins, M. A., "Pitfalls in Pipe Network Analysis Techniques," *ASCE Convention* (1979).

129. Rao, H. S. and D. W. Bree, "Extended Period Simulation of Water Systems—Part A," *Jour. Hyd. Div.*, ASCE, *103*: 97 (1977).

130. Bhave, P. R., "Node Flow Analysis of Serial Water Distribution Systems," *Jour. Ind. Wat. Wks. Assoc.*, *12*: 17 (1980).

131. Bhave, P. R., "Node Flow Analysis of Branching Water Distribution Systems," *Jour. Ind. Wat. Wks. Assoc.*, *13*: 25 (1981).

132. Bhave, P. R., "Node Flow Analysis of Water Distribution Systems," *Transp. Engrg. Jour.*, ASCE, *107*: 457 (1981).

133. Bhave, P. R., "Analysis of Water Distribution Networks: Part III," *Jour. Ind. Wat. Wks. Assoc.*, *13*: 305 (1981).

134. Shamir U. and C. D. D. Howard, "Engineering Analysis of Water Distribution System," *Jour. Am. Wat. Wks. Assoc.*, *69*: 510 (1977).

135. Walski, T. A., "Techniques for Calibrating Network Models," *Jour. Wat. Res. Planning and Management*, ASCE, *109*: 360 (1983).

136. Cesario, A. Lee and J. O. Davis, "Calibrating Water System Models," *Jour. Am. Wat. Wks. Assoc.*, *76*: 62 (1984).

137. Bhave, P. R., "Noncomputer Optimization of Single-Source Networks," *Jour. Env. Engrg. Div.*, ASCE, *104*: 799 (1978).

138. Bhave, P. R., "Optimality Criteria for Distribution Networks," *Ind. Jour. of Env. Hlth*, *19*: 120 (1977).

139. Bhave, P. R., "Critical Path Approach for the Design of Dead-End Gravity Distribution Systems," *Jour. Ind. Wat. Wks. Assoc.*, *8*: 41 (1976).

140. Bhave, P. R., "Economical Design of Single Source Distribution Networks by Critical Path Method," *Jour. Inst. of Engrs (India)*, *58* (EN1): 1 (1977).

141. Karmeli, D., Y. Gadish and S. Meyers, "Design of Optimal Water Distribution Networks," *Jour. Pipeline Div.*, ASCE, *94*: 1 (1968).

142. Gupta, I., "Linear Programming Analysis of a Water Supply System," *Trans. Am. Inst. of Industrial Engrs.*, *1*: 56 (1969).

143. Calhoun, C. A., "Optimization of Pipe Systems by Linear Programming," Proceedings on Control of Flow in Closed Conduits, J. P. Tullis, ed, Colorado St. Univ., Fort Collins, Colo., 175 (1970).

144. Bhave, P. R., "Selecting Pipe Sizes in Network Optimization by LP," *Jour. Hyd. Div.*, ASCE, *105*: 1019 (1979).

145. Gupta, I., M. Z. Hassan and J. Cook, "Linear Programming Analysis of a Water Supply System with Multiple Supply Points," *Trans. Am. Inst. of Industrial Engrs.*, *4*: 200 (1972).

146. Bhave, P. R., "Optimal Design of Dead End Water Distribution Systems for Variable Demand," *Jour. Ind. Wat. Wks. Assoc.*, *14*: 3 (1982).

147. Robinson, R. B. and T. A. Austin, "Cost Optimization of Rural Water Systems," *Jour. Hyd. Div.*, ASCE, *102*: 1119 (1976).

148. Cowan, J., "Checking Trunk Main Designs for Cost-Effectiveness," *Wat. and Wat. Engr.*, *75*: 385 (1971).

149. Deb, A. K., "Least Cost Design of Water Main System in Series," *Jour. Env. Engrg. Div.*, ASCE, *99*: 405 (1973).

150. Deb, A. K., "Least Cost Design of Branched Pipe Network System," *Jour. Env. Engrg. Div.*, ASCE, *100*: 821 (1974).

151. Perold, R. P., "Economic Pipe Sizing for Gravity Sprinkler Systems," *Jour. Irrig. and Drainage Div.*, ASCE, *100*: 107 (1974).

152. Perold, R. P., "Economic Pipe Sizing in Pumped Irrigation Systems," *Jour. Irrig. and Drainage Div.*, ASCE, *100*: 425 (1974).

153. Lekane, T. M., D. E. Hellemans and C. M. Whitwam, "Long-term Optimization Model of Tree Water Networks," *European Jour. of Opr. Res.*, *4*: 7 (1980).

154. Liang, T., "Design Conduit System by Dynamic Programming," *Jour. Hyd. Div.*, ASCE, *97*: 383 (1971).

155. Yang, K. P., T. Liang and I. P. Wu, "Design of Conduit System with Diverging Branches," *Jour. Hyd. Div.*, ASCE, *101*: 167 (1975).

156. Alperovits G. and U. Shamir, "Design of Optimal Water Distribution Systems," *Wat. Res. Research*, *13*: 885 (1977).

157. Cembrowicz, R. G. and J. J. Harrington, "Capital Cost Minimization of Hydraulic Networks," *Jour. Hyd. Div.*, ASCE, *99*: 431 (1973).

158. Tong, A. L., et al., "Analysis of Distribution Networks by Balacing Equivalent Pipe Lengths," *Jour. Am. Wat. Wks. Assoc.*, *53*: 192 (1961).

159. Raman V. and S. Raman, "New Method of Solving Distribution System Networks Based on Equivalent Pipe Lengths," *Jour. Am. Wat. Wks. Assoc.*, *58*: 615 (1966).

160. Deb, A. K. and A. K. Sarkar, "Optimization in Design of

Hydraulic Network," *Jour. San. Engrg. Div.*, ASCE, *97:* 141 (1971).

161. Kanga, A. R., "New Concepts in the Design and Analysis of Water Distribution Networks," *Env. Hlth, 5:* 320 (1963).

162. Liebman, J. C. and E. D. Brill, Discussion of "Optimization in Design of Hydraulic Network" by A. K. Deb and A. K. Sarkar, *Jour. San. Engrg. Div.*, ASCE, *97:* 786 (1971).

163. Swamee, P. K. and P. Khanna, "Equivalent Pipe Methods for Optimizing Water Networks—Facts and Fallacies," *Jour. Env. Engrg. Div.*, ASCE, *100:* 93 (1974).

164. Bhave, P. R., "Pressure Surface Approach in Network Optimization—Concept and Limitations," *Jour. Inst. of Pub. Hlth. Engrs., 4:* 15 (1980).

165. Kally, E., "Automatic Planning of the Least-Cost Water Distribution Network," *Wat. and Wat. Engrg, U.K., 75:* 148 (1971).

166. Kally, E., "Computerized Planning of the Least Cost Water Distribution Network," *Wat. and Sewage Wks.,* R. 121 (1972).

167. Quindry, G., et al., "Comments on ;osDesign of Optimal Water Distribution Systems' by Alperovits and Shamir," *Wat. Res. Research, 15:* 1651 (1979).

168. Lai, D. and J. Schaake, "Linear Programming and Dynamic Programming Applications to Water Distribution Network Design," *Report 116,* Dept. of Civil Engrg., Massachusetts Inst. of Tech., Cambridge, Mass., U.S.A. (1969).

169. Quindry, G., E. D. Brill and J. Liebman, "Water Distribution System Design Criteria," *Dept. of Civil Engrg.*, Univ. of Illinois at Urbana-Champaign, Urbana, Ill., U.S.A., (1979).

170. Quindry, G., E. D. Brill and J. C. Liebman, "Optimization of Looped Water Distribution Systems," *Jour. Env. Engrg. Div.*, ASCE, *107:* 665 (1981).

171. Pitchai, R., "A Model for Designing Water Distribution Pipe Networks," *Ph.D. Thesis,* Harvard Univ., Cambridge, Mass., U.S.A. (1966).

172. Jacoby, S. L. S., "Design of Optimal Hydraulic Networks," *Jour. Hyd. Div.*, ASCE, *94:* 641 (1968).

173. Lam, C. F., "Discrete Gradient Optimization of Water Systems," *Jour. Hyd. Div.*, ASCE, *99:* 863 (1973).

174. Watanatada, T., "Least Cost Design of Water Distribution Systems," *Jour. Hyd. Div.*, ASCE, *99:* 1497 (1973).

175. Rasmusen, H. J., "Simplified Optimization of Water Supply Systems," *Jour. Env. Engrg. Div.*, ASCE, *102:* 313 (1976).

176. Cembrowicz, R. G. and J. J. Harrington, "Capital Cost Minimization of Hydraulic Networks," *Jour. Hyd. Div.*, ASCE, *99:* 431 (1973).

177. Shamir, U., "Optimal Design and Operation of Water Distribution Systems," *Wat. Res. Research, 10:* 27 (1974).

178. Ridgik, T. A. and D. T. Lauria, discussion of "Optimization of Gravity-Fed Water Distribution Systems: Theory," by P. R. Bhave, *Jour. of Env. Engrg, 110:* 504 (1984).

179. Bhave, P. R., "Optimization of Gravity-Fed Water Distribution Systems: Theory," *Jour. of Env. Engrg.*, ASCE, *109:* 189 (1983).

180. Bhave, P. R., "Optimization of Gravity-Fed Water Distribution Systems: Application," *Jour. of Env. Engrg.*, ASCE, *109:* 383 (1983).

181. Rowell, W. F. and W. J. Barnes, "Obtaining Layout of Water Distribution Systems," *Jour. Hyd. Div.*, ASCE, *108:* 137 (1982).

182. Bhave, P. R., "Optimal Expansion and Strengthening of Dead-End Gravity Water Distribution Systems," *Jour. Ind. Wat. Wks. Assoc., 16:* 13 (1984).

183. Bhave, P. R., "Optimal Expansion of Water Distribution Systems," *Jour. of Env. Engrg.*, ASCE, *111:* 177 (1985).

184. "Inspection of Water Mains from Inside," *Jour. Ind. Wat. Wks. Assoc., 17:* 283 (1985).

185. "Water Losses in Distribution Systems," *Jour. Ind. Wat. Wks. Assoc., 17:* 280 (1985).

186. "Repairing Buried Pipelines by AMK Technology," *Jour. Ind. Wat. Wks. Assoc., 17:* 282 (1985).

CHAPTER 2

Water Management Using Interactive Computer Simulation

ALFRED B. CUNNINGHAM* AND JOHN R. AMEND**

INTRODUCTION

The purpose of this chapter is to provide a comprehensive summary of contemporary applications of interactive computer simulation to the field of water resource management. The chapter begins with an overview of simulator technology and hardware followed by discussion of applications for which this type of simulation is best suited. The first application discussed involves use of interactive simulators as part of education and demonstration programs. For this purpose generalized water resource system models are utilized to simulate system behavior which is representative of conditions within a given geographical region. Documentation of previous research is presented along with discussion of system models and hardware configurations utilized. The second major topic discussed is the role which interactive computer simulation is presently assuming as a contemporary water resource system analysis tool. Specific examples are provided which illustrate the use of interactive simulation to develop water resource system management guidelines along with similar applications in the training of system managers.

AN OVERVIEW OF INTERACTIVE COMPUTER SIMULATION TECHNOLOGY

Recent advances in microprocessor technology have made possible inexpensive and realistic computer simulation of complex natural resource and engineering systems. Main-frame computer simulation has traditionally been used for investigation of system response to external forces. The rapid response time now possible with high-capability, low cost dedicated microprocessor systems has made possible a new type of computer simulation in which the researcher is placed in the computer feedback loop, interacting with system operation problems as they occur. Such an approach leads rapidly to identification of optimum system operation parameters, since unreasonable alternatives are rejected as soon as they become visible in the simulation. The managerial nature of the simulator operator's role also lends itself well to illustration of system structure and interaction of variables, and therefore to the training of both technical and management personnel. The approach seems limited only by the availability of problems, algorithms and calibration data; we have successfully applied the technology discussed in this article to water resource management, hydropower system management, wastewater treatment plant operation, energy management, and management of grazing lands.

Interactive computer simulation makes some stringent demands on the computer-operator interface. The computer's task in interactive simulation is to compute and present the system condition parameters required for effective decision-making. Most computers—mainframe or personal size—communicate with their operators by keyboard and CRT or printed display. When decisions are made in a leisurely manner after the computer output is assembled and organized, this approach is satisfactory. However, in a "real-time" system, the pixel density on all but extremely high resolution monitors is too low to present the amount of needed information, and keyboards only permit interaction with one operator. We have found that few persons can handle more than three independent control variables in a "real-time" simulation; as in the real world, our systems use multiple operators with policy and management responsibility divided among them. Large system parameter displays are required to communicate with the operator groups.

*Department of Civil Engineering, Montana State University, Bozeman, MT
**Department of Chemistry, Montana State University, Bozeman, MT

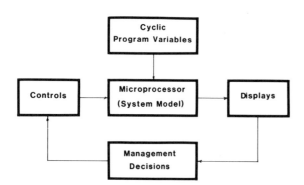

FIGURE 1. Conceptual features of interactive simulation.

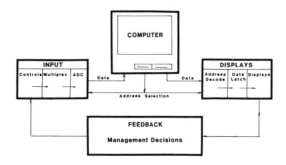

FIGURE 2. Computational steps in interactive simulation.

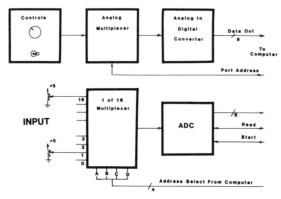

FIGURE 3. Input control components.

Figure 1 shows the conceptual features of an interactive simulation system. The key to this configuration is the system manager. The manager receives system condition information from the display block, compares this with predetermined "optimum conditions," and, within the social and economic constraints imposed upon him or her, makes a decision and acts to implement this decision.

The computer has four tasks. (1) It must accept control inputs and translate them into parameters in the system algorithm. (2) It must generate or look up external cyclic or random variables, and translate them into parameters in the system algorithm. (3) Using the system algorithm and updated parameters, it must compute the system condition. (4) Finally, it must display, in a manner readily understood by the operator, both the current system conditions and by means of graphs, the trend in specific conditions that brought us to the current situation.

The computational steps (2) and (3) may be handled by any personal computer with adequate speed and memory (we construct our own computers and are presently using 4 Mhz, 64K, Z-80 processors for algorithm computations). Steps (1) and (4) require external hardware. This situation is illustrated in Figure 2.

Input Controls

Input controls are of three types: (1) Initial condition controls, set once to determine static system parameters; (2) Simulator operator controls, such as start/stop, clock speed, etc.; and (3) control variables, such as reservoir discharge, irrigation withdrawal, and ground/surface water source.

Initial condition controls are usually thumbwheel switches which are read once at the beginning of the simulation. These are read as binary-coded numbers by a parallel input port to the computer. Operator controls are usually toggle switches or push-buttons, these are most conveniently read by an analog-to-digital converter which also reads the control potentiometers. All switches and potentiometers are read each pass through the program and observed values placed in a matrix which is referenced by the program algorithm (Figure 3).

Most of our systems use 8-bit analog-to-digital converters; units such as the National ADC-0804 convert in less than 200 microseconds and are accurate to one part in 255, or about 0.4%. Both the conversion time and accuracy is adequate for most interactive simulation. One can read 32 controls in about 6.4 milliseconds at this conversion speed; the simulator is computation limited rather than control limited. If the system completes at least three updates per second, the computer will appear transparent to the operator.

Display Technology

Four types of display are used in our simulator, depending on the type of information to be contributed to the decision

TABLE 1. Display Technology Must Match the Need for Information.

Display Type	Application
7-segment numeric	Where quantitative information is absolutely required for decision-making.
LED-bar graph	Where relative magnitude is sufficient for decision making.
Red/amber/green lamp banks	Where qualitative state is required.
Single colored lamps	Where on-off information is sufficient.
Color graphic display	Where trends and cause/effect variable interactions must be displayed. We put three graphs on a display, some units use three CRT's to display graphic information.
Sonalert Audio tone	When operator attention must be gained immediately. Usually used with a visual display.

process (Table 1). Numerical information is easiest to generate and hardest to assimilate; it must be used cautiously.

Displays involve three components. (1) An address decoder selects the display to be updated; (2) a data latch catches and holds this data, and (3) the display driver illuminates the display (Figure 4).

Graphic displays are driven by a separate 6809 processor; this unit uses 64 K of screen memory and can store and present ten different screens of either graphic data or alphanumeric message.

SIMULATION OF GENERALIZED WATER RESOURCES SYSTEMS

Microprocessor based water management simulators were first used to develop college level and public education programs dealing with water resource system operation problems Amend and Cunningham [1], Amend and Arnold [2]. These simulators have been calibrated to represent generalized hydrologic conditions in a chosen region and are, therefore, not site specific to any particular river basin. A schematic diagram of the front panel for a general water resources system simulator which models a water supply and demand situation for a high mountain sub-basin typical of intermountain regions is shown in Figure 5. A display panel for a simulator typical of conditions in the Great Plains and other areas with large groundwater reserves is shown in Figure 6. Simulators modeling regions exhibiting other hydrologic regimes have also been built.

The display panel for the simulator units shown in Figures 5 and 6 is approximately 24 in. × 35 in. × 8 in. (60.7 × 88.9 × 20.32 cm) and weighs about 45 lb (20.4 kg); it contains the computer, basin displays, and operator controls. Two processors are used which together access approximately 100K of memory. The simulator is controlled by several small control modules, each of which is operated by individuals participating in the simulation exercise. About two dozen water resources simulators are presently being used by universities and water planning and management agencies around the U.S.

System Model

A generalized hydrologic model around which the water resource management simulators have been developed is illustrated in Figure 7. This water balance model utilizes a monthly time step thereby eliminating the need to consider

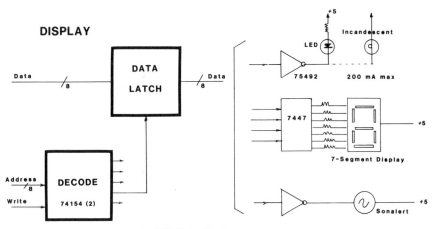

FIGURE 4. Display components.

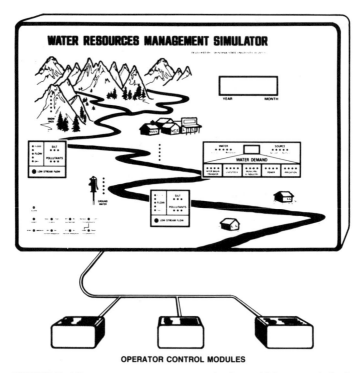

FIGURE 5. Water resources management simulator—high mountain basin.

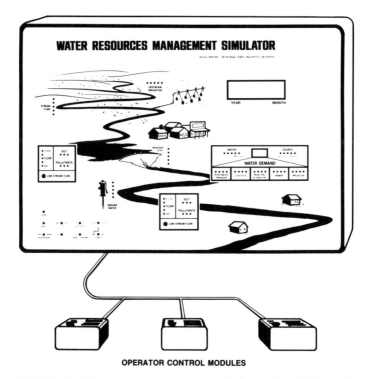

FIGURE 6. Water resources management simulator—Great Plains basin. The optional velcro-mounted reservoir has been added in this figure.

WATER RESOURCE MANAGEMENT MODEL

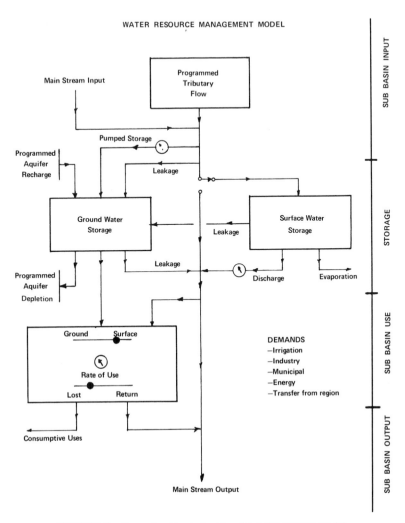

FIGURE 7. Generalized hydrologic/water management model.

hydraulic routing effects. The relative magnitude of ground-water and surface water resources can be defined for a given region in using field-programmable switches located on the back of the simulator cabinet. Evaporation of surface water, groundwater infiltration, and irrigation demand follow pre-programmed annual cycles which are representative of conditions in a given region.

Unregulated streamflows are generated by the system model at the beginning of each time step in the simulation. Equation 1 (after Fiering, [5]) is used to generate flow sequences which are statistically similar to recorded stream flow data from the study region.

$$q_{i+1} = q_{j+1} \, b_j (q_i - q_j) + t \sigma_{j+1} (1 - r_j^2)^{1/2} \qquad (1)$$

q_i and q_{i+1} are the flows in the ith and $(i + 1)$th months

from the start of the sequence, q_j and q_{j+1} are the mean monthly flows in the jth and $(j + 1)$th month of the annual cycle, and b_j is the regression coefficient for estimating flows in the $(j + 1)$th month from the flows in the jth month. t_i is a random normal deviate with a zero mean and unit variance, σ_{j+1} is the standard deviation of the flows for the $(j + 1)$th month, and r_j is the correlation coefficient between flows in the jth and $(j + 1)$th months. Equation 1 is "calibrated" for the flow regime in a particular region by choosing combinations of b_j, σ_j and r_j which are similar to those computed from historic stream flow records. Once values for q_j are assumed hypothetical monthly flow sequences are generated which exhibit similar statistical properties to observed region stream flows.

It is possible to specify surface diversions from the main stream to meet the demand of the water use area. Ground-

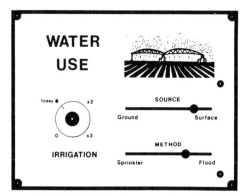

FIGURE 8. Simulator control model.

water supplies can also be developed and monitored during the simulation. Return flows from the use area are assumed to re-enter the stream. Water quality degradation resulting from return flows is computed based on observed relationships between flow and total dissolved solids (TDS) for the particular region. Mass balance computations allow water quality conditions to be monitored throughout the basin.

Stream Flow Displays

As stream flows are generated at the start of each time step, their values appear on displays provided for each tributary as well as selected points above and below the reservoir and water use area. These displays consist of a light-emitting-diode (LED) bar graph at each measurement point. Water quality (slit and total dissolved solid) is indicated by green, amber, and red lamps. Break points for the three qualitative indicators may be set by data statements in the computer program.

In-Stream Reservation

A large red lamp in each flow indicator warns when the flow level has fallen below that chosen as a reserve for wildlife or navigation. This low flow warning may be programmed as a percentage of the average monthly stream

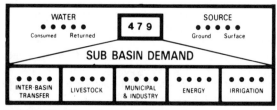

FIGURE 9. Water demand display on simulator front panel.

flow at the measurement point. The value is set by a switch located on the back of the display cabinet.

Water Storage

Surface water storage is provided by a surface reservoir which may be activated by a switch on the reservoir control panel. Lamps form a dam across the stream when a reservoir is turned on. Participants may elect to run with or without a dam. Reservoir capacity can be varied from zero up to a maximum of twice the annual flow. Reservoir level is indicated on the simulator display panel with a vertical bar graph of five LED's, and on a color graphic display. A programmable (and cyclic) amount of evaporation is withdrawn from the surface reservoir during each year.

A similar indicator is used for ground water reserve; it is possible to artificially recharge the ground water reservoir with surface water by use of a ground water control module. This control permits transfer of surface water to ground reserve through pumping or percolation from holding ponds. Ground water is also recharged by programmed infiltration from precipitation, and leakage from streams and surface water reservoirs. Permeability coefficients are set in the computer program. The size of the ground water reservoir in each sub-basin may be set by field-programmable controls on the headwaters display unit.

Five water use categories are provided in the model: (a) Irrigation, (b) Livestock, (c) Municipal and Industry, (d) Energy, and (e) Inter-basin transfer. For each of these uses the water may be drawn from either the ground water or surface water resource; the relative percentage from each resource is adjusted using control modules such as illustrated in Figure 8. Rate-of-use controls are calibrated in terms of the region's current use pattern thereby making the problem of projecting changes in use easier to understand. (The computer model, however, computes in terms of thousands of acre feet.) Irrigation demand is programmed to follow an annual cycle.

For all of the use sectors it is possible to decide the fate of the used water. Except for Inter-basin transfer, water may be either evaporated or consumed (lost) or returned to the river. Slide controls on each control module permit setting the recovery ratio for each of the uses; this may be changed at any time during the simulation to explore the effects of a change in the user's technology.

Relative demand is indicated on the main panel for each use (Figure 9). Four LED's comprise a simple horizontal bar graph for this indicator. The total sub-basin demand is indicated on a digital display to permit comparison of the relative use in each of the sub-basins. One may also determine the relative weighting of each demand by turning it off and observing the change in the total sub-basin demand.

Two horizontal five-LED indicators on the display panel

operate with one lighted LED – its position in the string indicates in 20% steps the mix of ground and surface water in use in that sub-basin, or the mix of consumptive vs non-consumptive use.

Operator Controls

Although the simulator is controlled using operator control modules, a number of functional controls are located in the lower-left corner of the display panel. These switches give the workshop instructor control of important operational characteristics of the simulator.

The *Sonalert* switch disables the audible alarm which sounds whenever stream flow falls below the preset instream reservation at any of the measurement stations. The function of the *Reset* and *Stand-by – Operate* switches is apparent. The *Test* switch activates a test cycle which will exercise all of the digital displays. The *Slow* switch doubles the number of sub-basin computer up-dates that occur each clock month, thus operating the simulator at half speed and permitting more time for discussion of the previous month's results and plans for corrective action. The *Random* switch generates a random seed for the random number generation used in the stream flow algorithm, making it impossible for even an experienced group to second guess the precipitation cycle represented by the tributary flow. The *Plot* switch activates the memory controls, which are discussed in the following paragraph.

Memory

As the simulator operates, important parameters are stored in the computer's random access memory as they are computed. Following a simulation, one can reconstruct the conditions and strategies used during any selected portion of the run. This data can be either manually plotted on an overhead projector or paper, or can be automatically presented with a color graphics display. The memory mode is accessed by turning on the *Plot* switch. The clock is then moved to the beginning of the year in question with the "up" and "dn" push-button switches on the Memory module. The desired parameter is selected with a push-button and displayed on the four-digit clock display by pressing the "read" push-button. (In the case of the graphic display, the entire annual hydrograph for the parameter in equation is presented on the display.) One can thus easily plot run-off hydrographs, and reservoir level hydrographs, permitting careful examination of the characteristics of the water resource system and discussion of alternate strategies for managing the water. The memory module is located on the left-hand end panel of the simulator, conveniently accessible to the operator.

SYSTEM ANALYSIS

In this section interactive computer simulation is discussed from the standpoint of enhancing performance of traditional water resource system analysis techniques including simulation modeling and optimization. An overview of system management policy development using interactive simulation is given along with discussion of the potential for introducing subjective judgement directly into the simulation process. A case study is also presented in which interactive simulation is used to develop operational guidelines for a three-reservoir system involving various physical, economic and legal constraints. Results of this case study indicate that interactive simulation is well suited for developing management policy for water resource systems involving intangible as well as tangible constraints. Through a process of trial and evaluation, system management policies are developed which are "acceptable" given the prevailing level of system uncertainty.

Simulation vs. Optimization

The traditional system analysis modeling techniques for optimization and simulation have different but related applications in water resources research and policy investigations. Optimization models are usually concerned with determining the "best" option from many system operation alternatives, given a clearly defined operator goal. Similarly, simulation involves mathematical modeling of system processes in order to gain insight into overall system behavior. Simulation models are particularly well suited for evaluating design alternatives and/or system operation policies. It is important to state that both of these approaches to system analysis are basically software oriented while interactive simulation, as previously discussed, is both hardware and software oriented. Thus interactive simulators can be developed for a particular system around *either* an optimization model or a simulation model. In either case it is apparent that interactive simulation is best suited for developing or analyzing system management policy for water resource systems involving both tangible and intangible constraints. Also, the ability to interact with a system model during the simulation process can make system analysis methods more usable by field operational personnel. Examples of resistance of operational staff to sophisticated computational methods are given by Helweg, et al [6].

Analysis of water resource systems (using either optimization or simulation) consists of similar functional elements. In either case, a mathematical model receives initial system parameter values along with sequences of input variables then computes variables describing the system's condition during each time step. Computed results are subsequently displayed or printed out for evaluation. This approach comprises a "single pass" analysis. One asks "what if" a certain set of conditions exists, sets the variables, runs the optimization or simulation model, and observes the projected result. After analyzing the results of the simulation, variables are changed and the process repeated. By repeated trial and error, it is possible to predetermine a set of optimum conditions for the management of a particular system.

An automated multiple-pass approach may be used to develop or analyze system management policy. A multiple-pass analysis uses a software and hardware feedback circuit to compare the computed output with the optimum conditions, automatically adjusts the control parameters for a more favorable comparison, then repeats the analysis. After a series of passes through the system model, the system condition indicators more and more closely approximate the desired optimum conditions, and an optimum set of control parameters is identified. This approach has been designated by Monro [7] for calibration of hydrologic system models. However, many water resource systems are simply too complex to permit the effective use of multiple-pass analysis. It is for such systems that the interactive simulation approach to management policy development is best suited.

Interactive Simulation Approach

Interactive simulation expands the scope of traditional system analysis techniques by placing the analyst in the computer's feedback loop as shown in Figure 10. After each time step the current state of the system is displayed for eval-

uation. Here the analyst is able to introduce subjective judgement regarding intangible goals and constraints for system operation. For example, during periods of water shortage, the system operator can make water allocation decisions considering the political, legal, and institutional aspects prevailing for the real system. If necessary, system parameters may be adjusted or new operational schemes added before continuing to the next time step. Through a process of successive trial and evaluation, analysts develop an understanding of the relationships between system variables along with identifying potential management alternatives. In the case of a simulation model, the result is efficient identification of system management policies which are "acceptable" (though probably not "optimal") given the existing level of system uncertainty. If an optimization model is used (i.e. linear or dynamic programming model) an interactive simulation would allow the development of optimal operation policies which include consideration of intangible aspects of system operation.

MADISON RIVER CASE STUDY

As an illustration of system analysis using interactive simulation, consider the simulator unit shown in Figures 11 and 12. This system is composed of three reservoirs located in a 50-mile (80 km) reach of the upper Madison River in southwestern Montana. These reservoirs provide flood control, water for irrigation, hydro-power, lake and stream recrea-

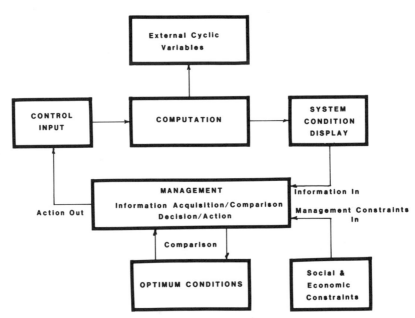

FIGURE 10. Management decisions incorporated in feedback loop.

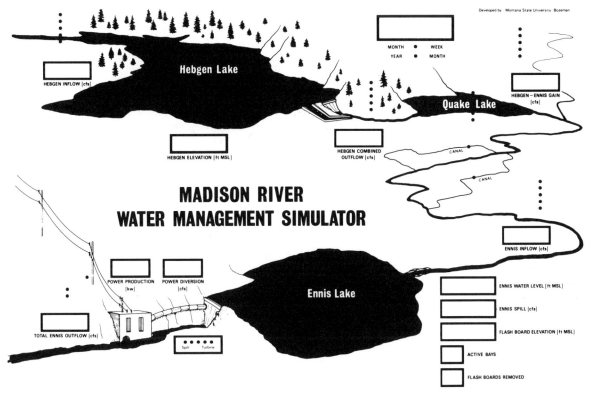

FIGURE 11. Main panel for Madison River water management simulator.

tion and wildlife management. As shown in Figure 11, the main simulator panel layout includes a schematic of the three reservoir system. Major system irrigation diversions, power production, and spillway flows. These system variables are displayed at the locations where they occur by means of lighted digital displays. The relative condition of selected variables such as reservoir water surface elevation and streamflow is also displayed using vertical light strings. These displays are updated after each time step; the user has the option to either run in a continuous mode, i.e. one time step each five seconds, or to use a manual advance switch.

System operation is subject to a variety of constraints including maximum and minimum lake levels, meeting sea-

sonal irrigation and power production demands, provision for minimum releases for trout spawning and wildlife management, and others. The status of all constraints is continuously displayed qualitatively (Figure 12) throughout the simulation (red lights indicating constraint violation, green lights indicating satisfactory condition, an amber light indicating impending violation).

System Model

The Madison River Simulator has been developed around a physically-based water balance model of the three-reservoir system. Major input variables include unregulated

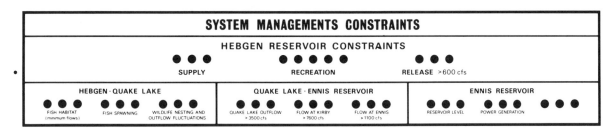

FIGURE 12. Constraint panel for Madison River water management simulator.

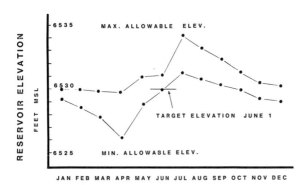

FIGURE 13. Guidelines for operation of Hebgen Reservoir developed using Madison River water management simulator.

stream flow values for each time step entering Hebgen Reservoir along with major tributary inflow entering the Madison River below Quake Lake. The user has the option of entering specific stream flow sequences from the input control panel or using random sequences generated by a stochastic stream flow generation routing incorporated in the system model. The time step between inflow values (as well as for operation of the entire system) may be chosen as either one week or one month. For either time step, it is unnecessary to consider channel routing in the water balance calculations.

Releases from Hebgen Reservoir are adjusted by the user as one of the major management decisions in the operation of the reservoir system. Any inflows to Hebgen Reservoir in excess of reservoir capacity are assumed to pass over the spillway. After exiting Hebgen Reservoir, stream flows next enter Quake Lake which is located approximately 10 miles down river. This natural lake was formed as a result of a large landslide in 1959 which completely blocked the original course of the Madison River. The lake outflow is unregulated and, therefore, has very little impact on the overall water supply picture for the system. No management decisions are made at Quake Lake. However, a maximum outflow constraint of 3,500 cfs must be met to prevent erosion. Following the addition of the tributary inflows the user next makes the decision regarding the magnitude of diversion for irrigation in the Upper Madison River Valley. Here flows are diverted up to the capacity of the canal system and excess flows are computed and returned to the Madison River at a point upstream from Ennis Reservoir.

Ennis Reservoir is operated so as to maintain a predetermined minimum water surface elevation. Releases at Ennis Dam are either in the form of minimum in-stream flows, diversion for hydropower generation, or excess spillway flow. In determining the release pattern the in-stream flow requirements are assumed to be met first. Additional inflow, up to the capacity of the penstock, is then diverted for hydropower generation and subsequently

returned at a point downstream from the dam. Any additional flow is spilled over the dam.

System Management Guidelines

The Madison River Simulator serves as an example of how interactive simulation can be used to develop system operation guidelines. The procedure used in this case consisted of analyzing the sensitivity of system response to a series of operation decisions. By employing stochastically generated system inflows and considering all constraints on system operation (both tangible and intangible), the simulator can be used to readily examine a wide range of possible operation decisions during each one-month time step of the simulation. When an acceptable system response condition is found, the corresponding reservoir release (Hebgen Reservoir) and power diversion (Ennis Reservoir) are noted for the particular time step and the simulation then advanced to the following time step. If an unacceptable condition develops that cannot be corrected in the present time step, then the operator must analyze how previous operation decisions should have been altered to avoid the present problem. For example, consistently meeting minimum reservoir release criteria was often difficult with the Madison River Simulator without lowering Hebgen Reservoir to an unacceptable level. When this occurred, it was necessary to reanalyze release decisions made in previous time steps to determine if the present problem could have been avoided.

After successive repetition of the analysis procedure described above, it is possible to develop system operation guidelines such as shown in Figure 13. Here guidelines have been expressed in terms of a range of "desirable" monthly water surface elevations for Hebgen Reservoir. In other words, the simulation has shown that by operating Hebgen Reservoir such that the level is kept within the bandwidth shown in Figure 13, the risk of an intolerable future condition developing is kept within acceptable limits.

In this example it is possible to express system operation guidelines in a reasonably straightforward manner. This may not be so easily accomplished in the case of system involving a larger number of decision variables or more complex hydraulic relationships. However, regardless of the degree of system complexity, an interactive approach to system analysis remains appealing because of the ability to incorporate both subjectivity and intangible constraints directly into the analysis procedure.

Training System Managers

Because of its ability to rapidly analyze and display system response, simulator technology has proven to be a useful tool for training system management personnel. Exemplifying the training function is the Power System Simulator developed for Electric Power Training Center of the Western Area Power Administration (WAPA) (Figure

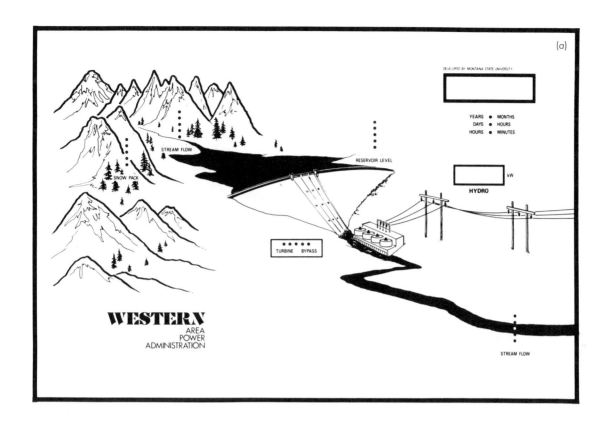

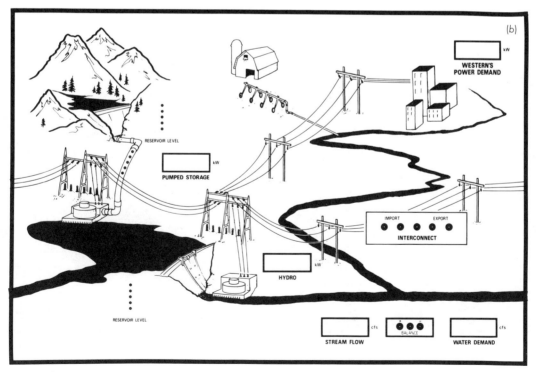

FIGURE 14. Schematic of power system simulator developed for Western Area Power Administration.

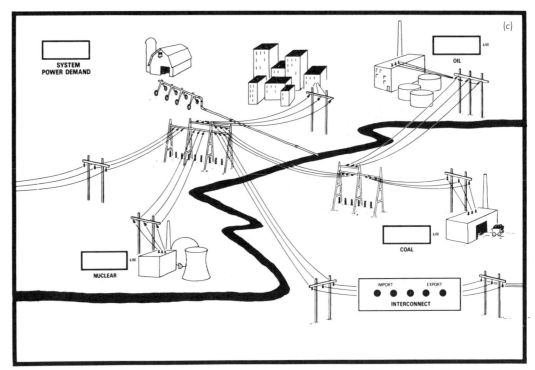

FIGURE 14 (continued). Schematic of power system simulator developed for Western Area Power Administration.

FIGURE 15a. Photograph of power system simulator.

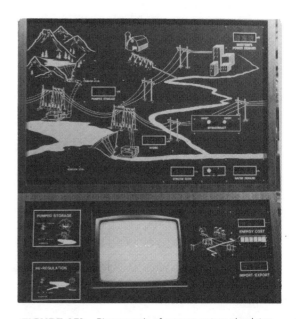

FIGURE 15b. Photograph of power system simulator.

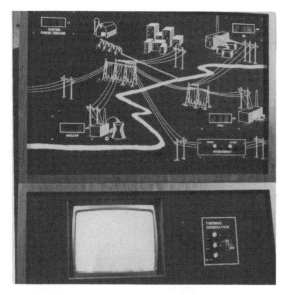

FIGURE 15c. *Photograph of power system simulator.*

14(a), (b) and (c)). Training on the simulator has been incorporated into classes and workshops for both management and technical personnel (Amend, Cunninghman et al., [3] and Cunningham and Amend, [4]). The major goal of the training program is to develop an understanding of the operational aspects of WAPA's electric power system and a realization of the impact of varying hydrologic conditions on power generation. The Power System Simulator is also being used as a tool for building public awareness of the basic principles of electric power generation, transmission and distribution systems and, likewise, serves to further understanding of WAPA's power management and economic policies.

The simulator deals with three conceptual problems. First is the generation of hydroelectric power and the associated hydrologic factors of snowpack prediction, reservoir management, streamflow regulation, and pumped-storage generating facilities. Next are the problems of seasonal and hourly variation in electrical load, base load and peaking facility, capacity, system reliability, and the problem of economic management of the power system. Costs for hydroelectric, coal-fired, oil-fired, and nuclear-generated electricity may be field programmed in the simulator. Third is the problem of transmission charges and import and export pricing.

The photographs in Figures 15a,b,c show that the simulator is constructed in six separate panels. Each panel has a digital display of information on top and a color graphic display on the bottom. The hydrologic stream-flow data currently stored by the computer is based on streamflow records from the Missouri River, the Colorado River, and the North Platte River and the Central Valley Project near Shasta Dam.

The first panel is concerned with the hydrology of power generation. A watershed forms the background; the source area for runoff is primarily snowfall in the mountains in the winter months. Snowpack is displayed by a vertical light emitting diode (LED) bar graph, and is based on historical snowfall records for the region. A vertical LED string shows the amount of inflow into the reservoir from melting snow; another series of lights shows the reservoir level. Reservoir size, head, and generating capacity may be varied by thumbwheel switches on the rear of one panel. A training instructor can also choose from among several river systems which have been programmed into the computer memory. On the lower panel is a television screen that plots hydrologic data in the form of inflow into the reservoir, reservoir level, and downstream riverflow. In the upper right hand corner is a display that shows how much power, in megawatts, is being generated at that dam. Just above that is a clock that displays time in years and months, days and hours, or hours and minutes.

The center panel simulates a load center which uses principally hydroelectric power. A re-regulation dam and power plant are depicted along with a display showing the amount of power being produced. In the upper left-hand corner is a pumped storage unit. Pumped storage is used mainly for peaking power. At offpeak times, power is used to pump water from the lower reservoir to the upper reservoir. During peak hours, the water is released through the generators to produce power. An interconnect provides for power transfer to and from the federal (hydroelectric) and private (thermal) generating systems. Streamflow and water demand are displayed in the lower-right-hand corner of this panel. If downstream water demand is met with streamflow, the system is in balance. However, if the streamflow is too high or too low, a series of lights are activated warning trainees that power plants along the hydro system should be operated to maintain proper streamflow. The center panel's television screen plots three variables: total hydro generation, total thermal generation, and total energy demand. To the right of these graphs is a series of buttons for display of energy cost for each type of generation, blended cost, and power being imported or exported.

The third panel on the right simulates a private power system and includes three thermal generating sources and a second load center. A coal-fired powerplant is represented in the lower, right-hand corner: its output is displayed in megawatts. Immediately above is an oil-fired plant, generally used as a peaking plant. In the lower left-hand corner is a nuclear plant, and the number of megawatts being generated is again indicated. In the lower right-hand corner of this panel are five lights labeled "interconnect." This again displays whether power is being exported, or imported to meet additional load. The lower portion of this panel has room for a video tape recorder, permitting an optional video introduction to the operation of the simulator. The video tape is used primarily for public display. When the video portion is not in use, the lower part of the third panel presents a graphic plot of the import and export information at the interconnect. The lower panel of this console also carries controls for the three thermal plants.

According to WAPA training program supervisors (WAPA press release [8]), the simulation training exercise confront the system operator with a realistic set of circumstances for decision making. WAPA requires that their simulation exercises 1) provide forecast data (streamflows, water demands, power demands) with a similar level of uncertainty to the real system, 2) create situations which occasionally require violation of certain system operating constraints in favor of meeting others and 3) encourage operators to make acceptable management decisions under varying levels of combined system uncertainty. The term "acceptable" refers here to the condition where system power demands are met without excessive violation of operation constraints (also the blended cost of the power supplied should be minimized). According to WAPA, interactive simulation represents a successful approach for modeling the physical as well as judgemental and intangible aspects of complex system operation.

REFERENCES

1. Amend, J. R. and A. B. Cunningham, "A Public Education Program in Water Resource Management," *Western Planner*, Missouri River Basin Commission (April 1980).

2. Amend, J. R. and Armold, Anita A., "A Public Education Program in Water Resources Management," *Journal of Geological Education,* V. 31, p. 362 (November 1983).

3. Amend, J. R., A. B. Cunningham, C. D. Council and G. B. Freeny, "Interactive Simulation of Hydropower Systems," *Proceedings of the International Conference on Hydropower.* 'Waterpower 85', Las Vegas (September 1985).

4. Cunningham, A. B. and J. R. Amend, "Water Management Using Interactive Simulation," *Journal of Water Resources Planning and Management*, ASCE, Vol. 110, No. 3 (July 1984).

5. Fiering, M. B. and B. B. Jackson, "Synthetic Streamflow," *American Geophysical Union, Water Resource Monograph 1*, Washington, D.C. (1971).

6. Helweg, O. T., R. R. Hinks and D. T. Ford, "Reservoir Systems Operation," *Journal of Water Resources Planning and Management*, ASCE, Vol. 108, No. WR2, pp. 169–179 (June 1982).

7. Monro, J. C., "Direct Search Optimization in Mathematical Modeling and a Watershed Model Application," *U.S. Natl. Weather Service, NOAA Tech. Mem. NWS HYDRO-12* (April 1971).

8. Western Area Power Administration press release, "New Power System Simulation Shows WAPA's Impact," *Closed Circuit*, Vol. 3, No. 2 (1982).

CHAPTER 3

Real-Time Reservoir Operations by Mathematical Programming

EMRE K. CAN* AND MARK H. HOUCK**

INTRODUCTION

Until recently, the focus of most modeling efforts on reservoir systems has been on the planning and design of those systems. The focus has now moved to the operation and management of those systems. There are several reasons for this move: most of the prime reservoir sites have already been developed; money for the design and construction of new dams is more difficult to obtain; and the potential benefits obtainable through increased efficiency of operation of existing reservoirs are great. In addition, the need for improved operations is enormous. The demands placed on reservoirs are increasing constantly. These increased demands are not only for more of the same output, but are also for different outputs. For example, the use of reservoirs for recreation purposes has increased tremendously yet many existing reservoir systems were not constructed for that purpose.

The uses of a reservoir system can be broken into three categories: flow attenuation, reservoir pool stabilization, and hydroelectric energy production. Flow attenuation is important in order to reduce flood damages due to high-flows; to increase the dependable supply of water for industrial, municipal, commercial, domestic, and agricultural uses; to enhance the navigatability of streams by increasing the number of days when minimum flows exist to permit ship and boat movements; to improve water quality through the dilution of pollutants; and to improve recreation downstream of a dam by increasing the low flows.

Pool stabilization is important for two reasons. Recreation at the reservoir is detrimentally affected by significant and rapid changes in the pool elevation. The facilities available for boating and swimming and picnicking at the reservoir may be unusable if the pool elevation is too high or too low. In addition, if the pool elevation changes rapidly, erosion of the banks can be significantly increased. This causes loss of property to land owners around the lake as well as increased sediment loading into the reservoir. Neither of these is desirable.

Hydroelectric energy production is a separate category because it cannot be categorized with any other reservoir use that involves only flows or only storages. Hydroelectric energy production is a function of the head acting on the turbines multipled by the release through the turbines.

The operation of a single reservoir with multiple purposes is difficult. If pool stabilization is desired, then any inflow should be exactly matched with an equivalent outflow. This will result in complete stabilization of the pool if losses through groundwater seepage and evaporation can be disregarded. If the purpose of the reservoir is flow attenuation, then the outflow should be kept as constant as possible and any excess or deficit inflow should be accommodated by adjustments to the storage volume. Clearly the flow attenuation and pool stabilization objectives are in conflict. If another purpose of the reservoir is hydropower production, then a third conflicting objective exists.

If a single reservoir is difficult to operate, multiple reservoir systems are extremely difficult to operate. In addition to all of the conflicts associated with a single reservoir system, now the interactions of several reservoirs must be accounted for. Therefore, mathematical models that can assist the reservoir operator in making real-time management decisions are extremely valuable [17,20,21].

REAL-TIME RESERVOIR OPERATIONS MODEL

The optimal, real-time operation of reservoir systems requires the determination of the most efficient release sched-

*School of Engineering, Lakehead University, Thunder Bay, Ontario, Canada
**School of Civil Engineering, Purdue University, W. Lafayette, IN

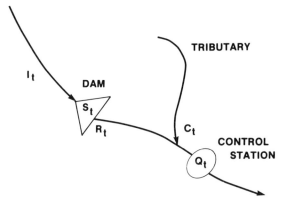

FIGURE 1. Sample reservoir system.

$$Q_t - \alpha_1 R_t - \alpha_2 R_{t-1} = \alpha_3 C_t \qquad t = 1, 2, \ldots, T \quad (4)$$

$$SMIN \leq S_{t+1} \leq CAP \qquad t = 1, 2, \ldots, T \quad (5)$$

$$RMIN \leq R_t \leq RMAX \qquad t = 1, 2, \ldots, T \quad (6)$$

$$Q_t \geq QMIN \qquad t = 1, 2, \ldots, T \quad (7)$$

$$R_t - R_{t-1} \leq RDIF \qquad t = 1, 2, \ldots, T \quad (8)$$

$$R_{t-1} - R_t \leq RDIF \qquad t = 1, 2, \ldots, T \quad (9)$$

where

$B_i(t)$ = benefits associated with objective i for day t, function of storages, releases, and downstream control station flows
λ_i = weighting factors between zero and unity
$\beta_i(.)$ = benefit function for i-th objective
S_t = expected reservoir storage volume at the beginning of day t
R_t = expected reservoir release during day t
Q_t = expected flow at a downstream control station during day t
I_t = forecasted effective reservoir inflow during day t
C_t = forecasted effective tributary flow during day t
CAP = storage capacity of the reservoir
$SMIN$ = minimum allowable storage volume
$RMAX$ = maximum allowable release rate
$RMIN$ = minimum allowable release rate
$QMIN$ = minimum allowable control station flow
$RDIF$ = maximum allowable fluctuation in releases
T = operating and forecast horizon
α_j = channel routing coefficients

ule to satisfy the given physical, environmental, and policy constraints. The reservoir pool levels and the flows at downstream control points are computed from releases. The decisions are based on limited forecast information. Inflows to the reservoirs, evaporations from the reservoir surfaces, seepage from the reservoirs and channels, uncontrolled tributary flows downstream of the reservoirs, and precipitation are forecasted. These forecasts are not one hundred percent reliable and they are only available for a limited time horizon. The time step used in the analysis depends on the nature of the problem. In general, short time steps such as daily, 6-hourly, or even hourly are used.

It is possible to express the general real-time operations problem as a mathematical programming model. The real-time reservoir operations model will be defined for an optimal daily operation of a simple single reservoir, one downstream control station system as shown in Figure 1. The decisions are based on the inputs of the model, i.e., initial storage of the reservoir, previous reservoir releases, and forecasted effective reservoir inflows (inflow + precipitation − evaporation − seepage) and effective tributary inflows for the operation horizon. Equations (1–9) define such a mathematical program.

Maximize $Z =$

$$\sum_{t=1}^{T} \lambda_1 B_1(t) + \lambda_2 B_2(t) + \ldots + \lambda_n B_n(t) \quad (1)$$

Subject to

$$B_i(t) = \beta_i(S_{t+1}, S_t, R_t, R_{t-1}, Q_t, Q_{t-1}, \ldots)$$
$$\qquad\qquad\qquad (2)$$
$$i = 1, 2, \ldots, n$$

$$S_{t+1} - S_t + R_t = I_t \qquad t = 1, 2, \ldots, T \quad (3)$$

Equations (3−9) are the constraints of the problem. Equation (3) represents the continuity of the reservoir storage volume from the beginning of the current day through the next T days. Equation (4) is the routing equation relating the reservoir releases, tributary flows, and the flows at the downstream control station. For simplicity, the routing equation is chosen to be a linear function of the current and the previous days' releases and current effective tributary inflow. For a given system any functional relationship can represent the routing equation; Muskingum routing or stochastic models (AR, MA, ARMA) can be used. However, most of the mathematical programming models require linear constraints to ensure optimality.

Equation (5) relates the physical restriction that storage volume in the reservoir may not exceed its capacity and may not be below a given quantity. Equation (6) limits the releases to discharge capacity and minimum legal or physical requirements (if any). In some cases total reservoir outflows are written as a sum of different releases, such as water supply release, directly pumped from the lake, release through penstocks for hydropower production, and controlled flow over the spillway. Equation (7) ensures that the flow at the downstream control point exceeds the required minimum flow.

Equations (8) and (9) indicate that rapid and drastic changes in the reservoir release from one day to the next are to be avoided. In this formulation, rate-of-increase and rate-of-decrease of releases are assumed to have the same restrictions. However, the allowable rate of changes could be different, as described by Yazicigil et al. [19].

The objective function, Equation (1), is the maximization of the weighted sum of anticipated benefits over the operating horizon. Equation (2) implies that the benefit functions are known mathematically. For example, the benefit function of hydroelectric power production may be expressed as:

$$B_1(t) = C_1 \cdot R_t \cdot H_t \qquad (10)$$

$$H_t = h(S_{t+1}, S_t) \qquad (11)$$

where

$B_1(t) =$ benefit for hydropower objective for day t
$C_1 =$ conversion constant
$H_t =$ productive storage head for day t
$h(.) =$ function relating available head to storage volume

Benefits associated with irrigation and urban−industrial water supply can directly be written as a function of amount used, i.e.,

$$B_2(t) = C_2 \cdot R_t \qquad (12)$$

where

$B_2(t) =$ irrigation and water supply objective for day t
$C_2 =$ unit cost of water

In some cases, it would be easier to express the objective function in terms of losses or penalties instead of benefits. Some targets or desired performance levels for the reservoir operations may be defined. For the recreational use the best reservoir pool level would be the target. For a flood control objective the target level could be the pool level corresponding to a sufficient empty storage or the target downstream control station flow would be the range that does not cause any flooding. Then deviations from the targets would result in losses or penalties. Figure 2 shows examples of such a penalty function. It can be observed that if the storage (or flow) deviates from the target storage (flow range), a penalty is associated.

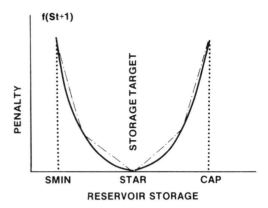

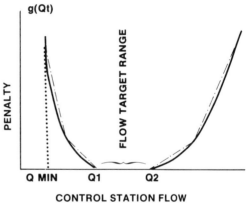

FIGURE 2. Typical penalty functions.

The shape of the penalty functions is convex implying that as operations deviate further from their targets or accepted ranges, the marginal penalty associated does not decrease, i.e., the slope of the penalty function increases as deviations increase. Penalty functions can be expressed as a single curve or as piecewise linear functions as shown in Figure 2.

The total penalty can then be expressed as a function of deviations from the desired target levels.

$$SPEN(t) = f(S_t - STAR) \qquad (13)$$

$$QPEN(t) = g(Q_t - Q_1) \qquad (14)$$

where $STAR$ and Q_1 (or Q_2) are storage and flow targets and $SPEN(t)$ and $QPEN(t)$ are the total storage and flow penalties corresponding to S_t or Q_t.

The relationship between the benefits and penalties can be expressed as:

$$B_i(t) = -PEN_i(t) \qquad (15)$$

Therefore, the objective function would be to minimize the total penalties minus the total benefits or to maximize the total benefits minus the total penalties. The weighting factor, λ_i, in the objective function, Equation (1), indicates the relative importance of each objective of the system. By varying these weights between zero and unity, different operations may be obtained for various conditions.

The performance of the model depends on the operating or forecast horizon, T. It is clear that if the forecast information is highly reliable or perfect, then the model should yield better operation as T increases. But, on the other hand, as T increases, reliability of forecasts decreases, and extending the operating horizon may not improve the operation [5,8]. Therefore, a trade-off has to be made between the forecast reliability and the operating or forecast horizon (T) to be used in the real-time operations model.

In actual operation, the mathematical program is solved at the beginning of each day with a limited forecast information. The solution of the model comprises suggested optimal releases and the corresponding anticipated reservoir storages and downstream control station flows. However, the only implemented portion of this solution is the suggested optimal release of the current day, R_c^*. At the beginning of the next day, the model is reconstructed with revised forecast information and the actual beginning storage for that day (which is not necessarily the same as model computes unless the forecasts have 100 percent reliability). Then, the model is resolved and optimal release for that day is made.

The mathematical model can also be used to simulate the operation of a reservoir system over a long period of time. The model is solved repeatedly while updating the forecast and past input data.

LINEAR PROGRAMMING MODELS

Linear programming (LP) models are widely used in real-time reservoir operations. An LP model requires the constraints of the problem to be expressed as linear equations or inequalities, and the objective function to be written as maximization or minimization of a linear function. In the real-time reservoir operations model given in Equations $(1-9)$, benefit and penalty functions (Equation (2)) may not be linear. In this case, they can not be incorporated into an LP model.

The penalty and/or benefit functions have to be given as linear or piecewise linear convex functions. An example may be used to illustrate mathematically how the piecewise linear convex penalty function is formulated. Let the objectives of the sample reservoir system shown in Figure 1 be flood control and recreation, and assume that the piecewise linear, convex function shown in Figure 3 represents the penalty function.

As the storage (or flow) deviates from the target level

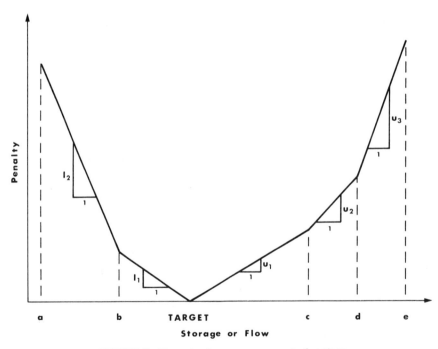

FIGURE 3. Piecewise linear convex penalty functions.

(TAR), penalty increases. Every unit of deviation results in a U_1 unit of penalty when storage (or flow) is higher than the TAR but less than c; U_2 units of penalty are assessed for each unit of storage (or flow) in the range c–d, etc.

Equations (16–21) define five new variables that subdivide the storage variable S_t. If the storage is greater than the target ($S_t >$ TAR), then: $A1_t$ is the volume of the storage in excess of the target that is in the first zone above the target (between TAR and c); $A2_t$ is the volume of the storage in excess of the target that is in the second zone above the target (between c and d); and $A3_t$ is the volume in excess of the target that is in the third zone above the target (between d and e). If the storage is below the target ($S_t <$ TAR), then, $B1_t$ is the portion of the storage deficit (TAR $- S_t$) that is in the first zone below the target (between TAR and b); and $B2_t$ is the portion of the storage deficit that is in the second zone below the target (between b and a).

$$S_t + B1_t + B2_t - A1_t - A2_t - A3_t = \text{TAR} \quad (16)$$

$$0 \le B1_t \le (\text{TAR} - b) \quad (17)$$

$$0 \le B2_t \le (b - a) \quad (18)$$

$$0 \le A1_t \le (c - \text{TAR}) \quad (19)$$

$$0 \le A2_t \le (d - c) \quad (20)$$

$$0 \le A3_t \le (e - d) \quad (21)$$

These zone constraints have to be written for every time step in the operating horizon. The total storage penalty to be minimized may be computed by the product of new variables and their corresponding unit penalties (slopes in Figure 3) as shown in Equation (22).

Minimize

$$Z1 = \sum_{t=2}^{T+1} (U_1 A1_t + U_2 A2_t + U_3 A3_t + L_1 B1_t + L_2 B2_t)$$

$$(22)$$

The complete LP model comprises an objective function [Equation (22)] and a set of constraints [Equations (3–9) and (16–21)]. An optimal solution to a model can be obtained by a well known simplex method [14,18]. Most algorithms only provide one optimal solution when there are alternate optima. Although each alternative will yield the same value for the objective function, intangible objectives may control the operation. Thus, an efficient model should indicate the existence of any alternate optima.

The LP model can also be used in simulation studies. The model will be solved repeatedly by updating the necessary information before each run. In this case, the only changes from one linear program to the next are on the right hand sides of the constraints. Therefore, the dual-simplex method described in Phillips et al. [14] may be used for computational efficiency. In this method, an initial basis drawn from the solution of the previous linear program eliminates the Phase I step (finding an initial feasible solution) because the solution of the previous problem is automatically feasible for the dual problem [19].

The linear programming model presented above is suitable for reservoir systems having flood control, recreation, and water supply as the primary objectives because the penalties and/or benefits can be expressed as linear functions. However, when hydropower production is one of the purposes of the reservoir system, any model requiring linearity may not be suitable. Hydropower is defined as being linearly related to the product of two decision variables, the reservoir head and the reservoir release. Furthermore, because the head is generally not a linear function of the reservoir storage (reservoir storage is used in the reservoir continuity equation in order to keep the equality linear), the nonlinearity is highly complex. Equations (10) and (11) define the benefit function due to hydropower production and the additional constraint which relates the reservoir storage volume and the productive storage head, respectively.

A successive linear programming (SLP) method may be used to overcome the nonlinearity due to maximization of hydropower production. The SLP method is based on the assumption that the available head does not change significantly over short periods. From Equation (10), it is apparent that once the value of head is treated as a constant, the benefit associated with the hydropower production is a linear function of reservoir releases. And because head is not a decision variable, the nonlinear constraint (Equation (11)) is not required in the optimization model. Therefore, an LP algorithm can be used to determine an optimal solution.

At the beginning of the iteration process, heads throughout the operating horizon are set to be equal to the head at the beginning of the analysis. The model is solved as an LP problem and an optimal solution is attained. This solution includes the suggested releases, reservoir storages, and downstream control station flows for the operating horizon. Then heads corresponding to these storages are computed from Equation (11) outside the optimization and are compared with the assumed heads. If the differences are not within allowable limits, the optimization program will be rerun after adjusting the assumed values of the heads. They are set equal to the average of the heads assumed in the previous step and the heads obtained from the optimization program. This process is repeated until convergence is reached.

This and other approximations for linearization of benefits and constraints due to hydropower production have been described in the literature [6,12,13]. Grygier and Stedinger [10] compare some of the algorithms for optimizing the operation of multi-reservoir hydrosystems and favor the SLP method.

The LP models reviewed in this section are deterministic. Streamflow forecasts are assumed to have a single value. Stochastic models, considering random streamflows, are discussed in the section on Chance Constrained Models.

GOAL PROGRAMMING MODELS

The linear programming models discussed in the previous section and quadratic programming models discussed in the next section require a set of mathematical functions that relate benefits from the system and penalties due to deviations from ideal operations to a common unit so that they can be added. An alternative approach to modelling the real-time reservoir operations that does not require the penalty functions is goal programming [4].

The constraints of the goal programming (GP) model are identical with the constraints of the LP model. However, the objective is expressed differently. Instead of a single equation that defines the objective in an LP model, a hierarchy of goals, expressed in terms of storages, flows, and releases, defines the operational objective of the GP model. The goals are ordered by priority with the highest priority goal first and the least important goal last.

The goal program is solved by considering the goals one at a time in order. First, a solution which maximizes attainment of goal 1 is found. If a unique optimum is obtained, then this is the optimal solution for the system and the remaining priority levels are computed using this solution. However, if alternate optima result, then for the same level of attainment of goal 1, different achievements of remaining goals may be obtained.

A constraint which ensures that goal 1 will always be satisfied to a level equal to the one just found is added to the constraint set. Goal 2 is now considered. A solution is found that maximizes the attainment of goal 2. If alternate optima exist, another constraint which ensures that goal 2 will always be satisfied to the level found is now added to the constraint set. This process continues until all of the goals have been considered, or a unique optimum is found after a goal.

This procedure is only appropriate if goal 1 is much more important than goal 2 and goal 2 is much more important than goal 3 and so on. It does not permit a reduction in the attainment of a higher priority goal in order to improve the attainment of a lower priority goal. For modelling real-time reservoir operations, goal programming may be appropriate.

Goals of the reservoir operations problem are related to physical phenomena, such as spillway levels, stream channel capacities, minimum storage requirements, minimum downstream flow requirements for fish life and/or navigation, and minimum release requirements for water supply. The first priority goals will have the broadest range and will include all the possible operations of the system. For example, only the minimum downstream control station flow and the minimum release requirements may be included as first priority goals.

The second priority goals will decrease the feasible region. Depending on the decision makers, the second priority goals may require all the system reservoirs to operate below their spillway crest elevations and above minimum elevations for water intakes, and require the flows at downstream control stations to be low enough to avoid major flooding. Similarly, each lower priority goal will force the operation to be closer to the targets for reservoir storages and target levels for downstream control station flows.

Consider the example given in the previous section. Figure 3 is the piecewise linear convex penalty function. Experience indicates that the points a, b, TAR, c, d, and e are relatively easily chosen. For example, the main purposes of the reservoirs in the Green River Basin in Kentucky are flood control and recreation, and these points correspond to physical performance levels such as:

Point	Storage	Flow
a	beaches unusable	fish just surviving
b	boat ramps unusable	no rafting possible
TAR	storage target	flow target
c	picnic tables underwater	minor flooding
d	spillway elevation	major flooding
e	top of dam	disaster

The selection of the penalty values (slopes in Figure 3) associated with the various storage and flow levels is more difficult. However, it is easier to list the priority of operating in different ranges. For example, operating in the range a−b may be the worst of all ranges. Operating in the d−e range may be next to worst. Operating in range c−d may be next. One way to summarize these observations is with a hierarchy of goals such as:

Goal 1. Operate in the range b−e.
Goal 2. Operate in the range b−d.
Goal 3. Operate in the range b−c.
Goal 4. Operate at the target TAR.

Any feasible solution to the constraint set ensures that all the possible operations are in the range a−e. The first goal eliminates those operations that are most detrimental. If this is possible, the remaining goals are considered starting from the second goal. Each goal limits the range of operations step by step and the last goal, if achieved, operates the system at the target.

Using the same notation as the LP models, it is possible to define the goal program for the same problem with four priority levels. The first priority goal is to minimize operations outside the range b−e. This objective may be stated as:

$$\text{Minimize } P1 = \sum_{t=2}^{T+1} (B2_t) \tag{23}$$

Solving the mathematical program comprising the same constraint set as the LP model (i.e., Equations (3−9) and 16−21)) and an objective function given in Equation (23) yields an optimal objective function $P1^*$.

The second priority goal is to minimize operations outside the region b−d while not affecting the first priority goals. This can be done by solving the program comprising Equations (3−9), (16−21), and (24−25). Equation (24) is the constraint restricting any impacts on the first priority.

$$\text{Minimize } P2 = \sum_{t=2}^{T+1} A3_t \tag{24}$$

$$\sum_{t=2}^{T+1} (B2_t) = P1^* \tag{25}$$

The third priority is to minimize operations outside the region b−c while not affecting the first and second priority goals. Equation (26) is added to the constraints of the previous problem to ensure the second priority goals are not affected and Equation (27) represents the objective function for the third priority.

$$\sum_{t=2}^{T+1} A3_t = P2^* \tag{26}$$

$$\text{Minimize } P3 = \sum_{t=2}^{T+1} A2_t \tag{27}$$

The fourth and last priority is to minimize the deviation of operation from the target while not affecting the first three priority goals. Equations (3−9), (16−21), (25), (26), (28), and (29) represent this problem. Equations (25), (26), and (28) restrict the effect of the fourth priority goals on the higher priority goals.

$$\sum_{t=2}^{T+1} A2_t = P3^* \tag{28}$$

Equation (29) is the objective of the last priority.

$$\text{Minimize } P4 = \sum_{t=2}^{T=1} (A1_t + B1_t) \tag{29}$$

Solution of this program yields an optimal objective function value of $P4^*$ and, among other variable values, the optimal value of R_1^*, the optimal release for the first day.

The GP model described above may be solved with the help of LP algorithms by solving successive linear programs for every priority after adding new constraints to ensure the level of higher priority goals. However, an efficient partitioning algorithm for solving linear goal programming problems is given in Arthur and Ravindran [1] and is recommended.

In the linear programming approach, two dimensional information is required. The determination of the penalties corresponding to unit changes in the storages or flows (i.e., slopes of the lines) may require detailed economic analysis. However, in the goal programming model, only one dimensional information, such as the location of points b, c, and d, is needed. These points are usually defined by physical phenomena such as storages corresponding to water intake levels or spillway elevations and flows corresponding to irrigation demands or flooding.

QUADRATIC PROGRAMMING MODELS

The quadratic programming model assumes that the nonlinear penalty and/or benefit functions are given as parabolic functions. Equations (30–32) represent parabolic, convex penalty functions for reservoir storages, releases, and flows at downstream control points.

$$\text{SPEN}_{t+1} = \alpha(S_{t+1} - \text{STAR})^2 \tag{30}$$

$$\text{RPEN}_t = \beta(T_t - \text{RTAR})^2 \tag{31}$$

$$\text{QPEN}_t = \gamma(Q_t - \text{QTAR})^2 \tag{32}$$

in which STAR, RTAR, QTAR are storage, release, and flow targets, respectively, and α, β, and γ are non-negative constants to be determined. Equations (30−32) indicate that no penalty is assessed if the operation is at the target and marginal penalty increases as deviations increase.

The QP model is subject to linear constraints as given in Equations (3−9). The objective function of the model is to minimize the total penalties, i.e.,

Minimize

$$Z2 = \sum_{t=1}^{T} \alpha(S_{t+1} - \text{STAR})^2 \qquad (33)$$

$$+ \beta(R_t - \text{RTAR})^2 + \gamma(Q_t - \text{QTAR})^2$$

A general quadratic programming (QP) problem is an optimization problem involving a quadratic objective function and linear constraints [14,18]. In vector form, it can be stated as:

Minimize

$$Z = YX + X' D X \qquad (34)$$

Subject to

$$AX \geq B \qquad (35)$$

$$X \geq 0 \qquad (36)$$

where A is an (mxn) matrix of constraint coefficients, B is an (mxl) column vector, Y is an (lxn) row vector, D is an (nxn) matrix of quadratic form, and X is an (nxl) vector of decision variables.

The region of feasible solutions is defined over a set of linear constraints, thus it is convex. Then, if the objective function, Z, is convex, it is known that if a local minimizing solution exists to the QP problem, it will also be a global minimizing solution. The quadratic objective function is convex, provided that the matrix D in the objective function is positive definite or positive semidefinite. In the QP model of real-time reservoir operations defined above, the constraints are linear and the matrix of quadratic form can be written as shown in Table 1.

TABLE 1. Matrix of Quadratic Form for QP Model.

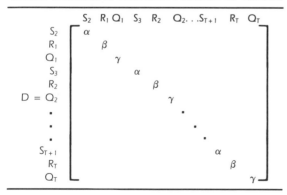

The matrix in Table 1 has zero off-diagonal and positive diagonal elements. One way to determine whether a symmetric matrix is positive (semi) definite is through the leading principal determinants. A symmetric matrix is positive definite if all the diagonal elements and all the leading principal determinants are positive (positive semidefinite if they are non-negative). It is clear that for the quadratic form matrix of the QP model all the leading principal determinants are positive. Hence, the matrix is positive definite and the objective function is convex. Therefore, an optimal solution for the model can be obtained by quadratic programming.

An efficient and simple method for solving the QP problems with convex objective functions is the complementary pivot method developed by Lemke which is based on solving the Kuhn-Tucker optimality conditions [14,18]. This method guarantees to find an optimal solution for quadratic programming problems with positive semidefinite quadratic form matrices. It is shown [16] that the complementary pivot method is computationally more attractive than most of the methods available for solving QP problems when the matrix of quadratic form is positive semidefinite. The complementary pivot method can easily be programmed in Fortran language within 320 executable statements. A computer program for the method is given in Ravindran [15].

The QP model has three advantages over the LP model. First, the size of the problem is significantly reduced [3]. Introduction of new variables and zone constraints (Equations (16−21)) are not required. Second, the information required to describe the penalty function is less. In the LP model, slopes of each zone need to be determined. On the other hand, only one parameter is required to define the penalty functions for the QP model. This simplification of the form of the objective function may, however, result in a less accurate description of the true objective because the true objective may be asymmetric or non-quadratic.

The third, but maybe the most important, advantage of the QP model is that it may offer a global optimal solution to a multiobjective system, including hydropower production. All the proposed models for maximizing hydropower production either require discretization (DP algorithms), linearization (LP models), or search methods. The difficulty arises due to the cross multiplication term in the hydropower benefits.

Suppose the objective of a multipurpose system is to minimize penalties due to deviations from storage, release, and flow targets for flood control, recreation, etc., and to maximize benefits due to hydropower production. It is assumed that the head-storage relationship is linear for the range that is possible for given forecasts, i.e.,

$$H_t = C_3 S_t + C_4 \qquad (37)$$

in which C_3 and C_4 are constants. Introducing this linear relationship in Equation (10), hydropower benefits can be

TABLE 2. Matrix of Quadratic Form ($\theta = C_1C_3/2$).

	S_2	R_1	S_3	R_2	Q_1	Q_2
S_2	α	$-\theta$	0	0	0	0
R_1	$-\theta$	β	0	0	0	0
S_3	0	0	α	θ	0	0
R_2	0	0	$-\theta$	β	0	0
Q_1	0	0	0	0	γ	0
Q_2	0	0	0	0	0	γ

expressed as:

$$B_1(t) = C_1C_3R_tS_t + C_1C_4R_t \qquad (38)$$

and the objective function for the system is written as shown in Equation (38).

Minimize

$$Z3 = \sum_{t=1}^{T} \alpha(S_{t+1} - STAR)^2$$

$$+ \beta(R_t - RTAR)^2 + \gamma(Q_t - QTAR)^2 \qquad (39)$$

$$- C_1C_3R_tS_t - C_1C_4R_t$$

For this model, the matrix of quadratic form can be written as shown in Table 2. For simplicity the operating horizon (T) is taken to be 2 days. The matrix in Table 2 has positive diagonal elements. The last two leading principal determinants are positive if the first four leading principal determinants are positive. However, the first four leading principal determinants are positive if and only if ($\alpha\beta - \theta^2$) is positive. It can easily be shown that the term ($\alpha\beta - \theta^2$) controls the positive definiteness of the matrix for any value of T. This term is related to constants in storage and release penalty functions and hydropower benefit function. Therefore, if this term is positive for a given multiobjective system, then the objective function in Equation (39) is convex and the optimal operation can be obtained via quadratic programming.

DYNAMIC PROGRAMMING MODELS

The real-time reservoir operations problem defined by Equations (1–9) can be solved by dynamic programming under certain conditions. As described in a previous section, linear programming could be used to solve the problem if the constraints and objective function are linear. If the problem is "separable" and satisfies a "monotonicity" criterion, then dynamic programming may be used as the solution procedure [2]. In general, these criteria are satisfied.

Instead of presenting the real-time operations problem in the format of Equations (1–9), the dynamic program formulation comprises a functional equation and boundary or initial conditions. For the case of a single reservoir, with penalties associated with deviations from storage and release targets, the functional equation is:

$$f_t(S_t) = \underset{S_{t+1}\epsilon\phi_{t+1}}{\text{Minimum}} [\{R(R_t - RTAR) + S(S_t - STAR)\}$$

$$+ f_{t+1}(S_{t+1})] \qquad (40)$$

for all values of $S_t\epsilon\phi_t$ where

R_t = $S_t - S_{t+1} + I_t$
RTAR = target release for day t
STAR = target storage for day t
ϕ_{t+1} = set of discrete storage volumes that are considered for day $t + 1$
$R(.)$ = penalty function associated with releases
$S(.)$ = penalty function associated with storages
$f_t(S_t)$ = minimum total penalty accrued from the beginning of day t to the end of the operating horizon
S_t = storage volume at the beginning of day t
R_t = release volume for day t
I_t = forecasted inflow for day t

The initial or boundary conditions are:

$$f_{T+1}(S_{T+1}) = 0 \qquad (41)$$

for all values of $S_{T+1}\epsilon\phi_{T+1}$.

This dynamic program is formulated as backward moving because the solution of Equation (40) occurs repeatedly, first for $t = T$, then for $T - 1$, and so on. The "stages" of the dynamic program are days or time periods and the "states" are storage volumes. This is a discrete dynamic program because the continuum of storage volumes (SMIN $\leq S_t \leq$ SMAX) is represented by discrete set of storages denoted by ϕ_t (i.e., $S_t\epsilon\phi_t$).

The formulation of the dynamic program above is only one example of how the real-time reservoir operations problem can be cast in dynamic programming form. Instead of backward moving, the problem could be stated as a forward moving dynamic program. In multiple reservoir cases, the functional equation has been formulated with the stages representing reservoirs and not time periods. Also, there are special techniques available for solving several classes of dynamic programs. For example, the dynamic program formulated in Equations (40) and (41) can be solved as it is shown or using a technique called discrete differential dynamic programming or using dynamic programming with successive approximations. Each of these techniques has advantages that reduce the computational effort for solution yet can guarantee global optimality of the solution if the problem meets certain conditions [13].

The important issue is: what advantages and disadvantages does the dynamic programming solution procedure have compared with the linear programming procedure? Both procedures are used to solve the real-time reservoir operations problem. Therefore, the selection of the procedure will depend upon specific characteristics of the problem and the ability of each procedure to address them.

The principal advantages of dynamic programming over linear programming are its abilities to handle nonlinearities. For example, in Equation (40), the objective function is denoted by the release and storage penalties. In linear programming, these penalty functions would have to be either linear or be able to be approximated by piecewise linear segments. In dynamic programming, that restriction does not exist. Clearly, those penalty functions should be convex or at least nondecreasing as the deviation from the targets increases. As long as that criterion is satisfied, dynamic programming can incorporate the nonlinear penalty function without approximation. Another example of the ability to handle nonlinearities is with the inclusion of hydropower in the problem. As mentioned earlier, hydropower is a complex nonlinear function of storage and release. In the linear programming models, it is essential to make assumptions about the behavior of the storage prior to solution of the model. In dynamic programming, this is avoided by directly incorporating the physics of the hydropower production.

There are several principal disadvantages of dynamic programming compared to linear programming. One disadvantage has already been mentioned: the storage in the reservoir must be represented by a set of discrete values. If that set is poorly chosen, then the discrete set of values may not adequately represent the continuous range of storage and may distort the solution obtained by solving the dynamic program. An obvious solution to this problem is to make the number of discrete storages very large. Unfortunately, this causes another problem. As the number of storages becomes larger, the computational requirements for solution increase. This increase is always faster than linear and may be quadratic with the number of discrete storage values considered. In addition, when the system is made up of more than one reservoir, the computational effort required for solution of the dynamic program again increases nonlinearly with the number of reservoirs. If these two effects are coupled, then very quickly, multiple reservoir systems become impossible to model with a dynamic program.

Much effort has been expended in overcoming this problem of rapidly increasing computational effort for multiple reservoir systems being modeled with dynamic programs. At one time, systems made up of two or three reservoirs represented the limit of dynamic programming ability. Recently, there have been reports in the literature of systems comprising up to twenty reservoirs as being successfully modeled with dynamic programs.

A related problem is the necessity to write the codes that are required to solve the mathematical programs. The codes for solving linear programs are highly reliable and thoroughly tested and are maintained usually by a computer systems staff. On the other hand, dynamic programs almost always must be coded specifically for each system under investigation. This can be an important distinction between the two procedures depending on the computer programming resources available.

Another major disadvantage of dynamic programming is the criterion called separability. This criterion requires that the optimization problem be capable of separation into stages. In the case of reservoir operations, the stages are typically time periods. Hence, any effect from one time period to the next is acceptable within the dynamic program; however, any effect that spans more than one time period cannot be included in the dynamic program. Therefore, routing of water from one dam downstream for two or three or four days to a point where flood damage may occur is very difficult to handle within the dynamic programming format. It would require the reformulation of the functional equation to include reservoir operations over that two- or three- or four-day period. This would mean an exponential increase in the computational effort required to solve the dynamic program. Typically, this increase in computational effort would make solution very expensive or prohibitive.

CHANCE CONSTRAINED MODELS

It is possible to use any of the approaches described previously to consider the effects of imperfect streamflow forecasts. One brute force way to measure the effects is to solve the optimization model many times under different streamflow forecast assumptions. A much more elegant procedure that explicitly considers the uncertainties in streamflow forecasts is chance constrained programming [7,9,11].

In problems that involve random variables, it is sometimes important to restrict the probability of occurrence of some events. For example, chance constrained programming has been used in long-term planning and operating models for reservoir systems. The random variable has been monthly or seasonal streamflow and the events that have been constrained have been storage levels, release rates, and hydroelectric power production. For example, a chance constraint on release may be:

$$Pr[R_t \leq RMAX] \geq 0.9 \qquad (42)$$

$$Pr[R_t \geq RMIN] \geq \alpha \qquad (43)$$

Because release is a function of the reservoir inflow and because inflow is a random variable, the release is a random variable. The selection of an appropriate reliability value such as 0.9 or α, however, is not necessarily easy.

Before a chance constraint can be included in an optimization model, it usually must be converted into a form accepted by the optimization algorithm. Typically, this conversion results in an equation that is called the deterministic equivalent (of the chance constraint) and that is in the proper form for inclusion in a linear program or some other optimization procedure.

One constraint and an objective function can be used to illustrate how the real-time reservoir operations problem can use chance constrained programming. The mass balance equation for reservoir storage from the beginning of the first day of operation to the beginning of the second day of operation is:

$$S_2 = S_1 + I_1 - R_1 \qquad (44)$$

In this equation, I_1 denotes the forecast inflow to the reservoir during the first day.

The error in the forecast is a random variable whose distribution may be determined. Therefore, the forecast is a random variable whose distribution may be determined. And through the mass balance equation, the storage at the beginning of the second day (S_2) becomes a random variable.

Chance constraints restricting the possible beginning storages for the second day may be formulated as:

$$Pr[S_2 \geq \text{SMIN}] \geq 0.9 \qquad (45)$$

or

$$Pr[S_2 \leq \text{SMAX}] \geq \gamma \qquad (46)$$

Appropriate objectives that correspond to these chance constraints (Equations (45) and (46)) are: to maximize the value of SMIN, which is the storage that is exceeded by at least 90 percent of all possible actual storages at the beginning of the second day; or to maximize the value of γ, which is the reliability of actual storage at the beginning of the second day not exceeding a specified value SMAX.

Other performance levels of the operation can be included in either the constraints or objective of the optimization model. For example, hydroelectric power production could be restricted to be within some designated range with high probability. The probability that the storage at the end of five days is within some specified range of the rule curve could be maximized. Or the probability of the flow at a downstream control station falling within a target range could be maximized.

Only limited experience testing chance constrained models for real-time, short-term reservoir operation is available. Datta [7] has described a single reservoir system whose operations was simulated using a chance constrained model. He also argued strongly in favour of the chance constrained approach applied to real-time reservoir operations.

REFERENCES

1. Arthur, J. L. and A. Ravindran, "A Partitioning Algorithm for (Linear) Goal Programming Problems," *ACM Transactions on Mathematical Software, 6* (3), pp. 378–386 (September 1980).
2. Bellman, R., *Dynamic Programming,* Princeton University Press, Princeton, N.J. (1957).
3. Can, E. K., "Real-Time Reservoir Operations by Quadratic Programming," *Proceedings, ASCE Specialty Conference Computer Applications in Water Resources,* Buffalo, New York, pp. 790–799 (June 1985).
4. Can, E. K. and M. H. Houck, "Real-Time Reservoir Operations by Goal Programming," *Journal of Water Resources Planning and Management,* ASCE, Vol. 110, No. 3, pp. 297–309 (July 1984).
5. Can, E. K. and M. H. Houck, "Problems with Modelling Real-Time Reservoir Operations," *Journal of Water Resources Planning and Management,* ASCE, Vol. 111, No. 4, pp. 367–381 (October 1985).
6. Can, E. K., M. H. Houck, and G. H. Toebes, "Optimal Real-Time Reservoir Systems Operation: Innovative Objectives and Implementation Problems," *Technical Report 150,* Purdue University Water Resources Research Center, West Lafayette, Indiana (November 1982).
7. Datta, B., "Stochastic Optimization Models for Long-Term Planning and Real-Time Operation of Reservoir Systems," Ph.D. Dissertation, Purdue University, West Lafayette, Indiana, U.S.A. (1981).
8. Datta, B. and S. Burges, "Short-Term, Single, Multiple-Purpose Reservoir Operation: Importance of Loss Functions and Forecast Errors," *Water Resources Research, 20* (9), pp. 1167–1176 (September 1984).
9. Datta, B. and M. H. Houck, "A Stochastic Optimization Model for Real-Time Reservoirs Using Uncertain Forecasts," *Water Resources Management, 20* (8), pp. 1039–1046 (August 1984).
10. Grygier, J. C. and J. R. Stedinger, "Algorithms for Optimizing Hydropower System Operation," *Water Resources Research, 21* (1) pp. 1–10 (January 1985).
11. Houck, M. H., "A Chance Constrained Optimization Model for Reservoir Design and Operation," *Water Resources Research, 15* (5), pp. 1011–1016 (October 1979).
12. Houck, M. H. and J. L. Cohen, "Sequential Explicitly Stochastic Linear Programming Models for Design and Management of Multipurpose Reservoir Systems," *Water Resources Research, 14* (2), pp. 161–169 (April 1978).
13. Loucks, D. P., J. R. Stedinger, and D. A. Haith, *Water Resources Systems Planning and Analysis,* Prentice-Hall, New York (1981).
14. Phillips, D. T., A. Ravindran, and J. J. Solberg, *Operations Research: Principles and Practice,* Wiley, New York (1976).
15. Ravindran, A., "A Computer Routine for Quadratic and Linear Programming Problems," Algorithm 431, *Communications of the Association for Computing Machines, 15* (9), pp. 818–820 (1972).

16. Ravindran, A. and H. Lee, "Computer Experiments on Quadratic Programming Algorithms," *European Journal of Operations Research, 8* (2), pp. 166–174 (1981).

17. Read, E. G. and R. E. Rosenthal, "Scheduling Reservoir Releases for Optimal Power Generation: Introduction and Survey," paper presented in A Special Energy Management Course offered during the 1982 Energy Exposition and World's Fair, Knoxville, Tennessee (June 1982).

18. Reklaitis, G. V., A. Ravindran, and K. M. Ragsdell, *Engineering Optimization: Methods and Applications*, Wiley, New York (1984).

19. Yazicigil, H., M. H. Houck, and G. H. Toebes, "Daily Operation of a Multipurpose Reservoir System," *Water Resources Research, 19* (1), pp. 1–13 (January 1983).

20. Yeh, W. W-G, "State of the Art Review: Theories and Applications of Systems Analysis Techniques to the Optimal Management and Operation of a Reservoir System," *Report UCLA-ENG-82-52,* University of California, Los Angeles (1982).

21. Yeh, W., "Reservoir Management and Operations Models: A State of the Art Review," *Water Resources Research,* Vol. 21, No. 12, pp. 1797–1818 (December 1985).

CHAPTER 4

Surge Shaft Stability for Pumped-Storage Schemes

ALEX ANDERSON*

INTRODUCTION – STABILITY AND SURGE SHAFT DESIGN

Surge Shafts and Their Function

A "surge shaft" in a pressure tunnel system is any shaft with an unbounded water surface free to move vertically (Figure 1). Such a shaft may occur because it has been introduced for another reason altogether (e.g., a gate slot, construction or access adit, additional water intake, etc.) but the term "surge shaft" is usually applied when it has been introduced specifically to provide a free water surface at that point in the pressure tunnel system (Figure 2). Nevertheless, surging of the free water surface may occur in either type of shaft and there is no physical difference in their response or mathematical difference in their analysis.

A free water surface is deliberately introduced into a pressure tunnel system (e.g., the surge tank in Figure 2a or the air chamber in Figure 2c) for one or both of the following reasons:

1. To protect a tunnel from high pressure water hammer waves by providing a reflecting free surface boundary.
2. To reduce the apparent inertia of the overall tunnel system in order to improve its response to load change transients (classically to improve the regulation of water turbines):
   - by storing/providing water during decreasing/increasing demand
   - while creating a negative/positive potential head (i.e., raising/lowering shaft free surface) in order to decelerate/accelerate the lower pressure side tunnel flow (Figure 1)

Whether a shaft has been introduced for these or for other reasons it will respond to transients in the pressure tunnel system and its design should have (Figure 3):

(a) Ability to contain surge extremes (prevention of overflow or complete drainage), usually in response to maximum changes of load on hydraulic machines (turbines, pumps).
(b) Ability to operate stably under steady state load conditions (i.e., oscillations in water surface level tend to damp out).
(c) In addition to these traditional requirements, modern pumped-storage plants may also require the ability to operate within an acceptable limiting amplitude under cyclic load conditions (e.g., frequency regulation duty).

Pumped-Storage Schemes and Their Surge Shaft Stability Problems

Conventionally, in surge shaft analysis, the first two requirements above (i.e., extreme surge response and stable operation) are considered as separate problems and this chapter discusses the stability problems and analyses specific to surge tanks in pumped-storage and other pumping systems.

Essentially these problems fall into one of three classes (for which the driving mechanisms will subsequently be illustrated by example) depending on the source of the excitation:

- Traditionally, surge shaft stability has been concerned with water turbines in hydro power plant governed to produce a constant power output. The governing action is the source of excitation.
- In a pumping system "pump surge" may occur. This arises from the interaction of the pump performance characteristic and the pressure tunnel system hydraulic characteristic.

*Department of Mechanical Engineering, University of Newcastle upon Tyne, United Kingdom

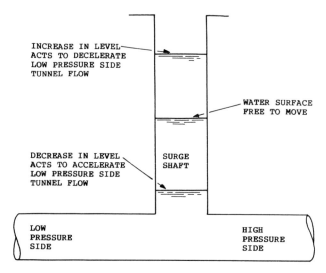

INCREASE IN LEVEL
ACTS TO DECELERATE
LOW PRESSURE SIDE
TUNNEL FLOW

WATER SURFACE
FREE TO MOVE

DECREASE IN LEVEL
ACTS TO ACCELERATE
LOW PRESSURE SIDE
TUNNEL FLOW

SURGE
SHAFT

LOW
PRESSURE
SIDE

HIGH
PRESSURE
SIDE

PART OF PRESSURE TUNNEL SYSTEM

FIGURE 1. Surge shaft.

— In multiple shaft systems (e.g., Figure 2) there is a possibility of unexpected instabilities arising from the interaction of the shafts.

For the first of these there is an extensive literature, admirably summarised in a series of excellent standard texts [16,19,29,42,62,74] and practical results and discussions of their applications are thus readily available (a brief review will be given). For the others, however, there is little guidance in the literature and in these cases new stability analyses may have to be carried out. Fortunately stability is a standard problem of dynamic systems so the general techniques are well described (though rarely in civil engineering or hydraulic textbooks) and are summarised for convenience in the next section.

In addition, some surge shaft stability problems peculiar to pumped-storage and other pumping systems are introduced because:

— Pumped-storage plants are increasingly being designed [49] for more rigorous operational requirements, including complex (e.g., pump to turbine, rapid load then unload, etc.) and cyclic loading sequences (Figure 4), leading to relatively larger surge tanks.
— Pumped-storage plants, for economic reasons, are being designed with increasingly long penstock and draft tube tunnels and with greater flexibility in the plant layout.

Thus, terms which have been neglected in traditional stability analyses may assume a new significance for pumped-storage schemes.

Objectives for Stability Analysis

The various methods for *analysis* of surge shaft stability are merely aids to *design* of surge shafts. In carrying out any stability analysis it should be borne in mind that in appropriate circumstances the design requirements may be:

— either to *avoid* instability altogether
— or to *control* any instability to acceptable locations and amplitudes.

Stability analysis may be used in two ways:

1. At the initial stages of the design some preliminary sizing and setting out is required—stability analysis can give rise to formulas which suggest a starting point for the design (e.g., the classic Thoma surge shaft area formula) or suggest the influence of variations in the sizing of particular components.
2. At subsequent stages of the design detailed stability analysis is required to confirm that the system continues to satisfy the criteria above—frequently the classic techniques are inadequate and more advanced methods are required to cope with the increased complexity of the final design.

Traditionally, stability analysis is chiefly concerned with the first objective but the demands of pumped storage plants mean that the second objective is becoming more important. Obviously different techniques of stability analysis are appropriate for each of these objectives, so a useful starting point is to review the techniques available.

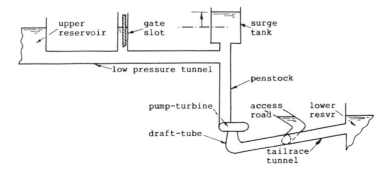

(a) TYPICAL PUMPED-STORAGE SCHEME WITH SURGE SHAFTS

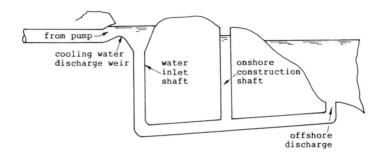

(b) TYPICAL OFFSHORE DISCHARGE TUNNEL WITH SURGE SHAFTS

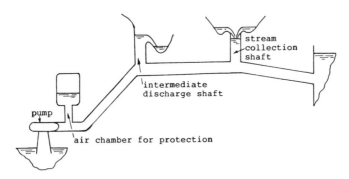

(c) TYPICAL WATER SUPPLY TUNNEL WITH SURGE SHAFTS

FIGURE 2. Surge shafts in pumping systems.

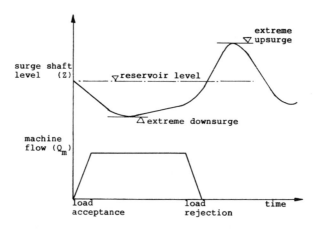

(a) CONTAINED SURGE SHAFT RESPONSE TO LARGE LOAD CHANGES

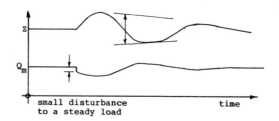

(b) STABLE SURGE SHAFT RESPONSE WITH STEADY-STATE LOADING

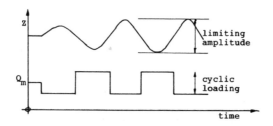

(c) LIMITED SURGE SHAFT RESPONSE TO CYCLIC LOADING

FIGURE 3. Design requirements for surge shaft.

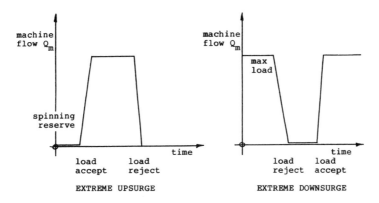

(a) COMPOSITE SEQUENCES FOR SURGE EXTREMES (Ref 49, 61, 67)

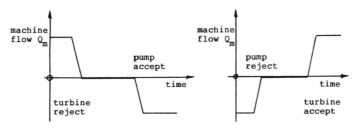

(b) 'HYDRAULIC REVERSAL' WITH PUMPED-STORAGE (Ref 56, 67)

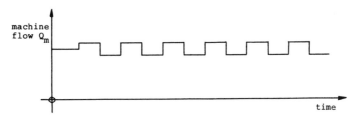

(c) CYCLIC LOADING, eg FREQUENCY REGULATION (Ref 49)

FIGURE 4. Pumped-storage scheme loading sequences.

(a) STABLE EQUILIBRIUM

(b) UNSTABLE EQUILIBRIUM

(c) SYSTEM WITH DIFFERENT EQUILIBRIUM STATES

FIGURE 5. *States of stable and unstable equilibrium.*

BACKGROUND—STABILITY OF SELF-EXCITED VIBRATIONS

Surge Shafts and General Dynamic Systems

For turbining flows in simple surge shaft systems there is an extensive literature which describes well-established techniques and many practical applications, e.g., [19,42]. This, however, is far from the case with:

— pumping flows, and
— multiple surge shaft systems (characteristic of pumping schemes)

so that it is likely that recourse will often have to be made to the basic techniques of analysis. The stability of surge shafts is merely a special case of the more general problem of the stability of dynamic systems, and those interested in the wider aspects of the problem will find useful discussions in standard textbooks on mechanical vibrations, e.g., [24,78], or on control, e.g., [8,26,58,64,65,66]. To assist in their application to surge shaft systems, the basic techniques of stability analysis will be described briefly.

Basic Concepts of Stability in Dynamic Systems

The idea of stability is readily expressed at an elementary level:

—stable —an equilibrium state is stable if the response to a small displacement from that state is an eventual return to the original position (Figure 5a)
—unstable —an equilibrium state is unstable if the response to a small displacement from that state causes the system to continue to move further away (Figure 5b)

Thus it can be seen that stability is a function

— not only of the dynamic system
— but also of particular equilibrium states of that system (Figure 5c).

An alternative definition, which can be mathematically formalised, e.g., [25,45], defines an equilibrium state as being stable if:

— either the response to an impulse away from the steady state tends to zero as time tends to infinity (Figure 6a),
— or the response to every bounded input (excitation) is a bounded output (Figure 6b).

While these definitions are apparently *objective*, in practice they are likely to be applied *subjectively*, i.e., the response should tend to zero "sufficiently rapidly" or should be bounded to a "sufficiently low" amplitude. Alternatively, a "safety margin" will always be applied to any formally derived result to avoid "undesirable" patterns of behaviour.

Origins of Instability

The physical basis behind the above ideas is best illustrated by example, e.g., the elementary lumped mass (m)/-spring (k)/damper (c) mechanical system (Figure 7) for which the standard dynamic equation is [24,78]:

$$m\ddot{x} + c\dot{x} + kx = F(t) \qquad (1)$$

or

$$\ddot{x} + 2\lambda\dot{x} + \omega_n^2 x = F/m \qquad (2)$$

where $2\lambda = (c/m)$ and $\omega_n^2 = k/m$

The type of vibration which this system undergoes is characterised by the nature of the exciting force $F(t)$

—free vibration —applied impulse $F(t) = \begin{cases} F_0(t=0) \\ 0(t>0) \end{cases}$

—forced vibration —explicit function of time (t), i.e., excitation $F(t)$ exists independently of the motion $x(t)$ and will always sustain it.

—self-excited vibration —function of displacement (x), velocity ($\dot{x}$) or acceleration ($\ddot{x}$), i.e., excitation $F(t) = F(x)$ or $F(\dot{x})$ or $F(\ddot{x})$

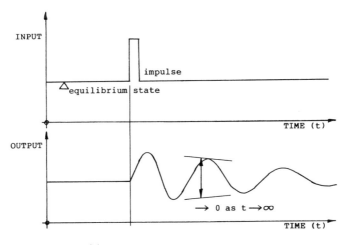

(a) STABLE RESPONSE TO AN IMPULSE

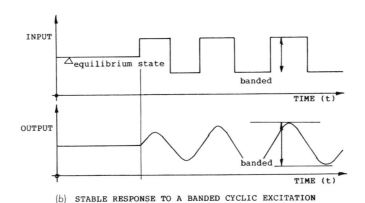

(b) STABLE RESPONSE TO A BANDED CYCLIC EXCITATION

FIGURE 6. Stable responses.

is created or controlled by the motion itself (so that when the motion disappears then the excitation also disappears).

Free and forced vibrations of this simple system are extensively discussed in standard textbooks, e.g., [24,78]. For the former, with $F = 0$ in Equations (1 and 2), the general solution is:

$$x = C_1 \exp(s_1 t) + C_2 \exp(s_2 t) \qquad (3)$$

where the constants C_1 and C_2 are determined by the initial conditions and s_1 and s_2 are given by the roots of the characteristic equation:

$$s^2 + 2\lambda s + \omega_n^2 = 0 \text{ i.e., } s = -\lambda \pm \sqrt{\lambda^2 - \omega_n^2} \quad (4)$$

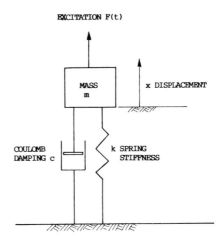

FIGURE 7. Elementary second order mechanical system.

Normally both roots (s_1, s_2) will have negative real parts as in Figure 8a [65], thus giving a solution (Equation (3)) which diminishes in amplitude with time.

It is *self-excited* vibration, however, that gives rise to stability problems. Consider each kind of self-excitation in turn.

EXCITING FORCE IS A FUNCTION OF DISPLACEMENT

This gives rise to a dynamic equation of the type:

$$m\ddot{x} + c\dot{x} + kx = F_x \cdot x \qquad (5)$$

which can be rewritten in the form of a free-vibration equation:

$$\ddot{x} + 2\lambda\dot{x} + \omega_n^2 = 0 \qquad (6)$$

in which the "apparent" damping and stiffness are:

$$2\lambda = (c/m) \text{ and } \omega_n^2 = (k - F_x)/m \qquad (7)$$

Depending on the relative magnitudes of k and F_x, this may give rise to negative values of ω_n^2, i.e., an apparently "negative stiffness," which will give the real part of one of the roots of Equation (4) positive as in Figure 8b. Thus the solution (Equation (3)) will contain a term which does not diminish with time but continues to increase, i.e., an instability by the definitions above. This case is often illustrated by means of the inverted pendulum [78] or feedback control proportional to position [66].

EXCITING FORCE IS A FUNCTION OF VELOCITY

Similarly:

$$m\ddot{x} + c\dot{x} + kx = F_v \cdot \dot{x} \qquad (8)$$

gives rise to Equation (6) but with the apparent damping and stiffness:

$$2\lambda = (c - F_v)/m \text{ and } \omega_n^2 = (k/m) \qquad (9)$$

In this case, depending on the relative magnitudes of c and F_v, there may be negative values of 2λ, i.e., an apparently "negative damping." Again this will cause instability due to positive real roots (Equations (3–4)) as in Figure 8c.

EXCITING FORCE IS A FUNCTION OF ACCELERATION

Finally, an excitation $F(t) = F_a \cdot \ddot{x}$ gives Equation (6) with

$$2\lambda = c/(m - F_a) \text{ and } \omega_n^2 = k/(m - F_a) \qquad (10)$$

in which there may be apparently "negative" values of stiffness depending on the relative values of m and F_a (Figure 8d).

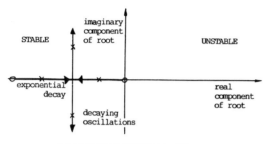

(a) POSITIVE STIFFNESS AND DAMPING

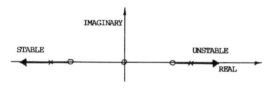

(b) NEGATIVE STIFFNESS, POSITIVE DAMPING

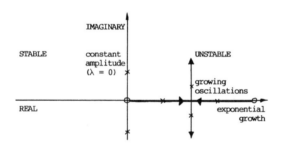

(c) NEGATIVE DAMPING, POSITIVE STIFFNESS

(d) NEGATIVE STIFFNESS AND DAMPING

FIGURE 8. Root-locus plot for the solution (Equation (3)) to Equation (6).

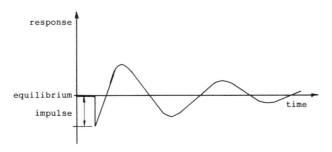

(a)   STATICALLY AND DYNAMICALLY STABLE

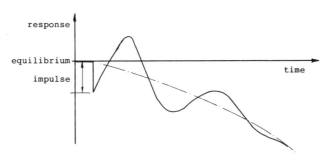

(b)   DYNAMICALLY STABLE BUT STATICALLY UNSTABLE

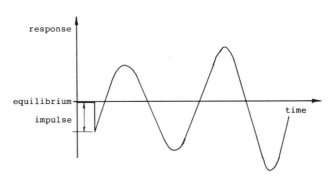

(c)   STATICALLY STABLE BUT DYNAMICALLY UNSTABLE

FIGURE 9. States of static and dynamic instability.

These cases illustrate two distinct types of instability (Figure 9):

—dynamic instability —a small displacement from an equilibrium position leads to increasingly large oscillations about that state due to apparently negative damping (Figure 8c).

—static instability —a small displacement from an equilibrium position sets up a force tending to drive the system away from the equilibrium position (Figure 5b) due to apparently negative stiffness (Figure 8b,d).

In either case, the self-excited vibration needs an energy source to drive it, but that need not in itself be an alternating source.

Determination of System Instability

Following from the definitions of stability above, a general approach to stability analysis is [78]:

1. Assume a deviation from the steady state (equilibrium) motion of the system.

2. Investigate the resulting vibrations with respect to the steady state motion.
3. If these vibrations tend to die out then assume that the system is stable for that particular steady state motion.

This approach gives rise to three general concepts of stability analysis, which are obviously closely interrelated:

(a) If the solution itself is available then it may be examined to see if it is of stable form.
(b) For linear systems, however, it is not even necessary to carry out the solution once the characteristic equation (e.g., Equation (4)) is determined, because the *nature* of the roots can be deduced directly from the coefficients of the characteristic equation.
(c) Finally, there exists a technique which does not require either the solution or even the characteristic equation — the Liapunov "direct" method (which obviously possesses great advantages for non-linear systems).

Each of these will be described briefly in turn.

Method (a) — Examination of Solution for Instabilities

The solution may be available:

— either in an explicit analytical form, thus allowing direct inspection of the roots to determine their stability or otherwise (cf Figure 8),
— or the form of the response may be available (e.g., from an analog computer or numerical integration on a digital computer), thus allowing inspection of the form of the response (cf Figure 9).

Each of these possibilities will be described in turn.

INSPECTION OF ROOTS

The general solution [10,26] to the non-homogeneous linear equation with constant coefficients a_i (of which Equations (1–2) above for Figure 7 are special cases):

$$a_n \frac{d^n x}{dt^n} + a_{n-1} \frac{d^{n-1} x}{dt^{n-1}} + --- + a_1 \frac{dx}{dt} + a_0 x = F(t)$$

(11)

possesses a particular integral depending on $F(t)$ as well as the complementary solution to the homogeneous equation:

$$(a_n D^n + a_{n-1} D^{n-1} + --- + a_1 D + a_0)x = 0 \quad (12)$$

where the differential operator $D \equiv d/dt$ in the usual way. By assuming a solution of the form (of which Equation (3) above for Equation (1–2) is a special case):

$$x = C_1 \exp(s_1 t) + C_2 \exp(s_2 t) + --- + C_n \exp(s_n t)$$

(13)

then back-substitution into Equation (12) shows that the exponents s_i are the roots of the "characteristic equation" (of which Equation (4) above is a special case):

$$a_n s^n + a_{n-1} s^{n-1} + --- + a_1 s + a_o = 0 \quad (14)$$

i.e., the impulse response of a linear autonomous (none of the coefficients a_i depend explicitly on the independent variable time t) system is a sum of exponential time functions whose exponents are the roots of the system characteristic equation. To ensure that the impulse response decays with time (stable), then the real parts of the roots must be negative.

In the case of $n = 2$ (e.g., Equations (1–2) for Figure 7) it is possible to obtain the explicit general solution as in Equations (3–4) above and stability depends entirely [26,65] on the location of the roots in the complex plane (Figure 8):

— right half plane (positive real parts):
  unstable (increasing exponential and/or sinusoid)
— on imaginary axis:
  constant amplitude oxcillations (distinct roots) or increasing exponential and/or sinusoid (repeated roots)
— left half plane:
  stable (decaying exponential and/or sinusoid).

To ensure stability, the roots should be excluded not only from the right half plane but also from the imaginary axis [65]. This approach is attractive for $n = 2$ because the existence of the general solution (Equations (3–4)) allows the effect on the solution of changing parameters to be seen directly (Figure 8). It is also possible (with difficulty) to obtain general solutions with $n = 3$ or 4 but with higher orders ($n > 4$) it is impossible to obtain an explicit general solution to Equation (14) [26] so that each individual case has to be examined separately. This is usually done by the "eigenvalue method." For this approach it is convenient to define first the "state variables" [58,64,66]:

$$x_1 = x, \ x_2 = \frac{dx_1}{dt} = \frac{dx}{dt}, \ --, \ x_n = \frac{dx_{n-1}}{dt} = \frac{d^{n-1} x}{dt^{n-1}} \quad (15)$$

(The state variables of a dynamic system are the smallest set of parameters which must be specified at any time t in order to predict its behaviour uniquely). Then Equation (11) can be rewritten as [58]:

$$\frac{dx_n}{dt} = \frac{-1}{a_n}(a_{n-1}x_n + --- + a_1 x_2 + a_0 x_1) + \frac{1}{a_n} F(t)$$

(16)

with

$$\frac{dx_{n-1}}{dt} = x_n, \ ---, \ \frac{dx_1}{dt} = x_2$$

i.e., in general any linear system can be represented in

matrix notation as [58]:

$$\dot{\underline{x}} = \underline{A}\,\underline{x} + \underline{b}.\,F(t)$$

with

$$\underline{x} = \begin{bmatrix} x_1 \\ x_2 \\ - \\ - \\ x_n \end{bmatrix} \quad \underline{A} = \begin{bmatrix} a_{11} & a_{12} & -- & a_{1n} \\ a_{21} & a_{22} & -- & a_{2n} \\ - & & & \\ - & & & \\ a_{n1} & a_{n2} & -- & a_{nn} \end{bmatrix} \quad \underline{b} = \begin{bmatrix} b_1 \\ b_2 \\ - \\ - \\ b_n \end{bmatrix} \quad (17)$$

where Equation (17) is the "state equation," x is the "state vector," $\underline{A}$ is the "system matrix" and $F(t)$ is the "system input." E.g., for the spring/mass/damper of Figure 7 then Equation (1–2) (with $x_1 \equiv x$, $x_2 \equiv \dot{x}$) can be rewritten as [64–66]:

$$\begin{bmatrix} \dot{x}_1 \\ \dot{x}_2 \end{bmatrix} = \begin{bmatrix} 0 & 1 \\ -\omega_n^2 & -2\lambda \end{bmatrix} \begin{bmatrix} x_1 \\ x_2 \end{bmatrix} + \begin{bmatrix} 0 \\ m^{-1} \end{bmatrix} \cdot F(t) \quad (18)$$

If, as in Equation (13) above, the solution for the free vibrations (i.e., $F(t) = 0$ in Equation (17)) is assumed to be of the form:

$$x_1(t) = C_1 \exp(st), \; ---, \; x_n(t) = C_n \exp(st) \quad (19)$$

then back substitution into Equation (17) yields:

$$\begin{aligned} (a_{11} - s)C_1 + a_{12}C_2 \quad &+ --- + a_{1n}C_n &= 0 \\ a_{21}C_1 + (a_{22} - s)C_2 \quad &+ --- + a_{2n}C_n &= 0 \\ - & & \\ - & & \\ a_{n1}C_1 + a_{n2}C_2 \quad &+ --- + (a_{nn} - s)C_n &= 0 \end{aligned}$$
$$(20)$$

and this Equation (20) has a nontrivial solution if [12]:

$$\begin{vmatrix} (a_{11} - s) & a_{12} & -- & a_{1n} \\ a_{21} & (a_{22} - s) & -- & a_{2n} \\ - & & & \\ - & & & \\ - & & & \\ a_{n1} & a_{n2} & -- & (a_{nn} - s) \end{vmatrix} = 0 \text{ or } |\underline{A} - s\underline{I}| = 0 \quad (21)$$

Equation (21) is the characteristic equation (it can be confirmed easily that for Equation (18) it gives the simple result of Equation (4) above) and the n roots s_i are the characteristic roots or "eigenvalues" [12]. The advantage of this formulation is that systematic methods of obtaining eigenvalues are described in standard textbooks, e.g., [12,58], and are available in standard program packages [57]. For such an autonomous linear system there is stability if and only if all the eigenvalues of $\underline{A}$ have negative real parts. For nonlinear autonomous systems:

$$\text{linear } \dot{\underline{x}} = \underline{A}\,\underline{x}, \quad \text{nonlinear } \dot{\underline{x}} = \underline{A}\,\underline{x} + \underline{f}(\underline{x})$$

there is stability if [27]:

1. Every solution of $\dot{\underline{x}} = \underline{A}\,\underline{x}$ is stable
2. $\underline{f}(\underline{x})$ is continuous in the region of interest
3. $|\underline{f}(\underline{x})|/|\underline{x}| \rightarrow 0$ as $|\underline{x}| \rightarrow 0$

so that the knowledge of the eigenvalues of $\underline{A}$ is still essential.

INSPECTION OF FORM OF RESPONSE

By whatever means a solution is arrived at, either analytically or numerically or by using an analog or digital [26,73] simulation, its form can be inspected immediately to see whether it is stable (Figures 6, 9a) or unstable (Figure 9b–c). In practice, however, examination of solutions in the time (t) domain may not be straightforward and it is easier to determine not only the presence of instability but also its likely type and thus source by presenting the solutions in the "phase plane."

Consider the autonomous state equations [45,59]:

$$\dot{x}_1 = F(x_1, x_2) \text{ and } \dot{x}_2 = G(x_1, x_2) \quad (22)$$

If these represent a system (cf Equation (18) above) in which $x_1 \equiv x$ (position) and $x_2 \equiv \dot{x}$ (velocity) then $(x_1, x_2) \equiv (x, \dot{x})$ is called the phase plane and:

$$\begin{bmatrix} x_1 \\ x_2 \end{bmatrix} \equiv \begin{bmatrix} x \\ \dot{x} \end{bmatrix} \quad (23)$$

is the phase vector. For a second order system ($n = 2$) the phase plane coordinates are position (x) and velocity ($\dot{x}$). For higher order systems ($n > 2$) then there will be a series of phase planes (with acceleration $\ddot{x}$, etc. as additional coordinates) which become difficult to plot so the method is rarely used for these [65]. However, using x and $\dot{x}$ as coordinates, then for any point on the $(x, \dot{x})$ plane there will be a gradient:

$$\frac{d\dot{x}}{dx} = \frac{G(x, \dot{x})}{F(x, \dot{x})} \quad (24)$$

Graphically these gradients build up into a series of "phase trajectories" which are not explicitly dependent on time (for F,G autonomous) but which have direction indicated by increasing time. These represent a family of solutions $x(t)$, $\dot{x}(t)$ to the differential equations (Equation (22)) traced out in the $(x, \dot{x})$ plane as t varies. Equilibrium points occur at the $\dot{x} = 0$ axis. Points at which both $F(x, \dot{x})$ and $G(x, \dot{x})$ vanish have a special significance (since Equation (24) is undetermined) and are known as "critical points" or "singularities."

The significance of the trajectories in the phase plane is illustrated by considering Equations (8) and (17) for $n = 2$ only:

$$\begin{bmatrix} \dot{x}_1 \\ \dot{x}_2 \end{bmatrix} = \begin{bmatrix} a_{11} & a_{12} \\ a_{21} & a_{22} \end{bmatrix} \begin{bmatrix} x_1 \\ x_2 \end{bmatrix} + \begin{bmatrix} b_1 \\ b_2 \end{bmatrix} F(t) \qquad (25)$$

This has the characteristic equation (Equation (21)):

$$\begin{vmatrix} (a_{11} - s) & a_{12} \\ a_{21} & (a_{22} - s) \end{vmatrix} = 0$$

or   (26)

$$s^2 - (a_{11} + a_{22})\, s + (a_{11}a_{22} - a_{12}a_{21}) = 0$$

which is identical to Equation (4) for the spring/mass/damper system (Figure 7) if:

sum of roots $(s_1 + s_2) = (a_{11} + a_{22}) \equiv -2\lambda$

product of roots $(s_1 s_2) = (a_{11}a_{22} - a_{12}a_{21}) \equiv + \omega_n^2$ (27)

discriminant $\Delta = (a_{11} - a_{22})^2 + 4a_{12}a_{21} \equiv 4(\lambda^2 - \omega_n^2)$

since the characteristic equation can also be written (for roots s_1, s_2):

$$(s - s_1)(s - s_2) = s^2 - (s_1 + s_2)s + (s_1 s_2) = 0 \quad (28)$$

The phase trajectories for each of the stable and unstable states of this system (Figure 8) can now be examined [8,9,59,66]:

– Stable System (Figure 8a) with $\lambda > 0$, $\omega_n^2 > 0$

(a) There are three possible cases depending on the value of the discriminant Δ. Starting with the case of $\Delta > 0$ then since:

$\Delta > 0 \rightarrow$ real roots
$\omega_n^2 = s_1 s_2 > 0 \rightarrow$ roots of same sign
$2\lambda = -(s_1 + s_2) > 0 \rightarrow$ roots negative

it can be seen that the solution (Equation (3)) is indeed stable and the phase trajectory (Figure 10a) is a "stable node."

(b) Similarly:

$\Delta = 0 \rightarrow$ equal real negative roots

and the phase trajectory (Figure 10b) is a "stable (proper) node."

(c) With:

$\Delta < 0 \rightarrow$ complex roots, negative real parts

giving a "stable focus" (Figure 10c) and a special case of this is given when:

(d) the system is undamped ($2\lambda = 0$) which gives rise to a constant amplitude oscillation (both roots imaginary) whose phase trajectory is a "centre" or "vortex" (Figure 10d).

– Dynamically unstable system (Figure 8c) with $\lambda < 0$, $\omega_n^2 > 0$:

(e) Again there are three possible cases depending on the value of the discriminant Δ. For $\Delta > 0$ then since:

$2\lambda = -(s_1 + s_2) < 0 \rightarrow$ real positive roots

the solution is now unstable and the phase trajectory (Figure 10e) is an "unstable node."

(f) Similarly:

$\Delta = 0 \rightarrow$ equal real positive roots

giving "(proper) node" as before but now unstable (Figure 10f).

(g) Finally:

$\Delta < 0 \rightarrow$ complex roots, positive real parts

giving an increasing oscillation with an "unstable focus" (Figure 10g).

– Statically unstable system (Figure 8b,d) with $\omega_n^2 < 0$:

(h) In this case (for $\lambda \gtrless 0$):

$\omega_n^2 < 0 \rightarrow \Delta > 0 \rightarrow$ real roots

$\omega_n^2 = s_1 s_2 < 0 \rightarrow$ roots have opposite signs.

The solution (Equation (3)) will have one root which increases in amplitude with time and this is represented by a "saddle" (Figure 10h).

In summary, phase trajectories of linear systems exhibit (Figure 11) one of the following types of behaviour (cf the original definitions of stability above):

1. Asymptotically stable:

   The trajectories all approach a critical point as $t \rightarrow \infty$ as in cases (a)–(c) above (Figure 12a).

2. Unstable:

   At least one of the trajectories $\rightarrow \infty$ as $t \rightarrow \infty$, i.e., cases (e)–(h) above (Figure 12b).

3. Cyclic oscillation:

   The closed trajectory neither approaches the critical point nor $\rightarrow \infty$ as $t \rightarrow \infty$, i.e., case (d) above. In addition there are "limit cycles" (Figure 12c) which are peculiar to nonlinear systems [8].

Method (b) – Examination of Characteristic Equation for Linearised Systems

For linear systems (Equation (11)) it is not even necessary to actually solve the equation or identify the eigenvalues in

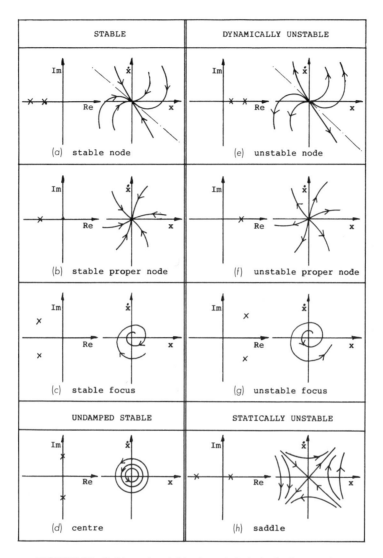

STABLE	DYNAMICALLY UNSTABLE
(a) stable node	(e) unstable node
(b) stable proper node	(f) unstable proper node
(c) stable focus	(g) unstable focus
UNDAMPED STABLE	STATICALLY UNSTABLE
(d) centre	(h) saddle

FIGURE 10. Stable and unstable phase trajectories for linear systems.

order to determine the nature of the roots. This can be done simply by examination of the coefficiencts (a_i) of the characteristic equation (Equation (14)) using criteria devised independently by Routh and Hurwitz [69].

E.g., the simplest case of the general characteristic equation (Equation (14)) is the second order:

$$a_2 s^2 + a_1 s + a_0 = 0 \qquad (29)$$

If s_1 and s_2 are roots of this then it can be written as Equation (28) above from which the results of Equation (27) can be deduced as before. Then, for roots of same sign:

$$s_1 s_2 > 0 \text{ i.e. } (a_0/a_2) = \omega_n^2 > 0 \qquad (30)$$

for these roots negative: $(s_1 + s_2) < 0$ i.e. $(a_1/a_2) = 2\lambda > 0$

from which it can be seen that the requirement is that all three coefficients (a_0, a_1, a_2) or, alternatively, both of 2λ and ω_n^2, must be positive, confirming the previous discussion.

In exactly the same way [24] conditions can be obtained for:

$$\text{cubic } (n = 3): \text{ all } a_i > 0 \text{ and } a_1 a_2 > a_0 \qquad (31)$$

$$\text{quartic } (n = 4): \text{ all } a_i > 0 \text{ and } a_1 a_2 a_3 > a_1^2 + a_3^2 a_0 \qquad (32)$$

These are known as the Routh-Hurwitz criteria. The general methods of Routh and Hurwitz are quite distinct but are

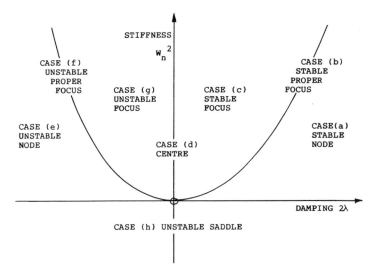

FIGURE 11. Influence of stiffness and damping on phase trajectories of linear system.

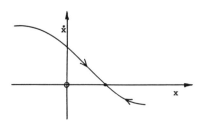

(a) STABLE (ARROWS TOWARDS CRITICAL POINT FROM BOTH SIDES)

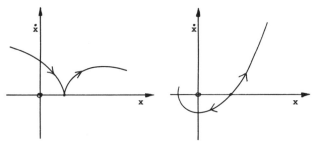

(b) UNSTABLE (ARROW AWAY FROM CRITICAL POINT EITHER/BOTH SIDES)

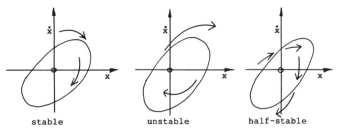

(c) STABLE AND UNSTABLE LIMIT CYCLES FOR NONLINEAR SYSTEMS

FIGURE 12. Stable and unstable phase trajectories for nonlinear systems.

easily shown to be exactly equivalent [58]. The algorithms are available in many standard texts, e.g., [25,58,65,66]. The Hurwitz form is very convenient for expressing the results of Equations (30–32) but for higher orders ($n > 4$) it is rather cumbersome and the Routh algorithm is preferable.

The general Routh algorithm is readily rearranged into a form suitable for use on, e.g., a programmable pocket calculator [36]:

$$\text{for } i = 1, n \text{ calculate } q_1 = b_0^{(i)}/b_0^{(i-1)} \tag{33}$$

$$\text{using } b_k^{(i+1)} = q_i b_{k+1}^{(i-1)} - b_{k+1}^{(i)}$$

$$\text{for } k = 0, (n/2) \text{ if } n \text{ even or } (n - 1)/2 \text{ if } n \text{ odd}$$

where $b_k^{(0)} = a_{2k}$, $b_k^{(1)} = a_{2k+1}$.

However, attractive as they seem, for higher orders ($n > 4$) considerable difficulties are often experienced with these and other methods based on the coefficients of the characteristic equation, e.g., continued fraction method [25], simply in actually obtaining the characteristic equation in the appropriate form (Equation (14)). Equation (21), which is suitable for the eigenvalue approach, is generally much simpler to generate.

The fundamental limitation of the Routh-Hurwitz and eigenvalue methods is that they apply only to linear systems. This means that real nonlinear systems have to be approximated by linear models. The physical justification for this is that the linear analysis, which is valid in the nonlinear case for *small* oscillations, indicates the *start* of instability and is thus often adequate for design purposes. The disadvantages are that:

— "small" is undefined and any inequalities obtained need a safety margin to ensure that they are acceptable, and
— nonlinear systems have forms of behaviour which are physically impossible for linear systems [8,26], in particular the existence of limit cycles (linear theory suggests that amplitudes can build up to ∞ which is clearly impossible in the real world).

There are two principal linearisation procedures [45] of which the second is more applicable in this field:

— Harmonic linearisation (or pseudolinear system):

It is assumed that any periodic output which can be represented by a Fourier series expansion:

$$x(t) = \sum_{k=1}^{n} f_k(t) \cong f_1(t) \tag{34}$$

is replaced by its first Fourier harmonic $f_1(t)$, i.e., it responds to a sinusoid with a sinusoid [4].

— Local linearisation:

Any function $y(x)$ is expanded locally as a power series in a small region $(x - x_0)$ about a steady state $y_0(x_0)$ and all terms of second and higher orders in $(x - x_0)$ are assumed to be negligible and are eliminated [26,65]:

$$y(x) \cong y_0(x_0) + \frac{dy}{dx}\bigg|_{x = x_0} \cdot (x - x_0) + \frac{d^2y}{dx^2}\bigg|_{x = x_0}$$

$$\times \frac{(x - x_0)^2}{2!} + \text{---} \tag{35}$$

For a function of several variables $y(x_1, x_2, \text{---}, x_n)$ with an equilibrium state $y_0(x_{01}, x_{02}, \text{---}, x_{0n})$ this becomes:

$$(y - y_0) \cong \frac{dy}{dx_1}\bigg|_{y = y_0} \cdot (x_1 - x_{01})$$

$$+ \text{---} + \frac{dy}{dx_n}\bigg|_{y = y_0} \cdot (x_n - x_{0n}) + \text{---} \tag{36}$$

Whether linearisation is acceptable in the region of a singularity depends on whether it is "simple" [11,19]. Considering Equations (22–24) above, substituting $x_1 = (x_{o1} + \Delta x_1)$ and $x_2 = (x_{02} + \Delta x_2)$ where (x_{01}, x_{02}) is a singularity and Δx_1 and Δx_2 are very small gives generally:

$$\frac{dx_2}{dx_1} \cong \frac{G(x_{01}, x_{02}) + (g_{11}\Delta x_1 + g_{21}\Delta x_2) + (g_{12}\Delta x_1^2 + g_{22}\Delta x_2^2) + \text{etc.}}{F(x_{01}, x_{02}) + (f_{11}\Delta x_1 + f_{21}\Delta x_2) + (f_{12}\Delta x_1^2 + f_{22}\Delta x_2^2) + \text{etc.}} \tag{37}$$

If the linear terms are present in both denominator and numerator then the very small higher order terms can be neglected giving in the region of the singularity (where F and G vanish):

$$\frac{dx_2}{dx_1} \cong \frac{g_{11}\Delta x_1 + g_{21}\Delta x_2}{f_{11}\Delta x_1 + f_{21}\Delta x_2} \tag{38}$$

On the other hand, if the linear terms are missing from Equation (37), then the singularity is "nonsimple" and the higher order terms cannot be neglected.

Method (c) – Liapunov Direct Method for Nonlinear Systems

The "Liapunov direct (or second) method," apparently has two significant advantages over the other techniques described above [8,9,27,58,64,65,66]:

— It does not require the form of the solution or even just the characteristic equation.
— It applies equally to nonlinear and linear systems.

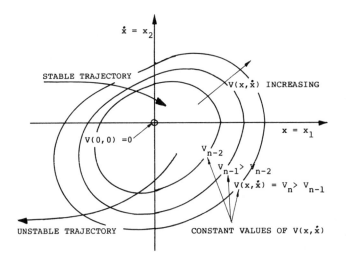

FIGURE 13. Liapunov function in the phase-plane.

The technique can be illustrated conveniently by the simple linear mass/spring/damper system (Figure 7), though it is by no means restricted to linear systems.

Consider the phase plane for such a system (Figure 13) and assume that some function $V(x_1,x_2) \equiv V(x,\dot{x})$ has been identified such that it is "positive definite," i.e., its equal value contours (Figure 13):

— do not intersect with each other
— increase in value the further they are from the origin
— are zero-valued at the origin

Then, for some stable phase trajectory which is moving towards the origin, it can be seen (Figure 13) that it cuts across successively decreasing contours of $V(x,\dot{x})$, cutting each contour only once, i.e., [9,65]:

$$\dot{V}(\underline{x}) = \frac{\partial V}{\partial x_1} \cdot \dot{x}_1 + \frac{\partial V}{\partial x_2} \cdot \dot{x}_2 + --- + \frac{\partial V}{\partial x_n} \cdot \dot{x}_n < 0$$

(39)

On the other hand, the unstable trajectory gives $\dot{V} > 0$ (Figure 13) and, by extension, $\dot{V} = 0$ implies a limit cycle (Figure 12c).

The method, then, requires only that a suitable positive definitive function $V(\underline{x})$ be identified. For a linear system (such as Equation (17) or (25)) this is straightforward because any quadratic function:

$$V(\underline{x}) = \sum_{i=1}^{n} \sum_{j=1}^{n} c_{ij}x_i x_j \text{ with } c_{ij} = c_{ji}$$ (40)

e.g.,

$$V(x_1,x_2) = c_{11}x_1^2 + (c_{12} + c_{21})x_1 x_2 + c_{22}x_2^2$$ (41)

can be shown to be positive definite by using Sylvester's Theorem [9,27,58,64]:

$$c_{11} > 0, \begin{vmatrix} c_{11} & c_{12} \\ c_{21} & c_{22} \end{vmatrix} > 0, ---, |\underline{A}| > 0$$ (42)

and a method for obtaining appropriate coefficients (c_{ij}) is given in many standard texts, e.g., [9,58,64,66].

E.g., for the spring/mass/damper system of Figure 7 and Equation (18) then:

$$V(x_1,x_2) = \left(\lambda^2 + \frac{1}{2} \omega_n^2 \right) x_1^2 + \lambda x_1 x_2 + \frac{1}{2} x_2^2$$ (43)

giving:

$$\dot{V}(x_1,x_2) = \left[2 \left(\lambda^2 + \frac{1}{2} \omega_n^2 \right) x_1 + \lambda x_2 \right] [x_2]$$

$$+ [\lambda x_1 + x_2][-\omega_n^2 x_1 - 2\lambda x_2]$$ (44)

$$= -2\lambda \left(\frac{1}{2} \omega_n^2 x_1^2 + \frac{1}{2} x_2^2 \right)$$

For positive stiffness with $\omega_n^2 > (2\lambda)^2$ (by Equation (42) to ensure that Equation (43) is positive definite) then Equation (44) gives:

stable if $\dot{V}(x) < 0$, i.e., $2\lambda > 0$ (positive damping)

limit cycle if $\dot{V}(x) = 0$, i.e., $2\lambda = 0$ (undamped) (45)

unstable if $\dot{V}(x) > 0$, i.e., $2\lambda < 0$ (negative damping)

However, the function chosen is not appropriate for negative stiffness ($\omega_n^2 < 0$). This indicates that Liapunov functions are not unique and the condition obtained is necessary but not sufficient so that failure to discover a function that gives clear guidance yields *no* information about stability or instability [27,58].

For nonlinear systems there is no general method available at present for constructing Liapunov functions, though a number of helpful techniques are available in the literature [8,9,27,58,65]. As with the linear system above, though there is considerable flexibility in choosing a Liapunov function $V(x)$, failure to find a suitable function does not imply instability but merely means that stability cannot be guaranteed [58] and that any domain of stability identified is dependent on the function used [38].

Summary—Methods for Examining Stability of Self-Excited Oscillations

In summary, a variety of methods for examining the stability of dynamic systems of various orders n are available and include:

—Methods for linearised equations (small oscillations using Equations (35–36)):

(a) Root-locus analysis for $n = 2$ (Figure 8) and also theoretically (though not always practically) $n = 3$ or 4 but not greater.

Eigenvalue determination (Equation (16–21)) for any n but in practice most useful for large $n \geq 4$.

Phase-plane topology (Equation (38)) by graphical or analytical means (e.g. Figure 10,11), for $n = 2$ in practice (though theoretically possible for any n).

(b) Routh-Hurwitz criteria (Equation (30–33)) for any n, though in practice the characteristic equation (Equation (14)) becomes tedious to obtain for large $n > 5$ or 6.

(c) Liapunov direct (or second) method using quadratic functions (Equations (40–42)) for any n but tedious for $n > 2$.

—Methods for nonlinear equations:

(a) Simulation (analog, digital, physical model) in time domain, suitable for any n though in practice it may be difficult to determine by inspection alone whether a solution exhibits instability, a limit cycle or just low damping.

Simulation in the phase-plane overcomes these difficulties of inspection but for a practical number of phase portraits $n \leq 4$.

Singularity analysis in the phase-plane allows the limits of validity of the linearised results above to be estimated for $n = 2$ in practice.

(c) Liapunov direct (or second) method, theoretically possible for any n if a positive definite Liapunov function can be found though in practice this is likely to be very difficult for $n > 2$.

Existing applications of these techniques to surge shaft stability problems will be reviewed, but, before this, it will be useful to establish first the physical basis of typical mechanisms for self-excitation in pumped-storage schemes (using Routh-Hurwitz criteria for illustration).

SOME FUNDAMENTAL MECHANISMS FOR SURGE SHAFT INSTABILITY

The System, Its Equilibrium State and the Source of Excitation

The surge tank stability problem originally arose due to governing of hydro turbines to give constant output power. This is, however, only one of the mechanisms by which instability may occur—pumped storage plant introduces others which unfortunately are not widely discussed in the literature.

It is invaluable, therefore, to develop an appreciation not only of the kind of excitation which may lead to instability but also of the different kinds of unstable behaviour that may be generated (e.g., dynamic or static instability as in Equations (5–10) and Figure 9 above). Three different instability mechanisms will be analysed and compared:

—the traditional and widely discussed governed constant power machine
—"pump surging" due to the shape of the pump performance characteristic
—interaction between shafts in multiple surge tank schemes.

To relate these to the standard mechanisms described above (Equations (5–10)) for the simple mass/spring/damper system of Figure 7, consider the standard single surge shaft hydraulic system of Figure 14. This can be represented [6] by the following equations for "rigid column" flow:

$$\frac{dZ}{dt} = \frac{Q_s}{A_s} \qquad (46)$$

$$\frac{L}{gA}\frac{dQ}{dt} = -Z - \frac{Q_s^2}{2gA_s^2}$$
$$(47)$$
$$- \frac{(L_s + Z)}{gA_s}\frac{dQ_s}{dt} - J_{cs} - \frac{K}{2gA^2}Q|Q|$$

SURGE SHAFT

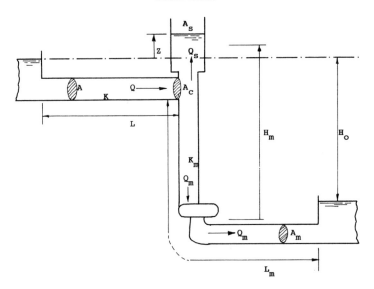

ARROWS INDICATE +VE DIRECTIONS OF Z, Q, Q_m, Q_s, H_m

FIGURE 14. Typical single surge shaft with turbine or pump-turbine.

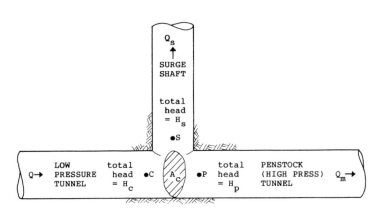

JUNCTION LOSSES:

$$J_{cs} = (\text{total head at c}) - (\text{total head at s}) = H_c - H_s$$

$$J_{cp} = H_c - H_p$$

$$J_{sp} = H_s - H_p = J_{cp} - J_{cs}$$

JUNCTION LOSS COEFFICIENTS:

$$J_{cs} = t_{cs} \cdot \{Q^2/(2gA_c^2)\} \quad \text{where } t_{cs} = t_{cs}(Q/Q_m)$$

etc

FIGURE 15. Definition of junction losses at surge shaft for Figure 14.

$$Q_s = Q - Q_m(H_m, t) \qquad (48)$$

$$H_m = H_0 + \left[Z + \frac{Q_s^2}{2gA_s^2} + J_{cs} \right] \qquad (49)$$

$$- \left[J_{cp} + \frac{K_m}{2gA_m^2} Q_m|Q_m| + \frac{L_m}{gA_m} \frac{dQ_m}{dt} \right]$$

in which time (t) is the independent variable, the principal (i.e., state) dependent variables are:

surge shaft level (Z), low pressure tunnel flow (Q)

with auxiliary variables for the surge tank flow (Q_s) and the head across the machines (H_m) and the flow $Q_m(H_m,t)$ is the excitation imposed by the hydraulic machine (Figure 14). The junction losses (J_{cs}, J_{cp}) are defined [3] in the usual way (Figure 15).

Stability, however, is a property not only of the system but also of its equilibrium states. An equilibrium state for any steady flow $\pm Q_0$ ($\pm$ for turbining/pumping in this case) through the system, i.e.:

$$Q = Q_m = Q_0 \qquad (50)$$

$$\frac{dZ}{dt} = Q_s = \frac{dQ}{dt} = \frac{dQ_m}{dt} = 0 \qquad (51)$$

is defined by (Z_0, Q_0) where:

$$Z_o = - \left(J_{cso} + \frac{K}{2gA^2} Q_0|Q_0| \right) \qquad (52)$$

$$H_{mo} = H_0 + [Z_0 + J_{cso}] - \left[J_{cpo} + \frac{K_m}{2gA_m^2} Q_0|Q_0| \right] \qquad (53)$$

$$= H_0 - \left[J_{cpo} + \left(\frac{K}{A^2} + \frac{K_m}{A_m^2} \right) \frac{Q_0|Q_0|}{2g} \right]$$

With the general Equations (1–10) discussed previously there were seen to be two types of instability (static and dynamic) depending on the nature of the excitation. In order to compare the surge tank Equations (46–49) with these [43] it is necessary to linearise them about the equilibrium position (Z_0, Q_0) defined by Equations (50–53). This local linearisation (cf Equations (35–36)) is done by assuming that a *small* excitation ΔQ_m about the steady flow $\pm Q_0$ ($\pm$ for turbining/pumping) flow:

$$Q_m(H_m, t) = (\pm Q_0) + \Delta Q_m(H_m, t) \qquad (54)$$

gives rise to *small* deviations from the steady state (Z_0, Q_0):

$$Z = Z_0 + \Delta Z \text{ and } Q = (\pm Q_0) + \Delta Q \qquad (55)$$

so that second and higher order terms in these deviations can be neglected, e.g.:

$$Q|Q| \cong \frac{|Q_0|}{Q_0} (Q_0 + \Delta Q)^2 \cong Q_0|Q_0| + 2|Q_0|\Delta Q \qquad (56)$$

In addition, since junction loss (J_{cs}, J_{cp}) coefficients vary as the flow ratio in the branches [3,34,80], it has also to be assumed that for small deviations as above, then the changes in the flow ratios at the junctions from the steady state (J_{cso}, J_{cpo}) are sufficiently small to allow changes in the junction loss coefficients (t_{cs}, t_{cp}) to be neglected, giving:

$$J_{cs} \cong t_{cs} \frac{Q|Q|}{2gA_c^2} \cong \frac{t_{cs}}{A_c^2} \frac{|Q_0|}{g} \Delta Q \qquad (57)$$

$$J_{cp} \cong t_{cp} \frac{Q|Q|}{2gA^2}, \qquad J_{sp} = J_{cp} - J_{cs}$$

With some manipulation, then, Equations (46–48) (for Figure 14) can be written in the form of Equations (1–2) (for Figure 7) with $\Delta Z \equiv x$ and

$$2\lambda = c/m \equiv |Q_0| \frac{A}{L} \left[\frac{K}{A^2} + \frac{t_{cs}}{A_c^2} \right] \bigg/ \left[1 + \frac{(L_s + Z_0)}{L} \frac{A}{A_s} \right] \qquad (58)$$

$$\omega_n^2 = k/m \equiv \left[\frac{gA}{LA_s} \right] \bigg/ \left[1 + \frac{(L_s + Z_0)}{L} \frac{A}{A_s} \right]$$

and an excitation given by:

$$F(t)/m \equiv - \left[\frac{2\lambda}{A_s} \right] \Delta Q_m - \left[\frac{\omega_n^2 L}{gA} \right] \frac{d(\Delta Q_m)}{dt} \qquad (59)$$

Obviously, since ΔQ_m is a function of H_m (Equation (54)) it must also be a function of (ΔZ), $d(\Delta Z)/dt$ and $d^2(\Delta Z)/dt^2$, thus giving rise to all three of the possibilities described by Equations (5–10) (and Figure 8) and thus to the danger of both static and dynamic instabilities. This will be illustrated for two particular cases:

— turbine flow governed to give constant power
— pump flow for constant speed operation

The Classic Instability—Turbine Governed for Constant Power

For constant power governing it is assumed [42] that the hydraulic machine changes the specific energy (i.e., energy/unit mass $\equiv$ power/ϱg) of the water flowing through it by a fixed amount E ($\pm Q_0$ for turbining/pumping):

$$Q_m H_m = (\pm Q_0) H_{mo} = E = \text{constant} \qquad (60)$$

Note that though the demand E itself remains invariant it provides the external energy source to drive the excitation created by the governor which is implicit in Equation (60).

With this exciting mechanism the classic Thoma-style surge shaft stability analysis [42,77] proceeds by the following stages.

LINEARISATION OF EXCITATION

Using Equation (49) then Equation (60) can be written as:

$$Q_m \left[H_0 + Z + \frac{Q_s^2}{2gA_s^2} + (J_{cs} - J_{cp}) \right. $$
$$\left. - \frac{K_m}{2gA_m^2} Q_m|Q_m| - \frac{L_m}{gA_m} \frac{dQ_m}{dt} \right] = (\pm Q_0)H_{mo} = E \tag{61}$$

which can be linearised [using Equations (54–57)] to ($\pm$ for turbining/pumping):

$$\Delta Q_m \cong - \left[\frac{\pm Q_0}{H_1} \right] \Delta Z + \left[\frac{2J_{spo}}{H_1} \right] \Delta Q$$
$$+ \left[\frac{\pm Q_0 L_m}{H_1 g A_m} \right] \frac{d(\Delta Q_m)}{dt} \tag{62}$$

in which (for convenience):

$$H_1 = H_{mo} - 2 \left[\frac{K_m}{A_m^2} \frac{Q_0|Q_0|}{2g} \right]$$
$$\text{and } J_{spo} = - \frac{(t_{cs} - t_{cp})}{A_c^2} \frac{Q_0|Q_0|}{2g} \tag{63}$$

In the past [42] it was usual to neglect the second and third terms on RHS of Equation (62) (representing the junction loss and penstock inertia) which would give:

$$\Delta Q_m \cong - \left[\frac{\pm Q_0}{H_1} \right] \Delta Z$$
$$\text{i.e., } \frac{d(\Delta Q_m)}{dt} \cong - \left[\frac{\pm Q_0}{H_1} \right] \frac{d(\Delta Z)}{dt} \tag{64}$$

Using this approximation in Equation (62) finally gives the linearised relationship:

$$\Delta Q_m \cong - \left[\frac{\pm Q_0}{H_1} \right] \Delta Z + \left[\frac{2J_{spo}}{H_1} \right] \Delta Q$$
$$- \left[\frac{Q_0^2 L_m}{H_1^2 g A_m} \right] \frac{d(\Delta Z)}{dt} \tag{65}$$

LINEARISED EQUATION FOR SELF-EXCITED OSCILLATIONS

Substituting Equations (65) into the linearised governing equation (Equations (1–2) with Equations (58–59)) gives a free vibration equation just like Equation (6) for the mass/spring/damper system previously, with:

$$2\lambda = \frac{(c - F_v)}{(m - F_a)} \equiv \frac{1}{A_s H_2} \left\{ |Q_0| \frac{A}{L} \left[\frac{K}{A^2} + \frac{t_{cs}}{A_c^2} \right] \right.$$
$$\times \left[A_s H_1 - \frac{Q_0^2}{gH_1} \frac{L_m}{A_m} \right] - (\pm Q_0) \right\} \tag{66}$$

$$\omega_n^2 = \frac{(k - F_x)}{(m - F_a)} \equiv \frac{gA}{H_2 L A_s} \left\{ H_0 - 3 \right.$$
$$\times \left[\frac{K}{A^2} + \frac{t_{cp}}{A_c^2} + \frac{K_m}{A_m^2} \right] \frac{Q_0|Q_0|}{2g} \right\} \tag{67}$$

in which (for convenience)

$$H_2 = H_1 - \frac{L_m}{L} \frac{A}{A_m} \frac{Q_0^2}{gH_1 A_s} + \frac{(L_s + Z_0)}{L} \frac{A}{A_s}$$
$$\times \left\{ H_0 - 3 \left[\frac{K}{A^2} + \frac{t_{cp}}{A_c^2} + \frac{K_m}{A_m^2} \right] \frac{Q_0|Q_0|}{2g} \right\} \tag{68}$$

Note that it has been usual in conventional stability analyses [42] to neglect:

—penstock inertia ($L_m = 0$) and losses ($K_m = 0$)
—surge shaft inertia ($L_s + Z_0 = 0$) and
—junction losses, representing only J_{cs} by the velocity head in the conduit, i.e., in place of Equation (57):

$$J_{cp} \cong 0 \text{ and } J_{cs} \cong Q_0^2/2gA_c^2 \tag{69}$$

With these simplifying assumptions then Equation (68) gives $H_2 = H_1 = H_{mo}$ and Equations (66–67) can be written simply as ($\pm$ for turbining/pumping):

$$2\lambda = \frac{(c - F_v)}{(m - F_a)} \equiv \frac{1}{A_s H_{mo}} \left\{ |Q_0| \frac{A}{L} \left[\frac{K}{A^2} \pm \frac{1}{A_c^2} \right] \right.$$
$$\times A_s H_{mo} - (\pm Q_o) \right\} \tag{66a}$$

$$\omega_n^2 = \frac{(k - F_x)}{(m - F_a)} \equiv \frac{gA}{H_{mo} L A_s} \left\{ H_{mo} - 2 \right.$$
$$\times \left[\frac{K}{A^2} \pm \frac{1}{A_c^2} \right] \frac{Q_0|Q_0|}{2g} \right\} \tag{67a}$$

Thus, by comparison of Equations (66–67) (or Equations

(66a–67a)) with Equation (58) it can be seen that with this particular form of excitation there is a possibility of apparently negative damping (leading to dynamic instability) or stiffness or mass (leading to static instability).

STABILITY CRITERIA

With the coefficients (Equations 66–67)) of Equation (6) now available it is possible to investigate the system parameters that may lead to different kinds of instability.

Dynamic Stability

For dynamic stability it is necessary to have positive damping, i.e., $2\lambda > 0$ (Equation (30)) or $(c - F_v) > 0$ (Equation (9)) and thus Equation (66) can be rearranged into the famous result:

$$A_s > A_{Th} \left[1 + \frac{2(-Z_0)}{H_1} \frac{L_m}{L} \frac{A}{A_m} \right] \qquad (70)$$

where

$$A_{Th} = \frac{LA\left(\dfrac{Q_0|Q_0|}{2gA^2}\right)}{|H_1| \, |Z_0|} \text{ with } |Z_0| = \left[\frac{K}{A^2} + \frac{t_{cs}}{A_c^2} \right] \frac{Q_0^2}{2g} \qquad (71)$$

or, using the simpler Equation (66a) [42,77], simply:

$$A_s > A_{Th}^1 = \frac{LA\left(\dfrac{Q_0|Q_0|}{2gA^2}\right)}{|H_{mo}| \, |Z_0^1|} \text{ with } |Z_0^1| = \left[\frac{K}{A^2} \pm \frac{1}{A_c^2} \right] \frac{Q_0^2}{2g} \qquad (71a)$$

Equation (71a) for turbining flow (Q_0 positive) gives the classic "Thoma surge shaft area" [42,77] and Equations (70–71) represent the additional effects on the dynamic stability criterion of penstock inertia and losses, surge shaft inertia and junction losses. For pumping flow (Q_0 negative) if penstock inertia is neglected then Equation (66a) suggests that there is unconditional dynamic stability [42]. However, the more general Equation (66) (including penstock inertia) shows that this is not necessarily the case (though in all practical cases it will be).

Static Stability

For static stability it is necessary to have positive stiffness, i.e., $\omega_n^2 > 0$ (Equation (30)) or $(k - F_x) > 0$ (Equation (7)) since the requirement for positive mass, i.e., $(m - F_a) > 0$ (Equation (10)) is that $H_2 > 0$ (Equation (68)) which is automatically satisfied if Equation (70) and Equation (72) (below) are satisfied. Thus static stability is guaranteed for pumping flows (Q_0 negative) and for turbin-

ing (Q_0 positive) Equation (67) requires that:

$$H_0 > 3 \left[\frac{K}{A^2} + \frac{t_{cp}}{A_c^2} + \frac{K_m}{A_m^2} \right] \frac{Q_0|Q_0|}{2g} = 3 \times \text{ net losses} \qquad (72)$$

or $(3H_{mo} - 2\,H_0) > 0$ i.e., $H_{mo} > \dfrac{2}{3}\,H_0$

or, from the simpler Equation (67a):

$$H_{mo} > 2|Z_0^1| \text{ or } H_0 > 3|Z_0^1| \qquad (72a)$$

Equation (72a) gives Thoma's second stability criterion [42,77] which is, however, less obviously physically meaningful than Equation (72).

The difference between pumping (which is effectively unconditionally stable) and turbining [which is only conditionally stable by Equations (70–72)] emphasises the point that stability is a property of a particular equilibrium state as well as of a system.

The stability criteria (Equations 70–72) are based on a linearised analysis which is valid only for small oscillations. In the case of the dynamic stability criterion, though this will not be valid for large oscillations, it will indicate incipient instability conditions and thus remains a useful guide in practice, even if the true behaviour of the actual non-linear system may be unstable or exhibit a stable limit cycle (which may or may not be "acceptable" depending on its magnitude).

For static stability, the physical significance of Equation (72) (i.e., of the so-called "negative stiffness") is readily visualized. From Equation (61) it can be seen that Equation (72) coincides with the condition for $dE/dQ_m = 0$, i.e. maximum possible specific energy (E) transmission through the tunnel system of Figure 14. If Equation (72) is not satisfied for turbining then $dE/dQ_m \leq 0$ (Figure 16) and thus if Q_0 is increased, the output E would have to decrease which would lead the governor to further increase Q_m in an attempt to restore E to its constant value (Equation (60)), leading ultimately to drainage of the surge shaft.

This interpretation, however, implies large deviations (ΔZ) in order to result in drainage though the analysis is valid only for small deviations. Thus conditions for static instabilities derived from linearised (i.e. small amplitude) studies have to be used with great caution, unlike the corresponding criteria for dynamic instabilities, i.e. an understanding of the physical basis of the instability is invaluable when subjectively interpreting the significance of any criterion obtained.

Instability Caused by Pump Surging

Pumps are usually operated with fixed geometry (guide vane opening) and speed with varying input power, rather

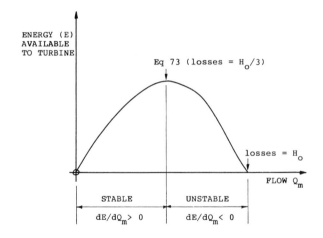

FIGURE 16. Energy available to turbine.

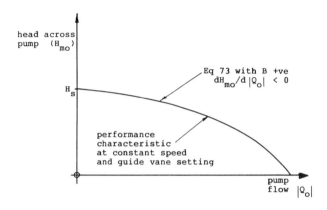

(a) STABLE PUMP CHARACTERISTIC WITH −VE SLOPE THROUGHOUT

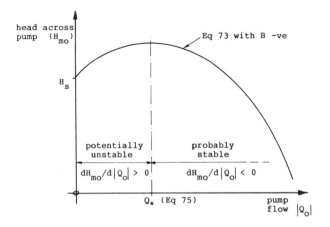

(b) PUMP CHARACTERISTIC WITH POTENTIAL FOR SURGING (+VE SLOPE)

FIGURE 17. Typical pump performance characteristics at constant speed and guide-vane setting.

than being governed at constant power as above. Under these conditions the well known phenomenon of "pump surging," i.e. unstable operation, may occur.

The likelihood of pump surging depends [1,68,72] on the shape of the pump performance characteristic (H_{mo} v $|Q_o|$). Where this exhibits a negative slope pump operation is usually stable but where it exhibits a positive slope the pump may operate unstably (Figure 17). Thus the actuator for the self-excitation is the pump steady-state performance characteristic rather than a governor (as implicit in Equation (60)).

While generally not a satisfactory representation of the pump performance over its full range, in a limited operating region an adequate description is (Appendix):

$$H_{mo} \cong H_s - B|Q_0| - CQ_0^2 \qquad (73)$$

in which H_s = pump shut valve head and B and C are approximately constants (at least over some portion of the characteristic). This may be used to consider small disturbances about a steady state.

The slope of this performance characteristic (Equation (73)) is:

$$\left(\frac{dH_{mo}}{d\,|Q_0|}\right)_{pump} = -(B + 2C|Q_o|) \qquad (74)$$

which will always be $-ve$ if the constant B is $+ve$, i.e., the characteristic is stable. If B is $-ve$, however, the characteristic will have a region of $+ve$ slope at low flows (Figure 17):

$$\left(\frac{dH_{mo}}{d\,|Q_0|}\right)_{pump} > 0 \text{ for } |Q_0| < Q* = -\frac{B}{2C} \qquad (75)$$

so that there is the potential for pump surging in this region.

Assuming that a linearised (small amplitude) stability analysis is satisfactory for pump surging [68], then the procedure is exactly as before except that Equation (60) is replaced by (with $Q_m - ve$ for pumping on Figure 14):

$$H_s - B|Q_m| - CQ_m^2 = H_m \text{ (Equation (49))} \qquad (76)$$

which gives results in a similar form to Equations (66–68) but with (in place of Equation (63) above):

$$H_1 = B|Q_0| + 2CQ_0^2 - 2\left[\frac{K_m}{A_m^2} \frac{Q_0|Q_0|}{2g}\right] \qquad (77)$$

Thus there are again both static and dynamic stability criteria.

Static Stability

From Equation (67) there is positive stiffness ($\omega_n^2 > 0$) if the gradient of the pump performance characteristic (Figure

17) satisfies the conditions (Figure 18):

$$|Q_0| \cdot \left(\frac{dH_{mo}}{d\,|Q_0|}\right)_{pump} < 2\left[\frac{K}{A^2} + \frac{t_{cp}}{A_c^2} + \frac{K_m}{A_m^2}\right]$$

$$\times \frac{Q_0^2}{2g} = 2 \times \text{net losses} \qquad (78)$$

i.e., $\left(\frac{dH_{mo}}{d\,|Q_0|}\right)_{pump} < \left(\frac{dH_{mo}}{d\,|Q_0|}\right)_{pipe\ system}$

and this is the case for flows (with coefficient $B - ve$):

$$|Q_0| > Q_*'' = \frac{-B}{\left\{2C + 2\left[\frac{K}{A^2} + \frac{t_{cp}}{A_c^2} + \frac{K_m}{A_m^2}\right]\frac{1}{2g}\right\}}$$

$$(79)$$

For a flow $|Q_0| < Q_*''$ then the system is statically unstable for any surge shaft size.

Dynamic Stability

The positive damping ($2\lambda > 0$) condition of Equation (66) with Equation (77) demonstrates one major problem with algebraic stability analysis—including a number of effects (penstock inertia, etc.) makes the resulting expressions rather indigestible. Simplification by ignoring the penstock (L_m) and surge shaft ($L_s + Z_0$) inertia terms allows a similar expression to that for static stability (Equations (78–79)) to be obtained (Figure 17) for unconditional dynamic stability:

$$|Q_0| \cdot \left(\frac{dH_{mo}}{d\,|Q_0|}\right)_{pump} < 2\left[\frac{K_m Q_0^2}{A_m^2 2g}\right] = 2 \times \begin{array}{l}\text{penstock}\\\text{losses}\end{array}$$

$$(80)$$

which applies for flows:

$$Q_0 > Q_*' = \frac{-B}{\left\{2C + 2\left[\frac{K_m}{A_m^2}\right]\frac{1}{2g}\right\}} \qquad (81)$$

For flows Q_*'' (Equation (79)) $< |Q_0| < Q_*'$ (Equation (81)) then there is conditional dynamic stability only if the surge shaft area A_s is made large enough to satisfy Equations (70–71) but using Equation (77) for H_1 (in place of Equation (63)).

In practice the values of $Q*$ (Equation (75)), Q_*' (Equation (81)) and Q_*'' (Equation (79)) are very close together, confirming the usual experience that between (Figure 18) the region of major surge excursions and the region of stable performance there is a small region where there may be dynamic instability [68]. In practice a real nonlinear system usually exhibits limit cycle oscillations (Figure 12c) in this small region [68] but it is unlikely that the differences between Q_*, Q'_* and Q''_* will be of any practical significance. Realistically the peak of the characteristic (Q_* by Equation (75)) would seem to be a sufficiently sensitive indicator of potentially unstable behaviour and the dynamic instability is unlikely to be of any importance—in contrast to the Thoma shaft area criterion for the previous excitation.

The physical significance of the static stability criterion (Equations (78–79)) is illustrated on Figure 19 which shows a pump characteristic and (a) stable and (b) unstable pipe system characteristics. For system (a) a small decrease in flow will give rise to an increased pump head which will tend to increase the flow back towards its equilibrium value, and vice versa. On the other hand, for system (b) any deviation will tend to drive the flow further away from the equilibrium point.

This is what is generally understood by the phenomenon of pump surging [44] and this static stability criterion (Equations 78–79)) is much more likely to be of practical in-terest than the tank drainage criterion (Equation (72)) for constant power operation (which will always be satisfied for any economically realistic scheme). However, like that condition, the limitations of the analysis to *small* deviations (in this case in the application of Equation (73) as well as in the linearisation process) mean that Equation (79) may not be an accurate delimiter of the unstable region.

**Higher Order Systems and Alternative
Forms of Instability**

The conventional single shaft system of Figure 14 and Equations (46–59) is readily visualised as an analog to the spring/mass/damper of Figure 7 and Equations (1–10) [43] with the result that forms of excitation leading to negative stiffness (static instability) and negative damping (dynamic instability) are readily identifiable. Unfortunately this is by no means the case for higher order, i.e. multiple shaft, systems [2].

This can be illustrated by the "tandem" double surge shaft arrangement (Figure 20) for which the mass oscillation

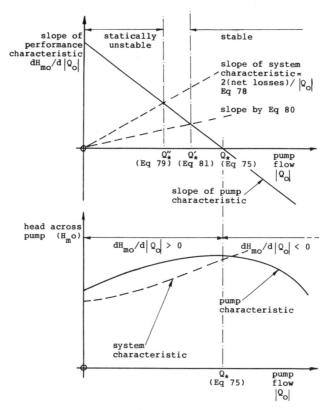

FIGURE 18. Relationship between pump operating characteristics and surge shaft stability—division of characteristic into stable and unstable regions.

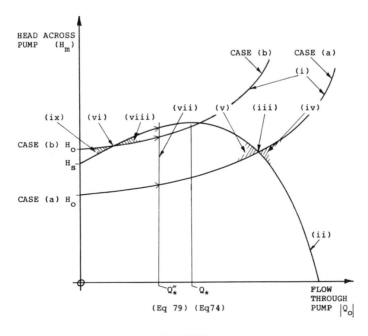

NOTES: (i)   SYSTEM CHARACTERISTICS
   (ii)   PUMP CHARACTERISTIC

   (iii)   STABLE EQUILIBRIUM POINT FOR CASE (a)
   (iv)   SYSTEM LOSSES ACT TO DRIVE FLOW TOWARDS TOWARDS
                       EQUILIBRIUM
   (v)   EXCESS PUMP HEAD ACTS TO DRIVE FLOW TOWARDS
                        EQUILIBRIUM

   (vi)   UNSTABLE EQUILIBRIUM POINT FOR CASE (b)
   (vii)   SYSTEM AND PUMP CHARACTERISTICS PARALLEL
   (viii) EXCESS PUMP HEAD ACTS TO DRIVE FLOW AWAY FROM
                        EQUILIBRIUM
   (ix)   SYSTEM LOSSES ACT TO DRIVE FLOW AWAY FROM
                        EQUILIBRIUM

FIGURE 19. *Pump surging by static instability.*

equations are (positive flows Q_m, Q_1, Q_2 only):

$$\frac{dZ_1}{dt} = \frac{Q_2 - Q_1}{A_{s1}} \quad \text{and} \quad \frac{L_1}{gA_1}\frac{dQ_1}{dt} = +Z_1$$

$$+ \frac{1}{2g}\left(\frac{dZ_1}{dt}\right)^2 + J_{c1} - \frac{K_1}{A_1^2}\frac{Q_1^2}{2g} \tag{82}$$

$$\frac{dZ_2}{dt} = \frac{Q_m - Q_2}{A_{s2}} \quad \text{and} \quad \frac{L_2}{gA_2}\frac{dQ_2}{dt}$$

$$= +\left[Z_2 + \frac{1}{2g}\left(\frac{dZ_2}{dt}\right)^2 + J_{c2}\right] \tag{83}$$

$$- \left[Z_1 + \frac{1}{2g}\left(\frac{dZ_1}{dt}\right)^2 + J_{b1}\right] - \frac{K_2}{A_2^2}\frac{Q_2^2}{2g}$$

where the junction losses (J) are as defined on Figure 20.

The corresponding steady state equations for the equilibrium point ($d/dt = 0$, $Q_m = Q_1 = Q_2 = Q_0$) are:

$$Z_{10} = +\left[\frac{K_1}{A_1^2} - \frac{t_{c1}}{A_{c1}^2}\right]\frac{Q_0^2}{2g} \quad \text{and}$$

$$Z_{20} = +\left[\frac{K_1}{A_1^2} + \frac{t_{a1}}{A_1^2} + \frac{K_2}{A_2^2} - \frac{t_{c2}}{A_{c2}^2}\right]\frac{Q_0^2}{2g} \tag{84}$$

in which the steady state values of the junction losses (J) have been defined as in Equation (57) (coefficients t +ve).

In addition it has been assumed, for illustrative purposes, that these are tailrace tunnel shafts with turbining flow (Figure 20), giving the head across the machine as:

$$H_m = H_0 - \left[Z_2 + \frac{1}{2g}\left(\frac{dZ_2}{dt}\right)^2 + J_{b2}\right] - \frac{K_m}{A_m^2}\frac{Q_0^2}{2g} \tag{85}$$

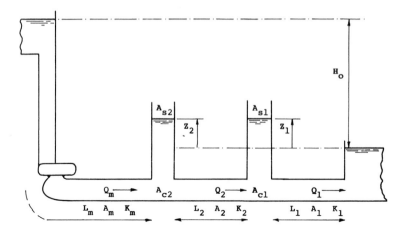

JUNCTION LOSSES:

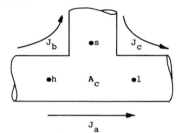

$$J_b = H_h - H_s \qquad J_c = H_1 - H_s \qquad J_a = H_h - H_1$$
$$\quad = t_b(Q_h^2/2gA_c^2) \qquad = t_c(Q_1^2/2gA_c^2) \qquad = t_a(Q_h^2/2gA_c^2)$$
$$\quad = J_a + J_c$$

FIGURE 20. "Tandem" tailrace tunnel surge shaft system.

with the steady state value:

$$H_{mo} = H_0 - \left[\frac{K_1}{A_1^2} + \frac{t_{a1}}{A_{c1}^2} + \frac{K_2}{A_2^2} + \frac{t_{a2}}{A_{c2}^2} + \frac{K_m}{A_m^2} \right] \frac{Q_0^2}{2g}$$
$$(86)$$

so that conventional constant power governing (Equation (60)) can be investigated.

The characteristic equation for higher order systems like this is obtained with simple but rather extensive algebraic manipulation. Linearisation of Equations (82) and (83) about the equilibrium point (Equation (84)) in the usual way (cf Equations (54–59) above) will give rise to two second order equations (one for each surge shaft):

$$D_{11}(\Delta Z_1) + D_{12}(\Delta Z_2) = 0$$
$$D_{21}(\Delta Z_1) + D_{22}(\Delta Z_2) = 0$$
$$(87)$$

where the D are differential operators in ΔZ, $d(\Delta Z)/dt$ and $d^2(\Delta Z)/dt^2$. By eliminating either ΔZ_1 or ΔZ_2 it can be shown [10] that Equation (87) reduces to:

$$\begin{vmatrix} D_{11} & D_{12} \\ D_{21} & D_{22} \end{vmatrix} \cdot \Delta Z_1 = 0 \text{ i.e. } (D_{11}D_{22} - D_{21}D_{12})\Delta Z_1 = 0$$
$$(88)$$

Thus the characteristic equation will be of fourth order:

$$S^4 + R_3S^3 + R_2S^2 + R_1S + R_0 = 0 \qquad (89)$$

in which the coefficients ($R_i = a_i/a_{i+1}$ in Equations (14) and (32)) are obtained (rather tediously) from the determinant in Equation (88) [2].

In this particular case then (with Z_{10} and Z_{20} as in Equation (84)):

$$R_3 = R_3^1 - \left(\frac{Q_0}{A_{s2}H_1}\right) \tag{90}$$

$$R_2 = R_2^1 + \left(\frac{gA_2}{L_2A_{s2}}\right)2\left[\frac{t_{c2}}{A_{c2}^2}\frac{Q_0^2}{2g}\right] \Big/ \left[H_2 - \left(\frac{Q_0}{A_{s2}H_1}\right)R_3^1\right] \tag{91}$$

$$R_1 = R_1^1 - 2g \cdot \frac{A_1A_2}{L_1L_2}\left\{\frac{Q_0}{A_{s2}H_1}\left[Z_{10} + \frac{t_{c2}}{A_{c2}^2}\frac{Q_0^2}{2g}\right]\right\} \tag{92}$$

$$R_0 = R_0^1 (3H_{mo} - 2H_0)/H_2 \tag{93}$$

with (for convenience):

$$H_1 = H_{mo} - 2\left[\frac{K_m}{A_m^2} + \frac{t_{a2}}{A_{c2}^2}\right]\frac{Q_0^2}{2g}$$

$$H_2 = H_1 - 2\left[\frac{t_{c2}}{A_{c2}^2}\frac{Q_0^2}{2g}\right] \tag{94}$$

and the following quantities have a special significance:

$$R_3^1 = \frac{g}{Q_0}\left\{\frac{2A_1}{L_1}\left[Z_{10} + \frac{L_1}{L_2}\frac{A_2}{A_1}(Z_{20} - Z_{10})\right]\right\} \tag{90a}$$

$$R_2^1 = g\left\{\frac{1}{A_{s1}}\left[\frac{A_1}{L_1} + \frac{A_2}{L_2}\right] + \frac{1}{A_{s2}}\cdot\frac{A_2}{L_2}\right. \tag{91a}$$
$$\left. + \frac{A_1A_2}{L_1L_2}\cdot\frac{4g}{Q_0^2}\cdot Z_{10}(Z_{20} - Z_{10})\right\}$$

$$R_1^1 = g\,Q_0\frac{A_1A_2}{L_1L_2}\left\{\frac{Z_{20}}{A_{s1}} + \frac{Z_{10}}{A_{s2}}\right\} \tag{92a}$$

$$R_0^1 = \frac{g^2A_1A_2}{L_1L_2A_{s1}A_{s2}} \tag{93a}$$

Their significance is this—the coefficients R (Equations 90–93) have been derived for the specific case of governed turbine operation with $\Delta Q_m \neq 0$ but, for a constant discharge $Q_m = Q_0$ with $\Delta Q_m = 0$, then the coefficients simplify to the R^1 given in Equations (90a–93a) [2].

For stability the Routh-Hurwitz criteria (Equation (32)) must be satisfied. With governed turbine flow ($\Delta Q_m \neq 0$) and the coefficients R (Equations (90–93)) it can be seen that there will be static and dynamic stability criteria analogous to those for the simple single shaft system discussed before.

E.g., the condition $R_0 > 0$ gives from Equation (93):

$$(3H_{mo} - 2H_0) > 0$$

or

$$H_0 > 3\left[\frac{K_1}{A_1^2} + \frac{t_{a1}}{A_{c1}^2} + \frac{K_2}{A_2^2} + \frac{t_{a2}}{A_{c2}^2} + \frac{K_m}{A_m^2}\right]$$
$$\times \frac{Q_0^2}{2g} = 3 \times \text{net losses} \tag{95}$$

which is exactly the same as the tank drainage (i.e., static stability) criterion of Equation (72) before. Similarly the remaining four criteria lead to dynamic stability conditions which may be rearranged into forms corresponding in appearance to the classic Thoma shaft area criterion of Equations (70–71). E.g., the condition $R_3 > 0$ gives from Equation (90):

$$A_{s2} > \frac{L_1A_1\left(\dfrac{Q_0^2}{2gA_1^2}\right)}{H_1\left|Z_{10} + \dfrac{L_1A_2}{L_2A_1}(Z_{20} - Z_{10})\right|} \tag{96}$$

However, these familiar conditions do not exhaust the possible types of instability in this situation. Looking instead at Equations (90a–93a) for the case with *no variation* in the turbine flow ($Q_m = Q_0$ with $\Delta Q_m = 0$) then, though static stability is guaranteed ($R_0^1 > 0$ unconditionally by Equation (93a)), other conditions cannot be, e.g., $R_3^1 > 0$ gives from Equation (90a):

$$\frac{L_2A_1}{L_1A_2} > \frac{Z_{10} - Z_{20}}{Z_{10}} \tag{97}$$

This condition could be violated [2,5] in a situation where L_1 is short and thus K_1 is small, especially if there is also some throttling below the shaft (i.e. $A_{c1} < A_1$). This situation will usually occur where the shaft A_{s1} is not intended primarily as a surge shaft but has been introduced to act as a gate, vent or access shaft (A_{s2} being the primary surge shaft close to the turbine).

The existence of a potential instability in the absence of an excitation ($\Delta Q_m = 0$) may seem surprising but the mechanism can be elucidated [5]. In this type of situation, digital simulation shows that surge shaft fluctuations have, as would be expected, both high and low frequency components but also that the high frequency component has a relatively much greater amplitude in the tunnel A_1L_1 and shaft A_{s1} due to the relatively low damping K_1 and inertia L_1. Thus a small increase in shaft level ΔZ_1, will, by continuity, be matched by a small decrease in conduit flow ΔQ_1, but this in turn will tend to cause ΔZ_1 to increase further (Equation (82)).

This, then, is a rather similar mechanism to the static instability described previously for the single shaft system

(Equation (72)) but, unlike that "tank drainage" instability, it seems physically reasonable that this one will be applicable for small flucutations as these are implicit to the exciting mechanism. This particular form of instability emphasises that the driving energy source need not itself be fluctuating.

Summary—Self-Excitation of Shaft Oscillations in Pumped-Storage Schemes

Traditional surge shaft stability analysis has been concerned with only one form of self-excitation, i.e., the action of a governor regulating a hydraulic turbine to operate at a constant load. The development of modern pumped-storage plant has altered this in three ways:

1. The design of plant with more flexible layout and longer penstock and draft-tube tunnels means that it may be necessary to introduce additional effects into the classic analysis, thus complicating the results obtained (e.g., Equations (70–72) cf Equations (71a–72a)).
2. Pumping plant often introduces multiple surge shafts and, though the stability of these has been extensively analysed in the past, the danger under certain circumstances of an excitation mechanism completely different in nature and additional to the governor is rarely recognised. The traditional concept of an element external to the hydaulic system (i.e. governor) creating the excitation has to be discarded, not only for multiple shaft schemes but also for:
3. Different kinds of machine operation may introduce new self-excitation mechanisms. In particular, constant speed and guide-vane setting operation of pumps leads to a very different form of instability (Equations (78–81)) even though the analysis itself is superficially similar. Previously the static instability has been less significant than the dynamic instability but this is not the case with "pump surging."

As more complex and demanding modes of operation are required of pumped-storage schemes it is possible that further self-excitation mechanisms will have to be considered.

To illustrate these points the analyses above have been carried out by:

- linearising (Equations (35–36)) the nonlinear governing equations
- forming the linear characteristic equation (Equation (14))
- applying the Routh-Hurwitz criteria (Equations (30–32)) to the coefficients of this

This approach, however, is by no means the only one of the general methods summarised previously which is applicable to surge shaft stability problems. Practical applications of these alternative methods, available in the literature, will now be reviewed.

REVIEW—APPLICATION OF STABILITY ANALYSIS TECHNIQUES TO SURGE SHAFTS

Use of Routh-Hurwitz Criteria

Most of the general techniques described previously have been used, in greater or lesser measure, for surge shaft stability problems and therefore examples of their application are available in the literature to be reviewed. There are four principal approaches, each of which will be discussed in turn.

Of all the existing techniques the most significant in surge shaft stability analsis has been the application of the Routh-Hurwitz criteria (Equations (30–33)) to the linearised characteristic equation (Equation (14)). Three practical applications of the technique have been illustrated in the preceeding section, but the classic surge shaft stability problem is that of a single surge shaft upstream of a turbine (Figure 14) which is governed to give constant power output (Equation (60)).

Following observed unstable behaviour of surge shafts, Thoma [77] was the first to analyse this problem by:

- linearising (Equation (35)) the equations of motion about a steady state
- obtaining the second order characteristic equation (Equation (4))
- applying the Routh-Hurwitz criteria (Equation (30)) to obtain dynamic and static criteria
- rearranging these into physically meaningful expressions [see Equations (71a), (72a)] which are appropriate for design purposes

It is this final step that was crucial to the success of the analysis and its practical acceptance and, since that time, the concept of the "Thoma surge shaft area" has dominated discussion of surge shaft stability and acted as an invaluable guide for designers [19,32,42].

The methodology has never been superseded but the analysis has been extensively refined. Its development can be followed through the standard textbooks on the subject from Calame and Gaden [16] to Jaeger, who gives the most comprehensive summary available [41,42]. The refinements have concentrated on two aspects:

- more precise modelling of the surge shaft/conduit system, introducing (as in Equations (46) through (72) above) e.g. surge tank throttle [31,81] and junction [3] losses, surge shaft inertia, penstock effects [4,41], etc.
- a less simplified approach to the modelling of the excitation, introducing, e.g., the effects of the location of the plant in an electrical system and of turbine efficiency characteristic [42], of governor action delay [7,20,32] and action limits [19,70], etc.

In addition the technique has been applied to other problems including:

—systems with two surge shafts (in various configurations) using the Routh-Hurwitz criteria Equation (32) with characteristic equations obtained as by Equations (87–89) [2,7,11,42,54] including as special cases, differential surge tanks [81] and a single surge shaft system [4] which includes penstock elasticity effects (lumped together to act effectively as a second small surge tank)
—forms of excitation other than ideal constant power governed turbine flow (Equation (60)), including valve flow (flow varies as root of head across machine) and constant power pumping [42], combined constant power and valve flow to represent the limits of governor action [19,70] constant speed pumping as in Equations (73–81) [3], no hydraulic machine excitation at all as in Equations (90a–93a) and (97) [2], etc.

However, while the original simple applications demonstrate the strengths of a method like this based on algebraic manipulations, the extensions summarised above expose its weaknesses. As has already been illustrated by example (Equations (46–72)), the introduction of additional terms (such as the junction losses, shaft inertia, penstock losses and inertia) results in:

—considerable algebraic labour
—more complicated final expressions (e.g. Equations (70–71) cf Equation (71a))

Indeed, with some forms of excitation it may not even be apparent how the criteria can be re-arranged into helpful forms without simplifiecation (e.g. Equations (80–81) above).

Increasing the order of the system, e.g. by introducing a second shaft as in Equations (82–97) above, leads not only to this problem (e.g. Equation (96)) but also makes the characteristic equation itself (Equations (89–93)) rather tedious to obtain. This is especially so for systems with more than two shafts and more often than not the coefficients R_i of the characteristic equation (Equation (89)) are evaluated numerically for particular cases [2,54,81] rather than algebraically for general expressions. Where algebraic expressions are obtained [2,62] these, as above with Equations (95–97), are only for some of the Routh-Hurwitz criteria and not all of them.

In practical terms, therefore, this method is appropriate only to one or two shaft systems in which it is reasonable to assume that many minor effects may be neglected and in which the excitation is relatively simple.

For such systems, as exemplified by the classic Thoma analysis [42], the method is very powerful, particularly as it does not just indicate instability but actually gives rise to a general expression showing the *influence* of scheme dimensions on stability.

To conclude, it is interesting to compare this algebraic approach with the graphical root-locus method (e.g., Figure 8). In practice the root-locus approach is only really appropriate for a second order ($n = 2$) system (e.g., Equations (2), (4)) corresponding to a single surge shaft scheme (with Equations (58) or (66–67)]. Where damping (2λ) and stiffness (ω_n^2) have a direct physical significance (as for Figure 7) then different values of these can be plotted to show their effect on stability (Figure 8). However, though the surge tank system (Figure 14) can be represented by a spring/mass/damper analogy [43], the "stiffness" and "damping" themselves depend on other parameters which have their own physical significance (e.g., surge shaft area A_s, etc.). In this case the values of 2λ and ω_n^2 in themselves do not contain the amount of information that can be obtained from derived expressions like the Thoma criteria (Equations (70–72)).

Use of the Phase-Plane Topology Approach

While application of the Routh-Hurwitz criteria dominated the early phase of surge shaft stability analysis, the next major developments in the subject were the use of:

—phase-plane topology investigation
—inspection of solutions arrived at by direction simulation

Ultimately the crucial limitation on the use of the Routh-Hurwitz criteria is that they are valid only for *small* oscillations. Typically, studies comparing linearised methods with full non-linear simulations [3] indicate that results from the former are vaild for variations in machine flow ($\Delta Q_m/Q_0$) $< 5\%$.

With advances in the subject, especially in examining experimental [11,28,46,50], site [20,42] and simulation [60] results, it was realised that the assumption of small oscillations led to limitations on the expressions obtained. Initially a factor of safety (N), typically $N = 1.2$ but often higher [40–42] was applied to the Thoma area A_{Th} (Equation (71a)). To quantify this Jaeger [42] developed a "time-average" approximation. The coefficients 2λ and ω_n^2 in Equation (6) are not constant in the nonlinear equations but have to be approximated as constants by, e.g. linearisation as in Equations (54–68). Improved time-average values for these allow the effect of large oscillations on the Routh-Hurwitz criteria to be estimated.

However, alternative methods of analysis were felt to be necessary and followed the two routes indicated above, each of which will be discussed in turn. The limitations imposed by the small oscillation requirement led investigators [11,38,41,46] to study the actual form of the solution using the phase-plane topology in the region of the singularities. Detailed descriptions and applications are available in the literature for the case of ideal governed constant turbine power [38] and also for other turbining flows [19].

The method, sometimes called the Poincaré orbital stability method [11], proceeds as follows:

— From the general governing equations (Equations (46–49)) coupled with the equation of excitation (e.g., Equations (60–61)) two phase-plane coordinates are selected (typically Z along with Q or Q/A or dZ/dt) and the governing equations are rearranged in terms of these in the form of Equation (24) or (37).
— From this equation the singularities (or critical points) can be identified (for the functions $F = 0$, $G = 0$). There are three singularities:
  1. one at the steady state equilibrium point of Equations (50–53)
  2. another complex conjugate pair, one of which occurs at low shaft levels, and
  3. the other of which can be disregarded since it has no physical significance (corresponds to $Z < -Ho$ which can never occur).
— The phase-plane equation [Equation (24) or (37)] can now be linearised about each individual singularity in turn to give the form of Equations (38), e.g. for the first singularity Equations (66–68) for 2λ and ω_n^2 could be used in the form of Equation (25):

$$\frac{d}{dt}\begin{bmatrix} Z \\ \left(\dfrac{dZ}{dt}\right) \end{bmatrix} = \begin{bmatrix} 0 & 1 \\ -\omega_n^2 & -2\lambda \end{bmatrix}\begin{bmatrix} Z \\ \left(\dfrac{dZ}{dt}\right) \end{bmatrix} \qquad (98)$$

—The phase trajectories can now be sketched in the regions close to (because the locally linearised equations still apply only to small oscillations) each individual singularity. The nature of the trajectories for each of the two physically real singularities will, of course, depend on the local values of 2λ and ω_n^2 in Equation (98) (cf Figure 10) but generally the steady state operating point will be a node or focus (dynamic stability) and the other will be a saddle (static instability).
—At large distances away from the singularities the details of the true (nonlinear) phase trajectories are not known. However if the separatrix between the phase trajectories converging to either singularity can be identified then (Figure 21a) oscillations starting on one side will travel towards the saddle (thus leading to tank drainage). Thus a qualitative appreciation of the behaviour of large (and thus nonlinear) oscillations is obtained.

In fact, the assumed stability region in Figure 21a still does not represent the true nonlinear behaviour of the system because this has features which linear systems do not. Evangelisti [38,46] showed that a limit cycle (Figure 12c) occurs around the focal singularity as in Figure 21b and attempted to investigate its size analytically. Development of this approach has led to an extremely elegant application of the Liapunov direct method [38] but in general the choice of an appropriate Liapunov function $V(x)$ is a daunting problem.

Considering practical applications of the phase-plane approach the following observations can be made:

—Though the linearised equations (e.g. Equation (98) with Equations (66–68)) are relatively straightforward to obtain and thus the nature of the singularities is readily identified (Figure 10), it is not necessarily easy to locate the singularities themselves. Existing studies [11,19,38,46] ignore, e.g. the junction and penstock losses and surge tank and penstock inertia terms in the head across the machines H_m (Equation (49)). Including these (as in Equations (47) and (49)) renders the analysis considerably less tractable.
—This is also the case if other forms of excitation are investigated (e.g. Equation (76) in place of Equation (60)) and, in addition, the multiple phase portraits occurring with multiple shaft schemes militates against the use of phase-plane topology with these.
—In practice extremely large oscillations cannot be sustained under governor action as this will usually reach limits of action beyond which stable behaviour is exhibited [19,60,70]. Thus the extent to which it is really necessary to investigate large amplitude oscillations can be questioned.
—As a general rule-of-thumb, for dynamic stability the small-amplitude linearised analysis is likely to be a reliable guide but for static instability (e.g. tank drainage) it may be wise to explore the effect of large oscillations.

Use of Direct Simulation

As with any other technical subject, the methods widely used for surge shaft stability analysis betray the historical context in which they were developed. The classic methods of analysis:

—use of Routh-Hurwitz criteria, and
—use of singularity analysis in the phase-plane

were introduced before reliable and convenient solution methods were widely available for nonlinear ordinary differential equations. As long as solution of the nonlinear equations could be carried out only by time-consuming manual graphical or finite-difference methods the classic techniques possessed considerable practical advantages [29].

Early analog devices such as the mechanical differential analyser had been applied to surge tank simulation [75] but when general purpose electronic analog and digital computers became available their many advantages were immediately perceived [60]:

—All nonlinear effects can be fully accounted for and there

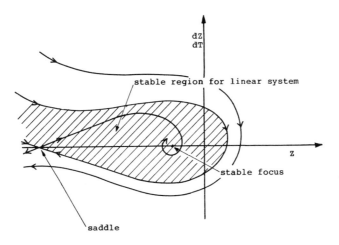

(a) **LINEAR OR QUASI-LINEAR SYSTEM**

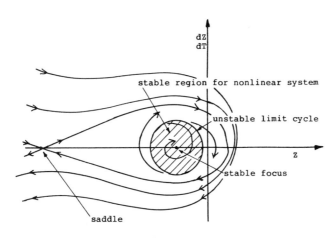

(b) **NONLINEAR SYSTEM WITH LIMIT CYCLE**

FIGURE 21. Phase-plane topology for single surge shaft.

is no restriction on the amplitude of oscillations that can be studied.

- The exact form of the solution is made available so that it becomes possible to form subjective judgements on the acceptability of the results, e.g. time taken to achieve a desired load change, rate of damping of oscillations, amplitude of any limit cycle, etc.

- Additional terms such as junction losses and penstock effects can be added to the governing equations simply so that detailed investigation of any particular design becomes possible.

- More complex (multiple shaft) models are easily assembled and it becomes relatively straightforward to ex-

tend the model to include, e.g. limits on the range of governor operation and governor models incorporating statism and droop, etc. In practice, however, the emphasis, especially with analog simulation [22,50,51,71], has been on the first two items above (i.e. reinforcing the techniques used in earlier studies) and few opportunities have been taken to extend models to embrace the overall electro-mechanical-hydraulic system [21,39] as had originally been envisaged [60].

The disadvantages of direct simulation for stability (as opposed to transient) studies are that:

- It does not lead to a generalized expression for stability showing the effect of various parameters (e.g. the Thoma criteria Equations (70–72) or Equations (78–81) or (90–97)) but will only establish the stability of individual particular cases. This applies not only to the system but also to the equilibrium points, i.e. it is possible to overlook a particular operating condition which might unexpectedly lead to instability.

— As a result of this, many individual cases will have to be tested separately throughout a study.
— With solutions in the time domain it may be difficult to decide whether and, in certain circumstances, where and why instability is occurring [5]. It is helpful in these cases to reform the solution in the phase-plane and simulations performed on digital computers should always be presented graphically as well as numerically.

Though analog computers have been widely used in the past [22,50,51,60,71], the balance of advantage now lies with digital computer [62,63] since:

— It is easier to cope with discontinuous models (e.g. limits to governor action, surge shafts with changes of cross section) and with tabulated information (e.g. machine performance characteristics which are not satisfactorily represented by formulas such as Equation (73) or junction loss data).
— It is easier to develop models of greater complexity.
— Digital computing power is cheaply available.

In using digital computer simulations, however, attention should be paid to:

— the careful formulation of the mass oscillation equations to ensure that all terms are represented consistently [6],
— especially with regard to the junction losses, since changes in junction layout can alter the Thoma shaft area by up to 25% [3], and
— the choice [15] of an accurate numerical method and appropriate step-size (especially as solutions will need to be carried on for a number of cycles in order to determine stability).

Use of Eigenvalue Determination

The fact that eigenvalue determination appears to have been applied to surge shaft stability problems only fairly recently [18] is probably due to:

— the previous success of the Routh-Hurwitz and phase-plane approaches, especially for single shaft schemes (Figure 14), and
— the relatively recent development of interest in the additional complexities introduced by pumped-storage plant and multi-shaft schemes.

Nevertheless, this is a technique which has considerable potential for surge shaft stability analysis with pumping plant.

The disadvantages of eigenvalue determination are that:

— it is necessary to linearise the equations, so that, as with the Routh-Hurwitz criteria, the method applies only to small amplitude oscillations and
— like direct simulation it will not lead to a generalized expression for stability but will only establish the stability or otherwise of individual particular cases (systems and equilibrium points).

On the other hand, the advantages of eigenvalue determination are [18]:

— Because, in effect, the complete linear solution is obtained then the results indicate not only the presence of instability (non-negative real parts of roots) but also the frequencies (very useful if a full numerical simulation is to be undertaken subsequently) and the amplitudes (or relative amplitudes) of all the components of oscillation. Thus the *significance* of any particular instability can be assessed.
— The surge shaft equations are naturally expressed in state-space form so systems of high order ($n \geq 4$) and complexity are easily analysed without the need to form the characteristic equation (Equation (14)) which may be tedious algebraically.
— Once the linearised equations are assembled in state-space form, standard matrix methods and program packages may be used to obtain the eigenvalues.
— Because of this it would also be relatively straightforward to extend the scope of the stability analysis in future so as to include further ordinary differential equations describing, e.g. pressure tunnel elasticity as well as inertia and losses [4], the action of the governor and mechanical system [19] and of the electrical system [7,21,39].

As one example, for the single shaft scheme (Figure 14) the linearised equations are given by Equation (18) with Equations (58–59) above. As another example of the formulation, consider the tandem surge shaft system (Figure 20, Equations (82–86)), previously analysed by the Routh-Hurwitz method (Equations (87–97)), for comparison.

The surge levels (Z_1, Z_2) and conduit flows (Q_1, Q_2) form natural state variables (x in Equation (17)) because, with appropriate representation of the junction losses, the governing equations (Equations (82–83)) are always in the state-space form [18]:

$$\frac{dZ_1}{dt} = f_1\,(Z_1,Z_2,Q_1,Q_2,Q_m)$$

$$\frac{dZ_2}{dt} = f_2\,(Z_1,Z_2,Q_1,Q_2,Q_m)$$

$$\frac{dQ_1}{dt} = f_3\,(Z_1,Z_2,Q_1,Q_2,Q_m)$$

$$\frac{dQ_2}{dt} = f_4\,(Z_1,Z_2,Q_1,Q_2,Q_m)$$

(99)

where, in addition to the state variables, Q_m is the system input ($F(t)$ in Equation (17)).

Equations (82–83) (or Equation (99)) are linearised about the equilibrium point (Equation (84)) exactly as before (for Equation (87)) but it is no longer necessary to form the determinant of Equation (88) in order to obtain the characteristic equation (Equation (89)) with Equations (90–94)). The resulting linearised equations are given simply by Equation (17), for any small excitation $F(t) = \Delta Q_m(t)$ about the steady state flow Q_o, where $\underline{x} = (\Delta Z_1, \Delta Z_2, \Delta Q_1, \Delta Q_2)$ and the coefficients are:

$$a_{11} = a_{12} = a_{21} = a_{22} = a_{23} = a_{32}$$

$$= a_{34} = a_{43} = b_1 = b_3 = b_4 = 0$$

$$a_{13} = -a_{14} = -1/A_{s1}$$

$$a_{24} = -b_2 = -1/A_{s2}$$

$$a_{31} = gA_1/L_1$$

$$a_{33} = \frac{-Q_oA_1}{L_1}\left[\frac{K_1}{A_1^2} - \frac{t_{c1}}{A_{c1}^2}\right]$$

$$a_{41} = -a_{42} = -gA_2/L_2$$

$$a_{44} = \frac{-Q_oA_2}{L_2}\left[\frac{t_{b1}}{A_{c1}^2} + \frac{K_2}{A_2^2} - \frac{t_{c2}}{A_{c2}^2}\right]$$

(100)

For the special case of no excitation ($\Delta Q_m = 0$) then the eigenvalues of $\underline{A}$ using the coefficients a_{ij} of Equation (100) have to be determined. On the other hand, if, as before, ideal constant power governed turbine flow is taken as the exciting mechanism, then (cf Equations (61–65) but using Equations (85–86) instead for the head across the machines H_m) the system input can be written as (H_1 as Equation (94)):

$$\Delta Q_m = \left[\frac{Q_0}{H_1}\right]\Delta Z_1 + \left[\frac{2}{H_1}\cdot\left(\frac{t_{c2}}{A_{c2}^2}\cdot\frac{Q_0^2}{2g}\right)\right]\Delta Q_2$$

(101)

which substituted into Equation (17) gives the free vibration form (cf Equation (6)):

$$\dot{\underline{x}} = \underline{A}\,\underline{x}$$

with a_{ij} as Equation (100) except:

$$a_{22} = \frac{Q_0}{A_{s2}H_1}, \quad a_{24} = \left[\frac{2}{H_1}\left(\frac{t_{c2}}{A_{c2}^2}\cdot\frac{Q_0^2}{2g}\right) - 1\right]\frac{1}{A_{s2}} \quad (102)$$

and the eigenvalues of the new system matrix $\underline{A}$ are determined. Any other form of linearised self-excitation can readily be accommodated in a similar fashion and its stability examined.

Similarly improvements to the model are also readily incorporated. E.g., the junction loss coefficients throughout the above have been represented by their steady state values at the equilibrium point flows, but it is well known that they vary substantially with the flow ratios at the branches [6,34,80]. These variations can be represented by polynomials fitted to the experimental data and incorporated into the linearised governing equations [18].

Though the determination of the eigenvalues of $\underline{A}$ is routine once numerical values of the coefficients a_{ij} are known, care has to be taken in the stability analysis of higher order systems such as the example above because of the dependence of stability not only on the system but also on the particular state of equilibrium being investigated (i.e., the values of Q_o, t_{c1}, t_{b1}, t_{c2} in Equations (100,102)).

This has important consequences because, though instability means that a system currently at a particular steady state operating point will not remain at that point given a slight disturbance, it is possible that the system might find another singularity, i.e. certain state variables may fluctuate between stable and unstable states [18], especially with static instabilities. Such behaviour has been observed from simulations of a tandem double shaft system [5,18] and contributes to the difficulty often noted in determining whether certain time-domain simulations actually do or do not exhibit instability.

Summary—Strategies for Surge Shaft Stability Analyses

For turbine operations, especially with the normal single shaft layout (Figure 14), there is an extensive literature of practical results, but with pumping plant it is possible that it will be necessary to undertake a new stability analysis. In choosing a method for any new stability problem the actual *purpose* of the analysis must be taken into account:

PRELIMINARY DESIGN STAGES

For optimisation of layout and initial sizing of components, stability analysis, in association with many other considerations, is used for general guidance. Thus a method should be chosen which gives a result of general applicability (the classic such result being, of course, that of Thoma Equations (70–72)) but it is not necessary at this stage to include much detail (e.g. "minor" terms such as penstock parameters, governor performance, etc.). Indeed it may be acceptable to simplify a system by neglecting smaller shafts or the form of the junction of the surge shaft and tunnel. For relatively simple schemes (i.e. low order $n \leq 4$ to 6) the Routh-Hurwitz approach is tried and tested. For multiple shaft schemes (i.e. higher order $n \geq 4$), however, though eigenvalue determination is straightforward it does not produce the desired general results.

DETAILED DESIGN CHECKING

At later stages the objective is to check the actual performance against the design intention and to "fine tune" the various components. Thus a generalised result as above is of less value for a particular system and the acceptability of linearised results and the nature of the model are changed. In conventional practice models at this stage take two forms (which may, however, be combined):

1. Full details of the hydraulic system (including junction losses, etc.) are included but with a relatively simple model of the machine behaviour (e.g. Equations (46–49) or Equations (82–83) above with the simple Equation (60) or (73)).
2. A relatively simple model of the hydraulic system but with more detailed modelling of the governor, mechanical and electrical systems [17,19,21].

For a single shaft ($n = 2$) system, phase-plane topology methods could be considered but generally a combination of direct simulation (for large oscillations and discontinuities) backed up by Routh-Hurwitz ($n \leq 4$) or eigenvalue ($n \geq 4$) analysis (to positively identify unstable components) offers the most flexible approach.

INVESTIGATION OF A STABILITY PROBLEM

Where an unexpected instability has occurred the resulting investigation has rather different objectives to the above. In particular, it is invaluable to identify the *mechanism* creating the self-excitation and hence the instability in order to establish a rational basis for controlling or curing it. Linearised small amplitude methods are usually acceptable for identifying dynamic instabilities but static instabilities may require large amplitude methods to define the stability boundaries sufficiently accurately:

– Routh-Hurwitz criteria will identify potential unstable behaviour (e.g. Equations (95–97) above) but in themselves contain little information about the *nature* of the solution [3] and thus further help is required to establish this, e.g. simulation [5].
– On the other hand phase-plane topology inspection yields comprehensive information about the nature of the problem [19] but may be difficult to apply to complex systems.
– With these direct simulation will illustrate the mechanism in action and, for high order ($n \geq 4$) systems, combination with eigenvalue analysis will help in locating instabilities.

SURGE SHAFT RESPONSE TO FORCED OSCILLATIONS

Forced Oscillations and Stability Analysis

To complete the consideration of surge shaft stability for pumped-storage schemes it is necessary to consider the response of the system to forced oscillations. By definition, stability analysis relates to self-excited oscillations (where the exciting mechanism is a function of the motion itself) and not to forced oscillations (where the exciting mechanism is independent of the motion). However, it may be possible to simplify an analysis considerably by representing it as a forced rather than a self-excited oscillation.

The overall dynamic system will generally consist of:

– the hydraulic system, i.e. tunnels, surge shafts, machine hydraulic characteristics, etc.
– the governor and hydraulic machine control system
– the mechanical system, i.e. rotating parts
– the external electrical network, including loads, other generating plant, etc.

In conventional surge shaft stablity analysis the overall system is generally *truncated* to consist only of the first of the above coupled with a simplified model of the second, e.g. the models of excitation mechanisms described by Equations (60) and (73). Thus assumptions are made about the behaviour of the overall system so that large parts of it can be omitted from the stability analysis.

These assumptions are frequently questioned, e.g. the Thoma criteria (Equations (70–72)) are based on the idealised model of Equations (60–61) and extensions and adaptations to this have already been reviewed. In the same way, if the overall system is too large and too complex to model conveniently, then it can be ignored and only its *effect* on the hydraulic system taken into account. This will appear, then, as a forced vibration, e.g. "flashy loading" or frequency regulation duty [49] may be represented by a cyclic variation in turbine demand (Figure 4).

Forced oscillation transients are, therefore, extensions of traditional stability studies in that they do not represent the conventional catastrophic modes of action but rather examples of normal operation which build up to give:

– large surge amplitudes, and
– ultimately a limit cycle (when the energy input is balanced by the energy dissipation occurring during the large amplitude oscillations).

Clearly linearised methods are inadequate to describe these. The loss terms, in particular, must be represented accurately in order to determine the final size of the limit cycle and in practice only direct transient simulation can achieve this. (For the inverse problem, however, where a design which deliberately minimises amplitudes is being sought [4], then a linearised analysis would be appropriate).

Complex and Cyclic Loading Sequences for Pumped-Storage Plant

Even with hydro turbine operation, practical experience has suggested that the extreme transient surges will be caused by a sequence of load changes [40,79] and these have become more complex for pumped-storage plants [23,48,52, 56,67] leading to the introduction of cyclic loading [49,55,61] to represent the effect of the worst possible self-excited behaviour of the overall system.

It is well-known that successive load change operations:

— may cause marked damping of surges (depending on their timing)
— but on the other hand, may cause a substantial increase in the surge amplitude.

In this latter case it is sometimes said that the surges are "superimposed," though strictly this is not an accurate terminology. To represent the worst possible case it is generally assumed that:

— The system imposes the cyclic load changes at instances which will lead to the maximum surge amplitudes (i.e. the hydraulic system is assumed to be in resonance with the remainder of the overall system). With multiple shaft schemes there is a question of which surge amplitude to maximise—most logically it is variations in the head across the machines (H_m).
— The loading sequence leading to the worst ultimate surge (size of the limit cycle) is the sequence of individual load changes each causing the maximum amplitude of the immediately following surge. This seems intuitively correct though it is by no means easy to show formally that it is so—this can be achieved in simple cases by the Principle of Optimality which shows that an optimal policy is made up only of optimal subpolicies [13].

Timing Load Changes to Maximise Surge Amplitudes

Ever since multiple operation sequences were first investigated there have been attempts to resolve the problem of timing the individual load changes so as to maximise the surge amplitude. Early contributions [14,30,61] suggested that:

— critical timing of instantaneous (i.e. idealised) load changes occurs at extreme (i.e. maximum and minimum) values of low pressure tunnel flow rather than surge level [76]
— this always leads to an increase in the surge amplitude
— where there are surge shaft throttle losses ideal instantaneous load-changes do not give the worst case.

For realistic load changes occurring over a finite time, there has been uncertainty over the appropriate starting time and form [48,53,56], though there have been some very elegant studies (restricted to very simple cases) which have applied optimisation theory to this problem [47].

However, for real complex systems which feature:

— discontinuities (e.g. in loss coefficients with flow direction, in shaft area, etc.), or

— graphical or tabular data (e.g. for machine characteristics, form of load change, junction loss coefficients, etc.), or
— governed (e.g. Equation (60)) or otherwise adjusted (e.g. Equation (73)) operation between imposed load changes

then direct simulation using a trial and error method to find the optimum timing is necessary [5,67]. This technique demonstrates that the critical load change time, while being close to the time of the flow extreme (and always after the surge extreme) is not necessarily exactly at it [67].

SUMMARY—THE POINTS TO WATCH

In considering surge shaft stability for pumped-storage schemes it is important to bear in mind that these may introduce a range of problems not extensively discussed in the existing literature. The first step is to decide whether any particular case is:

— described in the existing literature, as reviewed above or elsewhere [19,42,62 et al.], or
— not described in the literature and therefore requires a new analysis (as with two of the three exciting mechanisms discussed above).

In either case there are some important points to watch for.

When applying the existing literature to pumped-storage plants remember that:

— It is generally restricted to turbining flows only, particularly ideal governed constant power output, and other mechanisms for self-excitation (or, indeed, forced excitation) are rarely discussed.
— Important features of the hydraulic system are often oversimplified or ignored, especially the form of the surge shaft/conduit junction. It has been shown that the form of the junction has a large influence on both transient surges [37] and stability [3], with the Thoma area being underestimated by up to 25% for the latter.
— Conclusions applicable to one particular plant layout (typically Figure 14) should *not* be extended to other layouts. E.g. it is often suggested that the "venturi" surge shaft [40,62] improves stability and while this is true for the conventional layout of Figure 14 it has exactly the opposite effect [2,3] for the tailrace surge shaft (as shown, e.g., by the $-ve$ sign on the junction loss coefficients in Equation (84)).
— Simplifying some system so as to fit an existing analysis, e.g. omitting or combining together some minor shafts not primarily intended as surge shafts [2], may fundamentally alter the response of the system in unexpected ways.

If a new analysis is required because some new problem peculiar to pumped storage plants is introduced, then decide next on the objectives of the investigation because (as has been discussed) these will influence:

—the degree of system detail required,
—the choice of method for stability analysis and
—the required form of the solution.

While carrying out new stability analyses remember that:

—All self-excitation mechanisms should be identified, especially in complex schemes where there is a possibility (not covered in the existing literature) that there is more than one (e.g. two completely independent mechanisms were identified for the tandem double shaft scheme of Figure 20, Equations (82–97)).
—Static and dynamic instabilities should be distinguished as the former are less likely to be represented satisfactorily by linearised small amplitude analysis. The fact that a dynamic instability (Thoma area criterion) dominates much of the existing literature does not preclude static instabilities being of more importance for the other types of problem (e.g. pump surging as discussed above).
—The full range of stability analysis techniques (summarised above) should be exploited as appropriate. While the existing literature is dominated by the Routh-Hurwitz criteria and phase-plane singular point analysis, it is likely that techniques previously rarely used, such as eigenvalue determination, will prove useful for more complex pumping plants.

NOTATION

The following symbols have been used.

A, A_i = mean cross-sectional area of tunnel over full length L (Figure 14) or L_i ($i = 1, 2$ for double shaft system Figure 20)

A_c, A_{ci} = cross-sectional area of tunnel in vicinity of surge shaft (Figure 14) ($i = 1, 2$ for double shaft system Figure 20)

A_m = mean cross-sectional area of high-pressure tunnels (penstock and draft tube) over full length L_m (Figure 14, 20)

A_s, A_{si} = horizontal water free-surface area in surge shaft (Figure 14) ($i = 1, 2$ for double shaft system Figure 20)

A_{Th}, A_{Th}^1 = Thoma surge shaft area (Equations (70–72))

$\underline{A}$ = system matrix (coefficients a_{ij}) for general state-space representation (Equation (17))

a_i ($i = 0, n$) = coefficients of general linear equation (Equations (11–12)) or characteristic polynomial (Equation (14))

a_{ij} = elements of system matrix $\underline{A}$ (Equation 17))

B = coefficient ($\pm ve$) in pump characteristic polynomial (Equation (73))

$\underline{b}.F(t)$ = system input (excitation) for Equation (17)

C = coefficient ($+ ve$) in pump characteristic polynomial (Equation (73))

C_i ($i = 1, n$) = coefficients (Equations (3), (13)) in general linear solution for x

c = Coulomb damping coefficient (Equation (1), Figure 7)

c_{ij} = coefficients for quadratic Liapunov function V (Equations (40–42))

D = differential operator d/dt

D_{ij} = compound differential operators in ΔZ, $d(\Delta Z)/dt$, etc. (Equation (87))

E = specific energy (i.e. energy/unit mass) of water flowing through hydraulic machine

F, G = phase-plane gradient functions (Equations (22,24,37))

$F(t)$ = time-dependent exciting force (Equations (1–2), Figure 7)

F_x, F_V, F_a = self-excitation $F(t) = F_x \cdot x$ or $F_v \cdot \dot{x}$ or $F_a \cdot \ddot{x}$ for force dependent on displacement, velocity or acceleration

$\underline{f}(x)$ = nonlinear function

f_{ij}, g_{ij} = coefficient of small deviation terms in phase-plane equations (Equations (37–38))

$f_k(t)$ = Fourier harmonic of $x(t)$ (Equation (34))

G, F, g_{ij}, f_{ij} = see F, f_{ij}

H_1, H_2 = convenience terms (defined in Equations (63), (68), (77), (94) as appropriate)

H_m, H_{mo} = head across hydraulic machine (H_m) and steady-state value (H_{mo} for flow Q_o)

H_0 = difference between reservoir levels (Figures 14, 20)

H_s, B, C = coefficients in pump characteristic polynomial (Equation (73))

H_{sv} = true shut-valve-head of pump when flow $Q_m = 0$

I = unit matrix

J_{ai}, J_{bi}, J_{ci} = ($i = 1, 2$) junction head losses (Figure 20) from upstream pressure tunnel to surge shaft, shaft to downstream tunnel, upstream to downstream tunnel

J_{cp}, J_{cs}, J_{sp} = head losses occurring at surge shaft/tunnel junction (Figures 14–15), from tunnel to penstock, tunnel to shaft, shaft to penstock

J_{cpo}, J_{cso} = steady state values of J_{cp}, J_{cs} (Equations (50–51))

K, K_i = total loss coefficient for tunnel over full length L (Figure 14) or L_i ($i = 1, 2$ for Figure 20) based on velocity head ($Q^2/2gA^2$)

K_m = total loss coefficient for penstock and draft-tube tunnels over length L_m based on velocity head ($Q^2_m/2gA^2_m$)

k = spring stiffness (Equation (1), Figure 7)

L, L_i = total length of low pressure tunnel (Figure 14 or $i = 1, 2$ for Figure 20)

L_m = total length of penstock and draft-tube tunnels (Figure 14, 20)

L_s = height of surge shaft from base (at tunnel) to reservoir datum (i.e. zero flow steady state) level (Figure 14)

m = mass (Equation (1), Figure 7)

N = safety factor applied to Thoma shaft area (Equation (71a))

n = order of characteristic equation (Equation (14))

Q, Q_i = low pressure tunnel flow at any instant (positive direction defined on Figure 14 or Figure 20 for $i = 1, 2$)

Q_m = volumetric flow rate in penstock, draft tube and hydraulic machine at any instant (positive direction defined on Figure 14, 20)

Q_0 = steady state (Equations (50–51)) volumetric flow rate through whole system

Q_s = unsteady flow into shaft at any instant (Figure 14)

Q_* = volumetric flow rate at which pump performance characteristic peaks (Figure 17, Equation (75))

Q'_*, Q''_* = limits of dynamic (Equations (80–81)) and static (Equations (78–79)) stability regions for pump surging

R_i = general coefficients of 4th order characteristic equation (Equation (89), $i = 0, 3$)

R'_i = values of R_i (Equations (90–93)) for special case with no fluctuations in machine flow, i.e. $\Delta Q_m = 0$ (Equations (90a–93a))

s, s_i = roots of characteristic equation (Equations (13–14), (19–20))

t = time (independent variable)

t_{ai}, t_{bi}, t_{ci} = junction loss (J_{ai} etc.) coefficients for steady or very nearly steady flows (Figure 20), based on the velocity head ($Q_0^2/2gA_{ci}^2$)

t_{cp}, t_{cs}, t_{sp} = junction loss (J_{cp} etc.) coefficients for steady or very nearly steady flows (Figures 14–15), based on the velocity head ($Q_0^2/2gA_c^2$) (Equation (57))

$V(x)$ = Liapunov function (Figure 13)

$x, \dot{x}, \ddot{x}$ = displacement (Figure 7), velocity, acceleration (Equation (1))

$x_i(i = 1, n)$ = general state (Equations (15), (17)) or phase (Equations (22–24)) variable

x_0 = value of x at an equilibrium point of singularity

$\underline{x}$ = state vector ($x_1, x_2, \ldots, x_n$)

$y(x), y_0$ = any function of x and its value at a singularity ($x = x_0$)

Z, Z_i = free water surface level in surge shaft at any instant, measured positive above reservoir datum (Figure 14 or Figure 20 for $i = 1, 2$)

Z_0, Z_{i0} = steady state equilibrium point values of Z (Equation (52)) and Z_i (Equation (84))

Δ = discriminant of 2nd order characteristic equation (Equations (26–27))

ΔQ_m = fluctuation (self-excited or imposed) in machine flow Q_m about a steady state value Q_0 (Equation (54))

ΔX = small fluctuation in quantity ($X = x_i, Q, Q_i, Q_m, Z, Z_i$) about steady state value X_0

2λ = damping coefficient (Equation (2))

ω_n^2 = square of natural frequency (Equation (2))

REFERENCES

1. Allen, E. *Using Centrifugal Pumps.* Oxford University Press, London, Chap. 6 (1960).
2. Anderson, A., "A Novel Type of Surge Shaft Instability," *Proceedings, Institution of Civil Engineers,* London, Vol. 67, Part 2, pp. 695–706 (Sept. 1979).
3. Anderson, A., "Surge Shaft Stability with Pumped-Storage Schemes," *Journal of Hydraulic Engineering, ASCE,* Vol. 110, No. 6, pp. 687–706 (June 1984).
4. Anderson, A., "Interaction of Surge Shafts and Penstocks," *Proceedings, 4th International Conference on Pressure Surges, Bath, Sept. 1983,* Paper G2, BHRA Fluid Engineering, Cranfield (1983).
5. Anderson, A. and Robbie, J. F., "Some Novel Experiences in the Computer Aided Design of Dinorwic Pumped Storage Surge System," *Proceedings, 2nd International Conference on Pressure Surges, London, Sept. 1976,* Paper B2, BHRA Fluid Engineering, Cranfield (1977).
6. Anderson, A. and Robbie, J. F., "Effect of Equation Formulation on the Prediction of Mass Oscillations in Closed Conduits and Surge Tanks," *Proceedings, 3rd International Conference on Pressure Surges, Canterbury, March 1980,* Paper G4, BHRA Fluid Engineering, Cranfield (1981).
7. Aronovich, G. V. et al. *Water Hammer and Surge Tanks.* transl. Ledermann, D., Israel Program for Scientific Translation, Jerusalem (1970).
8. Atherton, D. P. *Nonlinear Control Engineering.* Van Nostrand Reinhold, New York (1975).
9. Atherton, D. P. *Stability of Nonlinear Systems.* Research Studies Press, John Wiley, Chichester (1981).
10. Ayres, F. *Theory and Problems of Differential Equations.* Schaum's Outline Series, McGraw-Hill, New York (1952).
11. Balint, E. et al., "Analyses of a Complex Surge Tank System (A Symposium of Three Papers)," *Journal of the Institution of Engineers, Australia,* pp. 155–166 (June 1955).
12. Bell, W. W. *Matrices for Scientists and Engineers.* Van Nostrand Reinhold, New York, Chap. 4 (1975).
13. Bellman, R. *Dynamic Programming.* Princeton University Press, Princeton, NJ (1975).
14. Bouvard, M. and Molbert, J., "Calcul de la cheminée à étranglement de la chute Isère-Arc," *La Houille Blanche,* No. 2, pp. 260–281 (May 1953).

15. Bullough, J. B. B. and Robbie, J. F., "Theoretical and Experimental Analysis of the Error in the Numerical Solution of the Equations of Mass Oscillation in Pumped Storage Schemes," *Proceedings, 2nd International Conference on Pressure Surges, London, Sept. 1976,* Paper H3, BHRA Fluid Engineering, Cranfield (1977).

16. Calame, J. and Gaden, D., *Théorie des Chambres d'Équilibre,* la Concorde, Lausanne and Gauthier-Villars, Paris, 1926 (*Theory of Surge Chambers,* transl. Elsden, O. and Whitaker, F. W. A., Sir Alexander Gibb and Partners, London, 1950).

17. Chaplin, R. A., "Simple Governing and Surging Simulation," *Water Power and Dam Construction,* Vol. 37, No. 5, May, pp. 25–28, No. 6, June, pp. 39–49 (1985).

18. Chapman, A. D. R. and Robbie, J. F., "Eigenvalue Stability Analysis of Surge Systems Involving Double Vented Shafts: Theory and Practice," *Proceedings, 4th International Conference on Pressure Surges, Bath, Sept. 1983,* Paper H3, BHRA Fluid Engineering, Cranfield (1983).

19. Chaudhry, M. H. *Applied Hydraulic Transients.* Van Nostrand Reinhold, New York, Chap. 5, 11 (1979).

20. Chevalier, J. and Hug, M., "Essais de la Cheminée d'Équilibre de Cordéac en ce qui concerne la Condition de Thoma," *La Houille Blanche,* Vol. 12, No. 6, pp. 888–902 (Dec. 1957).

21. Codrington, J. B. and Witherell, R. G., "The Use of Impedance Concepts and Digital Modelling Techniques in the Simulation of Pipeline Transients," *Proceedings, 2nd International Conference on Pressure Surges, London, Sept. 1976,* Paper A2, BHRA Fluid Engineering, Cranfield (1977).

22. Corbellini, G. and Silvestri, A., "Stabilità della Regolazione di Frequenza in Relazione al Dimensionamento della Camera Cilindrica del Pozzo Piezometrico," *L'Energia Elettrica,* Vol. 50, No. 11, pp. 653–660 (Nov. 1973).

23. Datta, O. P. and Mehra, J. M. L., "Rudiments of Surge Tank Design," *Indian Journal of Power and River Valley Development,* pp. 39–41 (Feb. 1973).

24. den Hartog, J. P. *Mechanical Vibrations.* 4th Edn., McGraw-Hill, New York (1956).

25. di Stefano, J. J. et al. *Theory and Problems of Feedback and Control Systems,* SI edn, Schaum's Outline Series, McGraw-Hill, New York (1976).

26. Doebelin, E. O. *System Modelling and Response: Theoretical and Experimental Approaches.* John Wiley, New York (1980).

27. Donaldson, D. D., "Liapunov's Direct Method in the Analysis of Nonlinear Control Systems," Chap. 5 of Leondes, C. T. (ed.), *Modern Control Systems Theory.* McGraw-Hill, New York (1965).

28. Duggins, R. K., "Hydraulic Stability of a Model Hydroelectric Installation," *Proceedings, Institution of Mechanical Engineers,* Vol. 182, Part 3M, Paper 3, pp. 20–27 (1967/68).

29. Escande, L. *Méthodes Nouvelles pour le Calcul des Chambres d'Équilibre.* Dunod, Paris (1950).

30. Escande, L., *Nouveau Compléments d'Hydraulique: Part 4,* Publications Scientifiques et Techniques du Ministère de l'Air, No. 395, Paris (July 1963).

31. Escande, L., "The Stability of Throttled Surge Tanks Operating with the Electric Power Controlled by the Hydraulic Power," *Journal of Hydraulic Research, IAHR,* Vol. 1, No. 1, pp. 4–13 (1963).

32. Forster, J. W., "Design Studies for Chute-des-Passes Surge-Tank System," *Journal of the Power Division, ASCE Proceedings,* Vol. 88, No. P01, May 1962, Paper No. 3142, pp. 121–152 (Discussion Vol. 88, No. P04, Dec. 1962, pp. 183–188, Vol. 89, No. P01, Sept. 1963, pp. 101–103).

33. Fox, J. A. *An Introduction to Engineering Fluid Mechanics.* 2nd edn., Macmillan, London, pp. 374–377 (1977).

34. Gardel, A. and Rechsteiner, G. -F., "Les pertes de charge dans les branchements en Té des conduites de section circulaire," *Bulletin Technique de la Suisse Romande,* Vol. 96, No. 25, pp. 363–391 (Dec. 1970).

35. Graeser, J. -E. and Hoffer, J. -P., "Deux exemples d'utilisation de petites calculatrices programmables pour la résolution de problèmes posés par les installations hydrauliques," *Bulletin Technique de la Suisse Romande,* Vol. 105, No. 6, pp. 53–58 (Mar. 1979).

36. Henrici, P. *Computational Analysis with the HP25 Pocket Calculator.* Wiley-Interscience, New York, pp. 101–106 (1977).

37. Hsu, S. T. and Elder, R. A., "Raccoon Mountain Pumped-Storage Plant Hydraulic Transient Studies," *Transactions, 6th Symposium, Section for Hydraulic Machinery Equipment and Cavitation, IAHR, Rome, 1972,* Part 1, Paper G1.

38. Infante, E. F. and Clark, L. G., "On the Large Oscillations of a Simple Surge Tank," *L'Energia Elettrica,* Vol. 41, No. 12, pp. 866–873 (Dec. 1964).

39. IEEE Power System Engineering Committee, "Dynamic Models for Steam and Hydro Turbines in Power System Studies," in Byerly, R. T. and Kimbark, E. W. (eds.), *Stability of Large Electric Power Systems.* IEEE Press, New York, pp. 128–139 (1974).

40. Jaeger, C., "Present Trends in Surge Tank Design," *Proceedings Institution of Mechanical Engineers,* Vol. 168, No. 2, pp. 91–124 (1954).

41. Jaeger, C., "A Review of Surge-Tank Stability Criteria," *Journal of Basic Engineering, ASME Transactions,* Series D, Vol. 82, pp. 765–783 (1960).

42. Jaeger, C. *Fluid Transients in Hydro-Electric Engineering Practice.* Blackie, Glasgow, Chap. 12, 26–27 (1977).

43. Koelle, E., "Hydraulic Transients in Pressure Conduits—Basic Equations," *Proceedings, International Institute on Hydraulic Transients and Cavitation, Sao Paulo, July 1982,* Associacao Brasileira de Hidrologia e Recursos Hidricos, Paper A 1.1 (1982).

44. Kolnsberg, A., "Reasons for Centrifugal Compressor Surging and Surge Control," *Journal of Engineering for Power, ASME Transactions,* Vol. 101, pp. 79–86 (Jan. 1979).

45. Kurman, K. J. *Feedback Control: Theory and Design.* Studies in Automation and Control, Vol. 4, Elsevier, Amsterdam (1984).

46. Marris, A. W., "Large Water-Level Displacements in the Simple Surge Tank," *Journal of Basic Engineering, ASME Transactions,* Series D, Vol. 81, pp. 446–454 (Dec. 1959).

47. Marro, G., "Manovre ottime negli impianti idroelettrici funzionanti a programma," "La manovre critica negli impianti idroelettrici con pozzo piezometrico," *L'Energia Elletrica*, Vol. 41, No. 8, Aug., pp. 556–566, No. 11, Nov., pp. 779–786 (1964).

48. Matthew, G. D. and Robbie, J. F., "A New Numerical Method Applied to the Study of Mass Oscillations in a Pumped-Storage Hydro-Electric Scheme," *Proceedings, Institution of Civil Engineers*, Vol. 41, pp. 499–521 (1968), Vol. 43, pp. 433–445 (1969).

49. Mawer, W. T. et al. "Dinorwic Pumped Storage Project—Pressure Surge Investigations," *Proceedings, 2nd International Conference on Pressure Surges, London, Sept. 1976*, Paper B3, BHRA Fluid Engineering, Cranfield (1977).

50. McCaig, I. W. and Jonker, F. H., "Applications of Computer and Model Studies to Problems Involving Hydraulic Transients," *Journal of Basic Engineering, ASME Transactions*, Vol. 81, pp. 433–445 (1959).

51. Mosonyi, E. and Nagy, L., "Stability Investigations by Computer," *Water Power*, Vol. 16, No. 7, pp. 312–314 (July 1964).

52. Mosonyi, E. and Seth, H. B. S., "The Surge Tank—A Device for Controlling Water Hammer," *Water Power and Dam Construction*, Vol. 27, No. 2, Feb., pp. 69–74; No. 3, Mar., pp. 119–123 (1975).

53. Murillo, J., "Application of a Digital Computer to a Surge Chamber Problem," *Proceedings, 9th Convention, International Association for Hydraulic Research, Dubrovnik, 1961*, pp. 926–930.

54. Natarajan, C. and Vittal, B. P. D., "Stability Criteria for a Double Surge Tank System," *Modern Trends in Hydraulic Engineering Research, Golden Jubilee Symposia, Poona, Jan. 1966*, Central Water and Power Resaerch Station, Vol. 2, pp. 135–143.

55. Nath, V. and Agrawal, B. G., "Electronic Digital Computer in Testing Hydraulic System of Yamuna Hydro-Electric Scheme Stage 2," Symposium on Electronic Techniques in Testing and Measurement in Research and Industry, Lucknow, Aug. 1970, *Journal of the Institution of Engineers, India*, Vol. 51, Part ET2, pp. 112–117 (Jan. 1971).

56. Noseda, G., "Sul funzionamento di un pozzo piezometrico inserito in un impianto idroelettrico con pompatura," *L'Energia Elettrica*, Vol. 44, No. 10, pp. 585–594 (Oct. 1967).

57. *NAG Library*, Fortran Mark 7, Routine No. F02AGF, NAGFLIB, Numerical Algorithms Ltd., Oxford, Vol. 3, MK5, 987/623 (Sept. 1975).

58. Ogata, K. *State Space Analysis of Control Systems*. Prentice Hall, Englewood Cliffs, NJ (1967).

59. O'Neil, P. V. *Advanced Engineering Mathematics*. Wadsworth, Belmont Calif. (1983).

60. Paynter, H. M., "Surge and Water Hammer Problems," Symposium on Electronic Computers, *Transactions ASCE*, Vol. 118, pp. 962–1009 (1953) (*Proceedings ASCE*, Separate No. 146, Aug. 1952).

61. Pistilli, G. and Savastano, G., "La risoluzione da problemi di oscillazione di massa con l'impiego delle calcolatrici elet-

troniche—Nota III," *L'Energia Elettrica*, Vol. 39, No. 6, pp. 472–482 (June 1962).

62. Popescu, M. *Probleme Actuale in Domeniul Hidraulicii Castelelor de Echilibru*, Studii de Hidraulica, Vol. 21, Institutul de Studii si Cercetari Hidrotechnice, Bucharest (1969).

63. Popescu, M. and Halanay, A., "A Computing Technique for Hydraulic Resonance in Hydropower Plants with Surge Tanks," *Revue Roumaine des Sciences Techniques, Série de Mécanique Appliquée*, Vol. 26, No. 3, pp. 421–431 (May-June 1981).

64. Porter, B. *Synthesis of Dynamical Systems*. Nelson, London (1969).

65. Raven, F. H. *Automatic Control Engineering*. 3rd edn., McGraw-Hill, New York (1978).

66. Richards, R. J. *An Introduction to Dynamics and Control*. Longman, London (1979).

67. Robbie, J. F. and Robson, F. M., "Computer Aided Design of Surge Chambers in Pumped Storage Conduit Systems," *Proceedings, International Conference on Pressure Surges, Canterbury, Sept. 1972*, Paper E2, BHRA Fluid Engineering, Cranfield (1973).

68. Rothe, P. H. and Runstadler, P.W., "First-Order Pump Surge Behaviour," *Journal of Fluids Engineering, ASME Transactions*, Series I, Vol. 100, pp. 459–466 (Dec. 1978).

69. Routh, E. J. *A Treatise on the Stability of a Given State of Motion, Particularly Steady Motion*. Macmillan, London (1877), reprinted in Fuller, A. T. (ed), *Stability of Motion: A Collection of Early Scientific Papers by Routh, Clifford, Sturm and Bocher*. Taylor and Francis, London (1975).

70. Ruus, E., "Stability of Oscillations in Simple Surge Tank," *Journal of the Hydraulics Division, ASCE Proceedings*, Vol. 95, No. HY5, Paper 6773, pp. 1577–1587 (Sept. 1969).

71. Sethuraman, V. and Meenakshisundaram, S., "Analog Computer Studies of the Stability of Orifice Surge Tanks," *Water Power*, Vol. 25, No. 9, pp. 336–341 (Sept. 1973).

72. Sliosberg, P., "Critères pour l'instabilité de fonctionnement des pompes centrifuges," *La Houille Blanche*, Vol. 8, No. 4, pp. 521–525 (Aug.–Sept. 1953).

73. Speckhart, F. H. and Green, W. L. *A Guide to Using CSMP—The Continuous Modelling Simulation Program*. Prentice-Hall, Englewood Cliffs, NJ (1976).

74. Stucky, A. *Chambres d'Équilibre*, Cours d'Aménagement des Chutes d'Eau, École Polytechnique de l'Université de Lausanne, Lausanne (1952).

75. Taylor, E. H., et al., "Unsteady Flow in Conduits with Simple Surge Tanks," *Journal of the Hydraulics Division, ASCE Proceedings*, Vol. 85, No. HY2, Part 1, Paper No. 1933, pp. 1–11 (Feb. 1959).

76. Thakar, V. S., "Digital and Analogue Computer Analysis of Mass Oscillations in Surge Systems," *Irrigation and Power Journal (India)*, Vol. 25, No. 2, pp. 193–202 (April 1968).

77. Thoma, D. *Zur Theorie des Wasserschlosses bie selbsttätig geregelten Turbinenlagen*, Oldenburg, Munich (1910).

78. Timoshenko, S. and Young, D. H. *Vibration Problems in Engineering*. 3rd edn., Van Nostrand, Princeton, NJ (1955).

79. Tonacca, E., "Ampliamento del pozzo piezometrico dell' im-

pianto di Cimego," *L'Energia Elettrica*, Vol. 40, No. 3, pp. 226–234 (1963).

80. Villemonte, J. R., "Some Basic Concepts on Flow in Branching Conduits," *Journal of the Hydraulics Division, ASCE Proceedings,* Vol. 103, No. HY7, Paper No. 13051, pp. 685–697 (July 1977).

81. Zienkiewicz, O. C., "Stability of Parallel-Branch and Differential Surge Tanks," *Proceedings, Institution of Mechanical Engineers,* Vol. 170, pp. 265–280 (1956).

APPENDIX—QUADRATIC PUMP CHARACTERISTIC REPRESENTATION

Use of Quadratic Polynomial to Represent Characteristic

A pump performance characteristic may exhibit the classic form showing potential for pump surging (Figure 17b) or, especially with pump-turbines, may be of generally stable form (Figure 17a) but exhibit an apparently discontinuous upward shift in the characteristic (Figure 22), which may cause pump surging if the operating point coincides with this feature. In either case [3] Equation (73) may be used as a linearised small amplitude stability analysis.

Equation (73) (Figure 17) is a widely quoted approximation to pump performance characteristics at constant speed [35]. The derivation, available in standard texts, e.g. [33], ignores leakage and off-design-point losses. As a result, any attempt to fit this type of curve to an actual pump characteristic will have only limited success (Figure 23a). It is for this reason that polynomial turbomachine characteristic representation is not recommended for transient studies. However, as Figure 23a suggests, over a *limited* region of applicability (small deviations) it may be an acceptable representation.

Evaluation of Coefficients (H_s, B, C) in Equation (73)

Given an actual performance graph (Figure 23a) there are three approaches to fitting the quadratic (Equation (73)), i.e. to obtaining values of the coefficients H_s, B and C.

LEAST SQUARES CURVE FITTING

The second order least squares algorithm can be used with any number of experimental points (≥ 3) picked off the curve but avoiding the shut-valve point ($Q_m = 0$, H_{sv}).

INTERPOLATING QUADRATIC

A very quick technique is to select carefully 3 points in the region of interest, e.g. (Q_1, H_1), (Q_2, H_2), (Q_3, H_3) not at ($Q_m = 0$, H_{sv}), and to look for the quadratic that actually passes through each of these, i.e. which satisfies

$$H_i = H_s - B|Q_i| - CQ_i^2 \text{ for } (Q_i, H_i) \text{ with } i = 1, 2, 3 \quad (103)$$

Then Equation (103) can be solved as three linear equations for the three coefficients (H_s, B, C):

$$\begin{bmatrix} 1 & -|Q_1| & -(Q_1)^2 \\ 1 & -|Q_2| & -(Q_2)^2 \\ 1 & -|Q_3| & -(Q_3)^2 \end{bmatrix} \begin{bmatrix} H_s \\ B \\ C \end{bmatrix} = \begin{bmatrix} H_1 \\ H_2 \\ H_3 \end{bmatrix} \quad (104)$$

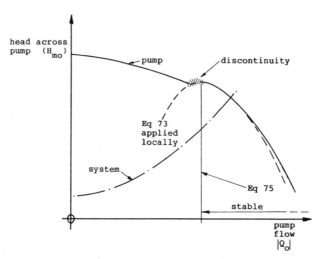

FIGURE 22. Representative pump-turbine pumping characteristic.

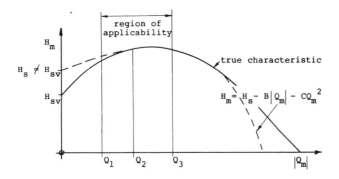

(a) PUMP CHARACTERISTIC AND POLYNOMIAL APPROXIMATION

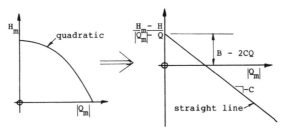

(b) 'RECTIFICATION' OF QUADRATIC TO PLOT AS A STRAIGHT LINE

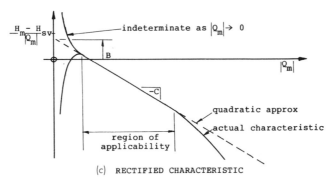

(c) RECTIFIED CHARACTERISTIC

FIGURE 23. Quadratic polynomial approximation to pump characteristic.

RECTIFICATION WITH STRAIGHT-LINE FITTING

Both of the above methods will give rise to the situation sketched in Figure 23a. With this type of result it is difficult to determine whether the fitted curve is quantitatively or qualitatively wrong, i.e. whether changes in the coefficients (H_s, B, C) would lead to a better fit or whether this is impossible because the actual characteristic is not of the form of Equation (73). To overcome this the curve can be "rectified" so that a quadratic would be represented by a *straight line* (Figure 23b).

If the actual characteristic is plotted to these coordinates (taking $Q_1 = 0$, $H_1 = H_{sv}$) then it will be seen clearly (Figure 23c) that overall it is not of the desired quadratic form of Equation (73) but that in a reasonably well delimited region (usually close to peak efficiency point) a straight line can be a reasonable approximation. The coefficients B and C are given by the intercept and gradient (Figure 23b,c) and H_s can be obtained by substituting these into Equation (103) evaluated at the equilibrium point under consideration.

CHAPTER 5

Water Use and Public Policy

ANDREW A. DZURIK*

INTRODUCTION

Historical Perspective

Historically water was consumed as an inexhaustible natural resource with little concern for costs, pollution, development of further resource bases, purification or transport of water. In colonial times water for residential use came from wells or other fresh water sources located nearby. Agricultural use of water was minimal; at that period much prime land existed which needed little, if any, irrigation. Irrigation was a more primitive affair—basically the diversion of small streams of water from creeks or rivers bordering the farmer's property. Similarly, industrial use was associated with nearby streams or lakes.

With the population relatively small and concentrated in water-abundant areas and with the existence of plentiful supplies, little need existed for conservation or careful water management. The small concern for water conservation and management was further due to the fact that many residents derived their water from wells they owned. In effect, there was no market since they represented both the producer and consumer of the product.

Even as the need for water in cities grew, little market could be found for such a commodity. Augmented by the fact that water is a necessity for life, human or otherwise, it easily fit under the category of a "public good". Municipalities and private water companies both produced and distributed water. Water was still usually seen as an inexhaustible resource and was priced accordingly. Through time those water-works that were privately owned came to be regulated by government, insuring that their product would be continuously supplied at a sufficient quality level as a necessary public good.

As this country expanded westward, population was introduced into arid and semi-arid areas. Often agricultural activity was impossible or greatly reduced if the water resources were not available. By the 1870s, irrigation through use of large ditches became increasingly prevalent in the West. Water still had a very low market price but it began to be recognized as a resource that did have costs (development of the resource base, transportation), however minimal.

The availability of water is generally taken for granted, especially in water-rich states. Whether for domestic, commercial, industrial, agricultural, or other purposes, it often seems that water can be found in endless supply. It is available in large quantities and in a variety of forms, although not distributed uniformly over time and geographic space. Water managers often say that there is not a shortage of water, but only difficulty in getting the available water cheaply to where it is needed.

Modern society has found a variety of solutions to make water widely available at low cost. Municipalities have spent millions of dollars in developing water supply and distribution systems to satisfy the consumptive requirements of domestic, commercial, and industrial users. With increases in population and urbanization, however, the demand for water begins to place financial burdens on suppliers.

Recent periods of low rainfall and high consumption in different parts of the nation have emphasized the delicate balance between liberal water use and severely constrained availability. Enlargement of water supply infrastructure is a costly venture, and may overburden readily available supplies. On the other hand, underestimating and ignoring increased future consumption can be costly to society in other ways. Water planners must strike an appropriate balance, and thus the estimation of future water requirements becomes a critical factor in planning for urban water supply.

*Department of Civil Engineering, Florida A&M University/
Florida State University, College of Engineering, Tallahassee, FL

Over four trillion gallons of rain and snow fall on the 48 contiguous states on an average day or about 18,000 gallons per person. Of that total, Americans use 450 to 700 billion gallons daily or a little over ten percent of this total. The majority (65%) returns to the atmosphere through evaporation and transpiration, but there are vast underground water supplies and extensive rivers and lakes that make the United States, on the average, a water-rich nation. Why, then, the frequent concern with water availability and the conflicts over the use of water?

Broadly speaking, supply and demand are not in equilibrium with regard to both space and time. Even within relatively small regions, some local areas have a relative abundance of water while others experience shortages. Much public concern focuses on current and projected shortages that result from increasing demands and a relatively fixed supply. At the same time, many regions occasionally face too much water in the form of flooding. The major problem is to assure the availability of adequate water of sufficient quality. Despite increasing demand, the United States Geological Survey (USGS) estimates that overall water supply in the U.S. is more than adequate to meet foreseeable needs, but the water resources must be carefully protected, conserved, and managed.

Purpose

This chapter addresses the important issues in water use, and provides information on public policies which have been adopted to deal with some of the issues. Two broad areas with regard to the physical resource are covered in detail: available water resources (or supply), and uses and requirements for water (or demand).

The policy issues that are discussed relate to demand and supply problems. Much of the focus is on federal law and policy in the United States, for these set the framework within which most water resource issues are addressed. Discussion of broader issues at the international level is also included. It is important to emphasize, however, that state and local laws and policies play an important role in U.S. water resources management and vary considerably from one jurisdiction to another.

Basic Concepts

Several concepts regarding water resources are important for gaining an understanding of the water resource systems within which we operate. The *hydrologic cycle* provides the basis for dealing with water resources by depicting a summary of flows in the earth's water system (see Figure 1). Precipitation falls to earth as rain, snow, or hail and may then follow several paths. Some will evaporate before reaching the ground, but a large portion will infiltrate the earth, and the remainder will enter surface or depression storage. With the accumulation of excess precipitation, water will overflow and move across the earth's surface. This runoff reaches streams, lakes, oceans, and other surface water bodies, much of which returns to the atmosphere through *evaporation*. The water infiltrating the ground enters the soil zone where it may be taken up by the soil and plants, part of which goes back to the atmosphere by the process of *evapotranspiration*. A portion may evaporate directly from the soil surface, or pass into the saturation zone and into *aquifers,* which are strata of sediments and rocks that store and transmit significant amounts of water.

Several concepts regarding water use should be clarified before proceeding to look at actual water use figures. Two common measures of use are *withdrawal* and *consumption.* Withdrawal is the process of taking water from a surface water or ground water source and conveying it to a place for a particular type of use. *Conveyance loss* is water that is lost in transit from a conveyance (pipe, channel, conduit, ditch) by leakage or evaporation and generally is not available for further use. Consumption takes place when water is removed from available supplies by evaporation or transpiration, by manufacturing and agriculture, or for food preparation and drinking. The more common term is *consumptive use* which implies that water is consumed. This concept applies best to a limited geographic scope, for broadly speaking, all water in the earth's system is in relatively fixed supply. It is consumed only in the sense that it is removed from a particular subsystem for a period of time.

Withdrawals and consumptive use apply only to *offstream uses.* In other words, this is where water is withdrawn from a surface or ground water source, and used in another place. *Instream uses* are a clear contrast, wherein the water is used without removing it from its sources for such purposes as navigation and recreation. Discussions of water use typically concentrate on offstream uses.

Safe yield is a concept that applies to the amount of water that can be withdrawn from surface- or ground-water sources. For surface water, especially reservoirs, this term refers to the amount that can be expected to be available, particularly during periods of low flow. The concept of safe yield more generally applies to groundwater. In the past, it was defined as the amount of water that could be withdrawn annually without ultimate depletion of the aquifer. Oftentimes, safe yield is viewed as the draft that results in no change in ground water level [1]. Perhaps the most useful definition of safe yield is "the annual draft of water that can be withdrawn without producing some undesirable results" [2]. It is important to keep in mind, however, that numerous undesirable results can be produced by withdrawals of ground water, and that there is no general agreement on exactly what constitutes safe yield.

The notion of the *water budget* or *water balance model* is helpful in understanding water supply and various forms of consumption, as well as the safe yield concept. The water budget for a specified water system, whether ground, surface, or the complete water system within a geographic area, is a summation of inputs and outputs to the system (see Figure 2). The net result is the change in storage. Some

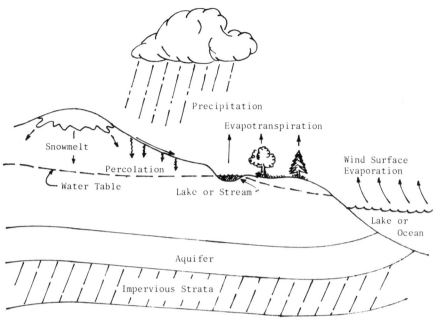

FIGURE 1. The hydrologic cycle.

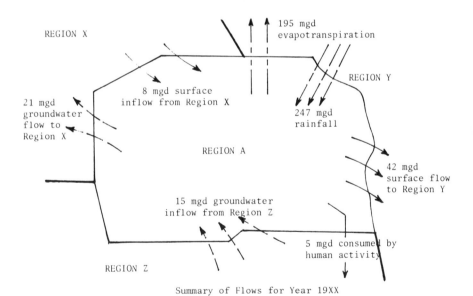

Summary of Flows for Year 19XX

Gains (Inflows)		Losses (Outflows)	
Inflow from X	8 mgd	Evapotranspiration	195 mgd
Inflow from Z	15 mgd	Surface flow	42 mgd
Rainfall	247 mgd	Groundwater flow	21 mgd
		Consumptive uses	5 mgd
TOTAL	270 mgd	TOTAL	263 mgd

Net Gain for Year = 7 mgd

FIGURE 2. Hypothetical water balance model.

years will show losses while other years may remain constant or show a net gain. The water budget concept is particularly useful in regions that show a continual loss, especially in areas where groundwater is being tapped heavily. Reviewing a number of annual water budgets in such a place will indicate the extent of groundwater depletion. Essentially the water budget is similar to a bank book that gives a record of deposits, withdrawals, and net balance.

WATER SOURCES AND SUPPLIES

Water sources and supplies are important in formulating any plans or policies for water use. As stated earlier, water resources may be plentiful on a national scale, but variations over time and among regions cause differing concerns regarding excessive or inadequate quantities at the regional scale. The Second National Water Assessment provides a summary of water supply conditions [3].

Above 40 billion gallons per day (bgd) pass over the

coterminous United States as water vapor, and about 10 percent of this total precipitates as rain, snow, sleet or hail at an average equivalent amount of 30 inches per year. Of the 4,200 bgd that precipitates, about ⅔ is evaporated from wet surfaces or transpired by vegetation back to the atmosphere. The remaining ⅓ or about 1,400 bgd accumulates in surface or ground storage, flows to the oceans or across international boundaries, or is used for consumption by domestic, agricultural, or industrial activities. Only a small percentage of this 1400 bgd can be actively used.

Approximately ⅔ of the combined fresh water (335 bgd) and saline water (57 bgd) withdrawals in the coterminous states comes from streamflow and surface storage, and 15 percent (61 bgd) comes from groundwater sources that are highly interactive with stream flows. The remaining sources are saline (15%) and ground water mining (5%).

The availability of surface water supplies from runoff and streamflow varies widely in accordance with variation in precipitation. Regions with wide annual variations in precipitation may show a high annual average, but still suffer

TABLE 1. Regional Streamflow Analysis.

Region	Mean	5%	50%	80%	95%	Mean Natural Runoff*
		Streamflow Analysis Percentage Exceedence				
		(billion gallons per day)				
New England	78.1	10.7.	77.4	62.7	48.3	78.6
Mid-Atlantic	79.2	115.1	77.8	61.2	48.4	81.0
South Atlantic-Gulf	228.0	356.6	219.2	164.1	121.8	232.5
Great Lakes	72.7	103.9	71.7	57.3	44.9	75.3
Ohio	178.0	254.0	178.0	141.0	105.0	180.1
Tennessee	40.8	57.9	40.8	35.9	31.4	41.1
Upper Mississippi	121.0	189.0	121.0	91.8	65.3	135.0
Lower Mississippi	433.0	757.0	433.0	282.0	202.0	455.4
Souris-Red-Rainy	6.0	11.4	5.6	3.4	18.1	6.1
Missouri	44.1	74.3	43.2	29.9	17.6	57.0
Arkansas-White-Red	62.6	120.7	59.1	37.4	21.6	65.2
Texas-Gulf	28.3	62.4	22.9	12.3	6.3	34.0
Rio Grande	1.2	4.4	.6	.3	.2	4.8
Upper Colorado	10.0	15.6	10.0	7.0	3.9	12.4
Lower Colorado	1.6	1.7	1.6	1.4	1.2	3.2
Great Basin	10.5	19.0	9.8	6.7	4.6	13.7
Pacific Northwest	255.3	344.7	254.3	213.3	179.7	266.5
California	48.2	88.7	45.2	30.4	20.0	72.9
Alaska	905.0	1003.0	898.0	795.0	705.0	905.1
Hawaii	6.7	10.3	6.3	4.9	3.8	7.4
Caribbean	4.8	7.1	4.5	3.3	1.6	5.2
Nation (Regions 1*18)	1242.1	1972.6	1219.1	895.1	679.3	1326.9
Nation (Regions 1*21)	2158.7	3020.1	2127.9	1698.3	1389.1	2044.0

*The difference between streamflow and natural runoff can be explained largely by consumptive uses, evaporation, and some deep percolation.
Source: [3]

TABLE 2. Ground Water Supplies—1975.

| Region | Current Withdrawals | | Current Ground Water in Storage | | |
	Total Withdrawals (BGD)	Amount Withdrawn in Excess of Natural or Man's Recharge (BGD)	Total Less Than 2500 Feet Deep (tril. gals.)	Amount That Can Be Feasibly Withdrawn (tril. gals.)	Length of Time to Deplete Half of Storage at Current Rates of Withdrawal* (years)
New England	0.6	0	N/A	N/A	OVER 2000
Mid-Atlantic	2.7	0.03	1,611	414	OVER 2000
South Atlantic-Gulf	5.5	0.3	N/A	4,460	OVER 2000
Great Lakes	1.2	0.03	N/A	261	OVER 2000
Ohio	1.8	0	1,399	383	OVER 2000
Tennessee	0.3	0	N/A	530	OVER 2000
Upper Mississippi	2.4	0	4,638	2,243	OVER 2000
Lower Mississippi	4.8	0.4	2,596	1,272	OVER 2000
Souris-Red-Rainy	0.1	0	703	172	OVER 2000
Missouri	10.4	2.5	1,118	445	235
Arkansas-White-Red	8.8	5.5	664	499	125
Texas-Gulf	7.2	5.6	2,803	1,434	350
Rio Grande	2.3	0.7	16,033	1,874	OVER 2000
Upper Colorado	0.1	0	1,054	78	OVER 2000
Lower Colorado	5.0	2.4	571	N/A	325
Great Basins	1.4	0.6	944	171	390
Pacific Northwest	7.4	0.6	1,377	180	410
California	19.2	2.2	327	82	51
Alaska	0.04	0	N/A	1,122	OVER 2000
Hawaii	0.9	0	N/A	N/A	OVER 2000
Caribbean	0.3	0.01	N/A	16	OVER 2000
Total, Regions 1*21	82.4	20.9	35,838	15,636	N/A

*Subregions may have critical problems that do not show up at the regional level.
Source: [3]

from drought conditions over parts of the year. Annual streamflow variations can be expressed as a probability of flow that is expected to be exceeded in that percentage of years. Table 1 provides a summary of streamflow analysis for the water resources regions of the U.S.

Reservoirs provide another important source of freshwater, with total storage capacity in the nation at about 224,600 billion gallons. About 35 percent of this capacity is for flood control, while the remainder is available for other uses.

Saline water is available for certain purposes in almost unlimited supply from the oceans. Conversion to freshwater, however, requires substantial expenditures and is not typically viewed as being a desirable alternative from a cost perspective.

Ground water is presently being mined, or depleted, at a rate of 21 bgd compared with total ground water withdrawals of 82 bgd. In an average year, approximately 30 percent of the nation's streamflow is supplied by ground water that reaches the surface through springs and seepage. This process has an inverse in that a principal source of ground water recharge is seepage from streams, rivers, canals, and reservoirs. Table 2 gives a summary of groundwater supplies by water resources regions.

The quantity of ground water far exceeds the available water in streams and lakes, but problem areas exist because of substantial ground water mining. Especially notable is the High Plains area which mines over 14 million acre-feet annually. The total ground water estimated to be available in the conterminous states within 2,500 feet of the earth's surface is 100 billion acre-feet, or 35×10^{12} gallons.

Detailed annual hydrologic data for each state are compiled and published by the USGS under the National Water Data System. They are readily available from district offices of the USGS and provide such information as stage, discharge, and water quality of streams; elevation and water quality of lakes and reservoirs; water levels and water quality of wells; and discharge and water quality of springs.

Further information on water resources is provided by the USGS National Water Summary series that was started in 1983 [4,5]. The series not only provides data on surface- and ground-water resources, but also discusses significant water-related events and water-quality trends.

TRENDS IN WATER USE

The USGS has compiled estimates of water withdrawals and uses in the United States every five years since 1950. The estimates were derived from a variety of sources with a wide range of accuracy. In order to improve the availability and quality of data, the Congress directed the USGS in 1977 to establish a National Water Use Information Program. The first major report under this program is the *Estimated Use of Water in the United States in 1980* [6]. The report presents summary information on various aspects of water use.

The estimated offstream use of water in the United States in 1980 was an average of 450 bgd, an increase of 8 percent over the 1975 estimate, and 22 percent more than 1970. In comparison, there was an increase of 37 percent from 1960 to 1970. Figure 3 shows the overall trend of water use in the U.S. from 1900 to 1980, while Figure 4 shows withdrawals by use for the period 1950–1980.

Offstream uses may be aggregated into four major categories: public supply, rural use, irrigation, and self-supplied industrial. Instream use consists primarily of hydroelectric power.

Agricultural Use

Agriculture is the largest user of water nationally. Of the 155.6 bgd used for agriculture in 1980, irrigation accounted for 150 bgd (97%) with the remainder going to rural domestic use and livestock production. A vast portion of the water used in irrigation was utilized in the arid western states, particularly California, Colorado, and Idaho.

Groundwater withdrawals for irrigation account for approximately 40 percent of all water used for irrigation. This heavy dependence on groundwater has led to serious depletion of regional aquifers. The huge Ogallala Aquifer, extending from northern Texas to southern South Dakota, was once thought to be inexhaustible. The aquifer, however, has dropped from an average saturated water thickness of 58 feet in 1930 to a present thickness of about 8 feet as a result of increased withdrawal for irrigation. Serious aquifer depletion is also occuring in the Texas-Gulf and Rio Grande Water Resources Regions.

Surface water depletion from irrigation is also a major concern in the western states. The Colorado River has been reduced to a mere trickle as it enters the Gulf of California as a result of appropriation of its waters by the states through which it flows. Much of this appropriated water is used for irrigation purposes in Central Arizona and Southern California, although a substantial portion is used for municipal supplies. Another example is Pyramid Lake in Nevada which has dropped by as much as 70 feet from its level in 1906, primarily a result of decreased inflow because of the diversion of water from the Truckee River for irrigation.

Demand for water for agricultural purposes is not expected to decline in spite of the depletion of major sources of supply. Agriculture will remain the leading consumptive user of the nation's water supply, with consumptive use of water for irrigation projected to increase from 83 bgd to 93 bgd in the year 2000. Competition for the use of the available ground and surface water supply is expected to be heavy, however, as population and industry continue to expand in arid regions such as the Southwest.

Because of its high withdrawal and consumption needs, the potential for water conservation in agriculture is very high. Supply augmentation is not very feasible in many places because of the economic and environmental constraints involved in creating new reservoirs or diverting

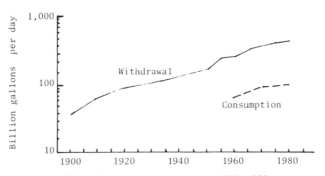

FIGURE 3. U.S. water use trends, 1900–1980.

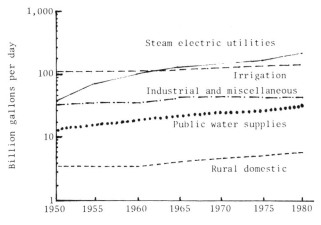

FIGURE 4. U.S. water withdrawals by use, 1950–1980.

water from distant sources (though recharge of groundwater deposits with waste agricultural water may be practical in specific areas). Conservation and reuse of irrigation waters appear more appropriate and would include: development of more productive and/or salt-tolerant crop species; increased use of chemical fertilizers, pesticides, and herbicides; lining of irrigation canals or the use of pipelines to prevent seepage loss; "trickle" or "drip" irrigation practices to reduce the quantity of water necessary for continued crop yield; and the reuse of municipal and/or agricultural wastewater for direct irrigation or for recharge of groundwater deposits. All of these practices involve conscientious management of irrigation water, and most require either significant economic expenditures or changes in traditional methods of crop management. The savings in the long-run, in terms of water supply and capital expenditures, may be worth the effort.

Industrial Use

Self-supplied industrial uses of water, excluding thermoelectric power plants, accounted for 45.3 bgd of the total water withdrawn in 1980. Of this total, 29 bgd was withdrawn from freshwater surface sources, 10 bgd from fresh ground water, and 6.3 bgd from saline water sources. A portion of industrial supplies are obtained from public water systems discussed in the following section.

The above figures do not provide the total picture of manufacturing use of water. Total gross water utilized by industries amounts to 2.2 times the amount withdrawn. This difference is accounted for by the fact that water withdrawn for industrial uses is recirculated extensively in the manufacturing process [3].

Demand projections for the industrial sector do not indicate an increase in the need for water. The WRC Second Assessment states that "combined fresh- and saline-water withdrawals are . . . projected to decrease from the present

60.9 bgd to about 29.1 in the year 2000" [3]. However, consumption in manufacturing processes in all water resources regions is expected to increase by the year 2000, the result of the perfection and widespread use of recycling and recirculation practices as well as a projected increase in manufacturing production.

According to the American Society of Civil Engineers' Task Committee on Water Conservation, federal water quality regulations regarding wastewater discharges have been the primary impetus for reduction of industrial demand [7]. This reduction in demand for water withdrawals has been brought about mainly through water reuse, recycling of internal wastewater, and water use reduction measures. Total industrial intake of water was reduced by 36 percent over the 1954 to 1968 period, much as a result of the implementation of these conservation practices.

The reuse and recirculation of wastewater is restricted by the economic feasibility of reuse (i.e., whether it is cheaper to recycle or to purchase from a public supply) and the quality of the recycled water for industrial processes. Overall, however, the outlook for continued conservation of industrial water supply appears promising.

Municipal Use

In 1980, 34 bgd of water was withdrawn for municipal use, with approximately 7.1 bgd of this amount being consumed. Domestic and public use accounted for 22 bgd, whereas industrial and commercial uses amounted to 12 bgd. Groundwater sources supplied approximately 35 percent of the total, while surface-water sources supplied the remainder.

Demand projections for municipal water use have been relatively consistent with projections for population increase. Though overall municipal withdrawal is expected to increase by 32 percent by the year 2000, certain regions, particularly Florida, California, Texas and Arizona, will experience a higher growth in demand for water than the rest of the nation because of increased population growth in those areas. Expansion of water supply sources in these and other regions, however, will be inhibited by increasing costs of expansion and competition among different users for available local and distant supplies. Municipal water conservation, then, presents itself as a major factor in future municipal water supplies.

Conservation of municipal water supplies can be attempted in a number of ways. First is maintenance and repair of existing facilities. Antiquated and deteriorated water supply systems are great wasters of water, yet with proper repair, they can continue to provide sufficient quantities of water for domestic and commercial use. Costs are high, however, for such maintenance and repair, thus illustrating the need for other conservation measures.

Residential water conservation methods can contribute significantly to overall municipal water conservation. Certain economic measures, such as water pricing policies, can be utilized to reduce the demand for water. Related to this is the need for metering water use. Water use restrictions (e.g., fines for illegal water use during times of drought, restrictions on outside water use, etc.) can be effective in reducing water consumption during emergency shortages or peak demand periods.

PLANNING FOR FUTURE WATER USE

Planning for future water supply must start with identifying uses and estimating demand for these uses. Estimating demand relationships is an important step in water resources planning and management. This section provides an overview of the major considerations in dealing with forecasting water demand and supply.

Demand

Most water supply agencies use relatively simple methods for determining the future demands on an area's water supply capabilities. Common approaches for projecting municipal demand focus on population size and number of households, while industrial demand forecasting may rely upon number of employees, and agricultural demand may relate to crop type and acreage. Such variables are used in simple mathematical calculations to project future demand, and to develop water supply management strategies. The realization has grown, however, that other less salient factors need to be examined in water use analysis.

Water use forecasts in most cases are long-range, for up to 50 years, and typically measure average daily use. This is because water use projections are usually used to plan major facilities, and these facilities are typically large projects such as dams, reservoirs and treatment plants. Short-range projections may be made for smaller facilities or selected

management strategies, and may deal with variations in water use by season, month, or week.

Even though most agencies use simple approaches, the available forecasting techniques are varied. The following sections describe the more important techniques. They are reviewed more fully in a report by the Institute for Water Resources, Corps of Engineers [8], and are explained in various textbooks on water supply.

TIME EXTRAPOLATION

This technique considers only past water use records, and extrapolates into the future using graphical or mathematical methods. It assumes a continuation of past trends over time, but may use a variety of functional forms such as linear, exponential, and logarithmic. Time and water use are the only variables considered in this technique. It is not highly reliable, especially for long-term projections.

SINGLE COEFFICIENT METHODS

The most commonly used technique estimates future water demand as a product of service area population and per capita water use. The per capita water use coefficient may be assumed constant, or it may be projected to increase (or decrease) over time. Population projections may be obtained for the service area through original work, from local sources such as local planning departments, or from higher level sources such as state agency projections, or federal level OBERS projections. The OBERS projections (U.S. Department of Commerce, Bureau of Economic Analysis/ U.S. Department of Agriculture, Economic Research Service) are used as standard practice in federal level water resources planning, and are also appropriate at lower levels. The per capita approach is usually applied to municipal water use. Many studies have shown population to be a reliable indicator of water use. The method may be refined by using separate per capita coefficients for different use categories such as residential, commercial, public, and industrial. The coefficients can also be disaggregated by geographic area and by season.

A variation of the per capita approach is to use the number of customers of the system within the study area. Single coefficient methods may also be applied to industrial use. For example, water use per employee may be used with projections of employment in the industrial sector. Such employment projections are often available from local or regional planning offices. Commercial forecasts may be done in terms of water use per employee, or per square foot. Similarly, agricultural forecasts may be done by projecting land area in specific crops and multiplying by irrigation requirements per unit of area for each crop type. In all cases, the single coefficient method relies entirely on projections of a key variable, and assumptions regarding future water use as a function of that variable. The method is reasonably reliable for short-range forecasts, but becomes increasingly questionable for long-term projections.

MULTIPLE COEFFICIENT METHODS

This approach defines future water use as a function of two or more variables associated with water use. Regression equations are typically developed as the statistical technique for estimating the relevant coefficients. This may be based on historic time series data for the study area, or on cross section data from a number of similar areas. To forecast water use, future values of the independent variables must be determined by other means. For example, average water use in a region (WU) may be specified as a function of the number of employees in manufacturing (Em) and number of households (H):

$$WU = a + b_1 Em + b_2 H + e$$

The values of a, b_1 and b_2 are parameters estimated statistically from the data set, while e is an "error term" to account for the unexplained variation in WU. To estimate the equation using time-series, historic data would be applied using regression analysis techniques. Future water use would then be determined by calculating the estimated equation with projected values of Em and H for the forecast period.

DEMAND MODELS

The term "water demand" is frequently used in water use projections. In economics, demand is a function of price, whereby a price increase is associated with a demand decrease. Water use projections in the absence of price considerations are sometimes termed "water requirements" by water resource economists [9]. If we consider price (P) and demand (Q) for water with other factors held constant, the "price elasticity of demand" can be defined as $-(dQ/Q)/(dP/P)$. Thus, if a price increase of 100% results in a demand decrease of 30%, the elasticity is -0.3. In economic theory, the elasticity is less for necessities than for luxuries; i.e., demand for a necessary item will be less influenced by price than will a luxury item. This concept is important in understanding the relative value of water and of pricing policies for water supply.

Water demand models are a subcategory of the multiple coefficient methods described above. The primary difference is that the price of water is included as an explanatory variable, and some measure of personal income is often included. One of the most significant studies of this type was by Linaweaver, et al. [10]. Their model showed that the most important variables in residential water use were climatic factors, economic levels of consumers, irrigable lawn areas, and number of homes. Howe and Linaweaver [11] demonstrated the potential effects of price on water demand by showing an average annual weighted price elasticity of -0.4 in a study of 21 metered areas; i.e., demand would drop by 40% if the unit price were doubled. A study by Kindler and Bower [12] showed that water demand relationships can be analyzed at four levels: national, regional, aggregated local, and individual. Each level is the aggregate of demand from

the next lower level. The primary advantage of demand models is that they contain more complete sets of explanatory variables and can reflect important policy considerations with respect to price. Substantial price increases may actually reflect changes in water supply costs as well as deliberate pricing policy.

PROBABILISTIC ANALYSIS

Models using regression analysis to explain variations in water use also include the "error term" (e) to account for unexplained variation. If the remaining variance is random and not accounted for by other variables, water use then can be identified as a stochastic process [8]. Probabilistic analysis includes not only independant variables to estimate future use, but also a probability distribution of that estimate. Probabilistic methods have not been highly successful in actual practice.

Supply

The supply side of water use forecasting requires information on the sources available to meet projected demand and the amount that can be provided from these sources, as well as the costs and environmental effects. In most places, water is available from ground water or surface water, or both. A few places have used desalinization plants to provide water, but the comparative costs are usually much higher than for other alternatives in most places.

Groundwater supplies are plentiful in many places and are frequently used for municipal, industrial and agricultural purposes. The significant information required is the depth to aquifers and the amount of water that can be withdrawn without impairing the quality or quantity of water in the aquifer, or safe yield. In many cases, ground water has been withdrawn to such an extent that a significant lowering of the water level and contamination by salt water intrusion has resulted.

Surface water provides naturally abundant supplies where lakes and rivers have sufficient capacity to meet demands. In such places, plans must be developed only for providing the necessary facilities for treatment, storage and distribution of water. In many cases, however, reservoirs are needed to augment natural systems in periods of low flow, and to regulate the distribution of surface water flows and volumes [13]. Essentially, reservoirs are for temporary storage of water over relatively long periods of dry weather and low flow. Many reservoirs, however, serve uses other than water supply, such as flood damage reduction, hydroelectric generation, and water-based recreational activity. There are three major elements of reservoir storage-capacity requirements: (1) active storage for firm and secondary yields, (2) Dead storage for sediment collection, hydropower production, and recreational purposes, and (3) flood storage for reduction of downstream flood damages [14]. Firm or safe yield can be defined as the maximum amount of water available from a

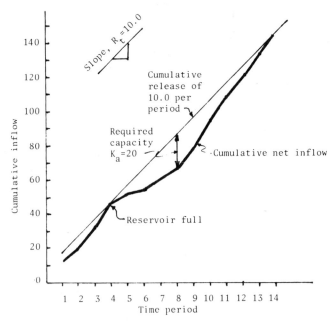

FIGURE 5a. Mass diagram.

reservoir based on historical streamflow records, whereas secondary yield is the amount greater than firm yield [14]. Dead storage is the amount available for purposes other than water supply.

MASS DIAGRAM ANALYSIS

A relatively simple method for determining reservoir storage requirements is a mass diagram, based on the assumption that past flows will be repeated in the future. The curve shows total cumulative inflow to a stream at the point of a proposed reservoir plotted against time. To determine required capacity, we find the maximum difference between cumulative inflows and cumulative demand. The example in Figure 5(a) is based on data in Table 3, showing cumulative monthly flow and assuming that average demand or required release per time period (R_t) is 10.0 units. The difference between total inflow and release is the quantity needed to meet demand. Demand is drawn as a sloped line and placed tan-

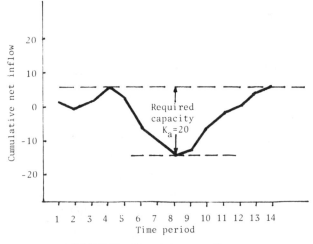

FIGURE 5b. Cumulative mass diagram.

TABLE 3. Calculation of Storage Capacity by the Sequent Peak Procedure.

Period t	Required Release (R_t)	Inflow (Q_t)	Previous Rqd. Capacity (K_{t-1})	Current Rqd. Capacity $(K_t^* = R_t - Q_t + K_{t-1})$
1	10.0	12	0	0
2	10.0	8	0	2
3	10.0	12	2	0
4	10.0	14	0	0
5	10.0	6	0	4
6	10.0	2	4	12
7	10.0	6	12	16
8	10.0	6	16	20**
9	10.0	12	20	18
10	10.0	16	18	12
11	10.0	14	12	8
12	10.0	12	8	6
13	10.0	14	6	2
14	10.0	12	2	0

* greater than or equal to zero
** maximum required storage capacity = 20

gent to the cumulative inflow curve. In months in which stream flow is less than demand, the demand slope is greater than the supply slope and the reservoir must make up the deficit. The maximum demand or required capacity (K_a) needed for this particular period of analysis is 20 units. The graphic approach can be done readily if demand in each time period is the same. If demand varies, the cumulative differences between inflow and demand can be plotted as in Figure 5(b). The maximum vertical distance between the highest point of the cumulative difference curve and the lowest point to its right represents the required capacity.

In order to determine storage capacity requirements over time, an estimate of the mean probability of unregulated streamflow makes it possible to define the probability of any particular reservoir yield. Thus, if we had a fifty-year record of streamflow, and the required reservoir capacity for each of those years, a frequency analysis of the record would allow selection based on recurrence intervals. Designing for the lowest flow of record in this case would be a 50-year recurrence interval, or two percent drought.

SEQUENT PEAK METHOD

The mass diagram method is still used often, but a more manageable modification is the sequent peak procedure [13]. Storage capacity required at the beginning of period t is K_t, and R_t is the required release in period t. Inflow is represented by Q_t. If we set $K_o = 0$, the procedure calls for calculating K_t for up to twice the total length of record according to the following:

$$\text{use } K_t = R_t - Q_t + K_{t-1} \quad \text{if positive}$$

$$\text{use } K_t = 0 \quad \text{if } R_t - Q_t + K_{t-1} \leq 0$$

If the critical sequence of flows occurs at the end of the record, this method assumes that the record repeats. The maximum of all K_t is the storage capacity needed to provide the required release, R_t. The method is demonstrated in Table 3 with data for 14 periods.

OPTIMIZATION MODELS

The mass diagram and sequent peak methods use few variables and cannot incorporate such considerations as evaporation losses, lake level regulation, or multiple reservoir systems. Mathematical models have been developed for such purposes, based on mass-balance equations which explicitly define inflows, outflows, and net storage.

A notable example of modeling for water supply is the Potomac River Basin model for the Washington, D.C. metropolitan area. Developed in the late 1970s after years of concern about water supply for the growing metropolitan area, the modeling approach has a number of important features, including the following [1]:

- combines optimization and simulation models to provide operating rules for the water supply system
- makes extensive use of the National Weather Service River Forecast System
- uses a technique for predicting water demand and applies the technique to system design and operation
- combines distribution analysis and hydrologic modeling for operating procedures for a complex, multiple water distribution system
- uses risk analysis to identify the start of potential droughts

Implementation of this systems analysis approach resulted

in a predominantly non-structural approach to a major water supply problem. Compared to other alternatives potential savings of $200 million to $1 billion were achieved [1].

WATER LAW AND PUBLIC POLICY

Framework of Water Law

Water law applicable to surface water generally has followed the *riparian rights* system in the eastern United States, while the western states, where water supply is more limited, have followed the *prior appropriation* system. Water law applicable to ground water has followed several different doctrines. Although laws are evolving among states to address contemporary issues, the two basic doctrines provide the framework for regulating water use. Beyond these basic concepts is a large set of laws relevant to water use and supply at all levels of government.

RIPARIAN RIGHTS

The system of riparian rights stems from English law under which the right to use lakes and streams was given to owners of riparian land, or land that borders on surface water. Use of the water was expected to be for beneficial purposes, and not to interfere with the use of water by downstream land owners. "Reasonable use" was established in the American adaptation to overcome what were viewed as overly restrictive provisions. The reasonable use doctrine allows riparian land owners to alter the quantity or quality of water within reason if the water is used for the purpose to which the land was dedicated [1]. Reasonable use is often subject to a balancing test by the courts in cases of conflict between riparian land owners. In theory, the standard of relative reasonableness facilitates a resolution of conflicts between uses based on the needs of each user and the general public interest [15].

Riparian rights generally go along with the purchase of riparian land, although deed provisions may specify that the water rights are retained by the seller. As water use continues to increase and riparian lands become more developed, many states have modified their water laws to incorporate permit systems for water use. The permits may regulate the amount of water withdrawn, the period of time when withdrawals are made, and the location of withdrawals. Whereas the basic riparian system grants equal rights to water use, the permit system may establish priorities for periods of limited supply.

PRIOR APPROPRIATION

The system of prior appropriation became the predominant water policy for the 17 western states in the mid-1800s. It evolved in a frontier society in an arid climate to meet the needs of the time. The system follows the basic concept that the first user has the superior right or priority in water use.

Thus, water rights are allocated on a time basis rather than in the public interest. The prior appropriation method of establishing rights and general rules for resolving disputes is not a fixed doctrine or law, but rather a general philosophy that the western states have followed in establishing water laws particular to their own situations.

Rights to water are generally obtained by applying to the appropriate state agency, and the date of approval becomes the significant date by which priority is established. A permit mechanism is used to grant rights to unappropriated water (if available) if not in conflict with existing users, and if the proposed use does not conflict with the public interest. The appropriation doctrine also follows a beneficial use determination in most western states. A beneficial use has been defined historically as having a clear economic value, such as domestic, agricultural, industrial, and municipal uses. Once a water right has been obtained, this right may be lost through nonuse. The right usually can be sold, but a new permit may be required.

Although the prior appropriation system may encourage the waste of water, it has several advantages over the riparian rights system. It offers flexibility with regard to place of use, certainty in the amount of use, elimination of unused rights, and more certainty in administration [15]. The absence of riparian restrictions allows for use of water where it will provide the most benefit to the individual user and to the public. Further, the system emphasizes beneficial use.

GROUNDWATER

Groundwater rights are distinct from the two surface water doctrines described above. They follow one of four doctrines dealing with groundwater rights: the absolute ownership doctrine, the reasonable use doctrine, the correlative rights doctrine, and the appropriation rights doctrine.

The absolute ownership doctrine treats percolating waters as part of the land in which they are found and belonging absolutely to the owners of the land. They can capture or divert this water for any purpose, and they are not liable for loss or injury to other landowners as long as the water is not used in a malicious manner. Furthermore, the owner is not restricted to using the water on the overlying land and the liability does not increase by using water away from the premises.

The reasonable use doctrine rectifies some of the problems with absolute ownership. The freedom from liability is possible only when the water use is reasonable with respect to the needs of the owner's land and will not extend to uses on other land.

The correlative rights doctrine is tied to the proportionate sharing of groundwater withdrawals among land owners. In other words, the rights to percolating water are proportional to the overlying land ownership. A person may not take more than the proportional share when the rights of other landowners will suffer.

The appropriation rights doctrine is similar to that for surface water. A permit is obtained from a state agency which sets a priority based on time. Most western states follow this procedure for groundwater, except Texas which follows the absolute ownership doctrine.

The entire system of water rights is in flux as increasing demands for water arise. There is wide variation among the states, and much debate over water rights and the most appropriate form of water law. It appears that the trend is toward a combination of the systems, with permits becoming a more common mechanism for administering water rights.

Legislation

The administration of water resources law and policy is divided among federal, state, and local governments. The federal government administers national water policies and legislation through various departments, councils, and commissions, and states typically interpret and apply these policies and laws to water within their boundaries. Further, the states enact legislation for the management of their water resources. Local governments enact their own water use and supply regulations in accordance with state and federal standards.

FEDERAL

Water resources legislation on a national scale has been extensive over the last thirty years and has strongly influenced the direction of water resources planning and engineering. The more recent legislation includes the Water Resources Planning Act (1965), the Federal Water Pollution Control Act Amendments (1972), the Safe Drinking Water Act (1974), the Resources Conservation and Recovery Act (1976), and the Clean Water Act (1977).

The Water Resources Planning Act established Congress' policies toward "the conservation, development, and utilization of water and related land resources on a comprehensive and coordinated basis by the Federal Government, states, localities, and private enterprise" [16]. The Act created the former Water Resources Council (WRC) whose main functions were to identify the nation's water resources and needs, and to oversee the implementation of federal water management plans by the various national river basin commissions (which could be established by the WRC or by appropriate states).

The Federal Water Pollution Control Act Amendments of 1972 are not concerned as much with water quantity as with water quality. The main directive of the Amendments was the elimination of pollutants into navigable waters by 1985, with interim goals of "fishable, swimmable" waters by 1983. Reduction and elimination of discharges were to be accomplished through federal financial assistance for the construction of wastewater treatment works and wastewater treatment management planning.

The Safe Drinking Water Act was a response to studies

starting in 1970 dealing with the deterioration of the nation's water supplies. The Act requires EPA to set national drinking water quality standards, with the states being responsible for enforcing those standards. EPA's National Interim Primary Drinking-Water Regulations and National Secondary Drinking-Water Regulations are summarized in Tables 4 and 5. Primary standards require public water systems to maintain a level of quality to protect public health and are legally enforceable, whereas secondary standards are to maintain satisfactory taste, odor and appearance and are intended to be guidelines for the states but are not legally enforceable. The Act establishes a program of state regulation and enforcement for protection of underground sources of drinking water.

The Resources Conservation and Recovery Act (RCRA) deals primarily with hazardous wastes, but is concerned

TABLE 4. National Interim Primary Drinking-Water Regulations.

Constituent	Maximum concentration*
Arsenic	0.05
Barium	1
Cadmium	0.010
Chromium	0.05
Lead	0.05
Mercury	0.002
Nitrate (as N)	10
Selenium	0.01
Silver	0.05
Fluoride	1.4–2.4
Turbidity	1–5 tu
Coliform bacteria	1/100 mL (mean)
Endrin	0.0002
Lindane	0.004
Methoxychlor	0.1
Toxaphene	0.005
2,4-D	0.1
2,4,5-TP Silvex	0.01
Total trihalomethanes [the sum of the concentrations of bromodichloromethane, dibromochloromethane, tribromomethane (bromoform) and trichloromethane (chloroform)]	0.10
Radionuclides:	
Radium 226 and 228 (combined)	5 pCi/L
Gross alpha particle activity	15 pCi/L
Gross beta particle activity	4 mrem/yr

*Data are given in milligrams per liter (mg/L) unless otherwise specified; mL = milliliters, tu = turbidity, pCi/L = picocurie per liter, mrem = millirem (one thousandth of a rem)

Data from U.S. Environmental Protection Agency, 1982, Maximum contaminant levels (subpart B of part 141, National interim primary drinking-water regulations): U.S. Code of Federal Regulations, Title 40, Parts 100 to 149, revised as of July 1, 1982 p. 315–318.

TABLE 5. National Secondary Drinking-Water Regulations.

Constituent	Maximum level*
Chloride	250
Color	15 color units
Copper	1
Corrosivity	Noncorrosive
Dissolved solids	500
Foaming agents	0.5
Iron	0.3
Manganese	0.05
Odor	3 (threshold odor number)
pH	6.5–8.5 units
Sulfate	250
Zinc	5

*Data are given in milligrams per liter (mg/L) unless otherwise specified.
Data from U.S. Environmental Protection Agency, 1982, Secondary maximum contaminant levels (section 143.3 of part 143, National secondary drinking-water regulations): U.S. Code of Federal Regulations, Title 40, Parts 100 to 149, revised as of July 1, 1982 p. 374.

with their effects upon water quality. The Act's main application is in regulating hazardous materials and preventing their introduction into the environment, particularly surface and ground water. The Act touches on water supply and wastewater treatment in that sludge from treatment plants can be included as a hazardous waste category. Waste materials from treatment plants other than sludge can also be regulated under the RCRA as hazardous wastes. Current penalties for non-compliance are quite strict and could have a definite impact on the future handling of wastes by water supply treatment facilities.

The Clean Water Act of 1977 consists of further amendments to the Federal Water Pollution Control Act. Broadly, the Clean Water Act can be viewed as the cumulative Federal Water Pollution Control Act because it amends earlier Acts and focuses on federal assistance for construction of public wastewater treatment plants and on coordination between federal, state, and local water agencies. The Act directs state and local authorities to develop area-wide wastewater management plans and to establish management agencies to implement the plans (specified in Section 208 of the Act).

It is safe to conclude that there is no single federal law or policy dealing with water resources and water use, but a number of individual acts plus extensive legislation by state and local governments.

STATE

Within the federal system, primary responsibility for regulation and control of water use generally lies with state government. States have initiated many of their water resources laws, policies, and regulations in coordination with

and as a result of federal water policies. State water resources legislation has long been influenced by federal statutes such as the federal reserved right to water use. Because of the predominance of the prior appropriations doctrine of water use in western water law, the western states have had to consider the reserved rights doctrine because of the possibility that an existing state water right may become subordinate to a federal reserved right [17]. Nevertheless, states have enacted legislation concerning the use and regulation of water within their boundaries in accordance with the laws of prior appropriation, riparian rights, or a water use permit system [18].

Aside from statutory regulation of water quantity and quality, state water management is usually accomplished by agencies which administer water rights by permit authorization or which administer and oversee water pollution regulations. In addition, there are a number of federal-state commissions, committees, or councils which coordinate water resources planning and management. Examples would include Federal-State Compact Commissions, which carry executive powers, and the Basin Inter-Agency Committees, composed of representatives from various state and federal water resources agencies. [17] Regional river basin commissions, such as the New England River Basin Commission, were created to plan and coordinate water resources activities and to integrate all three governmental levels of water management.

Four types of state legislation relate primarily to water supply distribution: (1) public utility acts, which are usually administered by public utility commissions and which set water service standards; (2) state water supply statutes, which are generally health and safety standards similar to federal enactments; (3) environmental statutes, usually similar to federal regulations and environmental impact assessments; and (4) water supply agreements, which are implemented primarily through local water authorities and usually applied in western states. The amount and degree of enforcement of these types of regulation varies from state to state.

Legislative authority for state water resource planning is similarly diverse. Two states (Delaware and Florida) require state-wide comprehensive water resources planning and management under the direction of a single state agency. Other states require either continuous comprehensive water planning (14 states), static comprehensive planning (7 states), or continuous comprehensive planning with a static water plan (4 states) [19].

LOCAL

Local water resources ordinances are usually implemented through municipal and county water authorities or districts, and deal primarily with drainage, water supply, or wastewater treatment. Much local water management is a result of federal and state delegation of powers. For example, Intra-State Special Districts are water management bodies that are "local units of government established by state law for planning, constructing, and ensuring the maintenance of local works" [16]. Most municipalities have their own water treatment or management authorities and most areas implement some type of water supply agreement to ensure provision of sufficient quantities of water.

Public Policy Issues

A 1983 report by the Conservation Foundation reviewed over two dozen national water policy studies that had been completed over the prior two decades [20]. The primary concerns were to identify the importance and consistency of issues among the many studies; to measure their long-term effects on water policy at all levels of government; and to evaluate their relevance to a contemporary water policy agenda. The Conservation Foundation report is an excellent summary of nationally significant water policy studies. A revealing commentary is this introductory statement:

> No subject in the field of natural resources has been studied so exhaustively and authoritatively, and yet been the subject of such vehement and continuing controversy as water policy. Since 1968 two full-fledged national commissions, two multi-year government assessments, several major national conferences, and a three-year presidential task force have focused on the issue, along with numerous other comprehensive studies. Despite this outpouring of impressive quantities of analysis, evaluation, and paper, we seem no closer today than 15 years ago to a "national water policy" (if such a beast is needed or even can be identified) [20].

Similarly, the Task Committee on Federal Policies in Water Resources Planning, American Society of Civil Engineers, reported that there is a myriad of organizations at all levels of government and in the private sector that are involved in managing the nation's water resources. In this organizational context, there is little clear definition of the nature and extent of federal interest. "There is no national strategy which (a) sets forth minimum goals to be met, (b) provides adequate definition of the responsibility to be exercised at different governmental levels and (c) provides adequate intergovernmental involvement" [21].

Although there is no single national water policy, there are a number of different water studies and issues identified by the Conservation Foundation [20] relative to water use and which show trends having implications for public policy:

Offstream Uses

- Municipal Supply: concern has shifted from the sufficiency of water for urban growth, to financing and rehabilitating older urban water systems, and to fiscal capabilities and responsibilities among levels of government.
- Industrial Supply: industrial withdrawals of fresh water are expected to continue to decline as recycling and use of lower quality and saline water increases.
- Irrigation: federal expansion of irrigation water supplies should be at an end, with attention shifting to problems of erosion, return flows, and possible socio-economic impacts of large-scale transfers of water out of agriculture.
- Electric Power Cooling/Energy Production: projections of water requirements for cooling and for synfuel plants have not materialized, allowing states a chance to review legal and administrative reforms for future periods of energy development.

Instream Uses

- Water Transport/Navigation: emphasis on navigation improvements has shifted to least-cost additions to a national transportation network with cost-sharing by beneficiaries.
- Hydropower: attention to hydropower has diminished as most suitable sites have already been developed, but an increasing interest is being shown in "low-head" hydro facilities.
- Fish and Wildlife/Recreation: increasing attention has been given to maintenance of adequate flow for fish, wildlife, and recreational uses.
- Flood Control/Floodplain Management: a clear shift has been made toward non-structural solutions, but the activity level of state and local governments in this regard is unclear.

Institutional Issues

- Federal-State Roles: a consistent preference is shown for increased state responsibility and authority, but questions of shared federal-state financial responsibility, state-local conflicts, and coordinated planning are important issues.
- General Planning Approaches: comprehensive, basin-wide planning and management are essential to dealing with water problems.

Another important discussion of water resource issues is given in the Conservation Foundation book on *State of the Environment—1984* [22]. The chapter on water resources provides considerable details on issues, conflicts, and policies regarding basic values pertaining to water. The conflicts in values apply to both water quantity and quality, but particular emphasis is given to conflicts in water use. The chapter concludes by noting that the U.S. has entered a period of fundamental change with regard to water policy, from a period of water development to one of water management.

To summarize the various studies, there have been a number of important trends in water use and public policy over the past two decades that have important implications for present and future water resources planning and management. New legislation and institutional reforms and new water management approaches at all levels are needed to address water policy problems such as cost-sharing, water transfers, inter-jurisdictional conflicts, and water quality-quantity issues. There will be an increasing concern for cost-effectiveness of new water projects, and a growing adoption of user fees.

Another issue that is expected to become increasingly significant is the planning crisis for small water users [23]. Whereas much financial and advisory help is available for planning large projects, the opposite situation exists for small projects. Emphasis will be increasingly placed on smaller water projects, plus maintenance and rehabilitation of existing projects. Therefore, smaller scale planning will become more important, and the need for information and financing from federal agencies will be significant to small water users.

In an effort to address many of the current water policy issues, the National Water Alliance was formed in 1983 as a broadly-based bi-partisan organization. Among the leadership of the Alliance are U.S. senators and congressmen from both parties, representing diverse geographic areas. The Alliance may play an important future role in bringing together political parties, urban and rural areas, and private sector participants to deal with water issues. It may also help to pull together the numerous federal agencies and private organizations that focus on water resources. Among the many issues already considered by the Alliance are the complex issues and sheer volume of information associated with water resources. The issues become further complicated because they are diffuse and involve many different perspectives. At one of its symposiums, the National Water Alliance dealt with the need for a national water policy, and with the multitude of public and private agencies and organizations involved with water resources. It also considered the general inadequacy of techniques for conflict resolution, and the need for an institutional mechanism to mediate water conflicts [24].

At the international level, water resources issues are becoming of increasing concern. For example, the Worldwatch Institute [25] reported that global trends in overuse and mismanagement of water threaten surface water quality and ground water supplies that could lead to shortages restricting food production and economic growth. The study cited the depletion of the Ogallala Aquifer on the high plain of the western United States as an example of mismanagement. It went on further to state that excessive withdrawals of ground and surface water threaten supplies worldwide. The author argued that governments are depending on large dams and river diversions while ignoring water management problems.

One focus of international attention has been the United Nation's International Drinking Water Supply and Sanitation Decade (1981–1990). A report released by the UN Secretary General in April 1985 indicated that as the decade reaches its halfway mark, an additional 530 million people will have received reasonable access to safe drinking water [26]. However, about 1.2 billion people will still be lacking a safe water supply, and 1.9 billion will be without adequate sanitation. Significant constraints include funding limitations, lack of trained personnel, and inadequate cost-recovery policies in most countries. The report emphasized the importance of using least-cost technologies.

Future worldwide water use and water supply issues came under investigation in *Global 2000* [27]. The various projections in the study (e.g., population, GNP) imply rapidly increasing freshwater demands, with increases in world water withdrawals of at least 200–300 percent expected between the years 1975 and 2000 (see Table 6). The largest increase, by far, is expected to be for irrigation, which already accounted for 70 percent of water uses in 1967. Much of the increased demand is expected to be in the developing countries, many of which are short in freshwater supplies currently. The study recognizes that the global supply of freshwater is large relative to current and projected demand, but the extreme seasonal and geographical variations in the distribution of water resources causes serious water shortages in many nations and regions, and conflicts among economic sectors and among nations drawing water from the same sources are anticipated.

TABLE 6. Projected Global Demand for Water by the Year 2000.

	Projected Demand (Km 3/yr)	
	Withdrawn	Consumed[a]
Irrigation	7,000	4,800
Domestic	600	100
Industrial	1,700	170
Waste dilution	9,000	
Other	400	400
Total	18,700	5,470

[a]Not returned to streams or rivers
Source: [27]

SUMMARY AND CONCLUSIONS

This chapter has provided an introduction to water use concepts, and an overview of water sources and supplies. It covered trends in water use by major sectors, and gave a summary of techniques for forecasting water demand and determining supply requirements. Finally, water law and public policy was discussed as it relates to various water use issues.

Water use and public policy are broad areas that touch on many aspects of water resources planning and engineering. As demand continues to increase for most uses of water, new approaches will have to be tried in order to find appropriate solutions. The nation and the world as a whole are, for the most part, beyond the old approach of providing more water to satisfy growing demands. New management techniques, innovative engineering including computerized modeling and systems analysis, greater emphasis on rational pricing policies, and increasing reliance on conservation and wise resource management are essential to meet future water needs.

GLOSSARY

Aquifer An underground bed of porous rock or soil that carries or holds water.

Artesian (aquifer or well) Water held under pressure in porous rock or soil confined by impermeable geologic formations. An artesian well is free-flowing.

Biochemical Oxygen Demand The amount of oxygen consumed by microorganisms (mainly bacteria) and by chemical reactions in the biodegradation process (BOD).

Brackish Mixed fresh and salt waters.

Consumptive Use Water removed from available supplies without direct return to a water resource system by use such as manufacturing, agriculture, and food preparation.

Conveyance Loss Water lost in conveyance (pipe, channel, conduit, ditch) by leakage or evaporation.

Dissolved Oxygen A measure of water quality indicating free oxygen dissolved in water.

Drainage Basin The area of land that drains water, sediment, and dissolved materials to a common outlet at some point along a stream channel.

Evaporation The process whereby water from land areas, bodies of water, and all other "moist" surfaces is absorbed into the atmosphere as a vapor.

Evapotranspiration The combined processes of evaporation and transpiration. It can be defined as the sum of water used by vegetation and water lost by evaporation.

Groundwater All water beneath the surface of the ground (whether in defined channels or not).

Hydrograph A graph of the rate of runoff plotted against time for a point on a channel.

Hydrologic Cycle Movement or exchange of water between the atmosphere and the earth.

Hydrology The study of the occurrence and distribution of the natural waters of the earth.

Infiltration Movement of water into the soil.

Infiltration Rate Quantity of water (usually measured in inches) that will enter a particular soil per unit time (usually one hour).

Instream Uses Water uses without removing it from its source, as in navigation and recreation.

Offstream Uses Water withdrawn from surface- or groundwater sources for use at another place.

Percolation The slow seepage of water into and through the ground.

Permeability Generally used to refer to the ability of rock or soil to transmit water.

Potable Water Safe and satisfactory drinking water.

Potentiometric Surface The level to which water will rise in cased wells or other cased excavations into aquifers, measured as feet above mean sea level.

Prior Appropriation A doctrine of water law that allocates the right to use water on a first-come first-serve basis.

Recharge Generally, the inflow to an aquifer and/or groundwater.

Recharge Area Generally, an area that is connected with the underground aquifer(s) by a highly porous soil or rock layer. Water entering a recharge area may travel for miles underground.

Recharge Rate The quantity of water per unit time that replenishes or refills an aquifer.

Riparian Rights A doctrine of water law under which the right to use lakes and streams rests with owners of riparian land, or land that borders on the surface water.

Runoff Generally defined as water moving over the surface of the ground, consisting of precipitation (rainfall) minus infiltration and evapotranspiration.

Safe Yield The amount of water that can be expected to be generally available from a source, particularly during periods of low flow.

Surface Water All water on the surface of the ground, including water in natural and man-made boundaries as well as diffused water.

Transmissivity The ability of an aquifer to transmit water.

Transpiration The process whereby water vapor is emitted or passes through plant leaf surfaces and is diffused into the atmosphere.

Unit Hydrograph The hydrograph of one inch of storm runoff generated by a rainstorm of fairly uniform intensity within a specific period of time.

Water Budget A summation of inputs, outputs, and net change to a particular water resource system over a fixed period. (Also, water balance model).

Watershed All land and water within the confines of a drainage divide.

Water Table The water level (or surface) above an impermeable layer of soil or rock (through which water cannot move). This level can be very near the surface of the ground or far below it.

Water Well An excavation where the intended use is for the location, acquisition, development, or artificial recharge of groundwater (excluding sandpoint wells).

Withdrawal The process of taking water from a source and conveying it to a place for a particular type of use.

REFERENCES

1. Viessman, W., Jr. and C. Welty, *Water Management: Technology and Institutions,* Harper and Row, New York (1985).
2. Dunne, T. and L. B. Leopold, *Water in Environmental Planning,* W. W. Freeman and Co., San Francisco (1978).
3. U.S. Water Resources Council, *Second National Assessment of the Nation's Water Resources,* U.S. Water Resources Council, Washington, D.C. (1978).
4. U.S. Geological Survey, 1984, *National Water Summary 1983–Hydrologic Events and Issues,* USGS Water-Supply Paper 2250.
5. U.S. Geological Survey, 1985, *National Water Summary 1984–Hydrologic Events, Selected Water-Quality Trends, and Ground-Water Resources,* USGS Water-Supply Paper 2275.
6. Solley, W. B., E. B. Chase, and W. B. Mann, *Estimated Use of Water in the United States in 1980,* Geological Survey Circular 1001, US Geological Survey, Washington, D.C., 1983.
7. American Society of Civil Engineers, Task Committee on Water Conservation, "Perspectives on Water Conservation," *Journal of the Water Resources Planning and Management Division,* ASCE, Vol. 107, No. WR1, pp. 225–238 (March, 1981).
8. Boland, J. D., D. D. Baumann, and B. Dziegielewski, *An Assessment of Municipal and Industrial Water Use Forecasting Approaches,* US Army Corp of Engineers, Institute for Water Resources, Ft. Beloir, Virginia (1981).
9. Basta, D. J. and B. T. Bower, *Analyzing Natural Systems,* Resources for the Future, Washington, D.C. (1982).
10. Linaweaver, F. P., J. C. Geyer, and J. B. Wolff, *A Study of Residential Water Use,* US Dept. of Housing and Urban Development, Washington, D.C. (1967)
11. Howe, C. H. and F. P. Linaweaver, "The Impact of Price on Residual Water Demand and its Relation to System Design and Price Structure," *Water Resources Research, Vol. 3,* No. 1 (February, 1967).
12. Kindler, J. and B. T. Bower,, "Modeling and Forecasting of Water Demands," presented at the November 28–29, 1978, Conference on Application of Systems Analysis in Water Management, held at Budapest, Hungary.
13. Loucks, D. P., J. R. Stedinger, and D. A. Haith, *Water Resource Systems Planning and Analysis,* Prentice-Hall, Inc., Englewood Cliffs, NJ (1981).
14. Loucks, D. P., "Surface-Water Quality Management Models," *Systems Approach to Water Management,* A. K. Biswas, ed., McGraw Hill, New York (1976).
15. Goodman, A. S., *Principles of Water Resources Planning,* Prentice-Hall, Englewood Cliffs, NJ (1984).
16. Cunha, L. V., et al., *Management and Law for Water Resources,* Water Resources Publications, Inc., Ft. Collins, Colorado (1977).
17. Reinke, C. E. and R. C. Allison, "State Water Laws: Effect on Engineering Solutions," in *Legal, Institutional and Social Aspects of Irrigation and Drainage and Water Resources Planning and Management,* American Society of Civil Engineers, New York, pp. 204–218 (1979).
18. Bird, J. W., "Origin and Growth of Federal Reserved Water Rights," *Journal of the Irrigation and Drainage Division, Proceedings of the American Society of Civil Engineers, Vol. 107,* No. IR1, pp. 11–24 (March, 1981).
19. U.S. Water Resources Council, *State of the States: Water Resources Planning and Management,* Washington, D.C. (April, 1980).
20. Metzger, P. C., "Nationally Significant Studies of Water Policy: A Review of Major Water Policy Studies and an Assessment of Present Issues," The Conservation Foundation, Washington, D.C. (1983).
21. American Society of Civil Engineers, Task Committee on Federal Policies in Water Resources Planning, "Federal Policies in Water Resources Planning," New York (1985).
22. The Conservation Foundation, *State of the Environment: An Assessment at Mid-Decade,* Washington, D.C. (1984).
23. Harris, S. and A. S. Maynard, "Small Water Users–Planning Crisis," *Civil Engineering, Vol. 55,* No. 8 (August, 1985).
24. *National Water Alliance Report, Vol. 1,* No. 3 (Summer, 984).
25. Postel, S., *Water: Rethinking Management in an Age of Scarcity,* Washington, D.C.: Worldwatch Institute (1985).
26. *World Water* (June, 1985).
27. Barney, G. O., *The Global 2000 Report to the President of the U.S., Volume 1: The Summary Report,* Report prepared by the Council on Environmental Quality and the Department of State, Pergamon Press, New York (1980).

APPENDIX : INFORMATION SOURCES

A wealth of information on water resources exists that is relevant to water use and public policy. At the local level the primary information is from water utilities that typically keep records of monthly use and charges for each customer as well as summary data. Depending on the structure of utility charges, many may also have information on sewer charges that are based on water use.

State level information varies, but usually includes data on water resources within the state and is kept within a department of natural resources or a similar type of agency. Some states have sub-state or regional authorities that have information on water issues relevant to their jurisdiction.

The federal government is the primary repository for water resources information, particularly through the U.S. Geological Survey (USGS) of the Department of the Interior. The USGS has been collecting water data for the U.S. since 1888 and today it maintains a national network of stream-gaging stations, ground-water observation wells, and water quality sampling sites for both surface- and groundwater. The U.S. Environmental Protection Agency (EPA) maintains substantial data on surface- and ground-water quality.

Relatively current information on water resources issues and policies can be obtained from the *U.S. Water News,* a monthly newspaper that started publication in 1984. This privately published newspaper was initiated because there was no single publication handling water news across the nation.

The following list provides a summary of specific information sources of the USGS and EPA.

USGS

WATER DATA PROGRAM

The USGS's Water Data Program provides basic information for hydrologic appraisals, environmental impact assessments, and energy-related studies. It also forms a basis for evaluation, development, and management of the nation's water resources. The field data are primarily collected by the USGS, but numerous other federal, state, and local agencies and private organizations contribute to the data collection effort.

Data are collected in three major categories: surface water, ground water, and water quality. Surface-water discharge data were collected at 11,076 stations in 1983, 7,152 of which monitored continuous flow. Information on ground-water levels was collected at 35,621 sites in 1983, 1,982 of which were collected continuously. Stream and lake samples were collected and tested for water quality characteristics at 4,610 stations throughout the nation, with a continuous record maintained at 784 of these sites. Water data for streamflow, water quality, and ground water are published in official USGS reports on a state-boundary basis. Additional information may be obtained from district offices of the USGS.

NATIONAL STREAM QUALITY ACCOUNTING NETWORK (NASQAN)

This network was established by the USGS to provide a uniform basis for assessing the quality of the nation's streams. It is made up of 501 stations with the same water quality characteristics being measured at each. In addition, 920 stations provided samples for short-term projects. The primary objectives of the program are to (1) depict areal variability of water-quality conditions nationwide on an annual basis, and (2) detect and assess long-term changes in stream quality. Further information may be obtained from:

> Chief, Operating Section, WRD
> U.S. Geological Survey
> 405 National Center
> Reston, Virginia 22092

WATER DATA STORAGE AND RETRIEVAL SYSTEM (WATSTORE)

WATSTORE was established in 1971 to modernize the USGS's existing water-data processing procedures. The system is maintained on the central computer facilities of the USGS National Center in Reston, VA. The WATSTORE system consists of several files having data grouped and stored by common characteristics. Files are presently maintained for the storage of data on (1) surface water, ground water, and water quality, (2) annual peak values for stream flow stations, (3) chemical analysis of surface and ground water, (4) water-data parameters, (5) geology and inventory of ground-water sites, and (6) water use. Additionally a file of monitoring sites is maintained. Further information on the availability of specific types of data can be obtained locally from each of the USGS Water Resources Division district offices. General information on WATSTORE can be obtained from:

> Chief Hydrologist
> U.S. Geological Survey
> 437 National Center
> Reston, VA 22092

NATIONAL WATER DATA EXCHANGES (NAWDEX)

With a vast amount of water resources data being collected nationwide, the potential user must determine if the specific information needed has been collected and is available. The NAWDEX system was established to help match user needs to available data. NAWDEX is a national confederation of water-oriented organizations whose primary objective is to assist users of water data to identify, locate, and acquire needed data. A central Program office is located under the Water Resources Division of the USGS. It maintains close working relationships with the USGS's Office of Water Data Coordination in updating the "Catalog of Information on Water Data", and it establishes working relationships with domestic and foreign organizations that maintain water-related data banks and information systems. The Program Office has four major areas of responsibility: (1) maintaining an internal data center; (2) indexing water data of participating organizations; (3) responding to requests for water data; and (4) formulating water-data handling and exchange standards. NAWDEX services are available to everyone through a nationwide network of 60 Assistance Centers which provide direct access to NAWDEX.

Although its function is not to provide data storage and retrieval services, NAWDEX does provide direct access to the following large water-data bases of members:

- WATSTORE (Water data Storage and Retrieval): USGS data file of streamflow, water quality, sediment discharge, and ground-water level.
- STORET (Storage and Retrieval system): U.S. Environmental Protection Agency file of over 40 million observations on water-quality parameters for surface- and

ground-water that have been collected by numerous federal and state agencies.

- EDIS (Environmental Data Information System): National Oceanic and Atmospheric Administration (NOAA) bibliographic data services on water-related subjects, and Environmental Data Index (ENDEX) service available under EDIS that provides references on more than 10,000 data files.

- Water Resources Scientific Information Center (WRSIC): Department of the Interior, Office of Water Research and Technology (OWRT) bibliographic data service with over 100,000 computerized abstracts available that relate to water resources.

- Several state government organizations provide extensive data, including the Texas Department of Water Resources, Iowa Geological Survey, Pennsylvania Department of Environmental Resources, Nebraska Natural Resources Commission, and Utah Division of Water Rights; several other state systems are under development. All other NAWDEX member data systems are available by referral from the Programs office and all Assistance Centers.

- Water Resources Document (WATDOC) Reference Center of the Inland Waters Directorate, Canadian Department of the Environment. A working relationship with WATDOC provides for a mutual exchange and processing of requests for water information or data under NAWDEX.

For further information about NAWDEX, contact:

National Water Data Exchange
U.S. Geological Survey
421 National Center
Reston, Virginia 22092

NATIONAL WATER-USE INFORMATION PROGRAM

This program under the USGS is a federal-state cooperative program to collect, store, and disseminate water-use information locally and nationally. The Program obtains data on 12 categories of water use: irrigation, agricultural nonirrigation, commercial, domestic, industrial, mining, public supplies, sewage treatment, and fossil-fuel, geothermal, hydroelectric, and nuclear power generation. Begun in 1978, it provides information that complements long-term USGS data on the availability and quality of U.S. water resources. Responsibility for disseminating raw data collected at the state level rests with each state, whereas the aggregated data are disseminated by the USGS. Additional information on the Program and its data bases are available from:

Program Manager
National Water Use Information Program

U.S. Geological Survey
440 National Center
Reston, Virginia 22092

NATIONAL WATER SUMMARY SERIES

Started in 1983, the initial volume of the series [4] introduced a chronology of hydrologic and water-related events to document their significance to human activity, and it also outlined a number of water issues of national concern. The second volume, *National Water Summary 1984 — Hydrologic Events, Selected Water-Quality Trends* and *Ground-Water Resources* [5], continues the chronology of events and presents additional information on several issues from the first volume. The *National Water Summary* series are broad in scope and are the work of many individuals and water resources organizations in the federal government and each state.

HYDROLOGIC DATA VIA SATELLITE

Such data is obtained by the USGS, which has developed automated earth-satellite telemetry for immediate transmission of data. More than 1,200 of the 8,000 automated hydrologic stations telemeter data via satellite. Data are processed, distributed, and stored as part of the USGS's hydrological information system. For further information, contact:

Chief Hydrologist
U.S. Geological Survey
409 National Center
Reston, Virginia 22092

EPA

NATIONAL GROUND-WATER INFORMATION CENTER

Under contract to the U.S. Environmental Protection Agency's National Center for Ground Water Research, the National Water Well Association's (NWWA) library has developed a computerized data base that includes 35,000 citations indexed in 23 fields of information. The data base is accessible worldwide using a computer terminal with a modem and telephone line. For further information on the NGWIC Data Base, contact:

National Ground Water Information Center
National Water Well Association
500 West Wilson Bridge Road
Worthington, Ohio 43085

STORAGE AND RETRIEVAL SYSTEM (STORET)

This computerized file contains over 40 million observations on surface- and ground-water quality parameters that have been collected by federal and state agencies.

CHAPTER 6

Accounting for Environmental Quality in Water Resources Planning

Leonard Ortolano*

INTRODUCTION

Environmental laws and regulations enacted during the 1970s in the United States required that environmental factors be accounted for in planning many types of facilities. New requirements led to much research on how environmental considerations could be effectively integrated into all stages of project planning. Elements of this research relevant to water resources planners and engineers are summarized below in the following topical areas:

- Impact identification typologies—categorizations of environmental impacts commonly associated with typical water resources projects.
- Forecasting procedures—techniques for predicting the magnitude and extent of a water project's environmental impacts.
- Evaluation methods—procedures for evaluating alternative project proposals in terms of environmental, social and economic impacts.
- Agency implementation surveys—analyses of environmental impact assessment (EIA) practices of water resources agencies.
- Policy enhancement studies—investigations of how the effectiveness of agency EIA's can be increased.

IMPACT IDENTIFICATION

Environmental and Social Effects

Much of the early literature on procedures for environmental assessment catalogued impacts typically associated with a particular type of project or activity. In the water resources literature, numerous "checklists" were developed indicating categories of impacts that might be associated with a particular project type. Examples of such listings are found in Canter (1985), Curran (1975, 1976), Hagan and Roberts (1973), Lohani (1984), Ortolano (1973), and U.S. Environmental Protection Agency (1973, 1975, 1977). These references focus on impoundments and channel modification works. Canter (1985), Lohani (1984), Ortolano (1973) and U.S. Environmental Protection Agency (1973, 1975) also consider impacts associated with dredging projects.

The checklists developed by Hagan and Roberts (1973) for water storage and diversion projects are illustrative. Tables 1, 2 and 3 are the listings they prepared for ecological impacts in the area of impoundment, downstream of the impoundment, and in the area where project water is used, respectively. Hagan and Roberts also provided numerous case examples for many of the listed items.

In a typical environmental impact assessment, the term "environment" is defined broadly to include social effects. These are mentioned only briefly in the tables prepared by Hagan and Roberts, since their focus was on changes affecting the natural environment. A classification of the types of social impacts observed in connection with water resources development projects in the United States has been prepared by Hitchcock (1981). His scheme, which is summarized in Table 4, is based on an analysis of 81 research studies of the social impacts of constructed water projects. The studies included impacts occurring prior to, during, and after construction. Hitchcock's categorization is notable because it is based on social impacts actually observed, not on conjectures about possible effects.

Three volumes on the social effects of water resource development projects have been prepared by the U.S. Army Corps of Engineers (1977). Information on the social impacts of large public works is given by Finsterbusch (1980).

*Department of Civil Engineering, Stanford University, Stanford, CA

TABLE 1. Ecological Impacts in Area of Impoundment.[a]

Disturbs natural state of area
 Desire to preserve natural conditions for present and future
  generations: "wild" rivers versus controlled rivers.
 Changes scenic values: conversion of rivers to lakes.
 Modifies micro-climate: temperature, humidity, wind.
 Alters land form, vegetation, wildlife, etc. (through
  construction and subsequent activities) and thus affects
  ecological diversity.
Increases evaporation loss
 Reduces water supply.
 Degrades water quality.
Changes water temperature
 Altered aquatic life.
 Effects on some water uses, primarily water sports.
Alters erosion and sedimentation
 Erodes reservoir banks, causes land slides.
 Deposits sediments in reservoirs: delta formation, loss of
  reservoir capacity.
Submerges land areas
 Affects scenic values: submerges scenic treasures; creates
  new scenic values; causes visual "pollution" from
  exposed banks during drawdown.
 Loss of historic sites.
 Displaces people (see economic and social effects).
 Loss of farm land (see economic and social effects).
 Alters habitat for fish and wildlife (see below): mitigation
  enhancement.
 Creates environments for new life forms: plants, insects,
  fish, and other wildlife.
 Possibly influences earthquake frequency.
Modifies fish production
 Substitution of lake for stream fishing: changes in fish
  species.
 Dam creates barrier for anadromous fish migrating to
  spawning grounds: fish ladders.

Still water and deeply submerged spawning beds affect
 reproduction and return of young fish to sea: fish
 hatcheries.
Alters wildlife production
 Submerges feeding areas.
 Substitutes other areas or more intensively managed areas.
 Provides new nesting and feeding areas for migrating
  birds.
 Denuded zone exposed during drawdown restricts access
  to water by timid animals.
 Reservoir brings "people pressures" which may upset
  delicate environmental balances, adversely affecting rare
  species.
Modifies recreation potential of area
 Alters opportunities for swimming, skiing, and boating.
 Modifies fishing and hunting.
 Creates or expands sites for camping and other
  recreational facilities.
 Increases people pressures on life of area with resultant
  ecological impacts.
 Increases penetration of adjacent and remote wilderness
  areas by hikers.
 Intensifies traffic, noise, and air pollution.
 Expands needs for pollution control and waste removal.
Increases development of surrounding lands for urban or
vacation housing
 Destroys native plant cover.
 Increases erosion from construction and loss of plant
  cover.
 Increases people pressure on surrounding areas.
 Intensifies pollution and waste disposal problems.
Alters economic, social, and political life of area with
resultant secondary ecological impacts

[a]From Environmental Quality and Water Development by C. R. Goldman, J. McEvoy, III and P. J. Richerson (editors). Copyright © 1973 by W. H. Freeman and Company. All rights reserved.

Environmental Impacts of the Aswan High Dam

Tables 1 through 4 are general and are intended to apply to a wide range of circumstances. In any particular situation, only some of the listed items will be important. This is illustrated by the environmental impacts associated with the Aswan High Dam constructed in Egypt during the 1960's and 1970's.

It is impractical to provide more than a brief overview of the types of impacts associated with the Aswan High Dam. Indeed, it is difficult even to provide a summary since the impacts reported in the literature are often inconsistent with each other. In some cases, the detailed field studies needed to document impacts have never been conducted. Notwithstanding these difficulties, the environmental impacts com-

monly discussed in connection with the Aswan High Dam are listed below and in Table 5. This information is compiled from reviews by Biswas (1978), Fahim (1981), Lohani (1984) and Waterbury (1979).

Evaporation and seepage losses—Although estimates of water lost from the Aswan High Dam Lake vary, Waterbury (1979, p. 125) asserts that "under the best of circumstances, more than half of Egypt's [anticipated] incremental gain [in water supply] from the High Dam will be lost in storage."

Resettlement—About 100,000 Sudanese and Egyptian Nubians were resettled from the area inundated by the Lake. The reservoir flooded their villages, agricultural lands, and historic remains. New communities were established for the uprooted people. In addition, new Lake-based settlements developed consisting of desert nomads, boat residents, re-

TABLE 2. Ecological Downstream from Impoundment and/or Diversion.[a]

A. In river channel and flood plain

Disturbs natural state of area

Modifies downstream hydrograph

Reduces peak flows.

Minimizes flood damage along channel and in flood plain.

Reduces channel scouring and increases sedimentation (affecting channel capacity, fish spawning ground).

Reduces opportunities for water spreading for ground water recharge.

Reduces capacity for flushing, diluting, and transporting wastes.

Increases minimum flows.

May weaken stream banks and levees, causing slumping.

May increase severity and duration of seepage and raise water table along channel and river basin.

Permits more adequate year-round waste dilution and transport.

Benefits navigation and power generation.

Increases water supply for expansion of agricultural, domestic, and industrial uses along river with resulting secondary impacts.

Introduces abnormal and variable flows caused by project operation.

High fows to create flood control space in reservoir.

Periodic discharges for peak power generation.

Mitigating effects and impacts of regulating reservoir.

Alters quality of river waters

Evaporation can increase salinity of stored waters.

Maintenance of higher minimum flows can reduce salinity of rivers affected by salty tributaries or return irrigation water.

Alters content of nitrogen and oxygen in discharged waters.

Alters river water temperature

Typically, lowers temperature during summer flow period.

Affects agricultural and industrial uses, types of fish and their production, and water sports.

Modifies sediment transport

Reduces peak flows lessening channel scouring, increasing sedimentation.

Traps sediment in reservoir.

Lowers downstream sediment load, affecting agricultural and other uses; increases channel scouring at given flows.

Muddy discharges can extend over greater portion of year where sediment remains in suspension in reservoir.

Changes aquatic and riparian vegetation

Increases encroachment on channels.

Influences ecological diversity.

Affects scenic values and recreation uses.

Modifies fish production

Changes water temperature altering fish production: may prevent survival of uniquely adapted species.

Dam interferes with migrating fish: mitigation by substitution of hatcheries and additional artificial spawning areas.

Dam reduces length of channel for stream fishing; increases people pressures on environment along shortened stream channel.

Changes recreational potential of river

Shortens channel available for river boating; may improve or worsen boating in remaining part.

Regulated flow: allows boating and navigation over greater part of year; affects other water sports; alters fishing.

Alters economic, social and political life of area with resultant secondary ecological impacts.

B. In delta, bay or ocean

Disturbs natural state of area

Alters patterns of water flows and possible effects

Reduces peak flows or reduces total flow due to diversion.

Decreases flood hazard to agriculture and cities in delta.

Reduces flooded areas available for bird resting and feeding.

Alters channel scouring and sedimentation in delta.

Affects commercial and recreational navigation in delta.

May alter salinity of inflow water.

Reduces capacity to flush pollutants and salts from delta and bay.

May reduce turbidity or receiving waters affecting light transmission and, in turn, algae production and estuarine life.

Affects land and estuarine plants and wildlife.

Reduces sediment supply to maintain beaches of ocean, and possibly increases wave erosion on beach-front lands.

Alters off-shore sandbars.

Increases minimum flows.

Alters sediment transport and delta formation.

May decrease salinity and pollutants in inflow water.

Increases capacity to dilute and transport pollutants from delta and bay.

Increases capacity to repel salt intrusion.

Improves navigation and water sports in delta.

Increases water supply for expansion of agricultural, domestic, and industrial uses in surrounding areas, with resulting secondary impacts.

Alters fish and wildlife potential of area.

Altered economic, social, and political life of area with resultant secondary ecological impacts.

TABLE 3. Ecological Impacts in Areas of Project Water Use.[a]

A. Agricultural and Rural Areas

Provides potential for changing natural state of area
Allows development of irrigated farming
  Causes visual changes in landscape.
  Alters environment for plants: native plants, crop plants and introduced weed species.
  Alters environment for wildlife.
    Area flooded during irrigation and leaching offers resting and feeding areas for birds.
    Replacement of natural vegetation by irrigated crops under intensive farming may restrict or encourage certain species of wildlife.
  Changes insect population.
    Allows breeding of insects in canals, ponds, poorly drained areas, and in wet fields.
    Allows breeding and development of insects on introduced crops and weeds.
  Alters incidence of plant, animal, and human diseases.
  Modifies local climate: increases humidity, moderates temperatures, and changes rainfall patterns (where large dry areas are under irrigation).
  Increased ground water recharge and high water tables create poorly drained and salinized areas.
  Increases pollution of surface and groundwater from return irrigation water; use of agricultural chemicals, and plant and animal wastes.
Supports increased population and altered economic, social, and political life of area, with resultant secondary ecological effects

B. In Urban Areas

Permits drastic changes in natural state of area
Permits expansion of cities to become vast urban areas
Provides water, permitting populations to exceed other physical or social resources of area.
Allows large industrial development including high water-use industries.
Need for flood control leads to channelization and levee construction along streams with changes in flow patterns, ground water recharge, and riparian vegetation.
Concentrates people, vehicles, and industry, leading to air and noise pollution.
Creates vast water pollution and waste dispoal requirements.
  Results in pollution of groundwater and surface waters, including bays and ocean shorelines.
  Affects recreation.
  Affects wildlife and especially growth of fish and shellfish and their suitability as food.
Increases power requirements in area, often leading to atmospheric and thermal pollution.
Increases people pressure on surrounding areas, particularly for recreation.
Increases social problems.
Increases availability of water which can be used to improve environment with parks, fountains, artificial ponds and gardened areas, and to develop water-based recreational facilities.
Permits introduction of urban areas into deserts and other water deficient locations
  Disturbs natural environment.
  Can spread populations over larger areas, reducing problems associated with urban crowding.
Affects land development, industry, and population, resulting in drastically altered economic, social, and political life, with resultant secondary ecological impacts

[a]From *Environmental Quality and Water Development* by C. R. Goldman, J. McEvoy, III and P. J. Richerson (editors). Copyright © 1973 by W. H. Freeman and Company. All rights reserved.

turning Nubians and people from the Wadi Halfa district. Fahim (1981) documents the traumas associated with Nubian adaptation to new government-planned communities and efforts made to establish new settlements along the Lake.

Sediment trapping and downstream erosion — Prior to the High Dam's construction, large quantities of silt were either deposited along the shore of the Nile or carried to the coastal delta and the Mediterranean Sea. After completion of the project, much silt settled out behind the dam, and reservoir releases were relatively silt-free. Downstream the river bed and banks have eroded as the Nile River attained its new equilibrium silt load. This erosion threatens to undermine the foundations of downstream bridges and dams. The trapping of silts behind the High Dam has also been linked to the increased rate of erosion of the Nile Delta, since silt is no longer deposited during the flood

season to offset the effects of wave-induced erosion of the Delta.

Sediment trapping and Nile Valley soil fertility — Prior to the construction of the Aswan High Dam, the nutrient-rich silts deposited on the Nile Valley during the annual flood provided a natural fertilizer. The contribution of these silts to soil fertility must now be provided using chemical fertilizers. Moreover, rich topsoil from the Nile Valley is being lost as farmers sell silt-rich topsoil for the manufacture of silt-bricks. Before the High Dam, brick manufacturers could obtain the needed silt by dredging local canals. Efforts are being made to persuade local brick manufacturers to make bricks using other materials, such as sand.

Sediment trapping and Mediterranean fish catches — The loss of nutrient-rich sediments, together with the decreased annual water outflow, has been a cause of dramatic reductions in Egypt's sardine catch from the Mediterranean Sea. Between 1965 and 1972, the annual catch fell from about 18,000 tons to 500 tons.

Waterlogging and soil salinization in newly farmed areas—Inadequate drainage facilities and the excessive use of readily available irrigation water have led to rapidly rising groundwater tables. In some cases, water tables rose from 15 m to the root level of plants. Improperly drained water often evaporates rapidly, leaving behind salt deposits. Salinized soil loses its agricultural productivity, and if salinization is extensive the soil can become unfit for cultivation. Some of the lands newly reclaimed as a result of the Aswan High Dam project are already facing a salination problem.

Poor quality irrigation return flows—Newly irrigated areas have relied on chemical fertilizers and pesticides. The drainage from these lands has contained notable concentrations of nitrates, phosphates and other chemicals associated with a reliance on chemical fertilizers and pesticides. Drainage from farmlands in the Nile Delta to lakes in northern Egypt (e.g., Lake Manzala) have caused deterioration in the quality of these lakes. In some cases, fish kills have been reported.

Increased incidence of debilitating schistosomiasis—Schistosomiasis is a parasitic disease caused by a tiny worm. The disease is acquired by humans through direct contact with water infected with the larva of the worm. Snails that serve as an intermediate host for the worm often proliferate with increases in the quantity of stagnant water and the amount of aquatic weeds to which snails cling. Changes in the Nile's flow pattern and the more intense reliance on irrigated agriculture in the Nile Valley have been linked to an increase in the rate of incidence of a severe variety of schistosomiasis.

The above review of environmental impacts associated with the Aswan High Dam is not intended as a condemnation of the project. Although the Dam has had numerous critics, many observers agree that it has contributed to Egypt's economic well being and political security. It has, for example, provided flood protection and a stable source of water to support perennial irrigation.

Rather than provide an assessment of the Aswan High Dam project as a whole, this summary serves to illustrate some of the items in Tables 1 through 4. The summary also provides a context for making additional observations about EIA. First, it demonstrates the indirect effects, such as the

TABLE 4. Hitchcock's Categories of Social Impacts Caused by Water Resources Projects.[a]

Distribution Impacts—changes in patterns of activity and status resulting from project actions.
- Population: change population density and population migration; e.g., increased in-migration to water project area.
- Land Use: change land uses and land values; e.g., increase in land values due to water projects.
- Distribution of Costs and Benefits: redistribute income; e.g., financial burdens on individuals forced to relocate by reservoir projects.

Opportunity impacts—changes which affect the ability of a member of a community to satisfy a range of needs and desires.
- Community Development: eliminate seasonal fluctuation in economic activity and increase diversity and stability of economic activity; sometimes community development is thwarted (e.g., reduction of tax base).
- Economic Opportunities: improve farming conditions, economic growth and investment security.
- Job Opportunities: enhance opportunities for employment.
- Amenities: improve "aesthetic quality" and recreational opportunities; decrease hunting and fishing opportunities.

Local Services Impacts—changes in delivery of community services.
- Local Finances: influence property tax (often reported as insignificant).

- Local Services: improve water systems and fire protection; strain local schools and law enforcement change local roads. (Both positive and negative changes in local roads have been reported.)
- Local Leadership: influence efficiency and responsiveness of local government and sructure of community (business, social and political) leadership. (Both positive and negative changes have been reported.)

Awareness—varying degrees of awareness of a project prior to its construction.
- Perception of Impacts: varying impressions (prior to project construction) of possible project effects such as changes in land use and school systems and the need to relocate families.
- Attitudes Toward Projects: varying levels of support for and opposition to projects over time.
- Level of involvement: varying citizen participation in project planning, ranging from a lack of involvement (stemming from a low sense of efficacy) to high degrees of involvement by groups supporting or opposing projects.
- Community Interactions: increasing community cohesion due to improved economic diversity and stability and pride in local area; decreasing cohesion resulting from animosity between groups supporting and opposing projects prior to construction and also due to changed social patterns (e.g., increased juvenile delinquency within the community).

[a]Based on Hitchcock's (1981) review of social impacts reported for water resources projects constructed in the United States.

TABLE 5. Principal Project Impacts at the Area of Impoundment—Aswan High Dam Lake.[a]

Disturbs natural state of area
  Modifies microclimate—increased humidity and
    precipitation.
Increases evaporation loss
  Estimates of annual losses due to evaporation and seepage
    range from 10 to 12.5 billion m³.
  Salinity concentration of outflows from the Lake are
    expected to be about ten percent higher than
    concentration of inflows.
Alters sedimentation
  Large quantities of sediments originating in the flood of the
    Blue Nile from Ethiopia are deposited behind the Aswan
    High Dam—storage allocated for these sediments is
    designed to be sufficient for 500 years.
Submerges land areas
  At maximum storage levels (reservoir height of 175 m),
    about 5200 km² are inundated.
  Inundated areas are rich in antiquities and monuments of
    past civilizations; 24 temples were salvaged and
    reconstructed (including temples of Nubia, Abu Simbel
    and Philae).
  Inundation of Nubian villages and relocation of about
    100,000 Nubians.
Modifies fish production
  Fish stocks in the reservoir resulted from species existing in
    the river prior to impoundment; while the Lake filled,
    catches increased by 1000 tons/yr., and for 1976 they
    reached 16,000 tons. Further increases were anticipated.
Other modifications in water quality
  High Aswan Dam Lake has supported a variety of
    phytoplankton, algae, and other plants that have
    interfered with navigation in the Lake. Releases of these
    organisms have caused problems with water supplies
    downstream at Cairo.

[a]Table entries organized in terms of topic headings in Table 1.

deterioration in water quality of Egypt's northern lakes, can be as important as the more obvious direct impacts of a water project. Second, the discussion indicates the uncertainties associated with impact assessment. For example, the increased incidence of a particularly severe form of schistosomiasis would have been difficult to predict with confidence. Indeed, according to Fahim (1981), there is still controversy regarding the extent to which the Aswan High Dam project is responsible for rising rates of schistosomiasis. Third, the review highlights the kinds of difficulties that can occur if adequate attention is not given to prior planning and impact analysis. The adverse social impacts caused by relocation of the Nubians is a case in point. Fourth, and finally, the summary reinforces the need for post-project follow-ups to reduce unanticipated adverse effects. This is demonstrated by ongoing efforts to offset the undermining of bridge foundations caused by erosion downstream of the Dam.

Checklists of environmental effects and studies of impacts caused by constructed water projects serve useful purposes. By consulting checklists (such as those in Tables 1 through 4) at the outset of project planning, water resources engineers and planners can avoid overlooking significant impacts. They can also learn important lessons from investigations of the environmental and social effects of completed projects.

FORECASTING ENVIRONMENTAL IMPACTS

Following passage of the U.S. National Environmental Policy Act of 1969 (NEPA), federal agencies were required to develop techniques for analyzing environmental impacts. As a result, government agencies, consultants, and university and private research groups produced a large number of what are termed "environmental assessment methods." The volume of literature is indicated by Canter's (1979) review of assessment procedures, which includes about 175 references to different approaches. An extensive bibliography is given by Clark, Bisset and Wathern (1980).

Impact assessment methods can be divided into two broad categories corresponding to basic planning activities: forecasting and evaluation. Forecasting consists of predicting the environmental impacts of alternative actions and is considered in this section. Evaluation, the subject of the next section, is the process of putting relative values on different impacts and establishing a preference ordering among alternative plans. Using these definitions, value judgments about whether predicted impacts are good or bad are part of evaluation, not forecasting.

Forecasting plays a central role in environmental impact assessment. Indeed, an environmental impact is generally defined as a projected change in one or more measures of environmental quality. Forecasts are typically made for a number of alternative plans, including the "no-action alternative," which consists of related events likely to occur if the agency pursues none of its alternatives. The no-action forecasts provide a base against which environmental conditions for other alternatives can be compared.

There are many ways of categorizing forecasting procedures. One approach classifies methods as judgmental (or expert-opinion), physical and mathematical. The discussion below uses this classification and summarizes a presentation by Ortolano (1984).

Judgmental Approaches to Forecasting

The most frequently used approaches to forecasting impacts rely heavily on expert judgment. In this context, an "expert" is someone with special knowledge useful in forecasting. Thus, for example, a real estate agent might be con-

sidered an expert in predicting land use changes induced by a proposed flood control project.

More generally, the term expert refers to a person with extensive academic training and practical experience relevant to the forecasting task. An expert's familiarity with impacts caused by similar projects in analogous settings commonly plays a key role in forecasting.

To illustrate how expert opinion is used in predicting environmental impacts, consider a proposal that involves destroying a marsh to create a fflood retention basin. Suppose a biologist with extensive knowledge of marsh ecosystems is asked to forecast the project's biological impacts. The biologist would employ standard scientific references and field investigation methods to characterize the project area from a biological perspective. If time were available, he or she might search for information on observed effects of related projects in similar settings. The forecast would consist of an opinion based on the above-mentioned information and the biologist's understanding of how marsh ecosystems function and respond to disturbances of the type proposed.

Sometimes individual specialists make forecasts of environmental impacts by working together in small groups. This has the advantage of allowing experts to build on each other's ideas and thereby provide a more useful prediction. However, there are some potential difficulties in using ordinary meetings as a format for group forecasting exercises. One is that a few vocal individuals may dominate the group's deliberations. Another is that some specialists may feel pressured to give opinions in conformity with others in the group.

The "Delphi method" is one of the many procedures developed to increase the effectiveness of experts in making forecasts as a group. It avoids the problems above by not relying on group meetings. The method obtains the opinions of experts by means of mail questionnaire surveys. Throughout the exercise, the anonymity of individuals providing responses to questionnaires is preserved. Several iterations (or "rounds") of the questionnaire survey are used to give individuals a chance to revise their previous forecasts based on the judgments of others participating in the Delphi process. Opinions are shared using statistical summaries of responses from preceding rounds. The summaries are mailed out with each successive round of the questionnaire. Statistical measures, such as the median and interquartile range of the forecasts, are employed in these summaries to reduce group pressure to arrive at a concensus forecast. Armstrong (1978) and Porter et al. (1980) discuss the Delphi method and numerous other techniques to elicit forecasts from groups of experts.

Physical Models in Forecasting

Physical models are small-scale three-dimensional representations of reality. They have been used to make predictions for thousands of years and are familiar, in some form, to almost everyone. A centuries-old example is the architect's model of a proposed building.

In addition to depicting individual buildings, physical models can show how the appearance of an entire landscape or cityscape would change with the addition of a new project. Examples of how physical models are used to predict visual impacts of projects are given by Appleyard et al. (1979) and Ortolano (1984).

A more common use of physical models in forecasting impacts of water projects involves predicting effects on water bodies. For example, models are useful in analyzing tidal estuaries. The three-dimensional mixing of fresh and salt water that occurs in estuaries makes it difficult to apply other forecasting techniques. Physical models of estuaries have been used to predict how navigation projects and port facilities influence water surface elevation, current velocity, salinity, and waste dispersion characteristics. Predictions have also been made of effects on estuaries of reducing fresh-water inflows, such as occurs when an upstream reservoir is developed for water supply purposes. Many authorities feel that physical models should not be relied on for quantitative predictions of water quality changes.

Estuary models generally include an array of electrical and mechanical devices to represent the effects of tides and the inflow of fresh waters. Instruments are included for measuring water depth, velocity, temperature, and the concentrations of salinity and various dyes used to simulate wastewater discharges. Figure 1 shows a portion of the U.S. Army Corps of Engineers' model of a complex estuarine system in northern California known as the San Francisco Bay—San Joaquin Delta. This model is built to a horizontal scale of 1 ft = 1000 ft and occupies about one acre. It is scaled such that 1 day of real time can be simulated in about 15 minutes.

The design of estuary models is based on scientific principles. However, these principles alone are not sufficient to guarantee an accurate representation of reality. Extensive measurements of water circulation and quality characteristics in a real estuary must be obtained to "verify" a model of that estuary. This is a time-consuming process in which model features, such as the roughness of the model's surface, are changed gradually until the behavior of the model replicates that of the estuary. Models can then be adjusted to provide forecasts of how physical changes within an estuary influence water surface elevation and other estuary characteristics. Additional information on physical models of estuaries is given by Tracor, Inc. (1971).

Forecasting with Mathematical Models

Mathematical models are also used to predict environmental impacts. These "models" are constructed from combinations of algebraic or differential equations. They are generally based on either scientific laws or statistical analyses of data, or both.

FIGURE 1. Portion of the U.S. Army Corps of Engineers' Model of the San Francisco Bay—San Joaquin Delta. (Courtesy of the U.S. Army Corps of Engineers.)

Since there are literally hundreds of mathematical models that can be used in forecasting environmental impacts of water projects, it is impractical to provide a survey here. Such surveys have been prepared by others, however. An example is the review by Brown et al. (1974).

As an alternative to a survey of the different mathematical models used in forecasting the impacts of water projects, an illustration is introduced: the Stanford Watershed Model (SWM), a widely used hydrologic simulation model developed by Crawford and Linsley (1966). Figure 2 delineates SWM's structure. Equations in the model provide a running account of how water entering a river basin moves via processes such as interception, infiltration, and runoff (see Figure 2). Formulas based on the law of conservation of mass keep track of all water entering and leaving the basin. Other theoretical and empirical equations route surface

runoff and groundwater flows into stream channels. Information on streamflow is used to develop hydrographs at selected points.

The SWM requires much data for calibration, the process of estimating model parameters. Between 20 and 40 parameters characterizing a river basin must be specified. Most of these can be determined from maps and hydrologic records. A small number are estimated using a trial-and-error procedure. Trial values of parameters are based on an analyst's judgment and past experience in using SWM. Once all parameter values are set, the model uses historical precipitation data to produce hydrographs at various locations. These "simulated hydrographs" are then compared with corresponding hydrographs constructed from field measurements ("observed hydrographs"). If there are significant differences between simulated and observed hydrographs, trial parameters are reestimated in an effort to reduce discrepancies. The model is then run again to produce another set of hydrographs and the comparison with measured hydrographs is repeated. This iterative process continues until the model is considered calibrated. Close agreement between simulated and measured hydrographs have been obtained in studies where extensive data was available for calibration.

The SWM and other hydrologic simulation models are used to predict how surface runoff and groundwater flows respond to water resources development projects and weather modification, for example, using cloud seeding. These models can also describe hydrologic impacts that accompany land surface modifications affecting infiltration and evapotranspiration. Additional examples of how mathematical models are used in forecasting hydrologic change are given by Linsley (1976).

The above discussion is based on a three part classification of methods useful in forecasting environmental impacts: judgmental, physical and mathematical. There are, however, other ways of categorizing forecasting procedures.

An Alternative Categorization of Forecasting Methods

An effort at categorizing prediction techniques that is relevant to water resources specialists is the one conducted by Mitchell et al. (1975) for the U.S. Army Engineer Institute for Water Resources. Mitchell and his colleagues were concerned with the forecasting tasks commonly faced by field level water resources planners. These tasks include making predictions of: economic and demographic trends; demands for project outputs (e.g., hydropower and flood control); and economic, social and environmental impacts. Thus the

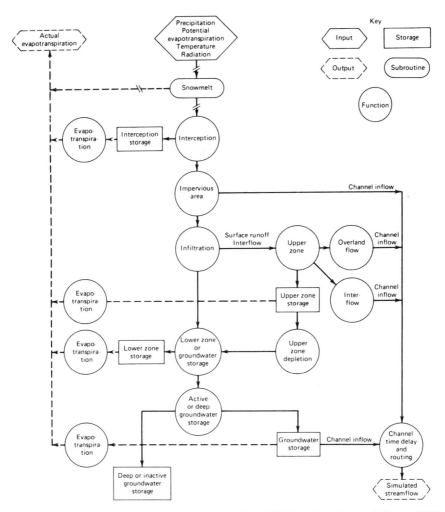

FIGURE 2. Flow chart for the Stanford Watershed Model IV. From Crawford and Linsley (1966).

TABLE 6. Categories of Forecasting Techniques Developed by Mitchell and Associates for Water Resources Planners.[a]

- Trend Extrapolation: extending time-series data using various curve-fitting procedures (see, also, Porter et al., 1980, pp. 115–122).
- Pattern identification: using statistical analysis techniques to identify patterns in observed data and relying on such patterns to make predictions.
- Probabilistic Forecasting: using methods of probability theory (e.g., Markov chains and queing theory) to make predictions (Porter et al., 1980, pp. 207–208).
- Dynamic Models: mathematical models that simulate a system's behavior over time, such as Forrester's systems dynamics models (Porter et al., 1980, pp. 204–207).
- Cross-Impact Analysis: means of studying interrelationships among events by evaluating changes in the likelihood of a set of events, given changes in the probability of some events within the set (Porter et al., 1980, pp. 190–195).
- KSIM: a process combining expert opinions and analytic methods in developing mathematical relationships among broadly defined environmental and socioeconomic parameters (Porter et al., 1980, pp. 196–203).
- Input-Output Analysis: representation of interdependencies among sectors of an economy that can be used to predict how a change in one sector's activity influences all other sectors (Porter et al., 1980, pp. 289–291).
- Policy Capture: use of statistical analysis techniques to idenitfy the relative weights an interest group places on various evaluative factors in judging alternative proposals; used to predict future judgments (Porter et al., 1980, pp. 371–373).
- Scenarios and Related Methods: use of plausible and imaginative conjectures to provide descriptions of potential "futures" (Porter et al., 1980, pp. 150–152).
- Expert Opinion Methods: systematically obtaining forecasts from groups of experts using expert panels, opinion surveys, the Delphi method and other approaches for deriving opinions from groups (Porter et al., 1980, pp. 122–130).
- Alternative Futures: techniques, such as "morphological analysis," that focus on defining conceivable, plausible and possible "futures," not just those that seem to be "without surprises" (Porter et al., 1980, pp. 174–175).
- Value Forecasting: predicting changes in people's values and attitudes in order to generate alternative futures (Porter et al., 1980, p. 143–146).

[a]Each of the twelve entries comprises a section in Mitchell et al., (1975). Many of the procedures are also described by Porter et al. (1980); page references are given in the body of the table.

set of forecasting problems examined by Mitchell was broader than the class of impact prediction problems considered here. Notwithstanding this difference in focus, the categories of techniques elaborated by Mitchell and his associates are noteworthy because they include prediction methods used in assessing environmental impacts.

After reviewing 150 different forecasting procedures, Mitchell *et al.* structured a *Handbook of Forecasting Techniques* around twelve basic procedures that were, in their view, "suitable for a wide range of technological, economic, social and environmental forecasting." Table 6 lists the types of forecasting methods they selected. Further information on the listed items is given by Mitchell *et al.* (1975) and Porter *et al.* (1980).

Additional information on methods for predicting environmental effects of water projects is available in handbooks and monographs prepared to assist in conducting environmental impact assessments. Examples of these materials include books by Canter (1985), Gordon (1985), McEvoy and Dietz (1977) and Rau and Wooten (1980).

Forecasting environmental impacts is only one aspect of an EIA. Another important element of the environmental assessment process is evaluation, the use of information on predicted impacts to reach decisions on which of several alternative projects is preferred.

METHODS FOR EVALUATING ALTERNATIVE PROJECTS

Since the 1960's there has been much emphasis on the systematic evaluation of alternative water resources projects in terms of economic, social and environmental factors. In this context, the term "evaluation" is used in the sense of Litchfield, Kettle and Whitebread (1975, p. 4), who define it as "the process of analyzing a number of plans or projects with a view of searching out their comparative advantages and disadvantages and the act of setting down the findings of such analysis in a logical framework." They stress that evaluation is not decision making. Instead it assists in decision making by highlighting the differences between alternatives and providing information for subsequent deliberation.

Extensions of Traditional Benefit-Cost Analysis

In traditional benefit-cost analysis (BCA), the basis for evaluating alternative water projects is maximization of the difference between national economic benefits and national economic costs. The economic benefits of a project are measured by the willingness of individuals to pay for the project's outputs. The costs of a project are the gains that must be sacrificed by using the "inputs" (resources such as land and labor) necessary to implement the project. Economists view costs as opportunities foregone, potential gains given up by not using the resources in other ways. If a project's inputs and outputs are traded in competitive markets, the market prices are used to compute the project's costs and benefits. Frequently, competitive markets do not exist. When suitable market prices are unavailable, costs and benefits are estimated using procedures that rely heavily on the professional judgments of individuals performing the BCA.

A common criticism of BCA is that it ignores significant effects on the natural and social environment that cannot be measured in monetary terms. One response to this consists of efforts to improve the economist's ability to place a monetary value on environmental impacts. Much research in this area has been conducted during the past two decades. It includes attempts to put a dollar value on changes in fisheries resources, decreased quality of water, and even the loss of human life. Freeman (1979, 1982) and Kneese (1984) review the tools that have been developed to make these types of economic evaluations.

Another response to the criticism that traditional benefit-cost analysis neglects environmental values involves extending BCA to include multiple objectives. The early work on multiple objective water resources project evaluation was done by Marglin (1962) and Maass (1966). Their illustrative cases considered two objectives: maximization of net income to a nation as a whole and maximization of net income to a particular region or group of citizens. The "national income net benefits" are defined as the net benefits included in a traditional BCA. Analysis of the second objective requires the introduction of "income redistribution net benefits," the net income that flows to the particular region or group singled out for special treatment. For each project under consideration, both national income and income redistribution net benefits are computed. Proposals may then be ranked using a weighted sum of the contributions to each objective.

During the past twenty years there has been substantial activity in the part of mathematical optimization theory that treats the multiobjective evaluation issues examined by Marglin and Maass. This branch of theory, known as "multiobjective programming," concerns the limited set of multicriteria evaluation problems that can be described adequately by mathematical equations and inequalities. Examples of how multiobjective programming methods have been used to consider environmental factors in water resources project evaluation are given by Major (1977) and Shamir (1983). For general reviews of multiobjective programming in water resources planning, see Cohon (1978) and Goicoechea, Hansen and Duckstein (1982).

Tabular Displays and the Sum of Weighted Factor Scores

A number of commonly used project evaluation methods rely on tabular displays of information. Table rows generally correspond to evaluative factors such as water quality, archaeological resources and construction cost. Columns correspond to the alternative projects under consideration. A few types of table entries are commonly used. In some cases, an entry consists of a brief description of how a particular project is likely to influence a given evaluative factor. In other cases, an entry is a numerical score characterizing the effects of a project on a factor.

Numerical table entries often result from comparative analyses of how alternatives influence an evaluative factor. Rank ordering is sometimes used as the basis for assigning scores. Suppose, for example, that five alternative water projects are ranked in terms of the disturbance they will cause to local trout and salmon fisheries. The project that causes the least disturbance is assigned a "1", the alternative causing the second to the least amount of disturbance is assigned a "2", and so on. Another approach to making comparative observations involves categories, for example, "positive effect, no impact, and negative effect." Sometimes these categories are assigned arbitrary numerical values, such as $+1$, 0, and -1, respectively.

Another common method for summarizing evaluative information uses the sum of weighted factor scores for each alternative. First, each proposal is assigned a score for each evaluative factor. All scores must be within the same numerical limits. For example, scores might vary from 1 to 10, with 10 representing the best alternative. Weights are assigned to indicate the relative importance of each factor and they are used in computing a weighted sum of factor scores. Numerous approaches of this type are discussed in Canter's (1979, 1985) reviews of water resources project evaluation methods.

Involving the Public in Evaluation

Public involvement in the process of evaluating alternative water projects is mandated by laws and regulations in the United States and many other countries. Some agencies use the term "public" broadly to include individual citizens and interest groups, elected and appointed officials, and governmental administrative units that may have an interest in the actions being planned.

The public's contribution to planning water resources projects is not limited to just evaluating alternatives. Many agencies involve the public in all planning tasks, including the formulation of alternatives and the prediction of impacts. No single approach is relied on for including the public in planning. Often combinations of public involvement techniques are tailored to meet the objectives of a particular planning study.

Most public involvement programs include meetings. Of the several commonly used meeting types, the public hearing is the most rigid (see Table 7). A hearing officer generally governs the proceedings and a stenographer makes a verbatim transcript. Presentations are formal and there is little interaction among participants. Large group meetings can be much less formal than hearings. However, it is difficult for many citizens to contribute directly in large

TABLE 7. Meeting Types Commonly Used to Include Citizens in Planning.[a]

Public hearings

Large public meetings
   Official presentation followed by question period
   Panelists offering alternative viewpoints
   Informal "town meeting" structure

Large public meetings utilizing small group discussions

Public workshops

Informal small group meetings

[a]Based on information in Creighton (1981). From Ortolano, L., *Environmental Planning and Decision Making.* Copyright © 1984 by John Wiley & Sons, Inc., New York, NY, Reprinted with permission.

assemblies unless provisions are made to break up into small groups for part of the time. Workshops are generally used to have individuals focus on a specific planning task. The least structured meeting types are small, informal get-togethers. These are sometimes held in private homes and provide an opportunity for citizens and agency planners to exchange ideas in a casual setting.

Public involvement programs rarely involve only meetings. Written materials and coverage in the "mass media" are useful in providing information to the public. Public displays and exhibits are valuable, especially when staffed by planners to answer questions. Interviews and mail questionnaires can be effective for obtaining information from the public. By randomly selecting persons to be surveyed, a representative sample of opinions and feelings can be obtained.

TABLE 8. Public Involvement Techniques Not Based on Meetings.[a]

Providing information to the public

   Reports, brochures, and information bulletins
   Mass media coverage
      Press releases
      Radio and television "talk shows"
      Documentary films
   Public displays and exhibits

Obtaining information from the public

   Interviews
   Mail questionnaires

Establishing two-way communications

   Advisory groups (also called task forces and citizens' committees)

[a]From Ortolano, L., *Environmental Planning and Decision Making,* Copyright © 1984 by John Wiley & Sons, Inc., New York, NY, Reprinted with permission.

Table 8 lists public involvement methods that do not rely on meetings. The last technique listed, advisory groups, is different from those discussed above. Agencies often form advisory groups to provide a convenient way of communicating informally with individuals representing a range of interests. Advisory groups can assist an agency in formulating and evaluating alternatives, and in designing other methods to include citizens in planning. For example, an advisory group might help prepare an information brochure that explains the motivations for an agency's planning study.

The design of an effective public involvement program requires both skill and effort. In the United States, many water resources agencies have developed sophisticated instruction manuals to help their staffs avoid common mistakes and take advantage of the experience gained by others. *The Public Involvement Manual* prepared by Creighton (1981) provides detailed instruction in the use of numerous techniques for including citizens in water resources planning. Creighton's position is consistent with that of many agencies in that he encourages public participation in all aspects of water resources planning, not just the evaluation of alternatives.

Environmental Mediation

In many cases, even when there is extensive citizen involvement in evaluating alternative projects, water resources agency decisions are objectionable to some groups. If there is great dissatisfaction, those adversely affected may sue in an effort to modify or halt the agency's final plan. In the United States during the 1970's, many citizens' groups used the courts for this purpose.

There are several reasons why judicial review is often inappropriate as a mechanism for resolving environmental conflicts between citizens' groups and water resources agencies. Court actions are costly and citizens adversely affected by an agency proposal may have difficulty raising the required funds. In addition, litigation can involve much time, sometimes years, before a settlement is reached. Furthermore, courts frequently make their decisions based on procedural (due process) questions, rather than the points of contention between citizens and an agency.

Even though "environmental mediation" is not a technique for evaluating alternative projects, it is included here because it represents an innovative means for resolving disputes that sometimes follow a water resources agency's evaluation. The following definition, used by the Office of Environmental Mediation at the University of Washington in Seattle, illuminates the mediation process:

> Mediation is a voluntary process in which those involved in a dispute jointly explore and reconcile their differences. The mediator has no authority to impose a settlement. His or her strength lies in the ability to assist the parties in resolving their own differences. The mediated dispute is settled when the parties themselves reach what they consider to be a workable solution (Cormick and Patton, 1980).

Since the mid-1970s, there have been dozens of applications of environmental mediation in the United States. Because many of them have involved citizens directly, mediation is sometimes confused with citizen participation in planning. Environmental mediation is not a public involvement technique; it is a process for making decisions. The goal is to resolve environmental conflicts by negotiating a formal agreement that can be implemented.

Environmental mediation is in its formative stages. There are many unsettled questions concerning the funding for and the qualifications of mediators and the selection of participants in the negotiations. In addition, there is a danger that the process will be abused, for example, by providing a delaying tactic for participants unwilling to engage in "good-faith bargaining." However, in light of the successful applications since 1975, mediation appears to be a promising supplement to court actions in resolving environmental disputes. Case studies of environmental mediation applied to water resources project planning are given by Talbot (1983).

The discussion in this section demonstrates that methods for evaluating alternative projects can take many forms. Some consist of computational procedures such as multiobjective programming algorithms. Others take the form of tabular displays of information. Still others revolve around processes for involving citizens in decision making. In any particular planning study, a combination of procedures may be useful in elucidating advantages and disadvantages of alternative projects and assisting decision makers in identifying a preferred plan.

ANALYSES OF AGENCY ENVIRONMENTAL IMPACT ASSESSMENT PRACTICES

Since the passage of the U.S. National Environmental Policy Act of 1969, there have been many surveys and analyses of how federal agencies have implemented the Act. Examples include the work by Andrews (1976), Caldwell (1982), Hill (1977), Mazmanian and Nienaber (1979), Taylor (1984) and the National Science Foundation (1982). An annotated bibliography of the literature on surveys of agency implementation of NEPA is included in Clark, Gisset and Wathern (1980). Reviews of the implementation of EIA legislation and regulations in countries other than the U.S. have also been conducted; see, for example, O'Riordan and Sewell (1981) and Clark *et al.* (1984).

The surveys cited above are *descriptive* in that they characterize actions taken by agencies in implementing NEPA and similar policy directives. Rather than suggest steps an agency *should* follow, the studies generally describe how an agency has responded to NEPA. Some investigations analyze agency responses using social science theories.

The above-noted studies clearly demonstrate the difficulty in reaching generalizable conclusions about what "causes" agencies to consider environmental factors in planning and

decision making. One problem is that, for each agency, a survey must be custom tailored to reflect agency-specific regulations and planning procedures. This makes it impractical to include more than a few agencies in any given research study. Moreover, even if conclusions regarding causal relations could be established rigorously for the particular agencies included in the research, they would soon be obsolete because the environment in which planning is carried out (e.g., the agency regulations applicable for a given study) is continually changing. Questions of rigor apart, however, descriptive studies can provide information useful in clarifying and appraising policy and in providing insights for those charged with issuing guidance for conducting environmental impact assessments.

The discussion below reports on research conducted at Stanford University on how U.S. water resources development agencies have responded to NEPA. The following theme has been dominant in much of the work at Stanford: if the goals of the U.S. National Environmental Policy Act are to be attained, the consideration given to environmental factors must be viewed as an integral part of planning, not as a series of separate activities carried out by environmental specialists to meet a set of laws and regulations. The remainder of this section is based on the Stanford University research and is organized in four parts:

1. Alternative roles of environmental specialists
2. Organizational location of environmental specialists within an agency's field office
3. Use of planning teams
4. Coordination with other agencies and the public

Alternative Roles of Environmental Specialists

Within a water resources planning agency a variety of professionals may be involved in the analysis of environmental issues (e.g., biologists, landscape architects, civil engineers, and sociologists); they are referred to herein as "environmental specialists." Following Jenkins and Ortolano (1978) and Ortolano *et al.* (1979), it is possible to distinguish the following roles for environmental specialists:

- *Planning*—a role as a member of a group involved in planning; this may include extensive coordination of environmental information with outside environmental interests (e.g., fish and wildlife agencies and citizens' groups) and with paid consultants;
- *Report preparation*—a role as the writer of a statement which fully discloses and assesses the environmental impacts of a proposed plan and alternatives to it;
- *Internal report review*—a role as an evaluator of environmental impact statements (EISs) or assessments prepared by other staff members within the agency;
- *Design*—a role as a member of a group involved in detailed design of a proposed plan;
- *System monitoring*—a role as a monitor of project opera-

tions to ensure that desired levels of environmental quality are maintained;
- *External report review* – a role as an evaluator of EISs or assessments for projects of other agencies whose work is related to the agency employing the environmental specialist.

Of the six roles listed above, the one that received the greatest attention in the early 1970's was report preparation. Jenkins (1977), whose research documents this focus on report preparation for the two water agencies he studied, observed: "several [participants in the research interviews] referred to the environmental unit as an 'EIS factory'. The pressure to prepare impact statements to meet legal requirements has kept environmental specialists from becoming very involved as members of project teams for planning studies."

If the role of environmental specialists is viewed narrowly in terms of EIS preparation, it seems to limit their influence on how planning study objectives are defined and alternative projects are formulated. The ability of environmental specialists to affect these two planning activities is of fundamental importance if environmental factors are to be integrated into planning. Ortolano *et al.* (1979) found evidence in support of this view and it is noted below. For water planning studies in their initial stages (e.g., prior to Congressional authorization of a proposed water project), the influence of environmental factors in formulating alternative plans was found to be high in Corps of Engineers and Bureau of Reclamation studies in which environmental specialists had a broad range of duties. This influence on the formulation of alternatives was not evident for Corps of Engineers and Bureau of Reclamation studies in which the role of environmental specialists was defined narrowly to emphasize environmental impact assessments.

A related observation (Ortolano *et al.*, 1979) concerns the importance of viewing the environmental specialist as an *active planner*. The consideration given to environmental factors is not enhanced by restricting the role of environmental specialists to "staff functions," i.e., providing information in response to specific requests made by a planning leader. Environmental information is likely to have a greater influence on planning outcomes if environmental specialists are active as planners, not just staff advisors.

Organizational Location of Environmental Specialists

In examining issues relating to alternative locations for environmental specialists within a field office of a water resources agency, it is useful to distinguish between the early and late stages of a planning investigation. Using the Bureau of Reclamation for illustrative purposes, the early stages of planning are exemplified by a "feasibility investigation," and the late stages are demonstrated by a "definite plan investigation."

For planning studies in their early stages, three of the above-listed roles for environmental specialists are relevant: planning, report preparation, and internal report review. Based on their analysis of water resources agencies, Jenkins and Ortolano (1978) conclude that the planning role of environmental specialists is highly interdependent with the engineering and other tasks (e.g., economic evaluations) associated with water resources planning. The placement of environmental specialists *in the same unit* as the engineers and planners (referred to as the "planning unit") facilitates task integration and the scheduling of work. In contrast, there is a high degree of "differentiation" between the planning task and the tasks of report preparation and internal report review. (Tasks are said to be differentiated if it is useful to have them performed by persons with different orientations or outlooks.) The report preparation and internal review roles require high objectivity since environmental assessments may reveal shortcomings of proposals formulated by the planning unit. Environmental specialists outside the planning unit can be more effective in providing objective analyses of proposed projects. Moreover, in instances where environmental specialists outside the planning unit differ with engineers and planners, the environmental specialists at least have the option of referring the differences to a higher level in the office hierarchy. While such a referral process may not be used frequently, the option of using it can be significant (Jenkins and Ortolano 1978). Thus, for planning investigations in their early stages, there is no single location for environmental specialists that increases the effectiveness of all of their roles.

For planning investigations in their late stages, the observations above must be modified because there is much less effort in formulating new plans and a much greater emphasis on report preparation and coordination with environmental agencies and groups. There are clear advantages for having these latter two activities performed by environmental specialists in a unit separate from the planning unit. In addition to the ability to make referrals to higher levels in the office hierarchy and increased objectivity in report preparation, a separate unit for environmental specialists can facilitate coordination with outside environmental spokespersons. These advantages provide a partial explanation for Price's (1979) finding that the influence of environmental information on the modification of a proposed plan was high in Bureau of Reclamation offices with environmental units separate from planning units.

The arguments presented above suggest that perhaps the best strategy for locating environmental specialists in an office of a water resources agency is to place them in more than one organizational unit. For example, environmental specialists engaged in the definition of objectives, the formulation of alternatives, and other basic planning activities could be located within the planning unit. Other environmental specialists could be located outside the planning unit for purposes of reviewing plans and reports and for providing an interface between engineers and planners and outside environmental spokespersons. This would make it possible to obtain good coordination and integration of environmental perspectives on routine planning activities *and* critical in-house reviews of the planning effort.

Use of Planning Teams

Although teams are often formed to carry out water resources planning, there is a question as to whether they are multidisciplinary or interdisciplinary. When multidisciplinary teams are used, different specialists contribute their expertise individually to a planning study. Interdisciplinary teams have different specialists working together and contributing to the team's overall conception of study objectives, alternative actions, and impact assessment.

Part of the Stanford University research examined whether or not interdisciplinary planning is being utilized by federal water resources agencies. Using Hill's (1977) criteria for deciding on whether planning is interdisciplinary or multidisciplinary, Ortolano *et al.* (1979) characterized planning in the 60 Corps of Engineers and Bureau of Reclamation studies they examined as multidisciplinary. This emphasis on multidisciplinary planning fails to capitalize on the principal advantage of the interdisciplinary team process: the synthesis of judgments and opinions of many individuals to yield an integrated perspective that is broader and more informed than the perspective of any single team member.

Aside from subtleties related to these distinctions, there is good reason to be concerned with how team planning is functioning at the field level. For one thing, NEPA and various planning regulations mandate interdisciplinary planning. There is also evidence suggesting that planning teams can be effective in integrating environmental factors into water resources planning.

Of the studies included in the research by Ortolano *et al.* (1979), planning teams were formed for many of the Bureau of Reclamation investigations and a few of the Corps of Engineers studies. For the most part, however, teams were only used on studies in the early stages of planning ("survey investigations" in the Corps and "feasibility investigations" in the Bureau). For such studies, there are indications that teams can be effective in enhancing the consideration given to environmental factors when:

1. One or more environmental specialists are team members.
2. Representatives of other agencies and environmental interest groups are included on the team.
3. The team is "active" in the sense that it meets frequently as a group.

An important aspect of the research results concerns the composition of planning teams. In many Bureau of Reclamation feasibility studies in which environmental factors in-

fluenced how alternatives were formulated, planning team members represented a broad range of interests such as other federal agencies, state and local agencies, environmental interest groups, and the general public. Frequently the team served as a principal mechanism for maintaining contact with these other agencies and groups. Another notable aspect of the Bureau's approach to planning is its use of subteams and task forces to assist a "main team" on specific issues such as impacts on fisheries. Specialized subteams can increase the effectiveness of planning teams by allowing participants with specific interests to concentrate on their areas of expertise. Representatives from subteams participate in the deliberations of a main team. This allows for synthesizing viewpoints expressed by subteam members with other perspectives represented on the main team.

Quite apart from whether the team approach can be shown to increase the influence of environmental information on planning outcomes, there are agencies that, for one reason or another, choose to adopt the team framework for planning. An analysis by Meersman and Ortolano (1980) of the use of teams in a transportation planning office provides ideas for increasing the effectiveness of planning teams. For example, they suggest expanding the basis for an individual's performance evaluations. If a person's professional efforts are appraised only within his or her organizational unit, the individual does not have strong incentive to participate in activities outside the unit. The extent to which a team member works cooperatively with other members and participates in team planning activities could be emphasized in evaluating the person's professional development and performance. Other suggestions for improving team performance include broadening the range of interests represented on teams and clarifying what is expected of team members.

The above discussion contains possible starting points in developing policies and procedures to increase planning team effectiveness. Organizational and procedural changes such as these will not produce more effective planning without the acceptance of, and commitment to, such changes by water agency staff members and their supervisors. Moreover, the use of an interdisciplinary orientation to water resources planning may involve increases in both time and personnel. It may also cause short term disruptions in customary work patterns as team members become accustomed to their new roles. Wagner and Ortolano (1976) discuss some of the additional difficulties that may be involved in trying to implement interdisciplinary water resources planning.

Team planning, which revolves largely around coordination of efforts and talents within an agency office, is important in integrating environmental factors into water planning. However, it is only one aspect of coordination. Also significant is "external coordination," the communication that occurs when a water agency tries to synchronize its planning with the interests of various outside agencies and groups.

Effectiveness of External Coordination

External coordination is of central importance in integrating environmental factors into all aspects of water planning. In the early stages of planning, external coordination can improve an agency's ability to identify environmental issues that will need to be investigated. As a study progresses, external coordination provides a continuing source of information on alternative actions, impacts, and values. The circulation of a draft EIS, which occurs relatively late in the planning process, allows for substantive public and interagency review of the water agency's judgments and analyses. Extensive coordination can yield useful information for agency decision making and increase both the range of environmental factors considered and the weight given to such factors in evaluating alternative plans.

The findings of Ortolano et al. (1979) confirm the importance of external coordination. They report that the influence of environmental information on decision making is enhanced when water resources agencies involve the public and other agencies in planning. Ortolano and his associates indicate that the role of the public may be especially important in early stages of planning, whereas the role of other agencies may be more influential in late stages. In the early stages of Corps of Engineers planning studies included in the research, other agencies often seemed hesitant to commit their limited resources toward making substantive contributions. Individual citizens and citizens' groups seemed more active than other agencies during these early stages. The opposite was often true for Corps of Engineers studies in their late stages. This was interpreted to indicate that as a Corps planning study draws to a close, other agencies have more opportunities to influence the outcome of the study than do citizens' groups. Sometimes so-called "conservation agencies" such as the U.S. Fish and Wildlife Service become strong advocates for the positions of citizens' groups that they view as being among their constituents.

The research by Ortolano et al. (1979) also provides information on specific coordination techniques that can be effective. For example, the discussion above mentioned that the influence of environmental information on the formulation of alternatives seems to be enhanced when planning teams meet frequently and have a broad membership. Planning workshops (sometimes called "public meetings") are also effective for coordinating interests. Results for all of the Corps of Engineers studies and many of the Bureau of Reclamation studies indicate that planning workshops facilitate communication between planners and concerned agencies and citizens. By avoiding the formality and authoritarian ambience typical of public hearings, workshops allow for more personal, informal communication among participants.

Another useful method of external coordination consists of environmental documents prepared and circulated for outside review before the draft EIS. In a survey of Corps of Engineers and Soil Conservation Service (SCS) planners, Hill and Ortolano (1975) reported that over half of the Corps respondents and about 70% of the SCS respondents felt the most useful review of environmental documents, in terms of improving the final plan, resulted from circulating "pre-EIS documents" (e.g., a preliminary draft EIS) before active planning was completed. The data of Ortolano et al. (1979) also supports the utility of a preliminary environmental assessment report that is circulated widely. Such a document, even one that is very modest in scope and detail, seems to stimulate communications with other agencies and segments of the public in the early stages of a planning investigation. The use of preliminary environmental assessment documents may be effective in overcoming the tendency of some reviewing agencies to put off substantive coordination with water resources agencies until late stages of planning.

This section draws heavily on Stanford University surveys of how federal water resources agencies are implementing NEPA. The findings from these surveys clarify the alternative roles played by environmental specialists and indicate that no single organizational location best serves all of these roles. The findings also emphasize the importance of both team planning and external coordination in integrating environmental factors into water resources planning. The descriptive surveys discussed above represent only part of the research at Stanford. There is also a *prescriptive* part.

MODIFYING PLANNING PROCESSES TO ENHANCE THE EFFECTIVENESS OF EIA

The prescriptive component of the Stanford University research rests on the premise that environmental factors should receive increased consideration in water resources planning, relative to the attention given environmental factors prior to 1970. An elaborate defense of the premise is not needed. Recent laws and regulations require federal water resources agencies to give added consideration to environmental factors. The most widely known of these is NEPA, but other mandates could also be cited.

Even presupposing that federal water resources agencies have a mandate to pay increased attention to environmental factors, there is still the question of how this requirement should be implemented. The prescriptive research at Stanford University employs the following reasoning. If environmental considerations are to be an integral part of water resources planning, then traditional agency planning processes may need to be restructured. Without such restructuring, there may be a tendency to simply "tack on" environmental impact assessment activities as an insignificant appendage to customary planning procedures. Moreover, environmental factors might tend to be considered only in the narrow context of impact assessment, not in relation to other planning activities such as the formulation of alternatives.

Research on restructuring planning processes was conducted at Stanford University in cooperation with the Army Corps of Engineers. The research contributed to the Corps' modification (during the 1970's) of the way it undertook its water resources planning. The modified process, which is reflected in Corps planning regulations (e.g., U.S. Army, 1975), is much more open and iterative than the process used by the Corps in the 1960's. The new planning process is "open" in that it relies on continual two-way communication between Corps planners and a wide range of interested citizens and government agencies beginning at the earliest stages of a planning study. The iterative nature of the process is reflected in its concurrent (as opposed to sequential) performance of the four traditional planning activities: identification of concerns, formulation of alternatives, impact assessment, and plan ranking (or evaluation). The research that contributed to the development of this process involved both conceptualization (Ortolano, 1974) and the subsequent field testing of concepts in a Corps planning study of the San Pedro Creek in Pacifica, California. The field test, which included an extensive follow-up evaluation, is described by Wagner and Ortolano (1976).

Another research project in which a Corps of Engineers planning process was modified to enhance the effectiveness of EIA involved the question of cumulative environmental impacts. The U.S. Council on Environmental Quality (1978) defines a cumulative impact as one "which results from the incremental impact of [a proposed] action when added to other past, present and reasonably foreseeable future actions." The Council indicates that cumulative impacts can result from individually minor but collectively significant actions taking place over a period of time.

An example of a cumulative impact is the effect of filling small portions of a bay to accommodate new shoreline development projects. A new project might cause only minor effects, such as reducing a very small fraction of the bay's surface area. Collectively, however, the impacts of many small development projects built on fill could decrease the bay's area significantly.

The San Francisco District Office of the Corps of Engineers wanted procedures for considering cumulative effects of its decisions to issue permits for projects involving dredging, filling or construction in navigable waters. Researchers at Stanford University proposed and tested an approach for assessing cumulative impacts of District Office permit decisions. The approach rests on an initial "carrying capacity" study that is updated using information generated during reviews of permit applications. In this context, carrying capacity represents the maximum number of boat berths in the study area consistent with maintaining acceptable levels of environmental quality and public welfare. Policies to guide shoreline development were identified during the carrying capacity study, and they provided a basis for ac-

tions that District Office staff took in reviewing new permit applications. The approach to cumulative impact assessment was implemented on a trial basis for proposed developments in the Oakland Bay Estuary in northern California. Key cumulative impact issues included water quality, boating congestion, and public access to the estuarine shoreline. Details of the approach and the case study application are given by Contant (1984).

In contrast to descriptive analyses of how water agencies are implementing NEPA, prescriptive studies rest on value judgments about what agencies *should be doing*. Consider the two studies above. One argued that the Corps could not effectively implement NEPA by simply appending EIA requirements to planning regulations of the 1960's. A revamping of traditional procedures made the Corps planning process more open and iterative. The other study presupposed that the Corps should account systematically for cumulative impacts of its permit decisions. Using carrying capacity concepts, the Corps' permit review process was changed so that cumulative effects could be considered.

SUMMARY AND CONCLUSIONS

Water resources professionals investigating the environmental impacts of a proposed project can draw on a voluminous literature summarizing past experience and recent research. Numerous procedures exist for identifying, predicting and evaluating the environmental effects of water projects.

Impact identification procedures generally take the form of checklists of environmental and social impacts that typically accompany water resources development. Case studies, as exemplified by analyses of the Aswan High Dam project, clarify the nature of impacts and how they have been assessed.

Impact forecasting consists of predicting the changes in environmental quality likely to result from a proposed development. Forecasting is discussed in terms of judgmental, physical and mathematical prediction methods.

Evaluation procedures account for environmental, social and economic effects in establishing a preference ordering among alternative projects. Evaluation methods range from computationally oriented schemes, such as multiobjective programming, to procedures for including various agencies and segments of the public in water resources planning.

Research on how federal water resources agencies have responded to NEPA identifies parameters affecting the way environmental factors influence decision making. Important parameters concern the roles assigned to environmental specialists and where those specialists are located within water resources agencies. Coordination issues are also important. The use of team planning and the ways in which water resources agencies interact with other agencies and citizens' groups have a bearing on how environmental factors affect decisions.

In addition to studies of agency implementation of NEPA, research has also explored what water resources agencies can do to more carefully account for environmental quality. An example is research to make Corps of Engineers water project planning more iterative and open to contributions from other agencies and the public. Another example uses carrying capacity analysis to account for cumulative environmental impacts of Corps permit decisions.

Much information exists to guide the consideration of environmental quality in water resources planning. However, the availability of information is not enough to assure that environmental factors will be given adequate attention. Even the existence of a well intentioned law such as the National Environmental Policy Act is not sufficient, since experience shows that meeting the Act's legal provisions does not assure attainment of its environmental quality goals. A critical factor is the commitment of water resources agencies to use existing information and innovative planning procedures to account for environmental quality in decision making.

ACKNOWLEDGEMENTS

The author acknowledges permission to reprint portions of his previously published works: *Environmental Planning and Decision Making*. Copyright ©1984 by John Wiley and Sons, Inc.; and "Integrating Environmental Considerations into Infrastructure Planning" in *Integrated Impact Assessment*, edited by F. A. Rossini and A. L. Porter. Copyright ©1983 by Westview Press, Inc. The author also thanks Robert Judd for helpful comments on an early draft of this chapter.

REFERENCES

1. Andrews, R. N. L., *Environmental Policy and Administrative Change*, Lexington Books/D.C. Heath, Lexington, Kentucky (1976).
2. Appleyard, D., P. Bosselmann, R. Klock and A. Schmidt, "Periscoping Future Scenes, How to Use an Environmental Simulation Lab," *Landscape Architecture*, Vol. 69, No. 5, pp. 487-510 (1979).
3. Armstrong, J. S., *Long-Range Forecasting: From Crystal Ball to Computer*, John Wiley & Sons, New York (1978).
4. Biswas, A. K., "Water Development and Environment," in Lohani, B. N. and N. C. Thanh (eds.), *Water Pollution Control in Developing Countries*, Vol. II, Asian Institute of Technology, Bangkok, Thailand (1978).
5. Brown, J. W., M. R. Walsh, R. W. McCarley, A. J. Green, Jr. and H. W. West, *Models and Methods Applicable to Corps of Engineers Urban Studies*, Miscellaneous Paper H-74-8, U.S. Army Engineer Waterways Experiment Station, Vicksburg, Mississippi (1974).
6. Caldwell, L. K., *Science and the National Environmental Policy Act, Redirecting Policy Through Procedure Reform*, University of Alabama Press, University, Alabama (1982).
7. Canter, L. W., *Water Resources Assessment—Methodology & Technology Sourcebook*, Ann Arbor Science, Ann Arbor, Michigan (1979).
8. Canter, L. W., *Environmental Impact of Water Resources Projects*, Lewis Publishers, Inc., Chelsea, Michigan (1985).
9. Clark, B. D., R. Bisset and P. Wathern, *Environmental Impact Assessment, A Bibliography with Abstracts*, Mansell Publishing, London (1980).
10. Clark, B. D., A. Gilad, R. Bisset and P. Tomlinson (eds.), *Perspectives on Environmental Impact Assessment*, D. Reidel Pub. Co., Dordrecht, The Netherlands (1984).
11. Cohon, J. L., *Multiobjective Programming and Planning*, Academic Press, New York (1978).
12. Contant, C. K., *Cumulative Impact Assessment: Design and Evaluation of an Approach for the Corps of Engineers Permit Program at the San Francisco District*, Ph.D. Dissertation, Stanford University, Stanford, California (1984).
13. Cormick, G. W. and L. Patton, "Environmental Mediation: Defining the Process through Experience," in L. M. Lake (ed.), *Environmental Mediation: The Search for Concensus*, Westview Press, Boulder, Colorado (1980).
14. Crawford, N. H. and R. K. Linsley, *Digital Simulation in Hydrology: The Stanford Watershed Model IV*, Technical Report No. 39, Department of Civil Engineering, Stanford University, Stanford, California (1966).
15. Creighton, J. L., *The Public Involvement Manual*, Abt Books, Cambridge, Massachusetts (1981).
16. Curran Associates, Inc., *Guidelines for EPA Review of Environmental Impact Statements on Projects Involving Impoundments*, report prepared for U.S. Environmental Protection Agency by Curran Associates, Inc., Northampton, Massachusetts (1975).
17. Curran Associates, Inc., *Guidelines for Review of Environmental Impact Statements—Vol. IV, Channelization Projects*, report prepared for U.S. Environmental Protection Agency by Curran Associates, Inc., Northampton, Massachusetts (1976).
18. Fahim, H. M., *Dams, People and Development: The Aswan High Dam Case*, Pergamon Press, New York (1981).
19. Finsterbusch, K., *Understanding Social Impacts, Assessing the Effects of Public Projects*, Vol. 110, Sage Library of Social Research, Sage Publications, Beverly Hills, Calif. (1980).
20. Freeman, A. M., III, *The Benefits of Environmental Improvement, Theory and Practice*, Johns Hopkins University Press for Resources for the Future, Inc., Baltimore, Maryland (1979).
21. Freeman, A. M., III, *Air and Water Pollution Control, A Benefit-Cost Assessment*, John Wiley & Sons, New York (1982).
22. Goicoechea, A., D. R. Hansen and L. Duckstein, *Multiobjec-

tive Decision Analysis with Engineering and Business Applications, John Wiley & Sons, New York (1982).

23. Gordon, S. I., *Computer Models in Environmental Planning*, Van Nostrand Reinhold Co., New York (1985).

24. Hagan, R. M. and E. B. Roberts, "Ecological Impacts of Water Storage and Diversion Projects" in Goldman, R., J. McEroy, III and P. J. Richerson (eds.), *Environmental Quality and Water Resources Development*, W. H. Freeman and Co., San Francisco, pp. 196–215 (1973).

25. Hill, W. W., *The National Environmental Policy Act and Federal Water Resources Planning: Effects and Effectiveness in the Corps and SCS*, Report IPM-4, Department of Civil Engineering, Stanford University, Stanford, California (1977).

26. Hill, W. W. and L. Ortolano, "Effects of NEPA's Review and Comment Process on Water Resources Planning: Results of a Survey of Planners in the Corps of Engineers and Soil Conservation Service," *Water Resources Research*, Vol. 12, No. 6, pp. 1093–1100 (1976).

27. Hitchcock, H., "SIA: On Leaving the Cradle" in Tester, F. J. and W. Mykes (eds.), *Social Impact Assessment: Theory, Method and Practice*, Kananaskis Centre for Environmental Research, University of Calgary, Alberta (1981).

28. Jenkins, B. R., *Changes in Water Resources Planning: An Organization Theory Perspective*, Ph.D. Dissertation, Stanford University, Stanford, California (1977).

29. Jenkins, B. R. and L. Ortolano, "Environmental Specialists in Water Agencies," *Proc. American Society of Civil Engineers, Journal of the Water Resources Planning and Management Division*, Vol. 104, No. WR1, pp. 61–74 (1978).

30. Kneese, A. V., *Measuring the Benefits of Clean Air and Water*, Resources for the Future, Inc., Washington, D.C. (1984).

31. Lichfield, N., P. Kettle and M. Whitbread, *Evaluation in the Planning Process*, Pergamon, Oxford (1975).

32. Linsley, R. K., "Rainfall–Runoff Models" in A. K. Biswas (ed.), *Systems Approach to Water Management*, McGraw-Hill, New York, pp. 16–53 (1976).

33. Lohani, B. N., *Environmental Quality Management*, South Asian Publishers, New Delhi, India (1984).

34. Maass, A., "Benefit-Cost Analysis: Its Relevance to Public Investment Decisions," in A. V. Kneese and S. C. Smith (eds.), *Water Research*, John Hopkins University Press for Resources for the Future, Inc., Baltimore, Maryland, pp. 311–328 (1966).

35. Major, D. C., *Multiobjective Water Resources Planning*, Water Resources Monograph 4, American Geophysical Union, Washington, D.C. (1977).

36. Marglin, S. A., "Objectives of Water Resource Development: A General Statement," in A. Maass *et al.* (eds.), *Design of Water-Resource Systems*, Harvard University Press, Cambridge, Massachusetts, pp. 17–87 (1962).

37. Mazmanian, D. A. and J. Nienaber, *Can Organizations Change? Environmental Protection, Citizen Participation and the Corps of Engineers*, The Brookings Institution, Washington, D.C. (1979).

38. Meersman, J. J. and L. Ortolano, "Environmental Considera-

tions in Highway Planning," *Proc. American Society of Civil Engineers, Journal of the Transportation Engineering Division*, Vol. 106, No. TE4, pp. 471–483 (1980).

39. Mitchell, A., B. H. Dodge, P. G. Kruzic, D. C. Miller, P. Schwartz and B. E. Suta, *Handbook of Forecasting Techniques*, IWR Report 75-7, U.S. Army Engineer Institute for Water Resources, Ft. Belvoir, Virginia (1975).

40. McEvoy, J., III and T. Dietz, (eds.) *Handbook for Environmental Planning, The Social Consequences of Environmental Change*, John Wiley & Sons, New York (1977).

41. O'Riordan, T. and W. R. D. Sewell (eds.), *Project Appraisal and Policy Review*, John Wiley & Sons, Chichester (1981).

42. Ortolano, L., "A Process for Federal Water Planning at the Field Level," *Water Resources Bulletin*, Vol. 10, No. 4, pp. 766–778 (1974).

43. Ortolano, L., *Environmental Planning and Decision Making*, John Wiley & Sons, New York (1984).

44. Ortolano, L. (ed.), *Analyzing the Environmental Impacts of Water Projects*, IWR Report 73-3, U.S. Army Engineer Institute for Water Resources, Ft. Belvoir, Virginia (1973).

45. Ortolano, L., C. M. Brendecke, J. E. Price and J. J. Meersman, *Environmental Considerations in Three Infrastructure Planning Agencies: An Overview of Research Findings*, Report IPM-6, Department of Civil Engineering, Stanford University, Stanford, California (1979).

46. Porter, A., F. A. Rossini, S. R. Carpenter and A. T. Roper, *A Guidebook for Technology Assessment and Impact Analysis*, Elsevier North Holland, New York (1980).

47. Price, J. E., *Consideration of Environmental Quality in Bureau of Reclamation Planning*, Ph.D. Dissertation, Stanford University, Stanford, California (1979).

48. Rau, J. G. and D. C. Wooten (eds.), *Environmental Impact Analysis Handbook*, McGraw-Hill, New York (1980).

49. Shamir, U., "Experiences in Multiobjective Planning and Management of Water Resources Systems," *Journal of Hydrological Sciences*, Vol. 28, No. 1, pp. 77–92 (1983).

50. Talbot, A. R., *Settling Things, Six Case Studies in Environmental Mediation*, The Conservation Foundation, Washington, D.C. (1983).

51. Taylor, S., *Making Bureaucracies Think, The Environmental Impact Statement Strategy of Administrative Reform*, Stanford University Press, Stanford, California (1984).

52. Tracor, Inc., *Estuarine Modeling: An Assessment*, report prepared for the U.S. Environmental Protection Agency by Tracor, Inc., Austin, Texas (1971).

53. U.S. Army Corps of Engineers, *Planning Process: Multiobjective Planning Framework*, ER1105-2-000, Office of the Chief of Engineers, Washington, D.C. (1975).

54. U.S. Army Corps of Engineers, *Proceedings of the Social Scientists Conference on Social Aspects of Comprehensive Planning*, Volumes 1–3, U.S. Army Engineer Institute for Water Resources, Ft. Belvoir, Virginia (1977).

55. U.S. Council on Environmental Quality, "Regulations for Implementing the Procedural Provisions of the National En-

vironmental Policy Act," *Federal Register*, 43:55978-56007 (November 28, 1978).

56. U.S. Environmental Protection Agency, *The Control of Pollution from Hydrographic Modification*, Report EPA-430-2F-9-73-017, Washington, D.C. (1973).

57. U.S. Environmental Protection Agency, *Impact of Hydrologic Modifications on Water Quality*, Report EPA-600/2-75-007, Washington, D.C. (1975).

58. U.S. Environmental Protection Agency, *Environmental Assessment of Water Quality Management Planning*, Technical Guidance Memorandum TECH-28, Washington, D.C. (1977).

59. U.S. National Science Foundation, *A Study of Ways to Improve the Scientific Content and Methodology of Environmental Impact Analysis*, Final Report on Grant PRA-79-10014, prepared by School of Public and Environmental Affairs, University of Indiana, Bloomington (1982).

60. Wagner, T. P. and L. Ortolano, *Testing an Iterative Open Process for Water Resources Planning*, Report 76-2, U.S. Army Engineer Institute for Water Resources, Ft. Belvoir, Virginia (1976).

61. Waterbury, J., *Hydropolitics of the Nile Valley*, Syracuse University Press, Syracuse, New York (1979).

CHAPTER 7

Evaluation of Runoff and Erosion from Surface-Mined Areas

DAVID M. HARTLEY* AND HELEN M. WEAGRAFF**

INTRODUCTION

The purpose of this chapter is to present and discuss methods for calculating runoff and soil loss from surface mined areas. The section will be of primary interest to the surface mine hydrologist because applications stress typical surface mine problems. However, the methods included herein are not exclusively surface mine methods and should be of interest to engineers, planners and concerned citizens who wish to evaluate the response of watersheds to agricultural, construction, or other disturbances.

One of the most difficult tasks facing the surface mine hydrologist is to determine the probable hydrologic consequences (PHC) of planned mining activities. The effect of these activities on surface runoff and sediment yield need to be evaluated for both PHC determination and for design of impact mitigation measures such as sediment ponds, diversion channels, culverts and surface treatments. Consequently, this chapter focuses on predictive methods which quantitatively estimate the surface runoff and erosion response to anticipated landscape disturbances caused by mining.

The methods included in this discussion were selected based on several criteria including their acceptability to regulatory authorities, predictive capability, adequacy of documentation and ease of use. Application of these criteria is somewhat subjective; therefore interested readers are encouraged to seek additional methods and more complete discussion of selected methods in the available literature. The reference list at the end of the chapter will provide a good place to begin this search. When alternate methods are

available it may be advantageous to compare estimates from different methods and select values which are most appropriate for the application under consideration. Regulatory authorities are often interested in "worst-case" scenarios and may require use of more than one method in order to establish a "worst-case" estimate of a given quantity. Also, the acceptability of methods depends both on region and regulatory agency. Local offices of pertinent authorities should be contacted to check their acceptance of particular methods.

The chapter applies two levels of treatment to selected methods depending on the method's complexity. Simple or moderately complex methods are presented in detail with all instructions, tables and figures necessary for using the method. More complex methods, such as computer programs which simulate watershed response would require too much space to describe completely. Their essential features are summarized and the means by which a prospective user obtains source codes, users manuals and advice on the programs are explained.

PHYSICAL SYSTEM

Surface runoff and erosion can be viewed as responses of a hydrologic system to driving forces of rainfall or snowmelt. The system of primary concern in this case is the soil surface with its topography and vegetation and the first meter or so of the soil profile below the surface. The primary input to this system is precipitation. Outputs include surface runoff with associated sediment transport, percolation to groundwater, and evapotranspiration to the atmosphere.

Runoff Processes

Vegetation intercepts a certain amount of precipitation on leaf and stem surfaces where it can evaporate and exit the

*Hydro-Ecosystems Research Group, U.S. Department of Agriculture, Agricultural Research Service, Fort Collins, CO
**Consultant in Environmental Hydrology, Fort Collins, CO

TABLE 1. Major Water Related Impacts of Surface Mining (Tourbier et al. [59]).

1. Water Quantity	Description of Impact	Major Operation Causing Impact	Remedial Measures
1:1 Alteration of flow patterns of streams.	Disturbing the surface during mining may cause increased infiltration of water. But often, consolidation causes increased runoff and reduced infiltration which can cause flooding and erosion problems, and may reduce recharge of aquifers and base flow of streams. Local increases in runoff also may originate from haul roads, etc. Runoff will increase due to excessive compaction during reclamation and the elimination of surface storage by creating smooth slopes.	Removal of vegetation, and all operations involving shifting and regrading and consolidation of overburden. All operations which increase the impermeability of the land surface.	Disturb smallest practicable area at any one time. Reclaim as contemporaneously as practicable. Design haul roads so as to minimize any increase in runoff.
1:2 Lowering of ground-water.	Dewatering the pit may cause a lowering of the ground-water. Deep exploratory bore-holes may also break through an impermeable stratum which confines an aquifer causing the aquifer to leak to lower strata.	Pit dewatering. Exploration boreholes. Mining through a stratum which previously confined an aquifer.	Casing and sealing of drilled holes. Plan mine excavation so as to prevent adverse impact.
1:3 Change in storage capacity and transmissibility of overburden.	Decrease in groundwater recharge may result from reduced permeability caused by the removal of vegetation. The removal and replacement of overburden will change both its storage capacity and transmissibility (often increasing both which can be a significant improvement). Vertical leakage to underlying aquifers can increase transmissibility.	Clearance of vegetation. Shifting, regrading and consolidation of overburden. Exploration boreholes. Blasting which causes fracturing and disturbance of basement rock.	Use straw dikes, riprap, check dams, etc. to reduce runoff volume. Minimize disturbance to prevailing hydrologic balance.
2. Water Quality			
2:1 Acidity.	Highly acidic runoff from mined sites results from the exposure of pyritic materials to air and water. Low pH tends to make some compounds toxic to plants, particularly Al and Mn. May cause local groundwater supply to become less than potable	Exposure of pyritic material, often lying in close proximity to coal, oxygen and water. The cause may be material exposed in exploration bore-holes, material in the pit bottom, material backfilled too close to the surface, or material used in road construction. Also, careless hauling of previously identified acid-producing materials causes this problem.	Conduct coal exploration in a manner which minimizes disturbance of hydrologic environment. Prevent or remove water from contact with acid-forming materials during mining operations. Bury acid-forming spoil. Correct pH before discharge of water from site. Acid-forming materials may not be used in construction of haul roads.

(continued)

TABLE 1 (continued).

2. Water Quality	Description of Impact	Major Operation Causing Impact	Remedial Measures
2:2 Sedimentation; Suspended solids.	Erosion of overburden materials may result in very high levels of sediment in runoff from mine sites, which causes a deterioration of stream health, silting of streambeds, etc. Loss of topsoil. Lessens the potential for post-mining use.	All mining operations involving earthmoving. Also haul roads may be serious sources of sediment.	Minimize erosion to the greatest extent possible. Reclaim as contemporaneously as practicable. Manage haul roads so as to cause no additional contribution of suspended solids to runoff flow. Provide sediment ponds.
2:3 Hardness; Deposit of iron hydroxide.	Hardness is rarely a serious problem. However, acidic drainage which is neutralized by treating with lime or limestone will increase in hardness. Neutralization will cause the deposit of iron hydroxide (Yellow Boy) and other compounds which may cause problems.	Operations involving the treatment of acid-forming materials.	Monitor surface water and groundwater. Treat acid water only as needed.
2:4 Groundwater pollution.	Groundwater pollution can result from acid water leaching into the groundwater. This This may be a problem when acid-producing material is placed so as not to prevent oxidation and leaching. Consolidation and in some cases sealing the acid-producing material should prevent this problem.	Results from placement of acid-forming materials during regrading where oxidation and leaching can take place.	Place backfill material so as to prevent groundwater pollution.

3. Other Water Related Problems

| 3:1 Instability. | Infiltration of water into the spoil may cause instability and slumping. Most reclamation measures seek to reduce run-of and increase infiltration but in cases where spoil has low shear strength the policy should be to prevent excessive seepage. A slide may have an adverse effect on public property, health, safety or the environment. | This problem occurs mostly on steep sites, particularly for large fills, Head of Hollow and Valley Fills. Providing bench or barrier on outslope. Backfilling and grading. | Provide barrier so as to assure stability. Backfill and grade so as to insure stability. Construct a subdrainage system. |
| 3:2 Erosion. | Besides giving rise to sedimentation problems, gully erosion may be so serious as to make it necessary to regrade the site. Careful attention to surface configuration and rapid protection with vegetation will avoid this problem. | Regrading operations. Revegetation operations. | Reclaim as contemporaneously as practicable. Perform regrading operations along contour. Regrade or stabilize rills or gullies. |

system. The amount evaporated is generally insignificant unless precipitation volume is low or vegetation density is high. Up to 100 percent of precipitation which is not intercepted infiltrates into the soil. The remainder becomes runoff. The amount infiltrated depends primarily on the characteristics of the top few centimeters of soil. The porosity, pore size distribution, and tortuosity of soil pores in this layer have a substantial effect on infiltration rates. Sands generally exhibit higher infiltration rates than loams or clays which have more porosity but much smaller pore sizes [20]. Other factors which affect the amount of infiltration include the level of saturation of the soil at the beginning of precipitation, the amount of depression storage on the soil surface, the vegetation density and the time distribution of precipitation intensities. All of these factors except for the precipitation pattern can be affected by mining and reclamation operations which may temporarily or permanently alter land surfaces, soil profile and vegetation characteristics.

Surface runoff begins after rainfall or snowmelt exceeds the infiltration rate and fills surface depressions. Surface runoff includes overland and channel type flow. Overland flow can occur as both sheet and rill flow. In sheet flow the water moves over the soil surface in thin broad layers while in rill flow it moves in numerous, small ephemeral channels. Rill flow tends to dominate on steep, eroded slopes. Overland flow velocities and runoff quantities can be reduced by the presence of vegetation or mulch ground cover. Ground cover increases hydraulic flow resistance and infiltration rates. Cover protects the soil surface from sealing by raindrop impact and vegetation opens up pores in surface soil and reduces soil saturation levels through evapotranspiration. Vegetation leaves, stems and litter increase hydraulic resistance and reduce velocity by obstructing the flow.

Erosion Processes

Erosion begins when raindrops strike the soil surface and detach particles from the soil mass. When surface runoff begins, these particles are transported downslope by overland and channel flow. In areas of overland type flow where the depth of water may be on the order of millimeters, raindrops continue to dislodge soil particles. On steeper and longer slopes, flowing water may also dislodge soil particles and cut small channels or rills in the soil surface. Erosion and soil transport in the rills is accelerated.

Erosion studies show that soil particle detachment is directly related to rainfall and runoff energy [16]. This underscores the importance of vegetation cover in protecting surfaces from erosion. As already mentioned, vegetation generally enhances infiltration and reduces both runoff volumes and runoff velocity. Additionally, vegetation reduces both raindrop and runoff energy by breaking up large raindrops and retarding flow velocity and shear stress at the soil surface. Thus vegetation functions in multiple

ways to reduce soil erosion. A comprehensive discussion of erosion phenomena as it relates to small watersheds in general is given by Foster [16].

Mining operations often result in compacted soil with reduced porosity as well as reduced vegetation cover. These disturbances are likely to reduce infiltration, increase the volume and peak rate of surface runoff and increase erosion and sediment yield unless conservation measures are incorporated into the operations. Tourbier et al. [59] associated specific surface mining operations with potential hydrologic effects including alteration of runoff characteristics and acceleration of soil erosion. Table 1 presents a summary of their findings.

METHODS OF SURFACE RUNOFF ESTIMATION

Depending on the situation, it may be desirable to estimate runoff volume, peak flow rates, or entire storm runoff hydrographs. Runoff hydrographs provide the most information and consequently are the most difficult to estimate. Often, estimation of runoff volumes or peak flow rates are sufficient and simpler methods can be used.

In the discussion of rainfall-runoff relationships, the terms "depth" and "volume" are used interchangeably when referring to amounts of rainfall, runoff, interception, depression storage and infiltration. Numerically, any depth can be converted to a volume by multiplying it by the watershed area.

Design Storms

In most cases surface mine hydrologists must make predictions of runoff or sediment yield in response to a design storm. Design storms are defined by a return period and duration (e.g., a 10-year, 24-hour precipitation event) which is specified by surface mine regulations or local regulatory authorities. Design storm rainfall depth and pattern of rainfall intensities are usually determined from historical records of precipitation at or near the mining site. Rainfall depths for various design storms are available from maps published by the National Weather Service [25,33]. Realistic and typical rainfall intensity patterns for design storms at a particular location are difficult to establish without considerable precipitation data. The U.S. Soil Conservation Service (SCS) analyzed available data in the United States and suggested two typical distributions of accumulated rainfall as shown in Figure 1. The "type I" distribution applies to states west of the Cascade and Sierra Nevada mountains as well as Alaska, Hawaii and North Dakota. The "type II" distribution applies to the remaining states. Rainfall intensity for any period in a storm is given by multiplying the average slope of the appropriate distribution during that period by the average rainfall rate of the entire storm. For example, if we wish to know the rainfall rate during the half hour spanning the temporal midpoint of a 6 hour storm with a total depth of 2.0 inches and a type II distribution, we first compute the average storm intensity of 2.0 ÷ 6 = 0.33 inches per hour. The slope of the type II distribution between 0.44 and 0.56 is 4.8. Hence the intensity is 4.8 × 0.33 = 1.6 inches per hour.

If historical rainfall data are available for locations near the mine site it may be possible to determine design storm characteristics more accurately than by using NWS and SCS methods. Analysis of historical precipitation data relies on statistical and probabilistic methods which are discussed by Barfield, et al. [2] and the OSM [39,40].

Snowmelt

Snowmelt rates are typically much smaller and less variable than rainfall rates in most surface mining areas. Consequently snowmelt is usually not a major consideration in quantifying peak flows or in estimating erosion from watersheds disturbed by surface mining. In some watersheds, however, snowmelt may represent a significant component of annual runoff volume. Simple methods for estimating snowmelt rates and snowmelt runoff are presented by the SCS [55].

Estimation of Runoff Volume

SCS CURVE NUMBER METHOD

The curve number method of estimating direct runoff volume uses three variables; the total rainfall volume, the antecedent moisture condition and the watershed soil-cover complex. Direct runoff is predominately surface runoff but in some watersheds may also include rapid subsurface flow. The Curve Number method is reported to be applicable to both small and large watersheds up to basin scale [55]. The following equation expresses the curve number (CN) relationship:

$$Q = \frac{(P - 0.2\ S)^2}{P + 0.8\ S} \qquad (1)$$

in which Q is the direct runoff in inches, P is the rainfall in inches and S is a parameter which combines the effect of antecedent moisture and watershed soil-cover conditions. The parameter S is usually expressed in terms of a curve number, CN:

$$S = \frac{1000}{CN} - 10 \qquad (2)$$

in which S is in inches. Once the curve number for the wa-

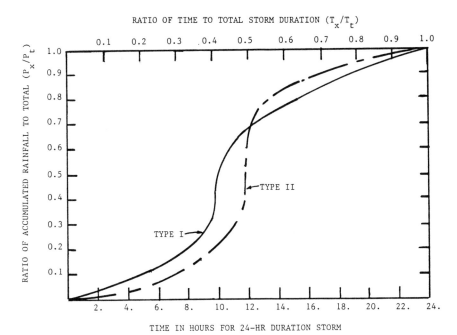

TYPE I Coastal side of Sierra Nevada and Cascade Mountains in California, Oregon and Washington, the Hawaiian Islands and Alaska.

TYPE II Remaining United States, Puerto Rico and Virgin Islands.

FIGURE 1. U.S. Soil Conservation Service design storms (SCS, [54]).

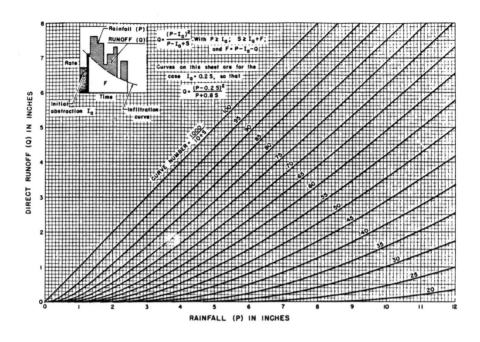

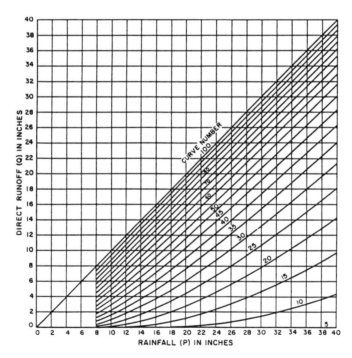

FIGURE 2. Solution to the runoff equation, $Q = (P - 0.2S)^2/(P + 0.8S)$ (SCS, [55]).

tershed is known, the direct runoff can be determined using Figure 2, or S and Q can be determined using Equations (2) and (1).

The major challenge in applying the curve number method is the selection of an appropriate curve number value. There is some subjectivity involved in this process, especially when dealing with severely disturbed watersheds. Hawkins [24] and Bondelid, et al. [6] have analyzed the effect of errors in curve number selection on runoff estimates. Appropriate regulatory authorities should be consulted for curve number selection procedures. The steps usually involved in curve number selection are as follows:

1. Determine the hydrologic soil group (A, B, C or D). For areas which have not been stripped, compacted or otherwise severely altered by mining activities, the soil group can be determined by looking up the soil name in the Appendix or by checking with a local Soil Conservation Service office. For severely disturbed areas, the hydrologic soils group can be chosen based on the definitions given in Table 2.
2. Determine the Curve Number (*CN*) for type II antecedent moisture conditions (AMC-II) based on land use and hydrologic soil group using Table 3.
3. Determine the appropriate AMC condition (I, II or III) based on the definitions in Table 4 and adjust the *CN* value from step 2 using Table 5.

If the watershed of interest contains areas of distinctly different soils or land use, the curve number procedure should be applied to each distinct subarea and runoff from all subareas should be summed to obtain the total runoff from the watershed.

The accuracy with which the *CN* method predicts runoff depths from rainfall depths depends to some extent on the ability of the user to choose an "optimal" curve number value. As pointed out earlier, there is some subjectivity involved in curve number selection, especially for disturbed watersheds where no rainfall-runoff data exists.

However, even if the "optimal" or "best" *CN* value were known for a watershed, accuracy is still limited because of the assumptions underlying the *CN* relationship given in Equation (1). As shown in Figure 2, the lines of Q vs. P tend to become parallel to the $Q = P$ lines as P increases. This trend is tantamount to assuming that as rainfall depth becomes large for any given *CN* value, the difference between rainfall and runoff approaches some constant value. In many actual cases, a plot of runoff vs. rainfall for a large storm on a watershed will show that the difference between P and Q increases; that is the Q vs. P line diverges from the $Q = P$ line. This situation would correspond to a deep soil with a saturated hydraulic conductivity in the same order of magnitude as the rainfall rate. In such a case, the concept embodied in the S parameter of a finite storage capacity which is exhausted as total rainfall depth increases is not realistic.

TABLE 2. Hydrologic Soil Group Definitions (SCS, [55]).

A. (Low runoff potential). Soils having high infiltration rates even when thoroughly wetted and consisting chiefly of deep, well to excessively drained sands or gravels. These soils have a high rate of water transmission.

B. Soils having moderate infiltration rates when thoroughly wetted and consisting chiefly of moderately deep to deep, moderately well to well drained soils with moderately fine to moderately coarse textures. These soils have a moderate rate of water transmission.

C. Soils having slow infiltration rates when thoroughly wetted and consisting chiefly of soils with a layer that impedes downward movement of water, or soils with moderately fine to fine texture. These soils have a slow rate of water transmission.

D. (High runoff potential). Soils having very slow infiltration rates when thoroughly wetted and consisting chiefly of clay soils with a high swelling potential, soils with a permanent, high water table, soils with a claypan or clay layer at or near the surface, and shallow soils over nearly impervious material. These soils have a very slow rate of water transmission.

Another problem with the *CN* method is that it ignores the time pattern of rainfall intensities. The method predicts the same amount of runoff for a short intense rainfall as for a long, low intensity rainfall as long as each contains the same rainfall volume. Infiltration may absorb up to 100 percent of the water from the low intensity storm and perhaps only a small percentage of the high intensity one.

These discrepancies between watershed behavior, and the form of the *CN* relationship explain much of the scatter of rainfall-runoff data points about the best fit line representing Equation (1). Further discussion of assumptions and limitations of this method are given by SCS [55], and Montgomery [34].

Despite its shortcomings, the work done by the SCS and others in relating *CN* (or *S*) values to different soils, antecedent moisture condition and land use activities make the curve number method a useful tool for the surface mine hydrologist.

Example: SCS Curve Number Method

Use the Curve Number (*CN*) method to estimate direct runoff from a watershed of 1000 acres located southwest of Steamboat Springs, Colorado. The soil is a complex of 30% Cochetopa loam and 70% Aaberg silty clay. The vegetation cover is 50% herbaceous and 50% sagebrush. Prior to mining, the hydrologic condition of the watershed is fair. After mining it is assumed that the condition will have deteriorated to poor. Determine runoff for the 10-year, 24-hour, rainfall both before and after mining.

TABLE 3. Runoff Curve Number (CN) for Hydrologic Soil—Cover Complexes.[1] (OSM, [38])

Land use or cover	Treatment[4] or practice	Hydrologic condition for infiltration	A	B	C	D
1. Agricultural Use						
Fallow	SR		77	86	91	94
Row crops	SR	Poor	72	81	88	91
	SR	Good	67	78	85	89
	C	Poor	70	79	84	88
	C	Good	65	75	82	86
	C&T	Poor	66	74	80	82
	C&T	Good	62	71	78	81
Small grain	SR	Poor	65	76	84	88
	SR	Good	63	75	83	87
	C	Poor	63	74	82	85
	C	Good	61	73	81	84
	C&T	Poor	61	72	79	82
	C&T	Good	59	70	78	81
Close-seeded legumes[2] or rotation meadow	SR	Poor	66	77	85	89
	SR	Good	58	72	81	85
	C	Poor	64	75	83	85
	C	Good	55	69	78	83
	C&T	Poor	63	73	80	83
	C&T	Good	51	67	76	80
Pasture or range		Poor	68	79	86	89
		Fair	49	69	79	84
		Good	39	61	74	80
	C	Poor	47	67	81	88
	C	Fair	25	59	75	83
	C	Good	6	35	70	79
Meadow (permanent)		Good	30	58	71	78
Woods (farm woodlots)		Poor	45	66	77	83
		Fair	36	60	73	79
		Good	25	55	70	77
Farmsteads			59	74	82	86
Roads (dirt)[3]			72	82	87	89
(hard surface)[3]			74	84	90	92
2. Surface Mine Use						
Paved road			98	98	98	98
Gravel			76	85	89	91
Dirt			72	82	87	89

(continued)

TABLE 3 (continued).

Land use or cover		Treatment[4] or practice	Hydrologic condition for infiltration		Hydrologic soil group			
					A	B	C	D
	Disturbed area (active mining)			72	81	88	91	
	Reclaimed spoil with vegetation cover			39–72	61–81	74–88	80–91	
3.	Range							
	Grassland		Poor	66	79	86	89	
			Fair	51	69	79	84	
			Good	37	61	74	80	
		C	Poor	47	67	78	83	
		C	Fair	38	61	74	80	
		C	Good	28	55	70	77	
	Herbaceous		Poor	68	80	87	90	
			Fair	55	71	81	85	
			Good	42	63	74	81	
	Desert shrub		Poor	68	80	87	90	
			Fair	52	70	80	85	
			Good	39	62	75	80	
	Sagebrush		Poor	47	67	78	83	
			Fair		48	65	74	
			Good		30	53	64	
	Pinyon-juniper		Poor	60	75	83	87	
			Fair	34	58	73	78	
			Good		41	61	70	
	Chaparral (Arizona)		Poor	68	80	87	93	
			Fair	32	57	71	83	
			Good		41	58	74	
	Oak-aspen		Poor	43	64	76	82	
			Fair		47	64	73	
			Good		30	53	64	
	Ponderosa pine		Poor	45	66	77	83	
			Fair	29	56	70	77	
			Good		46	64	72	
	Woods		Poor	45	66	77	83	
			Fair	36	60	73	79	
			Good	25	55	70	77	

[1]These CN values are for antecedent moisture condition II (AMC-II). To find the proper CN for AMC-III or AMC-I, use the CN from this table in Table 5.
[2]Close-drilled or broad cast.
[3]Including right-of-way.
[4]C = contoured, SR = straight row, T = terraced.

TABLE 4. Seasonal Rainfall Limits for AMC (SCS, [55]).

| AMC group | Total 5-day antecedent rainfall | |
	Dormant season (inches)	Growing season (inches)
I	Less than 0.5	Less than 1.4
II	0.5 to 1.1	1.4 to 2.1
III	Over 1.1	Over 2.1

BEFORE MINING

The runoff equation is:

$$Q = \frac{(P - 0.2\,S)^2}{P + 0.8\,S} \qquad (1)$$

1. Determine the precipitation depth, P: P values for different storm durations and return periods can be estimated from maps compiled by NOAA [33]. The 10-year, 24-hour rainfall for this particular location is 1.75 inches.

2. Determine the S parameter for the watershed. S is related to the curve number (CN) through Equation (2):

$$S = \frac{1000}{CN} - 10 \qquad (2)$$

The CN value depends on the hydrologic soil group (HSG) classification, the type of vegetation cover and the hydrologic condition of the watershed. HSG classifications are tabulated in the Appendix alphabetically by soil name.

Cochetopa, HSG = C

Aaberg, HSG = D

The curve number for each soil is obtained from Table 3 by taking a weighted average of the values for each cover type based on the cover percentages. For example, the Cochetopa soil with an HSG = C and fair hydrologic condition has a CN of 81 for the herbaceous cover and 65 for the sagebrush cover. Because each cover type occurs on 50% of the area, the simple average of these two values is the CN for the Cochetopa soil.

Cochetopa, $CN = (81.0 + 65.0)/2. = 73.0$

Similarly for the Aaberg, $CN = 79.5$

Because the two soils form a complex over a small area, an area weighted average CN value can be used to represent the entire watershed:

$$CN = 73.0(.30) + 79.5(.70) = 77.6$$

This CN value applies to the AMC-II antecedent moisture condition which is commonly used for design purposes. If another AMC condition is appropriate, the AMC-II CN value can be adjusted using Table 5.

Now the S parameter can be determined from Equation (2):

$$S = \frac{1000}{77.6} - 10 = 2.89 \text{ inches}$$

3. Determine the direct runoff, Q, from Equation (1):

$$Q = \frac{[1.75 - 0.2(2.89)]^2}{1.75 + 0.8(2.89)} = 0.34 \text{ inches}$$

The runoff volume in acre-feet is:

$$Q_{volume} = \frac{0.34 \text{ inches}}{12 \text{ inches per foot}} \times 1000 \text{ acres}$$

$$= 28.3 \text{ acre-feet}$$

AFTER MINING

1. The design storm is the same; $P = 1.75$ inches

2. Curve numbers are determined in the same way as for the "before mining" case except that values reflecting a poor hydrologic condition from Table 3 are used.

Cochetopa, $CN = 82.5$

Aaberg, $CN = 86.5$

The area weighted CN is:

$$CN = 82.5(.30) + 86.5(.70) = 85.3$$

The S value from Equation (2) is:

$$S = \frac{1000}{85.3} - 10 = 1.72 \text{ inches}$$

3. The direct runoff from Equation (1) is:

$$Q = \frac{[1.75 - 0.2(1.72)]^2}{1.75 + 0.8(1.72)} = 0.63 \text{ inches}$$

The runoff volume is:

$$Q_{volume} = \frac{0.63 \text{ inches}}{12 \text{ inches per foot}} \times 1000 \text{ acres}$$

$$= 52.5 \text{ acre-feet}$$

The assumed deterioration in hydrologic condition results in

TABLE 5. Curve Numbers (CN) and Constants for the Case $I_a = 0.2\ S$ (SCS, [55]).

1 CN for condition II	2 CN for conditions I	3 III	4 S values* (inches)	5 Curve* starts where P = (inches)	1 CN for condition II	2 CN for conditions I	3 III	4 S values* (inches)	5 Curve* starts where P = (inches)
100	100	100	0	0	60	40	78	6.67	1.33
99	97	100	.101	.02	59	39	77	6.95	1.39
98	94	99	.204	.04	58	38	76	7.24	1.45
97	91	99	.309	.06	57	37	75	7.54	1.51
96	89	99	.417	.08	56	36	75	7.86	1.57
95	87	98	.526	.11	55	35	74	8.18	1.64
94	85	98	.638	.13	54	34	73	8.52	1.70
93	83	98	.753	.15	53	33	72	8.87	1.77
92	81	97	.870	.17	52	32	71	9.23	1.85
91	80	97	.989	.20	51	31	70	9.61	1.92
90	78	96	1.11	.22	50	31	70	10.0	2.00
89	76	96	1.24	.25	49	30	69	10.4	2.08
88	75	95	1.36	.27	48	29	68	10.8	2.16
87	73	95	1.49	.30	47	28	67	11.3	2.26
86	72	94	1.63	.33	46	27	66	11.7	2.34
85	70	94	1.76	.35	45	26	65	12.2	2.44
84	68	93	1.90	.38	44	25	64	12.7	2.54
83	67	93	2.05	.41	43	25	63	13.2	2.64
82	66	92	2.20	.44	42	24	62	13.8	2.76
81	64	92	2.34	.47	41	23	61	14.4	2.88
80	63	91	2.50	.50	40	22	60	15.0	3.00
79	62	91	2.66	.53	39	21	59	15.6	3.12
78	60	90	2.82	.56	38	21	58	16.3	3.26
77	59	89	2.99	.60	37	20	57	17.0	3.40
76	58	89	3.16	.63	36	19	56	17.8	3.56
75	57	88	3.33	.67	35	18	55	18.6	3.72
74	55	88	3.51	.70	34	18	54	19.4	3.88
73	54	87	3.70	.74	33	17	53	20.3	4.06
72	53	86	3.89	.78	32	16	52	21.2	4.24
71	52	86	4.08	.82	31	16	51	22.2	4.44
70	51	85	4.28	.86	30	15	50	23.3	4.66
69	50	84	4.49	.90					
68	48	84	4.70	.94	25	12	43	30.0	6.00
67	47	83	4.92	.98	20	9	37	40.0	8.00
66	46	82	5.15	1.03	15	6	30	56.7	11.34
65	45	82	5.38	1.08	10	4	22	90.0	18.00
64	44	81	5.62	1.12	5	2	13	190.0	38.00
63	43	80	5.87	1.17	0	0	0	infinity	infinity
62	42	79	6.13	1.23					
61	41	78	6.39	1.28					

*For CN in column 1.

TABLE 6. Seasonal Rainfall Limits for S$_d$.

	Total 5 day antecedent rainfall in inches	
S$_d$	Dormant Season	Growing Season
0.85	<0.5	<1.4
0.50	0.5 to 1.1	1.4 to 2.1
0.15	>1.1	>2.1

TABLE 7. Infiltration Parameter Values Typical of Soil Textural Classes.*

Soil Texture Class	K$_s$ (inches/hour)	C (inches)
Sand	8.27	1.339
Loamy Sand	2.41	1.772
Sandy Loam	1.02	2.953
Loam	0.52	3.661
Silt Loam	0.268	5.709
Sandy Clay Loam	0.169	5.354
Silty Clay Loam	0.059	7.913
Clay Loam	0.091	6.102
Sandy Clay	0.047	7.402
Silty Clay	0.035	8.858
Clay	0.024	9.685

*After Rawls et al., [44].

TABLE 8. Depression Storage Depth, z(cm).*

	Soil Clay Content		
Surface Condition	5%	20%	>30%
Very rough, plowed on contour	0.6	1.2	2.5
Very rough, random clods or weathered after contour plowing	0.3	0.8	1.8
Slightly rough or harrowed along contour	0.0	0.5	0.5
Smooth, tilled up and down slope, or eroded by rilling	0.0	0.0	0.0

*Based on surface profile data of Evans [12] and assumption that effective depression storage depth equals approximately half the standard deviation of the soil surface height.

a post-mining runoff estimate which is 86% higher than the pre-mining estimate.

WATER BALANCE OR INFILTRATION-BASED METHODS

These methods are based on the principle of mass conservation. The conservation of rainfall on a watershed can be stated as:

$$P = I + F + D + Q \qquad (3)$$

in which P is rainfall, I is amount of rainfall trapped on vegetation leaves and stems, F is the amount of P which infiltrates, D is the amount of P trapped in surface depressions, and Q is the amount of P which becomes surface runoff. All of the components of Equation (3) can be expressed as a depth over the entire watershed area. If P is a known rainfall input, and I, F and D can be calculated, then Equation (3) can be solved for Q.

It is generally assumed that interception traps all rainfall at the beginning of a storm until interception capacity I_c of the vegetation canopy and ground cover has been exhausted. Interception depends on the type of vegetation, the percent cover and the rainfall intensity [32]. However, for the purposes of runoff estimation, the interception capacity can be estimated by

$$I_c = 0.2\,(GC + CC) \qquad (4)$$

in which I_c is in inches, GC is the fraction of ground cover at the soil surface and CC is the fraction of vegetation canopy cover.

After interception capacity has been satisfied, rainfall strikes the soil surface and infiltration begins. Water infiltrates the soil at a rate equal to the lesser of the potential infiltration rate and the rainfall rate. The potential infiltration rate depends on the antecedent soil moisture level, soil properties and the amount of water which has entered the soil since the initiation of infiltration. Smith [51] demonstrated the essential similarity of several modern infiltration equations (21,35,43,50). Smith and Parlange's potential infiltration rate equation is:

$$f_c = \left\{ \frac{e^{[F/(CS_d)]}}{e^{[F/(CS_d)]} - 1} \right\} K_s \qquad (5)$$

in which f_c is the potential infiltration rate when the total infiltrated depth is F, S_d is the saturation deficit reflecting the antecedent soil moisture condition, C is a soil capillary suction parameter, and K_s is the saturated hydraulic conductivity of the soil. S_d can be estimated from antecedent rainfall using Table 6, while the soil parameters K_s and C can be estimated based on the soil texture [23] as shown in Table 7.

The nature of Equation (5) requires that infiltration rates be determined in time increments beginning with the point at which the interception has been exhausted by the rainfall. Although this is a somewhat tedious way to obtain total infiltrated depth F and total runoff Q, it has the advantage of simultaneously determining the time distributions of F and Q.

At some point during a runoff-producing storm, the rainfall rate exceeds the infiltration rate and ponding occurs. Before any runoff is generated, it is assumed that depression storage capacity must be satisfied. This is similar to the assumed sequence of interception and infiltration discussed earlier. Depression storage capacity depends on management factors such as direction of tillage operations relative to slope, soil texture and erosion history [12,23,36]. Approximate surface depression storage capacities for a few surface conditions are given in Table 8.

As with the SCS-CN method, the primary factor affecting the accuracy of the water balance method is proper infiltration parameter selection. The parameter values of the soils in Table 7 can only be viewed as a rough approximation because of the variance of the data used to establish those values [44] and because other factors such as surface compaction, and type and density of vegetation cover, may affect infiltration characteristics. Thus, the accuracy of the water balance method can be expected to improve as more information on the effect of management activities on infiltration characteristics becomes available.

Despite difficulties in parameter estimation, the water balance method has the following advantages over the SCS-CN technique and other empirical approaches:

1. Interception and depression storage are treated as distinct processes; therefore the effect of anticipated changes in vegetation and surface configuration on runoff can be explicitly evaluated.
2. The effect of rainfall intensity patterns on runoff volume is predicted by the method. The method does not over-predict runoff volumes from long duration storms of low intensity.
3. The method provides a physically reasonable time distribution of rainfall excess which is a first step toward determining a complete storm hydrograph.

Example: Water Balance Method

Use the Water Balance method to estimate surface runoff for the same storm and watershed conditions described in the previous example illustrating the Curve Number method. Assume that the 10-year, 24-hour storm has a type II distribution. Prior to mining the watershed has an average ground cover (GC) density of 0.30 and a canopy cover (CC) density of 0.50. After mining, it is estimated that GC will be 0.10 and CC will be 0.25. The soil surface is slightly rough for both pre- and post-mining conditions.

PRE-MINING

Solving the Water Balance equation [Equation (3)] for the surface runoff Q results in:

$$Q = P - (I_c + D + F)$$

1. As determined from the NOAA atlas [33] in the CN example, the rainfall depth is:

$$P = 1.75 \text{ inches}$$

2. I_c is estimated using Equation (4):

$$I_c = 0.2(GC + CC)$$

$$I_c = 0.2(0.30 + 0.50) = 0.16 \text{ inches}$$

3. Depression storage, D is estimated using Table 8. For a slightly rough surface and soil which is more than 20% clay:

$$D = 0.5 \text{ cm} = 0.20 \text{ inches}$$

4. The infiltrated depth, F, is estimated using Equation (5):

$$f_c = \left\{ \frac{e^{[F/(CS_d)]}}{e^{[F/(CS_d)]} - 1} \right\} K_s$$

The saturation deficit, $S_d = 0.5$ is derived from Table 6 assuming antecedent conditions similar to an AMC II as defined in Table 4.

The soil capillary suction parameter, C can be estimated from Table 7 for each soil:

$$\text{Aaberg silty clay, } C = 8.86 \text{ inches}$$

$$\text{Cochetopa loam, } C = 3.66 \text{ inches}$$

The soil complex is 70% Aaberg silty clay and 30% Cochetopa loam; therefore the area weighted average C value is:

$$C = [0.70(8.86) + 0.30(3.66)] = 7.30 \text{ inches}$$

The saturated hydraulic conductivity, K_s can be estimated from Table 7 for each soil:

$$\text{Aaberg silty clay, } K_s = 0.035 \text{ inches/hour}$$

$$\text{Cochetopa loam, } K_s = 0.52 \text{ inches/hour}$$

The area weighted average K_s value is:

$$K_s = [0.70(0.0350 + 0.30(0.52)] = 0.18 \text{ inches/hour}$$

The total infiltrated depth, F is determined implicitly using Equation (5) on a time step by time step basis. These calculations are shown in Table WB-1. Column footnotes explain column operations. Infiltration begins after the interception capacity of 0.16 inches has been filled by rainfall. This takes a little less than 8.0 hours. As shown on the last line of Table WB-1, the final value of F at the end of the storm is:

$$F = 1.46 \text{ inches}$$

5. The runoff depth, Q is determined by:

$$Q = P - (I_c + F + D)$$

$$Q = 1.75 - (0.16 + 1.46 + 0.20) = -0.07$$

Thus,

$$Q = 0.0 \text{ inches}$$

No runoff is predicted for pre-mining conditions. This contrasts with the *CN* method estimate of 0.35 inches of runoff for the same conditions.

POST-MINING
1. The rainfall depth is the same:

$$P = 1.75 \text{ inches}$$

2. The interception capacity based on post-mining vegetation density estimates of $GC = 0.10$ and $CC = 0.25$ is:

$$I = 0.20(0.10 + 0.25) = 0.07 \text{ inches}$$

3. Depression storage is the same as for pre-mining conditions:

$$D = 0.20 \text{ inches}$$

4. Infiltration parameters are the same as for pre-mining conditions except that the saturated conductivity is expected to decrease as a result of compact of the surface soil. In order to estimate this effect, K_s values for each soil in the complex are reduced to values given for the next less permeable soil class listed in Table 7. That is, the Aaberg silty clay is estimated to have a post-mining conductivity typical of a clay soil, and the Cochetopa loam is estimated to have a conductivity typical of a silt loam. The resultant area weighted average conductivity is:

$$K_s = 0.10 \text{ inches/hour}$$

F is determined in the same manner as for pre-mining conditions. Calculations are shown in Table WB-2. The final F value at the end of the storm is:

$$F = 1.38 \text{ inches}$$

5. The estimated surface runoff is:

$$Q = P - (I_c + D + F)$$

$$Q = 1.75 - (0.07 + 0.20 + 1.38) = 0.10 \text{ inches}$$

A small amount of runoff is estimated to occur for post-mining conditions. This is only 16% of the amount estimated by the *CN* method for the same conditions. The main reason for the large difference in estimates by the two methods is that the infiltration-based Water Balance method is sensitive to the distribution of rainfall intensities in the storm. In this example the 24-hr, type II design storm includes intensities which are mostly lower than the infiltration capacity of the soil. If the same 1.75 inch storm were compressed into a 6 hour period, intensities would be 4 times greater and the Water Balance method would estimate a little less than 4 times as much runoff. In contrast, the *CN* method estimate would not change because the method does not distinguish between different rainfall intensities.

RUNOFF INDEX METHODS

These include the *Phi* and *W* index methods [36]. The *Phi* index represents an average rate of abstraction that includes infiltration, depression storage and interception which occurs during the period of direct runoff. It is subtracted from rainfall intensity to obtain an estimate of rainfall excess pattern and runoff volume (depth). The *Phi* index is determined by calibration using measured storm depths and runoff hydrographs. Sufficient rainfall and runoff data are needed to obtain an average *Phi* value which represents both watershed and rainfall characteristics. The *W* index is a refinement of the *Phi* index which separates infiltration from depression and interception storage. In the *W* index method, the latter two abstractions must be independently estimated.

Both methods require rainfall and runoff data to calibrate index values. Index values will change if watershed characteristics are modified. Consequently, they will only be useful to predict PHCs if sufficient rainfall-runoff data exists on neighboring watersheds which have been disturbed by similar mining activities. In this case, a *Phi* or *W* index can be calibrated from the data and used to estimate runoff volumes and rainfall excess patterns for design storms. The accuracy of these estimates will depend on the degree of watershed similarity and the quantity of available data. If data exist to relate *Phi* or *W* to antecedent rainfall or soil moisture, the effect of these conditions on runoff can be evaluated.

Index methods may also be used to characterize existing

TABLE WB-1. Pre-mining Water Balance Infiltration Calculations.

t hours (1)	Δt hours (2)	P inches (3)	ΔP inches (4)	i in/hr (5)	f_c in/hr (6)	ΔF (7)	F inches (8)	Comment (9)
6.00	6.00	0.16	0.16	0.03	∞	0.00	0.00	$P = I_c$, Interception filled
10.00	4.00	0.32	0.16	0.04	∞	0.16	0.16	$\Delta F = \Delta P$
11.00	1.00	0.41	0.09	0.09	4.18	0.09	0.25	$\Delta F = \Delta P$
11.50	0.50	0.53	0.12	0.24	2.71	0.12	0.37	$\Delta F = \Delta P$
11.75	0.25	0.70	0.17	0.68	1.86	0.17	0.54	$\Delta F = \Delta P$
12.00	0.25	1.16	0.46	1.84	1.31	0.33	0.87	$\Delta F = f_c (\Delta t)$, ponding
12.50	0.50	1.29	0.13	0.26	0.85	0.13	1.00	$\Delta F = \Delta P$
13.00	0.50	1.35	0.06	0.12	0.75	0.06	1.06	$\Delta F = \Delta P$
24.00	11.00	1.75	0.40	0.04	0.71	0.40	1.46	$\Delta F = \Delta P$

Column Notes:

(1) total time elapsed into storm, begins after interception, I_c, depth has been filled by precipitation

(2) time increment

(3) total precipitation that has fallen at time t, begins after interception, I_c, has been filled, calculated from type II distribution in Figure 1.

(4) precipitation increment that has fallen during Δt

(5) precipitation intensity during Δt, $i = \Delta P/\Delta t$

(6) infiltration capacity, f_c, Equation (5)

(7) incremental infiltration depth: $\Delta F = (i)(\Delta t)$ if $i \leq f_c$

$\Delta F = (f_c)(\Delta t)$ if $i > f_c$

(8) total infiltrated depth at time t

TABLE WB-2. Post-mining Water Balance Infiltration Calculations.

t hours (1)	Δt hours (2)	P inches (3)	ΔP inches (4)	i in/hr (5)	f_c in/hr (6)	ΔF (7)	F inches (8)	Comment (9)
3.00	3.00	0.07	0.07	0.02	∞	0.00	0.00	$P = I_c$, Interception filled
8.00	5.00	0.21	0.14	0.03	∞	0.14	0.14	$\Delta F = \Delta P$
10.00	2.00	0.32	0.11	0.06	2.66	0.11	0.25	$\Delta F = \Delta P$
11.00	1.00	0.41	0.09	0.09	1.51	0.09	0.34	$\Delta F = \Delta P$
11.50	0.50	0.53	0.12	0.24	1.12	0.12	0.46	$\Delta F = \Delta P$
11.75	0.25	0.70	0.17	0.68	0.85	0.17	0.63	$\Delta F = \Delta P$
12.00	0.25	1.16	0.46	1.84	0.63	0.16	0.79	$\Delta F = f_c(\Delta T)$, ponding
12.50	0.50	1.29	0.13	0.26	0.51	0.13	0.92	$\Delta F = \Delta P$
24.00	11.50	1.75	0.46	0.04	0.45	0.46	1.38	$\Delta F = \Delta P$

TABLE 9. Incremental W Values for Use in Cook's Method.

Watershed Characteristic	Extent or Degree	W
Relief	Steep rugged terrain with average slopes generally above 30%	40
	Hilly, with average slopes 10 to 30%	30
	Rolling, with average slopes 5 to 10%	20
	Relatively flat land, slopes 0 to 5%	10
Infiltration (I)	No effective cover; either rock or thin soil mantle of negligible infiltration capacity	20
	Slow to take up water; clay or other soil of low infiltration capacity	15
	Deep loams with infiltration about that of typical prairie soils	10
	Deep sand or other soil that takes up water readily and rapidly	
Vegetal cover (C)	No effective plant cover or equivalent	20
	Poor to fair cover; clean cultivated crops or poor natural cover; less than 10% of the watershed in good cover	15
	About 50% of watershed in good cover	10
	About 90% of watershed in good cover, such as grass, woodlands, or equivalent	5
Surface storage	Negligible; few surface depressions	20
	Well-defined system of small drainage channels	15
	Considerable depression storage with not more than 2% in lakes, swamps, or ponds	10
	Surface-depression storage high; drainage system poorly defined; large number of lakes, swamps or ponds	5

or "background" hydrologic conditions. Again, the quality of this characterization will depend on the quality and quantity of available rainfall and runoff data.

Estimation of Peak Runoff

THE COOK METHOD

The Cook method [41] estimates 10, 25 and 50-year return period peak discharges from small watersheds of up to 2500 acres in size. In this method peak discharges are determined by the drainage area, the geographical location of the watershed, and several watershed characteristics. The following steps illustrate how these factors are accounted for:

1. Factors representing watershed relief, infiltration characteristics, vegetation, and depression storage are determined based on descriptions in Table 9.

2. Factors determined in step 1 are summed to obtain a value for *W* which is then used in Figure 3 along with drainage area to obtain a preliminary discharge rate.

3. The discharge rate from step 2 is multiplied by a geographical (rainfall) factor obtained by locating the watershed in Figure 4. The result is an estimate of the 50-year frequency peak discharge.

4. If 10-year or 25-year frequency estimate is desired, Table 10 is entered with the sum of the vegetation and infiltration factors from step 1 to obtain a frequency factor. The 50-year discharge determined in step 3 is multiplied by this factor to obtain the desired peak discharge.

The Cook method's modular treatment of the most important factors controlling peak discharges from watersheds makes it easy to use to evaluate the affect of mining activities on peak discharge. Accuracy will be limited by uncertainty in factor selection and by the limited amount of data used to develop the tables and curves of the method. It should be emphasized that the Cook method predicts peak runoff for a

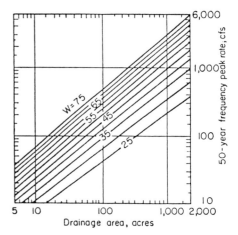

FIGURE 3. Cook's method for estimating 50-yr frequency peak rate of flow (after Ogrosky and Mockus, [41]).

FIGURE 4. Geographical (rainfall) factors for use with Cook's method (after Ogrosky and Mockus, [41]).

given return period, not peak runoff from a design storm of some return period and duration. If regulations require estimation of peak discharge resulting from a design storm, another method should be used.

THE RATIONAL METHOD

The Rational method estimates peak discharge from small watersheds. The use of a single parameter to describe watershed characteristics and the assumption of spatially uniform rainfall generally limits the use of the Rational method to areas smaller than 200 acres [10]. In the design of minor hydraulic structures such as road culverts, the method is quite commonly extended to areas as large as 2000 acres.

The Rational method for peak discharge is expressed

$$Q = C i A \qquad (6)$$

in which Q is the peak discharge in cubic feet per second (cfs), C is a coefficient varying from 0 to 1.0, i is a representative rainfall intensity in inches per hour and A is the watershed area in acres. The representative rainfall intensity, i, depends on both storm and watershed characteristics. To determine the proper value of i for Equation (6), the time of concentration (T_c) of the watershed must first be estimated. Then the rainfall intensity pattern of the storm of interest is

TABLE 10. Frequency Factors for Use with Cook's Method.

(I + C)	Average annual precipitation, in.					
	10	20	30	40	60	80
	Ratio: 25-year/50-year					
5	0.31	0.38	0.41	0.44	0.48	0.51
10	0.41	0.50	0.55	0.58	0.63	0.66
15	0.50	0.59	0.64	0.68	0.73	0.77
20	0.55	0.65	0.71	0.76	0.82	0.87
25	0.60	0.71	0.78	0.83	0.90	0.92
30	0.64	0.76	0.83	0.89	0.92	0.92
35	0.67	0.81	0.89	0.92	0.92	0.92
40	0.71	0.85	0.92	0.92	0.92	0.92
	Ratio: 10-year/50-year					
5	0.05	0.08	0.10	0.12	0.15	0.17
10	0.10	0.16	0.21	0.24	0.30	0.34
15	0.16	0.25	0.31	0.37	0.45	0.51
20	0.21	0.33	0.42	0.49	0.60	0.68
25	0.26	0.41	0.52	0.61	0.75	0.80
30	0.31	0.49	0.62	0.74	0.80	0.80
35	0.36	0.58	0.73	0.80	0.80	0.80
40	0.42	0.66	0.80	0.80	0.80	0.80

examined in order to find the time period equal to T_c with the largest average rainfall intensity. This average intensity is the correct value of i to use in Equation (6). Note that a short storm could have a duration which is smaller than T_c. In this case the proper value of i in Equation (6) will be equal to average rainfall intensity of the entire storm multiplied by the ratio of the storm duration to T_c; that is, i will be smaller than the average intensity of the entire storm. Strictly speaking, Equation (6) would be dimensionally consistent only if the units of Q were acre inches per hour (ac/in/hr); however, by coincidence an ac-in/hr and cfs are nearly equal.

The time of concentration, T_c, is defined as the time it takes for water to travel from the point in the watershed which is most remote in time to the downstream point where the peak discharge is to be estimated. Generally, T_c depends on surface slope and roughness, length of flow path and rainfall intensity. Kirpich [28] found that average slope and flow path length were more important than the other factors and proposed the following dimensional relationship:

$$T_c = \frac{L^{0.77}}{128 \ S^{0.38}} \qquad (7)$$

in which T_c is in minutes, L is the length of the most remote flow path in feet and S is the dimensionless ratio of the drop in elevation to the flow path length.

The Rational formula is essentially a statement of mass conservation at equilibrium conditions in which outflow, Q, equals inflow, iA. The coefficient, C, accounts mainly for the reduction in outflow caused by infiltration. The Rational method defines C as a watershed characteristic, independent of rainfall characteristics. Implicitly, the method assumes that the infiltration rate at the time of concentration is equal to a constant fraction $(1 - C)$ of the rainfall rate at the time of concentration. This is equivalent to assuming that higher rainfall intensities cause higher infiltration rates if all other factors are constant. This contradicts both theory and experience of infiltration phenomena, both of which indicate that over equal time periods, higher rainfall rates tend to reduce infiltration rates, not increase them. Despite these limitations, the Rational method continues to be popular because it requires little data or computational effort to obtain a peak discharge estimate.

MODIFIED RATIONAL METHOD

A modification of the Rational method which solves some of the problems discussed above can be expressed as

$$Q = i_e A \qquad (8)$$

in which i_e is maximum average rainfall excess rate which occurs over a period equal to the time of concentration of the watershed with area, A. Equation (8) requires knowl-

edge of the rainfall excess pattern. This can be established using the Water Balance method described previously, or other methods such as the SCS hydrograph method which is discussed below.

Time of concentration can be calculated using Equation (7) as described for the Rational method. This estimate of T_c can be improved by noting that T_c depends on i_e and on the average surface roughness as well as topographic factors. Hartley [23] showed that a more complete and dimensionally consistent relationship for T_c is

$$T_c = \frac{0.096 \ L^{0.59} \ K^{0.34} \ v^{0.08}}{(gS)^{0.34} \ i_e^{0.41}} \qquad (9)$$

in which i_e is the rainfall minus the infiltration rate at T_c, v is the kinematic viscosity of water, g is gravitational acceleration, and K is a surface roughness factor defined by

$$K = 60 + 3140 \ GC^{1.65} \qquad (10)$$

in which GC is the basal ground cover density. Noting that the viscosity and gravity factors of Equation (9) are nearly constant, the equation can be reduced to

$$T_c = \frac{C \ K^{0.34} \ L^{0.59}}{S^{0.34} \ i_e^{0.41}} \qquad (11)$$

in which $C = 0.012$ when the other variables are expressed in feet and seconds and $C = 0.015$ when they are expressed in meters and seconds. A first approximation of i_e using the T_c value from Equation (7) can be substituted into Equation (11) to determine a new value of T_c. A new value of i_e is then obtained from the known rainfall excess time distribution. In most cases it will be found that a couple of iterations of Equation (9) produce a virtually constant value of i_e which is then substituted into Equation (8) to estimate the peak discharge.

Use of the Modified Rational method in place of the Rational method is recommended whenever a reasonable approximation for the time distribution of the rainfall excess of the design storm is available.

SCS TIME OF CONCENTRATION (T_c) METHOD

This method [56] utilizes a graphical relationship between the peak discharge and time of concentration, T_c. It is restricted to homogeneous watersheds smaller than 2000 acres and 24-hr, type II design storms.

First the CN value and total runoff depth are estimated using the curve number method described previously. Second, the time of concentration in hours is determined by the following empirical relationship:

$$T_c = \frac{\ell^{0.8} \ (S + 1)^{0.7}}{1140 \ Y^{0.5}} \qquad (12)$$

in which ℓ is the hydraulic length in feet from the point which is most remote in time to the watershed outlet, S is the water storage factor in the Curve Number method [see Equation (2)] and Y is the average slope of the watershed expressed as a percentage. In this equation S is used as a surrogate parameter to represent surface roughness based on the assumption that surface roughness and water storage are positively correlated.

If the main channel has been partially channelized, or part of the watershed has been rendered impervious, the T_c value from Equation (12) is multiplied by factors derived from Figures 5 and 6 respectively. After T_c has been established, it is used in Figure 7 to obtain the peak discharge in CFS per inch of runoff and per square mile of watershed. This value is multiplied by the runoff depth in inches and the watershed area in sq mi to obtain the peak discharge rate in CFS.

The Time of Concentration method is simple to use and has the advantage of incorporating parameters which reflect topographic, hydrologic and hydraulic characteristics which are known to affect peak discharge. However, like many SCS methods, the accuracy of the estimate is very sensitive to the value of the curve number selected by the user. For example, for a 2 inch rainfall on a 2 square mile watershed with a hydraulic length of 3000 feet and an average slope of 7 percent, a 5 percent change in curve number from 80 to 84 results in a 45 percent change in estimated peak discharge. Thus, small errors in the estimation of the curve number can cause relatively large errors in estimated peak discharges using this method.

SCS TABULAR METHOD

This method [56] determines peak discharges for composite watersheds made up of subareas with different curve numbers. The method applies to watersheds of up to 20 sq mi subjected to type II, 24-hour design storms. The Tabular

method should not be used when large differences in CN values occur among subareas and when runoff volumes are less than 1.5 inches on areas with CN values greater than 60.

The following data for each subarea should be tabulated:

a. Curve number, CN.
b. Time of concentration, T_c, in hrs (use procedure described in SCS Time of Concentration Method above).
c. Travel time, T_t, in hours from subarea outlet to watershed outlet (use Manning equation (Equation 13) assuming bank-full conditions to obtain a mean velocity. Divide the channel length from subarea outlet to watershed outlet by the velocity to obtain time).
d. Drainage area in square miles.
e. The 24-hr, type II rainfall depth.
f. Runoff depth in inches [use Equation (1)].

The only item of input data which has not been previously discussed is item (c). The Manning equation is [9]:

$$V = \frac{1.49\ R^{0.67}\ S^{0.5}}{n} \qquad (13)$$

in which V is the average velocity in ft/sec, R is the hydraulic radius in feet which for bank-full conditions is simply the channel area divided by the channel perimeter, S is the friction slope which is approximated by the average ratio of elevation drop to channel length and n is the Manning roughness coefficient. Values of n for different channel materials are given in Table 11. Once T_c and T_t are known for a subarea, the contribution of that subarea to the total watershed discharge can be read from Table 12 in terms of a series of flow rates and corresponding times. Flow rates from subareas which reach the outlet at the same time are then summed. The largest sum is the peak discharge. It should be noted that the values in Table 12 are in terms of

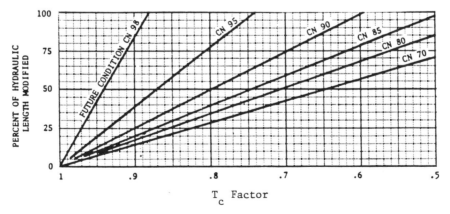

FIGURE 5. Factors for adjusting T_c when the main channel has been hydraulically improved (after SCS, [56]).

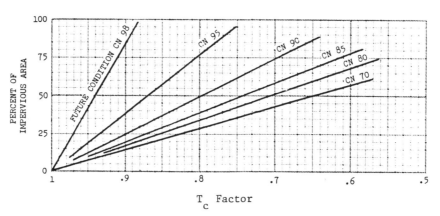

FIGURE 6. Factors for adjusting T_c when impervious areas occur in the watershed (after SCS, [56]).

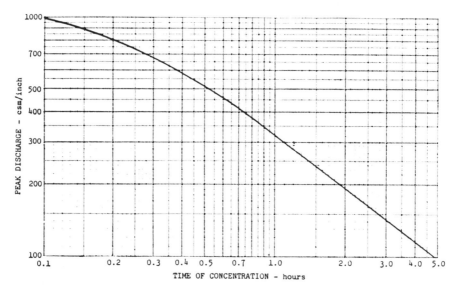

FIGURE 7. Peak discharge in csm per inch of runoff versus time of concentration (T_c for 24-hr, type-II storm distribution (SCS, [56]).

TABLE 11. Values of Manning's Roughness Coefficient, n (OSM, [40]).

Constructed Channel Condition	Values of n		
	Minimum	Maximum	Average
Earth channels, straight and uniform	0.017	0.025	0.0225
Dredged earth channels	0.025	0.033	0.0275
Rock channels, straight and uniform	0.025	0.035	0.033
Rock channels, jagged and irregular	0.035	0.045	0.045
Concrete lined, regular finish	0.012	0.018	0.014
Concrete lined, smooth finish	0.010	0.013	—
Grouted rubble paving	0.017	0.030	—
Corrugated metal	0.023	0.025	0.024

Natural Channel Condition	Value of n
Smoothest natural earth channels, free from growth with straight alignment.	0.017
Smooth natural earth channels, free from growth, little curvature.	0.020
Average, well-constructed, moderate-sized earth channels in good condition.	0.0225
Small earth channels in good condition, or large earth channels with some growth on banks or scattered cobbles in bed.	0.025
Earth channels with considerable growth, natural streams with good alignment and fairly constant section, or large floodway channels well maintained.	0.030
Earth channels considerably covered with small growth, or cleared but not continuously maintained floodways.	0.035
Mountain streams in clean loose cobbles, rivers with variable cross section and some vegetation growing on banks, or earth channels with thick aquatic growths.	0.050

discharge per unit area per unit of subarea runoff. These values must be multiplied by the subarea runoff depth in inches and area in square miles before they are summed with the contributions of the other subareas.

Example: SCS Tabular Method

Subareas 1, 2 and 4 in the watershed shown in Figure TM-1 will be mined. Estimate pre- and post-mining peak discharge from a 100-yr, 24-hr, type II rainfall of 6 inches at the watershed outlet given the data in Table TM-1.

1. Determine the time of concentration, T_c, for each subarea using Equation (11):

$$T_c = \ell^{0.8} (S + 1)^{0.7}/(1140) \, Y^{0.5} \qquad (11)$$

For example, for subarea 3:

$$T_c = [(3000^{0.8}) (3.3 + 1)^{0.7}]/(1140(15)^{0.5} = 0.38 \text{ hour}$$

2. Determine the travel time, T_t, from the subarea outlets to point of interest, here the watershed outlet:
* First, determine V, the mean velocity for each channel reach using Equation (13):
for example, V for stream reach 'A':

$$V = 1.49 \, (0.40)^{0.67} \, (0.05)^{0.5}/(0.04) = 4.5 \text{ fps.}$$

* Using V from above, determine T_t for reach "A":

$$T_t = CL/3600(V) = 1900/3600(4.5) = 0.12 \text{ hr.}$$

* Determine total T_t from each subarea outlet to watershed outlet:

Examination of Figure TM-1 shows that the total travel time, T_t, from outlets of subareas 1 and 2 to the watershed outlet is the sum of the T_t's for reaches "A", "B" and "C". Similarly, total travel time from the outlets of subareas 3 and 4 is the sum of the travel times for reaches "B" and "C". Note that the travel time from subarea 7 is zero. These results are summarized in Table TM-2.

3. Determine runoff, Q, from each subarea using Equations (2) and (1):

$$S = \frac{1000}{CN} - 10 \qquad (2)$$

$$Q = \frac{(P - 0.2 \, S)^2}{P + 0.8 \, S} \qquad (1)$$

Results are shown in Table TM-3.

TABLE 12. Tabular Discharges for Type-II Storm Distribution (csm/in)* (SCS, [56]).

Time of Concentration = 0.1 hours

T_t											Hydrograph Time in Hours													
	11.0	11.5	11.7	11.8	11.9	12.0	12.1	12.2	12.3	12.4	12.5	12.6	12.7	12.8	12.9	13.0	13.2	13.5	14.0	14.5	15.0	16.0	18.0	20.0
0	24	51	299	991	746	477	233	152	132	121	111	85	74	70	68	65	52	48	39	33	29	24	18	14
0.25	20	38	66	140	327	626	686	546	364	236	169	137	117	97	83	75	66	52	41	35	30	24	18	14
0.50	15	27	36	43	67	133	288	482	580	543	429	310	222	168	134	110	81	63	47	38	32	26	19	15
0.75	12	20	25	29	34	42	65	125	245	392	496	515	452	360	273	206	127	80	53	42	35	27	19	15
1.00	9	15	19	21	24	28	32	41	63	115	209	328	427	470	451	389	245	121	64	47	38	29	20	16
1.50	6	10	12	13	14	16	17	19	22	25	29	38	56	92	154	236	410	360	133	66	47	33	21	16
2.00	3	6	7	8	9	10	11	12	13	14	16	18	20	23	27	34	74	244	371	142	68	38	23	17
2.50	2	4	4	5	5	6	7	7	8	9	10	11	12	13	15	16	21	41	243	343	150	48	26	19
3.00	1	2	2	3	3	4	4	4	5	5	6	7	7	8	9	10	12	17	50	239	321	74	29	20
3.50	0	1	1	1	1	2	2	2	3	3	4	4	4	5	6	6	7	10	17	59	304	159	33	21
4.00	0	0	0	0	0	1	1	1	1	2	2	2	2	3	3	4	5	6	10	18	67	290	39	23

Time of Concentration = 0.2 hours

T_t											Hydrograph Time in Hours													
	11.0	11.5	11.7	11.8	11.9	12.0	12.1	12.2	12.3	12.4	12.5	12.6	12.7	12.8	12.9	13.0	13.2	13.5	14.0	14.5	15.0	16.0	18.0	20.0
0	23	47	208	509	796	641	424	245	170	138	121	104	85	75	71	68	56	49	40	34	29	24	18	14
0.25	18	34	49	91	196	419	603	627	486	341	235	173	138	114	96	83	70	55	43	36	31	25	18	15
0.50	14	24	32	37	50	87	181	341	490	545	497	397	296	219	167	133	92	67	49	39	33	26	19	15
0.75	11	18	23	26	30	36	49	84	161	284	409	491	481	422	340	263	157	89	56	43	36	27	19	15
1.00	9	14	18	20	22	25	29	35	48	79	143	240	347	426	452	427	299	147	69	49	39	29	20	16
1.50	5	9	11	12	13	14	16	18	20	23	26	32	43	67	110	176	330	399	159	72	50	33	22	17
2.00	3	6	7	7	8	9	10	11	12	13	15	16	18	21	24	29	56	192	363	168	75	40	24	18
2.50	1	3	4	5	5	6	6	7	7	8	9	10	11	12	13	15	19	33	200	337	174	51	26	19
3.00	0	2	2	2	3	3	4	4	5	5	6	6	7	8	8	9	11	15	40	203	316	82	29	20
3.50	0	0	1	1	1	2	2	2	2	3	3	4	4	5	5	6	7	9	16	46	300	180	34	22
4.00	0	0	0	0	0	1	1	1	1	1	2	2	2	3	3	3	4	6	9	16	53	286	41	24

(continued)

TABLE 12 (continued).

Time of Concentration = 0.3 hours

Hydrograph Time in Hours

T_t	11.0	11.5	11.7	11.8	11.9	12.0	12.1	12.2	12.3	12.4	12.5	12.6	12.7	12.8	12.9	13.0	13.2	13.5	14.0	14.5	15.0	16.0	18.0	20.0
0	21	43	141	324	586	658	535	372	251	184	148	124	102	86	77	71	61	51	41	34	30	24	18	14
0.25	17	31	43	67	134	279	461	559	530	428	318	234	179	143	116	97	76	59	45	37	32	25	18	15
0.50	13	22	29	34	42	65	124	238	378	479	499	447	363	281	216	168	110	74	51	41	34	26	19	15
0.75	10	17	21	24	27	32	41	63	114	203	316	413	457	443	389	319	198	105	60	45	37	28	20	15
1.00	8	13	16	18	20	23	26	31	40	60	103	176	269	358	415	426	344	182	77	51	41	30	20	16
1.50	5	8	10	11	12	13	15	16	18	21	24	28	36	52	82	132	272	382	192	81	52	34	22	17
2.00	3	5	6	7	8	8	9	10	11	12	14	15	17	19	21	25	44	151	351	198	85	41	24	18
2.50	1	3	4	4	5	5	6	6	7	8	8	9	10	11	12	14	17	28	162	328	200	54	27	19
3.00	0	1	2	2	3	3	3	4	4	5	5	6	6	7	8	9	10	14	33	169	309	94	30	20
3.50	0	0	1	1	1	1	2	2	2	3	3	3	4	4	5	5	6	9	14	38	172	294	35	22
4.00	0	0	0	0	0	0	1	1	1	1	2	2	2	3	3	4	5	9	15	43	281	281	42	24

Time of Concentration = 0.4 hours

Hydrograph Time in Hours

T_t	11.0	11.5	11.7	11.8	11.9	12.0	12.1	12.2	12.3	12.4	12.5	12.6	12.7	12.8	12.9	13.0	13.2	13.5	14.0	14.5	15.0	16.0	18.0	20.0
0	20	39	103	224	419	558	575	451	331	247	190	155	127	105	90	80	66	53	42	35	30	24	18	14
0.25	15	28	38	54	98	196	343	467	508	464	380	295	228	180	145	119	87	64	47	38	32	26	19	15
0.50	12	20	26	30	37	53	92	172	286	395	462	453	402	332	266	211	137	84	54	42	35	27	19	15
0.75	10	16	19	22	25	29	36	51	85	150	242	338	407	429	406	356	241	128	65	47	38	29	20	16
1.00	8	12	15	17	19	21	24	28	34	49	78	132	208	292	362	403	368	220	88	55	42	30	21	16
1.50	5	8	9	10	11	12	14	15	17	19	22	25	31	43	65	102	220	365	224	93	56	35	22	17
2.00	3	5	6	6	7	8	9	9	10	11	13	14	16	17	20	23	37	119	338	225	99	43	24	18
2.50	1	3	3	4	4	5	5	6	6	7	8	9	10	11	12	13	16	25	132	317	225	58	27	19
3.00	0	1	2	2	3	3	3	3	4	4	5	5	6	7	7	8	10	13	28	140	300	107	31	21
3.50	0	0	1	1	1	1	1	2	2	2	3	3	3	4	4	5	6	8	13	32	146	286	36	22
4.00	0	0	0	0	0	0	0	1	1	1	1	1	2	2	2	3	3	5	8	14	36	275	44	24

(continued)

TABLE 12 (continued).

Time of Concentration = 0.5 hours

Hydrograph Time in Hours

T_t	11.0	11.5	11.7	11.8	11.9	12.0	12.1	12.2	12.3	12.4	12.5	12.6	12.7	12.8	12.9	13.0	13.2	13.5	14.0	14.5	15.0	16.0	18.0	20.0
0	18	36	80	166	301	433	496	474	395	309	242	194	158	130	109	94	75	57	43	36	31	25	18	15
0.25	15	26	37	52	94	172	277	372	425	424	383	326	270	221	182	150	107	73	49	39	33	26	19	15
0.50	12	20	25	30	38	58	101	169	252	327	374	385	366	329	285	241	169	103	59	44	36	27	19	15
0.75	9	15	19	22	25	30	41	63	103	162	229	292	335	354	348	325	255	157	77	50	39	29	20	16
1.00	7	12	15	17	19	21	25	31	43	66	103	153	210	264	304	327	317	231	109	61	44	31	21	16
1.50	5	8	9	10	11	12	14	15	17	20	24	31	43	63	92	129	214	295	224	115	65	36	23	17
2.00	3	5	6	6	7	8	9	10	11	12	13	14	16	19	23	30	58	143	271	216	120	46	25	18
2.50	1	3	3	4	4	5	5	6	7	7	8	9	10	11	12	14	18	39	150	253	209	71	28	19
3.00	0	1	2	2	2	3	3	4	4	4	5	5	6	7	7	8	10	15	48	154	239	126	32	21
3.50	0	0	1	1	1	1	2	2	2	2	3	3	4	4	5	5	6	8	16	56	155	227	38	23
4.00	0	0	0	0	0	1	1	1	1	2	3	3	4	4	5	3	4	5	9	19	63	217	52	25

Time of Concentration = 0.75 hours

Hydrograph Time in Hours

T_t	11.0	11.5	11.7	11.8	11.9	12.0	12.1	12.2	12.3	12.4	12.5	12.6	12.7	12.8	12.9	13.0	13.2	13.5	14.0	14.5	15.0	16.0	18.0	20.0
0	15	29	57	98	163	248	329	375	388	369	325	276	232	195	165	142	107	76	51	39	33	26	19	15
0.25	12	21	29	39	61	100	158	227	291	336	355	348	321	285	247	212	156	103	62	44	36	27	19	15
0.50	10	16	21	24	29	41	63	100	150	208	263	305	327	329	314	288	226	147	79	52	40	29	20	16
0.75	8	13	16	18	20	24	30	43	65	98	142	192	239	278	303	311	286	208	107	63	45	31	21	16
1.00	6	10	13	14	15	17	20	24	31	44	65	95	134	177	220	256	294	264	149	81	53	33	21	16
1.50	4	6	8	9	10	11	12	13	14	16	19	23	31	42	60	83	147	269	248	152	85	40	23	17
2.00	2	1	5	5	6	7	7	8	9	10	11	12	14	16	18	23	39	97	251	235	153	56	26	19
2.50	1	2	3	3	4	4	4	5	5	6	7	7	8	9	10	11	15	28	107	218	236	91	29	20
3.00	0	1	1	2	2	2	2	3	3	4	4	5	5	6	6	7	8	12	33	113	225	153	34	22
3.50	0	0	0	1	1	1	1	1	1	2	2	3	3	3	4	4	5	7	13	39	117	215	44	24
4.00	0	0	0	0	0	0	0	1	1	1	1	1	1	2	2	2	3	4	7	15	45	207	63	26

(continued)

TABLE 12 (continued).

Time of Concentration = 1.0 hours

T_t	Hydrograph Time in Hours																							
	11.0	11.5	11.7	11.8	11.9	12.0	12.1	12.2	12.3	12.4	12.5	12.6	12.7	12.8	12.9	13.0	13.2	13.5	14.0	14.5	15.0	16.0	18.0	20.0
0	13	24	45	66	107	155	211	258	301	313	316	301	277	247	217	188	146	102	64	46	36	27	19	15
0.25	10	18	24	32	45	68	102	146	193	238	272	293	299	293	275	252	200	139	81	54	41	29	20	16
0.50	8	14	17	20	24	32	46	68	99	136	178	219	251	274	284	283	254	187	105	65	47	31	21	16
0.75	7	11	13	15	17	20	25	33	46	67	94	128	165	202	233	256	273	236	140	82	55	33	21	16
1.00	5	9	11	12	13	15	17	20	25	33	46	65	90	121	154	187	240	262	183	107	66	37	22	17
1.50	3	5	7	7	8	9	10	11	12	14	16	19	24	31	43	58	103	185	244	181	110	48	24	18
2.00	2	3	4	4	5	6	6	7	8	8	9	10	11	13	15	18	29	69	182	230	178	70	27	19
2.50	1	2	2	3	3	3	4	4	5	5	6	6	7	8	9	10	12	21	77	178	219	114	31	21
3.00	0	1	1	1	1	2	2	2	3	3	3	4	4	5	5	6	7	10	25	83	210	172	39	22
3.50	0	0	0	0	1	1	1	1	1	1	2	2	3	3	3	3	4	6	11	29	88	202	52	25
4.00	0	0	0	0	0	0	0	0	1	1	1	2	1	1	2	2	2	4	6	12	33	195	77	28

Time of Concentration = 1.25 hours

T_t	Hydrograph Time in Hours																							
	11.0	11.5	11.7	11.8	11.9	12.0	12.1	12.2	12.3	12.4	12.5	12.6	12.7	12.8	12.9	13.0	13.2	13.5	14.0	14.5	15.0	16.0	18.0	20.0
0	11	21	37	51	79	107	147	187	219	249	264	271	267	256	241	219	177	128	81	56	42	29	20	16
0.25	9	15	21	27	36	53	74	103	137	172	205	231	249	259	259	253	223	167	102	67	48	31	21	16
0.50	7	12	15	17	21	27	37	51	72	98	128	160	190	216	235	247	251	209	130	82	56	34	21	16
0.75	6	9	12	13	15	17	21	27	36	50	69	93	120	149	177	202	235	242	165	103	67	38	22	17
1.00	4	7	9	10	11	13	14	17	21	27	36	49	66	88	113	139	190	236	200	130	83	43	23	17
1.50	3	5	6	6	7	8	8	9	10	12	14	16	20	25	33	44	76	142	223	195	131	58	26	18
2.00	1	3	3	4	4	5	5	6	6	7	8	9	10	11	13	15	24	52	143	212	189	86	29	20
2.50	1	1	2	2	2	3	3	3	4	4	5	5	6	7	7	8	10	17	58	143	201	132	35	21
3.00	0	0	1	1	1	1	2	2	2	2	3	3	3	4	4	5	6	9	20	64	143	196	45	23
3.50	0	0	0	0	0	0	1	1	1	1	1	2	2	2	2	3	4	5	9	23	68	190	62	26
4.00	0	0	0	0	0	0	0	0	0	0	1	1	1	1	1	1	2	3	5	10	26	184	91	30

(continued)

TABLE 12 (continued).

Time of Concentration = 1.5 hours

T_t	11.0	11.5	11.7	11.8	11.9	12.0	12.1	12.2	12.3	12.4	12.5	12.6	12.7	12.8	12.9	13.0	13.2	13.5	14.0	14.5	15.0	16.0	18.0	20.0
0	10	18	31	42	57	81	105	133	164	192	209	227	235	236	236	225	201	153	99	68	50	32	20	16
0.25	8	13	17	22	30	41	57	76	99	125	153	178	199	215	225	230	224	188	122	82	58	36	21	16
0.50	6	10	13	15	18	22	30	40	54	72	94	118	143	167	188	204	224	214	152	99	68	39	22	17
0.75	5	8	10	11	13	15	18	22	29	39	52	69	89	111	134	157	194	219	182	122	82	44	23	17
1.00	4	6	8	9	10	11	12	14	17	22	29	38	50	66	84	105	148	198	214	150	100	50	24	18
1.50	2	4	5	5	6	7	7	8	9	10	12	14	17	21	26	34	58	109	191	204	149	70	28	19
2.00	1	2	3	3	4	4	4	5	5	6	7	8	8	10	11	13	19	40	112	184	197	102	33	20
2.50	0	1	1	2	2	2	3	3	3	4	4	5	5	6	6	7	9	14	45	114	190	147	40	22
3.00	0	0	0	1	1	1	1	1	2	2	2	3	3	3	4	4	5	7	16	49	115	184	53	25
3.50	0	0	0	0	0	0	0	1	0	0	1	1	2	2	2	2	3	4	8	18	53	178	74	28
4.00	0	0	0	0	0	0	0	0	0	0	0	1	1	1	1	1	2	2	4	8	21	174	105	34

Time of Concentration = 2.0 hours

T_t	11.0	11.5	11.7	11.8	11.9	12.0	12.1	12.2	12.3	12.4	12.5	12.6	12.7	12.8	12.9	13.0	13.2	13.5	14.0	14.5	15.0	16.0	18.0	20.0
0	7	14	22	30	38	49	64	80	95	114	133	152	165	175	184	192	190	176	129	93	68	41	23	17
0.25	6	10	13	17	22	28	37	47	61	75	91	108	126	143	157	168	185	189	153	109	79	46	24	17
0.50	5	8	10	11	13	17	21	27	35	45	57	71	86	103	119	135	162	186	172	129	92	52	26	18
0.75	4	6	8	8	10	11	13	16	21	26	34	43	55	67	82	97	129	166	183	149	109	59	27	18
1.00	3	5	6	7	7	8	11	11	13	16	20	26	33	42	52	64	92	136	180	167	127	68	29	19
1.50	1	3	3	4	4	5	5	6	7	8	9	10	12	15	18	23	37	68	135	175	163	93	34	21
2.00	1	1	2	2	3	3	3	4	4	5	5	6	6	7	8	10	14	26	71	133	170	127	42	23
2.50	0	1	1	1	1	1	2	2	2	3	3	3	4	4	5	5	7	11	29	74	132	166	53	26
3.00	0	0	0	0	1	1	1	1	1	1	2	3	2	4	3	3	4	5	12	32	76	162	71	30
3.50	0	0	0	0	0	0	0	0	0	1	1	1	1	1	1	2	2	3	6	13	35	158	95	35
4.00	0	0	0	0	0	0	0	0	0	0	0	0	0	1	1	1	1	2	3	6	14	80	155	43

–csm/in = cubic feet per second per square mile per inch of runoff.

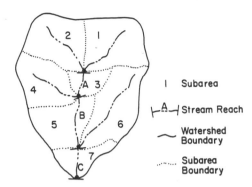

FIGURE TM-1. Watershed map for Tabular method example.

4. Determine partial hydrographs for each subarea based on T_c and T_t values. Note that Table 12 is actually composed of a series of tables corressponding to different T_c values and that each table has lines corresponding to different T_t values. For a given subarea, it may be necessary to interpolate both between tables and between lines in order to obtain discharges of the partial hydrograph from the subarea. For example, for subarea 4 ($T_c = 0.81$ and $T_t = 0.33$) it is necessary to interpolate between tables corresponding to $T_c = 0.75$ and $T_c = 1.00$ and lines corresponding to $T_t = 0.25$ and $T_t = 0.50$. Recall that discharge values from Table 12 must be converted as follows:

$$q = q_p(DA)Q$$

in which q is the subarea discharge in cfs, q_p is the value from Table 12 in cfs per sq. mi. per inch of runoff, DA is the subarea drainage area in sq. mi. and Q is the subarea runoff in inches.
Partial hydrographs for each subarea are tabulated in Table TM-3.

5. Determine the composite hydrograph at the watershed outlet. Sum partial hydrographs from each subarea. The results of this summation are shown on the last line of Table TM-3. The peak of 3693 cfs occurs at about $t = 12.6$ hours.

6. Determine post-mining peak and compare with pre-mining peak. Follow steps 1–5 using the post-mining CN values. Results are shown in Table TM-4. The post-mining peak of 4753 cfs is 29% higher than the pre-mining peak but the time to peak does not change.

REGIONAL METHODS

These are usually regression equations which relate peak flow rate to storm return periods and durations as well as selected watershed characteristics such as area and main channel slope. Applicability of the methods varies from small watersheds to basins of several thousand square miles. The United States Geological Survey (USGS) has been active for many years in collecting data and developing these relationships for regions throughout the United States. The Water Resources Division of local USGS offices should be contacted for further information about these methods. Additionally, regional and other methods of PHC determination have been collected, summarized and referenced in an annotated bibliography available from the U.S. Office of Surface Mining [39,40]. Regional methods are usually based on data which are collected from watersheds with similar land use such as "natural, undisturbed" or "agricultural" and they are not designed to predict changes in peak runoff resulting from land use changes. These methods, like index methods for runoff volume, may be useful for characterizing "background" hydrology prior to watershed disturbance.

Estimation of Runoff Hydrographs

Runoff hydrograph methods estimate the entire time distribution of runoff rates from a watershed. Once the runoff hydrograph has been established, such information as runoff volume, peak discharge and time to peak are immediately available or easily calculated.

TABLE TM-1. Example Watershed Data.

Subarea or (stream reach)	1	2	3 (A)	4	5 (B)	6	7 (C)
area, mi²	0.9	0.6	0.3	0.75	0.6	1.2	0.6
avg. % surface slope, Y	10	20	15	10	8	10	5
hydraulic length, ℓ, ft	7000	6500	3000	5000	5300	10,000	7000
channel slope, S	—	—	(0.05)	—	(0.04)	—	(0.02)
channel length, CL, ft	—	—	(1900)	—	(4300)	—	(2000)
avg. bankful hydraulic radius, R, ft	—	—	(0.40)	—	(0.55)	—	(0.65)
Manning roughness, n	—	—	(0.040)	—	(0.035)	—	(0.035)
Curve number, CN							
before mining	70	70	75	70	65	70	70
after mining	80	80	75	80	65	70	70

TABLE TM-2. Summary of Travel and Concentration Time for Subareas, (Pre-mining).

		V fps	T$_t$ hr	T$_c$ hr
Within subarea or (stream reach)	1	—	—	1.06
	2	—	—	0.71
	3 (A)	(4.5)	(0.12)	0.38
	4	—	—	0.81
	5 (B)	(5.7)	(0.21)	1.12
	6	—	—	1.41
	7 (C)	(4.51)	(0.12)	0.76
From subarea outlet to watershed outlet	1		0.45	
	2		0.45	
	3		0.33	
	4		0.33	
	5		0.12	
	6		0.12	
	7		0.0	

TABLE TM-3. Summary of Component and Composite Partial Hydrographs at the Watershed Outlet (Pre-mining).

Subarea	T$_c$ hr	T$_t$ hr	DA sq mi	Q in	Hydrograph time in hours						
					12.0	12.2	12.4	12.6	12.8	13.0	13.2
1	1.06	0.45	0.90	2.80	96	199	368	554	668	680	615
2	0.71	0.45	0.60	2.80	97	234	423	543	534	444	341
3	0.38	0.33	0.30	3.30	160	385	439	334	219	142	99
4	0.81	0.33	0.75	2.80	157	359	573	670	622	510	393
5	1.12	0.12	0.60	2.35	135	245	343	386	372	231	264
6	1.41	0.12	1.20	2.80	228	403	601	742	796	773	695
7	0.76	0.0	0.60	2.80	417	630	620	464	328	239	180
Total			4.95		1290	2096	3367	3693	3539	3109	2587

TABLE TM-4. Summary of Component and Composite Partial Hydrographs at the Watershed Outlet (Post-mining).

Subarea	T$_c$ hr	T$_t$ hr	DA sq mi	Q in	Hydrograph time in hours						
					12.0	12.2	12.4	12.6	12.8	13.0	13.2
1	0.79	0.45	0.90	3.78	170	398	742	1014	1061	932	745
2	0.53	0.45	0.60	3.78	184	476	784	846	696	506	356
3	0.38	0.33	0.30	3.30	160	385	439	334	219	127	99
4	0.60	0.33	0.75	3.78	323	734	1094	961	933	573	420
5	1.12	0.12	0.60	2.35	135	245	343	386	372	321	264
6	1.41	0.12	1.20	2.80	228	403	601	742	796	773	695
7	0.76	0.0	0.60	2.80	417	630	620	464	328	239	180
Total			4.95		1617	3271	4623	4753	4405	3471	2759

UNIT HYDROGRAPH CONCEPT

For several decades runoff hydrographs from watersheds have been determined using the unit hydrograph concept [47]. This concept holds that a constant excess rainfall uniformly distributed over a watershed produces a runoff hydrograph which is characteristic of the watershed and the duration of the rainfall excess. Further, instantaneous flow rates of the hydrograph are proportional to the rainfall excess rate. Thus, once a unit hydrograph for some convenient duration is known for a watershed, any pattern of excess rainfall can be subdivided into components of that duration and component hydrographs for each duration can be found and summed to determine the total hydrograph for the entire rainfall excess pattern. This process is shown in Figure 8. The unit hydrograph for the time duration, D is represented by the solid line. It contains 1.00 inches of excess rainfall. The unit hydrograph theory assumes that any rainfall excess amount which occurs during D time units produces a component hydrograph with ordinates equal to a factor times the ordinates of the unit hydrograph with duration, D. This factor is equal to the amount of the rainfall excess divided by the amount of rainfall excess in the unit hydrograph. For example, the first rainfall excess is 0.30 inches. Thus, the first component hydrograph of the rainfall excess pattern has ordinates which are 30 percent as large as those in the unit hydrograph. Similarly, the second and third component hydrographs have ordinates which are 50 percent and 20 percent of the unit hydrograph ordinates. Note, however, that component hydrographs 2 and 3 must begin D and $2D$ time units later than component 1 corresponding to the timing of their respective rainfall excesses. The total hydrograph for the entire rainfall excess pattern is estimated by summing the ordinates of the component hydrographs to determine a composite hydrograph as shown in Figure 8.

THE SCS TRIANGULAR UNIT HYDROGRAPH METHOD

Unit hydrographs are usually derived from rainfall-runoff data for the watershed of interest. Generally these data are not available for most watersheds and even if they were available they could not be used to anticipate changes in runoff which result from watershed disturbances. This has led the SCS to develop a synthetic unit hydrograph based on the time of concentration of the watershed of interest. Using data from many watersheds throughout the U.S., the SCS [55] has found that a universal unit hydrograph can be approximated by a triangle as shown in Figure 9. The peak discharge of the triangular unit hydrograph is given by

$$q_p = \frac{645 \; Q \; A}{0.89 \; T_c} \tag{14}$$

in which q_p is the peak discharge in cfs, Q is the excess rainfall depth in inches, A is the watershed area in sq mi, T_c is the time of concentration in hours and the constant is simply a conversion factor. It is customary to take Q equal to 1.0

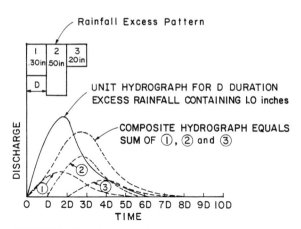

FIGURE 8. Development of a composite hydrograph using the unit hydrograph concept.

inches in which case q_p becomes simply

$$q_p = \frac{645 \; A}{0.89 \; T_c} \tag{15}$$

The base of the triangle is defined by

$$T_b = 1.78 \; T_c \tag{16}$$

and the time to peak by

$$T_p = 0.67 \; T_c \tag{17}$$

The duration, D, to which this unit hydrograph applies is

$$D = 0.133 \; T_c \tag{18}$$

In this case, the time of concentration T_c includes overland time as calculated by Equation (12) plus time of travel in any channels as determined using the Manning equation [Equation (13)].

The SCS Triangular method for estimation of a runoff hydrograph includes the following steps:

a. Estimate the time of concentration of the watershed using Equation (12) for the upland, overland flow portion and Equation (13) and channel length for the channel portion.
b. Estimate the rainfall excess pattern in increments of D hours using the Water Balance method or by applying the CN method in successive time increments (see example below).
c. Plot the individual component hydrographs for each rainfall excess of duration D making sure to begin each component hydrograph D time units later than the previous one.

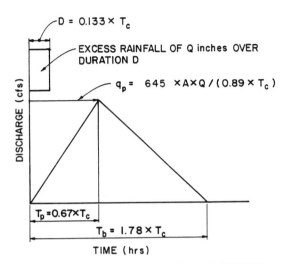

FIGURE 9. SCS triangular unit hydrograph (SCS, [55]).

d. Sum the component hydrographs to determine the total hydrograph.

The SCS approach to hydrograph synthesis relies on the same basic assumptions that apply to all unit hydrograph methods:

1. Rainfall excesses of a given duration produce component hydrographs with timing characteristics that are independent of rainfall excess amounts.
2. Component hydrograph ordinates are proportional to the rainfall excess amounts.
3. Component hydrographs are independent and can therefore be summed to determine a composite hydrograph.

These assumptions are inconsistent with the nonlinear process of unsteady rainfall excess and surface runoff on a watershed. The timing of peaks of component hydrographs do depend upon the size of rainfall excess as indicated by Equation (9). Also, components are not independent; there is an optimal pattern of rainfall excesses for which a particular unit hydrograph will most accurately estimate the runoff hydrograph. Departures from the optimal rainfall excess pattern reduce the accuracy of the method.

In addition to the assumptions associated with unit hydrograph methods in general, the SCS Triangular Hydrograph approach includes several other assumptions which allow development of unit hydrographs without rainfall-runoff data. These are outlined by the SCS [55]. Despite these assumptions, the method is useful for hydrograph estimation when computer facilities are limited or unavailable. The method can be used to develop hydrographs from large, complex watersheds by compositing several hydrographs from different subareas. For large watersheds, SCS [55] suggests that these subareas be hydrologically homogeneous and less than 20 sq mi in size. If SCS methods are used to calculate both rainfall excess patterns and time of concentration, then results will depend strongly on curve number selection.

Example: SCS Triangular Unit Hydrograph Method

Mining operations are planned for a 5.0 sq mi watershed which has a present curve number of 70. It is estimated that after mining and reclamation operations are completed the curve number will increase to 80. The watershed slope is 8.2 percent and the hydraulic length is approximately 18,000 ft. These values are expected to remain unchanged because planned reclamation will return the surface to its original contour. SCS methods will be used to determine the runoff hydrographs from the summer thunderstorm shown in Figure UH-1 for both pre-mining and reclaimed conditions.

PRE-MINING CONDITIONS

1. Determine the water storage parameter, S, using Equation (2):

$$S = (1000/CN) - 10$$

$$S = (1000/70) - 10 = 4.29 \text{ inches}$$

2. Determine the time of concentration, T_c, using Equation (12):

$$T_c = \ell^{0.8}(S+1)^{0.7}/(1140 \ Y^{0.5})$$

$$T_c = 18,000^{0.8}(4.29+1)^{0.7}/[(1140) \ 8.2^{0.5}] = 3.33 \text{ hours}$$

3. Determine the appropriate rainfall excess duration, D, from Equation (18):

$$D = 0.133 \ T_c$$

$$D = 0.133(3.33) = 0.44 \text{ hours}$$

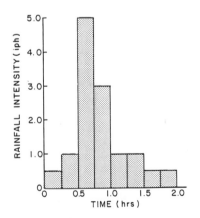

FIGURE UH-1. Unit hydrograph example rainfall pattern.

TABLE UH-1. Pre-Mining Conditions.

(i) N	(ii) i (in/hr)	(iii) $\Delta P = i(D)$ (in)	(iv) $t = N(D)$ (hr)	(v) $P = \Sigma\Delta P$ (in)	(vi) Q (in)	(vii) $\Delta Q = Q_N - Q_{N-1}$ (in)	(viii) q_p (cfs)
1	0.76	0.38	0.50	0.38	0.00	0.00	0
2	4.00	2.00	1.00	2.38	0.40	0.40	435
3	1.00	0.50	1.50	2.88	0.65	0.25	272
4	0.50	0.25	2.00	3.13	0.79	0.14	152

This can be rounded ± 15 percent for a more convenient duration. Let $D = 0.50$ hours.

4. Rearrange the rainfall excess hyetograph based on time increments of D hrs: Since the storm is already in the form of a series of 0.25 hours rates, conversion to 0.50 hour rates is easily accomplished by averaging pairs of consecutive rates. The results of this procedure are shown in column ii of Table UH-1.

5. Convert the series of rainfall intensities to a series of depths by multiplying each intensity from (4) by the duration, D. Results of this procedure are tabulated in column iii of Table UH-1.

6. Accumulate rainfall depths, P, corresponding to accumulated time (see column v of Table UH-1).

7. Determine successive accumulated runoff depths, Q, corresponding to accumulated rainfall depths, P: Input successive P values in column v of Table UH-1 into Equation (1). Note that P values and not ΔP values must be used in Equation (1). Negative Q results from this equation are set to zero. Tabulate the Q results in column vi of Table UH-1. For example, for the third accumulated time interval (time = 3(D) = 1.5 hours).

$$Q = (P - 0.2\ S)^2/(P + 0.8\ S)$$

$$Q = [2.88 - 0.2(4.29)]^2/[2.88 + 0.8(4.29)] = 0.65 \text{ inches}$$

8. Determine increments of Q: Disaggregate the successive accumulated Q values in column vi into separate ΔQ increments as shown in column vii of Table UH-1.

9. Determine peak discharge rates: Use each ΔQ increment in Equation (14) and tabulate results in column viii of Table UH-1. For example, for the second increment:

$$q_p = 645\ QA/(0.89\ T_c)$$

$$q_p = [645(0.4)5.0]/[0.89(3.33)] = 435 \text{ cfs}$$

10. Determine the time base of each component triangular hydrograph from Equation (16):

$$T_b = 1.78\ T_c$$

$$T_b = 1.78(3.33) = 5.93 \text{ hours}$$

11. Determine the time to peak for each component hydrograph from Equation (17):

$$T_p = 0.67\ T_c$$

$$T_p = 0.67(3.33) = 2.23 \text{ hours}$$

12. Plot component hydrographs: Use q_p values from column viii as vertices of successive component triangular hydrographs, each with base as determined in (10) and a time to peak as determined in (11). Begin each component at time = $(N - 1)$ − hours.

13. Determine the composite hydrograph: Sum ordinates of each component hydrograph as shown in Figure UH-2.

POST-MINING CONDITIONS

The same process is repeated for post-mining conditions with the following results:

1. $S = 2.50$ inches
2. $T_c = 2.50$ hours
3. $D = 0.33$ hours
4. through 9. i, ΔP, P, Q, ΔQ and q_p as in Table UH-2
10. $T_b = 4.45$ hours
11. $T_p = 1.67$ hours
12. and 13. component and composite hydrographs as shown in Figure UH-3.

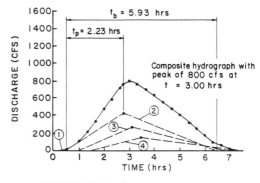

FIGURE UH-2. Pre-mining conditions.

TABLE UH-2. Post-Mining Conditions.

(i) N	(ii) i (in/hr)	(iii) ΔP (in)	(iv) t = N(D) (hr)	(v) P = ΣΔP (in)	(vi) Q (in)	(vii) ΔQ = $Q_N - Q_{N-1}$ (in)	(viii) q_p (cfs)
1	0.63	0.21	0.33	0.21	0.00	0.00	0
2	3.00	1.00	0.67	1.21	0.16	0.16	232
3	3.50	1.17	1.00	2.38	0.81	0.65	943
4	1.05	0.35	1.33	2.71	1.04	0.23	334
5	0.75	0.25	1.67	2.96	1.22	0.18	261
6	0.50	0.17	2.00	3.13	1.35	0.13	189

This example shows the sensitivity of the SCS procedure to the *CN* value. A 14 percent increase in *CN* from pre- to post-mining conditions results in an estimated increase in peak discharge of over 100 percent and an advance in time to peak of 0.5 hours. The procedure estimates that runoff volume increases by about 70 percent.

MECHANISTIC FLOW ROUTING METHOD

In the mechanistic flow routing approach, equations of mass and momentum conservation for the surface runoff water are written and combined to form a single, nonlinear, partial differential equation in both space and time for the flow rate. Solution of this equation is usually accomplished by numerical methods such as finite difference or finite element schemes which provide accurate results at a cost of considerable computational effort. Although it is possible to "step through" these methods using a calculator, a computer is essential for their practical use. A detailed description of mechanistic flow routing procedures is beyond the scope of

this discussion. The method is mentioned because:

1. It provides the most accurate mathematical representation of the surface runoff process and generally gives the most accurate results of any method.

2. Hydrographs, flow depths and velocities for virtually any point in the watershed are determined automatically. This is essential information for estimating the movement of sediment and other pollutants in the watershed.

3. The most advanced computer models of watershed hydrology utilize mechanistic flow routing methods. Some of these models which are currently available and useful to the surface mine hydrologist are discussed at the end of this chapter.

Further explanations of mechanistic flow routing in watersheds are given by Eagleson [11], Freeze [18], and Woolhiser [68].

ESTIMATION OF EROSION AND SEDIMENT YIELD

Erosion refers to the movement of soil from its original location or source area. A source may be a watershed surface or some part of the channel system. Some of the soil eroded from steep upstream sources often deposits in flatter downstream areas of a watershed. The amount of soil which passes a point of interest such as the watershed outlet during a specified time interval is referred to as the sediment yield. The sum of the sediment yield at a given location and the amount of soil deposited upstream of that location equals the total amount of soil eroded or "gross" erosion from all upstream areas. The sediment yield divided by the amount of gross erosion is referred to as the sediment delivery ratio (SDR). If all upstream source areas including channel segments are accounted for, the SDR will be less than or equal to 1.0.

Estimation of Erosion using the USLE

The Universal Soil Loss Equation (USLE) (Wischmeier and Smith, [67]) estimates soil loss caused by sheet and rill

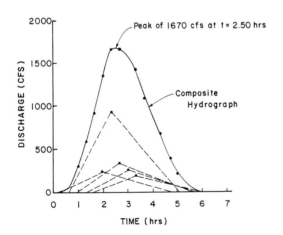

FIGURE UH-3. *Post-mining conditions.*

erosion from a field. The equation is

$$A = R K L S C P \qquad (19)$$

in which A is customarily the average annual mass of soil loss per unit area, R represents the average annual erosivity of rainfall and runoff with dimensions of energy-length per unit area-time per year and K is a soil erodibility parameter with dimensions of mass-time per energy-length. The remaining parameters are dimensionless. L represents the field length, S the slope, C the effect of canopy and ground cover and P the effect of conservation practices such as contouring and strip cropping.

Equation (19) is usually used to calculate average annual soil loss, however calculations for other time periods or even single storm events can be made if the magnitude and time units of the R parameter are suitably adjusted. In such cases R would be calculated per time period of interest or per storm instead of per year and time units of A would change accordingly.

Wischmeier and Smith [67] present the most comprehensive discussion of the theory, limitations and applications of the USLE. Prospective users of the soil loss equation should consult their work for a more complete discussion. The discussion below focuses on evaluation of USLE factors and use of the equation.

THE RAINFALL-RUNOFF FACTOR R

The R factor is the rainfall-runoff factor. In areas where snowmelt runoff is not a significant factor in soil loss, the R factor depends only upon rainfall intensities and amounts. The R factor for any storm can be determined by breaking the storm into a histogram of n components, each with its own intensity i_j in inches per hour and rainfall depth p_j in inches. The R factor for the storm is then defined by:

$$R = \sum_{j=1}^{n} \frac{e_j I_{30}}{100} \qquad (20)$$

in which I_{30} is the maximum 30-minute intensity during the storm in inches per hour and e_j is the energy of each rainfall component in hundreds of ft-tons per acre and is defined by

$$e_j = (916 + 331 \log_{10} i_j)p_j \qquad (21)$$

in which a maximum upper limit of 3.0 inches per hour is placed on i_j.

The R values for all "significant" storms in the period of interest are summed to determine an overall R value for the period. When calculating R values, successive rainfalls are considered to be individual storms if they are separated from one another by at least 6 hours without precipitation. Only storms which contain at least 0.5 inches of total rainfall or a 15-minute rainfall intensity of at least 1.0 iph are

used in the calculation of R values. At most mine sites there are usually insufficient data to determine average annual or monthly R values. Where data is unavailable, average annual R values for sites within the coterminous United States can be estimated from Figure 10. These values represent averages over a 22-year period which is considered long enough to incorporate major weather fluctuations at a given location. Table 13 presents useful data on expected deviations from the average annual R values for a range of probabilities for 181 locations. Table 14 presents R values for single storms with return periods ranging from 1 to 20 years.

It should be noted that the accuracy of R values determined from Figure 10 is suspect in areas where rainfall characteristics change rapidly with location such as in the mountainous West. More localized recording rain gauge data is needed to improve R estimates in these areas.

In areas where erosion from snowmelt runoff represents a significant portion of total annual erosion, USDA Handbook 537 [67] suggests that the average annual R value from Figure 10 should be incremented by an amount $R(s)$ defined as:

$$R(s) = 1.5 P_w \qquad (22)$$

in which P_w is the average annual precipitation depth in inches which occurs between December 1 and March 31. The handbook recommends that this snowmelt correction be applied to northern regions including the Pacific Northwest and areas in other northern states where experience indicates that erosion from snowmelt runoff is important.

Often it is desirable to estimate soil losses for specific seasons. Figure 11 partitions the coterminus United States into 33 regions. The cumulative distribution of R values within the year is given in Table 15 for each of these regions. Monthly distributions of the R factor for selected sites in the western states, Hawaii and Puerto Rico are given in Table 16. Distributions of R values in these tables do not include any snowmelt runoff component and must be corrected using $R(s)$ values for each month. The U.S. Soil Conservation Service (SCS) describes procedures for developing erosion index distribution curves [8]. Curves for specific regions within a given state can usually be obtained from the SCS state office.

THE ERODIBILITY FACTOR K

Ideally, K should be determined by setting up replicated field plots and collecting rainfall and soil loss data for many years. Recognizing that in most situations this would not be possible, Wischmeier and Smith [67] developed an alternative approach to estimating K for both surface and subsoils based on soil particle size distribution, organic matter content, soil structure and permeability class. These factors are incorporated into the nomograph shown in Figure 12.

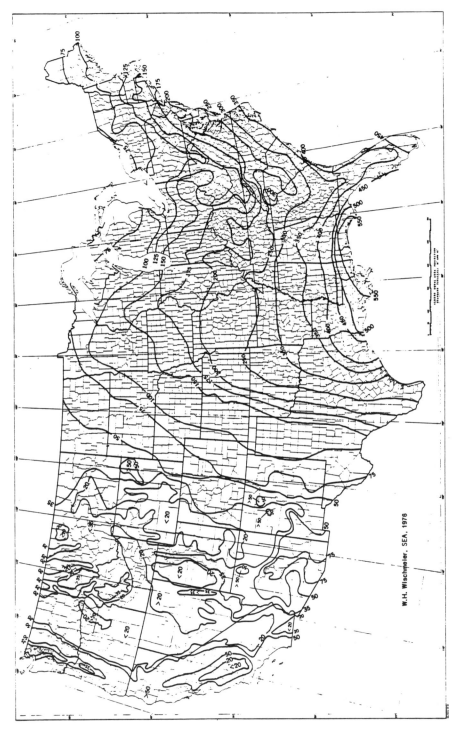

FIGURE 10. Average annual values of the rainfall erosion index (Wischmeier and Smith, [67]).

W.H. Wischmeier, SEA, 1976

TABLE 13. Observed 50-, 20-, and 5- Percent Probability Values of Erosion Index at Each of 181 Key Locations (Wischmeier and Smith, [67]).

Location	Observed 22-year range	Values of erosion index (EI) 50-percent probability	20-percent probability	5-percent probability
Alabama:				
Birmingham	179–601	354	461	592
Mobile	279–925	673	799	940
Montgomery	164–780	359	482	638
Arkansas:				
Fort Smith	116–818	254	400	614
Little Rock	103–625	308	422	569
Mountain Home	98–441	206	301	432
Texarkana	137–664	325	445	600
California:				
Red Bluff	11–240	54	98	171
San Luis Obispo	5–147	43	70	113
Colorado:				
Akron	8–247	72	129	225
Pueblo	5–291	44	93	189
Springfield	4–246	79	138	233
Connecticut:				
Hartford	65–355	133	188	263
New Haven	66–373	157	222	310
District of Columbia	84–334	183	250	336
Florida:				
Apalachicola	271–944	529	663	820
Jacksonville	283–900	540	693	875
Miami	197–1225	529	784	1136
Georgia:				
Atlanta	116–549	286	377	488
Augusta	148–476	229	308	408
Columbus	215–514	336	400	473
Macon	117–493	282	357	447
Savannah	197–886	412	571	780
Watkinsville[1]	182–544	278	352	441
Illinois:				
Cairo	126–575	231	349	518
Chicago	50–379	140	212	315
Dixon Springs[1]	89–581	225	326	465
Moline	80–369	158	221	303
Rantoul	73–286	152	201	263
Springfield	38–315	154	210	283
Indiana:				
Evansville	104–417	188	263	362
Fort Wayne	60–275	127	183	259
Indianapolis	60–349	166	225	302
South Bend	43–374	137	204	298
Terre Haute	81–413	190	273	389
Iowa:				
Burlington	65–286	162	216	284
Charles City	39–308	140	205	295
Clarinda[1]	75–376	162	220	295
Des Moines	30–319	136	198	284
Dubuque	54–389	175	251	356
Sioux City	56–336	135	205	308
Rockwell City	40–391	137	216	335

(continued)

213

TABLE 13 (continued).

Location	Observed 22-year range	Values of erosion index (EI) 50-percent probability	20-percent probability	5-percent probability
Kansas:				
Burlingame	57–447	176	267	398
Coffeyville	66–546	234	339	483
Concordia	38–569	131	241	427
Dodge City	16–421	98	175	303
Goodland	10–166	76	115	171
Hays[1]	66–373	116	182	279
Wichita	42–440	188	292	445
Kentucky:				
Lexington	54–396	178	248	340
Louisville	84–296	168	221	286
Middlesboro	107–301	154	197	248
Louisiana:				
Lake Charles	200–1019	572	786	1063
New Orleans	273–1366	721	1007	1384
Shreveport	143–707	321	445	609
Maine:				
Caribou	26–120	58	79	106
Portland	36–241	91	131	186
Skowhegan	39–149	78	108	148
Maryland:				
Baltimore	50–388	178	263	381
Massachusetts:				
Boston	39–366	99	159	252
Washington	65–229	116	153	198
Michigan:				
Alpena	14–124	57	85	124
Detroit	56–179	100	134	177
East Lansing	35–161	86	121	166
Grand Rapids	33–203	84	123	178
Minnesota:				
Alexandria	33–301	88	147	240
Duluth	7–227	84	127	189
Fosston	22–205	62	108	184
Minneapolis	19–173	94	135	190
Rochester	46–338	142	207	297
Springfield	37–290	96	154	243
Mississippi:				
Meridian	216–820	416	557	737
Oxford	131–570	310	413	543
Vicksburg	165–786	365	493	658
Missouri:				
Columbia	98–419	214	297	406
Kansas City	28–361	170	248	356
McCredie[1]	64–410	189	271	383
Rolla	105–415	209	287	387
Springfield	97–333	199	266	352
St. Joseph	50–359	178	257	366
St. Louis	59–737	168	290	488
Montana:				
Billings	2–82	12	26	50
Great Falls	3–62	13	24	44
Miles City	1–101	21	40	72

(continued)

TABLE 13 (continued).

Location	Values of erosion index (EI)			
	Observed 22-year range	50-percent probability	20-percent probability	5-percent probability
Nebraska:				
Antioch	18–131	60	86	120
Lincoln	44–289	133	201	299
Lynch	34–217	96	142	205
North Platte	14–236	81	136	224
Scribner	69–312	154	205	269
Valentine	4–169	64	100	153
New Hampshire:				
Concord	52–212	91	131	187
New Jersey:				
Atlantic City	71–318	166	229	311
Marlboro[1]	58–331	186	254	343
Trenton	37–382	149	216	308
New Mexico:				
Albuquerque	0–46	10	19	35
Roswell	5–159	41	73	128
New York:				
Albany	40–172	81	114	159
Binghamton	20–151	76	106	146
Buffalo	20–148	66	96	139
Geneva[1]	33–180	73	106	152
Marcellus[1]	24–241	74	112	167
Rochester	22–180	66	101	151
Salamanca	31–202	70	106	157
Syracuse	8–219	83	129	197
North Carolina:				
Asheville	76–238	135	175	223
Charlotte	113–526	229	322	443
Greensboro	102–357	184	244	320
Raleigh	152–569	280	379	506
Wilmington	196–701	358	497	677
North Dakota:				
Bismarck	9–189	43	73	120
Devils Lake	21–171	56	90	142
Fargo	5–213	62	113	200
Williston	4–71	30	45	67
Ohio:				
Cincinnati	66–352	146	211	299
Cleveland	21–186	93	132	185
Columbiana	29–188	96	129	173
Columbus	45–228	113	158	216
Coshocton[1]	72–426	158	235	343
Dayton	56–245	125	175	240
Toledo	32–189	83	120	170
Oklahoma:				
Ardmore	100–678	263	395	582
Cherokee[1]	49–320	167	242	345
Guthrie[1]	69–441	210	316	467
McAlester	105–741	272	411	609
Tulsa	19–584	247	347	478
Oregon:				
Pendleton	2–28	4	8	16
Portland	16–80	40	56	77

(continued)

TABLE 13 (continued).

Location	Observed 22-year range	Values of erosion index (EI)		
		50-percent probability	20-percent probability	5-percent probability
Pennsylvania:				
Erie	11–534	96	181	331
Franklin	50–228	97	135	184
Harrisburg	48–232	105	146	199
Philadelphia	72–361	156	210	282
Pittsburgh	43–201	111	148	194
Reading	84–308	144	204	285
Scranton	52–198	104	140	188
Puerto Rico:				
San Juan	203–577	345	445	565
Rhode Island:				
Providence	53–225	119	167	232
South Carolina:				
Charleston	174–1037	387	559	795
Clemson[1]	138–624	280	384	519
Columbia	81–461	213	298	410
Greenville	130–589	249	350	487
South Dakota:				
Aberdeen	19–295	74	129	219
Huron	18–145	60	91	136
Isabel	16–141	48	78	125
Rapid City	10–140	37	64	108
Tennessee:				
Chattanooga	163–468	269	348	445
Knoxville	64–370	173	239	325
Memphis	139–595	272	384	536
Nashville	116–381	198	262	339
Texas:				
Abilene	27–554	146	253	427
Amarillo	33–340	110	184	299
Austin	59–669	270	414	624
Brownsville	46–552	267	386	549
Corpus Christi	124–559	237	330	451
Dallas	93–630	263	396	586
Del Rio	19–405	121	216	374
El Paso	4–85	18	36	67
Houston	176–1171	444	674	1003
Lubbock	17–415	82	158	295
Midland	35–260	82	139	228
Nacogdoches	153–769	401	571	801
San Antonio	77–635	220	353	556
Temple[1]	81–644	261	379	542
Victoria	108–609	265	385	551
Wichita Falls	79–558	196	298	447
Vermont:				
Burlington	33–270	72	114	178
Virginia:				
Blacksburg[1]	81–245	126	168	221
Lynchburg	64–366	164	232	324
Richmond	102–373	208	275	361
Roanoke	78–283	129	176	237

(continued)

TABLE 13 (continued).

	Values of erosion index (EI)			
Location	Observed 22-year range	50-percent probability	20-percent probability	5-percent probability
Washington:				
Pullman[1]	1–30	6	12	21
Spokane	1–19	7	11	17
West Virginia:				
Elkins	43–223	118	158	209
Huntington	56–228	127	173	233
Parkersburg	69–303	120	165	226
Wisconsin:				
Green Bay	17–148	77	107	147
LaCrosse[1]	61–385	153	228	331
Madison	38–251	118	171	245
Milwaukee	31–193	93	139	202
Rice Lake	24–334	122	202	327
Wyoming:				
Casper	1–24	9	15	26
Cheyenne	8–66	28	43	66

[1]Computations based on SEA rainfall records. All others are based on Weather Bureau records.

TABLE 14. Expected Magnitudes of Single-Storm Erosion Index Values (Wischmeier and Smith, [67]).

	Index values normally exceeded once in—				
Location	year 1	years 2	years 5	years 10	years 20
Alabama:					
Birmingham	54	77	110	140	170
Mobile	97	122	151	172	194
Montgomery	62	86	118	145	172
Arkansas:					
Fort Smith	43	65	101	132	167
Little Rock	41	69	115	158	211
Mountain Home	33	46	68	87	105
Texarkana	51	73	105	132	163
California:					
Red Bluff	13	21	36	49	65
San Luis Obispo	11	15	22	28	34
Colorado:					
Akron	22	36	63	87	118
Pueblo	17	31	60	88	127
Springfield	31	51	84	112	152
Connecticut:					
Hartford	23	33	50	64	79
New Haven	31	47	73	96	122
District of Columbia	39	57	86	108	136
Florida:					
Apalachicola	87	124	180	224	272
Jacksonville	92	123	166	201	236
Miami	93	134	200	253	308

(continued)

217

TABLE 14 (continued).

Location	Index values normally exceeded once in—				
	year 1	years 2	years 5	years 10	years 20
Georgia:					
Atlanta	49	67	92	112	134
Augusta	34	50	74	94	118
Columbus	61	81	108	131	152
Macon	53	72	99	122	146
Savannah	82	128	203	272	358
Watkinsville	52	71	98	120	142
Illinois:					
Cairo	39	63	101	135	173
Chicago	33	49	77	101	129
Dixon Springs	39	56	82	105	130
Moline	39	50	89	116	145
Rantoul	27	39	56	69	82
Springfield	36	52	75	94	117
Indiana:					
Evansville	26	38	56	71	86
Fort Wayne	24	33	45	56	65
Indianapolis	29	41	60	75	90
South Bend	26	41	65	86	111
Terre Haute	42	57	78	96	113
Iowa:					
Burlington	37	48	62	72	81
Charles City	33	47	68	85	103
Clarinda	35	48	66	79	94
Des Moines	31	45	67	86	105
Dubuque	43	63	91	114	140
Rockwell City	31	49	76	101	129
Sioux City	40	58	84	105	131
Kansas:					
Burlingame	37	51	69	83	100
Coffeyville	47	69	101	128	159
Concordia	33	53	86	116	154
Dodge City	31	47	76	97	124
Goodland	26	37	53	67	80
Hays	35	51	76	97	121
Wichita	41	61	93	121	150
Kentucky:					
Lexington	28	46	80	114	151
Louisville	31	43	59	72	85
Middlesboro	28	38	52	63	73
Louisiana:					
New Orleans	104	149	214	270	330
Shreveport	55	73	99	121	141
Maine:					
Caribou	14	20	28	36	44
Portland	16	27	48	66	88
Skowhegan	18	27	40	51	63
Maryland:					
Baltimore	41	59	86	109	133
Massachusetts:					
Boston	17	27	43	57	73
Washington	29	35	41	45	50

(continued)

TABLE 14 (continued).

Location	Index values normally exceeded once in—				
	year 1	years 2	years 5	years 10	years 20
Michigan:					
Alpena	14	21	32	41	50
Detroit	21	31	45	56	68
East Lansing	19	26	36	43	51
Grand Rapids	24	28	34	38	42
Minnesota:					
Duluth	21	34	53	72	93
Fosston	17	26	39	51	63
Minneapolis	25	35	51	65	78
Rochester	41	58	85	105	129
Springfield	24	37	60	80	102
Mississippi:					
Meridian	69	92	125	151	176
Oxford	48	64	86	103	120
Vicksburg	57	78	111	136	161
Missouri:					
Columbia	43	58	77	93	107
Kansas City	30	43	63	78	93
McCredie	35	55	89	117	151
Rolla	43	63	91	115	140
Springfield	37	51	70	87	102
St. Joseph	45	62	86	106	126
Montana:					
Great Falls	4	8	14	20	26
Miles City	7	12	21	29	38
Nebraska:					
Antioch	19	26	36	45	52
Lincoln	36	51	74	92	112
Lynch	26	37	54	67	82
North Platte	25	38	59	78	99
Scribner	38	53	76	96	116
Valentine	18	28	45	61	77
New Hampshire:					
Concord	18	27	45	62	79
New Jersey:					
Atlantic City	39	55	77	97	117
Marlboro	39	57	85	111	136
Trenton	29	48	76	102	131
New Mexico:					
Albuquerque	4	6	11	15	21
Roswell	10	21	34	45	53
New York:					
Albany	18	26	38	47	56
Binghamton	16	24	36	47	58
Buffalo	15	23	36	49	61
Marcellus	16	24	38	49	62
Rochester	13	22	38	54	75
Salamanca	15	21	32	40	49
Syracuse	15	24	38	51	65

(continued)

TABLE 14 (continued).

Location	Index values normally exceeded once in—				
	year 1	years 2	years 5	years 10	years 20
North Carolina:					
Asheville	28	40	58	72	87
Charlotte	41	63	100	131	164
Greensboro	37	51	74	92	113
Raleigh	53	77	110	137	168
Wilmington	59	87	129	167	206
North Dakota:					
Devils Lake	19	27	39	49	59
Fargo	20	31	54	77	103
Williston	11	16	25	33	41
Ohio:					
Cincinnati	27	36	48	59	69
Cleveland	22	35	53	71	86
Columbiana	20	26	35	41	48
Columbus	27	40	60	77	94
Coshocton	27	45	77	108	143
Dayton	21	30	44	57	70
Toledo	16	26	42	57	74
Oklahoma:					
Ardmore	46	71	107	141	179
Cherokee	44	59	80	97	113
Guthrie	47	70	105	134	163
McAlester	54	82	127	165	209
Tulsa	47	69	100	127	154
Oregon:					
Portland	6	9	13	15	18
Pennsylvania:					
Franklin	17	24	35	45	54
Harrisburg	19	25	35	43	51
Philadelphia	28	39	55	69	81
Pittsburgh	23	32	45	57	67
Reading	28	39	55	68	81
Scranton	23	32	44	53	63
Puerto Rico:					
San Juan	57	87	131	169	216
Rhode Island:					
Providence	23	34	52	68	83
South Carolina:					
Charleston	74	106	154	196	240
Clemson	51	73	106	133	163
Columbia	41	59	85	106	132
Greenville	44	65	96	124	153
South Dakota:					
Aberdeen	23	35	55	73	92
Huron	19	27	40	50	61
Isabel	15	24	38	52	67
Rapid City	12	20	34	48	64
Tennessee:					
Chattanooga	34	49	72	93	114
Knoxville	25	41	68	93	122
Memphis	43	55	70	82	91
Nashville	35	49	68	83	99

(continued)

TABLE 14 (continued).

Location	Index values normally exceeded once in—				
	year 1	years 2	years 5	years 10	years 20
Texas:					
Abilene	31	49	79	103	138
Amarillo	27	47	80	112	150
Austin	51	80	125	169	218
Brownsville	73	113	181	245	312
Corpus Christi	57	79	114	146	171
Dallas	53	82	126	166	213
Del Rio	44	67	108	144	182
El Paso	6	9	15	19	24
Houston	82	127	208	275	359
Lubbock	17	29	53	77	103
Midland	23	35	52	69	85
Nacogdoches	77	103	138	164	194
San Antonio	57	82	122	155	193
Temple	53	78	123	162	206
Victoria	59	83	116	146	178
Wichita Falls	47	63	86	106	123
Vermont:					
Burlington	15	22	35	47	58
Virginia:					
Blacksburg	23	31	41	48	56
Lynchburg	31	45	66	83	103
Richmond	46	63	86	102	125
Roanoke	23	33	48	61	73
Washington:					
Spokane	3	4	7	8	11
West Virginia:					
Elkins	23	31	42	51	60
Huntington	18	29	49	69	89
Parkersburg	20	31	46	61	76
Wisconsin:					
Green Bay	18	26	38	49	59
LaCrosse	46	67	99	125	154
Madison	29	42	61	77	95
Milwaukee	25	35	50	62	74
Rice Lake	29	45	70	92	119
Wyoming:					
Casper	4	7	9	11	14
Cheyenne	9	14	21	27	34

FIGURE 11. Key map for selection of applicable EI-distribution data from Table 15 (Wischmeier and Smith, [67]).

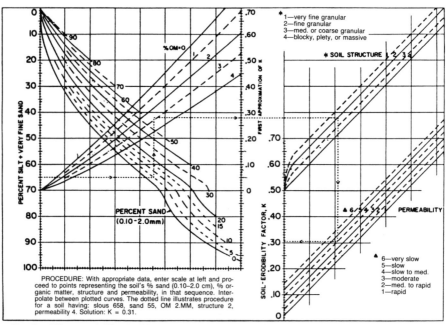

PROCEDURE: With appropriate data, enter scale at left and proceed to points representing the soil's % sand (0.10–2.0 cm), % organic matter, structure and permeability, in that sequence. Interpolate between plotted curves. The dotted line illustrates procedure for a soil having: slous 658, sand 55, OM 2.MM, structure 2, permeability 4. Solution: K = 0.31.

—The soil erodibility nomograph. Where the silt fraction does not exceed 7 percent, the equation is 100 K = 2.1 M (10^{-9}) (12 – a) + 3.25 (b – 2) + 2.5 (c – 3) where M = (percent si + vfs) (100 – percent c), a = percent organic matter, b = structure code, and c = profile permeability class.

FIGURE 12. USLE soil erodibility factor nomograph (Wischmeier and Smith, [67]).

TABLE 15. Percentage of the Average Annual EI Which Normally Occurs Between January 1 and the Indicated Dates.* Computed for the Geographic Areas Shown in Figure 11 [Wischmeier and Smith, [67]].

Area No.	Jan.		Feb.		Mar.		Apr.		May		June		July		Aug.		Sept.		Oct.		Nov.		Dec.	
	1	15	1	15	1	15	1	15	1	15	1	15	1	15	1	15	1	15	1	15	1	15	1	15
1	0	0	0	0	0	0	1	2	3	6	11	23	36	49	63	77	90	95	98	99	100	100	100	100
2	0	0	0	0	1	1	2	3	6	10	17	29	43	55	67	77	85	91	96	98	99	100	100	100
3	0	0	0	0	1	1	2	3	6	13	23	37	51	61	69	78	85	91	94	96	98	99	99	100
4	0	0	1	1	2	3	4	7	12	18	27	38	48	55	62	69	76	83	90	94	97	98	99	100
5	0	1	2	3	4	6	8	13	21	29	37	46	54	60	65	69	74	81	87	92	95	97	98	99
6	0	0	0	0	1	1	1	2	6	16	29	39	46	53	60	67	74	81	88	95	99	99	100	100
7	0	1	1	2	3	4	6	8	13	25	40	49	56	62	67	72	76	80	85	91	97	98	99	99
8	0	1	3	5	7	10	14	20	28	37	48	56	61	64	68	72	77	81	86	89	92	95	98	99
9	0	2	4	6	9	12	17	23	30	37	43	49	54	58	62	66	70	74	78	82	86	90	94	97
10	0	1	2	4	6	8	10	15	21	29	38	47	53	57	61	65	70	76	83	88	91	94	96	98
11	0	1	3	5	7	9	11	14	18	27	35	41	46	51	57	62	68	73	79	84	89	93	96	98
12	0	0	0	0	1	1	2	3	5	9	15	27	38	50	62	74	84	91	95	97	98	99	99	100
13	0	0	0	1	1	2	3	5	7	12	19	33	48	57	65	74	82	88	93	96	98	99	100	100
14	0	0	0	1	2	3	4	6	9	14	20	28	39	52	63	72	80	87	91	94	97	98	99	100
15	0	0	1	2	3	4	6	8	11	15	22	31	40	49	59	69	78	85	91	94	96	98	99	100
16	0	1	2	3	4	6	8	10	14	18	25	34	45	56	64	72	79	84	89	92	95	97	98	99
17	0	1	2	3	4	5	6	8	11	15	20	28	41	54	65	74	82	87	92	94	96	97	98	99
18	0	1	2	4	6	8	10	13	19	26	34	42	50	58	63	68	74	79	84	89	93	95	97	99
19	0	1	3	6	9	12	16	21	26	31	37	43	50	57	64	71	77	81	85	88	91	93	95	97
20	0	2	3	5	7	10	13	16	19	23	27	34	44	54	63	72	80	85	89	91	93	95	96	98
21	0	3	6	10	13	16	19	23	26	29	33	39	47	58	68	75	80	83	86	88	90	92	95	97
22	0	3	6	9	13	17	21	27	33	38	44	49	55	61	67	71	75	78	81	84	86	90	94	97
23	0	3	5	7	10	14	18	23	27	31	35	39	45	53	60	67	74	80	84	86	88	90	93	95
24	0	3	6	9	12	16	20	24	28	33	38	43	50	59	69	75	80	84	87	90	92	94	96	98
25	0	1	3	5	7	10	13	17	21	24	27	33	40	46	53	61	69	78	89	92	94	95	97	98
26	0	2	4	6	8	12	16	20	25	30	35	41	47	56	67	75	81	85	87	89	91	93	95	97
27	0	1	2	3	5	7	10	14	18	22	27	32	37	46	58	69	80	89	93	94	95	96	97	99
28	0	1	3	5	7	9	12	15	18	21	25	29	36	45	56	68	77	83	88	91	93	95	97	99
29	0	1	2	3	4	5	7	9	11	14	17	22	31	42	54	65	74	83	89	92	95	97	98	99
30	0	1	2	3	4	5	6	8	10	14	19	26	34	45	56	66	76	82	86	90	93	95	97	99
31	0	0	0	1	2	3	4	5	7	12	17	24	33	42	55	67	76	83	89	92	94	96	98	99
32	0	1	2	3	4	5	6	8	10	13	17	22	31	42	52	60	68	75	80	85	89	92	96	98
33	0	1	2	4	6	8	11	13	15	18	21	26	32	38	46	55	64	71	77	81	85	89	93	97

*For dates not listed in the table, interpolate between adjacent values.

TABLE 16. Monthly Distrubtion of EI at Selected Raingage Locations (Wischmeier and Smith, [67]).

| Location[1] | Average percentage of annual EI occurring from 1/1 to: | | | | | | | | | | | |
	2/1	3/1	4/1	5/1	6/1	7/1	8/1	9/1	10/1	11/1	12/1	12/31
California												
Red Bluff (69)	18	36	47	55	62	64	65	65	67	72	82	100
San Luis Obispo (51)	19	39	54	63	65	65	65	65	65	67	83	100
Colorado												
Akron (91)	0	0	0	1	18	33	72	87	98	99	100	100
Pueblo (68)	0	0	0	5	14	23	40	82	84	100	100	100
Springfield (98)	0	0	1	4	26	36	60	94	96	99	100	100
Hawaii												
Hilo (770)	9	23	34	44	49	51	55	60	65	72	87	100
Honolulu (189)	19	33	43	51	54	55	56	57	58	62	81	100
Kahului (107)	14	32	49	62	67	68	69	70	71	76	86	100
Lihue (385)	19	29	36	41	44	45	48	51	56	64	80	100
Montana												
Billings (18)	0	0	1	6	22	49	86	88	96	100	100	100
Great Falls (17)	1	1	2	6	20	56	74	93	98	99	100	100
Miles City (28)	0	0	0	1	10	32	65	93	98	100	100	100
New Mexico												
Albuquerque (15)	1	1	2	4	10	21	52	67	89	98	99	100
Roswell (52)	0	0	2	7	20	34	55	71	92	99	99	100
Oregon												
Pendleton (6)	8	12	15	22	56	64	67	67	74	87	96	100
Portland (43)	15	27	35	37	40	45	46	47	54	65	81	100
Puerto Rico												
Mayaguez (600)	1	2	3	6	15	31	47	63	80	91	99	100
San Juan (345)	5	8	11	17	33	43	53	66	75	84	93	100
Washington												
Spokane (8)	5	9	11	15	25	56	61	76	84	90	94	100
Wyoming												
Casper (11)	0	0	1	6	32	44	70	90	96	100	100	100
Cheyenne (32)	0	1	2	5	17	42	73	90	97	99	100	100

[1]Numbers in parentheses are the observed average annual EI.

Local offices of the U.S. Soil Conservation Service should be helpful in estimating K values for soils for pre-mining conditions. For disturbed soils, which have been stripped, stockpiled, mixed with other soils or compacted, K values may change. The extent of this change is unknown, though it is likely that K will increase because mechanical handling breaks down soil structure and compaction reduces infiltration rates. Because of the uncertainty involved in determining K for post-mining conditions, the following procedure is recommended for estimating values for the first few years following soil placement:

1. Determine a "first approximation" of K using estimates of soil particle size distribution and organic matter content from the left half of the nomograph in Figure 12.
2. Assume the most erodible soil structure class (line 4) and a permeability class which is at least two classes higher (i.e. lower permeability) than the class of the original un-

disturbed soil and refine the value from step 1 using the right hand side of the nomograph.

Although this procedure is "conservative" in that it results in higher post-mining K values, mine hydrologists should consult with appropriate regulatory authorities before using it.

THE LENGTH FACTOR L

The length factor L for surfaces with uniform slopes is defined as:

$$L = \left(\frac{\lambda}{72.6}\right)^m \tag{23}$$

in which λ is the length in feet from the top of the slope where overland flow begins to the point where slope decreases enough to cause deposition or where the runoff enters a channel, and m is an exponent dependent on percent slope as follows:

slopes steeper than 5% $m = 0.5$
between 3.5 and 5% $m = 0.4$
between 1.0 and 3.5% $m = 0.3$
less than 1.0% $m = 0.2$

THE SLOPE FACTOR S

The slope factor is defined by

$$S = 65.41 \sin^2\theta + 4.56 \sin\theta + 0.065 \quad (24)$$

in which θ is the angle of inclination of the sloping surface.

The relationships for L and S expressed by Equations (23) and (24) are based on field data where λ ranged from 30 to 300 feet and slopes from 3 to 18 percent. Use of the relationships for slopes and lengths far outside this range leads to results of unknown accuracy.

CURVED AND IRREGULAR SLOPES

Equations (23) and (24) apply to uniform slopes. Soil loss from concave slopes tends to be less than uniform slopes while it tends to be greater for convex slopes. Foster and Wischmeier [17] presented the following relationship for the product of L and S for nonuniform slopes which can be broken into a series n uniform slopes on which no deposition occurs:

$$LS = \frac{\sum_{j=1}^{n} S_j(\lambda_j^{l+m} - \lambda_{j-l}^{l+m})}{\lambda_e \, 72.6^m} \quad (25)$$

in which j refers to the slope segment number beginning with l at the top of the slope, S_j is the slope factor for the segment defined by Equation (24), λ_j is the length measured from the top of the slope to the bottom of the jth segment, m is as defined previously, and λ_e is the overall length of the nonuniform slope.

COVER AND MANAGEMENT FACTOR C

The C factor ranges from 0.0 to 1.0 and represents the ratio of soil loss from a surface of given cover and management characteristics to the soil loss from a bare surface which has been tilled up and down slope. This factor accounts for erosion reduction by vegetation, mulches and management practices not specifically included in the P factor described below. The C factor changes with time as a result of changes in vegetation and surface litter. The average value of C over any time period of n intervals must be weighted by the R factor for those intervals as follows:

$$C = \frac{\sum_{i=1}^{n} C_i R_i}{\sum_{i=1}^{n} R_i} \quad (26)$$

Equation (26) demonstrates that when C values vary over a period, the average C value for the same vegetation system may vary geographically because the temporal variation of R can vary geographically.

EVALUATION OF C FOR PRE-MINING CONDITIONS

Values of C for premining conditions will depend upon premining land use and variation in vegetation and mulch conditions over the period of interest. For woodland areas, Table 17 gives estimates of C as a function of forest canopy and ground cover densities. These C factors will exhibit seasonal variation with changes in canopy. For permanent pasture and rangeland, Table 18 can be used. It should be noted that C values in these tables reflect effects of the long-term build up of roots, organic matter and soil structure which must be assumed to be absent following the severe disturbance caused by surface mining activities such as soil removal, stockpiling and replacement.

Values of the C factor for premining conditions on agri-

TABLE 17. "C" Factors for Woodland (EPA, [61]).

1/ Tree Canopy % of Area	2/ Forest Litter % of Area	3/ Undergrowth	"C" Factor
100–75	100–90	Managed 4/	.001
		Unmanaged 4/	.003–.011
70–40	85–75	Managed	.022–.004
		Unmanaged	.01–.04
35–20	70–40	Managed	.003–.009
		Unmanaged	5/

1) When tree canopy is less than 20%, the area will be considered as grassland for estimating soil loss. See Table 18.
2) Forest litter is assumed to be at least two inches deep over the percent ground surface area covered.
3) Undergrowth is defined as shrubs, weeds, grasses, vines, etc., on the surface area not protected by forest litter. Usually found within canopy openings.
4) Managed—grazing and fires are controlled. Unmanaged—stands that are overgrazed or subjected to fires from natural causes.
5) For unmanaged woodland with litter cover of less than 75%, C values should be derived by taking 0.7 of the appropriate values in Table 18. The factor of 0.7 adjusts for the much higher soil organic matter on permanent woodland.

TABLE 18. "C" Factors for Permanent Pasture and Rangeland (EPA, [61]).

Vegetative Canopy			Cover that Contacts the Surface 1/						
Type and Height of Raised Canopy 2/	Canopy Cover 3/	Type 4/	Percent Ground Cover						
			0	10	20	40	60	80	95–100
No appreciable canopy		G	1.0	.45	.20	.10	.042	.013	.003
		W	1.0	.45	.24	.15	.090	.043	.011
Canopy of tall forbs or short brush (0.5 m fall ht.)	25	G	1.0	.36	.17	.09	.038	.012	.003
		W	1.0	.36	.20	.13	.082	.041	.011
	50	G	1.0	.26	.13	.07	.035	.012	.003
		W	1.0	.26	.16	.11	.075	.039	.011
	75	G	1.0	.17	.10	.06	.031	.011	.003
		W	1.0	.17	.12	.07	.067	.038	.011
Appreciable brush or brushes (2 m fall ht.)	25	G	1.0	.40	.18	.09	.040	.013	.003
		W	1.0	.40	.22	.14	.085	.042	.011
	50	G	1.0	.34	.16	.085	.038	.012	.003
		W	1.0	.34	.19	.13	.081	.041	.011
	75	G	1.0	.28	.14	.08	.036	.012	.003
		W	1.0	.28	.17	.12	.077	.040	.011
Trees but no appreciable low brush (4 m fall ht.)	25	G	1.0	.42	.19	.10	.041	.013	.003
		W	1.0	.42	.23	.14	.087	.042	.011
	50	G	1.0	.39	.18	.09	.040	.013	.003
		W	1.0	.39	.21	.14	.085	.042	.011
	75	G	1.0	.36	.17	.09	.039	.012	.003

1/ All values shown assume: (1) random distribution of mulch or vegetation, and (2) mulch of appreciable depth where it exists.
2/ Average fall height of waterdrops from canopy to soil surface: m = meters.
3/ Portion of total-area surface that would be hidden from view by canopy in a vertical projection, (a bird's-eye view).
4/ G: Cover at surface is grass, grasslike plants, decaying compacted duff, or litter.
   W: Cover at surface is mostly broadleaf herbaceous plants (as weeds with little lateral-root network near the surface, and/or undecayed residue).

cultural land will depend upon the particular cropping system and management practices employed by the farmer. Readers interested in C factors for cropland should consult *ARS Handbook 537* [67].

EVALUATION OF C FOR MINING AND POST-MINING CONDITIONS

Denuded surfaces such as regraded mine spoil or freshly placed topsoil will have C values of 1.0 unless surface mulch has been applied. Table 19 gives C values for different amounts and types of mulches. Note that the values are somewhat dependent upon surface slope and that the mulches become ineffective when lengths of denuded surfaces exceed a given limit. Reclamation sometimes requires that excessively long slopes be partitioned into a series of shorter ones by constructing various types of terraces. Information on terrace design should be obtained from appropriate regulatory authorities.

After topsoiling and seeding, C will depend on ground and canopy cover. If mulch has been applied, its contribution to ground cover may diminish with time because of decomposition, wind or water erosion. Percentages of ground cover, including residual mulch and live vegetation, and canopy cover can be used to estimate C values using Figures 13 and 14. In most cases, use of Figure 14 for canopy cover with an average height lower than 20 inches will not result in serious error.

THE SUPPORT PRACTICE FACTOR P

The P factor is usually 1.0 for premining conditions unless pre-mined land is in crop production involving contour tillage, strip-cropping or terracing techniques. It is unusual for such land to be subjected to surface mining; consequently assessment of P factors for premining conditions will not be discussed herein.

Reclamation operations conducted on the contour can be expected to result in reduced soil loss because contour tillage increases surface depression storage and surface roughness and reduces runoff volumes and velocities. Table 20 gives estimates of P factors for basic contour tillage and contour tillage with furrows or pits. As in the case of mulching systems described previously, long slopes may have to be partitioned using diversion terraces in order for contour practices to be effective in reducing erosion.

LIMITATIONS OF THE USLE

Accuracy of the USLE is unknown when it is used to predict soil loss for conditions which are substantially different from the range of conditions used to establish factor values in the equation. This can be a serious problem in surface mine areas where slopes are often steeper or longer than the ones which were used to establish the *S* and *L* factors. The USLE does not explicitly account for infiltration and runoff. Consequently, although it can be used to estimate soil losses from single storms, results will be insensitive to such factors as antecedent soil moisture and rainfall intensity pattern. It should be recognized that sediment yield from a watershed may include material from sources which are ignored by the USLE. These include gully erosion, channel bed degradation, mass failure of channel banks or mud slides. Conversely, USLE will not account for sediment deposition which can occur on flatter downstream areas or in channels. When deposition or erosion processes not included in the USLE occur, they must be estimated by other techniques in order to determine watershed sediment yield. The problem of quantifying deposition was studied by Neibling and Foster [37] and their work was utilized by Barfield et al. [2] in combination with the USLE to estimate sediment yield. However, their methods do not estimate such channel erosion processes as head-cut migration or channel bank failure. Unfortunately, reliable techniques for estimating the contribution of these channel erosion processes to watershed sediment yield are not yet available.

Example: USLE Soil Loss Estimate

It is proposed that an area southwest of Streamboat Springs, Colorado which includes three distinct soils and two vegetation communities should be mined. Vegetation, soils and topographic data indicate that the area can be divided into 5 erosion response units (ERUs) which can be considered homogeneous for the purpose of estimating soil erosion as shown in Figure US-1. Use the USLE to estimate average annual soil loss from the area for pre-mining, active mining and reclamation, and post-mining conditions. The following soils, vegetation and topographic data are available:

a. Soils Data

Ascalon Fine Sandy Loam—56% silt and fine sand, 30% sand > 0.1 mm, 1.0% organic matter, fine granular structure, moderate to rapid permeability.

Heldt Clay Loam—40% silt and fine sand, 25% sand > 0.1 mm, 1.5% organic matter, medium to coarse granular structure, moderate permeability.

Potsl Loam—40% silt and fine sand, 40% sand > 0.1 mm, 0.5% organic matter, fine granular texture, moderate to rapid permeability.

b. Vegetation Data

1. *Pre-mining conditions*—Range areas have a low-lying brush canopy with a density of 30% and a grass ground

TABLE 19. Mulch Factors and Length Limits for Construction Slopes (EPA, [61]).

Type of mulch	Mulch Rate	Land Slope	Factor C	Length Limit[1]
	Tons per acre	Percent		Feet
None	0	all	1.0	—
Straw or hay,	1.0	1–5	0.20	200
tied down by	1.0	6–10	0.20	100
anchoring and				
tacking	1.5	1–5	0.12	300
equipment[2]	1.5	6–10	0.12	150
Do.	2.0	1–5	0.06	400
	2.0	6–10	0.06	200
	2.0	11–15	0.07	150
	2.0	16–20	0.11	100
	2.0	21–25	0.14	75
	2.0	26–33	0.17	50
	2.0	34–50	0.20	35
Crushed stone,	135	<16	0.05	200
½ to 1½ in	135	16–20	0.05	150
	135	21–33	0.05	100
	135	34–50	0.05	75
Do.	240	<21	0.02	300
	240	21–33	0.02	200
	240	34–50	0.02	150
Wood chips	7	<16	0.08	75
	7	16–20	0.08	50
Do.	12	<16	0.05	150
	12	16–20	0.05	100
	12	21–33	0.05	75
Do.	25	<16	0.02	200
	25	16–20	0.02	150
	25	21–33	0.02	100
	25	34–50	0.02	75

[1]Maximum slope length for which the specified mulch rate is considered effective. When this limit is exceeded, either a higher application rate or mechanical shortening of the effective slope length is required.
[2]When the straw or hay mulch is not anchored to the soil, C values on moderate or steep slopes of soils having K values greater than 0.30 should be taken at double the values given in this table.

cover with a density of 20%. Shrub areas have a low-lying canopy with a 50% density and grass ground cover of 30%.

2. *Active mining and reclamation*—It is projected that active mining will take approximately one year and that it will be composed of two periods. The first period includes 5 months beginning on May 1 during which existing vegetation will be stripped off and the area will be bare. In early October the area will be mulched with straw at a rate of 2 tons/acre and seeded with a revegetation mix. Straw will be anchored to the soil by crimping.

3. *Post-mining conditions*—It is planned to reclaim the entire area for range use. Estimates indicate that revegeta-

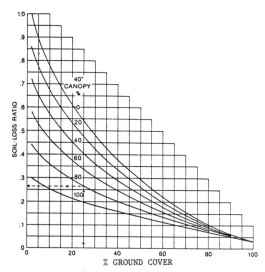

FIGURE 13. Combined ground and canopy effects when average fall distance of drops from canopy to the ground is about 40 inches (1 m) (after Wischmeier and Smith, [67]).

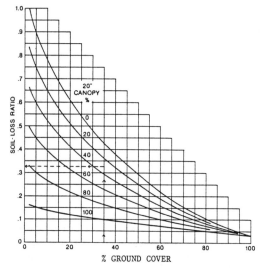

FIGURE 14. Combined ground and canopy effects when average fall distance of drops from canopy to ground is about 20 inches (0.5 m) (after Wischmeier and Smith, [67]).

tion will result in a low-lying canopy cover with a density of 40% and a grass and residual mulch ground cover with a density of 35%.

c. Topographic Data

The average percent slope, runoff length and area of each ERU are as follows:

ERU	Slope (%)	Length (feet)	Area (acres)
R-A	0.5	150	2000
R-B	1.5	200	1000
R-C	3.0	300	600
S-A	5.0	300	600
S-B	5.0	400	800

PRE-MINING

The soil loss equation is:

$$A = RKLSCP$$

The rainfall factor, R, can be estimated from Figure 10. Interpolation between R values on the map in the figure can be used to define a more precise value for the site. For this area, the average annual value is:

$$R = 30$$

The erodibility factor, K, is estimated using the soil information and the K-factor nomograph (Figure 12). The K values for each ERU are tabulated in Table US-1.

The slope-length factor, L, is estimated using the topo-

**TABLE 20. Determining Factor "P"
(After EPA, [61]).**

Land Slope %	Contouring[1]	Contour[2] Furrows or Pits
2.0 to 7	0.50	0.25
8.0–12	0.60	0.30
13.0–18	0.80	0.40
19.0–24	0.90	0.45
25.0–30	1.0	0.65

[1]Topsoil spreading, tillage, and seeding on the contour. Contour Limits—2 percent 400 feet, 8 percent 200 feet, 10 percent 100 feet, 14–30 percent 60 feet. The effectiveness beyond these limits is speculative.
[2]Estimating values for surface manipulation of reclaimed land disturbed by surface mining. Furrows or pits installed on the contour. Spacing between furrows 40–60 inches with a minimum 6 inch depth. Pits equal or exceed 12 inch width 36 inch length and 6 inch depth. Pit spacing is dependent on pit size, but generally the pits should occupy 50% of the surface area.

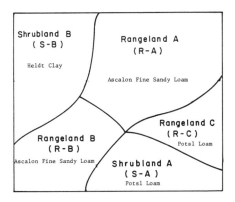

FIGURE US-1. *Soils and erosion response units (ERUs).*

graphic data for each ERU in Equation (23). For example, for ERU R-A,

$$L = \left(\frac{\lambda}{72.6}\right)^m$$

and *m* for a 0.5% slope is 0.2; therefore:

$$L = \left(\frac{150}{72.6}\right)^{0.20} = 1.16$$

The slope-steepness factor, *S*, is estimated using Equation (24):

$$S = 65.41\sin^2\theta + 4.56\sin\theta + 0.065$$

For example, for ERU S-B the average slope is 5%. Note that the sine of the slope angle is approximately equal to the percentage slope expressed as a decimal; therefore:

$$S = 65.41(0.05)^2 + 4.56(0.05) + 0.065 = 0.46$$

The *S* values for each ERU are tabulated in Table US-1.

The cover and management factor, *C*, can be estimated using the canopy and ground cover data and Table 18. Note that values in this table apply to permanent pasture and rangeland which have a well developed and distributed system of roots in the surface soil. Values in this table can be used for pre-mining soil loss estimates, but should not be used for active mining or post-mining conditions in the first years following reclamation because these root systems are unlikely to exist during these periods. *C* values for each ERU are tabulated in Table US-1.

The *P* factor is 1.0 for all ERUs as no support practices are present.

Once the USLE factors are tabulated for each ERU, soil loss calculations can easily be made. See Table US-1 for results, and column notes for explanations of calculations.

ACTIVE MINING AND RECLAMATION

Two *R* factors must be determined for the year which encompasses active mining and reclamation because two distinct periods with different cover conditions have been identified for this year. The first or "phase 1" period includes the 5 months beginning on May 1. The percentage of average annual *R* which occurs during this period can be estimated using Tables 15 and 16 or distribution curves developed by the SCS such as the one in Figure US-2. To determine percentage of average annual *R* which occurs during phase 1 from Figure US-2, the cumulative percentage for May 1 is read from the curve and subtracted from the percentage for October 1. The result shows that 69% of the average annual *R* occurs during phase 1 when it is projected that the area will be bare, and the remaining 31% must therefore occur during phase 2. As shown in Table US-2, USLE factors and soil loss estimates should be established for each phase when distinctly different factor values apply to different parts of a year.

As discussed in the text, it is conservative to assume that compaction and soil handling causes erodibility to increase over pre-mining conditions. Pre-mining *K* values are adjusted for disturbed conditions by using the most erodible soil structure and a permeability class which is two levels more erodible than the one used for the undisturbed soil. Adjusted *K* values are tabulated in Table US-2.

L factors are the same as for pre-mining conditions assuming that the area is returned to its original contour.

S factors are also the same as for pre-mining conditions.

C factors must be estimated for each phase of the year. For phase 1, it is assumed that the surface will be bare; therefore *C* = 1.0. As described above, during phase 2, 2 tons/acre of straw mulch will be in place and vegetation will be emerging. Table 19 indicates that with this mulching rate and relatively flat slopes, *C* = 0.06. Assuming that emerging vegetation compensates for losses of mulch with time, this value of *C* can be used to represent cover conditions during phase 2.

As before, *P* = 1.0.

Soil loss estimates for the year which includes active mining and reclamation are summarized in Table US-2. Column notes explain the calculations.

POST-MINING

The *R* factor is the same as for pre-mining conditions.

During the first few years following reclamation, it is assumed that *K* factors will reflect land disturbance. Consequently they are the same as for the active mining and reclamation period.

The topographic factor *L* is the same as for pre-mining conditions.

The topographic factor *S* is the same as for pre-mining conditions.

The *C* factor for post-mining conditions reflects the projected ground and canopy cover densities and the recent dis-

TABLE US-1. USLE Factors and Erosion Calculations for Pre-Mining Conditions.

ERU (1)	R (2)	K (3)	L (4)	S (5)	C (6)	P (7)	A t/ac/yr (8)	area ac (9)	soil loss t/yr (10)
R-A	30.0	0.37	1.16	0.09	0.16	1.00	0.19	2000.	380
R-B	30.0	0.37	1.36	0.15	0.16	1.00	0.36	1000.	360
R-C	30.0	0.25	1.53	0.46	0.16	1.00	0.48	600.	288
S-A	30.0	0.25	1.76	0.46	0.10	1.00	0.61	600.	366
S-B	30.0	0.21	1.98	0.46	0.10	1.00	0.57	800.	456

Estimated total average annual soil loss: 1850 tons

Column notes
(1) Erosion response unit with homogeneous soil, topography and vegetation properties.
(2) Average annual erosivity factor from Figure 10.
(3) Erodibility factor from Figure 12.
(4) Length factor from Equation (23).
(5) Steepness factor from Equation (24).
(6) Cover factor from Table 18.
(7) P = 1.0 for no support practices.
(8) Average annual soil loss per acre for the ERU = (2) × (3) × (4) × (5) × (6) × (7).
(9) Area of the ERU.
(10) Average annual soil loss for the ERU = (8) × (9).

TABLE US-2. USLE Factors and Erosion Calculations for Active Mining and Reclamation Conditions.

ERU (1)	R (2)	K (3)	L (4)	S (5)	C (6)	P (7)	A t/ac (8)	area ac (9)	soil loss t (10)
Phase 1									
R-A	20.7	0.47	1.16	0.09	1.00	1.00	1.01	2000.	2036
R-B	20.7	0.47	1.36	0.15	1.00	1.00	1.98	1000.	1980
R-C	20.7	0.33	1.53	0.46	1.00	1.00	2.72	600.	1632
S-A	20.7	0.33	1.76	0.46	1.00	1.00	5.53	600.	3318
S-B	20.7	0.29	1.98	0.46	1.00	1.00	5.68	800.	4547

Estimated phase 1 total average soil loss: 13513 tons

ERU (1)	R (2)	K (3)	L (4)	S (5)	C (6)	P (7)	A t/ac (8)	area ac (9)	soil loss t (10)
Phase 2									
R-A	9.3	0.47	1.16	0.09	0.06	1.00	0.02	2000.	40
R-B	9.3	0.47	1.36	0.15	0.06	1.00	0.06	1000.	60
R-C	9.3	0.33	1.53	0.46	0.06	1.00	0.08	600.	48
S-A	9.3	0.33	1.76	0.46	0.06	1.00	0.15	600.	90
S-B	9.3	0.29	1.98	0.46	0.06	1.00	0.14	800.	112

Estimated phase 2 total average soil loss: 350 tons

Total estimated average annual (phase 1 + phase 2) soil loss: 13863 tons

Column notes
(1) Erosion response unit.
(2) 69% of annual avg. R in phase 1 and 31% in phase 2.
(3) K values for undisturbed soils adjusted upwards by 2 permeability classes as defined in Figure 12.
(4) L-factor, same as for pre-mining conditions.
(5) S factor, same as for pre-mining conditions.
(6) Phase 1 value for bare surface. Phase 2 value from Table 19.
(7) P = 1.0; no support practices.
(8) Average soil loss per acre for the ERU during phase 1 or phase 2.
(9) Area of ERU.
(10) Average soil loss for the ERU.

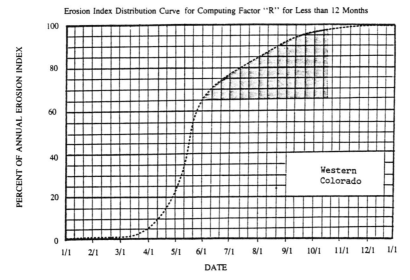

Erosion Index Distribution Curve for Computing Factor "R" for Less than 12 Months

FIGURE US-2. Erosion index distribution curve for computing EI values for periods of less than 12 months (EPA, [61]).

turbance of the area by mining operations. For the years immediately following reclamation, C should be estimated using Figures 13 and 14. Assuming a low-lying canopy of 40% and a ground cover of 35%, Figure 14 indicates that $C = 0.29$.

As before, $P = 1.0$.

Soil loss estimates for post-mining conditions are summarized in Table US-3. Note that ERUs which were previously designated as *S-A* and *S-B* are now designated as *R-D* and *R-E* reflecting their conversion from shrubland to rangeland. The estimate of post-mining, average, annual soil loss from the entire area (5606 tons or 1.12 tons/ac) is 3.0 times as large as the pre-mining estimate and 2.5 times smaller than the active mining estimate. The post-mining estimate uses a C factor (from Figure 14) which applies to a recently disturbed soil without a well developed root system. A "permanent" range with the same amount of canopy and ground cover would have much smaller C values (see Table 18). If revegetation continues to be successful, post-mine conditions can with time converge on pre-mine conditions. The amount of time this process takes will depend on local soil, vegetation and climate conditions as well as how the reclaimed area is managed.

Channel Erosion and Stability

QUALITATIVE ANALYSIS OF CHANNEL RESPONSE.

Channels draining areas disturbed by mining activities may become sources of additional sediment if water or sediment discharges deviate significantly from background rates. Natural channels respond to changes in water or sedi-

ment discharge or to local alterations of channel geometry by changing shape and gradient. Depending on the nature and location of the disturbance, deposition or erosion may result. Erosion may take the form of bed scour, bank scour, head-cut migration, channel-bank mass failure, or change in channel alignment. Examples of typical stream disturbances and stream responses are outlined by Simons and Senturk [48]. They present the following guidelines for the qualitative prediction of channel responses due to changes in flow and sediment load characteristics:

1. Depth of flow tends to increase if water discharge increases.
2. Channel width tends to increase if water discharge or sediment discharge increases.
3. The ratio of channel width to channel depth tends to increase if sediment discharge increases.
4. Channel gradient tends to decrease if water discharge increases, but it tends to increase if either sediment discharge or sediment size increases.
5. Channel sinuosity tends to increase if valley gradient increases but it tends to decrease if sediment discharge increases.
6. Transport of channel bed material tends to increase as average flow velocity and concentration of fine sediment increase, but it tends to decrease if the fall diameter of the bed material increases.

DESIGN OF STABLE CHANNELS

The preceding rules of thumb provide a means of making qualitative predictions of how surface mining operations can stimulate changes in channel characteristics which can in

TABLE US-3. USLE Factors and Erosion Calculations for Post-Mining Conditions.

ERU (1)	R (2)	K (3)	L (4)	S (5)	C (6)	P (7)	A t/ac/yr (8)	area ac (9)	soil loss t/yr (10)
R-A	30.0	0.47	1.16	0.09	0.29	1.00	0.43	2000.	860
R-B	30.0	0.47	1.36	0.15	0.29	1.00	0.83	1000.	830
R-C	30.0	0.33	1.53	0.46	0.29	1.00	1.14	600.	684
R-D	30.0	0.33	1.76	0.46	0.29	1.00	2.32	600.	1392
R-E	30.0	0.29	1.98	0.46	0.29	1.00	2.30	800.	1840

Estimated total average annual soil loss: 5606 tons

Column notes
(1) Same as pre-mining ERUs except shrub areas S-A and S-B have been converted to range areas R-D and R-E.
(2) Same as for pre-mining conditions.
(3) Same as for active mining conditions.
(4) Same as for pre-mining conditions.
(5) Same as for pre-mining conditions.
(6) Based on post-mining ground and canopy cover estimates and Figure 14.
(7) P = 1.0 for no support practices.
(8) Average annual soil loss per acre for the ERU = (2) × (3) × (4) × (5) × (6) × (7).
(9) Same as for pre-mining conditions.
(10) Average annual soil loss for the ERU = (8) × (9).

turn cause erosion both upstream and downstream. Estimations of sediment yield caused by channel changes are more difficult to make. While field surveys and historical data can be used to establish background quantities of channel erosion [57] no reliable and comprehensive method exists to quantitatively predict changes in sediment yield due to channel processes which may be set in motion by future mining activities. Due to the uncertainties involved, channel disturbances should be minimized and stabilization measures should be planned when they are unavoidable. The design of stable diversion channels and stabilization measures for natural channels are beyond the scope of this discussion and have been comprehensively treated by OSM [38] and others [13,15,29,48,52].

Estimation of Watershed Sediment Yield

As stated earlier, methods such as the USLE estimate erosion from a field but cannot account for reductions in sediment yield due to deposition or increases due to channel erosion. Methods which attempt to estimate sediment yield from watersheds are discussed below.

USLE WITH SEDIMENT DELIVERY RATIO.
This method determines watershed sediment yield using the following relationship:

$$SY = SDR\ A \qquad (26)$$

in which SY is the sediment yield from the watershed in the same units as A, A is soil loss predicted by the USLE, and SDR is the sediment delivery ratio which reflects the amount

of deposition which has occurred upstream of the point of interest. The accuracy of Equation (26) is contingent upon several items. First, if A is as defined for the USLE, erosion from gullies, mudslides, and channels must be insignificant, or estimates of these must be available and added to A. Second, the SDR must be quantified with a reasonable degree of accuracy. The second requirement is difficult to fulfill because deposition is a complex phenomenon which varies in time and space with hydraulic, topographic and sediment properties. Strictly speaking there can be no single characteristic SDR for a watershed because time and space distributions of these parameters vary from storm to storm. Nevertheless, researchers have proposed empirical equations relating SDR to watershed characteristics such as drainage area [7] and relief ratio [46]. Boyce's relationship is shown in Figure 15. The plotted data points indicate that errors in SDR using the curve may be more than 100 percent. More recently, the U.S. Forest Service [14] presented a method for SDR determination which includes effects of hydraulic, ground cover and sediment as well as geomorphic characteristics. However, the accuracy of this method is unknown. If sediment yield data is available on the watershed of interest and similar watersheds, it may be possible to arrive at a more accurate SDR estimate. In any case, uncertainty in the estimation of the sediment delivery ratio may cause regulatory authorities to require high SDR values regardless of watershed characteristics.

MODIFIED USLE (MUSLE)
The Modified USLE (63) can be written

$$SY = R_m\ K\ L\ S\ C\ P \qquad (27)$$

in which SY is watershed sediment yield in tons per acre K, L, S, C and P are as previously defined for the USLE except that in this case they are area weighted averages taken over the watershed; and R_m is a modified erosivity factor defined as:

$$R_m = \frac{23.6(Q\,q_p)^{0.56}}{DA^{0.44}} \qquad (28)$$

in which Q is the runoff amount in inches, q_p is the peak discharge in cubic feet per second, and DA is the area of the watershed in acres. Equation (28) is used to represent both the erosivity of rainfall and runoff as well as the efficiency with which sediment is transported downstream in the watershed. Q and q_p can be determined using techniques described in the runoff section of this chapter.

The MUSLE procedure appears to predict single storm values better than the USLE because it uses runoff instead of rainfall energy parameters. Although erosion correlates strongly with both rainfall and runoff, runoff is a better predictor of soil loss for single storms because without it soil loss cannot occur. Unlike the USLE, soil loss estimates made using the MUSLE will be sensitive to antecedent soil moisture conditions and rainfall intensity patterns provided that an adequate hydrologic procedure is used to estimate Q and q_p. Additionally, use of peak discharge as well as drainage area to represent what is effectively a storm-dependent SDR makes sense from a physical point of view. Unfortunately, the generality of the MUSLE is questionable because of its empirical nature. Equation (27) is based on a regression analysis of data from 18 small watersheds in Texas and Nebraska [63]. The watersheds were predominantly agricultural with average slope ranging from 1 to 6

percent and drainage areas of 3 to 4400 acres. Although the MUSLE explained 92 percent of the variation in single storm, sediment yield data, its predictive ability on watersheds with characteristics different than those upon which the equation is based is unknown. All the limitations associated with USLE parameters K, L, S, C and P also apply to the MUSLE. Williams [62,63] has addressed some of these limitations by modifying the original MUSLE to represent a complex watershed composed of heterogeneous sub-areas and to include channel erosion and deposition processes.

Despite its limitations, the MUSLE will probably provide more accurate estimates of single storm sediment yields, than the USLE with a single storm R factor and a sediment delivery ratio (SDR). This increased accuracy is achieved at the expense of the added computational effort involved in estimating Q and q_p.

THE PSIAC PROCEDURE

The PSIAC procedure [42] estimates average annual sediment yield for watersheds larger than 10 square miles. Use of the method is restricted to California, Nevada, Arizona, Utah, New Mexico and western Wyoming and Colorado. The sediment yield estimate is based on nine factors which describe, geologic, soil, climate, hydraulic, topographic, vegetation, land use and erosion characteristics. As shown in Table 21, each factor has three descriptive levels with three corresponding numerical values. A user of the PSIAC method must choose the level for each factor which most closely describes conditions on the watershed of interest. The appropriate numerical values for each of the nine factors are summed and the total is used to determine average annual sediment yield as shown in Table 22.

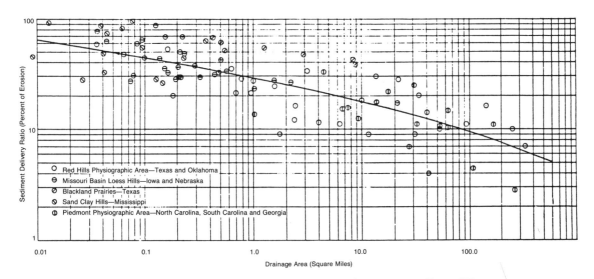

FIGURE 15. Sediment-delivery ratio versus size of drainage area (Boyce, [7]).

TABLE 21. PSIAC Sediment Yield Factors (PSIAC, [42]).

Sediment Yield Levels	A Surface Geology	B Soils	C Climate	D Runoff
High	(10)* a. Marine shales and related mud-stones and slit stones.	(10) a. Fine textured; easily dispersed; saline-alkaline; high shrink-swell characteristics. b. Single grain silts and fine sands.	(10) a. Storms of several days duration with short periods of intense rain-fall. b. Frequent intense convective storms. c. Freeze-thaw occurrence.	(10) a. High peak flows per unit area. b. Large volumes of flow per unit area.
**				
Moderate	(5) a. Rocks of medium hardness. b. Moderately weathered. c. Moderately frac-tured.	(5) a. Medium textured soil. b. Occasional rock fragments. c. Caliche layers.	(5) a. Storms of moder-ate duration and intensity. b. Infrequent con-vective storms.	(5) a. Moderate peak flows. b. Moderate volume of flow per unit area.
**				
Low	(0) a. Massive, hard formations.	(0) a. High percentage of rock fragments. b. Aggregated clays c. High in organic matter.	(0) a. Humid climate with rainfall of low intensity. b. Precipitation in form of snow. c. Arid climate, low intensity storms. d. Arid climate; rare convective storms.	(0) a. Low peak flows per unit area. b. Low volume of runoff per unit area. c. Rare runoff events.

(continued)

TABLE 21 (continued).

E Topography	F Ground Cover	G Land Use	H Upland Erosion	I Channel Erosion & Sediment Transport
(20) a. Steep upland slopes (in excess of 30%). High relief; little or no floodplain development.	(10) Ground cover does not exceed 20%. a. Vegetation sparse; little or no litter. b. No rock in surface soil.	(10) a. More than 50% cultivated. b. Almost all of area intensively grazed. c. All of area recently burned.	(25) a. More than 50% of area characterized by rill and gully or landslide erosion.	(25) a. Eroding banks continuously or at frequent intervals with large depths and long flow duration. b. Active headcuts and degradation in tributary channels.
**				
(10) a. Moderate upland slopes (less than 20%). b. Moderate fan or floodplain development.	(0) Cover not exceeding 40%. a. Noticeable litter. b. If trees present understory not well developed.	(0) a. Less than 25% cultivated. b. 50% or less recently logged. c. Less than 50% intensively grazed. d. Ordinary road and other construction.	(10) a. About 25% of the area characterized by rill and gully or landslide erosion. b. Wind erosion with deposition in stream channels.	(10) a. Moderate flow depths, medium flow duration with occasionally eroding banks or bed.
**				
(0) a. Gentle upland slopes (less than 5%). b. Extensive alluvial plains.	(−10) a. Area completely protected by vegetation, rock fragments, litter Little opportunity for rainfall to reach erodible material	(−10) a. No cultivation. b. No recent logging. c. Low intensity grazing.	(0) a. No apparent signs of erosion	(0) a. Wide shallow channels with flat gradients, short flow duration. b. Channels in massive rock, large boulders or well vegetated. c. Artificially controlled channels.

*The numbers in specific boxes indicate values to be assigned appropriate characteristics. The small letters a, b, c, refer to independent characteristics to which full value may be assigned.

**If experience so indicates, interpolation between the 3 sediment yield levels may be made.

TABLE 22. PSIAC Sediment Yield Estimates (PSIAC, [42]).

Sum of 9 Factors from Table 21	PSIAC Classification	Average Annual Sediment Yield (acre-ft/sq. mi.)	Average Annual Sediment Yield (tons/acre)*
>100	1	>3.0	>10.2
75–100	2	1.0–3.0	3.4–10.2
50–75	3	0.5–1.0	1.7–3.4
25–50	4	<0.2	<0.7

*Assuming soil specific weight of 100 lbs/cu-ft.

The major problem with the PSIAC method is the subjective interpretation necessary to establish numerical values for the climate, runoff and erosion factors. Different users may decide on different values for these factors when considering the same watershed and arrive at substantially different sediment yield estimates. As indicated by Table 22, the method cannot predict sediment yield from watersheds which deliver more than 3.0 ac-ft/sq mi of sediment. Additionally, the method has limited use as a predictive tool because the user must have foreknowledge of the severity of upland and channel erosion in order to obtain a sediment yield estimate. These considerations coupled with the geographical and watershed size restrictions noted above severely limit the utility of the PSIAC procedure for making absolute sediment yield estimates. However, the procedure is very easy to use and can provide a rough, relative index for comparing the sediment yield potential of different watersheds.

RESERVOIR SURVEY METHOD

The SCS measures the annual sediment build-up in several thousand reservoirs across the U.S. If the watershed of interest has similar steepness, drainage area, vegetative cover and land use to one for which survey data are available, average annual sediment yield may also be similar on the two watersheds. Barfield et al. [2] pointed out that it is often difficult to find surveyed reservoirs which can be used in surface mine sediment pond design because reservoir drainage areas are usually much larger and contain far less disturbed land than the surface mined areas which are expected to drain to the pond. On the other hand, sediment pond survey data from neighboring mined watersheds may be very useful in estimating sediment yields from the watershed of interest.

MECHANISTIC SEDIMENT ROUTING

The mechanistic sediment routing approach employs a partial differential equation for the conservation of transported soil which is directly analogous to the mass conservation relationship for surface water routing described earlier. In addition to the conservation relationship, this approach requires equations which determine the rate of

supply of soil particles to the flow stream from raindrop impact, runoff shear and other soil detachment mechanisms. Additional equations which describe soil particle deposition may also be included.

The mechanistic sediment routing method is too complex to describe in detail. It is mentioned herein because of its use in conjunction with the mechanistic water routing technique in computer models which provide the most accurate and detailed predictions of erosion, deposition and sediment yields from single storm events. Some of these computer models estimate the selective transport and deposition of different sizes of sediment particles which is very useful in water quality assessment and the design of sediment ponds and other sediment control measures. Some of the more accessible, better documented computer models which include these capabilities are discussed below.

SELECTED COMPUTER MODELS OF WATERSHED RESPONSE

Surface mine hydrologists may sometimes find it necessary or convenient to use computer models which estimate watershed runoff and sediment yield. A few selected models are briefly described herein. Two criteria were used in selecting the models: their potential utility in surface mine applications and the availability of computer source codes, documentation and expert assistance to prospective users. Readers interested in exploring these and other models which treat hydrology, erosion and sediment yield are urged to consult Renard et al. [45] who provide abstracts, references and tabular summaries of 80 watershed models. Foster [16] also abstracts 13 models which treat erosion and sediment yield processes. More recently, an ASCE task committee (Task Committee on Quantifying Land-Use Change Effects, 1985) evaluated 28 hydrologic models with potential use in quantifying hydrologic effects of land-use changes.

HEC-1 Watershed Runoff Model

The HEC-1 model [60] is a single event watershed runoff simulation model. It includes no erosion or sediment transport routines. Similar to many other watershed models, the user must subdivide the watershed of interest into a network of channels with tributary overland flow surfaces. Each channel segment and overland flow surface can have its own hydraulic, hydrologic and topographic characteristics. The model also simulates the effect of reservoirs on storm runoff. The HEC-1 model considers the following processes which control storm runoff: interception and infiltration losses, overland flow, channel flow and reservoir storage effects. One of the distinct features of this model is that it provides the user with several optional methods which simulate these processes. Initial abstractions and infiltration are determined either by the SCS Curve Number method or by an empirical infiltration equation known as the Holtan Loss Rate method. Overland flow routing can be determined using one of several different unit hydrograph methods, or by mechanistic flow routing techniques. There are several channel routing options including the Muskingum, kinematic and storage routing methods. Reservoirs are simulated using a level-pool routing method. The availability of these options may or may not be an advantage depending on the training and interests of the user. Other capabilities of HEC-1 include generation of various types of design storms based on historical data and simulation of snowmelt, flow diversions, channel infiltration and pumping station effects.

DATA REQUIREMENTS

These vary depending on the options selected by the user. Generally, topographic data, rainfall or snowmelt data, water loss/infiltration parameters, overland and channel routing parameters and reservoir characteristics must be input. If rainfall-runoff data is available, built-in optimization routines can be used to determine some parameters.

SOURCE CODE AVAILABILITY

Users manuals, Fortran tapes or IBM-PC compatible diskettes can be obtained from:

U.S. Army Corps of Engineers
Hydrologic Engineering Center
609 Second Street
Davis, California 95616

FURTHER INFORMATION
U.S. Army Corps of Engineers
Hydrologic Engineering Center
609 Second Street
Davis, California 95616

SCS TR-20 Watershed Runoff Model

TR-20 [53] is a single event, runoff model which includes no erosion or sediment transport. The model uses SCS curve number, dimensionless hydrograph and routing methods to compute runoff hydrographs and water surface

elevations in complex watersheds which can include up to 99 reservoirs and 200 stream reaches. It is specifically designed to predict the effect of alternative reservoir designs and watershed management options on the hydrologic response of the watershed to any user-specified storm.

DATA REQUIREMENTS

Input data include curve numbers, drainage areas, and times of concentration for each subarea of the watershed, stream cross section and length data, characteristics of proposed and existing hydraulic structures, dimensionless hydrograph specifications (default is SCS standard dimensionless hydrograph), and initial reservoir water surface elevations. Methods for obtaining all of these data are described in the *SCS National Engineering Handbook*, Section 4 [55] and in SCS technical Release No. 20 (1965) which is the users' manual for the model.

SOURCE CODE AVAILABILITY

The TR-20 program is written in Fortran and requires a computer with 400 K words of memory. The TR-20 users' manual and a tape which includes the TR-20 Fortran source code and example data sets can be purchased from:

National Technical Information Service
U.S. Department of Commerce
P.O. Box 1553
Springfield, Virginia 22151

The U.S. Soil Conservation Service anticipates development of TR-20 software which will be compatible with micro and personal computers; however, it is not available at the time of this writing. Persons interested in PC or micro software should check with software companies which specialize in civil engineering and water resources programs. One such company which offers an IBM-PC compatible version of TR-20 is:

Civilsoft
290 South Anaheim Blvd.
Anaheim, California 92805

FURTHER INFORMATION
Additional information on TR-20 can be obtained from SCS state offices and regional technical centers.

ANSWERS Watershed Runoff and Sediment Yield Model

ANSWERS [3,4] is a single event (rainfall only, no snowmelt) surface runoff, erosion and sediment transport model. The model determines water abstractions by interception, depression storage and infiltration. Infiltration is calculated using a modified Holtan equation. The model uses a kinematic approximation to mechanistically route rainfall excesses overland and in channels and to generate surface runoff hydrographs. The watershed is represented by a square grid and channel network which easily accommodates the spatial variability of infiltration, surface roughness and erosion characteristics. Erosion and sediment yield are determined by a set of relationships which describe raindrop detachment, runoff detachment and flow transport capacity. The detachment relationships use USLE erodibility and ground cover parameters. The model does not predict channel erosion, but does allow resuspension of sediment which may be deposited early in a runoff event. ANSWERS does not explicitly simulate sediment ponds or terraces. Their effect on sediment is approximated by assuming a constant sediment trapping efficiency.

DATA REQUIREMENTS

Time-intensity data are required to define the storm of interest. Antecedent moisture, infiltration and drainage parameters and USLE K factors must be input for each of the soils in the watershed. Interception, vegetation cover, depression storage, hydraulic roughness and USLE C factors must be input for each land use. Topographic data, land use and soil identifiers must be defined for each surface element. Channel cross section and roughness data for each channel element is also required. Existing soils surveys and topographical maps provide much of the information necessary to set up ANSWERS data sets. Methods for estimating model parameters are detailed in the model users manual.

SOURCE CODE AVAILABILITY

ANSWERS Fortran source code is available on tape or on diskette for IBM-PC or compatible computers. Codes and users' manuals can be obtained from:

Department of Agricultural Engineering
Purdue University
West Lafayette, Indiana 47907

FURTHER INFORMATION

Professors David B. Beasley or Larry F. Huggins
Department of Agricultural Engineering
Purdue University
West Lafayette, Indiana 47907

SEDIMOT II Watershed Runoff and Sediment Yield Model

This model [65,66] is designed to evaluate the runoff and sediment yield response of watersheds to surface management and structural options including detention ponds, grass filters and check dams. The model performs short-term, storm simulations only; it includes no long-term, con-

tinuous hydrology components. The user divides the watershed into subwatersheds of relatively uniform land use. The model determines hydrographs, sediment graphs and particle size distributions for each subwatershed, routes the water and sediment down the channel system and composites the contributions from each subwatershed to determine the total hydrograph, sediment graph and particle size distribution for the entire watershed. Users can opt for SCS design storms or user-specified storms. Rainfall excesses are generated on each subwatershed using the SCS curve number method. Excesses are routed over the surface using the unit hydrograph method. Runoff is routed through channels using the Muskingum procedure. Sediment yield is predicted using two optional methods, one based on the Modified Universal Soil Loss Equation (MUSLE) and the other based on the detachment vs. transport limiting concept. Sediment graphs are determined by assuming that sediment load is directly proportional to flow rate. Particle size distributions are determined by assuming selective deposition of the largest sediment particles.

DATA REQUIREMENTS

Rainfall data include specification of depth, duration and type of SCS design storm or accumulated depth versus time data for other storms. For surface runoff calculations, curve numbers and times of concentration for each subwatershed are required. Channel routing requires user specification of the Muskingum routing coefficients for each channel segment. Erosion and sediment routing calculations require USLE factors and the particle size distribution of eroded sediment for each subwatershed. Additional inputs are required to describe detention ponds, porous check dams and grass filters if present in the watershed.

SOURCE CODE AVAILABILITY

Users' manuals and tapes which include Fortran source codes, example data sets and outputs which are compatible with most main-frame computers can be ordered from:

Department of Agricultural Engineering
University of Kentucky
Lexington, Kentucky 00751

A version of the program which will run on an IBM-PC compatible, personal computer can be obtained from:

Oklahoma Technical Press
815 Hillcrest
Stillwater, Oklahoma 74074

FURTHER INFORMATION

Dr. B. J. Barfield or Dr. R. C. Warner
Department of Agricultural Engineering
University of Kentucky
Lexington, Kentucky 00751

USGS Hydrologic Modeling System

The United States Geological Survey maintains several watershed models of surface runoff, water quality and sediment transport. Many of these models are summarized by Lorens [31]. Three of the models supported by the USGS have been integrated with a special interactive program which aids the user in the preparation, checking, updating and reformatting of input data sets and the analysis of program output. This interactive program is known as ANNIE. Users requesting copies of source codes for PRMS, HSPF, and DR3M (described below) can get them with the ANNIE program. The USGS has also completed a version of ANNIE which is IBM-PC compatible. It will be available for use with PC codes of hydrologic and hydraulic models developed by the USGS. For further information contact:

Chief, Surface Water Branch
U.S. Geological Survey
415 National Center
Reston, Virginia 22092

DR3M Watershed Runoff and Sediment Yield Model

DR3M (Distributed routing rainfall-runoff model, version II) [49] is predominantly an urban watershed hydrology model although it includes routines for pervious areas, detention storage and erosion and sediment transport which may make it useful in surface mine or other disturbed land hydrology analyses. The model tracks soil moisture continuously, routes surface runoff from individual storms, but includes no subsurface or base flow components. Rainfall excess is determined by a physically based infiltration equation. Excesses are mechanistically routed overland and in channels using the kinematic wave approximation. Sediment yield is determined by the Modified Universal Soil Loss Equation (MUSLE). Sedimentation in reservoirs is estimated by assuming plug flow and settlement in accordance with Stokes' Law. The model predicts the sedimentation by particle size, but the user must input the influent particle size distribution to each reservoir. Influent distributions are assumed to be constant throughout the runoff event. DR3M can be applied to watersheds which range from a few acres to over 10 square miles. On larger watersheds, input data should reflect spatial variations in rainfall patterns.

DATA REQUIREMENTS
DR3M includes an optimization procedure which uses a short record of rainfall and runoff data to calibrate hydrologic parameters. Additionally, topographical and hydraulic data are required to describe watershed surfaces, reservoirs, conduits and channels. As mentioned above, the model requires influent particle size distributions for each reservoir. Additionally, USLE parameters are required for each watershed subarea. Limited availability of calibration data and reliance on users to input influent particle size distributions may limit the utility of this model in surface mine applications.

SOURCE CODE AVAILABILITY
Users' manuals and tapes of the Fortran source code for DR3M are available from the Surface Water Branch of the USGS. The program comes with the ANNIE data management system. Contact:

Chief, Surface Water Branch
U.S. Geological Survey
415 National Center
Reston, Virginia 22092

FURTHER INFORMATION

P. E. Smith
U.S. Geological Survey
Gulf Coast Hydroscience Center, Building 1100
NSTL Station, MS 39529

HSPF Watershed Hydrology and Water Quality Model

HSPF [26] is a comprehensive hydrologic model which simulates all the significant hydrologic components operative within a watershed as well as the transport of sediment and the temporal and spatial distribution of nutrients, pesticides, dissolved oxygen, temperature, biological oxygen demand, organics, pH and microorganisms. The model is capable of single storm (short) as well as continuous (long-term) simulations. The watershed is represented in the model as a network of surfaces, channels and reservoirs. An infiltration capacity curve is used to generate rainfall excesses. These excesses are routed using the Manning equation and an empirical relationship between surface storage volume and downstream flow depth. Water is routed through channels and reservoirs using a storage routing technique. Sediment detachment by overland flow and by rainfall are simulated separately. The model separates the transport and deposition of fine (silts and clays) and coarse (sands and gravels) materials but the user must supply the settling rate for fine material. Many of the relationships in the HSPF runoff, erosion and sediment transport routines are based on approximate conceptualizations of the processes involved. Some of the empirical relationships utilize parameters which are difficult to define, measure or estimate for different watershed conditions. Consequently, the model requires calibration. These factors may limit the value of HSPF as a truly predictive watershed model. However, the model may be quite useful in making relative comparisons of the effects of different surface treatment options, reservoir and diversion designs, etc. on both water quality and quantity.

DATA REQUIREMENTS

Meteorological data include precipitation and potential evapotranspiration. Snowmelt simulation requires air and dew point temperature, wind and solar energy data. Water quality simulation requires humidity, cloud cover, and tillage practice data as well as identification of point source loadings and pesticide and fertilizer application schedules. Watershed geometry, reservoir, vegetation, and soil and sediment parameters must also be specified. The assembly of data sets for HSPF can be difficult and time consuming. This task can be significantly streamlined by the USGS AN-NIE data management program.

SOURCE CODE AVAILABILITY

Fortran tapes of this program along with the ANNIE data management program can be obtained from the USGS by contacting:

Director
U.S. Environmental Protection Agency
Environmental Research Laboratory
Athens, GA 30605

FURTHER INFORMATION

Alan M. Lumb
U.S. Geological Survey
National Center, M.S. 415
Reston, Virginia 22092

or

Tom Barnwell
U.S. Environmental Protection Agency
Environmental Research Laboratory
Athens, GA 30605

PRMS Watershed Runoff and Sediment Yield Model

The Precipitation-runoff modeling system (PRMS) [30] divides complex watersheds into a group of homogeneous plane, channel and reservoir subunits. The model uses mechanistic runoff and sediment routing to determine watershed hydrographs and sediment discharges from single rainfall events with intensities which can vary in time and space over the watershed. The model can continuously track snowmelt, soil moisture, evapotranspiration, percolation and runoff on a daily basis. When simulating storm runoff, the model uses a Green-Ampt type infiltration equation to generate rainfall excesses. These excesses are routed overland and in channels using the kinematic flow approximation. Flow is routed through reservoirs using a storage routing technique. Erosion is determined using separate expressions for raindrop and runoff detachment. The amount of sediment in suspension is limited by either the erosion rate or the transport capacity of the flow. Sediment is routed using the equation of sediment continuity. The model does not simulate erosion or deposition in channels, nor does it include calculations of the sediment trapping efficiency of reservoirs. For the purposes of continuous daily hydrologic simulation PRMS conceptualizes the watershed as a network of storages representing interception, snowpack, surface depression, root zone, subsurface and groundwater. Water is transferred down-gradient, from one storage to another and to the stream by processes of through-fall, infiltration, surface runoff, subsurface runoff, percolation and groundwater runoff. Water is lost from the system by evaporation, transpiration and groundwater leakage. Estimation of some erosion and water balance parameters may be difficult. PRMS includes a built-in parameter optimization routine which may be useful in estimating some parameters if sufficient data are available. The program is compatible with the ANNIE data management system.

DATA REQUIREMENTS

Data requirements for the daily hydrology option include daily precipitation and temperature extremes. Snowmelt computations are improved if daily short-wave radiation data is available. If no snowmelt computations are needed, daily pan evaporation data can be substituted for temperature data. Additionally, topographic, soil and vegetation data are needed for each watershed subunit. For single storm hydrograph and sediment yield computations, breakpoint precipitation and sediment data are needed.

SOURCE CODE AVAILABILITY

Tapes of Fortran source codes and users' manuals are available from:

Chief, Surface Water Branch
U.S. Geological Survey
415 National Center
Reston, Virginia 22092

FURTHER INFORMATION

George H. Leavesley
U.S. Geological Survey, WRD
Box 25046, M.S. 412
Denver Federal Center
Lakewood, Colorado 80225

HYSIM Watershed Runoff and Sediment Yield Model

HYSIM [5] is predominantly an empirically based model which uses relationships which apply to the Tennessee Valley or other areas of similar climate and geography. The model estimates daily stream flow volumes for a watershed as well as storm hydrographs and sediment graphs. The model includes a stochastic rainfall generator which simulates rainfall amounts for time periods as small as 5 minutes. The continuous, daily simulation algorithm considers interception, evapotranspiration and watershed losses to deep seepage. Runoff water is partitioned as impervious area surface runoff, pervious area direct runoff or groundwater runoff. Impervious area runoff is assumed to reach the stream on the same day as rainfall occurs. Pervious area and groundwater runoff are routed to the stream using empirical relationships. When daily runoff exceeds a user-specified limit, the model generates a storm hydrograph using a modified curve number approach, and the TVA double triangle unit hydrograph method. Unlike some of the other models discussed in this section, HYSIM treats the entire watershed as a spatially lumped system. It performs no channel routing and does not simulate the effects of ponds or reservoirs. The model does not simulate snowfall or snowmelt. Soil loss amounts are determined using the MUSLE. There is also a procedure for determining the amount of dirt and dust washed off from impervious, urbanized areas. Sediment graphs are generated using a unit sediment graph routing technique.

DATA REQUIREMENTS

Data input is facilitated by an interactive system which prompts the user for the necessary information. This includes watershed topography, vegetation cover, geomorphic parameters, soil permeability and water holding capacity and USLE factors. Meteorological data include long-term average rainfall, monthly rainfall and evapotranspiration statistics.

SOURCE CODE AVAILABILITY

Users' manuals and tapes of Fortran source codes can be obtained from:

Water Systems Development Branch
Tennessee Valley Authority
P.O. Drawer E
Norris, Tennessee 37828

FURTHER INFORMATION

Roger P. Betson or Michael L. Poe
Tennessee Valley Authority
P.O. Drawer E
Norris, Tennessee 37828

REFERENCES

1. Alley, W. M., and P. E. Smith, "Distributed Routing Rainfall-runoff Model: Version II," U.S. Geological Survey Open File Report No. 82-0344 (1982).
2. Barfield, B. J., R. C. Warner, and C. T. Haan, *Applied Hydrology and Sedimentology for Disturbed Areas,* Oklahoma Technical Press, Stillwater, Oklahoma (1983).
3. Beasley, D. B., "ANSWERS: A Mathematical Model for Simulating the Effects of Land Use and Mangement on Water Qual-

ity," Ph.D. Thesis, Purdue University, West Lafayette, Indiana, 226 p. (1977).
4. Beasley, D. B. and L. F. Huggins, "ANSWERS–Users Manual," EPA-905-82-001, U.S. Environmental Protection Agency, 54 p. (1982).
5. Betson, R. P., J. Bales and H. E. Pratt, "User's Guide to TVA-HYSIM a Hydrologic Program for Quantifying Land-use Change Effects," EPA-600/70-80-048, 81 p. (1980).
6. Bondelid, T. R., R. H. McCuen and T. J. Jackson, "Sensitivity of SCS Models to Curve Number Variation," *Water Resources Bulletin, Vol. 18,* No. 1, pp. 111–116 (1982).
7. Boyce, R. C., "Sediment Routing with Sediment-delivery Ratios," In: *Present and Prospective Technology for Predicting Sediment Yields and Sources,* U.S. Department of Agriculture ARS-S-40 (1975).
8. Brooks, F. L. and J. W. Turelle, "Universal Soil Loss Equation," Technical Note No. 32, West Technical Service Center, U.S. Soil Conservation Service, Portland, Oregon (1974).
9. Chow, V. T., *Open-channel Hydraulics,* McGraw-Hill, New York, New York (1959).
10. Dunne, T. and L. B. Leopold, "Water in Environmental Planning," W. H. Freeman and Co., San Francisco, California (1978).
11. Eagleson, P. S., *Dynamic Hydrology,* McGraw-Hill Book Company (1970).
12. Evans, R., "Mechanics of Water Erosion and Their Spatial and Temporal Controls; An Empirical Viewpoint," In: *Soil Erosion,* John Wiley and Sons (1980).
13. Federal Highway Administration, "Design of Stable Channels with Flexible Linings," Hydraulic Engineering Circular No. 15, Federal Highway Administration, Washington, D.C. (1975).
14. Forest Service, "An Approach to Water Resources Evaluation of Non-point Silvicultural Sources (A Procedural Handbook)," EPA-600/8-80-012, Environmental Protection Agency, Washington, D.C. (1980).
15. Fortier, S. and F. C. Scobey, "Permissible Canal Velocities," *Transactions of the ASCE, Vol. 89,* pp. 940–956 (1926).
16. Foster, G. R., "Modeling the Erosion Process," In: *Hydrologic Modeling of Small Watersheds,* ASAE Monograph No. 5, American Society of Agricultural Engineers, St. Joseph, Michigan (1982).
17. Foster, G. R. and W. H. Wischmeier, "Evaluating Irregular Slopes for Soil Loss Prediction," *American Society of Agricultural Engineers Transactions, Vol. 17,* pp. 305–309 (1974).
18. Freeze, R. A., "Mathematical Models of Hillslope Hydrology," In: *Hillslope Hydrology* by M. J. Kirkby, John Wiley and Sons (1978).
19. Fullerton, W. T., "Water and Sediment Routing from Complex Watersheds: An Example Application to Surface Mining," In: *Proceedings of the D. B. Simons Symposium on Erosion and Sedimentation,* in cooperation with Colorado State University and Waterway, Port, Coastal and Ocean Division of the American Society of Civil Engineers, Simons, Li and Associates, Inc., Fort Collins, Colorado (1983).
20. Gardener, H. R. and D. A. Woolhiser, "Hydrologic and Cli-

matic Factors," In: *Reclamation of Drastically Disturbed Lands,* edited by Frank W. Schaller and Paul Sutton, American Society of Agronomy, Crop Science Society of America and Soil Science Society of America, Madison, Wisconsin (1978).

21. Green, W. A. and G. A. Ampt, "Studies on Soil Physics, I, The Flow of Air and Water Through Soils," *J. Agr. Sci., Vol. 4,* pp. 1–14 (1911).

22. Hartley, D. M., "Runoff and Erosion Response of Reclaimed Surfaces," *Journal of Hydraulic Engineering, Vol. 110,* No. 9, pp. 1181–1199 (1984).

23. Hartley, D. M., "Simplified Process Model for Water and Sediment Yield from Single Storms," *Transactions of the American Society of Agricultural Engineers,* Vol. 30, No. 3, pp. 710–717 (1987).

24. Hawkins, R. H., "The Importance of Accurate Curve Numbers in the Estimation of Storm Runoff," *Water Resources Bulletin, Vol. 11,* No. 5, pp. 887–891 (1975).

25. Herschfield, D. M., "Rainfall Frequency Atlas of the United States," Technical Paper 40, U.S. Department of Commerce, Weather Bureau, Washington, D.C. (1961).

26. Johanson, R. C., J. C. Imhoff and H. H. Davies, Jr., "Users Manual for Hydrologic Simulation Program–Fortran (HSPF): Environmental Research Laboratory," EPA-600/9-80-015, Athens, GA. (1980).

27. Kent, K. M., "A Method for Estimating Volume and Rate of Runoff in Small Watersheds," SCS-TP149, Soil Conservation Service, U.S. Department of Agriculture (1968).

28. Kirpich, P. Z., "Time of Concentration of Small Agricultural Watersheds," *Civil Engineering, Vol. 10,* No. 6 (1940).

29. Lane, E. W. and E. J. Carlson, "Some Factors Affecting the Stability of Canals Constructed on Coarse Granular Materials," *Proceedings of the Minnesota International Hydraulics Convention; joint meeting of IAHR and ASCE Hydraulics Division,* September (1953).

30. Leavesley, G. H., R. W. Lichty, B. M. Troutman, and L. G. Saindon, "Precpitation-runoff Modeling System: Users Manual," U.S. Geological Survey Water Resources Investigations Report 83-4238 207 p. (1983).

31. Lorens, J. A., "Computer Programs for Modeling Flow and Water Quality of Surface Water Systems," U.S. Geological Survey Open-fil Report 81-430, U.S. Geological Survey Gulf Coast Hydroscience Center, NSTL Station, MS 39529 (1982).

32. Lull, H. W., "Ecological and Silvicultural Aspects" Section 6. In: *Handbook of Applied Hydrology* by Ven Te Chow, McGraw-Hill, Inc. (1964).

33. Miller, J. F., R. H. Frederick, and R. J. Tracy, "Precipitation-frequency Atlas of the Western United States," Volume III-Colorado. NOAA Atlas 2, U.S. Department of Commerce (1973).

34. Montgomery, R. J., "A Data-based Evaluation of the SCS Curve Number Method for Runoff Prediction," Master's Thesis, Colorado State Univesity, Fort Collins, Colorado (1980).

35. Morel-Seytoux, H. J., "Derivation of Equations for Rainfall Infiltration," *J. of Hydrol., Vol. 31,* pp. 203–219 (1976).

36. Musgrave, G. W. and H. N. Holtan, "Infiltration," In:

Handbook of Hydrology by Ven Te Chow, McGraw-Hill Book Company (1964).

37. Neibling, W. H. and Foster, G. R., "Estimating Deposition and Sediment Yield from Overland Flow Processes," *Proceedings International Symposium on Urban Hydrology,* Hydraulics and Sediment Control, UK BU 114, College of Engineering, University of Kentucky, Lexington, Kentucky (1977).

38. Office of Surface Mining, "Surface Water Hydrology and Sedimentology Manual," Prepared by J. F. Sato and Associates, Simons, Li and Associates and Shepherd Water Geologists for OSM Region V, Denver, Colorado (1982).

39. Office of Surface Mining, "Methods Applicable to Determination of Probable Hydrologic Consequences–Appendix to Methods for PHC Determination," Prepared by J. F. Sato and Associates for OSM Western Technical Center, Denver, Colorado (1982).

40. Office of Surface Mining, "Methods for PHC Determination–Final Report," Prepared by J. F. Sato and Associates for OSM Western Technical Center, Denver, Colorado, U.S. Department of Interior (1984).

41. Ogrosky, H. O. and V. Mockus, "Hydrology of Agricultural Lands," In: *Handbook of Hydrology* by Ven Te Chow, McGraw-Hill Book Company (1964).

42. Pacific Southwest Inter-agency Committee, "Report on the Watershed Management Subcommittee on Factors Affecting Sediment Yield in the Pacific Southwest Area and Selection and Evaluation of Measures for Reduction of Erosion and Sediment Yield," (1968).

43. Philip, J. R., "The Theory of Infiltration, 5. Sorptivity and Algebraic Infiltration Equations," *Soil Sci., Vol. 84,* No. 3, pp. 257–264 (1957).

44. Rawls, W. J., D. L. Brakensiek, and K. E. Saxton, "Estimation of Soil Water Properties," *Transactions of the ASAE, Vol. 25,* No. 5, pp. 1316–1320, 1328 (1982).

45. Renard, K. G., W. J. Rawls, and M. M. Fogel, "Currently Available Models," In: *Hydrologic Modeling of Small Watersheds* by C. T. Haan, H. P. Johnson and D. L. Brakensiek, ASAE Monograph Number 5, St. Joseph, Michigan (1982).

46. Renfro, G. W., "Use of Erosion Equations and Sediment-delivery Ratios for Predicting Sediment Yield," In: *Present and Prospective Technology for Predicting Sediment Yields and Sources,* U.S. Department of Agriculture ARS-S-40 (1975).

47. Sherman, L. K., "Streamflow from Rainfall by the Unit-graph Method," *Eng. News-rec., Vol. 108,* pp. 501–505 (1932).

48. Simons, D. B. and F. Senturk, "Sediment Transport Technology," Water Resources Publications, Fort Collins, Colorado.

49. Smith, P. E. and W. M. Alley, "Rainfall-runoff-quality Model for Urban Watersheds," *Proceedings, International Symposium on Rainfall-Runoff Modeling,* Mississippi State University, pp. 421–442 (1981).

50. Smith, R. E. and J.-Y. Parlange, "A Parameter-efficient Hydrologic Infiltration Model," *Water Resources Res., Vol. 14,* No. 3, pp. 533–538 (1978).

51. Smith, R. E., "Flux Infiltration Theory for Use in Watershed Hydrology," *Proceedings of the Conference on Advances in In-*

filtration. December 12–13, 1983, American Society of Agricultural Engineers, ASAE Publication 11-83 (1983).

52. Soil Conservation Service, "Hydraulics" Section 5, *Soil Conservation Service National Engineering Handbook,* U.S. Department of Agriculture, Washington, D.C. (1954).

53. Soil Conservation Service, "Computer Program for Project Formulation Hydrology," U.S. Department of Agriculture, Technical Release No. 20 (1965).

54. Soil Conservation Service, "Estimating Peak Discharges for Watershed Evaluation Storms and Preliminary Designs," U.S. Department of Agriculture, Technical Service Center Technical Note Hydrology-PO-2, 6 p. (1970).

55. Soil Conservation Service, "Hydrology" Section 4, *Soil Conservation Service National Engineering Handbook,* U.S. Department of Agriculture, Washington, D.C. (1972).

56. Soil Conservation Service, "Urban Hydrology for Small Watersheds," Technical Release No. 55, U.S. Department of Agriculture, Washington, D.C. (1975).

57. Soil Conservation Service, "Sedimentation" Section 3, *Soil Conservation Service National Engineering Handbook,* U.S. Department of Agriculture, Washington, D.C. (1983).

58. Task Committee on Quantifying Land-Use Change Effects on the Watershed Management and Surface Water Committees of the Irrigation and Drainage Division, "Evaluation of Hydrologic Models Used to Quantify Major Land-use Change Effects," *J. of Irrigation and Drainage Engineering, ASCE, Vol. III,* No. 1, pp. 1–17 (1985).

59. Tourbier, J., T. Westmacott and R. Westmacott, *A Handbook for Small Surface Coal Mine Operators,* Water Resources Center, University of Delaware, for Office of Surface Mining, United States Department of the Interior (1980).

60. U.S. Army Corps of Engineers, "HEC-1 Flood Hydrograph Package Users Manual," Hydrologic Engineering Center, Davis, California, 192 p. (1981).

61. U.S. Environmental Protection Agency, "Preliminary Guidance for Estimating Erosion on Areas Disturbed by Surface Mining Activities in the Interior Western United States," Interim final report, July, EPA-908/4-77-005 (1975).

62. Williams, J. R., "Sediment Routing for Agricultural Watersheds," *Water Resources Bulletin, Vol. 11,* No. 5, pp. 965–974 (1975).

63. Williams, J. R., "Sediment-yield Prediction with Universal Equation Using Runoff Energy Factor," In: *Present and Prospective Technology for Predicting Sediment Yields and Sources,* U.S. Department of Agriculture, ARS-S-40 (1975).

64. Williams, J. R., "A Sediment Yield Routing Model," Verification of Mathematical and Physical Models in Hydraulic Engineering, ASCE, New York (1978).

65. Wilson, B. N., et al., "A Hydrology and Sedimentology Watershed Model. Part I: Operational Format and Hydrologic Component," *Transactions of the ASAE, Vol. 25,* No. 5, pp. 1370–1377 (1984).

66. Wilson, B. N., et al., "A Hydrology and Sedimentology Watershed Model. Part II: Sedimentology Component," *Transactions of the ASAE, Vol. 25,* No. 5, pp. 1378–1384 (1984).

67. Wischmeier, W. H. and D. D. Smith, "Predicting Rainfall Erosion Losses–A Guide to Conservation Planning," U.S. Department of Agriculture, *Agriculture Handbook No. 537,* (1978).

68. Woolhiser, D. A., "Simulation of Unsteady Overland Flow," In: *Unsteady Flow in Open Channels, Volume II,* by K. Mahmood and V. Yevjevich, Water Resources Publications, Fort Collins, Colorado (1975).

APPENDIX—SOIL NAMES AND HYDROLOGIC CLASSIFICATIONS (SCS, 1972)

Soil	HSG	Soil	HSG	Soil	HSG	Soil	HSG	Soil	HSG
AABERG	C	AHL	C	ALMY	B	ANLAUF	C	AROOSTOOK	
AASTAD	B	AHLSTROM	C	ALOHA	C	ANNABELLA	B	AROSA	C
ABAC	D	AHMEEK	B	ALONSO	B	ANNANDALE	C	ARP	C
ABAJO	C	AHOLT	D	ALOVAR	C	ANNISTON	B	ARRINGTON	B
ABBOTT	D	AHTANUM	C	ALPENA	B	ANOKA	A	ARRITOLA	D
ABBOTTSTOWN	C	AHWAHNEE	C	ALPHA	C	ANONES	C	ARROLIME	B
ABCAL	D	AIBONITO	C	ALPON	B	ANSARI	D	ARRON	D
ABEGG	B	AIKEN	B/C	ALPOWA	B	ANSEL	B	ARROW	B
ABELA	B	AIKMAN	D	ALPS	C	ANSELMO	A	ARROWSMITH	B
ABELL	B	AILEY	B	ALSEA	B	ANSON	B	ARROYO SECO	B
ABERDEEN	D	AINAKEA	B	ALSPAUGH	C	ANTELOPE SPRINGS	C	ARTA	C
ABES	D	AIRMONT	C	ALSTAD	B	ANTERO	B	ARTOIS	C
ABILENE	C	AIROTSA	B	ALSTOWN	B	ANT FLAT	C	ARVADA	D
ABINGTON	B	AIRPORT	D	ALTAMONT	D	ANTHO	B	ARVANA	B
ABIQUA	C	AITS	B	ALTAVISTA	C	ANTHONY	B	ARVESON	D
ABO	B/C	AJO	C	ALTDORF	D	ANTIGO	B	ARVILLA	B
ABOR	D	AKAKA	A	ALTMAR	B	ANTILON	B	ARZELL	B
ABRA	C	AKASKA	B	ALTO	C	ANTIOCH	D	ASA	B
ABRAHAM	B	AKELA	C	ALTOGA	C	ANTLER	C	ASBURY	B
ABSAROKEE	C	ALADDIN	B	ALTON	B	ANTOINE	C	ASCALON	B
ABSCOTA	B	ALAE	A	ALTUS	B	ANTROBUS	B	ASCHOFF	B
ABSHER	D	ALAELOA	B	ALTVAN	B	ANTY	B	ASHBY	C
ABSTED	D	ALAGA	A	ALUM	B	ANVIK	B	ASHCROFT	B
ACACIO	C	ALAKAI	D	ALUSA	D	ANWAY	B	ASHDALE	B
ACADEMY	C	ALAMA	B	ALVIN	B	ANZA	B	ASHE	B
ACADIA	D	ALAMANCE	B	ALVIRA	C	ANZIANO	C	ASHKUM	C
ACANA	D	ALAMO	D	ALVISO	D	APACHE	D	ASHLAR	B
ACASCO	D	ALAMOSA	C	ALVOR	C	APAKUIE	A	ASHLEY	A
ACEITUNAS	B	ALAPAHA	D	AMADOR	D	APISHAPA	C	ASH SPRINGS	C
ACEL	D	ALAPAI	A	AMAGON	D	APISON	B	ASHTON	B
ACKER	B	ALBAN	B	AMALU	D	APOPKA	A	ASHUE	B
ACKMEN	B	ALBANO	D	AMANA	B	APPIAN	C	ASHUELOT	C
ACME	C	ALBANY	C	AMARGOSA	D	APPLEGATE	C	ASHWOOD	C
ACO	B	ALBATON	D	AMARILLO	B	APPLETON	C	ASKEW	C
ACOLITA	B	ALBEE	C	AMASA	B	APPLING	B	ASO	C
ACOMA	C	ALBEMARLE	C	AMBERSON		APRON	B	ASOTIN	C
ACOVE	C	ALBERTVILLE	C	AMBOY	C	APT	C	ASPEN	B
ACREE	C	ALBIA	C	AMBRAW	C	APTAKISIC	B	ASPERMONT	B
ACRELANE	C	ALBION	B	AMEDEE	A	ARABY		ASSINNIBOINE	
ACTON	B	ALBRIGHTS	C	AMELIA	B	ARADA	C	ASSUMPTION	B
ACUFF	B	ALCALDE	C	AMENIA	B	ARANSAS	D	ASTATULA	A
ACWORTH	B	ALCESTER	B	AMERICUS	A	ARAPIEN	C	ASTOR	A/D
ACY	C	ALCOA	B	AMES	C	ARAVE	D	ASTORIA	B
ADA	B	ALCONA	B	AMESHA	B	ARAVETON	B	ATASCADERO	B
ADAIR	D	ALCOVA	B	AMHERST	C	ARBELA	C	ATASCOSA	D
ADAMS	A	ALDA	C	AMITY	C	ARBONE	B	ATCO	B
ADAMSON	B	ALDAX	D	AMMON	B	ARBOR	B	ATENCIO	B
ADAMSTOWN		ALDEN	D	AMOLE	C	ARBUCKLE	B	ATEPIC	D
ADAMSVILLE	C	ALDER	B	AMOR	B	ARCATA	B	ATHELWOLD	B
ADATON	D	ALDERDALE	C	AMOS	C	ARCH	B	ATHENA	B
ADAVEN	D	ALDERWOOD	C	AMSDEN	B	ARCHABAL	B	ATHENS	B
ADDIELOU	C	ALDINO	C	AMSTERDAM	B	ARCHER	C	ATHERLY	B
ADDISON	D	ALDWELL	C	ANTOFT	D	ARCHIN	C	ATHERTON	B/D
ADDY	C	ALEKNAGIK	B	AMY	D	ARCO	B	ATNMAR	C
ADE	A	ALEMEDA	C	ANACAPA	B	ARCOLA	C	ATHOL	B
ADEL	A	ALEX	B	ANAHUAC	D	ARD	C	ATKINSON	B
ADELAIDE	D	ALEXANDRIA	C	ANAMITE	D	ARDEN	B	ATLAS	D
ADELANTO	B	ALEXIS	B	ANAPRA	B	ARDENVOIR	B	ATLEE	C
ADELINO	B	ALFORD	B	AMASAZI	B	ARDILLA	C	ATMORE	B/D
ADELPHIA	C	ALGANSEE	B	ANATONE	D	AREDALE	B	ATOKA	B
ADENA	C	ALGERITA	B	ANAVERDE	B	ARENA	C	ATON	B
ADGER	D	ALGIERS	C/D	ANAWALT	D	ARENALES	A	ATRYPA	C
ADILIS	A	ALGOMA	B/D	ANCHO	B	ARENDTSVILLE	B	ATSION	C
ADIRONDACK		ALHAMBRA	B	ANCHOR BAY	D	ARENOSA	A	ATTERBERRY	B
ADIV	B	ALICE	A	ANCHOR POINT	D	ARENZVILLE	B	ATTEWAN	A
ADJUNTAS	C	ALICEL	B	ANCLOTE	D	ARGONAUT	D	ATTICA	B
ADKINS	B	ALICIA	B	ANCO	C	ARGUELLO	B	ATTLEBORO	
ADLER	C	ALIDA	B	ANDERLY	C	ARGYLE	B	ATWATER	B
ADOLPH	D	ALIKCHI	B	ANDERS	C	ARIEL	C	ATWELL	C/D
ADRIAN	A/D	ALINE	A	ANDERSON	B	ARIZO	A	ATWOOD	B
AENEAS	B	ALKO	D	ANDES	C	ARKABUTLA	C	AUBBEENAUBBEE	B
AETNA	B	ALLAGASH	B	ANDORINIA	C	ARKPORT	B	AUBERRY	B
AFTON	D	ALLARD	B	ANDOVER	D	ARLAND	B	AUBURN	C/D
AGAR	B	ALLEGHENY	B	ANDREEN	B	ARLE	B	AUBURNDALE	D
AGASSIZ	D	ALLEMANDS	D	ANDREESON	C	ARLING	D	AUDIAN	
AGATE	D	ALLEN	B	ANDRES	B	ARLINGTON	C	AU GRES	C
AGAMAN	B	ALLENDALE	C	ANDREWS	C	ARLOVAL	C	AUGSBURG	C
AGENCY	C	ALLENS PARK	B	ANED	D	ARMAGH	D	AUGUSTA	C
AGER	D	ALLENSVILLE	C	ANETH	A	ARMIJO	D	AULD	D
AGNER	B	ALLENTINE	D	ANGELICA	D	ARMINGTON	D	AURA	B
AGNEW	B/C	ALLENWOOD	B	ANGELINA	B/D	ARMO	B	AURORA	C
AGNOS	B	ALLESSIO	B	ANGELO	C	ARMOUR	B	AUSTIN	C
AGUA	B	ALLEY	C	ANGIE	C	ARMSTER	C	AUSTWELL	D
AGUADILLA	A	ALLIANCE	B	ANGLE	A	ARMSTRONG	D	AUXVASSE	D
AGUA DULCE	C	ALLIGATOR	D	ANGLEN	B	ARMUCHEE	C	AUZQUI	C
AGUA FRIA	B	ALLIS	D	ANGOLA	C	ARNEGARD	B	AVA	C
AGUALT	B	ALLISON	C	ANGOSTURA	B	ARNHART	C	AVALANCHE	C
AGUEDA	B	ALLOUEZ	C	ANHALT	D	ARNHEIM	C	AVALON	B
AGUILITA	B	ALLOWAY		ANIAK	D	ARNO	D	AVERY	B
AGUIRRE	D	ALMAC	B	ANITA	D	ARNOLD	B	AVON	C
AGUSTIN	B	ALMENA	C	ANKENY	A	ARNOT	C/D	AVONBURG	D
AHATONE	D	ALMONT	A			ARNY	A	AVONDALE	E

NOTES   A BLANK HYDROLOGIC SOIL GROUP INDICATES THE SOIL GROUP HAS NOT BEEN DETERMINED
TWO SOIL GROUPS SUCH AS B/C INDICATES THE DRAINED/UNDRAINED SITUATION

244

Name	Group
AMBREY	D
AXTELL	D
AYAR	D
AYCOCK	B
AYON	B
AYR	B
AYRES	B
AYRSHIRE	C
AYSEES	B
AZAAR	C
AZARMAN	C
AZELTINE	B
AZFIELD	B
AZTALAN	B
AZTEC	B
AZULE	C
AZWELL	B
BABB	A
BABBINGTON	B
BABCOCK	C
BABYLON	A
BACA	C
BACH	D
BACHUS	C
BACKBONE	A
BACULAN	A
BADENAUGH	B
BADGER	C
BADGERTON	B
BADO	D
BADUS	C
BAGARD	C
BAGDAD	B
BAGGOTT	D
BAGLEY	B
BAHEM	B
BAILE	D
BAINVILLE	C
BAIRD HOLLOW	C
BAJURA	D
BAKEOVEN	D
BAKER	C
BAKER PASS	B
BALAAM	A
BALCH	D
BALCOM	B
BALD	C
BALDER	C
BALDOCK	B/C
BALDWIN	D
BALDY	B
BALE	C
BALLARD	B
BALLER	D
BALLINGER	C
BALM	B/C
BALMAN	B/C
BALON	B
BALTIC	D
BALTIMORE	B
BALTO	D
BAMBER	B
BAMFORTH	B
BANGAS	B
BANCROFT	B
BANDERA	B
BANGO	C
BANGOR	B
BANGSTON	A
BANKARD	A
BANKS	A
BANNER	C
BANNERVILLE	C/D
BANNOCK	B
BANQUETE	D
BARABOO	B
BARAGA	C
BARBARY	D
BARBOUR	B
BARBOURVILLE	B
BARCLAY	C
BARCO	B
BARCUS	B
BARD	D
BARDEN	C
BARDLEY	C
BARELA	C
BARFIELD	D
BARFUSS	B
BARGE	C
BARISHMAN	C
BARKER	C
BARKERVILLE	C
BARKLEY	B
BARLANE	D
BARLING	C
BARLOW	B
BARNARD	D
BARNES	B
BARNESTON	B
BARNEY	A
BARNHARDT	B
BARNSTEAD	
BARNUM	B
BARRADA	D
BARRETT	D
BARRINGTON	B
BARRON	B
BARRONETT	C
BARROWS	D
BARRY	D
BARSTOW	B
BARTH	C
BARTINE	C
BARTLE	D
BARTLEY	C
BARTON	B
BARTONFLAT	B
BARVON	C
BASCOM	B
BASEHOR	D
BASHAW	D
BASHER	B
BASILE	D
BASIN	C
BASINGER	C
BASKET	C
BASS	A
BASSEL	B
BASSETT	B
BASSFIELD	B
BASSLER	D
BASTIAN	D
BASTROP	B
BATA	A
BATAVIA	B
BATES	B
BATH	C
BATTERSON	D
BATTLE CREEK	C
BATZA	D
BAUDETTE	B
BAUER	C
BAUGH	B/C
BAXTER	B
BAXTERVILLE	B
BAYAMON	B
BAYBORO	D
BAYERTON	C
BAYLOR	D
BAYSHORE	B/C
BAYSIDE	C
BAYUCOS	D
BAYWOOD	A
BAZETTE	C
BAZILE	B
BEAD	C
BEAGLE	C
BEALES	A
BEAR BASIN	B
BEAR CREEK	C
BEARDALL	C
BEARDEN	C
BEARDSTOWN	C
BEAR LAKE	D
BEARMOUTH	A
BEARPAW	B
BEAR PRAIRIE	B
BEARSKIN	D
BEASLEY	C
BEASON	C
BEATON	C
BEATTY	C
BEAUCOUP	B
BEAUFORD	D
BEAUMONT	D
BEAUREGARD	C
BEAUSITE	B
BEAUVAIS	B
BEAVERTON	A
BECK	C
BECKER	B
BECKET	C
BECKLEY	B
BECKTON	D
BECKWITH	C
BECKWOURTH	B
BECREEK	B
BEDFORD	C
BEDINGTON	B
BEDNER	C
BEEBE	A
BEECHER	C
BEECHY	
BEEHIVE	B
BEEK	C
BEENOM	D
BEEZAR	B
BEGAY	B
BEGOSHIAN	C
BEHANIN	B
BEHEMOTOSH	B
BEHRING	D
BEIRMAN	D
BEJUCOS	B
BELCHER	D
BELDEN	D
BELDING	B
BELEN	C
BELFAST	B
BELFIELD	B
BELFORE	B
BELGRADE	B
BELINDA	D
BELKNAP	C
BELLAMY	C
BELLAVISTA	D
BELLE	B
BELLEFONTAINE	
BELLICUM	B
BELLINGHAM	C
BELLPINE	C
BELMONT	B
BELMORE	B
BELT	D
BELTED	D
BELTON	C
BELTRAMI	B
BELTSVILLE	C
BELUGA	D
BELVOIR	C
BENCLARE	C
BENEVOLA	C
BENEWAH	C
BENFIRLD	C
BENGE	B
BEN HUR	B
BENIN	D
BENITO	D
BENJAMIN	D
BEN LOMOND	B
BENMAN	A
BENNDALE	B
BENNETT	C
BENNINGTON	D
BENOIT	D
BENSUN	C/D
BENTEEN	B
BENTONVILLE	C
BENZ	D
BEOTIA	B
BEOWAWE	D
BERCAIL	C
BERDA	B
BEREA	C
BERENICETON	B
BERENT	A
BERGLAND	D
BERGSTROM	B
BERINO	B
BERKELEY	D
BERKS	C
BERKSHIRE	B
BERLIN	C
BERMESA	C
BERMUDIAN	B
BERNAL	D
BERNALDO	B
BERNARD	D
BERNARDINO	C
BERNARDSTON	C
BERNHILL	B
BERNICE	A
BERNING	C
BERRENDOS	D
BERRYLAND	D
BERTELSON	B
BERTHOUD	B
BERTIE	C
BERTOLOTTI	B
BERTRAND	B
BERVILLE	D
BERYL	B
BESSEMER	B
BETHANY	C
BETHEL	D
BETTERAVIA	C
BETTS	B
BEULAH	B
BEVENT	B
BEVERLY	B
BEW	D
BEWLEYVILLE	B
BEWLIN	D
BEXAR	C
BEZZANT	B
BIBB	B/D
BIBON	A
BICKELTON	B
BICKLETON	C
BICKMORE	C
BICONDOA	C
BIDDEFORD	D
BIDDLEMAN	C
BIDMAN	C
BIDWELL	B
BIEBER	D
BIENVILLE	A
BIG BLUE	D
BIGEL	A
BIGELOW	C
BIGETTY	C
BIGGS	A
BIGGSVILLE	B
BIG HORN	C
BIGNELL	B
BIG TIMBER	D
BIGWIN	D
BIJOU	A
BILLETT	A
BILLINGS	C
BINDLE	B
BINFORD	B
BINGHAM	B
BINNSVILLE	D
BINS	B
BINTON	C
BIPPUS	B
BIRCH	A
BIRCHWOOD	C
BIRDOW	B
BIROS	C
BIRDSALL	D
BIRDSBORO	B
BIRDSLEY	D
BIRKBECK	B
BISBEE	A
BISCAY	C
BISHOP	B/C
BISPING	B
BISSELL	B
BISTI	C
BIT	D
BITTERON	A
BITTERROOT	C
BITTER SPRING	C
BITTON	B
BIXBY	B
BJORK	C
BLACHLY	C
BLACK BURN	B
BLACK BUTTE	C
BLACK CANYON	D
BLACKCAP	A
BLACKETT	B
BLACKFOOT	B/C
BLACKHALL	D
BLACKHAWK	D
BLACKLEAF	B
BLACKLEED	A
BLACKLOCK	D
BLACKMAN	C
BLACK MOUNTAIN	C
BLACKOAR	C
BLACKPIPE	C
BLACK RIDGE	D
BLACKROCK	B
BLACKSTON	B
BLACKTAIL	B
BLACKWATER	D
BLACKWELL	B/D
BLADEN	D
BLAGO	D
BLAINE	B
BLAIR	C
BLAIRTON	C
BLAKE	A
BLAKELAND	A
BLAKENEY	C
BLAKEPORT	B
BLALOCK	D
BLAMER	C
BLANCA	B
BLANCHARD	B
BLANCHESTER	B/D
BLAND	C
BLANDFORD	C
BLANDING	B
BLANEY	B
BLANKET	C
BLANTON	A
BLANYON	C
BLASDELL	A
BLASINGAME	C
BLAZON	D
BLENCOE	C
BLEND	D
BLENDON	B
BLETHEN	B
BLEVINS	B
BLEVINTON	B/D
BLICHTON	D
BLISS	D
BLOCKTON	C
BLODGETT	A
BLOMFORD	B
BLOOM	C
BLOOMFIELD	A
BLOOMING	B
BLOOR	D
BLOSSOM	C
BLOUNT	C
BLOUNTVILLE	C
BLUCHER	C
BLUEBELL	C
BLUE EARTH	D
BLUEJOINT	B
BLUE LAKE	A
BLUEPOINT	B
BLUE STAR	B
BLUEWING	B
BLUFFDALE	C
BLUFFTON	D
BLUFORD	D
BLY	B
BLYTHE	D
BOARDTREE	C
BOBS	D
BOBTAIL	B
BOCK	B
BODELL	D
BOENBURG	B
BODINE	B
BOEL	A
BOELUS	A
BOESEL	B
BOETTCHER	C
BOGAN	C
BOGART	B
BOGUE	D
BOHANNON	C
BOHEMIAN	B
BOISTFORT	C
BOLAR	C
BOLD	B
BOLES	C
BOLIVAR	B
BOLIVIA	B
BOLTON	B
BOMBAY	B
BON	B
BONACCORD	D
BONAPARTE	A
BOND	D
BONDRANCH	D
BONDURANT	B
BONE	D
BONG	B

Soil	HSG	Soil	HSG	Soil	HSG	Soil	HSG	Soil	HSG	Soil	HSG
BONMAN	C	BRANDON	B	BROOKLYN	D	BUSTER	D	CAMPSPASS	C	CANEADEA	D
BONIFAY	A	BRANDYWINE	C	BROOKSIDE	C	BUTANO	C	CAMPUS	B	CANEEK	C
BONILLA	B	BRANFORD	B	BROOKSTON	B/D	BUTLER	D	CAMRODEN	C	CANEL	B
BONITA	D	BRANTFORD	B	BROOKSVILLE	D	BUTLERTOWN	D	CANA	C	CANELO	D
BONN	D	BRANYON	D	BROOMFIELD	D	BUTTE	C	CANAAN	C/D	CANEY	C
BONNER	B	BRASHEAR	C	BROSELEY	B	BUTTERFIELD	C	CANADIAN	B	CANEYVILLE	C
BONNET	B	BRASSFIELD	B	BROSS	B	BUTTON	C	CANADICE	D	CANEZ	B
BONNEVILLE	B	BRATTON	B	BROUGHTON	D	BUXIN	D	CANANDAIGUA	D	CANFIELD	C
BONNICK	A	BRAVANE	D	BROWARD	C	BUXTON	C	CANASERAGA	C	CANISTEO	C
BONNIE	D	BRAXTON	C	BROWNELL	B	BYARS	D	CANAVERAL	C	CANNINGER	D
BONO	D	BRAYMILL	B/D	BROWNFIELD	A	BYNUM	C	CANBURN	D	CANNON	B
BONSALL	D	BRAYS	D	BROWNLEE	B	BYRON	A	CANDELERO	C	CANOE	B
BONTA	C	BRAYTON	C	BROYLES	B			CANE	C	CANONCITO	B
BONTI	C	BRAZITO	A	BRUCE	D	CABALLO	B	CANEADEA	D	CANOVA	B/D
BOOKER	D	BRAZOS	A	BRUFFY	C	CABARTON	D			CANTALA	B
BOOMER	B	BREA	B	BRUIN	C	CABBA	C			CANTON	B
BOONE	A	BRECKENRIDGE	D	BRUNEEL	B/C	CABBART	D			CANTRIL	B
BOONESBORO	B	BRECKNOCK	B	BRUNO	A	CABEZON	D			CANTUA	B
BOONTON	C	BREECE	B	BRUNT	C	CABIN	C			CANUTIO	B
BOOTH	C	BREGAR	D	BRUSH		CABINET				CANYON	D
BORACHO	C	BREMEN	B	BRUSSETT	B	CABLE	D			CAPAC	B
BORAH	A/C	BREMER	B	BRYAN	A	CABO ROJO				CAPAY	D
BORDA	D	BREMO	C	BRYCAN	B	CABOT	D			CAPE	D
BORDEAUX	B	BREMS	A	BRYCE	D	CACAPON	B			CAPE FEAR	D
BORDEN	B	BRENDA	C	BUCAN	D	CACHE	D			CAPERS	D
BORDER	B	BRENNAN	B	BUCHANAN	C	CACIQUE	C			CAPILLO	C
BORNSTEDT	C	BRENNER	C/D	BUCHENAU	C	CADDO	D			CAPLES	C
BORREGO	C	BRENT	C	BUCHER	C	CADEVILLE	C			CAPPS	B
BORUP	B	BRENTON	B	BUCKHOUSE	A	CADMUS	A			CAPSHAW	B
BORVANT	D	BRENTWOOD	B	BUCKINGHAM		CADOMA	D			CAPULIN	B
BORZA	C	BRESSER	B	BUCKLAND	C	CADOR	C			CAPUTA	C
BOSANKO	D	BREVARD	B	BUCKLEBAR	B	CAGEY	C			CARACO	C
BOSCO	B	BREVORT	B	BUCKLEY	B/C	CAGUABO	D			CARALAMPI	B
BOSKET	B	BREWER	C	BUCKLON	D	CAGWIN	B			CARBO	C
BOSLER	B	BREWSTER	D	BUCKNER	A	CAHABA	B			CARBOL	D
BOSQUE	B	BRENTON	C	BUCKNEY	A	CAHILL	B			CARBONDALE	D
BOSS	D	BRICKEL	C	BUCKS	B	CAHONE	C			CARBURY	B
BOSTON	C	BRICKTON	C	BUCKSKIN	C	CAHTO	B			CARCITY	D
BOSTWICK	B	BRIDGE	C	BUCODA	C	CAID	D			CARDIFF	C
BOSWELL	D	BRIDGEHAMPTON	B	BUDD	B	CAIRO	D			CARDINGTON	C
BOSWORTH	D	BRIDGEPORT	B	BUDE	C	CAJALCO	C			CARDON	D
BOTELLA	B	BRIDGER	A	BUELL	B	CAJON	A			CAREY	B
BOTHWELL	C	BRIDGESON	B/C	BUENA VISTA	B	CALABAR	D			CAREY LAKE	B
BOTTINEAU	C	BRIDGET	B	BUFFINGTON	B	CALABASAS	B			CAREYTOWN	D
BOTTLE	A	BRIDGEVILLE	B	BUFFMEYER	B	CALAIS	C			CARGILL	C
BOULDER	B	BRIDGPORT	B	BUFF PEAK	C	CALAMINE	D			CARIBE	B
BOULDER LAKE	D	BRIEDWELL	B	BUICK	C	CALAPOOYA	C			CARIBEL	B
BOULDER POINT	B	BRIEF	B	BUIST	B	CALAWAH	B			CARIBOU	B
BOULFLAT	D	BRIENSBURG		BUKREEK	B	CALCO	C			CARLIN	D
BOURNE	C	BRIGGS	A	BULLION	D	CALDER	D			CARLINTON	B
BOW	C	BRIGGSDALE	C	BULLREY	B	CALDWELL	B			CARLISLE	A/D
BOWBAC	C	BRIGGSVILLE	C	BULL RUN	B	CALEAST	C			CARLOTTA	B
BOWBELLS	B	BRIGHTON	A/D	BULL TRAIL	B	CALEB	B			CARLOW	D
BOWDOIN	D	BRIGHTWOOD	C	BULLY	B	CALERA	C			CARLSBAD	D
BOWDRE	C	BRILL	B	BUMGARD	B	CALHI	A			CARLSBORG	A
BOWERS	C	BRIN	C	BUNCOMBE	A	CALHOUN	D			CARLSON	C
BOWIE	B	BRIMFIELD	C/D	BUNDO	B	CALICO	B			CARLTON	B
BOWMAN	B/D	BRIMLEY	B	BUNDYMAN	C	CALIFON	C			CARMI	B
BOWMANSVILLE	C	BRINEGAR	B	BUNEJUG	C	CALIMUS	B			CARNASAW	C
BOXELDER	C	BRINKERT	C	BUNKER	D	CALITA	B			CARNEGIE	C
BOXWELL	C	BRINKERTON	D	BUNSELMEIER	C	CALIZA	B			CARNERO	C
BOY	A	BRISCOT	B	BUNTINGVILLE	B/C	CALKINS	C			CARNEY	D
BOYCE	B/D	BRITE	C	BUNYAN	B	CALLABO	C			CAROLINE	C
BOYD	D	BRITTON	C	BURBANK	A	CALLAHAN	C			CARR	B
BOYER	B	BRIZAN	A	BURCH	B	CALLEGUAS	D			CARRISALITOS	D
BOYNTON		BROAD	C	BURCHARD	B	CALLINGS	C			CARRIZO	A
BOYSAG	D	BROADALBIN	C	BURCHELL	B/C	CALLOWAY	C			CARSITAS	A
BOYSEN	D	BROADAX	B	BURDETT	C	CALMAR	C			CARSLEY	C
BOZARTH	C	BROADBROOK	C	BUREN	C	CALNEVA	C			CARSO	D
BOZE	B	BROAD CANYON	B	BURGESS	C	CALOUSE	C			CARSON	D
BOZEMAN	A	BROADHEAD	C	BURGI	B	CALPINE	B			CARSTAIRS	B
BRACEVILLE	C	BROADHURST	D	BURGIN	D	CALVERT	D			CARSTUMP	C
BRACKEN	D	BROCK	D	BURKE	C	CALVERTON	C			CART	B
BRACKETT	C	BROCKLISS	C	BURKHARDT	B	CALVIN	C			CARTAGENA	D
BRAD	D	BROCKMAN	C	BURLEIGH	D	CALVISTA	D			CARTECAY	C
BRADDOCK	C	BROCKO	B	BURLESON	D	CAM	B			CARUSO	C
BRADENTON	B/D	BROCKPORT	D	BURLINGTON	A	CAMAGUEY	D			CARUTHERSVILLE	B
BRADER	D	BROCKTON	D	BURMA		CAMARGO	D			CARVER	A
BRADFORD	B	BROCKWAY	B	BURMESTER	D	CAMARILLO	B/C			CARWILE	D
BRADSHAW	B	BRODY	C	BURNAC	C	CAMAS	A				
BRADWAY	D	BROE	B	BURNETTE	B	CAMASCREEK	B/D				
BRADY	B	BROGAN	B	BURNHAM	D	CAMBERN	C				
BRADYVILLE	C	BROGDON	B	BURNSIDE	B	CAMBRIDGE	B				
BRAHAM	B	BROLLIAR	D	BURNSVILLE	B	CAMDEN	B				
BRAINERD	B	BROMO	B	BURNT LAKE	B	CAMERON	B				
BRALLIER	D	BRONAUGH	B	BURRIS	D	CAMILLUS	B				
BRAM	B	BRONCHO	B	BURT	D	CAMP	B				
BRAMARD	B	BRONSON	B	BURTON	B	CAMPBELL	B/C				
BRAMBLE	C	BRONTE	C	BUSE	B	CAMPHORA	B				
BRAMWELL	C	BROOKE	C	BUSH	B	CAMPIA	B				
BRAND	D	BROOKFIELD	B	BUSHNELL	C	CAMPO	C				
BRANDENBURG	A	BROOKINGS	B	BUSHVALLEY	D	CAMPONE	B/C				

Name	Grp	Name	Grp	Name	Grp	Name	Grp	Name	Grp
CARYVILLE	B	CENTRAL POINT	B	CHILGREN	C	CLARESON	C	COKEDALE	B/C
CASA GRANDE	C	CERESCO	A	CHILHOWIE	C	CLAREVILLE	C	COKEL	B
CASCADE	C	CERRILLOS	B	CHILI	B	CLARINDA	D	COKER	D
CASCAJO	B	CERRO	C	CHILKAT	C	CLARION	B	COKESBURY	D
CASCILLA	B	CHACRA	C	CHILLICOTHE	C	CLARITA	D	COKEVILLE	B
CASCO	B	CHAFFEE	C	CHILLISQUAQUE		CLARK	B	COLBATH	C/D
CASE	B	CHAGRIN	B	CHILLUM	B	CLARK FORK	A	COLBERT	D
CASEBIER	D	CHAIX	B	CHILMARK	B	CLARKSBURG	C	COLBURN	B
CASEY	C	CHALFONT	C	CHILO	B/D	CLARKSDALE	C	COLBY	B
CASHEL	C	CHALMERS	C	CHILOQUIN	B	CLARKSON	B	COLCHESTER	B
CASHION	D	CHAMA	B	CHILSON	D	CLARKSVILLE	B	COLDCREEK	B
CASHMERE	B	CHAMBER	C	CHILTON	B	CLARNO	B	COLDEN	D
CASHMONT	B	CHAMBRINO	C	CHIMAYO	D	CLARY	B	COLD SPRINGS	C
CASINO	A	CHAMISE	B	CHIMNEY	B	CLATO	B	COLE	B/C
CASITO	D	CHAMOKANE	B	CHINA CREEK	B	CLATSOP	D	COLEBROOK	B
CASPAR	B	CHAMPION	B	CHINCHALLO	B/D	CLAVERACK	C	COLEMAN	C
CASPIANA	B	CHANCE	B/D	CHINIAK	A	CLAWSON	C	COLEMANTOWN	D
CASS	A	CHANDLER	B	CHINO	B/C	CLAYBURN	B	COLETO	A
CASSADAGA		CHANEY	C	CHINOOK	B	CLAYSPRINGS	D	COLFAX	C
CASSIA	C	CHANNAHON	B	CHIPETA	D	CLAYTON	B	COLIBRO	B
CASSIRO	C	CHANNING	B	CHIPLEY	C	CLEARFIELD	C	COLINAS	B
CASSOLARY	B	CHANTA	B	CHIPMAN	D	CLEAR LAKE	D	COLLAMER	C
CASSVILLE		CHANTIER	D	CHIPPENY	D	CLEEK	C	COLLARD	B
CASTAIC	C	CHAPIN	C	CHIPPEWA	B/D	CLE ELUM	B	COLLBRAN	C
CASTALIA	C	CHAPMAN	B	CHIQUITO	C/D	CLEGG	B	COLLEEN	C
CASTANA	B	CHAPPELL	B	CHIRICAHUA	D	CLEMAN	B	COLLEGIATE	C
CASTELL	C	CHARD	B	CHISPA	B	CLEMS	B	COLLETT	C
CASTILE	B	CHARGO	D	CHITINA	B	CLEMVILLE	B	COLLIER	A
CASTINO	C	CHARITON	D	CHITTENDEN	C	CLEORA	B	COLLINGTON	B
CASTLE	D	CHARITY	D	CHITWOOD	C	CLERF	C	COLLINS	C
CASTLEVALE	D	CHARLEBOIS	C	CHIVATO	D	CLERMONT	D	COLLINSTON	C
CASTNER	C	CHARLESTON	C	CHIWAWA	B	CLEVERLY	B	COLLINSVILLE	C
CASTO	C	CHARLEVOIX	B	CHO	C	CLICK	A	COLMA	B
CASTRO	C	CHARLOS	A	CHOBEE	D	CLIFFDOWN	B	COLMOR	B
CASTROVILLE	B	CHARLOTTE	A/D	CHOCK	B/D	CLIFFHOUSE	C	COLO	B
CASUSE	D	CHARLTON	B	CHOCOLOCCO	B	CLIFFORD	B	COLOCKUM	B
CASWELL	D	CHASE	C	CHOPAKA	C	CLIFFWOOD	C	COLOMA	A
CATALINA	B	CHASEBURG	B	CHOPTANK	A	CLIFTERSON	B	COLOMBO	B
CATALPA	C	CHASEVILLE	A	CHOPTIE	D	CLIFTON	C	COLONA	C
CATANO	A	CHASKA	C	CHORALMONT	B	CLIFTY	B	COLONIE	A
CATARINA	D	CHASTAIN	D	CHOSKA	B	CLIMARA	D	COLORADO	B
CATAULA	C	CHATBURN	B	CHOTEAU	C	CLIMAX	D	COLOROCK	D
CATAMBA	B	CHATFIELD	C	CHRISTIAN	C	CLIME	C	COLOSO	D
CATH	D	CHATHAM	B	CHRISTIANA	B	CLINTON	B	COLOSSE	A
CATHCART	C	CHATSWORTH	D	CHRISTIANBURG	D	CLIPPER	B/C	COLP	D
CATHEDRAL	D	CHAUNCEY	C	CHRISTY	B	CLODINE	D	COLRAIN	B
CATHERINE	B/D	CHAVIES	B	CHROME	C	CLONTARF	B	COLTON	A
CATHRO	D	CHAWANAKEE	C	CHUALAR	B	CLOQUALLUM	C	COLTS NECK	B
CATLETT	C/D	CHEADLE	C	CHUBBS	C	CLOQUATO	B	COLUMBIA	B
CATLIN	B	CHECKETT	D	CHUCKAWALLA	B	CLOQUET	B	COLUMBINE	A
CATNIP	D	CHEDEHAP	B	CHUGTER	B	CLOUD	D	COLUSA	C
CATOCTIN	C	CHEEKTOWAGA	D	CHULITNA	B	CLOUDCROFT	D	COLVILLE	B/C
CATOOSA	B	CHEESEMAN	C	CHUMMY	C/D	CLOUD PEAK	C	COLVIN	C
CATSKILL	A	CHEHALEM	C	CHUMSTICK	C	CLOUD RIM	B	COLWOOD	B/D
CATTARAUGUS	C	CHEHALIS	B	CHUPADERA	C	CLOUGH	D	COLY	B
CAUDLE	B	CHEHULPUM	D	CHURCH	D	CLOVERDALE	D	COLYER	C/D
CAVAL	B	CHELAN	B	CHURCHILL	D	CLOVER SPRINGS	B	COMER	B
CAVE	D	CHELSEA	A	CHURCHVILLE	D	CLOVIS	B	COMERIO	B
CAVELT	D	CHEMAWA	B	CHURN	B	CLUFF	C	COMETA	D
CAVE ROCK	A	CHEMUNG	B	CHURNDASHER	B	CLUNIE	D	COMFREY	C
CAVO	D	CHEN	D	CHUTE	A	CLURDE	C	COMITAS	A
CAVODE	C	CHENA	A	CIALES	D	CLURO	C	COMLY	C
CAVOUR	D	CHENANGO	A	CIBEQUE	B	CLYDE	D	COMMERCE	C
CAWKER	B	CHENEY	B	CIBO	D	CLYMER	B	COMO	A
CAYAGUA	C	CHENNEBY	C	CIBOLA	B	COACHELLA	B	COMODORE	B
CAYLOR	B	CHENOWETH	B	CICERO	D	COAD	B	COMORO	B
CAYUGA	C	CHEQUEST	C	CIDRAL	C	COAL CREEK	D	COMPTCHE	B
CAZADERO	C	CHEREETE	A	CIENEBA	C	COALMONT	C	COMPTON	C
CAZADOR	B	CHERIONI	D	CIMA	C	COAMO	C	COMSTOCK	C
CAZENOVIA	B	CHEROKEE	D	CIMARRON	C	COARSEGOLD	B/C	COMUS	B
CEBOLIA	C	CHERRY	C	CINCINNATI	C	COATICOOK	C	CONALB	B
CEBONE	C	CHERRYHILL	B	CINCO	A	COATSBURG	D	CONANT	C
CECIL	B	CHERRY SPRINGS	C	CINDERCONE	B	COBB	B	CONASAUGA	C
CEDA	B	CHESAW	A	CINEBAR	B	COBEN	D	CONATA	D
CEDARAN	D	CHESHIRE	B	CINTRONA	D	COBEY	B	CONBOY	D
CEDAR BUTTE	C	CHESHNINA	C	CIPRIANO	D	COBURG	C	CONCHAS	C
CEDAREDGE	B	CHESNIMNUS	B	CIRCLE	C	COCHETOPA	C	CONCHO	C
CEDAR MOUNTAIN	D	CHESTER	B	CIRCLEVILLE	C	COCOA	C	CONCONULLY	B
CEDARVILLE	B	CHESTERTON	C	CISNE	D	COCOLALLA	C	CONCORD	D
CEDONIA	B	CHETCO	D	CISPUS	A	CODORUS	C	CONCREEK	B
CEDRON	C/D	CHETEK	B	CITICO	B	CODY	A	CONDA	C
CELAYA	B	CHEVELON	C	CLACKAMAS	C	COEBURN	B	CONDIT	D
CELETON	D	CHEWACLA	C	CLAIBURNE	B	COEROCK	D	CONDON	C
CELINA	C	CHEWELAH	B	CLAIRE	A	COFF	D	CONE	A
CELIO	A/D	CHEYENNE	B	CLAIREMONT	B	COFFEE	B	CONEJO	C
CELLAR	D	CHIARA	D	CLALLAM	C	COGGON	B	CONESTOGA	B
CENCOVE	B	CHICKASHA	B	CLAM GULCH	D	COGSWELL	C	CONESUS	B
CENTER	C	CHICOPEE	B	CLAMO	C	COHASSET	B	CONGAREE	B
CENTER CREEK	B	CHICOTE	D	CLANTON	C	COHOCTAH	D	CONGER	B
CENTERFIELD	B	CHIGLEY	C	CLAPPER	B	COHOE	B	CONI	D
CENTERVILLE	D	CHILCOTT	D	CLAREMORE	D	COIT	C	CONKLIN	B
CENTRALIA	B	CHILDS	B	CLARENCE	D			CONLEN	B

Name	Grp	Name	Grp	Name	Grp	Name	Grp	Name	Grp	Name	Grp
CONLEY	C	COURT	B	CROWLEY	D	DANSKIN	B	DELLROSE	B		
CONNEAUT	C	COURTHOUSE	D	CROWN	B	DANT	D	DELM	D		
CONNECTICUT		COURTLAND	B	CROWSHAW	B	DANVERS	C	DELMAR	D		
CONNERTON	B	COURTNEY	D	CROZIER	C	DANVILLE	C	DELMITA	C		
CONOTTON	B	COURTROCK	B	CRUCES	D	DANZ	B	DELMONT	B		
CONOVER	B	COUSE	C	CRUCKTON	B	DARCO	A	DELNORTE	C		
CONOWINGO	C	COUSHATTA	B	CRUICKSHANK	C	DARGOL	D	DELPHI	B		
CONRAD	B	COVE	D	CRUME	B	DARIEN	C	DELPHILL	C		
CONROE	B	COVEILO	B	CRUMP	D	DARNELL	C	DELPIEDRA	C		
CONSER	C/D	COVELAND	C	CRUTCH	C	DARNEN	B	DELPINE	D		
CONSTABLE	A	COVELLO	B/C	CRUTCHER	D	DARR	A	DELRAY	A/D		
CONSTANCIA	D	COVENTRY	B	CRUZE	C	DARRET	C	DEL REY	C		
CONSUMO	B	COVEYTOWN	C	CRYSTAL LAKE	B	DARROCH	C	DEL RIO	B		
CONTEE	D	COVINGTON	D	CRYSTAL SPRINGS	D	DARROUZETT	C	DELSON	E		
CONTINE	C	COWAN	A	CRYSTOLA	B	DART	A	DELTA	B		
CONTINENTAL	C	COWARTS	C	CUBA	B	DARVADA	D	DELTON	B		
CONTRA COSTA	C	COWDEN	D	CUBERANT	B	DARWIN	D	DELWIN	A		
CONVENT	C	COWDREY	C	CUCHILLAS	D	DASSEL	D	DELYNDIA	A		
COOK	D	COWEEMAN	D	CUDAHY	D	DAST	C	DEMAST	B		
COOKPORT	C	COWERS	B	CUERO	B	DATEMAN	C	DE MASTERS	B		
COOLBRITH	B	COWETA	C	CUEVA	D	DATINO	C	DE MAYA	C		
COOLIDGE	B	COWICHE	B	CUEVITAS	D	DATWYLER	C	DEMERS	D		
COOLVILLE	C	COWOOD	C	CULBERTSON	B	DAULTON	D	DEMKY	D		
COOMBS	B	COX	D	CULLEN	C	DAUPHIN		DEMONA	C		
COONEY	B	COXVILLE	D	CULLEOKA	B	DAVEY	A	DEMOPOLIS	C		
COOPER	C	COY	D	CULLO	C	DAVIDSON	B	DEMPSEY	B		
COOTER	C	COYATA	C	CULPEPER	C	DAVIS	B	DEMPSTER	B		
COPAKE	B	COZAD	B	CULVERS	C	DAVISON	B	DENAY	B		
COPALIS	B	CRABTON	B	CUMBERLAND	B	DAVTONE	B	DENHAWKEN	D		
COPELAND	B/D	CRADDOCK	B	CUMLEY	C	DAWES	C	DENISON	C		
COPITA	B	CRADLEBAUGH	D	CUMMINGS	B/D	DAWHOO	B/D	DENMARK	D		
COPLAY		CRAFTON	C	CUNDIYO	B	DAWSON	D	DENNIS	C		
COPPER RIVER	D	CRAGO	B	CUNICO	C	DAXTY	C	DENNY	D		
COPPERTON	B	CRAGOLA	D	CUPPER	B	DAY	D	DENROCK	D		
COPPOCK	B	CRAIG	C	CURANT	B	DAYBELL	A	DENTON	D		
COPSEY	D	CRAIGMONT	C	CURDLI	C	DAYTON	D	DENVER	C		
COQUILLE	C/D	CRAIGSVILLE	A	CURECANTI	B	DAYVILLE	B/C	DEODAR	D		
CORA	D	CRAMER	D	CURHOLLOW	D	DAZE	D	DEPEW	C		
CORAL	C	CRANE	B	CURLEW	C	DEACON	B	DEPOE	D		
CORBETT	B	CRANSTON	B	CURRAN	C	DEADFALL	B	DEPORT	D		
CORBIN	B	CRARY	C	CURTIS CREEK	D	DEAMA	C	DERA	B		
CORCEGA	C	CRATER LAKE	B	CURTIS SIDING	A	DEAN	C	DERINDA	C		
CORD	C	CRAVEN	C	CUSHING	B	DEAN LAKE	C	DERR	C		
CORDES	B	CRAWFORD	D	CUSHMAN	C	DEARDURFF	B	DERRICK	B		
CORDOVA	C	CREAL	D	CUSTER	C	DEARY	C	DESAN	A		
CORINTH	C	CREBBIN	C	CUTTER	D	DEARYTON	B	DESART	C		
CORKINDALE	B	CREDO	C	CUTZ	D	DEATMAN	C	DESCALABRADO	D		
CORLENA	A	CREEDMAN	D	CUYAMA	B	DEAVER	C	DESCHUTES	C		
CORLETT	B	CREEDMOOR	C	CUYON	A	DEBENGER	C	DESERET	C		
CORLEY	C	CREIGHTON	B	CYAN	D	DEBORAH	D	DESERTER	B		
CORMANT	C	CRELDON	B	CYLINDER	B	DECAN	D	DESHA	D		
CORNHILL	B	CRESBARD	C	CYNTHIANA	C/D	DECATHON	D	DESHLER	C		
CORNING	D	CRESCENT	B	CYPREMORT	C	DECATUR	B	DESOLATION	C		
CORNISH	B	CRESCO	C	CYRIL	B	DECCA	B	DESPAIN	B		
CORNUTT	C	CRESPIN	C			DECKER	C	DETER	C		
CORNVILLE	B	CREST	C	DABOB	B	DECKERVILLE	C	DETLOR	C		
COROZAL	C	CRESTLINE	B	DACONO	C	DECLO	B	DETOUR	C		
CORPENING	D	CRESTMORE		DACOSTA	D	DSCORRA	B	DETRA	B		
CORRALITOS	A	CRESTON	A	DADE	A	DECROSS	B	DETROIT	C		
CORRECO	C	CRESWELL	C	DAFTER	B	DGE	C	DEV	B		
CORRERA	D	CRETE	D	DAGFLAT	C	DEEPWATER	C	DEVILS DIVE	D		
CORSON	C	CREVA	D	DAGGETT	A	DEER CREEK	C	DEVOE	D		
CURTADA	B	CREVASSE	A	DAGLUM	D	DEERFIELD	B	DEVOIGNES	C/D		
CORTEZ	D	CREWS	D	DAGOR	B	DEERFORD	D	DEVOL	B		
CORTINA	A	CRIDER	B	DAGUAO	C	DEERING	B	DEVON	B		
CORUNNA	D	CRIM	B	DAGUEY	C	DEERLODGE	D	DEVORE	B		
CORVALLIS	B	CRISFIELD	B	DAHLQUIST	B	DEER PARK	A	DEVOY	D		
CORWIN	B	CRITCHELL	B	DAIGLE	C	DEERTON	B	DEWART	B		
CORY	C	CRIVITZ	A	DAILEY	A	DEERTRAIL	C	DEWEY	B		
CORYDON	C	CROCKER	A	DAKOTA	B	DEFIANCE	D	DEWVILLE	B		
COSAD	C	CROCKETT	D	DALBO	B	DEFORD	D	DEXTER	B		
COSH	C	CROESUS	C	DALBY	D	DEGARMO	B/C	DIA	C		
COSHOCTON	C	CROFTON	B	DALCAN	B	DEGNER	C	DIABLO	C		
COSKI	B	CROGHAN	B	DALE	B	DE GREY	D	DIAMOND	D		
COSSAYUNA	C	CROOKED	C	DALHART	B	DEJARNET	B	DIAMOND SPRINGS	C		
COSTILLA	A	CROOKED CREEK	D	DALIAN	B	DEKALB	C	DIAMONDVILLE	C		
COTACO	C	CROOKSTON	B	DALLAM	B	DEKOVEN	D	DIANEV	C		
COTATI	C	CROOM	B	DALTON	C	DELA	B	DIANOLA	D		
COTITO	C	CROPLEY	D	DALUPE	B	DELAKE	B	DIAZ	C		
COTO	C	CROSBY	C	DAMASCUS	D	DELANCO	C	DIBBLE	C		
COTOPAXI	A	CROSS	D	DAMON	D	DELANEY	A	DICK	A		
COTT	B	CROSSVILLE	B	DANA	B	DELANO	B/C	DICKEY	A		
COTTER	B	CROSWELL	A	DANBURY	C	DELECO	D	DICKINSON	A		
COTTERAL	B	CROT	D	DANBY		DELENA	D	DICKSON	C		
COTTIER	B	CROTON	D	DANDREA	C	DELFINA	B	DIGBY	C		
COTTONWOOD	C	CROUCH	B	DANDRIDGE	D	DELHI	A	DIGGER	C		
COTTRELL	C	CROW	C	DANGBERG	D	DELICIAS	B	DIGHTON	B		
COUCH	C	CROW CREEK	B	DANIC	C	DELKS	B/D	DILL	B		
COUGAR	D	CROWFOOT	B	DANIELS	B	DELL	C	DILLARD	C		
COULSTONE	B	CROWHEART	D	DANKO	D	DELLEKER	B	DILLDOWN			
COUNTS	C	CROW HEART	D	DANLEY	C	DELLO	A/C	DILLINGER	B		
COUPEVILLE	C	CROW HILL	C	DANNEMORA	D			DILLON	D		

Name	Group	Name	Group	Name	Group	Name	Group	Name	Group
DILLWYN	A	DOUGHTY	A	DU PAGE	B	EGBERT	B/C	EMILY	B
DILMAN	C	DOUGLAS	B	DUPEE	C	EGELAND	B	EMLIN	B
DILTS	D	DOURO	B	DUPLIN	C	EGGLESTON	B	EMMA	C
DILWORTH	D	DOVER	B	DUPO	C	EGNAR	C	EMMERT	A
DIMAL	D	DOVRAY	D	DUPONT	D	EICKS	C	EMMET	B
DIMYAW	C	DOW	B	DUPREE	D	EIFORT	C	EMMONS	C
DINGLE	B	DOWAGIAC	B	DURALDE	C	EKAH	C	EMORY	B
DINGLISHMA	D	DOWDEN	C	DURAND	B	EKALAKA	B	EMPEDRADO	B
DINKELMAN	B	DOWELLTON	D	DURANT	D	ELAM	A	EMPEY	B
DINKEY	A	DOWNER	B	DURELLE	B	ELBERT	D	EMPEYVILLE	C
DINNEN	B	DOWNEY	B	DURHAM	B	ELBURN	B	EMPIRE	C
DINSDALE	B	DOWNS	B	DURKEE	C	ELCO	B	EMRICK	B
DINUBA	B/C	DOXIE	C	DUROC	B	ELD	B	ENCE	B
DINZER	B	DOYCE	C	DURRSTEIN	D	ELDER	B	ENCIERRO	D
DIOXICE	B	DOYLE	A	DUSTON	B	ELDER HOLLOW	D	ENCINA	B
DIPMAN	D	DOYLESTOWN	D	DUTCHESS	B	ELDERON	B	ENDERS	C
DIQUE	B	DOYN	C	DUTSON	D	ELDON	B	ENDERSBY	B
DISABEL	D	DRA	C	DUTTON	D	ELDORADO	C	ENDICOTT	C
DISAUTEL	B	DRACUT	C	DUVAL	B	ELDRIDGE	C	ENET	B
DISCO	B	DRAGE	B	DUZEL	B	ELEPHANT	D	ENFIELD	B
DISHNER	D	DRAGOON	B	DWIGHT	D	ELEROY	B	ENGLE	B
DISTERHEFF	C	DRAGSTON	C	DWYER	A	ELFRIDA	B	ENGLESIDE	B
DITCHCAMP	C	DRAHAT	D	DYE	D	ELIJAH	C	ENGLEWOOD	C
DITHOD	C	DRAIN	D	DYER	B	ELIOAK	C	ENGLUND	D
DIVERS	B	DRAKE	B	DYKE	B	ELK	B	ENNIS	B
DIVIDE	B	DRANYON	B	DYRENG	D	ELKADER	B	ENOCHVILLE	B/D
DIX	A	DRAPER	C			ELKCREEK	C	ENOLA	B
DIXIE	C	DRESDEN	B	EACHUSTON	D	ELK HOLLOW	B	ENON	C
DIXMONT	C	DRESSLER	C	EAD	C	ELKHORN	B	ENOREE	D
DIXMORE	B	DREWS	B	EAGAR	B	ELKINS	D	ENOS	B
DIXONVILLE	C	DREXEL	B	EAGLECONE	B	ELKINSVILLE	B	ENOSBURG	D
DIXVILLE	A	DRIFTON	C	EAKIN	B	ELKMOUND	C	ENSENADA	B
DOAK	B	DRIGGS	B	EAMES	B	ELK MOUNTAIN	B	ENSIGN	D
DOBBS	C	DRUM	C	EARLE	D	ELKOL	D	ENSLEY	D
DOBEL	D	DRUMMER	B	EARLMONT	B/C	ELKTON	D	ENSTROM	B
DOBROW	D	DRUMMOND	D	EARP	B	ELLABELLE	B/D	ENTENTE	B
DOBY	D	DRURY	B	EASLEY	D	ELLEDGE	C	ENTERPRISE	B
DOCAS	B	DRYAD	C	EAST FORK	C	ELLERY	D	ENTIAT	D
DOCKERY	C	DRYBURG	B	EAST LAKE	A	ELLETT	D	ENUMCLAW	C
DOCT	B	DRY CREEK	C	EASTLAND	C	ELLIBER	A	EPHRAIM	C
DODGE	B	DRYDEN	B	EASTON	C	ELLICOTT	A	EPHRATA	B
DODGEVILLE	B	DRY LAKE	C	EASTONVILLE	A	ELLINGTON	B	EPLEY	B
DODSON	C	DUANE	B	EAST PARK	D	ELLINOR	C	EPOUFETTE	D
DOGER	A	DUART	C	EASTPORT	A	ELLIOTT	C	EPPING	D
DOGUE	C	DUBAKELLA	C	EATONTOWN	C	ELLIS	D	EPSIE	D
DOLAND	B	DUBAY	D	EAUGALLIE	B/D	ELLISFORDE	C	ERA	B
DOLE	C	DUBBS	B	EBA	C	ELLISON	B	ERAM	C
DOLLAR	B	DUBOIS	C	EBBERT	D	ELLOAM	D	ERBER	C
DOLLARD	C	DUBUQUE	B	EBBS	B	ELLSBERRY	C	ERIC	B
DOLORES	B	DUCEY	B	EBENEZER	C	ELLSWORTH	C	ERIE	C
DOLPH	C	DUCHESNE	B	ECCLES	B	ELLUM	B	ERIN	B
DOMEZ	C	DUCKETT	C	ECHARD	C	ELMA	B	ERNEST	C
DOMINGO	C	DUCOR	D	ECHLER	B	ELMDALE	B	ERNO	B
DOMINGUEZ	C	DUDA	A	ECKERT	D	ELMENDORF	D	ERRAMOUSPE	C
DOMINIC	A	DUDLEY	D	ECKLEY	B	ELMIRA	A	ESCABOSA	C
DOMINO	C	DUEL	B	ECKMAN	B	ELMO	C	ESCAL	B
DOMINSON	A	DUELM	C	ECKRANT	D	ELMONT	B	ESCALANTE	B
DONA ANA	B	DUFFAU	B	ECTOR	D	ELMORE	B	ESCAMBIA	C
DONAHUE	C	DUFFER	D	EDALGO	C	ELMWOOD	C	ESCONDIDO	C
DONALD	B	DUFFIELD	B	EDDS	B	ELNORA	B	ESMOND	B
DONAVAN	B	DUFFSON	B	EDDY	C	ELOIKA	B	ESPARTO	B
DONEGAL		DUFFY	B	EDEN	C	ELPAN	D	ESPIL	D
DONERAIL	C	DUFUR	B	EDENTON	C	EL PECO	C	ESPINAL	A
DONEY	C	DUGGINS	D	EDENVALE	D	EL RANCHO	B	ESPLIN	D
DONICA	A	DUGOUT	D	EDGAR	B	ELRED	B/D	ESPY	C
DONLONTON	C	DUGWAY	D	EDGECUMBE	B	ELROSE	B	ESQUATZEL	B
DONNA	D	DUKES	A	EDGELEY	C	ELS	A	ESS	B
DONNAN	C	DULAC	C	EDGEMONT	B	ELSAH	B	ESSEN	C
DONNAROO	B	DUMAS	B	EDGEWATER	C	ELSINBORO	B	ESSEX	C
DONNYBROOK	D	DUMECQ	C	EDGEWICK	B	ELSINORE	A	ESSEXVILLE	D
DONOVAN	B	DUMONT	B	EDGEWOOD	A	ELSMERE	A	ESTACADO	B
DOOLEY	A	DUNBAR	D	EDGINGTON	C	ELSO	D	ESTELLINE	B
DOONE	B	DUNBARTON	C	EDINA	D	EL SOLYO	C	ESTER	D
DOOR	B	DUNBRIDGE	B	EDINBURG	C	ELSTON	B	ESTERBROOK	B
DORA	D	DUNCAN	D	EDISON	B	ELTOPIA	B	ESTHERVILLE	B
DORAN	C	DUNCANNON	B	EDISTO	C	ELTREE	B	ESTIVE	C
DORCHESTER	B	DUNCOM	D	EDITH	A	ELTSAC	D	ESTO	B
DOROSHIN	D	DUNDAS	C	EDLOE	B	ELWHA	B	ESTRELLA	B
DOROTHEA	C	DUNDAY	A	EDMORE	D	ELWOOD	C	ETHAN	B
DOROVAN	D	DUNDEE	C	EDMUND	C	ELY	B	ETHETE	B
DORS	B	DUNELLEN	B	EDNA	C	ELYSIAN	B	ETHRIDGE	C
DORSET	B	DUNE SAND	A	EDNEYVILLE	B	ELZINGA	B	ETIL	A
DOS CABEZAS	C	DUNGENESS	B	EDOM	C	EMBDEN	B	ETNA	B
DOSS	C	DUN GLEN	C	EDROY	D	EMBRY	C	ETOE	B
DOSSMAN	B	DUNKINSVILLE	B	EDSON	B	EMBUDO	B	ETOWAH	B
DOTEN	D	DUNKIRK	B	EDWARDS	B/D	EMDENT	C	ETOWN	B
DOTHAN	B	DUNLAP	B	EEL	C	EMER	C	ETSEL	D
DOTTA	B	DUNMORE	B	EFFINGTON	D	EMERALD	B	ETTA	C
DOTY	B	DUNNING	C	EFWUN	A	EMERSON	B	ETTER	B
DOUBLETOP	B	DUNPHY	C	EGAM	C	EMIDA	D	ETTERSBURG	B
DOUDS	B	DUNUL	A	EGAN	B	EMIGRANT	B	ETTRICK	D
DOUGHERTY	A	DUNVILLE	B			EMIGRATION	D	EUBANKS	B

Name	Group	Name	Group	Name	Group	Name	Group	Name	Group
EUDORA	B	FE	D	FLOWELL	C	FRENCH	C	GARLOCK	C
EUFAULA	A	FEDORA	B	FLOWEREE	B	FRENCHTOWN	D	GARMON	C
EUREKA	D	FELAN	A	FLOYD	B	FRENEAU	C	GARMORE	B
EUSTIS	A	FELDA	B/D	FLUETSCH	C	FRESNO	C/D	GARNER	D
EUTAW	D	FELIDA	B	FLUSHING		FRIANA	D	GARO	D
EVANGELINE	C	FELKER	D	FLUVANNA	C	FRIANT	D	GARR	D
EVANS	B	FELT	B	FLYGARE	B	FRIDLO	C	GARRARD	B
EVANSTON	B	FELTA	C	FLYNN	D	FRIEDMAN	B	GARRETSON	B
EVARO	A	FELTHAM	A	FOARD	D	FRIENDS	D	GARRETT	B
EVART	D	FELTON	B	FOGELSVILLE	B	FRIES	D	GARRISON	B
EVENDALE	C	FELTONIA	B	FOLA	B	FRINDLE	B	GARTON	C
EVERETT	B	FENCE	B	FOLEY	D	FRIO	B	GARWIN	C
EVERGLADES	A/D	FENDALL	C	FONDA	D	FRIZZELL	C	GASCONADE	D
EVERLY	B	FENWOOD	B	FONDIS	C	FROBERG	D	GAS CREEK	C
EVERMAN	C	FERA	C	FONTAL	D	FROHMAN	C	GASKELL	C
EVERSON	D	FERDELFORD	C	FONTREEN	B	FRONDORF	C	GASS	D
EVESBORO	A	FERDIG	C	FOPIANO	D	FRONHOFER	C	GASSET	D
EWA	B	FERDINAND	C	FORBES	B	FRONTON	D	GATESBURG	A
EWAIL	A	FERGUS	B	FORD	D	FROST	D	GATESON	C
EWALL	A	FERGUSON	B	FORDNEY	A	FRUITA	B	GATEVIEW	B
EWINGSVILLE	B	FERNANDO	B	FORDTRAN	C	FRUITLAND	B	GATEWAY	C
EXCELSIOR	B	FERN CLIFF	B	FORDVILLE	B	FRYE	C	GATEWOOD	D
EXCHEQUER	D	FERNDALE	B	FORE	D	FUEGO	C	GAULDY	B
EXETER	C/D	FERNLEY	C	FORELAND	D	FUERA	C	GAVINS	C
EXLINE	D	FERNOW	B	FORELLE	B	FUGAWEE	B	GAVIOTA	D
EXRAY	D	FERNPOINT	C	FORESMAN	B	FULCHER	C	GAY	D
EXUM	C	FERRELO	B	FORESTDALE	D	FULDA	C	GAYLORD	B
EYEBROW	D	FERRIS	D	FORESTER	C	FULLERTON	B	GAYNOR	C
EYRE	B	FERRON	D	FORESTON	C	FULMER	B/D	GAYVILLE	B
		FERTALINE	D	FORGAY	A	FULSHEAR	C	GAZELLE	D
FABIUS	B	FESTINA	B	FORMAN	B	FULTON	D	GAZOS	B
FACEVILLE	B	FETT	D	FORNEY	D	FUQUAY	B	GEARHART	A
FAHEY	B	FETTIC	D	FORREST	C	FURNISS	B/D	GEARY	B
FAIM	C	FIANDER	C	FORSEY	C	FURY	B/D	GEE	B
FAINES	A	FIBEA	D	FORSGREN	C	FUSULINA	C	GEEBURG	C
FAIRBANKS	B	FIDALGO	C	FORT COLLINS	B			GEER	C
FAIRDALE	B	FIDDLETOWN	C	FORT DRUM	C	GAASTRA	C	GEFO	A
FAIRFAX	B	FIDDYMENT	C	FORT LYON	B	GABALDON	B	GELKIE	B
FAIRFIELD	B	FIELDING	B	FORT MEADE	A	GABBS	D	GEM	C
FAIRHAVEN	B	FIELDON	B	FORT MOTT	A	GABEL	C	GEMID	C
FAIRMOUNT	D	FIELDSON	A	FORT PIERCE	C	GABICA	D	GEMSON	C
FAIRPORT	C	FIFE	B	FORT ROCK	C	GACEY	D	GENESEE	B
FAIRYDELL	C	FIFER	D	FORTUNA	D	GACHADO	D	GENEVA	C
FAJARDO	C	FILLMORE	D	FORTWINGATE	C	GADDES	C	GENOA	D
FALAYA	C	FINCASTLE	C	FORWARD	C	GADES	G	GENOLA	B
FALCON	D	FINGAL	C	FOSHOME	B	GADSDEN	D	GEORGEVILLE	B
FALFA	C	FINLEY	B	FOSSUM	B	GAGE		GEORGIA	B
FALFURRIAS	A	FIRESTEEL	B	FOSTER	B/C	GAGEBY	B	GERALD	D
FALK	B	FIRGRELL	B	FOSTORIA	B	GAGETOWN	C	GERBER	D
FALKNER	C	FIRMAGE	B	FOUNTAIN	D	GAHEE	B	GERIG	B
FALL	B	FIRO	D	FOURLOG	D	GAINES	C	GERING	B
FALLBROOK	B/C	FIRTH	B/C	FOURMILE	B	GAINESVILLE	A	GERLAND	C
FALLON	C	FISH CREEK	B	FOUR STAR	B/C	GALATA	D	GERMANIA	
FALLSBURG	C	FISHERS	B	FOUTS	B	GALE	B	GERMANY	B
FALLSINGTON	D	FISHHOOK	D	FOX	B	GALEN	B	GERRARD	D
FANCHER	C	FISHKILL		FOXCREEK	B/D	GALENA	C	GESTRIN	B
FANG	B	FITCH	A	FOXMOUNT	C	GALEPPI	C	GETTA	C
FANNIN	B	FITCHVILLE	C	FOXOL	D	GALESTOWN	A	GETTYS	C
FANNO	C	FITZGERALD	B	FOXPARK	D	GALETON	D	GEYSEN	D
FANU	C	FITZHUGH	B	FOX PARK	D	GALEY	B	GHENT	C
FARADAY	B	FIVE DOT	B	FOXTON	C	GALISTEO	C	GIBBLER	C
FARALLONE	B	FIVEMILE	B	FRAILEY	B	GALLAGHER	B	GIBBON	B
FARAWAY	D	FIVES	B	FRAM	B	GALLATIN	A	GIBBS	D
FARB	D	FLAGG	B	FRANCIS	A	GALLEGOS	B	GIBBSTOWN	A
FARGO	D	FLAGSTAFF	C	FRANCITAS	D	GALLINA	C	GIFFIN	C
FARISITA	C	FLAK	B	FRANK	D	GALLION	B	GIFFORD	C
FARLAND	B	FLAMING	B	FRANKFORT	D	GALVA	B	GILA	B
FARMINGTON	C/D	FLAMINGO	D	FRANKIRK	C	GALVESTON	A	GILBY	B
FARNHAM	B	FLANAGAN	B	FRANKLIN	B	GALVEZ	C	GILCHRIST	B
FARNHAMTON	B/C	FLANDREAU	B	FRANKSTOWN	B	GALVIN	C	GILCREST	B
FARNUF	B	FLASHER	A	FRANKTOWN	D	GALWAY	B	GILEAD	C
FARNUM	B	FLATHEAD	A	FRANKVILLE	B	GAMBLER	A	GILES	B
FARRAGUT	C	FLAT HORN	D	FRATERNIDAD	D	GAMBOA	B	GILFORD	B/D
FARRAR	B	FLATTOP	D	FRAZER	C	GANNETT	D	GILHOULY	B
FARRELL	B	FLATWILLOW	B	FRED	C	GANSNER	D	GILISPIE	C
FARRENBURG	B	FLAXTON	A	FREDENSBORG	C	GAPO	D	GILLIAM	C
FARROT	C	FLEAK	A	FREDERICK	B	GAPPMAYER	B	GILLIGAN	B
FARSON	B	FLECHADO	B	FREDON	C	GARA	B	GILLS	C
FARWELL	C	FLEER	D	FREDONIA	C	GARBER	A	GILLSBURG	C
FASKIN	B	FLEETWOOD		FREDRICKSON	C	GARBUTT	B	GILMAN	B
FATIMA	B	FLEISCHMANN	D	FREEBURG	C	GARCENO	C	GILMORE	C
FATTIG	C	FLEMING	C	FREECE	D	GARDELLA	D	GILPIN	C
FAUNCE	A	FLETCHER	B	FREEDOM	C	GARDENA	B	GILROY	C
FAUQUIER	C	FLOKE	D	FREEHOLD	B	GARDINER	A	GILSON	B
FAUSSE	D	FLOM	C	FREEL	B	GARDNER'S FORK	B	GILT EDGE	D
FAWCETT	C	FLOMATION	A	FREEMAN	C	GARDNERVILLE	D	GINAT	D
FAWN	B	FLOMOT	B	FREEMANVILLE	B	GARDONE	A	GINGER	C
FAXON	D	FLORENCE	B	FREEON	B	GAREY	C	GINI	B
FAYAL	C	FLORESVILLE	C	FREER	C	GARFIELD	C	GINSER	C
FAYETTE	B	FLORIDANA	B/D	FREESTONE	C	GARITA	C	GIRARDOT	D
FAYETTEVILLE	B	FLORISSANT	C	FREEZENER	C	GARLAND	B	GIRD	A
FAYWOOD	C			FREMONT	C	GARLET	A	GIVEN	C

Soil	Group	Soil	Group	Soil	Group	Soil	Group	Soil	Group
GLADDEN	A	GOTHARD	D	GROWDEN	B	HAMBRIGHT	D	HASTINGS	B
GLADE PARK	C	GOTHIC	C	GROWLER	B	HAMBURG	B	HAT	D
GLADSTONE	B	GOTHO	C	GRUBBS	D	HAMBY	C	HATBORO	C
GLADWIN	A	GOULDING	D	GRULLA	D	HAMEL	C	HATCH	C
GLAMIS	C	GOVAN	C	GRUMMIT	D	HAMERLY	C	HATCHERY	C
GLANN	B/C	GOVE	B	GRUNDY	C	HAMILTON	A	HATFIELD	C
GLASGOW	C	GOWEN	B	GRUVER	C	HAMLET	B	HATHAWAY	B
GLEAN	B	GRABE	B	GRYGLA	C	HAMLIN	B	HATTIE	C
GLEASON	C	GRABLE	B	GUADALUPE	B	HAMMONTON	C	HATTON	C
GLEN	B	GRACEMONT	B	GUAJE	A	HAMPDEN	A	HAUBSTADT	C
GLENBAR	B	GRACEVILLE	A	GUALALA	D	HAMPSHIRE	C	HAUGAN	B
GLENBERG	B	GRADY	D	GUAMANI	B	HAMPTON	C	HAUSER	D
GLENBROOK	D	GRAFEN	B	GUANABANO	B	HANTAH	C	HAVANA	B
GLENCOE	D	GRAFTON	B	GUANAJIBO	C	HANA	A	HAVEN	B
GLENDALE	B	GRAHAM	D	GUANICA	D	HANALEI	A	HAVERLY	B
GLENDIVE	B	GRAIL	C	GUAYABO	B	HANAMAULU	A	HAVERSON	B
GLENDORA	D	GRAMM	B	GUAYABOTA	D	HANCEVILLE	B	HAVILLAH	C
GLENELG	B	GRANATH	B	GUAYAMA	D	HANCO	D	HAVINGDON	D
GLENFIELD	D	GRANBY	A/D	GUBEN	B	HANO	B	HAVRE	B
GLENFORD	C	GRANDE RONDE	D	GUCKEEN	C	HANDRAN	B	HAVRELON	B
GLENHALL	B	GRANDFIELD	B	GUELPH	B	HANDSBORO	D	HAW	B
GLENHAM	B	GRANDVIEW	C	GUENOC	C	HANDY	D	HAWES	A
GLENMORA	C	GRANER	C	GUERNSEY	C	HANEY	B	HAWI	B
GLENNALLEN	C	GRANGER	C	GUERRERO	C	HANFORD	B	HAWKEYE	A
GLENOMA	B	GRANGEVILLE	B/C	GUEST	D	HANGAARD	C	HAWKSELL	A
GLENROSE	B	GRANILE	B	GUIN	A	HANGER	B	HAWKSPRINGS	A
GLENSTED	D	GRANO	D	GULER	B	HANIPOE	B	HAXTUN	A
GLENTON	B	GRANT	B	GULKANA	B	HANKINS	C	HAYBOURNE	B
GLENVIEW	B	GRANTSBURG	C	GUMBOOT	C	HANKS	B	HAYBRO	C
GLENVILLE	C	GRANTSDALE	A	GUNBARREL	A	HANLY	A	HAYDEN	B
GLIDE	B	GRANVILLE	B	GUNN	B	HANNA	B	HAYESTON	B
GLIKON	B	GRAPEVINE	C	GUNNUK	C	HANNUM	D	HAYESVILLE	B
GLORIA	C	GRASMERE	B	GUNSIGHT	B	HANOVER	C	HAYFIELD	B
GLOUCESTER	A	GRASSNA	B	GUNTER	A	HANS	C	HAYFORD	C
GLOVER	C/D	GRASSY BUTTE	A	GURABO	D	HANSEL	C	HAYMOND	B
GLYNDON	B	GRATZ	C	GURNEY	B	HANSKA	C	HAYNESS	B
GLYNN	C	GRAVDEN	C	GUSTAVUS	D	HANSON	A	HAYNIE	B
GOBLE	C	GRAVE	B	GUSTIN	C	HANTHO	B	HAYPRESS	A
GODDARD	B	GRAVITY	C	GUTHRIE	D	HANTZ	D	HAYSPUR	B/D
GODDE	D	GRAYCALM	A	GUYTON	D	HAP	B	HAYTER	B
GJOECKE	D	GRAYFORD	B	GWIN	D	HAPGOOD	B	HAYTI	D
GODFREY	C	GRAYLING	A	GWINNETT	B	HAPNEY	C	HAYWOOD	B
GODWIN	D	GRAYLOCK	B	GYMER	C	HARBORD	B	HAZEL	C
GOEGLEIN	C	GRAYPOINT	B	GYPSTRUM	B	HARBOURTON		HAZELAIR	D
GOESSEL	D	GRAYS	B			HARCO	B	HAZEN	B
GOFF	C	GREAT BEND	B	HACCKE	C	HARDEMAN	B	HAZLEHURST	B
GOGEBIC	B	GREELEY	B	HACIENDA	D	HARDESTY	B	HAZLETON	B
GOLBIN	C	GREEN BLUFF	B	HACK	B	HARDING	D	HAZTON	D
GOLCONDA	D	GREENBRAE	C	HACKERS	B	HARDSCRABBLE	B	HEADLEY	B
GOLD CREEK	D	GREEN CANYON	B	HACKETTSTOWN	B	HARDY	D	HEADQUARTERS	B
GOLDENDALE	B	GREENCREEK	B	HADAR	A	HARGREAVE	B	HEAKE	D
GOLDFIELD	B	GREENDALE	B	HADES	C	HARKERS	C	HEATH	C
GOLDHILL	B	GREENFIELD	B	HADLEY	B	HARKEY	B	HEATLY	A
GOLDMAN	C	GREENHORN	D	HADO	B	HARLAN	B	HEBBRONVILLE	B
GOLDRIDGE	B	GREENLEAF	B	HAGEN	B	HARLEM	C	HEBER	B
GOLDRUN	A	GREENOUGH	C	HAGENBARTH	B	HARLESTON	C	HEBERT	A
GULDSBORO	C	GREENPORT		HAGENER	A	HARLINGEN	D	HEBGEN	C
GOLDSTON	C	GREEN RIVER	B	HAGER	C	HARMEHL	C	HEBO	D
GOLDSTREAM	D	GREENSBORO		HAGERMAN	C	HARMONY	C	HEBRON	C
GOLDVALE	C	GREENSON	C	HAGERSTOWN	C	HARNEY	C	HECHT	C
GOLDVEIN	C	GREENTON	C	HAGGA	B	HARPER	D	HECKI	B
GOLIAD	C	GREENVILLE	B	HAGGERTY	B	HARPETH	B	HECLA	D
GOLLAHER	A	GREENWATER	A	HAGSTADT	C	HARPS	B	HECTOR	C
GOLTRY	A	GREENWICH	B	HAGUE	A	HARPSTER	C	HEDDEN	B
GOMEZ	B	GREENWOOD	D	HAIG	C	HARPT	B	HEDRICK	B
GOMM	D	GREER	C	HAIKU	B	HARQUA	C	HEDVILLE	D
GONVICK	B	GREGORY	A	HAILMAN	B	HARRIET	D	HEGNE	D
GOOCH	D	GREHALEM	B	HAINES	B/C	HARRIMAN	B	HEIDEN	D
GOODALE	C	GRELL	D	HAIRE	C	HARRIS	D	HEIDTMAN	C
GOODING	C	GRENADA	C	HALAWA	B	HARRISBURG	D	HEIL	D
GOODINGTON	C	GRENVILLE	B	HALDER	C	HARRISON	C	HEIMDAL	B
GOODLOW	B	GRESHAM	C	HALE	B	HARRISVILLE	C	HEISETON	B
GOODMAN	B	GREWINGK	D	HALEDON	D	HARSTENE	B	HEISLER	B
GOODRICH	B	GREYBACK	B	HALEIWA	B	HARSTINE	C	HEIST	C
GOODSPRINGS	D	GREYBULL	C	HALEY	B	HART	D	HEITT	C
GOOSE CREEK	B	GREYCLIFF	C	HALF MOON	B	HART CAMP	C	HEITZ	D
GOOSE LAKE	D	GREYS	B	HALFORD	A	HARTFORD	A	HEIZER	D
GOOSMUS	B	GRIFFY	B	HALFWAY	D	HARTIG	B	HELDT	C
GORDO	B	GRIGSTON	B	HALGAITOH	B	HARTLAND	B	HELEMANO	C
GORDON	D	GRIMSTAD	B	HALII	B	HARTLETON	B	HELENA	C
GORE	D	GRISWOLD	B	HALIIMAILE	B	HARTLINE	B	HELMER	C
GORGONIO	A	GRITNEY	C	HALIS	C	HARTSBURG	B	HELVETIA	B
GORHAM	B	GRIVER	C	HALL	B	HARTSELLS	B	HELY	B
GORIN	C	GRIZZLY	C	HALLECK	B	HARTSHORN	B	HEMBRE	C
GORING	C	GROGAN	B	HALL RANCH	C	HARVARD	B	HEMMI	
GORMAN	B	GROSECLOSE	C	HALLVILLE	B	HARVEL	B	HEMPFIELD	
GORUS	A	GROSS	C	HALSEY	D	HARVEY	C	HEMPSTEAD	C
GORZELL	B	GROTON	A	HAMACER	A	HARWOOD	C	HENCRATT	B
GOSHEN	B	GROVE	A	HAMAKUAPOKO	B	HASKI	B	HENDERSON	B
GOSHUTE	D	GROVELAND	B	HAMAN	B	HASKILL	A	HENDRICKS	
GOSPORT	C	GROVER	B	HAMAR	B	HASKINS	C	HENEFER	C
GOTHAM	A	GROVETON	B	HAMBLEN	C	HASSELL	C	HENKIN	B

Name		Name		Name		Name		Name	
HENLEY	C	HOBOG	D	HORD	B	HYAT	A	IZAGORA	C
HENLINE	C	HOBSON	C	HOREB	B	HYATTVILLE	C	IZEE	C
HENNEKE	D	HOCHHEIM	B	HORNE	D	HYDABURG	D		
HENNEPIN	B	HOCKING	B	HORNELL	D	HYDE	D	JABU	C
HENNINGSEN	C	HOCKINSON	C	HORNING	A	HYDRO	C	JACAGUAS	B
HENRY	D	HOCKLEY	C	HORNITOS	D	HYMAS	D	JACANA	B/D
HENSEL	B	HODGE	B	HORROCKS	B	HYRUM	B	JACINTO	B
HENSHAW	C	HODGINS	C	HORSESHOE	B	HYSHAM	D	JACK CREEK	A
HENSLEY	D	HODGSON	C	HORTON	B			JACKLIN	B
HEPLER	D	HOEBE	B	HORTONVILLE	B	IAO	C	JACKNIFE	C
HERBERT	B	HOELZLE	C	HOSKIN	C	IBERIA	D	JACKPORT	D
HEREFORD	B	HOFFMAN	C	HOSKINNINI	D	ICENE	C	JACKS	C
HERKIMER	B	HOFFMANVILLE	C	HOSLEY	D	IDA	B	JACKSON	B
HERLONG	D	HOGANSBURG	B	HOSMER	C	IDABEL	B	JACKSONVILLE	C
HERMISTON	B	HOGELAND	B	HOTAW	C	IDAK	B	JACOB	D
HERMON	A	HOGG	C	HOT LAKE	C	IDANA	C	JACOBSEN	C
HERNDON	B	HOGRIS	B	HOUDEK	B	IDEON	D	JACOBY	C
HERO	B	HOH	B	HOUGHTON	A/D	IDMON	B	JACQUES	C
HERRERA	A	HOHMANN	C	HOUK	C	IGNACIO	C	JACQUITH	C
HERRICK	C	HOKO	C	HOULKA	D	IGO	D	JACWIN	B
HERRON	B	HOLBROOK	B	HOULTON	C/D	IGUALDAD	D	JAFFREY	A
HERSH	A	HOLCOMB	D	HOUNDBY	D	IHLEN	D	JAGUEYES	B
HERSHAL	B/D	HOLDAWAY	D	HOURGLASS	B	IJAM	D	JAL	A
HESCH	B	HOLDEN	A	HOUSATONIC	D	ILDEFONSO	B	JALMAR	A/C
HESPER	C	HOLDER	B	HOUSE MOUNTAIN	C	ILKA	B	JAMES CANYON	B/C
HESPERIA	B	HOLDERMAN	C	HOUSEVILLE	C	ILLION	B/D	JAMESTOWN	C
HESPERUS	B	HOLDERNESS	C	HOUSTON	D	IMA	B	JANE	C
HESSE	C	HOLDREGE	B	HOUSTON BLACK	D	IMBLER	B	JANISE	C
HESSEL	D	HOLLAND	B	HOVDE	A/C	IMLAY	C	JANSEN	A
HESSELBERG	D	HOLLINGER	B	HOVEN	D	IMMOKALEE	B/D	JARAB	D
HESSELTINE	B	HOLLIS	C/D	HOVENWEEP	C	IMPERIAL	B	JARBOE	C
HESSLAN	C	HOLLISTER	D	HOVERT	D	INAVALE	A	JARITA	C
HESSON	C	HOLLOMAN	C	HOVEY	C	INDART	B	JARRE	B
HETTINGER	D	HOLLOWAY	A	HOWARD	B	INDIAHOMA	D	JARVIS	B
HEXT	B	HOLLY	B	HOWELL	C	INDIAN		JASPER	B
HEZEL	B	HOLLY SPRINGS	D	HOWLAND	C	INDIAN CREEK	D	JAUCAS	A
HIALEAH	D	HOLLYWOOD	D	HOYE	B	INDIANO	C	JAVA	B
HIAWATHA	A	HOLMDEL	C	HOYLETON	C	INDIANOLA	A	JAY	C
HIBBARD	D	HOLMES	B	HOYPUS	A	INDIO	B	JAYEM	B
HIBBING	C	HOLOMUA	B	HOYTVILLE	D	INGA	B	JAYSON	D
HIBERNIA	C	HOLOPAW	B/D	HUBBARD	A	INGALLS	B	JEAN	C
HICKORY	C	HOLROYD	B	HUBERLY	D	INGARD	B	JEANERETTE	D
HICKS	B	HOLSINE	B	HUBERT	B	INGENIO	C	JEAN LAKE	B
HIDALGO	B	HOLST	B	HUBLERSBURG	C	INGRAM	D	JEDD	C
HIDEAWAY	D	HOLSTON	B	HUCKLEBERRY	C	INKLER	B	JEDDO	D
HIDEWOOD	C	HOLT	B	HUDSON	C	INKS	D	JEFFERSON	B
HIERRO	C	HOLTLE	B	HUECO	C	INMACHUK	D	JEKLEY	C
HIGHAMS	D	HOLTVILLE	C	HUEL	A	INMAN	C	JELM	D
HIGHFIELD	B	HOLYOKE	C/D	HUENEME	B/C	INMO	A	JENA	B
HIGH GAP	C	HOMA	C	HUERHUERO	D	INNESVALE	D	JENKINS	B
HIGHLAND	B	HOME CAMP	C	HUEY	D	INSKIP	C	JENKINSON	D
HIGHMORE	B	HOMELAKE	B	HUFFINE	A	INVERNESS	D	JENNESS	B
HIGH PARK	B	HOMER	C	HUGGINS	C	INVILLE	B	JENNINGS	C
HIHIMANU	A	HOMESTAKE	D	HUGHES	B	INWOOD	C	JENNY	D
HIIBNER	C	HOMESTEAD	B	HUGHESVILLE	B	IO	B	JERAULD	D
HIKO PEAK	B	HONAUNAU	C	HUGO	B	IOLA	A	JERICHO	C
HIKO SPRINGS	D	HONCUT	B	HUICHICA	C/D	IOLEAU	C	JEROME	C
HILDRETH	D	HONDALE	D	HUIKAU	A	IONA	B	JERRY	C
HILEA	D	HONDO	C	HULETT	B	IONIA	B	JESBEL	D
HILES	B	HONDOHO	B	HULLS	C	IOSCO	B	JESSE CAMP	C
HILGER	B	HONEOYE	B	HULLT	B	IPAVA	B	JESSUP	C
HILGRAVE	B	HONEY	D	HULUA	D	IRA	C	JETT	B
HILLEMANN	C	HONEYGROVE	C	HUM	B	IREDELL	D	JIGGS	C
HILLERY	D	HONEYVILLE	C	HUMACAO	B	IRETEBA	C	JIM	C
HILLET	D	HONN	B	HUMATAS	C	IRIM	C	JIMENEZ	C
HILLFIELD	B	HONOKAA	A	HUMBARGER	B	IROCK	B	JIMTOWN	C
HILLGATE	D	HONOLUA	B	HUMBIRD	C	IRON BLOSSOM	C	JOB	C
HILLIARD	B	HONOMANU	B	HUMBOLDT	D	IRON MOUNTAIN	D	JOBOS	C
HILLON	B	HONOULIULI	D	HUMDUN	B	IRON RIVER	B	JOCITY	A
HILLSBORO	B	HONUAULU	A	HUME	C	IRONTON	C	JOCKO	B
HILLSDALE	B	HOOD	B	HUMESTON	C	IRRIGON	C	JODERO	B
HILMAR	C/D	HOODLE	B	HUMMINGTON	C	IRVINGTON	C	JOEL	B
HILO	A	HOODSPORT	C	HUMPHREYS	B	IRWIN	D	JOES	C
HILT	B	HOODVIEW	B	HUMPTULIPS	B	ISAAC	C	JOHNS	C
HILTON	B	HOOKTON	C	HUNSAKER	B/C	ISAAQUAH	B/C	JOHNSBURG	D
HINCKLEY	A	HOOLEHUA	B	HUNTERS	B	ISAN	D	JOHNSON	B
HINDES	C	HOOPAL	D	HUNTING	C	ISANTI	D	JOHNSTON	B/D
HINESBURG	C	HOOPER	D	HUNTINGTON	B	ISBELL	C	JOHNSWOOD	B
HINKLE	D	HOOPESTON	B	HUNTSVILLE	B	ISHAM	C	JOICE	D
HINMAN	C	HOOSIC	A	HUPP	B	ISHI PISHI	C	JOLAN	C
HINSDALE		HOOT	D	HURDS	B	ISLAND	B	JOLIET	C
HINTZE	D	HOOTEN	D	HURLEY	D	ISOM	B	JONESVILLE	A
HIPPLE	C	HOOVER	B	HURON	C	ISSAQUAH	B/C	JONUS	B
HISLE	D	HOPEKA	D	HURST	D	ISTOKPOGA	D	JOPLIN	B
HITT	B	HOPETON	C	HURWAL	B	ITCA	D	JOPPA	B
HI VISTA	C	HOPEWELL	C	HUSE	C	ITSWOOT	B	JORDAN	D
HIWASSEE	B	HOPGOOD	C	HUSSA	B/D	IUKA	D	JORGE	B
HIWOOD	A	HOPKINS	B	HUSSMAN	D	IVA	C	JORNADA	C
HIXTON	B	HOPLEY	B	HUTCHINSON	C	IVAN	B	JORY	C
HOBACKER	B	HOPPER	B	HUTSON	B	IVES	B	JOSE	C
HOBAN	C	HOQUIAM	B	HUXLEY	D	IVIE	A	JOSEPHINE	B
HOBBS	B	HORATIO	D	HYAM	D	IVINS	C	JOSIE	B

Soil	Grp	Soil	Grp	Soil	Grp	Soil	Grp	Soil	Grp
JOY	B	KARNAK	D	KEOWNS	D	KIPP	C	KOVICH	D
JUANA DIAZ	B	KARNES	B	KEPLER	C	KIPPEN	A	KOYEN	B
JUBILEE	C	KARRO	B	KERBY	B	KIPSON	A	KOYUKUK	B
JUDD	D	KARS	A	KERMEL	B	KIRK	B/D	KRADE	B
JUDITH	B	KARSHNER	D	KERMIT	A	KIRKHAM	C	KRANZBURG	B
JUDKINS	C	KARTA	C	KERMO	A	KIRKLAND	D	KRATKA	C
JUDSON	B	KARTAR	B	KERR	B	KIRKTON	B	KRAUSE	A
JUDY	C	KASCHMIT	D	KERRICK	B	KIRKVILLE	C	KREAMER	C
JUGET	D	KASHWITNA	B	KERRTOWN		KIRTLEY	C	KREMLIN	B
JUGHANDLE	B	KASILOF	A	KERSHAW	A	KIRVIN	C	KRENTZ	C
JULES	B	KASKI	B	KERSICK	D	KISRING	D	KRESSON	C
JULESBURG	A	KASOTA	C	KERSTON	A/D	KISSICK	D	KRUM	D
JULIAETTA	B	KASSLER	A	KERT	C	KISTLER	C/D	KRUSE	B
JUMPE	B	KASSON	C	KERWIN	C	KITCHELL	B	KRUZOF	B
JUNCAL	C	KATAMA	B	KESSLER	C	KITCHEN CREEK		KUBE	B
JUNCOS	D	KATENCY	C	KESWICK	D	KITSAP	C	KUBLER	C
JUNCTION	B	KATO	C	KETCHLY	B	KITTANNING		KUBLI	C
JUNEAU	B	KATRINE	B	KETTLE	B	KITTITAS	D	KUCERA	B
JUNIATA	B	KATULA	B	KETTLEMAN	B	KITTREDGE	C	KUCK	C
JUNIPERO	B	KATY	C	KETTNER	C	KITTSON	C	KUGRUG	D
JUNIUS	C	KAUFMAN	D	KEVIN	C	KIUP	B	KUHL	D
JUNO	B	KAUPO	A	KEWAUNEE	C	KIVA	B	KUKAIAU	A
JUNQUITOS	C	KAVETT	D	KEWEENAW	A	KIWANIS	A	KULA	B
JURA	C	KAWAIHAE	C	KEYA	B	KIZHUYAK	B	KULAKALA	B/C
JUVA	B	KAWAIHAPAI	B	KEYES	D	KJAR	D	KULLIT	B
JUVAN	D	KAWBAWGAM	C	KEYNER	B	KLABER	C	KUMA	B
		KAWICH	A	KEYPORT	C	KLAMATH	B/D	KUNIA	B
KAALUALU	A	KAWKAWLIN	C	KEYSTONE	A	KLAUS	A	KUNUWEIA	C
KACHEMAK	B	KEAAU	D	KEYTESVILLE	D	KLAWASI	D	KUPREANOF	B
KADAKE	D	KEAHUA	B	KEZAR	B	KLEJ	B	KUREB	A
KADASHAN	B	KEALAKEKUA	C	KIAWAH	C	KLICKER	C	KURO	D
KADE	C	KEALIA	D	KIBBIE	B	KLICKITAT	C	KUSKOKWIM	D
KADIN	B	KEANSBURG	D	KICKERVILLE	B	KLINE	B	KUSLINA	D
KADOKA	B	KEARNS	B	KIDD	D	KLINESVILLE	C/D	KUTCH	D
KAENA	D	KEATING	C	KIDMAN	B	KLINGER	B	KUTZTOWN	B
KAHALUU	D	KEAUKAHA	D	KIEHL	A	KLONDIKE	D	KVICHAK	B
KAHANA	B	KEAWAKAPU	B	KIETZKE	D	KLONE	B	KWETHLUK	A
KAHANUI	B	KEBLER	B	KIEV	B	KLOOCHMAN	C	KYLE	D
KAHLER	B	KECH	D	KIKONI	B	KLOTEN	B	KYLER	D
KAHOLA	B	KECKO	B	KILARC	D	KLUTINA	B		
KAH SHEETS	D	KEDRON	C	KILAUEA	B	KNAPPA	B	LA BARGE	B
KAHUA	D	KEEFERS	C	KILBOURNE	A	KNEELAND	C	LABETTE	C
KAIKLI	D	KEEGAN		KILBURN	B	KNIFFIN	C	LABISH	D
KAILUA	A	KEEI	D	KILCHIS	D	KNIGHT	C	LABOU	D
KAIMU	A	KEEKEE	B	KILDOR	C	KNIK	B	LABOUNTY	C
KAINALIU	A	KEELDAR	B	KILGORE	B/D	KNIPPA	D	LA BOUNTY	C
KAIPOIOI	B	KEENE	C	KILKENNY	B	KNOB HILL	B	LA BRIER	C
KAIWIKI	A	KEENO	C	KILLBUCK	C/D	KNOWLES	B	LABSHAFT	B
KALAE	B	KEESE	D	KILLEY	D	KNOX	B	LACAMAS	C/D
KALALOCH	B	KEG	B	KILLINGWORTH		KNULL	C	LA CASA	C
KALAMA	C	KEHENA	C	KILLPACK	C	KNUTSEN	B	LACITA	B
KALAMAZOO	B	KEIGLEY	C	KILMERQUE	C	KOBAR	C	LACKAWANNA	C
KALAPA	B	KEISER	B	KILN	D	KOBEH	B	LACONA	C
KALAUPAPA	D	KEITH	B	KILOA	A	KOCH	C	LACOTA	D
KALIFONSKY	D	KEKAHA	B	KILOHANA	A	KODAK	C	LACY	D
KALIHI	D	KEKAKE	D	KILWINNING	C	KODIAK	B	LADD	B
KALISPELL	A	KELLER	C	KIM	B	KOEHLER	C	LADDER	D
KALKASKA	A	KELLY	D	KIMAMA	B	KOELE	B	LADELLE	B
KALMIA	B	KELN	C	KIMBALL	C	KOEPKE	B	LADOGA	C
KALOKO	D	KELSEY	D	KIMBERLY	B	KOERLING	B	LADUE	B
KALOLOCH	B	KELSO	C	KIMBROUGH	D	KOGISH	D	LADYSMITH	D
KALSIN	D	KELTNER	B	KIMMERLING	D	KOHALA	A	LA FARGE	B
KAMACK	B	KELVIN	C	KIMMONS	C	KOKEE	B	LAFE	D
KAMAKOA	A	KEMMERER	C	KIMO	C	KOKERNOT	C	LAFITTE	D
KAMAOA	B	KEMOO	B	KINA	D	KOKO	B	LA FONDA	B
KAMAOLE	B	KEMPSVILLE	B	KINCO	A	KOKOKAHI	D	LAFONT	B
KAMAY	D	KEMPTON	B	KINESAVA	C	KOKOMO	B/D	LAGLORIA	B
KAMIE	B	KENAI	C	KINGFISHER	B	KOLBERG	B	LAGONDA	C
KAMRAR	B	KENANSVILLE	A	KINGHURST	B	KOLEKOLE	C	LA GRANDE	C
KANABEC	B	KENDAIA	C	KINGMAN	D	KOLLS	D	LAGRANGE	D
KANAKA	B	KENDALL	B	KINGS	C/D	KOLLUTUK	D	LAHAINA	B
KANAPAHA	A/D	KENDALLVILLE	B	KINGSBURY	C	KOLOA	C	LA HOGUE	B
KANDIK	B	KENESAW	B	KINGSLEY	B	KOLOB	C	LAHONTAN	D
KANE	B	KENMOOR	B	KINGS RIVER	C	KOLOKOLO	B	LAHRITY	A
KANEOHE	B	KENNALLY	B	KINGSTON	B	KONA	D	LAIDIG	C
KANEPUU	B	KENNAN	B	KINGSVILLE	C	KONAWA	B	LAIDLAW	B
KANIMA	C	KENNEBEC	B	KINKEAD	C	KONNER	D	LAIL	C
KANLEE	B	KENNEDY	B/C	KINKEL	B	KONOKTI	C	LAIRDSVILLE	C
KANOSH	C	KENNER	D	KINKORA	D	KOOGLAU	C	LAIREP	D
KANZA	D	KENNEWICK	B	KINMAN	C	KOOSKIA	C	LAJARA	D
KAPAA	A	KENNEY	A	KINNEAR	B	KOOTENAI	A	LAKE	A
KAPAPALA	B	KENNEY LAKE	C	KINNEY	B	KOPIAH	D	LAKE CHARLES	D
KAPOD	B	KENO	D	KINNICK	C	KOPP	B	LAKE CREEK	B
KAPOWSIN	C	KENOMA	D	KINREAD	D	KOPPES	B	LAKEHELEN	B
KAPUHIKANI	D	KENSAL	B	KINROSS	D	KORCHEA	B	LAKEHURST	A
KARAMIN	B	KENSPUR	A	KINSTON	D	KORNMAN	B	LAKE JANEE	B
KARDE	B	KENT	D	KINTA	D	KOSMOS	D	LAKELAND	A
KARHEEN	D	KENYON	C	KINTON	C	KOSSE	D	LAKEMONT	D
KARLAN	C	KEO	B	KINZEL	B	KUSTER	C	LAKEPORT	B
KARLIN	A	KEOLDAR	B	KIOMATIA	A	KUSZTA	B	LAKESHORE	D
KARLO	D	KEOMAH	C	KIONA	B	KOTEDO	D	LAKESOL	B
KARLUK	D	KEOTA	C	KIPLING	D	KOUTS	B	LAKETON	B

Name	Group	Name	Group	Name	Group	Name	Group	Name	Group
LAKEVIEW	C	LATAH	C	LENAWEE	B/D	LINVILLE	B	LORADALE	C
LAKEWIN	B	LATAHCO	C	LENNEP	D	LINWOOD	A/D	LORAIN	C/D
LAKEWOOD	A	LATANG	B	LENOIR	D	LIPAN	D	LORDSTOWN	C
LAKI	B	LATANIER	D	LENOX	B	LIPPINCOTT	B/D	LOREAUVILLE	C
LAKIN	A	LATENE	B	LENZ	B	LIRIOS	B	LORELLA	D
LAKOMA	D	LATHAM	D	LEO	B	LIRRET	D	LORENZO	A
LALAAU	A	LATHROP	C	LEON	A/D	LISADE	B	LORETTO	B
LA LANDE	B	LATINA	D	LEONARD	C	LISAN	D	LORING	C
LALLIE	D	LATON	D	LEONARDO	B	LISBON	B	LOS ALAMOS	B
LAM	B/D	LATONIA	B	LEONARDTOWN	D	LISMAS	D	LOS BANOS	C
LAMAR	B	LATTY	D	LEONIDAS	B	LISHORE	B	LOSEE	B
LAMARTINE	B	LAUDERDALE	B	LEOTA	C	LITCHFIELD	A	LOS GATOS	B/C
LAMBERT	B	LAUGENOUR	B/D	LEPLEY	D	LITHGOW	C	LOS GUINEOS	C
LAMBETH	C	LAUGHLIN	B	LERDAL	C	LITHIA	C	LOSHMAN	D
LAMBORN	D	LAUMAIA	B	LEROY	B	LITIMBER	C	LOS OSOS	C
LAMINGTON	D	LAUREL	C	LESAGE	B	LITLE	C	LOS ROBLES	B
LAMO	B	LAURELHURST	C	LESHARA	B	LITTLEBEAR	A	LOS TANOS	B
LAMONI	D	LAURELWOOD	B	LESHO	C	LITTLEFIELD	D	LOST CREEK	B
LAMONT	A	LAUREN	B	LESLIE	D	LITTLE HORN	C	LOST HILLS	B
LAMONTA	D	LAVALLEE	B	LESTER	B	LITTLE POLE	D	LOS TRANCOS	B
LAMOURE	C	LAVATE	B	LE SUEUR	B	LITTLETON	B	LOSTWELLS	B
LAMPHIER	B	LAVEEN	B	LETA	C	LITTLE WOOD	B	LOTHAIR	C
LAMPSHIRE	D	LAVELDO	D	LETCHER	D	LITZ	C	LOTUS	B
LAMSON	D	LAVERKIN	C	LETHA	D	LIV	C	LOUDON	C
LANARK	B	LA VERKIN	C	LETHENT	C	LIVERMORE	A	LOUDONVILLE	C
LANCASTER	B	LAVINA	C	LETORT	B	LIVIA	D	LOUIE	C
LANCE	C	LAWAI	B	LETTERBOX	B	LIVINGSTON	D	LOUISA	B
LAND	D	LAWET	C	LEVAN	A	LIVONA	A	LOUISBURG	B
LANDES	B	LAWLER	B	LEVASY	C	LIZE	C	LOUP	D
LANDISBURG	C	LAWRENCE	C	LEVERETT	C	LIZZANT	B	LOURDES	C
LANDLOW	C	LAWRENCEVILLE	C	LEVIATHAN	B	LLANOS	C	LOUVIERS	D
LANDUSKY	D	LAWSHE	C	LEVIS	C	LOBDELL	C	LOVEJOY	C
LANE	C	LAWSON	B	LEWIS	D	LOBELVILLE	C	LOVELAND	C
LANEY	C	LAWTHER	D	LEWISBERRY	B	LOBERG	B	LOVELL	C
LANG	B/D	LAWTON	C	LEWISBURG	C	LOBERT	B	LOVELOCK	C/D
LANGFORD	C	LAX	C	LEWISTON	C	LOBITOS	C	LOWELL	C
LANGHEI	B	LAXAL	B	LEWISVILLE	C	LOCANE	D	LOWRY	B
LANGLEY	C	LAYCOCK	B	LEX	B	LOCEY	C	LOWVILLE	B
LANGLOIS	D	LAYTON	A	LEXINGTON	B	LOCHSA	B	LOYAL	B
LANGOLA	B	LAZEAR	D	LHAZ	D	LOCKE	B	LOYALTON	D
LANGRELL	B	LEA	C	LIBBINGS	D	LOCKERBY	C	LOYSVILLE	D
LANGSTON	C	LEADER	B	LIBBY	B	LOCKHARD	B	LOZANO	B
LANIER	B	LEADPOINT	B	LIBEG	A	LOCKHART	B	LOZIER	D
LANIGER	B	LEADVALE	C	LIBERAL	D	LOCKPORT	D	LUALUALEI	D
LANKBUSH	B	LEADVILLE	B	LIBERTY	C	LOCKWOOD	B	LUBBOCK	C
LANKIN	C	LEAF	D	LIBORY	A	LOCUST	C	LUBRECHT	C
LANKTREE	C	LEAHY	C	LIBRARY	D	LODAR	D	LUCAS	C
LANOAK	B	LEAL	B	LIBUTTE	D	LODEMA	A	LUCE	C
LANSDALE	B	LEAPS	C	LICK	B	LODI	C	LUCEDALE	B
LANSDOWNE	C	LEATHAM	C	LICK CREEK	D	LODO	D	LUCERNE	B
LANSING	B	LEAVENWORTH	B	LICKDALE	D	LOFFTUS	C	LUCIEN	C
LANTIS	B	LEAVITT	B	LICKING	C	LOFTON	D	LUCILE	D
LANTON	D	LEAVITTVILLE	B	LICKSKILLET	D	LOGAN	D	LUCILETON	B
LANTONIA	B	LEBANON	C	LIDDELL	D	LOGDELL	D	LUCKENBACH	C
LANTZ	D	LEBAR	B	LIEBERMAN	C	LOGGERT	A	LUCKY	B
LAP	D	LE BAR	B	LIEN	D	LOGHOUSE	B	LUCKY STAR	B
LA PALMA	C	LEBEC	B	LIGGET	B	LOGY	B	LUCY	A
LAPEER	B	LEBO	C	LIGHTNING	D	LOHLER	C	LUDDEN	D
LAPINE	A	LEBSACK	C	LIGNUM	C	LOHMILLER	C	LUDLOW	C
LAPLATTA	C	LECK KILL	B	LIGON	D	LOHNES	A	LUEDERS	C
LAPON	D	LEDBEDER	B	LIHEN	A	LOIRE	B	LUFKIN	D
LAPORTE	C	LEDGEFORK	A	LIHUE	B	LOLAK	D	LUHON	B
LA POSTA	A	LEDGER	D	LIKES	A	LOLALITA	B	LUJANE	C
LA PRAIRIE	B	LEDRU	D	LILAH	A	LOLEKAA	B	LUKIN	C
LARABEE	B	LEDY		LILLIWAUP	A	LOLETA	C/D	LULA	B
LARAND	B	LEE	D	LIMA	B	LOLO	A	LULING	D
LARCHMOUNT	B	LEEDS	C	LIMANI	B	LOLON	A	LUMBEE	D
LARDELL	C	LEEFIELD	C	LIMBER	B	LOMA	C	LUMNI	B/C
LAREDO	B	LEELANAU	A	LIMERICK	C	LOMALTA	D	LUN	C
LARES	C	LEEPER	D	LIMON	C	LOMAX	B	LUNA	C
LARGENT	D	LEESVILLE	B/C	LIMONES	B	LOMIRA	B	LUNCH	C
LARGO	B	LEETON	C	LIMPIA	C	LOMITAS	D	LUNDIMO	C
LARIN	A	LEETONIA	C	LINCO	B	LONDO	C	LUNDY	D
LARIMER	B	LEFOR	B	LINCOLN	A	LONE	C	LUNT	C
LARKIN	B	LEGLER	B	LINCROFT	A	LONEPINE	C	LUPPINO	C
LARKSON	C	LEGORE	B	LINDLEY	C	LONERIDGE	B	LUPTON	D
LA ROSE	B	LEHEW	C	LINDSEY	D	LONE ROCK	A	LURA	D
LARRY	D	LEHIGH	C	LINDSIDE	C	LONETREE	A	LURAY	C/D
LARSON	D	LEHMANS	D	LINDSTROM	B	LONGFORD	C	LUTE	D
LARUE	A	LEHR	B	LINDY	C	LONGLOIS	B	LUTH	C
LARVIE	D	LEICESTER	C	LINEVILLE	C	LONGMARE	D	LUTHER	B
LAS	C	LEILEHUA	B	LINGANORE	B	LONGMONT	C	LUTIE	B
LAS ANIMAS	C	LELA	D	LINKER	B	LONGRIE	C	LUTON	D
LASAUSES	C	LELAND	D	LINKVILLE	B	LONGVAL	B	LUVERNE	B
LAS FLORES	D	LENETA	D	LINNE	C	LONG VALLEY	B	LUXOR	D
LASHLEY		LEMING	C	LINNET	D	LONGVIEW	C	LUZENA	D
LASIL	D	LEMM	B	LINNEUS	B	LONOKE	B	LYCAN	B
LAS LUCAS	C	LEMONEX	D	LINO	C	LONTI	C	LYCOMING	C
LAS POSAS	C	LEMPSTER	C/D	LINOYER	B	LOOKOUT	C	LYDA	D
LASSEN	D	LEN	C	LINSLAW	D	LOON	B	LYDICK	B
LASTANCE	B	LENA	A	LINT	B	LOPER	B	LYFORD	C
LAS VEGAS	D	LENAPAH	D	LINTON	B	LOPEZ	D	LYLES	B

Soil	Group
LYMAN	C/D
LYMANSON	C
LYNCH	D
LYNCHBURG	B/D
LYNDEN	A
LYNNDYL	A
LYNN HAVEN	B/D
LYNNVILLE	C
LYNX	B
LYONMAN	C
LYONS	D
LYONSVILLE	B
LYSINE	D
LYSTAIR	B
LYTELL	B
MABANK	D
MABEN	C
MABI	D
MABRAY	D
MACAR	B
MACEDONIA	C
MACFARLANE	B
MACHETE	C
MACHIAS	B
MACHUELO	D
MACK	C
MACKEN	D
MACKINAC	B
MACKSBURG	B
MACOMB	B
MACOMBER	B
MACON	B
MACY	B
MADALIN	D
MADAWASKA	B
MADDOCK	A
MADDOX	
MADELIA	C
MADELINE	D
MADERA	D
MADISON	B
MADONNA	C
MADRAS	C
MADRID	B
MADRONE	C
MADUREZ	B
MAFURT	B
MAGALLON	B
MAGENS	B
MAGGIE	D
MAGINNIS	C
MAGNA	D
MAGNOLIA	B
MAGNUS	C
MAGOTSU	D
MAGUAYO	D
MAHAFFEY	C/D
MAHAFFY	C/D
MAMALA	C
MAMALASVILLE	B/D
MAMANA	B
MAMASKA	B
MAHER	C
MAHONING	D
MAHUKONA	B
MAIDEN	B
MAILE	A
MAINSTAY	D
MAJADA	B
MAKAALAE	B
MAKALAPA	D
MAKAPILI	A
MAKAWAO	B
MAKAWELI	B
MAKENA	B
MAKIKI	B
MAKLAK	A
MAKOTI	C
MAL	B
MALA	B
MALABAR	A/D
MALABON	C
MALACHY	B
MALAGA	B
MALAMA	A
MALAYA	D
MALBIS	B
MALCOLM	B
MALETTI	C
MALEZA	B
MALIBU	D
MALIN	C/D
MALJAMAR	B
MALLOT	A
MALM	C
MALO	B
MALONE	B
MALOTERRE	D
MALPAIS	C
MALPOSA	C
MALVERN	C
MAMALA	D
MAMOU	C
MANAHAA	C
MANALAPAN	
MANANA	C
MANASSA	C
MANASSAS	B
MANASTASH	C
MANATEE	B/D
MANAWA	C
MANCELONA	A
MANCHESTER	A
MANDAN	B
MANDERFIELD	B
MANDEVILLE	B
MANFRED	D
MANGUM	D
MANHATTAN	A
MANHEIM	C
MANI	C
MANILA	C
MANISTEE	B
MANITOU	C
MANLEY	B
MANLIUS	C
MANLOVE	B
MANNING	B
MANOGUE	D
MANOR	B
MANSFIELD	D
MANSIC	B
MANSKER	B
MANTACHIE	C
MANTEO	C/D
MANTER	B
MANTON	B
MANTZ	B
MANU	C
MANVEL	C
MANWOOD	D
MANZANITA	C
MANZANO	C
MANZANOLA	C
MAPES	C
MAPLE MOUNTAIN	B
MAPLETON	C/D
MARAGUEZ	B
MARATHON	B
MARBLE	A
MARBLEMOUNT	B
MARCELINAS	D
MARCETTA	A
MARCIAL	D
MARCUM	B
MARCUS	C
MARCUSE	D
MARCY	D
MARDEN	C
MARDIN	C
MARENGO	C/D
MARESUA	B
MARGERUM	B
MARGUERITE	B
MARIA	B/C
MARIANA	C
MARIAS	D
MARICAO	B
MARICOPA	B
MARIETTA	C
MARILLA	C
MARINA	A
MARION	D
MARIPOSA	C
MARISSA	C
MARKES	D
MARKEY	D
MARKHAM	C
MARKLAND	C
MARKSBORO	C
MARLA	A
MARLBORO	B
MARLEAN	B
MARLETTE	B
MARLEY	C
MARLIN	D
MARLOW	C
MARLTON	C
MARMARTH	B
MARNA	D
MARPA	B
MARPLEEN	D
MARQUETTE	A
MARR	B
MARRIOTT	B
MARSDEN	C
MARSELL	B
MARSHALL	B
MARSHAN	D
MARSHDALE	C
MARSHFIELD	C
MARSING	B
MART	C
MARTELLA	B
MARTIN	C
MARTINA	A
MARTINECK	D
MARTINEZ	D
MARTINI	B
MARTINSBURG	B
MARTINSDALE	B
MARTINSON	D
MARTINSVILLE	B
MARTINTON	B
MARTY	B
MARVAN	C
MARVELL	B
MARVIN	C
MARY	C
MARYDEL	B
MARYSLAND	D
MASADA	C
MASCAMP	D
MASCHETAH	B
MASCOTTE	D
MASHEL	C
MASHULAVILLE	B/D
MASON	B
MASONVILLE	B
MASSACK	B
MASSENA	C
MASSILLON	B
MASTERSON	D
MATAGORDA	D
MATAMOROS	C
MATANUSKA	C
MATANZAS	B
MATAPEAKE	B
MATAMAN	C
MATCHER	A
MATFIELD	C
MATHERS	B
MATHERTON	B
MATHESON	B
MATHEWS	A
MATHIS	A
MATHISTON	C
MATLOCK	D
MATMON	D
MATTAPEX	C
MATTOLE	C
MAU	D
MAUDE	C/D
MAUGHAN	B
MAUKEY	C
MAUMEE	A/D
MAUNABO	D
MAUPIN	C
MAUREPAS	D
MAURICE	A
MAURINE	D
MAURY	B
MAVERICK	C
MAVIE	D
MAWAE	A
MAX	B
MAXEY	C
MAXFIELD	C
MAXSON	A
MAXTON	B
MAXVILLE	A
MAXWELL	D
MAY	B
MAYBERRY	C
MAYBESO	D
MAY DAY	D
MAYER	D
MAYES	D
MAYFIELD	B
MAYFLOWER	C
MAYHEW	D
MAYLAND	C
MAYMEN	D
MAYNARD LAKE	B
MAYO	B
MAYODAN	B
MAYOWORTH	B
MAYSDORF	B
MAYSVILLE	B
MAYTOWN	C
MAYVILLE	B
MAYWOOD	B
MAZEPPA	B
MAZON	C
MAZUMA	C
MCAFEE	C
MCALLEN	B
MCALLISTER	C
MCALPIN	C
MCBEE	B
MCBETH	D
MCBRIDE	B
MCCABE	B
MCCAFFERY	A
MCCAIN	C
MCCALEB	B
MCCALLY	D
MCCAMMON	D
MCCANN	C
MCCARRAN	D
MCCARTHY	B
MCCLAVE	C
MCCLEARY	C
MCCLELLAN	B
MCCLOUD	C
MCCOIN	D
MCCOLL	D
MCCONNEL	B
MCCOOK	B
MCCORNICK	C
MCCOY	C
MCCREE	B
MCCRORY	D
MCCROSKIE	D
MCCULLOUGH	C
MCCULLY	C
MCCUNE	D
MCCUTCHEN	C
MCDOLE	B
MCDONALD	B
MCDONALDSVILLE	C
MCEWEN	B
MCFADDEN	B
MCFAIN	C
MCFAUL	C
MCGAFFEY	B
MCGARR	C
MCGARY	C
MCGEHEE	C
MCGILVERY	D
MCGINTY	B
MCGIRK	C
MCGOWAN	B
MCGRATH	B
MCGREW	A
MCHENRY	B
MCILWAINE	A
MCINTOSH	B
MCINTYRE	B
MCKAMIE	D
MCKAY	D
MCKENNA	C/D
MCKENZIE	D
MCKINLEY	B
MCKINNEY	D
MCLAIN	C
MCLAURIN	B
MCLEAN	C
MCLEOD	B
MCMAHON	C
MCMEEN	C
MCMULLIN	D
MCMURDIE	C
MCMURPHY	B
MCMURRAY	D
MCNARY	C
MCPAUL	B
MCPHERSON	C
MCPHIE	B
MCQUARRIE	D
MCQUEEN	C
MCRAE	B
MCTAGGART	B
MCVICKERS	C
MEAD	D
MEADIN	A
MEADOWVILLE	B
MEADVILLE	C
MEANDER	D
MECAN	B
MECCA	B
MECKESVILLE	C
MECKLENBURG	C
MEDA	B
MEDANO	C
MEDARY	C
MEDFORD	B
MEDFRA	D
MEDICINE LODGE	C
MEDINA	B
MEDLEY	B
MEDWAY	B
MEEKS	A
MEETEETSE	D
MEGGETT	D
MEGON	C
MEHL	C
MEHLHORN	C
MEIGS	B
MEIKLE	D
MEISS	D
MELBOURNE	B
MELBY	C
MELITA	B
MELLENTHIN	D
MELLOR	D
MELLOTT	B
MELOLAND	C
MELROSE	C
MELSTONE	A
MELTON	B
MELVILLE	B
MELVIN	D
MEMALOOSE	D
MEMPHIS	B
MENAHGA	A
MENAN	B
MENARD	C
MENCH	C
MENDEBOURE	C
MENDOCINO	B
MENDON	B
MENDOTA	B
MENEFEE	D
MENFRO	B
MENLO	D
MENO	C
MENOKEN	C
MENOMINEE	C
MENTO	B
MENTOR	B
MEQUON	C
MERCED	C/D
MERCEDES	D
MERCER	C
MERCEY	C
MEREDITH	B
MERETA	C
MERGEL	B
MERIDIAN	B
MERINO	D
MERKEL	B
MERLIN	D
MERMILL	B/D
MERNA	D
MEROS	A
MERRIFIELD	B
MERRILL	C
MERRILLAN	C
MERRIMAC	A
MERRITT	B/C
MER ROUGE	B
MERTON	B
MERTZ	B
MESA	B
MESCAL	B
MESCALERO	C
MESITA	C
MESKILL	C

Name	Group
MESMAN	C
MESPUN	A
MESSER	C
MET	D
METALINE	B
METAMORA	B
METEA	B
METHOW	B
METIGOSHE	A
METOLIUS	B
METRE	D
METZ	A
MEXICO	D
MHOON	D
MIAMI	B
MIAMIAN	C
MIGCO	A/D
MICHELSON	B
MICHIGAMME	C
MICK	B
MIDAS	D
MIDDLE	C
MIDDLEBURY	B
MIDESSA	B
MIDLAND	D
MIDNIGHT	D
MIDVALE	C
MIDWAY	D
MIFFLIN	B
MIFFLINBURG	B
MIGUEL	D
MIKE	D
MIKESELL	C
MILACA	B
MILAN	B
MILES	B
MILFORD	C
MILHAM	C
MILHEIM	C
MILL	B
MILLARD	B
MILLBORO	D
MILLBROOK	B
MILLBURNE	B
MILLCREEK	B
MILLER	D
MILLERLUX	D
MILLERTON	D
MILLETT	B
MILLGROVE	B/D
MILL HOLLOW	B
MILLICH	D
MILLIKEN	C
MILLINGTON	B
MILLIS	C
MILLRACE	B
MILLSAP	C
MILLSDALE	B/D
MILLSHOLM	C
MILLVILLE	B
MILLWOOD	D
MILNER	C
MILPITAS	C
MILROY	D
MILTON	C
MIMBRES	C
MIMOSA	C
MINA	C
MINAM	B
MINATARE	D
MINCHEY	B
MINCO	B
MINDALE	B
MINDEGO	B
MINDEMAN	B
MINDEN	C
MINE	B
MINEOLA	
MINER	D
MINERAL	A
MINERAL MOUNTAIN	C
MINERVA	B
MING	B
MINGO	B
MINIDOKA	C
MINNEISKA	C
MINNEOSA	B
MINNEQUA	B
MINNETONKA	D
MINNEWAUKAN	B
MINNIECE	D
MINOA	C
MINORA	C
MINTO	C
MINU	D
MINVALE	B
MIRA	D
MIRABAL	C
MIRACLE	B
MIRAMAR	B
MIRANDA	D
MIRES	B
MIRROR	B
MIRROR LAKE	A
MISSION	B
MITCH	B
MITCHELL	B
MITIWANGA	C
MITRE	C
MIZEL	D
MIZPAH	C
MOANO	D
MOAPA	D
MOAULA	A
MOBEETIE	B
MOCA	D
MOCHO	D
MODA	D
MODALE	C
MODEL	C
MODENA	B
MODESTO	C
MODOC	C
MOENKOPIE	D
MOEPITZ	B
MOFFAT	B
MOGOLLON	B
MOGUL	B
MOHALL	B
MOHAVE	B
MOHAWK	B
MOIRA	C
MOKELUMNE	D
MOKENA	C
MOKIAK	B
MOKULEIA	B
MOLAND	B
MOLCAL	B
MOLENA	A
MOLINOS	B
MOLLVILLE	D
MOLLY	B
MOLOKAI	B
MOLSON	B
MOLYNEUX	B
MONAD	A
MONAHAN	D
MONAHANS	B
MONARDA	C
MONCLOVA	B
MONDAMIN	C
MONDOVI	B
MONEE	D
MONICO	B
MONIDA	B
MONITEAU	D
MONMOUTH	C
MONO	D
MONOLITH	C
MONONA	B
MONONGAHELA	C
MONROE	B
MONROEVILLE	C/D
MONSE	B
MONSERATE	C
MONTAGUE	D
MONTALTO	C
MONTARA	D
MONTAUK	C
MONTCALM	A
MONTE	B
MONTE CRISTO	D
MONTEGRANDE	D
MONTELL	D
MONTELLO	C
MONTEOLA	D
MONTEROSA	D
MONTEVALLO	D
MONTGOMERY	D
MONTICELLO	B
MONTIETH	A
MONTHORENCI	B
MONTOSO	B
MONTOUR	D
MONTOYA	D
MONTPELLIER	C
MONTROSE	B
MONTVALE	B
MONTVERDE	A/D
MONTWEL	C
MONUE	B
MOODY	B
MOOHOO	B
MOOSE RIVER	D
MORA	B
MORADO	C
MORALES	D
MORD	C
MOREAU	D
MOREHEAD	C
MOREHOUSE	C
MORELAND	D
MORELANDTON	A
MORET	D
MOREY	D
MORFITT	B
MORGANFIELD	B
MORGNEC	D
MORIARTY	D
MORICAL	C
MORLEY	C
MORMON MESA	D
MOROCCO	A/C
MORONI	D
MOROP	C
MORRILL	B
MORRIS	C
MORRISON	B
MORROW	C
MORSE	D
MORTENSON	C
MORTON	B
MORVAL	B
MOSBY	C
MOSCA	A
MOSCOW	C
MOSEL	C
MOSHANNON	B
MOSHER	D
MOSHERVILLE	C
MOSIDA	B
MOSQUET	D
MOSSYROCK	B
MOTA	B
MOTLEY	B
MOTOQUA	D
MOTTSVILLE	A
MOULTON	B/D
MOUND	C
MOUNTAINBURG	D
MOUNTAINVIEW	B/D
MOUNTAINVILLE	B
MOUNT AIRY	A
MOUNT CARROLL	B
MOUNT HOME	B
MOUNT HOOD	B
MOUNT LUCAS	C
MOUNT OLIVE	D
MOUNTVIEW	B
MOVILLE	C
MOWATA	D
MOWER	C
MOYERSON	D
MOYINA	D
MUCARA	D
MUCET	C
MUDRAY	D
MUD SPRINGS	C
MUGHOUSE	C
MUIR	B
MUIRKIRK	B
MUKILTEO	D
MULCROW	D
MULKEY	C
MULLINS	D
MULLINVILLE	B
MULT	C
MULTORPOR	A
MUMFORD	B
MUNDELEIN	B
MUNDOS	B
MUNISING	B
MUNK	C
MUNSON	D
MUNUSCONG	D
MURDO	B
MURDOCK	C
MUREN	B
MURRILL	B
MURVILLE	D
MUSCATINE	B
MUSE	C
MUSELLA	B
MUSICK	B
MUSINIA	B
MUSKINGUM	C
MUSKOGEE	C
MUSQUIZ	C
MUSSEL	B
MUSSELSHELL	B
MUSSEY	D
MUSTANG	A/D
MUTNALA	B
MUTUAL	B
MYAKKA	A/D
MYATT	B/D
MYERS	D
MYERSVILLE	B
MYLREA	B
MYRICK	D
MYRTLE	B
MYSTEN	A
MYSTIC	D
MYTON	B
NAALEHU	B
NABESNA	D
NACEVILLE	C
NACHES	B
NACIMIENTO	C
NACOGDOCHES	B
NADEAN	B
NADINA	D
NAFF	B
NAGEESI	B
NAGITSY	C
NAGLE	B
NAGOS	D
NAHATCHE	C
NAHMA	C
NAHUNTA	C
NAIWA	B
NAKAI	B
NAKNEK	D
NALDO	B
NAMBE	B
NAMON	C
NANAMKIN	A
NANCY	B
NANNY	B
NANNYTON	B
NANSENE	B
NANTUCKET	C
NANUM	C
NAPA	D
NAPAISHAK	D
NAPAVINE	B
NAPIER	B
NAPLENE	B
NAPLES	B
NAPPANEE	D
NAPTOWNE	B
NARANJITO	C
NARANJO	C
NARCISSE	B
NARD	B
NARLON	C
NARON	B
NARRAGANSETT	B
NARROWS	D
NASER	B
NASH	B
NASHUA	A
NASHVILLE	B
NASON	C
NASSAU	C/D
NASSET	B
NATALIE	C
NATCHEZ	B
NATHROP	B
NATIONAL	B
NATRONA	B
NATROY	D
NATURITA	B
NAUKATI	D
NAUMBURG	C
NAVAJO	D
NAVAN	B
NAVARRO	B
NAVESINK	
NAYLOR	
NAYPED	C
NAZ	B
N-BAR	B
NEAPOLIS	B/D
NEBEKER	C
NESGEN	D
NEBISH	B
NEBO	
NECHE	C
NEDERLAND	B
NEEDHAM	D
NEEDLE PEAK	C
NEEDMORE	C
NEELEY	B
NEESOPAH	C
NEGITA	C
NEGLEY	B
NEHALEM	B
NEHAR	B
NEILTON	A
NEISSON	B
NEKIA	C
NELLIS	B
NELMAN	B
NELSCOTT	B
NELSON	B
NEMAH	C
NEMOTE	A
NENANA	B
NENNO	B
NEOLA	D
NEOTOMA	B
NEPALTO	A
NEPESTA	B
NEPHI	B
NEPPEL	B
NEPTUNE	A
NERESON	B
NESDA	A
NESHAMINY	B
NESIKA	B
NESKAHI	B
NESKOWIN	C
NESPELEM	B
NESS	D
NESSEL	B
NESSOPAH	B
NESTER	C
NESTUCCA	C
NETARTS	A
NETCONG	B
NETO	B
NETTLETON	C
NEUBERT	B
NEUNS	B
NEUSKE	B
NEVADOR	C
NEVILLE	B
NEVIN	C
NEVINE	B
NEVKA	C
NEVOYER	D
NEVTAH	C
NEVU	D
NEWARK	C
NEWART	B
NEWAYGO	B
NEWBERG	B
NEWBERRY	C
NEWBY	B
NEW CAMBRIA	C
NEWCASTLE	B
NEWCOMB	A
NEWDALE	B
NEWELL	B
NEWELLTON	D
NEWFANE	
NEWFORK	D
NEWKIRK	D
NEWLANDS	B
NEWLIN	B
NEWMARKET	B
NEWPORT	C
NEWRUSS	B
NEWRY	B
NEWSKAH	B
NEWSTEAD	D
NEWTON	A/D
NEWTONIA	B

Name		Name		Name		Name		Name	
NEWTOWN	C	NORTON	C	OKAW	D	ORELLA	D	PACK	C
NEWVILLE	C	NORTONVILLE	C	OKAY	B	OREM	A	PACKARD	B
NEZ PERCE	C	NORTUNE	D	OKEECHOBEE	A/D	ORESTIMBA	C	PACKER	C
NIAGARA	C	NORWALK	B	OKEELANTA	A/D	ORFORD	C	PACKHAM	B
NIART	B	NORWAY FLAT		OKEMAH	C	ORIDIA	C	PACKSADDLE	B
NIBLEY	C	NORWELL	C	OKLARED	B	ORIF	A	PACKWOOD	D
NICHOLSON	C	NORWICH	D	OKLAHAMA	A/D	ORIO	C	PACOLET	C
NICHOLVILLE	C	NORWOOD	B	OKMOK	B	ORION	B	PACTOLUS	C
NICKEL	B	NOTI	D	OKO	D	ORITA	D	PADEN	C
NICODEMUS	B	NOTUS	A/C	OKOBOJI	C	ORLANO	B	PADRONI	B
NICOLAUS	C	NOUQUE	D	OKOLONA	D	ORLANDO	A	PADUCAH	B
NICOLLET	B	NOVARA	B	OKREEK	D	ORMAN	B	PADUS	C
NIELSEN	D	NOVARY	B	OKTIBBEHA	D	ORMSBY	B/C	PAESL	B
NIGHTHAWK	B	NOWOOD	C	OLA	C	ORODELL	C	PAGET	B
NIHILL	B	NOYO	C	OLAA	A	ORO FINO	B	PAGODA	C
NIKABUNA	D	NOYSON	C	OLALLA	C	ORO GRANDE	C	PAHRANAGAT	C
NIKEY	B	NUBY	C/D	OLANTA	B	ORONO	C	PAHREAH	D
NIKISHKA	A	NUCKOLLS	C	OLATHE	C	OROVADA	B	PAHROC	C
NIKLASON	B	NUCLA	B	OLD CAMP	D	ORPHANT	D	PAIA	C
NIKOLAI	D	NUECES	C	OLDHAM	C	ORR	C	PAICE	C
NILAND	C	NUGENT	A	OLDS	D	ORRVILLE	C	PAINESVILLE	C
NILES	C	NUGGET	C	OLDSMAR	B/D	ORSA	A	PAINTROCK	B
NIMROD	C	NUMA	C	OLDWICK	B	ORSINO	A	PAIT	B
NINCH	C	NUNDA	C	OLELO	B	ORTELLO	A	PAJARITO	B
NINEMILE	D	NUNICA	C	OLENA	B	ORTIGALITA	C	PAJARO	B
NINEVEH	B	NUNN	C	OLEQUA	B	ORTING	C	PAKA	B
NINIGRET	B	NUSS	D	OLETE	C	ORTIZ	C	PAKALA	B
NININGER	B	NUTLEY	C	OLEX	B	ORTLEY	B	PAKINI	B
NINNESCAH	E	NUTRAS	C	OLGA	C	ORWET	A	PALA	B
NIOBELL	C	NUTRIOSO	B	OLI	B	ORWOOD	B	PALACIO	B
NIOTA	D	NUVALDE	C	OLIAGA	B/D	OSAGE	D	PALAPALAI	B
NIPE	B	NYALA	D	OLINDA	B	OSAKIS	B	PALATINE	B
NIPPERSINK	B	NYMORE	A	OLIPHANT	B	OSCAR	D	PALESTINE	B
NIPPT	A	NYSSA	C	OLIVENHAIN	B	OSCURA	C	PALISADE	B
NIPSUM	C	NYSSATON	B	OLIVER	C	OSGOOD	B	PALMA	B
NIRA	B	NYSTROM	C	OLIVIER	C	OSHA	B	PALMAREJO	C
NISHNA	C			OLJETO	A	OSHAWA	D	PALM BEACH	A
NISHON	D	OAHE	B	OLMITO	D	O'SHEA	C	PALMER	D
NISQUALLY	A	OAKDALE	B	OLMITZ	B	OSHKOSH	D	PALMER CANYON	B
NISSWA	B	OAKDEN	D	OLMOS	C	OSHTEMO	B	PALMICH	B
NIU	B	OAKFORD	B	OLMSTFO	B/D	OSIER	B/D	PALMS	D
NIULII	C	OAK GLEN	B	OLNEY	B	OSKA	C	PALMYRA	B
NIVLOC	D	OAK GROVE	C	OLOKUI	D	OSMUND	B	PALO	B
NIWOT	C	OAK LAKE	B	OLPE	C	OSO	B	PALODURO	B
NIXA	C	OAKLAND	C	OLSON	D	OSOBB	D	PALOMAS	B
NIXON	B	OAKS RIDGE	C	OLTON	C	OSORIDGE	D	PALOMINO	B
NIXONTON	B	OAKVILLE	A	OLUSTEE	B/D	OSOTE	B	PALOS VERDES	B
NIZINA	A	OAKWOOD	D	OLYIC	B	OSSIAN	C	PALOUSE	B
NOBE	D	OANAPUKA	B	OLYMPIC	B	OST	B	PALSGROVE	B
NOBLE	B	OASIS	B	OMADI	B	OSTRANDER	B	PAMLICO	D
NOBSCOTT	A	OATMAN	B	OMAHA	B	OTERO	B	PAMOA	C
NOCKEN	C	OBAN	C	OMAK	C	OTHELLO	D	PANSDEL	D
NODAWAY	B	OBARC	B	OMEGA	A	OTIS	C	PANUNKEY	C
NOEL	D	OBEN	C	OMENA	B	OTISCO	A	PANA	B
NOHILI	D	OBRAST	D	OMNI	C	OTISVILLE	A	PANACA	D
NOJASIPPI	D	OBRAY	D	ONA	A/D	OTLEY	B	PANAEWA	D
NOKAY	C	OBURN	D	ONALASKA	B	OTSEGO	C	PANASOFFKEE	D
NOKOMIS	B	OCALA	B	ONAMIA	B	OTTER	B/D	PANCHERI	B
NOLAM	B	OCEANET	D	ONARGA	B	OTTERBEII	C	PANCHUELA	C
NOLICHUCKY	B	OCEANO	A	ONAWA	D	OTTERHOLT	B	PANDO	C
NOLIN	B	OCHEYEDAN	B	ONAWAY	B	OTTOKEE	A	PANDOAH	C
NOLO	B	OCHLOCKONEE	B	ONDAWA	B	OTWAY	D	PANDORA	D
NOME	D	OCHO	D	ONEIDA	B	OTWELL	C	PANDURA	D
NONDALTON	B	OCHOCO	C	O'NEILL	B	OUACHITA	C	PANE	B
NONOPAMU	D	OCHOPEE	B/D	ONEONTA	B/D	OURAY	A	PANGUITCH	C
NOOKACHAMPS	C/D	OCILLA	C	ONITA	C	OUTLET	C	PANHILL	C
NOOKSACK	B	OCKLEY	B	ONITE	B	OVALL	C	PANIOGUE	B
NOONAN	D	OCOEE	A/D	ONOTA	A/D	OVERGAARD	C	PANKY	C
NORA	B	OCONEE	C	ONOVA	C	OVERLAND	D	PANOCHE	B
NORAD	B	OCONTO	B	ONRAY	B	OVERLY	C	PANOLA	C
NORBERT	D	OCOSTA	D	ONSLOW	D	OVERTON	D	PANSEY	D
NORBURNE	B	OCQUEOC	B	ONTARIO	B	OVID	C	PANTEGO	D
NORBY	B	OCTAGON	B	ONTKO	B/D	OVINA	B	PANTHER	D
NORD	B	ODEE	D	ONTONAGON	D	OWEGO	D	PANTON	D
NORDBY	B	ODELL	B	ONYX	B	OWEN CREEK	C	PAOLA	A
NORDEN	B	ODEM	A	OOKALA	A	OWENS	D	PAOLI	B
NORDNESS	B	ODERMOTT	C	OPAL	D	OWHI	B	PAONIA	C
NORFOLK	B	ODESSA	D	OPEQUON	C/D	OWOSSO	B	PAPAA	D
NORGE	B	ODIN	C	OPHIR	C	OWYHEE	B	PAPAI	A
NORKA	B	ODNE	C	OPIHIKAO	D	OXALIS	C	PAPAKATING	D
NORMA	B/C	O'FALLON	D	OPPIO	D	OXBOW	D	PAPOOSE	C
NORMANGEE	D	OGDEN	D	OQUAGA	D	OXERINE	C	PARADISE	C
NORREST	C	OGEECHEE	C	ORA	C	OXFORD	D	PARADOX	B
NORRIS	C	OGEMAN	C	ORAN	B	OZAMIS	B/D	PARALOMA	C
NORRISTON	B	OGILVIE	C	ORANGE	D	OZAN	D	PARAMORE	D
NORTE	B	OGLALA	B	ORANGEBURG	B	OZAUKEE	C	PARASOL	B
NORTHDALE	C	OGLE	B	ORCAS	D			PARCELAS	D
NORTHFIELD	B	OHAYSI	D	ORCHARD	B	PAAIKI	B	PARDEE	C
NORTHMORE	C	OHIA	A	ORD	A	PAALOA	B	PAREHAT	C
NORTHPORT	C	OJAI	B	ORDNANCE	C	PAAUHAU	A	PARENT	C
NORTH POWDER	C	OJATA	D	ORDWAY	D	PACHAPPA	B	PARIETTE	
NORTHUMBERLAND	C/D	OKANOGAN	B	ORELIA	D	PACHECO	B/C	PARIS	

Name		Name		Name		Name		Name		Name	
PARISHVILLE	C	PELIC	D	PICAYUNE	B	PLEASANT VIEW	B	POS.	D
PARKAY	B	PELLA	D	PICKAWAY	C	PLEDGER	D	POTAMO	D
PARKDALE	B	PELLEJAS	B	PICKENS	D	PLEEK	C	POTH	C
PARKE	B	PELONA	C	PICKET.	B	PLEINE	D	POTLATCH	C
PARKER	B	PELUK	D	PICKFORD	D	PLEVNA	D	POTRATZ	C
PARKFIELD	C	PEMBERTON	A	PICKRELL	D	PLONE	B	POTSDAM	B
PARKHILL	D	PEMBINA	C	PICKWICK	B	PLOVER	B	POTTER	B
PARKHURST		PEMBROKE	B	PICO	B	PLUMAS	B	POTTS	B
PARKINSON	B	PENA	B	PICOSA	C	PLUMMER	B/D	POUDRE	B
PARKVIEW	B	PENCE	A	PICTOU	B	PLUSH	B	POULTNEY	B
PARKVILLE	C	PENDEN	B	PIE CREEK	D	PLUTH	B	POUNCEY	D
PARKWOOD	A/D	PEND OREILLE	B	PIERIAN	A	PLUTOS	C	POVERTY	A
PARLEYS	B	PENDROY	D	PIERPONT	C	PLYMOUTH	A	POWDER	B
PARLIN	C	PENELAS	D	PIERRE	D	POALL	C	POWDERHORN	C
PARLO	B	PENISTAJA	B	PIERSONTE	B	POARCH	B	POWELL	C
PARMA	C	PENITENTE	B	PIIHONUA	A	POCALLA	A	POWER	B
PARNELL	D	PENLAW	C	PIKE	B	POCATELLO	B	POWHITE	C/D
PARR	B	PENN	C	PILCHUCK	A	POCKER	D	POWLEY	C
PARRAN	D	PENNEL	C	PILGRIM	B	POCOMOKE	D	POWMATKA	C
PARRISH	C	PENNINGTON	B	PILOT	B	POOO	D	POY	D
PARSHALL	B	PENO	C	PILOT ROCK	C	PODUNK	B	POYGAN	D
PARSIPPANY	D	PENOYER	C	PIMA	B	POE	B/C	POZO	C/D
PARSONS	D	PENROSE	D	PIMER	B	POEVILLE	D	POZO BLANCO	B
PARTRI	C	PENSORE	B	PINAL	D	POGAL	B	PRAG	C
PASAGSHAK	D	PENTHOUSE	D	PINALENO	B	POGANEAB	D	PRATHER	B
PASCO	B/C	PENTZ	D	PINANT	B	POGUE	B	PRATLEY	C
PASO SECO	D	PENWELL	A	PINATA	C	POHAKUPU	A	PRATT	A
PASQUETTI	C/D	PENWOOD	A	PINAVETES	A	POINDEXTER	C	PREACHER	B
PASQUOTANK	B/D	PEOGA	C	PINCHER	C	POINSETT	B	PREAKNESS	D
PASSAR	C	PEOH	C	PINCKNEY	C	POINT	B	PREBISH	D
PASS CANYON	D	PEONE	B/C	PINCONNING	D	POINT ISABEL	C	PREBLE	C
PASSCREEK	C	PEORIA	C	PINCUSHION	B	POJOAQUE	B	PRENTISS	C
PASTURA	D	PEOTONE	C	PINEDA	B/D	POKEGEMA	B	PRESQUE ISLE	B
PATAHS	B	PEPOON	B	PINEDALE	B	POKEMAN	B	PRESTO	A
PATENT	C	PEQUEA	C	PINEGUEST	B	POKER	C	PRESTON	A
PATILLAS	B	PERCHAS	D	PINELLOS	A/D	POLAND	B	PREWITT	B
PATILO	C	PERCIVAL	C	PINETOP	C	POLAR	B	PREY	D
PATIT CREEK	B	PERELLA	C	PINEVILLE	B	POLATIS	C	PRICE	C
PATNA	B	PERHAM	C	PINEY	C	POLE	A	PRIDA	D
PATOUTVILLE	C	PERICO	B	PINICON	B	POLEBAR	C	PRIDHAM	D
PATRICIA	B	PERITSA	C	PINKEL	C	POLELINE	B	PRIETA	D
PATRICK	B	PERKINS	C	PINKHAM	C	POLEO	B	PRIMEAUX	C
PATROLE	C	PERKS	A	PINKSTON	B	POLEY	C	PRIMGHAR	B
PATTANI	D	PERLA	C	PINNACLES	C	POLICH	B	PRINCETON	B
PATTENBURG	B	PERMA	A	PINO	C	POLLARD	C	PRINEVILLE	B
PATTER	C	PERMANENTE	C	PINOLA	C	POLLASKY	C	PRING	B
PATTERSON	C	PERRIN	B	PINOLE	B	POLLY	B	PRINS	C
PATTON	B/D	PERRINE	D	PINON	C	POLO	B	PRITCHETT	C
PATWAY	C	PERROT	D	PINONES	D	POLSON	C	PROCTOR	B
PAUL	B	PERRY	D	PINTAS	D	POLVADERA	B	PROGRESSO	C
PAULDING	D	PERRYPARK	B	PINTLAR	A	POMAT	C	PROMISE	D
PAULINA	D	PERRYVILLE	B	PINTO	C	POMELLO	C	PROMO	D
PAULSELL	D	PERSANTI	C	PINTURA	A	POMPANO	A/D	PROMONTORY	B
PAULSON	B	PERSAYO	D	PINTWATER	D	POMPONIO	C/D	PRONG	C
PAULVILLE	B	PERSHING	C	PIOCHE	D	POMPTON	B	PROSPECT	B
PAUMALU	B	PERSIS	B	PIOPOLIS	D	POMROY	B	PROSPER	B
PAUNSAUGUNT	D	PERT	D	PIPER	B/C	PONCA	B	PROSSER	C
PAUSANT	B	PERU	C	PIROUETTE	D	PONCENA	D	PROTIVIN	C
PAUWELA	B	PESCADERO	C/D	PIRUM	B	PONCHA	A	PROUT	C
PAVANT	D	PESET	C	PISGAH	C	POND	B/C	PROVIDENCE	C
PAVILLION	B	PESHASTIN	B	PISHKUN	B	POND CREEK	B	PROVO	D
PAVOHROO	B	PESO	C	PISTAKEE	B	PONDILLA	A	PROVO BAY	D
PAWCATUCK	D	PETEETNEET	D	PIT	D	PONIL	D	PROWERS	B
PAWLET	B	PETERBORO	B	PITTMAN	D	PONTOTOC	B	PTARMIGAN	B
PAWNEE	D	PETERS	D	PITTSFIELD	B	PONZER	D	PUAULU	A
PAXTON	C	PETOSKEY		PITTSTOWN	C	POOKU	A	PUCHYAN	A
PAXVILLE	D	PETRIE	D	PITTWOOD	B	POOLE	B/D	PUDDLE	D
PAYETTE	B	PETROLIA	D	PITZER	C	POOLER	D	PUERCO	D
PAYMASTER	B	PETTONS	C	PIUTE	D	POORMA	B	PUERTA	D
PAYNE	C	PEWAMO	B/D	PLACEDO	D	POPE	B	PUETT	D
PAYSON	D	PEYTON	B	PLACENTIA	D	POPPLETON	A	PUGET	B/C
PEACHAM	D	PFEIFFER	B	PLACERITOS	C	POQUONOCK	C	PUGSLEY	B
PEARL HARBOR	D	PHAGE	B	PLACID	A/D	PORRETT	B/D	PUHI	A
PEARMAN		PHANTOM	C	PLACK	D	PORT	B	PUHIMAU	D
PEARSOLL	D	PHARO	B	PLAINFIELD	A	PORTAGEVILLE	D	PULASKI	B
PEAVINE	C	PHAROLIO	D	PLAINVIEW	C	PORTALES	B	PULEHU	B
PECATONICA	B	PHEBA	C	PLAISTED	C	PORTALTO	B	PULLMAN	D
PECOS	D	PHEENEY	B	PLANO	B	PORT BYRON	B	PULS	D
PEDEE	C	PHELAN	B	PLASKETT	D	PORTERS	B	PULSIPHER	D
PEDERNALES	C	PHELPS	B	PLATA	B	PORTERVILLE	D	PULTNEY	C
PEDIGO	B/C	PHIFERSON	B	PLATEA	C	PORTHILL	C	PUMEL	C
PEDLAR	D	PHILBON	B/D	PLATEAU	B	PORTINO	C	PUMPER	C
PEDOLI	C	PHILIPSBURG	B	PLATNER	C	PORTLAND	D	PUNA	A
PEDRICK	B	PHILLIPS	C	PLATO	C	PORTNEUF	B	PUNALUU	D
PEEBLES	C	PHILO	B	PLATORO	B	PORTOLA	C	PUNOHU	A
PEEL	C	PHILOMATH	D	PLATTE	D	PORTSMOUTH	D	PURDAM	C
PEELER	B	PHIPPS	C	PLATTVILLE	B	PORUM	C	PURDY	C
PEEVER	C	PHOEBE	B	PLAZA	B/C	POSANT	C	PURGATORY	D
PEGLER	D	PHOENIX	D	PLEASANT	C	POSEY	B	PURNER	D
PEGRAM	B	PIASA	C	PLEASANT GROVE	B	POSITAS	D	PURSLEY	B
PEKIN	C	PICACHO	C	PLEASANTON	B	POSKIN	C	PURVES	D
PELHAM	B/D			PLEASANT VALE	B	POSOS	C	PUSTOI	A

Name		Name		Name		Name		Name	
PUTNAM	D	RANDMAN	D	REELFOOT	C	RIFFE	B	ROLETTE	C
PUUKALA	D	RANDOLPH	D	REESER	C	RIFLE	A/D	ROLFE	C
PUUONE	C	RANDS	C	REESVILLE	C	RIGA	D	ROLISS	D
PUU OO	A	RANGER	D	REEVES	C	RIGGINS	A	ROLLA	C
PUU OPAE	B	RANIER	C	REFUGE	C	RIGLEY	B	ROLLII	D
PUU PA	B	RANKIN	C	REGAN	B	RILEY	C	ROLOFF	C
PUYALLUP	B	RANTOUL	D	REGENT	C	RILLA	B	ROMBERG	B
PYLE	A	RANYHAN	B	REHM	C	RILLITO	B	ROMBO	C
PYLON	D	RAPELJE	C	REICHEL	B	RIMER	C	ROMEO	C
PYOTE	A	RAPHO	B	REIFF	B	RIMINI	A	ROMNEY	A
PYRAMID	D	RAPIDAN	B	REILLY	A	RIMROCK	D	ROMULUS	D
PYRMONT	D	RAPLEB	C	REINACH	B	RIN	B	ROND	C
		RARDEN	C	REKOP	D	RINCON	C	RONNEBY	B
QUACKENBUSH	C	RARICK	B	RELAN	A	RINCONADA	C	RONSON	B
QUAKER	C	RARITAN	C	RELAY	B	RINDGE	D	ROOSE?	B
QUAKERTOWN	B	RASBAND	B	RELIANCE	C	RINGLING	C	ROOTEL	D
QUAMSA	D	RASSET	B	RELIZ	D	RINGO	D	ROSACHI	C
QUANON	A	RATAKE	C	RELSE	B	RINGOLD	B	ROSAMOND	B
QUANAH	B	RATHBUN	C	REMBERT	D	RINGWOOD	B	ROSANE	C
QUANDAHL	B	RATLIFF	B	REMMIT	A	RIO	D	ROSANKY	C
QUARLES	D	RATON	D	REMSEN	D	RIO ARRIBA	D	ROSARIO	C
QUARTZBURG	C	RATTLER	B	REMUDAR	B	RIOCONCHO	C	ROSCOE	D
QUATAMA	C	RATTO	D	REMUNDA	C	RIO GRANDE	B	ROSCOMMON	D
QUAY	B	RAUB	B	RENBAC	D	RIO KING	C	ROSEBERRY	B/D
QUAZO	D	RAUVILLE	C	RENCALSON	C	RIO LAJAS	A	ROSEBLOOM	D
QUEALY	D	RAUZI	B	RENCOT	A	RIO PIEDRAS	B	ROSEBUD	B
QUEBRADA	C	RAVALLI	C	RENFROW	D	RIPLEY	B	ROSEBURG	B
QUEENY	D	RAVENDALE	D	RENICK	D	RIPON	B	ROSE CREEK	C
QUEETS	B	RAVENNA	C	RENNIE	C/D	RIRIE	B	ROSEGLEN	B
QUENADO	C	RAVOLA	B	RENO	D	RISBECK	B	ROSEHILL	D
QUENZER	D	RAWAH	B	RENOHILL	C	RISLEY	D	ROSELAND	D
QUICKSELL	D	RAWHIDE	D	RENOVA	D	RISTA	C	ROSELLA	D
QUIETUS	C	RAWSON	B	RENOX	D	RISUE	D	ROSELMS	D
QUIGLEY	B	RAY	B	RENSHAW	B	RITCHEY	B	ROSEMOUNT	B
QUILCENE	C	RAYADO	C	RENSLOW	B	RITNER	C	ROSENDALE	B
QUILLAYUTE	B	RAYENOUF	B	RENSSELAER	C	RITO	B	ROSE VALLEY	C
QUIMBY	B	RAYMONDVILLE	D	RENTIDE	C	RITTER	B	ROSEVILLE	B
QUINCY	A	RAYNE	C	RENTON	B/C	RITTMAN	C	ROSEWORTH	C
QUINLAN	C	RAYNESFORD	B	RENTSAC	C	RITZ	B/D	ROSHE SPRINGS	D
QUINN	D	RAYNHAM	C	REPARADA	D	RITZCAL	B	ROSITAS	A
QUINNEY	C	RAYNOR	D	REPP	A	RITZVILLE	B	ROSLYN	B
QUINTON		RAZOR	C	REPPART	B	RIVERHEAD	B	ROSMAN	B
QUITMAN	C	RAZORT	B	REPUBLIC	B	RIVERSIDE	A	ROSNEY	C
QUONSET	A	READING	C	RESCUE	C	RIVERTON	C	ROSS	B
		READINGTON	C	RESERVE	B	RIVERVIEW	B	ROSS FORK	C
RABER	C	READLYN	B	RESNER	B	RIVRA	A	ROSSI	C
RABEY	A	REAGAN	B	RET	B/C	RIXIE	C	ROSSMOYNE	C
RABIDEUX	B	REAKOR	B	RETRIEVER	D	RIXON	C	ROSS VALLEY	C
RABUN	B	REAL	C	RETSOF	C	RIZ	D	ROTAN	C
RACE	D	REAP	D	RETSOK	B	ROANOKE	D	ROTHIEMAY	B
RACHERT	D	REARDAN	C	REXBURG	B	ROBANA	B	ROTHSAY	B
RACINE	B	REAVILLE	C	REXFORD	C	ROBBINS	B	ROTTULEE	B
RACOON	D	REBA	C	REXOR	A	ROBBS	D	ROUBIDEAU	C
RAD	C	REBEL	B	REYES	C/D	ROBERTS	D	ROUEN	C
RADERSBURG	B	REBUCK	C	REYNOLDS		ROBERTSDALE	C	ROUND BUTTE	D
RADFORD	B	RECAL	B	REYNOSA	B	ROBERTSVILLE	D	ROUNDLEY	C
RADLEY	C	RECLUSE	C	REYWAT	D	ROBIN	B	ROUNDTOP	C
RADNOR	D	REDBANK	B	RHAME	B	ROBINSON	B	ROUNDUP	C
RAFAEL	D	RED BAY	B	RHEA	B	ROBINSONVILLE	B	ROUNDY	C
RAGER	B	RED BLUFF	C	RHINEBECK	D	ROBLEDO	D	ROUSSEAU	A
RAGLAN	C	RED BUTTE	B	RHOADES	D	ROB ROY	C	ROUTON	D
RAGNAR	B	REDBY	C	RHOAME	C	ROBY	C	ROUTT	C
RAGO	C	REDCHIEF	C	RIB	C	ROCA	D	ROVAL	D
RAGSDALE	B/D	REDCLOUD	B	RICCO	D	ROCHE	C	ROWE	D
RAGTOWN	D	REDDICK	C	RICETON	B	ROCHELLE	C	ROWENA	C
RAHAL	C	REDDING	D	RICEVILLE	C	ROCHEPORT	C	ROWLAND	C
RAHM	C	REDFIELD	B	RICHARDSON	B	ROCKAWAY	C	ROWLEY	B
RAIL	C/D	RED HILL	C	RICHEAU	C	ROCKCASTLE	D	ROXAL	D
RAINBOW	C	RED HOOK	C	RICHEY	C	ROCK CREEK	D	ROXBURY	B
RAINEY	B	REDLAKE	D	RICHFIELD	C	ROCKFORD	B	ROY	B
RAINS	B/D	REDLANDS	B	RICHFORD	A	ROCKHOUSE	A	ROYAL	B
RAINSBORO	C	REDLODGE	D	RICHLIE	A	ROCKINGHAM	C/D	ROYALTON	C
RAKE	D	REDMANSON	B	RICHMOND	D	ROCKLIN	C/D	ROYCE	B
RALSEN	B/C	REDMOND	C	RICHTER	B	ROCKLY	D	ROYSTONE	B
RAMADA	C	REDNUN	C	RICHVALE	B	ROCKPORT	C	ROZA	C
RAMADERO	B	REDOLA	B	RICHVIEW	C	ROCK RIVER	B	ROZELLVILLE	B
RAMBLER	B	REDONA	B	RICHWOOD	B	ROCKTON	B	ROZETTA	B
RAMELLI	C	REDRIDGE	B	RICKMORE	B	ROCKWELL	B	ROZLEE	C
RAMIRES	D	REDROB	D	RICKS	A	ROCKWOOD	B	RUARK	C
RAMMEL	C	RED ROCK	B	RICO	C	ROCKY FORD	B	RUBICON	A
RAMO	C	RED SPUR	B	RICREST	B	RODDY	C	RUBIO	C
RAMONA	B	REDSTOE	B	RIDO	C	RODMAN	A	RUBY	B
RAMPART	B	REDTHAYNE	B	RIDGEBURY	C	ROE	B	RUBYHILL	C
RAMPARTAR	A	REDTOM	C	RIDGECREST	C	ROEBUCK	D	RUCH	B
RAMPARTER	A	REDVALE	C	RIDGEDALE	B	ROELLEN	D	RUCKLES	D
RAMSEY	D	REDVIEW	B	RIDGELAND	D	ROEMER	C	RUCLICK	D
RAMSHORN	B	REE	B	RIDGELAWN	A	ROESIGER	B	RUDD	D
RANCE	C	REEBEX	C	RIDGELY	B	ROGERT	D	RUDEEN	B
RANCHERIA	B	REED	D	RIDGEVILLE	B	ROHNERVILLE	B	RUDOLPH	C
RANO	B	REEDER	B	RIDGEWAY	D	ROHRERSVILLE	C	RUDYARD	D
RANDADO	C	REEDPOINT	C	RIDIT	C	ROIC	D	RUELLA	B
RANDALL	D	REEDY	D	RIETBROCK	C	ROKEBY	D	RUGGLES	B

Name		Name		Name		Name		Name	
RUIDOSO	C	SALVISA	C	SAUK	B	SEDAN	B	SHELBY	B
RUKO	D	SALZER	D	SAULICH	D	SEDILLO	B	SHELBYVILLE	B
RULE	B	SAMBA	D	SAUM	C	SEDOWELL	C	SHELDON	C
RULICK	C	SAMISH	C/D	SAUNDERS	C	SEEDSKADEE	D	SHELIKOF	D
RUMBO	C	SAMMAMISH	C	SAUVIE	C/D	SEES	C	SHELLABARGER	B
RUMFORD	B	SAMPSEL	D	SAUVOLA	C	SEGMEE	B	SHELLDRAKE	A
RUMNEY	C	SAMPSON	B	SAVAGE	C	SEGAL	D	SHELLROCK	A
RUMPLE	C	SAMSIL	D	SAVANNAH	C	SEGNO	C	SHELMADINE	D
RUM RIVER	C	SAN ANDREAS	C	SAVENAC	C	SEHORN	D	SHELOCTA	B
RUNE	C	SAN ANTON	B	SAVO	C	SEITZ	C	SHELTON	C
RUNGE	B	SAN ANTONIO	C	SAVOIA	B	SEJITA	D	SHENA	C
RUNNELLS	C	SAN ARACIO	B	SAWABE	D	SEKIL	C	SHENANDOAH	C
RUNNYMEDE	B	SAN BENITO	B	SAWATCH	C	SEKIU	D	SHEP	B
RUPERT	A	SANCHEZ	D	SAWCREEK	B	SELAH	C	SHEPPARD	A
RUSCO	C	SANDALL	C	SAWMILL	C	SELDEN	C	SHERANDO	C
RUSE	D	SANDERSON	B	SAWYER	C	SELEGNA	D	SHERAR	A
RUSH	C	SANDLAKE	C	SAXBY	D	SELFRIDGE	C	SHERBURNE	B
RUSHTOWN	A	SANDLEE	A	SAXON	B	SELKIRK	D	SHERIDAN	B
RUSHVILLE	D	SANELI	D	SAYBROOK	B	SELLE	B	SHERLOCK	B
RUSS	B	SAN EMIGDIO	B	SAYLESVILLE	C	SELLERS	A/D	SHERM	D
RUSSELL	B	SANFORD	A	SAYLOR	A	SELMA	B	SHERRYL	B
RUSSELLVILLE	C	SANGER	B	SCALA	B	SEMIAHMOO	D	SHERWOOD	B
RUSSLER	C	SAN GERMAN	D	SCAMMAN	C	SEMIHMOO	D	SHIBLE	B
RUSTON	B	SANGO	C	SCANDIA	B	SEMINARIO	D	SHIELDS	C
RUTLAND	C	SANGREY	A	SCANTIC	C	SEMIX	C	SHIFFER	B
RUTLEGE	D	SANILAC	C	SCAR	A	SEN	B	SHILOH	C
RYAN	D	SAN ISABEL	B	SCARBORO	D	SENECAVILLE	C	SHINAKU	D
RYAN PARK	B	SAN JOAQUIN	D	SCAVE	B	SEQUATCHIE	B	SHINGLE	C
RYDE	B/D	SAN JON	C	SCHAFFENAKE?	A	SEQUIM	A	SHINGLETOWN	C
RYDER	C	SAN JOSE	B	SCHAMBER	A	SEQUIN	B	SHINN	B
RYEGATE	B	SAN JUAN	A	SCHAMP	C	SEQUOIA	C	SHINROCK	C
RYELL	A	SAN LUIS	B	SCHAPVILLE	C	SERENE	D	SHIOCTON	B
RYEPATCH	D	SAN MATEO	B	SCHEBLY	C	SERNA	D	SHIPLEY	B
RYER	C	SAN MIGUEL	C	SCHERRARD	D	SEROCO	A	SHIPROCK	B
RYORP	C	SANPETE	B	SCHLEY	B	SERPA	C/D	SHIRAT	B
RYUS	C	SAN POIL	B	SCHMUTZ	B	SERVOSS.	D	SHIRK	C
		SAN SABA	D	SCHNEBLY	D	SESAME	C	SHOALS	C
SABANA	D	SAN SEBASTIAN	B	SCHNEIDER	C	SESPE	C	SHOEBAR	C
SABANA SECA	D	SANTA	C	SCHNOORSON	B/D	SESSIONS	C	SHOEFFLER	B
SABENYO	B	SANTA CLARA	C	SCHNORBUSH	C	SESSUM	D	SHONKIN	D
SABINA	C	SANTA FE	D	SCHODACK	D	SETTERS	C	SHOOFLIN	C
SABINE	A	SANTA ISABEL	D	SCHOOSON	C	SETTLEMEYER	D	SHOOK	A
SABLE	D	SANTA LUCIA	C	SCHOFIELD	B	SEVAL	D	SHOREWOOD	C
SAC	B	SANTA MARTA	C	SCHOHARIE	C	SEVERN	B	SHOREY	B
SACO	D	SANTANA	C	SCHOLLE	B	SEVILLE	D	SHORN	B
SACRAMENTO	C/D	SANTAQUIN	A	SCHOOLEY	C/D	SEVY	C	SHORT CREEK	D
SACUL	D	SANTA YNEZ	C	SCHOONER	D	SEWARD	B	SHOSHONE	D
SADDLE	B	SANTEE	D	SCHRADER	D	SEWELL	B	SHOTWELL	D
SADDLEBACK	B	SANTIAGO	B	SCHRAP	D	SEXTON	D	SHOUNS	B
SADER	D	SANTIAM	C	SCHRIER	B	SEYMOUR	C	SHOWALTER	C
SADIE	B	SAN TIMOTEO	C	SCHROCK	B	SHAAK	D	SHOWLOW	C
SADLER	C	SANTONI	D	SCHUMACHER	B	SHADELAND	C	SHREWSBURY	D
SAFFELL	B	SANTOS	C	SCHUYLKILL	C	SHAFFER	A	SHRINE	B
SAGANING	D	SANTO TOMAS	B	SCIO	B	SHAKAN	B	SHROE	D
SAGE	D	SAN YSIDRO	D	SCIOTOVILLE	C	SHAKESPEARE	C	SHROUTS	D
SAGEHILL	B	SAPP	D	SCISM	B	SHAKOPEE	C	SHUBUTA	C
SAGEMOOR	C	SAPPHIRE	B	SCITUATE	C	SHALCAR	D	SHULE	B
SAGERTON	C	SAPPHO	B	SCOBEY	C	SHALET	D	SHULLSBURG	C
SAGINAW		SAPPINGTON	B	SCOOTENEY	B	SHAM	D	SHUMWAY	D
SAGO	D	SARA	C	SCORUP	C	SHAMBO	B	SHUPERT	C
SAGOUSPE	C	SARALEGUI	B	SCOTT	D	SHAMEL	B	SHUWAH	B
SAGUACHE	A	SARANAC	D	SCOTT LAKE	B	SHANAHAN	B	SI	B
SAHALIE	B	SARAPH	D	SCOUT	B	SHANDON		SIBLEYVILLE	B
SAINT HELENS	A	SARATOGA	B	SCOWLALE	C	SHANE	D	SIBYLEE	B
SAINT MARTIN	C	SARATON	B	SCRANTON	B/D	SHANO	B	SICILY	B
SALADO	B	SARBEN	A	SCRAVO	A	SHANTA	B	SICKLESTEETS	B
SALADON	D	SARCO	B	SCRIBA	C	SHAPLEIGH	C/D	SIDELL	C
SALAL	B	SARDINIA	C	SCRIVER	B	SHARATIN	B	SIEANCIA	B
SALAMATOF	D	SARDO	B	SCROGGIN	C	SHARKEY	D	SIEBER	A
SALAS	C	SARGEANT	D	SCULLIN	C	SHARON	B	SIELO	C
SALCHAKET	B	SARITA	A	SEABROOK	C	SHARPSBURG	B	SIEROCLIFF	C
SALEM	B	SARKAR	D	SEAMAN	C	SHARROTT	D	SIERRA	B
SALEMSBURG	B	SARPY	A	SEAQUEST	C	SHARVANA	C	SIERRAVILLE	B
SALGA	C	SARTELL	A	SEARCHLIGHT	B	SMASKIT	B/C	SIESTA	D
SALIDA	A	SASKA	B	SEARING	B	SHASTA	A	SIFTON	B
SALINAS	C	SASPANCO	B	SEARLA	B	SHAVANO	B	SIGNAL	C
SALISBURY	D	SASSAFRAS	B	SEARLES	C	SHAVER	B	SIGURD	B
SALIX	B	SASSER	B	SEATON	B	SHAMA	B	SIKESTON	D
SALKUM	C	SATANKA	C	SEATTLE	D	SHAMANO	A	SILCOX	B
SALLISAW	B	SATANTA	B	SEAWILLOW	B	SHAMMUT	B	SILENT	D
SALLYANN	C	SATELLITE	C	SEBAGO	D	SHAY	D	SILER	B
SALMON	B	SATT	D	SEBASTIAN	D	SHEAR	C	SILERTON	B
SALOL	D	SATTLEY	B	SEBASTOPOL	C	SHECKLER	C	SILI	D
SALONIE	D	SATTRE	B	SEBEKA	D	SHEDADO	B	SILSTID	A
SALREE	C/D	SATURN	B	SEBEWA	B/D	SHEDD	C	SILVER	C
SALTAIR	D	SATUS	B	SEBREE	D	SHEEGE	D	SILVERADO	C
SALT CHUCK	A	SAUCIER	B	SEBRING	D	SHEEP CREEK	C	SILVERBOW	D
SALTER	B	SAUDE	B	SEBUD	B	SHEEPHEAD	C	SILVER CREEK	D
SALTERY	D	SAUGATUCK	C	SECATA	C/D	SHEEPROCK	A	SILVERTON	C
SALT LAKE	D	SAUGUS	B	SECCA	C	SHEET IRON	B	SILVIES	D
SALUDA	C			SECRET	C	SHEFFIELD	D	SIMAS	C
SALUVIA				SECRET CREEK	B	SHELBURNE	C	SIMCOE	C

Name	Group	Name	Group	Name	Group
SIMEON	A	SNOW	B	SQUALICUM	B
SIMMLER	D	SNOWDEN	C	SQUAM	B
SIMMONT	C	SNOWLIN	B	SQUILCHUCK	B
SIMNER	A	SNOWVILLE	D	SQUIMER	B
SIMON	C	SNOWY	A	SQUIRES	B
SIMONA	D	SOAKPAK	B	ST. ALBANS	B
SIMOTE	C	SOAP LAKE	B	ST. CHARLES	B
SIMPERS	B	SOBOBA	A	ST. CLAIR	D
SIMPSON	C	SOBRANTE	C	ST. ELMO	A
SIMS	D	SODA LAKE	B	ST. GEORGE	C
SINAI	C	SODHOUSE	D	ST. HELENS	A
SINCLAIR	C	SODUS	C	ST. IGNACE	C
SINE	C	SOELBERG	C	ST. JOE	B/D
SINGLETREE	C	SOFIA	B	ST. JOHNS	B/D
SINGSAAS	B	SOGN	D	ST. LUCIE	A
SINNIGAM	C	SOGZIE	B	ST. MARTIN	C
SINOMAX	B	SOKOLOF	B	ST. MARYS	B
SINTON	B	SOLANO	D	ST. NICHOLAS	D
SINUK	D	SOLDATNA	B	ST. PAUL	B
SION	B	SOLDIER	C	ST. THOMAS	D
SIOUX	A	SOL DUC	B	STAATSBURG	
SIPPLE	A	SOLDUC	B	STABLER	B
SIRI	B	SOLLEKS	C	STACY	B
SISKIYOU	B	SOLLER	D	STADY	B
SISSETON	B	SOLOMON	D	STAFFORD	C
SISSON	B	SOLONA	B	STAGECOACH	B
SITES	C	SOMBRERO	D	STAHL	C
SITKA	B	SOMERS	B	STALEY	C
SIXMILE	B	SOMERSET	D	STAMBAUGH	B
SIZEMORE	B	SOMERVELL	B	STAMFORD	D
SIZER	B	SOMSEN	C	STAMPEDE	D
SKAGGS	B	SONOITA	B	STAN	B
SKAGIT	B/C	SONOMA	D	STANDISH	C/D
SKAMA	A	SONTAG	D	STANEY	D
SKALAN	C	SOPER	B/C	STANFIELD	C
SKAMANIA	B	SOQUEL	B	STANLEY	C
SKAMOKAWA	B	SORDO	C	STANSBURY	D
SKANEE	C	SORF	C	STANTON	D
SKELLOCK	B	SORRENTO	B	STAPLETON	B
SKERRY	C	SORTER	B/D	STARBUCK	D
SKIDMORE	B	SOSA	C	STARGO	B
SKILLET	C	SOTELLA	C	STARICHKOF	D
SKINNER	C	SOTIN	B	STARKS	C
SKIYOU	C	SOUTHFORK	D	STARLEY	D
SKOKOMISH	B/C	SOUTHGATE	D	STARR	B
SKOOKUMCHUCK	B	SOUTHWICK	C	STASER	B
SKOWHEGAN	D	SPAA	D	STATE	B
SKULL CREEK	D	SPACE CITY	A	STATEN	D
SKUMPAH	D	SPADE	B	STATLER	B
SKUTUM	C	SPALDING	D	STAVE	D
SKYBERG	C	SPAN	D	STAYTON	D
SKYHAVEN	D	SPANAWAY	B	STEAMBOAT	D
SKYKOMISH	B	SPANEL	D	STEARNS	D
SKYLICK	C	SPARTA	A	STECUM	A
SKYLINE	D	SPEARFISH	B	STEED	A
SKYWAY	B	SPEARMAN	C	STEEDMAN	D
SLAB	D	SPEARVILLE	C	STEEKEE	C
SLATE CREEK	C	SPECK	D	STEELE	B
SLAUGHTER	C	SPECTER	D	STEESE	C
SLAVEN	D	SPEELYAI	C	STEFF	C
SLAWSON	B	SPEIGLE	B	STEGALL	C
SLAYTON	D	SPENARD	D	STEIGER	A
SLEETH	C	SPENCER	B	STEINAUER	B
SLETTEN	D	SPENLO	B	STEINBECK	B
SLICKROCK	B	SPERRY	C	STEINMETZ	D
SLIGHTS	D	SPICER	C	STEINSBURG	C
SLIGO	B	SPILLVILLE	B	STEIWER	C
SLIKOK	D	SPINKS	A	STELLAR	C
SLIP	B	SPIRES	D	STEMILT	C
SLIPMAN	B/C	SPIRIT	B	STENDAL	C
SLOAN	D	SPIRO	B	STEPHEN	B
SLOCUM	B	SPLENDORA	C	STEPHENSBURG	B
SLODUC	C	SPLITRO	D	STEPHENVILLE	B
SLOSS	C	SPOFFORD	C	STERLING	B
SLUICE	B	SPOKANE	B	STERLINGTON	B
SMARTS	B	SPOUNSELLER	B	STETSON	B
SMITH CREEK	A	SPOON BUTTE	C	STETTER	D
SMITHDALE	B	SPOONER	C	STEUBEN	B
SMITHNECK	B	SPOTTSWOOD	B	STEVENS	B
SMITHTON	D	SPRAGUE	B/C	STEVENSON	B
SMOLAN	C	SPRECKELS	C	STEWART	D
SMOOT	D	SPRING	C/D	STICKNEY	C
SNAG	B	SPRING CREEK	C	STIDHAM	A
SNAHOPISH	B	SPRINGDALE	B	STIGLER	C
SNAKE	C	SPRINGER	B	STILLMAN	A
SNAKE HOLLOW	B	SPRINGERVILLE	D	STILLWATER	D
SNAKELUM	B	SPRINGFIELD	D	STILSON	D
SNEAD	D	SPRINGMEYER	C	STIMSON	B/C
SNELL	C	SPRINGTOWN	C	STINGAL	B
SNELLING	B	SPROUL	D	STINSON	C
SNOHOMISH	D	SPUR	B	STIRK	D
SNOQUALMIE	B	SPURLOCK	B	STIRUM	B

Name	Group	Name	Group	Name	Group
STISSING	C	SURGH	B	TABERNASH	B
STIVERSVILLE	B	SURPRISE	B	TABIONA	B
STOCKBRIDGE	B	SURRENCY	B/D	TABLE MOUNTAIN	B
STOCKLAND	B	SURVYA	C	TABLER	D
STOCKPEN	D	SUSIE CREEK	D	TABOR	D
STOCKTON	D	SUSITNA	B	TACAN	B
STODICK	D	SUSQUEHANNA	D	TACOMA	D
STOKES	D	SUTHER	C	TACODSH	D
STOMAR	C	SUTHERLIN	C	TAFT	C
STONER	B	SUTLEW	B/C	TAGGERT	C
STONEWALL	A	SUTPHEN	A	TAHOMA	B
STONO	B/D	SUTTLER	B	TAHQUAMENON	D
STOOKEY	B	SUTTON	B	TAHQUATS	C
STORDEN	B	SVEA	B	TAINTOR	C
STORLA	B	SVERDRUP	B	TAJO	C
STORMITT	B	SVOLD	C	TAKEUCHI	C
STORM KING	D	SWAGER	C	TAKILMA	B
STORY	C	SWAKANE	C	TAKOTNA	B
STOSSEL	C	SWAN	C	TALAG	D
STOUGH	C	SWANBOY	D	TALANTE	C
STOWELL	D	SWANNER	B	TALAPUS	B
STOY	C	SWANSON	B	TALBOTT	C
STRAIGHT	C	SWANTON	B/D	TALCOT	C
STRAIN	B	SWANTOWN	C	TALIHINA	D
STRASBURG	C	SWAPPS	C	TALKEETNA	C
STRATFORD	B	SWARTSWOOD	C	TALLAC	B
STRAUSS	C	SWARTZ	D	TALLADEGA	C
STRAW	B	SWASEY	D	TALLAPOOSA	C
STRAWN	B	SWASTIKA	C	TALLEYVILLE	B
STREATOR	C	SWATARA	A	TALLS	B
STROLE	B	SWAUK	C	TALLULA	B
STRONGHURST	B	SWAWILLA	A		
STRONTIA	B	SWEATMAN	C		
STROUPE	C	SWEDE	B		
STRYKER	B	SWEDEN	B		
STUBBS	C	SWEEN	C		
STUCKCREEK	B	SWEENEY	B		
STUKEL	D	SWEET	C		
STUKEY	B	SWEETGRASS	B		
STUMBLE	A	SWEETWATER	C		
STUMPP	D	SWENODA	B		
STUMP SPRINGS	B	SWIFTCREEK	B		
STUNNER	B	SWIFTON	A		
STUTTGART	D	SWIMS	A		
STUTZMAN	B	SWINGLER	C		
STUTZVILLE	B/C	SWINK	D		
SUBLETTE	B	SWISBOB	D		
SUDBURY	B	SWITCHBACK	C		
SUDDUTH	C	SWITZERLAND	B		
SUFFIELD	C	SWOPE	C		
SUGARLOAF	B	SWYGERT	C		
SUISUN	D	SYCAMORE	B/C		
SULA	B	SYCAN	A		
SULLY	B	SYLACAUGA	B/D		
SULPHURA	D	SYLVAN	B		
SULTAN	B	SYNERTON	B		
SUMAS	B/C	SYNAREP	B		
SUMDUM	D	SYRACUSE	B		
SUMMA	B	SYRENE	D		
SUMMERFIELD	C	SYRETT	C		
SUMMERS	B				
SUMMERVILLE	C				
SUMMIT	C				
SUMMITVILLE	B				
SUMTER	C				
SUN	D				
SUNBURST	C				
SUNBURY	B				
SUNCOOK	A				
SUND	C				
SUNDELL	C				
SUNDERLAND	C/D				
SUNDOWN	B				
SUNFIELD	B				
SUNNILAND	C				
SUNNYHAY	D				
SUNNYSIDE	B				
SUNNYVALE	C				
SUNRAY	B				
SUNRISE	D				
SUNSET	B				
SUNSHINE	C				
SUNSWEET	C				
SUNUP	D				
SUPAN	B				
SUPERIOR	C				
SUPERSTITION	A				
SUPERVISOR	C				
SUPPLEE	B				
SUR	B				
SURGEM	C				

Name	Grp	Name	Grp	Name	Grp	Name	Grp	Name	Grp
TALLY	B	TENINO	B	TIGERON	A	TOMERA	D	TRENTON	D
TALMAGE	A	TENNO	D	TIGIWON	B	TOMICHI	A	TREP	B
TALMO	B	TENORIO	B	TIGRET1		TOMOKA	A/D	TRES HERMANOS	B
TALOKA	D	TENOT	C	TIGUA	D	TONASKET	B	TRETTEN	C
TALPA	D	TENRAG	B	TIJERAS	B	TONATA	C	TREVINO	D
TAMA	B	TENSAS	D	TILFORD	B	TONAWANDA	C	TREXLER	C
TAMAHA	C	TENSED	C	TILLEDA	B	TONEY	D	TRIAMI	C
TAMALCO	D	TENSLEEP	B	TILLICUM	B	TONGUE RIVER	C	TRIASSIC	
TAMBA	C/D	TEOCULLI	B	TILLMAN	C	TONINI	B	TRICON	C
TAMELY	:	TEPEE	D	TILMA	C	TONKA	C	TRIDELL	B
TAMMANY CREEK	B	TEPETE	B/D	TILSIT	C	TONKEY	D	TRIDENT	D
TAMMANY RIDGE	B	TERBIES	C	TILTON	B	TONKIN	C	TRIGO	C
TAMMS	C	TERESA	C	TIMBERG	C	TONKS	B/D	TRIMBLE	B
TAMPICO	B	TERINO	D	TIMBERLY	B	TONOPAH	B	TRIMMER	B
TANAMA	D	TERMINAL	D	TIMBLIN	D	TONOR	C	TRINCHERA	C
TANANA	D	TERMO	C	TIMENTWA	B	TONOWEK	B	TRINITY	D
TANBERG	D	TEROUGE	D	TIMKEN	D	TONRA	A	TRIOMAS	B
TANDY	C	TERRA CEIA	A/D	TIMMERMAN	B	TONSINA	B	TRIPIT	C
TANEUM	C	TERRAD	D	TIMMONS	B	TONUCO	C	TRIPLEN	B
TANEY	C	TERRERA	C	TIMPAHUTE	B	TOOLE	D	TRIPOLI	C
TANGAIR	C	TERRETON	C	TIMPANOGOS	B	TOOMES	D	TRIPP	B
TANNA	C	TERRIL	B	TIMPER	D	TOP	C	TRITON	C
TANNER	C	TERRY	B	TIMPOONEKE	B	TOPIA	D	TRIX	B
TANSEM	B	TERWILLIGER	C	TIMULA	B	TOPPENISH	B/C	TROJAN	C
TANTALUS	A	TESAJO	A	TINA	C	TOPTON		TROMMALD	D
TANWAX	D	TESCOTT	C	TINDAHAY	A	TOQUERVILLE	C	TROMP	C
TAOPI	C	TESUQUE	B	TINE	A	TOQUOP	A	TRONSEN	B
TAOS	D	TETON	A	TINGEY	B	TORBOY	B	TROOK	B
TAPIA	C	TETONIA	B	TINSLEY	A	TORCHLIGHT	C	TROPAL	D
TAPPEN	D	TETONKA	C	TINTON	A	TORDIA	D	TROSI	D
TARA	B	TETOTUM	C	TINYTOWN	B	TORHUNTA	C	TROUP	A
TARKIO	D	TEW	B/D	TIOCANO	D	TORNING	B	TROUT CREEK	C
TARKLIN	C	TEX	B	TIOGA	B	TORODA	B	TROUTDALE	B
TARPO	C	TEXLINE	B	TIPPAH	C	TORONTO	C	TROUT LAKE	C
TARRANT	D	TEZUMA	C	TIPPECANOE	B	TORPEDO LAKE	D	TROUT RIVER	A
TARRETE	D	THACKERY	B	TIPPER	A	TORREON	C	TROUTVILLE	B
TARRYALL	B	THADER	C	TIPPERARY	A	TORRES	B	TROXEL	B
TASCOSA	B	THAGE	C	TIPPIPAH	D	TORRINGTON	B	TROY	C
TASSEL	D	THANYON	A	TIPPO	C	TORRO	C	TRUCE	C
TATE	B	THATCHER	B	TIPTON	B	TORSIDO	D	TRUCKEE	C
TATIYEE	C	THATUNA	C	TIPTONVILLE	B	TORTUGAS	D	TRUCKTON	B
TATU	C	THAYNE	B	TIRO	C	TOSTON	D	TRUEFISSURE	A
TATUM	C	THEBES	B	TISBURY	B	TOTELAKE	A	TRUESDALE	C
TAUNTON	C	THEBO	D	TISCH	C	TOTEM	B	TRULL	C
TAVARES	A	THEODALUND	C	TISH TANG	B	TOTTEN	B	TRULON	B
TAWAS	A/D	THENAS	C	TITUSVILLE	C	TOUCHET	B	TRUMAN	B
TAWCAW	C	THEO	C	TIVERTON	A	TOUHEY	B	TRUMBULL	D
TAYLOR	C	THERESA	B	TIVOLI	A	TOULON	B	TRUMP	D
TAYLOR CREEK	D	THERIOT	D	TIVY	C	TOURN	C	TRYON	D
TAYLORSFLAT	D	THERMAL	C	TOA	C	TOURNQUIST	B	TSCHIGOMA	B
TAYLORSVILLE	C	THERMOPOLIS	D	TOBICO	D	TOURS	B	TUB	C
TAYSON	B	THESS	B	TOBIN	B	TOUTLE	A	TUBAC	C
TAZLINA	A	THETFORD	A	TOBISH	C	TOWER	D	TUCANNON	C
TEAL	D	THIEL	A	TOBLER	B	TOWHEE	D	TUCKERMAN	D
TEALSON	C	THIGKOL	C	TOBOSA	D	TOWNER	B	TUCSON	B
TEALWHIT	C	THOENY	D	TOBY	B	TOWNLEY	C	TUCUMCARI	B
TEANAWAY	C	THOMAS	D	TOCCOA	B	TOWNSBURY	B	TUFFIT	D
TEAPO	B	THORNDALE	D	TODD	B	TOWNSEND	C	TUGHILL	D
TEAS	C	THORNDIKE	C/D	TODDLER	B	TOWSON	B	TUJUNGA	A
TEASDALE	B	THORNOCK	D	TODDVILLE	B	TOXAWAY	D	TUKEY	C
TEBO	B	THORNTON	D	TOEHEAD	C	TOY	D	TUKWILA	C
TECHICK	B	THORNWOOD	B	TOEJA	C	TOYAH	B	TULA	C
TECOLOTE	B	THOROUGHFARE	B	TOEM	C	TOZE	B	TULANA	C/D
TECUMSAH	B	THORP	C	TOGO	B	TRABUCO	C	TULARE	C/D
TEDROW	B	THORR	B	TOGUS	D	TRACK	B/C	TULAROSA	B
TEEL	B	THORREL	B	TOHONA	C	TRACY	B	TULIA	B
TEHACHAPI	D	THOW	B	TOINE	C	TRAER	C	TULLAHASSEE	C
TEHAMA	C	THREE MILE	D	TOISNOT	D	TRAIL	A	TULLER	D
TEJA	[	THROCK	C	TOIYABE	C	TRAIL CREEK	B	TULLOCK	B
TEJON	B	THUNDERBIRD	D	TOKEEN	B	TRAM	B	TULLY	C
TEKOA	C	THURBER	C	TOKUL	C	TRANSYLVANIA	B	TULUKSAK	D
TELA	B	THURLONI	C	TOLBY	A	TRAPPER	A	TUMBEZ	B
TELEFONO	C	THURLOW	C	TOLEDO	D	TRAPPIST	C	TUMEY	D
TELEPHONE	D	THURMAN	A	TOLICHA	D	TRAPPS	B	TUMITAS	B
TELFER	A	THURMONT	B	TOLKE	B	TRASK	C	TUMWATER	A
TELFERNER	D	THURSTON	B	TOLL	A	TRAVELERS	D	TUNEHEAN	D
TELIDA	D	TIAGOS	B	TOLLGATE	B	TRAVER	B/C	TUNICA	D
TELL	B	TIAK	C	TOLLHOUSE	D	TRAVESSILLA	D	TUNIS	D
TELLER	B	TIBAN	B	TOLMAN	D	TRAVIS	C	TUNITAS	B
TELLICO	B	TIBBITTS	B	TOLNA	B	TRAWICK	B	TUNKHANNOCK	A
TELLMAN	B	TICA	D	TOLO	B	TRAY	C	TUNNEL	B
TELSTAD	B	TICE	C	TOLSONA	D	TREADWAY	D	TUPELO	D
TEMESCAL	D	TICHIGAN	C	TOLSTOI	D	TREASURE	B	TUPUKNUK	D
TEMPLE	B/C	TICHNOR	D	TOLT	D	TREBLOC	D	TUQUE	B
TEMVIK	B	TICKAPOO	D	TOLTEC	C	TREGO	C	TURBEVILLE	C
TENABO	D	TICKASON	B	TOLUCA	B	TRELONA	D	TURBOTVILLE	C
TENAHA	B	TIDWELL	D	TOLVAR	B	TREMANT	B	TURBYFILL	B
TENAS	C	TIERRA	D	TOMAH	C	TREMBLES	B	TURIN	B
TENCEE	D	TIETON	B	TOMAS	B	TREMPE	A	TURK	D
TENERIFFE	C	TIFFANY	C	TOMAST	C	TREMPEALEAU	B	TURKEYSPRINGS	C
TENEX	A	TIFTON	B	TOME	B	TRENARY	B	TURLEY	C
TENIBAC	B	TIGER CREEK	B	TOMEL	D	TRENT	B	TURLIN	B

TURNBOW	C	USINE	B	VERDUN	D	WADDELL	B	WARDEN	B
TURNER	B	USKA	D	VERGENNES	D	WADDOUPS	B	WARDWELL	C
TURNERVILLE	B	UTALINE	B	VERHALEN	D	WADELL	B	WARE	B
TURNEY	B	UTE	C	VERMEJO	D	WADENA	B	WAREHAM	C
TURRAH	D	UTICA	A	VERNAL	B	WADESBORO	B	WARMAN	D
TURRET	B	UTLEY	B	VERNALIS	B	WADLEIGH	D	WARM SPRINGS	C
TURRIA	C	UTUADO	B	VERNIA	A	WADMALAW	A	WARNERS	A/D
TURSON	B/C	UVADA	D	VERNON	D	WADSWORTH	C	WARREN	B
TUSCAN	D	UVALDE	C	VERONA	C	WAGES	B	WARRENTON	B/D
TUSCARAWAS	C	UWALA	B	VESSER	C	WAGNER	D	WARRIOR	
TUSCARORA	C			VESTON	D	WAGRAM	A	WARSAW	B
TUSCOLA	B	VACHERIE	C	VETAL	A	WAHA	C	WARSING	B
TUSCUMBIA	D	VADER	B	VETERAN	B	WAHEE	D	WARWICK	A
TUSEL	C	VADO	B	VEYO	D	WAHIAWA	B	WASATCH	A
TUSKEEGO	C	VAIDEN	D	VIA	B	WAHIKULI	B	WASEPI	B
TUSLER	B	VAILTON	B	VIAN	B	WAHKEENA	B	WASHBURN	
TUSQUITEE	B	VALBY	C	VIBORAS	D	WAHKIACUS	B	WASHINGTON	B
TUSTIN	B	VALCO	C	VIBORG	B	WAHLUKE	B	WASHOE	C
TUSTUMENA	B	VALDEZ	B/C	VICKERY	C	WAHMONIE	D	WASHOUGAL	B
TUTHILL	B	VALE	B	VICKSBURG	B	WAHPETON	C	WASHTENAW	C/D
TUTNI	B	VALENCIA	B	VICTOR	A	WAHTIGUP	B	WASIELA	D
TUTWILER	B	VALENT	A	VICTORIA	D	WAHTUM	D	WASIOJA	B
TUXEDO		VALENTINE	A	VICTORY	B	WAIAHA	D	WASSAIG	B
TUXEKAN	B	VALERA	C	VICU	D	WAIAKOA	C	WATAB	C
TWIN CREEK	B	VALKARIA	B/D	VIDA	B	WAIALEALE	D	WATAUGA	B
TWINING	C	VALLAN	D	VIDRINE	C	WAIALUA	B	WATCHAUG	B
TWISP	B	VALLECITOS	C/D	VIEJA	D	WAIAWA	D	WATCHUNG	D
TWO DOT	C	VALLEONO	B	VIENNA	B	WAIHUNA	D	WATERBORO	
TYBO	D	VALLERS	C	VIEQUES	B	WAIKALOA	B	WATERBURY	D
TYEE	D	VALMONT	C	VIEW	C	WAIKANE	B	WATERINO	C
TYGART	D	VALMY	B	VIGAR	C	WAIKAPU	B	WATERS	C
TYLER	D	VALOIS	B	VIGO	D	WAIKOMO	D	WATKINS	B
TYNDALL	B/C	VAMER	D	VIGUS	C	WAILUKU	B	WATKINS RIDGE	B
TYNER	A	VANAJO	D	VIKING	D	WAINEA	B	WATO	B
TYRONE	C	VANANOA	D	VIL	D	WAINEE	B	WATOPA	B
TYSON	C	VAN BUREN		VILAS	A	WAINOLA	A	WATROUS	B
		VANCE	C	VILLA GROVE	B	WAIPAHU	C	WATSEKA	C
UANA	D	VANDA	D	VILLARS	B	WAISKA	B	WATSON	C
UBAR	C	VANDALIA	C	VILLY	D	WAITS	B	WATSONIA	D
UBLY	B	VANDERDASSON	D	VINA	B	WAKE	D	WATSONVILLE	D
UCOLA	D	VANDERGRIFT	C	VINCENNES	C	WAKEEN	B	WATT	D
UCOLO	C	VANDERHOFF	D	VINCENT	C	WAKEFIELD	B	WATTON	C
UCOPIA	B	VANDERLIP	A	VINEYARD	C	WAKELAND	B/D	WAUBAY	B
UDEL	D	VAN DUSEN	B	VINGO	B	WAKONDA	C	WAUBEEK	B
UJOLPHO	C	VANET	D	VINING	C	WAKULLA	A	WAUBONSIE	B
UFFENS	D	VANG	B	VINITA	C	WALCOTT	B	WAUCHULA	B/D
UGAK	D	VANHORN	B	VINLAND	C	WALDECK	C	WAUCOMA	B
UHLAND	B	VAN NOSTERN	B	VINSAD	C	WALDO	D	WAUCONDA	B
UHLIG	B	VANNOY	B	VINT	B	WALDRON	D	WAUKEE	B
UINTA	B	VANOSS	B	VINTON	B	WALDROUP	D	WAUKEGAN	B
UKIAH	C	VANTAGE	C	VIRA	C	WALES	B	WAUKENA	D
ULEN	B	VAN WAGONER	D	VIRATON	C	WALFORD	C	WAUKON	B
ULLOA	B	VARCO	C	VIRDEN	C	WALKE	C	WAUMBEK	B
ULM	B	VARELUM	C	VIRGIL	B	WALL	B	WAURIKA	D
ULRICHER	B	VARICK	D	VIRGIN PEAK	D	WALLACE	B	WAUSEON	B/D
ULUPALAKUA	B	VARINA	C	VIRGIN RIVER	D	WALLA WALLA	B	WAVERLY	B/D
ULY	B	VARNA	C	VIRTUE	C	WALLER	B/D	WAWAKA	C
ULYSSES	B	VARRO	B	VISALIA	B	WALLINGTON	C	WAYCUP	B
UMA	A	VARYSBURG	B	VISTA	C	WALLIS	B	WAYDEN	D
UMAPINE	B/C	VASHTI	C	VIVES	B	WALLKILL	C/D	WAYLAND	C/D
UMIAT	D	VASQUEZ	B	VIVI	B	WALLMAN	C	WAYNE	B
UMIKOA	B	VASSALBORO	D	VLASATY	C	WALLOWA	C	WAYNESBORO	B
UMIL	D	VASSAR	B	VOCA	C	WALLPACK	C	WAYSIDE	
UMNAK	B	VASTINE	C	VODERMAIER	B	WALLROCK	B/C	WEA	B
UMPA	B	VAUCLUSE	C	VOLADORA	B	WALLSBURG	D	WEAVER	C
UMPQUA	B	VAUGHNSVILLE	C	VOLCO	D	WALLSON	B	WEBB	C
UNA	D	VAYAS	D	VOLENTE	C	WALPOLE	C	WEBER	B
UNADILLA	B	VEAL	B	VOLGA	D	WALSH	B	WEBSTER	C
UNAWEEP	B	VEAZIE	B	VOLIN	B	WALSHVILLE	D	WEDEKIND	D
UNCOM	B	VEBAR	B	VOLINIA	B	WALTERS	A	WEDERTZ	C
UNCOMPAHGRE	D	VECONT	D	VOLKE	C	WALTON	C	WEDGE	A
UNEEDA	B	VEGA	C	VOLKMAR	B	WALUM	B	WEDOWEE	D
UNGERS	B	VEGA ALTA	C	VOLMER	D	WALVAN	D	WEED	B
UNION	C	VEGA BAJA	C	VOLNEY	B	WAMBA	B/C	WEEDING	A/C
UNIONTOWN	B	VEKOL	D	VOLPERIE	C	WAMIC	B	WEEDMARK	B
UNIONVILLE	C	VELDA	B	VOLTAIRE	D	WANATAH	B	WEEKSVILLE	B/D
UNISUN	C	VELMA	B	VOLUSIA	C	WANBLEE	D	WEEPON	D
UPDIKE	D	VELVA	B	VONA	B	WANDO	A	WEHADKEE	D
UPSAL	C	VENA	C	VORE	B	WANETTA	A	WEIKERT	C/D
UPSATA	A	VENANGO	C	VROGMAN	B	WANILLA	C	WEIMER	D
UPSHUR	C	VENATOR	D	VULCAN	C	WANN	A	WEINBACH	C
UPTON	C	VENETA	C	VYLACH	D	WAPAL	B	WEIR	D
URACCA	B	VENEZIA	D			WAPATO	C/D	WEIRMAN	B
URBANA	C	VENICE	D	WABANICA	D	WAPELLO	B	WEISER	C
URBO	D	VENLO	D	WABASH	D	WAPINITIA	B	WEISHAUPT	D
URICH	D	VENUS	B	WABASHA	D	WAPPING	B	WEISS	A
URNE	B	VERBOORT	D	WABASSA	B/D	WAPSIE	B	WEITCHPEC	B
URSINE	D	VERDE	C	WABEK	B	WARBA	B	WELAKA	A
URTAH	C	VERDEL	D	WACA	C	WARD	D	WELBY	B
URWIL	D	VERDELLA	D	WACOTA	B	WARDBORO	A	WELCH	C
USAL	B	VERDICO	D	WACOUSTA	C	WARDELL	D	WELD	C
USHAR	B	VERDIGRIS	B	WADAMS	B			WELDA	C

Name	Group	Name	Group	Name	Group	Name	Group	Name	Group
WELDON	D	WICKIUP	C	WISNER	D	YALMER	B	ZUNDELL	B/C
WELDONA	B	WICKLIFFE	D	WITBECK	D	YAMAC	B	ZUNHALL	B/C
WELLER	C	WICKSBURG	B	WITCH	D	YAMHILL	C	ZUNI	D
WELLINGTON	D	WIDTSOE	C	WITHAM	D	YAMPA	C	ZURICH	B
WELLMAN	B	WIEHL	C	WITHEE	C	YAMSAY	D	ZWINGLE	D
WELLNER	B	WIEN	D	WITT	B	YANA	B		
WELLSBORO	C	WIGGLETON	B	WITZEL	D	YANCY	C		
WELLSTON	B	WIGTON	A	WODEN	B	YARDLEY	C		
WELLSVILLE	B	WILBRAHAM	C	WODSKOW	B/C	YATES	D		
WELRING	D	WILBUR	C	WOLCOTTSBURG		YAUCO	C		
WEMPLE	B	WILCO	C	WOLDALE	C/D	YAWDIM	D		
WENAS	B/C	WILCOX	D	WOLF	B	YAWKEY	C		
WENATCHEE	C	WILCOXSON	C	WOLFESEN	C	YAXON	B		
WENDEL	B/C	WILDCAT	D	WOLFESON	C	YEARY	C		
WENHAM		WILDER	B	WOLFORD	B	YEATES HOLLOW	C		
WENONA	C	WILDERNESS	C	WOLF POINT	D	YEGEN	B		
WENTWORTH	B	WILDROSE	D	WOLFTEVER	C	YELM	B		
WERLOW	C	WILDWOOD	D	WOLVERINE	A	YENRAB	A		
WERNER	B	WILEY	C	WOODBINE	B	YEOMAN	B		
WESO	C	WILKES	C	WOODBRIDGE	C	YESUM	B		
WESSEL	B	WILKESON	C	WOODBURN	C	YETULL	A		
WESTBROOK	D	WILKINS	D	WOODBURY	D	YODER	B		
WESTBURY	C	WILL	D	WOODCOCK	B	YOKOHL	D		
WESTCREEK	B	WILLACY	B	WOODENVILLE	C	YOLLABOLLY	D		
WESTERVILLE	C	WILLAKENZIE	C	WOODGLEN	D	YOLO	B		
WESTFALL	C	WILLAMAR	D	WOODHALL	B	YOLOGO	D		
WESTFIELD		WILLAMETTE	B	WOODHURST	A	YOMBA	C		
WESTFORD		WILLAPA	C	WOODINVILLE	C/D	YOMONT	B		
WESTLAND	B/D	WILLARD	B	WOODLY	B	YONCALLA	C		
WESTMINSTER	C/D	WILLETTE	A/D	WOODLYN	C/D	YONGES	D		
WESTMORE	B	WILLHAND	B	WOODMANSIE	B	YONNA	B/D		
WESTMORELAND	B	WILLIAMS	B	WOODMERE	B	YORDY	B		
WESTON	D	WILLIAMSBURG	B	WOOD RIVER	D	YORK	C		
WESTPHALIA	B	WILLIAMSON	C	WOODROCK	C	YORKVILLE	D		
WESTPLAIN	C	WILLIS	C	WOODROW	C	YOST	C		
WESTPORT	A	WILLITS	B	WOODS CROSS	D	YOUGA	B		
WESTVILLE	B	WILLOUGHBY	B	WOODSFIELD	C	YOUMAN	C		
WETHERSFIELD	C	WILLOWCREEK	B	WOODSIDE	A	YOUNGSTON	B		
WETHEY	B/C	WILLOWDALE	B	WOODSON	D	YOURAME	A		
WETTERHORN	C	WILLOWS	D	WOODSTOCK	C/D	YOVIMPA	D		
WETZEL	D	WILLWOOD	A	WOODSTOWN	C	YSIDORA	D		
WEYMOUTH	B	WILMER	C	WOODWARD	B	YTURBIDE	A		
WHAKANA	B	WILPAR	D	WOOLMAN	B	YUBA	D		
WHALAN	B	WILSON	D	WOOLPER	C	YUKO	C		
WHARTON	C	WILTSHIRE	C	WOOLSEY	C	YUKON	D		
WHATCOM	C	WINANS	B/C	WOOSLEY	C	YUNES	D		
WHATELY	D	WINBERRY	D	WOOSTER	C	YUNQUE	C		
WHEATLEY	D	WINCHESTER	A	WOOSTERN	B				
WHEATRIDGE	C	WINCHUCK	C	WOOTEN	A	ZAAR	D		
WHEATVILLE	B	WINDER	B/D	WORCESTER	B	ZACA	D		
WHEELER	B	WINDHAM	B	WORF	D	ZACHARIAS	B		
WHEELING	B	WINDMILL	B	WORK	C	ZACHARY	D		
WHEELON	D	WINDOM	B	WORLAND	B	ZAFRA	B		
WHELCHEL	B	WIND RIVER	B	WORLEY	C	ZAHILL	B		
WHETSTONE	B	WINDSOR	A	WORMSER	C	ZAHL	B		
WHIDBEY	C	WINDTHORST	C	WOROCK	B	ZALESKI	C		
WHIPPANY	C	WINDY	C	WORSHAM	D	ZALLA	A		
WHIPSTOCK	C	WINEG	C	WORTH	C	ZAMORA	B		
WHIRLO	B	WINEMA	C	WORTHEN	B	ZANE	C		
WHIT	B	WINETTI	B	WORTHING	D	ZANEIS	B		
WHITAKER	C	WINFIELD	C	WORTHINGTON	C	ZANESVILLE	C		
WHITCOMB	C	WING	D	WORTMAN	C	ZANONE	C		
WHITE BIRD	C	WINGATE	B	WRENTHAM	C	ZAPATA	C		
WHITECAP	D	WINGER	C	WRIGHT	C	ZAVALA	B		
WHITEFISH	B	WINGVILLE	B/D	WRIGHTMAN	C	ZAVCO	C		
WHITEFORD	B	WINIFRED	C	WRIGHTSVILLE	D	ZEB	B		
WHITEHORSE	B	WINK	B	WUNJEY	B	ZEESIX	C		
WHITE HOUSE	C	WINKEL	D	WURTSBORO	C	ZELL	C		
WHITELAKE	B	WINKLEMAN	C	WYALUSING	D	ZEN	C		
WHITELAW	B	WINKLER	A	WYARD	B	ZENDA	C		
WHITEMAN	D	WINLO	D	WYARNO	B	ZENIA	B		
WHITEROCK	D	WINLOCK	C	WYATT	C	ZENIFF	B		
WHITESBURG	C	WINN	C	WYEAST	C	ZEONA	C		
WHITE STORE	D	WINNEBAGO	B	WYEVILLE	C	ZIEGLER	A		
WHITE SWAN	C	WINNEMUCCA	B	WYGANT	C	ZIGWEID	C		
WHITEWATER	B	WINNESHIEK	B	WYKOFF	B	ZILLAH	B/C		
WHITEWOOD	C	WINNETT	D	WYMAN	B	ZIM	D		
WHITLEY	B	WINONA	D	WYMORE	C	ZIMMERMAN	A		
WHITLOCK	B	WINOOSKI	B	WYNN	B	ZING	C		
WHITMAN	D	WINSTON	A	WYNOOSE	D	ZINZER	B		
WHITNEY	B	WINTERS	C	WYO	B	ZION	C		
WHITORE	A	WINTERSBURG	C	WYOCENA	B	ZIPP	C/D		
WHITSOL	B	WINTERSET	C			ZITA	B		
WHITSON	D	WINTHROP	A	XAVIER	B	ZOAR	C		
WHITWELL	C	WINTONER	C			ZOATE	D		
WHOLAN	C	WINU	C	YACOLT	B	ZOHNER	B/D		
WIBAUX	C	WINZ	C	YAHARA	B	ZOOK	C		
WICHITA	C	WIOTA	B	YAHOLA	B	ZORRAVISTA	A		
WICHUP	D	WISHARD	A	YAKI	D	ZUFELT	B/D		
WICKERSHAM	B	WISHEYLU	C	YAKIMA	B	ZUKAN	D		
WICKETT	C	WISHKAH	C	YAKUS	D	ZUMBRO	B		
WICKMAN	B	WISKAH	C	YALLANI	B	ZUMWALT	C		

CHAPTER 8

Sediment Transport and Unit Stream Power

CHIH TED YANG*

INTRODUCTION

The subject of sediment transport has been studied by both hydraulic engineers and geologists because of its importance to the understanding of river hydraulics, river morphology, and river engineering. Due to its complex nature, sediment transport is often subject to semiempirical or empirical treatment. Because of the large number of variables which may effect the rate of sediment transport and the interdependence among them, theoretical treatment is often based on simplified and idealized assumptions. It is often assumed that the rate of sediment transport can be determined by one or two dominant variables, such as water discharge, average flow velocity, water surface or energy slope, shear stress, etc. Numerous equations have been developed based on these assumptions. Each equation is supported by limited laboratory data and, occasionally, by field data. However, when different equations are applied to a given river, the computed results may differ drastically from each other and from the measurements. This is an indication of the lack of generality of the assumptions used in the derivation of these equations.

This chapter provides a brief review of the basic approaches used in the study of incipient motion and sediment transport. With the exception of unit stream power equations, mathematical derivations of transport equations are not included. Comparisons between measured and computed results from different equations are made to determine their accuracies. Recommendations are made on the selection of sediment transport equations under different flow and sediment conditions.

INCIPIENT MOTION

The flow condition under which sediment particles on the bed start to move, i.e., the condition of incipient motion, is important to the study of sediment transport. Criteria for incipient motion can be obtained from the balance of forces acting on a sediment particle. The magnitudes of these forces can be determined from either a shear stress or a velocity approach.

Shear Stress Approach

One of the most widely used criterion for incipient motion is the Shields' diagram. Shields' [24] applied dimensional analysis to determine some dimensionless parameters and established his well-known diagram for incipient motion. The factors which are important to the determination of incipient motion are shear stress τ, the difference in density between sediment and fluid ($\varrho_s - \varrho_f$), the diameter of the particle d, the kinematic viscosity ν, and gravitational acceleration g. These five quantities can be grouped into two dimensionless quantities, i.e.,

$$\frac{d \sqrt{\tau_c/\varrho_f}}{\nu} = \frac{d \, U_*}{\nu} \tag{1}$$

and

$$\frac{\tau_c}{d(\varrho_s - \varrho_f)g} = \frac{\tau_c}{d\gamma[(\varrho_s/\varrho_f) - 1]} \tag{2}$$

where

ν = kinematic viscosity
ϱ_s, ϱ_f = density of sediment and fluid, respectively
γ = the specific weight of water
U_* = the shear velocity
τ_c = critical shear stress at initial motion

*U.S. Department of the Interior, Bureau of Reclamation, Engineering and Research Center, Denver, CO

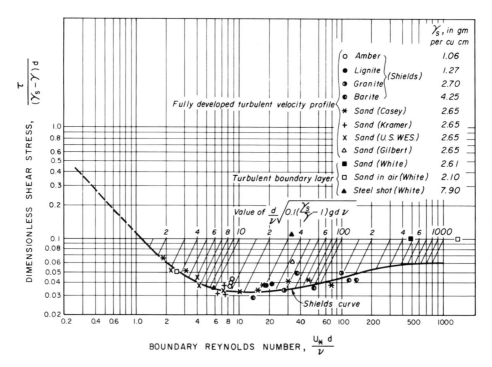

FIGURE 1. Shield's diagram for incipient motion [32].

The relationship between these two parameters was then determined from experiments. Figure 1 shows the experimental results obtained at incipient motion. At points above the line, the particle will move. At points below the line, the flow is unable to move the particle.

Although Shields' diagram has been widely used by engineers as a criterion for incipient motion, considerable dissatisfaction can be found in the literature. Yang [35] pointed out several deficiencies of the Shields' diagram and suggested that it may not be the most desirable criterion for incipient motion. One of the objections to the use of Shields' diagram is that the dependent variables appear in both ordinate and abscissa parameters because $U_* = \sqrt{\tau/\varrho_s}$. Depending on the nature of the problem, the dependent variable can be critical shear stress or grain size. The American Society of Civil Engineers Task Committee for the Preparation of Manual on Sedimentation [32] uses a third parameter

$$\frac{d}{\nu}\sqrt{0.1\left(\frac{\gamma_s}{\gamma}-1\right)gd}$$

as shown in Figure 1 to alleviate the problem. The use of this parameter enables us to determine its intersection with Shields' diagram and its corresponding values of shear stress.

Velocity Approach

There are several incipient motion criteria based on velocity approach. One of the more recent criteria is that proposed by Yang [35]. Yang's approach is based on established theories in fluid mechanics. Unlike the Shields' diagram which neglects the effect of lift force, Yang considered the effect of lift, drag, and resistance forces acting on a sediment particle at the state of incipient motion. The drag force acting on a sediment particle resting on the bed is

$$F_D = C_D \frac{\pi d^2}{4} \frac{\varrho}{2} V_d^2 \qquad (3)$$

where

C_D = drag coefficient at velocity V_d
d = particle diameter
ϱ = density of water
V_d = local velocity at a distance, d, above the bed

The terminal fall velocity of a spherical particle is reached when there is a balance between drag force and submerged weight of the particle, i.e., when

$$C_D' \frac{\pi d^2}{4} \frac{\varrho}{2} \omega^2 = \frac{\pi d^3}{6} (\varrho_s - \varrho)g \qquad (4)$$

where

C'_D = drag coefficient at ω
ω = terminal fall velocity of sediment
ϱ_s = density of sediment particle

By substituting C'_D with $\psi_1 C_D$, and eliminating C_D from Equations (3) and (4), the drag force becomes

$$F_D = \frac{\pi d^3}{6\,\psi_1\omega^2}\,(\varrho_s - \varrho)gV_d^2 \qquad (5)$$

where ψ_1 = coefficient.

If we assume that the logarithmic law for velocity distribution can be applied in this case, then

$$\frac{V_y}{U_*} = 5.75 \log \frac{y}{d} + B \qquad (6)$$

where

V_y = local velocity at distance y above the bed
B = roughness function

Based on the logarithmic velocity distribution, it can be

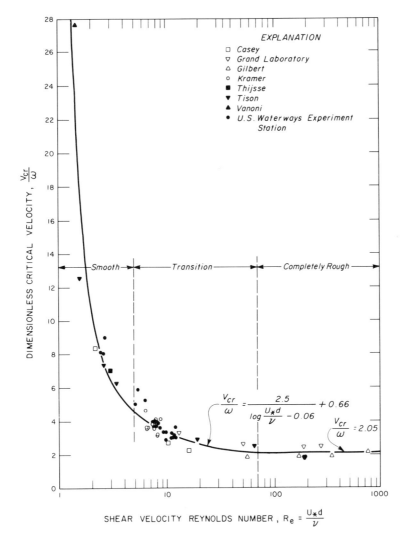

FIGURE 2. Relationships between dimensionless critical average velocity and Reynolds number [35].

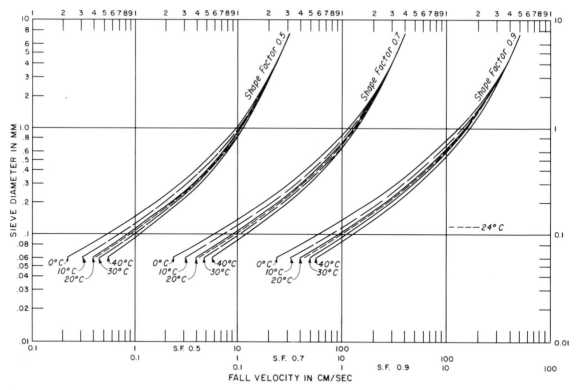

FIGURE 3. Relation of sieve diameter and fall velocity for naturally worn quartz particles falling alone in quiescent distilled water of infinite extent [31].

shown that

$$F_D = \frac{\pi d^3}{6\psi_1} (\varrho_s - \varrho)g \left(\frac{V}{\omega}\right)^2$$
$$\times \left[\frac{B}{5.75 \left(\log \dfrac{D}{d} - 1 \right) + B} \right]^2 \quad (7)$$

where D = water depth.

Similarly, the lift force can be expressed by

$$F_L = \frac{\pi d^3}{6\psi_1\psi_2} (\varrho_s - \varrho)g \left(\frac{V}{\omega}\right)^2$$
$$\times \left[\frac{B}{5.75 \left(\log \dfrac{D}{d} - 1 \right) + B} \right]^2 \quad (8)$$

where ψ_2 = coefficient.

The submerged weight of the particle is

$$W_s = \frac{\pi d^3}{6} (\varrho_s - \varrho)g \quad (9)$$

Then the resistant force becomes

$$F_R = \psi_3(W_s - F_L)$$
$$= \frac{\psi_3 \pi d^3}{6} (\varrho_s - \varrho)g \left\{ 1 - \frac{1}{\psi_1\psi_2}\left(\frac{V}{\omega}\right)^2 \right.$$
$$\left. \times \left[\frac{B}{5.75 \left(\log \dfrac{D}{d} - 1 \right) + B} \right]^2 \right\} \quad (10)$$

where ψ_3 = the friction coefficient.

Assume that the incipient motion occurs when $F_D = F_R$. From Equations (7) and (10),

$$\frac{V_{cr}}{\omega} = \left[\frac{5.75 \left(\log \dfrac{D}{d} - 1 \right)}{B} + 1 \right] \sqrt{\frac{\psi_1\psi_2\psi_3}{\psi_2 + \psi_3}} \quad (11)$$

where

V_{cr} = the average critical velocity at incipient motion
V_{cr}/ω = the dimensionless average critical velocity at incipient motion

Equation (11) is the basic equation which specifies the flow condition when a sediment particle is ready to move on the bed of an open channel. The values of ψ_1, ψ_2, and ψ_3 have to be determined from experiments. The roughness function, B, depends on whether the boundary is in a hydraulically smooth, transition, or completely rough regime, i.e.,

$$B = 5.5 + 5.75 \log \frac{U_* d}{\nu}, \, 0 < \frac{U_* d}{\nu} < 5 \quad (12)$$

$$B = 8.5, \frac{U_* d}{\nu} > 70 \quad (13)$$

Substituting Equations (12) and (13) into Equation (11) and calibrating the coefficients against laboratory data, Yang's incipient motion criteria becomes [35]

$$\frac{V_{cr}}{\omega} = \frac{2.5}{\log \frac{U_* d}{\nu} - 0.06} + 0.66, \, 1.2 < \frac{U_* d}{\nu} < 70 \quad (14)$$

and

$$\frac{V_{cr}}{\omega} = 2.05, \, 70 \leq \frac{U_* d}{\nu} \quad (15)$$

Comparison between Equations (14) and (15) and measured data by different investigators are shown on Figure 2. The application of Equations (14) and (15) are easy and straight-forward. The accuracy of these criteria have been more recently verified by laboratory tests [27] and field data [26]. Because most natural river sand has a shape factor of 0.7 and its size is outside the Stokes' range, engineers should use the measured fall velocity shown on Figure 3 [31] for the computation of dimensionless critical velocity.

BASIC APPROACHES IN SEDIMENT TRANSPORT

The basic approaches used in the development of sediment transport equations are deterministic, probabilistic, and regression. This section provides a brief review of the advantages and disadvantages of developing a transport equation from these approaches.

A. Deterministic Approach

Equations derived from deterministic approach assume that there is a one-to-one correlation between dependent and independent variables. The advantage of this approach is that once the values of independent variables are given, sediment transport rate or concentration can be computed directly. The disadvantage is that if the assumed one-to-one relationship does not exist, or only exists under certain special conditions, the computed results could be far from

reality. Most sediment transport equations were developed from deterministic approach and can be classified into one of the following basic forms.

$$Q_s = A_1 (Q - Q_c)^{B_1} \quad (16)$$

$$Q_s = A_2 (V - V_c)^{B_2} \quad (17)$$

$$Q_s = A_3 (S - S_c)^{B_3} \quad (18)$$

$$Q_s = A_4 (\tau - \tau_c)^{B_4} \quad (19)$$

$$Q_s = A_5 (\tau V - \tau_c V_c)^{B_5} \quad (20)$$

$$Q_s = A_6 (VS - V_c S_c)^{B_6} \quad (21)$$

where

Q, Q_s = water and sediment discharge, respectively
V = average flow velocity
S = energy or water surface slope
τ = shear stress
τV = stream power
VS = unit stream power
$A_1, A_2, A_3, A_4, A_5, A_6, B_1, B_2, B_3, B_4, B_5, B_6$ = parameters related to flow and sediment characteristics
c = subscript denotes the critical condition at incipient motion

Laboratory data collected by Guy et al. [14] with 0.93-mm sand, as shown on Figure 4, were used by Yang [34,38] as an example to examine the validity of these assumptions.

Figure 4a shows the relationship between total sediment discharge and water discharge. For a given value of Q, two different values of q_t can be obtained. Apparently, different sediment discharges can be transported by the same water discharge, and a given sediment discharge can be transported by different water discharges. The same sets of data shown on Figure 4a are plotted on Figure 4b to show the relationship between total sediment discharge and average velocity. Although q_t increases steadily with increasing V, it is apparent that for approximately the same value of V the value of q_t can differ considerably due to the steepness of the curve. Figure 4c indicates that there is no single relationship between total sediment discharge and slope. Figure 4d shows that there is a one-to-one correlation between total sediment discharge and shear stress when total sediment discharge is in the middle range of the curve. For either higher or lower sediment discharge, the curve becomes vertical, which means for the same shear stress numerous values of sediment discharge can be obtained.

It is apparent from Figures 4a, b, c, and d that more than one value of total sediment discharge can be obtained for the same value of water discharge, velocity, slope, or shear stress. The validity of the assumption that sediment discharge of a given particle size could be determined from water discharge, velocity, slope, or shear stress is questiona-

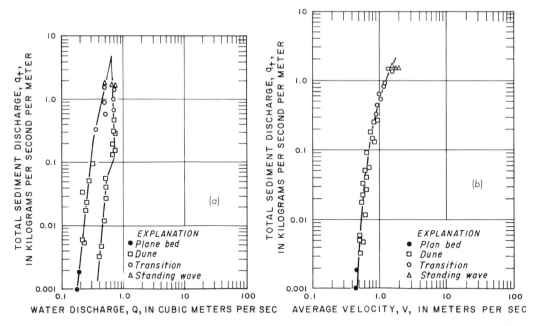

FIGURE 4a–4b. Relationship between (a) total sediment discharge and water discharge, (b) total sediment discharge and velocity for 0.93-mm sand in 8-ft-wide flume [38].

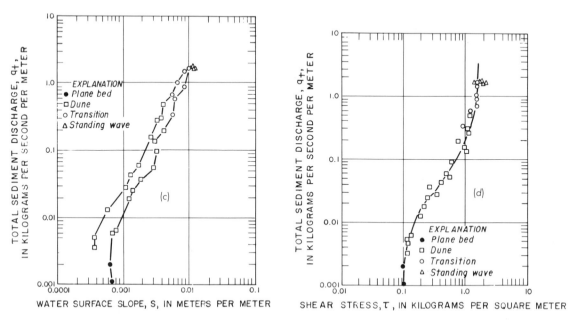

FIGURE 4c–4d. Relationship between (c) total sediment discharge and slope, (d) total sediment discharge and shear stress for 0.93-mm sand in 8-ft-wide flume [38].

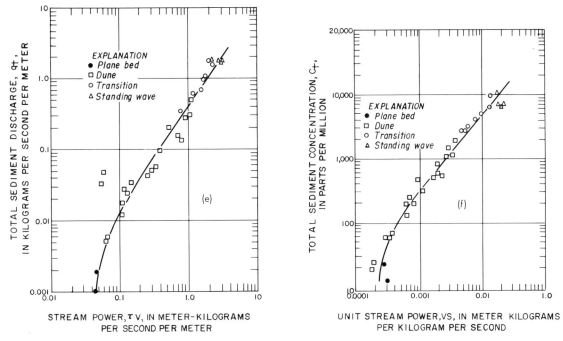

FIGURE 4e–4f. Relationship between (e) total sediment discharge and stream power, (f) total sediment discharge and unit stream power for 0.93-mm sand in 8-ft-wide flume [38].

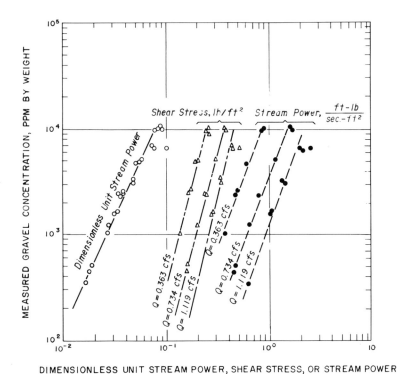

FIGURE 5. Relationship between dimensionless unit stream power, shear stress, stream power, and 4.94-mm gravel concentration measured by Gilbert from 0.3-m flume [39].

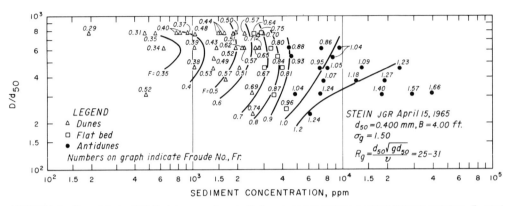

FIGURE 6. Plot of Stein (1965) data as sediment discharge concentration against Froude number, F_r, and relative roughness D/d_{50} [33].

ble. Because of the basic weakness of these assumptions, the generality of an equation which was derived from one of these assumptions is also questionable. When the same sets of data are plotted on Figure 4e, with stream power as the independent variable, the correlation between total sediment concentration and stream power is improved. Further improvement can be made by using unit stream power as the dominant variable which is shown on Figure 4f. This close correlation exists in spite of the presence of different bed forms such as plane bed, dune, transition, and standing wave.

Gilbert's [13] bedload gravel data were used by Yang [39] to test the relative merit between unit stream power, stream power, and shear stress. Figure 5 clearly indicates that the one-to-one relationship between gravel concentration and unit stream power still exists while multiple values of gravel

concentration can be obtained from a given value of shear stress or stream power. Besides the six independent variables shown in Equations (16) through (21), some engineers also tried to relate sediment concentration to relative roughness and Froude number. An example of this is shown on Figure 6 [33]. The same data can be plotted by using the unit stream power as shown on Figure 7. The improvement is remarkable. Schumm and Khan [25] collected sediment data during channel development from straight to meandering and to braided. Yang [36] reanalyzed their data. Figure 8 shows that the strong correlation between unit stream power and sediment concentration is being maintained during the process of changing channel pattern. The superiority of using unit stream power over other parameters in predicting sediment transport rate or concentration is apparent.

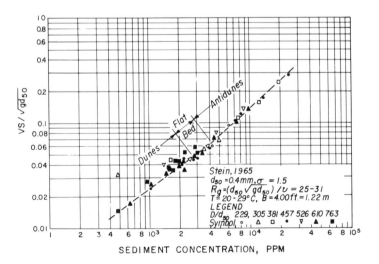

FIGURE 7. Plot of Stein (1965) data as sediment discharge concentration against dimensionless unit stream power $VS/\sqrt{gd_{50}}$ [33].

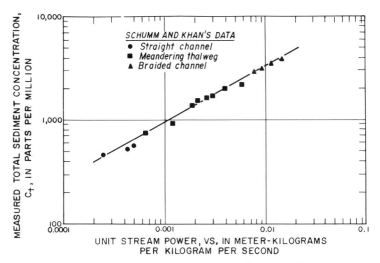

FIGURE 8. Relationship between total concentration and unit stream power during process of channel pattern development from straight to meandering and to braided [36].

B. Probabilistic Approach

The criteria for incipient motion and the rate of sediment transport are related to the instantaneous flow condition and turbulent flow fluctuation. Consequently, the beginning and ceasing of sediment motion as well as the average rate of transport can be expressed with the probability concept.

Einstein's [10] bedload transport function is the most prominent one based on the probabilistic approach. The mathematical formulation of Einstein's transport function is very sophisticated and complicated. As a result, many of the assumptions and coefficients used in his transport function could not be easily and thoroughly verified by laboratory and field tests. Many of his coefficients and functional relationships were developed from limited observations. The accuracy obtained from probabilistic approach does not seem to justify the vast amount of data and time required for its application. This is especially true in solving practical engineering design problems with limited field data.

C. Regression Approach

Frustrated by the inaccurate results obtained from deterministic or probabilistic approaches, many engineers use regression analyses alone to obtain empirical relationships between sediment transport rate and some flow and sediment parameters. The advantage of using regression approach is that it can give us a quick site-specific solution where reliable data are available for regression analyses. The disadvantage is that it does not provide any physical meaning or explanation of the sediment transport process. An equation obtained purely from regression analysis can be applied only to those conditions which are similar to those from where the data were used in obtaining the regression equation.

UNIT STREAM POWER THEORY AND EQUATION

The rate of sediment transport should be directly related to the rate of energy dissipation of fluid used in transporting sediment [4,5,34,35,41]. For open channel flow under steady equilibrium condition, there is no change of kinetic energy. The total rate of energy dissipation is completely due to potential energy dissipation. The rate of potential energy dissipation per unit weight of water in a reach of open channel with total fall of Y is

$$\frac{dY}{dt} = \frac{dx}{dt}\frac{dY}{dx} = VS \tag{22}$$

Yang [34,35] defined the velocity-slope product, VS, as the unit stream power. The fact that rate of sediment transport or concentration should be directly related to unit stream power has been demonstrated by figures 4f, 5, 7, and 8. Sediment transport mainly occurs under turbulent flow conditions. Only turbulent flow can keep sediment particles in suspension is a strong indication that the relationship between sediment concentration and unit stream power can be obtained from basic turbulent flow theories.

The basic form of Yang's [35,39] unit stream power equation is

$$\log C_t = M_1 + N_1 \log\left(\frac{VS}{\omega}\right) \tag{23}$$

where

M_1 and N_1 = dimensionless parameters related to flow and sediment characteristics

C_t = total sediment concentration with wash load excluded

Hinze [15] expressed the velocity and pressure as the sum of their mean and fluctuating parts, i.e.,

$$U_i = \bar{U}_i + u_i \qquad (24)$$

$$P_i = \bar{P}_i + p \qquad (25)$$

$$U_i U_i = \bar{U}_i \bar{U}_i + 2\bar{U}_i u_i + u_i u_i = \bar{U}_i \bar{U}_i + 2\bar{U}_i u_i + q^2 \qquad (26)$$

where

U_i and P_i = velocity and pressure, respectively

$\bar{U}$ and $\bar{P}$ = mean parts of the velocity and pressure, respectively

u_i and p = fluctuating parts of the velocity and pressure, respectively

$q^2 = u_i u_i$

Then the energy or power equation for turbulent flows becomes

$$\underbrace{\frac{d}{dt} \frac{q^2}{2}}_{(I)} = \underbrace{- \frac{\partial}{\partial x_i} \overline{u_i \left(\frac{p}{\varrho} + \frac{q^2}{2} \right)}}_{(II)}$$

$$\underbrace{- \overline{u_i u_j} \frac{\partial \bar{U}_j}{\partial x_i}}_{(III)} + \underbrace{\nu \frac{\partial}{\partial x_i} \overline{u_j \left(\frac{\partial u_i}{\partial x_j} + \frac{\partial u_j}{\partial x_i} \right)}}_{(IV)} \qquad (27)$$

$$\underbrace{- \nu \overline{\left(\frac{\partial u_i}{\partial x_j} + \frac{\partial u_j}{\partial x_i} \right) \frac{\partial u_j}{\partial x_i}}}_{(V)}$$

Hinze [15] explained the physical meaning of Equation (27) as follows:

> The change (I) in kinetic energy of turbulence per unit of mass of the fluid is equal to (II), the convective diffusion by turbulence of the total turbulence energy, plus (III), the energy transferred from the mean motion through the turbulence shear stresses or the production of turbulence energy, plus (IV) the work done per unit of mass and of time by the viscous shear stresses of the turbulent motion, plus (V) the dissipation per unit of mass by turbulent motion.

It should be noted that the term "energy" as quoted is ac-

tually the rate of energy or power which is apparent by the dimension of (I).

The second term on the right-hand side of Equation (27) is of particular interest to the study of sediment transport. For uniform turbulent flows with high Reynolds numbers flowing in the longitudinal direction ($\bar{U}_y = \bar{U}_z = 0$), the expression for the turbulence energy production rate becomes [41]

$$- \overline{u_i u_j} \frac{\partial \bar{U}_j}{\partial x_i} = - \overline{u_x u_y} \frac{d \bar{U}_x}{dy} \qquad (28)$$

Because at high Reynolds numbers, the shear stresses due to viscosity can be neglected,

$$\tau_{xy} = - \varrho \, \overline{u_x u_y} \qquad (29)$$

and

$$- \overline{u_x u_y} \frac{d \bar{U}_x}{dy} = \frac{\tau_{xy}}{\varrho} \frac{d \bar{U}_x}{dy} \qquad (30)$$

Equation (30) enables us to express the rate of turbulence energy production for an open channel flow in terms of shear stress and velocity gradient. The shear stress near the bed of an open channel flow can be expressed by [22]

$$\tau_{xy} = \varrho k^2 \, y^2 \left(\frac{d \bar{U}_x}{dy} \right)^2 \qquad (31)$$

where

k = von Karman's universal coefficient

y = distance above the bed

Assuming the vertical distribution of shear stress in a wide open channel flow can be approximated by

$$\tau_{xy} = \tau_0 \left(1 - \frac{y}{D} \right) \qquad (32)$$

where

τ_0 = shear stress at the bottom of the channel

D = depth of flow

y = the distance above the bed

then

$$\tau_0 = \int_0^D \varrho g \, S \, dy = gS \int_0^D \varrho \, dy = \bar{\varrho} \, g \, DS \qquad (33)$$

and

$$\tau_{xy} = \bar{\gamma}(D - y)S \qquad (34)$$

where

ϱ and $\bar{\varrho}$ = local and depth averaged water density, respectively

S = channel slope

$\bar{\gamma}$ = average specific weight of water

Assuming the velocity can be approximated by the universal logarithmic type velocity distribution

$$\frac{\bar{U}_x}{U_*} = \frac{1}{k} \ln \left(\frac{y}{k_s} \right) + \text{constant} \tag{35}$$

or

$$\frac{d\bar{U}_x}{dy} = \left(\frac{1}{yk} \right) U_* \tag{36}$$

where

$U_* = \sqrt{gDS}$ = shear velocity

k_s = average height of roughness element

D = depth of flow

Then the product of shear stress and velocity gradient becomes

$$\tau_{xy} \frac{d\bar{U}_x}{dy} = \left(\frac{D - y}{y} \right) \left(\frac{\varrho g U_* S}{k} \right) \tag{37}$$

At a small distance $y = a$ above bed, Equation (37) becomes

$$\left(\tau_{xy} \frac{d\bar{U}_x}{dy} \right)_{y=a} = \left(\frac{D - a}{a} \right) \left(\frac{\varrho g U_* S}{k} \right) \tag{38}$$

Based on logarithmic velocity distribution, the vertical sediment concentration distribution derived by Rouse [21] is

$$\frac{\bar{C}}{\bar{C}_a} = \left(\frac{D - y}{y} \frac{a}{D - a} \right)^z \tag{39}$$

where $\bar{C}$ and $\bar{C}_a$ = time-averaged sediment concentration at distance of y and a above the bed, respectively; and

$$Z = \frac{\omega}{kU_*} \tag{40}$$

Einstein and Chien [11] replaced Z with Z_1, i.e.,

$$Z = Z_1 \beta \tag{41}$$

where β = a coefficient which can be determined graphically [10].

Replacing Z with Z_1, Equation (39) can be modified to

$$\frac{\bar{C}}{\bar{C}_a} = \left(\frac{D - y}{y} \frac{a}{D - a} \right)^{Z_1} \tag{42}$$

Substituting Equations (37) and (38) into (42)

$$\frac{\bar{C}}{\bar{C}_a} = \left[\frac{\tau_{xy} \dfrac{d\bar{U}_x}{dy}}{\left(\tau_{xy} \dfrac{d\bar{U}_x}{dy} \right)_{y=a}} \right]^{Z_1} \tag{43}$$

Equation (43) indicates that the vertical distribution of sediment concentration in an open channel turbulent flow is directly related to the distribution of the rate of turbulence energy production of the flow given in Equation (30). Coleman [8] measured vertical sediment concentration distributions in a laboratory flume. His data are reanalyzed here in accordance with Equation (43). Figure 9 shows an example comparison between the experimental and theoretical results stated in Equation (43). The P in Figure 8 stands for the rate of turbulence energy production shown in Equation (43). Figure 9 indicates that the measured suspended sediment concentration distribution follows the relationship given by Equation (43). This relationship is not sensitive to the selection of reference depth a.

Suspended sediment discharge per unit width through a small element perpendicular to the main flow is the product of the time-averaged local concentration and local velocity. The total suspended sediment discharge per unit width can be obtained through integration, i.e.,

$$q_s = \int_a^D \frac{\bar{C}_a}{\left(\tau_{xy} \dfrac{d\bar{U}_x}{dy} \right)_{y=a}^{Z_1}} \left(\tau_{xy} \frac{d\bar{U}_x}{dy} \right)^{Z_1} \bar{U}_x \, dy \tag{44}$$

Based on logarithmic velocity distribution, it can be shown that

$$\frac{d\bar{U}_x}{dy} = \left(\frac{V}{A_2' \ln D + B_2'} \right) \left(\frac{A_1'}{y} \right) \tag{45}$$

where A_2', B_2' = coefficients,

and

$$\left(\tau_{xy} \frac{d\bar{U}_x}{dy} \right)_{y=a} = \varrho k^2 a^2 \left(\frac{d\bar{U}_x}{dy} \right)_{y=a}^3 = \varrho k^2 a^2 \left(\frac{U_*}{ka} \right)^3 \tag{46}$$

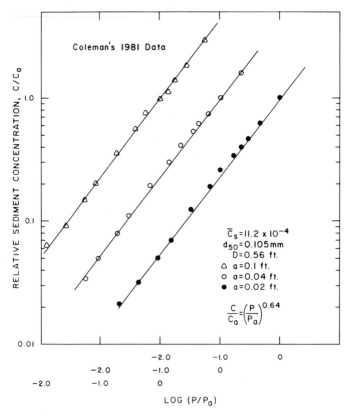

FIGURE 9. Relationship between vertical distribution of sediment concentration and rate of turbulence energy production.

From Equations (44), (45), and (46), the average suspended sediment concentration becomes

$$C_s = \frac{q_s}{q_w} = \frac{\bar{C}_a}{\left(\frac{\varrho U_*^3}{ka}\right)^{z_1} (A_2' \ln D + B_2')^{z_1+1}} \frac{(A_1'\bar{\gamma})^{z_1}}{D}$$

$$\times \ (VS)^{z_1} \int_a^D \left(\frac{D-y}{y}\right)^{z_1} (A_1' \ln y + B_1') \, dy$$

(47)

where

q_w = water discharge per unit channel width
A_1', A_2', B_1', B_2' = coefficients

The integral term at the right-hand side of Equation (47) does not have a closed-form solution. However, a numerical solution similar to that used by Einstein [10] can be applied

for various a/D and Z_1 values to evaluate the integrals

$$K_1 = \int_{a/D}^1 \left(\frac{1-y'}{y'}\right)^{z_1} \ln y' \, dy';$$

(48)

$$K_2 = \int_{a/D}^1 \left(\frac{1-y'}{y'}\right)^{z_1} dy'$$

where $y' = y/D$.

The resulting solution has the form

$$\int_a^D \left(\frac{D-y}{y}\right)^{z_1} (A_1' \ln y + B_1') \, dy = (P_E K_1 + K_2)$$

(49)

where P_E = parameter related to constants A_1' and B_1'.

Then

$$\log C_s = \log \left[\frac{\bar{C}_a}{D \, (A_2' \ln D + B_2')^{z_{,+1}}} \right.$$
$$\left. \times \left(\frac{kaA_1' \, \bar{\gamma}}{\varrho U_\ast^\frac{1}{3}} \right)^{z_,} (P_E K_1 + K_2) \right] + Z_1 \log (VS) \qquad (50)$$

The resulting expression for suspended sediment concentration can be reduced to the general form of Equation (23), i.e.,

$$\log C_s = M_2 + N_2 \log (VS) \qquad (51)$$

or

$$\log C_s = M_3 + N_3 \log \frac{VS}{\omega} \qquad (52)$$

where M_2, N_2, M_3, N_3 = parameters related to flow and sediment characteristics.

It is reasonable for us to assume that there is a continuity of sediment concentration at the interface between suspended and bedload regimes. This in return would required that the bedload be related to suspended load. Einstein [10] assuming the concentration in the bed layer of thickness $a = 2d_{65}$ remained constant, related the reference concentration C_a to bedload by

$$C_a = \frac{\beta_1 \, i_B \, q_B}{a V_b} \qquad (53)$$

where

$i_B \, q_B$ = bedload discharge for a size fraction, i_B
V_b = the average velocity of bedload
β_1 = a correction factor to compensate for the assumption of uniform bed layer concentration
q_B = the bedload discharge
d_{65} = sediment particle diameter for which 65 percent of the sediment mixture is finer

Relating V_b to the shear velocity due to grain roughness that Einstein [10] defined, and using the experimental result, Equation (53) becomes

$$\bar{C}_a = \frac{1}{11.6} \left(\frac{i_B \, q_B}{a U_\ast'} \right) \qquad (54)$$

where $U_\ast'$ = the shear velocity due to grain roughness.

According to Einstein [10], the expression relating suspended and bedload discharges is

$$i_s \, q_s = i_B \, q_B \, (P_E \, I_1 + I_2) \qquad (55)$$

where I_1 and I_2 = integrals used in Einstein's sediment transport equation which are similar to K_1 and K_2 in Equation (48).

The procedure outlined herein follows similar assumptions as Einstein's [10]. However, Yang and Molinas [41] relate suspended sediment concentration to the rate of turbulence energy production, $\tau_{xy} \, (d\bar{U}_x/dy)$, instead of using Rouse's equation, i.e., Equation (39) directly. Total load can be obtained as the sum of bed load and suspended load from Equation (55), i.e.,

$$q_t = q_s + q_B = \left(\frac{1 + P_E \, I_1 + I_2}{P_E \, I_1 + I_2} \right) q_s \qquad (56)$$

and

$$C_t = \left(\frac{1 + P_E \, I_1 + I_2}{P_E \, I_1 + I_2} \right) \left(\frac{q_s}{q_w} \right) \qquad (57)$$

Combining Equations (52) and (57), the basic form of Yang's equation, i.e., Equation (23) is obtained. Equation (23) can be rewritten as

$$\log C_t = M + N \log \left(\frac{VS}{\omega} - \frac{V_{cr}S}{\omega} \right) \qquad (58)$$

where M, N = dimensionless parameters related to flow and sediment characteristics.

to include the critical dimensionless unit stream power required at incipient motion.

The derivations have shown that sediment concentration can be expressed as a function of the rate of turbulence energy production in open channel flow. By the application of the logarithmic velocity distribution, Rouse's equation and a procedure similar to that used by Einstein, sediment concentration can be obtained as a function of unit stream power through the integration of turbulence energy production over the depth of flow. Thus, the basic form of unit stream power equations proposed by Yang can be derived from well-established theories in fluid mechanics and turbulence. Strictly speaking, M and N in Equation (58) should be related to the coefficients shown in Equations (50) and (57). However, analytical solution which relates all these coefficients to different flow and sediment characteristics is extremely difficult and may not be necessary. The approach used by Yang [35] was the dimensional analysis which related M and N in Equation (58) to some dimensionless parameters. The constants associated with the dimensionless parameters were determined by Yang [35,39] through multiple regression analysis of laboratory data.

Yang developed two dimensionless unit stream power equations for sand and gravel transport, respectively. They are [35,39]

$$\log C_{ts} = 5.435 - 0.286 \log \frac{\omega d}{\nu} - 0.457 \log \frac{U_*}{\omega}$$

$$+ \left(1.799 - 0.409 \log \frac{\omega d}{\nu} - 0.314 \log \frac{U_*}{\omega} \right)$$

$$\times \log \left(\frac{VS}{\omega} - \frac{V_{cr}S}{\omega} \right) \quad (59)$$

and

$$\log C_{tg} = 6.681 - 0.633 \log \frac{\omega d}{\nu} - 4.816 \log \frac{U_*}{\omega}$$

$$+ \left(2.784 - 0.305 \log \frac{\omega d}{\nu} - 0.282 \log \frac{U_*}{\omega} \right)$$

$$\times \log \left(\frac{VS}{\omega} - \frac{V_{cr}S}{\omega} \right) \quad (60)$$

where

C_{ts}, C_{tg} = total sediment concentration, with wash load excluded, in parts per million by weight for sand and gravel, respectively
VS = unit stream power
U_* = shear velocity
ν = kinetic viscosity
ω = fall velocity of sediment
d = median particle diameter

The incipient motion criteria given in Equations (14) and (15) should be used for Equations (59) and (60). Because of the range of data used in calibrating the coefficients, Equation (59) should be applied to sand transport and Equation (60) should be applied to gravel transport with median particle size from 2 to 10 mm. Equations (59) and (60) were developed for fairly uniform materials. When they are applied to highly nonuniform materials, the total sediment concentration should be computed by

$$C_t = \sum_{i=1}^{n} p_i C_i \quad (61)$$

where

n = total number of size fractions used in the computation
p_i = percent of material in size i
C_i = computed concentration for size i by either Equation (59) or (60)

EVALUATION OF TRANSPORT EQUATIONS

Numerous sediment transport equations have been proposed in the literature for the prediction of sediment load or concentration in natural rivers. Engineers often face the problem of selecting a reliable equation for engineering application. The fact that computed results from different equations may differ drastically from each other and from measurements further complicates the selection. The accuracy of an equation can be judged from the generality of basic assumptions used, theoretical basis of the equation, direct comparison with measurements and other equations, and indirect comparison based on computer model simulation. Section III already compared the validity and generality of the basic assumptions used in deriving different sediment transport equations. Those comparisons indicate that unit stream power is a reliable parameter for the determination of sediment concentration or load. Other parameters may be used only under certain special flow and sediment conditions. Most sediment transport equations are empirical or semiempirical in nature. The relationship between unit stream power and sediment concentration can be derived directly from well-established theories in fluid mechanics and turbulence. This gives the unit stream power equations a clear advantage from theroetical point of view. The sim-

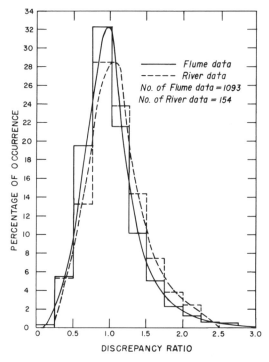

FIGURE 10. Distribution of discrepancy ratio of Yang's (1973) equation [36].

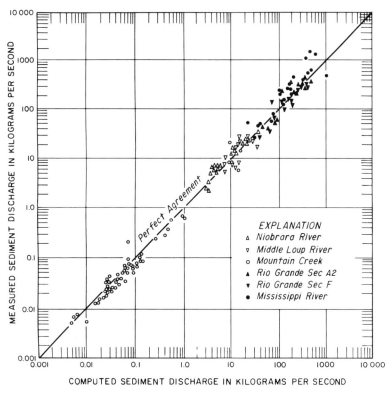

FIGURE 11. Comparison between measured total bed-material discharge from six river stations and computed results from Yang's (1973) equation [36,37].

plicity of the unit stream power equations and the small amount of data required for its application further enhance their acceptability by engineers. Examples of direct and indirect comparisons made by different investigators are used in this section to assist engineers in making their selections.

A. Direct Comparison

Yang [36] made detailed comparisons between computed results from Equation (59) and 1,247 sets of measured results from laboratory flumes and natural rivers. A discrepancy ratio, defined as the ratio between computed and measured results, is used as an indicator of the accuracy of an equation. Figure 10 shows that the discrepancy ratio of Equation (59) follows the normal distribution with an overall mean value of 1.03 and a very small standard deviation. This is a clear indication of the accuracy and consistency of Equation (59). The reason that the mean discrepancy ratio for river data is slightly higher than that of the flume data is because some river data used in the comparison did not include bedload transported in the unmeasured zone near the river bottom. Consequently, computed total load, with wash load excluded, should be higher than the measured results.

Figure 11 shows very good agreement between measured river data and computed results from Equation (59). These river data cover a wide range of river size, flow, and sediment conditions [42].

Yang and Molinas [41] made detailed comparison between measured and computed results from different equations. Table 1 shows that Equation (59) is the most accurate one for both laboratory flume and river data in the sand size range. It also shows that equations proposed by Shen and Hung [23] and Maddock [17] are also fairly accurate for laboratory data. Because of the lack of physical meaning and theoretical basis of these two empirical regression or regime equations, their applications should be limited to those conditions which are similar to where the data were obtained and used in their analyses.

Alonso [2] studied the accuracy and applicability of 31 sediment transport equations. Only 8 among the 31 equations received detailed comparison and evaluation. Some of the equations were not included for detailed evaluation by Alonso because they have not received extensive applications. Others, such as Toffaleti [28] and Modified Einstein [7] methods, are too complicated or require knowledge of the measured concentration of the suspended load, and

TABLE 1. Summary of Comparisons of Accuracies of Different Equations [41].

Author of equations (1)	Discrepancy ratio					Standard deviation (7)	Number of data (8)
	Mean (2)	0.75–1.25 (3)	0.5–1.5 (4)	0.25–1.75 (5)	0.5–2.0 (6)		
(a) Laboratory data							
Colby	0.31	4%	10%	29%	10%	0.64	865
Yang (1973)	1.01	55%	85%	95%	92%	0.44	1,093
Shen and Hung	0.91	46%	84%	96%	89%	0.39	1,093
Engelund and Hansen	0.88	26%	59%	91%	65%	0.72	1,093
Ackers and White	1.28	37%	68%	84%	86%	0.69	1,093
Maddock	0.99	46%	72%	85%	82%	0.51	1,093
(b) River data							
Colby	0.61	13%	29%	71%	33%	0.66	102
Yang (1973)	1.13	48%	77%	92%	90%	0.43	166
Shen and Hung	1.18	43%	71%	80%	81%	0.61	166
Engelund and Hansen	1.51	34%	58%	72%	79%	0.75	166
Ackers and White	1.50	31%	61%	75%	80%	0.80	166
Maddock	0.49	24%	43%	56%	45%	0.48	166
(c) All data							
Colby	0.34	5%	12%	33%	12%	0.64	967
Yang (1973)	1.03	54%	84%	95%	92%	0.44	1,259
Shen and Hung	0.95	46%	82%	94%	88%	0.42	1,259
Engelund and Hansen	0.96	27%	59%	88%	67%	0.72	1,259
Ackers and White	1.31	36%	67%	83%	85%	0.71	1,259
Maddock	0.92	43%	68%	81%	78%	0.51	1,259

TABLE 2. Analysis of Discrepancy Ratio Distribution of Different Transport Formulas [2].

| Formula | No. of tests | Ratio between predicted and measured load | | | | Percentage of tests with ratio between ½ and 2 |
		Mean	95%-confidence limits of the mean		Standard deviation	
		Field data				Percent
Ackers and White	40	1.27	1.05	1.48	0.68	87.8
Engelund and Hansen	40	1.46	1.28	1.64	0.56	82.9
Laursen	40	0.65	0.49	0.80	0.48	56.1
MPME	40	0.83	0.50	1.15	1.02	58.5
Yang (1973)	40	1.01	0.89	1.13	0.39	92.7
Bagnold	40	0.39	0.31	0.47	0.26	32.0
Meyer-Peter & Muller	40	0.24	0.22	0.27	0.09	0
Yalin	40	2.59	2.08	3.11	1.62	46.3
		Flume data with D/d ≥ 70				
Ackers and White	177	1.34	1.24	1.54	1.29	73.0
Engelund and Hansen	177	0.73	0.63	0.83	0.68	51.1
Laursen	177	0.81	0.73	0.88	0.51	71.4
MPME	177	3.11	2.95	3.52	2.75	42.1
Yang (1973)	177	0.99	0.93	1.08	0.60	79.8
Bagnold	177	0.85	0.81	1.22	2.50	20.8
Meyer-Peter & Muller	177	0.40	0.39	0.47	0.49	18.5
Yalin	177	1.62	1.38	2.23	4.08	32.6
		Flume data with D/d < 70				
Ackers and White	48	1.12	0.93	1.28	0.52	89.6
Engelund and Hansen	48	0.75	0.59	0.90	0.50	66.7
Laursen	48	1.04	0.76	1.32	0.99	79.2
MPME	48	1.34	1.04	1.64	1.04	66.7
Yang (1973)	48	0.90	0.79	1.05	0.51	85.4
Bagnold	48	1.53	1.46	1.87	1.14	45.8
Meyer-Peter & Muller	48	1.03	1.00	1.27	0.83	72.9
Yalin	48	1.92	1.45	2.41	1.65	64.6

TABLE 3. Summary of Rating of Selected Sediment Transport Formulas [3].

Formula number (1)	Reference (2)	Type (3)	Comments (4)
1	Ackers and White (1973)	Total load	Rank[1] = 3
2	Engelund and Hansen (1967)	Total load	Rank = 4
3	Laursen (1958)	Total load	Rank = 2
4	MPME[2] (1948,1950)	Total load	Rank = 6
5	Yang (1973)	Total load	Rank = 1, Best overall predictions
6	Bagnold (1956)	Bedload	Rank = 5
7	MPM[3] (1948)	Bedload	Rank = 7
8	Yalin (1963)	Bedload	Rank = 8

[1]Based on mean discrepancy ratio (calculated over observed transport rate) from 40 tests using field data and 165 tests using flume data.
[2]MPME = Meyer-Peter and Müller (1948) formula for bedload and modified Einstein (1950) formula for suspended load.
[3]MPM = Meyer-Peter and Müller (1948) formula.

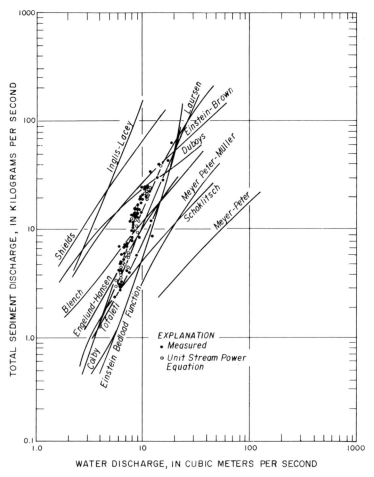

FIGURE 12. Comparison between measured total sediment discharge of the Niobrara River near Cody, Nebraska, and computed results of various equations [36,42].

therefore are not suitable for hydrologic simulation. The result of comparison by Alonso [2] for sand transport is shown in Table 2. The MPME method, shown in Table 2, estimates the total load by adding the bedload predicted by the Meyer-Peter and Müller [19] formula to the suspended load computed by the Einstein procedure [10].

Alonso limited his comparisons of field data to those where the total bed-material load can be measured by special facilities. Thus, the uncertainty of unmeasured load does not exist. Table 2 indicates that Equation (59) has an average error of only 1 percent for both field and flume data. When the relative depth D/d is less than 70, the flow is shallow and surface wave effect becomes important. In this range, most sediment formulas may fail because their formulations do not account for interactions with free surface waves.

The ASCE Task Committee on Relations Between Mor-

phology of Small Streams and Sediment Yield of the Committee on Sedimentation of the Hydraulic Division [3] considered the results obtained by Alonso [2] "to be representative inasmuch as they reflect the magnitude of prediction errors likely to result in comparison with field data assumed to be correct." The ASCE Committee also stated that "when we consider errors likely to be present in field data, these results probably underestimate the probable magnitude of the errors." The relative rating of accuracy of eight formulas made by the ASCE Committee [3] is shown in Table 3.

The ASCE Manual on Sedimentation Engineering [32] compared the computed sediment discharges from different equations with the measured results from natural rivers. Those comparisons were replotted by Yang and Stall [42] and Yang [36] to include Equation (59). The total measured sediment load does not include wash load. These compari-

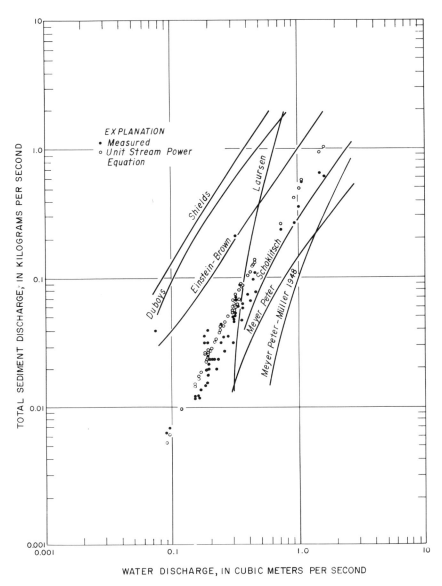

FIGURE 13. Comparison between measured total sediment discharge of the Mountain Creek at Greenville, South Carolina, and computed results of various equations [36,42].

sons are shown on Figures 12 and 13. With the exception of Yang's [35] unit stream power equation, i.e., Equation (59), computed results from different equations differ drastically from each other and from the measurements.

Gravels are transported mainly as bedload. Becuase there is no reliable instrument for bedload measured under field conditions, the accuracy of Equation (60) has to be judged from laboratory data only. The Meyer-Peter and Müller equation [19] is one of the most widely used bedload equations for gravel transport, especially in Europe. Figure 14 indicates that there is no correlation between measured and computed results from the Meyer-Peter and Müller equation. This may be caused by the lack of generality of the assumption used in deriving their equation [39]. Figure 15 summarizes the comparison between measured and computed results from Equation (60). Comparisons made by Yang [39] are also summarized in Table 4. Because of the amount of laboratory data used in calibrating the coefficients in Equation (60) is rather limited and the lack of reliable field data for verification, further testing and verification of Equation (60) may be needed.

B. Indirect Comparison

One of the most widely used computer models for the simulation of scour and deposition processes of natural rivers is the HEC-6 model developed by the U.S. Army Corps of Engineers [29]. The sediment transport equations included in HEC-6 which can be selected by users are those by Yang [35], Toffaleti [28], Laursen [16], and DuBoy [9]. The HEC-6 model was applied by the Corps of Engineers Los Angeles District [30] to the study of scour and deposition process along several rivers due to engineering constructions. They found that the computed results are sensitive to the selection of equations. After a thorough comparison of all the transport equations in HEC-6, Yang's [35] equation was selected because it clearly yielded the most reasonable results.

Figure 16 shows an example of comparison between actually surveyed and computed river bed profiles based on Equation (59). Comparisons betwen Equation (59) and other equations were also made by the U.S. Geological Survey [18] in a computer model for the simulation of bed degradation below Cochiti Dam, New Mexico. Computed bed profiles were again in good agreement with field survey where the bed-material is in the sand size range. Molinas, et al. [20] applied both Equations (59) and (60) in their stream tube computer model to simulate the development of local scour at the Mississippi River Lock and Dam No. 26 replacement site near St. Louis, Missouri. The computed results agree very well with those from actual field survey.

C. Selection of Equations

Sediment transport rate in natural rivers depends not only on the independent variables mentioned in previous sec-

tions, but also on the gradation and shape factor of sediment, percentage of bed surface covered by coarse material, or armor layer, variation of hydrological circle, rate of supply of fine material or wash load, water temperature, channel pattern and bed configuration, strength of turbulence, etc. Because of the tremendous uncertainties involved in estimating sediment discharge at different flow and sediment conditions under different hydrologic, geologic, and climatic constraints, it is extremely difficult, if not impossible to recommend one equation for engineers to use in the field under all circumstances. The agreements between measured and computed results from Equations (59) and (60) only indicate that, on the average, they are more accurate than other equations. Engineers should not be surprised to find out that one equation may be more accurate than others when applied to a particular river but may not be true for other rivers. The following procedures were based on Yang's [40] recommendation with minor modifications to assist engineers in their selection of sediment transport equations.

1. Determine the type of field data available or measurable within the time, budget, and manpower limitations.
2. Examine all the formulas and select those with measured values of independent variables determined from step 1.
3. Compare the field situation and the limitations of formulas selected in step 2. If more than one formula can be used, calculate the rate of sediment transport by these formulas, and compare the results.
4. Decide which formulas can best agree with the measured sediment load and use these formulas to estimate the rate of sediment transport at those flow conditions when actual measurements are not available.
5. In the absence of measured sediment load for comparison, the following formulas should be considered:
   a. Use Meyer-Peter and Müller's [19] formula when the bed material is coarser than 5 mm.
   b. Use Einstein's [10] procedure when bedload is a significant portion of the total load.
   c. Use Toffaleti's [28] formula for large sand-bed rivers.
   d. Use Colby's [6] formula for rivers with depth less than 10 feet.
   e. Use Shen and Hung's [23] formula for laboratory flumes and very small rivers.
   f. Use Yang's [35] sand formula for sand bed laboratory flumes and natural rivers with wash load excluded. Use Yang's [39] gravel formula for gravel transportation when the bed material is between 2 and 10 mm.
   g. Use Ackers and White's [1], or Engelund and Hansen's [12] equation for subcritical flow condition in the lower flow regime.
   h. Use Laursen's [16] formula for laboratory flumes and shallow rivers with fine sand or coarse silt.
   i. A regime or regression equation can be applied to a river only if the flow and sediment conditions are similar to that from where the equation was derived.

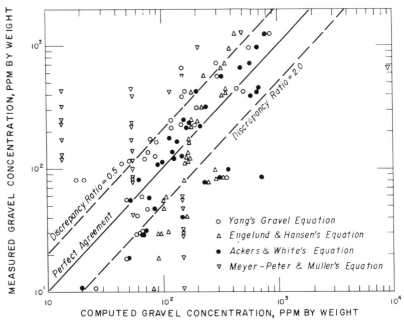

FIGURE 14. Comparison between measured 2.46-mm gravel concentration by Casey and computed results by different equations [39].

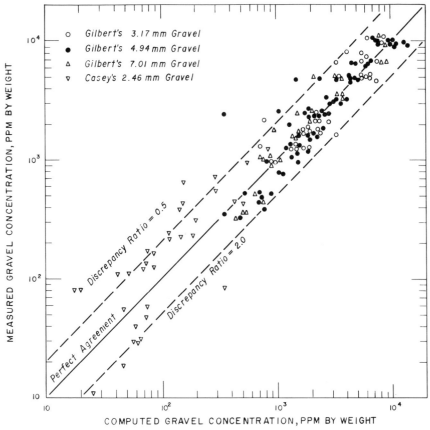

FIGURE 15. Comparison between measured and computed gravel concentrations by Yang's (1984) gravel equation [39].

TABLE 4. Summary of Comparisons Between Computed and Measured Gravel Concentrations [39].

Equations (1)	Discrepancy ratio				Correlation coefficient (6)	Standard deviation (7)	Mean error (8)	Number of data (9)
	Mean (2)	0.75–1.25 (3)	0.5–1.5 (4)	0.25–1.75 (5)				
Yang (1984)	1.05	47%	75%	92%	0.92	0.51	5%	166
Engelund and Hansen	0.85	13%	40%	86%	0.93	0.99	−15%	167
Ackers and White	1.21	44%	76%	83%	0.93	0.89	21%	167
Meyer-Peter & Müller	1.86	61%	81%	87%	0.91	8.98	86%	167

j. Select an equation according to its degree of accuracy shown in Table 3.

6. In case none of the existing sediment transport equations can give satisfactory results, use the existing data collected from a river station and plot sediment load or concentration against water discharge, velocity, slope, depth, shear stress, stream power, and unit stream power. The least scattered curve without systematic deviation from a one-to-one correlation between dependent and independent variables should be selected as the sediment rating curve for the station.

CONCLUSIONS

This chapter reviews and evaluates the basic concepts and approaches used in the development of sediment transport equations. Laboratory and field data are used to determine the validity of the assumptions used and the accuracy of the equations derived from these assumptions. The review and evaluation reach the following conclusions:

1. It is questionable that the rate of sediment transport or concentration can be consistently determined from water

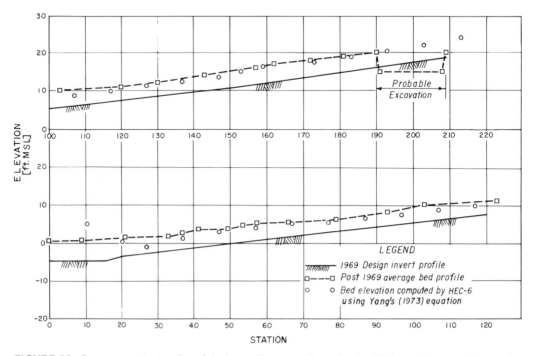

FIGURE 16. Reconstituted bed profiles of the Lower Santa Ana River after the 1969 flood by using Yang's (1973) sand equation [30].

discharge, average flow velocity, energy slope, or shear stress. The lack of generality of the one-to-one correlation between sediment load or concentration and those parameters may be the basic reason of inaccuracy and inconsistency for most sediment transport equations.

2. The strong correlation between sediment concentration and unit stream power is not affected by the variation of bed forms, relative roughness, Froude number, and channel pattern.

3. The fact that sediment concentration should be related to unit stream power can be derived directly from well-established theories in fluid mechanics and turbulence.

4. Comparison between measured and computed results indicates that, on the average, the unit stream power equations are more accurate than others.

5. Due to the complex nature of field conditions, it is impossible to recommend one equation for universal application. Recommendations are made on the procedures for engineers to follow in their selection of sediment transport equations.

LIST OF SYMBOLS

$A_1, A_2, A_3, A_4, A_5, A_6, B_1, B_2, B_3, B_4, B_5, B_6$ = parameters
A'_1, A'_2, B'_1, B'_2 = coefficients
$\bar{C}$ and $\bar{C}_a$ = time-averaged sediment concentration at distance of y and a above the bed, respectively
C'_D, C_D = drag coefficient at ω and V_d, respectively
C_i = computed sediment concentration in size i
C_s, C_t = suspended and total sediment concentration with wash load excluded, respectively
C_{ts}, C_{tg} = total sediment concentration, with wash load excluded, in parts per million by weight for sand and gravel, respectively
D = depth of flow
d = sediment particle diameter
d_{65} = sediment particle diameter for which 65 percent of the sediment mixture is finer
F_D, F_L, F_R = drag, lift, and resistance force, respectively
g = gravitational acceleration
I_1, I_2 = integrals used in Einstein's sediment transport equation
$i_B q_B$ = bedload discharge for a size fraction, i_B
K_1, K_2 = integrals
k = the von Karman's universal constant
k_s = the average height of roughness element
$M, N, M_1, N_1, M_2, N_2, M_3, N_3$ = parameters related to flow and sediment characteristics
n = total number of sediment size fractions used in the computation
P = rate of turbulence energy production

$\bar{P}$ = mean part of pressure
P_i = pressure in ith direction
p = fluctuating part of pressure
p_i = percent of material in size i
P_E = parameter related to A'_1 and B'_1
Q, Q_s = water and sediment discharge, respectively
$q^2 = u_i u_i$
q_B, q_s, q_t, q_w = bedload, suspended sediment, total sediment, and water discharge per unit channel width, respectively
S = energy or water surface slope
U_i = velocity in the ith direction
$\bar{U}_i, u_i$ = mean and fluctuating parts of velocity, respectively
U_*, U'_* = shear velocity and shear velocity due to grain roughness, respectively
V = average flow velocity
V_b = the average velocity of bedload
V_d = local velocity at a distance d above the bed
VS = unit stream power per unit weight of water
W_s = submerged weight of sediment particle
X, Y = horizontal and vertical distance, respectively
y = distance above the bed
$Z = \dfrac{\omega}{kU_*}$
$Z_1 = Z/\beta$
β = coefficient
β_1 = a correction factor to compensate for the assumption of uniform bed layer concentration
ψ_1, ψ_2, ψ_3 = coefficients
$\varrho_s, \varrho_f, \varrho$ = density of sediment, fluid, and water, respectively
$\gamma, \bar{\gamma}$ = specific and average specific weight of water, respectively
γ_s = specific weight of sediment
τ = shear stress
τ_c = critical shear stress at initial motion
τ_{xy} = turbulent shear stress
τV = stream power per unit bed area
ν = kinetic viscosity
ω = fall velocity of sediment

REFERENCES

1. Ackers, P. and W. R. White, "Sediment Transport: New Approach and Analysis," *Journal of the Hydraulics Division,* ASCE, Vol. 99, No. HY11, Proceeding Paper 10167, pp. 2041–2060 (November 1973).

2. Alonso, C. V., "Selecting a Formula to Estimate Sediment Transport Capacity in Nonvegetated Channels," Chapter 5, *CREAMS (A Field Scale Model for Chemicals, Runoff, and*

Erosion from Agricultural Management System), W. G. Knisel, editor, U.S. Department of Agriculture, Conservation Research Report No. 26, pp. 426–439 (May 1980).

3. ASCE Task Committee on Relations Between Morphology of Small Streams and Sedimentation of the Hydraulics Division, "Relationships Between Morphology of Small Streams and Sediment Yields," *Journal of the Hydraulics Division*, ASCE, Vol. 108, No. HY11, Proceedings Paper 17450, pp. 1328–1365 (November 1982).

4. Bagnold, R. A., "An Approach to the Sediment Transport Problem from General Physics," U.S. Geological Survey Professional Paper A-22-J (1966).

5. Bagnold, R. A., "An Empirical Correlation of Bedload Transport Rates in Flumes and Natural Rivers," *Proceedings of Royal Society of London*, Vol. 372A, pp. 453–473 (1980).

6. Colby, B. R., "Practical Computations of Bed-Material Discharge," *Journal of the Hydraulics Division*, ASCE, Vol. 90, No. HY2 (1964).

7. Colby, B. R. and C. H. Hembree, "Computation of Total Sediment Discharge, Niobrara River near Cody, Nebraska," U.S. Geological Survey Water Supply Paper 1357 (1955).

8. Coleman, N. L., "Velocity Profiles with Suspended Sediment," *Journal of Hydraulic Research*, Vol. 19, No. 3, pp. 211–229 (1981).

9. DuBoy, M. P., "Le Rhone et les Rivieres a Lit affouillable," *Annals de Pont et Chausses*, Ser. 5, Vol. 18, pp. 141–195 (1879).

10. Einstein, H. A., "The Bedload Function for Sediment Transportation in Open Channel Flows," U.S. Department of Agriculture Soil Conservation Service Technical Bulletin, No. 1026 (1950).

11. Einstein, H. A. and N. Chien, "Second Approximation to the Solution of the Suspended Load Theory," Research Report No. 3, University of California Institute (1954).

12. Engelund, F. and E. Hansen, "A Monograph on Sediment Transport in Alluvial Streams," Teknisk Forlag, Copenhagen (1972).

13. Gilbert, K. G., "The Transportation of Debris by Running Waters," U.S. Geological Survey Professional Paper 86 (1914).

14. Guy, H. P., D. B. Simons, and E. V. Richardson, "Summary of Alluvial Channel Data from Flume Experiment, 1956–1961," U.S. Geological Survey Professional Paper 462-I (1966).

15. Hinze, J. O. *Turbulence*, first edition, McGraw-Hill Book Company, Inc., New York, NY, pp. 64–66 (1959).

16. Larsen, E. M., "The Total Sediment Load of Streams," *Journal of The Hydraulics Division*, ASCE, Vol. 84, No. HY 1 (1958).

17. Maddock, Jr., T., "Equations for Resistance to Flow and Sediment Transport in Alluvial Channels," *Water Resources Research*, Vol. 12, No. 1, pp. 11–12 (1976).

18. Mengis, R. C., "Modeling of a Transient Streambed in the Rio Grande, Cochiti Dam to near Albuquerque, New Mexico," U.S. Geological Survey Open File Report 82-106, 99 pp., Denver, Colorado (1981).

19. Meyer-Peter, E. and R. Müller, "Formulas for Bedload Transport," *Proceedings of the Third Meeting of International Association for Hydraulic Research*, pp. 39–64, Stockholm (1948).

20. Molinas, A., C. W. Denzel and C. T. Yang, "Application of the Stream Tube Computer Model," *Proceedings of the Fourth Interagency Sedimentation Conference*, Las Vegas, Nevada (1986).

21. Rouse, H., "Modern Concept of the Mechanics of Turbulence," *Transactions*, ASCE, Vol. 102, Paper No. 1965, p. 4630 (1937).

22. Shames, I. H. *Mechanics of Fluid*, McGraw-Hill Book Company, Inc., New York, NY, p. 332 (1962).

23. Shen, H. W. and C. S. Hung, "An Engineering Approach to Total Bed Material Load by Regression Analysis," *Proceedings of the Sedimentation Symposium*, Chapter 14, pp. 14-1–14-7 (1972).

24. Shields, A., "Application of Similarity Principles and Turbulence Research to Bedload Movement," Translated from German to English by W. P. Ott and J. C. Van Uchelen, California Institute of Technology, Pasadena, California (1936).

25. Shumm, S. A. and H. R. Khan, "Experimental Study of Channel Patterns," *Geological Society of America*, Bulletin 83, p. 407 (1972).

26. Strand, R. I. and E. L. Pemberton, "Reservoir Sedimentation," Technical Guideline for Bureau of Reclamation, U.S. Bureau of Reclamation, Denver, Colorado (1982).

27. Talapatra, S. C. and S. N. Ghosh, "Incipient Motion Criteria for Flow Over a Mobile Bed Sill," *Proceedings of the Second International Symposium on River Sedimentation*, Nanjing, China, pp. 459–471 (October 1983).

28. Toffaleti, F. B., "Definitive Computations of Sand Discharge in Rivers," *Journal of the Hydraulics Division*, ASCE, Vol. 95, No. HY 1, pp. 225–246 (January 1969).

29. U.S. Army Corps of Engineers, The Hydrologic Engineering Center, "Generalized Computer Program, HEC-6, Scour and Deposition in Rivers and Reservoirs," Users Manual, Davis, California (March 1977).

30. U.S. Army Corps of Engineers, Los Angeles District, "General Design Review Conference," Los Angeles California (December 15–16, 1982).

31. U.S. Interagency Committee on Water Resources, Subcommittee on Sedimentation, Report No. 12, "Some Fundamentals of Particle Size Analysis," (1957).

32. Vanoni, V. A., Editor, *Sedimentation Engineering*, ASCE Task Committee for the Preparation of the Manual on Sedimentation of the Sedimentation Committee, Committee of the Hydraulics Division, 1975 (Reprinted 1977).

33. Vanoni, V. A., "Predicting Sediment Discharge in Alluvial Channels," *Water Supply and Management*, Pergamon Press, pp. 399–417 (1978).

34. Yang, C. T., "Unit Stream Power and Sediment Transport," *Journal of the Hydraulics Division*, ASCE, Vol. 18, No. HY 10, Proceeding Paper 9295, pp. 1805–1826 (October 1972).

35. Yang, C. T., "Incipient Motion and Sediment Transport," *Journal of the Hydraulics Division*, ASCE, Vol. 99, No. HY 10, Proceeding Paper 10067, pp. 1679–1704 (October 1973).

36. Yang, C. T., "The Movement of Sediment in Rivers," *Geophy-*

sical Survey 3, D. Reidel Publishing Company, pp. 39–68 Dordrecht, Holland (1977).

37. Yang, C. T., "Sediment Transport and River Engineering," *Proceedings of the International Symposium on River Sedimentation,* Vol. 1, pp. 350–386, Beijing, China (March 1980).

38. Yang, C. T., "Rate of Energy Dissipation and River Sedimentation," *Proceedings of the Second International Symposium on River Sedimentation,* pp. 575–585, Nanjing, China (October 1983).

39. Yang, C. T., "Unit Stream Power Equation for Gravel," *Journal of the Hydraulic Engineering,* ASCE, Vol. 110, No. HY 12, pp. 1783–1797 (December 1984).

40. Yang, C. T., *Lecture Notes on Sediment Transport,* U.S. Department of the Interior, Bureau of Reclamation, Engineering and Research Center, Denver, Colorado (November 1984).

41. Yang, C. T. and A. Molinas, "Sediment Transport and Unit Stream Power Function," *Journal of the Hydraulics Division,* ASCE, Vol. 108, No. HY 6, Proceeding Paper 17161, pp. 776–793 (June 1982).

42. Yang, C. T. and J. B. Stall, "Applicability of Unit Stream Power Equation," *Journal of the Hydraulics Division,* ASCE, Vol. 102, No. HY 5, Proceeding Paper 12103, pp. 559–568 (May 1976).

CHAPTER 9

Estimating Extreme Events: Log Pearson Type 3 Distribution

DONTHAMSETTI VEERABHADRA RAO*

INTRODUCTION

A knowledge of extreme hydrologic events associated with certain probability level (or return period/recurrence interval) is essential in evaluation and design of water control projects and in dealing with several other aspects of water environment. These events may be flood flows, droughts, rainfalls, storages, water levels, etc. Throughout the world, rivers have been harnessed by building dams, and waterfront properties have been protected by levees. Failure of these structures can result in catastrophic losses to life and property. In the United States, major hydraulic structures are designed to be safe against probable maximum flood [7,8,16] which is determined from the meteorological considerations. In most other countries, the magnitude of these design events is related to some rare probability of occurrence; e.g., in the U.S.S.R., depending on the hazard the failure of a hydraulic facility may pose, the project is designed for 100 yr to 10,000 yr flood event, and in India, for large- and medium-sized dams, a design exceedance probability of not less than 0.1 percent ($T = 1,000$ yr) is used [25]. Water resources projects also are designed to provide assured irrigation and municipal water supplies and instream river flows for droughts of certain severity. Rainfall data are analyzed to determine frequencies of both maximum events and drought rainfall.

The common technique for estimating extreme hydrologic events consists of choosing an appropriate probability distribution and fitting it to sample data by one of the available statistical procedures. Sample data comprise observed annual maximum or minimum values. If P is the exceedance probability of the annual events, the return period T is given by $1/P$ for maximum values (e.g., floods) and $1/(1-P)$ for minimum values (e.g., low flows).

Several probability distributions have been used to determine frequencies of hydrologic variables [5,17,18]. These vary in complexity from the simple two-parameter extreme value distribution [10] to the five-parameter Wakeby distribution [12]. A distribution widely recommended for use is log Pearson [2]. This chapter describes some theoretical properties of log Pearson type 3 distribution and illustrates its application by two methods for parameter estimation.

LOG PEARSON TYPE 3 (LP) DISTRIBUTION

The *LP* is a three-parameter probability distribution. Let a, b, and c be the parameters of *LP* and X the *LP* variable and $Y = \ln X$. The logarithmic variable Y is distributed as Pearson type 3 (*P3*) and the parameters a, b, and c are common for both *P3* and *LP*. The probability density function of *P3*, $f(y)$, is given by:

$$f(y) = \frac{|a|}{\Gamma(b)} [a(y - c)]^{b-1} \exp[-a(y - c)] \quad (1)$$

The density function of *LP* distribution was derived by Bobee [3] as:

$$f(x) = \frac{|a|}{\Gamma(b)} \frac{\exp(ac)}{x^{1+a}} [a(\ln x - c)]^{b-1} \quad (2)$$

If $a > 0$, the *P3* distribution is positively skewed, and $c \le y \le +\infty$. In this case, the *LP* distribution is also positively skewed, and $\exp(c) \le x < +\infty$.

If $a < 0$, the *P3* distribution is negatively skewed and $-\infty < y \le c$. In this case, the *LP* distribution is either positively or negatively skewed depending on the values of a and b and $0 < x \le \exp(c)$. At $x = 0$, $f(x)$ may be defined as zero for this case.

---

*St. Johns River Water Management District, Palatka, FL

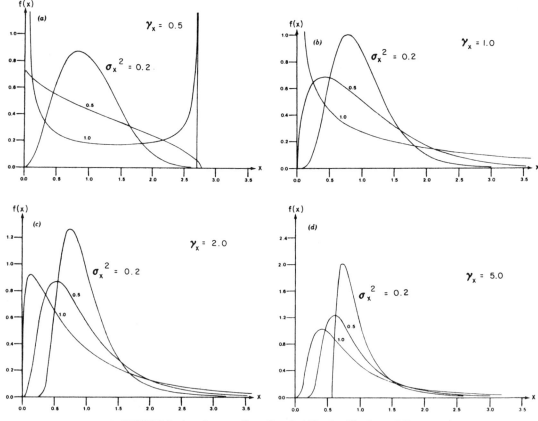

FIGURE 1. Log Pearson Type 3 probability densities ($\mu_x = 1.0$).

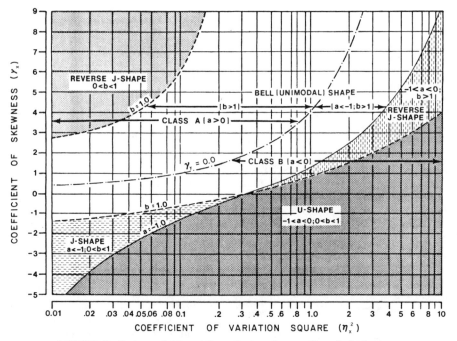

FIGURE 2. Regions of different forms for Log Pearson Type 3 distribution.

The distribution parameters (a, b and c) are related to the statistical parameters (SP) mean, variance and skew by the following equations:

1. $P3$ distribution:

$$\text{Mean: } \mu_y = c + ba^{-1} \qquad (3)$$

$$\text{Variance: } \sigma_y^2 = ba^{-2} \qquad (4)$$

$$\text{Skew: } \gamma_y = \frac{a}{|a|} 2b^{-1/2} \qquad (5)$$

2. LP distribution:

$$\text{Mean: } \mu_x = \frac{\exp(c)}{\left(1 - \dfrac{1}{a}\right)^b} \qquad (6)$$

$$\text{Variance: } \sigma_x^2 = \exp(2c) \left[\frac{1}{\left(1 - \dfrac{2}{a}\right)^b} - \frac{1}{\left(1 - \dfrac{1}{a}\right)^{2b}} \right] \qquad (7)$$

Skew: $\gamma_x = \exp(3c)$

$$\frac{\left\{ \dfrac{1}{\left(1 - \dfrac{3}{a}\right)^b} - \dfrac{3}{\left[\left(1 - \dfrac{2}{a}\right)\left(1 - \dfrac{1}{a}\right)\right]^b} + \dfrac{2}{\left(1 - \dfrac{1}{a}\right)^{3b}} \right\}}{\sigma_x^3} \qquad (8)$$

The forms taken by LP probability density function, $f(x)$, and its bounds are of some concern in fitting hydrologic data. Parameters a and b, known as scale and shape parameters, govern the overall geometric form of LP density, which takes four basic forms, viz., J, reverse J, bell (unimodal) and U. In addition, several transitional shapes varying from one basic form to the other occur (Figure 1). Figure 2 shows the regions of SP, which produce different forms of LP. Note that the coefficient of variation $\eta_x = \sigma_x/\mu_x$. The forms which need particular attention are the J, reverse J, and U. If the solution indicates one of these forms, it should be ensured that data are really distributed accordingly. With regard to bounds, a comparative study of different three-parameter probability distributions shows that for a given hydrologic sample, in general, the LP fit will have better applicable properties or its estimates of design events compare well with other distributions [21,23].

THE SAMPLE MOMENTS

For a sample of size n ($X_1 \ldots X_i \ldots X_n$, in which i is the ith item of data), the sample moments or SP are commonly calculated by:

$$\text{Mean: } \bar{X} = \frac{1}{n} \sum_{i=1}^{n} X_i \qquad (9)$$

$$\text{Variance: } S_x^2 = \frac{1}{(n-1)} \sum_{i=1}^{n} (X_i - \bar{X})^2 \qquad (10)$$

$$\text{Skew: } CS_x = \frac{n}{S_x^3(n-1)(n-2)} \sum_{i=1}^{n} (X_i - \bar{X})^3 \qquad (11)$$

Likewise, the logarithmic moments (for data $Y = \ln X$), $\bar{Y}$, S_y^2, and CS_y are obtained from Equations (9–11), respectively, by substituting Y for X. The sample moments are the estimates of the corresponding population SP, i.e., $\bar{X}$ for μ_x, etc.

EVALUATION OF LP QUANTILES

For given exceedance probability P, the log Pearson quantile (design value), X_p, is given by:

$$X_p = \exp(\mu_y + K'\sigma_y) \qquad (12)$$

in which $K' = $ Pearson frequency factor (Table 1). Value of K' depends upon P and γ_y. The simplest approach for solving Equation (12) is through the logarithmic sample moments (log moments method), i.e.,

1. Set $\mu_y = \bar{Y}$; $\sigma_y = S_y$; and $\gamma_y = CS_y$.
2. Obtain K' values from Table 1 for P values of interest. Calculate X_p values by Equation (12).

Calculation of LP quantiles is somewhat indirect in other methods [3,6,11,19,20], which solve for distribution parameters (a, b, and c). Then the log SP are obtained by Equations (3–5) and X_p are calculated by Equation (12).

BIAS IN ESTIMATED QUANTILES

In 1967, the United States Water Resources Council recommended the use of LP distribution (by log moments method) for estimating frequencies of flood flows [1]. Intensive research by numerous investigators since then contributed to both development of alternative fitting methods for LP [3,20] and greater understanding of the statistics of fre-

TABLE 1. Pearson Frequency Factors (K' Values) for Use in Equation (12).

LOGARITHMIC SKEWNESS COEFFICIENT (γ_y)

P	0.0	0.1	0.2	0.3	0.4	0.5	0.6	0.7	0.8	0.9	1.0	1.1	1.2	1.3
0.995	-2.5758	-2.4819	-2.3880	-2.2942	-2.2009	-2.1083	-2.0164	-1.9258	-1.8366	-1.7492	-1.6639	-1.5811	-1.5011	-1.4244
0.990	-2.3264	-2.2526	-2.1784	-2.1039	-2.0293	-1.9547	-1.8803	-1.8062	-1.7327	-1.6600	-1.5884	-1.5181	-1.4494	-1.3827
0.980	-2.0538	-1.9997	-1.9450	-1.8896	-1.8336	-1.7772	-1.7203	-1.6633	-1.6060	-1.5489	-1.4919	-1.4353	-1.3793	-1.3241
0.960	-1.7507	-1.7158	-1.6800	-1.6433	-1.6057	-1.5674	-1.5283	-1.4885	-1.4481	-1.4072	-1.3658	-1.3241	-1.2823	-1.2403
0.900	-1.2816	-1.2704	-1.2582	-1.2452	-1.2311	-1.2162	-1.2003	-1.1835	-1.1657	-1.1471	-1.1276	-1.1073	-1.0861	-1.0641
0.800	-0.8416	-0.8461	-0.8499	-0.8529	-0.8551	-0.8565	-0.8572	-0.8570	-0.8561	-0.8543	-0.8516	-0.8481	-0.8437	-0.8384
0.500	0.0000	-0.0166	-0.0333	-0.0499	-0.0665	-0.0830	-0.0995	-0.1158	-0.1320	-0.1481	-0.1640	-0.1797	-0.1952	-0.2104
0.200	0.8416	0.8364	0.8304	0.8238	0.8164	0.8083	0.7995	0.7900	0.7799	0.7690	0.7575	0.7454	0.7326	0.7192
0.100	1.2816	1.2918	1.3011	1.3094	1.3167	1.3231	1.3285	1.3329	1.3364	1.3389	1.3404	1.3409	1.3405	1.3390
0.040	1.7507	1.7846	1.8176	1.8495	1.8804	1.9102	1.9390	1.9666	1.9931	2.0185	2.0427	2.0657	2.0876	2.1082
0.020	2.0538	2.1070	2.1594	2.2108	2.2613	2.3108	2.3593	2.4067	2.4530	2.4981	2.5421	2.5848	2.6263	2.6666
0.010	2.3264	2.3996	2.4723	2.5442	2.6154	2.6857	2.7551	2.8236	2.8910	2.9574	3.0226	3.0866	3.1494	3.2110
0.005	2.5758	2.6697	2.7632	2.8564	2.9490	3.0410	3.1323	3.2228	3.3124	3.4011	3.4887	3.5753	3.6607	3.7450
0.002	2.8782	2.9998	3.1217	3.2437	3.3657	3.4874	3.6087	3.7296	3.8498	3.9693	4.0880	4.2058	4.3226	4.4384
0.001	3.0902	3.2332	3.3770	3.5214	3.6661	3.8109	3.9557	4.1002	4.2444	4.3881	4.5311	4.6734	4.8149	4.9555

LOGARITHMIC SKEWNESS COEFFICIENT (γ_y)

P	1.4	1.5	1.6	1.7	1.8	1.9	2.0	2.1	2.2	2.3	2.4	2.5	2.6	2.7
0.995	-1.3511	-1.2817	-1.2162	-1.1548	-1.0975	-1.0443	-0.9950	-0.9495	-0.9074	-0.8686	-0.8328	-0.7997	-0.7691	-0.7407
0.990	-1.3182	-1.2561	-1.1968	-1.1404	-1.0871	-1.0370	-0.9900	-0.9461	-0.9052	-0.8672	-0.8320	-0.7992	-0.7688	-0.7405
0.980	-1.2700	-1.2172	-1.1658	-1.1163	-1.0686	-1.0231	-0.9798	-0.9388	-0.9001	-0.8637	-0.8296	-0.7977	-0.7678	-0.7399
0.960	-1.1984	-1.1568	-1.1157	-1.0751	-1.0354	-0.9967	-0.9592	-0.9230	-0.8881	-0.8549	-0.8232	-0.7931	-0.7646	-0.7377
0.900	-1.0414	-1.0181	-0.9942	-0.9698	-0.9450	-0.9199	-0.8946	-0.8694	-0.8442	-0.8193	-0.7947	-0.7706	-0.7471	-0.7242
0.800	-0.8322	-0.8252	-0.8172	-0.8084	-0.7987	-0.7882	-0.7769	-0.7648	-0.7521	-0.7388	-0.7250	-0.7107	-0.6960	-0.6811
0.500	-0.2254	-0.2400	-0.2542	-0.2681	-0.2815	-0.2944	-0.3069	-0.3187	-0.3300	-0.3406	-0.3506	-0.3599	-0.3685	-0.3764
0.200	0.7051	0.6905	0.6753	0.6596	0.6434	0.6266	0.6094	0.5918	0.5738	0.5555	0.5368	0.5179	0.4987	0.4793
0.100	1.3367	1.3333	1.3290	1.3238	1.3176	1.3105	1.3026	1.2938	1.2841	1.2737	1.2624	1.2504	1.2377	1.2242
0.040	2.1277	2.1459	2.1629	2.1787	2.1933	2.2067	2.2189	2.2299	2.2397	2.2483	2.2558	2.2622	2.2674	2.2716
0.020	2.7056	2.7433	2.7796	2.8147	2.8485	2.8809	2.9120	2.9418	2.9703	2.9974	3.0233	3.0479	3.0712	3.0932
0.010	3.2713	3.3304	3.3880	3.4444	3.4994	3.5530	3.6052	3.6560	3.7054	3.7535	3.8001	3.8454	3.8893	3.9318
0.005	3.8280	3.9097	3.9902	4.0693	4.1470	4.2234	4.2983	4.3719	4.4440	4.5147	4.5839	4.6518	4.7182	4.7831
0.002	4.5530	4.6665	4.7788	4.8897	4.9994	5.1077	5.2146	5.3201	5.4243	5.5269	5.6282	5.7280	5.8263	5.9232
0.001	5.0951	5.2335	5.3709	5.5070	5.6419	5.7755	5.9078	6.0387	6.1682	6.2963	6.4229	6.5481	6.6719	6.7942

LOGARITHMIC SKEWNESS COEFFICIENT (γ_y)

P	0.0	-0.1	-0.2	-0.3	-0.4	-0.5	-0.6	-0.7	-0.8	-0.9	-1.0	-1.1	-1.2	-1.3
0.995	-2.5758	-2.6697	-2.7632	-2.8564	-2.9490	-3.0410	-3.1323	-3.2228	-3.3124	-3.4011	-3.4887	-3.5753	-3.6607	-3.7450
0.990	-2.3264	-2.3996	-2.4723	-2.5442	-2.6154	-2.6857	-2.7551	-2.8236	-2.8910	-2.9574	-3.0226	-3.0866	-3.1494	-3.2110
0.980	-2.0538	-2.1070	-2.1594	-2.2108	-2.2613	-2.3108	-2.3593	-2.4067	-2.4530	-2.4981	-2.5421	-2.5848	-2.6263	-2.6666
0.960	-1.7507	-1.7846	-1.8176	-1.8495	-1.8804	-1.9102	-1.9390	-1.9666	-1.9931	-2.0185	-2.0427	-2.0657	-2.0876	-2.1082
0.900	-1.2816	-1.2918	-1.3011	-1.3094	-1.3167	-1.3231	-1.3285	-1.3329	-1.3364	-1.3389	-1.3404	-1.3409	-1.3405	-1.3390
0.800	-0.8416	-0.8364	-0.8304	-0.8238	-0.8164	-0.8083	-0.7995	-0.7900	-0.7799	-0.7690	-0.7575	-0.7454	-0.7326	-0.7192
0.500	0.0000	0.0166	0.0333	0.0499	0.0665	0.0830	0.0995	0.1158	0.1320	0.1481	0.1640	0.1797	0.1952	0.2104
0.200	0.8416	0.8461	0.8499	0.8529	0.8551	0.8565	0.8572	0.8570	0.8561	0.8543	0.8516	0.8481	0.8437	0.8384
0.100	1.2816	1.2704	1.2582	1.2452	1.2311	1.2162	1.2003	1.1835	1.1657	1.1471	1.1276	1.1073	1.0861	1.0641
0.040	1.7507	1.7158	1.6800	1.6433	1.6057	1.5674	1.5283	1.4885	1.4481	1.4072	1.3658	1.3241	1.2823	1.2403
0.020	2.0538	1.9997	1.9450	1.8896	1.8336	1.7772	1.7203	1.6633	1.6060	1.5489	1.4919	1.4353	1.3793	1.3241
0.010	2.3264	2.2526	2.1784	2.1039	2.0293	1.9547	1.8803	1.8062	1.7327	1.6600	1.5884	1.5181	1.4494	1.3827
0.005	2.5758	2.4819	2.3880	2.2942	2.2009	2.1083	2.0164	1.9258	1.8366	1.7492	1.6639	1.5811	1.5011	1.4244
0.002	2.8782	2.7571	2.6367	2.5174	2.3994	2.2831	2.1688	2.0570	1.9481	1.8424	1.7406	1.6431	1.5502	1.4623
0.001	3.0902	2.9483	2.8079	2.6692	2.5326	2.3987	2.2678	2.1405	2.0174	1.8989	1.7857	1.6783	1.5770	1.4822

(continued)

TABLE 1 (continued).

LOGARITHMIC SKEWNESS COEFFICIENT (γ_y)

P	-1.4	-1.5	-1.6	-1.7	-1.8	-1.9	-2.0	-2.1	-2.2	-2.3	-2.4	-2.5	-2.6	-2.7
0.995	-3.8280	-3.9097	-3.9902	-4.0693	-4.1470	-4.2234	-4.2983	-4.3719	-4.4440	-4.5147	-4.5839	-4.6518	-4.7182	-4.7831
0.990	-3.2713	-3.3304	-3.3880	-3.4444	-3.4994	-3.5530	-3.6052	-3.6560	-3.7054	-3.7535	-3.8001	-3.8454	-3.8893	-3.9318
0.980	-2.7056	-2.7433	-2.7796	-2.8147	-2.8485	-2.8809	-2.9120	-2.9418	-2.9703	-2.9974	-3.0233	-3.0479	-3.0712	-3.0932
0.960	-2.1277	-2.1459	-2.1629	-2.1787	-2.1933	-2.2067	-2.2189	-2.2299	-2.2397	-2.2483	-2.2558	-2.2622	-2.2674	-2.2716
0.900	-1.3367	-1.3333	-1.3290	-1.3238	-1.3176	-1.3105	-1.3026	-1.2938	-1.2841	-1.2737	-1.2624	-1.2504	-1.2377	-1.2242
0.800	-0.7051	-0.6905	-0.6753	-0.6596	-0.6434	-0.6266	-0.6094	-0.5918	-0.5738	-0.5555	-0.5368	-0.5179	-0.4987	-0.4793
0.500	0.2254	0.2400	0.2542	0.2681	0.2815	0.2944	0.3069	0.3187	0.3300	0.3406	0.3506	0.3599	0.3685	0.3764
0.200	0.8322	0.8252	0.8172	0.8084	0.7987	0.7882	0.7769	0.7648	0.7521	0.7388	0.7250	0.7107	0.6960	0.6811
0.100	1.0414	1.0181	0.9942	0.9698	0.9450	0.9199	0.8946	0.8694	0.8442	0.8193	0.7947	0.7706	0.7471	0.7242
0.040	1.1984	1.1568	1.1157	1.0751	1.0354	0.9967	0.9592	0.9230	0.8881	0.8549	0.8232	0.7931	0.7646	0.7377
0.020	1.2700	1.2172	1.1658	1.1163	1.0686	1.0231	0.9798	0.9388	0.9001	0.8637	0.8296	0.7977	0.7678	0.7399
0.010	1.3182	1.2561	1.1968	1.1404	1.0871	1.0370	0.9900	0.9461	0.9052	0.8672	0.8320	0.7992	0.7688	0.7405
0.005	1.3511	1.2817	1.2162	1.1548	1.0975	1.0443	0.9950	0.9495	0.9074	0.8686	0.8328	0.7997	0.7691	0.7407
0.002	1.3798	1.3028	1.2313	1.1653	1.1047	1.0490	0.9980	0.9513	0.9085	0.8693	0.8332	0.7999	0.7692	0.7407
0.001	1.3941	1.3128	1.2381	1.1697	1.1074	1.0507	0.9990	0.9519	0.9089	0.8695	0.8333	0.8000	0.7692	0.7407

TABLE 2. Effect of Bias in Estimate of γ_y on Log Pearson Quantiles by Logarithmic Moments Method.

Return period, in years, T (1)	γ_y Positive; Bias Negative		γ_y Negative; Bias Algebraically Positive	
	Annual smallest events (2)	Annual largest events (3)	Annual smallest events (4)	Annual largest events (5)
2	Quantiles will: Increase for γ_y = 0 to 3.7; decrease for γ_y > 3.7	Quantiles will: Increase for γ_y = 0 to 3.7; decrease for γ_y > 3.7	Quantiles will: Decrease for $|\gamma_y|$ = 0 to 3.7; increase for $|\gamma_y|$ > 3.7	Quantiles will: Decrease for $|\gamma_y|$ = 0 to 3.7; increase for $|\gamma_y|$ > 3.7
5	Decrease	Increase	Decrease	Increase
10	Decrease	Decrease for γ_y = 0 to 1.1; increase for γ_y > 1.1	Increase for $|\gamma_y|$ 0 to 1.1; decrease for $|\gamma_y|$ > 1.1	Increase
25	Decrease	Decrease for γ_y = 0 to 3.1; increase for γ_y > 3.1	Increase for $|\gamma_y|$ = 0 to 3.1; decrease for $|\gamma_y|$ > 3.1	Increase
50	Decrease	Decrease for γ_y = 0 to 4.8; increase for γ_y > 4.8	Increase for $|\gamma_y|$ = 0 to 4.8; decrease for $|\gamma_y|$ > 4.8	Increase
100	Decrease	Decrease for γ_y = 0 to 7.1; increase for γ_y > 7.1	Increase for $|\gamma_y|$ = 0 to 7.1; decrease for $|\gamma_y|$ > 7.1	Increase
≥200	Decrease	Decrease	Increase	Increase

quency analysis [13,14,26,27]. Bobee [3] showed that the *LP* distribution can also be fitted through the moments of real data. Studies by Wallis, et al. [27] and Kirby [13] showed that *CS* (Equation (11)) on the average, is an underestimate of γ for sample sizes normally found in practice. Table 2 summarizes the effect of such bias on *LP* quantiles. Note that when γ_y is negative, CS_y will be an overestimate of γ_y algebraically. Reference [9] suggests development and use of regional log skew coefficient for improving the accuracy of flood flow estimates. Others have proposed bias correction factors for *CS* [4,15].

METHOD OF MIXED MOMENTS

Since parameters *a*, *b*, and *c* are common for *LP* [Equation (2)] and *P3* [Equation (1)] distributions, it is possible to solve for these values by mixing the log and real data moments and choosing any three moments. This approach gives rise to methods which eliminate the use of *CS* from calculations, e.g., values of *a*, *b*, and *c* can be solved from: (1) $\bar{X}$, S_x^2, and $\bar{Y}$; (2) $\bar{Y}$, S_y^2, and $\bar{X}$; (3) $\bar{X}$, S_x^2, and S_y^2; and (4) $\bar{Y}$, S_y^2, and S_x^2. However, simulation experiments have shown that Mixed Moments–I (MXM1) method which is based on mean and variance of real data, and the mean of log data (i.e., $\bar{X}$, S_x^2, and $\bar{Y}$), has generally superior statistical properties compared to logarithmic moments as well as other mixed moments methods [20,24].

Estimating *LP* parameters by MXM1 method involves an iterative process in which either parameter *a* or parameter *b* is optimized [22]. The method which optimizes parameter *b* is described below as it is computationally efficient.

Parameters *a*, *b*, *c*, μ_x, and σ_x^2 are related as follows [19,22]:

$$b = \frac{\ln\left(\frac{\mu_x^2 + \sigma_x^2}{\mu_x^2}\right)}{\ln\left[\frac{\left(1 - \frac{1}{a}\right)^2}{\left(1 - \frac{2}{a}\right)}\right]} \tag{13}$$

$$c = \ln \mu_x + b \ln\left(1 - \frac{1}{a}\right) \tag{14}$$

Equation (13) can be rearranged in the form of a quadratic equation in *a* as:

$$a^2 - 2a - \frac{1}{Z} = 0 \tag{15}$$

in which *Z* is given by

$$Z = \left(\frac{\mu_x^2 + \sigma_x^2}{\mu_x^2}\right)^{b^{-1}} - 1 \tag{16}$$

Equation (15) yields:

$$a = 1 \pm \left(1 + \frac{1}{Z}\right)^{0.5} \tag{17}$$

Parameter *a* has the same sign as γ_y and *b* is related to γ_y by Equation (5). Thus, if an initial estimate of γ_y is available, it would provide an initial value for *b*, and the sign of *a*. Then the procedure consists of: (1) calculating *Z*, *a*, *c*, and μ_y by Equations (16), (17), (14), and (3), respectively; and (2) adjusting *b* until $\mu_y \cong \bar{Y}$. However, optimization of parameters is greatly facilitated if calculations are performed by converting real data into dimensionless variates by the transform $K_i = X_i/\bar{X}$. For the transformed data K_i, the mean $\bar{K} = 1.0$ and the final results (e.g., flood flows for different return periods) are obtained as ratios to mean. Substituting $\mu_x = \mu_k = 1.0$ and $\sigma_x^2 = \sigma_k^2$ (in which μ_k and σ_k^2 are the population mean and variance of *K*, respectively) in Equations (14) and (16), a Newton-Raphson iteration scheme can be formulated for optimizing parameter *b* (see Appendix). Also, for choosing an appropriate initial value for γ_y, Tables 3, 4, and 5 present $\mu_y - \sigma_k^2 - \gamma_y$ relations for the variate *K*.

The following algorithm may be used to obtain *LP* quantiles by MXM1 method for a given sample.

Let the given data (e.g., annual flood flows) be represented by $X_1, \ldots X_i, \ldots X_n$.

Step 1. Calculate the mean of data, $\bar{X}$ [Equation (9)].
Step 2. Convert X_i into dimensionless variates by the transform, $K_i = X_i/\bar{X}$. Calculate the variance of K_i, S_k^2 (Equation (10)).
Step 3. Obtain logarithmic data, $Y_i = \ln K_i$. Calculate $\bar{Y}$.
Step 4. For $\sigma_k^2 = S_k^2$ and $\mu_y = \bar{Y}$, obtain an approximate value of γ_y from Tables 3–5. Calculate the initial value of $b = 4/\gamma_y^2$ [Equation (5)]. Note: Column 8, Table 4 ($\gamma_y = 0$) represents the two-parameter log normal distribution (*LN2*) for which the logarithmic mean is given by:

$$\mu_{yLN2} = -\frac{1}{2} \ln (1 + \sigma_k^2) \tag{18}$$

If S_k^2 and $\bar{Y}$ indicate that γ_y may be close to zero, the sign of γ_y can be readily determined on the basis of μ_{yLN2}. At given σ_k^2, μ_y increases (algebraically) with γ_y (see Tables 3–5). Thus, if $\bar{Y} < \mu_{yLN2}$, parameter *a* and γ_y will be negative; and if $\bar{Y} > \mu_{yLN2}$, parameter *a* and γ_y will be positive.
Step 5. Calculate $Z = (1 + S_k^2)^{1/b} - 1$, and then the two values for *a* from Equation (17). Accept the value of *a*, which has the same sign as γ_y in Step 4.

TABLE 3. Logarithmic Mean (μ_y) Values for Different Variance and Logarithmic Skewness Coefficients of Dimensionless Variate, K ($\gamma_y = -3.0 - -0.7$).

	LOGARITHMIC SKEWNESS COEFFICIENT (γ_y)												
VARIANCE OF K	-3.0	-2.6	-2.4	-2.2	-2.0	-1.8	-1.6	-1.4	-1.2	-1.0	-0.9	-0.8	-0.7
(1)	(2)	(3)	(4)	(5)	(6)	(7)	(8)	(9)	(10)	(11)	(12)	(13)	(14)
0.03	-0.0210	-0.0200	-0.0196	-0.0191	0.0187	-0.0182	-0.0178	-0.0174	-0.0170	-0.0166	-0.0164	-0.0162	-0.0160
0.04	-0.0295	-0.0279	-0.0271	-0.0264	-0.0257	-0.0250	-0.0243	-0.0236	-0.0230	-0.0224	-0.0221	-0.0218	-0.0215
0.05	-0.0385	-0.0362	-0.0350	-0.0340	-0.0329	-0.0320	-0.0310	-0.0301	-0.0292	-0.0283	-0.0279	-0.0275	-0.0271
0.06	-0.0481	-0.0448	-0.0433	-0.0419	-0.0405	-0.0392	-0.0379	-0.0366	-0.0354	-0.0343	-0.0337	-0.0332	-0.0326
0.07	-0.0581	-0.0539	-0.0519	-0.0501	-0.0483	-0.0465	-0.0449	-0.0433	-0.0418	-0.0403	-0.0396	-0.0389	-0.0382
0.08	-0.0686	-0.0633	-0.0609	-0.0585	-0.0563	-0.0541	-0.0521	-0.0501	-0.0482	-0.0464	-0.0455	-0.0447	-0.0438
0.09	-0.0796	-0.0731	-0.0701	-0.0672	-0.0644	-0.0618	-0.0593	-0.0570	-0.0547	-0.0525	-0.0515	-0.0505	-0.0495
0.10	-0.0911	-0.0832	-0.0796	-0.0761	-0.0728	-0.0697	-0.0668	-0.0639	-0.0613	-0.0587	-0.0575	-0.0563	-0.0551
0.15	-0.1547	-0.1382	-0.1308	-0.1238	-0.1172	-0.1111	-0.1053	-0.0999	-0.0948	-0.0900	-0.0878	-0.0855	-0.0834
0.20	-0.2283	-0.2001	-0.1876	-0.1760	-0.1652	-0.1553	-0.1460	-0.1374	-0.1294	-0.1219	-0.1183	-0.1149	-0.1116
0.25	-0.3114	-0.2682	-0.2493	-0.2321	-0.2162	-0.2017	-0.1883	-0.1759	-0.1645	-0.1539	-0.1490	-0.1442	-0.1395
0.30	-0.4034	-0.3420	-0.3156	-0.2916	-0.2698	-0.2500	-0.2319	-0.2153	-0.2001	-0.1861	-0.1796	-0.1733	-0.1673
0.35	-0.5043	-0.4212	-0.3859	-0.3543	-0.3257	-0.2999	-0.2765	-0.2553	-0.2359	-0.2183	-0.2101	-0.2022	-0.1946
0.40	-0.6138	-0.5054	-0.4602	-0.4198	-0.3836	-0.3512	-0.3220	-0.2957	-0.2719	-0.2504	-0.2403	-0.2308	-0.2217
0.45	-0.7318	-0.5946	-0.5380	-0.4879	-0.4434	-0.4038	-0.3683	-0.3366	-0.3080	-0.2823	-0.2704	-0.2591	-0.2483
0.50	-0.8582	-0.6884	-0.6192	-0.5584	-0.5048	-0.4574	-0.4153	-0.3777	-0.3442	-0.3141	-0.3002	-0.2871	-0.2746
0.55	-0.9929	-0.7866	-0.7036	-0.6312	-0.5678	-0.5120	-0.4627	-0.4191	-0.3803	-0.3457	-0.3298	-0.3147	-0.3005
0.60	-1.1358	-0.8892	-0.7911	-0.7061	-0.6321	-0.5674	-0.5106	-0.4606	-0.4163	-0.3770	-0.3590	-0.3420	-0.3260
0.65	-1.2869	-0.9960	-0.8815	-0.7830	-0.6977	-0.6236	-0.5589	-0.5022	-0.4522	-0.4081	-0.3880	-0.3690	-0.3511
0.70	-1.4462	-1.1069	-0.9748	-0.8617	-0.7645	-0.6805	-0.6075	-0.5438	-0.4880	-0.4390	-0.4166	-0.3956	-0.3758
0.75	-1.6137	-1.2218	-1.0707	-0.9423	-0.8324	-0.7380	-0.6564	-0.5855	-0.5237	-0.4695	-0.4450	-0.4219	-0.4002
0.80	-1.7893	-1.3405	-1.1692	-1.0245	-0.9014	-0.7961	-0.7055	-0.6273	-0.5592	-0.4999	-0.4730	-0.4478	-0.4242
0.85	-1.9730	-1.4630	-1.2703	-1.1083	-0.9713	-0.8547	-0.7549	-0.6689	-0.5946	-0.5299	-0.5007	-0.4734	-0.4479
0.90	-2.1648	-1.5893	-1.3737	-1.1937	-1.0421	-0.9138	-0.8044	-0.7106	-0.6297	-0.5597	-0.5281	-0.4987	-0.4712
0.95	-2.3647	-1.7191	-1.4796	-1.2805	-1.1138	-0.9733	-0.8540	-0.7521	-0.6647	-0.5892	-0.5553	-0.5237	-0.4941
1.00	-2.5728	-1.8525	-1.5877	-1.3687	-1.1863	-1.0331	-0.9037	-0.7936	-0.6994	-0.6184	-0.5821	-0.5483	-0.5168
1.25	-3.7350	-2.5709	-2.1604	-1.8291	-1.5592	-1.3373	-1.1533	-0.9996	-0.8702	-0.7605	-0.7119	-0.6670	-0.6253
1.50	-5.1020	-3.3702	-2.7820	-2.3176	-1.9466	-1.6472	-1.4031	-1.2023	-1.0356	-0.8962	-0.8350	-0.7787	-0.7269
1.75	-6.6761	-4.2453	-3.4477	-2.8301	-2.3456	-1.9609	-1.6520	-1.4013	-1.1958	-1.0259	-0.9520	-0.8843	-0.8223
2.00	-8.4600	-5.1924	-4.1537	-3.3637	-2.7540	-2.2770	-1.8992	-1.5964	-1.3511	-1.1503	-1.0635	-0.9844	-0.9123
2.25	-10.4568	-6.2082	-4.8969	-3.9159	-3.1701	-2.5946	-2.1445	-1.7877	-1.5016	-1.2696	-1.1700	-1.0796	-0.9975
2.50		-7.2902	-5.6749	-4.4848	-3.5927	-2.9131	-2.3874	-1.9752	-1.6477	-1.3844	-1.2720	-1.1704	-1.0784
2.75		-8.4360	-6.4856	-5.0691	-4.0209	-3.2318	-2.6280	-2.1590	-1.7896	-1.4950	-1.3699	-1.2573	-1.1556
3.00		-9.6437	-7.3272	-5.6673	-4.4540	-3.5506	-2.8662	-2.3393	-1.9277	-1.6018	-1.4641	-1.3405	-1.2293
3.25		-10.9115	-8.1982	-6.2784	-4.8913	-3.8691	-3.1019	-2.5162	-2.0622	-1.7050	-1.5549	-1.4205	-1.2998
3.50			-9.0973	-6.9016	-5.3323	-4.1871	-3.3351	-2.6899	-2.1932	-1.8050	-1.6426	-1.4975	-1.3676
3.75			-10.0233	-7.5361	-5.7765	-4.5045	-3.5660	-2.8606	-2.3211	-1.9020	-1.7273	-1.5718	-1.4327
4.00				-8.1811	-6.2238	-4.8212	-3.7945	-3.0283	-2.4460	-1.9961	-1.8094	-1.6435	-1.4955
4.25				-8.8361	-6.6737	-5.1370	-4.0208	-3.1932	-2.5680	-2.0877	-1.8890	-1.7128	-1.5561
4.50				-9.5005	-7.1259	-5.4521	-4.2448	-3.3554	-2.6874	-2.1767	-1.9662	-1.7800	-1.6147
4.75				-10.1739	-7.5804	-5.7662	-4.4666	-3.5151	-2.8043	-2.2635	-2.0413	-1.8452	-1.6714
5.00					-8.0369	-6.0793	-4.6863	-3.6723	-2.9188	-2.3481	-2.1144	-1.9085	-1.7263
6.00					-9.8795	-7.3221	-5.5452	-4.2789	-3.3552	-2.6672	-2.3887	-2.1449	-1.9307
7.00					-11.7435	-8.5491	-6.3749	-4.8537	-3.7620	-2.9602	-2.6388	-2.3591	-2.1147
8.00						-9.7604	-7.1787	-5.4014	-4.1438	-3.2317	-2.8692	-2.5553	-2.2823
9.00						-10.9568	-7.9593	-5.9254	-4.5044	-3.4852	-3.0832	-2.7367	-2.4365
10.00							-8.7190	-6.4286	-4.8468	-3.7234	-3.2834	-2.9056	-2.5796

TABLE 4. Logarithmic Mean (μ_y) Values for Different Variance and Logarithmic Skewness Coefficients of Dimensionless Variate, K (γ_y = -0.6 - 0.6).

VARIANCE OF K (1)	LOGARITHMIC SKEWNESS COEFFICIENT (γ_y)												
	-0.6 (2)	-0.5 (3)	-0.4 (4)	-0.3 (5)	-0.2 (6)	-0.1 (7)	0.0 (8)	0.1 (9)	0.2 (10)	0.3 (11)	0.4 (12)	0.5 (13)	0.6 (14)
0.03	-0.0158	-0.0157	-0.0155	-0.0153	-0.0151	-0.0149	-0.0148	-0.0146	-0.0144	-0.0143	-0.0141	-0.0140	-0.0138
0.04	-0.0212	-0.0210	-0.0207	-0.0204	-0.0201	-0.0199	-0.0196	-0.0194	-0.0191	-0.0189	-0.0186	-0.0184	-0.0181
0.05	-0.0267	-0.0263	-0.0259	-0.0255	-0.0251	-0.0248	-0.0244	-0.0240	-0.0237	-0.0233	-0.0230	-0.0227	-0.0223
0.06	-0.0321	-0.0316	-0.0311	-0.0306	-0.0301	-0.0296	-0.0291	-0.0287	-0.0282	-0.0278	-0.0273	-0.0269	-0.0265
0.07	-0.0376	-0.0369	-0.0363	-0.0356	-0.0350	-0.0344	-0.0338	-0.0332	-0.0327	-0.0321	-0.0316	-0.0310	-0.0305
0.08	-0.0430	-0.0422	-0.0414	-0.0407	-0.0399	-0.0392	-0.0385	-0.0378	-0.0371	-0.0364	-0.0357	-0.0351	-0.0345
0.09	-0.0485	-0.0475	-0.0466	-0.0457	-0.0448	-0.0439	-0.0431	-0.0423	-0.0414	-0.0406	-0.0399	-0.0391	-0.0383
0.10	-0.0540	-0.0529	-0.0518	-0.0507	-0.0497	-0.0486	-0.0477	-0.0467	-0.0457	-0.0448	-0.0439	-0.0430	-0.0422
0.15	-0.0813	-0.0792	-0.0773	-0.0753	-0.0735	-0.0716	-0.0699	-0.0682	-0.0665	-0.0649	-0.0633	-0.0618	-0.0603
0.20	-0.1084	-0.1052	-0.1022	-0.0993	-0.0965	-0.0938	-0.0912	-0.0886	-0.0861	-0.0837	-0.0814	-0.0792	-0.0770
0.25	-0.1351	-0.1308	-0.1267	-0.1227	-0.1189	-0.1151	-0.1116	-0.1081	-0.1048	-0.1016	-0.0985	-0.0955	-0.0926
0.30	-0.1615	-0.1559	-0.1506	-0.1454	-0.1405	-0.1357	-0.1312	-0.1268	-0.1226	-0.1185	-0.1146	-0.1108	-0.1071
0.35	-0.1874	-0.1805	-0.1739	-0.1676	-0.1615	-0.1556	-0.1501	-0.1447	-0.1395	-0.1346	-0.1298	-0.1253	-0.1209
0.40	-0.2130	-0.2046	-0.1967	-0.1891	-0.1818	-0.1749	-0.1682	-0.1619	-0.1558	-0.1499	-0.1443	-0.1390	-0.1338
0.45	-0.2381	-0.2283	-0.2190	-0.2101	-0.2016	-0.1935	-0.1858	-0.1784	-0.1713	-0.1646	-0.1582	-0.1520	-0.1461
0.50	-0.2627	-0.2514	-0.2407	-0.2305	-0.2208	-0.2115	-0.2027	-0.1943	-0.1863	-0.1787	-0.1714	-0.1644	-0.1577
0.55	-0.2870	-0.2741	-0.2620	-0.2504	-0.2395	-0.2290	-0.2191	-0.2097	-0.2007	-0.1922	-0.1840	-0.1762	-0.1688
0.60	-0.3108	-0.2964	-0.2828	-0.2699	-0.2576	-0.2460	-0.2350	-0.2245	-0.2146	-0.2051	-0.1961	-0.1876	-0.1794
0.65	-0.3342	-0.3182	-0.3031	-0.2888	-0.2753	-0.2625	-0.2504	-0.2389	-0.2280	-0.2176	-0.2078	-0.1984	-0.1895
0.70	-0.3572	-0.3396	-0.3230	-0.3074	-0.2926	-0.2786	-0.2653	-0.2528	-0.2409	-0.2296	-0.2190	-0.2088	-0.1992
0.75	-0.3798	-0.3606	-0.3425	-0.3254	-0.3094	-0.2942	-0.2798	-0.2662	-0.2534	-0.2412	-0.2297	-0.2188	-0.2085
0.80	-0.4020	-0.3812	-0.3616	-0.3431	-0.3257	-0.3094	-0.2939	-0.2793	-0.2655	-0.2525	-0.2401	-0.2285	-0.2174
0.85	-0.4239	-0.4014	-0.3803	-0.3604	-0.3417	-0.3242	-0.3076	-0.2920	-0.2772	-0.2633	-0.2502	-0.2377	-0.2260
0.90	-0.4454	-0.4212	-0.3986	-0.3773	-0.3574	-0.3386	-0.3209	-0.3043	-0.2886	-0.2738	-0.2599	-0.2467	-0.2343
0.95	-0.4665	-0.4407	-0.4166	-0.3939	-0.3726	-0.3527	-0.3339	-0.3163	-0.2997	-0.2840	-0.2693	-0.2554	-0.2422
1.00	-0.4874	-0.4599	-0.4342	-0.4101	-0.3876	-0.3664	-0.3466	-0.3279	-0.3104	-0.2939	-0.2784	-0.2637	-0.2499
1.25	-0.5867	-0.5509	-0.5176	-0.4865	-0.4576	-0.4307	-0.4055	-0.3819	-0.3599	-0.3393	-0.3200	-0.3018	-0.2848
1.50	-0.6791	-0.6349	-0.5940	-0.5562	-0.5211	-0.4885	-0.4581	-0.4299	-0.4036	-0.3791	-0.3562	-0.3349	-0.3149
1.75	-0.7653	-0.7130	-0.6647	-0.6202	-0.5791	-0.5410	-0.5058	-0.4731	-0.4428	-0.4146	-0.3884	-0.3640	-0.3412
2.00	-0.8463	-0.7859	-0.7304	-0.6794	-0.6325	-0.5892	-0.5493	-0.5124	-0.4782	-0.4466	-0.4172	-0.3899	-0.3646
2.25	-0.9226	-0.8543	-0.7918	-0.7346	-0.6821	-0.6338	-0.5893	-0.5484	-0.5105	-0.4756	-0.4433	-0.4133	-0.3856
2.50	-0.9949	-0.9189	-0.8495	-0.7862	-0.7283	-0.6752	-0.6264	-0.5815	-0.5403	-0.5022	-0.4671	-0.4346	-0.4046
2.75	-1.0635	-0.9799	-0.9039	-0.8347	-0.7716	-0.7138	-0.6609	-0.6123	-0.5677	-0.5267	-0.4889	-0.4541	-0.4219
3.00	-1.1288	-1.0379	-0.9555	-0.8805	-0.8123	-0.7500	-0.6931	-0.6411	-0.5933	-0.5495	-0.5092	-0.4721	-0.4379
3.25	-1.1912	-1.0931	-1.0044	-0.9239	-0.8508	-0.7842	-0.7235	-0.6680	-0.6172	-0.5706	-0.5279	-0.4887	-0.4526
3.50	-1.2509	-1.1458	-1.0509	-0.9651	-0.8872	-0.8165	-0.7520	-0.6933	-0.6396	-0.5905	-0.5455	-0.5042	-0.4663
3.75	-1.3082	-1.1962	-1.0954	-1.0043	-0.9219	-0.8471	-0.7791	-0.7172	-0.6607	-0.6091	-0.5619	-0.5187	-0.4791
4.00	-1.3632	-1.2446	-1.1379	-1.0418	-0.9548	-0.8761	-0.8047	-0.7398	-0.6806	-0.6267	-0.5774	-0.5323	-0.4910
4.25	-1.4163	-1.2911	-1.1787	-1.0776	-0.9864	-0.9039	-0.8291	-0.7612	-0.6995	-0.6433	-0.5920	-0.5451	-0.5022
4.50	-1.4674	-1.3358	-1.2179	-1.1120	-1.0165	-0.9303	-0.8524	-0.7817	-0.7175	-0.6590	-0.6058	-0.5572	-0.5128
4.75	-1.5168	-1.3790	-1.2556	-1.1450	-1.0454	-0.9557	-0.8746	-0.8012	-0.7345	-0.6740	-0.6189	-0.5686	-0.5228
5.00	-1.5646	-1.4206	-1.2920	-1.1767	-1.0732	-0.9800	-0.8959	-0.8198	-0.7508	-0.6882	-0.6313	-0.5795	-0.5323
6.00	-1.7417	-1.5743	-1.4256	-1.2930	-1.1746	-1.0684	-0.9730	-0.8870	-0.8095	-0.7393	-0.6758	-0.6182	-0.5658
7.00	-1.9001	-1.7110	-1.5438	-1.3954	-1.2633	-1.1453	-1.0397	-0.9449	-0.8597	-0.7829	-0.7135	-0.6508	-0.5940
8.00	-2.0437	-1.8344	-1.6499	-1.4869	-1.3422	-1.2135	-1.0986	-0.9958	-0.9036	-0.8208	-0.7462	-0.6789	-0.6181
9.00	-2.1753	-1.9469	-1.7464	-1.5697	-1.4134	-1.2747	-1.1513	-1.0412	-0.9426	-0.8543	-0.7750	-0.7036	-0.6392
10.00	-2.2968	-2.0504	-1.8348	-1.6453	-1.4782	-1.3303	-1.1989	-1.0820	-0.9777	-0.8843	-0.8007	-0.7255	-0.6579

TABLE 5. Logarithmic Mean (μy) Values for Different Variance and Logarithmic Skewness Coefficients of Dimensionless Variate, K (γy = −0.7 − 3.0).

VARIANCE OF K (1)	LOGARITHMIC SKEWNESS COEFFICIENT (γ_y)												
	0.7 (2)	0.8 (3)	0.9 (4)	1.0 (5)	1.2 (6)	1.4 (7)	1.6 (8)	1.8 (9)	2.0 (10)	2.2 (11)	2.4 (12)	2.6 (13)	3.0 (14)
0.03	−0.0136	−0.0135	−0.0133	−0.0132	−0.0129	−0.0126	−0.0123	−0.0121	−0.0118	−0.0115	−0.0113	−0.0110	−0.0105
0.04	−0.0179	−0.0177	−0.0174	−0.0172	−0.0168	−0.0163	−0.0159	−0.0155	−0.0151	−0.0147	−0.0144	−0.0140	−0.0133
0.05	−0.0220	−0.0217	−0.0214	−0.0211	−0.0205	−0.0199	−0.0193	−0.0188	−0.0183	−0.0177	−0.0173	−0.0168	−0.0158
0.06	−0.0261	−0.0256	−0.0252	−0.0248	−0.0241	−0.0233	−0.0226	−0.0219	−0.0212	−0.0206	−0.0200	−0.0194	−0.0182
0.07	−0.0300	−0.0295	−0.0290	−0.0285	−0.0275	−0.0266	−0.0257	−0.0249	−0.0241	−0.0233	−0.0225	−0.0218	−0.0204
0.08	−0.0338	−0.0332	−0.0326	−0.0320	−0.0309	−0.0298	−0.0288	−0.0278	−0.0268	−0.0258	−0.0249	−0.0241	−0.0224
0.09	−0.0376	−0.0369	−0.0362	−0.0355	−0.0342	−0.0329	−0.0317	−0.0305	−0.0294	−0.0283	−0.0273	−0.0263	−0.0244
0.10	−0.0413	−0.0405	−0.0397	−0.0389	−0.0374	−0.0359	−0.0345	−0.0331	−0.0319	−0.0306	−0.0295	−0.0283	−0.0262
0.15	−0.0588	−0.0574	−0.0560	−0.0547	−0.0521	−0.0496	−0.0473	−0.0451	−0.0430	−0.0410	−0.0391	−0.0373	−0.0340
0.20	−0.0749	−0.0728	−0.0708	−0.0689	−0.0652	−0.0617	−0.0584	−0.0554	−0.0524	−0.0497	−0.0471	−0.0447	−0.0402
0.25	−0.0897	−0.0870	−0.0844	−0.0819	−0.0771	−0.0725	−0.0683	−0.0643	−0.0606	−0.0571	−0.0539	−0.0508	−0.0452
0.30	−0.1036	−0.1002	−0.0970	−0.0938	−0.0879	−0.0823	−0.0771	−0.0723	−0.0678	−0.0636	−0.0597	−0.0560	−0.0494
0.35	−0.1166	−0.1126	−0.1087	−0.1049	−0.0978	−0.0912	−0.0851	−0.0794	−0.0742	−0.0693	−0.0647	−0.0605	−0.0529
0.40	−0.1289	−0.1241	−0.1196	−0.1152	−0.1070	−0.0994	−0.0924	−0.0859	−0.0799	−0.0743	−0.0692	−0.0644	−0.0559
0.45	−0.1404	−0.1350	−0.1298	−0.1248	−0.1155	−0.1069	−0.0990	−0.0917	−0.0850	−0.0789	−0.0731	−0.0679	−0.0585
0.50	−0.1514	−0.1453	−0.1395	−0.1339	−0.1235	−0.1139	−0.1052	−0.0971	−0.0897	−0.0829	−0.0767	−0.0709	−0.0608
0.55	−0.1617	−0.1550	−0.1486	−0.1424	−0.1309	−0.1204	−0.1108	−0.1020	−0.0940	−0.0866	−0.0799	−0.0737	−0.0628
0.60	−0.1716	−0.1642	−0.1572	−0.1504	−0.1379	−0.1265	−0.1161	−0.1066	−0.0979	−0.0900	−0.0828	−0.0761	−0.0646
0.65	−0.1811	−0.1730	−0.1654	−0.1581	−0.1445	−0.1322	−0.1210	−0.1108	−0.1015	−0.0931	−0.0854	−0.0784	−0.0661
0.70	−0.1901	−0.1814	−0.1731	−0.1653	−0.1507	−0.1375	−0.1256	−0.1147	−0.1049	−0.0959	−0.0878	−0.0804	−0.0675
0.75	−0.1987	−0.1894	−0.1805	−0.1722	−0.1566	−0.1426	−0.1299	−0.1184	−0.1079	−0.0985	−0.0900	−0.0822	−0.0688
0.80	−0.2069	−0.1970	−0.1876	−0.1787	−0.1622	−0.1473	−0.1339	−0.1218	−0.1108	−0.1009	−0.0920	−0.0839	−0.0699
0.85	−0.2149	−0.2044	−0.1944	−0.1849	−0.1675	−0.1518	−0.1377	−0.1250	−0.1135	−0.1032	−0.0938	−0.0854	−0.0709
0.90	−0.2225	−0.2114	−0.2009	−0.1909	−0.1726	−0.1561	−0.1413	−0.1280	−0.1160	−0.1052	−0.0955	−0.0868	−0.0719
0.95	−0.2298	−0.2181	−0.2071	−0.1966	−0.1774	−0.1602	−0.1447	−0.1308	−0.1184	−0.1072	−0.0971	−0.0881	−0.0727
1.00	−0.2369	−0.2247	−0.2131	−0.2021	−0.1820	−0.1640	−0.1479	−0.1335	−0.1206	−0.1090	−0.0986	−0.0893	−0.0735
1.25	−0.2689	−0.2539	−0.2398	−0.2265	−0.2023	−0.1809	−0.1618	−0.1449	−0.1299	−0.1166	−0.1047	−0.0942	−0.0765
1.50	−0.2962	−0.2787	−0.2623	−0.2470	−0.2191	−0.1946	−0.1730	−0.1539	−0.1371	−0.1223	−0.1092	−0.0976	−0.0785
1.75	−0.3200	−0.3002	−0.2818	−0.2645	−0.2333	−0.2060	−0.1821	−0.1611	−0.1428	−0.1267	−0.1126	−0.1002	−0.0800
2.00	−0.3411	−0.3191	−0.2987	−0.2797	−0.2455	−0.2157	−0.1897	−0.1671	−0.1474	−0.1302	−0.1153	−0.1023	−0.0811
2.25	−0.3599	−0.3360	−0.3138	−0.2931	−0.2561	−0.2241	−0.1963	−0.1722	−0.1513	−0.1331	−0.1174	−0.1038	−0.0819
2.50	−0.3768	−0.3511	−0.3272	−0.3051	−0.2655	−0.2314	−0.2019	−0.1765	−0.1545	−0.1356	−0.1192	−0.1051	−0.0825
2.75	−0.3922	−0.3648	−0.3394	−0.3159	−0.2739	−0.2379	−0.2069	−0.1802	−0.1573	−0.1376	−0.1207	−0.1061	−0.0830
3.00	−0.4064	−0.3773	−0.3504	−0.3256	−0.2815	−0.2436	−0.2113	−0.1835	−0.1597	−0.1393	−0.1219	−0.1070	−0.0834
3.25	−0.4194	−0.3888	−0.3606	−0.3345	−0.2883	−0.2488	−0.2152	−0.1864	−0.1618	−0.1408	−0.1230	−0.1077	−0.0837
3.50	−0.4315	−0.3994	−0.3699	−0.3427	−0.2945	−0.2535	−0.2187	−0.1889	−0.1637	−0.1422	−0.1239	−0.1083	−0.0839
3.75	−0.4427	−0.4092	−0.3785	−0.3502	−0.3002	−0.2578	−0.2218	−0.1912	−0.1653	−0.1433	−0.1247	−0.1089	−0.0842
4.00	−0.4531	−0.4184	−0.3865	−0.3572	−0.3055	−0.2617	−0.2247	−0.1933	−0.1668	−0.1443	−0.1254	−0.1093	−0.0843
4.25	−0.4630	−0.4270	−0.3940	−0.3637	−0.3104	−0.2654	−0.2273	−0.1952	−0.1681	−0.1452	−0.1260	−0.1097	−0.0845
4.50	−0.4722	−0.4350	−0.4010	−0.3698	−0.3149	−0.2687	−0.2297	−0.1969	−0.1693	−0.1461	−0.1265	−0.1101	−0.0846
4.75	−0.4809	−0.4426	−0.4076	−0.3755	−0.3191	−0.2718	−0.2320	−0.1985	−0.1704	−0.1468	−0.1270	−0.1104	−0.0848
5.00	−0.4892	−0.4498	−0.4138	−0.3808	−0.3231	−0.2747	−0.2340	−0.1999	−0.1714	−0.1475	−0.1274	−0.1107	−0.0849
6.00	−0.5182	−0.4749	−0.4354	−0.3995	−0.3367	−0.2845	−0.2410	−0.2047	−0.1746	−0.1496	−0.1288	−0.1115	−0.0851
7.00	−0.5425	−0.4957	−0.4533	−0.4147	−0.3477	−0.2922	−0.2463	−0.2084	−0.1770	−0.1511	−0.1297	−0.1121	−0.0853
8.00	−0.5632	−0.5134	−0.4683	−0.4275	−0.3568	−0.2986	−0.2507	−0.2112	−0.1789	−0.1523	−0.1304	−0.1125	−0.0854
9.00	−0.5811	−0.5287	−0.4813	−0.4384	−0.3645	−0.3039	−0.2542	−0.2136	−0.1803	−0.1532	−0.1309	−0.1128	−0.0855
10.00	−0.5970	−0.5421	−0.4926	−0.4479	−0.3711	−0.3084	−0.2572	−0.2155	−0.1815	−0.1539	−0.1314	−0.1130	−0.0856

```
      SUBROUTINE MXMPAR(VAK,YB,GY,A,B,C,ITR)
C--------------------------------------------------------------------------
C    PARAMETERS:
C    VAK        - VARIANCE OF DIMENSIONLESS DATA (K)
C    YB         - MEAN OF NATURAL LOGARITHMS OF DIMENSIONLESS DATA
C    GY         - SAMPLE LOGARITHMIC SKEW
C    A,B,C      - ESTIMATES OF LP PARAMETERS BY MXM1 METHOD
C    ITR        - NUMBER OF ITERATIONS REQUIRED FOR OPTIMIZATION (IF ITR=50,
C                   CALCULATIONS ARE ABANDONED AND THE VALUES OF A,B, & C
C                   OBTAINED AT 50TH ITERATION RETURNED)
C--------------------------------------------------------------------------
      DOUBLE PRECISION A,B,C,FB,FD,Z,GY,VAK,BSAVE
C DOUBLE PRECISION STATEMENT IS OPTIONAL DEPENDING ON THE TYPE OF COMPUTER USED
      GY=DABS(GY)
      CC=DLOG(1.+VAK)
      CC1=-CC/2.
C ** CC1=LOG MEAN OF 2-PARAMETER LOGNORMAL DISTRIBUTION
C    DETERMINE THE SIGN OF LOG SKEW FOR MXM1 SOLUTION
      IF(YB.LT.CC1) GY=-GY
      B=4./GY**2
      ITR=0
   20 Z=(1.+VAK)**(1./B)-1.
      IF(GY.GT.0.) A=1.+DSQRT(1.+1./Z)
      IF(GY.LT.0.) A=1.-DSQRT(1.+1./Z)
      C=B*DLOG(1.-1./A)
      CPMY=C+B/A
      FB=CPMY-YB
      ITR=ITR+1
      IF(DABS(FB).LE.0.00001) GO TO 40
      IF(ITR.EQ.50) GO TO 40
      FD=DLOG(1.-1./A)+1./A+DLOG(1.+VAK)/(2.*Z*B*A**2)
      DB=-FB/FD
C IF THE INITIAL B IS MUCH GREATER THAN OPTIMAL B, A NEGATIVE B IS GENERATED.
C (DB WORKS OUT AS A LARGE NEGATIVE QUANTITY > INITIAL B)
C  IN THIS SITUATION REVISED B IS ASSUMED AS ONE HALF OF THE INITIAL B
C INSTEAD OF B+DB. THIS OPERATION LEADS TO CONVERGENCE OF SOLUTION.
      BSAVE=B
      B=B+DB
      IF(B.LE.0.) B=BSAVE/2.
      GO TO 20
   40 RETURN
      END
```

FIGURE 3. Subroutine MXMPAR.

Step 6. Calculate the value of $c = b \ln (1 - 1/a)$.

Step 7. Calculate $F(b) = \mu_y - \bar{Y} = c + b/a - \bar{Y}$. If $|F(b)| \leq 10^{-5}$, go to the next step. Otherwise, calculate:

$$F'(b) = \ln\left(1 - \frac{1}{a}\right) + \frac{1}{a} + \frac{\ln(1 + S_k^2)}{2Za^2b} \quad (19)$$

and

$$\Delta b = -\frac{F(b)}{F'(b)} \quad (20)$$

Calculate the new value of b by applying the correction Δb to the old value, i.e., $b_{new} = b_{old} + \Delta b$. Then repeat Steps 5–7.

Step 8. Calculate the estimates of σ_y and γ_y from Equations (4) and (5), respectively. Obtain the LP quantiles from Equation (12).

Choosing γ_y from Tables 3–5 in Step 4 considerably reduces the number of iterations required for optimization; this should be done if calculations are performed by pocket-type electronic calculators. However, when developing a program for the high speed or personal computers, the subroutine MXMPAR given in Figure 3 may be used for

TABLE 6. Sequence of Iterations in Optimizing Log Pearson Parameters by MXM1 Method.

Iteration Number (1)	b (2)	Z (3)	a (4)	c (5)	F(b) (6)	F'(b) (7)	Δb (8)
1	11.11111	0.03046	−4.81632	2.09620	0.002774	0.002263	−1.22598
2	9.88513	0.03430	−4.49114	1.98720	−0.000278	0.002742	0.10146
3	9.98659	0.03395	−4.51878	1.99647	−0.000002	0.002696	0.00086

Steps 4–7. This subroutine has S_k^2, $\bar{Y}$, and CS_y as input parameters. The value of CS_y should be calculated in Step 3.

EXAMPLE

The following statistical parameters have been computed for the annual flood flow data (1903–1972), Cedar River at Cedar Rapids, Iowa. Calculate log Pearson flood estimates by the MXM1 and logarithmic moments methods.

Sample size, $n = 70$; $\bar{X} = 27{,}558$ cfs (771.62 m³/sec). For the dimensionless data (K_i): $S_k^2 = 0.39570$; $\bar{Y} = -0.21355$; $S_y^2 = 0.48770$; and $CS_y = -0.50800$.

From Table 4, for $\mu_y = -0.21355$, and $\sigma_k^2 = 0.3957$, an approximate value of $\gamma_y = -0.6$. Table 6 presents the sequence of iterations made and different parameter estimates in each iteration. (For the same example, using subroutine MXMPAR with trial $\gamma_y = -0.508$ took 5 iterations.)

The estimates of LP parameters are: $a = -4.51878$; $b = 9.98659$; and $c = 1.99647$. The estimates of logarithmic standard deviation and skewness coefficient may be calculated as:

$$\hat{\sigma}_y = \frac{\sqrt{b}}{|a|} = \frac{\sqrt{9.98659}}{4.51878} = 0.69934$$

$$\hat{\gamma}_y = \frac{a}{|a|} \cdot \frac{2}{\sqrt{b}} = -\frac{2}{\sqrt{9.98659}} = -0.63288$$

For $P = 0.01$, $K' = 1.8559$ (Table 1). The 100 yr flood estimate is given by:

$$
\begin{aligned}
X_{100\ yr} &= \exp\left(\hat{\mu}_y + K'\,\hat{\sigma}_y\right) \times \bar{X} \\
&= \exp\left(-0.21355 + 1.8559 \times 0.69934\right) \times 27{,}558 \\
&= 81{,}500 \text{ cfs (rounded to three significant figures)} \\
&\quad (2{,}272 \text{ m}^3/\text{sec}).
\end{aligned}
$$

By logarithmic moments method,

$$\hat{\sigma}_y = S_y = \sqrt{0.48770} = 0.69836; \quad \hat{\gamma}_y = CS_y = -0.50800$$

For $P = 0.01$, $K' = 1.9487$. The 100 yr flood estimate is

TABLE 7. Log Pearson Annual Flood Estimates for Cedar River at Cedar Rapids, Iowa.

Exceedance Probability, P (1)	Return Period (T) Years (2)	Flow, cfs	
		MXM1 Method (3)	Logarithmic Moments Method (4)
0.995	1.005	2,440	2,650
0.990	1.010	3,190	3,400
0.980	1.020	4,230	4,420
0.960	1.042	5,700	5,850
0.900	1.111	8,780	8,830
0.800	1.250	12,800	12,700
0.500	2	24,000	23,600
0.200	5	40,500	40,500
0.100	10	51,300	52,000
0.040	25	64,200	66,400
0.020	50	73,200	76,800
0.010	100	81,500	86,800
0.005	200	89,300	96,500
0.002	500	98,900	109,000
0.001	1,000	106,000	118,000

Note: 1 cfs = 0.028 m³/sec.

TABLE 8. Flood Estimates (in cfs) in the Presence of a Low Outlier.
North Prong St. Marys River at Moniac, Georgia.

Return Period Years (1)	MXM1 Method			Logarithmic Moments Method			
	Case 1 (2)	Case 2 (3)	Case 3 (4)	Case 1 (5)	Case 2 (6)	Case 3 (7)	Case 4 (8)
10	4,290	4,030	3,960	3,240	4,070	3,990	3,970
100	9,690	9,750	9,670	3,350	9,220	9,150	8,850
500	14,400	16,000	16,000	3,360	14,300	14,300	13,600
1,000	16,700	19,500	19,500	3,360	17,000	17,000	16,000

Notes:
1 cfs = 0.028 m³/sec
Case 1 = From observed data
Case 2 = Lowest value (7.8 cfs) deleted.
Case 3 = Lowest value deleted and a conditional probability adjustment applied (Reference [9]).
Case 4 = Case 3 + weighted log skew was computed based on the generalized skew map for U.S.A. given in Reference [9].

given by

$$X_{100 \ yr} = \exp \left(-0.21355 + 1.9487 \times 0.69836\right) \times 27,558$$
$$= 86,800 \text{ cfs } (2,430 \text{ m}^3/\text{sec}).$$

Flood estimates for different significant P values are shown in Table 7.

OUTLIERS IN OBSERVED DATA

Sometimes hydrologic data may contain observations which are markedly different in magnitude from the remaining data. These observations, known as outliers, may occur among the highest, or the lowest, or both magnitudes in a given sample of data. They may result from an unusual combination of natural phenomena, or due to the consequence of sample analyzed being small or simply due to an error in recording.

Results given by the logarithmic moments method are significantly affected by the outliers in data. Reference [9] recognizes that the presence of low outliers in a sample leads to an underestimation and high outliers to overestimation of flood flows (i.e., maximum events) by this method. Reference [9] also suggests treatment procedures which revise flood estimates suitably. The following examples illustrate that the results of MXM1 method are not unduly affected by the presence of outliers whereas some outlier treatment will be necessary, especially when low outliers are present in data, if logarithmic moments method is used.

Example 1: Annual peak flow (1922–23, 28–30, 33–34, 52–83) data in cfs (1 cfs = 0.028 m³/sec) are presented below for North Prong St. Marys River at Moniac, Georgia.

DATA IN DESCENDING ORDER

11600	6060	4590	4080	3110	2760	2730	2650	2640	2200
2180	2160	2140	2020	1870	1870	1800	1770	1640	1500
1430	1410	1360	1340	1330	1150	1080	1050	894	890
828	821	764	729	670	429	390	244	7.8	

The lowest annual peak flow of 7.8 cfs (0.218 m³/sec) was produced during an extreme drought in 1954–1955. For improving results of logarithmic moments method, Reference [9] suggests: (1) deleting low outliers from the

TABLE 9. Flood Estimates (in cfs) in the Presence of a High Outlier.
Oconee River near Greensboro, Georgia.

Return Period Years (1)	MXM1 Method (2)	Logarithmic Moments Method	
		From Observed Data (3)	Using the Weighted Skew (Ref. [9]) (4)
10	27,100	27,300	27,100
100	55,400	57,700	54,500
500	83,100	89,100	80,800
1,000	97,700	106,100	94,500

Note: 1 cfs = 0.028 m³/sec.

record; (2) modification of sample statistics (obtained from the truncated data) by a conditional probability adjustment; and (3) weighting station skew.

For the observed data, the following SP are calculated: Sample size, $n = 39$; $\bar{X} = 2{,}005$ cfs (56.14 m³/sec). For the dimensionless data (K_i), $S_k^2 = 0.98317$; $\bar{Y} = -0.39825$; $S_y^2 = 1.26851$; and $CS_y = -2.46655$. The MXM1 solution gives, $\hat{\sigma}_y^2 = 0.86275$; and $\hat{\gamma}_y = -0.27143$.

Deleting the lowest flow value gives the following SP: Sample size, $n = 38$; $\bar{X} = 2{,}057$ cfs (57.60 m³/sec).

For the dimensionless data: $S_k^2 = 0.93268$; $\bar{Y} = -0.28857$; $S_y^2 = 0.56684$; and $CS_y = 0.06673$.

The MXM1 solution gives $\hat{\sigma}_y^2 = 0.54193$; and $\hat{\gamma}_y = 0.24676$.

Table 8 presents flood estimates for different cases (see Reference [9] for details of conditional probability adjustment).

Example 2: For Oconee River near Greensboro, Georgia, the following data represent annual peak flows in cfs (1 cfs = 0.028 m³/sec) in descending order.

```
66800 44000 41100 34100 31800 29400 28800 26100 22700 22400
20100 19800 19200 18200 18100 17800 17200 16400 15400 15400
15300 15200 15200 15100 14500 14300 14200 13800 13300 13300
13300 13000 13000 12800 12700 12200 12000 11900 11900 11000
10700  9800  9500  9290  8910  8910  8260  8210  8100  7840
 7820  7580  7480  6960  6800  6520  6330  6180  5990  5690
 5440  5260  3610
```

The highest observation (66,800 cfs) is considered as an outlier. If high outliers are detected, Reference [9] suggests to examine historic data to determine whether the observations are maximum in an extended period of record. If so, an adjustment is made in the sample SP [9]. Otherwise, they are retained as part of the record. For the present example, it will be assumed that no historic data information is available.

The following SP are calculated for the observed data: Sample size, $n = 63$; $\bar{X} = 15{,}300$ cfs (428.4 m³/sec). For the dimensionless data (K_i), $S_k^2 = 0.48512$; $\bar{Y} = -0.17226$; $S_y^2 = 0.32336$; and $CS_y = 0.42918$. The MXM1 solution gives, $\hat{\sigma}_y^2 = 0.32268$; and $\hat{\gamma}_y = 0.33194$.

Table 9 presents flood estimates by the MXM1 and logarithmic moments methods.

NOTATION

The following symbols are used in this chapter:

a,b,c = parameters of Pearson type 3 or log Pearson type 3 distribution
CS = sample skewness coefficient
$f(\ )$ = probability density function of variate in parenthesis
K,k = dimensionless variate
$\bar{K}$ = sample mean of K
K' = Pearson frequency factor
$LN2$ = two-parameter lognormal distribution
LP = log Pearson type 3 distribution
$MXM1$ = method of mixed moments-I
n = sample size
P = exceedance probability
$P3$ = Pearson type 3 distribution
S^2 = sample variance
SP = statistical parameters
T = return period
X,x = real data (log Pearson type 3 variate)
$\bar{X}$ = sample mean of X
Y,y = logarithmic data (Pearson type 3 variate, ln X)
$\bar{Y}$ = sample mean of Y
Z = constant
γ = population skewness coefficient
η = population coefficient of variation
μ = population mean
σ = population standard deviation; and
σ^2 = population variance
$\hat{\ }$ = symbol to denote estimate

Subscripts:

i = ith value (for variates).

REFERENCES

1. "A Uniform Technique for Determining Flood Flow Frequencies," *Bulletin No. 15 of the Hydrology Committee*, United States Water Resources Council, Washington, D.C. (December 1967).
2. Benson, M. A., "Uniform Flood-Frequency Estimating Methods for Federal Agencies," *Water Resources Research*, Vol. 4, No. 5, pp. 891–908 (October 1968).
3. Bobee, B., "The Log Pearson Type 3 Distribution and its Application in Hydrology," *Water Resources Research*, Vol. 11, No. 5, pp. 681–689 (October 1975).
4. Bobee, B., and Robitaille, R., "Correction of Bias in the Estimation of the Coefficient of Skewness," *Water Resources Research*, Vol. 11, No. 6, pp. 851–854 (December 1975).
5. Chow, V. T., "Statistical and Probability Analysis of Hydrologic Data, Part I, Frequency Analysis," *Hand Book of Applied Hydrology*, McGraw-Hill Book Co., New York, N.Y., pp. 8-1 to 8-42 (1964).
6. Condie, R., "The Log Pearson Type 3 Distribution: The T-Year Event and its Asymptotic Standard Error by Maximum Likelihood Theory," *Water Resources Research*, Vol. 13, No. 6, pp. 987–991 (December 1977).
7. Dalrymple, T., "Hydrology of Flow Control, Part I. Flood Characteristics and Flow Determination," *Hand Book of Applied Hydrology*, McGraw-Hill Book Co., New York, N.Y., pp. 25-1 to 25-33 (1964).
8. Gilman, C. S., "Rainfall," Section 9, *Hand Book of Applied Hydrology*, McGraw-Hill Book Co., New York, N.Y., pp. 9-62 to 9-65 (1964).

9. "Guidelines for Determining Flood Flow Frequency," *Bulletin #17B of the Hydrology Subcommittee,* Interagency Advisory Committee on Water Data, Office of Water Data Coordination, U. S. Geological Survey, Reston, Virginia (March 1982).

10. Gumbel, E. J., *Statistics of Extremes,* Columbia University Press, New York, N.Y. (1958).

11. Hoshi, K., and Burges, S. J., "Sampling Properties of Parameter Estimates for the Log Pearson Type 3 Distribution, Using Moments in Real Space," *Journal of Hydrology, Vol. 53,* No. 3/4, pp. 305–316 (October 1981).

12. Houghton, J. C., "Birth of a Parent: The Wakeby Distribution for Modeling Flood Flows," *Water Resources Research, Vol. 14,* No. 6, pp. 1105–1115 (December 1978).

13. Kirby, W., "Algebraic Boundedness of Sample Statistics," *Water Resources Research, Vol. 10,* No. 2, pp. 220–222 (April 1974).

14. Landwehr, J. M., Matalas, N. C., and Wallis, J. R., "Some Comparison of Flood Statistics in Real and Log Space," *Water Resources Research, Vol. 14,* No. 5, pp. 902–920 (October 1978).

15. Lettenmaier, D. P., and Burges, S. J., "Correction for Bias in Estimation of the Standard Deviation and Coefficient of Skewness of the Log Pearson 3 Distribution," *Water Resources Research, Vol. 16,* No. 4, pp. 762–766 (August 1980).

16. Linsley, R. K., and Franzini, J. B., "Probability Concepts in Design," *Water Resources Engineering,* McGraw-Hill Book Co., New York, N.Y., pp. 119–147 (1972).

17. Markovic, R. D., *Probability Functions of Best Fit to Distributions of Annual Precipitation and Runoff,* Hydrology Papers No. 8, Colorado State University, Fort Collins, Colorado (August 1965).

18. Matalas, N. C., *Probability Distribution of Low Flows,* U.S. Geological Survey Professional Paper 434-A, Washington, DC (1963).

19. Rao, D. V., "Log Pearson Type 3 Distribution: A Generalized Evaluation," *Journal of the Hydraulics Division, ASCE, Vol. 106,* No. HY5, pp. 853–872 (May 1980).

20. Rao, D. V., "Log Pearson Type 3 Distribution: Method of Mixed Moments," *Journal of the Hydraulics Division, ASCE, Vol. 106,* No. HY6, pp. 999–1019 (June 1980).

21. Rao, D. V., "Three-Parameter Probability Distributions," *Journal of the Hydraulics Division, ASCE, Vol. 107,* No. HY3, pp. 339–358 (March 1981).

22. Rao, D. V., "Estimating Log Pearson Parameters by Mixed Moments," *Journal of Hydraulic Engineering, ASCE, Vol. 109,* No. 8, pp. 1118–1132 (August 1983).

23. Rao, D. V., "Three-Parameter Probability Distribution of Best Hydrologic Bounds," *Frontiers in Hydraulic Engineering,* Proceedings of the ASCE Hydraulics Division Conference held at Massachusetts Institute of Technology, pp. 486–491 (August 9–12, 1983).

24. Rao, D. V., *Magnitude and Frequency of Flood Discharges in Northeast Florida,* Publication No. SJ 86-2, St. Johns River Water Management District, Palatka, Florida (1986).

25. Sokolov, A. A., Rantz, S. E., and Roche, M., *Floodflow Computation,* The Unesco Press, Paris, pp. 22–25 (1976).

26. Tasker, G. D., "Flood Frequency Analysis with a Generalized Skew Coefficient," *Water Resources Research, Vol. 14,* No. 2, pp. 373–376 (April 1978).

27. Wallis, J. R., Matalas, N. C., and Slack, J. R., "Just a Moment!," *Water Resources Research, Vol. 10,* No. 2, pp. 211–219 (April 1974).

APPENDIX

Newton-Raphson iteration scheme for optimizing parameter b in estimating log Pearson parameters by MXM1 method. For the dimensionless variate K, the following equations can be written:

$$Z = (1 + \sigma_k^2)^{1/b} - 1 \tag{21}$$

$$a = 1 + \epsilon \left(1 + \frac{1}{Z} \right)^{1/2} \text{ with } \epsilon = \frac{|a|}{a} \tag{22}$$

$$c = b \ln \left(1 - \frac{1}{a} \right) \tag{23}$$

$$\mu_y = c + \frac{b}{a} \tag{24}$$

Consider function $F(b)$ given by

$$F(b) = c + \frac{b}{a} - \bar{y} \tag{25}$$

in which c and a = functions of b. The MXM1 method requires that $\mu_y = \bar{Y}$ or $F(b) = 0$.

It is possible to show that

$$\frac{dc}{db} = \ln \left(1 - \frac{1}{a} \right) + \frac{b}{a(a-1)} \frac{da}{db} \tag{26}$$

$$\frac{da}{db} = \frac{\epsilon}{2Z^{3/2}} \frac{(Z+1)^{1/2}}{b^2} \ln (1 + \sigma_k^2) \tag{27}$$

Finally

$$\frac{dF(b)}{db} = F'(b) = \ln \left(1 - \frac{1}{a} \right) + \frac{1}{a} + \frac{\ln (1 + \sigma_k^2)}{2Za^2 b} \tag{28}$$

By the method of Newton-Raphson the correction, Δb, should be applied to parameter b at each iteration. This is given by

$$\Delta b = - \frac{F(b)}{F'(b)} \tag{29}$$

CHAPTER 10

Arid Lands: Evapotranspiration Under Extremely Arid Climates

ABDIN M. A. SALIH*

INTRODUCTION

Evapotranspiration, which combines the evaporation of water from land and water surfaces together with transpiration by vegetation, is a very important element of the hydrological cycle. A rational evaluation of this element is a key factor for estimating irrigation water requirement and is also useful in other demand and supply aspects of water resources (i.e. losses in open water systems, aquifers yield estimation, rainfall disposition, municipal and industrial demands, etc.). This evaluation is needed at planning, designing and operation levels in water resources development aspects, particularly under extremely arid climates which prevail in a considerable portion of the world's area. Such severe climates are usually linked with the scarcity of conventional water supplies such as surface and rechargeable groundwater. Accelerating demand of water must, hence, be met from expensive sources such as deep, nonrenewable groundwater acquifers, re-use and desalination schemes. The allocation of this precious water to present and anticipate future demands (i.e. irrigation, domestic, industrial, etc.) must be reached through rational methods of evaluations in order to avoid any future disappointments.

Irrigation demands usually require the major allocation of that limited water supplies. Hence, rational estimation of evapotranspiration and consequently of that demand should have been a mandatory requirement preceding the establishment of such projects. Unfortunately, this is not done, since the thrust for food production has recently accelerated huge irrigation agriculture in many of these desert type regions. In the absence of direct measurement of evapotranspiration, experts tend to adopt empirical formulas, based on one or more climatic parameters, for predicting irrigation water requirement. These formulas, probably being calibrated for other regions, may fail to give realistic values of the actual need and hence lead to future problems.

The problem is also complicated further by the vague definition of aridity. The current definition of semi-arid, arid and extremely arid, as devised by climatologists, could serve the purposes of many specializations, yet its broad divisions are far from describing successfully evapotranspiration changes as a function of climate. Hence, these estimation methods are greatly affected by local conditions, siting of climatic stations and the effect of advection on oasis-type agriculture pertinent to extremely arid zone.

All of these aspects shall be considered in this section which is divided into five topics. The first of these topics is this introduction which will be followed by a short review on the concept of aridity. The third topic will deal with the history of irrigation water requirement in a few extremely arid locations. It inclusion here is meant to give the reader a few examples of the wisdom of the old inhabitants of these regions in rationally managing their scarce irrigation water supplies. The fourth topic will deal with the question of evapotranspiration under extremely arid climates, which is the main purpose of this work. Thus greater space and material is devoted to this part. The chapter will be ended by a brief summary to the aspects that need further studies in future. It is hoped that by this presentation the reader will be provided with a broad introcution to this very important subject which is sparsely documented under extremely arid conditions.

THE QUESTION OF ARIDITY

In spite of the long history of worldwide inspirations for reaching a standard definition for aridity, a consensus on a

*Khartoum University, Khartoum, Sudan

single approach is far from being attained. The main difficulties stem from the fact that studies on arid zones involve
a broad spectrum of specializations such as climate, soil,
vegetation, human settlement, land use, etc. Hence, each
sector will tend to address the question with somewhat
different techniques and with views that suit their end objectives [25,33,54]. In a recent publication by Heathcote [25]
an excellent review has been given to the evolving definitions of arid lands. Two approaches have been identified: the
first is descriptive in nature, while the other is a rather quantitative approach based on measurable parameters (i.e. precipitation, temperature, evaporation, potential evapotranspiration, etc.).

For the descriptive approach, which is based on the
changing literal meanings of the word "desert," Heathcote
[25] listed the definitions that evolved through the period
1225–1968. According to Heathcote, the parameters used
for the quantitative approach were very much influenced by
the objectives of the enquiry. He then reviewed the contributions of: Albrecht Penck (1894) whose interest was based on
global landforms and hence he used evaporation and precipitation as parameters; Martonne and Aufrere (1927) whose
aim was to prepare a global map of the areas of interior
basin drainage; Shantz (1956) whose works included both
global soils and vegetations; and Koppen (1900–31), Thornthwaite (1948) and Meigs (1953) whose approaches were
based on climatic parameters. Due to the greater relevance
of the work of the last group of climatologists to the objectives of this chapter, brief elaborations will be given to some
of the terms and parameters used in their approaches.

The Koppen Classification

It consists of five major climatic categories identified as:

A—Tropical forest; hot all season
B—Dry climates
C—Warm temperature rainy climates; mild winter
D—Cold forest climates; severe winters
E—Polar climates

Each of these five categories is further divided into
various subdivisions (see Critchfield [12]). For example arid
climates under B are identified as:

BWh—Tropical deserts—Arid; hot
BWk—Mid latitude deserts, Arid; cool or cold

The letter symbols are usually defined. For example, for
arid ranges the corresponding symbols can be explained as
follows:

1. B = 70% or more of annual precipitation falls in
warmer six months (April through September in Northern Hemisphere) and r less than $2t + 28$. (r is average
annual ppt. in cm.; t = average annual temperature in
°C.) or 70% or more of annual precipitation falls in

cooler six months and r less than $2t$ or neither half of the
year with more than 70% of precipitation and r less than
$2t + 14$.
2. W stands for r less than ½ the upper limit of applicable
requirement for B.
3. h stands for t greater than 18°C.
   k stands for t less than 18°C.

The Thornwaite Classification (from Critchfield [12])

The main features of this system are:

1. A precipitation of effectiveness index, which relates precipitation to evaporation; and
2. A temperature efficiency index (based on mean monthly
temperatures).

Thornthwaite proposed at a later stage the concept of
potential evapotranspiration.

The new classification is expressed as follows:

$$I_m = \frac{(S - D)}{ETP} \times 100$$

where

I_m = monthly moisture index
ETP = potential evapotranspiration
S = monthly surplus ($P - ETP$)
D = monthly deficit ($P - ETP$)
P = precipitation

or if the soil moisture is assumed to be constant, the above
equation can be simplified to:

$$I_m = 100 \left(\frac{P}{ETP} - 1 \right)$$

The sum of the 12 months gives the annual moisture index. Arid climatic types (symbol E) are defined for I_m in the
range −100 to 66.7.

The classifications also include thermal efficiency (potential evapotranspiration in cms), seasonal distribution of
moisture adequacy (i.e. Aridity index = water deficit/ETP
× 100; humidity index = water surplus/ETP × 100), and
summer concentration of thermal efficiency (accumulation
in three summer months).

Meigs Contribution [12,25,34]

Based on Thornthwaite, and to some extent Koppen
classifications, Meigs [34] has developed a classification, as
a request from UNESCO, that has been recognized in somewhat a global way. Compared to the previous approaches,
Meigs added a new classification, named extremely arid,
which is defined as an area in which there is not a regular
seasonal rhythm of rainfall and at least 12 consecutive

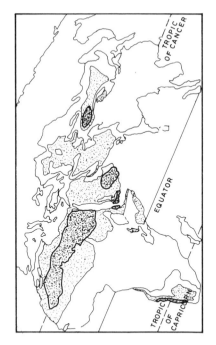 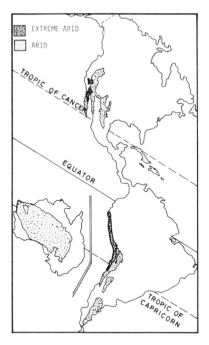

FIGURE 1. Arid and extreme arid areas of the world according to Meigs classification (prepared from Heathcote [25]).

months have elapsed without recorded rainfall. Figure 1 represents a sketch showing arid and extremely arid lands in the world as recommended by Meigs in 1953. From this figure the following zones can be identified as containing some arid and extreme arid areas.

A. In the African continent such areas can be identified in both North and South Africa.
   (1) For North Africa—arid areas include most of the land between latitudes 12° to 35°N all through the continent (including the vast Sahara Desert).
   (2) For South Africa—arid areas include a narrower strip on the western coast located between latitudes 10° to 32°S (including Kalahari Desert).
B. Most of central and western parts of the Australian continent.
C. Arid areas in the Asian continent include most of the Arabian peninsula, parts of Persia, parts of Pakistan and India, and the deserts of U.S.S.R., China, Turkestan, and Mongolia.
D. Arid areas in North America include much of Mexico, the Gulf of California, Parts of California, Arizona, Nevada, New Mexico, etc. The South American dry areas, on the other hand, consist of a narrow strip along the western coast almost between latitudes 5° to 35°S. They then switch to the eastern coast approximately between latitudes 38° to 50°S.

Comparison Between Classification Techniques

Table 1 gives a broad comparison between global arid lands estimated from vegetative and climatic classifications. The results reflect an acceptable agreement between the total and extreme-arid areas, yet it demonstrates large variations on the extents of arid and semi-arid types. Further variations have been reported by Shantz [45] who estimated the total arid soils as 43% of the global land area (estimated as 135 million square kilometers). This clearly demonstrates that the so-called quantitative approach depends greatly on

TABLE 1. Comparison of Classification Methods (% World Area).

Classification Type of Aridity	Vegetative Shantz, 1956)*	Climatic (Meigs, 1953)*	Meigs/ Shantz
Semi-arid	5.2	15.8	3.04
Arid	24.8	16.2	0.65
Extremely arid	4.7	4.3	0.91
Total	34.7	36.3	1.05

*Numerical figures are taken from Heathcote [25].

TABLE 2. Location and Climatic Parameters for Selected Stations.

Station Parameters	Riyadh[a]	Kharj[a]	Dirab[a]	Hofuf[b]	Aflaj[a]	Shagra[c]	Hutat[c] Sudair	Zilfi[c]	Sakaka[d]
Latitude	24° 39′ N	24° 10′ N	24° 25′ N	25° 17′ N	22° 15′ N	25° 15′ N	25° 32′ N	26° 18′ N	29° 58′ N
Longitude	46° 43′ E	47° 34′ E	46° 34′ E	49° 35′ E	46° 44′ E	45° 15′ E	45° 48′ E	40° 12′ E	40° 12′ E
Altitude, m	564	430	630	160	539	730	665	506	574
M. annual Temp., °C	24.6	24.5	26.0	25.1	25.3	24.9	24.2	23.4	22.5
M. annual Radiation, cal $cm^{-2} d^{-1}$	477	480	504	501	458	487	465	427	407
M. annual Wind Speed, Km d^{-1}	123	73	112	221	117	123	168	108	139
Mean annual Relative Humidity, %	34.0	32.3	30.0	45.2	34.4	33.7	33.5	33.7	40.1
M. annual* Precip., mm	88	68	47	68	54	120	119	130	60
M. annual class A Pan Evap., mm	2938	3398	3907	3205	3854	3468	4391	4380	3128
M. July Temp., °C	33.3	33.3	34.8	34.1	33.8	33.1	33.1	32.0	31.7
M. July Radiation, cal $cm^{-2} d^{-1}$	576	593	600	598	507	606	595	549	545
M. July Wind Speed, km d^{-1}	160	92	148	264	104	160	197	121	170
M. July Relative Humidity, %	19.6	17.8	18.0	29.9	18.7	19.9	18.7	19.9	27.0
M. July* Precip., mm			N	O		N	E		
M. Class A Pan Evap; July, mm	402	486	572	437	488	451	605	591	434
Minimum Length of Record, years	12	9	5	11	4	6	4	3	20

*Precipitation is averaged for longer periods than shown in the table.
Sources of data: a = Sendil, et al. [44];   b = Hydrological Pub. [28]
            c = Khowaiter   [33];   d = AlMansour [6]

the aim of the enquiry (being climatic, botanic, soil oriented, etc.).

In this work the climatic approach is preferred in spite of the different values given by Koppen, Thornthwaite, Meigs and Rogers [38]. The Meigs approach is currently more acceptable and has received wider international recognition and publicity. However, the distinction between Arid and Extreme Arid Zones shall not be strictly adhered to in this study when quoting typical personal experiences in arid lands. Table 2 gives the location, altitude and mean annual climatic parameters for some of the sites being referred to in the course of this section. They are included here in order to give the reader a sense of acquaintance with the climate of the region often being referred to in this work.

HISTORY OF IRRIGATION WATER MANAGEMENT UNDER ARID CONDITIONS

As mentioned in the previous section, the degree of aridity is very much linked with the scarcity of rainfall coupled with high evaporation rates. Hence, for any worthwhile food production, irrigation would be the only way for agriculture and consequently intelligent management of its water demand should be expected to be mandatory. Other than a few rivers (i.e. Nile, Colorado, Euphrates, Tigris . . .) that traverse part of the world's arid lands, groundwater is the only possible resource. It is thus the purpose of this section to present to the reader a few examples of the ingenious old techniques that had been used by the inhabitants of the arid lands of Arabia and North Africa in managing their irrigation water. Elaborations shall be confined to experiences from water resources other than those from the great rivers of the Nile, Tigris and Euphrates. This is because literature on irrigation from these famous rivers is relatively well documented.

Irrigation Water Requirement

Though terms such as "evapotranspiration" and "consumptive use" are relatively new, the literature on the history of irrigation water requirement in arid lands demonstrates that similar general concepts were used by the old farmers in places such as Al-Hasa (eastern Saudi Arabia), Tadmur (Syria), Nafr (Iraq), Egypt and North Africa.

EXPERIENCE FROM AL-HASA

Irrigation in Al-Hasa oasis is thousands of years old. It consists of a system of canals receiving their water from many natural springs. This water flows by gravity and is distributed through smaller branch canals to the different farms. Part of the water delivered to the first farm could be re-used for many times in other farms of lower grade till it finally drained to a salty depression (Sabkha) when it became very saline [53]. The system as a whole consists of a network of canals and drains crossing through aqueducts and other implements. This system is recently modernized, keeping almost the same intelligent features that existed all through successful history.

The irrigation water application, at the rate of every two days in summer and 4–5 days in winter, was criticized by a FAO expert [51] as being 2¼ times that in similar arid environments of the United States. The expert, however, was unjustified in his criticism since two recent studies [8,11,37] on actual evapotranspiration in this area have both independently confirmed the rates that were used for thousands of years as a result of shear experience. This is also an indication that aridity is a term of wide range that should be taken cautiously in transferring irrigation water consumption rates from one arid area to another.

EXAMPLE FROM TADMUR [40]

There is a group of farms in this very old city irrigated since old times by a natural local spring. In order to measure and control the flow from that spring, a local expert had made an unwise decision of fixing a Parshall flume near the outlet of the spring, across the rectangular cross section of a built canal (0.7 × 0.3m) that conveys the spring water to the farms. The farmers sabotaged the flume and removed it from the canal claiming that it reduced their irrigated area by about 10%. While calculating the effect of the back water created by the flume on the water source, it was found that the reduction in the spring discharge was almost equal to 10% and thus confirmed the farmers claim. A closer investigation to the empirical methods they used locally for providing their irrigation water requirement and its scheduling reflected an impressive rational judgement based on their drive for maximum yield and the limited supply of water.

NAFAR OF IRAQ

Al-Hashmi [2] has reported a document, 3500 years old, describing the methods of irrigation water scheduling in old Iraq in the words of a father advising his son on when to irrigate and how much to apply. The document also mentioned the effect of field preparation on efficient irrigation and the effect of the number of irrigations on crop yield.

This reference also included various other aspects related to irrigation at that time. These aspects reflect a great awareness on the necessity of wise management of their irrigation water requirement.

EXAMPLE FROM EGYPT

Al-Shami [7] has reviewed irrigation water practices in Egypt as documented in Arab writings. He showed that agricultural yield was very much related to a particular height of the Nile water. An increase or decrease from that height meant a reduction in yield. However, that height, which could be related to the irrigated area and the corresponding heights of dykes, was only 16 arm lengths in the first Arabic century, and gradually increased to 20 arm lengths by the 11th Arabic century. The document has also included review of other non-Nilotic systems of irrigation especially in the oases where groundwater was the only source.

DRIP IRRIGATION IN NORTH AFRICA [1]

Drip irrigation is described in modern text books as being a new recent method of irrigation accredited for its highest irrigation efficiency compared to all other methods of irrigation. Al-Amami [1] has, however, reported from a book published in the 12th century on agriculture by an Arabic scientist called Ibn Al-Awam. In that book, Ibn Al-Awam described a form of drip irrigation as follows, "Let us install two pots filled with water within the root system of a tree. Make a small hole in the pot and protect that hole from be-

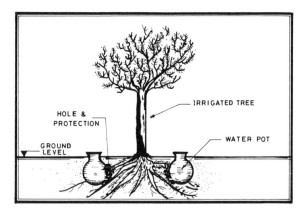

FIGURE 2. Sketch of old drip irrigation arrangements.

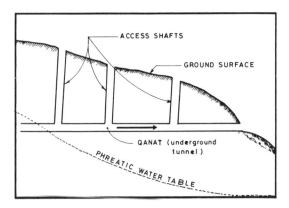

FIGURE 3. Sketch of a qanat (underground water tunnel).

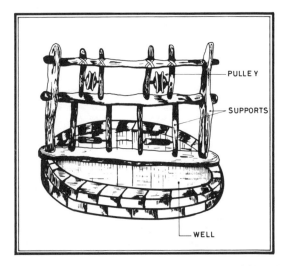

FIGURE 4. Sketch showing the fixed components of Aswani. (The pulley connects the water buckets to the animal.)

ing blocked by the neighbouring soil. Add water manually to the pot when its water is reduced." Figure 2 demonstrates this method which was historically used in many parts of North Africa and Arabia and is at present receiving admirable renovation by a research center in Tunisia [1].

Management of Irrigation Water

Other than the above few examples on irrigation water requirements, the old inhabitants of the Arid lands of North Africa, Arabia and Asia had produced great innovations for a better management of their irrigation water. Due to the limitation of space only a few examples shall be cited in this section.

Al-Khaldi [3] has reported the wonderful contributions of the Nabatans, who lived in the deserts of Northern Arabia centuries before Christ's time, in water management. They were able to turn the desert into green belts through an efficient system of water conservation, storage and management.

To abstract groundwater, intelligent methods such as the famous underground tunnels (Qanats) and lifting of water from Dug wells by Animals (Asawani) were largely used [41]. The Qanats (Figure 3) were used in many parts of Arabia (i.e. Hijax, Najd, Aflaj, Al-Hasa, Oman, Bahrain, etc.), Persia and other arid parts of the world. Asawani systems (Figure 4) were mostly used in Arabia through lifting irrigation water from dug wells using the force of camels, oxen or donkeys. To avoid or reduce losses of such precious water through evaporation or seepage, canals were normally lined and the times of irrigation were carefully selected.

Use was also made of the intermittent flash floods from sparse rainfall by building dikes and larger dams. Salih and Al-Farra [41] have reported different types of such dams which were built across Arabian wadis thousands of years ago.

An important element for efficient irrigation in Arid zones is a thorough knowledge of local agricultural almanac. Theodoratus [50] and Varisco [52] in different publications have reviewed and praised the contributions of old Arab farmers in this direction. Al-Maktari [5], however, has reported the legal aspects of water and irrigation systems in the Arabian peninsula. He demonstrated the strict adherence of individuals to the traditional regulations for irrigation water distribution in spite of the limited supply.

DETERMINATION OF EVAPOTRANSPIRATION UNDER EXTREMELY ARID CLIMATES

Introduction

Some of the major problems encountered in reaching sound estimates of evapotranspiration under extremely arid climates are: the vagueness with which aridity is currently defined, the scarcity of direct measurements of evapotranspiration in such zones, and the relatively inferior quantity and quality of climatic data recorded from these regions. As demonstrated earlier in this study, extreme arid lands are expressed in the literature as making between 4 to 6% of the world's area [25,38]. A distinct demarkation line between this category and the so-called arid land is rather hypothetical. Both categories often overlap since the climatic parameters that define the degree of aridity could vary greatly even within the same category in time and space horizons.

It is also clear from the literature that very little direct measurements of ETR are available from extreme arid areas. Such measurements are very important for evaluating and calibrating the vast number of climatic methods usually published with the general label of being suitable for arid climates. A further deficiency in extremely arid areas is related to the quantity and quality of climatic data which can be used in the above climatic methods. The range of this problem could extend from total lack of such data, incompleteness of climatic parameters, shortness of time of record, poor siting of measuring equipments to observation and record-keeping problems.

It must also be remembered that a complete picture of evapotranspiration should include, in addition to these climatic parameters, inputs related to the type of crop and its stage of growth as well as soil characteristics. These elements are especially important in time scheduling of consumptive use. Due to the above difficulties, it is hence decided to keep the emphasis on this section towards estimation of evapotranspiration of longer time intervals (i.e. monthly) that could be useful for water resources planning and management purposes. Irrigation scheduling purposes can be better dealt with at later stages when more information is available to the literature from direct measurement experiments. Management should, however, play a key role in controlling this last aspect as a function of yield, other needs and availability of water in time and space.

Basic Definitions

The literature is rich with various definitions related to evapotranspiration. These definitions are unfortunately far from being internationally standardized and hence they appear in different forms as noted in the reviewed wide literature. Hence the following meanings and abbreviations shall be adhered to in the course of this section.

EVAPOTRANSPIRATION (ET)
This is a combination of transpiration and evaporation processes. Transpiration is meant, here, to include the water that enters the plant roots, with a little part of it being used for building plant tissue and the major part passes through the plant leaves to the atmosphere. Evaporation, on the other hand, includes the water evaporating from adjacent soil and plant surfaces.

CONSUMPTIVE USE (CU)
From a practical point of view the term consumptive use has the same meaning as that of evapotranspiration. Academically, however, consumptive use is equal to evapotranspiration plus the little water used for building the plant tissue.

POTENTIAL EVAPOTRANSPIRATION (ETP)
It is the rate of evapotranspiration from a crop community which is never short of water.

REFERENCE CROP EVAPOTRANSPIRATION (ETR)
It is the rate of evapotranspiration from an extensive surface of a green crop (usually alfalfa or grass) of specified range of height and is never short of water. The terms ETR_a and ETR_g shall be used in this work for alfalfa and grass, respectively, when used as reference crops. The concept of reference is relatively new in the literature, yet it is quickly replacing the older concept of potential evapotranspiration. With proper standardization, this concept can be quite useful in comparative studies.

CROP EVAPOTRANSPIRATION (ETC)
This term is meant to give the actual consumptive use, or evapotranspiration, for any particular crop at any stage of growth.

$$ETC = K_c ETP \text{ or } ETC = K_c ETR \qquad (1)$$

The numerical values of K_c, crop coefficient, depend on the corresponding definition of *ETP* and *ETR*. Empirical tables of K_c that suit each of these cases are published in many textbooks and reports [i.e. 15,31 . . .].

Methods of Determination

Evapotranspiration is usually determined through direct measurements, by using empirical formulas built on climatic observations, or by utilizing evaporimeter measurements as indices to evapotranspiration.

Direct Measurements

Direct measurements of consumptive use are usually achieved by: weighing and nonweighing lysimeters; soil moisture depletion; water balance techniques; energy balance method; mass balance; or by the use of the combination equation. An elaboration of each of these methods, together with further references on their use, can be found in Jensen, 1974 [30]. Direct measurements, if available would naturally give the best evaluation of evapotranspiration, but unfortunately such measurements would most probably be deficient within most of extremely arid areas. Estimation is, thus, obtained from climatic or evapometric methods whose calibration might have been achieved under different climatic parameters, yet they may carry a generalized label of being suitable for arid areas. It is, hence, very important to encourage the establishment of direct measurements experiments for the calibration of these methods under various conditions of aridity in order to fill the present gap in extremely arid areas where irrigation is, nowadays, expanding at a frightening rate.

Climatic Methods

These are empirical or semi-empirical methods based on historical climatic data (i.e. temperature, radiation, windspeed, relative humidity, etc.) and used for estimating potential or reference evapotranspiration. They are usually calibrated by direct measurement data obtained from the area at which they were originally calibrated. Hence, their use for other areas of different climatic conditions could be unwarranted.

The lack or poor quality of climatic data can also form problems in using some of these methods in many zones of extremely arid conditions. A recent work by Camilo [9] has reported a sensitivity analysis which indicated that measurement biases of climatic data produces larger errors in evaporation than random errors. The siting of the weather station and the selection of the most suitable method can also affect the results obtained from these methods.

SELECTION OF ESTIMATION METHODS

It is not easy, at this stage of knowledge, to select a particular method or group of methods that leads to reliable estimation of evapotranspiration under extremely arid conditions. Yet, a trial shall be made, in here, to inspect few of the publications that dealt with a sort of rating or modifications to selected group of these methods. These publications include the work of Jensen [30], Doorenbos and Pruitt [15], Hargreaves and Samani [22], Salih and Sendil [42], and few others [18,35].

Jensen's Rating (1974)

Jensen [30] had used available data from 10 sites selected at Aspendole (Australia), Brawley (California), Copenhagen (Denmark), Coshocton (Ohio), Dauis (California), Kimberly (Idaho), Lompoc (California), Ruzizi Valley (Zaire), Seabrook (New Jersey) and South Park (Colorado). He then selected 15 methods of estimation and subjected them to a rational evaluation according to their accuracy for seasonal estimates and root mean square of the monthly differences between the measured and estimated values. According to that evaluation the groups in Table 3 of five methods are rated in the shown order of merits as being the best for coastal and inland-semiarid to arid zones.

The excellent work of Jensen is however limited by the selection of sites—number-wise and area-wise. None of these sites can be taken as good representative of extreme arid conditions which is the purpose of this section. Another limitation is the absence of any rating for evaporation methods due to the unavailability of the corresponding data.

Doorenbos and Pruitt Selection (1977)

The report prepared by Doorenbos and Pruitt [15] to the Food and Agriculture Organization of the United Nations included three methods for estimating reference crop evapotranspiration (*ETR*) at various extremes of climatic conditions. The chosen methods are Blaney-criddle, Modified Penman and Pan Evaporation. Elaborations on these methods shall be given in later parts of this section.

Doorenbos and Puritt did not, however, give explicit rating or evaluation criteria for the selection of these particular methods.

Hargreaves and Samani (1982)

Hargreaves and Samani [22] reported another evaluation for four estimation methods, based upon smallness of intercept, a, coefficient of determination, R^2, and standard deviation of ratio of measured to estimated values. As a result of the rating, the four selected methods are ranked in the order of merits shown in Table 4.

TABLE 3. Jensen's Rating.

Coastal		Inland-Semiarid to Arid	
Order of merit	Name of method	Order of merit	Name of method
1	Christiansen Rs	1	Jensen-Haise
2	Turc	1	Van Bavel-Businger, 0.25
3	Kohler, et al. lake	3	Penman
4	Blaney-Criddle	4	Kohler, et al. lake
4	Ivanov	5	Van Bavel-Businger, 0.5

TABLE 4. Rating of Hargreaves and Samani.

Order of merit	Name of Method
1	Class A evaporation pan sited in an irrigated grass pasture
2	Hargreaves
3	Jensen-Haise
4	Blaney-Criddle

In a discussion of the above paper, Hasan [23] indicated that he had tested Hargreaves' equations with two forms of Penman equation using a year data obtained for an arid site in the Middle East. The four tested equations resulted in relatively comparable results. In a closure to the above discussion, Hargreaves [20] stressed his reservation on the capability of Penman equation, in spite of its sound physical rationale, in estimating transpiration which is more important than evaporation.

Salih and Sendil Evaluation (1984)

Salih and Sendil [42] have recently reported a study from four sites (Riyadh, Kharj, Dirab and Hofuf) with extreme arid climates for most of the year (see Table 2). The work was built on a previous study by Sendil et al. [43] where four of the five top methods recommended by Jensen [30] for inland-semiarid to arid, were tested using climatic data from the above four sites. These four methods included Jensen-Haise, van Bavel-Businger (0.25), Penman, and van Bavel-Businger (0.5). The results indicated that the two van Bavel-Businger's methods greatly underestimate reference evapotranspiration under extremely arid climates as those prevailing in the tested region. For this reason van Bavel-Businger's methods were dropped from any further analysis in this area. Figure 5 shows a sample diagram typical to the theme of values obtained from that test.

Hence finally Salih and Sendil [42] had selected Jensen-Haise and Modified Penman methods (from Jensen [30]) and the three additional methods (class A Pan, Hargreaves and Blaney-Criddle) that were evaluated by Hargreaves and tested them by using the corresponding climatic data from the above four sites. The results of that test showed that for all stations the common trend is that Class A Pan data (unadjusted) gave the highest values, then followed by Jensen-Haise as second, Hargreaves third, Modified Penman forth, while all tested versions of Blaney-Criddle (local, FAO and USDA-SCS) indicated the lowest values of evapotranspiration. The FAO version is a modified form of Blaney-Criddle recommended by Doorenbos and Pruitt [15], while the local version is a form recommended by the authority in this region. This form can be reached by substituting the monthly coefficient (k) of Blaney-Criddle formula by 0.60 and 0.85 to obtain evapotranspiration for alfalfa in winter months (December to February) and the rest of the months of the year, respectively. Figure 6 gives a sample diagram showing typical monthly values estimated by six of these methods using climatic data taken from one of the four sites.

Similar results to the above have been reported by Khowaiter [33] from five other sites (Aflaj, Zilfi, Hutat Sudair and Shagra) in the same region. Table 5 shows a summary of the mean annual values obtained from this investigation using modified Penman, Jensen-Haise and Hargreaves methods. The same trend has been reported by Al-Mansour [6] for Sakake station further to the North from the

previous stations (see Table 2). Typical values from this station can be seen in Table 5 and Figure 7. A noticeable observation from Figure 7 is that the values estimated by modified Penman are now closer to Jensen-Haise's estimation if compared with previous values estimated for station further to the south. This trend is currently under study in more than six sites in the area of Sakaka. Another observation is, that all versions of Blaney-Criddle indicate lower values compared to the other methods.

Salih and Sendil [42] evaluated Jensen-Haise, Class A Pan, Hargreaves, Modified Penman, and three versions of Blaney-Criddle (local, F.A.O., and USDA-SCS) according to the criteria of Jensen's [30], Hargreaves and Samani's [22] in addition to the closeness of the average ratio to unity. As a reference to that rating, data of direct measurements at Hofuf area [8,11,37] was used. The result of that rating in order of merits is shown in Table 6.

In addition to the above rating it is worth mentioning that all of the climatic methods have generally given lower values than those of direct measurements, while Class A Pan data was always the highest (see Figure 8). Of all estimation methods, Jensen-Haise was the closest to the actual values, followed by the class A Pan data when adjusted according to Doorenbos and Pruitt Criterion [15]. Figure 9 indicates a striking closeness between the results from Jensen-Haise, adjusted Class A Pan and the direct measurements.

Other Evaluations

The above are only but a few of the evaluations of some selected methods which are widely used for estimating evapotranspiration under extremely arid climates. This work is by no means comprehensive, since the literature is voluminous with methods that carry the general label of being suitable for arid climates, yet one or only a few of them were evaluated in large areas of extreme aridity. However, two of the works which came across the author's search and deserve to be noted in here are the good review by Gay [18] and the evaluation by Olaniran [36]. Both publications have made references to other interesting contributions and contributers, from arid regions, that deserve to be quoted if not for the limited space.

Except for the Thornthwaite, Priestly and Taylor, and Monteith-Penman methods, which did not add much to the present aim, the rest of the methods reviewed in Gay's publication have already been mentioned in this work. An interesting estimate of potential evapotranspiration was obtained from several methods using climatic data from Tucson, Arizona (1966–1979). Of these methods, Jensen-Haise, adjusted Pan, and Penman showed consistency and predicted maximum estimates of 9.9 mm/day, 8.8 mm/day, 8.0 mm/day, respectively. In the absence of direct measurements, these methods were not rated in a similar fashion to the above. However, the results of monthly variations from these methods have great resemblance to Figures 5 and 6.

Olaniran's work [36] included an evaluation of seven cli-

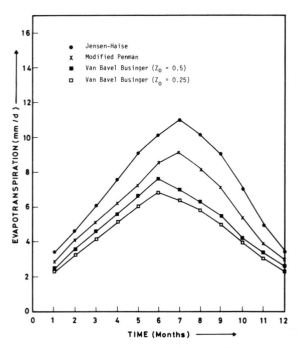

FIGURE 5. Reference evapotranspiration at Riyadh (1970–1981).

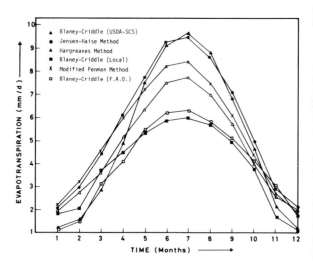

FIGURE 7. Comparison between different methods for estimating evapotranspiration (mm/day) at Sakaka for the period 1970–1983.

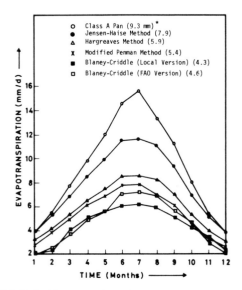

FIGURE 6. Mean monthly reference evapotranspiration for alfalfa at Kharj (1973–1981). [( )* number between brackets represents the mean annual ETR_a in mm/d.]

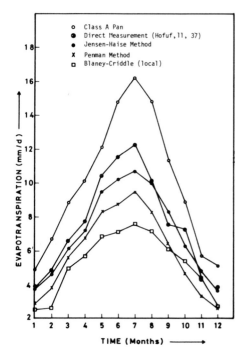

FIGURE 8. Mean monthly reference evapotranspiration for alfalfa at Hofuf (1970–71).

TABLE 5. Mean Annual Reference (Alfalfa) Evapotranspiration Estimated from Three Climatic Methods for Central Saudi Arabia.

	Estimated ETR_a (mm/d)			
Station	Jen-Haise	Modified Penman	Hargreaves	Reference
Aflaj	7.7	5.8	5.8	
Zilfi	7.4	5.9	5.8	Khowaiter
Hutat Sudair	7.3	6.7	5.9	[33]
Shagra	8.0	6.2	6.2	
Sakaka	5.7	5.2	4.8	Al-Mansour [6]

matic methods using data from Nigeria. In the absence of the corresponding climatic data it is very difficult to check the degree of aridity of his chosen area. The conclusions reached from his evaluation are that the temperature and radiation methods tend to underestimate potential evapotranspiration at small time intervals. The accuracy of the radiation methods improve most significantly when the time interval is increased.

SITING OF CLIMATIC STATION

As demonstrated from the previous review, climatic data could play an important role in estimating potential or reference evapotranspiration in the absence of direct measurements. However, the climatic data can be affected significantly by the siting of the weather station, a matter that leads to erroneous estimation of evapotranspiration. Allen et al. [4] have recently published an article that contained a good review of relevant literature to this problem as well as results obtained from a study performed in arid rangeland of southern Idaho. The article revealed an increase in temperature and decrease in vapor pressure due to the location of the station in non-irrigated site. It recommended a procedure for adjusting historical temperature data to reflect irrigation conditions.

TABLE 6. Salih and Sendil Rating.

Order of merit	Name of Method
1	Jensen-Haise
1	Class A Pan
3	Hargreaves
4	Modified Penman
5	Blaney-Criddle (USDA-SCS)
5	Blaney-Criddle (FAO)
7	Blaney-Criddle (Local)

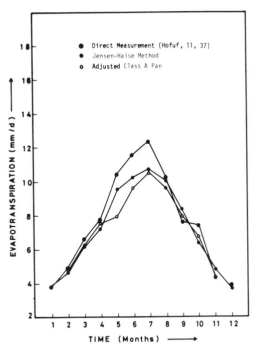

FIGURE 9. Mean monthly reference evapotranspiration for alfalfa at Hofuf (1970–71).

Gay [18] and Morton [35] have also reviewed some publications that dealt with the effect of siting of climatic station under arid climates (Oasis effect). An extreme has been reported from the Sudan where evaporation was reduced to 50% between the edge of a cotton field and a position 300 m to the inside of the field. The effect of siting on Pan readings and the corresponding adjusting coefficients that relate it to evapotranspiration has been reported by Doorenbos and Pruitt [15]. Pan coefficient (kp) is affected by climatic factors as well as the position and type of upwind fetch; whether being green crop or dry fallow.

The above are only a few examples of the influence of site on the value of climatic parameters and consequently on the estimation of evapotranspiration. These effects are greater under extremely arid conditions where irrigated agriculture is made in most cases of scattered oases amid vast desert areas. The effect of advection in these situations can be enormous, especially in mid-summer.

ADVECTION EFFECT

The transfer of sensible heat, from dry non-irrigated desert around irrigated land, and its conversion to latent heat is termed advection. In spite of its importance in extremely arid areas, it is unfortunately the least documented. A brief description to this phenomenon has been included in Chow's handbook of applied hydrology [10].

Jury and Tanner [32] claimed that by modifying the

Priestly and Taylor's equation, by adding a saturation deficit term to meet high local advection, they obtained better results for daily evapotranspiration from irrigated potatoes. However, their modification was not better than the Priestly-Taylor method, calibrated in a period of high advection, when used for alfalfa.

Shouse et al. [46] reported results from a field experiment conducted fro three successive summers in the arid southwestern United States with the aim of testing suitability of five formulas under extreme advective conditions. From all of these methods, Pan evaporation was found the poorest indicator of full-cover crop water use. Solar radiation methods correlated with functions of humidity and wind speed, advection-modified Priestly and Taylor equation, and especially Penman combination equation worked quite well.

The work of Sumayao et al. [49] was concerned with the assessment of evapotranspiration advection by using leaf temperature. They reported that at midday when the air temperature exceeded 33°C the leaves were cooler than the air and sensible heat flux was toward the canopy. As a result of that the rate of evapotranspiration exceeded net radiation during these periods.

The Hofuf study [11,37] recorded a daily evapotranspiration of 15 mm from Alfalfa and 21 mm for the corresponding Class A Pan as measured in August 1971. These values almost correspond to 900 and 1260 cal. cm^{-2} d^{-1}, respectively. These values were reached, while the actual global radiation was only about 500 cal. cm^{-2} d^{-1}. Hence, these results indicate an extreme high advection in that region which is of extreme aridity for most of the year, particularly summer months. In spite of this high effect, the previous evaluation reported by Salih and Sendil [42] was carried out with data from this particular region including the period during which the above results were collected. Salih and Sendil rating, however, has recommended almost opposing results to these recommended by Shouse et al. [45].

Similar extreme advection to that of the above region was reported by Rosenberg and Verma [39] from a study in Nebraska during the midwestern drought of 1976. Using a precision weighing lysimeter in a field of alfalfa, evapotranspiration rate had reached up to 14.22 mm/day and exceed 10.0 mm/day for a third of the study days. The net radiation was found to provide only up to 7 mm/day of that rate with the sensible heat advection catering for the balance.

There is no doubt that advection plays a significant role under extremely arid climates. Though the present review is by no means complete, yet it can be concluded that more intensive researches are certainly required under these extreme conditions before the problem is reasonably understood.

Popular Climatic Methods

The previous review has demonstrated the difficulty of specifying particular climatic method or methods as being most suitable under extreme conditions, yet one can list a few of the many reasons that restrain one from making such general recommendations. These reasons include the poor state of knowledge on evapotranspiration under extremely arid conditions, the current vague definition of aridity, the lack of data on direct measurements in both space and time, doubts on the quality and length of record of climatic data, and the poor documentation of advection effects. In spite of these effects many worldwide authorities believe that climatic data can be used successfully for prediction of ETP or ETR [15,35, . . .]. A brief review shall be given here for a few of the methods that received some popularity and wider acceptance under arid conditions. These methods shall be classified under three main categories; namely, simple, radiation and combination methods.

SIMPLE METHODS

This category includes all methods that basically utilize one climatic parameter such as temperature or relative humidity. Two of the famous temperature based methods are Thornthwaite and Blaney-Criddle methods, while Halstead, Ostromecki, Papadakis are examples of humidity methods [30]. Tempeature-methods are poorly rated for arid and extremely arid climates [22,24,30,42], while no firm recommendations are reported in favor of humidity methods. In spite of its poor rating, the parameters of different versions of Blaney-Criddle shall be presented here, only because of its wide popularity in many arid zones.

The Blaney-Criddle Method

The general expression for the Blaney-Criddle method can be written as [19]

$$U = 25.4 \; KF = 25.4 \; K \; \Sigma \; \frac{tP}{100} \qquad (2)$$

where

U = evapotranspiration of crop in mm for a given time period

K = empirical crop coefficient (annual, irrigation season, or growing season)

F = sum of evapotranspiration factors ($\Sigma \; tP/100$)

t = mean temperature in °F

P = percentage of day-time hours of the year, occurring during the period (tables can be found in standard textbooks and references [15,19,30, . . . etc.])

For monthly calculations, the following form is used,

$$u = 25.4 \; kf = 25.4 \; k \; \frac{tP}{100} \qquad (3)$$

where

u = monthly evapotranspiration, in mm

k = monthly crop coefficient

f = monthly evapotranspiration factor = $txP/100$

The soil conservation service of the United States Department of Agriculture [29] reported a modified version of Blaney-Criddle by replacing the monthly crop coefficient, k, by the following two coefficients,

$$k = k_c \, k_t \qquad (4)$$

where

k_c = monthly crop coefficient [29]
k_t = a climatic coefficient = $0.0173t - 0.314$ for $(k_t \geq 0.3$ and t is in °F)

Typical monthly values of k_c for alfalfa was reported in the USDA-SCS report [29] as follows:

Month	1	2	3	4	5	6	7	8	9	10	11	12
k_c	0.63	0.73	0.86	0.99	1.08	1.13	1.11	1.06	0.99	0.91	0.78	0.64

Doorenbos and Pruitt [15] have also suggested further refinement for u by entering f values in curves that involve wind speed, relative humidity and hours of sunshine to obtain monthly reference evapotranspiration.

RADIATION METHODS

This group consists of all climatic formulas, other than the combination method, that include radiation as the basic climatic parameter. Vast numbers of such methods have been reported in the literature with few of them given high rating when evaluated under arid and extreme arid climates [22,42,46], but only three of them shall be presented in here. These methods are: Jensen-Haise which was highly rated by Salih and Sendil [42] as well as Jensen [30]; Hargreaves which was highly rated by Hargreaves and Samani [22]; and the advection modified Priestley-Taylor which was recommended by Shouse, et al. [46].

Jensen-Haise Method [30]

This method was developed by its authors using observations in western United States. It can be written as;

$$ETR_a = C_t \, (T - T_x) \, R_s \qquad (5)$$

where

ETR_a = reference crop evapotranspiration, well watered alfalfa in langleys/day
C_t = a temperature coefficient = $1/(C_1 = C_2 C_H)$
T = average daily temperature in °C
T_x = the intercept on the temperature axis
R_s = incident solar radiation in langleys/day
C_H = 50 mbar/$(e_2 - e_1)$
C_1 = $38 - (2°C \times EL/305)$
C_2 = 7.6 °C

e_2, e_1 = saturation vapor pressure of water at the mean maximum and mean minimum temperatures, respectively, for the warmest month of the year in a given area
EL = Elevation above sea level in meters

and

$$T_x = -2.5 - 0.14 \, (e_2 - e_1) \text{ °C/mbar} - EL/550 \qquad (6)$$

A modified version of this method has been reported by Salih and Sendil [42] in the following form;

$$ETR_a = 1.16 \, ETR_{J\text{-}H} - 0.37 \qquad (7)$$

where

ETR_a = reference crop actual evapotranspiration, taken for well watered alfalfa, in mm/d.
$ETR_{J\text{-}H}$ = computed reference crop potential evapotranspiration, taken for well watered alfalfa using Equation (5), in mm/d.

Equation (7) was derived for conditions of extreme aridity, taken from central Saudi Arabia.

Hargreaves Method

This method was based on data from grass lysimeters and has recently been strongly recommended by Hargreaves and Samani [22]. It can be written as;

$$ETR_g = 0.0135 \, (T + 17.78) \, R_s \qquad (8)$$

where

ETR_g = reference crop potential evapotranspiration, well watered grass in langleys/day
T = average daily temperature in °C
R_s = incident solar radiation in langleys/day

For alfalfa as a reference crop, potential evapotranspiration is suggested by Hargreaves as follows:

$$ETR_a = 1.2 \, ETR_g \qquad (9)$$

In places where no measured values of R_s are available, Hargreaves and Samani [22] suggested the following equation;

$$ETR_g = 0.0075 \times T°F \times KT \times RA \times TD^{1/2} \qquad (10)$$

TABLE 7. Determination of Vapor Pressure Deficit (Δe) [13,30].

Method No.	Procedure
1	$\Delta e = \bar{e}_s - e_{dP_{min}}$
	Saturation vapor pressure at average temperature minus vapor pressure at minimum dewpoint temperature.
2	$\Delta e = \bar{e}_s - e_{dP_{av}}$
	Saturation vapor pressure at average temperature minus vapor pressure at average dewpoint temperature.
3	$\Delta e = \bar{e}_s (1 - R.H.)$
	Saturation vapor pressure at average temperature times one minus average relative humidity
4	$\Delta e = \frac{1}{2} (e_{s_{max}} + e_{s_{min}}) - e_{dP_{av}}$
	Average saturation vapor pressure based on maximum and minimum temperatures minus vapor pressure at average dewpoint temperature
5	$\Delta e = \frac{1}{2} [(e_{s_{max}} - e_{max}) + (e_{s_{min}} - e_{min})]$
	Average saturation vapor pressure deficit at maximum and minimum temperature
6	$\Delta e = \bar{e}_s - e_{air}$
	Saturation vapor pressure at average temperature minus actual vapor pressure (e_{air}) which is computed from wet bulb depression (ΔT_{wet}) measured by a psychromater

where

T = average daily temperature in °F
RA = extraterrestrial radiation in langleys/day
KT = a coefficient of temperature = $0.035 \times (100 - RH)^{1/2}$
RH = mean monthly relative humidity in percent for values of $R.H. \geq 54\%$
TD = mean maximum minus mean minimum temperature

Hargreaves and Samani [22] claimed that Equation (10) may give better estimates than those given by Equation (8).

The Advection—Modified Priestly-Taylor Equation

Jury and Tanner [32] have modified the Priestly and Taylor formula to include a saturation deficit term to account for high local advection. This equation was tested by Shouse, et al. [46] who recommended its adequacy in advective environments. The recommended form of the equation is written by Shouse, et al. [46] as follows:

$$ETP = \left[1 + \frac{(\alpha - 1)}{\Delta e_A} \Delta e \right] \frac{S}{S + \gamma} R_N \quad (11)$$

where

ETP = potential evapotranspiration
R_N = net radiation (cal/cm/day)
α = a constant to be calibrated locally
Δe = vapor pressure deficit
Δe_A = average vapor pressure deficit for crop cycle, to be obtained from local calibration

THE COMBINATION METHODS

The main advantage of such methods is their sound physical base where the formulas describing evapotranspiration are derived from a combination of energy balance and an aerodynamic term to cater for mass transport. These methods have witnessed vast developments since Penman introduced its first version in 1948.

In spite of the modest rating given to the Penman approach under arid conditions [20,21,22,30,42], yet its components represent significant potential for reaching a generalized approach to predict evapotranspiration under various climatic conditions.

The Modified Penman Method

The original Penman equation has been slightly remodeled in various recent publications [15,19,30,31]. Most of these changes were related to the aerodynamic term which is quite important under arid conditions [15,20,21]. Another noticeable change is related to the utilization of the equation for estimating evapotranspiration for a specific reference crop, usually being alfalfa or grass. To cater for all these forms this equation will be defined, here, in a general form. With the aid of the information compiled in Tables 7 and 8, various corresponding components of the aerodynamic terms that suit each of the four versions, implied in Table 8, can be selected. The general form can be written as;

$$ETR = \frac{\Delta}{\Delta + \gamma} (R_n + G) + \frac{\Delta}{\Delta + \gamma} \times M(W_1 + W_2 U_2) \Delta e \quad (12)$$

where

ETR = reference crop evapotranspiration, well watered alfalfa or grass
Δ = slope of saturation vapor pressure-temperature curve de/dT in mbar/°C (see Jensen [30], Hansen, et al. [19])

TABLE 8. Various versions for M, W₁, and W₂ of Equation (12).

Reference, Author	Reference crop	M	W_1	W_2	Δe, Method No., Table 8	ETR, units
Jensen [30]	Grass	15.36	1.0	0.0062	2	Langleys d^{-1}
Doorenbos and Pruitt [15]	Grass	0.27	1.0	0.01	1, 3, 6	mmd^{-1}
Hansen, et al. [19]	Alfalfa	15.36	1.10	0.0062	2	Langleys d^{-1}
Jensen [31]	Alfalfa	15.36	0.75	0.0115	4	Langleys d^{-1}

*The terms $\Delta/\Delta + \gamma$ and $\gamma/(\Delta + \gamma)$ in Equation (12) have been substituted, in Doorenbos and Pruitt's [15] equation by the terms W and (1 − W) respectively. W and (1 − W) are temperature-altitude related terms, prepared in tabular form. Also the soil heat flux G is dropped.

γ = psychrometric constant
R_n = net radiation in cal/cm² per day
G = soil heat flux in cal/cm² per day
U_2 = wind movement in km/day at 2 m above ground level
Δe = vapor pressure deficit in mbar (see Table 8 for methods of determination)
M = constant of proportionality (see Table 8 for typical values)

and

$$\gamma = C_p \frac{P}{(0.622\lambda)} \quad (13)$$

where

C_p = 0.24
P = 1013−0.1055 EL
λ = latent heat of water in cal/g = 595.9 − 0.55T, (T in °C)

R_n = 0.77 R_s − R_b (assuming an average albedo of 0.23)

$$(14)$$

$$R_b = R_{bo} ([a \, R_s/R_{so}] + b) \quad (15)$$

$$R_{bo} = (a_a + b_1 \sqrt{e_a}) \, 11.71 \times 10^{-8} \, (T_a^4 + T_b^4)/2 \quad (16)$$

where

R_s = incident solar radiation in cal/cm² per day (measured)
R_{so} = clear day solar radiation in langleys/day (tables are found in standard text books, see Hansen, et al. [19], Jensen [30], . . . etc.)
a,b,a_1,b_1 = empirical constants that differ with locations (see Jensen [30,31], Hansen, et al. [19])
e_a = mean actual vapor pressure in mbar

T_a = maximum daily temperature in °K
T_b = minimum daily temperature °K

The Aerodynamic Term

Cuenca and Nicholson [13] have published an article describing a methodology for relating the proper wind coefficient with the matching vapor pressure deficit procedure. The importance of their work lies on the fact that there are almost six methods for computing the pressure deficit (see Table 7). Each of these methods is tied to specific wind coefficients and reference crops provided that the units of vapor pressure, wind speed and evapotranspiration are strictly noted. For example Table 8, clearly demonstrates the varying values of these parameters for four methods reported with common wind speed and vapor pressure units (km/day and mbar, respectively).

In addition to the procedure recommended for the application of Penman's wind function, Cuenca and Nicholson have praised a new wind function that was recommended by Wright in the form of polynomials which are a function of time during the growing season. In a discussion of the above paper, Stigter [48] has criticized Cuence and Nicholson empirical approach for solving the controversies related to the wind function. He called for a sound physical approach based on the following equation, which was reported previously by Stigter [47].

$$Ea = [30 \, f(u) \, \Delta e] \left/ \left[\ln \frac{z - d}{z_0} \right]^2 \right. \quad (17)$$

where

Ea = aerodynamic term in mm/d
u = wind speed at 2m, kmd^{-1}
Δe = vapor pressure deficit mbar
d = displacement height, same unit as z
z_0 = roughness length, same unit as z

z = measured from soil surface upwards

$f(u)$ = same as $f(u)$ reported by Doorenbos and Pruitt [15] (or the modified Thom and Oliver [46]; $f(u) = 0.37 \ (1 + u/160)$)

In a closure to the above discussion, Cuenca and Nicholson [14], have appreciated Stigter's recommendation, yet they doubted its practicality in determining the zero plane displacement, d, and roughness length, z_0.

To cater for the effect of wind, Doorenbos and Pruitt [15] have recommended various adjustments, (based on the ratio of the wind speed during the day and night together with relative humidity) to be applied for the version of their equation implied in Table 8. For seven classifications, based mainly on wind variations, relative humidity, and sometimes on radiation and season of the year, the adjustment coefficient could vary between a minimum of 0.3 to a maximum of 1.2 of ETR_g calculated from the unadjusted formula (Equation (12)).

The difficulty of applying Doorenbos and Pruitt approach stems from the fact that day and night variation of wind speed is not readily available in standard hydrometeorological data books and also the longer time needed for entering these adjustments into the computer. As an effort to solve the latter problem, Frevert, et al. [17] recommended regression equations for faster determination of these adjusting factors.

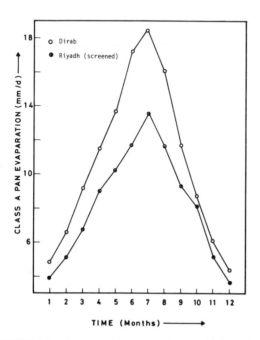

FIGURE 10. Mean monthly evaporation (mm/d) from Dirab and Riyadh Stations (1979-1982).

Evaporimeters

Vast publications have been devoted, in the literature, to the use of evaporimeter records as indices to estimates of evapotranspiration. A simple relation can be developed as follows:

$$ETR = K_e \times E_e \qquad (18)$$

where K_e is a coefficient based on the type of evaporimeter and reference crop; E_e is evaporation from evaporimeter and ETR is reference evapotranspiration.

It is not intended, nor even possible at this occasion, to review comprehensively the numerous pros and cons on the use of this method for estimating ETR, yet reference will be given to few contributions that were at hand during the preparation of this section. Though Jensen [30] was not able to use Class A pan as one of the methods that rated for suitability in estimating evapotranspiration in his excellent work, yet he included a good review of previous contributions as well as the limits within which this method can be used. He concluded that, with careful attention, the method could be used for monthly ETR estimate within an accuracy of $\pm 10\%$ or even better for most climates.

The Hofuf study [11,37] had the advantage of being performed under extremely arid climates. In that study direct measurement of evapotranspiration for alfalfa, using lysimeter technique, was compared with the corresponding evaporation from standard Class A pan and Czeratzki disc placed 2.0 m above soil surface. Both measurements had shown high correlation to direct measurements. When the evaporation measurements of Czeratzki and Pan, together with estimates from Blaney-Criddle and Penman methods, were compared with direct measurements, they respectively gave the following coefficients of proportionality: 0.67, 0.75, 1.36 and 1.27. The corresponding variations were 0.06, 0.06, 0.27 and 0.11, respectively, a result that puts these evaporimers at a greater advantage over Blaney-Criddle and Penman methods.

A better rationalization for the use of pans as indices for reference evapotranspiration, was recommended by Doorenbos and Pruitt [15]. In this last work the pan coefficients K_e (K_p) were tabulated as a function of wind speed, relative humidity, pan environment and type of pan. Two sets of tables were prepared, separately, for estimating K_p values for Class A pan and Colorado Sunken Pan. These values of K_p were qualified with further considerations well identified in that publication. For easier computation of K_p, Frevert, et al. [17], suggested a regression equation for predicting K_p values.

Doorenbos and Pruitt [15] suggested that a screen mounted over pans will lead to up to 10% reduction in evaporation rate. This figure has been confirmed by Howell, et al. [27] for a pan in a semi-arid climate. This value seems to be low compared to the mean annual reduction of about

FIGURE 11. Class A Pan at Dirab Station.

23% given by comparing Riyadh's screened pan to neighbouring Dirab's pan (see Table 2). Figure 10, which gives a comparison of the mean monthly evaporations at both stations, indicates greater reduction in the summer months. The environments and level of maintenance (Figures 11 and 12) are quite different for the two stations, hence it is possible that this larger reduction is due to additional local causes. This problem is at present receiving a closer investigation.

It is also worth mentioning that Class A pan method has received very high rating by Salih and Sendil [42] for an ex-

tremely arid climate. Similar rating for this method was also given by Hargreaves and Samani [22]. The detail of these ratings have already been given earlier in this section.

Crop Evapotranspiration (ETC)

According to the previous definitions, crop evapotranspiration (or consumptive use) should be calculated by multiplying reference evapotranspiration by a crop coefficient in the following form;

$$ETC = K_c \, ETR \qquad (19)$$

where

ETC = estimated crop evapotranspiration
ETR = reference crop evapotranspiration (alfalfa or grass)
K_c = crop coefficient

The crop coefficient is a function of many factors such as the type of crop, stage of growth, etc., while ETR depends on the type of reference crop and its environments. These aspects are unfortunately far from being standardized at present. Different textbooks and references [15,30,55, etc.] contain tables of experimental values of K_c. Hence it is warned, that great attention must be paid to the selection of the most relevant of these tables.

CONCLUDING REMARKS

The state of knowledge on evapotranspiration under extremely arid climates is relatively poor, at present, in spite

FIGURE 12. Screened Class A Pan at Riyadh Station.

of the worthwhile efforts scattered here and there. The main impediments are: vagueness in the definitions of aridity, shortage in climatic and direct measurements data, little knowledge on the effect of siting and advection in climatic measurements, and the controversial definitions of evapotranspiration terms (i.e. potential, reference, crop, etc.). It is hence recommended that these problems should be thoroughly investigated and clarified in future. A generalized approach utilizing the basic components of the combination equation is advisable and appears to be possible if better understanding is developed on the effects of the aerodynamic term.

The current climatological definition of aridity, as reviewed earlier in this section, seems to be too broad for meeting any worthwhile generalization on evapotranspiration under arid and extremely arid climates. There is certainly a need for a more detailed definition of aridity based on evapotranspiration concept and more oriented towards irrigation water requirements.

Awareness on the importance of climatic data for many purposes has recently led to the establishment of a vast number of climatic stations in many areas under extremely arid climates. The effect of siting of these stations on the quality of data and the type of adjustments to be effected on the data under various levels of aridity are matters that need further investigations.

In spite of the huge agricultural expansion in many extremely arid regions, irrigation is actually practiced through a large number of isolated small holdings instead of a single expanding area due to water, soil quality and ownership restrictions. The effect of advection on this Oasis-type system of irrigation, especially under these extreme conditions, is poorly understood and requires special attention.

In spite of the importance of direct measurements of evapotranspiration for calibration of climatical and theoretical methods, yet very little on such lines has been reported from extremely arid areas. It is, hence, recommended that more direct measurement experiments should be encouraged.

The concept of reference crop evapotranspiration (*ETR*) seems to overshadow gradually the older term of potential evapotranspiration. However, it seems that a standardization of the reference crop (or crops) and the corresponding crop factor is a matter that needs urgent contribution from one of the concerned international authorities (i.e. ICID, FAO, ASCE, ASAE, . . . etc.) to eliminated the current controversies.

It has been demonstrated earlier, from few examples presented in this chapter, that worthwhile information on irrigation water requirement under extremely arid conditions can be gathered from old experiences. These experiences have been reported in the literature mainly from historical or archaeological point of view. It seems that a closer engineering look is needed for better understanding of these very old systems.

REFERENCES

1. Al-Amami, S., "Drip Irrigation by Ibn Al-Awam," presented at the December 10–14, 1983, Third International Symposium on the History of Arab Science, National Council for Culture and Arts, Kuwait, Kuwait.

2. Al-Hashmi, R. J., "History of Irrigation in Old Iraq," presented at the December 10–14, 1983, Third International Symposium on the History of Arab Science, National Council for Culture and Arts, Kuwait, Kuwait.

3. Al-Khaldi, S., "Arab's Methods of Lifting Control and Distribution of Groundwater in Al-Ghizwini and Al-Damonhori Writings," presented at the December 10–14, 1983, Third International Symposium on the History of Arab Science, National Council for Culture and Arts, Kuwait, Kuwait.

4. Allen, R. G., C. E. Brockway, and J. L. Wright, "Weather Station Siting and Consumption Use Estimates," *Journal of Water Resources Planning and Management Division, ASCE, Vol. 109* (No. 2), pp. 134–146 (April, 1983).

5. Al-Maktari, A. M. A., "Legal Aspects of Water and Irrigation Systems in the Arabian Peninsula," presented at the December 10–14, 1983, Third International Symposium on the History of Arab Science, National Council for Culture and Arts, Kuwait, Kuwait.

6. Al-Mansour, A. E., "Evapotranspiration at Sakaka Area," thesis submitted to King Saud University, at Riyadh, Saudi Arabia, in 1985, in partial fulfillment of the requirement of the degree of B.Sc. in Civil Engineering.

7. Al-Shami, A. A. M., "Arab Writings on Irrigation Agriculture System in Egypt," presented at the December 10–14, 1983, Third International Symposium on the History of Arab Science, National Council for Culture and Arts, Kuwait, Kuwait.

8. Asseed, M., A. Turjoman, and H. Eteway, "Water Use for Agriculture in Al-Hassa Area in Eastern Province," presented at the April 17–20, 1983, Symposium on Water Resources in Saudi Arabia; Its Management, Treatment and Utilization, College of Engineering, King Saud University, Riyadh, Saudi Arabia.

9. Camillo, P. J., and R. J. Gurney, "A Sensitivity Analysis of a Numerical Model for Estimating Evapotranspiration," *Water Resources Research, Vol. 20* (No. 1), pp. 105–112 (January, 1984).

10. Chow, V. T., *Handbook of Applied Hydrology*, McGraw-Hill Book Company, New York, pp. 24–1–24–46 (1964).

11. *Consumptive Use of Alfalfa in Al-Hassa Region*, Publication No. 8, Hofuf Agricultural Research Centre, Hofuf, Saudi Arabia, 75 pp. (September, 1973).

12. Critchfield, H. J., *General Climatology*, 3rd ed., Prentice Hall Inc., New Jersey, pp. 145–148 (1974).

13. Cuenca, R. H. and M. T. Nicholson, "Application of Penman Equation Wind Function," *Journal of the Irrigation and Drainage Division, ASCE, Vol. 108* (No. 1R1), pp. 13–23 (March, 1982).

14. Cuenca, R. H. and M. T. Nicholson, closure to "Application of

Penman Equation Wind Function," *Journal of the Irrigation and Drainage Engineering, ASCE, Vol. 109* (No. 2), pp. 283–287 (June, 1983).

15. Doorenbos, J. and W. O. Pruitt, "Crop Water Requirements," Irrigation and Drainage Paper 24, Food and Agriculture Organization of the United Nations, Revised, Rome, Italy, 144 pp. (1977).

16. Faulkner, R. D. and T. E. Evans, "The Measurement of Evapotranspiration in an Arid Climate Using the Energy Balance Method," *Proc. Instn. Civ. Engrs., Vol. 71* (Part 2), pp. 51–61 (March, 1981).

17. Frevert, D. K., R. W. Hill, and B. C. Braaten, "Estimation of FAO Evapotranspiration Coefficient," *Journal of the Irrigation and Drainage Engineering, Vol. 109* (No. 2), pp. 265–270 (June, 1983).

18. Gay, L. W., "Potential Evapotranspiration for Deserts," in D. D. Evans and J. L. Thames, editors, *Water in Desert Ecosystems*, School of Renewable Natural Resources, Tuscon, Arizona, U.S.A., pp. 172–194 (1981).

19. Hansen, V. E., O. W. Isrealson, and G. E. Stringham, *Irrigation Principles and Practices*, 4th ed., John Wiley and Sons, Inc., New York, N.Y., 417 pp. (1980).

20. Hargreaves, G. H., closure to "Estimating Potential Evapotranspiration," *Journal of the Irrigation and Drainage Engineering, ASCE, Vol. 109* (No. 3), pp. 343–344 (September, 1983).

21. Hargreaves, G. H., Discussion of "Application of Penman Equation Wind Function," by R. H. Cuenca and M. T. Nicholson, *Journal of the Irrigation and Drainage Engineering, ASCE, Vol. 109* (No. 2), pp. 277–278 (June, 1983).

22. Hargreaves, G. H., and Z. A. Samani, "Estimating Potential Evapotranspiration," *Journal of the Irrigation and Drainage Division, ASCE, Vol. 108* (No. 1R3), pp. 225–230 (September, 1982).

23. Hasan, M. R., Discussion of "Estimating Potential Evapotranspiration," by G. H. Hargreaves and Z. A. Samani, *Journal of Irrigation and Drainage Engineering, ASCE, Vol. 109* (No. 3), pp. 341–342 (September, 1983).

24. Hashemi, F. and M. T. Habibian, "Limitations of Temperature-Based Methods in Estimating Crop Evapotranspiration in Arid-Zone Agricultural Development Projects, *Agricultural Meteorology* (20), pp. 237–247 (1979).

25. Heathcote, R. L., *The Arid Lands: Their Use and Abuse*, Longman, London & New York, pp. 12–21 (1983).

26. Howard, K. W. F. and J. W. Lloyd, "The Sensitivity of Parameters in the Penman Evaporation Equations and Direct Recharge Balance," *Journal of Hydrology* (41), pp. 329–344 (1979).

27. Howell, T. A., C. J. Phene, D. W. Meek, and R. J. Miller, "Evaporation from Screened Class A Pans in a Semi-Arid Climate," *Agricultural Meteorology, Vol. 29*, pp. 111–124 (1983).

28. Hydrological Publications Nos. 45, 53, 61, 67, 74, 82, 89, and 90–95, Ministry of Agriculture and Water, Riyadh, Saudi Arabia (1970–1981).

29. "Irrigation Water Requirement," Technical Release No. 21,

United States Department of Agriculture (SCS), U.S.A., April, 1967, 88 pp. (Revised September, 1970).

30. Jensen, M. E., ed., *Consumptive Use of Water and Irrigation Water Requirements,* ASCE, New York, N.Y., 215 pp. (1974).

31. Jensen, M. E., ed., "Design and Operation of Farm Irrigation System," Monograph No. 3, American Society of Agricultural Engineers, St. Joseph, Mich., 829 pp. (1980).

32. Jury, W. A. and C. B. Tanner, "Advective Modification of the Priestley and Taylor Evapotranspiration Formula," *Agronomy Journal, Vol. 67* (No. 6), pp. 840–842 (1975).

33. Khowaiter, A. H., "Estimated Consumptive Use Requirement of Crops in Saudi Arabia," Thesis presented to King Saud University, at Riyadh, Saudi Arabia in 1983, in partial fulfillment of the requirement of the degree of B.Sc. in Civil Engineering.

34. Meigs, P., "World Distribution of Arid and Semi-Arid Homoclimates," in UNESCO Arid Zone Res. series No. 1, *Arid Zone Hydrology*, pp. 203–209 (1953).

35. Morton, F. I., "Estimating Evapotranspiration from Potential Evapotranspiration: Practicality of an Iconoclastic Approach," *Journal of Hydrology, 38*, pp. 1–32 (1978).

36. Olaniran, O. J., "Empirical Methods of Computing Potential Maximum Evapotranspiration," *Arch. Met. Geoph. Biokl., Ser. A., Vol. 30*, pp. 369–381 (1981).

37. *Problems Related to the Estimation of Consumption Use Under Extremely Arid Conditions*, Publication No. 14, Hofuf Agricultural Research Center, Hofuf, Saudi Arabia, 105 pp. (June, 1976).

38. Rogers, J. A., "Fools Rush In, Part 3: Selected Dryland Areas of the World," *Arid Lands Newsletter,* Arizona, *Vol. 14*, pp. 24–5 (1981).

39. Rosenberg, N. J. and S. B. Verma, "Extreme Evapotranspiration by Irrigated Alfalfa: A Consequence of the 1976 Midwestern Drought," *Journal of Applied Meteorology, Vol. 17* (No. 7), pp. 934–941 (July, 1978).

40. Salih, A. M. A., "A Visit to Al-Daw Basin," Unpublished Report to the Arab Centre for the Studies of Arid Zones and Dry Lands, Damascus, Syria (1976).

41. Salih, A. M. A. and T. O. Al-Farra, "Old Irrigation Systems in the Kingdom of Saudi Arabia," presented at the December 10–14, 1983, Third International Symposium on the History of Arab Science, National Council for Culture and Arts, Kuwait, Kuwait.

42. Salih, A. M. A., and U. Sendil, "Evapotranspiration under Extremely Arid Climates," *Journal of Irrigation and Drainage Engineering, ASCE, Vol. 110* (No. 3), pp. 289–303 (September, (1984).

43. Sendil, U., A. M. A. Salih, and S. H. Alvi, "Hydrological Characteristics of Wadi Hanifa," Presented at the April 17–20, 1983, Symposium on Water Resources in Saudi Arabia; Its Management, Treatment, and Utilization, College of Engineering, King Saud University, Riyadh, Saudi Arabia.

44. Sendil, U., et al., "Hydrological Characteristics of Wadi

Hanifa," Research Report No. CE-2/1404, College of Engineering, King Saud University, p. 239 (1983).

45. Shantz, H. L., "History and Problems of Arid Lands Development," in G. F. White (ed.), *The Future of Arid Lands*, Amer. Assoc. Adv. Sc. Pub. No. 43, Washington (1956).

46. Shouse, P., W. A. Jury, and L. H. Stolzy, "Use of Deterministic and Empirical Models to Predict Potential Evapotranspiration in an Advective Environment," *Argronomy Journal, Vol. 72,* pp. 994–998 (November/December, 1980).

47. Stigter, C. J., "Comparison and Combination of Two Recent Proposals for a Generalized Penman Equation," *Quarterly Journal of the Royal Meterological Society, Vol. 105,* pp. 1071–1073 (1979).

48. Stigter, C. J., Discussion of "Application of Penman Equation Wind Function," by R. H. Cuenca and M. T. Nicholson, *Journal of the Irrigation and Drainage Engineering, ASCE, Vol. 109* (No. 2), pp. 278–281 (June, 1983).

49. Sumayao, C. R., E. T. Kanemasu, and T. W. Brakke, "Using Leaf Temperature to Assess Evapotranspiration and Advection," *Agricultural Meteorology, Vol. 22,* pp. 153–166 (1980).

50. Theodoratus, R. J., "Traditional Agriculture in the Lower Elevations of Southwest Arabia," Presented at the April, 1982, Meeting of the Association for Arid Lands Studies, Denver, Colorado.

51. Twitchel, K. S., *Saudi Arabia,* 3rd ed., Greenwood Press Publishers, New York (1958).

52. Varisco, D. M., Arab Classical Writings and Agriculture: The Agricultural Almanac, Presented at the December 10–14, 1983, Third International Symposium on the History of Arab Science, National Council for Culture and Arts, Kuwait, Kuwait.

53. Vidal, F. S., The Oasis of Al-Hasa, Arabian American Oil Company, Arabian Research Division, Dahaharan, Saudi Arabia (1955).

54. Walton, K., The Arid Zones, Hutchinson University Library, London, pp. 8–17 (1969).

55. Wright, J. L., "New Evapotranspiration Crop Coefficient," *Journal of the Irrigation and Drainage Division, ASCE, Vol. 108* (No. 1R2), pp. 57–74 (1982).

CHAPTER 11

The IUH of General Hydrologic System Model

V. C. Kulandaiswamy*

1. INTRODUCTION

1.1 General

The concept of basin modelling has been the prime source of inspiration to various investigators working in the area of hydrology. The basin models developed over the years ranged over a wide spectrum from near physical scale models of the basin with artificial rain to pure mathematical abstraction while passing through the zone of analogs.

The general objectives of all these models have been:

1. To establish the relationship between the rainfall and runoff, both of them observed and available from the records
2. To use this relationship to predict or forecast the runoff from known sequence of rainfall data.

From the spectrum of the models, one could identify certain specific groups like those which (i) create analogs for a basin, (ii) develop simple empirical relationships between rainfall and runoff, and (iii) develop mathematical relationship on a conceptual replacement of the basin with varying degrees of generalization of the basin characteristics.

1.2 Rainfall-Runoff Models

The analogy of tanks in series or series-parallel to simulate the basin transformation of rainfall into runoff has been first attempted. Subsequently, in this category, the basin is replaced through several storages to obtain the observed hydrograph of runoff.

The digital simulation [19] of the watershed though supposed to follow the hydrological cycle has enormous data re-

quirements that would hardly be available. The phases of the runoff cycle are simulated through appropriately considered storages, their coefficients requiring calibration. This category of models requires expert manpower, large variety of data requirement, and computer facility of adequate capability. The model calibration is tedious and rather arbitrary.

The practising professionals, having been not either adequately equipped with advanced methods of watershed modelling or not being able to wait that long to have the basin calibrated and resort to forecasting, have evolved their own empirical rainfall-runoff models.

1.3 Rainfall-Runoff Relationships

These relationships using mathematical transformations like Laplace, Fourier, Z etc. [4,21,30,31] yield the runoff hydrograph for a given rainfall function. Their capability of predicting the runoff hydrographs has not been to the required degree of satisfaction. They are specific to the task of reproducing the observed hydrograph without going into the characteristics of the watershed mathematically reflected in the procedure.

1.4 Hydrologic System

Preservation of the mathematical rigour, *generalisation of the approach* and capability to predict the runoff hydrographs are all possible only when the watershed characteristics are mathematically identified and modelled without imposing any assumption purely for the purpose of simplification of solution procedure. The watershed is to be approached as a system with its mutually interacting dynamic elements like interception, evaporation, transpiration, infiltration, detention, retention, surface runoff, sub-surface runoff, etc., appropriately integrated, equally satisfying the physics of the process and the postulated mathematical inter-relationship.

*Vice Chancellor, Anna University, Madras 600 025, India

The general hydrologic system model [14-17,29] has shown its potential to remain general, with other models being special cases of themselves, and has proved its applicability to a variety of hydrological problems like flood routing, infiltration, basin transformation, etc., to name a few of the well investigated situations which are presented in the subsequent sections.

2. EVOLUTION OF THE GENERAL HYDROLOGIC SYSTEM MODEL

2.1 Characteristics of Drainage Basins and Storms

The factors that influence the abstractions and runoff in a basin are [16]:

1. State of the basin
   a. Soil moisture
   b. Density of growth of vegetation
   c. Land use
   d. Temperature
   e. Wind
   f. Atmospheric pressure
   g. Humidity
   h. Solar radiation
2. Basin characteristics
   a. Area
   b. Shape
   c. Slope
   d. Soil type
   e. Geological conditions
   f. Land use
   g. Stream density
   h. Soil surface condition
   i. Size, shape, slope and roughness of the channel
   j. Storage capacity

3. Storm characteristics
   a. Type of precipitation
   b. Intensity
   c. Duration
   d. Time distribution
   e. Areal distribution

The list is neither exhaustive, nor are the factors enumerated independent. Many of them are in fact interdependent. The variables are too many and most of them are not susceptible to quantitative evaluation. Each one of the factors listed above have some bearing on the runoff characteristics of the basin. However they do not still lend themselves to analytic treatment of any significance, though empirical and semi-empirical methods have been developed for runoff prediction.

NONLINEAR ELEMENTS IN A BASIN SYSTEM

In general all physical systems are nonlinear. They are linear only by assumption. In many systems the assumption of linearity may be satisfactory. In some, the assumption of linearity may hold good over certain ranges of operation. In certain others, nonlinear effects may be quite large. The runoff process in a basin system is indicated in Figure 1a. In the case of a basin system, surface runoff studies as now made involve three major operations [29,38,40]:

1. Determination of rainfall excess from total rainfall
2. Separation of the hydrograph of runoff into surface hydrograph and base flow hydrograph
3. Establishing a relationship between surfac runoff hydrograph obtained in step 2 and the rainfall excess determined in step 1

The procedures used in 1 and 2 are empirical; nevertheless they are nonlinear in nature. It is in step 3 that an assumption is made that the process by which rainfall excess

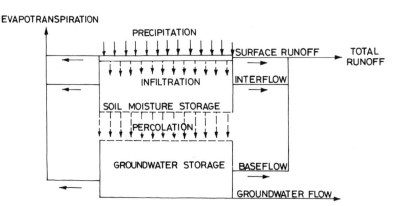

FIGURE 1a. Runoff cycle.

is converted into surface runoff is linear. The validity of this assumption has not been conclusively proved. If the results obtained on the basis of a linear theory is uniformly satisfactory, the validity of linearity could perhaps be established by inductive reasoning. If, on the other hand, the results show variations, they could be due to many reasons, and nonlinearity may or may not be one of them. But supposing the variations show a trend that could be explained by nonlinear elements in the process under consideration, it may be reasonable to conclude that the conversion of rainfall excess into surface runoff is apparently not a linear transformation. Whether it could be linearised or not, is another problem, and depends on the degree of nonlinearity and the accuracy of results expected. The possible errors in baseflow separation and rainfall excess determination, inaccuracies in the original data and nonuniform areal distribution of rainfall, do not fully account for the variation. In the application of unithydrograph theory [70], it has been consistently observed that for the same basin the unithydrographs derived from minor storms differ significantly from those obtained from major storms. It can be seen from Paynter's studies [62] that the instantaneous unithydrographs derived from a number of hydrographs from the same basin differ considerably. He attributed the variations to the seasonal effect and the magnitude of flood size.

A systematic study by Singh [71], using various methods for deriving instantaneous unithydrograph, showed that the parameters of the instantaneous unithydrograph are not constant for a basin, and their variation shows a dependence on factors involving inflow and outflow. Instantaneous unithydrographs derived by Ramaseshan [65] for 26 storms over a 17,570 sq km basin, using Nash method [59] indicated that the parameters vary considerably and the variation shows dependence on rainfall excess. Studies by Minshall [58] for deriving unithydrographs for small watersheds, have indicated that the unithydrograph parameters vary with the storm characteristics. Amorocho [1], in his discussion on Minshall's paper, has shown that unithydrograph parameters for a given basin are functions of storm characteristics. Later, Amorocho and Orlob [2] have proposed the use of multiple system functions representing the complex nonlinear runoff process in a drainage basin. But, so far, techniques are not yet developed for solving the multiple system functions by direct inversion using complex input and output data. Studies by Diskin [20] have shown that nonlinearity in the runoff process can be investigated in terms of the properties of the kernel function in the Duhamel convolution integral. But he has not given any simple method of arriving at the values of the parameters in the model. From the foregoing review, it is clear that hydrologists have been aware of the fact that nonlinear elements in runoff process need a detailed investigation. Using continuity equation and storage equation a systematic procedure for the study of nonlinearity is described below.

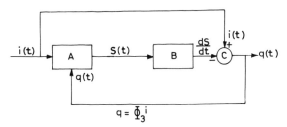

FIGURE 1b. Rainfall excess–surface runoff relationship.

THE EQUATION OF CONTINUITY

Assuming that the basin system is lumped, the continuity equation can be written as:

$$i(t) - q(t) = \frac{dS}{dt} \qquad (2.1)$$

where $i(t)$ is the inflow, $q(t)$ the outflow and S the storage at time t.

A simple block diagram for Equation (2.1) is given in Figure 1b. The block diagram consists of three elements, namely A, B and C. The element A transforms quantities involving $i(t)$ and $q(t)$ into $S(t)$. The element $S(t)$ is transformed into dS/dt by the element B, i.e., $S(t)$ is differentiated with respect to time t. The element C merely adds $i(t)$ and dS/dt to yield $q(t)$:

$$q(t) = i(t) - \frac{dS}{dt} \qquad (2.2)$$

Element B is linear as it is a differential operator. Element C is also linear as it represents mere summation. Nonlinearity, if any, is in the element A; in other words, the storage in the basin may be nonlinearly related to $q(t)$ and/or $i(t)$. The problem of conversion of rainfall excess into surface runoff reduces to one of deriving an expression for storage.

STORAGE CHARACTERISTICS OF DRAINAGE BASINS

A careful study of the various methods to represent rainfall-runoff relationship reveals that a satisfactory equation representing the relationship between storage and discharge would help considerably an understanding of the basin system. Hydrologists working on rainfall-runoff relationship, have attempted to develop expressions for storage in terms of discharge at the gauging station and other variables. The following are some of the equations proposed:

Zoch [76]: $\qquad\qquad S = Kq \qquad (2.3)$

Many authors: $\qquad\qquad S = Kq^n \qquad (2.4)$

Meyer and
Muskingum River
Basin Board [27,7,8]: $\quad S \text{ Vs } \sqrt{q(t) \cdot q(t + \Delta t)}$ (2.5)
(Graphical relationship)

Laurenson [51]: $\qquad\qquad S = K(q) \cdot q$ (2.6)

McCarthy [5,8,50,55]: $\quad S = K[Xi + (1 - X)q]$ (2.7)

Kulandaiswamy [14,16,17,29,46,47]:

$$S = \sum_{n = 0}^{N} a_n(q,i) \frac{d^n q}{dt^n}$$

(2.8)

$$+ \sum_{m = 0}^{M} b_m(q,i) \frac{d^m i}{dt^m}$$

Prasad [63] $\qquad S = K_1 q^n + K_2 \frac{dq}{dt}$ (2.9)

Where K, $K(q)$, K_1, K_2, $a_n(q,i)$ and $b_m(q,i)$ are coefficients, which depend upon the outflow q and/or i and their derivatives. Δt is a time constant, t is the time, n is an exponent in Equation (2.4) and variable in Equation (2.8); m is a variable, N and M are constants and X is a weighting factor. The merits of the above equations have been discussed in many publications [7,9,15,16,18,32,51–54,56,61,66,72,77, 79]. Equation (2.3) is a straight line approximation of the loop formed by S vs. q plot (Figure 2) and Equation (2.4) is a curvilinear approximation of the same. In both the cases storage is assumed to be a single valued function of q. The relationships in Equations (2.5) and (2.6) are also similar to Equation (2.4). Equation (2.7) is also a single valued function of the weighted flow. Equation (2.9) can be shown as a simple case of Equation (2.8).

A preliminary investigation was made by Kulandaiswamy [29] using an equation of the following form for storage:

$$S = a_o q + a_1 \frac{dq}{dt} + b_o i$$ (2.10)

The study indicated the possible presence of nonlinearity in the runoff process. He concluded that it is necessary to analyse the basin behaviour on the basis of a general theory which will constitute a medium for examining whether or not the system can be treated as linear. The basin system was represented by means of a nonlinear differential equation, in which a parameter provides the method for testing the nonlinearity of the system [35,73]. This study was further extended to cover a large number of storms over basins of different areas [3,38,40].

Taking into consideration the nature of the curves repre-

senting $q(t)$ and $i(t)$ and the magnitude of the error likely to be introduced by numerical differentiation, the values of $N = 1$ and $M = 0$ have been adopted [3,29,38,40] in Equation (2.8) and the following equation is obtained:

$$S = a_0(q,i)q + a_1(q,i) \frac{dq}{dt} + b_0(q,i)i$$ (2.11)

Whether the system is linear or nonlinear will depend on whether the coefficients a_0, a_1 and b_0 are constants or functions of q and/or i.

$$\text{When } q = q_p, \frac{dq}{dt} = 0$$

For those cases where rainfall excess ends before the peak, $i = 0$. Thererfore Equation (2.8) reduces to:

$$S_p = a_o(q_p) q_p$$ (2.12)

where q_p is the peak discharge and S_p, the corresponding storage.

Using the actual storage at peak (S_p), the value of a_0 is computed in the above equation for a set of storms in a basin. The value of a_0 is found to decrease with q_p when plotted against q_p. A plot of a_0 vs. q_p is prepared for six large basins (covering areas from 41 to 301 sq km) and five small basins (covering areas from 1/2 to 850 hectares). This is subsequently treated as a plot of a_0 vs. q, since a range of storms with different magnitudes of q_p is taken for the preparation of the graph of a_0 vs. q_p. The details of the basins studies are given in Table 1. The plots of a_0 vs. q_p for North Creek River basin and for Nebraska watershed-3 which are representative of large and small basins studied are given in Figures 3a and 3b, respectively. It can be seen that a_0 approaches a constant value with the increase in discharge. This may be taken to indicate that nonlinear effects predominate in the case of minor storms and for major storms they are not so pronounced.

Plots of S_p vs. q were also made for six large basins and five small basins. One of them for each case is given in Figures 4 and 5. It can be seen from the graphs that S_p and q are linearly related in higher range and curvilinearly in the lower range. The values of a_1 and b_o are determined from the parts of the surface runoff hydrograph; a_1 is from recession limb of the hydrograph, and b_o is from the hydrograph and hyetograph up to the end of rainfall excess and these coefficients are treated as constants for the basin.

The storage equation can now be written as:

$$S = a_o(q)q + a_1 \frac{dq}{dt} + b_o i$$ (2.13)

The combination of the above storage equation with con-

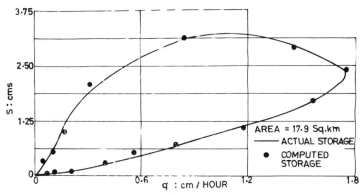

(a) WATERSHED –97, COSHOCTON, OHIO; STORM OF JUNE 28, 1957.

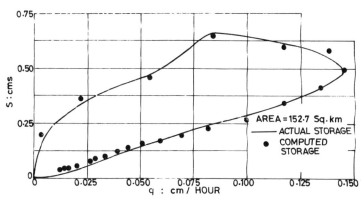

(b) FARM CREEK BASIN; STORM OF MAY 17–18, 1943.

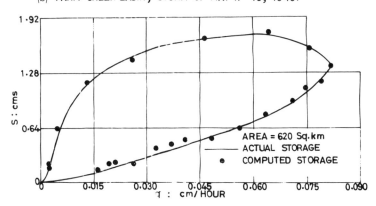

(c) WILLS CREEK BASIN; STORM OF APRIL 4–5, 1941.

FIGURE 2. Storage vs. discharge.

TABLE 1. Details of Basins.

Sl. No.	Drainage basin	Area	No. of recording rain-gauges	No. of storms	Baseflow recession constant	Source of data
1	2	3	4	5	6	7
1.	Oxford, Mississippi Watershed-WC2, North Mississippi Branch Experiment Station, Holly Springs, Mississippi	0.587 ha	1	11	—	U.S. Dept. of Agr., Washington, DC
2.	Riesel (Waco) Watershed-Y4, Texas	32.34 ha	2	7	—	U.S. Dept. of Agr., Washington, DC
3.	Riesel (Waco) Watershed-W1, Texas	71.00 ha	4	10	—	U.S. Dept. of Agr., Washington, DC
4.	Hastings, Nebraska Watershed-3	194.7 ha	6	11	—	U.S. Dept. of Agr., Washington, DC
5.	Safford, Arizona Watershed-W11	276.0 ha	2	8	—	U.S. Dept. of Agr., Washington, DC
6.	Riesel (Waco) Watershed-D, Texas	449.2 ha	4	8	—	U.S. Dept. of Agr., Washington, DC
7.	Hastings, Nebraska Watershed-8	844.2 ha	8	9	—	U.S. Dept. of Agr., Washington, DC
8.	Coshocton Watershed-97, Ohio	17.9 sq km	5	4	0.0554	U.S. Dept. of Agr., Washington, DC
9.	Beech River Basin near Lexington, Tennessee	41.2 sq km	1	33	0.0014	Tennessee Valley Authority, Knoxville, Tennessee
10.	North Creek Basin near Jacksboro, Texas	56.2 sq km	2	13	0.0785	USGS through Colorado State University
11.	West Fork, Deep River Watershed-W1, High Point, North Carolina	83.2 sq km	2	9	0.0204	U.S. Dept. of Agr., Washington, DC
12.	Farm Creek, Illinois River, East Peoria, Illinois	152.7 sq km	2	6	—	Dept. of Public Works, Illinois
13.	Vero Beach, Watershed-W1, Indian River Farms Drainage District, Indian River Country, Florida	202.0 sq km	5	21	0.0119	U.S. Dept. of Agr., Washington, DC
14.	Chestuee Creek, Basin, Dentiville	296.0 sq km	5	53	0.0015	Tennessee Valley Authority, Knoxville, Tennessee
15.	Mississippi Watershed-34, Oxford	304.0 sq km	29	7	0.0064	U.S. Dept. of Agr., Washington, DC
16.	Wills Creek Basin, Potomac River, Cumberland, MD	543.0 sq km	4	5	0.0179	U.S. Army Corps of Engrs., Washington, DC
17.	Ash Brook, United Kingdom	620.0 sq km	—	1	—	Proc. Int. Ass. Sci. Hydr. General Assembly of Toronto, No. 45, 1958. pp. 114–118.
18.	N. Br. Potomac River, near Cumberland, MD	2237.0 sq km	6	5	—	U.S. Army Corps of Engrs., Washington, DC
19.	Little Red River, White River, Heber Springs, AK	2852.5 sq km	4	6	—	U.S. Army Corps of Engrs., Washington, DC

tinuity equation [Equation (2.1)] gives:

$$a_1 \frac{d^2q}{dt^2} + \Phi(q) \frac{dq}{dt} + q = i - b_o \frac{di}{dt} \quad (2.14)$$

where

$$\Phi(q) = a_o(q) + q \frac{da_o(q)}{dq} \quad (2.15)$$

Equation (2.14) is a second order nonlinear differential equation. The degree of nonlinearity and the nature of the solution of Equation (2.14) depends upon the function $\Phi(q)$, as other coefficients a_1 and b_o are treated as constants for the basin.

The relationship of $\Phi(q)$ with q is to be determined from the analysis of the rainfall and runoff data. If the varation of $\Phi(q)$ with q is very pronounced, a strictly nonlinear system is to be considered. If the variation is not significant, $\Phi(q)$ may be approximated to a constant value, resulting in a linear differential equation. From the graph of a_o vs. q_p, (treated as a graph of a_o vs. q), the value of $\Phi(q)$ is computed for various values of q and a plot of $\Phi(q)$ vs. q is made and drawn on the corresponding plot of S_p vs. q. It can be seen that there are three regions (Figures 4 and 5):

1. $\Phi(q)$ varies linearly with $q-$ the system is nonlinear.
2. $\Phi(q)$ varies with q but in an irregular way—transition region.
3. $\Phi(q)$ is relatively independent of $q-$ the system can be treated as linear.

From the results of the analysis made, the following conclusions can be drawn [3,38,40]:

1. Existence of the nonlinearity in the runoff process could be revealed by studying the variation of outflow with the storage in the basin at peak flow.
2. Nonlinear effects are observed to be predominant in the basin system only in the case of minor storms, whatever be the size of the basin.
3. In certain basins where the nonlinear effects are observed to be mild, linear analysis of such basin systems could be possible.
4. The rainfall excess–surface runoff relationship can be represented by the differential equation, Equation (2.14).

An approximate expression for $\Phi(q)$ in Equation (2.14) will be as follows:

$$\Phi(q) = C + mq \qquad q < q_{lnr}$$
$$\qquad = C_1 \qquad q \geq q_{lnr} \qquad (2.16)$$

In the nonlinear and transitional ranges, the first equation could be used. For $q \geq q_{lnr}$ the system can be treated as linear.

For solving the differential equation [Equation (2.14)] for

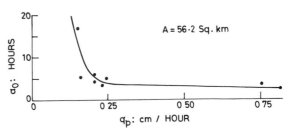

FIGURE 3a. North Creek basin: graph of a_0 vs. q_p.

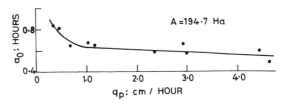

FIGURE 3b. Nebraska watershed-3: graph of a_0 vs. q_p.

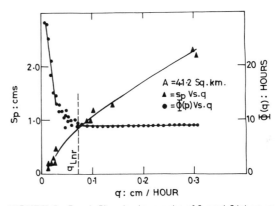

FIGURE 4. Beech River basin: graphs of S_p and $\Phi(q)$ vs. q.

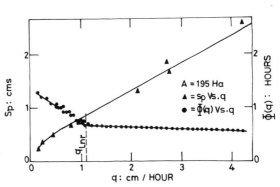

FIGURE 5. Nebraska watershed-3: graphs of S_p and $\Phi(q)$ vs. q.

the nonlinear surface runoff model, two methods can be used:

1. Runge-Kutta method
2. Regular perturbation technique, where nonlinearity is mild

The performance of nonlinear surface runoff model [3,43], i.e., Equation (2.14), is found good in a few cases but in most cases the results are not satisfactory. The unsatisfactory results may possibly be due to treating the parameters a_1 and b_0 as constants or the method of their determination. At present Equations (2.13) and (2.14) are useful only for a study of the presence and the degree of nonlinearity in the surface runoff process. Their use for surface runoff prediction can be thought of, if a satisfactory method of determining the parameters especially a_1 and b_0 could be found.

In the light of the above discussion, it is necessary to look for an approach which considers the nonlinearity in the rainfall excess–surface runoff relationship, at least indirectly. Formulation of the general hydrologic system (GHS) model, study of variation of the model parameters with the known characteristics and study of the variation of the instantaneous unithydrograph (IUH) from storm to storm and from season to season making use of the linear systems (IUH) approach were attempted [3,13,14,29] and they are presented below:

2.2 Formation of the GHS Model

Eliminating S from Equations (2.2) and (2.8) and assuming coefficients are constant for a particular storm, gives:

$$q(t) = \frac{-b_M D^{M+1} - b_{M-1} D^M - \ldots - b_o D + 1}{a_N D^{N+1} + a_{N-1} D^N + \ldots + a_o D + 1} i(t)$$

$$(2.17)$$

in which $D^M = d^M/dt^M$, and $D^N = d^N/dt^N$.

Equations (2.1) and (2.8) together or Equation (2.17) is called the general hydrologic system model, or can be abbreviated as the GHS model. Because the coefficients, a_n and b_m, are functions of i and q or both, the differential equation, Equation (2.8) or (2.17) is nonlinear, and therefore the system being modelled is nonlinear. Both i and q are functions of the independent variable of time. The coefficients also may vary with long-term seasonal and yearly changes in hydrologic phenomena, making the system time-variant. If the coefficients are treated, for approximation, as constant and independent of i and q and of long-term changes, then the model becomes linear and time-invariant and its solution will be greatly simplified. For practical purposes, the coefficients may be assumed as functions of certain characteristic values of i and q, such as the average input and output or the peak input and output of a hydrologic

event or as a parameter having a constant value for a given hydrologic event. Thus, for a particular hydrologic event in a particular period of time, the coefficients may be treated as constants, and the model in this case is linear and time-invariant. It can readily be solved for that event. After the model is solved linearly and as time-invariant for a number of events, the coefficients may be expressed as functions of the characteristic values of i and q in magnitude (or at different times of occurrence for the events). Alternately, the coefficients may be expressed as functions of a parameter which may have a definite value for each storm. In this manner, the application of the GHS model may be extended to a nonlinear and time-variant hydrologic system.

Analysis of the data listed in Table 1 showed that the values of derivatives of order higher than two are comparatively small and it is likely that the errors involved in obtaining higher derivatives may offset the refinement achieved by considering them. It may therefore be possible to drop one or more terms in Equation (2.8) without appreciable loss of accuracy. From the investigation made by Kulandaiswamy [17,29], the choice of the storage equation is presented in the following form:

$$S = a_o q + a_1 \frac{dq}{dt} + a_2 \frac{d^2 q}{dt^2} + b_o i + b_i \frac{di}{dt} \quad (2.18a)$$

$$S = a_o q + a_1 \frac{dq}{dt} + b_o i + b_1 \frac{di}{dt} \quad (2.18b)$$

$$S = a_o q + a_1 \frac{dq}{dt} + a_2 \frac{d^2 q}{dt^2} \quad (2.18c)$$

$$S = a_o q + a_1 \frac{dq}{dt} + b_o i \quad (2.18d)$$

$$S = a_o q + a_1 \frac{dq}{dt} \quad (2.18e)$$

The coefficients in the above equations are determined from the known rainfall excess and surface runoff data, by multiple linear regression. By a comparison between the observed storage and the computed storage values, obtained from the above storage equations, it was found that in all cases the order of preference for the storage equations for accuracy of results is exactly as presented earlier by Kulandaiswamy [29]. The computed storage using Equation (2.18a) is presented in Figure 2. The following two simplified cases [3,14,17,29] of the general storage equation are used for analysis:

Case 1. Five coefficient storage equation

$$S = a_o q + a_1 \frac{dq}{dt} + a_2 \frac{d^2 q}{dt^2} + b_o i + b_i \frac{di}{dt} \quad (2.18a)$$

Case 2. Three coefficient storage equation

$$S = a_o q + a_1 \frac{dq}{dt} + b_o i \qquad (2.18d)$$

Substituting the storage equations (Equations (2.18a) and (2.18d)) in the continuity equation (Equation (2.1)), or by taking $M = 1$, $N = 2$ and $M = 0$, $N = 1$ in Equation (2.17), the following models for surface runoff are obtained:

Case 1. Five coefficient surface runoff model

$$q(t) = \frac{-b_1 D^2 - b_o D + 1}{a_2 D^3 + a_1 D^2 + a_o D + 1} i(t) \qquad (2.19)$$

Case 2. Three coefficient surface runoff model

$$q(t) = \frac{-b_o D + 1}{a_1 D^2 + a_o D + 1} i(t) \qquad (2.20)$$

where D is the differential operator d/dt.

Recently Singh and McCann [54,72] made a mathematical analysis of the GHS model. By considering mathematical properties of the delta function, the GHS model was simplified. In their study, it amounts to making $b_o = b_1 = 0$ in Equation (2.19) and $b_o = 0$ in Equation (2.20) resulting in the following equation.

$$q(t) = \frac{1}{a_2 D^3 + a_1 D^2 + a_o D + 1} i(t) \qquad (2.21)$$

and

$$q(t) = \frac{1}{a_1 D^2 + a_o D + 1} i(t) \qquad (2.22)$$

which indicate that the IUH will be independent of b_m. By this, the inconsistencies in the initial stage like $u(0)$, a positive value and a few negative values in the initial period of IUH, which might appear in some cases of storms, are removed straight away. In order to determine the coefficients in the GHS model, the method of moments and the method of cumulants are proposed in their study. From the results of their study, it is observed that there is a loss in the reproduction of surface runoff hydrograph indicating a consistently lower surface runoff peak for 5 hydrographs studied by them. Further, the method of moments yielded too low values of the flood peaks.

The possibility that an IUH is characterized by a positive instantaneous response at $t = 0$ and negative ordinates at small t is quite realistic. In the hydrologic analysis, the effective rainfall hyetograph is expressed by a bar diagram or histogram as the rainfall record is commonly reported at time intervals. In the hydrologic cycle, the effective rainfall seems to begin somewhat suddenly rather than gradually

once the potential infiltration rate and the depression and retention storages of the ground surface are satisfied. Because of these realistic reasons, the hydrologic system may give a positive instantaneous response to a positive impulse, thus $u(0) \geq 0$. This initial impulse may soon be absorbed by the system, resulting in an immediate decrease in its response or even to a negative value slightly afterwards before the system recovers to a positive state.

It must be recognised that the IUH is only a theoretical concept and it cannot be real in nature. Although it may exhibit negative ordinates and oscillatory graphs, it will produce a normal positive direct runoff hydrograph after it convolutes with the effective rainfall hyetograph which was used in computing the IUH. In a paper by Boneh and Golan [6] titled "Instantaneous Unit Hydrograph with Negative Ordinates – Possible?", it is postulated as follows:

> The answer to the question posed in the title of this paper is yes, provided that some restrictions are imposed on the rainfall excess which is the input to the surface runoff system.

However, the restrictions are purely theoretical and cannot be imposed on the rainfall excess in the realistic situation. Thus, the negative ordinates may arise because of unavoidable errors in data and inexactness, and the lacking of reality in the theory of analyais. After a lengthy analysis of IUH, Boneh and Golan [6] concluded:

> The concept of an instantaneous unit hydrograph with negative ordinates (IUHWNO) was considered in this paper. It may be concluded that such an IUH is often capable of representing a real-world system.

The IUHs of 5-coefficient and 3-coefficient GHS models [Equations (2.19) and (2.20)] have been derived and their comparison has been examined [3,14,29,46,47]. The instantaneous unithydrographs derived using Equation (2.19) for some of the basins listed in Table 1 are reproduced in Figures 6a to 6c. The 5-coefficient model may produce negative ordinates of the IUH, but this is often capable of representing a real world system and it was found to reproduce the surface runoff hydrograph very well with positive ordinates (Figures 7a to 7c). In Figures 7a and 7b, the surface runoff hydrographs reproduced using the 3-coefficient GHS model are also presented.

Following the work on the GHS model by Kulandaiswamy [29] and the publication of the paper on the GHS model by Chow and Kulandaiswamy [14], the model has been applied to a large number of basins by investigators [3,10–13,22, 24–26,28,33–35,37–44,46–49,54,61,64,67–69,72,73,75,76] in this field. The following additional refinements made on the IUH from the 5-coefficient model [Equation (2.19)] yield extremely satisfactory results [13,17]:

1. $u(0)$ is set to zero, i.e., $u(0) = 0$.
2. If $u(t) < 0$, set those values of $u(t)$ to zero.

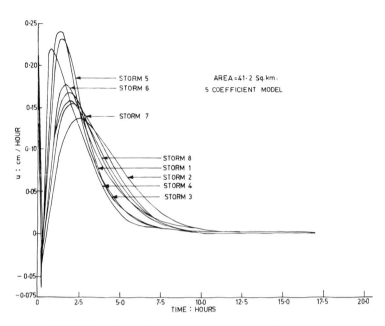

FIGURE 6a. Beech River basin: instantaneous unithydrographs.

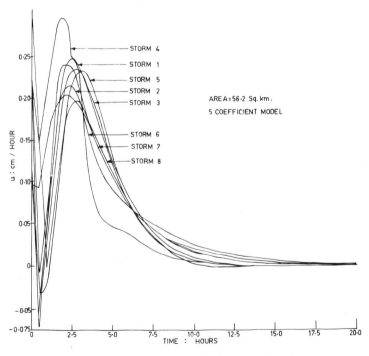

FIGURE 6b. North Creek basin: instantaneous unithydrographs.

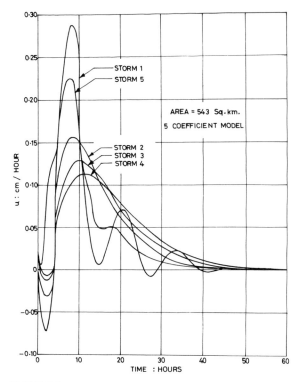

FIGURE 6c. Willscreek basin: instantaneous unithydrographs.

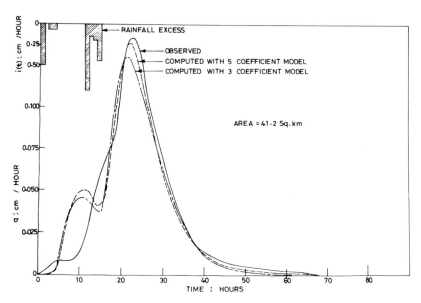

FIGURE 7a. Beech River basin: storm of March 23, 1953.

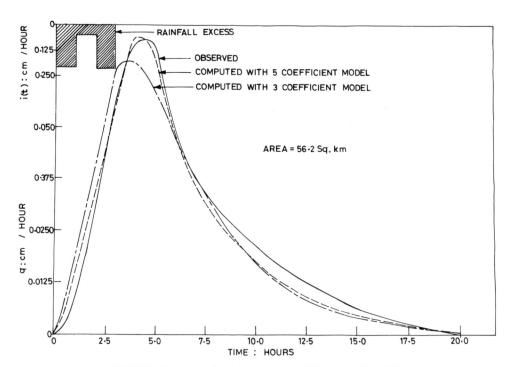

FIGURE 7b. North Creek basin: storm of November 26, 1962.

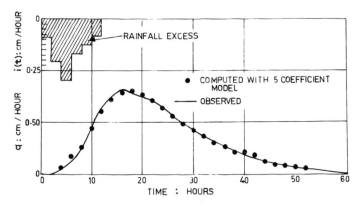

FIGURE 7c. Willscreek basin: storm of April 4–5, 1941.

3. After making the changes in 1 and 2, adjust if necessary the $u(t)$ ordinates so that the resulting $u(t)$ will satisfy the following condition:

$$\int_0^\infty u(t)\, dt = 1 \qquad (2.23)$$

Conditions 1 and 2 aforementioned are not inevitable requirements as far as the definition of IUH is concerned. It is more to conform to the general concept of unithydrograph, since IUH is thought of as a limiting case of the unithydrograph though this is not exactly the case.

If for any reason non-zero values for $u(0)$ and negative values for $u(t)$ are to be avoided, it would be better to use these refinements rather than eliminating the coefficients b_0 and b_1, which are found relevant from the studies on storage. If the IUH ordinates should meet certain requirements at values of time, close to zero, it is in this case reasonable to deal with it as a local problem. It is inadvisable to impose alteration that would affect adversely the overall performance of the model.

Because the three-coefficient model contains less parameters and produces IUH of positive oridinates, it is slightly easier to use. This advantage becomes insignificant since the compilation is now often made by electronic computer. The five coefficient model is therefore superior to the three coefficient model, more realistic to the real world system and from a practical viewpoint, it fits the data better.

The parameters a_0, a_1, a_2, b_0 and b_1 in Equation (2.19) are treated as constants for a given storm and the differential equation [Equation (2.19)] is solved for a given input rainfall excess (Figures 7a to 7c). Similarly, the parameters a_0, a_1 and b_0 in Equation (2.20) are treated as constants and the differential equation [Equation (2.20)] is solved for a given input rainfall excess (Figures 7a and 7b).

2.3 Study of Model Parameters

It is seen that the parameters in the surface runoff model vary from storm to storm and these variations are quite considerable. This aspect has been investigated by Kulandaiswamy [29]. He has shown that the parameters a_0, a_1 and a_2 in Equation (2.19) are functions of q_p. Since b_0 and b_1 did not vary with any dependable trend with q_p, he treated them as constants for a given basin. For a new storm for which the surface runoff is to be predicted, a trial procedure is involved for obtaining the values of a_0, a_1 and a_2. A plot of q_p vs. R_{ev}, the rainfall excess volume is made use of, in obtaining the first approximate value of q_p. Thus Kulandaiswamy [29] indirectly considered the variation of parameters a_0, a_1 and a_2 with the volume of rainfall excess R_{ev}. Later [3,46,47] the parameters a_0, a_1, a_2, b_0 and b_1 were treated as functions of the storm characteristics, R_{ev}, t_r and t_e as follows:

$$a_0 = f_1(R_{ev}, t_r, t_e)$$

$$a_1 = f_2(R_{ev}, t_r, t_e)$$

$$a_2 = f_3(R_{ev}, t_r, t_e) \qquad (2.24)$$

$$b_o = f_4(R_{ev}, t_r, t_e)$$

$$b_1 = f_5(R_{ev}, t_r, t_e)$$

The variables R_{ev}, t_r and t_e are combined into a single paramater η [3,46,47]:

$$\eta = R_{ev}\frac{t_r}{t_e} \qquad (2.25)$$

and the coefficients in both the models [Equations (2.19) and (2.20)] are expressed as functions of η, (Figures 8a to 8c). The correlation coefficient, R, between the computed and the observed hydrographs varies between 0.9017 and 0.9968 for 97 storms studied in 12 basins and it is above 0.99 for 222 storms in two watersheds of Bhavani basin in India. A study of the multiple correlation relationships among the parameters a_0, R_{ev} and t_r/t_e, a_1, R_{ev} and t_r/t_e and b_0, R_{ev} and t_r/t_e has also been made and they are expressed in non-dimensional form as follows (Figure 9) [48]:

$$a_{0n} = f(R_{en}, \tau_{gr})$$

$$a_{1n} = f(R_{en}, \tau_{gr}) \qquad (2.26)$$

$$b_{0n} = f(R_{en}, \tau_{gr})$$

where $a_{0n} = a_0/a_{0min}$, $a_{1n} = a_1/a_{1min}$, $b_{0n} = b_0/b_{0max}$, $R_{en} = R_{ev}/R_{evmax}$ and $\tau_{gr} = t_r/t_e$. For Nebraska watershed, from these plots of a_0, a_1 and b_0 vs. R_{ev}. t_r/t_e, a_{0min}, a_{1min}, b_{0max} and R_{emax} are found as 0.588 hours 0.04 hours², 0.08 hours and 5 cm, respectively. Using the graphs in Figure 9, the runoff for a storm is predicted (Figure 10).

Also a regression analysis has been made for three sub-basins of Bhavani basin among the coefficients and the known input characteristics using 119 storms of 5 years (1933 to 1938) as below [22,68]:

$$a_0 = f(X_1, X_2, X_3, X_4, X_5)$$

$$a_1 = f(X_1, X_2, X_3, X_4, X_5) \qquad (2.27)$$

$$b_0 = f(X_1, X_2, X_3, X_4, X_5)$$

where

X_1 = rainfall excess volume, R_{ev}
X_2 = time distribution of rainfall excess, t_r/t_e
X_3 = η as in Equation (2.25)
X_4 = API at the start of the storm
X_5 = ratio of the mean rainfall to the maximum rainfall

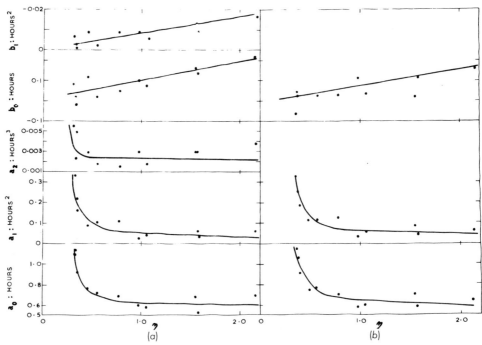

FIGURE 8a. Nebraska watershed-3: (a) graphs of a_0, a_1, a_2, b_0, b_1, vs. η; (b) graphs of a_0, a_1, b_0 vs. η.

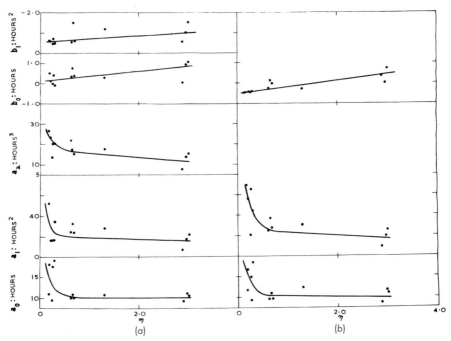

FIGURE 8b. Beech River basin: (a) graphs of a_0, a_1, a_2, b_0, b_1, vs. η; (b) graphs of a_0, a_1, b_0 vs. η.

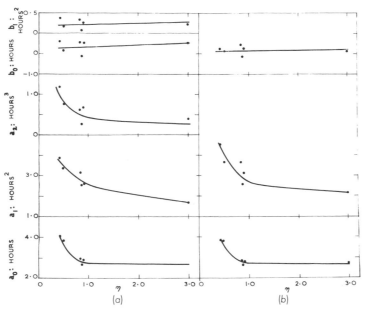

FIGURE 8c. Farm Creek basin: (a) graphs of a_0, a_1, a_2, b_0, b_1, vs. η; (b) graphs of a_0, a_1, b_0 vs. η.

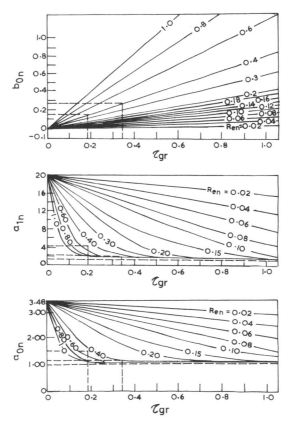

FIGURE 9. Nebraska watershed-3: multiple correlations among coefficients a_{0n}, a_{1n} and b_{0n} vs. τ_{gr} and R_{en}.

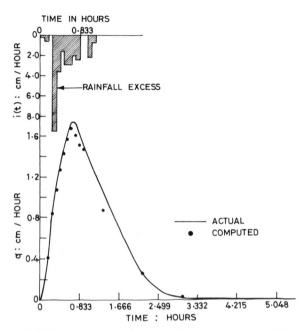

FIGURE 10. Nebraska watershed-3: storm of June 7, 1953.

For Mukurthy sub-basin, the regression equations using 29 storms were obtained as follows.

For monsoon period:

$$a_o = 3.39 - 0.00024X_1 - 1.57X_2$$
$$\quad - 0.01X_3 - 0.04X_4 - 0.47X_5$$
$$a_1 = 3.02 - 0.0058X_1 - 1.31X_2$$
$$\quad - 0.0026X_3 - 0.05X_4 - 1.80X_5 \qquad (2.28a)$$
$$b_o = 0.20 - 0.004X_1 - 0.95X_2$$
$$\quad + 0.02X_3 - 0.01X_4 + 1.23X_5$$

For non-monsoon period:

$$a_o = 1.17 + 5.87X_1 + 2.53X_2$$
$$\quad - 9.52X_3 + 0.05X_4 - 1.41X_5$$
$$a_1 = 0.11 - 0.78X_1 - 1.95X_2$$
$$\quad + 8.14X_3 + 0.84X_4 + 0.99X_5 \qquad (2.28b)$$
$$b_o = -0.70 - 0.67X_1 - 0.29X_2$$
$$\quad + 0.23X_3 + 0.38X_4 + 1.32X_5$$

Using the above equations for the parameters in the 3-coefficient model [Equation (2.20)], the correlation coefficient (R) between computed and the observed surface runoff hydrographs is about 0.94.

The model parameters in the case of varied land uses are studied separately (Section 3).

3. APPLICATION OF THE GHS MODEL FOR SOME HYDROLOGIC PROBLEMS

3.1 General

The evolution of the GHS model was discussed in the previous section and it was found to represent the rainfall excess-surface runoff relationship very satisfactorily. Its applicability to a variety of hydrologic problems like infiltration in drainage basins, rainfall-runoff relationship, flood routing, interbasin flow estimation in large drainage basins, land use effect on streamflows, was investigated [22,28,33, 34,37,41,46,47,67–69,75,76]. A brief account of these investigations is presented in this section.

3.2 Infiltration in Drainage Basins

Hydrologists are often required to make a reasonably accurate estimate of rainfall excess for a given storm, for making forecast of floods, for the estimation of total yield from a basin and for computing design discharge for projects. All studies involving the unithydrograph concept [70] require the determination of effective hyetograph after making allowances for all abstractions and it is the first step to estimate precipitation losses or abstractions. An accurate determination of various abstractions is rather difficult and for practical purposes, mostly abstraction indices are used. They have come to be designated as infiltration indexes, even though infiltration is not the only form of abstraction accounted for in the index approach. The most widely used forms of abstraction rate are Φ-index and f_{av}-index [3,45,47] and if carefully estimated, they may be satisfactory for most purposes. In the following, a method is presented for the estimation of average abstraction rate, Φ-index.

PREDICTION OF Φ-INDEX

A procedure for estimating the total abstractions, F_v is given below. The Φ-index is to be arrived at from abstractions F_v by a trial and error procedure. From the method of investigation proposed by Kulandaiswamy and Babu Rao [37,45], the abstractions, F_v can be expressed as a function of initial conditions and storm characteristics as follows (Figures 11a and 11b):

$$F_v = f_3(R_v, T, T_g, I_{ap}) \qquad (3.1)$$

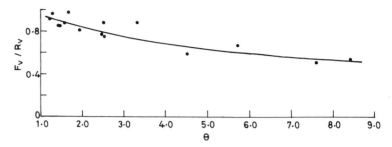

FIGURE 11a. Beech River basin: graph of F_v/R_v vs. Θ.

where R_v is the total volume of rainfall, T, the duration of rainfall, T_g, the time to the centre of area of the rainfall from its start and

$$I_{ap} = \frac{P_1}{1} + \frac{P_2}{2} + \frac{P_3}{3} + \cdots + \frac{P_n}{n} \qquad (3.2)$$

where P_n is the rainfall during 24 hour period, recorded on the nth day preceding the storm. The value of n varies from 10 to 30. Through the process of dimensional analysis and treating R_v/T as average rainfall rate, r_a, the variables can be grouped as follows:

$$F_v/R_v = F_4(T_g/T, I_{ap}/r_a) \qquad (3.3)$$

In Equation (3.3), the parameter T_g/T represents the storm pattern, while I_{ap}/r_a gives a measure of the initial condition of the precipitation going as abstraction. If the total abstractions F_v is known, Φ-index can be computed by a trial and error procedure with the condition that F_v equals the area below Φ-index.

In order to study the relationship between F_v/R_v and the parameters T_g/T and I_{ap}/r_a, 126 storms were chosen from 6 basins (Basin Nos. 9, 10, 11, 13, 14 and 15 in Table 1). The parameters I_{ap}/r_a and T_g/T were combined into a single parameter θ:

$$\theta = \frac{I_{ap} T_g}{R_v} \qquad (3.4)$$

The plots of F_v/R_v vs. θ are presented for two basins in Figures 11a and 11b. The parameter F_v/R_v correlated fairly well with I_{ap}/r_a and T_g/T for all the basins studied and hence their random variation is ignored. The method for prediction of Φ-index yields satisfactory results and is recommended for the prediction of average abstraction rate.

DETERMINATION OF ABSTRACTION RATE

It is known from a detailed discussion [3,33], that an improved relationship between detention-storage and discharge, that is applicable for the entire hydrograph, will improve the accuracy of the abstraction rate curve obtained through the detention flow relationship method. A detailed discussion of the merits of the storage equations used in infiltration studies was made and a review of these equations is given in References [3,17,33,37]. Two methods have been proposed for obtaining the abstraction rates during a storm over a basin:

1. By using 5-coefficient storage equation [Equation (2.18a)] and 5-coefficient surface runoff model [Equation (2.19)]

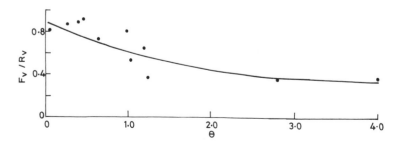

FIGURE 11b. North Creek basin: graph of F_v/R_v vs. Θ.

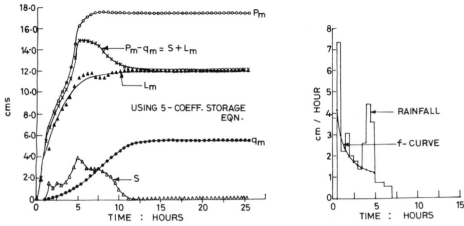

FIGURE 12. North Creek basin: storm of October 3, 1959.

2. By using 3-coefficient storage equation [Equation (2.18d)] and 3-coefficient surface runoff model [Equation 2.20)]

These methods are verified using a large number of storms in a number of basins. In the following, only the first procedure for obtaining abstraction rate curve using 5-coefficient storage equation and 5-coefficient surface runoff model is described. The use of the second method is similar in principle and in details.

METHOD USING 5-COEFFICIENT MODEL

For a storm where runoff is to be predicted, only the time distribution of precipitation, the mass precipitation and the rainfall excess predicted using the infiltration index described above are known. From the past rainfall and runoff records from the basin under study, the curves relating the coefficients a_0, a_1, a_2, b_0 and b_1 in Equation (2.24) to a parameter η [Equation (2.25)] describing rainfall characteristics can be prepared as per procedure indicated in Section 2.3. Having obtained the foregoing information, the abstraction rate for a particular storm over the basin can be determined using 5-coefficient storage equation [Equation (2.18a)] and 5-coefficient surface runoff model [Equation (2.19)] as follows [3,47].

Step 1. Mass curve of precipitation P_m for the storm is plotted. If q_m is the mass curve of surface runoff, F_m is the infiltration mass curve, V_d the depression storage, S the detention-storage and increment in channel storage due to surface runoff and E_m is the mass curve of interception and evaporation losses, then

$$P_m - q_m = S + F_m + E_m + V_d \qquad (3.5a)$$

or

$$P_m - q_m = S + L_m \qquad (3.5b)$$

where L_m represents the mass curve of losses or abstractions given by:

$$L_m = F_m + E_m + V_d \qquad (3.6)$$

Step 2. The parameter η [Equation (2.25)] is computed from the rainfall excess diagram as the values of volume of rainfall excess R_{ev}, the time to the centre of area of the rainfall excess t_r, and the duration of rainfall excess t_e are known.

Step 3. The values of coefficients a_0, a_1, a_2, b_0 and b_1 are read from the tables: a_0 vs. η, a_1 vs. η, a_2 vs. η, b_0 vs. η and b_1 vs. η.

Step 4. As the coefficients a_0, a_1, a_2, b_0 and b_1 and the input rainfall excess rate $i(t)$ are known, the time distribution of surface runoff $q(t)$ is obtained using the 5-coefficient surface runoff model [Equation (2.19)].

Step 5. As the surface runoff is known in step 4 the mass curve of surface runoff q_m can also be prepared. At any point of time, the values P_m and q_m are known. If S is known, the L_m curve can be plotted. In cases where L_m curve is not monotonically increasing with time, the following assumptions are made: The final value of L_m, i.e., the value of L_m corresponding to the end of surface runoff, is assumed as the maximum value and when this final value is exceeded in the range lying between the beginning and the end of surface runoff, such exceeded values are brought down to a final value. An average L_m-curve is drawn. If from the L_m-curve, losses due to depression storage, evaporation and interception storage are subtracted, the F_m-curve is obtained. If it can be assumed that losses due to evaporation and interception during the storm are negligible, the total loss rate or abstraction rate can be approximated to infiltration rate. Then the slope of L_m-curve gives the infiltration rate. Since V_d can be assumed constant, its inclusion in L_m does not affect its slope.

Step 6. As the coefficients a_0, a_1, a_2, b_0 and b_1 and the

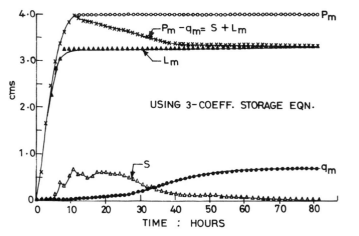
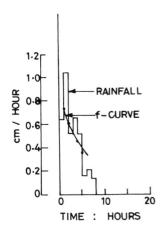

FIGURE 13. Beech River basin: storm of November 22, 1961.

rainfall excess rate $i(t)$ are known, the value of $S(t)$ is computed using the 5-coefficient storage equation [Equation (2.18a)].

Step 7. The abstraction rate curve is obtained by considering the slope of L_m-curve. If the evaporation and interception losses are negligible during the period of storm, the abstraction rate can be approximated to infiltration rate.

In order to verify the method explained in the foregoing, 96 storms in 9 basins (Basin Nos. 4, 8, 9, 10, 11, 14, 15, 16 and 19 in Table 1) were analysed. A plot for a storm in a basin is presented in Figure 12.

It can be seen that the abstraction rate curves which are treated as infiltration rate curves very closely resemble those obtained from experimental studies [23,60]. The results are consistent and for each storm the points plotted yield a well defined infiltration rate curve. Since the storage equation used in this method yields farily satisfactory results, in estimating the storage corresponding to the rising as well as falling stage of the surface runoff hydrograph, the method is relatively accurate.

METHOD USING 3-COEFFICIENT MODEL

Similar study is also made using 3-coefficient storage equation [Equation (2.18d)] and 3-coefficient surface runoff model [Equation (2.20)] for infiltration rate curve and the plots mentioned in steps 1 to 7 above were obtained for the same 96 storms. A plot for a storm in a basin is presented in Figure 13.

3.3 Baseflow

The falling limb of a streamflow hydrograph during a period of drought was a depletion curve, having the characteristic exponential decay equation [16]:

$$q_b = q_i e^{-Kt} \qquad (3.7)$$

where q_b is the baseflow at time t, q_i is the initial baseflow, K is the recession constant and e is the Napierian base. The following model based on Equation (3.7) was used for the baseflow hydrograph during the existence of surface runoff and thereafter [3,47]:

$$
\begin{aligned}
q_b &= q_i \cdot e^{-Kt} & O < t &\le t_p \\
&= q_{bp} & t_p < t &\le t_s \qquad (3.8) \\
&= q_{bs} \cdot e^{-Kt} & t &> t_p
\end{aligned}
$$

where q_{bp} is the baseflow at time t_p, q_{bs} is the baseflow at t_s, the end of surface runoff.

3.4 Rainfall–Runoff Relationship

The investigation of rainfall–runoff relationship consists mainly of three components:

1. Estimation of abstractions and consequent rainfall excess
2. Conversion of rainfall excess into surface runoff
3. Determination of baseflow contribution

The combination of surface runoff with baseflow gives runoff. For a given time distribution of rainfall and antecedent conditions of the drainage basin, the GHS model is used to convert rainfall into abstractions and runoff. A brief description of the model with a block diagram representing the basin system is given below.

BLOCK DIAGRAM REPRESENTATION OF RAINFALL–RUNOFF RELATIONSHIP

A block diagram representing the basin system [3,29,47,79] that converts rainfall into abstractions and runoff is given in Figure 14. The operators Φ_1' and Φ_1'', Φ_2 and Φ_3 in the figure represent the operations performed by

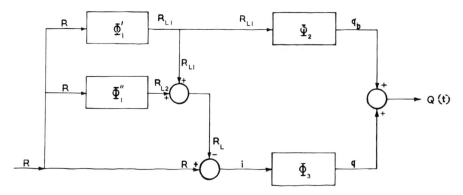

FIGURE 14. Block diagram representation of a basin system.

the elements to convert rainfall rate into: abstractions (R_L), baseflow (q_b) and surface runoff (q). The surface runoff is added to baseflow to get hydrograph of runoff (Q). The following equations can be readily written for the model shown in Figure 14:

$$R_L = R_{L1} + R_{L2} \tag{3.9}$$

$$R_L = \Phi_1' R + \Phi_1'' R \tag{3.10}$$

$$i = R - R_L \tag{3.11}$$

$$i = R - \Phi_1' R - \Phi_1'' R \tag{3.12}$$

$$Q(t) = \Phi_3(R - \Phi_1' R - \Phi_1'' R) + \Phi_2(\Phi_1' R) \tag{3.13}$$

In the above equations, R_{L1} is the infiltration abstraction, Φ_1' is the operator that converts rainfall into infiltration abstraction and Φ_1'' is the operator that converts rainfall into evaporation abstraction. The depression storage ultimately is depleted partly by infiltration and partly by evaporation. The intercepted water ultimately evaporates and gets depleted from the intercepting surfaces. Hence, for all practical purposes, it would be adequate if losses due to infiltration and evaporation are considered. Among the abstractions again, infiltration constitutes the major component during the period of precipitation, and evaporation during the period of precipitation can be treated as negligible. Hence, the estimation of total abstractions during precipitation, by and large, amounts to the determination of infiltration abstraction. Now Equation (3.13) can be written in the following form:

$$Q(t) = \Phi_3(R - \Phi_1' R) + \Phi_2(\Phi_1' R) \tag{3.14}$$

where $\Phi_1' R$ represents the infiltration rate, which includes all other abstractions also. In Equation (3.14), the first term $\Phi_3(R - \Phi_1' R)$ indicates the surface runoff and the second

term $\Phi_2(\Phi_1' R)$ indicates the baseflow. The combination of surface runoff with baseflow gives the total runoff $Q(t)$ and Equation (3.14) can also be written as follows:

$$Q(t) = q + q_b \tag{3.15}$$

where q is the surface runoff and q_b is the baseflow, at any time t.

The model described in the foregoing consists of three components or sub-systems, each of which has been studied in detail in Sections 2.2, 3.2 and 3.3. These subsystems can now be combined and an integrated model for basin runoff can be arrived at. The integrated model will now consist of:

1. A model for the determination of the average infiltration rate, Φ-index and the infiltration rate curve using a 5-coefficient storage equation
2. A 5-coefficient model for the determination of surface runoff from the rainfall excess
3. A baseflow model

The successive steps involved in the use of the model are discussed in the following:

RUNOFF MODEL WITH 5-COEFFICIENT SYSTEM FUNCTION

Input Data Needed
The following information about the basin must be available for predicting the runoff using 5-coefficient runoff model for a given storm:

1. A graph of F_v/R_v vs. θ
2. Graphs connecting a_0, a_1, a_2, b_0 and b_1 with η
3. Values of the delay time, τ and the baseflow recession constant, K

The value of τ can be determined by analysing the available rainfall excess–surface runoff data. The values of K can be computed by studying the baseflow characteristic over a period of time. Having obtained the foregoing information, the runoff for any known storm real or hypothetical can be predicted as follows (Figure 15).

Computational Procedure

Step 1.

- For the given storm, the values of T_g and R_v are computed. Antecedent precipitation index, I_{ap}, is computed using Equation (3.2). From these values, Θ is computed using Equation (3.4).
- The value of F_v/R_v is read from the plot of F_v/R_v vs. Θ and F_v is computed as R_v is known.
- The value of average abstraction rate Φ-index is computed by a trial procedure (Section 3.2).

Step 2. Abstraction rate curve is determined, using the method based on 5-coefficient storage equation, proposed in Section 3.2.

Step 3. Rainfall excess volume R_{ev} is determined subtracting abstractions from rainfall rate. The ratio t_r/t_e is calculated from rainfall excess diagram and the parameter η is calculated from Equation (2.25).

Step 4. The values of coefficients a_0, a_1, a_2, b_0, b_1 are read from the graphs or tables of a_0 vs. η, a_1 vs. η, a_2 vs. η, b_0 vs. η and b_1 vs. η.

Step 5. The three roots of the characteristic polynomial in Equation (2.19) are evaluated and the instantaneous unithydrograph is obtained using the appropriate value of the root.

Step 6. A delay time of τ hours is used, if necessary in certain basins and surface runoff is computed using Equation (2.19).

Step 7. Baseflow is computed using Equation (3.8).

Step 8. The total runoff, $Q(t)$, is obtained by adding surface runoff to baseflow [Equation (3.15)].

RUNOFF MODEL WITH 3-COEFFICIENT SYSTEM FUNCTION

The model proposed above, with the 5-coefficient system function, will involve a certain amount of tedious computations. These can be done with ease if a digital computer is available. Where calculations are to be made with the help of a calculator, a simpler model will be more convenient. Investigations have shown that a simpler runoff model with a 3-coefficient system function is quite satisfactory (Figure 16). The procedure for the use of this simpler form of the model is the same as in the case of determination with 5-coefficient model.

For the investigation of the 5-coefficient and 3-coefficient runoff models for rainfall–runoff relationship 96 storms in 9 basins (Basin Nos. 4, 8, 9, 10, 11, 14, 15, 16 and 19 in Table 1) were used and based on the study the following recommendations were made: The runoff predictions made by using 5-coefficient runoff model and 3-coefficient runoff model are both satisfactory. But the runoff predictions made by using 5-coefficient runoff model are more satisfactory when compared with the results obtained by using the 3-coefficient runoff model.

As the best results are obtained using the 5-coefficient runoff model, this model is recommended for practical use where a computer facility is available. The 3-coefficient runoff model is simple and hand computations can be made quickly. The results obtained by using this model are also

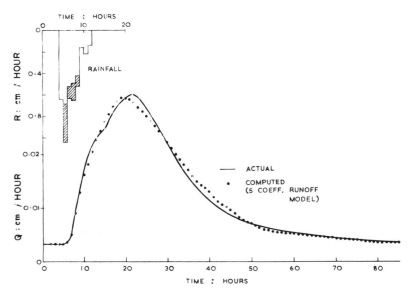

FIGURE 15. Beech River basin; storm of November 22, 1961.

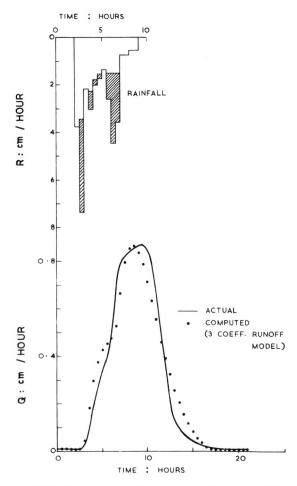

FIGURE 16. North Creek basin; storm of October 3, 1959.

very good. Therefore, the 3-coefficient runoff model is recommended where a computer is not available, and the computations have to be made by using a calculator.

3.5 Flood Routing Through Channels

In the case of rivers whose stages rise above the normal channel bank, damages occur and protection measures are needed to regulate the floods and to eliminate or minimize losses. These measures are in the form of flood control reservoirs, levees, flood walls and channel improvement. The determination of the river stage at a given location requires the prediction of the flow rate, based on known inflow at specified upstream point or points.

There are a number of methods that are in vogue for flood routing through channels. Among them, methods of hydrologic routing based on storage equations are widely used in field practice.

The procedure is based on the establishment of a relationship between stage and storage or discharge and storage. Often, this relationship is obtained on the basis of an analysis of past floods in a channel reach and an assumption is made that the relationship so established for a known flood is valid for future floods in the reach. It may be relevant to observe at this point that the above assumption is not justifiable. It is a fairly severe approximation. In this section a method based on storage discharge relationship, known as "General sotrage equation method," is presented and used to carry out a nonlinear analysis of channel routing.

GENERAL STORAGE EQUATION METHOD

The simpler version of the general storage equation [Equation (2.18d)] has been used to develop a method of flood routing. The continuity equation [Equation (2.1)] and the storage equation [Equation (2.18d)] are rewritten for the channel routing as below:

$$I - Q = \frac{dS}{dt} \tag{3.16}$$

$$S = a_0 Q + a_1 \frac{dQ}{dt} + b_0 I \tag{3.17}$$

where Q is the outflow from the channel and I is the inflow into the channel and S is the storage in the channel. Using these relationships and finite difference technique, Kulandaiswamy et al. [34] developed the following inflow–outflow relationship for a channel reach:

$$Q_j = A\,Q_{j-1} + BQ_{j-2} + CI_j + DI_{j-1} \tag{3.18}$$

where

$$A = \frac{2a_1 + a_0\Delta t - \tfrac{1}{2}(\Delta t)^2}{a_1 + a_0\Delta t + \tfrac{1}{2}(\Delta t)^2} \tag{3.19a}$$

$$B = \frac{-a_1}{a_1 + a_0\Delta t + \tfrac{1}{2}(\Delta t)^2} \tag{3.19b}$$

$$C = \frac{-b_0\Delta t + \tfrac{1}{2}(\Delta t)^2}{a_1 + a_0\Delta t + \tfrac{1}{2}(\Delta t)^2} \tag{3.19c}$$

$$D = \frac{b_0\Delta t + \tfrac{1}{2}(\Delta t)^2}{a_1 + a_0\Delta t + \tfrac{1}{2}(\Delta t)^2} \tag{3.19d}$$

For a given inflow and outflow, the storage equation can be written at intervals of Δt. To solve for the three unknowns, namely a_0, a_1 and b_0, the method of least squares can be used. With the coefficients known, the values of A, B, C and D can be computed from Equations (3.19a), (3.19b), (3.19c), and (3.19d). For a given inflow into the

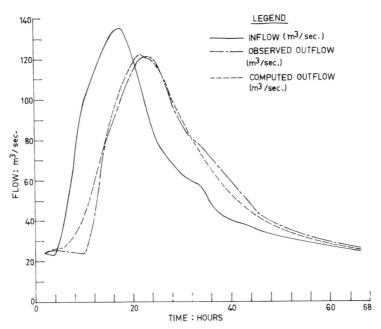

FIGURE 17. Toboso Dillon Reach, Licking River, Ohio: channel routing.

channel, the outflow can be computed, using the relationship in Equation (3.18).

The method explained in the foregoing has been applied to 3 flood hydrographs [3,28,64]. A plot of observed and computed flood hydrograph using the simpler version of the general storage equation is presented in Figure 17. It is found to reproduce the observed hydrograph fairly well.

NONLINEAR CHANNEL ROUTING

If the coefficients in the storate equation [Equation (3.17)] are constants, the equation is linear; on the other hand if they are functions of Q and/or I, then the equation becomes nonlinear. The most dominant parameter in the storage equation is a_0. It is treated as a function of the outflow from the channel, Q. The coefficients a_1 and b_0 are treated as constants for the channel. Under these conditions for the nonlinear analysis of the channel, the following equation is written [74]:

$$S = a_0 (Q) Q + a_1 \frac{dQ}{dt} + b_0 I \qquad (3.20)$$

From the given inflow and outflow data for the channel, the mass balance is maintained and the parameters a_0, a_1 and b_0 are determined by least squares and a relation between a_0 and Q_p, the observed peak outflow, of the following form is fitted:

$$a_0 = a + bQ_p + cQ_p^2 \qquad (3.21)$$

Average values of a_1 and b_0 are considered for the reaches. The relation between a_0 and Q_p is treated as the relation for a_0 and Q. The relation between a_0 and Q was studied for channel reaches Shelbyville to Columbia, Columbia to Centerville, and Centerville to Hurricane Mills in Duck River (Figure 18). A total of 12 events have been studied. It was found that a_0 can be expressed as a function of Q for the channel reach, Columbia to Centerville, as below:

$$a_0 = 21.482 + 0.018886Q + 0.000011Q^2 \qquad (3.22)$$

where Q is the outflow in m³/sec from the channel.

The coefficients a_1 and b_0 from the 12 flood events have been taken and averaged; they are 97.1 hours² and 4.0 hours, respectively. The inflow flood hydrographs for the reach have been routed by solving Equations (3.16), (3.20) and (3.22) by iteration. The result for a flood event in the reach from Columbia to Centerville is presented in Figure 19, where interbasin contribution was also modeled as described below.

3.6 Interbasin Flow Estimation in Large Drainage Basins

In many river basins, the contribution of runoff from sub-basins which are generally ungauged assumes importance in the water resources planning. For ungauged sub-basins, the flows are estimated either by using the nearby sub-basin flows on an area basis or by synthetic unit hydrograph methods. Both the methods are based on certain hydrological homogeneity and uniform distribution of rainfall.

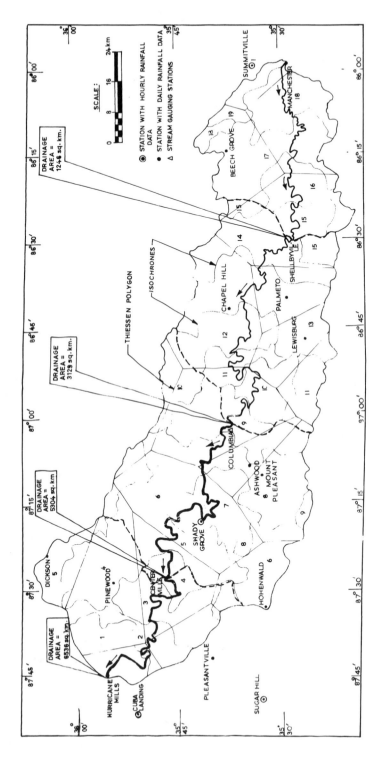

FIGURE 18. Duck River basin.

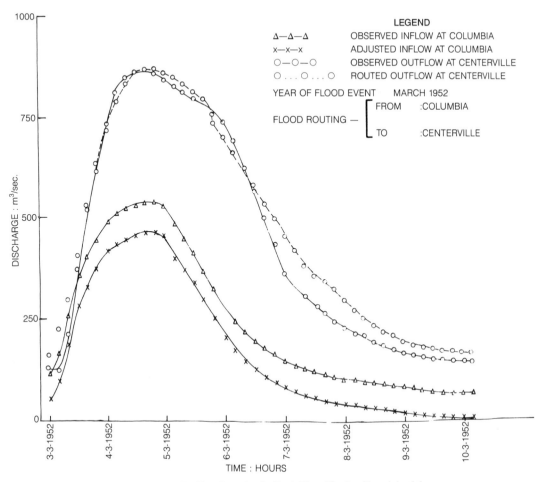

FIGURE 19. Flood routing in Duck River (Centerville sub-basin).

Methods of estimating sub-basin contribution based on the main-stem flow data have not received much attention.

A procedure for identifying the sub-basin contribution, knowing the flow at a downstream point, based on Clark's time-area diagram and the instantaneous unithydrograph [8] is given. A model based on a combination of rainfall–runoff relationship and channel routing using the 3-coefficient model for predicting stream flow of large basin for a given rainfall pattern is also presented. The model parameters are related to input characteristics of the basin. Using these relationships in the basin model, a flood hydrograph is predicted for a known storm.

INTERBASIN FLOW ESTIMATION

In Figure 18 the rainfall recording stations in and around the Duck River basin and the stream gauging stations along the river are indicated. The Duck River basin up to Hurricane Mills is considered for modelling the runoff process. The Shelbyville, Columbia, Centerville and Hurricane

Mills sub-basins are referred to as sub-basins-A, -B, -C and -D, respectively. For 8 storm events, the rainfall as well as stream flow data are available.

The time–area concentration diagram represents the basin translation and rainfall characteristics. It involves the basin area A, sub-area ΔA, between consecutive isochrones, the time of concentration T, unit period ΔT and the rainfall excess rate i_i (Figure 18). The time of concentration primarily depends upon the drainage basin characteristics such as relief, area, vegetal growth, length of basin, length of stream, soil condition, etc., and the storm characteristics such as direction of movement of storm, its intensity, areal distribution, duration, etc. Isochrones are drawn considering channel length of $\Delta t = 5$ hr interval. Thiessen weights are found for each raingauge station and average rainfall excess between consecutive isochrones is determined, and time–area diagram is constructed. When the time–area diagram is routed through an assumed storage $S = Kq$ where K is a storage coefficient and q is the surface runoff, instan-

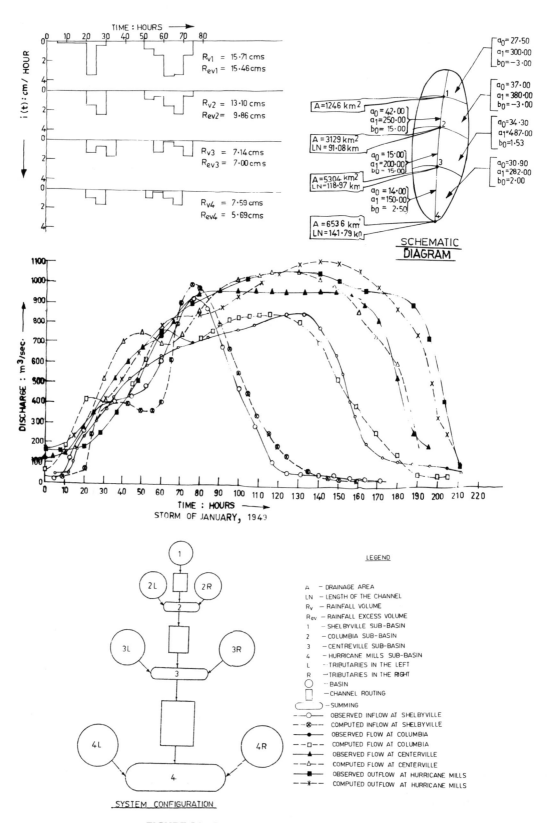

FIGURE 20. Duck River basin: system configuration.

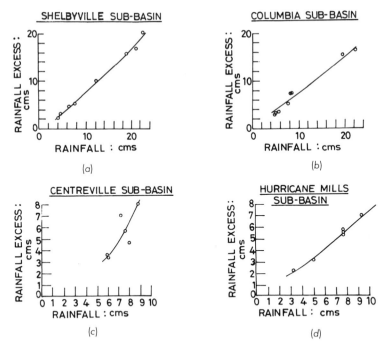

FIGURE 21. Duck River basin: rainfall vs. rainfall excess volume.

taneous unithydrograph is obtained. The average value of K is 32 hours for each of the four sub-basins. The rainfall excess is convolved with IUH and the surface runoff is obtained, e.g., one for the entire basin gauged at Columbia (Q_i) and the other for the sub-basin B (q_i) with reference to the same gauging stations. The ratio $R_i = q_i/Q_i$ gives the fraction of the discharge at Columbia contributed by the sub-basin-B. For 7 storm events, the baseflow was separated from the total runoff at Columbia, the ordinates of surface runoff were multiplied with the ratio R_i and the surface runoff contribution from the sub-basin-B estimated. The sub-basin-B contribution was subtracted from the outflow hydrograph at Columbia; then the attenuated flow from basin-A at Columbia was obtained and this procedure was repeated to other sub-basins-C and -D.

MODEL DESCRIPTION

Basin system configuration for Duck River basin is indicated in Figure 20. Basin runoff routing and channel flow routing under study are represented in the system configuration. For sub-basins-A, -B, -C and -D, the rainfall excess surface runoff-relationships were studied (Figure 21). The flow (Q_s) from Shelbyville basin was routed through the channel up to Columbia to get the channel routed flow Q_{sc} at Columbia and added to the sub-basin-B contribution and the runoff hydrograph at Columbia was obtained. The flow at Columbia was routed through the channel up to Centerville to get the channel routed flow, Q_{cct} at Centerville and added

to the sub-basin-C contribution and the runoff hydrograph at Centerville was obtained. The flow at Centerville was routed through the channel up to Hurricane Mills to get the channel routed flow Q_{cth} at Hurricane Mills and added to the sub-basin-D contribution and the stream flow at Hurricane Mills was obtained. A study of the volume of rainfall excess (R_{ev}) with the rainfall volume (R_v) (Figure 21) and initial base-flow (q_{bo}) with soil moisture index (SMI) for the sub-basins has been made and the equations are given below:

For sub-basin-A

$$q_{bo} = 10.2\text{SMI} + 2.66 \qquad (3.22a)$$

For sub-basin-B

$$q_{bo} = 40\text{SMI} + 22 \qquad (3.22b)$$

For sub-basin-C

$$q_{bo} = 48\text{SMI} + 28.70 \qquad (3.22c)$$

For sub-basin-D

$$q_{bo} = 64\text{SMI} + 3.72 \qquad (3.22d)$$

The value of SMI was calculated based on daily losses for a period of 15 days before the storm. For studying basin

TABLE 2. Basin Routing Parameters

Parameters	Shelbyville sub-basin	Columbia sub-basin	Centerville sub-basin	Hurricane Mills sub-basin
a_0 = (Hours)	$23.6 -$ $19R_{ev} -$ $5.06 T_c -$ $1.11 \Theta_1$ $R = 0.944$	$23.38 +$ $4.24R_{ev} -$ $3.7 \Theta_1$ $R = 0.93$	$26.33 +$ $1.64R_{ev}$ $R = 0.8$	$20.16 +$ $4.39 \Theta_1$ $R = 0.75$
a_1 = (Hours^2)	300.0	300.0	600.0	325.0
b_0 = (Hours)	-2.4	-2.4	1.0	1.0

R_{ev} = volume of rainfall excess (cm)
T_c = duration of rainfall excess
Θ_1 = $R_{ev}.T_g/T_c$
T_g = time to the centre of gravity of rainfall excess
R = correlation coefficient

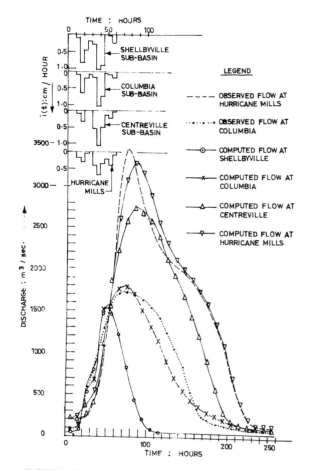

FIGURE 22. Duck River basin: storm of February, 1948.

routing, Equation (2.20) and for channel routing Equations (3.18), (3.19a) to (3.19d) were used. The coefficients were determined by least squares using known input and output. Similarly, the channel routing equations for Q_{cct} and Q_{cth} were studied. In all 8 storm data were available. Seven of these storms were used for the calibration phase of the model. The coefficients used in Equations (2.20), (3.18) and (3.19a) to (3.19d) are presented in Tables 2 and 3.

For the prediction of outflow hydrograph at Hurricane Mills due to a given storm in the basin, the following information is required. These relationships can be prepared from the past storm data:

1. Graphs of R_{ev} vs. R_v for the sub-basins-A, B, C and D (Figure 21).
2. Coefficients for basin routing and channel routing parameters a_0, a_1 and b_0 (Tables 2 and 3).
3. Relationships of q_{bo} vs. SMI for the sub-basins-A, -B, -C and -D [Equations (3.22a) to (3.22d)].

Having obtained the foregoing information, the outflow hydrograph can be predicted (Figure 22) using Equations (2.20), (3.15), (3.18) and (3.19a) to (3.19d) due to a given storm over a basin. For 7 of the storms studied in the calibration phase in Shelbyville, Columbia, Centerville and Hurricane Mills sub-basins, the correlation coefficient (R) varied between 0.94 and 0.99 for the predicted flood hydrographs. In the verification phase one storm data available was used. For this the percentage error in computed peak at Columbia is 2.9 percent and at Hurricane Mills is 4.9 percent and the correlation coefficient is 0.96. Similar studies were made for Sigur lower in Bhavani basin with 8 storm events and the results are presented for a storm in Figure 23.

3.7 Effect of Land Use on Stream Flows

In many developing countries, the need to produce more food, fibre and shelter has caused many watersheds to undergo rapid land use changes altering drastically its hydrologic characteristics such as peak and low flows. Predicting the hydrologic characteristics of these transitional

TABLE 3. Channel Routing Parameters.

Parameters	Shelbyville to Columbia	Columbia to Centerville	Centerville to Hurricane Mills
a_0 (Hours)	42.0	15.0	14.0
a_1 (Hours)^2	250.0	200.0	150.0
b_0 (Hours)	15.0	15.0	2.5

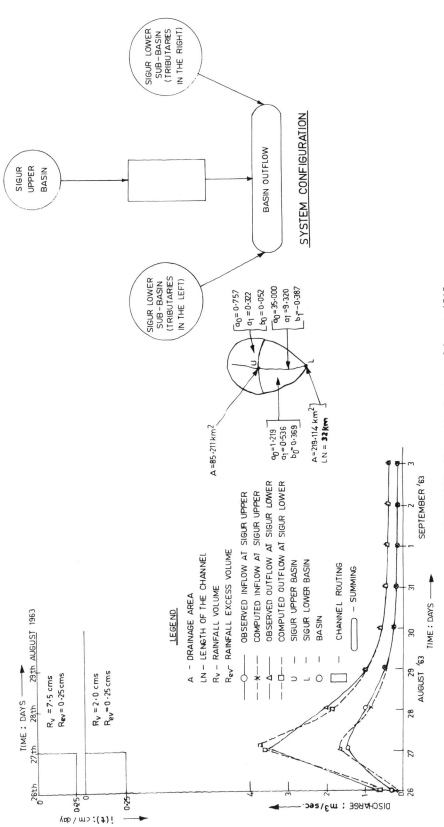

FIGURE 23. Bhavani basin: storm of August, 1963.

watersheds has always posed problems mainly because of the difficulties in including the land use either directly or indirectly in the rainfall-runoff modelling. The land use changes alter mainly the volume of runoff as well as its time distribution. Bhavani river (catchment area 6730 sq km, the average runoff 2300 Mm³ per annum), an important tributary of the river Cauvery in South India, has undergone significant changes with regard to land use pattern, resulting in perceptible alteration in its hydrologic characteristics. During the last 25 years, several parts of the basin have experienced floods of magnitude 2 to 4 folds higher than what it was previously, causing large scale land slides. Various aspects of the land use effect studied using the general hydrologic system model are presented below.

WATERSHEDS IN TRANSITION

Hydrologic data collected from two small experimental watersheds lying in the high hills of Nilgiris (latitude 11°13′30″N, longitude 76°39′50″E) were analysed for the effect of land use on hydrologic characteristics of these watersheds. The two watersheds (watershed-A and watershed-B) selected have approximately equal areas (Figure 24), are lying side by side and have the same geologic, geomorphologic and vegetative characteristics. Watershed-A and -B consist of respectively 15.67 percent and 8.34 percent of shola, 7.53 percent and 7.52 percent of swamps and 76.8 percent and 84.14 percent of grassland. Watershed-B was subjected to transition by clearing of trees and the grass in pockets and planting with bluegum, while watershed-A was kept in its natural state. Hydrologic measurements in these

watersheds were started in the year 1970. Three years (1970, 1973 and 1976) of hydrologic data were selected for analysis for both the watersheds. They represent one year of pretransition period (1970) and two years of transition period (1973 and 1976) for the watershed-B.

A simpler version of the GHS model, namely the 3-coefficient surface runoff model was selected for modelling. A total of 230 storms in the years 1970, 1973 and 1976 in both the watersheds were studied by Equation (2.20). The correlation coefficient for flood hydrographs determined using the 3-coefficient model was 0.99 for 222 storms and varied between 0.95 and 0.99 for the balance 8 events (Figure 25).

The model parameters a_0, a_1 and b_0 werre assumed to have the following functional relationships [10,11,26,68]:

$$a_0 = f(\Theta, t_r, t_{da})$$

$$a_1 = f(\Theta, t_r, t_{da}) \qquad (3.23)$$

$$b_0 = f(\Theta, t_r, t_{da})$$

where $\Theta = t_g \cdot R_{ev}$ representing input characteristics, t_{da}, the delay time, from the end of effective rainfall to the end of surface runoff, t_r a transition factor for the watershed-B and its computation is explained below. The coefficients in the model have been found insensitive to the land use changes and storm characteristics (Figure 26); but they vary with the delay time, τ (Figure 27). To eliminate the effect of delay time on model parameters, storms having a fixed delay time (1 day) (185 storms) were chosen from among the 230

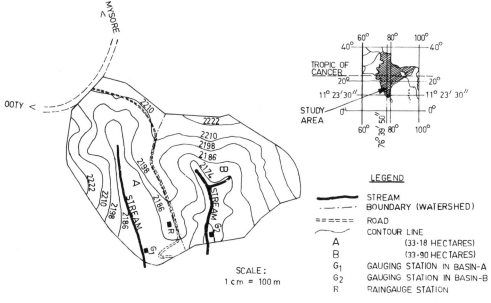

FIGURE 24. Location map of Glenmorgan.

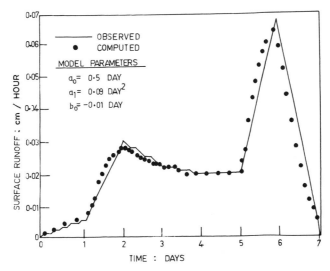

FIGURE 25. Watershed-A, storm of June 10, 1973: observed and computed hydrographs.

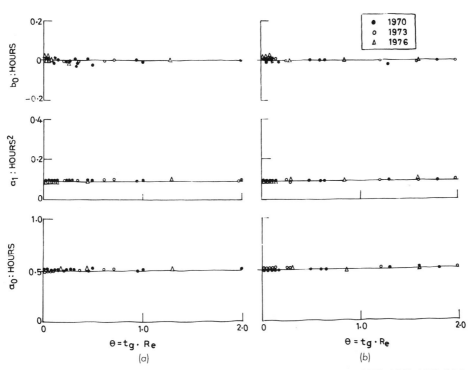

FIGURE 26. Graph of a_0, a_1, b_0 vs. Θ (a) for storms in watershed A—during 1970, 1973, 1976; (b) for storms in watershed B—during 1970, 1973, 1976.

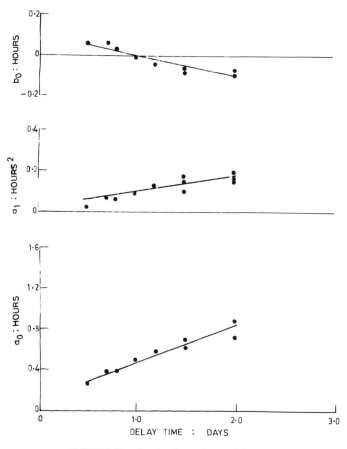

FIGURE 27. Graph of a_0, a_1, b_0 vs. delay time.

storms for further analysis. For these storms, the model parameters can be considered as $a_0 = 0.5$ day, $a_1 = 0.09$ days^2 and $b_0 = -0.01$ day.

COMPUTATION OF TRANSITION FACTOR

The annual rainfall excess in cm in watershed-A and watershed-B were respectively 8.070 and 5.83 for 1970, 10.68 and 10.16 for 1973 and 5.53 and 4.7 for 1976. The rainfall excess in the year 1970 in watershed-B is 72.25 percent of that in watershed-A in the same year. The increase in annual rainfall excess in watershed-B in 1973 and 1976 over that in 1970 was 32 percent and 18 percent, respectively, indicating that the watershed-B was in various stages of transition. For the prediction of rainfall excess, quantification of various stages of transition is required. For this purpose a transition factor t_r, in terms of the rainfall excess, is introduced. For the watershed-B in 1970 the value of t_r is 1.00, and in 1973 and 1976, it is 1.32 and 1.18, respectively.

The rainfall excess as a percentage of rainfall (y) is ex-
pressed as a function of antecedent precipitation index (x) and the transition factor, t_r. Assuming the functional relationship to be of the form $y = a x^n$, the values of the parameters a and n were determined for different values of t_r. From the log–log plot of y vs. x using the data from watershed-B (Figure 28) the values of the parameters a and n were determined. The values of a and n were respectively 0.26 and 1.55 (1970), 0.32 and 1.57 (1973) and 0.29 and 1.56 (1976). From the plots of y vs. x, the rainfall excess was estimated knowing the antecedent precipitation index and the value of t_r. Knowing the rainfall excess and the model coefficients for the watershed, the flood hydrograph can be predicted.

There is considerable increase in rainfall excess immediately after the land preparation for afforestation and this reduces with the growth of vegetation. The rainfall excess volume in a watershed under transition can be predicted as a function of antecedent precipitation index and transition factor. The reproduction of runoff hydrographs by the 3-coefficient model is excellent even though its coefficients are insensitive to the land use changes.

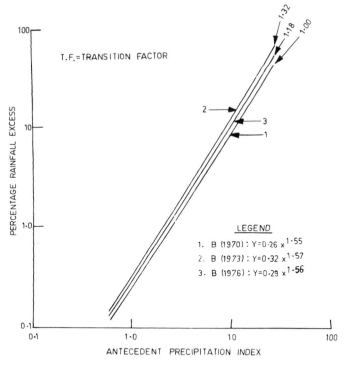

FIGURE 28. Log–log plot of percentage rainfall excess vs. API for various transition factors for watershed-B.

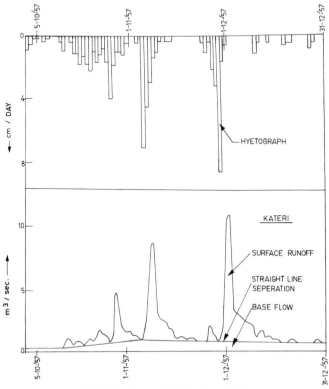

FIGURE 29. Separation of baseflow.

AFFORESTATION AND URBANIZATION OF AN AGRICULTURAL WATERSHED

In the above study the model parameters a_0, a_1 and b_0 are found to vary with the delay time, t_{da}, and independent of the land use changes. This study was then extended to actual watersheds in the same hydrologic environ consisting of forest watershed (Mukurthy—25.25 Km² EL = 2150m − 2700 m), agricultural watershed (Kateri— 54.755 Km², EL = 1800 m − 2200 m) and an urban watershed (Coonoor−44.03 Km², EL = 1600 m − 2150 m). The effect of urbanization and afforestation of the agricultural

watershed was investigated on the rainfall excess volume and time distribution of surface runoff, namely peak q_p and time to peak, t_p.

Five years of daily rainfall and flow data (119 storms) were subjected to analysis. The separation of baseflow is shown in Figure 29. Reproduction of flood hydrographs was made using 3-coefficient runoff model (Sec. 3.4) for three watersheds, as shown in Figure 30. Based upon the hydrologic features of the Bhavani basin the year was divided into workable periods as follows: (a) Non-monsoon period (January to May), (b) Monsoon period (June to December).

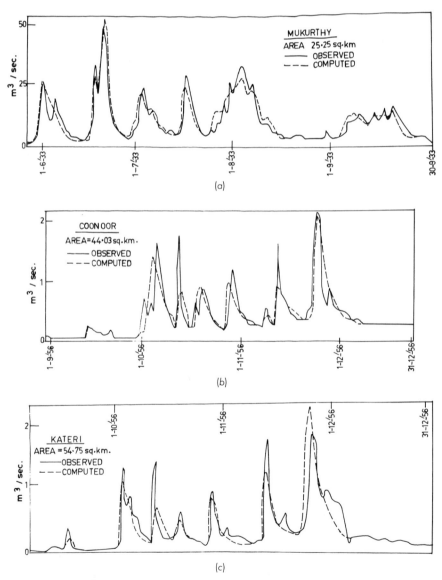

FIGURE 30. Plotting of observed and computed runoff for three watersheds.

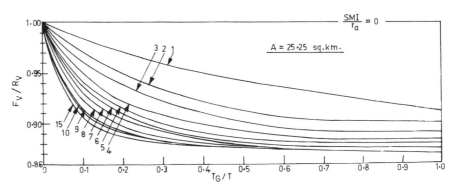

FIGURE 31. Forest watershed (Mukurthy): graph of F_v/R_v vs. SMI/r_a and T_G/T.

For the storms studied, the correlation coefficient (R) varied between 0.88 and 0.99 for predicted flood hydrograph for the three watersheds.

The rainfall excess of a given storm is intimately related to the storm characteristics as well as the infiltration characteristics of the watershed. The infiltration characteristics are mainly related to the wetness of the soil prior to the storm, the type of soil and the land use to which the watershed is subjected to. This concept can be expressed by the following relationship [10,11,68]:

$$\frac{F_v}{R_v} = \left(\frac{SMI}{r_a} , \frac{T_G}{T} , \text{type of land use} \right) \quad (3.24)$$

where F_v represents abstractions, R_v is the volume of rainfall, SMI is the soil moisture index, r_a is the average rainfall rate, T_G is the time to the centre of gravity of the rainfall area and T is the duration of rainfall. From the observed data, curves were drawn between F_v/R_v vs. T_G/T with SMI/r_a as parameters for all the three watersheds. Subtracting F_v from R_v, the volume of rainfall excess was arrived at. In the case of forest watershed (Mukurthy) a single set of curves was obtained (Figure 31), while in the case of urban (Coonoor) and rural (Kateri) watersheds, two sets of curves, for mon-

soon and non-monsoon separately, were prepared (Figures 32 to 35). Assuming that the rainfall excess curves obtained for one watershed can be interchanged with the others, depending on the land use to which the watershed is subjected, the rainfall excess of the rural watershed when it is completely urbanised or afforested can be computed. For an assumed-rainfall excess of 5 cm in the rural watershed (Figure 36), it was observed that there was an increase of 49 percent in rainfall excess, if the watershed was urbanised and a decrease of 20 percent in rainfall excess if the watershed was afforested as indicated in Figure 36.

For obtaining the distribution of surface runoff, the 3-coefficient model (Figures 37 and 38) was used. The coefficients a_0, a_1 and b_0 were estimated for the rural watershed for a number of storms and average values of these coefficients were computed. The model parameters a_0, a_1 and b_0 were assumed to have the following functional relationships:

$$a_0 = f(A, L, \sqrt{S}, R_e, t_{da})$$

$$a_1 = f(A, L, \sqrt{S}, R_e, t_{da}) \quad (3.25)$$

$$b_0 = f(A, L, \sqrt{S}, R_e, t_{da})$$

where A is the area of the watershed, L the length of the

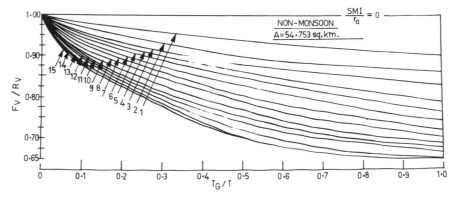

FIGURE 32. Agricultural watershed (Kateri): graph of F_v/R_v vs. SMI/r_a and T_G/T.

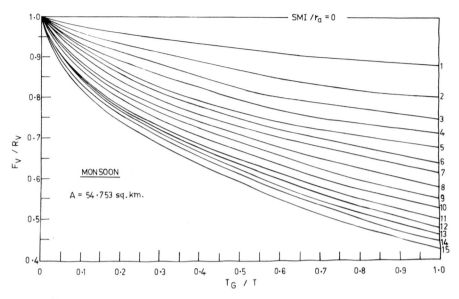

FIGURE 33. Agricultural watershed (Kateri): graph of F_V/R_V vs. SMI/r_a and T_G/T.

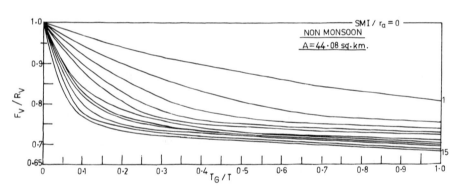

FIGURE 34. Urban watershed (Coonoor): graph of F_V/R_V vs. SMI/r_a and T_G/T.

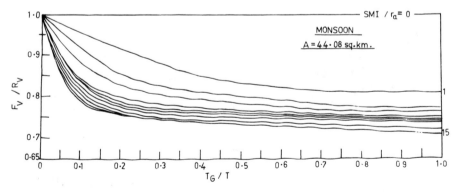

FIGURE 35. Urban watershed (Coonoor): graph of F_V/R_V vs. SMI/r_a and T_G/T.

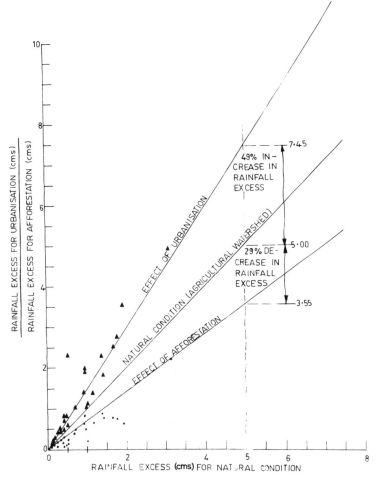

FIGURE 36. Kateri: effect of afforestation and urbanisation on the rainfall excess.

main channel, S the average slope of the main channel, R_e the elongation ratio of the basin and t_{da} the average delay time of the basin. By a study of a_0, a_1 and b_0 with the watershed characteristics and the time delay, the following relationships for the three watersheds were obtained [10,11,68]:

$$a_0 = K_0 \frac{AL}{R_e\sqrt{S}} t_{da}$$

$$a_1 = K_1 \frac{AL}{R_e\sqrt{S}} t_{da} \qquad (3.26)$$

$$b_0 = K_2 \frac{AL}{R_e\sqrt{S}} t_{da}$$

Knowing the values of A, L, R_e and S, t_{da} and a_0, a_1 and b_0,

the coefficients K_0, K_1 and K_2 in the above equations can be computed for different types of land uses. Using the values K_0, K_1 and K_2 of the urban watershed (Coonoor) the values of the parameters a_0, a_1 and b_0 were computed for the rural watershed (Kateri) in order to study the effect of urbanisation on the time distribution of surface runoff and time to peak. Similarly using the values of K_0, K_1, and K_2 of the forest watershed (Mukurthy) the values of the parameters a_0, a_1 and b_0 were computed for the rural watershed (Kateri) to study the effect of afforestation. The model parameters so evaluated were used in the 3-coefficient models (Figures 37 and 38) and the time distribution of surface runoff was computed. A total of 30 storms, 18 from the monsoon period and 12 from the non-monsoon period of the rural watershed were used. From these studies, it was found that there was an increase of 107 percent in the surface runoff peak if the watershed was urbanised and 80 percent decrease in the sur-

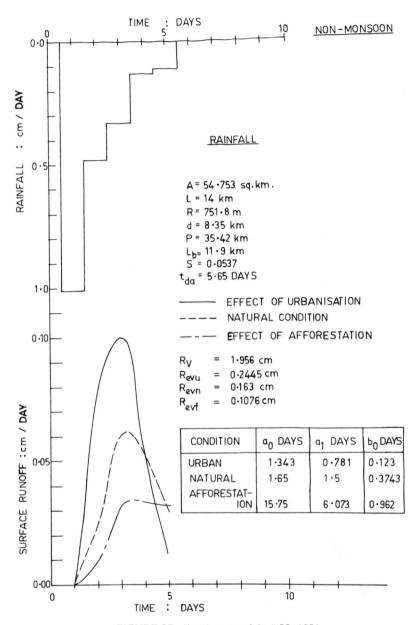

FIGURE 37. Kateri: storm of April 22, 1956.

362

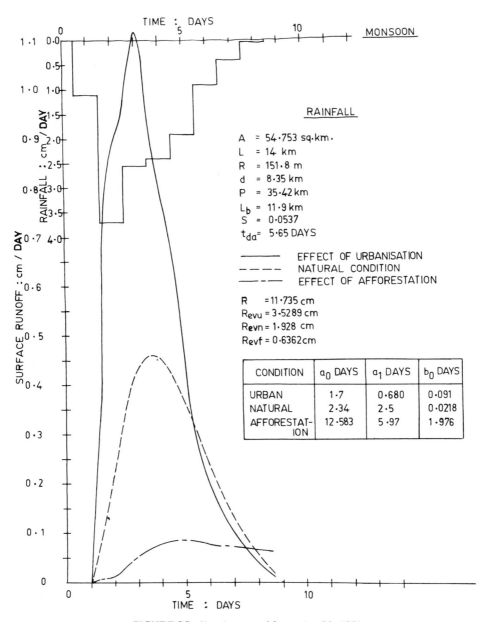

FIGURE 38. Kateri: storm of September 30, 1956.

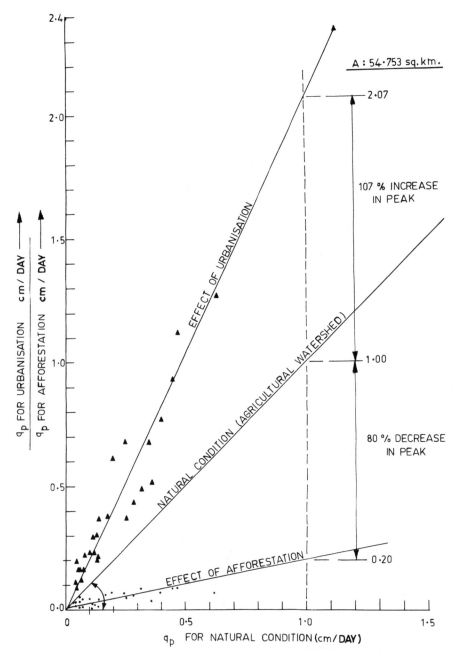

FIGURE 39. Kateri: effect of afforestation and urbanisation on the peaks of surface runoff.

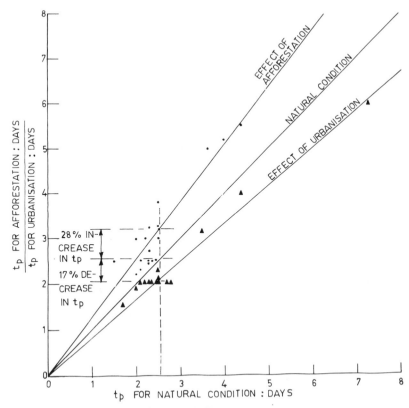

FIGURE 40. Kateri: effect of afforestation and urbanisation on t_p.

face runoff peak if the watershed was afforested for an assumed peak of 1.0 cm/day in the natural condition (rural) (Figure 39). Further it was found that there was a decrease of 17 percent in t_p, time to peak if the watershed was urbanised and an increase of 28 percent in t_p if it was affortested, for an assumed value of 2.5 days in the natural condition (Figure 40). By using the above procedure, any mixed watershed consisting of urban, rural and forest can be modelled and the effect of land use changes on the streamflow can be predicted [25].

4. CONCLUSIONS

The general hydrologic system model since its formulation two decades ago is still being actively pursued by researchers and several results are getting added to the existing knowledge of the model and to its applicability. Only a few remarks in the nature of overall review are given as conclusions.

The outstanding merit of the model has been its general storage relationship which could model the transformation in almost all phases of the runoff cycle like infiltration, surface runoff, routing through the stream, etc. Hence, it could derive the capability to provide a powerful alternative to model the rainfall-runoff relationship for the basin itself. On a long term basis, the basin transformation could also be detected and modelled to be of use in prediction.

The hesitation of its adoption in view of the requirement of fast computational facility is taken into account and a simpler version of the same model capable of being tackled by calculators is provided and recommended as three parameter model.

Method for estimation of average abstraction rate through Φ-index is given for the completeness of the topic. The use of GHS model for obtaining the infiltration rate curve during a storm over a basin is demonstrated in Section 3.2.

The rainfall excess–surface runoff relationship has been established either as nonlinear case or as a linear model.

A simple version of the model with only three coefficients is presented as:

$$q(t) = \frac{-b_0 D + 1}{a_1 D^2 + a_0 D + 1} \, i(t) \qquad (2.20)$$

The model parameters have been shown to be the functions of storm characteristics. This is true for the 5 parameter model and to somewhat a lower degree to three parameter model also. These models are shown to produce the abstraction rate curve satisfactorily.

The GHS model helps to route the flood through a channel reach and reproduces the outflow hydrograph very satisfactorily. Its use in river forecasting is thus demonstrated.

The continuation of basin routing and channel routing both based on general storage equation is found to forecast the streamflow accurately enabling the flow prediction in large drainage basins.

The three coefficient model on application to various watersheds in transitional stages is seen to predict the altered hydrologic characteristics with changed land use effects. The model parameters are regressed with catchment characteristics.

The limitations faced by this model are not unique to itself but commonly valid for any hydrologic model. The subbasin variability of storm and basin characteristics are at best approximately accounted. Accounting the nonlinearity to the required rigour is still a formidable problem. Its impact, even if it is included in the solution, is difficult to justify the effort.

REFERENCES

1. Amorocho, J., "Predicting Storm Runoff on Small Experimental Watershed, Discussion," *Trans. Am. Soc. Civ. Engrs.*, V-127, Part-I (1962).

2. Amorocho, J. and G. T. Orlob, "Nonlinear Analysis of Hydrologic Systems," University of Calif., Water Resources Centre, Contribution No. 40 (1961).

3. Babu Rao, T., "A Mathematical Model for Basin Runoff," Ph.D. thesis, University of Madras, Madras (1973).

4. Blank, D. and J. W. Delleur, "A Program for Estimating Runoff from Indiana Watersheds, Part-I Linear System Analysis in Surface Hydrology and its Application to Indiana Watersheds," Tech. Rep. No. 4 (1968).

5. Bilcrest, B. R. and L. E. Marsh, "Channel Storage and Discharge Relations in the Lower Ohio River Valley," *Trans. Am. Geophysics, Uni.*, 637–649 (1941).

6. Boneh, A. and A. Golan, "Instantaneous Unit Hydrograph with Negative Ordinates – Possible?" *Water Resources Research*, Vol. 15, No. 1, 121–130 (1979).

7. Butler, S. S. *Engineering Hydrology*. Prentice Hall, Inc., Anglewood Cliffs, NJ (1957).

8. Clark, C. O., "Storage and the Unit Hydrograph," *Trans. Am. Soc. Civ. Engrs.*, Vol. 110, 1419–1446 (1945).

9. Clarke, R. T., "Mathematical Models in Hydrology," FAO, U.N., Rome, Irri. and Drain. Paper 19 (1973).

10. Chinnamani, S., "An Integrated Study of Hydrology of the Bhavani Basin," Ph.D. thesis, Anna University, Madras (1982).

11. Chinnamani, S. and R. Sakthivadivel, "An Integrated Study of Hydrology of the Bhavani Basin," Final Report, Centre for Water Resources, Anna University, Madras (1982).

12. Choudhary, F. H., "General Linear Hydrologic Response Models," *Proc. Third Ann. Symp. Waterways, Harbours and*

13. Choudhary, F. H., M. A. Simoes and W. M. Ferreira Filho, "Chow's General Hydrologic System Model," *Jour. Hyd. Div., Proc. Am. Soci. Civ. Engrs.*, 102 (HY9), 1387–1390 (1976).

14. Chow, V. T. and V. C. Kulandaiswamy, "General Hydrologic System Model," *Jour. Hyd. Div. Proc. Am. Soc. Civ. Engrs.*, 97 (HY6), 791–804 (1971).

15. Chow, V. T. and V. C. Kulandaiswamy, "Reply to Discussion by Franz on General Hydrolgic System Model," *Proc. Am. Soc. Civ. Engrs., Jour. Hyd. Div.*, Vol. 98, No. HY.10 (1972).

16. Chow, V. T. *Handbook of Applied Hydrology*. McGraw-Hill Book Co., Inc., New York (1964).

17. Chow, V. T. and V. C. Kulandaiswamy, "The IUH of General Hydrologic System Model," *Jour. Hyd. Div., Proc. Am. Soc. Civ. Engrs.*, HY7, 830–844 (1982).

18. Chow, V. T. *Open-Channel Hydraulics*. McGraw-Hill Book Co., Inc., New York (1959).

19. Crawford, N. H. and R. K. Linsley, "Digital Simulation in Hydrology – Stanford Watershed Model IV," Tech. Rep. No. 39, Dept. Civ. Engg., Stanford (1966).

20. Diskin, M. H., "A Basic Study of the Linearity of Rainfall–Runoff Process in Watersheds," Ph.D. thesis, University of Ill., Urbana, IL (1964).

21. Eagelson, P. S., "Deterministic Linear Hydrologic Systems," *Proc. First Int. Symp. Hydro. Professors*, Vol. I, Dept. of Civil Engg., University of Illinois, IL (1969).

22. Gupta, C. V. S., "A Study of Mathematical Models for the Subwatersheds of the Bhavani Basin," M.E. thesis, Anna University, Madras (1981).

23. Horton, R. E., "Approach Towards a Physical Interpretation of Infiltration Capacity," *Soc. Sc. Soc. Am., Proc.*, 5, 399–417 (1940)

24. Ilangovan, M., "A Comparative Study of Mathematical Models for Rainfall–Runoff Relationship," M.Sc.(Eng.) thesis, University of Madras, Madras (1966).

25. Jayaraman, A., T. Babu Rao and R. Sakthivadivel, "Effect of Land Use on Surface Runoff Characteristics," 50th Annual R and D Session, CBIP, New Delhi, Tech. Sess. held at Simla, Paper-1 (1983).

26. Jebaraj, S., "Mathematical Modelling of Watersheds in Transition," M.E. thesis, Anna Univ., Madras (1980).

27. Johnstone, D. and W. P. Cross. *Elements of Applied Hydrology*. The Ronald Press Co., New York (1949).

28. Krishnasami, M., "A Study on the Uses of Storage Equation in Some Phases of Runoff Process," University of Madras, Madras (1965).

29. Kulandaiswamy, V. C., "A Basic Study of the Rainfall Excess–Surface Runoff Relationship in a Basin System," Ph.D. thesis presented to the University of Ill., Urbana, IL (1964).

30. Kulandaiswamy, V. C., "Linear Analysis of Rainfall–Runoff Relationship," *Jour. Inst. Engrs.* (India) (1966).

31. Kulandaiswamy, V. C., "Derivation of IUH Using Z-transformation," *Jour. Cent. Bo. Irr. Pow.*, Vol. 23, No. 3, 251–255 (1966).

32. Kulandaiswamy, V. C., "A Note on Muskingum Method of Flood Routing," *Jour. Hydr.*, Holland, No. 4, 273–276 (1966).

33. Kulandaiswamy, V. C. and T. Babu Rao, "An Analytical Approach to the Determination of Infiltration Rate," *Jour. Inst. Civil Engrs.* (London), 137–144 (1967).

34. Kulandaiswamy, V. C., M. Krishnasami, and R. M. Ramalingham, "Flood Routing Through Channels," *Jour. Hydr.*, Holland, No. 5, 279–285 (1967).

35. Kulandaiswamy, V. C. and C. V. Subramanian, "A Nonlinear Approach to Runoff Studies," *Int. Hydr. Symp. IASH*, Fort Collins, 72–79 (1967).

36. Kulandaiswamy, V. C., "Discussion on the Paper by Ramanand Prasad on Nonlinear System Response Model," *Jour. Hyd. Div. Proc. Am. Soc. Civ. Engrs.*, Vol. 94, No. HY3, Proc. Paper 5920, 800–803 (1968).

37. Kulandaiswamy, V. C. and T. Babu Rao, "Infiltration in Drainage Basins," Report-1, Hydraulics and Water Resources Dept., College of Engg., Madras (1969).

38. Kulandaiswamy, V. C. and T. Babu Rao, "An Investigation of Non-Linearity in Storage–Discharge Relationship for Watersheds," Report 4, Hydraulics and Water Resources Department, College of Engg., Madras (1970).

39. Kulandaiswamy, V. C. and T. Babu Rao, "Reply to the Discussion of Dr. Painter on An Analytical Approach to the Determination of Infiltration Rate," *Jour. Inst. Civ. Engrs.* (London), 395–396 (1970).

40. Kulandaiswamy, V. C. and T. Babu Rao, "An Investigation of Non-linearity in the Runoff Process," Symp. Wat. Res., I.I.Sc., Bangalore (1971).

41. Kulandaiswamy, V. C. and T. Babu Rao, "Infiltration in Drainage Basins," Symp. Wat. Res., I.I.Sc., Bangalore (1971).

42. Kulandaiswamy, V. C. and T. Babu Rao, "A Mathematical Model for Rainfall–Runoff Relationship, 41st Ann. Res. Ses., Cen. Bo. Irr. Pow., New Delhi held at Jaipur (1971).

43. Kulandaiswamy, V. C. and T. Babu Rao, "Digital Simulation of a Drainage Basin," *Int. Symp. Math. Mod. Hydro.*, IASH, Warsaw, Poland, 19–24 (1971).

44. Kulandaiswamy, V. C. and T. Babu Rao, "Reply to Discussion of Dr. Majumdar on A Mathematical Model for Rainfall–Runoff Relationship," *Proce. 41st Ann. Res. Sess., Cen. Bo. Irr. Pow.*, New Delhi held at Jaipur (1971).

45. Kulandaiswamy, V. C. and T. Babu Rao, "A Method for the Prediction of Φ-Index," *Irr. and Power Jour.*, Cen. Bo. Irr. Pow., New Delhi, 135–142 (1971).

46. Kulandaiswamy, V. C. and T. Babu Rao, "A Mathematical Model for Basin Runoff," Report 10, Hydraulics and Water Resources Dept., College of Engg., Madras (1974).

47. Kulandaiswamy, V. C. and T. Babu Rao, "A Mathematical Model for Basin Runoff," *Second World Congress on Water Resources, I.W.R.A., Proc. Vol. V*, New Delhi (1975).

48. Kulandaiswamy, V. C. and T. Babu Rao, "A Mathematical Model for Simulation of Drainage Basins," *Proc. Dia. Jub. Symp.*, CWPRS, Pune (1976).

49. Lakshmanan, P., "Modelling for Flood Routing in Channels," M.E. thesis, University of Madras, Madras (1980).

50. Langbein, W. B., "Channel – Storage and Unit Hydrograph Studies," *Trans. Am. Geophys. Uni.*, 620–627 (1940).

51. Laurenson, E. M., "A Catchment Storage Model for Runoff Routing," *Jour. Hydr.* Vol. II, No. 2, 141–163 (1964).

52. Linsley, R. K. and J. B. Franzini. *Elements of Hydraulics Engineering.* McGraw-Hill Book Co., Inc., New York (1955).

53. Linsley, R. K., M. A. Kohler, and J. L. H. Paulhus. *Applied Hydrology.* McGraw-Hill Book Co. Inc., New York (1949).

54. McCann, R. C. and V. P. Singh, "An Analysis of the Chow-Kulandaiswamy GHS Model," *Advances in Water Resources,* Vol. 3, No. 4 (1980).

55. McCarthy, G. R., "The Unit Hydrograph and Flood Routing," U.S.A. Engrs. Office, Providence, R.I., A Paper Presented at the Conference of the North Atlantic Div., U.S. Engineers Dept. at New London, CT, June 24, 1938, revised March 21, 1939.

56. Mein, R. G., E. M. Laurenson and T. A. McMohan, "Simple Nonlinear Model for Flood Estimation," *Jour. Hydr. Div., Proc., Am. Soc. Civ. Engrs.*, Vol. 100, No. HY11, 1507–1518 (1974).

57. Meyer, A. F. *Elements of Hydrology.* 2nd Ed., Wiley, New York (1928).

58. Minshall, N. E., "Predicting Storm Runoff on Small Experimental Watersheds," *Trans. Am. Soc. Civ. Engrs.*, Vol. I, 127. Part I, 625 (1962).

59. Nash, J. E., "The Form of the Instantaneous Unit Hydrograph," *Proc. IASH, Gen. Ass.*, Toronto, No. 45, 114–118 (1958).

60. Neal, T. H., "The Effect of the Degree of Slope and Rainfall Characteristics on Runoff and Soil Erosion," Missouri University, Agr. Dept. Sta., Resource Bul No. 280 (1938).

61. Padmanabhan, G., M. Narayanan, R. Sakthivadivel and V. C. Kulandaiswamy, "A Review of Deterministic Mathematical Models for Rainfall–Runoff Relationship," *Proc. Dia. Jub. Symp.*, CWPRS, Pune (1976).

62. Paynter, H. M., "Methods and Results from M.I.T. Studies Unsteady Flow," *Jour. Bost. Soc. Civ. Engrs.*, Vol. 39, No. 2, 120–165 (1952).

63. Prasad, R., "A Nonlinear Hydrologic System Response Model," *Jour. Hyd. Div. Proc. Am. Soc. Civ. Engrs.*, Vol. 93, No. HY4, Proc. Pap. 5350, 201–221 (1967).

64. Ramalingam, T. N., "A Study on the Uses of General Storage Equation in Runoff Process, "M.Sc. (Eng.) thesis, University of Madras, Madras (1965).

65. Ramaseshan, S., "Synthetic Hydrology and Simulation for the Analysis of Rainfall-Runoff Characteristics," Ph.D. thesis, University of Ill., Urbana, IL (1964).

66. Raudkivi, A. J. *Hydrology—An Advanced Introduction to Hydrological Processes and Modelling.* Pergamon Press (1979).

67. Sakthivadivel, R., T. Babu Rao, P. Lakshmanan and K. Venugopal, "Runoff Modeling of an Ungauged Sub-Basin," 49th Annual Res. and Dev. Sess., Cen. Bo. Irri. Pow., New Delhi, Tech. Sess. held at Ooty, Pap.1 (1981).

68. Sakthivadivel, R., T. Babu Rao, S. Chinnamani and C. V. S. Gupta, "Runoff Prediction of High Mountain Watersheds with

Varied Landuses," *Proc., Int. Symp. Hydro. Asp. Mount. Watersheds,* School of Hydrology, University of Roorkee, Roorkee, Vol. 1, VI-1–VI-6 (1982).

69. Sathiamoorthy, K. K., "Streamflow Modelling for Large Drainage Basins," M.E. thesis, Anna University, Madras (1982).

70. Sherman, L. K., "Streamflow from Rainfall by the Unitgraph Method," *Engg. News Rec.,* Vol. 108, 501–505 (1932).

71. Singh, K. P., "Nonlinear Instantaneous Unit Hydrograph Theory," *Jour. Hyd. Div., Proc. Am. Soc. Civ. Engrs.,* 313–347 (1964).

72. Singh, V. P. and R. C. McCann, "A Mathematical Study of the General Hydrologic System Model," NSF ENG 79-05560, Mississippi State University, Mississippi (1979).

73. Subramanian, C. V., "A Nonlinear Approach to Runoff Studies," M.Sc. (Eng.) thesis, University of Madras, Madras (1965).

74. Sundaramahalingam, M., "A Non-linear Analysis of Channel Routing," M.E. thesis, Anna University, Madras (1983).

75. Venugopal, K., T. Babu Rao and R. Sakthivadivel, "A Conceptual Model for Hydrologic Forecasting of a Large River System," *Proc., Int. Symp. Hydro. Asp. Mount. Wat.,* School of Hydrology, Univ. of Roorkee, Roorkee, Vol. 1, V-37–V-41 (1982).

76. Zoch, R. T., "On the Relation Between Rainfall and Streamflow," *Monthly Weather Review,* Vol. 62, 315–322 (1934), Vol. 64, 105–121 (1936), and Vol. 65 135–147 (1937).

77. ICE (LONDON), Engineering Hydrology To-day, IHD Committee and the Institute of Hydrology (1975).

78. Muskingum River Basin Board, "Report on Analysis of Hydrologic Data for Index Area," Ohio, U.S. Engineers Office, Huntington, W.Va. (1943).

79. USDA Tech. Bul. No. 1468, "Linear Theory of Hydrologic Systems," ARS, Washington, D.C. (1973).

CHAPTER 12

Piezometer Installation and Monitoring Under Artesian Conditions

KULBHUSHAN L. LOGANI*

INTRODUCTION

A surface or subsurface facility is designed based on a set of assumed design parameters, which at best represents a range of most probable numerical values of the engineering properties of naturally occurring materials: soil and rock. Uncertainties in assigning engineering properties necessitate performance-monitoring during and after construction. The actual performance is then compared with the design-predicted performance to evaluate whether a change is required in the design, construction, or operation of the facility.

Pore pressure measurement is one of the most important of the design parameters. The pore pressure, under ordinary conditions, can easily be measured by installing pore pressure measuring devices (piezometers) by known conventional techniques. Although the principle of installation and monitoring piezometers, under ordinary conditions, is simple, reliable, and generally well understood, it is still difficult to install piezometers under artesian conditions. For monitoring the artesian pressure in a particular location rather than a zone, the conventional method does not assure an accurate placement of the piezometer cell, a water tight seal, and an effective backfilling of the borehole. The piezometer installation is still more difficult when the artesian pressure and flow are relatively high—pressure head of more than 5 meters above the ground surface and flow of more than a few gallons per minute.

ARTESIAN CONDITIONS

The artesian conditions—artesian flow and pore pressure—are found in pervious strata that are confined between impervious strata and are connected to a water source at a higher elevation (artesian source). The confined stratum can consist of sands and gravels, porous sandstones, highly fractured rocks, or cavernous limestones. A borehole (Figure 1) drilled to such a stratum, which has a pressure head greater than its depth, will flow freely under artesian pressure.

Ground treatment, sometimes is applied either to reduce the permeability or compressibility or to increase strength or stability of soils or rocks. This is usually achieved by filling cavities, fissures, and porespaces in the rock or soil by pressure injected cement-grout. The ground treatment thus alters the artesian conditions either by reducing the stratum permeability, or by disconnecting the stratum from the artesian source, or by both. A borehole in such a stratum will experience extreme variations of artesian conditions; from high pressure and flow before the treatment to no pressure and flow after the treatment. On the other hand, a borehole in a stratum which has not been subjected to any ground treatment, will experience constant steady state artesian conditions.

PRINCIPLE OF INSTALLATION AND MONITORING

Extremely variable artesian conditions are monitored by locating the piezometer within the pervious stratum in the borehole. The installation is completed under temporary no-flow-conditions, which are induced by the use of special equipment and techniques.

The constant artesian conditions, with at least a few gallons per minute flow, is monitored by locating a wellpoint within the pervious stratum. The piezometer is then attached to the riser pipe sticking out either at the ground surface or, in the case of an offshore operation, at the mudline; about 10 feet (3 meters) below still water level at the shoreline.

At the completion of the installation, the piezometers are either read manually or automatically (punched tapes or

*Civil and Geotechnical Engineering Consultant (Formerly of Harza Engineering Company), Glenview, IL

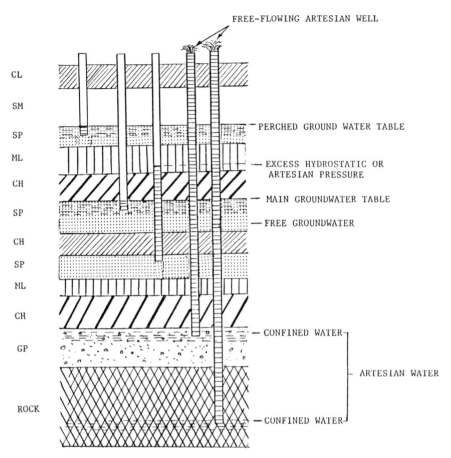

FIGURE 1. Artesian conditions.

digital outputs) at the piezometer locations or at a central terminal. The raw data is then converted into pore pressure units, summarized, compared with the previous data, and plotted in a form that can be quickly reviewed by an engineer. Complete data processing can now be conducted by a computer base data management system.

Piezometers can be installed under most of the artesian conditions likely to be encountered in the field, and the artesian conditions can be monitored by utilizing procedures compatible with the field conditions.

BOREHOLE DRILLING AND STABILIZATION

The borehole for the piezometer is advanced by an appropriate boring method. Selection of a boring method in a given situation depends upon economics; composition, physical properties, and natural conditions of soil or rock; relative efficiency of the boring procedure in the material

encountered; possible disturbance of material sampled; and depth of the artesian stratum (Table 1). In general, the quickest, cheapest, and most commonly used is the rotary drilling method which has the capabilities of drilling to depths of several hundred feet.

Common to all boring methods is the problem of caving-in of the sides and blowing-up of the borehole bottom. Uncased dry boreholes are generally stable when they are shallow and above the ground water table, but danger of caving increases rapidly with depth and presence of free ground water. Clean and uncemented sands and gravels, especially when located below the water table, are more susceptible to caving. Soft cohesive or organic soils obstruct boreholes by squeezing in or reducing the diameter of the borehole. The squeezing of soft soils often occurs at greater depths. In firm cohesive soils a borehole may remain open for a considerable and often impractical depth.

Several techniques are used to stabilize (prevent caving and squeezing) a borehole.

TABLE 1. Methods of Borings.

Boring Method	Procedure Utilized	Applicability
Displacement type	Repeatedly driving or pushing tube or spoon sampler into soil and withdrawing recovered materials. Changes indicated by examination of materials and resistance to driving or static force for penetration. No casing required.	Used in loose to medium compact sands above water table and soft to stiff cohesive soils. Economical where excessive caving does not occur. Limited to holes <3" in diameter.
Auger boring	Hand or power operated augering with periodic removal of material. In some cases continuous auger may be used requiring only one withdrawal. Changes indicated by examination of material removed. Casing generally not used.	Ordinarily used for shallow explorations above water table in partly saturated sands and silts, and soft to stiff cohesive soils. May be used to clean out hole between drive samples. Very fast when power-driven. Large diameter bucket auger permits examination of hole.
Wash type boring for undisturbed or dry samples	Chopping, twisting and jetting action of a light bit as circulating drilling fluid removes cuttings from hole. Changes indicated by rate of progress, action of rods and examination of cuttings in drilling fluid. Casing used as required to prevent caving.	Used in sands, sand and gravel without boulders and soft to hard cohesive soils. Most common method of subsoil exploration. Usually can be adapted for inaccessible locations, such as over water, in swamps, on slopes or within buildings.
Rotary drilling	Power rotation of drilling bit as circulating fluid removes cuttings from hole. Changes indicated by rate of progress, action of drilling tools and examination of cuttings in drilling fluid. Casing usually not required except near surface.	Applicable to all soils except those containing much large gravel, cobbles and boulders. Difficult to determine changes accurately in some soils. Not practical in inaccessible locations because of heavy truck mounted equipment, but applications are increasing since it is usually most rapid method of advancing bore hole.
Percussion drilling (Churn drilling)	Power chopping with limited amount of water at bottom of hole. Water becomes a slurry which is periodically removed with bailer or sand pump. Changes indicated by rate of progress, action of drilling tools and composition of slurry removed. Casing required except in stable rock.	Not preferred for ordinary exploration or where undisturbed samples are required because of difficulty in determining strata changes, disturbance caused below chopping bit, difficulty of access, and usually higher cost. Sometimes used in combination with auger or wash borings for penetration of coarse gravel, boulders and rock formations.
Rock core drilling	Power rotation of a core barrel as circulating water removes ground-up material from hole. Water also acts as coolant for core barrel bit. Generally hole is cased to rock.	Used alone and in combination with boring types to drill weathered rocks, bedrock and boulder formations.

NAVDOCK DM-7, 1962

Stabilization with Water

Boreholes are often filled with water to stabilize the hole. The water in the hole counteracts soil and pore water pressures and thus provides limited borehole support. Water alone can neither prevent caving of borings in soft or cohesionless soils nor a gradual squeezing in of a borehole in plastic or organic soils and is therefore of limited use as a means of borehole support. When water is reduced or lost, as in highly permeable strata, any effect gained by water as a borehole stabilizer is lost. Uncased boreholes filled with water are generally used in rock and often in stiff, cohesive soils.

Stabilization with Drilling Fluid

A borehole can often be stabilized by filling it with a properly proportioned drilling fluid or "mud" which when circulated also serves to remove ground-up material from the bottom of the hole. Table 2 summarizes the approximate proportion of mud mixtures. A 5 to 10 foot long steel casing is grouted in place to stabilize the top of the hole and to hold the piezometer installation equipment. Satisfactory drilling fluid can occasionally be obtained by mixing locally available fat clays with water, but it is usually advantageous and often necessary to add commercially prepared products such as volclay or Aquagel. These mud forming products

TABLE 2. Approximate Proportions of Mud Mixtures.

Purpose of drilling mud	Approximate proportion of material per barrel of water	Viscosity	Descriptive Consistency
For lifting cuttings from borehole	10 to 15 pounds of bentonite for fine-grained soils	Slightly higher than water	Thin cream
	30 pounds of bentonite for coarse-grained soils	About 1.3* times the viscosity of water	Very thick cream
For Supporting drill hole	30 pounds of bentonite and chemical additives and wall stabilizers, as directed by the manufacturer	About 1.3* times the viscosity of water	Very thick cream
	When barite is used about 5 pounds is recommended.		

*Viscosity is measured by a Marsh Funnel which is calibrated with water at 72°F. The time required for a given amount of water to flow through the funnel is considered as 1.0. The value listed above is the relative time for the same amount of mud mixture to flow through the funnel.

consist of highly colloidal, gel-forming, thixotropic clays—primarily bentonite and preferably sodium montmorillonite—with various chemicals added to control dispersion, thixotropy, viscosity, and gel strength. Special chemicals must be added to prevent flocculation when foundations containing salt or sodium sulfate are encountered.

The stabilizing effect of the drilling fluid, in comparison with water alone, is caused in part by its specific gravity and in part by the formation of a relatively impervious lining or "mud cake" on the side walls of the borehole. This mud cake prevents sloughing of cohesionless soils and decreases the rate of swelling of cohesive soils. The drilling fluid also facilitates removal of cuttings from the hole on account of its greater specific gravity and viscosity. Drilling fluid may be lost when cavities or highly permeable strata, such as clean gravels are encountered, especially, where there is also a strong ground water flow. This loss, when it occurs before the artesian stratum is reached, can often be stopped by adding cement, straw, corn stalk, cotton seed hulls, or special commercially prepared fibrous materials to the drilling fluid. These materials deposit in, and seal off, the pervious strata.

A recent development in drilling fluid is a material called Revert. This is a biodegradable mud which behaves very similar to drilling mud, but which reverts back to the viscosity of water after 72 hours. The biodegradable mud is composed of an organic polymer which self-destructs through enzyme breakdown with time, leaving only the water used to mix the mud in the borehole. The boreholes in which wellpoints are to be placed can use Revert for stabilization. However, a minimum period of 72 hours is required between the completion of the borehole and the installation of a wellpoint or piezometer. Johnson Chemical Company has a material called Fastbreak which increases the rate of the reversion process.

Stabilization with Casing

Casing, or the lining of the borehole with steel pipe, provides the safest and cleanest, though relatively expensive (steel pipes may have to be left in the borehole) method of stabilizing the borehole. The casing is normally driven into place with a 250 lb. to 400 lb. drop hammer. It can also be advanced by rotation using special adaptors for the rotary drill with cuttings washed to the surface using water or drilling fluid. After driving or rotating and advancing the casing to the desired depth, the material within the casing is removed by washing or drilling.

Casing is generally installed in 5 foot to 10 foot (1.5 to 3 meters) long sections and may have to be left in place. Where coring is to be continued in rock, the casing is extended through the overburden and socketed two feet into the rock.

While casing prevents caving or squeezing-in of the sides of a borehole, it does not prevent upward movement of soil into the casing under hydrostatic/artesian and insitu pressures. The upward movement of soil into the casing is especially pronounced in cohesionless, organic, and very plastic soils. The insitu pressures must be counteracted by filling the casing with water or drilling fluid and keeping the casing full at all times.

Stabilization by Grouting

A borehole passing through a troublesome zone in rock-cavities, faults, fissures, and broken rock, etc. may be stabilized by filling that zone of the hole with cement grout and thereafter redrilling the hole through the concrete plug. This method can be used only when the hole remains open until the grouting is completed and setting started. Grouting is often preferred to the use of casing, since the diameter of the hole can be maintained, whereas it must be reduced for any extension of the hole below the casing.

PIEZOMETER TYPES

During the last two decades, the instrumentation manufacturers have developed a large variety of versatile pore pressure measuring devices. There are five types of piezometers available in the market: 1) vibrating wire, 2) pneumatic, 3) hydraulic, 4) open standpipe, and 5) bonded resistance strain gage. The first three types are considered suitable for installation under artesian conditions.

Vibrating Wire Piezometer

A vibrating wire piezometer (Figure 2) consists of a porous stone, a metallic disc diaphragm, a prestressed wire attached to the center of the metallic disc, an electric coil, a permanent magnet, and a sensor body. The diaphragm separates the pore water chamber from the measuring system so that a change in pore pressure deflects the diaphragm and changes the tension in the wire. The change in tension is measured by plucking the wire, using the electric coil, and measuring the frequency of the vibration. The wire vibrates in the magnetic field of the permanent magnet causing an alternating voltage to be induced in the plucking coil. The frequency of the output voltage is identical to the frequency of the wire vibration and is transmitted to a frequency-counting device. A calibration curve or table is then used to calculate the pore pressure from the measured frequency changes. The fundamental frequency of vibration of a wire is related to its tension, length, and mass. The wire tension can be expressed in terms of change of strain in the wire or deflection of the metallic diaphragm.

$$f = \tfrac{1}{2}\, \ell \sqrt{t/m}$$
$$t/m = \epsilon E g / \varrho$$
$$\epsilon = \Delta / \ell$$

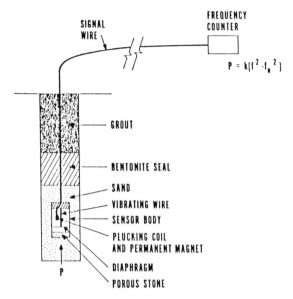

FIGURE 2. Schematic of vibrating wire piezometer.

$$f = \tfrac{1}{2}\, \ell \sqrt{(\Delta E G)/(\ell \varrho)}$$
$$\Delta = f^2 \cdot \tfrac{1}{4}\ell \cdot \varrho / E g$$
P is directly proportional to Δ
$$P = K\,[f^2 - f_0^2]$$

where

f = fundamental frequency of vibration
f_0 = observed frequency of vibration
l = wire length
t = wire tension
m = wire mass per unit length
ϵ = wire strain
E = wire material modulus
ϱ = wire material density
g = acceleration due to gravity
Δ = diaphragm deflection
K = instrument calibration constant
P = pore water pressure

Advantages and disadvantages associated with vibrating wire piezometers:

Advantages:

• adaptable to automatic reading
• suitable for long distance data transmission
• accommodates central terminal observation
• easy and quick to read
• minimum interference with construction activities
• short time lag
• no freezing problems

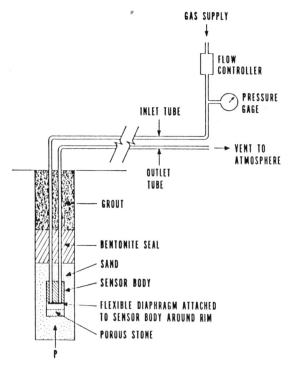

FIGURE 3. Schematic of pneumatic piezometer.

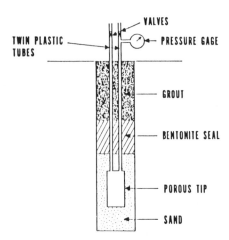

FIGURE 4. Schematic of hydraulic piezometer.

- can read negative pressure
- levels of wire and readout independent of level of tip
- in-place calibration checkup

Disadvantages:

- zero drift problem
- sensitive to temperature, barometric pressure, and electromagnetic changes
- not suitable for dynamic readings
- sensitive to rough handling

Pneumatic Piezometer

A pneumatic piezometer (Figure 3) consists of a porous stone, a flexible diaphragm, a sensitive check valve, a sensor body, an inlet tube, an outlet tube, a pressure gage, and a flow controller. The diaphragm separates the pore water chamber from the measuring system. Pressurized gas applied to the inlet tube causes the check valve to open and lets out the gas from the outlet tube when the applied gas pressure equals the pore water pressure. The pressure is read at the pressure gage on the inlet tube.

Advantages and disadvantages associated with pneumatic piezometers:

Advantages:

- adaptable to continuous recording
- accommodates central terminal observation
- easy to read
- minimum interference with construction activities
- short time lag
- no freezing problems
- level of tubes and readout independent of level of tip

Disadvantages:

- check valve sensitive to corrosion and electrolytic problems
- periodic purging required to remove condensed water in tubings

Hydraulic Piezometer

A hydraulic piezometer (Figure 4) consists of a porous tip, a plastic standpipe, a pressure gage, and a valve. A twin-tube piezometer, also called a closed hydraulic piezometer, consists of two standpipes and two valves. The porous tip is connected to two plastic pipes with the pressure gage on the upper end of one standpipe. The second standpipe is used for periodic flushing of gas that may enter the piezometer tip. The standpipes can be extended to locate the pressure gage at a convenient place. The standpipes are controlled by the valves. Great care is needed to exclude air bubbles by using high quality de-aired water.

There are advantages and disadvantages associated with hydraulic piezometers.

TABLE 3. Piezometer Sensors.

Type	Sensor	Comments on Longevity	Type	Sensor	Comments on Longevity
Hydraulic	Bourdon tub pressure gage	Must be high quality sensor without risk of galvanic corrosion. Gages should be installed in series with a shut-off valve for easy replacement. Check periodically against master gage. Oil filled versions are available and perform well. Best reported longevity appears to be an acetate copolymer and stainless steel gage developed in Australia.	Vibrating Wire	Vibrating Wire Transducer	are made while gas is flowing, a constant flow rate is essential, preferably maintained by an automatic constant volume flow controller. Cause of failure will usually be either zero drift or corrosion. Zero drift will be minimized by annealing and load cycling the wire prior to assembly (most manufacturers insist both are essential for long-term stability, but others report excellent results when neither precaution is taken), by using wire attachment procedures that do not weaken the wire (squeezed capillary tube clamps and swaged pins appear to be preferable), and by keeping wire tension less than 10% of yield. Corrosion can be minimized by selection of materials that are not subject to galvanic corrosion and by drying and hermetically sealing the space around the wire. Attempts to minimize corrosion by venting transducer to atmosphere and using a continuous flow of dry nitrogen have not been successful. Pressure transducer versions are available for checking zero drift during the life of the transducer.
	Manometer	Longevity better than all but acetate copolymer and stainless steel Bourdon tube pressure gages.			
Pneumatic	Pressure Transducer	Diaphragm should be resilient so that a permanent distortion will not be created if sensor is accidentally over-pressurized: nitrile rubber preferable to steel if over-pressurization is possible. A dry gas should be used rather than air. Accessible filter required in inlet line. Periodic flushing may be necessary to remove moisture. Tubing should be impermeable to both air and water: most suitable materials appear to be nylon II with a polyethylene or PVC jacket, or polyethylene with a PVC jacket. Nylon 66 (trade name "type H") is not the best choice, due to its water absorption characteristics. If readings			

Dunnicliff, 1981.

TABLE 4. System Performance.

Type of Piezometer	Comments on Performance	Type of Piezometer	Comments on Performance
Hydraulic (without diaphragm)	Periodic flushing necessary to remove air. However, if system is properly designed, flushing is very seldom necessary unless sub-atmospheric pore water pressures are being read. Use back pressure during flushing (Vaughan, 1973). Use high air entry filters, approx. 0.1 inch (3 mm) i.d. tubing, that is impermeable to both air and water. Do not use nylon 66 tubing (Type H) with high water absorption characteristics. Polyethylene coated nylon 11 has been used with wide success in England and by some U.S.A. public agencies. Use de-aired water with less than 1 ppm dissolved oxygen (Walter Nold Co.). Cannot obtain reading if any part of tubing is significantly (10 to 15 ft., 3 to 5 m) above piezometric level. Can be subjected to falling or rising head test to verify functioning, as for open standpipe piezometer. If Bourdon gages are used, check periodically against master gage. Avoid overpressurizing during flushing. Can use electrical transducer readout system, with transducer calibrator. Use one manometer, Bourdon gage, or transducer for each tube. Cap tubing prior to installation to keep insects out. Operation requires highly trained conscientious personnel. Avoid couplings in tubing (if unavoidable, seal and strengthen using electrical cable splicing kit). Protect all parts of system from freezing. Select terminal house paints carefully: certain paint solvents dissolve polyethylene and nylon tubing.		Check valve displacement caused by the reading operation should be a minimum. (Commercial versions are available with displacement as low as 0.002 cc). Types having one or two additional tubes to the soil side of the diaphragm must be installed such that tubes do not rise more than 10 to 15 ft. (3 to 5 m) above minimum piezometric level of interest (as for twin tube hydraulic piezometers). Cap tubing prior to installation to keep insects out. Avoid couplings in tubing (if unavoidable, seal and strengthen using electrical cable splicing kit). Select terminal house paints carefully: certain paint solvents dissolve polyethylene and nylon tubing.
		Vibrating Wire pressure transducer	Used extensively in Europe, South America, Canada and elsewhere for up to 25 years. Two versions available with check-in-place features. Geonor version (DiBiagio, 1974) has method of applying gas backpressure to diaphragm to check present pressure reading, initial (zero) reading and calibration. Ingenjorsfirman Geotech version has internal method of returning vibrating wire to initial (zero) position by use of an electromagnet. Most common cause of failure of vibrating wire piezometers appears to be sheared cables. Hence cables should, if possible, not be routed through zones of large differential settlement, and cushioning and slack should be provided at any such transitions. Avoid cable splices (if unavoidable, seal and strengthen using electrical cable splicing kit).
Pneumatic pressure transducer	Used in U.S.A. for nearly 20 years, extensively and successively for the past 10 years.		

Dunnicliff, 1981.

Advantages:

- long performance record
- simple and reliable
- accommodates central terminal observation

Disadvantages:

- periodic de-airing required
- freezing problem
- tubing must not be significantly above piezometric level
- long time lag
- other advantages associated with vibrating wire or pneumatic piezometers are not available

PIEZOMETER PERFORMANCE CRITERIA

The selection of a piezometer for installation under artesian conditions must be based on its long-term performance. The nine desirable features that must be addressed during a selection process are: reliability, longevity, simplicity, self verification, connecting linkage, terminal facility, chemical durability, calibration, and data management. Dunnicliff (1981) has discussed the long-term performance of the instruments in great detail.

Reliability

Reliability is a characteristic that assures that the observed reading is either an accurate reading or very close to it. This is an ideal assumption which often cannot be achieved, but should be kept in mind when comparing different types of piezometers.

Longevity

If a piezometer has been properly installed, read, and maintained, yet it fails, the failure is generally associated with its sensor. The sensor of a piezometer must have a proven longevity. Various sensors are listed in Table 3 together with comments on estimated longevity. The longevity estimates are made based on the simplicity and conservatism incorporated into the piezometer design, proven performance, the extent of proof testing under field condition if it is a new design, and the reputation of the manufacturer. The system performance is discussed in Table 4.

Simplicity

A simple piezometer design with minimum moving parts and minimum complex electronics has been found to be more reliable.

Self-Verification

A characteristic whereby the reading is verified in-place. Vibrating wire piezometers are available in which the vibrating wire can be returned to its zero position as a check on zero drift.

Connecting Linkage

Routing of connecting cables, tubes, and pipes needs special attention. Linkages should be free of splices and strong enough to withstand unavoidable and accidental differential strain.

Sufficient slack in the linkage should be allowed in zones suspected of differential settlements.

Terminal Facility

An above ground facility should be able to survive climatic and moisture changes.

Chemical Durability

All components should be designed to survive chemical deterioration: corrosion and electrolytic breakdown.

Calibration Check

Accessible readout instruments should be regularly check-calibrated with standards acceptable in the industry.

Data Management

Data monitoring, computation, and interpretation should be straightforward and simple.

PIEZOMETER SELECTION

The objective is to select a piezometer that provides the value of the pore pressure to an acceptable accuracy over a specific range under specific artesian conditions and does this reliably during the specified observation period.

From the information available to the author and from his experience with the installation and monitoring of several instrumentation programs, there appears to be no basic long-term performance difference between vibrating wire and pneumatic piezometers. Both the vibrating wire and the pneumatic piezometer can provide reliable data over long periods of time, but a substantial number of instruments can fail. The following suggestions, if implemented, can improve overall system longevity and performance:

1. Provide redundancy by installing an instrument package which contains several types of piezometers.

2. Purchase good quality piezometers, cables, and tubing that can withstand rough handling under field conditions, and that have demonstrated good performance at other similar projects. The cost of the piezometer hardware should not be a major factor in the selection of the right instrument.

3. Adhere to the criteria for piezometer long-term performance.

4. Entrust the installation, monitoring, and maintenance of the piezometers to a well qualified staff that is interested in the program and its result and evaluation.

PIEZOMETER CALIBRATION AND INITIAL READING

Piezometer Calibration

Before installation, the piezometers are calibrated. The piezometer calibration serves several important purposes:

1. Verifies that the piezometer and the readout devices are working properly after their delivery from the manufacturer.

The instruments generally undergo a rough handling during shipment, therefore they are checked in accordance with the step-by-step acceptance tests supplied by the manufacturer.

2. Establishes the accuracy of the piezometer and a true relationship between the value measured and the true value of the quantity being measured. The calibration curves or factors—linear or non-linear relationships which are applied to measured values to obtain true values—are checked by immersing piezometers under known depths of water. The zero reading must agree with the reading obtained from the manufacturer's calibration curves.

Where an absolute standard of accuracy is required, certificates of traceability are obtained from the manufacturer. The certificates should indicate that the calibrations were performed using instruments whose accuracy is traceable to the National Bureau of Standards.

3. Verifies correction for field variation: such as temperature and other factors that influence the measurement.

A field calibration is an excellent method to assure an absolute accuracy against field variability.

4. Establishes a correction for long-term drift of the instrument.

Drift can result from deterioration and aging of electrical components, connections, or insulation; from creep and shippage of mechanical anchors; and from build up of corrosion or dirt on operating surfaces.

Readout devices are also recalibrated frequently by using acceptable standards. The recalibration is either performed at the field or at commercial calibration houses, using equipment and techniques certified by the National Bureau

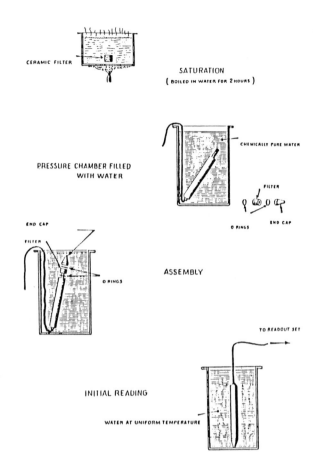

FIGURE 5. Initial reading of piezometer.

of Standards. A sticker on the instrument should indicate the last and next calibration dates.

Piezometer Initial Reading

After checking the piezometer for proper functioning, the instrument is dismantled and the porous element is saturated with water.

Sintered bronze porous elements are merely immersed in chemically pure water, for a certain time, for a complete saturation.

Ceramic porous elements are de-aired and saturated by boiling them in water for two hours. After the element is saturated, it is reassembled under water, upside down (Figure 5). This eliminates the risk of trapping air in the pressure chamber. The cell is then left in the normal position for two to three hours and preferably overnight for temperature stabilization. The initial reading is then taken. The readings sometimes are influenced by a change in the barometric pressure. For accurate measurement, barometric pressure is taken into consideration.

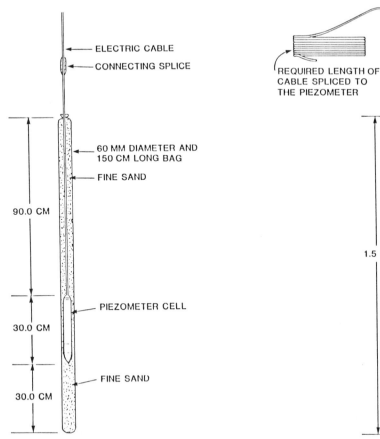

FIGURE 6. Piezometer assembly.

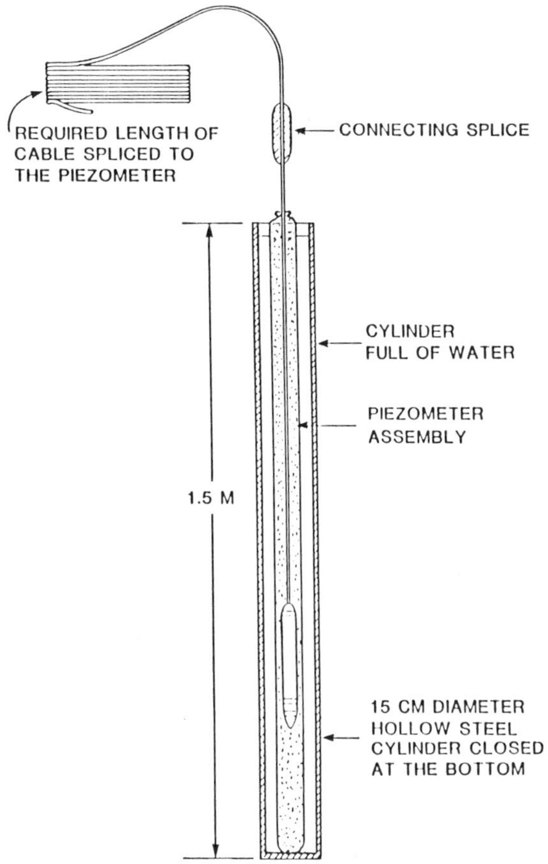

FIGURE 7. Piezometer assembly in carrying-cylinder.

PIEZOMETER INSTALLATION

Piezometers are installed in accordance with the procedures applicable to the artesian conditions encountered in the field:
- variable artesian head and flow
- constant artesian head and flow

Installation Under Variable Head and Flow

PIEZOMETER ASSEMBLY

The piezometer assembly consists of a 2.5 inch (6.0 cm) diameter and 5 feet (150 cm) long bag, fine sand, and a piezometer cell (Figure 6). The bag is sewn from a highly permeable polyester fabric. The bottom one foot (30 cm) of the bag is filled with thoroughly washed and compacted fine sand which is compacted under running water by tamping with a half inch (1.25 cm) diameter steel rod. After taking the initial reading, the piezometer cell is transferred to the bag and placed directly over the sand column. Before the cell transfer, the cell alone; but after the cell transfer, both

the cell and the bag are kept immersed in water to keep the cell filter element saturated and water chamber filled with water. The space around and on the top of the cell is filled with clean, washed, and saturated sand; and is compacted gradually until the whole bag is firm and compact. The bag is then tied with heavy twine to avoid loss of sand.

The piezometer assembly is prepared in the field laboratory and is transferred to the installation site, fully immersed in a carrying cylinder consisting of a five foot (150 cm) long and six inch (15 cm) diameter closed length of casing (Figure 7).

MATERIALS

Sand—ASTM designation C-33 sand (Figure 8) is used to prepare the piezometer assembly and to provide the bottom sand pad and top sand column in the borehole (Figure 12). Sand for piezometer assembly is selected from the fine, and for the borehole from the coarse gradation limits. The fine sand is thoroughly washed with clean water to remove silt size particles, and to avoid contamination of the filter element. The coarse sand gradation is selected for the bore-

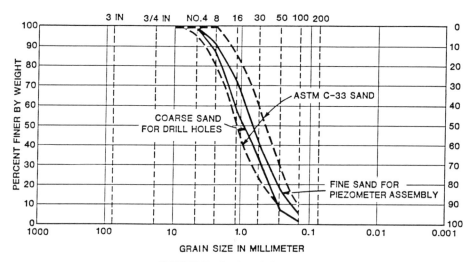

FIGURE 8. Sand gradation curve.

hole bottom pad and top sand column to partially overcome artesian flow resistance against settlement of the sand under gravity, should such flow exist due to any equipment leakage at the ground surface.

Bentonite Pellets—Commercially prepared pre-formed bentonite pellets (Figure 9) have been available for a long time. The pre-formed pellets provide a unique method of placing a bentonite seal at a required location in a borehole. The pellets are placed in the borehole with the same ease as pea gravel and can be used in dry or water-filled holes. The commercially available pellets are manufactured in accordance with the following specifications:

Material Contents: Highly compressed bentonite granular particles of sizes between ASTM No. 20 and No. 70 standard seives.

Composition: Bentonite—a hydrous silicate of alumina comprised essentially of 90 percent montmorillonite.

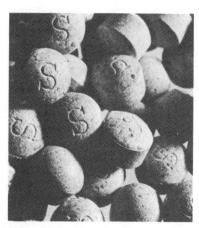

FIGURE 9. Commercially prepared bentonite pellets.

Chemical Composition: Silica 63 percent, alumina 21 percent, iron 3 percent, magnesium 3 percent, sodium and potassium 3 percent, and calcium 1 percent.

Moisture Content: Approximately 10 percent.

pH: 8.5 to 10.5

Size: 1/2″ and 7/8″ diameter

Standard Packing: 25 lbs. protected in waterproof polyethelene bag and sturdy cardboard carton.

Relative Density (to water): 1.8

Dry Hole Density: 70 lb. per cubic foot (Weight required to backfill a borehole volume of one cubic foot).

In case commercially pre-formed pellets are not available in the field, then the half inch (1.25 cm) size pellet can be prepared manually. A thick paste of bentonite is prepared by mixing 0.22 pound (100 gm) of bentonite with one pound (480 gm) of potable water. The paste is mixed in a laboratory mixer and rolled by hand into half inch (1.25 cm) size spherical pellets. The pellets are air-dried before use.

Cement-Bentonite Slurry—The cement-bentonite slurry used to backfill the piezometer borehole is prepared using a cement-water ratio of 1.0 and bentonite-water ratio of 0.08 (all by weights).

INSTALLATION EQUIPMENT

For artesian head ≤ about 15 feet (5 meters), the installation equipment (Figure 10) consists of a steel cylinder, coupling connections, pressure valves, a pressure gage, a fine wire screen, and leak-proof stuffing boxes. The equipment can be fabricated in the field or in a workshop. The dimensions of the installation equipment and accessories are determined by the maximum artesian flow. A high flow requires large sized equipment to facilitate free water flow through the pressure regulating valve, and to allow for easy introduction of material and cell by opening the upper coupling connection.

For high artesian pressure the above equipment (Figure 10) is modified to include a pressure loading chamber with valves on both ends and a material loading hopper (Figure 11). The modified installation equipment is suitable for use under most of the artesian conditions and allows all the installation operations to be performed under no-flow conditions.

INSTALLATION PROCEDURE

For artesian head ≤ about 15 feet (5 meters), the installation equipment is connected to the piezometer borehole by a coupling and the artesian pressure is recorded to determine grout pressure for backfilling the hole. The pressure regulating valve is opened to release artesian pressure and allow free water flow. A precalculated amount of coarse sand for the bottom sand pad (Figure 12) is poured by removing the upper coupling. The bottom sand pad is provided to locate the piezometer cell within a previous zone in the artesian stratum. The upper coupling is then replaced and the pressure regulating valve is closed to stop the free upward flow of water, thus allowing the sand to settle by gravity, under no-flow conditions, to a level below the pervious zone. After sufficient time is allowed to settle, water is allowed to flow again through the pressure regulating valve. The upper coupling is removed, the second time, to introduce the piezometer assembly and upper sand column in the borehole. The upper coupling is replaced and the pressure regulating valve is closed again to allow the sand to settle under gravity and no-flow conditions. Bentonite pellets are similarly introduced into the borehole to form the impervious seal. Lastly, the flexible PVC grout pipe is introduced into the borehole through a side mounted stuffing box to a depth corresponding to the top of the bentonite seal. The borehole is backfilled with cement-bentonite slurry under pressure to counter the artesian head. In order to maintain the pressure for 24 hours to allow the cement-bentonite slurry to set, a calculated volume of water is injected to avoid the slurry set up in the equipment itself.

For high artesian pressure, installation equipment shown in Figure 11 is connected to the piezometer borehole and the artesian pressure is determined. The pressure regulating valve is opened to release artesian pressure and allow free water flow. The upper coupling is removed to introduce the piezometer assembly (Figure 6) with its cable passed through one of the stuffing boxes connected to the coupling. The upper coupling is placed back with the piezometer assembly hanging, with its bottom just above the junction of the vertical pipe and the pressure loading chamber, inside the installation equipment. The PVC grout pipe, with its lower end barely sticking out into the vertical pipe, is introduced through the side stuffing box. The pressure regulating valve is closed to induce no-flow conditions. A precalculated amount of coarse sand for the bottom sand pad is poured through the loading hopper into the pressure loading chamber by opening the upper valve. The upper valve is closed and the lower one is opened to introduce the material

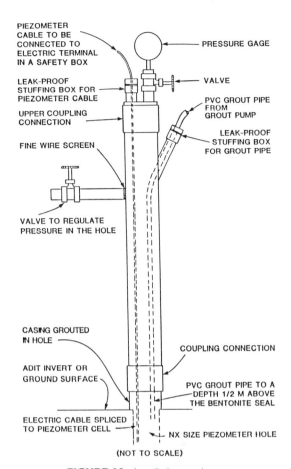

FIGURE 10. Installation equipment.

to settle into the borehole under no-flow conditions. After allowing sufficient settling time, the upper stuffing box is loosened a little to let the cable or tubes slide and piezometer assembly move to its intended position. The stuffing box is tightened and the upper sand column and bentonite pellets are similarly introduced. Lastly, the side stuffing box is loosened to let the PVC pipe reach the top of the bentonite seal, and backfilling the borehole with cement-bentonite slurry is started under pressure. As the borehole gets backfilled, the PVC pipe is withdrawn simultaneously. The borehole is finished in accordance with the procedure discussed above.

Installation Under Constant Head and Flow

PIEZOMETER ASSEMBLY

The piezometer assembly consists of a piezometer head, holding one or more pressure transducers (Figure 13) connected to a pipe nipple. A gate valve at the top and a floor flange at the bottom are threaded to the nipple. A steel plate is welded to the nipple to protect the piezometer head. Two

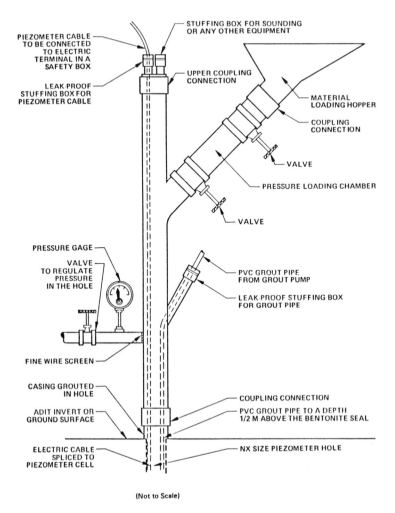

FIGURE 11. Installation equipment.

different types of pore pressure sensing transducers are included in the piezometer head to incorporate redundancy. The piezometer assembly is either installed at the ground surface or at the mudline in an offshore operation. This allows the piezometer assembly to be replaced rapidly and inexpensively, in case it malfunctions or gets damaged.

INSTALLATION EQUIPMENT

The installation equipment (Figure 14) consists of a stainless steel wellpoint of an appropriate length, a steel riser pipe, a riser pipe flange, a rubber gasket, and a pneumatic packer.

INSTALLATION PROCEDURE

After the borehole is advanced to the required depth, the wellpoint, riser pipe, and packer are installed. The packer is inflated, and the borehole is grouted with a cement-bentonite slurry or a cement grout. The grout is allowed to set up for 24 hours and the wellpoint and riser pipe are purged with water until the artesian flow is restored. The restored artesian flow also cleans the drilling mud out of the artesian stratum if drilling fluid had been used for borehole stabilization.

At the completion of the purging, the riser pipe sticking up at the ground surface or at the mudline is threaded for connecting the riser pipe flange. The piezometer assembly, with its gate valve opened to facilitate assembly under flowing condition, is then bolted to the riser pipe flange. The gate valve is closed. The piezometer head is connected to the terminal facility by a readout cable which is encased in an armored conduit and laid on the ground surface or mudline. All the underwater operations are conducted by divers.

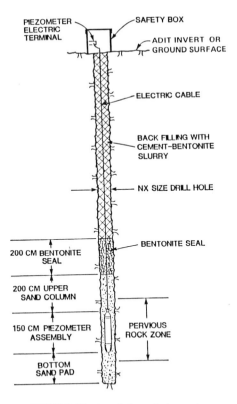

FIGURE 12. Installation of piezometer.

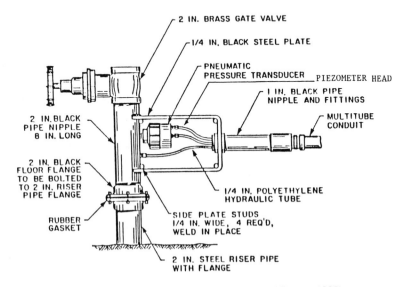

FIGURE 13. Piezometer assembly (Kinner and Dugan, 1985).

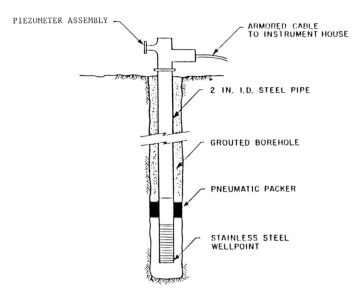

PIEZOMETER ASSEMBLY

ARMORED CABLE
TO INSTRUMENT HOUSE

2 IN. I.D. STEEL PIPE

GROUTED BOREHOLE

PNEUMATIC PACKER

STAINLESS STEEL
WELLPOINT

FIGURE 14. *Installation equipment (Kinner and Dugan, 1985).*

LIGHTNING PROTECTION

Many cases of damage to instruments by electrical disturbance or lightning have been reported. In one of the author's instrumentation programs in Argentina, 25 percent of the instruments (vibrating wire) were demaged by one electrical storm. A well advanced technology to provide protection against lightning damage is now available in the telephone industry. The lightning protection is provided by the use of a high voltage discharge path, a voltage limiting circuit, and a diode limiting technique. Surge arrestors have been introduced into instrument's electrical circuits as a protection against lightning, but surge arrestors have been frequently damaged by transient electromagnetic disturbance. Burkitt (1980) reports that instrument damage due to lightning can be eliminated by providing adequate surge arrestors and good grounding. Burkitt has suggested a surge arrestor that is believed to provide lightning protection. The surge arrestor consists of a spark-gap and gas discharge tubes, Zener diodes, and an automatic disconnect and reset circuit. He also recommends the following rules for a good grounding technique:

1. Connect the surge arrestor via a low-resistance and low-inductance cable to a well-defined ground;

2. Connect the ground termination of the protected equipment directly to the ground terminal on the surge arrestor;

3. Remove all other grounds not complying with the above rule from the protected equipment;

4. Locate the surge arrestor as close as possible to the instrument it is protecting.

Further information on lightning protection is available from General Electric (1976).

PIEZOMETER MAINTENANCE

Maintenance requirements vary with each instrument. The maintenance of a piezometer should regularly be conducted in accordance with the manufacturer's requirements. However, as a general guide, the following provisions are usually applicable:

- Maintain a protective enclosure, guard posts, terminal facility, barricades, identification markings, and instrument approaches for accessibility.
- Refer to the manufacturer's instructions for routine maintenance and trouble-shooting guide in the event of malfunction.
- Check lightning protection provisions.
- Paint carrying cases and protective enclosure with highly visible bright orange color.
- Check and service batteries frequently and follow recommended battery charging procedures.
- Check readout devices against accepted standards and monitor zero drift.
- Clean and dry, periodically, all electronic readout devices. Some readout devices need periodic lubrication.
- Watch for visible sign of damage, deterioration, and malfunction, and repair or replace without delay.
- Protect delicate instruments from shock and rough handling.

- Protect electrical plugs with dust caps and plastic bags under highly humid and wet conditions.
- Check the continuity of signal cables and their leakage to ground characteristics in accordance with the manufacturer's recommended test procedures.
- Dry periodically pneumatic piezometer lines and de-air periodically hydraulic piezometer lines.
- Flush-out periodically wellpoint riser pipe.

ARTESIAN CONDITIONS MONITORING

Data Recording

The piezometers are monitored under the supervision of a geotechnical engineer who coordinates the efforts of one or more competent data-collection and data-processing technicians. The data-collection technicians must understand how the instruments work, what they are supposed to measure, and why they are necessary. They must be able to recognize a faulty instrument so that it can be rapidly corrected. They must know how to install, maintain, and read the instruments properly. They should be instructed on what to look for in the way of changes in the data so that significant changes can be rapidly brought to the attention of the engineer.

Data is recorded either manually (in a field book or on specially prepared field data sheets), or automatically (on punched tapes), or some combination thereof. In any case, it is important that the latest reading be compared immediately with previous reading so that changes can be verified as real or as errors caused by misreading or instrument malfunction.

Manual Data Recording

The data is recorded manually either at each individual piezometer location or at a central terminal. At the central terminal, output signals are transmitted by wireless telemetry, or over electric cables, or by instrument lead pipe/cable, respectively, to long, medium, and short distance locations. In the simplest form, the technician activates one instrument at a time and records the data manually. The data is recorded either in a field book or on specially prepared field data sheets. If data sheets are used, they are prepared especially for each piezometer. All the important information including project name, piezometer number, readings, remarks, weather, temperature, electric storms, severe climatic changes, construction activity, and any other factor that might possibly influence the readings are recorded either on each individual sheet or in the field book.

Automatic Data Recording

The data is recorded automatically at the individual piezometer location, if the location is accessible by a panel truck (containing automatic recording equipment) or at the central terminal. The central terminal is equipped with a completely automatic system, varying from simple continuous strip-chart records to a high speed chart recording. The data, sometimes, also is recorded on punched tape. There are both advantages and disadvantages associated with the automatic recording system, but the disadvantages often overweigh its advantages.

Advantages:

- reduced manpower cost
- instantaneous data transmittal and use of telemetry
- more frequent or continuous data monitoring
- measurement of rapid and instant change in data
- electronic data storage, management, and processing
- increased reading sensibility and accuracy
- data retrieval from inaccessible locations

Disadvantages:

- entail high initial cost
- complexity and lack of reliability, require initial debugging period
- needs a reliable and continuous source of power
- susceptible to weather and construction damage
- unable to record factors that can influence reading
- encourages "file and forget" data attitude
- replaces knowledgeable data collection technician by a computer programmer

The automatic recording system should be used only where it is absolutely necessary. The system is associated with many pitfalls.

Data Processing

The data processing provides a rapid assessment of data to detect sudden changes requiring immediate action. It summarizes and presents the data in graphical plots to show trends and predict behavior. The graphical plots are necessary to:

- distinguish between trends, real changes, erratic readings, miscalculations, or noise in the data due to lack of reading precision
- easily compare data of various locations and times
- ensure that the important data is not overlooked
- pinpoint location variation in general trends for further investigation
- understand and digest the mass observation data

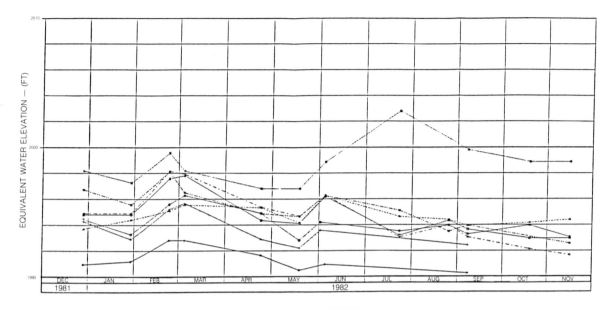

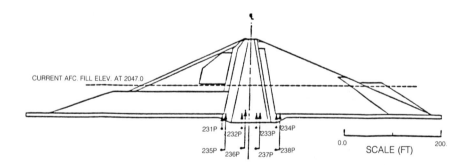

CURRENT AFC. FILL ELEV. AT 2047.0

231P 232P 233P 234P
235P 236P 237P 238P

SCALE (FT)
0.0   200.

EQUIVALENT WATER ELEVATIONS FOR 7 SEP 1982

FIGURE 15. Computer data plotting.

- display mass data clearly and precisely
- make decision making processes simple and quick

The data processing can either be done manually or by computer data-base system.

Manual Data Processing

The raw field data from the field books, field data sheets, or punched tapes is transcribed from digital number into known engineering units by using calibration charts, formulas, and correction factors. Next, the data is compared with the previous data and the difference or changes computed, summarized, and plotted. The transcribed data is then plotted and updated every so often. Somewhere in this chain of events erroneous data from instrument malfunctions or from unexplained causes are identified and rejected. Some guidelines on plotting are discussed by Hansmire (1980) and Lambe (1970).

- All work is done with lead pencils. Use symbols to distinguish between different instruments or different times.
- Choose scales so that the data fills the space available, but do not exaggerate scales so that minor movements appear alarmingly large.
- Draw plots slightly darker than the underlying grid so that when copied both are visible. Make sure data points are visible.
- Plot elevations, depths, and pressure heads on the vertical axes.
- Plots should stand by themselves, that is, they should be self explanatory. Show project name, the type of instrument, the scale and units of measurement, and time of measurement.
- Use sketches on the plot to show the location of the instrument relative to the construction activity and the geology.
- Maintain consistency of scales in a given document so that comparisons of one plot with another can be made easily.

- Time plots showing the progressive change in any parameter that can influence the instrument should be given.
- Plot predicted-behavior on the same chart as the field readings.

Computer Data Processing

The raw field data is transcribed from digital form to computer language on punched cards or magnetic tapes. Once this is done, the tapes can be handled by the computer data-base system with any desired degree of sophistication. The computer data-base system capabilities vary from simple reduction of data with digital printout of results to extremely sophisticated data reduction. It can also be used for data comparison, data plotting (Figure 15), comparison of changes with previous changes, and issuing warning when the boundary between tolerable and intolerable changes are encountered. The engineering profession feels rather strongly that no computer or computer warning system can ever replace the element of engineering judgment in the evaluation of field data.

DATA INTERPRETATION

The data is analyzed and interpreted by intelligent and experienced technical personnel. The first step is to assess the correctness of the reading and to decide if any adverse situation exists that calls for immediate attention. The second step is to establish a cause and effect relationship to study the deviation of the reading from predicted behavior. Sometimes a real data, at first sight, does not appear to be reasonable, and there is a temptation to reject the genuine data which may in fact be carrying an important message. In such cases, it is important to come up with a hypothesis that is consistent with the data. The resultant discussion, together with the use of appropriate procedures for ensuring reading correctness, often leads to an assessment of data validity. Without a clear sense of purpose for an instrumenta-

TABLE 5. Instrument Pre-Installation Data.

Piezometer Number	Date Installed	Cell Elevation (Meter)	Ground Level Elevation (Meter)	Artesian Head Above Ground Level (Meter)	Position From Grout & Drain Curtain
PZ-1	04/20/77	345.8	380.8	10	Downstream
PZ-3	03/07/77	343.1	421.8	12	Upstream
PZ-4	03/10/77	344.8	374.2	7	Downstream
PZ-6	04/30/77	343.8	424.0	0	Downstream

1 meter = 3.28 foot.

tion program, there can be no interpretation. Once a purpose is established, the method of data interpretation follows.

CASE HISTORIES

Arch Dam Foundation Monitoring

Vibrating wire piezometers were installed (Logani, 1983) at several critical locations in the abutment of a 650 foot (200 meter) high arch dam in Iran. The piezometers were installed to monitor the increasing foundation uplift pressure during the first reservoir filling. The dam is founded on sedimentary rocks: predominantly dolomites and cavernous limestones. The foundation contained a very pervious zone which was subjected to a widely fluctuating reservoir water head. In addition, a provision was made for the future treatment of the zone for excessive foundation pore pressure. Consequently, extreme changes in pore water pressure and flow were anticipated. The artesian pressure of more than 35 feet (10 meters) of water head above the

ground surface was recorded during the piezometer installation. The installation of piezometers was completed by the "Variable Head and Flow" method (Figure 10). The piezometer record for the four instruments (Table 5) during the first three years is presented in Figure 16. The reservoir and tail water levels are also included. The data indicates the instruments responded accurately to the reservoir water level fluctuations.

Aquifer Depressurization Monitoring

Hydraulic piezometers were installed (Reyes, 1985) in the Swan River aquifer at the Nipawin dam site, in central Saskatchewan, Canada. The piezometers were installed to obtain information for design and later to monitor depressurization of the aquifer. The dam is founded on impervious glacial till underlain by the Swan River Formation: an artesian siltstone-sandstone aquifer with a 42.5 foot (13 meter) water head above the ground surface and a variable permeability. A thin shale zone, the Ashville Formation separates the till from the Swan River. Extreme changes in pore water pressure and flow were anticipated. The installation of

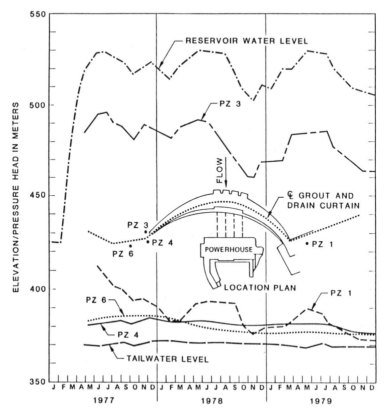

FIGURE 16. Foundation pressure and reservoir filling.

INVENTORY OF PIEZOMETERS—Pore Pressure Measuring Devices.

Category	Type of Instrument	Manufacturers Or Service Agents
Piezometers	Open standpipe	Borros ELE Geonor Hall Landtest Piezometer R&D Roctest Slope Indicator Soil Instruments Soiltest Terrametrics Westbay Instruments
	Open standpipe, drive-in type	Borros ELE Geonor Landtest Roctest Soil Instruments Terrametrics
	Open standpipe, multiple point type	Westbay Instruments
	Closed hydraulic without diaphragm	Landtest Piezometer R&D Soil Instruments Terrametrics Westbay Instruments
	Hydraulic diaphragm	Gloetzl Terrametrics
	Pneumatic transducer	Apparatus Specialties Gloetzl Petur Instruments Slope Indicator Soil Instruments Terrametrics Terra Technology Westbay Instruments
	Closed hydraulic and pneumatic combination	Geonor Petur Instruments Soil Instruments Terrametrics Terra Technology
	Vibrating wire strain gage transducer	Geonor Ing. Geotech Irad Gage Maihak Soil Instruments Telemac

(continued)

INVENTORY OF PIEZOMETERS (continued).

Category	Type of Instrument	Manufacturers Or Service Agents
	Semi conductor strain gage transducer	Geotechniques International Huggenberger Linden-Alimak Slope Indicator Terrametrics Terra Technology
	Semi conductor strain gage transducer, multiple point type	Westbay Instruments
	Detachable semi conductor strain gage transducer	Linden-Alimak
	Unbonded resistance strain gage	Carlson Huggenberger Terrametrics

Dunnicliff, 1980.

piezometer was completed by a procedure very similar to the "Variable Head and Flow" method.

Offshore Cofferdam Foundation Monitoring

Pneumatic piezometers were installed (Kinner and Dugan, 1985) from a drill barge at 300 feet to more than 1,000 feet offshore in water depths up to more than 100 feet. The piezometers were installed to monitor artesian pore pressure required for the design studies of an offshore Cellular Cofferdam. The cofferdam is founded on (from top to bottom) a 10 foot thick alluvial layer commencing at the mudline, a 20 to 40 feet thick deposit of glacial till, and a very thick and pervious sand and gravel aquifer exhibiting artesian pressure equal to 25 to 35 foot water head above mean tide level. A flow of 1 to 100 gallons per minute was measured from 6-inch (15 cm) diameter casing. Due to complex and costly drilling operation, a replaceable piezometer assembly was recommended—an assembly which could be rapidly and inexpensively replaced in case of damage. The assembly contained a pneumatic pressure transducer and a simple hydraulic tube connection. No change in the artesian conditions was expected. The installation of the piezometer was completed by the "Constant Head and Flow" method. Eleven such piezometers were installed and operated satisfactorily throughout a 5-1/2 month period.

MANUFACTURERS' ADDRESSES

Apparatus Specialties Co., Box 122, Saddle River, N.J. 07458 USA.

Borros Co., Ltd., Box 3063, S-17103 Solna 3, Sweden (Contact Roctest).

Carlson Instruments, 1190-C Dell Avenue, Campbell, CA 95008 USA.

Engineering Laboratory Equipment (ELE) Ltd., Durrants Hill Trading Estate, Apsley, Hemmel Hempstead, Hertfordshire, UK.

Geonor A/S P.O. Box 99, Roa, Oslo 7, Norway (Contact ELE, Roctest, or Slope Indicator).

Geotechnik 3 Hannover, Gabelborgerstrasse 5, W. Germany.

Geotechniques International, Inc., P.O. Box E, Middleton, MA. 01949 USA.

Franz Gloetzl, D-7501 Forchheim, Baumesstechnik, W. Germany (Contact Terrametrics or Roctest).

Hall, Inc., 1050 Northgate Drive, San Rafael, CA. 94903 USA.

Huggenberger AG, Hohlstrasse 176, CH-8004 Zurich, Switzerland.

Irad Gage, Etna Road, Lebanon, NH. 03766 USA.

Landtest Ltd., 43 Raywood Road, Rexdale, Ontario, Canada M9V 3Y8.

Linden—Alimak AB, S-93103 Skelleftea, Sweden.

Maihak—H. Maihak AG, 2000 Hamburg, P.O. Box 39, Samperstrausse 38, W. Germany (Contact Roctest).

Petur Instrument Company, Inc., 11300 25th Avenue N.E., Seattle, WA. 98125 USA.

Piezometer R&D Inc., 33 Magee Avenue, Stamford, CT. 06902 USA.

Roctest Ltd., 665 Pine, St. Lambert (Montreal), Quebec, Canada J4P 2P11 or

Roctest, Inc., 7 Pond Street, Platsburg, NY. 12901 USA.

Slope Indicator—Slope Indicator Co., 3668 Albion Place, N. Seattle, WA. 98103 USA.

Soil Instruments, Ltd., Bell Lane, Uckfield, East Sussex, TN22101, England.

Soiltest International, Inc., 2205 Lee Street, Evanston, Illinois 60202 USA.

Telemac, s.r.l., 17 Rue Alfred Roll, Paris 17e, France (Contact Roctest).

Terrametric—Terrametrics, 16027 W. 5th Avenue, Golden, Colorado 80401 USA.

Terra Technology (Division) Corporation, 3018, Western Avenue, Seattle, WA 98121 USA.

Westbay Instruments, Ltd., Suite 1B, 265-25th Street, W. Vancouver, B.C., Canada V7V 4H9.

REFERENCES

1. Acker, W. L., "Basic Procedures for Soil Sampling and Core Drilling," Acker Drill Co., Inc., Scranton, PA., USA (1974).
2. American Association of State Highway and Transportation Officials, Standard Method for Measurement of Pore Pressure in Soils, Designation T252-76, Washington, D.C. (1976).
3. American Association of State Highway and Transportation Officials, Standard Method for Installing, Monitoring, and Processing Data of the Traveling Type Slope Inclinometer, Designation T 254-78, Washington, D.C., pp. 941−950 (1978).
4. American Society for Testing and Materials, "Measurement of Pore-Fluid Pressures in Earth Dams, Field Testing of Soils," Special Technical Publication 322, Philadelphia, PA., pp. 8−18 (1962).
5. Bishop, A. W., M. F. Kennard and P. R. Vaughn, "Developments in the Measurement and Interpretation of Pore Pressure in Earth Dams," *Transactions, Eighth International Congress on Large Dams*, Edinburgh, Vol. 2, pp. 47−72 (1964).
6. Boiten, R. G. and G. Plantema, "An Electrically Operating Pore Water Pressure Cell," *Proceedings, Second International Conference on Soil Mechanics and Foundation Engineering*, Rotterdam, Vol. 1, pp. 306−309 (1948).
7. Bozzouk, M., "Description and Installation of Piezometers for Measuring Pore Water Pressures in Clay Soils," Division of Building Research, National Research Council of Canada, Building Research Note No. 37 (1960).
8. British Geotechnical Society, *Field Instrumentation in Geotechnical Engineering*, John Wiley & Sons, Inc., New York (1974).
9. Burkitt, C. J., "Lightning Protection," *Measurements and Control*, MTL Inc., 7541 Gary Road, Virginia 22110, pp. 128−132 (October 1980).
10. Casagrande, A., "Non-Metallic Piezometer for Measuring Pore Pressures in Clay, Appendix to Soil Mechanics in Design and Construction of Logan Airport," *Journal, Boston Society of Civil Engineers*, Vol. 36, No. 2, pp. 214−221 (1949).
11. Cooling, L. F., "Second Rankine Lecture: Field Measurements in Soil Mechanics," *Geotechnique*, Vol. 12, No. 2, pp. 77−104 (June 1962).
12. Cording, et al., "Methods for Geotechnical Observations and Instrumentation in Tunnelling," Report No. UILU-Eng 75 2002, Department of Civil Engineering, University of Illinois at Urbana-Champaign, Urbana, Illinois (1975).
13. Corps of Engineers, *Instrumentation of Earth and Rockfill Dams, Part I, Groundwater and Pore Pressure Observations*, Engineer Manual 1110-2-1908, Department of the Army, Office of the Chief of Engineers (1971).
14. Corps of Engineers, *Instrumentation of Earth and Rockfill Dams, Part II, Earth-Movement and Pressure Measuring Devices*, Engineer Manual 1110-2-1908, Department of the Army, Office of the Chief of Engineers (1976).
15. Cumming, J. D., *Diamond Drill Handbook*, J. K. Smit & Sons Diamond Products Ltd., Toronto, Canada (1975).
16. Dachn, W. W., "Development and Installation of Piezometers for Measurement of Pore-Fluid Pressure in Earth Dams, Field Testing of Soils," Special Technical Publication 322, ASTM, pp. 8−18 (1962).
17. DiBiagio, E., "Discussion," *Proceedings, Symposium on Field Instrumentation in Geotechnical Engineering*, British Geotechnical Society, John Wiley & Sons Inc., New York, pp. 565−566 (1974).
18. Dunnicliff, J., "Equipment for Field Deformation Measurements," *Proc. Fourth Panamerican Conference on Soil*

Mechanics and Foundation Engineering, San Juan, Vol. 2, pp. 319 – 332 (1971).

19. Dunnicliff, J., "Schematic Arrangements of Various Types of Soil Mechanics and Rock Mechanics Measuring Instruments," *Highway Focus*, Vol. 4, No. 2, pp. 134 – 144 (1972).

20. Dunnicliff, J., "Notes for U.S. Department of Transportation," Federal Highway Administration Training Course on Geotechnical Instrumentation (1980).

21. Dunnicliff, J., D. Hampton and E. T. Selig, "Tunnel Instrumentation: Why and How," *Proc. Rapid Excavation and Tunneling Conference*, San Francisco, pp. 1455 – 1472(1981).

22. Dunnicliff, J., "Long-Term Performance of Embankment Dam Instrumentation," *Proc. Symposium on Recent Developments in Geotechnical Engineering for Hydro Projects, Instrumentation Reliability and Long-Term Performance Monitoring of Embankment Dams*, American Society of Civil Engineers, New York, pp. 1 – 22 (1981).

23. Dunnicliff, J., "Geotechnical Instrumentation for Monitoring Field Performance, NCHRP Synthesis 89," Transportation Research Board, NRC, Washington, D.C. (1982).

24. Forsyth, R. A. and M. McCauley, *Monitoring Devices to Control Embankment Construction on Soft Foundations*, Course Notes Chapter 9, "Slope Stability and Foundation Investigation," Institute of Transportation and Traffic Engineering, University of California, Berkeley (1973).

25. General Electric, *Transient Voltage Suppression Manual*, G. E. Semiconductor Products Department, Electronics Park, Syracuse, New York 13201 (1976).

26. Gibson, R. E., "An Analysis of System Flexibility and Its Effect on Time-Lag in Pore-Water Pressure Measurements," *Geotechnique*, Vol. 13, No. 1, pp. 1 – 11 (1963).

27. Hanna, T. H., *Foundation Instrumentation*, Trans Tech Publications, 21330 Center Ridge Road, Cleveland, OH. 44116 (1973).

28. Hansmire, W. H., "Collection, Processing, and Interpretation of Field Measurements," Rolla Short Course (1978).

29 Heinz, R. A., "In Situ Soils Measuring Devices," *Civil Engineering*, Vol. 45, No. 10, pp. 62 – 65 (1975).

30. Hvorslev, J., "Time Lag and Soil Permeability in Groundwater Observations," Bulletin No. 36, U.S. Army Engineer Waterways Experiment Station, Corps of Engineers, Vicksburg (1951).

31. Kinner, E. B. and J. P. Dugan, "Discussion on Piezometer Installation Under Artesian Conditions," *Journal of Geotechnical Engineering*, Vol. 111, No. 11, ASCE, New York (1985).

32. Kleiner, D. E. and K. L. Logani, "Discussion on Long-Term Performance of Embankment Dam Instrumentation," *Journal of the Geotechnical Engineering Division*, Vol. 102, No. GT8, ASCE, New York (1982).

33. Lambe, T. W., "Interpretation of Field Data," *Proc. Lecture Series on Observational Methods in Soil & Rock Engineering*, Chicago, pp. 116 – 148 (1970).

34. Lindberg, D. A., "Comparative Aspects of Five Piezometer Designs," M. S. Thesis, Department of Civil Engineering,

University of Alberta, Edmonton, Alberta, Canada (May 1965).

35. Little, A. L., et al., "Some Developments in the Measurement of Pore Pressure," *Proceedings, Conference on Pore Pressure and Suction in Soils*, Butterworth, London, pp. 75 – 80 (1961).

36. Logani, K. L., "Piezometer Installation Under Artesian Conditions," *Journal of Geotechnical Engineering*, Vol. 109, No. 8, ASCE, New York (1983).

37. Logani, K. L., "Closure on Piezometer Installation Under Artesian Conditions," *Journal of Geotechnical Engineering*, Vol. 111, No. 11, ASCE, New York (1985).

38. Peck, R. B., "Observation and Instrumentation: Some Elementary Considerations," *Highway Focus*, Vol. 4, No. 2, pp. 1 – 5 (June 1972).

39. Penman, A. D. M., "A Field Piezometer Apparatus," *Geotechnique*, Vol. 6, No. 2, pp. 57 – 65 (1956).

40. Penman, A. D. M., "A Study of the Response Time of Various Types of Piezometers," *Proceedings, Conference on Pore Pressure and Suction in Soils*, Butterworth, London, pp. 53 – 58 (1961).

41. Penman, A. D. M. and M. F. Kennard, "Long-Term Monitoring of Embankment Dams in Britain," *Proc. Symp. Recent Developments in Geotechnical Engineering for Hydro Projects*, ASCE, New York (1981).

42. Peters, N. and W. C. Long, "Performance Monitoring of Dams in Western Canada," *Proc. Symp. Recent Developments in Geotechnical Engineering for Hydro Projects*, ASCE, New York (1981).

43. Plantema, G., "Electrical Pore Water Pressure Cells: Some Designs and Experiences," *Proceedings, Third International Conference on Soil Mechanics and Foundation Engineering*, Switzerland, Vol. 1 (1953).

44. Reyes, M. S., "Discussion on Piezometer Installation Under Artesian Conditions," *Journal of Geotechnical Engineering*, Vol. 111, No. 11, ASCE, New York (1985).

45. Scott, J. D. and J. Kilgour, "Experience with Some Vibrating Wire Instruments," paper presented at the 19th Canadian Soil Mechanics Conference (1966).

46. Shannon, W. L., S. D. Wilson and R. H. Meese, "Field Problems: Field Measurements," in *Foundation Engineering* (Leonards, G. A., ed.), McGraw-Hill, New York, pp. 1025 – 1080 (1962).

47. Sherard, J. L., et al., *Earth and Earth-Rock Dams*, Wiley, New York (1963).

48. Snyder, J. W., "Pore Pressures in Embankment Foundations," Technical Report S-68-2, U.S. Army Engineer Waterways Experiment Station, CE, Vicksburg, Miss. (1968).

49. Terzaghi, K. and R. B. Peck, *Soil Mechanics in Engineering Practice*, John Wiley & Sons, Inc., New York, p. 672 (1967).

50. U.S. Department of the Interior, Bureau of Reclamation, *Earth Manual*, 2nd Ed., pp. 243 – 253, 650 – 699 (1974).

51. Vaughan, P. R., "A Note on Sealing Piezometers in Boreholes," *Geotechnique*, Vol. 19, No. 3, pp. 405 – 413 (1969).

52. Vaughan, P. R., "The Measurement of Pore Pressures with Piezometers," *Field Instrumentation in Geotechnical Engineer-*

ing (British Geotechnical Society), Wiley, New York, pp. 411–422 (1974).

53. Vaughan, P. R., "Discussion on Pore Pressure Reading Equipment for Hydraulic Piezometers," *Proc. Symp. Field Instrumentation in Geotechnical Engineering*, British Geotechnical Society, John Wiley & Sons, Inc., New York, pp. 559–560 (1974).

54. Wilson, S. D. and R. Squier, "Earth and Rockfill Dams," *7th International Conference on Soil Mechanics and Foundation Engineering*, Mexico City, State-of-the-Art, Vol., pp. 137–223 (1969).

55. Wilson, S. D. and D. E. Hilts, "Application of Instrumentation to Highway Stability Problems," *Proc., Joint ASCE-ASME National Transportation Engineering Meeting*, Seattle, American Society of Civil Engineers, New York, and American Society of Mechanical Engineers, New York (1971).

56. Wilson, S. D., "Instrumentation for Dams," *Bulletin of the Association of Engineering Geologist*, Vol. 9, No. 3 (1972).

57. Wilson, S. D. and P. E. Mikkelsen, "Field Instrumentation, Landslide Analysis and Control," Special Report 176 Transp. Res. Board Commission on Sociotechnical Systems, Nat. Res. Council National Academy of Sciences, Washington, D.C. (1978).

CHAPTER 13

Economic Evaluation of Urban Flood Damage Reduction Plans

RALPH A. WURBS*

INTRODUCTION

Due to their flatness, fertility, and esthetics and their accessibility to water supply, navigation, hydropower, and recreational opportunities, flood plains provide advantageous locations for urban and agricultural development. Unfortunately, the same rivers and streams which attract development periodically overflow their banks causing loss of life and property. People can develop the flood plain as if the flood threat did not exist and suffer periodic damages, or they can implement measures to reduce the damages. Externalities, economies of scale, and technical and financial capabilities required to implement flood damage reduction programs necessitate collective action. Consequently, control or alleviation of flooding has been viewed as a public function to be handled primarily by governmental agencies and programs.

One of the major accomplishments of the federal program for water resources development has been the introduction of economic criteria into government decision making. The Flood Control Act of 1936 initiated the use of benefit–cost analysis as a basis for evaluation of proposed federal flood control projects. The act states that federal flood control improvements are justified "if the benefits to whomsoever they may accrue are in excess of the estimated costs, and if the lives and social security of people are otherwise adversely affected." Since 1936, benefit–cost analysis has been applied to essentially all federal flood control projects. The method has also been extended to the other water resources development purposes and has been incorporated into state and local water resources planning and management programs.

Economic evaluation consists of estimating and compar-

ing the benefits and costs, expressed in dollars, which would result from implementation of alternative plans of action. Thus, a quantitative basis for comparing alternative plan configurations and scales of development is provided as well as a criterion for judging whether public expenditures are justified. Although many other important considerations influence the public decision-making process, economic analyses can significantly aid in determining optimal courses of action. Policies, procedures, and practices for evaluating flood damage reduction plans have evolved and developed over the past several decades, primarily in conjunction with the programs of the federal water agencies. The purpose of this chapter is to outline state-of-the-art procedures for economic evaluation of plans for reducing flood damages in urban areas.

FLOOD DAMAGE REDUCTION MEASURES

Economic evaluation of urban flood damage reduction plans involves development and computational manipulation of certain hydrologic, hydraulic, and economic relationships. Flood damage reduction measures are considered here within the framework of a classification scheme based upon which relationship a measure is primarily intended to alter and how it alters this relationship. Thus, the system for categorizing measures adopted below is based upon basic functional relationships used in the economic evaluation procedures discussed later.

The classification system outlined in Figure 1 consists of four major categories of measures: (1) measures which primarily alter the frequency versus discharge relationship, (2) measures which primarily alter the discharge versus stage relationship, (3) measures which primarily alter the stage versus damage relationship, and (4) measures which primarily redistribute the costs related to flood damages. The first two categories consist of measures which are designed

*Civil Engineering Department, Texas A&M University, College Station, TX

393

Measures which primarily alter the frequency versus discharge relationship:

> watershed management
> dams and reservoirs
> diversions

Measures which primarily alter the discharge versus stage relationship:

> levees and floodwalls
> channel improvements

Measures which primarily alter the stage versus damage relationship:

> floodproofing
> flood plain regulation
> public acquisition
> flood forecasting, warning, and emergency measures
> postflood recovery operations
> information and education

Measures which primarily redistribute the costs related to flood damages:

> financial relief
> flood insurance

No action.

FIGURE 1. Classification of flood damage reduction measures.

to partially control the flood water. The third category consists of measures which are designed to reduce the susceptibility of property to damage. Measures in the fourth category are not intended to control the water or reduce damage susceptibility but rather redistribute the monetary losses due to flooding. The "no action" alternative, which means continuing to bear flood losses without attempting to implement damage reduction measures, is a fundamental alternative to be considered in formulating and evaluating flood damage reduction plans.

Measures are also commonly divided into the two broad categories of structural and nonstructural. Structural refers to major public works projects, such as dams, diversions, levees, flood walls, and channel improvements, constructed to partially control flood waters. Nonstructural refers to all other measures. Nonstructural measures include: increasing the capability of individual buildings or other facilities to withstand the damaging effects of flood waters; modifying the way people occupy or use flood plain land; increasing the effectiveness of attempts to deal with a flood emergency situation; and partially controlling flood waters by methods

other than major constructed works, such as watershed land use practices. Programs such as flood insurance, which are designed to redistribute costs, rather than actually prevent damages, are also termed "nonstructural measures."

Measures Which Primarily Alter the Frequency Versus Discharge Relationship

Stream discharge as a function of the probability that a given discharge will be equalled or exceeded in any year is a basic hydrologic relationship used in flood control planning and management. Peak flow at a damage prone location is usually the discharge of greatest concern. An exceedence frequency versus peak discharge function can be altered by four subcategories of measures: (1) weather modification, (2) watershed management, (3) reservoirs, and (4) diversions. The first category modifies the precipitation, the second modifies the runoff hydrograph prior to the flow reaching the stream, and the last two directly modify streamflow. The first two subgroups of measures can be termed "nonstructural" and the last two "structural." Weather modification research related to flood damage reduction has not advanced to the stage of routine practical application and is not addressed further. The other three types of measures discussed below have been extensively used to reduce flood damages.

WATERSHED MANAGEMENT

Watershed management measures reduce runoff by increasing the detention storage capacity of the soil or land surface and by increasing the infiltration rate. The idea is to control the precipitation where it falls. Any approach designed to reduce flood flows by improving conditions of vegetative cover or soil structure or by altering the slope, travel distance, hydraulic roughness, or detention storage related to overland, gully, or small tributary flow is included in this category.

Typical land treatment measures, usually associated with rural watershed areas, include terracing, land leveling, contour plowing, farm ponds, vegetative cover control, fire control, and forest management. Although land treatment measures can be implemented solely for flood control, the usual situation is to use these measures primarily for erosion control with flood control being an incidental benefit. Flood control and erosion control are often closely related. Most erosion occurs during very high flow conditions, and the same measures reduce both sediment and water discharge. Also flood flows usually are made up largely of sediment as well as water. Flood damages are greatly increased by the deposition of sediment in structures. Stream bank erosion during floods is also related to this problem. Land treatment for erosion control also indirectly reduces the potential for downstream flooding by reducing channel and reservoir aggradation.

Snow management for flood damage reduction purposes

is also related to land treatment. Patterns of harvesting forests and other schemes have been proposed to alter snow pack accumulation and melting rates in areas subject to floods from melting snow.

Roofs, streets, and parking areas associated with urban development decrease concentration time and infiltration and increase volumes and peak rates of flood flows. The acceleration and concentration of runoff by construction of storm sewers and other drainage facilities also contribute to increased flood flows. Therefore, control of the manner in which urban areas are developed from the point of view of effects of flood runoff is another flood damage reduction measure which fits under the category of watershed management. A growing number of urban communities are including onsite detention requirements in land development ordinances. Onsite detention measures provide temporary storage of runoff and can be used to offset increased concentration and decreased infiltration caused by urbanization. Any number of approaches, including earthen or paved holding areas, can be used for onsite detection storage. Schemes are also being developed for a municipal government to tax or otherwise require from developers monetary mitigation for increased potential for downstream flood damages, thus encouraging consideration of the effects new residential, commercial, or industrial development has on runoff patterns and characteristics and, consequently, on downstream flooding.

Quantitative and often even qualitative analysis of the hydrologic effects of a particular watershed management measure can be extremely difficult. Disagreement concerning the effects of vegetal cover on floods has existed for many years. Additional research is much needed concerning the effects of urbanization on the hydrology of an area. However, in general, most of the measures considered under the watershed management category would be expected to result in a reduction in the frequency and severity of minor floods but would have relatively little effect as the floods become very large.

DAMS AND RESERVOIRS

Dams and reservoirs also modify the exceedence frequency versus peak discharge relationship for the downstream flood plain. However, the hydrologic and economic impacts of a flood control reservoir typically can be quantified more precisely than the impacts of watershed management measures. Temporary storage of flood waters reduces downstream peak flows. The magnitude of the flow reduction depends upon the storage capacity and release policy of the reservoir, the location of the reservoir in relation to the damage area, and the magnitude and centering of the storm. The degree of protection provided by a reservoir is greatest immediately below the dam and decreases with distance downstream as inflows from uncontrolled portions of the watershed become more dominate.

Reservoirs can be either controlled or uncontrolled. Uncontrolled reservoirs are provided with fixed, ungated outlets which automatically regulate the outflow in accordance with the volume of water in storage. Gated outlet works and spillways provide greater flexibility in operating to maintain allowable nondamaging flow rates at specified downstream control points.

Multiple purpose reservoirs contain storage capacity for flood control as well as conservation purposes such as municipal, industrial, and agricultural water supply, hydroelectric power, and recreation. Flood flows can be converted to beneficial uses by reservoir storage. Multiple purpose development has been found often to offer significant economic advantages over single purpose projects.

DIVERSIONS

Where urban or other improvements are located on the banks of a flood-producing stream such that adequate space is not available for channel enlargement or levees, diversion works may prove economical. Diversions lower the downstream discharge versus frequency function. Diverted flows may be returned directly to the stream downstream from the protected area, temporarily stored in off-channel reservoirs before returning the stream, or diverted to another watershed. The discharge versus frequency function downstream of the point of diversion return will be altered by both the diversion magnitude and timing of return flow. The amount of flow diverted can be controlled by structures such as gated or ungated weirs. Pumping plants are also sometimes used. A floodway is a special type of diversion facility in which flows are diverted through a large depression. The floodway provides both increased flow capacity and detention storage. Opportunities for the construction of floodways are usually limited by the topography of the valley and the availability of low-value land which can be used for the floodway. Physical opportunities for use of all types of flow diversion facilities are limited and consequently used much less frequently than other types of structural flood control measures.

Measures Which Primarily Alter the Discharge Versus Stage Relationship

The stream discharge versus water surface elevation is a basic hydraulic relationship used for many planning and management purposes. Measures in this category lower the water surface profile in the protected flood plain area by increasing the flow capacity of the stream. This can be accomplished by either levees and floodwalls or channel improvements. This is the only one of the four main classification categories which consists entirely of structural measures.

LEVEES AND FLOODWALLS

One of the oldest and most widely used methods of providing protection from floods is to construct barriers preventing stream overflow. Levees are continuous earthen embankments placed parallel to and at varying distances from the natural banks of the stream to serve as an artificial floodway during high flows. In urban areas where adequate space for constructing levees is not available, concrete, masonry, or sheet pile floodwalls are often constructed.

The confinement of flood waters between levees or floodwalls significantly alters the hydrologic characteristics of the flow. Since the water is prevented from flowing through portions of the natural flood plain, the stage and velocity in the stream must be increased in order to carry the same volume of water within the constricted channel section. The downstream discharge is increased, since the valley storage is reduced in the leveed reaches. Thus, while levees and floodwalls have the local effect of increasing the channel capacity and reducing flooding adjacent to the levee, flooding can be increased downstream.

The consequences of overtopping and failure of a levee can be catastrophic. The water stage at the protected area is essentially nondamaging up to the design discharge, but floods above the design discharge can cause much greater damages than if the levee did not exist. Consequently, levees in urban areas are almost always designed to pass at least the standard project flood with a safe freeboard. Tributary streams entering the stream in a leveed or walled reach and interior drainage must be accommodated by ponding areas with gravity drains or pumps. Flood damages sometimes occur due to ponding behind the levees.

CHANNEL IMPROVEMENTS

Channel improvements involve removal of brush and debris, straightening of bends, or channel enlargement in order to increase the discharge capacity of a reach of stream. An increase in slope or cross-sectional area or decrease in hydraulic roughness permits a given discharge to be accommodated through the reach at a lower water surface elevation. Unlike levees and floodways, channel improvements reduce the stage in the protected damage area for all flood discharges including those larger than design channel capacity. Like levees and floodways, the reduction in channel attenuation accompanying channel improvements can significantly increase flooding downstream. Like all other types of structural flood control measures, channel improvements alter the natural sediment transport, ecological, and aesthetic characteristics of the stream.

Measures Which Primarily Alter the Stage Versus Damage Relationship

The stage versus damage function is the economic counterpart to the hydraulic function of stage versus discharge and represents the damages which will occur in a river reach if flood waters reach various levels. Measures in this category either modify the susceptibility of property to

damage, guide the use of flood plain land, or increase the effectiveness of emergency actions taken during a flood event. This and the next main classification categories are composed entirely of nonstructural measures.

FLOODPROOFING

Floodproofing consists of modifications to buildings, contents, grounds, or other facilities to make them less vulnerable to flood damage. Floodproofing may be permanent or may be contingent on some action at the time of flooding. Existing structures can be altered to make them less susceptible to damage or measures can be installed during construction of new structures. Floodproofing measures are probably more likely to be economically justified for commercial than for residential structures. These measures are most applicable where only isolated units of high value are threatened by flooding. However, floodproofing can be beneficially utilized, either alone or in combination with other types of measures, under a wide range of flooding conditions.

Many varied floodproofing approaches and techniques can be devised depending on the individual situation. New structures can be built with floor elevations above a specified flooding level or existing structures can be raised in place. Another approach is to prevent water from entering a building by sealing walls to control seepage; permanently or temporarily closing doors, windows and other openings; and related techniques. Floodproofing includes water resistant building materials, installation of devices such as valves and drains, and construction methods which decrease the susceptibility of buildings and facilities to flooding. Small flood walls designed to protect one or a few structures can also be classified as floodproofing measures. Susceptibility to flood damage can also be reduced by sensible location and arrangement of building contents.

FLOOD PLAIN REGULATIONS

Flood plain regulations modify future susceptibility to damages on flood plains that are not fully developed or where older structures are being rehabilitated. Local communities and state agencies use zoning, land use restrictions, and urban planning to direct the type and location of activities in the flood plain. Subdivision regulations and building codes provide the legal tools for controlling the manner in which buildings are constructed. Wise development policies and action decisions by local governments to prevent construction of street improvements, schools, and other public facilities in locations that would encourage unwise use of flood plain lands can also significantly reduce flood losses.

Under the impetus provided by the National Flood Insurance Program, flood plain regulation became a major component of flood damage reduction programs throughout the nation during the 1970s and 1980s. To be eligible for participation in the federally sponsored and subsidized National Flood Insurance Program, a community must adopt and enforce flood plain management regulations consistent with program criteria. This includes requiring new buildings or substantial improvements to old buildings to be constructed with floor elevations above the 100-year return period base flood level or to be floodproofed. Encroachments within a designed regulatory floodway are not permitted if increases in the base flood level would result.

PUBLIC ACQUISITION

Areas adjacent to streams have a natural attraction and are readily adaptable to recreation and open areas. Acquisition of flood plains can result in beneficial uses compatible with the flood hazard and also reduce flood damage potential by preventing development of uses highly susceptible to flood damages. Public acquisition of land and relocation of occupants is most appropriate for sparsely developed areas. In some instances permanent evacuation of flood hazard areas may be the only economically feasible alternative.

FLOOD FORECASTING, WARNING, AND EMERGENCY MEASURES

Flood damages may be significantly reduced by reliable, accurate, and timely flood forecasts coupled with an effective system to warn floodplain occupants and take emergency actions. An emergency plan of action to respond to a flood threat includes: a system for early recognition and evaluation of potential floods; a warning system; a plan for temporary evacuation of people and property; provisions for installation of temporary protective measures; means to maintain vital services; and a plan for postflood reoccupation and rehabilitation of the flooded area.

POSTFLOOD RECOVERY OPERATIONS

Effective postflood recovery operations can significantly reduce the overall impact of a flood. Postflood recovery usually occurs in two stages, the initial rescue and cleanup operations which occur immediately following the flood and the rebuilding which requires many months or years of effort. The first stage is a continuation of the emergency actions which are discussed in the previous paragraph. The latter stage is closely related to financial relief which is discussed later.

INFORMATION AND EDUCATION

Preparing and disseminating flood hazard information is prerequisite to successful implementation of the various other types of flood damage reduction measures. Vital information includes delineation of flood hazard areas, probabilities of various flooding levels, impacts of flooding, alternative flood damage reduction measures which might be implemented, and potential impacts of land use decisions on flood potential. Based on this information, government

officials, existing and potential property owners, and other floodplain occupants and managers can better make decisions concerning courses of action in dealing with the flood hazard.

Measures Which Primarily Redistribute the Costs Related to Flood Damages

Certain measures do not modify hydrologic or economic relationships but rather redistribute the cost of flood damages. Governmental financial relief to flood victims and flood insurance are common types of measures which fit into this category.

FINANCIAL RELIEF

Financial relief is closely related to postflood recovery operations. Although relief does not directly reduce flood losses, it does reduce the overall loss impact by shortening the period of disruption and by accelerating the return to normalcy. Thus, relief could possibly also be considered under the category of measures which alter the stage versus damage relationship. Relief may be in the form of federal or state grants, loans, or tax adjustments. Aid may also be provided by public and quasipublic agencies to flood victims in the form of donations of food and clothing; temporary shelter and living facilities; and money, materials, and manpower for rebuilding.

FLOOD INSURANCE

Insurance lessens the financial burden of flood losses to the individual and community by spreading the cost over time and over a relatively large number of similarly exposed risks. Property owners in eligible communities may purchase flood insurance through private insurance companies under the federally sponsored and subsidized National Flood Insurance Program. A community is required to adopt and enforce flood plain regulations to maintain eligibility for participation in the program.

Historical Policy Perspective

The nation's early flood control policy was basically every man for himself. Individuals using the flood plain could accept periodic flood damages or use a limited range of nonstructural measures. As communities developed, construction of levees and other protective structures also became feasible. However, cooperative efforts by local groups to protect flood prone lands generally proved inadequate. Thus, pressure for the federal government to do something about flooding problems began to emerge in the mid-19th Century. Prior to 1936, federal responsibility for flood control evolved gradually. Sometimes done under the guise of aiding navigation, it tended to focus on the lower Mississippi River Valley. After a disastrous series of floods in the early 1930s, Congress passed the Flood Control Act of 1936 by which the federal government assumed responsibility for flood control and initiated a nationwide program of constructing structural improvements.

The flood damage reduction activities of the U.S. Army Corps of Engineers, the principal federal flood control agency, are carried out primarily under the authorities made available by the Flood Control Act of 1936 and a subsequent series of flood control acts. The Corps of Engineers has constructed numerous dam, levee, channel improvement, and diversion projects. The Watershed Protection and Flood Prevention Act of 1954 and subsequent amendments thereto established the flood control program of the Soil Conservation Service. Soil Conservation Service projects are generally similar to but smaller than Corps of Engineers projects and are limited to headwater watersheds. The Tennessee Valley Authority, created in 1933, has constructed extensive flood control improvements in the Tennessee River Basin. The Reclamation Act of 1939 authorized the inclusion of flood control storage in Bureau of Reclamation reservoirs. States, cities, drainage and flood control districts, and other nonfederal entities also plan, finance, and construct flood control improvements.

Although numerous flood control projects have been constructed, annual flood damages continue to increase. With few exceptions, flood control structures have adequately performed the services for which they were designed. The increases in flood losses are caused primarily by intensified development of urban flood plain lands and by the increased values of buildings and contents. In small urbanizing watersheds, increased runoff due to urbanization also contributes to downstream flooding problems. The dominant theme of the flood control planning and management literature of the past 20 years has been the necessity for supplementing structural measures with nonstructural measures in order to restrain the increasing flood losses.

The Flood Plain Management Services Program of the Corps of Engineers was one of the first major federal actions related to nonstructural measures. Section 206 of the Flood Control Act of 1960 authorized the Corps of Engineers, when requested by a responsible nonfederal entity, to provide the information needed by state and local governments to guide and regulate the use of a flood plain. This includes identification of flood prone areas and provisions of engineering advice to the nonfederal entities for their use in planning to reduce flood hazards. According to the National Water Commission in 1973, requests for services under this program had far exceeded the expectations of the drafters of the legislation [1]. In recent years, federal responsibility for providing flood plain information has shifted largely to the National Flood Insurance Program.

By 1965, concern by the Bureau of the Budget over an increasing percentage of project benefits that were being claimed for protection of future development in flood plains led to the appointment of the Task Force on Federal Flood Control Policy. The findings and recommendations of the task force, published in 1966, marked a major turning point in the nation's approach to flood problems [2]. A broad approach for managing flood losses based on a comprehensive mix of nonstructural and structural programs was urged. The necessity for planning and regulating land use to reduce flood losses was emphasized. The recommendations of the Task Force on Federal Flood Control Policy were strongly endorsed and reemphasized by the later recommendations of the National Water Commission [1].

The recommendations of the Task Force on Federal Flood Control Policy resulted in creation of a federal flood insurance program. The National Flood Insurance Program was established by the National Flood Insurance Act of 1968, expanded by the Housing and Urban Development Act of 1969 and Flood Protection Act of 1973, and further modified by the Housing and Community Development Act of 1977. The program is a joint effort of the federal government, local communities, and the insurance industry. The Federal Emergency Management Agency is responsible for administration of the program. Program purposes are basically to provide flood insurance at rates made affordable through a federal subsidy and to require local governments to adopt and administer flood plain regulation measures to restrict future development. The flood plain management aspects of the National Flood Insurance Program provide a major thrust toward implementation of a program of nonstructural measures as a basis for avoiding flood losses.

Section 73(a) of the Water Resources Development Act of 1974 requires federal agencies to consider nonstructural alternatives in the planning and design of any project involving flood protection. The Water Resources Council in 1976 laid out a conceptual framework articulating the responsibilities of the federal, state, and local levels of government in responding to the nation's flood problem [3]. Again the necessity for a broader, more comprehensive mix of nonstructural and structural measures was emphasized. Executive Order 11988, signed in May 1977, directed implementation of the Water Resources Council's recommendations. The 1979 revisions to the "Principles and Standards for Water and Related Land Resources Planning" required that a primarily nonstructural alternative plan be formulated whenever structural measures were being considered.

The emphasis on nonstructural measures has significantly affected the economic feasibility of structural measures. Whereas structural measures control the flood water and thus reduce damages related to essentially everything located within the protected flood plain throughout the life of a project, nonstructural measures can be used to protect selected improvements. For example, zoning prevents damage to future development without protecting existing activities. Floodproofing can be applied to individual structures without protecting neighboring structures. Flood plains can be partially evacuated. In many cases, if some of the flood plain occupants are protected by nonstructural measures, benefits to be derived from protection of the remaining occupants do not justify additional structural improvements even though structural improvements would be justified without the nonstructural measures. Whereas the economic efficiency criterion has been stringently applied in the justification of structural measures, certain types of nonstructural measures, most notably flood plain regulation, typically are implemented without performing an economic evaluation.

The most significant implementation of nonstructural measures to date has been flood plain zoning and regulation accomplished under the impetus of the National Flood Insurance Program. The primary purpose of this program is to prevent unwise use of flood plain lands. In the past, benefits related to future development were typically a large percentage of the total benefits included in an economic evaluation. For example, the Task Force on Federal Flood Control Policy pointed out that over 40 percent of the total benefits included in the justification of the 59 Corps of Engineers projects authorized by the Flood Control Act of 1965 were derived from protecting future development [2]. Without flood plain regulation, continued development in a flood plain could very well be expected. Flood control improvements were relied upon to protect the future development. Since the mid-1970s most of the future benefits formerly credited to structural projects in economic studies are eliminated by regulation, or assuming future regulation, of the 100-year flood plain. Economic justification now depends much more heavily on benefits accruing to existing development located in the flood plain.

BASIC ECONOMIC EVALUATION CONCEPTS

Literature

A thorough knowledge of the physical, hydrologic, and economic characteristics of various types of flood mitigation measures combined with a general understanding of analysis techniques from the disciplines of hydrology, hydraulics, civil design, urban planning, and engineering economics is a prerequisite to performing an economic evaluation. An enormous amount of experience in the planning, design, and operation of flood control reservoirs, levees, and channel improvements has been acquired, particularly by the federal agencies since 1936. Economic analyses have been a central thrust of this work. Large collections of planning and design documents related to specific structural projects have been accumulated by the Corps of Engineers, Soil Conservation Service, and other federal and nonfederal agencies and consulting firms. Although practitioners and researchers have published extensively on topics related to both structural and nonstructural measures, the major emphasis in the published literature of the past two decades has been on nonstructural measures.

A bibliography on flood damage prevention published in 1976 lists 800 references [4]. Early general works on flood damage reduction include books by Barrows [5] and Hoyt and Langbein [6]. The more recent water resources planning and management textbooks by James and Lee [7], Linsley and Franzini [8], Petersen [9], Goodman [10], and Viessman and Welty [11] cover benefit–cost analysis and other topics pertinent to the present discussion. Whimple, et al. [12] addresses flood damage reduction from the perspective of urban stormwater management. Grigg, et al. [13,14] outline procedures for evaluation of urban drainage and flood control projects. The Baltimore District of the Corps of Engineers provides a concise overview of structural and nonstructural flood damage reduction measures [15]. Several recent Hydrologic Engineering Center and Institute for Water Resources reports deal specifically with estimating benefits and costs for nonstructural measures [16–19]. Penning-Rowsell and Chatterton outline flood alleviation benefit assessment techniques used in England [20].

Procedures for evaluating flood damage reduction plans have evolved and developed primarily within the federal water resources development community. Both flood control and economic evaluation have been major focuses of the federal water program. The "Green Book" adopted in 1950 and re-issued with revisions in 1958 provided guidance for many early economic analyses [21]. Procedures for evaluating water resources development projects were revised with adoption of Senate Document 97 in 1962 [22]. The Principles and Standards (P&S) became the basic evaluation guidelines in 1973 and were revised in 1979 [23]. A manual of economic evaluation procedures adopted in 1979 supplemented the revised P&S [24]. The Principles and Guidelines (P&G) replaced the Principles and Standards in 1983 [25]. The economic evaluation procedures outlined in the P&G are essentially identical to the P&S. The primary difference is that whereas the P&S were a set of rules that the federal agencies were required to follow, the P&G allow the agencies greater flexibility in determining the extent to which the guidelines are actually followed by the various organizations on each specific study.

The Corps of Engineers developed ER 1105-2-351 [26] to provide more detailed guidance under the general framework of the Principles and Standards. Corps of Engineers regulations were subsequently revised in accordance with the Principles and Guidelines. A Planning Guidance Notebook was developed which includes Engineering Regulation 1105-20-20 which addresses flood damage reduction in general and Engineering Regulation 1105-2-40 and Engineering Pamphlet 1105-2-45 on economic considerations [27–29]. An emphasis in the revisions was in providing greater flexibility in adapting procedures to specific studies.

Economic Objectives

Economic evaluations address both objectives and constraints, which can be formulated in various forms. For a plan to be economically feasible, the benefits must exceed the costs. Stated another way, the ratio of benefits to costs must exceed unity. In comparing alternatives, a typical objective function is to maximize net benefits, which are total benefits minus total costs. In federal programs, net benefits rather than benefit-to-cost ratios are maximized. Net benefit and benefit-to-cost ratio maximization are not the same. For example, in sizing projects based on maximizing net benefits the scale of development will be increased until incremental benefits no longer exceed incremental costs, which tends to drive the benefit-to-cost ratio toward unity.

Another approach is to minimize cost subject to providing a judgementally or institutionally specified degree of protection. For example, levees in urban areas are designed to contain the standard project flood while minimizing the costs for construction and maintenance. The net benefit maximization and specified degree of protection approaches are often combined. A flood control reservoir or channel improvement project may be designed to maximize net benefits subject to the constraint of containing at least the 50-year return period flood.

In some cases, flood damage reduction programs are based on providing a specified degree of protection without consideration of economics at all. The flood plain regulation requirements of the National Flood Insurance Program is a major example of this approach. The 100-year return period flood is the institutionally established base for regulating flood plains.

Economic benefits and costs are those impacts of a proposed action which can be quantified in dollars. Intangible benefits related to the well being of people, such as reducing loss of life, suffering, anxiety, inconvenience, and disruption of economic and social activities, were recognized by the Flood Control Act of 1936 and essentially all pertinent subsequent statements of policy. Environmental quality has been strongly emphasized since enactment of the National Environmental Policy Act of 1969. Impacts of a project on a region's income, employment, population, economic base, and social and cultural development are also assessed. However, these considerations are not reflected in a project's benefit-to-cost ratio. Non-economic impacts are extremely difficult to quantify in common units of measurement for comparison of adverse and beneficial consequences for a given plan of improvement and comparison of the relative merits of different plans of improvement. Consequently, these crucial non-economic consequences must be considered on a largely subjective basis. Although estimates of economic benefits and costs, particularly benefits, are difficult and necessarily approximate, reasonably meaningful numbers can be developed. Consequently, economic analysis plays an important role in providing objective information which can be systematically applied in a complex decision-making process. National policy emphasis during the 1980s on reduced spending and economic efficiency can

be expected to result in continued reliance on economic criteria in project evaluation and selection.

Benefits and Costs

Flood control benefits are derived from reducing damages and permitting more efficient use of land resources. Benefits are conceptually considered in three categories: inundation reduction, intensification, and location. Inundation reduction benefits apply in a situation in which an activity uses the flood plain exactly the same with and without a flood damage reduction plan. The benefit is the increase in net income to the flood plain activity. Intensification benefits occur when a commercial, industrial, or agricultural activity on the flood plain modifies its operation because the reduction in flood damages makes it profitable to do so. Location benefits occur when an activity uses the flood plain with a flood control project but uses a site out of the flood plain if there is not a flood control project. The location and intensification benefits are the increased net income to the activity and land owner comparing the method of operation without a flood control project to that with a project.

Reduction in flood damages represents an increase in net income or an inundation reduction benefit. Flood damages include physical damages or losses, income losses, and emergency costs. Physical damages include damages to buildings and contents and other facilities such as roads, bridges, and utilities. Loss of wages or net profits to business over and above physical flood damages usually results from a disruption of normal activities. Emergency costs include expenses resulting from a flood that would not otherwise be incurred, such as the costs of evacuation and reoccupation, flood fighting, and disaster relief, and the increased costs of normal operations during the flood.

The benefits of a flood damage reduction plan are determined by a comparison of the with and without project conditions. This refers to the land use and related conditions projected to occur in the future if the proposed plan is not implemented compared to those which would occur if the plan is implemented.

The benefits to be derived from a proposed plan of action must be compared with the opportunity cost of the various resources required for plan implementation. The opportunity costs of resource use are usually reflected in the marketplace. When market prices do not adequately reflect resource value, surrogate values may be used to adjust or replace market prices. However, costs are based essentially on the market prices of goods and services.

Cost estimates for structural and nonstructural flood damage reduction plans are developed in essentially the same manner as for other types of public works projects. Many items are typically estimated in terms of quantities and unit prices. Typical cost categories include: land acquisition and severance damages; relocation of highways and utility lines, business operations, and displaced persons; construction; planning and design; administration; and operation, maintenance, and replacement costs.

Discounting

Benefits and costs are discounted to a common time base using interest or equivalence formulas to account for the time value of capital. The time sequence of benefits and costs are converted to either equivalent annual or present worth values for purposes of comparison. A period of analysis of 50 to 100 years is typically used in an economic analysis. Discounting techniques are described by James and Lee [7] as well as a number of engineering economics textbooks.

In an economic analysis using traditional discounting procedures, the time value of money is completely specified by one number, the discount rate. The discount rate is a measure of willingness to sacrifice present consumption in order to produce capital for future use [7]. Selection of an appropriate discount rate for use in public works programs has been a controversial topic. The various arguments for using higher or lower rates are summarized by James and Lee [7] and Steinberg [30]. The current procedure for annually updating the discount rate used in federal programs was established by the Water Resources Development Act of 1974. Quoting from the act, the rate is established as follows:

> The interest rate to be used in plan formulation and evaluation for discounting future benefits and computing costs, or otherwise converting benefits and costs to a common time basis, shall be based upon the average market yield during the preceding fiscal year on interest-bearing marketable securities of the United States which, at the time the computation is made, have terms of 15 years or more remaining to maturity. Provided, however, that in no event shall the rate be raised or lowered more than one-quarter of 1 percent for any year.

The previous discount rate formulas established by Senate Document 97 in 1962 and the Principles and Standards in 1973 varied in detail but were also based on the yield of government bonds.

From 1965 through 1968, the federal discount rate was 3.25 percent. The rate was increased to 4.625 percent in 1969 and has progressed through a series of annual increases to 8.375 percent in 1985. During the last decade, the discount rate formula provision which limits increases to 0.25 percent per year has been the controlling factor in setting the rate each year.

The benefits, costs, and discount rate used in an economic evaluation should all be inflation free or all be equally affected by inflation in a consistent manner [30]. Traditional procedures use a common price level date for benefits and costs without considering inflation. Consequently, the dis-

count rate should also be unaffected by inflation. The 0.25 percent change per year restriction has the effect of somewhat limiting the impact of inflation on the discount rate [30].

The discount rate adopted can significantly influence the results of an economic analysis. Although operation, maintenance, and replacement costs are distributed approximately uniformly over the period of analysis, the bulk of the total costs for a typical structural flood control project are construction related costs occurring at the beginning of the period of analysis. Benefits are distributed relatively uniformly over time. For nonstructural measures, benefits and costs are typically more evenly distributed over time. A high discount rate is unfavorable to proposals involving investment of large amounts of capital now in order to achieve benefits in the future. For example, assuming benefits evenly distributed over a 100-year period of analysis and the 3.25 percent discount rate used in 1968, total undiscounted benefits of $3.39 are required to justify an initial investment of $1.00. If the discount rate is increased to the 1985 value of 8.375 percent, $8.38 in benefits are needed to justify the $1.00 initial investment. With an 8.375 percent discount rate, a $1.00 benefit occurring 50 years and 100 years from now justify investments today of 1.8 cents and 0.03 cents, respectively.

Flood Damage Reduction Versus Stormwater Management

Flood control (or flood damage reduction) and urban stormwater management are two closely associated, yet somewhat different activities. Flood control deals with streams and rivers overflowing their banks. Flood control normally involves relatively large watersheds and well-defined streams. Urban stormwater management is concerned with drainage and detention storage facilities for controlling excessive rainfall on a watershed which has not yet collected as runoff into a major stream. Of course, stormwater management practices affect the frequency and magnitude of flooding on downstream major streams. Stormwater management has traditionally been the responsibility of local communities. The federal government has played a dominant role in flood control. The large federal flood control programs have fostered development of hydrologic and economic analysis techniques. Since stormwater management responsibilities involve a multitude of separate local entities, coordinated funding of research and development efforts have been somewhat hampered compared to the federal programs. Hydrologic engineering, and to a lesser extent economic evaluation, techniques developed by the federal agencies are often applied by local governmental entities and consulting firms doing stormwater management work.

The economic evaluation objectives for flood control and

urban stormwater management are somewhat different. All streams naturally overflow their banks periodically, temporarily inundating their flood plains. Control of all overflows on all streams is impractical and undesirable. Procedures are required for determining on a case by case basis when flood control improvements are warranted and what degree of protection should be provided. To what extent should the flood water be controlled and to what extent should the use of the flood plain be controlled? Benefit–cost analysis provides quantitative information to help make the decisions typically associated with flood control programs. Stormwater management is viewed more as a public service, like police and fire protection, garbage and sewage disposal, and water supply, which should be provided without requiring a benefit–cost justification. The objective of an economic evaluation of urban drainage and detention storage facilities is usually to minimize the costs required to provide a judgementally or institutionally determined level of service or degree of protection. Economic benefits are typically not computed. However, flood control and stormwater management are overlapping and closely related activities. Defining the line of demarcation between the two activities is often difficult. Comprehensive flood damage reduction planning and management dictates an integration of stormwater management and flood control programs. The economic evaluation concepts outlined here have been traditionally associated primarily with flood control but are, in many respects, applicable to urban stormwater management as well.

BENEFIT EVALUATION PROCEDURES

Each flooding situation is different. Numerous approaches and combinations of various types of measures can be formulated to reduce damages. Economic analysis is an interdisciplinary undertaking involving urban planning, hydrology, hydraulics, civil design, and engineering economics. Benefit and cost estimates must necessarily be based largely on judgement and experience. Consequently, an economic evaluation of alternative flood damage reduction plans requires considerable professional skill and ingenuity. A rigid set of detailed economic evaluation rules is not possible or necessarily desirable. A general framework for performing an economic evaluation is outlined here which can be adapted as necessary to each particular study. These general procedures have evolved within the federal program for water resources development and are documented in governmental publications primarily as guidance for the federal agencies. However, the procedures are equally applicable for nonfederal studies.

Steps in Evaluating Benefits

The process of estimating urban flood damage reduction benefits is organized into ten steps which are outlined by the

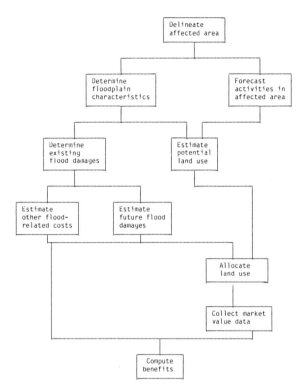

FIGURE 2. Flowchart of urban flood damage benefit evaluation procedures.

planning manual [24] developed by the Water Resources Council in 1979 to supplement the Principles and Standards, repeated again in the Principles and Guidelines [25], incorporated into agency regulations [28], and discussed by others [30]. The ten steps are delineated in Figure 2 and summarized below. The procedure is designed primarily to determine land use and to relate use to flood hazard from a national economic development perspective. The level of detail and effort expended on each step depends upon the nature of the study and the sensitivity of results to further refinements.

STEP 1 – DELINEATE AFFECTED AREA

The area affected by a proposed plan consists of the flood plain plus all other nearby areas likely to serve as alternative sites for any major type of activity that might use the flood plain if it were protected.

STEP 2 – DETERMINE FLOOD PLAIN CHARACTERISTICS

The flood plain is inventoried to determine those characteristics that make it attractive or unattractive for the land use demands established in steps 3 and 4, with emphasis on those characteristics that distinguish the flood plain from other portions of the affected area. Characteristics of a flood plain include: flooding conditions and flood characteristics; areas which would affect flood flows elsewhere if urbanized or structurally protected; natural attributes such as open space, wildlife, and recreation potential; physical characteristics such as slope, soil types, and water table; resources such as water supply or sand, gravel, and mineral deposits; transportation and other services; and existing activities.

STEP 3 – PROJECT ACTIVITIES IN AFFECTED AREA

Economic and demographic projections normally include: population, personal income, recreation demand, manufacturing, and employment. Additional projections may be necessary for any given area, depending on the potential uses of the flood plain and the sensitivity of the plan to these projections. Projections should be based on the most recent available studies, assessments of trends in larger areas and the relationship of historical data for the affected area to trends projected for the larger areas; and consultations with local officials and planners.

STEP 4 – ESTIMATE POTENTIAL LAND USE

Potential land use within the affected area is estimated by converting demographic projections to acres. The conversion factors can normally be derived from published sources, studies of similar areas, or from empirical data for the affected area.

STEP 5 – PROJECT LAND USE

Land use demand is allocated to flood plain and non-flood plain lands for the without project condition and for each alternative flood damage reduction plan. The allocation is based on a comparison of flood plain characteristics, the characteristics sought by potential occupants, and the availability of sought-after characteristics in the non-flood plain portions of the affected area.

STEP 6 – DETERMINE EXISTING FLOOD DAMAGES

Existing flood damages are potential losses to activities affected by flooding at the time of the study. Average annual flood damages are estimated using damage-frequency techniques that relate hydrologic flood variables, such as discharge and stage, and probability of occurrence to damage.

STEP 7 – PROJECT FUTURE FLOOD DAMAGES

Future flood damages are the potential losses to activities that are projected to use the flood plain in the future in the absence of a plan. Future is any time period after the year in which the study is completed. Projections developed in steps 3 and 5 are used to estimate how the existing flood damages estimated in step 6 will change in the future in response to changing hydrologic, economic, and physical conditions.

STEP 8 – DETERMINE OTHER COSTS OF USING THE FLOOD PLAIN

Economic impacts other than flooding on existing and potential future flood plain occupants include floodproofing costs, administrative costs incurred in servicing flood insurance policies, and losses due to a flood hazard causing facilities to be used less efficiently than they would be if protected.

STEP 9 – COLLECT LAND MARKET VALUE AND RELATED DATA

If land use is different with and without a flood damage reduction plan, the difference in income for the land is computed using land market value data.

STEP 10 – COMPUTE BENEFITS

The first nine steps provide the information needed to compute the benefits for structural and nonstructural measures. Table 1 displays the types of benefits claimable for three example types of measures and the steps in the procedure that provide the necessary data. The table is generally applicable, but specific cases may vary.

To the extent that step 5 indicates that land use is the same with and without the project, the inundation reduction benefit is the difference in flood damages with and without the project (step 7), plus the reduction in floodproofing costs (step 8), plus the reduction in insurance overhead (step 8), plus the restoration of land values in certain circumstances (step 9).

If step 5 indicates that land uses are the same with and without the project but activity is more intense with the project, an intensification benefit is applicable. The benefit should be measured as the increase in market value of land from step 9 or changes in direct income from step 6, exercising caution to avoid double counting. The intensification benefit cannot exceed the increased flood damage potential when the existing activity is compared to the intensified activity without the proposed plan. Although intensification benefits are theoretically applicable to urban situations, few case studies have been documented in which this type of benefit was actually included.

If step 5 indicates that land use is different with and without the project, the location benefit is measured by the change in the net income or market value of the floodplain land and certain adjacent land where, for example, the plan creates open space (step 9).

TABLE 1. Guide to Types of Benefits.

Type of Benefit (and step)	Structural	Floodproofing	Evacuation
Inundation:			
Incidental flood damages (step 6)	Claimable	Claimable	Claimable
Primary flood damages (step 6)	Claimable	Claimable	Not claimable
Floodproofing costs reduced (step 7)	Claimable	Not claimable	Not claimable
Reduction in insurance overhead (step 7)	Claimable	Claimable	Claimable
Restoration of land value (step 9)	Claimable	Claimable	Not claimable
Intensification (steps 7 and 9)	Claimable	Claimable	Not claimable
Location:			
Difference in use (step 9)	Claimable	Claimable	Not claimable
New use (step 9)	Not claimable	Not claimable	Claimable
Encumbered title (step 9)	Not claimable	Not claimable	Claimable
Open space (step 9)	Not claimable	Not claimable	Claimable

Average Annual Damage

The economic evaluation process outlined in the previously cited federal planning guidance documents and summarized in the previous paragraphs provides a general framework for analysis but does not prescribe the numerous detailed data collection, modeling, and evaluation tasks which are encompassed within the overall process. Likewise, the present discussion does not attempt to describe in detail all the complex urban planning, hydrology, hydraulics, civil design, and engineering economics work required to perform an economic evaluation of urban flood damage reduction plans. Rather, the discussion below focuses on one particular set of computations which typically provides a key central thrust of the overall economic evaluation effort.

The estimation of inundation reduction benefits attributable to reducing physical flood damages is the heart of the economic evaluation process. This benefit is defined as the difference in average annual damages without and with a proposed plan. Stated another way, the benefit is the reduction in average annual damages which would result from implementation of a particular course of action.

Average annual damage computations are based on the concept of expected value from statistics. The terms "average annual damage" and "expected annual damage" are used interchangeably. Expected or average annual damage is computed as the integral of the damage versus exceedence frequency function. Exceedence frequency versus peak discharge, discharge versus stage, and stage versus damage relationships are combined to develop a damage versus frequency relationship. A fundamental assumption of the procedure is that damages can be estimated as a function of peak discharge or stage. Additional analyses are required to show how damages change with variations in flow velocity, duration, and sediment content.

The magnitude of a flooding problem without implementation of damage reduction measures as well as the reductions in flood hazard achieved by various alternative plans of improvement can be quantified in various ways. Discharges, stages, and damages at specified locations can be estimated for historical storms (such as the most severe flood on record), statistical floods (such as the 50-year and 100-year recurrence interval floods), and/or hypothetical floods (such as the standard project flood). Expected or average annual damage is actually a frequency weighted sum of damage for the full range of damaging flood events and can be viewed as what might be expected to occur, on the average, in any present or future year. Additional meaningful information, including discharges, stages, and damages associated with a range of statistical storms, are generated in the process of computing average annual damages. Historical and hypothetical storms can be simulated using similar techniques.

Computational Procedures

The basic functional relationships used in computing expected annual damages are illustrated in Figure 3. The discharge–frequency, stage–discharge, and stage–damage relationships are computed from field data. The damage–frequency relationship is derived from the other three functions. Expected annual damage is the integral of the damage–frequency function.

Frequency, discharge, stage, and damage functions vary along a river. Consequently, a river system must be divided into reaches for analysis purposes. Expected annual damages are computed for each reach and summed to get the total. The reaches can be handled by either index locations or zones. The analysis is conceptually identical either way, varying only in computational detail. Subsequent discussions here address the computations from the perspective of using index locations. This approach involves selecting an index location within each reach to represent the reach. The functional relationships are developed for each index location and represent the variables for the entire

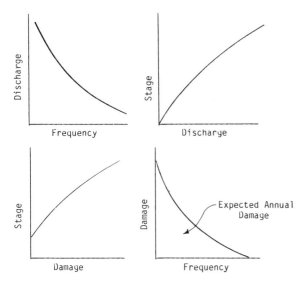

FIGURE 3. Evaluation relationships.

reach. In the alternative approach, water surface profile computations are used to delineate areas of the flood plain inundated along the reach for specified discharges. Damages are estimated corresponding to zones on a flood plain map associated with each discharge. Reach and index location selection is based on experience and judgement and depends upon flood and flood plain characteristics, damage reduction measures being considered, and the level of detail of the study.

Since watershed and flood plain conditions change over time due to urbanization and other factors, average annual damages are computed assuming conditions expected to occur at a particular point in time. Average annual damage computations are typically repeated for present and several alternative future conditions of development such as the years 1990, 2000, 2010, etc. If future changes in watershed and flood plain conditions are considered negligible, the computations are performed for a single set of present conditions. The exact number of discrete changes in conditions for which the computations are repeated is a matter of judgement and depends upon the level of study detail and the extent to which conditions are expected to change in the future. The average annual damages computed for the alternative future years are converted to an equivalent value using discounting techniques and an appropriate discount rate and period of analysis. Thus, a nonuniform distribution of expected values over time is converted to an equivalent uniform distribution for purposes of comparing alternatives and comparing benefits with discounted annual costs.

BASIC RELATIONSHIPS

The peak discharge versus exceedence frequency relationship describes the probabilistic nature of flood flows and

is developed using standard hydrologic engineering techniques. Exceedence frequency is the probability that a given discharge level will be equalled or exceeded in any year. The exceedence probability is the reciprocal of the recurrence interval. Often a single discharge–frequency relationship is considered representative of several reaches or the entire flood plain being studied. Discharge–frequency functions are commonly computed either from a statistical analysis of gaged streamflow data or through rainfall-runoff modeling. Watershed or rainfall-runoff modeling consists of computing the discharge hydrograph resulting from a given rainfall input over a watershed. Rainfall probabilities are obtained from a statistical analysis of gaged data or from published intensity–duration–frequency information. Watershed modeling is often necessitated by the lack of adequate measured streamflow data. Watershed modeling is also required whenever watershed management measures are being considered in order to analyze the change in expected annual damage to result from measure implementation.

Stage versus discharge is a basic hydraulic relationship that relates stage or water surface elevation to discharge and is commonly referred to as a rating curve. It is usually developed from water surface profile computations. A stage at an index location corresponds to a water surface profile along the river reach.

The stage versus damage relationship represents the damages which would occur along a river reach if flood waters reach various levels. Three alternative approaches which have been taken in developing stage versus damage relationships involve using: historical flood damage data for the study area; synthetic data for the study area; or generalized local, regional, or national inundation depth versus percent damage functions.

A historical stage-damage curve can be developed if post-flood damage surveys have been made for several major floods which have occurred in the flood plain in the past. Damages, with price level corrections for inflation, are plotted against stage for the historical floods. Although numerous detailed post-flood surveys have been made at various locations, adequate historical data is still not available for most flood plains. Consequently, generalized depth versus percent damage functions and/or synthetically developed damage data must be used for most studies.

Synthetic damage data can be developed based on engineering estimates of damages which would be sustained by specific structures as a result of various depths of inundation. This approach is used primarily for facilities for which the generalized damage versus percent damage functions discussed below are not pertinent.

The most common approach for developing a stage versus damage relationship for a river reach involves the use of generalized flooding depth versus percent damage functions which have been previously developed from flood data for a specific river basin or region or the nation as a whole. Damage expressed as a percentage of the market value of

TABLE 2. Example FIA Depth Versus Percent Damage Function.

Water Depth in Feet	Damage as a Percent of Property Value			
	Building		Contents	
	1973 Study	1985 Update	1973 Study	1985 Update
0	7	7.7	10	14.4
1	10	11.9	17	21.6
2	14	19.7	23	30.3
3	26	22.5	29	33.4
4	28	25.4	35	35.6
5	29	29.3	40	39.9
6	41	39.6	45	44.6
7	43	42.6	50	49.8
8	44	43.7	55	54.8
9	45	45.0	60	59.9
10	46	46.0	60	59.9
11	47	47.1		
12	48	48.1		
13	49	49.0		
14	50	50.0		
15	50	50.0		

Non-Velocity Zones
One Floor—No Basement

TABLE 3. Example of Integration of Damage–Frequency Function.

Exceedence Probability	Damage ($1,000)	Probability Interval	Average Damage ($1,000)	Weighted Damage ($1,000)
0.000	4,820			
0.001	4,820	0.001	4,820	4.82
0.002	3,620	0.001	4,220	4.22
0.005	2,970	0.003	3,295	9.89
0.01	2,390	0.005	2,680	13.40
0.02	1,510	0.010	1,950	19.50
0.03	810	0.010	1,160	11.60
0.05	220	0.020	515	10.30
0.10	10	0.050	115	5.75
0.20	0	0.100	5	0.50
				79.98

Expected Annual Damages = $79,980.

property is a function of depth of inundation. The depth versus percent damage curves are applied to an inventory of the properties located in the flood plain to develop a stage versus damage relationship. A field survey of the flood plain is made to tabulate pertinent data including the location, floor elevation, type of structure, size, and market value of each building located along the river reach. Location and floor elevation are referenced to the water surface profiles associated with the stage versus discharge function. Market values of structures and contents are estimated. Damages to individual or aggregated groups of structures corresponding to a given flood stage are then determined from the generalized depth versus percent damage curves for different types of damageable property.

The approach just outlined is dependent upon the availability of generalized depth versus percent damage relationships for various types of damageable property. The various organizations which routinely conduct economic evaluations have compiled depth versus percent damage data from a variety of sources. With the implementation of the Federal Flood Insurance Program, the Federal Insurance Administration has become the primary source for this type of information.

Grigg and Helweg [31] investigated the availability of depth versus percent damage data for residential buildings and contents in 1974. They compared curves developed by the Federal Insurance Administration [32], Corps of Engineers [33], Soil Conservation Service [34], and Tennessee Valley Authority [35]. Considerable variation was found to exist between the different data sources. The Federal Insurance Administration curves were in the middle range and were concluded to be probably the most reasonable for estimation purposes.

Davis [36] recently conducted a survey of damage evaluation procedures used in each of the various Corps of Engineers district offices as well as several other organizations. A number of different sets of depth versus damage data were found to be in use. Federal Insurance Administration data were found to be used more than all other residential damage functions combined. A lack of flood damage data was concluded to be particularly evident for commercial, industrial, and institutional property. Federal Insurance Administration data combine all business in the same damage functions.

The Federal Insurance Administration (FIA) is the organization within the Federal Emergency Management Agency (FEMA) which is responsible for administration of the Federal Flood Insurance Program. FIA published depth versus damage relationships developed from a variety of sources in 1970 [32] and revised the data based on a 1973 study [37]. The damage functions were updated in 1985 based on actual claims data for the period 1978 through 1984 [38]. Damage as a percent of property value is tabulated as a function of depth of water above the first floor elevation of the building. Separate functions are provided for structures and their contents. Data sets for structures and contents are provided for the following categories of buildings: one floor with no basement, one floor plus a basement, two floors with no basement, split level with no basement, split level with basement, and mobile homes. Additional contents functions are provided for commercial buildings. The data usually used is for low flow velocities. Limited additional damage data is available for high flow velocities. Table 2 illustrates the data format. Depth versus percent damage functions are provided for a one story building with no basement. The data from the 1973 study is shown along with the 1985 revisions for purposes of comparison. As an example of interpreting the table, if flood waters rise three feet above the floor elevation of a one story building with no basement, based on the 1985 data, the damage is estimated to be 22.5 percent of the market value of the building plus 33.4 percent of the market value of its contents.

INTEGRATION OF THE DAMAGE-FREQUENCY FUNCTION

The damage versus exceedence frequency relationship is derived by combining the three basic functions discussed above. For an assumed frequency, the corresponding peak discharge is determined from the first function, stage from the second, and damage from the third. Expected annual damage is then computed by numerical integration of the damage–frequency relationship. An example of numerically integrating or computing the area under a damage–frequency curve is presented in Table 3. The curve is plotted in Figure 4. The exceedence probability versus damage function is tabulated at discrete intervals in Table 3. The average damage for each interval is multiplied by the probability interval to obtain a weighted damage. The expected annual damage is the sum of the weighted damages. This is the expected annual damage for a given reach corresponding to a specified set of watershed and flood plain conditions.

Modeling Effects of Flood Damage Reduction Measures

Expected annual damages are computed for without project conditions. The computations are then repeated for each of the flood damage reduction plans being considered. The effects of the various types of measures are reflected in the computation of the basic functional relationships. Watershed management, reservoirs, and diversions modify the frequency–discharge relationships at downstream locations. Levees, flood walls, and channel improvements change the discharge–stage function. Nonstructural measures are reflected in the stage–damage function. Any change in these three basic relationships results in a corresponding change in the frequency–damage function and thus in expected annual damages. Benefits are the reduction in expected annual damages resulting from a plan.

Figures 5 and 6 illustrate the changes in expected annual damages resulting from implementation of damage reduc-

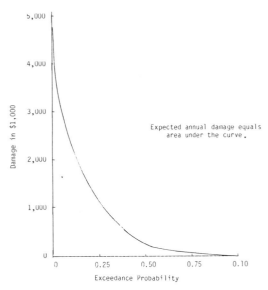

FIGURE 4. Damage–frequency curve. The damage–frequency curve is numerically integrated in Table 3 to obtain an expected annual damage of $79,980.

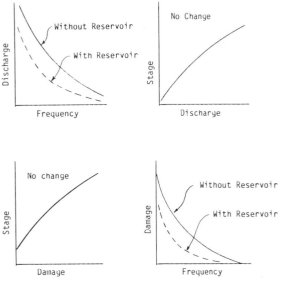

FIGURE 5. Effect of reservoir.

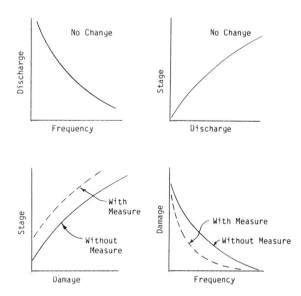

FIGURE 6. Effect of nonstructural measures.

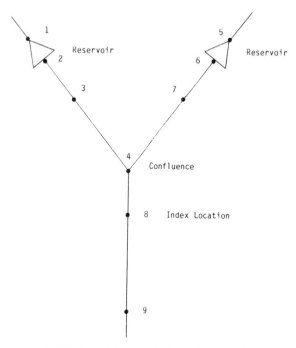

FIGURE 7. Schematic for illustrative example.

tion measures. The expected annual damages for the with and without project conditions are represented by the areas under the corresponding damage–frequency curves. The expected annual damages prevented, or benefit, is the area between the two damage–frequency curves.

INFLOW HYDROGRAPHS

In order to model the effects of structural measures, a series of flood hydrographs representing a broad range of magnitudes must be routed through the stream system. Each flood provides one point on the basic relationships. Each flood consists of a set of inflow hydrographs to the stream system. Hydrographs are included on each tributary at a location upstream of all damage areas and damage reduction measures. Additional hydrographs are included at downstream locations to reflect incremental lateral inflows. The locations of the inflow hydrographs are determined based on engineering judgement considering the watershed and stream configuration and the locations of damage areas and damage reduction measures. Since each flood will create one point on the frequency–damage curve which is to be integrated to obtain expected annual damages, an adequate number and range of magnitude of floods are needed to properly define the frequency–damage curve at each of the index damage locations.

Several alternative approaches can be taken in developing sets of hydrographs corresponding to several floods representing a range of magnitudes. Gaged streamflow data from historical floods is used whenever available. Watershed modeling and related hydrologic engineering methods can be used to generate flood hydrographs from rainfall data. The relationship between the hydrographs at various locations in the stream system for a given storm depends upon storm characteristics as well as watershed characteristics. Maintaining a representative relationship between hydrographs at various locations is a key factor in this type of analysis. Hydrographs can be developed for each flood in the series using historical streamflow and/or rainfall data. An alternative simplified approach, which is incorporated into the Hydrologic Engineering Center computer programs discussed later, involves a set of pattern hydrographs combined with a set of multiflood ratios. One set of pattern hydrographs representative of the relationship between flows at different locations is developed from historical or synthetic data. The pattern hydrographs consist of a hydrograph for each location. The pattern hydrographs are all multiplied by a constant ratio to obtain a given flood. A flood series is represented by a set of ratios with each ratio representing one point on the damage–frequency curve being developed. The ratios are selected such that the resulting range of flood magnitudes will adequately define the damage–frequency curves at each damage location.

ILLUSTRATIVE EXAMPLE

A flood damage reduction plan consisting of two reservoirs is used to illustrate the procedure for computing inundation reduction benefits. The system is shown schematically in Figure 7. The dams are located at points 2 and 6 on two tributaries which confluence at point 4. The flood plain has been divided into four reaches which are represented by index locations 3, 7, 8, and 9. Index damage location 3 represents the flood plain reach between points 2 and 4. Index damage location 7 represents the flood plain reach between points 6 and 4. Index location 8 represents the flood plain reach between point 4 and a point somewhere between points 8 and 9. Index location 9 represents the most downstream reach of the flood plain. Water flows from the flood control reservoirs through ungated outlet structures. Discharge from the reservoirs is a function of the volume of water in storage. The system will be simulated with and without the reservoirs to obtain expected annual damages for with and without project conditions. The inundation reduction benefits are the reduction in expected annual damages. The simulation will be for year 1990 conditions of watershed and flood plain development.

Peak discharge versus exceedence frequency relationships are developed for the index locations 3, 7, 8, and 9 for without project conditions. Water surface profiles are developed for a range of discharges. Discharge versus stage and stage versus damage relationships are developed for index locations 3, 7, 8, and 9. Since the reservoirs will not change these relationships, they can be combined to form a discharge versus damage relationship or can be viewed as separate functions depending on the preference of the analyst.

A set of pattern hydrographs combined with a set of flood ratios will be used to provide inflow hydrographs. (An alternative approach would be to use a series of hydrographs from historical storms.) The pattern hydrographs consist of a representative hydrograph at points 1, 5, 3, 7, 4, 8, and 9. The hydrographs at points 3, 4, 7, 8, and 9 are incremental local flows entering the stream between that point and the next upstream point.

The system is first simulated without the reservoirs. The simulation begins by multiplying each pattern hydrograph by the first flood ratio, thus creating the first flood to be routed through the system. Since the reservoirs do not exist, the flood hydrograph at point 1 is routed to point 2 using channel routing techniques. The resulting hydrograph at point 2 is routed to point 3 and added to the lateral inflow hydrograph at point 3. The peak discharge is read from the resulting hydrograph. The peak discharge is related to the discharge versus frequency curve to assign a frequency for the first flood at index location 3. The peak discharge is related to the discharge–damage function (or discharge–stage and stage–damage functions) to determine the damage at index location 3 resulting from the first flood. The hydrograph at location 3 is then routed to location 4. Next, the hydrograph at location 5 is routed to location 6 using channel routing techniques. The resulting hydrograph at location 6 is routed to location 7 and combined with the lateral inflow hydrograph there. The peak discharge at loca-

tion 7 is used to tabulate the corresponding exceedence frequency and damage for the first flood. The hydrograph is then routed to the confluence of the two tributaries at location 4. The hydrographs from the two tributaries and the lateral inflow hydrograph at location 4 are added. The resulting hydrograph is routed to location 8, the lateral inflow hydrograph added, peak discharge determined from the combined hydrograph, exceedence frequency determined from the frequency–discharge relationship, and damage determined from the discharge–damage relationship. The hydrograph is then routed to location '9 and the process repeated.

At this stage in the computations, the first flood, corresponding to the first flood ratio, has been simulated without the reservoirs. The results include assigning both an exceedence frequency and damage for the first flood for each of the index damage locations. The entire simulation is repeated for each flood. The simulations begin by multiplying the pattern hydrographs by a flood ratio to obtain a different magnitude flood. The computations for each flood are identical to those described above for the first flood. Therefore, at the completion of the without project simulation, the exceedence frequency and damage for each index damage location for a range of flood magnitudes have been determined. The frequency versus damage relationship is numerically integrated to obtain expected annual damages. The exceedence frequencies assigned to each flood (or multiflood ratio) at each index damage location are saved for use in the with project simulation discussed below.

Since frequency–discharge, discharge–stage, and stage–damage relationships for each index damage location were furnished as input to the simulation process described above, these relationships could be combined to compute expected annual damages for the without project condition without routing hydrographs through the system. However, the without project computations were necessary to assign frequencies to the floods. The original frequency–discharge functions are developed for without project conditions and used to assign probabilities to the flood events. Expected annual damages were computed along with the assigning of exceedence frequencies to the floods at each index location.

Next, the simulation is repeated assuming the two-reservoir plan has been implemented. The simulation begins by multiplying the pattern hydrographs by the first flood ratio to recreate the first flood event. The hydrograph at location 1 is now routed through the reservoir. The physical characteristics of the reservoir are reflected in the reservoir routing computations. The outflow hydrograph at point 2 has been modified by the reservoir storage. The hydrograph at location 2 is routed to location 3 using channel routing techniques and combined with the lateral inflow hydrograph at location 3. The peak discharge from the resulting hydrograph is combined with the discharge–damage relationship at location 3 to determine the damage associated with the first flood. This damage value corresponds to the frequency of the flood event at that location which was previously determined during the without project simulation. The hydrograph is then routed to location 4 and so forth. The with project simulation is identical to the without project simulation with two important exceptions. Reservoir routing computations have modified the reservoir outflow hydrographs at points 2 and 6. The impact of the reservoirs on the hydrographs are propagated on downstream. The exceedence frequency assigned to each flood at each index location is that computed for without project conditions.

Reiterating the general procedure, floods are created by multiplying pattern hydrographs by multiflood ratios. The system is simulated for a range of flood magnitudes. Each flood represents a point on the frequency–damage curves developed at each index location. The expected annual damages are computed by numerical integration of the frequency–damage relationship at each index damage location and summed to obtain the total. Expected annual damages are computed for with and without project conditions. The benefit is the reduction in expected annual damages resulting from the project.

The analysis just described was for year 1990 conditions of watershed and flood plain development. The entire procedure can be repeated for future conditions of development, such as 2000, 2010, and 2020. Urbanization and other changes in watershed conditions are reflected in the frequency–discharge relationships which are provided as input to the above described simulation for each index location. Rainfall-runoff modeling is required to simulate the impact of watershed changes on the runoff hydrograph. Changes in flood plain use will be reflected in the stage–damage relationships developed for each index location. After expected annual damages have been computed reflecting watershed and flood plain conditions expected to occur at various times in the future, discounting techniques can be used to compute an equivalent expected annual damage.

OTHER TYPES OF MEASURES

The general procedure for computing inundation reduction benefits, as illustrated by the reservoir example, is modified in various ways to model the effects of other types of measures.

The reservoirs in the example above had ungated outlet structures which meant that outflow was a function of reservoir water surface elevation or the volume of water in storage. A gated reservoir is typically operated in such a manner that releases are dependent upon flow conditions at downstream control points. This complicates the flood routing portion of the simulation, but the general economic evaluation approach is the same.

Diversions remove water from the river system at a certain location. The water may be returned at another location or may be permanently lost to the system being modeled. In computing expected annual damages, a diversion is reflected in the routing computations. The diversion reduces peak discharges at downstream index damage locations.

Channel improvements are reflected primarily in the stage–discharge function for the index damage location at which the channel improvement is located. The channel lowers flood stages in the adjoining flood plain. The stage–discharge functions are typically developed using water surface profile computations. In computing inundation reduction benefits for a plan including a channel improvement, the water surface profile computations are repeated with and without the project. The channel improvement also has an incidental impact of reducing natural flood attenuation and increasing downstream discharges. This can be reflected in the expected annual damage determination through the channel routing computations.

Levees and floodwalls in urban areas are typically designed for at least the standard project flood. Consequently, essentially the entire flood plain behind the levees or floodwalls are protected for the entire range of floods. This effect is reflected in the stage–damage relationships. Levees and floodwalls, like channel improvements, can significantly increase downstream flooding which can be modeled by changing the routing parameters for the reach.

Watershed management measures for reducing flood damages, like the impacts of urbanization in increasing runoff, are reflected in the discharge–frequency relationships. Changes in watershed runoff conditions are typically analyzed using rainfall-runoff modeling techniques.

With the exception of certain watershed management measures and measures which redistribute costs rather than reduce damages, non-structural measures are reflected in the stage–damage function. Various nonstructural measures change the stage–damage function in different ways. For example, floodproofing can be considered from the perspective of the resulting changes in the depth versus percent damage curves used in developing stage–damage functions for the flood plain. Flood plain regulations will not affect the expected annual damages for present conditions but will be reflected in the projected changes in future flood plain conditions. Permanent evacuation decreases the number of structures included in the stage–damage relationship. Implementation of a flood warning system can be reflected in the stage–damage curve as a combination of floodproofing and removing a portion of the building contents and other property.

Computer Programs

Numerous computer programs for performing hydrologic and economic analysis of flood damage reduction plans have been developed at the various offices of the Corps of Engineers, Soil Conservation Service, Tennessee Valley Authority, and other federal and state agencies, private engineering firms, and universities. Viessman, et al. [39] summarize the availability of hydrologic simulation models. A number of computer models are described, many of which can be used for various tasks related to economic evaluation of flood damage reduction plans.

The present discussion is limited to a series of computer programs developed by the Hydrologic Engineering Center (HEC) of the U.S. Army Corps of Engineers. These generalized models are well documented, extensively tested, and readily available from the HEC. HEC computer programs are widely used by other agencies, consulting firms, and universities as well as by the district offices of the Corps of Engineers. Detailed documentation of each computer program is available from the HEC. Feldman [40] provides a general overview summary of the generalized water resources system simulation computer models available from the HEC. The HEC [41] summarizes the capabilities of models available for evaluating nonstructural flood damage reduction measures. Several HEC programs which are particularly pertinent to economic analysis of flood damage reduction plans are cited below.

HEC-1 FLOOD HYDROGRAPH PACKAGE

The "HEC-1 Flood Hydrograph Package" computer program was first published in 1968. Expanded versions with major revisions were released in 1973, 1981, and 1985 [42]. HEC-1 is a general purpose hydrologic model for simulating flood events, which provides a number of options including economic flood damage analysis capabilities. The model provides single-event rainfall-runoff and hydrologic flood routing capabilities. Various options are available for performing each of the tasks involved in the hydrologic simulation of a flood event, including handling of precipitation data, computing infiltration and other losses, developing and applying unit hydrographs, hydrologic reservoir and channel routing, and kinematic routing. Special options are included for calibrating unit hydrograph and loss rate parameters and routing parameters, precipitation depth–area relationship simulation, and dam safety analysis. Uncontrolled reservoirs, diversions, and pumping plants can be included in a system simulation.

The flood damage analysis capabilities of HEC-1 include computing expected annual damages. The computations are performed in the manner outlined in the previous section using pattern flood hydrographs and multiflood ratios. Inundation reduction benefits for up to four plans can be computed in a single run of the program.

Optimal capacities for flood control measures are typically determined by simulating a number of alternative capacities in a trial-and-error manner. System costs and benefits are computed for the alternative capacities. HEC-1 also contains an optional routine in which this process is automated. The flood control system optimization option automatically sizes components of a flood control system based

on the objective of maximizing net benefits. Net benefits are inundation reduction benefits minus discounted annual costs associated with constructing and operating plans of improvement. Reservoirs, diversions, pumping plants, and channel improvements can be optimally sized. The program uses an univariate gradient search algorithm combined with the system simulation.

HEC-2 WATER SURFACE PROFILES

HEC-1 has the capability to compute uniform flow depths using the Manning equation but does not have the capability for gradually varied water surface profile computations. Consequently, HEC-2 is often used in combination with HEC-1 to develop water surface profiles. HEC-2 simulates steady, gradually varied flow in natural and improved open channels, using the standard step method to solve the energy equation [43]. Options are provided for conveniently changing channel improvement and levee configurations and simulating flood plain encroachments.

HEC-5 SIMULATION OF FLOOD CONTROL AND CONSERVATION SYSTEMS

The HEC-5 computer program was designed to simulate the operation of multipurpose, multireservoir systems. The initial version of the program was developed for simulating reservoir operation for a single flood event and was released as "HEC-5 Reservoir System Operation for Flood Control" in 1973. The program was subsequently expanded to include operation for conservation purposes and for period-of-record routings and has undergone several major revisions [44].

HEC-5 has the capability to compute expected annual damages using pattern hydrographs and multiflood ratios in the same manner as HEC-1. The HEC-5 model has the additional optional capability to use a number of historical floods instead of pattern hydrographs and multiflood ratios for computing expected annual damages. Whereas HEC-1 is limited to reservoir routing with a fixed storage versus outflow relationship, HEC-5 incorporates various reservoir operating criteria including making releases based on flows at downstream control points. Unlike HEC-1, HEC-5 has no rainfall-runoff modeling capabilities. Inflow hydrographs, rather than rainfall, must be input to the model.

EXPECTED ANNUAL FLOOD DAMAGE COMPUTATION (EAD)

The Expected Annual Flood Damage Computation computer program [45] uses essentially the same damage–frequency integration procedures as HEC-1 and HEC-5 but has no hydrologic simulation capabilities. Discharge–frequency or stage–frequency, discharge–stage, and stage–damage relationships are provided as input data. Sets of input data can be provided for various future years along with a discount rate and period of analysis, and the program will compute the equivalent expected annual damages. Other computer programs interface with this program for expected annual damage computations.

INTERACTIVE NONSTRUCTURAL ANALYSIS PACKAGE (INA)

The Interactive Nonstructural Analysis Package provides a convenient means to investigate nonstructural measures for particular structures [45]. The package consists of two parts: a preprocessor program that creates a data file containing necessary information for a nonstructural analysis and an interactive analysis program that allows the user to selectively access data for evaluation of nonstructural measures. Damageable property in the floodplain is identified by spatial coordinates or an identification number. Four types of data are input: structure data, hazard data, economic data, and environmental data. The structure data describes the type of structure and its present condition. The hazard data identifies location in the flood-plain, depth of flooding, and the stage versus frequency relationship. Economic data describe the depth versus damage relationship for the structure and its contents. Environmental data indicate features of special significance.

The program is designed to be executed in an interactive mode. The program provides the capability to quickly and efficiently evaluate alternative types of nonstructural measures applied to specific types of structures. Nonstructural flood damage reduction plans are specified and the program interacts with the EAD program to compute expected annual damage.

STRUCTURE INVENTORY FOR DAMAGE ANALYSIS (SID)

The SID computer program is designed to aid in the systematic and expeditious collection and management of data related to structures subjected to flooding. Its basic function is to process structure inventory data to develop aggregate elevation-damage functions by damage categories and location. Because of the capability to develop and manipulate elevation-damage relationships, the program can evaluate the modifications to those functions for nonstructural measures such as flood proofing, relocation and raising. The evaluations are based on individual structural data and user specifications [47].

DAMAGE REACH STAGE-DAMAGE CALCULATION (DAMCAL)

The DAMCAL computer program performs similar evaluations as the SID program except the analyses procedures are based on area concepts (grid cells) of damage potential instead of individual structures. The program accesses geographic information stored in a grid cell data bank for the evaluation (elevation–damage functions by category and damage reach) of existing and future land use patterns for with and without conditions. The non-structural analytical capabilities include: floodplain regulation, flood proofing, permanent relocation, and removal of contents in response to flood warnings. The measures may be evaluated in terms of providing a uniform level of protection, such as the 100-year flood, or protection to specific heights above ground or first floor elevations. The resulting elevation-damage functions are interfaced with other evaluation tool results to perform desired analyses [48].

REFERENCES

1. National Water Commission, "Water Policies for the Future" (June 1971).
2. Task Force on Federal Flood Control Policy, "A Unified National Program for Managing Flood Losses," House Document No. 465, 89th Congress, 2nd Session (August 1966).
3. Water Resources Council, "A Unified National Program for Flood Plain Management," Washington, DC (July 1976).
4. Weathers, J. W., editor, *Flood Damage Prevention, An Indexed Bibliography,* Tennessee Valley Authority and Water Resources Research Center of the University of Tennessee, 8th edition (October 1976).
5. Barrows, H. K., *Floods, Their Hydrology and Control,* McGraw-Hill (1948).
6. Hoyt, W. G. and W. B. Langbein, *Floods,* Princeton University Press, Princeton, New Jersey (1955).
7. James, L. D. and R. R. Lee, *Economics of Water Resources Planning,* McGraw-Hill (1971).
8. Linsley, R. K. and J. B. Franzini, *Water Resources Engineering,* McGraw-Hill, third edition (1979).
9. Petersen, M. S., *Water Resource Planning and Development,* Prentice-Hall (1984).
10. Goodman, A. S., *Principles of Water Resources Planning,* Prentice-Hall (1984).
11. Viessman, W., Jr. and C. Welty, *Water Management Technology and Institutions,* Harper and Row (1985).
12. Whipple, W., N. S. Grigg, T. Grizzard, C. W. Randall, R. P. Shubinski, and L. S. Tucker, *Stormwater Management in Urbanizing Areas,* Prentice-Hall (1983).
13. Grigg, N. S., L. R. Rice, L. H. Botham, and W. J. Shoemaker, "Evaluation and Implementation of Urban Drainage and Flood Control Projects," Colorado State University, Environmental Resources Center, Completion Report No. 56 (June 1974).
14. Grigg, N. S., L. H. Botham, L. Rice, W. J. Shoemaker, and L. S. Tucker, "Urban Drainage and Flood Control Projects: Economic, Legal, and Financial Aspects," Colorado State University, Hydrology Paper 85 (February 1976).
15. U.S. Army Corps of Engineers, Baltimore District, "Flood Damage Reduction Manual," DP800-1-80 (May 1984).
16. Carson, W. D., "Estimating Costs and Benefits for Nonstructural Flood Control Measures," U.S. Army Corps of Engineers, Hydrologic Engineering Center (October 1975).
17. Johnson, W. K., "Physical and Economic Feasibility of Nonstructural Flood Plain Management Measures," U.S. Army Corps of Engineers, Hydrologic Engineering Center and Institute for Water Resources (March 1978).
18. Carson, W. D., "National Economic Development Benefits for

Nonstructural Measures," U.S. Army Corps of Engineers, Hydrologic Engineering Center (October 1980).

19. Moser, D. A., "Assessment of the Economic Benefits From Flood Damage Mitigation by Relocation and Evacuation," U.S. Army Corps of Engineers, Institute for Water Resources, Research Report 85-R-1 (February 1985).

20. Penning-Roswell, E. C. and J. B. Chatterton, *The Benefits of Flood Alleviation, a Manual of Assessment Techniques,* Saxon House, Teakfield Limited, England (1977).

21. Federal Inter-Agency River Basin Committee, Subcommittee on Evaluation Standards, "Proposed Practices for Economic Analysis of River Basin Projects," Government Printing Office (May 1958).

22. President's Water Resources Council, "Policies, Standards, and Procedures in the Formulation, Evaluation, and Review of Plans for Use and Development of Water and Related Land Resources," Senate Document No. 97, 87th Congress, 2nd Session (May 29, 1962).

23. Water Resources Council, "Principles and Standards for Planning Water and Related Land Resources," *Federal Register* (September 10, 1973), Revised, *Federal Register, 44* (242), Part X (December 14, 1979).

24. Water Resources Council, "Procedures for Evaluation of National Economic Development (NED) Benefits and Costs in Water Resources Planning," *Federal Register, 44* (242), Part IX (December 14, 1979).

25. Water Resources Council, "Economic and Environmental Principles and Guidelines for Water and Related Land Resources Implementation Studies," Government Printing Office (March 19, 1983).

26. U.S. Army Corps of Engineers, Office of the Chief of Engineers, "Evaluation of Beneficial Contributions to National Economic Development for Flood Plain Management Plans," ER 1105-2-351 (June 13, 1975).

27. U.S. Army Corps of Engineers, Office of the Chief of Engineers, "Chapter 3—Flood Damage Reduction" of ER 1105-20-20 entitled "Project Purpose Planning Guidance" (May 15, 1985).

28. U.S. Army Corps of Engineers, Office of the Chief of Engineers, "Economic Considerations," ER 1105-2-40, Reprint with Change 1 dated January 8, 1982, Change 2 dated July 9, 1983, and Change 3 dated December 23, 1983.

29. U.S. Army Corps of Engineers, Office of the Chief of Engineers, "Economic Considerations," EP 1105-2-45 (August 6, 1984).

30. Steinberg, B., "Flood Damage Prevention Services of the U.S. Army Corps of Engineers: An Evaluation of Policy Changes and Program Outcomes During 1970–1983 Measured Against Criteria of Equity, Efficiency, and Responsiveness," U.S. Army Corps of Engineers, Institute for Water Resources, Dissertation 84-D-2 (February 1984).

31. Grigg, N.S. and O. J. Helweg, "State-of-the-Art of Estimating Flood Damage in Urban Areas," *AWRA, Water Resources Bulletin, 11* (2) (April 1975).

32. Federal Insurance Administration, "Flood Hazard Factors, Depth Damage Curves, Elevation Frequency Curves, Standard Rate Tables" (September 1970).

33. U.S. Army Corps of Engineers, "Guidelines for Flood Studies" (1970).

34. U.S. Department of Agriculture, Soil Conservation Service, South Regional Technical Service Center, "Floodwater Damage Estimates for Residential and Commercial Property," EWP Technial Guide No. 21, Supplement 21 (December 1970). (Reprinted as Technical Note No. 603 dated October 1978 and supplemented by Bulletin No. S200-0-1 dated August 1980).

35. Tennessee Valley Authority, "TVA Research on Flood Loss Rates" (1969).

36. Davis, S. A., "Business Depth-Damage Analysis Procedures," U.S. Army Corps of Engineers, Institute for Water Resources, Research Report 85-R-5 (September 1985).

37. Federal Insurance Administration, "Depth-Percent Damage: Structure, Residential Contents, Commercial Contents" (January 1974).

38. Federal Insurance Administration, "Flood Insurance Rate Review—1985, Depth Percent Damage, Non-Velocity Zones," unpublished set of tables (1985).

39. Viessman, W., Jr., J. W. Knapp, G. L. Lewis, and T. E. Harbaugh, *Introduction to Hydrology,* Harper and Row, second edition (1977).

40. Feldman, A. D., "HEC Models for Water Resources System Simulation: Theory and Experience," in *Advances in Water Science,* Volume 12, edited by Yen Te Chow, Academic Press (1981).

41. Hydrologic Engineering Center, "Analytical Instruments for Formulating and Evaluating Nonstructural Measures," Training Document No. 16 (January 1982).

42. Hydrologic Engineering Center, "HEC-1 Flood Hydrograph Package, Users Manual" (January 1985).

43. Hydrologic Engineering Center, "HEC-2 Water Surface Profiles, Users Manual" (September 1982).

44. Hydrologic Engineering Center, "HEC-5 Simulation of Flood Control and Conservation Systems, Users Manual" (May 1983).

45. Hydrologic Engineering Center, "Expected Annual Flood Damage Computation (EAD), Users Manual" (February 1984).

46. Hydrologic Engineering Center, "Interactive Nonstructural Analysis Package, Problem Users Manual" (1980).

47. Hydrologic Engineering Center, "Structure Inventory for Damage Analysis (SID), Users Manual" (January 1982).

48. Hydrologic Engineering Center, "Damage Reach Stage-Damage Calculation (DAMCAL), Users Manual (February 1979).

CHAPTER 14

Multi-Objective Approaches to River Basin Planning

L. DUCKSTEIN* AND I. BOGARDI**

This chapter consists of five sections. The first section describes principles of multicriterion decision making (MCDM), gives basic definitions and underlines the need of using MCDM to river basin planning.

The second section defines river basin planning as an MCDM problem by considering objectives, specifications, criteria, generation of alternatives and the alternatives-criteria array. The example of the Santa Cruz River Basin Planning illustrates the methodology.

The third section contains the description of three selected types of MCDM techniques amenable to river basin planning.

The fourth section provides an example of application of MCDM techniques to the Santa Cruz River Basin Planning.

In the fifth section the selected MCDM methods are compared according to viewpoints which can serve as guidelines for choosing among MCDM techniques.

PRINCIPLES OF MCDM

Traditionally, river basin planning has considered a single, mostly economic objective such as expected net benefit. In that case all benefits stemming from the multiple uses or purposes of river basin development has been endeavored to be expressed in monetary terms. For some purposes such as hydropower generation, irrigation, water supply, the consideration of economic benefit is realistic. But for other benefits such as those stemming from flood control, water quality management or recreation the use of monetary term is often artificial and cannot express the real content of the benefit. In such cases the use of physical units such as the number of people protected from floods, amount of sedi-

ment or dissolved oxygen, visitor day seems to be more preferable.

Even within economic benefit, several economic efficiency criteria such as expended net benefit or benefit-cost ratio can be used.

Multiobjective analysis has developed in explicit form largely through the work of the Harvard Water Program (Maass, et al. 1962). Prior to that, the problem of the formation of a single optimality criterion from a number of noncommensurable elementary criteria was treated by Pareto (1896). Since then the concept of "Pareto optimality" slowly found its way into several MCDM techniques.

In reference to multiobjective planning within the USA, the National Environmental Policy Act (NEPA) of 1979, as amended (42 U.S.C. 4321 et seq.) deserves mention. More than any other single piece of legislation, perhaps, it has directed planners to consider more than the narrow criteria used in the part to justify projects. An excerpt from that act reads: "It is the continuing responsibility of the Federal Government to use all practical means . . . to improve and coordinate Federal plans, functions, programs, and resources to the end that the Nation may

1. Fulfill the responsibilities of each generation as trustee of the environment for succeeding generations;
2. Assure for all Americans safe, healthful, productive, and esthetically and culturally pleasing surroundings;
3. Attain the widest range of beneficial uses of the environment without degradation, risk to health and safety, or other undesirable and unintended consequences;
4. Preserve important historic, cultural, and natural aspects of our national heritage, and maintain, wherever possible, an environment which supports diversity, and variety of individual choice;
5. Achieve a balance between population and resource use which will permit high standards of living and a wide sharing of life's amenities; and

*Systems & Industrial Engineering, University of Arizona, Tucson, AZ
**University of Nebraska, Lincoln, NE

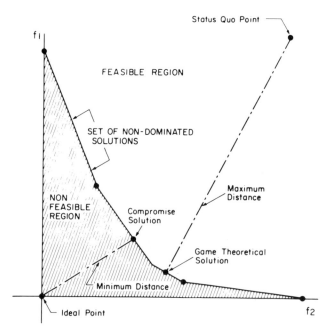

FIGURE 1. Principles of multiple objective optimization: minimize the distance from an ideal point or maximize the distance from a status quo point or trade-off along the Pareto-optimum boundary.

6. Enhance the quality of renewable resources and approach the maximum attainable recycling of depletable resource."

The July 1970 report (United States Water Resources Council, 1973, p. 16) represented a full committment to the use of multiple objectives in the process of project and program formulation: "No one objective has any inherently greater claim on water and land use than any other."

On the international scene, the United Nations Industrial Development Organization (UNIDO, 1972) has issued guidelines for project evaluation that take into account multiple objectives: employment, redistribution of income, balance of payments, self-reliance, and aggregate consumption. These guidelines are addressed primarily to government project evaluators and represent a determined commitment to multiobjective analysis for developing nations.

The single objective programming problem consists of optimizing one objective subject to a constraint set. On the other hand, a multiobjective programming problem is characterized by a p-dimensional vector of objective functions

$$z(\underline{x}) = [z_1(\underline{x}), z_2(\underline{x}), \ldots , z_p(\underline{x})] \qquad (1)$$

and the feasible region is denoted X and defined as follows:

$$X = \{x{:}x \in R^n, \, g_i(x) \le 0, \, x_j \ge 0, \text{ for all } i \text{ and } j\} \qquad (2)$$

In using multiobjective techniques instead of seeking a single optimal solution, a set of "nondominated" solutions is sought. This set of nondominated solutions is a subset of the feasible region. The main characteristic of the non-dominated set of solutions is that for each solution outside the set (but still within the feasible region), there is a non-dominated solution for which all objective functions are unchanged or improved and at least one which is strictly improved. The concept of a nondominated solution also appears in the literature under the terms of Pareto optimum and efficient solution. Figure 1 shows the main elements of multiple objective optimization technique for two objectives such as hydropower generation and amount of water supply. The set of nondominated solution separates the region of feasible and infeasible solutions. The ideal point represents the individual maxima of objective functions, evidently a non-feasible point. On the other hand, a "status quo" point referring to the worst individual values of the objective functions is a feasible solution that may sometimes "be achieved" as a consequence of incorrect river basin planning (poor economic efficiency and poor environmental quality).

MCDM PROBLEM FORMULATION

The procedure of MCDM to river basin planning can be formulated in five steps: (1) objectives, (2) specifications, (3) criteria, (4) generation of alternatives, and (5) evaluation matrix (criteria-alternatives tableau).

Objectives

At this stage, the objectives that the river basin management plan is to fulfill or approach are formulated in general verbal terms. In this respect the objective is the direction of change of state in the river basin to be pursued (or desired). This may call for a better management of an existing system or a new development of river basin. The definitions of objectives requires an overview of the river basin considered.

As for our example, river basin planning in the Upper Santa Cruz Basin, Arizona (Figure 2) is a complex issue involving urban, agricultural, and mining interests. These users account for 29, 41, and 27 percent, respectively, of total water depletion in 1975. Competition for water is extremely intense and the basin tends to be one of the most critically overdrafted areas in the state. Roughly 236,000 acre feet of water were used in 1975 compared with only 74,000 acre feet of water supplied by imports or natural

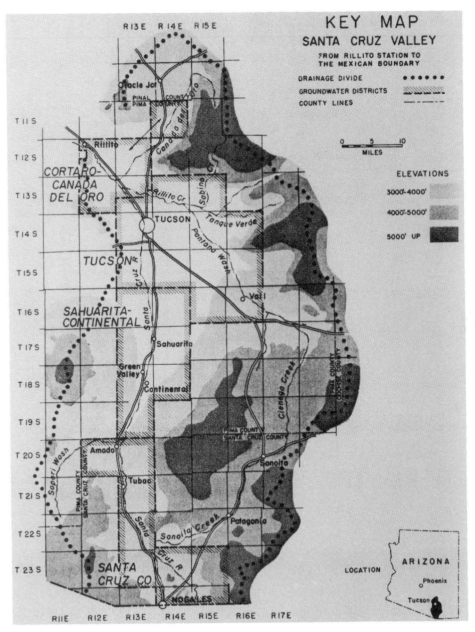

FIGURE 2. Map of the upper Santa Cruz river basin.

recharge. This represents a depletion-supply ratio of 3.3 to
1. Agricultural water demands are projected to decline
slightly over the next thirty years, but both urban and mining demands are expected to more than offset any savings
from agriculture.

Planning problems in the basin are exacerbated by the
rapid growth in population of Tucson and its surrounding
areas. The demand for flood protection, recreational opportunities, and high environmental quality, as well as adequate
water supplies, are of importance to persons living in the
basin.

In developing this river basin planning within a multiobjective context, the following system objectives are given.

1. *Water Requirements:* The region's supply of water (both
   quality and quantity) should be regulated in order to
   meet both present and future demands without seriously
   affecting the region's present ground water level.
2. *Flood Protection:* Flood protection should be provided
   along the main stem of the Upper Santa Cruz River
   while maintaining environmental quality.
3. *Enhancement of Environment:* The effects on wildlife,
   vegetation, and historical or anthropological sites should
   be kept to a minimum.
4. *Utilization of Resources:* The physical and socio-
   economic resources required to implement the alternatives should be kept to a minimum.
5. *Recreational Enhancement:* Recreation opportunities
   should be provided wherever possible along the Santa
   Cruz River Basin.

Specifications

To be useful to the decision maker (DM), the objectives of
river basin planning should be quantified to the fullest extent
possible. A method to achieve this quantification is to express each objective in terms of a set of specifications. By
specification, it is meant the clarification of the objective by
subdividing it into one or more measurable (if quantifiable)
or comparable (if not quantifiable) detailed characteristics
of the problem under consideration. A planning horizon of
40 years in the present example has been adopted to take
into account the long range implications of any development
strategy. The following specifications can be given to the
Santa Cruz planning objectives.

WATER REQUIREMENTS
In defining the specifications with respect to water requirements, a distinction is made between urban and agricultural uses. The specifications of each of these categories
is the same but the criteria (standards) defining them may
differ.

Aquifer Level
Given present supply and demand conditions, there has
been a net decrease in the underground aquifer level. This

can be looked at as a reduction in the supply of water
available for future use. As such, the ground water level
should be stabilized.

Water Quality
The quality of water for urban and agricultural uses
should be maintained in accordance with the State of Arizona Water Quality Standards (Bureau of Water Quality
Control, 1978).

FLOOD PROTECTION
As the population of the study area continues to grow,
flood protection increasingly becomes an issue of importance. Two specifications with respect to this objective will
be considered.

Expected Losses Due to a Flood
Both the social and economic losses due to floods should
be minimized. Resettlement of persons already living in
flood-prone areas should be avoided.

Expected Frequency of Floods
The expected frequency or expected number of floods per
period of time should be minimized. Flood occurrence is a
random event and may be specified in terms of a probability
distribution function.

PRESERVATION OF ENVIRONMENT
While none of the proposed basin plans will be
specifically designed for environmental enhancement, no
system can be developed without taking this objective into
account. Two specifications have been established for this
objective.

Effects on Native Wildlife and Vegetation
The natural environment of the study area should be
altered as little as possible. Any potential environmental
damage or dislocation should be minimized.

*Preservation of Historical, Archeological or
Anthropological Sites*
Those sites designated as having either local or national
value in any of the above categories should be preserved.

UTILIZATION OF RESOURCES
The costs of the alternative plans must be considered in
defining the overall system's effectiveness. Costs are defined
in a general way to express both the direct and indirect
economic impact of the different strategies. Indirect costs
can be thought of in terms of benefits foregone by choosing
one plan over another. Direct costs refer to the cost of capi-
talization as well as annual operation and maintenance expenses.

Capital Costs

This is a measure of all capital outlays up until the completion of the system.

Operation and Maintenance Expenses

These costs are incurred after the completion of the system. Included are yearly operation expenditures as well as annual maintenance expenses.

Indirect Benefits and Costs

Indirect benefits and costs measure secondary spin-off effects of various basin plans. For example, certain strategies necessarily will promote more economic activity or displace more people than other plans. These effects will be estimated qualitatively and included in the analysis.

Natural Resources

The alternative systems should make optimal use of land, water and manpower.

RECREATIONAL ENHANCEMENT

Additional recreational opportunities are desirable to maintain the quality of life for a growing population. This objective can be measured as follows:

1. *Preservation of existing facilities:* Existing recreation facilities should be maintained to the greatest possible extent.
2. *Creation of new facilities:* Where possible, the construction of the system should include the creation of new recreational opportunities.

Criteria

In this step, system evaluation criteria are established that resolve system capabilities to specifications and hence to objectives. A criterion is a measure of performance, effectiveness or weakness, or system capacity and is used as the basis for evaluation. For the Santa Cruz River Basin planning the following criteria are used to quantify the degree to which each of the alternative systems meet the specifications. One criterion is provided for each specification. These criteria, in many cases, are noncommensurable. For example, water quality cannot be compared directly with losses due to floods. A systems versus criteria array is provided as a summary of the material to follow.

WATER REQUIREMENTS

Underground Aquifer Level

The effect of the underground aquifer level is measured by the net change in water level in feet over the years.

Water Quality

Quality of the water is measured against standards set (State of Arizona, Water Quality Standards) for each of the user groups. An ordinal scale assigning ratings a, b, c, d, e for the best to the worst case, respectively, is used.

FLOOD PROTECTION

Expected Flood Losses

The direct flood losses should be measured in expected dollars. The losses are based on the probabilities of given size floods occurring and the losses to be expected if such a flood did occur.

Expected Frequency of Floods

Protection can be geared to the level of a flood that is expected to occur once in a given time interval based on a probability distribution of the river's peak levels. The criterion is the number of years making up this time interval, a greater number of years between floods being preferred. Federal projects are designed to the 100-year flood. The number of years between floods is translated into a probability of a flood occurring in a given year.

ENHANCEMENT OF ENVIRONMENT

Preservation of Historical, Archeological or Anthropological Sites

The degree of preservation of these designated areas is measured on a scale (a,e) with "a" representing "all areas saved" and "e" representing "all areas destroyed."

Effect on Native Wildlife and Vegetation

The system's effects on the native wildlife and vegetation is also measured on a scale (a,e). In this case "a" represents "enhancement," "c" represents "no effect," and "e" represents "extreme environment impacts."

UTILIZATION OF RESOURCES

Implementation Costs

These costs are measured in present value dollars. Included are all costs incurred from the planning stages until the project is complete and the chosen system is operable.

Maintenance and Operating Costs

These costs are also measured in present value dollars. All costs necessary to keep the system operational after its completion are included.

Indirect Benefits and Costs

The indirect benefits and costs resulting from the implementation of each system are evaluated and rated on a scale (a,e) where "a" represents the best and "e" represents the worst.

Natural Resources

A subjective scale (a,e) is again used to evaluate the utilization of the available natural resources.

RECREATION ENHANCEMENT

Preservation of Existing Facilities

At present, recreational facilities include trails for horseback riding or hiking. A scale (a,e) will be used to evaluate this criterion.

Creation of New Opportunities

This criterion is measured by using a scale (a,b,c,d,e) as has been done earlier with "a" being the best case and "e" the worst case.

Table 1 summarizes objectives, specifications and criteria for the Santa Cruz (Arizona) example.

Generation of Alternatives

The set of objectives of river basin planning can be approached by several alternative systems, based on engineering interventions (management and/or development), and economic and social measures, such as water pricing policy or hydrological forecasting.

In general, the set of alternative systems for a given river basin planning problem may be discrete or continuous. However, the planner should try to consider the largest possible number and/or types of alternative systems, as the chance of finding the best system will also be higher.

The generation of alternatives can utilize the fact, that river basin plans are generally developed for the purpose of reducing the risk of floods or increasing water supplies. Keeping in this spirit, several alternative systems will be evaluated. Although these systems are designed with flood protection and water requirements in mind, they can and will be evaluated in terms of all objectives. This spirit is followed in the Santa Cruz River Basin example.

In an MCDM problem, the decision maker must, in some manner, accept tradeoffs among the objectives. A number of alternative development strategies are defined which embody these trade-offs.

Alternative actions for flood control include:

- levee construction
- channelization
- construction of dams and multipurpose reservoirs
- flood plain management, including floodproofing of existing structures
- no action

Alternatives for water supply include:

- waste water reclamation
- new groundwater development
- the Central Arizona Project that is, large-scale water transfer
- conservation and education program
- no action

In the present example, a discrete number of alternatives are available, since there are five actions under flood control and five for water supply; thus there are twenty-five different alternative systems to be evaluated. Table 2 defines these systems in tabular form. A short description of these systems follows.

FLOOD CONTROL SYSTEM

Levee Construction

Levee construction entails the construction of barriers, usually of earth or concrete to contain the overflow during time of flooding. Levees are built outside of the river channel and therefore do not interfere with the natural flow of the river.

Channelization

Channelization is the modification of the river channel so that it can contain the design water flow (equivalent to 100-year flood) without having water overflow the channel. It may include deepening the channel or creating a larger channel from concrete through bottom stabilization or bank stabilization, or both.

Dams and Reservoirs

Dams and reservoirs involve the construction of a dam and a reservoir to control the flows of the river. One possible result of this alternative would be to store stormwaters and increase the water supply or keep water in the river at all times.

Floodplain Management

Floodplain management includes nonstructural solutions such as floodplain land purchases, strict control of new development in the floodplain, raising of bridges or floodproofing individual structures.

No action keeps the present system maintained and operating. No changes would be implemented.

WATER SUPPLY SYSTEM

Waste water reclamation processes municipal wastes so that the water can be reused. It can be used to keep a flow in the river or put right back into urban use.

New groundwater development includes recharge of the underground water levels and expanding the number of wells now in use. The first will help the long term water supply, while the second will help the short term supply at the expense of the future.

Central Arizona Project (CAP) is a project of the Depart-

TABLE 1. Objectives, Specifications and Criteria Summary Sheet for the Santa Cruz River Basin Planning.

Objective	Specifications	Criteria
Water Supply	Aquifer Level	Net change in ft/year
	Water Quality—Urban	Ordinal
	Water Quality—Agriculture	Ordinal
Flood Protection	Expected Flood Losses	Expected dollars
	Expected Frequency	Number of years between expected floods
Environment	Preservation of Designated Areas	Ordinal
	Effect on Wildlife and Vegetation	Ordinal
Utilization of Resources	Implementation	Present dollars
	Operation & Maintenance	Present dollars
	Indirect Costs	Ordinal
	Natural Resources	Ordinal
Recreation	Preservation of Existing Facilities	Ordinal
	Creation of New Opportunities	Ordinal

Note: "Ordinal" means five-point ordinal scale a, b, c, d, e, with a = best and e = worst.

TABLE 2. Definition and Numbering of Alternative Systems for the Santa Cruz River Basin.

Alternatives	Levee Construction	Channelization	Reservoirs and Dams	Flood Plain Management	No Action
1. Waste Water Reclamation	1	2	3	4	5
2. Groundwater Development	6	7	8	9	10
3. Central Arizona Project	11	12	13	14	15
4. Conservation and Education	16	17	18	19	20
5. No Action	21	22	23	24	25

TABLE 3. Water Supply Action Versus Criteria.

	C.A.P.	Wastewater Reclamation	Groundwater Development	Conservation	No Action
Aquifer Level	2.0	2.7	1.6	2.4	3.5
Urban	b	e	a	a	b
Agriculture	b	a	b	b	b

TABLE 4. Number of Structures and Residents in Various Basin Floodplains.

Area	Structures	Residents
Green Valley	60	120
Marana	20	50
Canada del Oro	1,000	4,000
Rillito	1,500	6,000
Tanque Verde	70	180
Agua Caliente	130	400
Pantano	130	650
Rodeo	520	2,100

TABLE 5. Total Flood Losses for the 25-, 50- and 100-Year Floods.

Area	25-year	50-year	100-year
		(In Millions)	
Santa Cruz	.5	10.3	12.4
Tributaries	21	63	84

TABLE 6. Expected Flood Frequency Versus Preventive Action.

	Levees	Chan.	Dams, Reservoirs	Floodplain Management	No Action
Probability of flooding in a given year	.01	.01	.003	.02	.04

ment of the Interior which will bring new supplies of water into the basin from the Colorado River.

Conservation and education would invoke a large scale advertising campaign in combination with a series of programs designed to help the population to reduce water consumption voluntarily.

No action continues operating with the supplies available.

Alternative Systems Versus Criteria Array

The last step of formulating a river basin planning as an MCDM problem consists of the preparation of the alternative systems versus criteria array. This work includes the engineering calculation or estimation of each criterion value pertaining to the different alternatives. Methods to perform this work range from detailed computer-based physical modeling, such as regional groundwater prediction to rough engineering judgement. The importance of the planning problem and data available would determine the actual methods in a given area. For illustration purposes, methods to estimate criterion values for the Santa Cruz River Basin alternatives are briefly reviewed.

WATER SUPPLY

Aquifer Level
The effect of the water supply alternatives on the aquifer level is calculated in feet decreases per year. The no action alternative represents the worst case where the drawdown rate is approximately 3.5 feet/yr. The system which will provide the best results is new groundwater development. In this case the corresponding figure for the rate of drawdown is estimated at 1.6 feet/yr.

Water Quality
As the water is pumped from deeper levels, the amount of solids in the water increases (Davidson, 1973), thus reducing the quality of the water. If no action is taken, the quality of the water will deteriorate with time since the aquifer level will drop. Additional pumping will increase the rate of deterioration. However, water pumped for the first time from new areas will be of a higher quality.

Water from the Central Arizona Project (CAP) will not be suitable for urban use, but the present plan is to blend it with the water pumped from the ground to meet state drinking water regulations. Although acceptable, this will result in a lower quality of the water. Water obtained as a result of waste water recycling will not be suitable for drinking, but it may actually be advantageous for agricultural use.

The evaluation of the alternative water supply actions with respect to the water supply criteria follow.

TABLE 7. Environmental Specifications Versus Criteria.

	Waste water Reclam.	Ground water Develop.	C.A.P.	No Action	Levees	Chan.	Cons. and Educ.	Dams and Res.	Floodplain Manag.
Preservation of designated areas	a	a	b	a	d	d	a	c	a
Effect on wildlife & vegetation	a	d	c	b	d	c	a	e	b

FLOOD PROTECTION

Flood Losses

The number of structures in the floodplain is given in Table 4.

Table 5 illustrates total flood losses based upon an average flood loss per structure at $5,000, $15,000 and $25,000 for the 25-, 50-, and 100-year flood, respectively, for the affected areas given in Table 4. In addition, agricultural losses along the Santa Cruz, based upon actual figures of the 1977 flood, are included.

Expected Frequency of Floods

None of the alternative actions suggested for the water supply objective will have an effect on this criterion. The expected frequency of floods is just a function of what flood control action is selected, as shown in Table 6.

These figures are based on the fact that federal law requires federal projects to protect against the 100-year flood (prob = .01). Levees and channelization would be designed to this level. Dams and reservoirs would actually do better than the design level because there are other functions they must fulfill. Floodplain management, while providing some protection, would not be designed to protect against the 100-year flood. No action would preserve the status quo. At this time the banks can protect against the 25-year flood.

ENVIRONMENT

Preservation of Designated Areas and Effect on Native Wildlife and Vegetation

For criteria under the environmental objective, subjective ratings are given in Table 7.

UTILIZATION OF RESOURCES

Implementation, Operation and Maintenance Costs

Cost estimates for the flood control actions were obtained from discussions with the Corps of Engineers (1978). Costs for the water supply alternatives were obtained from discussions with members of the Tucson City Planning Department. These costs are given as gross figures and will be combined and converted into present values in the criteria versus systems array. Costs are given in Table 8.

TABLE 8. Costs of Alternative Actions.

System	Capital Costs $ (million)	Operation & Maintenance $ (million)
Waste Water	.18	2.7
Groundwater Development	20.0	.2
C.A.P.	15.6	.16
Conservation & Education	.05	.05
Levee Construction	12.5	.08
Channelization	16.6	.08
Dams—Reservoirs	12.1	.12
Floodplain Management	1.8	.02

TABLE 9. Indirect Costs and National Resources Utilization Versus Actions.

	Waste water Reclam.	Ground water Develop.	C.A.P.	No Action	Levees	Chan.	Cons. and Educ.	Dams and Res.	Floodplain Manag.
Indirect costs	c	e	a	d	c	c	c	b	e
Natural Resources	b	e	e	a	a	c	c	c	b

Indirect Costs and Natural Resources

Subjective evaluations are again developed for each of the alternative actions and combined to evaluate the alternative systems for the criteria versus systems array. The development is shown in Table 9.

RECREATION

Preservation of Existing Facilities and Creation of New Facilities

Subjective evaluations are performed as described previously and shown in Table 10 for this objective.

SELECTION AMONG ALTERNATIVES MCDM TECHNIQUES

The Task

Once the river basin planning has been properly defined as a MCDM problem, the alternative systems-criteria array is available either in a tabular form as shown in Table 11 for discrete alternatives, or as a generator of alternatives in the case of continuous alternatives (e.g., amount of water treated or stored in a reservoir).

The selection among alternatives would be relatively easy if one could find a single alternative which performs best in view of all objectives. However, in river basin planning generally no such overall optimum solution can be found, and the selection among alternatives would be possible only by considering trade-offs among the achievement of objectives.

This type of selection is called multicriterion decision making (MCDM) and would range from simple engineering judgement, which in the case of a few alternatives and few criteria is a quite realistic approach, to sophisticated multiobjective programming methods.

Both the theory and water resources application of MCDM can be found in several excellent sources including Haimes, et al. (1975), Keeney and Raiffa (1976), Major (1977), Cohon (1978), Goicoechea, et al. (1982), Zeleny (1982), Voogd (1983), Fraser and Hipel (1984), Zoints and Wallenius (1984), Teghem, et al. (1984), Szidarovszky, et al. (1986), Tecle, et al. (1985), Massam (1984). A Tchebycheff metric was applied by Greis, et al. (1983) to a water allocation problem in a river basin.

A goal programming management model was proposed by Moosburner and Wood (1980) for controlling contamination in a New Jersey aquifer. Multiobjective programming techniques were reviewed and evaluated by Cohon and Marks (1975). Similar reviews have been made on analytical MCDM techniques from time to time (Macrimmon, 1973, Gershon, 1981, Evans, 1984 and Khairullah and Zionts, 1979, Szidarovszky and Duckstein, 1986).

As a result of the abundant literature on MCDM, this chapter will not give an exhaustive review of available MCDM techniques. Rather, main types of the techniques will be specified, their applications will be illustrated and guidelines for selecting proper MCDM techniques will be provided.

Three types of MCDM techniques may be distinguished:

1. Outranking—types such as ELECTRE

TABLE 10. Recreation Facilities Versus Alternative Actions.

	Waste water Reclam.	Ground water Develop.	C.A.P.	No Action	Levees	Chan.	Cons. and Educ.	Dams and Res.	Floodplain Manag.
Existing facilities	b	a	a	a	d	b	b	d	a
New facilities	b	e	e	e	c	c	c	a	a

2. Distance–based such as compromise programming or composite programming
3. Utility–types such as multiattribute utility functions

Outranking–Types of MCDM techniques

ELECTRE I

This methodology, developed by Benayoun, et al. (1966), has been used for river basin planning by David and Duckstein (1976). The main idea in ELECTRE I is to choose those systems which are preferred for most of the criteria and yet do not cause an unacceptable level of discontent for any one criteria. Three concepts are developed in this methodology: concordance, discordance and threshold values.

The concordance of any two actions i and j is a weighted measure of the number of criteria for which action i is preferred or indifferent to action j and is given as:

$$C(i,j) = \frac{\text{Sum of weights for criteria where } i \geq j}{\text{Total sum of weights}}$$

where the weights are elicited from the decision maker. Concordance can be thought of as the weighted percentage of criteria for which one action is preferred to another.

To compute the discord matrix, an interval scale common to each criterion is first defined. The scale is used to compare the discomfort caused between the "worst" and the

TABLE 11. Systems Versus Criteria Array.

Objective	Criteria	W	1	2	3	4	5	6	7	8	9	10	11	12
							Alternatives							
Water Supply	Aquifer Level	9	2.7	1.6	2.0	2.4	3.5	2.7	1.6	2.0	2.4	3.5	2.7	1.6
	Water Quality Urban	3	e	a	d	a	b	e	a	d	a	b	e	a
	Water Quality Agric.	3	a	b	b	b	b	a	b	b	b	b	a	b
Flood Protection	Expect. Flood Losses	4	7.72	7.72	0	19.45	26.33	7.72	7.72	0	19.45	26.33	7.72	7.72
	Expect. Freq.	5	.01	.01	.003	.02	.04	.01	.01	.003	.02	.04	.01	.01
Environment	Pres. Desig. Areas	5	d	d	c	a	a	d	d	c	a	a	d	d
	Effect on Wild. Veg.	5	c	b	d	a	a	d	c	e	c	c	c	c
Utilizaton Of Resource	Implem. Costs	2	12.7	16.8	12.3	1.9	0.2	32.5	36.6	32.1	21.8	20	28.1	32.2
	O&M Costs	2	37.6	37.8	38.2	37.2	37.0	2.6	2.8	3.2	2.2	2.0	2.2	2.4
	Indirect Costs	2	c	c	b	d	c	d	d	d	e	e	b	b
Recreation	Natural Resource	2	c	c	c	b	a	d	d	d	c	c	e	e
	Pres. of Exist. Fac.	1.5	c	b	c	b	b	c	b	c	a	a	c	b
	Creation New Oppor.	1.5	b	b	a	a	c	d	d	b	c	e	d	d

(continued)

TABLE 11 (continued).

Objective	Criteria	W	13	14	15	16	17	18	19	20	21	22	23	24	25
								Alternatives							
Water Supply	Aquifer Level	9	2.0	2.4	3.5	2.7	1.6	2.0	2.4	3.5	2.7	1.6	2.0	2.4	2.5
	Water Quality Urban	3	d	a	b	e	a	d	a	b	e	a	d	a	b
	Water Quality Agric.	3	b	b	b	a	b	b	b	b	a	b	b	b	b
Flood Protection	Expect. Flood Losses	4	0	19.45	26.33	7.72	7.72	0	19.45	26.33	7.72	7.72	0	19.45	26.33
	Expect. Freq.	5	.003	.02	.04	.01	.01	.003	.02	.04	.01	.01	.003	.02	.04
Environment	Pres. Desig. Areas	5	c	a	a	d	d	c	a	a	d	d	c	a	a
	Effect on Wild. Veg.	5	d	d	b	c	c	d	b	b	c	c	d	b	b
Utilizaton Of Resource	Implem. Costs	2	27.6	17.3	15.6	12.6	16.7	12.2	1.8	.01	12.5	16.6	12.1	1.8	0
	O&M Costs	2	2.8	1.8	1.6	1.1	1.3	1.7	.6	.5	.6	.0	1.2	.2	0
	Indirect Costs	2	a	c	b	c	c	b	d	d	c	c	c	e	d
Recreation	Natural Resource	2	e	d	c	c	b	e	a	a	c	b	c	b	a
	Pres. of Exist. Fac.	1.5	c	b	b	c	b	c	a	a	c	b	c	b	b
	Creation New Oppor.	1.5	b	c	e	d	d	b	c	e	d	d	b	c	e

"best" of each criterion. For example, a range of 1–150 might be used where the best value would be assigned the highest value of the range and the "worst" value would receive the lowest value of the range. Each criterion, however, can have a different range.

Given this information, the discord index is defined to be:

$$D(i,j) = \frac{\text{Max. interval where } i < j}{\text{maximum range of scale}}$$

For example, suppose two systems are being compared with regard to two criteria. The maximum range is given as 150. Let System 2 be preferred to System 1 only on the first crite-rion and assume the difference between the two systems is 75 in terms of range. The discordance $D(1,2)$ is given as $75/150 = .5$ for this simplistic example.

To synthesize both the concordance and discordance matrices, threshold values (p,q) are defined by the decision maker (DM). One must have p less than 1 since no action dominates all other actions for 100 percent of the criteria. If this were not the case, the set of non-dominated solutions would be reduced to one action. Similarly, q must be greater than 0, since no action is strictly dominated for all criteria. By choosing a value of p, the DM specifies how much "concordance" he wants; by specifying q, he specifies the amount of "discordance" he is willing to tolerate. It is possible that for some choices of p and q, there may not be any action which fulfills such a choice. If this is the case, the DM is asked to restate his values of p and q. It is also possible for cycles to occur in ELECTRE I; that is, if System 1 is preferred to System 2, which is preferred to System 3, which in turn is preferred to System 1, then there exists a cycle and the three nodes in question are collapsed into one new node.

The result of ELECTRE I is a preference graph which presents a partial ordering of the alternative systems. ELECTRE II (Roy and Bertier, 1971; Duckstein and Gershon, 1984) is then used to obtain a complete ordering.

ELECTRE II

Two preference graphs must be generated by ELECTRE I for use as input to the ELECTRE II procedure. These graphs represent the strong and the weak preference structures of the decision maker. The strong preference graph results from the use of the stringent threshold values; that is, the decision maker is asked to select a high level of concordance and a low level of discordance. For the weak preference graph, the decision maker is asked to relax his threshold values (lower p, higher q). These relaxed threshold values can be thought of as lower bounds on the system performance that the decision maker is willing to accept. A precise algorithm is presented after the following general discussion.

The graph representing the strong preference structure is defined as the graph G_F. The graph representing weak preferences is defined as G_f. The graph G_F is defined by the pair (Y_F, V_F) and the graph G_f by (Y_f, V_f) where Y and V, respectively, define the sets of nodes and arcs in the graphs. For the river basin study, Y is the set of alternative systems $(1, 2, \ldots 25)$ and V is the set of directed arcs showing preferences among these twenty-five systems.

Forward Ranking

The first step in the forward ranking is to identify all nodes in the graph G_F which have no precedents. This set is defined as the set C. Next, the nodes in set C having no precedents in graph G_f are identified; that is, a subset C is defined where the elements of this subset have no precedents in either graph G_F or G_f. This set is defined as set A, and the elements of this set are assigned the rank of 1.

The next step consists of reducing G_F and G_f by eliminating all nodes contained in the set A and all arcs originating at these nodes. The reduced graph G_F is again examined to identify all nodes having no precedents. These nodes comprise a new set C and the procedure outlined above is repeated. The next set of nodes in set A receives the rank of 2. This iterative procedure is continued until all nodes of G_F and G_f have been eliminated and all systems are ranked.

Reverse Ranking

The first step in the reverse ranking is to reverse the direction of all arcs in G_F. If System i is strongly preferred to

System j in the forward ranking, System j is now weakly preferred to System i in the reverse ranking. By reversing the arcs, high concordance becomes low concordance and low discordance becomes high discordance. The remaining steps are identical to the steps outlined in the forward ranking with one exception. The system which is ranked last is ranked first and the remaining systems are ranked in reverse order. This reestablishes the correct direction of the ranking process.

Average Ranking

Upon completion of the forward and reverse rankings, an average of the two is taken for each node. Thus, if System i is ranked first in the forward ranking and second in the reverse ranking, its average ranking is 1.5. The final stage in ELECTRE II is to order the systems with respect to their average rankings. This establishes a complete ranking among the systems. A concise algorithm follows.

FORWARD RANKING
1. Set $K = 0$
2. Select all nodes in $G_F(K)$ having no precedent. Denote this set as C.
3. Select all nodes in C having no precedent in $G_f(K)$. Denote this set as A.
4. Assign rank $V'(y) = K + 1$ to all nodes in set A.
5. Reduce G_F and G_f. If all nodes are eliminated, stop.
6. Set $K = K + 1$ and return to Step 2.

REVERSE RANKING
1. Reverse direction of all arcs V_F of G_F and V_f of G_f.
2. Obtain ranking $\alpha(y)$ analogous to $v'(y)$ above.
3. Reestablish correct direction of the ranking by setting $v'(y) = 1 + \alpha_{max}(y) - \alpha(7)$ for all nodes y.

AVERAGE RANKING
1. $v(y) = \{v'(y) + v'(y)\}/2$.

Distance-Based Techniques: Compromise Programming (C.P.)

In compromise programming, the "best" solution is that point which minimizes the distance from an "ideal" point to the set of non-dominated solutions. This concept is illustrated with the aid of Figure 1 which corresponds to a two-criteria analysis. Let Z_1 and Z_2 denote the first and second criterion, respectively. The set C represents the feasible region (mapping of the set of all feasible solutions into the decision space) and is bounded by the curve *abcdef*.

In the case where the set of nondominated solutions is defined by a set of discrete points (i.e., alternative river basin plans), the curve *ab* would be reduced to a set of points such as *a, b, c, d, e* and *f*. If the decision maker were to maximize only the first criterion, point *(f,o)* would be achieved. Likewise, point (o,a) would be attained given maximization of only the second criterion. The point (a,f) represents an "ideal" point (Zeleny, 1977), or the point where both criteria are maximized simultaneously. This point, however, is not in the feasible region and therefore a compromise solution is necessary. This solution may be at point c, d, or any other point which is an element of the set of nondominated solutions.

As suggested, the "best" solution is that point which minimizes the distance from the ideal point to the Pareto optimal frontier.

The distance measure used in compromise programming is an 1_p metric distance. This is given as:

$$\text{Min}\left\{ 1_p(x) = \left[\sum_{i=1}^{n} \alpha_i^p \left(\frac{|f_1^* - f_i(x)|}{\text{Range }(i)} \right)^p \right]^{1/p} \right\} \quad (3)$$

where α_i is the weight associated with criterion $;$; f_i^* is best outcome among all systems being evaluated with respect to criterion i (ith component of the ideal point) and $f_j(x)$ is the actual outcome of system (j) with respect to criterion i. The range (i) is the difference from worst to best of criterion i. Dividing by range (i) helps to scale the solution to values ranging from 0 to 1. The value associated with p defines the 1_p metric measure being used. For $p = 1$, all deviations from f_i^* and $F_i(x)$ are given equal importance; for $p = 2$, the largest deviations exert the most influence. As p increases, more weight is given to the largest deviations, and when $p = \infty$, a mini-max situation exists.

In order to use compromise programming where qualitative data are present, it is necessary to scale the systems versus criteria array. This scaling proceeds in the following manner. First, a range associated with each criterion is defined where the "best" value is assigned the highest value and the "worst" value is given the lowest value. Next the values of each criterion for each system are scaled proportionally to the range associated with that criterion. This idea will become clearer in the application section. The qualitative data are scaled such that the letter a is given the highest value of the range, e is given the lowest value, c is given the mid-point value and b and d are scaled in a similar manner. This scaling procedure is analogous to the scaling procedure used in the discord matrix of ELECTRE. To test the robustness of this heuristic scaling procedure, a sensitivity analysis is performed.

Objectives of river basin planning can often be divided into two or three groups such as economic and environmental objectives. Then to each group several criteria pertain. It has been recognized that (a) criteria pertaining to different groups have different features such as economic related criteria (cost, income, cost/benefit) versus environmental criteria (water quality, nitrate loading, etc.); (b) DM's would prefer a final trade-off graph such as the one shown in

Figure 1 between two groups of objectives instead of considering each criterion separately.

These two reasons have led to the development of *composite programming* which is an extension of compromise programming (Bardossy, 1984). Composite programming is also a distance-based technique, but the distance measured from the idea point is defined stepwise. In each group of objectives a compromise solution, and then the overall compromise are successively sought. For this purpose, in each group j, one selects a value of the compromise programming parameter p_j and a set of weights $\{\alpha\}$. Then the overall objective function with parameter q and weight set $\{\beta\}$ is defined. This procedure yields the objective function as a composite distance between sections x and y:

$$f(x,y) = \left[\sum_j \beta(j)^q \left(\sum_{i \in I} \alpha_{(i)}^{p(j)} \mid \gamma(i) \right. \right.$$
$$\left. \left. - y(i) \mid^{p(j)} \right)^{q/p(j)} \right]^{1/q} \quad (4)$$

where $\alpha(i)$, $\beta(j)$ are weights and $p(j)$, q exponents: $i = 1 \ldots I$, $j = 1 \ldots J$. The choice of the composite programming parameters p and q should reflect the nature of the criteria; thus p can be, say, 3 to emphasize the limiting character of the "worst" pollution element.

The mode of finding the minimum of Equations (3) and (4) depends on the number of alternatives and the type of the objective functions. In the case of a relatively small number of discrete alternatives, such as 25 (Santa Cruz River Basin) total enumeration can be used. In other cases standard mathematical programming algorithms can be applied. For discrete alternatives a branch and bound algorithm or, specifically, 0–1 programming algorithm (Bogardi, et al., 1984) can be used. For continuous linear objective functions linear programming, and for nonlinear objective functions other standard methods such as the gradient method and quadratic programming can be applied.

MULTIATTRIBUTE UTILITY FUNCTIONS (MAUT)

In the context of river basin planning, utility is defined as the subjective benefit(s) derived by the decision maker from the achievement of the stated objectives. The motivation factor of multiattribute utility theory is that the DM's utility function can be specified numerically. This is accomplished by eliciting the DM's utility for each criterion and then combining these single utilities into one overall utility function. The system which provides the highest degree of utility with respect to all the objectives is defined to be the preferred alternative.

Several axioms (von Neumann and Morgenstern, 1947,

and Owen, 1982) insure the existence of a well-behaved utility function. Given that these axioms hold, the following steps, as outlined in Keeney and Wood (1977), must be considered in evaluating alternative systems via multiattribute utility theory.

1. Check for utility independence among criteria.
2. Check for preferential independence among criteria.
3. Check for additive independence.
4. Decide on form of utility function.
5. Assess utility functions.
6. Assess scaling factors.

These steps are explored in more detail below.

Utility Independence

One of the fundamental concepts of multiattribute utility theory is that of utility independence. It is a necessary and sufficient condition for having a single utility function over one of the criteria (Keeney and Raiffa, 1976). To define utility independence, consider for the two criteria case, $\{X,Y\}$ where $x_1, x_2 \in X$ and $y_1, y_2 \in Y$. In addition, assume that:

$$x_1 \le x \le x_2 \text{ and } y_1 \le y \le y_2$$

in which the symbol $<$ indicates a preference relationship and stands for indifference. To better describe the concept a 50–50 gamble (see Gamble 1 below) is presented and the DM is asked to determine x such that he/she is indifferent between certainty equivalent (x, x_1) and the gamble. The value of y is then changed

Gamble 1

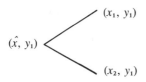

to some other value, say y_2, and the DM again is asked to state the certainty equivalent $(\hat{x}, y_2)$. If the value of x remains the same for all $y \epsilon Y$, then the criterion X is said to be utility independent of criterion Y. If the converse is also true, that is, that Y is also utility independent of X, the condition is defined as mutual utility independence.

Preferential Independence

Consider the two criteria case, $\{X,Y\}$ where $x \in X$ and $y \in Y$. Further consider the ordered pair $<x_1, y_1>$ which represents an attainment level of criteria x and y. In this case, the necessary and sufficient condition for criterion X to be preferentially independent of criterion Y is the case in which $(x_1, y_1) \ge (x_2, y_2) \Rightarrow (x_1, y_2) \ge (x_2, y_2)$ holds for all y, x_1 and x_2. This does not necessarily mean that y is also preferentially independent of x. However, if the latter is true then the condition is called mutual preferential inde-

pendence. This is a necessary condition for the existence of an additive value function such as

$$v(x_1, x_2, \ldots , x_n) = \sum_{i=1}^{n} v(x_i)$$

where $n \ge 3$, and v_i is a value function over the mutually preferentially independent criteria X_i).

Additive Independence

For $x \in X$ and $y \in Y$, criteria X and Y are said to be additive independent if and only if the paired preference comparison of any two gambles involving x and y depends only upon their marginal probability distributions and not on their joint probability distribution. In two dimensions, criteria X and Y are additive independent if and only if the lotteries (see Gamble 2) are equally preferable.

Gamble 2

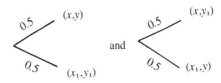

(i.e., indifferent) for all (x,y) given a specific x_1 and y_1. Additive independence is a necessary and sufficient condition for the existence of additive utility function (Keeney, 1973 and Pollak, 1967). The latter allows the addition of the separate contributions of the two criteria to obtain a unidimensional utility, such as:

$$u(x,y) = k_x u_x (x) + k_y u_y (y)$$

where u is total utility scaled from 0 to 1, u_x and u_y are the utilities, and k_x and k_y are the scaling constants for criteria X and Y, respectively.

Form of Utility Function

A general type of utility function for which utility independence is the necessary and sufficient condition is multilinear utility function. The mathematical formulation and its derivation are discussed in Keeney and Raiffa (1976) and will not be presented here. However, special cases of the multilinear utility function which are more handy to use in practice are presented. These particular multiattribute utility functions are the additive and multiplicative forms. A sufficient condition for the existence of both forms of utility functions is mutual utility independence, say between criteria X_1 and X_2 in a two-criterion space.

The general form of the multiplicative utility function

under this condition is given by

$$u(x_1,x_2) = k_1u_1(x_1) + k_2u_2(x_2)$$
$$+ kk_1k_2u_1(x_1)u_2(x_2) \tag{5}$$

where $u(\cdot)$ is the total utility function for each system under consideration and is scaled from 0 to 1; $u_i(x_i)$ is the utility function, and k_1 is the scaling factor of the ith criterion; and K is an additional scaling constant. Now, by multiplying each side of Equation (5) by k, adding 1 and factoring, it can be present in short form as

$$ku(x_1,x_2) + 1 = \prod_{i=1}^{n} [kk_iu(x_i) + 1] \tag{6}$$

where n in this case is 2 and the other symbols are as described above. The multiplicative utility function is reduced to an additive form if both criteria of the two utilities under consideration show both mutual utility independence and additive utility independence. The latter implies $k = o$. As a result Equation (5) becomes

$$u(x_1, x_2) = k_1u_1(x_1) + k_2u_2(x_2) \tag{7}$$

which can be represented in short form as

$$u(x_1,x_2) = \sum_{i=1}^{n} k_iu_i(x_i)$$

where the u, $u_i(x_i)$ and k_i are as defined above and $n = 2$.

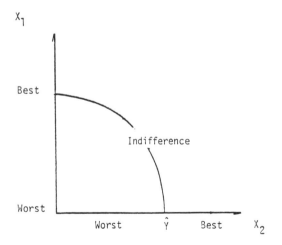

FIGURE 3. Indifference curve.

Assess Utility Functions

Along with additive, multiplicative and multilinear utility functions, preferential independence, additive independence and utility independence have to be used to reduce the assessment of an n-criteria utility function to the assessment of n one-criterion utility functions. This can be done through a sequence of assessment of the utility functions which may proceed along the following lines:

1. Specify the range from worst to best for each criterion.
2. Perform two gambles, one involving the best value and the midpoint value of each criterion, the other involving the midpoint and the worst value.
3. In the gamble involving the midpoint value and the best, let the utility of the midpoint be defined as U_n (midpoint) $= 0$ and define U_n (best) $= 1$. Similarly in the gamble between the worst and midpoint, define $U_0(X_j)$.
4. Perform a linking gamble. This involves a gamble between $U_n(X_i)$ and $U_0(X_j)$.
5. Use a linking technique (see Krzysztofowicz and Duckstein 1980) to link $U_0(X_i)$ and $U_n(X_i)$ to form $U_i(X_i)$.

These steps are elaborated in more detail in the application section.

Assess Scaling Factors

Keeney and Wood (1977) suggest that the first step in assessing the scaling factor k_i is to determine the relative magnitude. This involves asking the DM which criteria were at their worst level. This question is asked iteratively until all the k_i are ranked.

To establish the actual magnitude of the scaling factors, tradeoffs between criteria are established. For example, consider Figure 3. The DM is indifferent between $X_1 =$ best, $X_2 =$ worst and $X_1 =$ worst, $X_2 = y$ where y is some value less than the best value of X_2. Utilizing the multiplicative utility function given in Equation (6) and equating the two indifference pairs yields:

$$k_i = k_2 u_2(y)$$

In a similar fashion, $n - 1$ other equations are generated. These equations are given in the application section.

Finally, the constant k must be evaluated. To assess the constant k, another gamble is introduced involving a probability assessment between the two highest ranked criteria (i.e., with respect to the k_i).

For illustrative purposes consider the following situation:

Gamble 3

The DM is asked to specify a value of p, such that the above indifference relationship holds. Note that the gamble involves setting the two criteria at their best and worst levels. Utilizing Equation (7), we have:

$$k_0 = p(k_0 + k_1 + k_0 k_1 k)$$

At this point of the analysis there are $n + 1$ unknowns and n equations. By directly assessing the highest ranked k_i, the system of equations reduces to n equations and n unknowns.

APPLICATION

In the following section, ELECTRE, Compromise Programming and Multiattribute Utility Theory are applied to the Santa Cruz River Basin Study. First the use of the techniques is presented, then the results are given.

The Use of MCDM

ELECTRE

In order to apply ELECTRE, the DM is asked to indicate which one of the criteria is most important. This criterion then receives the highest weight. Next, that criterion which is least important is selected. This, in turn, receives the lowest weight. This process is continued until all the weights are generated (see Table 12).

In specifying the range for each criterion the decision maker is asked to specify the relative difference between best and worst for each criterion. If there was a small difference between best and worst, then the range for this criterion should reflect this fact.

Recall that the range values are used in the calculation of the discordance matrix. It might be noted that the range associated with each criterion is quite similar to that of scaling the k_i; that is, we not only want to determine which criteria are most important, but also take into account the difference between best and worst of each criterion. The scales for each criterion are given in Table 12.

The threshold values (p,q) are chosen somewhat arbitrarily. Ideally we want p to be close to 1, and q to be close to 0. For this study, p and q are defined to be .9 and .2 for the strong preference graph and are relaxed to .7 and .5 for the weak graph. When ELECTRE is applied to the systems versus criteria array of Table 11, the strong and weak preference graphs of Figures 4 and 5 are obtained. Working from these graphs, the ELECTRE II algorithm proceeds as shown in Table 13.

COMPROMISE PROGRAMMING

The first step in applying compromise programming is to scale the systems versus criteria array, as shown for our example in Table 14. Notice the qualitative rankings have been assigned numerical values. For example, for Criterion 2

TABLE 12. Weights and Criterion Range for ELECTRE Applications.

	Criterion	Weight	Scale
X1	Aquifer Level	9	150
X2	Water Quality—Urban	3	110
X3	Water Quality—Agriculture	3	130
X4	Expected Flood Losses	4	160
X5	Expected Frequency	5	160
X6	Preservation Designated Areas	5	150
X7	Effect on Wildlife & Vegetation	5	200
X8	Implementation Costs	2	200
X9	Operation & Maintenance Costs	2	110
X10	Indirect Costs	2	110
X11	Natural Resources	2	100
X12	Preservation of Existing Facilities	1.5	100
X13	Creation of New Opportunities	1.5	76

(water quality-urban), the letter "e" is now 0, whereas the letter "a" is assigned a value of 110. The other letters are assigned numerical values as well. In scaling the systems versus criteria array, the values associated with the range of ELECTRE have been used so that the comparison between results is consistent.

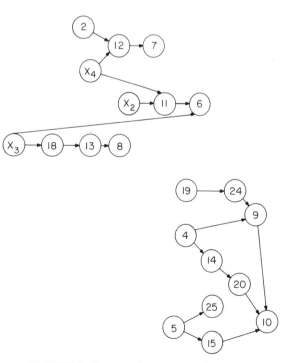

FIGURE 4. Strong preference graph G_F (.9, .2).

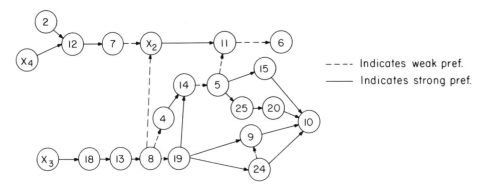

FIGURE 5. Weak preference graph G_f (.7, .5): solid line indicates weak preference; dotted line indicates strong preference.

Next, ideal values of each criterion must be provided. An ideal point can be thought of as the "best" value of each criterion; alternatively, it may be defined without regard to the actual systems in question. In this case study, the "best" values have been chosen as ideal points.

Finally the weights α_i and the 1_p distance measure p are defined. The weights correspond to those elicited earlier in the ELECTRE anaïysis. The value of p is taken to be 1, so that all deviations from the ideal point are weighted equally. The values of p and α are modified in the sensitivity analysis.

APPLICATION OF MULTIATTRIBUTE UTILITY THEORY

In order to apply multiattribute utility theory to the systems versus criteria array of Table 11, the steps of Keeney

(1979) which have been presented earlier in the methodology section, are followed.

To show utility independence, the decision maker is asked to state his indifference point in each of several lotteries. If the indifference point for each situation remained the same regardless of the levels of the other criteria, then utility independence is assumed. A sampling of the lotteries as presented to the DM is shown in Figure 6. In order to avoid gambles between extreme events, two gambles for each tradeoff situation are provided. The first gamble involves a tradeoff between the worst value of the criterion and the midpoint value, whereas the second gamble involves the midpoint and the best value. For example, consider Figure 6a and 6b. The decision maker responded that a "sure thing" of 2.0 was as equally satisfactory as the risky option yielding

TABLE 13. ELECTRE II Rankings with Sensitivity Analysis.

	Forward Ranking				Reverse Ranking			
Iteration K	Set C	Set A	Rank v^I	Iteration K	Set C	Set A	Rank α	Reverse Rank v^{II}
0	$(X_2,X_3,X_4,2,4,5,19)$	$(X_3,X_4,2)$	1	0	(6,7,8,10,14)	(6,10)	1	10
1	$(X_2,4,5,12,18,19)$	(12,18)	2	1	(7,8,9,11,14,15,20)	(9,11,15,20)	2	9
2	$(X_2,4,5,7,13,19)$	(7,13)	3	2	$(X_2,7,8,14,24,25)$	$(X_2,24,25)$	3	8
3	$(X_2,4,5,8,19)$	(8)	4	3	(5,7,8,14)	(5,7)	4	7
4	$(X_2,4,5,19)$	$(X_2,4,19)$	5	4	(8,12,14)	(12,14)	5	6
5	(5,14,24)	(14,24)	6	5	$(X_4,2,4,8,19)$	$(X_4,2,4,19)$	6	5
6	(5,9)	(5,9)	7	6	(8)	(8)	7	4
7	(11,15,25)	(11,15,25)	8	7	(13)	(13)	8	3
8	(6,20)	(6,20)	9	8	(18)	(18)	9	2
9	(10)	(10)	10	9	(23)	(23)	10	1

Set C = All nodes having no precedents in strong preference graph
Set A = All nodes in set C having no precedents in weak preference graph
X_2 = systems (1,16,21)
X_3 = systems (3,23)
X_4 = systems (17,22)

a value of 1.6 or 2.55 with probability $p = .5$. This is found to be the case regardless of the level of the other criteria. Similar results have been obtained with respect to other lotteries. Utility independence is therefore assumed.

To test for preferential independence, the decision maker is asked to state an indifference point in a two-criteria trade-off situation. These tradeoffs are presented in Figure 7 and involve the pairwise coupling of those criteria which appear most likely to violate the assumption of preferential independence. Consider Figure 7c. In this example, the decision maker is indifferent between the outcome represented by $X_2 = 1$, $X_1 = 3.5$, and the outcome defined by $X_2 = 4$, $X_1 = 6$. This was found to be the case in all the tradeoff situations, regardless of the levels of the other criteria. Thus, preferential independence is assumed. It might be noted that, if preferential independence does not exist, alternative proxy criteria may be introduced so as to circumvent this problem.

In checking for additive independence it has been found that there is at least one violation of this assumption, so that an additive utility function does not seem appropriate. The specific violation occurred with respect to Criteria 1 and 7; that is, the DM is not indifferent between the following gambles:

Gamble 4:   Gamble 5:

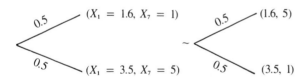

In the above case, Gamble 4 was preferred to Gamble 5. Because of this violation, a multiplicative utility function appears appropriate.

In order to assess the utility functions of each criterion (i.e., $U_i(X_i)$), gambles with respect to each criterion are performed and the linking technique as presented in Krzysztofowicz (1978) is used. An illustrative example defining the utility function for Criterion 2 (water quality-agriculture) is presented below.

The DM is asked to give his certainty equivalents in the following lotteries:

Gamble 6:   Gamble 7:

Recall that Number 1 defines high quality, 3 represents acceptable quality and 5 is unacceptable. In Gamble 6, the DM responded with the value of 1.4; in Gamble 7, the response is 4.2. Equating the certainty equivalents with the

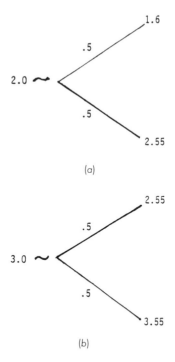

FIGURE 6. Sample lotteries for checking utility independence.

expected value of the gambles implies that $U_n(1.4) = .5$ and $U_0(4.2) = .5$. Summarizing this information, we have the following:

$$U_n(1) = 1 \qquad\qquad U_0(3) = 1$$

$$U_n(3) = 0 \qquad\qquad U_0(5) = 0$$

$$U_n(1.4) = .5 \qquad\qquad U_0(4.2) = .5$$

Next, the DM was presented with the linking gamble. This consisted of a gamble between the two certainty equivalents of Gambles 6 and 7. This lottery is given in Gamble 8.

Gamble 8:

The certainty equivalent response for Gamble 8 was 3.6. The utility of this value (i.e., $U_0(3.6)$) was directly assessed by the DM to be .8.

The purpose of the linking gamble is to provide a bridge between $U_0(X_i)$ and $U_n(X_i)$ such that the utility function

TABLE 14. Rescaled Systems vs. Criteria Array.

Objective	Criteria	W	1	2	3	4	5	6	7	8	9	10	11	12
Water Supply	Aquifer Level		63	150	118	87	0	63	150	118	87	0	63	150
	Water Quality Urban		0	110	27.5	110	82.5	0	110	27.5	110	82.5	0	110
	Water Quality Agric.		130	104	104	104	104	130	104	104	104	104	130	104
Flood Protection	Expect. Flood Losses		113	113	160	42	0	113	113	160	42	0	113	113
	Expect. Freq.		120	120	160	80	0	120	120	160	80	0	120	120
Environment	Pres. Desig. Areas		37.5	37.5	75	150	150	37.5	37.5	75	150	150	37.5	37.5
	Effect on Wild. Veg.		100	150	50	200	200	50	100	0	100	100	100	100
Utilizaton Of Resource	Implem. Costs		113	109	134	191	199	23	0	25	81	91	47	24
	O&M Costs		1.7	1.2	0	2.8	3.5	102.5	102	101	104	103	104	103
	Indirect Costs		50	50	75	25	50	25	25	65	0	0	75	75
Recreation	Natural Resource		50	50	50	75	100	25	25	25	50	50	0	0
	Pres. of Exist. Fac.		50	75	50	75	75	50	75	75	100	100	50	75
	Creation New Oppor.		57	57	76	30	38	19	19	57	38	0	19	19

$U_i(X_i)$ can be scaled from 0 to 1. The linking technique is then nothing more than a scaling technique which insures that the utility function $U_i(X_i)$ be evaluated from 0 to 1 while allowing the DM to avoid gambles involving extreme consequences. To find the linking factor (defined as h), the unknown in Equation (8a or 8b) is solved for; that is,

$$2U_0(y^{111}) = U_0(y^1) + U_0(y^{11}) \qquad (8a)$$

$$2U_n(y^{111}) = U_n(y^1) + U_n(y^{11}) \qquad (8b)$$

where y^1, y^{11}, and y^{111} are the certainty equivalents for Gambles 6, 7 and 8, respectively. Equation (8a) is used when the certainty equivalent of Gamble 8 is greater than the midpoint value; Equation (8b) is used if it is less than the midpoint value. For the present example, we have

$$2U_0(3.6) = U_0(1.4) + U_0(4.2)$$
$$U_0(1.4) = 1.1$$

Two utility values of 1.4 are now apparent, namely $U_0(1.4) = 1.1$ and $U_n(1.4) = .5$. The linking factor h used to take into account these differences is defined as:

$$h = \frac{U_0(y) - 1}{U_n(y)} \qquad (9)$$

and for this example,

$$h = \frac{U_0(1.4) - 1}{U_n(1.4)} = .2 \qquad (10)$$

The exact manner in which h is used will differ depending upon whether $U_0(y)$ or $U_n(y)$ is being scaled. If $U_0(y)$ is being scaled, then:

$$U_i(y) = \frac{U_0(y)}{1 + h} \qquad (11)$$

whereas if U_n is being scaled, then:

$$U_i(y) = \frac{(h\, U_n(y) + 1}{1 + h} \qquad (12)$$

ASSESS SCALING FACTORS (k_i)

In this task a procedure similar to that suggested by Keeney and Wood (1977) is followed. The response for this study was X_7 (effect on wildlife and vegetation). This implied that k_7 must be the largest of the k_i. The DM was asked this question iteratively and the final order resulted in the following:

$$k_7 > k_2 > k_9 > k_8 > k_4 > k_1 =$$

$$k_2 > k_5 > k_{12} > k_3 = k_{12} > k_{13}$$

To establish actual magnitude of the scaling factors, k_i, trade-offs between the criteria were established. To provide a consistency check, X_1 was pairwise coupled with the other criteria. These twelve tradeoff situations are given in Figure 8.

Consider the first graph in Figure 7 (i.e., tradeoff between X_1 and X_2). The DM is indifferent between ($X_1 = 1.6$, $X_2 = 5$) and ($X_1 = 3.5$, $X_2 = 3.6$). Utilizing the multiplicative utility function given in Equation (7) and equating the above two indifference pairs yields:

$$k_1 = k_2\, U_2\, (3.6) \qquad (13)$$

In a similar fashion, eleven other equations are generated. The twelve equations are summarized in Table 15.

Next, the constant k is introduced into a gamble involving a probability assessment between criteria X_7 and X_2 (those criteria with highest k_i). The gamble is the following.

Gamble 9:

$$(X_7 = 1, X_2 = 5) \sim \overset{\displaystyle p\ (X_7 = 5, X_2 = 5)}{\underset{\displaystyle (1-p)\ (X_7 = 1, X_2 = 1)}{<}}$$

The DM is asked to specify the value of p. Recall that when $p = .5$, the above gamble is essentially a check for additive

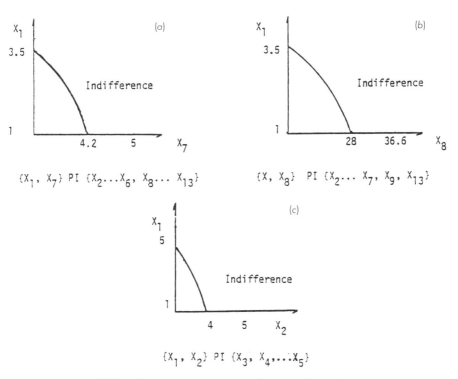

FIGURE 7. Sample gambles for preferential independence.

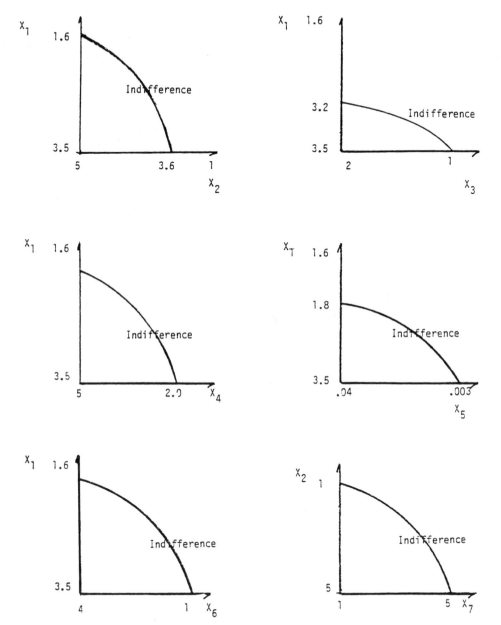

FIGURE 8. Indifference curves for assessment of k_i.

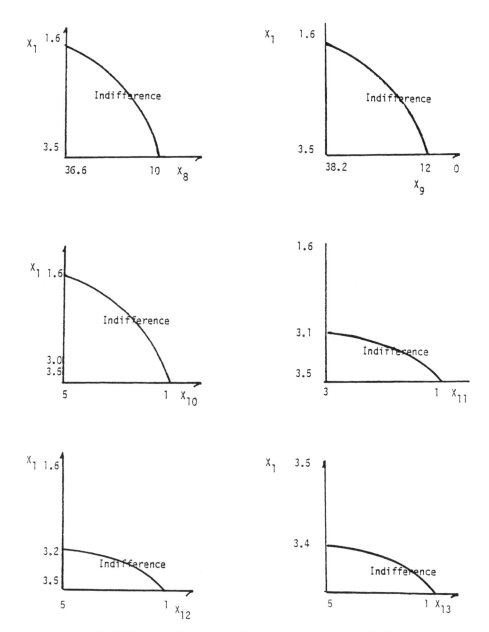

FIGURE 8 (continued). Indifference curves for assessment of k_i.

TABLE 15. Tradeoff Equations.

Tradeoff 1	Tradeoff 2	Equation
$X_1 = 1.6$, $X_2 = 5$	$X_1 = 3.5$, $X_2 = 3.6$	$k_1 = k_2 U_2$ (3.6)
$X_1 = 3.5$, $X_3 = 1$	$X_1 = 3.2$, $X_3 = 5$	$k_3 = k_1 U_1$ (3.2)
$X_1 = 1.6$, $X_4 = 26.33$	$X_1 = 3.5$, $X_4 = 2$	$k_4 = k_1/U_4$ (2)
$X_1 = 3.5$, $X_5 = .003$	$X_1 = 1.8$, $X_5 = .04$	$k_5 = k_1 U_1$ (1.8)
$X_1 = 3.5$, $X_6 = 1$	$X_1 = 1.6$, $X_6 = 4$	$k_1 = k_6$
$X_2 = 1$, $X_7 = 5$	$X_2 = 5$, $X_7 = 1.5$	$k_2 = k_7 U_7$ (1.5)
$X_1 = 1.6$, $X_8 = 36.6$	$X_2 = 3.5$, $X_8 = 10$	$k_8 = k_1 U_8$ (10)
$X_1 = 1.6$, $X_9 = 38.2$	$X_1 = 3.5$, $X_9 = 12$	$k_9 = k_1 U_9$ (12)
$X_1 = 3.5$, $X_{10} = 1$	$X_1 = 3.0$, $X_{10} = 5$	$k_{10} = k_1 U_1$ (3.0)
$X_1 = 3.5$, $X_{11} = 1$	$X_1 = 3.1$, $X_{11} = 5$	$k_{11} = k_1 U_1$ (3.1)
$X_1 = 3.5$, $X_{12} = 1$	$X_1 = 3.2$, $X_{12} = 5$	$k_{12} = k_1 U_1$ (3.2)
$X_1 = 3.5$, $X_{13} = 1$	$X_1 = 3.4$, $X_{13} = 5$	$k_{13} = k_1 U_1$ (3.4)

independence. In this case the DM responded with a value of $p = .57$. Once again utilizing Equation (4), we have:

$$(1 + k\,k_2\,U_2(X_2))(1 + k_7\,U_7(X_7)) =$$

$$.57(1 + k\,k_2(U_2))(1 + k\,k_7\,(U_7)) \qquad (14)$$

$$+ .43(1 + k_2(U_2))(1 + k\,k_2(U_2))$$

therefore:

$$k_7 = .57(k_2 + k_7 + k\,k_7\,k_2) \qquad (15)$$

There are now thirteen equations and fourteen unknowns. By directly assessing k_7, this system of equations reduces to thirteen equations and thirteen unknowns. The assessment of k_7 involves one further tradeoff situation. The tradeoff involves Criterion 7, and Criterion 1. At one extreme, the ordered pair representing the best of both criteria (i.e., $X_7 = 1$, $X_1 = 1.6$) receives a value of 1, whereas the number pair presenting the worst of both criteria ($X_7 = 5$, $X_1 = 3.5$) receives a value of 0. The decision maker was asked to evaluate the pair which represented the best of Criterion 7, and the worst of Criterion 1. This essentially

amounts to the assessment of the corner utility of X_7. The response of the DM was .28 (see Figure 9).

To establish the final values of the k_i, equations of Table 15 are solved. The value of the k_i are given in Table 16.

A brief diversion at this point is made in regard to the interpretation of the k_i. The k_i do not indicate which criteria are most important in a weighting sense, rather the k_i reflect the DM's preference to move criteria away from their worst levels. This conceptual distinction between weights and the k_i is best illustrated with an example. In the systems versus criteria array (Table 11), aquifer level received a weight of 9, whereas effect on wildlife and vegetation (X_7) received a weight of 5. This suggests that Criterion 1 is just less than twice as important as Criterion 7. On the other hand, k_7 is higher than k_1. The reason for this is because the difference between the best outcome and the worst outcome of Criterion 1 is only 1.9 feet per year. In contrast, the difference between the best outcome and the worst outcome of Criterion 7 is the preservation of the status quo versus near disaster. As such, the utility associated with the worst level of this criterion should be less than the utility assigned to the worst level of Criterion 1. The values of k_i confirm that this is indeed the case.

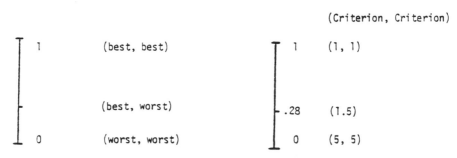

FIGURE 9. Assessment of corner utility (X_1).

TABLE 16. The Values of k_i.

i	1	2	3	4	5	6	7	8	9	10	11	12	13
k_i	.17	.26	.017	.19	.14	.17	.28	.20	.22	.049	.047	.017	.008

k = .65

Application Results

The ELECTRE method, compromise programming and multiattribute utility theory have been applied to the systems versus criteria array of Table 3.

RESULTS OF ELECTRE

The results of ELECTRE I indicate that waste water reclamation combined with channelization, reservoirs or levee construction are the alternatives most often preferred. While ELECTRE I determines that the above systems are preferable to other systems, it does not provide a complete ranking among these preferred alternatives. The application of the ELECTRE II methods yields such a ranking. In general, the final ranking groups the systems according to the flood control actions. Among these groups reservoirs rank highest followed by channelization, floodplain management, levees and no action. The fact that reservoirs ranked highest can be seen by observing that these alternatives [3,23,18, 13,8] are ranked within the top nine of twenty-five systems. At the bottom of the ranking, alternatives [10,20,15,25,5] are dominated.

Within each flood control grouping, a trend among the water supply actions emerges. Waste water reclamation is preferred to all other water supply actions within four of the five flood control groups. It is followed by no action, conservation and education, Central Arizona Project and new groundwater development. It should be noted that the new groundwater alternative ranks last within every flood control group.

The system selected as best by the ELECTRE II methodology is reservoirs combined with waste water reclamation (System 3). It should be noted that public response to reservoir construction has been negative in the past. Such a consideration has not been included explicitly in the analysis, but this could easily be done. The complete ranking is given in Figure 10.

Sensitivity Analysis—To test the robustness of the model with respect to changes in the scales and weights, sensitivity analysis was performed. The scales are changed such that the same scale interval is used for all criteria. Likewise, the weights are changed so that all criteria are given the same weight. These changes, shown in Table 17, are designed to provide insight as to the relative importance of correctly specifying the scales and weights.

Effect of Scale Changes

Reservoirs combined with waste water reclamation remain the best choice upon implementation of the scale changes. However, the scale changes do have an impact on the overall ranking. Among the flood control alternatives, channelization replaces reservoirs as the most preferred choice and levees replace no action as the least preferred choice. Observe that the channelization alternatives [2,22,17,12,7] are ranked within the top eight system choices, whereas the levee construction alternatives are ranked in four of the last six positions.

Within the flood control groups, waste water reclamation remains the most preferred water supply alternative, while new groundwater development remains last. No further trends are apparent.

Effect of Weighting Changes

When all criteria receive the same weights, channelization combined with no action is the most preferred choice. Although some trends appear evident (for example, levees are consistently ranked lowest), it is difficult to rank systems according to flood control or water supply groupings.

The complete rankings after sensitivity analysis are shown in Figure 10. A closer inspection of Figure 10 indicates that waste water reclamation and reservoirs (System 3) should be implemented. However, waste water reclamation with floodplain management (System 2) and reservoirs with no action (System 23) would also constitute good compromises.

COMPROMISE PROGRAMMING RESULTS

The best compromise programming solution is channelization combined with the no action water supply alternative. This is immediately followed by channelization combined with conservation and education, and then by floodplain management with conservation and education. In general, the results are grouped according to the flood control alternatives. The channelization alternatives (2, 7, 12, 17, 22) ranked highest (i.e., 1st, 2nd, 3rd, 8th, 11th) followed by floodplain management systems (4, 9, 14, 19, 24 ranked 3rd, 5th, 6th, 13th and 14th), reservoirs and dams (Systems 3, 8, 13, 18, 23 ranked 7th, 9th, 10th, 12th, and 16th), levees (1, 6, 11, 16, 21 ranked 17th, 18th, 19th, 20th, and 24th), and finally no action (5, 10, 15, 20, 25 ranked 14th, 16th, 22nd, 23rd, and 25th). Within the flood control groups, the no ac-

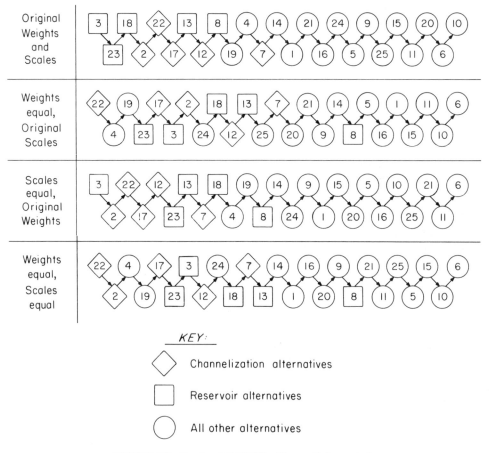

Original
Weights
and
Scales

Weights
equal,
Original
Scales

Scales
equal,
Original
Weights

Weights
equal,
Scales
equal

KEY:

◇ Channelization alternatives

□ Reservoir alternatives

○ All other alternatives

FIGURE 10. Results of ELECTRE with sensitivity analysis.

tion alternatives are most often preferred, whereas those strategies which include new groundwater development are most often dominated. The complete ranking of the twenty-five alternative systems is provided in Figure 11.

Sensitivity Analysis—To test the robustness of the compromise programming solutions to shifts in parameter values, three types of changes have been made. First, the weights α_i are all set equal. Next, the value of p, the l_p metric, is varied from $p = 1$ through $p = 10$. Finally, the scales associated with the systems vs. criteria array are changed. These changes are recorded in Table 18.

Effect of Weight Changes

When the weights α_i are set equal to one another the most preferred system becomes floodplain management combined with conservation and education (System 19). Again, the systems are ranked according to the flood control alternatives. In this case, however, the floodplain management systems replace the channelization alternatives as the most preferred strategies. It is noted that the floodplain manage-

ment systems ranked 1st, 2nd, 3rd, 14th, and 18th, whereas the channelization alternatives ranked 4th, 5th, 6th, 17th, and 19th. Both of these systems are preferred to dams and reservoirs which placed 7th, 8th, 10th, 13th, and 23rd; no action which ranked 9th, 11th, 12th, 21st, and 24th; and levees which ranked 15th, 16th, 20th, 22nd, and 25th.

Within the flood control groupings, the no action water supply alternative is most often preferred, followed by conservation and education, waste water reclamation, Central Arizona Project and new groundwater development. These groupings are similar to those obtained with the original scales, with the exception that the trend appears more predominant in the latter case.

Effect of Distance Metric Changes

As the metric distance parameter p is given the values $p = 1, 2, \ldots 10$, the groupings among the flood control actions become even more obvious. Given the original set of weights and scales, the channelization alternatives are still ranked highest (1st, 2nd, 3rd, 4th, and 5th) but now they dominate all other systems no matter which water supply alternative is considered. Interestingly enough, channelization combined with waste water has replaced channelization with no action as the highest ranked system. Reservoirs and dams (3, 8, 13, 18, 23) are now ranked second within the flood control groups (6th, 7th, 8th, 9th, and 15th), floodplain management (4, 9, 14, 19, and 25 ranked 10th, 11th, 12th, 13th, and 14th); levees (1, 6, 11, 16, and 21 ranked 17th, 18th, 19th, 20th, and 21st); and finally, no action (5, 10, 15, 20, and 25 ranked 16th, 22nd, 23rd, 24th, and 25th).

With respect to the water supply alternatives it appears that both the Central Arizona Project and new groundwater development are the least preferred systems. It is difficult to draw trends among the three other alternatives other than that they are preferred to the CAP and groundwater development alternatives.

Effect of Scale Changes

Two scale changes have been made. First, the scales are redefined such that all scales are set equal and then the scales are made proportional to the values of the k_i used in multiattribute utility theory; that is, if k_1 is twice that of k_2, then the range of Criterion 1 is defined to be double that of Criterion 2 (see Figure 11).

Equal Scales—When the scales are set equal, channelization combined with no action is the most preferred system. This is followed by channelization with conservation and education, and then by floodplain management with conservation and education. Once again, the systems are ranked according to the flood control alternatives. Channelization ranked highest (1, 2, 4, 9 and 11), followed by floodplain management (3, 6, 7, 12 and 13), dams and reservoirs (5, 8, 10, 15, 22), levees (16, 17, 18, 21 and 23), and finally no action (14, 19, 20, 24 and 25).

A trend among the water supply alternatives emerges with respect to the dominated solutions; that is, the Central Arizona Project, and new groundwater development programs are most often dominated.

Scales Proportioned to the k_i—Given the original set of weights and the scales proportioned to the k_i, the results change very little. The most preferred system is once again channelization combined with no action. For a complete analysis, the reader is referred to Figure 11. More interesting are the results generated when the weights are equal. In this case the most preferred system is floodplain management combined with conservation and education (System 19). The complete set results is recorded in Figure 11c. In general, the floodplain management alternatives tend to rank highest (1, 2, 3, 9, and 11), followed by channelization (4, 5, 7, 13 and 16), reservoirs and dams (6, 8, 14, 16, and 23), then by the no action alternatives (10, 12, 15, and 20), and finally by levees (18, 19, 21, 22 and 24). The interesting point worth noting is that this ranking is close to the ranking obtained in multiattribute utility theory.

In general, the compromise results are fairly robust with

TABLE 17. Sensitivity Changes.

Criteria	Original Scales	Equal Scales	Original Weights	Equal Weights
1	150	100	9	1
2	110	100	3	1
3	130	100	3	1
4	160	100	4	1
5	160	100	5	1
6	150	100	5	1
7	200	100	2	1
8	200	100	2	1
9	110	100	2	1
10	100	100	2	1
11	100	100	2	1
12	100	100	1.5	1
13	76	100	1.5	1

respect to changes in parameter values. Through the sensitivity analysis, the results remained grouped according to the flood control alternatives, with the floodplain management systems and channelization alternatives being the most preferred system most of the time. The no action alternatives and levee construction were consistently dominated.

Within the flood control groupings, the compromise programming results are more sensitive to changes in parameter values. In general, however, it can be safely stated that the Central Arizona Project and new groundwater development are dominated most of the time.

RESULTS OF MAUT

The results of multiattribute utility theory indicate that floodplain management combined with conservation and education is the most preferred system. In general, the

TABLE 18. Sensitivity Changes for Compromise Programming.

Criteria	Original Scales	Equal Scales	Scale prop. to K_i
1	150	100	60.7
2	110	100	92.8
3	130	100	60.0
4	160	100	67.8
5	160	100	50.0
6	150	100	60.7
7	200	100	100.0
8	200	100	67.9
9	110	100	77.2
10	100	100	17.2
11	100	100	16.8
12	100	100	6.0
13	76	100	2.3

Original Weights and Scales	22, 17, 19, 2, 4, 24, 23, 12, 18, 13, 7, 3, 9, 14, 25, 8,
	21, 16, 1, 11, 5, 20, 15, 6, 10

Original Weights and Scales proportional to the k_i	22, 17, 19, 2, 18, 4, 24, 3, 12, 7, 14, 9, 13, 25, 23, 21,
	16, 1, 5, 20, 11, 8, 6, 15, 10

Equal Weights, Scales proportional to the k_i	19, 4, 24, 22, 17, 18, 2, 3, 14, 25, 9, 20, 12, 23, 5, 7, 13,
	21, 16, 15, 1, 11, 3, 6, 10

Original Weights, Equal Scales	22, 17, 19, 2, 18, 4, 24, 3, 12, 13, 7, 14, 9, 25, 23, 21, 16,
	1, 5, 20, 11, 8, 6, 15, 10

Equal Weights, Equal Scales	19, 4, 24, 22, 17, 18, 2, 3, 14, 25, 13, 9, 12, 20, 23, 5, 7,
	21, 16, 15, 1, 8, 11, 6, 10

Equal Weights, Original Scales	19, 4, 24, 22, 17, 2, 23, 3, 9, 14, 18, 5, 13, 12, 20, 25, 7,
	15, 8, 21, 16, 1, 11, 10, 6

FIGURE 11. Results of compromise programming.

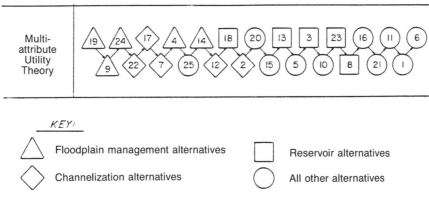

FIGURE 12. Results of MAUT.

floodplain management alternative is the most preferred system, regardless of the water supply action being considered (see Figure 12). Note the fact that Systems 19, 9 and 24 are ranked the top three positions. The results are again grouped according to the flood control alternatives. Floodplain management ranks highest, followed by channelization, no action, dams and reservoirs, and levees. With regard to the water supply alternatives, trends are not as readily apparent. It is safe to say, however, that conservation and education and no action are preferred to the other three alternatives. Given the above information, one concludes that the best systems are combinations of floodplain management or channelization combined with conservation and education and no action.

Sensitivity Analysis—To test the robustness of the model to variations in parameter values, three types of changes are made. First, the assessment of the scaling factor k is challenged. Next, the assessment of the corner utility of k_7 is altered. This in turn has the effect of changing the values associated with the other k_i. The final sensitivity change involves shifting the utilities associated with the individual criteria. This provides information with respect to importance of accurately assessing the DM's utility functions.

Effects on Change in k

Recall that the purpose of the scaling factor k is to insure that the overall utility function is properly scaled between 0 and 1. In varying k, we wish to determine the effect that a misspecification of this parameter has upon the final ranking. For this purpose, k is varied from $-.1653$ to $-.999$. The effect of these changes was slight with regard to the final outcomes (i.e., rankings). However, for $k \leq .6$, some system utility values increased above the value of 1. This can be attributed to the fact that, in order to arrive at a value of $k \geq .6$, the decision maker would have to be inconsistent in the assessment of p in Gamble 9. It is noted that the assessment of k is not critical if we are only interested in the final

ranking and not the utility values associated with this ranking.

Effects on Changes in Corner Utility of k_7

Two changes with respect to the direct assessment of k_7 were made. These involved assessing k_7 to be .06 and .4. The other k_i have been recalculated as a consequence of these changes. These new values are recorded in Table 19. There is little effect on the final outcome as a result of changing the k_i in the above manner. This outcome is expected since only the relative values of the k_i changed; that is, although the absolute values were varied, the k_i are still ranked in the same position as in the original formulation. Once again, however, the utilities associated with the ranking changed as a result of the variation in the value of the k_i.

Effect on Changes in Utility Values

Three changes in utility values have been made. The first change involved increasing the utilities associated with the

TABLE 19. Recalculated Values of k_i.

Criteria	Original k_i	$k_7 = .06$	$k_7 = .4$
1	.17	.037	.2455
2	.26	.056	.372
3	.017	.004	.0245
4	.19	.041	.279
5	.14	.03	.1989
6	.17	.037	.2455
7	.28	.06	.40
8	.20	.043	.288
9	.22	.047	.314
10	.049	.011	.0417
11	.047	.01	.0394
12	.017	.004	.0245
13	.017	.004	.0245

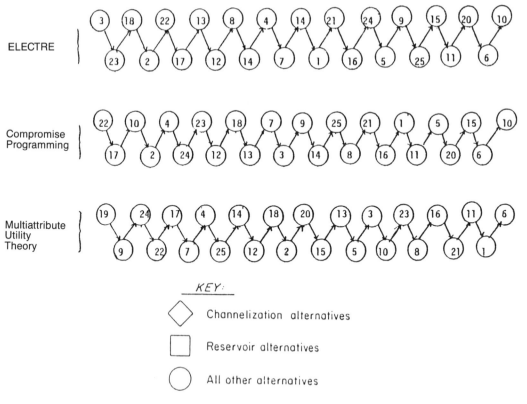

FIGURE 13. Comparison of results.

second highest ranked system (i.e., System 9) until this system is slightly preferred to System 19 (i.e., highest ranked). This required adding a value of .02 to each of the criterion utility values.[1] As such it is concluded that System 19 is only slightly better than System 9, and this may be entirely due to random noise in the evaluation of the original utility functions.

The next change focused upon System 22. It is desired to see how much better the floodplain management alternatives are with respect to the highest ranked channelization system. By adding .05 to each of the utility values associated with System 22, the floodplain management systems still predominated. When .10 was added, however, the channelization alternative ranked first. This suggests that the floodplain management alternatives and the channelization systems are very closely ranked.

The final utility change involved changing the utility values associated with the most preferred dams and reservoirs alternative. This alternative is System 18. When .25 was added to this system's utility value, it became the most preferred system. It is noted, however, that adding .25 to each utility value is a significant parameter change.

Discussion

Changing parameter values in the multiattribute formulation did not have a major impact on the final solutions. The floodplain management alternatives were generally preferable to the other systems. Although this is the case, the channelization alternatives were ranked closely behind these preferred systems.

Figure 13 illustrates this comparison, and Table 20 provides this information in tabular format.

MCDM MODEL CHOICE

In this section, the ELECTRE methodology, compromise programming, and multiattribute utility theory are compared based on a set of specified viewpoints. These viewpoints may be used for selecting the proper MCDM method

---

[1]Note that if $U_i(X_i) > .98$, then $U_i(X_i) + 0.2 > 1$ which is not allowed. In that case, $U_i(X_i)$ is assumed to be 1.0. The same reasoning is used for other utility changes.

to a given river basin planning problem. The viewpoints are:

1. Type of data present
2. Nature of alternative systems under investigation
3. Consistency of results between methodologies
4. Robustness of results with respect to changes in parameter values
5. Ease of computation
6. Amount of interaction required between the decision maker and the systems analyst
7. The type of decision maker (i.e., group or single)

These points will be explored in more detail below.

TYPE OF DATA PRESENT

Data can be classified as being either qualitative or quantitative. The ELECTRE methodology can be used for ranking alternative systems which are evaluated with regard to either of these types of data. The other two example methodologies can handle only quantitative data. However, if a way can be found to quantify the qualitative data, then these methods are appropriate. Recall that in compromise programming a heuristic scaling procedure is used to convert qualitative data. In multiattribute utility theory, a function defining quantified utilities is specified. The advantage of using a method such as ELECTRE over the other two methods is that no bias is encountered through this quantification.

NATURE OF ALTERNATIVE SYSTEMS

Sets of systems can be classified as being either discrete or continuous. An example of the latter might be the presence of an infinite number of reservoir systems corresponding to continuously varied reservoir sizes. A discrete

systems analysis involves a finite number of alternative systems.

The ELECTRE methodology can only handle discrete sets of systems, whereas C.P. and MAUT can effectively deal with either case. MAUT requires, however, interpolation between values of the independent variables.

CONSISTENCY OF RESULTS

In the present example all three methods yield similar rankings. By this it is meant that the same groups of systems consistently are ranked near the top, and another group of systems are ranked at the bottom. Two discrepancies within the results are apparent. First, reservoirs and dams are ranked highest in the ELECTRE method while they are placed third and fourth in C.P. and MAUT, respectively. The other inconsistency is that channelization ranked above floodplain management in the ELECTRE and compromise programming rankings, whereas this order is switched in MAUT.

The reasons for these differences may lie in how the concept of discordance is used. In C.P. and MAUT, all discordances (i.e., distances in C.P., disutilities in MAUT), no matter how small, are included in the analysis. In ELECTRE the threshold value for discordance is either satisfied or not. If it is satisfied, the value is not used again. For example, reservoirs had a high discordance on the environmental criterion, but this was ignored since that value was less than the threshold value chosen. Thus, in comparing two systems, say System A and System B, it is possible for System A to be preferred to System B, even if System A caused more discomfort than System B on the majority of criteria. It is difficult to say which measure of discordance is better; rather, one should be aware of the different ways discordance can be measured.

TABLE 20. Comparison of Results.

	ELECTRE	Compromise Programming	Multiattribute Utility Theory
Flood Control Alternatives	Dams and Reservoirs Channelization Floodplain Management Levees No Action	Channelization Floodplain Management Dams and Reservoirs Levees No Action	Floodplain Management Channelization No Action Dams and Reservoirs Levees
	Waste Water No Action Conservation and Education Central Arizona Project (CAP) New Ground	No Action Conservation and Education Waste Water Central Arizona Project (CAP) New Ground	No Action Conservation and Education ------------

ROBUSTNESS OF RESULTS TO CHANGES IN PARAMETER VALUES

In general, all the methodologies considered are fairly robust with respect to changes in parameter values. In ELECTRE, changing the weights had more of an effect than changing the scales. In compromise programming, the results consistently were ranked according to the flood control alternatives, with the channelization alternative most often being preferred. This was followed closely by the floodplain management systems. In none of the parameter changes did this basic ranking change. In multiattribute utility theory, the results also appear to be robust. Changing the values of k and the k_i had little effect on the final ranking. Varying the utility values did change the final outcome, but this is more of an indication that there was little difference between the top systems rather than an indication of a sensitive model.

EASE OF COMPUTATION

Ease of computation can be thought of in two ways. First, how much learning is required by the systems analyst to master the technique, and secondly, once mastered, how much time is required to implement the technique and analyze the results. The time required for implementation refers to person hours, and not computer time.

With regard to the first point, multiattribute utility theory requires the most time for the systems analyst to learn; the ELECTRE technique follows, and then compromise programming. It is noted that C.P. is extremely easy for discrete analysis and can be understood by persons having little or no background in multiobjective planning.

With regard to the second point, MAUT also requires the most time with regard to implementation. Several assumptions must be checked and then the assessment of individual utility functions must be performed. Analyzing the results, however, is rather straightforward. Both compromise programming and ELECTRE are fairly easy to implement. The major difference in time is in the analysis of results. Compromise programming is quite simple, whereas ELECTRE is tedious and cumbersome. This is because the results of ELECTRE are a set of preference graphs in which 25 systems are entangled. From these graphs, preferences must be drawn and the systems ranked. It is not out of the ordinary to find cycles which in turn must be broken, which necessarily takes time.

AMOUNT OF INTERACTION BETWEEN DM AND SYSTEM ANALYST

Multiattribute utility theory requires an extensive amount of interaction between the decision maker and the systems analyst. In fact, the amount of interaction time required in MAUT is so extensive that it seems farily impractical for real world applications. ELECTRE and compromise programming, on the other hand, do not require a significant amount of the DM's time. This is because these latter techniques only require the specification of weights and scales. As already noted, these weights and scales need not be pinpoint precise. The dominant compromise solution surfaces as long as the weights and scales are close to the desired values.

TYPE OF DECISION MAKER

Although all three methodologies can be implemented with either a single or a group decision maker, it seems that C.P. and ELECTRE would do better than MAUT with respect to the group DM. This is because of the DM time required in this latter technique. A group of individuals sitting down to define a set of weights is one thing; the same group to evaluate their collective utility for 13 criteria is quite another.

REFERENCES

1. Bardossy, A., The mathematics of composite programming, working paper. No. 84-18, Dept. of Systems Eng., Univ. of Ariz., Tucson, Ariz. 85721 (1984).
2. Benayoun, R., B. Roy and B. Sussman, ELECTRE: Une methode pour guider le choix en presence de points de vue multiples. Direction Scientifique, Note de Travail No. 49, SEMA, Paris (1966).
3. Bogardi, I., L. Duckstein and A. Bardossy, Trade-off between cost and efficiency of pollution control. *Proceedings, Sixth Int. Conf. on MCDM*, Cleveland, Ohio (1984).
4. Bureau of Water Quality Control, State of Arizona, Water Quality Standards (1978).
5. Cohon, H. L., *Multiobjective Programming and Planning*. Academic Press, New York, 333 p. (1978).
6. Cohon, J. and D. Marks, A review and evaluation of multiobjective programming techniques. *Water Resources Research*, 11(2), April, 208–220 (1975).
7. Corps of Engineers, Urban Study for the Metropolitan Area of Tucson, Arizona, No. 12192, Tucson, Arizona (December, 1978).
8. David, L. and L. Duckstein, Multicriterion ranking of alternative long-range water resources systems. *Water Resources Bulletin*, 12(4), August, pp. 731–734 (1976).
9. Davidson, E. S., Geohydrology and water resources of the Tucson Basin, Arizona. Geological Survey Water Supply Paper 1939-E (1973).
10. Duckstein, L. and M. Gershon, "Multiobjective analysis of a vegetation management problem using ELECTRE II," *Applied Mathematical Modeling*, Vol. 7, pp. 254–261 (1984).
11. Evans, G. W., "An Overview of Techniques for Solving Multiobjective Mathematical Programs." *Management Science*, Vol. 30, No. 11 (1984).
12. Fraser, N. M. and K. W. Hipel, *Conflict Analysis: Models and Resolutions*. Elsevier Science Publishing Co., Inc. New York, NY (1984).

13. Gershon, M. E., "Model Choice in Multiobjective Decision Making in Water and Mineral Resource Systems," Natural Resource Systems Technical Report Series #37, Department of Hydrology and Water Resources, University of Arizona, Tucson, AZ 85721 (1981).

14. M. Gershon and L. Duckstein, A procedure for selection of a multiobjective technique with application to water and mineral resources. *Applied Mathematics and Computation,* 14(3), pp. 245–271 (April 1984).

15. Goicoechea, A., D. Hansen and L. Duckstein, *Introduction to Multiobjective Analysis with Engineering and Business Applications,* John Wiley & Sons, New York, N.Y., p. 511 (1982).

16. Greis, N. P., E. F. Wood, and R. E. Steuer, Multicriteria analysis of water allocation in a river basin: the Tchebycheff approach, *Water Resources Research,* Vol. 19, No. 4, pp. 865–875 (1983).

17. Haimes, Y., W. Hall, and H. Freedman, *Multiobjective optimization in water resources systems: The surrogate worth trade-off method.* Elsevier Scientific Publishing Company (1975).

18. Keeney, R. L., "Evaluation of Proposed Storage Sites," *Operations Research,* Vol. 27, No. 1, pp. 48–64 (January–February 1979).

19. Keeney, R. L., "Concepts of Independence in Multiattribute Utility Theory." In *Multicriteria Decision Making,* J. Cochrane and M. Zeleny, eds. University of South Carolina Press, Columbia, SC (1973).

20. Keeney, R. L. and E. F. Wood, An Illustrative Example of the Use of Multiattribute Utility Theory for Water Resources Planning, *Water Resources Research,* Vol. 13, No. 4, pp. 705–712 (August 1977).

21. Keeney, R. L. and H. Raiffa, *Decisions with Multiple Objectives: Preferences and Value Tradeoffs,* John Wiley & Sons, New York, p. 569 (1976).

22. Khairullah, Z. and S. Zionts, "An Experiment with Some Approaches for Solving Problems with Multiple Criteria." Working Paper No. 442, Third International Conference on Multiple Criteria Decision Making, Konigswinter, West Germany (1979).

23. Krzysztofowicz, R., Preference Criterion and Group Utility Model for Reservoir Control under Uncertainty. Reports on Natural Resource Systems, No. 30, Department of Hydrology and Water Resources, University of Arizona, Tucson, Arizona 85721 (1978)

24. Maas, A., M. M. Hufschmidt, R. Dorfman, H. A. Thomas, S. A. Manglin, and G. N. Fair, *"Design of Water Resource Systems",* Harvard University Press, Cambridge, Mass. (1962).

25. MaCrimmon, K. R., "An Overview of Multiple Objective Decision Making." In *Multiple Criteria Decision Making,* J. Cochrane and M. Zeleny (eds.), University of Southern Carolina Press, Columbia, South Carolina, pp. 18–44 (1973).

26. Major, D., *Multiobjective Water Resource Planning,* Water Resources Monograph 4, American Geophysical Union, Washington, D.C. (1977).

27. Massam, G. H., "The Central Arizona Water Control Study: A Comparison of Alternate Plans Using Concordance Analysis and Multidimensional Scaling." *Water Resources Bulletin,* Vol. 20, No. 4, pp. 483–492 (1984).

28. Moosburner, G. J., and E. F. Wood, Management model for controlling nitrate contamination in the New Jersey Pine Barrens aquifer, *Water Resources Bulletin,* 16/5, p. 971 (1980).

29. Owen, G., *Game Theory,* 2nd ed. Academic Press, Inc., New York, NY (1982).

30. Pareto, V., Cours d' Economie Politique, Lausanne, Rouge (1896).

31. Pollak, R. A., "Additive von Neumann-Morgenstern Utility Functions." *Econometrica,* Vol. 35, pp. 485–494 (1967).

32. Roy, B. and P. Bertier, "La Methode ELECTRE II," Note de Travail No. 142, Direction Scientifique, Groupe Metra, Paris, April (1971).

33. Szidarovszky, F. and L. Duckstein, "A Framework for Dynamic Multiobjective Optimization with Application to Water and Mineral Resources Management." EURO Sixth European Congress on Operations Research, Vienna, July 19–22, 1983. *Eur. J. Oper. Res.,* Vol. 24, pp. 305–307 (1986).

34. Szidarovszky, F., M. Gershon and L. Duckstein, *Techniques for Multiobjective Decision-Making in Systems Management,* Elsevier Amsterdam, 506 p. (1986).

35. Tecle, A., L. Duckstein and M. Fogel, "Regional Water Resources Management: A Multiobjective Case Study of the Upper Santa Cruz River Basin." Western Regional Science Association Meeting, San Diego, CA (1985).

36. Teghem, J., Jr. and P. L. Kunsch, "Application of Multiobjective Stochastic Linear Programming to Power Systems Planning." Working Paper. Fac-Polytechn. Mons. (1984).

37. UNIDO, Guidelines for Project Evaluation, New York (1972).

38. von Neumann, J. and O. Morgenstern, *Theory of Games and Economic · Behavior,* Second Edition, Princeton University Press, Princeton, New Jersey (1947).

39. Voogd, H., *Multicriteria Evaluation for Urban and Regional Planning.* Plon Limited, London (1983).

40. U.S. Water Resources Council, "Principles and Standards for Planning and Related Land Resources," *Federal Register,* Vol. 38, No. 174, Part III, pp. 24779–24869 (September 10, 1973).

41. U.S. Water Resources Council, "Economic and Environmental Principles and Guidelines for Water and Related Land Resources Implementation Studies," *Federal Register,* p. 10259 (March 10, 1983).

42. Zeleny, M., "Adaptive Displacement of Preference in Decision Making," in: *Multiple Criteria Decision Making,* M. K. Starr and M. Zeleny, eds., North-Holland Publishing Co., Amsterdam-New York, pp. 147–157 (1977).

43. Zeleny, M., *Multiple Criteria Decision Making,* McGraw Hill Book Co., New York, p. 563 (1982).

44. Zionts, S. and J. Wallenius, "Recent Developments in Our Approach to Multiple-Criteria Decision Making." *Interactive Decision Analysis,* M. Grauer and A. Wierczbicki (eds.), Springer-Verlag, Vienna, Austria (1984).

APPENDIX

For Further Sources of Information

Bardossy, A., I. Bogardi and L. Duckstein, "Analyse Multi-critère 'Floue' de la Gestion d'une Nappe Karstique Regionale: II. Application Numérique," *Revue Internationale des Sciences de l'eau*, Vol. 2, No. 1 (February 1986).

Bogardi, I., A. Bardossy and L. Duckstein, "Multicriterion Network Design Using Geostatistics," *Water Resources Research*, Vol. 21, No. 2, pp. 199–208 (February 1985).

Bogardi, I., L. Duckstein and A. Bardossy, "An Efficient Solution of Multiobjective Compromise Optimization in Water Resources by Differential Dynamic Programming," Working Paper No. 81-25, Department of Systems & Industrial Engineering, University of Arizona (1981).

Bogardi, I., L. Duckstein and A. Bardossy, "Regional Management of An Aquifer for Mining Under Fuzzy Environmental Objectives," *Water Resour. Res.*, Vol. 19, No. 6, pp. 1394–1402 (December 1983).

Bowman, V. J., "On the Relationship of the Tchebycheff Norm and the Efficient Frontier of Multiple-Criteria Objectives;" and Shapiro, J. F., "Multiple-Criteria Public Investment Decision Making by Mixed-Integer Programming," *Lecture Notes on Economics and Mathematical Systems*, Vol. 130 (May 1975).

Cohon, J. L., C. S. ReVelle and J. Current, "Application of a Multi-objective Facility Location Model to Power Plant Sitting in a Six-State Region of the U.S.," *Computers and Operations Research*, 7, pp. 107–123 (1980).

Dalkey, N.C., "Group Decision Analysis," in *Multiple Criteria Decision Making*, M. Zeleny, ed., Kyoto (1975).

Davis, D. and R. Krysztofowicz, "The Performance of Flood Forecast-Response Systems," Paper WMO-NOAA Int. Conf. on Mitigation of Natural Hazards Through Real-Time Data Collection Systems, Sacramento, CA (Sept. 1983).

Davis, D. R., L. Duckstein and F. Szidarovszky, "Opportunity Losses Due to Uncertainty in Multiobjective Decision Problems," *Applied Math. and Computations* (to appear).

Deason, J. P. and K. P. White, "Specification of Objectives by Group Processes in Multiobjective Water Resources Planning," *Water Resources Research*, Vol. 20, No. 2, pp. 189–196 (1984).

Despontin, M. and J. Spronk, "Comparison of Multiple Criteria Decision Models, First Results of an International Investigation," Presented at the 10th Meeting of the EURO Working Group on MCDM, Liege, Report 7923A, Centre for Research in Business Economics, Erasmus University, Rotterdam (October 1979).

Despontin, M., J. Moscarola and J. Spronk, "Tutorial Paper XVI," *Revue Belge de Statistique, D'Informatique et de Recherche Operationnelle*, Vol. 23, No. 4: A User-Oriented Listing of Multiple Criteria Decision Methods (1983).

Duckstein, L., "General Report: Systems Approach to Groundwater Resources," *Proceedings, Int'l. Conf. on Modern Approaches to Groundwater Resources Management*, Capri, Italy, Vol. 11, pp. 13–15 (October 25–27, 1982).

Duckstein, L., "Imbedding Uncertainties into Multiobjective Decision Models in Water Resources," Keynote Paper, Session T1, International Symposium on Risk & Reliability in Water Resources, University of Waterloo, Waterloo, Canada (June 1978).

Duckstein, L., "Multiobjective Optimization in Structural Design: The Model Choice Problem," *Proceedings, International Symposium on Optimum Structural Design*, pp. 10/1–10/4 (October 1981).

Duckstein, L. and J. Bernier, "Risk-Related Criteria in Water Resources: A Mathematical System Framework with Applications," *Proceedings, Workshop on Risk Analysis in Hydrological System Operation*, Ecole Polytechnique, Montreal, Quebec, Canada (November 1985).

Duckstein, L. and J. Kempf, "Multicriteria Q-Analysis for Plan Evaluation," Paper presented at 9th Meeting of EURO Working Group on MCDM, Amsterdam, April, also available as Report No. 79-15, Department of Systems & Industrial Engineering, University of Arizona (1979).

Duckstein, L., D. Davis, and E. Plate, "Use of Climatological Information in River Basin Planning Under Risk," *Proceedings, Int'l. Symp. on Statistical Climatology*, Lisbon, Portugal, pp. 9.3.1–9.3.10 (September 26–30, 1983).

Duckstein, L., H. Hiessl and M. Becker, "Multicriterion Q-Analysis with a Discordance Concept: Application to River Basin Management," EURO VII, Multicriterion Decision Making Session, Bologna, Italy (16–20 June, 1985).

Duckstein, L., I. Bogardi and A. Bardossy, "Analyse Multi-critère 'Floue' de la Gestion d'une Nappe Karstique Regionale: I. Théorie et Cas de la Transdanubie (Hongrie)," *Revue Internationale des Sciences de l'eau* (1980).

Duckstein, L., J. Kempf and J. Casti, "Design and Management of Regional Systems by Fuzzy Ratings and Polyhedral Dynamics," Chapter II in *Macro-Economic Planning with Conflicting Goals*, P. Nijkamp, M. Despontin and J. Spronk, eds., Springer-Verlag, Berlin, Vol. 230, pp. 223–238 (1984).

Duckstein, L., S. Ambrus and D. Davis, "Management Forecasting Requirements," Ch. 16 in *Hydrological Forecasting*, MN. G. Anderson and T. P. Burt, eds., Wiley, pp. 559–585 (1985).

Dyers, J.S. and R. K. Sarin, "Measurable Multiattribute Value Functions," Discussion Paper No. 66, Management Science Study Center, University of California, Los Angeles, California (1978).

Engineering Reliability and Risk in Water Resources, Pro-

ceedings of the NATO Advanced Study Institute, held in Tucson, Arizona, 19–31 May 1985, L. Duckstein and E. J. Plate, eds., Martinus Nijhoff, The Hague Netherlands (1987).

Fishburn, M. L., "Methods of Estimating Additive Utilities," *Management Science*, *13*, No. 7, pp. 435–453 (1969).

Fishburn, P. C., "A Survey of Multiattribute/Multicriterion Evaluation Theories," in *Multiple Criteria Problem Solving*, S. Zionts, ed., *Proceedings, Buffalo, New York, Lecture Notes in Econ, and Math. Systems*, No. 155, Springer Verlag, Heidelberg, Germany (1978).

Fishburn, P. C., "Lexicographic Orders, Utilities and Decision Rules: A Survey," *Management Sciences*, Vol. 20, pp. 1442–1471 (1976).

Fishburn, P. C. *Utility Theory for Decision Making*. John Wiley & Sons, New York (1970).

Goicoechea, A., L. Duckstein and M. Fogel, "Multiobjective Programming in Watershed Management: A Study of the Charleston Watershed," *Water Resources Research*, Vol. 12, No. 6, pp. 1085–1092 (December 1976).

Goicoechea, A., L. Duckstein and M. Fogel, "Multiple Objectives Under Uncertainty: An Illustrative Application of PROTRADE," *Water Resources Research*, Vol. 15, No. 2, pp. 203–210 (April 1979).

Haimes, Y. Y., "The Surrogate Worth Trade-Off (SWT) Method and Its Extensions," *Proceedings of the Third Conference on Multiple Criteria Decision Making – Theory and Application*, Konigswinter, West Germany (August 1979).

Haimes, Y. Y., B. Das and K. Sung, "Multipleobjective Analysis in the Maumee River Basin: A Case Study," Report to National Science Foundation No. AEN75015820 and OWRT No. 14-34-0001-6221 (October 1979).

Haimes, Y. Y., K. A. Loparo, S. C. Olenik and S. K. Nanda, "Multiobjective Statistical Method for Interior Drainage Systems," *Water Resources Research*, Vol. 16, No. 3, pp. 465–475 (1980).

Haimes, Y. Y., D. A. Wismer and L. S. Lasdon, "On Bicriterion Formulation of the Integrated System Identification and System Optimization," *IEEE Transactions on Systems, Man and Cybernetics*, Vol. SMC-1, pp. 296–297 (July 1971).

Heidel, K. and L. Duckstein, "Extension of ELECTRE Technique to Group Decision Making: An Application to Fuel Emergency Control," Paper TIMS/ORSA Joint National Meeting, Chicago; Systems & Industrial Engineering Working Paper 83-014 (April 1983).

Hiessl, H., L. Duckstein and E. J. Plate, "Multiobjective Analysis with Concordance and Discordance Concepts," *Applied Math. & Computation*, Vol. 17, pp. 107–122 (1985).

Hobbs, B. F., "Experiments in Multicriteria Decision-Making and What We Can Learn from Them," in *Deci-sion-Making with Multiple Objectives*, J. J. Haimes and V. Chankong, eds., Springer-Verlag, New York (1985).

Hobbs, B. F., "Multiobjective Power Plant Siting Methods," *Journal of the Energy Division, ASCE*, Pro. No. 15745, Vol. EY2, pp. 187–200 (October 1980).

Hwang, C. I. and A. S. M. Masud, "Multiple Objective Decision Making – Methods and Applications," *Lecture Notes in Econ. and Math. Systems*, No. 164, Springer Verlag, New York (1979).

Keeney, R. L., "Evaluation of Proposed Storage Sites," *Operations Research*, Vol. 27, No. 1, pp. 48–64 (1979).

Keeney, R. L., "Multidimensional Utility Functions: Theory, Assessment and Applications," Technical Report No. 43, Operations Research Center, Massachusetts Institute of Technology, Cambridge, MA (1969).

Krzysztofowicz, R., "Strength of Preference and Risk Attitude in Utility Measurements," *Organizational Behavior and Human Performance*, Vol. 31, pp. 88–113 (1983).

Krzysztofowicz, R. and L. Duckstein, "Preference Criterion for Flood Control under Uncertainty," *Water Resources Research*, Vol. 15, No. 3, pp. 513–520 (June 1979).

Krzysztofowicz, R., "Preference Criterion and Group Utility Model for Reservoir Control Under Uncertainty," *Natural Resource Systems Technical Report Series #30*, Department of Hydrology and Water Resources, University of Arizona, Tucson, AZ (1978).

Lee, S. M., G. E. Green and C. S. Kim, "A Multi-Criteria Model for the Location Allocation Problem," *Computers and Operations Research*, 8, pp. 1–8 (1981).

Lord, W. B., "Objectives and Constraints in Federal Water Resources Planning," *Water Resour. Bull.*, Vol. 17, No. 6, pp. 1060–1065 (Dec. 1981).

Lovell, R. E., "Hydrologic Model Selection in a Decision-Making Context," *Natural Resources Systems Technical Report Series No. 26*, Department of Hydrology and Water Resources, University of Arizona, Tucson, AZ (1975).

Maddock, T. and L. Duckstein, "Multicriterion Groundwater Management," 11th IFIP Conference on System Modelling and Optimization, Copenhagen, Denmark (July 25–29 1983).

Monarchi, S., C. Kisiel and L. Duckstein, "Interactive Multiobjective Programming in Water Resources: A Case Study," *Water Resour. Res.*, Vol. 9, No. 4, pp. 837–850 (1973).

Nakamura, M. and J. M. Riley, "A Multiobjective Branch and Bound Method for Network-Structured Water Resources Planning Problems," *Water Resources Research*, Vol. 17, No. 5, pp. 1349–1359 (October 1981).

Opricovic, S., "An Extension of Compromise Programming to the Solution of Dynamic Multicriteria Problems," Paper, Ninth IFIP Conference on Optimization Techniques, Warsaw, Poland (September 1979).

Osyzka, A. *Multicriterion Optimization in Engineering*. Ellis-Orwood, Chichester, United Kingdom (1984).

Rarig, H. M. and Y. Y. Haimes, "Risk/Dispersion Index Method," *IEEE Transactions on Systems, Man and Cybernetics,* Vol. SMC 13, No. 3, pp. 317–318 (May-June 1983).

Roy, B., "Problems and Methods with Multiple Objective Functions," *Mathematical Programming,* Vol. 1, No. 2, pp. 239–268 (1971).

Salis, M. and L. Duckstein, "Mining Under a Limestone Aquifer in Southern Sardinia: A Multiobjective Approach," *Intern. J. of Mining Eng.,* Vol. 1, No. 2, pp. 357–374 (1983).

Starr, M. K. and M. Zeleny, "MCDM—State and Future of the Arts," in *Multiple Criteria Decision Making,* North-Holland Publishing Co., New York (1977).

Szidarovszky, F., I. Bogardi and L. Duckstein, "Use of Cooperative Games in a Multiobjective Analysis of Mining and Environment," *Proceedings, Second International Conference on Applied Numerical Modeling,* Madrid, Spain (September 1978).

Szidarovszky, F., L. Duckstein and I. Bogardi, "Multiobjective Management of Mining Under Water Hazard by Game Theory," *Eur. J. Oper. Res.,* Col. 15, No. 2, pp. 251–258 (1984).

Tauxe, G. W., R. R. Inman and D. M. Mades, "Multiobjective Dynamic Programming: A Classic Problem Redressed," *Water Resources Research,* 15, No. 6, pp. 1398–1402 (1979a).

Tauxe, G. W., R. R. Inman and D. M. Mades, "Multiobjective Dynamic Programming with Application to a Reservoir," *Water Resources Research,* 15, No. 6, pp. 1403–1408 (1979b).

Van de Nes, T. J., "Applicability of Multicriteria Decision Models with Multiple Decision Makers in Water Resources Systems," Seminar on Economic Instruments for Rational Utilization of Water Resources, Committee on Water Problems, Economic Commission for Europe (1980).

Wallenius, J., "Comparative Evaluation of Some Interactive Approaches to Multicriterion Optimization," *Management Science,* 21, No. 12, pp. 1387–1396 (1975).

Wallenius, J. and S. Zionts, "Some Tests of an Interactive Programming Method for Multicriteria Optimization and an Attempt at Implementation," in *Multiple Criteria Decision Making,* H. Thiriez and S. Zionts, eds., Jouyen-Josas, France, 1975, Springer-Verlag, Berlin, pp. 318–330 (1976).

White, C. C., A. P. Sage and S. Dozono, "A Model of Multiattribute Decision-Making and Trade-Off Weight Determination under Uncertainty," *IEEE Transactions on Systems, Man, and Cybernetics,* Vol. SMC-14, No. 2 (March/April 1984).

Yu, P. L., "Dynamic Programming in Finite Stage Multicriteria Decision Problems," Working Paper No. 118, School of Business, University of Kansas, Lawrence, KS (October 1978).

Zeleny, M., "Adaptive Displacement of Preferences in Decision Making," in *Multiple Criteria Decision Making,* M. K. Starr and M. Zeleny, eds., North-Holland Publishing Co., New York, NY, pp. 147–157 (1977).

Zeleny, M. *Linear Multiobjective Programming.* Springer-Verlag, Berlin, West Germany (1974).

Zionts, S., "Integer Linear Programming with Multiple Objectives," *Annals of Discrete Mathematics,* Vol. 1, pp. 551–562 (1977).

Zionts, S. and D. Deshpande, "A Time-Sharing Computer Programming Application of a Multiple Criteria Decision Method to Energy Planning—A Progress Report," *Multiple Criteria Problem Solving,* S. Zionts, ed., Proceedings 1977, Buffalo, New York, Springer-Verlag, Berlin (1978).

Zionts, S. and J. Wallenius, "An Interactive Multiple Objective Linear Programming Method for a Class of Underlying Nonlinear Utility Functions," *Management Science,* Vol. 29, pp. 519–520 (1983).

Zionts, S. and J. Wallenius, "An Interactive Programming Method for Solving Multiple Criteria Problem," *Management Science,* Vol. 22, No. 6, pp. 652–663 (1976).

SECTION TWO
Irrigation

CHAPTER 15

Crop Yield: Economic Considerations of Deficit Irrigation

George H. Hargreaves* **and Zohrab A. Samani****

INTRODUCTION

Deficit irrigation has been used as an approximate method for maximizing profit from the very beginnings of irrigation. The concept needs to be further developed and the economic considerations more fully defined. Stressing the crop may produce many advantages including:

1. A larger yield of mature fruit or grain at harvest time.
2. An increased yield of sugar content in sugar cane or rubber content in guayule.
3. Improved quality and/or uniformity.
4. Better root development and improved drought resistance.
5. A reduction in crop waterlogging due to rainfall.
6. Increased profits from limited water supplies.

The prosperity of an individual or of a country depends upon the cost and availability of the factors of production and on how efficiently these factors are used. Access to water, technology and stable markets are frequently more important than land ownership. Water rights associated with land ownership may be more valuable than the land. Careful water control combined with optimum use of fertilizers may double or quadruple water use efficiencies and profits per unit of water or land.

Maximum yield per unit land may or may not represent the maximum benefit. Figure 1 shows a typical yield-water relationship. As is shown in Figure 1, the maximum yield is obtained at maximum applied water, but the optimum water level for maximum profit is somewhat lower than maximum yield and maximum water use efficiency (WUE) is obtained at a water level less than those of maximum profit or maximum yield.

An analysis is presented of many of the factors that influence yields and profits under irrigated and rainfed agriculture.

In order to analyze the economics of deficit irrigation, it is useful to calculate the amount of water required for maximum yields and the desirable frequency and method of application. This chapter, therefore, presents equations for calculating crop water requirements for maximum yields, describes how other factors influence yields and profits and discusses modeling of some of the more important relationships.

An attempt is made to present the modeling process in its simplest terms and concepts and also to make available some advanced and technical methods of analysis.

CROP WATER REQUIREMENTS

If all other factors of production remain constant the crop yields increase with increasing water availability, reach a maximum, and then decline as excessive water is applied. There are several special cases of water tolerant crops such as rice and Californiagrass, but in general, the exponential relationship of yield as a function of water availability is similar for most commercial crops.

In order to determine the degree of deficit in irrigation applications, it is necessary to first define the crop water requirements for maximum yields, Y_m. The crop water use or crop evapotranspiration, ET_m, associated with the maximum crop yield is given by the equation:

$$ET_m = ET_o \times Kc \tag{1}$$

in which ET_o is reference crop evapotranspiration and Kc is a crop coefficient for the specific crop under consideration. Values of Kc are presented in Table 1.

*International Irrigation Center, Utah State University, Logan, UT

**Department of Civil, Agricultural and Geotechnical Engineering, New Mexico State University, Las Cruces, NM

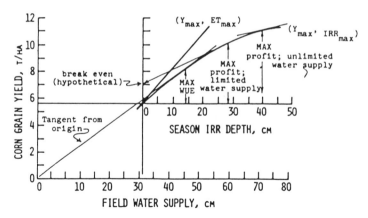

FIGURE 1. Graphic analysis of a yield vs seasonal irrigation [IRR] function to determine IRR levels that are likely to maximize water use efficiency [WUE] and profit [from Stewart and Hagan, 1973].

Evapotranspiration, ET, is the process by which water is transferred from the plant and soil into the atmosphere. It includes evaporation from plant and soil surfaces and transpiration of water through the plant tissue. The rate of *ET* is usually expressed as equivalent depth of water per unit time (e.g., mm/day or mm/month).

Reference crop evapotranspiration, ET_o, is the rate of *ET* from an extensive surface of 8 to 15 cm tall, green, cool-season (C3) grass cover of uniform height, actively growing, completely shading the ground and not short of water. Perennial ryegrass has been commonly used as the reference crop. The equations presented herein for *ET_o* were derived using Alta fescue grass, one of the tall fescues, as the reference crop. Kentucky bluegrass has also sometimes been used as the reference crop. The tall fescues, perennial ryegrass and Kentucky bluegrass have very similar rates of ET.

Maximum crop evapotranspiration, ET_m, is the potential *ET* of an agricultural crop grown in large fields under disease free and non-restricting soil and other conditions. The temperature and climate should be favorable for good levels of commercial crop production. *ET_m* may be measured under field conditions or calculated by multiplying *ET_o* times the appropriate crop coefficients, *Kc,* as indicated in Equation (1).

Reference crop evapotranspiration, *ET_o,* also known as potential evapotranspiration, PET, or ETP, is best estimated using one of the following equations:

$$ET_o = 0.00094 \times Ra \times T°F \times TD^{0.50} \quad (2)$$

$$ET_o = 0.0023 \times Ra \times (T°C + 17.8) \times TD^{0.50} \quad (3)$$

in which *Ra* is extraterrestrial radiation (the amount of solar radiation received at the top of the atmosphere) expressed in the same units of equivalent water evaporation as *ET_o,* and

TD is mean maximum minus mean minimum temperature (in °F for Equation (2) and in °C for Equation (3)). *Ra* is dependent only upon latitude and the time of year. Mean monthly values of *Ra* are given in Table 2 in mm/day. Daily values can be obtained by graphing the monthly means and connecting by a smooth curve. Fortran computer equations for monthly and daily values of *Ra* or *RA* are presented in the Appendix of this chapter.

Equations (2) and (3) are simpler and easier to use than most of the other methods for estimating *ET_o* and require only measured values of maximum and minimum daily air temperatures. They can, however, be considered as combination equations as the daily temperature range compensates for differences in water vapor in the atmosphere and for advective energy.

Worldwide cool-season grass lysimeter *ET* was compared with values of *ET_o* from Equations (2) and (3) and with values from 18 other estimating methods including the Penman and other combination methods. Equations (2) and (3) proved superior to the other methods evaluated and are therefore, the only procedures presented herein.

WATER-FERTILITY–YIELD INTERACTIONS

If the other factors of production or yield are constant, then maximum yield, *Y_m,* can be shown as a function of one variable. The principal variables to be considered are water, *ET_o,* fertility, crop variety, and actual crop evapotranspiration, *ET_a.* Various yield models relate production to *ET_a.* The Stewart yield model (Stewart et al., 1977) has been adopted by Doorenbos and Kassam (1979) and provides the basic criteria used by the Food and Agriculture Organization of the United Nations, FAO, for estimating the crop

TABLE 1. Crop Coefficients (Kc) for Use With Equations Based on Cool Season Grasses.

Crop	Crop Development Stages					Total Growing Period
	Initial	Crop Development	Mid-Season	Late Season	At Harvest	
Banana						
tropical	0.4 −0.5	0.7 −0.85	1.0 −1.1	0.9 −1.0	0.75−0.85	0.7 −0.8
subtropical	0.5 −0.65	0.8 −0.9	1.0 −1.2	1.0 −1.15	1.0 −1.15	0.85−0.95
Bean						
green	0.3 −0.4	0.65−0.75	0.95−1.05	0.9 −0.95	0.85−0.95	0.85−0.9
dry	0.3 −0.4	0.7 −0.8	1.05−1.2	0.65−0.75	0.25−0.3	0.7 −0.8
Cabbage	0.4 −0.5	0.7 −0.8	0.95−1.1	0.9 −1.0	0.8 −0.95	0.7 −0.8
Cotton	0.4 −0.5	0.7 −0.8	1.05−1.25	0.8 −0.9	0.65−0.7	0.8 −0.9
Grape	0.35−0.55	0.6 −0.8	0.7 −0.9	0.6 −0.8	0.55−0.7	0.55−0.75
Groundnut	0.4 −0.5	0.7 −0.8	0.95−1.1	0.75−0.85	0.55−0.6	0.75−0.8
Maize						
sweet	0.3 −0.5	0.7 −0.9	1.05−1.2	1.0 −1.15	0.95−1.1	0.8 −0.95
grain	0.3 −0.5	0.7 −0.85	1.05−1.2	0.8 −0.95	0.55−0.6	0.75−0.9
Onion						
dry	0.4 −0.6	0.7 −0.8	0.95−1.1	0.85−0.9	0.75−0.85	0.8 −0.9
green	0.4 −0.6	0.6 −0.75	0.95−1.05	0.95−1.05	0.95−1.05	0.65−0.8
Pea, fresh	0.4 −0.5	0.7 −0.85	1.05−1.2	1.0 −1.15	0.95−1.1	0.8 −0.95
Pepper, fresh	0.3 −0.4	0.6 −0.75	0.95−1.1	0.85−1.0	0.8 −0.9	0.7 −0.8
Potato	0.4 −0.5	0.7 −0.8	1.05−1.2	0.85−0.95	0.7 −0.75	0.75−0.9
Rice	1.1 −1.15	1.1 −1.5	1.1 −1.3	0.95−1.05	0.95−1.05	1.05−1.2
Safflower	0.3 −0.4	0.7 −0.8	1.05−1.2	0.65−0.7	0.2 −0.25	0.65−0.7
Sorghum	0.3 −0.4	0.7 −0.75	1.0 −1.15	0.75−0.8	0.5 −0.55	0.75−0.85
Soybean	0.3 −0.4	0.7 −0.8	1.0 −1.15	0.7 −0.8	0.4 −0.5	0.75−0.9
Sugarbeet	0.4 −0.5	0.75−0.85	1.05−1.2	0.9 −1.0	0.6 −0.7	0.8 −0.9
Sugarcane	0.4 −0.5	0.7 −1.0	1.0 −1.3	0.75−0.8	0.5 −0.6	0.85−1.05
Sunflower	0.3 −0.4	0.7 −0.8	1.05−1.2	0.7 −0.8	0.35−0.45	0.75−0.85
Tobacco	0.3 −0.4	0.7 −0.8	1.0 −1.2	0.9 −1.0	0.75−0.85	0.85−0.95
Tomato	0.4 −0.5	0.7 −0.8	1.05−1.25	0.8 −0.95	0.6 −0.65	0.75−0.9
Watermelon	0.4 −0.5	0.7 −0.8	0.95−1.05	0.8 −0.9	0.65−0.75	0.75−0.85
Wheat	0.3 −0.4	0.7 −0.8	1.05−1.2	0.65−0.75	0.2 −0.25	0.8 −0.9
Alfalfa	0.3 −0.4				1.05−1.2	0.85−1.05
Citrus						
clean weeding						0.65−0.75
no weed control						0.85−0.9
Olive						0.4 −0.6

First figure: Under high humidity (RHmin >70%) and low wind (U <5 m/sec).
Second figure: Under low humidity (RHmin <20%) and strong wind (>5 m/sec).
Source: Doorenbos and Kassam, 1979.

TABLE 2. Extraterrestrial Radiation (Ra) Expressed in Equivalent Evaporation in mm/day.

Northern Hemisphere												Lat.	Southern Hemisphere											
Jan.	Feb.	Mar.	Apr.	May	June	July	Aug.	Sept.	Oct.	Nov.	Dec.		Jan.	Feb.	Mar.	Apr.	May	June	July	Aug.	Sept.	Oct.	Nov.	Dec.
3.8	6.1	9.4	12.7	15.8	17.1	16.4	14.1	10.9	7.4	4.5	3.2	50°	17.5	14.7	10.9	7.0	4.2	3.1	3.5	5.5	8.9	12.9	16.5	18.2
4.3	6.6	9.8	13.0	15.9	17.2	16.5	14.3	11.2	7.8	5.0	3.7	48	17.6	14.9	11.2	7.5	4.7	3.5	4.0	6.0	9.3	13.2	16.6	18.2
4.9	7.1	10.2	13.3	16.0	17.2	16.6	14.5	11.5	8.3	5.5	4.3	46	17.7	15.1	11.5	7.9	5.2	4.0	4.4	6.5	9.7	13.4	16.7	18.3
5.3	7.6	10.6	13.7	16.1	17.2	16.6	14.7	11.9	8.7	6.0	4.7	44	17.8	15.3	11.9	8.4	5.7	4.4	4.9	6.9	10.2	13.7	16.7	18.3
5.9	8.1	11.0	14.0	16.2	17.3	16.7	15.0	12.2	9.1	6.5	5.2	42	17.8	15.5	12.2	8.8	6.1	4.9	5.4	7.4	10.6	14.0	16.8	18.3
6.4	8.6	11.4	14.3	16.4	17.3	16.7	15.2	12.5	9.6	7.0	5.7	40	17.9	15.7	12.5	9.2	6.6	5.3	5.9	7.9	11.0	14.2	16.9	18.3
6.9	9.0	11.8	14.5	16.4	17.2	16.7	15.3	12.8	10.0	7.5	6.1	38	17.9	15.8	12.8	9.6	7.1	5.8	6.3	8.3	11.4	14.4	17.0	18.3
7.4	9.4	12.1	14.7	16.4	17.2	16.7	15.4	13.1	10.6	8.0	6.6	36	17.9	16.0	13.2	10.1	7.5	6.3	6.8	8.8	11.7	14.6	17.0	18.2
7.9	9.8	12.4	14.8	16.5	17.1	16.8	15.5	13.4	10.8	8.5	7.2	34	17.8	16.1	13.5	10.5	8.0	6.8	7.2	9.2	12.0	14.9	17.1	18.2
8.3	10.2	12.8	15.0	16.5	17.0	16.8	15.6	13.6	11.2	9.0	7.8	32	17.8	16.2	13.8	10.9	8.5	7.3	7.7	9.6	12.4	15.1	17.2	18.1
8.8	10.7	13.1	15.2	16.5	17.0	16.8	15.7	13.9	11.6	9.5	8.3	30	17.8	16.4	14.0	11.3	8.9	7.8	8.1	10.1	12.7	15.3	17.3	18.1
9.3	11.1	13.4	15.3	16.5	16.8	16.7	15.7	14.1	12.0	9.9	8.8	28	17.7	16.4	14.3	11.6	9.3	8.2	8.6	10.4	13.0	15.4	17.2	17.9
9.8	11.5	13.7	15.3	16.4	16.7	16.6	15.7	14.3	12.3	10.3	9.3	26	17.6	16.4	14.4	12.0	9.7	8.7	9.1	10.9	13.2	15.5	17.2	17.8
10.2	11.9	13.9	15.4	16.4	16.6	16.5	15.8	14.5	12.6	10.7	9.7	24	17.5	16.5	14.6	12.3	10.2	9.1	9.5	11.2	13.4	15.6	17.1	17.7
10.7	12.3	14.2	15.5	16.3	16.4	16.4	15.8	14.6	13.0	11.1	10.2	22	17.4	16.5	14.8	12.6	10.6	9.6	10.0	11.6	13.7	15.7	17.0	17.5
11.2	12.7	14.4	15.6	16.3	16.4	16.3	15.9	14.8	13.3	11.6	10.7	20	17.3	16.5	15.0	13.0	11.0	10.0	10.4	12.0	13.9	15.8	17.0	17.4
11.6	13.0	14.6	15.6	16.1	16.1	16.1	15.8	14.9	13.6	12.0	11.1	18	17.1	16.5	15.1	13.2	11.4	10.4	10.8	12.3	14.1	15.8	16.8	17.1
12.0	13.3	14.7	15.6	16.0	15.9	15.9	15.7	15.0	13.9	12.4	11.6	16	16.9	16.4	15.2	13.5	11.7	10.8	11.2	12.6	14.3	15.8	16.7	16.8
12.4	13.6	14.9	15.7	15.8	15.7	15.7	15.7	15.1	14.1	12.8	12.0	14	16.7	16.4	15.3	13.7	12.1	11.2	11.6	12.9	14.5	15.8	16.5	16.6
12.8	13.9	15.1	15.7	15.7	15.5	15.5	15.6	15.2	14.4	13.3	12.5	12	16.6	16.3	15.4	14.0	12.5	11.6	12.0	13.2	14.7	15.8	16.4	16.5
13.2	14.2	15.3	15.7	15.5	15.3	15.3	15.5	15.3	14.7	13.6	12.9	10	16.4	16.3	15.5	14.2	12.8	12.0	12.4	13.5	14.8	15.9	16.2	16.2
13.6	14.5	15.3	15.6	15.3	15.0	15.1	15.4	15.3	14.8	13.9	13.3	8	16.1	16.1	15.5	14.4	13.1	12.4	12.7	13.7	14.9	15.8	16.0	16.0
13.9	14.8	15.4	15.4	15.1	14.7	14.9	15.2	15.3	15.0	14.2	13.7	6	15.8	16.0	15.6	14.7	13.4	12.8	13.1	14.0	15.0	15.7	15.8	15.7
14.3	15.0	15.5	15.5	14.9	14.4	14.6	15.1	15.3	15.1	14.5	14.1	4	15.5	15.8	15.6	14.9	13.8	13.2	13.4	14.3	15.1	15.6	15.5	15.4
14.7	15.3	15.6	15.3	14.6	14.2	14.3	14.9	15.3	15.3	14.8	14.4	2	15.3	15.7	15.7	15.1	14.1	13.5	13.7	14.5	15.2	15.5	15.3	15.1
15.0	15.5	15.7	15.3	14.4	13.9	14.1	14.8	15.3	15.4	15.1	14.8	0	15.0	15.5	15.7	15.3	14.4	13.9	14.1	14.8	15.3	15.4	15.1	14.8

Source: Doorenbos and Pruitt, 1977.

yield response to ET_a. The Stewart equation can be written:

$$\left(1 - \frac{Y_a}{U_m}\right) = Ky\left(1 - \frac{ET_a}{ET_m}\right) \qquad (4)$$

in which Y_a = actual harvested yield; Y_m = maximum or potential harvested yield; Ky is a crop yield response factor that relates the decline or decrease in Y_a to the unit decrease in ET_a; and ET_a and ET_m are as defined above.

Values of Ky given by Doorenbos and Kassam (1979) are given in Table 3. There is some evidence that these values may be somewhat optimistic. In general, most yield models show crop production to be a linear function of ET_a. This is valid over a range of conditions. However, as indicated by various models, the crop growth stage during which water deficits occur has a large influence on yield. The timing of water stress also influences crop quality, uniformity of ripening and profits.

The relationship of irrigation amount to crop yield is much more complex. Consideration must be given to initial soil moisture, utilizable rainfall, efficiency and uniformity of irrigation applications, and the influence of reduced aeration due to soil saturation after irrigation or rainfall on crop yield reductions. As irrigation approaches the amount required for maximum yield, it becomes increasingly difficult to maintain high efficiencies. The possibility of waterlogging by rain following too soon after irrigation, particularly on heavy soils, is also increased.

Hargreaves (1975) used yield data for several crops at various locations to derive a yield function for relative yield (Y) as a function of relative total water (initial soil moisture plus rain, plus irrigation). The range in the data was from 0.33 to 1.10 times full water adequacy. (Y) was assigned a value of 1.00 for maximum yield and (X) a value of 1.00 for the amount required to produce maximum yield. The best fit equation is:

$$Y = 0.8X + 1.3X^2 - 1.1X^3 \qquad (5)$$

TABLE 3. Yield Response Factor (Ky).

Crop	Vegetative Period (1) Early (1a)	Late (1b)	Total	Flowering Period (2)	Yield Formation (3)	Ripening (4)	Total Growing Period
Alfalfa			0.7–1.1				0.7–1.1
Banana							1.2–1.35
Bean		0.2		1.1	0.75	0.2	1.15
Cabbage	0.2				0.45	0.6	0.95
Citrus							0.8–1.1
Cotton		0.2		0.5		0.25	0.85
Grape							0.85
Groundnut		0.2		0.8	0.6	0.2	0.7
Maize		0.4		1.5	0.5	0.2	1.25
Onion			0.45		0.8	0.3	1.1
Pea	0.2			0.9	0.7	0.2	1.15
Pepper							1.1
Potato	0.45	0.8			0.7	0.2	1.1
Safflower		0.3		0.55	0.6		0.8
Sorghum		0.2		0.55	0.45	0.2	0.9
Soybean		0.2		0.8	1.0		0.85
Sugarbeet							
beet							0.6–1.0
sugar							0.7–1.1
Sugarcane			0.75		0.5	0.1	1.2
Sunflower	0.25	0.5		1.0	0.8		0.95
Tobacco	0.2	1.0				0.5	0.9
Tomato		0.4		1.1	0.8	0.4	1.05
Watermelon	0.45	0.7		0.8	0.8	0.3	1.1
Wheat							
winter		0.2		0.6	0.5		1.0
spring		0.2		0.65	0.55		1.15

Source: Doorenbos and Kassam, 1979.

This results in the equation:

$$\frac{dY}{dX} = 0.8 + 2.6X - 3.3X^2 \qquad (6)$$

Data were not available for (X) less than about 0.33 so the lower portion of the curve is undefined. As intercept of zero was used for convenience.

When yield is shown as a function of irrigation water applied, there is frequently some yield due to rainfall or rain plus stored soil moisture when the irrigation amount is zero. The yield equation can then frequently be expressed as:

$$Y = a + bX - cX^2 \qquad (7)$$

The yield response to applied nitrogen in kg/ha can frequently be expressed by a similar equation in the form of

$$Y = a + bN - cN^2 \qquad (8)$$

Yield data can also be expressed in one equation as a function of both water and nitrogen. However, there is a simpler methodology that seems more practical.

Hargreaves, 1983, proposed the use of isoquant curves (lines of equal yields) as functions of ET_a and N. A straight line through the approximate points of maximum curvature of the isoquants defines the optimum combinations of ET_a and N. Figure 2 is based upon research data from Israel and the United States for high yielding corn hybrids. The actual values will depend upon several factors including the carryover of fertility from previous years. Similar isoquant curves have been developed by plotting the percentage of available soil water remaining when irrigation is applied on the X axis and applied nitrogen on the Y axis. Isoquants of yield can be drawn as functions of several variables. Pasture production can be shown as a function of ET_a and N or as a function of solar radiation, RS, and leaf area index, LAi, or of ET_a and LAi.

Equation (8) can be used as an approximation of an

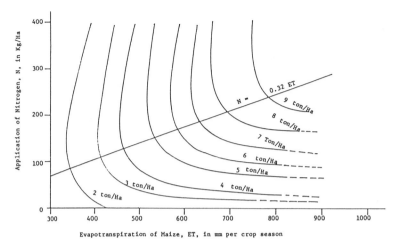

FIGURE 2. Isoquants of maize production. Source: Hargreaves, 1981.

economic model. If a = yield in kg of grain with no application of N, b = the average increase in yield per kg of N for the first 20 to 30 kg of N applied and $b - 2C \times N_{mx} = 0$ or $c = b/2N_{mx}$ for N_{mx} = the kg of N for maximum yield. Then if N_p = the price of a kg of N and Y_p = the price or value of a kg of grain and $Z = 2.5\ Y_p/N$ the value of dY/dN can be set equal to Z for the point on the yield curve where one kg of N will produce Z kg of grain worth 2.5 times the cost of the required kg of N. The equation

$$\frac{dY}{dN} = b - 2CN = Z \qquad (9)$$

will then indicate a desirable upper limit for the application

of N since the production of grain worth 2.5 times the cost of a kg of N should be adequate to pay for transportation, application and harvesting costs and still yield a reasonable profit to the farmer.

Table 4 presents recommendations for fertilizer applications. Requirements for P and K need to be determined based on soil conditions. Requirements for N are those associated with good yields as shown in Table 5.

ET_o is an index of energy available for plant growth. The potential yield depends upon the available energy, climatic conditions, fertility, variety, and management as well as upon water availability. The better farmers will develop sufficient knowledge of the other factors of production so that an approximate optimization will result. Attention can then be given to the maximization of benefits from that frequently scarce or expensive resource input, water.

TABLE 4. FAO Fertilizer Recommendations in Kg per Ha.

Crop	N	P	K	Crop	N	P	K
Alfalfa	0–40	55–65	75–100	Pineapple	230–300	45–65	110–120
Banana	200–300	45–60	240–480	Potato	80–120	50–80	125–160
Bean	20–40	40–60	50–120	Rice	100–150	20–40	80–120
Cabbage	100–150	50–65	100–130	Safflower	60–110	15–30	25–40
Citrus	100–200	35–45	50–160	Sorghum	100–180	20–45	35–80
Cotton	100–180	20–60	50–80	Soybean	10–20	15–30	25–60
Grape	100–160	40–60	160–230	Sugarbeet	150	50–70	100–160
Groundnut	10–20	15–40	25–40	Sugarcane	100–200	20–90	125–160
Maize	100–200	50–80	60–100	Sunflower	50–100	20–45	60–125
Olive	200–250	55–70	160–210	Tobacco	40–80	30–90	50–110
Onion	60–100	25–45	45–80	Tomato	100–150	65–110	160–240
Pea	20–40	40–60	80–160	Watermelon	80–100	25–60	35–80
Pepper	100–170	25–50	50–100	Wheat	100–150	35–45	25–50

1 kg of P = 2.4 kg of $P_{20}5$, 1 kg of K = 1.2 kg of K20.
Source: Doorenbos and Kassam, 1979.

TABLE 5. Good Yields of High-Producing Varieties Adapted to the Climatic Conditions of the Available Growing Season under Adequate Water Supply and High Level of Agricultural Inputs under Irrigated Farming Conditions (ton/ha).

Crop		Tropics[1] <20°C[4]	Tropics[1] >20°C	Subtropics[2] <20°C	Subtropics[2] >20°C	Temperate[3] <20°C	Temperate[3] >20°C
Alfalfa	hay	15		25		10	
Banana	fruit		40–60		30–40		
Bean: fresh	pod	6–8		6–8		6–8	
dry	grain	1.5–2.5		1.5–2.5		1.5–2.5	
Cabbage	head	40–60		40–60		40–60	
Citrus:							
grapefruit	fruit		35–50		40–60		
lemon	fruit		25–30		30–45		
orange	fruit		20–35		25–40		
Cotton	seed cotton		3–4		3–4.5		
Grape	fruit	5–10		15–30		15–25	
Groundnut	nut		3–4		3.5–4.5		1.5–2
Maize	grain	7–9	6–8	9–10	7–9		4–6
Olive	fruit			7–10			
Onion	bulb	35–45		35–45		35–45	
Pea: fresh	pod	2–3		2–3		2–3	
dry	grain	0.6–0.8		0.6–0.8		0.6–0.8	
Fresh pepper	fruit	15–20		15–25		15–20	
Pineapple	fruit		75–90		65–75		
Potato	tuber	15–20		25–35		30–40	
Rice	paddy		6–8		5–7		4–6
Safflower	seed			2–4			
Sorghum	grain	3–4	3.5–5	3–4	3.5–5		2–3
Soybean	grain	2.5–3.5		2.5–3.5			
Sugarbeet	beet			40–60		35–55	
Sugarcane	cane		110–150		100–140		
Sunflower	seed	2.5–3.5		2.5–3.5		2–2.5	
Tobacco	leaf		2–2.5		2–2.5		1.5–2
Tomato	fruit	45–65		55–75		45–65	
Watermelon	fruit		25–35		25–35		
Wheat	grain	4–6		4–6		4–6	

[1]Semi-arid and arid areas only.
[2]Summer and winter rainfall areas.
[3]Oceanic and continental areas.
[4]Mean temperature.
Source: Doorenbos and Kassam, 1979.

Nitrogen is essential for protein production and the larger the yield the more nitrogen is required. Hargreaves, 1983, and Hargreaves and Samani, 1984, analyzed considerable yield data by using isoquant graphs and developed an equation for the optimum requirements for nitrogen, N, in kg per ha. The equation is:

$$N = KN \times ET_a \qquad (10)$$

in which KN = a constant for each crop and variety. The yield data evaluated indicated values of KN for corn (maize), wheat, cotton, and sugar beets averaging approximately 0.32, 0.12, 0.10, and 0.15, respectively, for ET_a in mm of total depth for the crop growing season.

Crop varieties differ in their response to nitrogen. In some circumstances the optimum response for one will occur at twice the nitrogen application associated with the optimum for another variety. If the variety has been well selected and other factors of production are favorable, the optimization of yield depends largely upon finding the most suitable management practices related to water and fertility consistent with the available energy for crop production. The costs associated with precisely managing these factors may be quite high. However, an approximate optimization is frequently possible without excessive management input.

DEFICIT IRRIGATION AND DROUGHT

Agricultural drought has been defined and evaluated in many different ways. For rainfed agriculture, drought could be defined as that condition when rainfall occurrence falls below the amount required for economical crop production. A drought for one crop may not be a drought for another. Rice, corn and sorghum are all important food grains, but differ widely in their water requirements for economical production. Drought may occur during a part of the growing season or throughout the entire vegetative period.

The definition of drought should be related to ET_m for the particular crop under consideration. In general, when the water supply to the crop is less than one-third of ET_m, or when ET_a is less than one third of ET_m for a significant period of time, crop production is almost always uneconomical. However, some subsistence crops may be produced when somewhat less water is available.

In years of low rainfall the water supply for irrigation may be limited. Land may not be limited. During drought an attempt should be made to either optimize profit or maximize food production from the limited water supply. For the data analysed by Hargreaves, 1975, the maximum contribution of water to crop yield occurred at about 40 percent of full water adequacy. At 85 percent of full adequacy, the contribution of water to yield is less than half that at 40 percent adequacy decreasing to zero contribution as the requirement for maximum yield is approached.

During drought years rainfall or rain plus stored soil moisture may be only slightly below the amount required for subsistence or for economical agriculture. In these circumstances, it may be desirable to irrigate for much less than maximum yield in order to maximize total production and to also benefit as many farmers as is practical without reducing total productivity.

THE EFFECT OF GOVERNMENTAL POLICIES

Throughout much of the world, water is limiting and the good lands potentially well suited for irrigation development significantly exceed the available developed water supply. Yet on many large and important projects, considerable yield reduction occurs due to over irrigation. Governments may spend large sums each year on new construction yet fail to appropriate funds or delegate responsibility for improving irrigation scheduling and for increasing yields on the lands placed under irrigation. More emphasis needs to be given to irrigating the maximum area with available water supplies and in increasing yields consistent with optimization of profits or of food production.

If water is in short supply, irrigating for maximum yield is seldom in the best interests of the state or nation, unless some excess irrigation is required to control or reduce soil salinity. Research, extension and various training programs should be developed to promote greater emphasis on deficit irrigation.

THE INFLUENCE OF SOILS AND MANAGEMENT

Soil texture and structure influence the desirability of deficit irrigation. Most crops require a continuously adequate supply of oxygen in the root zone. If irrigation saturates the entire root zone for a few days, this may result in significant reductions in yield and/or quality. On heavy textured soils in Israel, drip and sprinkle irrigation were compared for cotton production. The drip system applied water between every other row, used 80 percent as much water as sprinkle and applied fertilizer and herbicide in the water. Yields were increased by 88 percent on the drip irrigated area as compared with those on lands irrigated by sprinkle.

Crop yield data from black cotton soils (heavy clay or clay loam soils) in Maharashtra, India, indicate the desirability of scheduling irrigation so as to avoid saturating the soils in the crop root zone, particularly during the rainy months. If too much water is applied there is a probability of waterlogging due to rain following irrigation. For rainfed agriculture on heavy soils, sometimes the years of most rain result in the lowest yield and the highest yields may occur during low rainfall years.

Some farmers have increased their profits by irrigating in alternate furrows for each successive water application.

Others have reported improved profits from skip furrow irrigation with decreased water use. Alfalfa will scald due to decreased respiration if flooded for a significant period during hot weather. Saturating the entire surface and the presence of some standing water reduces aeration resulting in a decrease in crop evapotranspiration, a lack of adequate cooling and permanent damage to the plants.

YIELD RESPONSE TO REDUCED SOIL AERATION

The production of plant biomass results from net photosynthesis or from gross photosynthesis minus respiration. If all other factors are equal, gross photosynthesis can be shown as a function of temperature. The rate increases with increasing temperature and then declines. The relationship can be approximated by an equation in the form of

$$GP = a + bT - cT^2 \qquad (11)$$

in which GP = gross photosynthesis and T = leaf temperature.

Respiration, R, is also a function of temperature and can be approximated by an equation of the form

$$R = K \times T^n \qquad (12)$$

Net photosynthesis is indicated by the area between these two curves. At the temperature where the curves meet, net photosynthesis declines very rapidly as respiration is then using up more plant material than is being produced through gross photosynthesis.

Photosynthesis requires energy to combine CO_2 and H_2O to form carbohydrates. Respiration requires oxygen to reverse the reaction and release energy to maintain normal plant functions. The requirements for soil aeration therefore vary exponentially with temperature. Inadequate amounts of oxygen in the soil results in corresponding yield reductions.

Translocation and conversion of the first products of photosynthesis requires the energy available from respiration. Attempts to irrigate for maximum yield sometimes result in enough over irrigation so as to interfere with the full potential of the benefits from respiration.

AGRONOMIC ADVANTAGES OF DEFICIT IRRIGATION

Various crops are less affected by moisture stress later in the growth cycle providing they have been stressed during the initial or vegetative stage. Initial root development may be too shallow if soil moisture levels are too high.

Withholding water is used to improve crop quality, control disease, and regulate the maturing of the crop. Stressing sugar cane prior to harvest increases sugar content and im-

proves the quality of the sugar. Stressing the guayule plant decreases total biomass but increases the production of rubber.

The period of flowering and fruit setting is usually the most critical in that moisture stress reduces yields more than stress during other vegetative stages. However, mildly stressing a crop in its flowering stage may be economically advantageous. The crop consists of thousands of plants with some more mature than others. A low level of water stress can be used to force less mature plants out of the vegetative stage and produce flowering and fruit setting so that all plants can be harvested at the same time. Withholding water is used to produce a smaller total yield, but a larger yield of mature fruit or grain at harvest time. For example, mechanically harvested tomatoes are stressed prior to harvest in order to increase the percentage of mature fruit.

Cotton is a perennial plant. If there is too much water at harvest time, new vegetative growth will start. Early water cut off usually increases the profitability of cotton production. Low moisture stress in the early vegetative stage of some crops may promote lodging or result in shallow rooting to the degree that subsequent water stress will have more serious effects resulting in greater yield reductions.

Guayule is native to Texas and Mexico and grows with limited water supplies. For the yield data used to develop Figure 3 the maximum production of biomass occurred with about 900 mm of water. However, the optimum net income from rubber production was from an application of 500 mm of irrigation. The analysis by Fangmeyer, et al., 1984, was made assuming that the land area was limited. Under conditions of limited water and unlimited land the optimum water application should be significantly lower.

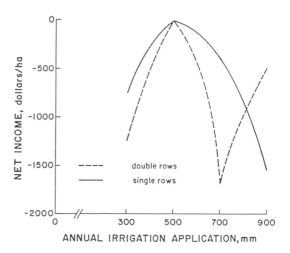

FIGURE 3. Effect of annual irrigation application on net income as lowest breakeven rubber price. Source: Fangmeier, et al., 1984 [3].

DATA REQUIREMENTS FOR EVALUATING RISK

A principle consideration in determining potential yield is whether water will be deficit, adequate, or excessive. Usually some form of probability analysis is desirable both for the water supply and for rainfall. Crop water requirements can also be analyzed for variability as the potential crop water use is usually higher during dry years due to higher radiation and temperature. However, the variability in potential crop water use from year to year is usually not great.

The selection of the method for estimating probabilities is not very important since there is a large degree of uncertainty in the estimation of future rainfall and future water supplies. The gamma probability distribution is often preferred for analyzing rainfall data and is usually recommended when computer facilities are readily available. If the rainfall records are fairly long, the ranking probability

distribution produces results very similar to those of the gamma distribution except for very low and very high probabilities. When the two distributions are graphed for comparison, the gamma distribution has the advantage that it tends to smooth irregularities.

The ranking method for determining the probability of assured rainfall or assured water supply can be expressed as follows:

$$F = \frac{m}{n + 1} \tag{13}$$

$$P = 100 - 100\,F \tag{14}$$

in which F = a frequency number; m = the order varying from 1 to n ($n = 1$ for the smallest value in the series); and P = the probability of the assured amount corresponding to the value in the data series associated with the order number m.

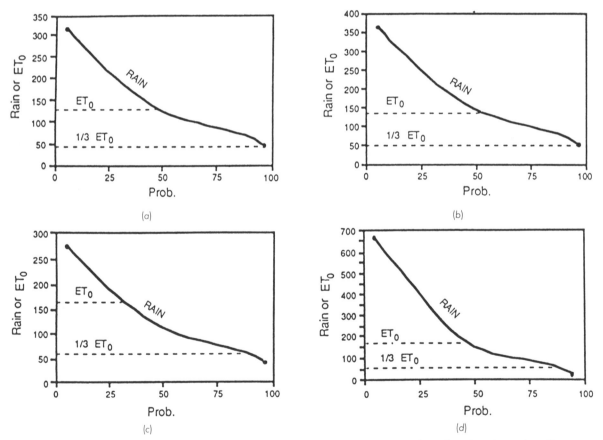

FIGURE 4. Probabilities of assured rainfall compared with estimated values of ET_o and $1/3$ ET_o for selected months at Begampet and Vishakhapatnam, Andhra Pradesh, India. Source: Hargreaves, et al., 1985. (a) Begampet/August; (b) Begampet/September; (c) Vishakhapatnam/August; and (d) Vishakhapatnam/October.

For planning purposes, an assured water supply with a 75 percent probability of occurrence is desired. An assured amount of rain corresponding to the 75 percent probably is defined as dependable rainfall.

The entire data series does not have to be ranked in order to determine the 75 percent probability. If $P = 75$, then $100 F$ will be 25 and $m/n + 1$ will equal 0.25. The data value corresponding to m is then the 75 percent probability. The 5, 50, and 95 percent probabilities can be calculated in a similar manner and connected by smooth curves and the values of ET_o and $1/3\ ET_o$ placed on the graph. Figure 4 presents computer graphs comparing the range of rainfall probabilities with ET_o and $1/3\ ET_o$. The probabilities were calculated by computer using the gamma distribution.

Equation (4) indicates how yields are reduced with increasing water deficits. Equation (2) or (3) and Table 1 present the method for calculating the potential crop water use. However, allowance needs to be made for irrigation or use efficiencies. The data requirements consist of rainfall, maximum and minimum temperatures and the water supply data. If daily data are recorded, summaries and evaluations can be made for any selected time period. In most cases, soil moisture storage will buffer water availability to the degree that a ten-day time period will be adequate. If this period is too long, 5-day or weekly periods can be selected.

IRRIGATION EFFICIENCIES

The economics of irrigation depends to a significant degree upon efficiencies. Table 6 presents a summary of the expected range of conveyance, distribution and application efficiencies.

A precise economic analysis of deficit irrigation must consider irrigation efficiencies. As maximum yields are approached, the efficiency of irrigation normally declines. A reduction of a given percentage in the irrigation application produces a somewhat lower reduction in the water available to the plant. Reported data on yields and profits related to irrigation applications are usually deficient of adequate information on initial soil moisture, effective rainfall and irrigation efficiencies.

MAXIMIZING ECONOMIC BENEFITS FROM RAINFALL

Irrigation is often supplemental to rainfall. The value of scarce water can be significantly increased by attempts to obtain maximum benefits from rain. Equation (5) is a best fit of the relationship of total water (soil moisture plus rain plus irrigation) to yield. The yield data are from research sites, principally in the western United States. The shape of the curve depends significantly on the irrigation efficiency.

Equation (6) indicates that at about 0.4 times full water adequacy, a small increase in water makes a maximum con-

tribution to yield. If rainfall amounts were more consistent from year to year, it would be much easier to plan for the maximization of economic benefits. Figure 4 as indicated above presents typical curves showing the probabilities of various rainfall amounts compared with ET_o and $1/3\ ET_o$.

In several countries and in a worldwide study by Hargreaves, 1977, a moisture adequacy or availability index, MAI, is used as a planning tool to indicate the potential agricultural benefits from rainfall. The values of MAI are calculated from dependable rainfall, PD, (the 75 percent probability of assured rainfall amount) and ET_o. The equation is:

$$MAI = \frac{PD}{ET_o} \tag{15}$$

When monthly values of MAI are less than 0.33, there is usually little probability of developing a profitable rainfed agricultural system. However, with MAI values of 0.45 or 0.55, there is often a marked response to the application of fertilizers and an opportunity for profitable agriculture. By optimizing both water and fertilizer, the spreading of scarce irrigation water over fairly large areas offers a potential for maximum benefits from the limited water supplies.

Benefits from rainfall can also be maximized on irrigation projects if steps are taken to minimize losses resulting from unforeseen or extreme rainfall. On heavy textured soils in areas where rainfall is significant during the crop growing season, rains following too soon after irrigation may reduce soil aeration, cause waterlogging of the soils and reduce crop production to unacceptable levels. The probability of such waterlogging can be reduced through deficit irrigation leaving capacity in the soil to store a significant portion of the rain in the event that this becomes necessary.

Hourly infiltration rates of water into agricultural soils usually vary from about 5 to 35 mm per hour. Hourly rainfall intensities exceeding these rates produce runoff unless provision is made to increase the opportunity time for the water to enter the soil. Frequently, in the developing countries hourly depth-duration rainfall data are not available. More often records can be obtained of extreme daily rainfall amounts. Hourly amounts can be estimated with reasonable accuracy from the daily records. Twenty-four hour extreme values average 1.13 times extreme daily amounts. For return periods, T, in years from 5 to 100 and durations, t, in hours, from 0.5 to 100 or more the depth-duration-frequency amounts of rain, D, are given by the equation:

$$D = K\ (Txt)^{0.25} \tag{16}$$

in which K is a constant requiring calibration. K can be calculated from daily rainfall records and used to estimate hourly or other time period amounts.

A common cause of reduced benefits on irrigation proj-

TABLE 6. Conveyance (Ec), Field Canal (Eb), Distribution (Ed) and Field Application Efficiency (Ea).

Conveyance Efficiency (Ec)			ICID/ILRI
Continuous supply with no substantial change in flow			0.9
Rotational supply in projects of 3,000–7,000 ha and rotation areas of 70–300 ha, with effective management			0.8
Rotational supply in large schemes (> 10,000 ha) and small schemes (< 1,000 ha) with respective problematic communication and less effective management:			
based on predetermined schedule			0.7
based on advance request			0.65
Field Canal Efficiency (Eb)			
Blocks larger than 20 ha: unlined			0.8
lined or piped			0.9
Blocks up to 20 ha: unlined			0.7
lined or piped			0.8
Distribution Efficiency (Ed = Ec. Eb)			
Average for rotational supply with management and communication adequate			0.65
sufficient			0.55
insufficient			0.40
poor			0.30

Field Application Efficiency (Ea)	USDA	US(SCS)	
Surface methods			
light soils	0.55		
medium soils	0.70		
heavy soils	0.60		
graded border		0.60–0.75	0.53
basin and level border		0.60–0.80	0.58
contour ditch		0.50–0.55	
furrow		0.55–0.70	
corrugation		0.50–0.70	
Subsurface		up to 0.80	
Sprinkler, hot dry climate		0.60	
moderate climate		0.70	0.67
humid and cool		0.80	
Rice			0.32

Source: Doorenbos and Pruitt, 1977.

ects is damage to canals, culverts, roads or other features or structures from extreme and unforeseen rainfall events. If an irrigation development is financed with a 50-year repayment period, it then seems logical that the design should be such that there will be little probability of serious damage from rain during the 50-year period. Equation (16) can be used to estimate risk.

Comparisons of long records with shorter duration series estimates indicate that the probable error in estimating values of K from 30-year records is about ± 30 percent.

A comparison was made between the probability distribution given by Equation (16) and those of several other distributions. All gave similar values for $T = 10$ and $t = 24$. Many long records exceeding 90 years were used to compare Equation (16) with the censored log normal probability distribution for a range of T from 2 to 200 years, and of t from 1 to 7 days. The maximum difference between the two distributions were generally less than ± 16 percent and the regression correlation between the two sets of probabilities yielded very high R^2 values.

Due to the large uncertainties in predicting the future extreme events, it seems probable that Equation (16), although simple and easy to use, may be about as good as any of the other probability distributions.

SIMULATING DAILY CLIMATIC DATA FROM MONTHLY

Many crop growth and crop yield models have been developed for the use of daily climatic data in evaluating the potential for crop production. In the developing countries, daily records over significant periods of time are often difficult to obtain and where available may not receive much use due to the lack or shortage of adequate computer facilities.

Considerable effort has been made towards the simulation of daily data from monthly values and some useful relationships developed. In many large areas, sometimes encompassing several countries or a very large region, climate may be uniform enough so that a useful equation can be developed for the number of rainy days in a month or week or other time period. The relationship is often simple and the R^2 values high. The equation for the number of rainy days in a given period as a function of the amount of precipitation, prec., can be written:

$$\text{Days} = K \times \text{Prec.} \qquad (17)$$

For Africa, the number of rainy days was taken as those days with either 0.1 mm or 0.2 mm of rain. Some of the countries with fairly uniform relationships are listed below together with the factors to be multiplied times the monthly rainfall amounts in mm.

Country	Factor	Country	Factor
Botswana	0.105	Malawi	0.090
Cameroon	0.085	Nambia	0.125
Central African Rep.	0.085	Upper Volta	0.085
Chad	0.085	Senegambia	0.075
Congo	0.075	Sadan	0.080
Guinea	0.085	Tunesia	0.115
Ivory Coast	0.085	Zaire	0.085
Kenya	0.090	Zambia	0.095
		Zimbawi	0.105

Figure 5 illustrates how weekly probabilities of useful rainfall amounts compare with the monthly *MAI* index. In locations where this relationship is reliable and consistent, monthly rainfall adequacies can be used to simulate weekly dependability of rain. Comparisons can then be made with the probable number of days or rain in the various time periods.

Simplicity often enhances utility. There is sometimes a tendency to have more confidence in the more complicated and difficult procedures, as illustrated by the widespread use of the Penman equation, when simple methods, such as Equations (2) and (3), produce more consistent, accurate and reliable results. The rooting depths and soil conditions are such that the crop water requirements need to be supplied only periodically. The required interval is seldom less than one week and may be as long as 10, 15, or even 30 days. Except for unusual events, weekly or 10-day average data can be used to develop crop growth models that are nearly as accurate as those using daily data. It is therefore recommended that efforts be concentrated on simulating 10-day data from monthly and in modifying the crop growth models to accommodate 10-day data.

FORECASTING SEASONAL RAINFALL ADEQUACY

The forecasting of seasonal rainfall is difficult and subject to large uncertainties. However, some relationships are useful. The yearly rainfall amounts seem to follow cycles in that several years in succession may be above normal or below normal. If one year is drier than normal, the probability that next year will be below normal is increased.

In bimodal climate the first rainy period may be followed by drought and then a more pronounced period of rain. If the first rainy month or months produce less rain than normal, then the probability that rains in the later rainy period will be below normal is significantly increased. A similar relationship exists in many locations where rainfall has a single mode. When rains are later in onset or when the rainfall in the first rainy month is significantly less than average, there is a higher probability that the seasonal rainfall will be below normal.

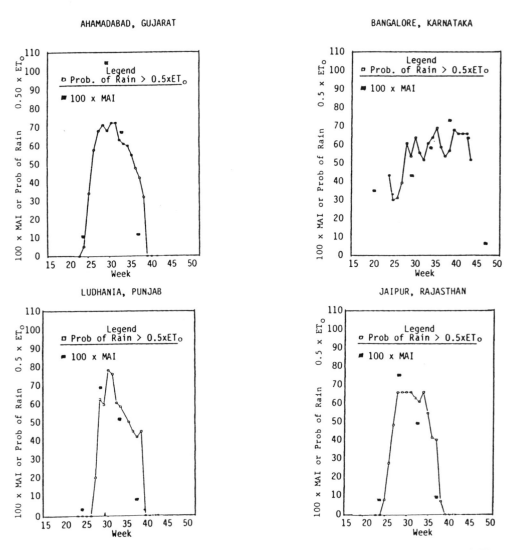

FIGURE 5. Comparison of MAI × 100 with weekly probability of rain exceeding 50 percent of ET_o. Source: Hargreaves, et al., 1985.

The movement of the sun's zenith from north to south and vice versa creates conditions for the onset of the rainy season. Differences between the surface water temperature of the ocean and the land have large influences on rainfall amounts. A knowledge of shifting ocean currents and of mass movements of warm or cold bodies of ocean water can be used to predict whether a year's rainfall will be significantly more or less than normal.

Unfortunately, over some fairly large areas of relatively uniform topography, rainfall is usually a localized and casual event that is much less predictable than in other areas. However, even in these very unpredictable areas there is a strong tendency for rainfall amounts to be cyclical.

THE USE OF CROP GROWTH MODELS

The principal factors of production are water, fertility, and energy. Other very important considerations include variety, planting density and soil aeration. The relationships of water and nitrogen are shown above. Within the range of optimum temperature for growth, several crops have a linear yield response to temperature. The maximum yield of maize or alfalfa that can be achieved (providing other factors are optimum) is a function of energy as measured by mean daily Class A pan evaporation during the crop growing season or by mean daily ET_o.

Curves showing the growth response to temperature are

shown as Figure 6 for 4 of the five crop groups. The crops in the various groups are given in Table 7. Temperature and radiation are the principle sources of energy for crop growth. Equations 2 and 3 integrate the effects of mean air temperature, solar radiation and advective energy (sensible heat transfer through air mass movements) on crop growth. If fertility, variety and drainage are optimum, then the principle consideration that drives the yield model is the ratio of ET_a/ET_m as illustrated by the Stewart Model (Equation (4)).

Some authorities prefer to separate evaporation, E, from transpiration, T, and use T_a/T_m (actual transpiration/maximum or potential transpiration) as the principle factor driving the yield model.

In the United States frost and/or low temperatures restrict the yields of various crops. One cold night can reduce ET_a or potential ET (ET_m) for a period of several days or more after the daily event. Probably because of this, many yield models place undue emphasis on the daily contribution of climate to yield. In the developing world, emphasis needs to be placed upon simplicity and models that operate with a minimum of data. For that reason, crop growth models need to be developed and calibrated for use with weekly or 10-day rainfall sums and mean maximum and mean minimum temperatures for these time periods.

A MODEL FOR ECONOMIC EVALUATION

Hargreaves and Samani, 1984, used an equation to vary yields as total water applied or available to the crop was varied. Total water included initial soil moisture, rainfall and irrigation. Equation (5) was used in a computer model to vary yields as irrigation was varied under various climatic and rainfall conditions. The model generates curves that

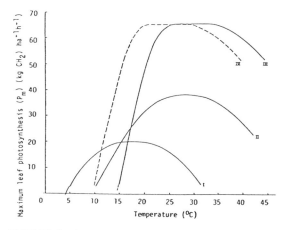

FIGURE 6. *Average relationship between maximum leaf photosynthesis rate and temperature for crop groups I, II, III and IV. Source: FAO, 1978.*

TABLE 7. Crop Groups for Temperature Requirements.

Group No.		Crops
Group No. 1		
Opt.	15–20°C	field mustard, potato, chickpea (gram)
Rg.	5–30°C	lentil, French bean, rape, cabbage, sunflower, barley, wheat, linseed, tomato, oat, rye, grape, pyrethrium, sugar beet, olive, Arabica coffee.
Group No. 2		
Opt.	25–30°C	groundnut, cowpea, soybean, French
Rg.	10–35°C	bean, tobacco, sunflower, sesame, tomato, safflower, rice, hyacinth, bean, roselle, kenaf, fig, cotton, castor bean, sweet potato, rape casava, white yam, olive, sweet orange, lemon, mango, cocoa, Robusta coffee, greater yam, banana, avocado, coconut, Para rubber, oil palm.
Group No. 3		
Opt.	30–35°C	Japanese barnyard millet, foxtail millet,
Rg.	15–45°C	finger millet, common millet, pearl millet, Hungry rice, sorghum, maize, sugar cane.
Group No. 4		
Opt.	20–30°C	Japanese barnyard millet, foxtail millet,
Rg.	10–35°C	common millet, sorghum, maize.
Group No. 5		
Opt.	25–30°C	Sisal and pineapple.
Rg.	10–45°C	

Source: FAO 1978.

show how crop yields and net benefits vary with different amounts of applied irrigation water.

In developing the model, data were used showing water costs, costs of applying water, other production costs, typical yields and crop prices. For purposes of analyzing different levels of economic returns, including net profits or net benefits, different price levels were assumed. The model provided knowledge relative to the following questions:

1. What factors influence the relationships of yields to profits?
2. When and under what conditions does maximum yield or near maximum yield produce maximum net profits?
3. How does crop value and selling price interact with irrigation amount to influence net profits?
4. When water is scarce or in limited supply and land abundant, how can profit be optimized?

5. When water is scarce or expensive is it profitable to invest in improved technology by changing from furrow irrigation to sprinkle or drip? Under what conditions?
6. What is the interaction of the price of water with irrigation amount on net profits?

In using a model, it is difficult to obtain fully adequate data that accurately define the influence of a single variable on profit. Data are available for current farm practices and for research conditions, but even on research plots, other conditions may differ significantly from optimum. Changes in water adequacy, for example, require corresponding changes in availability of nitrogen and in the planting density in order to produce optimum conditions for yield and profits. The model does, however, provide answers to some of the questions listed above.

Figures 7 and 8 compare yields and profits for wheat and alfalfa irrigated with different seasonal total amounts of water. Figure 7 is for wheat under low rainfall conditions with low water costs. Figure 8 is for alfalfa with an effective rainfall of 660 mm and water pumped from a deep well. For wheat with low rainfall and low cost of water, maximum net profits are with nearly as much water as that required for maximum yield. For alfalfa with more rainfall and higher water costs, the maximum profit is indicated to be when the applied irrigation water is about 30 percent less than required for maximum yield.

Figures 9 and 10 compare net profits from the production of grain hay at two price levels. The amount of water required to produce maximum net benefits is shown to vary with the price of grain hay.

Figures 11 and 12 compare total benefits from a limited water supply as a fixed amount of water is spread over an increasing land area with the possible benefits of varying the amount of water applied to a limited land area. From Equation (6), the contribution to yield of an incremental increase in available water is maximum when water is about 40 percent adequate for maximum yield. Maximum profit considers both the unit increase of yield from an incremental increase in water and the costs of spreading water over an increasingly large area.

Figure 13 compares net benefits from furrow, sprinkle and drip irrigation for grapes in the San Joaquin Valley of California. From the available data, drip irrigation slightly improved net profits but required only about half as much irrigation water as was applied in furrow irrigation. In the Coachella Valley of California, on sandy soils, drip irrigation is increasing profits significantly over those from furrow irrigation of grapes while requiring only one-third the amount of water. Not only is the farmer's profit increased, but the irrigation district costs are reduced due to the decreased requirements for drainage.

Figure 15 indicates how the allowable or desirable irrigation deficit varies as rainfall increases when land is limited and water is adequately available and when water is limited. The data used are from a crop of wheat in Bangladesh.

When water is limited and rainfall is significant it is often desirable to spread the limited water over a relatively large land area thereby increasing the irrigation deficit.

Figure 16 shows how the nitrogen level in the soil influences the optimum amount of water to be applied for maximum yield and for optimum profit at two crop selling prices with limited land. In this model it was assumed that the nitrogen was already in the soil and the cost of nitrogen was not considered. The data are for corn production in Kansas. With more nitrogen it becomes profitable to apply more irrigation.

Figure 17 is similar to Figure 16 but in this case it was assumed that water was limiting. The yield increases as more water is applied but this decreases the profit since less land can be irrigated with the available water. These data are from the same experiment as for Figure 16.

The graphs are for the conditions reported and demonstrate typical relationships. Significant differences in yield can be produced with the same water deficits due to variability of weather and to differences in sequencing of water stress or deficits. Stress during a critical growth period, usually flowering, fruit setting, or grain formation, has a larger influence on yield reduction than a deficit in other growth stages.

A comprehensive economic analysis requires fairly complete cost data and reasonably accurate yield models. Such an analysis can accurately define the merits of deficit irrigation. However, the above graphs demonstrate that deficit irrigation is less desirable when:

1. Water costs are low.
2. Irrigation efficiencies and uniformities are high.
3. Rainfall is low and provides little useful water.
4. Yields and crop values are high.
5. Leaching by irrigating is required to control soil salinity.
6. Land is limited and water less limited.

If rainfall is negligible, and water costs are about one-third of production costs, a 10 percent reduction in water applied would reduce production costs by about 3.3 percent. The model indicates that gross crop value would be reduced about 3 percent. At this point an increment of water costs produces approximately a corresponding increment in crop value. If farmers can estimate this relationship accurately, they may wish to irrigate with a small deficit so as to reduce the possibility of further reducing yields due to over-irrigation resulting from overestimation of crop water requirements.

DEFICIT IRRIGATION WHEN LAND IS LIMITED

There are many situations in which land is a major limiting factor and water is less limiting. This may be a seasonal or a year round condition. The water supply may be almost entirely from irrigation or principally from rainfall.

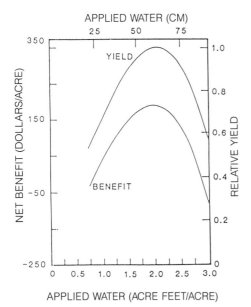

FIGURE 7. Comparison of yields and net benefits at one price level for wheat with 4″ (102 mm) effective rain.

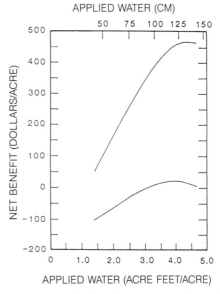

FIGURE 9. Net benefits for grain hay at two price levels with no effective rainfall.

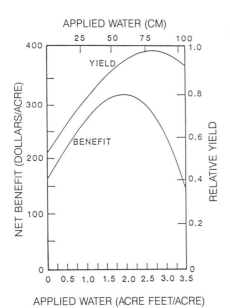

FIGURE 8. Comparison of yields and net benefits at one price level for alfalfa hay with 26″ (660 mm) effective rain.

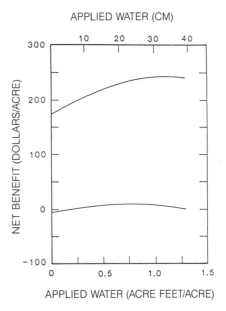

FIGURE 10. Net benefits for grain hay at two price levels with effective rainfall of 26″ (660 mm).

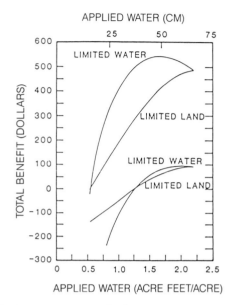

FIGURE 11. Total benefits from limited land and limited water for wheat at two price levels with 3″ (76 mm) of effective rain.

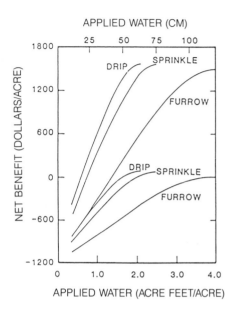

FIGURE 13. Comparison of net benefits for two price levels from drip, sprinkle, and furrow irrigation for grapes with 5″ (127 mm) of effective rain.

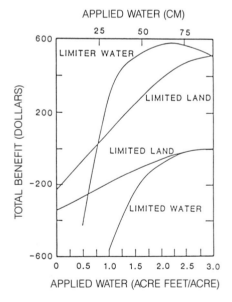

FIGURE 12. Total benefits for corn with limited land and limited water at two price levels with 10″ (245 mm) of effective rain.

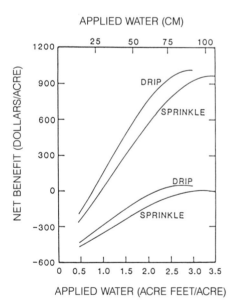

FIGURE 14. Comparison of net benefits for two price levels from drip and sprinkle irrigation for oranges with 4″ (102 mm) effective rain.

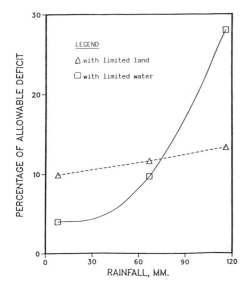

FIGURE 15. Allowable irrigation deficits with limited land and with limited water based on wheat production in Bangladesh.

If the amount of available land is less than the amount that would require the full water supply the benefit function for positive benefits, B, or net profits can be written:

$$B = Y_w (P_y - P_F) - (W \times P_w) - P_{cu} \qquad (18)$$

in which Y_w = yield per unit of land as a function of the irrigation water applied, W = depth of irrigation water applied, P_y = the price per unit of crop yield, P_F = cost of production per unit of land, P_w = the unit cost of water, and P_{cu} = the initial cost of irrigation per unit of land area. If

$$\frac{\partial B}{\partial W} = 0 \qquad (19)$$

then

$$\frac{\partial B}{\partial W} = \left(\frac{\partial Y_w}{\partial W} \times P_y \right) - \left(\frac{\partial P_F}{\partial Y_w} \times \frac{\partial Y_w}{\partial W} \right) - P_w = 0 \qquad (20)$$

or

$$\frac{\partial Y_w}{\partial W} = \frac{P_w}{P_y} \left[\frac{1}{1 - \left(\frac{\partial P_F}{\partial Y} \times \frac{1}{P_y} \right)} \right] \qquad (21)$$

The function $\partial P_F / \partial Y$ indicates the increase in cost per

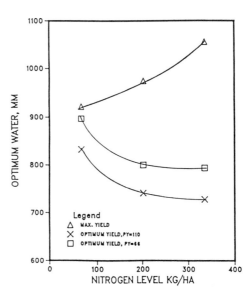

FIGURE 16. Optimum irrigation amounts for maximum yield and optimum yields at two crop price levels for corn in Kansas with limited land. (Py = 110 is a selling price for corn of $110 per ton.)

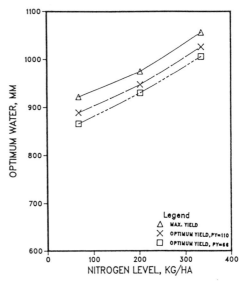

FIGURE 17. Optimum irrigation amounts for maximum yield and optimum yields at two price levels for corn in Kansas with limited water. Py = 110 is a selling price for corn of $110/ton.

unit increase in yield and can be very small when divided by
Py. In this analysis it is assumed to be zero.

A fixed cost of production per unit of land is assumed in
order to analyze the parameters of deficit irrigation. The approximation:

$$\frac{\partial Y_w}{\partial W} = \frac{P_w}{Py} \qquad (22)$$

provides a method for calculating the optimum amount of irrigation to be applied when land is limited (Steman, et al.
1980).

In rainfed agriculture, management has no control over
the amount of rain but can significantly control its utility.
Runoff can be reduced and surface drainage can be improved. Sometimes both can be accomplished simultaneously. As indicated previously, nitrogen, N, is often the
limiting factor on grain and/or other crop yields. Equation
(22) can be modified to indicate the desirable level of nitrogen to apply as follows:

$$\frac{\partial Y_N}{\partial N} = \frac{P_N}{Py} \qquad (23)$$

in which Y_N = yield per unit of N, and P_N = the cost per
unit of N including the application and associated costs.

If both water and fertilizer are variable, then optimum
conditions can be approximated by using Equations (22) and
(23), simultaneously.

Some authors have developed production functions for
yield, Y, as a function of water, W, and nitrogen, N, that can
be written in the form:

$$Y = + a + bW + CN - dW^2 - eN^2 + f(W \times N) \qquad (24)$$

Attempts to apply these equations to other conditions than
those for the specific research site may lead to unrealistic
results.

Although there are various ways of showing the nitrogen
and water yield functions, relationships similar to Equations
(7) and (8) seem the most practical. Research results from
various countries have been analyzed in this manner and
used successfully in technology transfer.

Hargreaves and Samani, 1984, concluded that for conditions of limited land the allowable irrigation deficit will increase when:

1. The price of water increases.
2. The crop price decreases.
3. The effective rainfall increases.

However, under normal economic conditions when land is
limited, the optimum amount of water and fertilizer will

usually be close to the amounts required to produce maximum yield.

DEFICIT IRRIGATION WHEN WATER IS LIMITED

Often in irrigated agriculture, water becomes the limiting
factor. The water may be limited either during some stages
of the crop or during the entire growing season. The question in this situation is, how much of the available land
should be irrigated. The optimum level of water per unit
land under limited water condition depends on whether the
rainfed agriculture is economical.

If the rainfed agriculture is not economical and the production depends mainly on irrigation, then the optimum
level of water can be calculated by the following equation:

$$\frac{\partial}{\partial W}\left(\frac{Yw}{W}\right) = \left(\frac{PF}{Py}\right)\left(\frac{-1}{W^2}\right) \qquad (25)$$

in which PF = production cost per unit land including the
cost of land preparation for irrigation.

If the rainfed agriculture is economical and irrigation is
supplementary, then the optimum level of irrigation per unit
land can be calculated by

$$\frac{\partial}{\partial W}\left(\frac{Yw}{W}\right) = \left(Yr + \frac{PF - PF_v}{Py}\right)\left(\frac{-1}{W^2}\right) \qquad (26)$$

in which Yr = the yield per unit land without irrigation,
PF_v is production costs for rainfed units.

For an irrigation project with limited water supplying
farmers who own small parcels (have limited land), the
farmers would benefit from irrigating for maximum yield.
This can create some social problems. Increasing the cost of
irrigation can promote deficit irrigation and increase the irrigation efficiency. On the other hand, if the government
subsidizes agriculture and artificially raises prices, the
farmers will tend to irrigate for maximum yield with less irrigation efficiency. With a limited supply of water, this will
result in less total food production and probably an increase
in social tension.

If the water supply in an area is limited and the individual
farmers cannot get enough water to irrigate all their irrigable lands, then the policies could be to reduce the fixed
cost of production per unit of land. The fixed initial cost of
irrigation and the cost of irrigation can be decreased
(through subsidizing) and the market price of crops can be
increased. These policies would encourage the farmers economically to stretch their allocated (limited) water by irrigating more lands and irrigating more efficiently.

The policy of stretching the water to irrigate more lands
cannot be promoted, however, without providing the farm-

ers with required facilities to reduce the conveyance loss of water, and without a significant improvement in social administration. The local politics and power struggles among the farmers can easily dominate the economic factors of deficit irrigation and result in an inequitable distribution and inefficient use of water.

The allowable deficit in irrigation under conditions of limited water will increase if:

1. The fixed cost of production, or fixed initial cost of irrigation, decreases.
2. The value of the yield produced is increased.

Deficit irrigation has long been practiced as a method to increase the irrigation efficiency. It can also be used to reduce erosion, maintain the water table, promote aeration in heavy soils, and increase the crop quality.

PRACTICAL APPLICATIONS

An underlying theme of this discussion is "keep it simple" (KIS). The developed nations frequently have problems from overproduction of agricultural commodities while in a significant number of developing nations the population is increasing faster than agricultural production. Simple yet effective methodologies that maximize the use of available data can in many situations have greater impact on crop production than more sophisticated technology. If the method is too complicated or requires much data, the usefulness is limited in that more training is required, more time is consumed in computation, and there are more chances for errors in the data used.

The equation for reference crop evapotranspiration, ET_o [Equations (2) and (3)], the form of the Stewart equation used [Equation (4)], the yield equations [Equations (6) through (10)] and the depth-duration rainfall equation [Equation (16)] are some of the simplest, requiring a minimum of data, and product results that are as good or better than those from more complicated methodologies.

The cardinal factors of production are water, energy and fertility. Et_o is an index of energy and can be accurately evaluated. Water and fertility can be controlled so as to produce maximum yields or maximum net profits commensurate with the maximum yield potential as determined by climate (principally ET_o or by pan evaporation). The need for various fertilizer elements such as P, K, S, etc., can be determined by laboratory analysis, POT tests and field trials. The requirements for nitrogen, when other conditions are optimum, can largely be determined from well established relationships. The amount of N required for a given yield has been fairly well defined for many crops. In the tropics soils laboratories rarely test the soil for N as the amounts prior to fertilization are universally quite low.

The farmer frequently has limited land and may wish to maximize the productivity of the land. An irrigation project often has a limited water supply and it may be in the best interests of the state and nation to maximize food production with the limited water supply. The concepts presented herein provide an indication as to when and how much irrigation deficit is desirable.

An irrigation system designed and operated so as to supply water at a small deficit will reduce the probability of over irrigation which reduces yields due to decreased soil aeration and the consequent reduction in plant respiration and photosynthesis. Waterlogging increases denitrification and the leaching of plant nutrients. Frequently, the yield curve declines more rapidly from over irrigation than it does from deficit irrigation. Too much may be significantly worse than too little.

Deficit irrigation can produce significant benefits under favorable circumstances. The magnitude of these benefits depends upon the interaction of the various factors that influence production as well as crop value, water costs and other production costs. The graphs indicate the relative significance of several factors, but only a comprehensive economic analysis can accurately assess the merits of a deficit irrigation system or program.

REFERENCES

1. Doorenbos, J. and A. H. Kassam, *Yield Response to Water*, FAO (Food and Agriculture Organization of the United Nations) Irrigation and Drainage Paper 33, 193 pages (1979).
2. Doorenbos, J. and W. O. Pruitt, *Crop Water Requirements*, FAO Irrigation and Drainage Paper 24, (Rev.) 156 pages (1977).
3. Fangmeier, D. D., Z. Samani, D. Garrat, Jr. and D. T. Ray, Water Requirements for Guayale for Rubber Production, Paper No. 512 presented at the Winter Meeting of ASCE in New Orleans, Louisiana, 14 pages (1984).
4. FAO, *Report on the Agri-Ecological Zones Project, Vol. 1*, Methodology and Results for Africa, 158 pages, plus tables (1978).
5. Hardee, J. E., *Analysis of Colombian Precipitation to Estimate Irrigation Requirements*, Utah Water Research Laboratory, College of Engineering, Utah State University, Logan, Utah 84322, 67 pages (1971).
6. Hargreaves, G. H., "Moisture Availability and Crop Production," *Transactions of the ASAE*, Vol. 18, No. 5, pp. 980–984 (1975).
7. Hargreaves, G. H., *World Water for Agriculture*, Utah State University, Logan and Agency for International Development (AID), Washington, D.C., 177 pages (1977).
8. Hargreaves, G. H., "Responding to Tropical Climates—An Approach," *Food and Climate Review, 1980–81*, Queenstown, MD. The Food and Climate Forum of the Aspen Institute, pp. 29–38 (1981).

9. Hargreaves, G. H. and Z. A. Samani, "Economic Considerations of Deficit Irrigation," *Journal of Irrigation and Drainage Engineering*, Vol. 110, No. 4, ASCE, Paper No. 198367, pp. 343–358 (1984).

10. Hargreaves, et al., *A Crop Water Evaluation Manual for India*, International Irrigation Center, Utah State University, Logan, Utah 84322 (1985).

11. Stegman, E. C., et al., "Irrigation Water Management," *Design and Operation of Farm Irrigation System*, ASAE, pp. 763–817 (1980).

12. Stewart, et al., *Optimizing Crop Production Through Control of Water and Salinity Levels in the Soil*. Utah Water Research Laboratory, Utah State University, Logan, Utah, 191 pages (1977).

APPENDIX

Fortran Computer Equations for Estimating Daily RA(Ra)

DEFINITION OF TERMS

ACOS	=	Arc cosine
D	=	Julian Date (January 1 = 1)
DER	=	Declination (angle of the sun) in radians
ES	=	Distance of the sun to the earth divided by the mean distance
LD	=	Latitude in degrees
LDM	=	Minutes of latitude
RA	=	Extraterrestrial radiation in mm/day
RLD	=	Extraterrestrial radiation in langleys/day
TM	=	Mean daily temperature in degrees Celcius = (Mean maximum plus mean minimum)/2
XLR	=	Latitude in radians
D	=	0

DO 8 K = 1,365

C 8 IS THE END OF THE DO LOOP

$$D = D + 1$$
$$Y = \cos(0.0172142 * (D + 192))$$
$$DER = 0.40876 * Y$$
$$ES = 1.00028 + 0.03269 * Y$$
$$XLR = (\text{FLOAT}(LD) + \text{FLOAT}(LDM)/60.)/57.2958$$

$$Z = -\text{TAN}(XLR) * \text{TAN}(DER)$$
$$OM = \text{ACOS}(Z)$$
$$DL = OM/.1309$$
$$RLD = 120. *(DL*\text{SIN}(XLR)*\text{SIN}(DER) + 7.639* \cos(XLR)* \cos(DER)*\text{SIN}(OM))/ES$$
$$RA = 10. * RLD/(595.9 - 0.55* TM)$$

Fortran Computer Equation for Estimating Monthly RA

DEFINITION OF TERMS

ACOS	=	Arc cosine
DEC	=	Declination of the sun in radians
DL	=	Day length in hours (sunrise to sunset)
DM	=	Number of days in a month
ES	=	Mean monthly distance of the sun to the earth divided by the mean annual distance
LD	=	Latitude in degrees
LDM	=	Minutes of latitude
RA	=	Extraterrestrial radiation in mm per month
RAL	=	Mean monthly extraterrestrial radiation in langleys/day
TM	=	Mean temperature in degrees Celcius

DATA (DM(M), M = 1, 12)/31., 28., 31., 30., 31., 30., 31., *31., 30., 31., 30., 31./

DATA (DEC(M), M = 1, 12)/−.3656, −.2365, −.04682, *.1607, .3247, .4017, .3699, .2360, .03995, −.1669, −.3291, *−.4021/

DATA (ES(M), M = 1, 12)/.97104, .98136, .99653, 1.01313, *1.02625, 1.03241, 1.02987, 1.01916, 1.00347, .98693, .97369, *.96812/

C CONVERT LAT TO RADIANS

$$XLR = (\text{FLOAT}(LD) + \text{FLOAT}(LDM)/60.)/57.2958$$
$$Z = -\text{TAN}(XLR) * \text{TAN}(DCE(M))$$
$$OM = \text{ACOS}(Z)$$
$$DL = OM/.1309$$
$$RAL = 916.732 * (OM * \text{SIN}(XLR) * \text{SIN}(DEC(M)) + *\cos(XLR) * \cos(DEC(M))* \text{SIN}(OM))/ES (M)$$
$$RA = DM(M) * 10. * RAL/(595.9 - 0.55 * TM (M))$$

CHAPTER 16

Modelling the Effect of Depth on Furrow Infiltration

Francisco de Souza*

INTRODUCTION

It has long been recognized that the variation of flow depth with time and distance is a key factor determining the rate at which water infiltrates into a furrow during irrigation [1,8,14–16]. The depth of water in the furrow is influenced by furrow shape, slope, roughness as well as inflow rate. Consequently, all these parameters play a role in the furrow intake rate.

To study the infiltration of water into furrows two important aspects have to be considered. One of them is related to the measurement of infiltration; the other one refers to modeling furrow infiltration.

Mathematical models of furrow irrigation call for more complex descriptions of infiltration than in border-irrigation models [21]. In borders, the effect of flow depth is minimal [11,12]; with a given soil structure, changes in surface-water depth affect only the hydraulic gradient driving the water into the soil. Since most of this gradient stems from the high negative pore pressures in the unsaturated soil ahead of the wetting front, even large variations in surface depths affect infiltration rates by but a few percent. Consequently, cumulative infiltration, in terms of volume per unit plan area of border (i.e., *depth*) z, is generally considered a function of infiltration time τ alone:

$$z = k(\tau) \qquad \text{(borders)} \qquad (1)$$

A common empirical formula describing this function is the Kostiakov power law, suitable in many soils at small to moderate τ,

$$z = k\tau^a \qquad \text{(borders)} \qquad (2)$$

Here, k and a are constants depending upon soil type and initial water content.

In furrows, the depth has a first-order influence on the surface area over which infiltration is occurring, and hence upon the volume A_z infiltrated per unit length of furrow.

In this chapter a mathematical approach to modeling the effect of depth on furrow infiltration will be presented as it was first postulated by Strelkoff and Souza [21]. In addition, considering that a problem of major practical significance, of course, is the "a priori" determination of the appropriate k and a for the mathematical model used, a discussion on the techniques available for measuring furrow infiltration will also be presented.

MEASURING FURROW INFILTRATION

The methods for determining the infiltration constants k and a can be divided into infiltration tests and volume-balance techniques.

Several kinds of infiltration tests have been developed and used to measure infiltration in furrow irrigation. These are: ring or cylinder infiltrometer; Bondurant, furrow, or blocked-furrow infiltrometer; inflow-outflow; bypass infiltrometer, and variations or modifications of any of these methods.

The ring-infiltrometer method utilizes two concentric cylinders. The area inside the smaller cylinder is the test area and the annular space is a buffer area [10]. Sometimes furrow infiltration characteristics are determined by unbuffered cylinder infiltrometers [8]. Total depth of infiltration is

*Department of Agricultural Engineering, University of Ceara, P.O. Box 3038, Fortaleza, CE, 60.000, Brazil

determined by measuring the lowering of the surface water ponded in the cylinder [5]. Karmeli et al. [8] stated that the advantages of using a cylinder infiltrometer are the small amount of water required for a test and the ease of transportation and installation. However, the use of a cylindrical infiltrometer to measure furrow intake can be meaningless for several reasons: (a) effects of flowing water over the surface are not accounted for [8]; (b) water movement from a cylinder is restricted to the downward direction until the water passes beyond the lower edge of the cylinder, while in a furrow water infiltrates downward and laterally in a direction perpendicular to the furrow wall [1,8]; (c) entrapped air below the ponded surface and between cylinder walls may significantly reduce infiltration [8]. Davis and Fry [5] compared four methods for determining infiltration rates in furrows: cylinder infiltrometer, inflow-outflow, blocked-furrow infiltrometer, and a volume-balance equation. They found in all cases that the infiltration rate from cylinder infiltrometers was one-half to one-fourth that determined by any of the other methods. Karmeli et al. [8] used a cylinder infiltrometer to measure furrow intake rate, but they discarded the data taken due to the variability and extremely low infiltration rate predicted for the soil under study.

The blocked-furrow infiltrometer [1] was designed to approximate the conditions which exist in a furrow during irrigation, and directly measure the rate of infiltration in the furrow. It consists of isolating a cross section of the furrow by driving two metal plates into the soil, placed one foot (0.3048 m) apart. A float-valve arrangement provides a constant water level within the test area and in buffer areas surrounding the infiltrometer. The Bondurant infiltrometer differs from the cylinder infiltrometer in that the test area is an actual furrow and the test accounts for lateral water infiltration through the sidewalls. However, it does not account for variability of soil characteristics along the furrow or the effects of flowing water [5,8,10,15]. In addition, the water depth is held constant during the test, and consequently the constants found apply only to depth values similar to the ones in the test conditions.

The inflow-outflow method [3] determines the intake of water over a length of furrow by measuring the inflow and outflow of the test area. It has the advantages of measuring infiltration of flowing water and providing a large sample area of infiltration [8]. Shull [5] stated that the inflow-outflow method, although capable of good measurement accuracy, measures infiltration over a length of furrow, and thus does not necessarily represent infiltration as it occurs in each small longitudinal furrow segment. The disadvantages of this method are that most outflow-measuring devices generally obstruct the flow and cause a consequent increase in the depth of flow, while the buildup of surface storage between measuring devices is neglected by assuming that the difference between inflow and outflow is the intake rate [8]. Bondurant [1] compared furrow-infiltrometer data with infiltration data obtained from field irrigation trials using inflow-outflow measurements. He found that basic infiltration rates obtained by the two methods showed very close agreement in most of the trials. However, he stated that the furrow-infiltrometer should be used whenever field measurements by irrigation trials is not feasible. Davis and Fry [5] also agree that infiltration rates from furrows should be determined with flowing water, whenever possible, i.e., by using an inflow-outflow method or a volume-balance equation.

At this point it must be emphasized that the results described in the literature of infiltration tests in furrows are somewhat contradictory. As an example, Davis and Fry [5] stated that furrow-infiltrometers may overestimate furrow infiltration in medium-to-fine soils. The claim that furrow flow may have some tendency to seal over the surface and cause a relative decrease in infiltration rate in sandy-textured soils, whereas the infiltrometer with ponded water does not have this effect; in the case of fine-textured soils, there is a tendency to place the infiltrometer in regions without cracks, which gives a smaller infiltration rate than the inflow-outflow method or the volume-balance equation. Ramsey [14] also compared results of infiltration tests from ponded furrows with the volume-balance equation, and for each case studied he observed that results from the ponded test underestimated infiltration during the entire irrigation. He concluded that this underestimation could be due to the fact that, when water is moving over the soil surface, some of the smaller particles are kept in suspension and do not block the pore spaces. Ramsey's conclusions are thus exactly opposite those of Davis and Fry. Smerdon and Hohn [18] studied infiltration rates by a volume-balance equation and the blocked-furrow infiltrometer for two different sites. For one site (Pecos) they observed that most infiltration rates in the furrow infiltrometers were higher, and they concluded that the water movement in the irrigated furrow causes agitation of the very fine soil particles which results in greater surface sealing; for the other site (College Station) the infiltrometer gave smaller infiltration rates than the volume-balance equation; the authors explained that the soil at College Station was severely cracked and the furrow infiltrometer was purposely not placed on these cracks. Observe now the agreement between the Davis and Fry [5] and the Smerdon and Hohn [18] conclusions. These authors [18] stated that no broad conclusions are warranted from the data presented. Karmeli et al. [8] observed that infiltration rates determined by inflow-outflow measurements extremely overpredicted the actual infiltration rate of the furrows under study. They explained that the reason for this overprediction was the effect the trapezoidal measuring flumes had on increasing the depth of flow and hence the wetted perimeter. These same authors [8] found that blocked-furrows also overpredicted infiltration rate, but that the results were within the range in which field values of infiltration can be obtained.

Nance and Lambert [10] developed a modified inflow-outflow method and compared infiltration results obtained by this method with results obtained by a modified Bondurant infiltrometer. This modified inflow-outflow method differed from the original inflow-outflow method in that it did not require the entire field to be set up for normal field tests. The equipment was designed to measure the intake rate of the soil in a 15-ft (4.57 m) length of furrow under normal irrigation conditions; the depth of water flowing in the furrow was adjustable up to a depth of 6 inches (15.23 cm). The Bondurant infiltrometer was modified so that the water depth and the length of furrow blocked were the same as in the modified inflow-outflow method. Comparing results of three tests, Nance and Lambert [10] observed that in two of the three tests the modified inflow-outflow method gave intake rates higher than the modified Bondurant method. In the other test, the infiltration rates were almost the same. They concluded that the flow of water over the soil surface increased the infiltration rate in the furrow, but they did not attempt to explain the increase.

Shull [15] designed the "by-pass furrow infiltrometer" to measure infiltration during irrigation over a small longitudinal furrow segment under conditions more nearly the same as those that occur in an irrigated furrow than is possible by other methods. The by-pass infiltrometer consists of a sheet metal unit that blocks one-half of the furrow width for a length of 3 ft (0.91 m). Water is supplied to the infiltrometer from a portable tank. The water inflow to the infiltrometer is controlled by two floats and a pivoting float-controlled valve. One of the floats is located in the furrow outside the infiltrometer; the other is located inside the infiltrometer. The water-level controlling mechanism is constructed so that the control valve is opened when the water level in the infiltrometer is lower than in the furrow. The control valve is closed when the water in the infiltrometer rises to the same level as in the furrow. In operation, the by-pass infiltrometer simulates actual irrigation conditions by providing a water level that changes at the same rate as that in the adjoining furrow. The presence of the infiltrometer may affect the depth of water in the furrow at the infiltrometer location, but Shull claims that such effect may be smaller than variations in depth due to the unevenness of the furrow bottom. From photographs of the water level at the ends of the infiltrometer it is possible to calculate the varying volume of surface storage, which is then used to find the actual accumulated furrow infiltration. Shull [15] found that accumulated infiltration as a function of time, as measured with the by-pass infiltrometer, does not plot as a straight line on a log-log graph paper and suggested another form of equation to fit the data. Close agreement between results obtained with the by-pass infiltrometer and a volume-balance equation was observed in three of the six field trials.

The volume-balance equation has been extensively used to determine infiltration characteristics in surface irrigation. It requires measurements of the advancing wetting front, flow depth at various stations along the run and the corresponding cross-sectional flow areas. The border or furrow is not disturbed, and conditions under normal irrigating practice are well accounted for, if a reasonable approximation of surface storage can be obtained [8]. However, the surface storage is difficult to estimate accurately.

Christiansen et al. [2] developed a volume-balance equation under the assumptions that advance is described by a power-law, that the infiltration rate of a soil can be expressed by the Kostiakov equation or a modification thereof, and that the average depth of surface water is constant.

Davis and Fry [5] used the Christiansen equation to determine the constants a and k in the Kostiakov formula. Smerdon and Hohn [18] also used Christiansen equation to find the values of the Kostiakov constants. However, they assumed that surface storage is not a function of advance but is dependent only upon the depth of flow, the furrow shape, and the shape of the surface profile of the advancing water front. Wilke [22] estimated the Kostiakov constants a and k by using the Philip and Farrell [13] solution of the Lewis-Milne volume-balance equation [9].

Ramsey [14] used the Gilley method for borders to determine the intake characteristics of furrows after attempting, without success, to modify the method for furrows. In the original Gilley method, the fundamental assumption was made that advance obeys a power-law, say, $x_A = f\,t^h$; the Kostiakov constants a and k are thus dependent upon f and h. Ramsey applied the Gilley method to the advance and other phases of the irrigation and found that this method is accurate for the advance phase; for the other phases, the method badly underestimates the actual intake. He did not explain, however, how he could apply a method based on the power-law advance assumption to the other phases of the irrigation.

Davis and Fry [5] concluded that among the aforementioned four methods of determining infiltration rates, the volume-balance equations may represent the potentially most accurate, because this is the only method that avoids sources of inaccuracy such as ponded water and measurement devices that influence the flow depth. Karmeli et al. [8] also used the Christiansen equation [2] to predict infiltration rates in furrows, using a modified Kostiakov equation. In contrast to Davis and Fry [5], they found that the method is characterized by a gross inability to represent adequately the infiltration rate of the soil under study.

Singh and Chauhan [17] stated that the disadvantage of Christiansen et al. [2] method in the determination of the intake function, is that both the intake and advance rates are forced to follow assumed empirical functions. They improved existing methods for estimating intake rate in surface irrigation based on the solution of Lewis and Milne [9] volume-balance equation. These authors found that cylinder infiltrometers under-estimate intake rate at small values of time and over-estimate the intake at large values of time, while inflow-outflow method also may not represent actual intake rate because of errors due to surface storage. They recommended the method based on volume balance equation for good estimation of intake rate in surface irrigation.

Elliott and Walker [6] studied the adequacy of various infiltration and advance relations based on a volume-balance approach. They found that the use of the Kostiakov-Lewis function with an additional term for the asymptotic long-time infiltration rate proved highly effective in simulating infiltrated volumes when reliable estimates of the steady infiltration rate could be obtained.

As can be seen, the problem of determining an infiltration function for furrows is still open. As Shull [16] pointed out, "it is probable that a more general furrow infiltration equation would also include water depth in the furrow as a variable." Up to the present time no method has been presented to account for the effect of depth changes on the infiltration rate of furrows. Any attempt to represent the intake rate of furrows by an expression having only time as an independent variable has the consequence of incorporating the effects of depth change with distance into the constants of the equation. As a result, the equation will only be reasonably valid under conditions where flow depths are approximately the same as in the test.

MODELING FURROW INFILTRATION

To study the effect of changing flow depth on the empirical constants of the Kostiakov infiltration function, Ramsey and Fangmeier [7,14] conducted a series of tests in the precision-furrow facilities at the University of Arizona. They postulated that if the flow depth in a furrow were to remain constant, a smooth monotonic infiltration curve would result; however, if the flow depth decreases with time and distance along the furrow, an anomaly in the slope of the curve on log-log paper would result, due to the decrease in soil contact area. They plotted cumulated intake volume against time on log-log graph paper and noticed that the curve deviated from a straight line. However, when the cumulative intake volume was divided by a characteristic length transverse to the flow (either the top width or the wetted perimeter) and this cumulative intake was plotted versus time, the curves remained straight. The conclusion was that either the top width or the wetted perimeter could be used to characterize the furrow intake rate. Another important conclusion of the study was that when the empirical constants of the Kostiakov equation were determined using top width or wetted perimeter as a characteristic transverse length, the "a" value remained nearly unchanged, while the "k" value changed in direct proportion to the ratio of wetted perimeter to top width. These results were partially confirmed by a theoretical, numerical study of Souza [19].

As stated before, in furrows the depth has a first-order influence on the volume A_z infiltrated per unit length of furrow. According to Strelkoff and Souza [21] if the symbol z is retained for the concept of volume infiltrated per unit of infiltrating-surface area (in borders, a unit of infiltrating-surface area is the same as a unit of plan area, i.e., a unit length of a strip of border of unit width), then, in general,

$$A_z(\tau, y) = \int_{w_L}^{w_R} z \, ds \qquad \text{(furrows)} \qquad (3)$$

In Equation (3) s = arc length along the wetted perimeter of the furrow, taken positive up one side of the furrow and negative up the other side; w_L and w_R are the values of arc length at the free surface on the left and right sides of the furrow, respectively. The total wetted perimeter for the depth y is then given by

$$w = -w_L + w_R \qquad (4)$$

In the general case, z must be considered a function of $z(\tau_s, s, y)$ of the time τ_s that a particular point on the currently wetted perimeter has been in contact with water, and also of its position s in the furrow perimeter, especially relative to the point s_y, that value of s at which the furrow perimeter intersects the free surface of the water in the furrow. For at the free surface particularly, the infiltration is probably not taking place perpendicular to the infiltrating surface; there would be some attraction for the water upward, as well as inward, into the soil. These effects would cause z to depend, in addition to its dependence on τ_s, upon s and y as well, even if the soil structure were identical in different portions of the furrow perimeter.

However, in the interests of simplification, it can be assumed that z is a function of τ_s alone:

$$z = z(\tau_s) \qquad (5)$$

For example, the same Kostiakov power law as used in borders, Equation (2), can be postulated, in a localized sense, for every point in the furrow perimeter

$$z = k\tau_s^a \qquad \text{(furrows)} \qquad (6)$$

in which

$$\tau_s = \tau_s(s, \tau) \qquad (7)$$

Where τ, to correspond to the case of borders, represents the time that the furrow *bottom* has been wetted, i.e.,

$$\tau = t - t_A \qquad (8)$$

t being current time, and $t_A(x)$ representing the time stream reached station x in the furrow. Thus,

$$\tau_s(0, \tau) = \tau \qquad (9)$$

For a given τ, τ_s varies from this maximum at the furrow bottom to zero at the free surface. The specific function describing this variation depends upon the shapes of the stream profile and furrow cross section, and upon the speed with which the profile advances in the furrow.

Per unit plan area of field, then, given furrows at the spacing W, the volume infiltrated Z is

$$Z(\tau, y) = \frac{1}{W} \int_w z(\tau_s)ds \qquad (10)$$

Earlier works (e.g., [4]) have ignored the dependence of τ_s on s and hence of Z on y, and tended to use formulas of the type

$$Z = k\tau^a \qquad (11)$$

just as in the case of borders. Such a formulation should work well if (a) the depth of flow does not vary a great deal over the length of the stream for most of the period of irrigation, and (b) the constants k and a were determined for flow at about the same depth as that encountered in the predicted case.

Equation (10), with Equation (6) in force, should in general be a more accurate representation of infiltration in furrows than Equation (11). On the other hand, it is a great deal more complicated than Equation (11). To apply it, a mathematical model of furrow irrigation must keep track of the growth of depth at each station in order to construct the function given, schematically, by Equation (7). That is, the wetted perimeter at any station under calculation at any instant must be broken up into a series of segments, then τ_s and z calculated for each segment, and finally the integration of Equation (10) or Equation (3) must be performed numerically.

In an effort to find a suitable formulation for Z, more accurate than Equation (11) and less complicated than Equation (10), a series of approximations were tested by Strelkoff and Souza [21] in a mathematical model of furrow irrigation and compared to each other, and to measurements in a real furrow. These approximations, in decreasing order of complexity were

MOD_{AZ}

1 $\qquad A_z = \int_w k_1 \tau_s^{a_1} ds \qquad (12)$

2 $\qquad A_z = k_2 \tau^{a_2} w(y) \qquad (13)$

3 $\qquad A_z = k_3 \tau^{a_3} B(y) \qquad (14)$

4 $\qquad A_z = k_4 \tau^{a_4} w(y_0) \qquad (15)$

5 $\qquad A_z = k_5 \tau^{a_5} B(y_0) \qquad (16)$

6 $\qquad A_z = k_6 \tau^{a_6} B(y_n) \qquad (17)$

The index MOD_{AZ} identifies the formula defining the volume infiltrated per unit length of furrow. The rationale behind Equation (12) has been presented in detail earlier. Equation (13) is based on the hope that appropriate, constant, k_2 and a_2 can be found for any given furrow, so that $k_2 \tau^{a_2}$ multiplied by the wetted perimeter, extant at any station at a depth y will adequately represent the volume infiltrated per unit length of furrow there. Equation (14) is based on the similar hope that suitable k_3 and a_3 exist so that $k_3 \tau^{a_3}$ multiplied by the top width B pertinent to a given depth y will adequately represent the volume infiltrated per unit length at a station wetted to that depth. The constants k_4 and a_4 are similarly associated with the time varying wetted perimeter at the head end of the furrow, at the depth $y_0(t)$; k_5 and a_5 relate to the top width at the upstream end, i.e., imply that the effect of depth is adequately embodied in the effect of the upstream depth. The crudest approximation, Equation (17), assumes that A_z, for any given discharge into any given furrow, is dependent only upon the wetting time of the furrow bottom at any station, and that the effect of depth is provided for by normal depth for the given furrow at the given discharge. This approximation is just one step removed from Equation (11), in which the depth does not appear explicitly at all.

The k and a, constant in each of the approximations, would be expected, for any given furrow irrigation, to be different for each approximation. Strelkoff and Souza's [21] aim was to establish which models, 1–6, adequately define infiltrated volume per unit length. This was determined by comparing irrigation results computed by each of the six models with measured field data, taken from a furrow used, in effect, in its entirety as an infiltrometer. Specifically, the six infiltration models were inserted, in turn, in a mathematical model of stream advance in an irrigation furrow. This computer model solved locally linearized equations of mass conservation and force equilibrium in each of a series of water cells comprising the surface stream and subsurface, infiltrated water in a scheme similar to that described by Strelkoff and Katopodes [20] for borders. In differential form, these equations are

$$\frac{\partial Q}{\partial x} + \frac{\partial A}{\partial t} + \frac{\partial A_z}{\partial t} = 0 \qquad (18)$$

and

$$\frac{\partial y}{\partial x} = S_0 - \frac{Q^2}{A^2 C_h^2 R} \qquad (19)$$

Here x and t are distance down the furrow and time, respectively, Q is discharge at any point in the furrow, A is cross-sectional area of the surface stream there, and A_z is defined through one of Equations (12–17); y is the depth of furrow flow, S_0 is the bottom slope, C_h is the Chezy C, expressed in terms of the Manning formula

$$C_h = \frac{C_u}{n} R^{1/6} \qquad (20)$$

with C_u a units coefficient ($C_u = 1.468 \ldots$ ft$^{1/2}$/sec in the English system, $C_u = 1$ m$^{1/2}$/s in the metric system), n is the Manning n and R is the hydraulic radius

$$R = \frac{A}{w} \qquad (21)$$

In numerically solving Equations (18) and (19), the length of the surface stream was divided up into a sufficient number of cells so that the variation of any of the variables over the length of a cell could be assumed linear (except at the very tip of the stream front, where a power-law variation was assumed). Similarly, time steps for solution were chosen small that the time variation of variables at any station could be similarly assumed linear over the duration of a step. To avoid iterative solutions, the equations were locally linearized about values on the current time line.

The independent variables governing the behavior of the advancing stream is a prismatic furrow of given shape are

Q_{in}	the (constant) inflow at the head end of the furrow
S_0	the bottom slope
k_j	$j = 1, 2, \ldots, 6$ – the infiltra-
a_j	tion-formula constants
n	the Manning n

The dependent variables chosen for comparison with measured data were

$x_A(t)$	the advance function
$V_z(t)$	the total volume infiltrated over the length of the stream
$y_o(t)$	the upstream depth of flow

The inflow and bottom slope were measured in the field tests and applied as input to the mathematical model. The field advance $x_A(t)$ was measured, as was the depth profile in the furrow at successive instants of time. The latter values, together with a known furrow shape yielded the area profile; numerical integration of the latter provided the volume of water $V_y(t)$ in the furrow at any of the given instants. Subtraction of this figure from the volume of inflow leads to the infiltrated volume,

$$V_z(t) = Q_{in} \cdot t - V_y(t) \qquad (22)$$

Values of k, a, and n for mathematical-model input were determined by trial and error for each of the infiltration models, $j = 1$–6, and each of the field tests. In each case, the values of k, a, and n which gave the best agreement between computation and measurement of both dependent functions $x_A(t)$ and $V_z(t)$ were selected. Furthermore, the Manning n had to be consistent with the measurement of depth at the upstream end of the furrow where normal-depth conditions were approached.

Procedure—Measurement of Infiltration Parameters Using an Entire Furrow as an Infiltrometer

As the basis of comparison, a total of four physical tests were chosen from amongst those reported by Ramsey and Fangmeier [7,14]. These test runs were simulated by the aforementioned zero-inertia model. The input for the model was the furrow shape, a constant inflow rate, bottom slope, furrow length, Manning n, and the Kostiakov constants a and k (see Table 1). Ramsey's data were the only ones available that were sufficiently detailed to make comparison. The results of the four trials reported, however, were too similar to represent a significant range of conditions. In addition, they are not typical in that advance is completed in less than 30 minutes. Therefore, it should be clear that the results presented should not be taken as conclusive, and that the problem of which geometrical parameter to use in the description of furrow intake is not totally solved.

In his dissertation [14], and subsequent paper with Fangmeier [7], Ramsey presented computed values of n, a, and k. Values for a and k were obtained by the volume-balance method, the Gilley method, and blocked-furrow infiltrometer. The computer model used all the input data from Ramsey's work except the values of a, k, and n. Instead, trial values of a, k, and n were used for each different MOD_{AZ} until the model output best approximated the Ramsey-measured values of advance, infiltration volume, and depth at the upstream boundary. It was assumed that the rate of advance is the best field-measured value among the three under study; therefore, in the trial process an attempt was made to obtain the best match between the computed and measured advance curves while maintaining the computed upstream depth within the range of measured values. Then model-computed infiltrated volumes $V_z(t)$ were compared with the values Ramsey calculated by use of the volume-balance method. In this way the importance of top width and wetted perimeter as a characteristic transverse length could be assessed.

The model was given input data from four of the seven irrigations conducted by Ramsey, namely, irrigations IF-2, IF-3, IF-4, and IF-5 (see Table 1). The best combination of input a, k, and n is presented in Table 2 for for each irrigation and each of the six different numerical ways of computing the furrow intake ($MOD_{AZ} = 1,6$).

TABLE 1. Values of a, k, n; Inflow Rate Q_0; Bottom Slope S_0; and Field Length L for Four Irrigations (Data from Ramsey, 1976).

Irrigation	a	k		n	Q_0		S_0	L	
		(in/hr^a)	(mm/hr^a)		(ft^3/s)	(l/s)	(ft/ft)	(ft)	(m)
IF-2	0.497	3.34	84.8	0.022–0.024	0.047	1.3	0.001032	330	100
IF-3	0.664	5.75	146.0	0.022–0.024	0.065	1.8	0.000996	330	100
IF-4	0.659	4.03	102.0	0.022–0.024	0.060	1.7	0.001068	330	100
IF-5	0.630	5.21	132.0	0.025–0.027	0.063	1.8	0.001029	330	100

Comparison of the results obtained by multiple use of the zero-inertia model (Table 2) and Ramsey's results (Table 1) indicates good agreement for values of a and n. In addition, it was observed that the "best" values of a and n remain unchanged for any given irrigation when changing from one MOD_{AZ} to another (Table 2). This is in agreement with Ramsey's observations. These results suggest that a and n are independent of the characteristic transverse dimension used to compute the infiltrated volume per unit furrow length, in Equations (13–17). However, the values of the Kostiakov k are not only different from the values reported by Ramsey, but they also vary significantly with different characteristic lengths (different MOD_{AZ}).

During the trial-and-error procedure used to find the best combination of a, k, and n (which affect the rate of advance, the furrow intake rate, and the upstream surface-water depth), several features of those constants were assessed.

It was observed that the Manning resistance factor n plays a major role in determining the surface-water depth. Figure 1 shows (for IF-2, $MOD_{AZ} = 2$) the variation of the upstream water depth with time for three different values of Manning n; it is apparent that the values of depth are higher for larger values of n. It was also observed that once the computed values of upstream depth were within the range of field-measured values for one value of MOD_{AZ}, the value of n need not be changed for other values of MOD_{AZ}. This fact can be observed in Figure 2 in which the computed depth versus time curves for various MOD_{AZ} do not differ from each other or from the curve measured by Ramsey; in this case the value of $n = 0.022$ was used for every MOD_{AZ} (see also Table 2). The infiltration constants a and k have a very small effect on the variation of upstream depth. Figure 3 shows that the curves for the three different values of k are almost the same. Figure 4 presents the variation of depth with time when three different values of a are considered; in both figures the value $n = 0.02$ was used.

The main effect of the Kostiakov constant a is on the shape of the cumulative infiltration curve V_z, and thus on the shape of the advance curve. In Figure 5, the logarithm of the total infiltrated volume was plotted against the logarithm of time. It can be observed that as a increases, the slope of the lines also increases. This variation of slope with a is an indication of the effect of a on the shape of the infiltration and advance

curves. During the trial-and-error process used in this study, the following approach was used: (1) first, the value of n which gave a match between measured and computed upstream depth was established (as in Figure 2); (2) as soon as this value of n was found, the value of a that would generate an advance (and infiltration) curve with the same shape (same slope on log-log paper) as the measured advance (and infiltration) was sought (see Figures 6, 7 and 8); (3) from then on, the value of a and n would remain unchanged for different MOD_{AZ}; only the value of k would be changed until the best agreement between computed and measured advance and infiltrated volume were reached (see Figures 7 and 8).

An increase or reduction of k, with a and n fixed for the

TABLE 2. Values of a, k, and n for Four Irrigations by Using the Zero-Inertia Furrow-Advance Model.*

Irrigation		MOD_{AZ}					
		1	2	3	4	5	6
IF-2	k	3.0	2.95	3.10	2.60	2.80	2.60
		(76.2)	(74.9)	(78.7)	(66.0)	(71.1)	(66.0)
	a	0.50	0.50	0.50	0.50	0.50	0.50
	n	0.022	0.022	0.022	0.022	0.022	0.022
IF-3	k	4.7	4.5	4.90	3.90	4.20	3.93
		(119)	(114)	(124)	(99.1)	(107)	(99.8)
	a	0.58	0.58	0.58	0.58	0.58	0.58
	n	0.022	0.022	0.022	0.022	0.022	0.022
IF-4	k	4.0	3.95	4.10	3.30	3.55	3.25
		(102)	(100)	(104)	(83.8)	(90.2)	(82.6)
	a	0.60	0.60	0.60	0.60	0.60	0.60
	n	0.022	0.022	0.022	0.022	0.022	0.022
IF-5	k	4.65	4.60	4.85	3.55	4.10	3.85
		(118)	(117)	(123)	(90.2)	(104)	(97.8)
	a	0.63	0.63	0.63	0.63	0.63	0.63
	n	0.026	0.026	0.026	0.026	0.026	0.026

*units of k—in/hr^a (mm/hr^a)

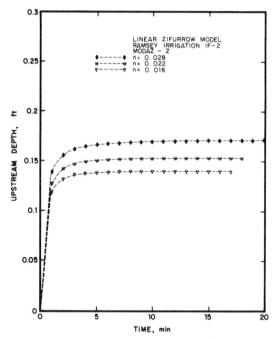

FIGURE 1. Effect of Manning n on the variation of upstream depth with time as computed by the zero-inertia approach.

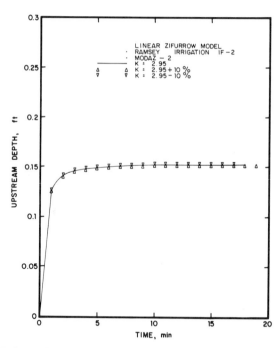

FIGURE 3. Effect of Kostiakov k on the variation of upstream depth with time as computed by the zero-inertia approach.

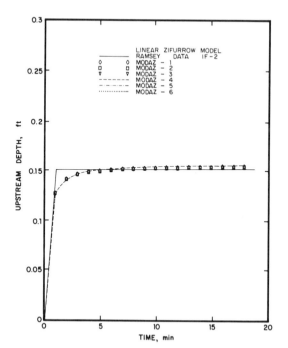

FIGURE 2. Comparison between measured and computed upstream depth.

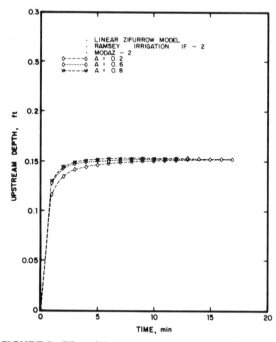

FIGURE 4. Effect of Kostiakov a on the variation of upstream depth with time as computed by the zero-inertia approach.

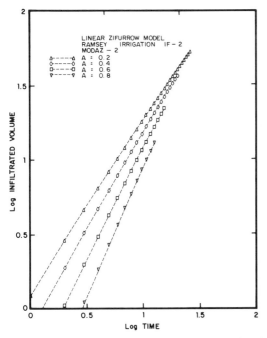

FIGURE 5. Effect of Kostiakov a on the cumulative infiltration in furrows.

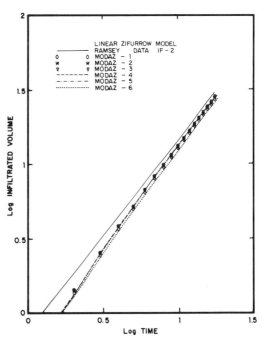

FIGURE 7. Comparison between measured and computed furrow cumulative intake.

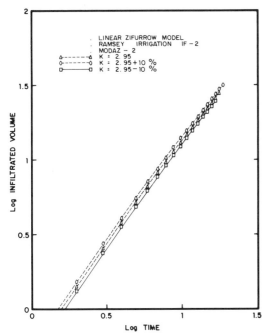

FIGURE 6. Effect of Kostiakov k on the furrow infiltrated volume.

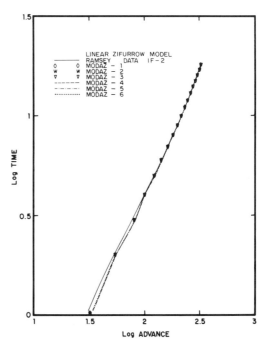

FIGURE 8. Comparison between observed and computed advance.

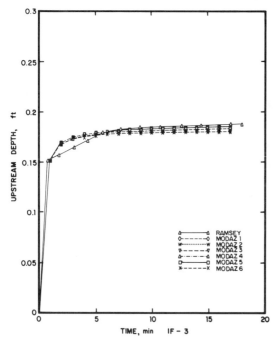

FIGURE 9. Time variation of upstream depth with "best" n, a, k. Test IF-3.

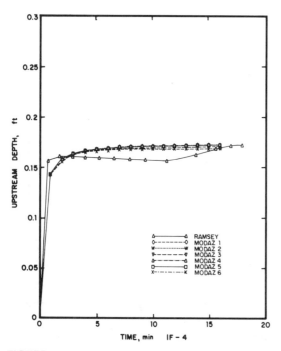

FIGURE 10. Time variation of upstream depth with "best" n, a, k. Test IF-4.

same MOD_{AZ}, would have the effect of increasing or decreasing the cumulative infiltration; as a consequence, the advance would be slower or faster. This effect can be observed in Figure 6, in which the logarithm of infiltrated volume was plotted against the logarithm of time for three different values of k (2.95, 2.95 + 10% = 3.25 and 2.95 − 10% = 2.66 in/hr[a] (74.9, 82.4, 67.5 mm/hr[a], respectively). Observe that the curves are moved up or down as k changes but the slope is the same for each case.

Figure 7 is a log-log plot of infiltrated volume versus time for irrigation IF-2. It should be noted that the infiltrated volume computed with the "best" n, a, and k values is smaller than the measured infiltrated volume (as reported by Ramsey) for all six numerical approaches (all MOD_{AZ}).

Results of Test Series

Figures 9–11 show the time variation of upstream depth in the Ramsey tests IF-3, IF-4, and IF-5, respectively. The computed results with the "best" n value (as well as "best" a and k values) and each of the six MOD_{AZ} options are shown on a background of measured data.

Figures 12–14 show the corresponding match of advance curves, while Figures 15–17 portray the accumulated infiltrated volume.

For three of the four irrigations studied (IF-2, IF-3, and IF-4) the best approximation to measured values was obtained when characteristic transverse length is the wetted perimeter computed with the local depth ($w(y)$, $MOD_{AZ} = 2$). In the case of irrigation IF-5, the best performance was obtained with τ_s taken as a function of the arc length along the wetted perimeter, s ($MOD_{AZ} = 1$). Figures 7 and 15–17 also show that there is not a great deal of difference in the infiltrated volume if $B(y)$, $WP(y_0)$ or $B(y_0)$ were used as the characteristic transverse length to compute the infiltration ($MOD_{AZ} = 3$, 4, and 5, respectively). On the other hand, the use of the top width computed with the normal depth ($MOD_{AZ} = 6$) as the characteristic transverse length gives a relatively poor match between computed and measured infiltrated volume, in all cases studied.

CONCLUSIONS

The quantity A_z is defined as infiltrated volume per unit length, i.e., is given by the following definition expression,

$$V_z(t) = \int_0^{x_A(t)} A_z(y,t)\,dx \qquad (23)$$

in which $V_z(t)$ = the infiltrated volume at a given time t, x_A = distance along the furrow, and y = the flow depth. With τ = the time duration that a given station, x, in the

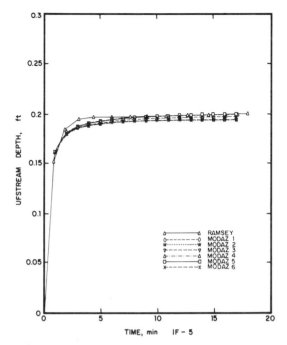

FIGURE 11. Time variation of upstream depth with "best" n, a, k. Test IF-5.

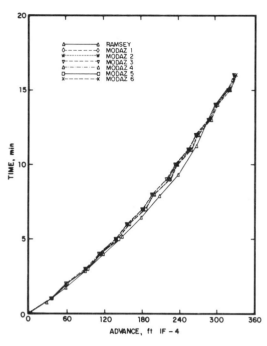

FIGURE 13. Advance curves with "best" n, a, k. Test IF-4.

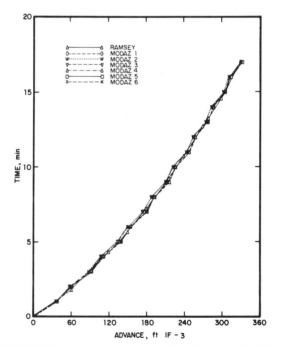

FIGURE 12. Advance curves with "best" n, a, k. Test IF-3.

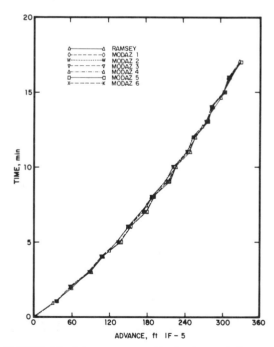

FIGURE 14. Advance curves with "best" n, a, k. Test IF-5.

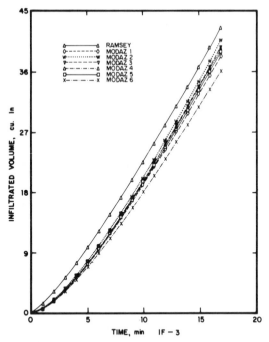

FIGURE 15. Total infiltrated volume with "best" n, a, k. Test IF-3.

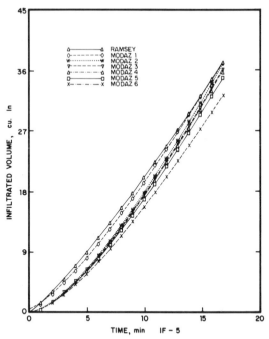

FIGURE 17. Total infiltrated volume with "best" n, a, k. Test IF-5.

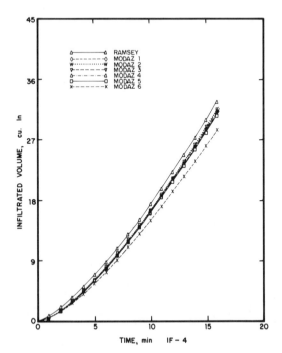

FIGURE 16. Total infiltrated volume with "best" n, a, k. Test IF-4.

length of the furrow has been wetted, A_z is approximated by a formula of the type

$$A_z(y,t) = C(y)\,z(\tau) \qquad (24)$$

in which C = some characteristic transverse dimension of the furrow; z is given by the Kostiakov formulation

$$z = k\tau^a \qquad (25)$$

in which the constants k and a depend upon the choice for C. Comparisons of six different choices for C were made with experimental data from a precision field furrow, which was used in this study as an infiltrometer. Measurement of a sucession of depth profiles in the furrow yielded the data necessary to determine appropriate values of k and a. An attempt was made to find which choice of characteristic transverse dimension yielded the best match for computed and measured time variations of stream-advance, upstream depth, and total cumulative infiltration volume. With the warning that the results presented are not at all exhaustive, it is concluded that wetted perimeter based on local depth ($MOD_{AZ} = 2$) is the best choice of transverse length to characterize furrow intake in mathematical models of furrow irrigation. A close second is the local top width ($MOD_{AZ} = 3$). Perhaps surprisingly, the most complex formulation, $MOD_{AZ} = 1$ (Equation 12), did not yield the best results in three out of the four cases tested. A possible explanation for this is that z is not truly a function of τ_s alone, as in Equation (6), but depends on s and y as well. Finally, little difference was noted amongst the performance of $MOD_{AZ} = 1$, 2, or 3, as long as the appropriate value of k was used. Even use of upstream values of wetted perimeter or top width ($MOD_{AZ} = 4$, 5) as the characteristic transverse dimension did not affect results too much. Use of constant top width, based on normal depth, however, ($MOD_{AZ} = 6$) does not yield very good results.

It should be emphasized the need for further research in this subject, with more reliable data covering a broader range of field conditions. In other words, the choice of transverse length to characterize furrow intake in mathematical models of furrow irrigation is still open to question.

On the other hand, the problem of determining an appropriate infiltration function to account for the effect of depth changes on the infiltration rate of furrows is also not concluded.

REFERENCES

1. Bondurant, J. A., "Developing a Furrow Infiltrometer," *Agric. Engineering, 38* (8), 602–604 (1957).
2. Christiansen, J. E., A. A. Bishop, F. W. Kiefer, and Yu-Si Fok, "Evaluation of Intake Rate Constants as Related to Water in Surface Irrigation," *Trans. of ASAE,* 671–674 (1966).
3. Criddle, W. D., S. Davis, C. H. Pair, and D. G. Schokley, "Method for Evaluating Irrigation Systems," *Agricultural Handbook No. 82,* USDA, Soil Conservation Service (1956).
4. Davis, J. R., "Estimating Rate of Advance for Irrigation Furrows," *Proceedings of the ARS-SCS Workshop in Hydraulics of Surface Irrigation,* Denver, Colorado, 67–86 (1960).
5. Davis, J. R. and Fry, A. W., "Measurement of Infiltration Rates in Irrigated Furrows," *Trans. of ASAE, 6* (4), 318–319 (1963).
6. Elliott, R. L. and W. R. Walker, "Field Evaluation of Furrow Infiltration and Advance Functions," *Transactions of the ASAE, 25* (2), 396–400 (1982).
7. Fangmeier, D. and M. K. Ramsey, "Intake Characteristics of Irrigation Furrows," *Transactions of the ASAE, 21,* Sw No. 4, 696–700 (1978).
8. Karmeli, D., L. J. Salazar, and W. R. Walker, "Assessing the Spatial Variability of Irrigation Water Applications," Report EPA-600/2-78-041. Office of Research and Development, U.S. Environmental Protection Agency, Ada, Oklahoma (1978).
9. Lewis, M. R. and W. E. Milne, "Analysis of Border Irrigation," *Agricultural Engineering, 19,* 267–272 (1938).
10. Nance, L. A., Jr. and J. R. Lambert, "A Modified Inflow-Outflow Method of Measuring Infiltration in Furrow Irrigation," *Trans. ASAE,* 792–798 (1968).
11. Parlange, Jean-Yves, "Theory of Water Movement in Soils: 6. Effect of Water Depth Over Soil," *Soil Science, 113* (5), 308–312 (1972).
12. Philip, J. R., "The Theory of Infiltration: 6. Effect of Water Depth Over Soil," *Soil Science, 85,* 278–286 (1958).
13. Philip, J. R. and D. A. Farrel, "General Solution of the Infiltration-Advance Problem in Irrigation Hydraulics," *J. of Geophysical Research, 69* (4), 621–631 (1964).
14. Ramsey, M. K., "Intake Characteristics and Flow Resistance in Irrigation Furrows," Thesis presented to the University of Arizona in partial fulfillment of the requirements for the degree of Master of Science (1976).
15. Shull, H., "A Bypass Furrow Infiltrometer," *Trans. of the ASAE, 4* (1), 15–17 (1961).
16. Shull, H., "Furrow Hydraulics at the Southwestern Irrigation Field Station," *Proceedings of the ARS-SCS Workshop on Hydraulics of Surface Irrigation,* Denver, Colorado, 55–62 (1960).
17. Singh, P. and H. S. Chauhan, "Determination of Water Intake Rate from Rate of Advance," *Transaction of ASAE, 16* (6), 1081–1084 (1973).
18. Smerdon, E. T. and C. M. Hohn, "Relationship Between the Rate of Advance and Intake Rate in Furrow Irrigation," Miscellaneous Publications No. 509. Texas Agr. Exp. Sta., College Station, Texas (April 1961).
19. Souza, Francisco, "Nonlinear Hydrodynamic Model of Furrow Irrigation," dissertation submitted in partial satisfaction of the requirements for the degree of Doctor of Philosophy in Engineering, University of California at Davis (1981).
20. Strelkoff, T. and N. D. Katopodes, "Border Irrigation Hydraulics with Zero Inertia," *Journal of the Irrigation and Drainage Division, ASCE, 103* (IR3), Proc. Paper 13189, 325–342 (September 1977).
21. Strelkoff, T. and F. Souza, "Modeling Effect of Depth on Furrow Infiltration," *Journal of Irrigation and Drainage Division, ASCE, 110* (4), Paper 19366, 375–387 (Dec. 1984).
22. Wilke, O. C., "A Hydrodynamic Study of Flow in Irrigation Furrows," Tech. Report No. 13, Water Resources Institute, Texas A&M University (1968).

CHAPTER 17

Principles and Design of Furrow Irrigation Systems

Yu-Si Fok*

INTRODUCTION

Furrow irrigation is the most common surface irrigation method in practice. As reported by Musick et al. (1973), in the U. S. southern high plains alone, there are 5 million acres under furrow irrigation. In fact, furrows are naturally created for row crops. Perhaps, in history, the practice of furrow irrigation can be traced as early as the use of the plow for tillage. Furrow irrigation has many advantages. It is widely adaptable to various terrain and climate with relative low cost in land preparation; it uses gravitational flowing water to distribute soil-moisture in the furrow and thus the energy requirement for furrow irrigation is lower than other irrigation methods. Furrows are suitable for a wide range of cultivation tools and implements. On the other hand, furrow irrigation has the soil erosion problems due to too steep a furrow slope and too big an irrigation stream flow. It requires higher irrigation labor cost to regulate the furrow irrigation water flows. It also has the tendency to over-irrigate the upper ends of the furrows and under-irrigate the lower ends of the furrows, whenever the furrows are too long for uniform distribution of water. This is due to the fact that water has to advance from the upper end of a furrow to the lower end which will need a longer time in a long furrow. Thus the upper end of the furrow will have excessive opportunity to absorb irrigation water while the lower end of the furrow may have to wait for the arrival of the advancing irrigation water. A properly designed furrow irrigation system will minimize those aforementioned disadvantages of the furrow irrigation systems so that soil and water can be conserved under this popular surface irrigation method. This chapter will present the design factors and principles related to furrow irrigation systems and show how well the design factors should be fitted to each other under the guidance of the physical principles. Since the problem of furrow irrigation is very complex, the approach of solutions is purposely tailored with engineering means for problem solution. In other words, aside from the mathematical solutions being used in this chapter, graphical, empirical and approximate solutions are also used. Equations presented in this chapter are in the form of algebraic equations so that the design factors can be evaluated for their relationship to each other.

Physical Nature of Furrow Water Flow

The physical nature of the flowing water in a furrow is unsteady, nonuniform three-dimensional open-channel flow on a sloped porous soil bed with a varying infiltration rate. It is a very complex hydrodynamic problem. In order to describe this furrow water flow problem, a set of partial differential equations which involve the principle of motion and the principle of conservation of mass is generally required. Interested readers can refer to Hansen (1958), Su (1961), Chen and Hansen (1966) and Tinney and Bassett (1961). At any cross section of a furrow, one can observe the rise of flow depth when the water advances and passes that cross section. After some time the flow depth will reach a constant depth since the inflow to the furrow is constant and outflow and the infiltration amount plus other fluid flow elements may reach an equilibrium. This situation is analogical to a flood flow in an open-channel with a rising and then a steady limb of the hydrograph. When the inflow water in a furrow is turned off, one can observe recession of water depth at that cross section too, because there is no incoming water to maintain the water depth. This situation is again analogical to the recession limb of the flood hydrograph.

In the viewpoint of furrow irrigation, the arrival and the departure of irrigation water at a point is the main interest,

*Professor, Department of Civil Engineering and Researcher, Water Resources Research Center, University of Hawaii at Manoa, Honolulu, HI

because the duration between the water arrival and departure is the time of irrigation for that location. This time of irrigation is also the infiltration opportunity time at that location. Consequently, the irrigation water depth applied to that point can be evaluated by means of an infiltration equation.

The Coverage of this Chapter

This chapter covers the land preparation for furrow irrigation systems; the maximum allowable non-erosive irrigation stream size; the irrigation, application and distribution efficiencies; the water advance and recession in furrows; the infiltration of water into the furrows and infiltration equations; the design of proper length of furrows and directions of improvements in furrow irrigation.

LAND PREPARATION FOR FURROW IRRIGATION

In order to have better control of the water, soil and the use of farming equipments for furrow irrigation, land shaping and grading are of prime importance. The furrow slope is an important factor for efficient distribution of irrigation water in furrows and also a limiting factor for the non-erosive irrigation stream. Although the non-erosive irrigation stream flow in a furrow is also constrained by the soil type, the soil type of the furrows is more or less a fixed factor. Therefore, the furrow slope is the only factor that can be modified and designed. The modern machines for land shaping and grading can provide fast and precise land preparation for furrow irrigation at a moderate cost. According to Jacobson (1966) and Houston (1966), the unit cost to move a cubic yard (0.764 m³) of earth on a farm is about $0.15. Houston (1966) also reported that the per acre earth movement to shape the land for irrigation in California is generally less than 500 cubic yards (382 m³) and the recommended furrow slope is from 0.1 to 0.2 percent where contour furrows may be used successfully on a sloped land such that the furrow slope can be controlled. Land grading operation is recommended to be conducted during the dry seasons. In general, most of the land grading for surface irrigation in California is done by contractors who follow farm-land grading as a full-time business. Farmers should be forewarned about the suitability of their lands for land grading. Farm lands with high infiltration rate, shallow top soil layer or the topography that requires more than 1,000 cubic yards per acre (309 m³ per hectare) of earth movement should use other methods of irrigation such as sprinkler or drip irrigation in order to conserve water and soil. Houston (1966) reported that the estimated land-grading costs in most irrigated areas of California in dollars per acre are: surveying and staking $4-7; earth moving 500 cubic yards $50-100; and land planning $4-6. With the advances in surveying and computer technology, the costs of surveying and land planning are expected to be less in today's market

value. However, the cost for earth movements may be higher because of higher energy and labor costs. Additional information for land shaping and grading for furrow irrigation can be found in Mickelson (1966), Buchta et al. (1966), Frey et al. (1966), Harris (1966), Dominy and Worley (1966) and *Agricultural Engineers Yearbook of Standards* (1983).

MAXIMUM ALLOWABLE NON-EROSIVE FURROW STREAM SIZE

Criddle et al. (1956) recommended the maximum allowable furrow stream to be estimated from the empirical relationship $Q = 10/S$, in which Q is the maximum allowable non-erosive furrow stream size in gallons per minute (1 gallon = 3.7854 liters) and S is the slope of the furrow in percent. Of course, this simple relation between the allowable furrow stream size and the furrow slope does not

TABLE 1. Values of Coefficients α and β for Different Soil Groups using Furrow Irrigation.*

	$q = \alpha S^\beta$		
Soil group (1)	α, in liters per second (gallons per minute) (2)	β (3)	Correlation coefficient (4)
I[a]	0.892 (14.144)	−0.937	0.891
II[b]	0.988 (15.666)	−0.550	0.724
III[c]	0.613 (9.712)	−0.733	0.800
IV[d]	0.644 (10.216)	−0.704	0.729
V[e]	1.111 (17.605)	−0.615	0.731
VI[f]	0.665 (10.543)	−0.548	0.921

*After Hamad and Stringham (1978) and with permission from American Society of Civil Engineers for reproduction.
[a]Heavy textured soil with very slow permeable subsoil and substratum. Depth to impermeable layer is more than 914 mm (36 in.).
[b]Moderately heavy textured soil with slow permeable subsoil and substratum. Depth to impermeable layer is 508 mm – 914 mm (20 in. – 36 in.).
[c]Medium textured soil with moderately slow permeable subsoil and substratum. Depth to impermeable layer is 508 mm – 914 mm (20 in. – 36 in.).
[d]Medium textured soil with moderately slow permeable subsoil and substratum. Depth to impermeable layer is 254 mm – 508 mm (10 in. – 20 in.).
[e]Light textured soil with moderately permeable subsoil and substratum. Depth to impermeable layer is 254 mm – 508 mm (10 in. – 20 in.).
[f]Very light textured soil with moderately rapid permeable subsoil and substratum. Depth to impermeable layer is less than 254 mm (10 in.).

include the other important factor, the soil type, in this empirical relationship. Hamad and Stringham (1978) reported a refinement of the Criddle et al. (1956) empirical development by including the soil type as the third parameter and used $Q = \alpha S^\beta$ as the general empirical equation with numerical values for α and β evaluated from experimental data obtained from six soil groups. Their study results are presented in Table 1.

IRRIGATION EFFICIENCIES

In a broad sense, irrigation efficiency considers many aspects of input factors to irrigation; these factors are land, labor, farming implements, seeds, fertilizer, etc. All of these input factors may be expressed in terms of capital which are impacted by the availability of irrigation water for crop productions as the outputs. Thus there are many ways to analyze the irrigation efficiency with the input factors versus the availability of irrigation water to crop production.

If the duty of irrigation water is the major concern in agricultural production, then, as reported by Jensen (1967), the U.S. agricultural industry has improved the efficiency of food production in the past decades. This is partly due to improvements in irrigation. He also suggested that irrigation efficiency should be considered as a measure of the control of irrigation water physically and chemically in the irrigated soils. Israelsen (1950) considered irrigation efficiency as the ratio of the water consumed by the crops of an irrigation farm to the water diverted from a natural water source into the farm. Although a large percentage of water is lost between the points of diversion and delivery, farmers are responsible to apply their irrigation water to their farms efficiently. In this special concern, Israelsen (1950) introduced the water-application efficiency for irrigation farmers to measure how well they apply the delivered water to the crops' root zone. Water-application efficiency is defined as the ratio of irrigation water stored in the root zone of the farm to irrigation water delivered to the farm. High water application efficiency ensures that water will be economically used.

Another important consideration is the uniformity of irrigation water being distributed in the soil profile; this concern was introduced by Hansen (1960). He defined the water distribution efficiency as the ratio of the difference between average depth of water stored in soil profile and average deviation from the mean depth of stored water in soil to the average depth of water stored in soil profile. Hansen (1960) pointed out that uniformly distributed irrigation water can produce crop stands in uniform height and the quality of crop yield is good for better market value. The water application efficiency and the water distribution efficiency are the measures of irrigation operation which stress that irrigation water is an important input for crop production and can be controlled for beneficial uses by the irrigation farmers.

In order to attend high water-application and -distribution

efficiencies, irrigation engineers can use their knowledge of the water advance and recession in the furrows so that proper length of furrows can be designed and suitable furrow irrigation operation can be scheduled. At this stage of irrigation furrows design, the land planning, shaping and grading are considered to be completed for furrow irrigation farming. However, before the presentation of the water flow advance and recession in furrows, the factor of water infiltration into soils has to be investigated.

WATER INFILTRATION INTO SOILS

Theoretically, water infiltration into the soil of the furrows is three-dimensional. However, because three-dimensional infiltration studies have just started being introduced recently into surface irrigation as shown by Fok and Chiang (1983), Fok and Chiang (1984) and Clothier et al. (1985), the most popular infiltration equation for furrow irrigation is the Lewis (1937)-Kostiakov (1932) empirical equation which takes the form of Equation (1):

$$d = kt^n \qquad (1)$$

in which d = cumulative infiltration depth; t = time of infiltration; and k and n = the empirical constants. The Lewis-Kostiakov equation has been shown to relate experimental data very well as reported in literature, among others, Criddle et al. (1956), Fok and Bishop (1965), Fangmeier and Ramsey (1978) and Clemmens (1983). Although it is evident that the empirical constants k and n in Equation (1) are dependents of soil hydrological properties and the duration of inltration time, the physical meanings of k and n have not been fully explored since they were postulated in the 1930s. The only investigation of the physical meanings of k and n was reported by Fok (1967). He has shown that Equation (1) can be represented by a set of four power-law equations of which each equation is valid within a specific time zone; and each time zone can be computed from a set of soil hydrological properties with known values. There are other infiltration equations, such as the Philip two-term equation (1957) and the modified Lewis-Kostiakov equation as stated in Fok and Bishop (1965), which have been used in furrow irrigation studies; however, since these two equations have two terms in them, they are not widely used in furrow irrigation analysis, although the Philip two-term equation has been derived from theoretical one-dimensional equations.

Furrow Infiltration Measurements

There are two major methods for the furrow infiltration measurements. They are:

1. Cylinder Infiltrometer Method
2. Inflow-Outflow Method

CYLINDER INFILTROMETER METHOD

Traditionally, infiltration measurements from cyclinder infiltrometers are obtained from the inner cylinder infiltrometer to measure the dropping of the water level versus the infiltration time. The outer cylinder infiltrometer is installed and filled with water to act as a buffer zone to prevent the effects of lateral soil-water movement on the observed data. In other words, the objective of the measurements is to measure one-dimensional infiltration. This method has been used since the 1930s as reported by Lewis (1937), and evidently had been used in France and other countries before then. In literature, this method is also called double-ring infiltrometers method. Sometimes a single cylinder infiltrometer is installed and surrounded by soil dike (10-20 cm in height) to create a buffer zone to hold ponding water during infiltration measurement.

The cylinder infiltrometer method is very popular, because it is simple to use and does not require larger quantity of water supply to support the field measurement, and can be operated by one person to measure four or five sites simultaneously within a distance of 100 m. The cylinders are usually made of 14 gage cold rolled sheet steel; the inside diameter of the cylinder should be at least 23 cm; the minimum height should be 30 cm. For convenience in storage and transportation, cylinders can be made with different and gradual diameters so that they can be nested one within another. A set of six cylinders having inside diameters of 24, 25.5, 27, 28.5, 30 and 31.5 cm with the same height of 32 cm would fit each other nicely as a set of infiltrometers. When in use, each cylinder is driven into the soil for a depth of 9 cm or more. A plastic bag is then placed inside of the cylinder to hold water enough to fill the cylinder to a depth of at least 15 cm. When the infiltration experiment is to be started, the water bag can be opened to let water contact the soil and at the same time to prevent the disturbance of the soil surface. The starting water depth in the cylinder can be measured by a hook gage or other marked scale or tape. The falling water depths versus clock times should be read frequently at the beginning of the experiment, at every one or two minutes for 10 to 20 minutes. The water depth readings should be read to the accuracy of mm unit.

The major disadvantages of using cylinder infiltrometers to measure furrow infiltration are that this method provides only point measurements and the observed data represent one-dimensional infiltration. Detailed information for the use of cylinder infiltrometers can be found in Haise et al. (1956) while the simple ones are given by Criddle et al. (1956).

INFLOW-OUTFLOW METHOD

The inflow-outflow method for furrow infiltration measurements applies a constant flow of water into a selected section of an experimental furrow. To prevent the effects of lateral flows to the adjacent furrows, waters are also applied to the adjacent furrows. Usually, the length of the experimental furrow section is not more than 10 m. The water inflow and outflow rates to this furrow section are recorded for the infiltration computations. The advantages of using the inflow-outflow method for furrow infiltration measurements are: the two- or even three-dimensional fluid flow conditions in the furrow are simulated, and therefore the infiltration measurements are close to the actual field conditions. The disadvantages are: the infiltration data base can not be established at the early stage of the experiment when there are no outflow waters at the outflow control section; and this method requires large quantity of water supply to maintain furrow flows in not only the furrow selected for the experiment but also to keep running water in the two adjacent furrows to prevent experimental error created from lateral flows to the adjacent furrows. Infiltration measurements using the inflow-outflow method had been reported in Criddle et al. (1956) and recently in Walker and Willardson (1983). Data analysis for this method was also presented by Criddle et al. (1956). However, Walker and Willardson (1983) suggested that the surface flow modeling techniques should be used to extract furrow infiltration data from inflow-outflow method's observations. Perhaps, computer aided infiltration data extraction and processes are needed when the surface flow modeling techniques are used.

In summary, infiltration measurements can be obtained by using the cylinder infiltrometer method and the inflow-outflow method; users can select the appropriate method to gather furrow infiltration information for furrow irrigation design and operation planning. Tabulations of observed data and data analyses for these two methods are given in Tables 2 and 3.

Furrow Infiltration

Furrow infiltration is not one-dimensional. The studies of two- and three-dimensional infiltration using highly mathematical analysis were presented by Philip (1966). The use of applied mathematics and field experiments to investigate two-dimensional infiltration was presented by Toksoz et al. (1965). Following this initial attempt, Fok (1970) introduced the component infiltration approach to evaluate two-dimensional infiltration for furrows. In 1982, Fok et al. (1982) utilized the Fok power-law infiltration equations (1967) and the component infiltration approach to develop a set of two-dimensional algebraic infiltration equations. This development provides the tool for the design of furrow irrigation because Fok and Chiang (1984) were able to present a set of two-dimensional infiltration algebraic equations for furrow irrigation. The Fok-Chiang (1984) Two-Dimensional Infiltration Equations for Furrow Irrigation are explicit. In other words, this set of equations is expressing infiltration volume or depth as a function of infiltration time when related soil hydrological properties (soil moisture content, hydraulic conductivity and capillary head) are known. As stated by Fok and Chiang (1984), there is no general infiltration equation for furrow irrigation because the cross-sectional area of a furrow may take many geometric shapes.

TABLE 2. Ring-Intake Data.

Observed time 24-hr. clock	Elapsed time		Distance to water surface from reference point		Intake during period		Accumulated intake during test	
	Since last reading	Since beginning of test	Before filling	After filling	Depth	Average rate per hour	Depth	Average rate per hour
	Minutes	Minutes	Inches	Inches	Inches	Inches	Inches	Inches
8:00			Fill	0.0				
					1.72	10.3		
8:10	10	10	1.72				1.72	10.3
					1.02	6.1		
8:20	10	20	2.74				2.74	8.2
					.75	4.5		
8:30	10	30	3.49				3.49	7.0
					.90	3.6		
8:45	15	45	4.39	.0			4.39	5.9
					.80	3.2		
9:00	15	60	.80				5.19	5.2
					1.40	2.8		
9:30	30	90	2.20				6.59	4.4
					1.15	2.3		
10:00	30	120	3.35	.0			7.74	3.9
					1.90	1.9		
11:00	60	180	1.90				9.64	3.2
					3.00	1.5		
13:00	120	300	4.90	.0			12.64	2.5
					2.40	1.2		
15:00	120	420	2.40				15.04	2.2
					2.20	1.1		
17:00	120	540	4.60				17.24	1.9

After Criddel et al. (1956) U.S. Department of Agriculture—SCS.

In their paper, the furrow cross-sectional area was in rectangular shape to serve as an example to illustrate the use of the method of components to account for the 1-D (one-dimensional) and 2-D (two-dimensional) portions of infiltration in a furrow. Since the Fok-Chiang Furrow Infiltration Equations (1984) are tailored for furrow irrigation design and operation, the derivations of this set of equations are presented briefly in the following.

DEVELOPMENT OF FOK-CHIANG FURROW INFILTRATION EQUATIONS

Downward Infiltration—Along y-axis

By following a similar approach developed by Green and Ampt (1911), Fok and Hansen (1966) presented the Fok-Hansen dimensionless infiltration equation as

$$\frac{d}{\Delta\theta \, \Sigma h} - \ln \left(1 + \frac{d}{\Delta\theta \, \Sigma h} \right) = \frac{Kt}{\Delta\theta \, \Sigma h} \quad (2)$$

in which $\Sigma h = h_c + h_o$ = capillary head at wetting front plus constant water depth on soil surface; $\Delta\theta = \theta_o - \theta_n$ = final soil moisture content after infiltration minus initial soil moisture content before infiltration; K = hydraulic conductivity in the transmission zone; and d and t have the same meaning as defined in Equation (1).

Since Equation (2) is a dimensionless equation, Fok and Hansen (1966) presented graphical solutions of Equation (2) by plotting the dimensionless parameters $d/\Delta\theta \, \Sigma h$ and $Kt/\Delta\theta \, \Sigma h$ on log-log paper and obtained a unique curve relating these two parameters as shown in Figure 1. Subsequently, Fok (1967) used four consecutive straight lines to make an approximation for this unique curve. From this linear approximation of Equation (2), a set of power-law infiltration equations was developed according to their respective time zones as defined by the following time equations [Equations (3) to (6)]:

$$t_1 = 0.005 \, \Delta\theta\Sigma h/K \quad (3)$$

$$t_2 = 0.337 \ \Delta\theta\Sigma h/K \tag{4}$$

$$t_3 = 3.367 \ \Delta\theta\Sigma h/K \tag{5}$$

$$t_4 = 85.15 \ \Delta\theta\Sigma h/K \tag{6}$$

$$\text{For } t < t_1, \ d = 1.41(\Delta\theta\Sigma h)^{0.50} \ (Kt)^{0.50} \tag{7}$$

$$\text{For } t_1 \leq t < t_2, \ d = 1.82(\Delta\theta\Sigma h)^{0.45} \ (Kt)^{0.55} \tag{8}$$

$$\text{For } t_2 \leq t < t_3, \ d = 2.19(\Delta\theta\Sigma h)^{0.32} \ (Kt)^{0.68} \tag{9}$$

$$\text{For } t_3 \leq t < t_4, \ d = 1.83(\Delta\theta\Sigma h)^{0.15} \ (Kt)^{0.85} \tag{10}$$

Equations (7) to (10) are a set of downward infiltration equations which have the same form as Equation (1) and are called Fok power-law infiltration equations.

Horizontal Infiltration—Along x-axis

The horizontal infiltration can be derived with the same approach developed by Green and Ampt (1911) and Fok et al.

(1982) as

$$d = 1.41(\Delta\theta\Sigma h)^{0.50} \ (Kt)^{0.50} \tag{11}$$

Equation (11) has the same form as Equation (7). This is expected to happen when the soil is relatively dry and infiltration time is shorter than t_1. When the infiltration time is prolonged into other time zones such as t_2, t_3, and t_4, the downward infiltration equations will change their forms respectively. This situation leads Fok et al. (1982) to use half-ellipses as the loci of infiltration wetting patterns to approximate the two-dimension infiltration.

Formation of Fok-Chiang Furrow Inltration Equations

Figure 2 shows the components of the one- and two-dimensional infiltration elements in a rectangular furrow section such that the formation of the Fok-Chiang Furrow inltration equations can be followed.

By using y to represent the downward infiltration and x to represent the horizontal infiltration, the rectangular furrow section's infiltration, I, can be expressed as the sum of the composite portions of the one- and two-dimension infiltrations (as shown in Figure 2 designated as 1-D and 2-D), in

TABLE 3. Inflow-Outflow Method: Measurement of Furrow-Intake Rate.

Clock time (24-hr.)	Elapsed time Station 0+00	Elapsed time Station 1+00[1]	Average	Inflow	Outflow	Loss in furrow	Intake per 100 ft.
FURROW NO. 2							
	Minutes	Minutes	Minutes	G.p.m.	G.p.m.	G.p.m.	G.p.m.
8:02	Start			4.0			
8:24	22	0					
8:27	25	3	14		0.60	3.40	3.40
8:50	48	26	37		1.90	2.10	2.10
9:20	78	56	67		2.44	1.56	1.56
10:00	118	96	107	Constant	2.80	1.20	1.20
11:10	118	166	177	Flow	3.00	1.00	1.00
12:30	268	246	257		3.12	.88	.88
14:00	358	336	347		3.30	.70	.70
16:00	478	456	467	4.0	3.40	.60	.60
FURROW NO. 4							
8:10	Start			8.0			
11:32	202	0					
11:40	210	8	109		0.20	7.80	1.30
12:00	230	28	129	Constant	1.40	6.60	1.10
12:40	270	68	169	Flow	2.30	5.70	.95
14:10	360	158	259		3.20	4.80	.80
16:15	485	283	384	8.0	4.16	3.84	.64

[1]For furrow No. 4 station 6 + 100.
After Criddle et al. (1956) USDA—SCS.

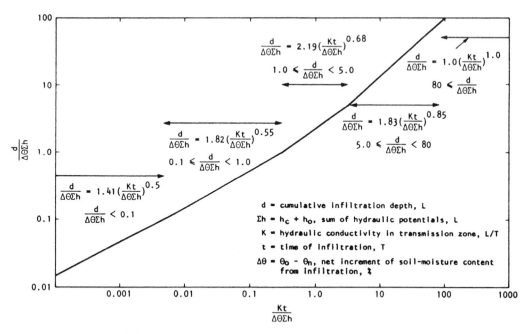

FIGURE 1. Power relationships between d/$\Delta\theta\Sigma$h and Kt/$\Delta\theta\Sigma$h.

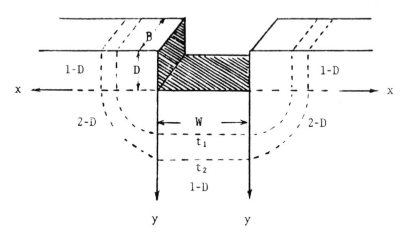

FIGURE 2. Dimensions of rectangular furrow composite infiltration [after Fok and Chiang (1984) and with permission from American Society of Civil Engineers for reproduction].

which D = furrow depth, W = furrow width and B = furrow length.

$$I = \left[2Dx + Wy + \left(\frac{\pi}{2} \right) xy \right] B \qquad (12)$$

As reported by Fok and Chiang (1984), there is no general infiltration equation for furrow irrigation because the cross-sectional area of a furrow may take many geometric shapes. Equation (12) serves as an example to illustrate the methodology to determine furrow infiltration by method of components. With measured or evaluated data substituted in Equation (12) as shown by Fok and Chiang (1984), the measured and calculated infiltrations have been in good agreement. Furthermore, Equation (12) can take the form of Equation (1) — the Lewis-Kostiakov intake equation — without introducing much error. Perhaps this is the reason for the popularity enjoyed by the Lewis-Kostiakov intake equation in irrigation, and this is the reason for using Equation (1) for subsequent analyses in this chapter. However, readers should remember that Equation (12) is fundamental and has physical meaning to summarize the 1-D and 2-D infiltration components in a furrow section to produce the infiltration, I, as related to time, t, so that I and t can be expressed in a relation as Equation (1) by means of approximation. Equation (1) takes the simplest mathematical form in which it provides great convenience in mathematical analyses of furrow irrigation. Detailed expansions of Equation (12) in different time zones have been presented by Fok and Chiang (1984) and therefore are not presented in this chapter.

WATER ADVANCE AND RECESSION IN FURROWS

In order to be able to present the analyses of water advance and recession in furrows by algebraic equations, the derivation of water advance and recession in furrows is presented by means of the principle of conservation of mass only. In other words, the analyses use the inflow-outflow water balance method to account for water volumes flowing into a furrow and to keep tracking the portions of the water on the soil surface and into the soil at a given time. The neglecting the energy loss during the furrow flow is assumed very mild and perhaps less than 10% in error term. In other words, without using the principle of motion, the analyses are much simpler and have a solvable ordinary differential equation to produce an algebraic water advance equation and an approximate water recession equation for furrow irrigation. Readers of this chapter are reminded that irrigation water flow in a furrow is much more complex than water flow in a concrete lined open channel. The design of a concrete lined open channel is still using semi-empirical hydraulic equations such as Manning or Chezy formula, and for the engineering practice, this is acceptable. Fok and Bishop (1965), Hart et al. (1968) and Singh and Chauhan

(1972) have shown how well the computed results from furrow advance equations derived from water-balance methods compare with measured data.

Fok-Bishop Furrow Water Advance Equation

Fok and Bishop (1965) used inflow-outflow method in which the inflow water volume is equal to the outflow volume at any given time as expressed in Equation (13):

$$Q T = A D \qquad (13)$$

in which Q = inflow water rate, T = furrow water application time, A = area covered by water = $W L$ = width of the water-covered furrow times length of the water-covered furrow, D = water depth in the furrow, and $D = D_s + D_a$ = average water depth of furrow plus average water depth infiltrated into soil of the furrow. Solving the furrow water advance length, L, from Equation (13), Fok and Bishop (1965) presented the inflow-outflow water-balance water advance equation for surface irrigation:

$$L = \frac{Q T}{(D_s + D_a) W} \qquad (14)$$

According to Equation (14), if D_s and D_a are properly derived from mathematical analyses, Equation (14) can be an exact solution for furrow water advance; however, D_s and D_a can only be approximated because they are dependent on each other physically, and each of them has its own hydrodynamic constraints to defy theoretical solution. According to many field observations the furrow water advance length versus time can be empirically related as:

$$L = a T^b \qquad (15)$$

in which a and b = empirical constants of the advance empirical equation. Fok and Bishop (1965), using Equations (1), (14) and (15), derived the expressions for D_s and D_a as:

$$D_s = \frac{A_o}{(1 + b) W} = \frac{u D_o^m}{(1 + b) W} \qquad (16)$$

and

$$D_a = \frac{F k T^n}{n (1 + n)} \qquad (17)$$

in which A_o = flow cross-sectional area; D_o = normal depth at furrow inlet; u and m are empirical constants evaluated from Manning Formula when shape and the slope of the furrow are given; and F = the Kiefer factor which can be approximated by:

$$F = \frac{(n - 1)(1 - b) + 2}{1 + b} \qquad (18)$$

according to Christiansen et al. (1966). According to Fok and Bishop (1965), the *b* value can be approximated by:

$$b = e^{-0.6n} \qquad (19)$$

in which e = the base of natural logarithms. According to Fok (1969):

$$\log_e b = -(k)^{0.29} \, n^{1.29} \, e^{(0.86-0.06q)} \qquad (20)$$

in which k = the empirical constant of Equation (1) in feet per minute; n = the empirical exponent of Equation (1); $q = Q/W$ = cubic feet per min. per foot width of the water covered furrow. The other terms have been defined previously. Equations (19) and (20) are empirical equations; as shown by Hart et al. (1968), Equation (19) can be regarded as a rough estimation of the value *b* by just knowing the value of *n*. Equation (20) relates *b* as a function of *k, n* and *q,* and it provides additional refinements for the estimation of *b* as reported by Fok (1969). The *b* value estimated from Equations (19) and (20) is needed for the evaluation of D_a in Equation (17) through the finding of the *F* factor from Equation (18). Again, the same value of *b* is used in Equation (16) for the estimation of D_s. With values of D_a and D_s obtained, Equation (14) can be solved numerically to obtain values of *L* versus *T*. As a check of how well the estimated value of *b* has been made, one can use the computed values of *L* versus *T* to obtain the computed *b* using Equation (15). If the estimated and computed values of *b* are close enough, there is no need to repeat the processes just described; otherwise, the evaluation of the *b* value should be repeated once again.

Furrow Water Recession

Furrow water recession is much more difficult to obtain an algebraic equation for than furrow water advance because of the lack of information on surface roughness and infiltration, as pointed out by Wu (1972). Su (1962) considered the recession flow has two parts, vertical recession and horizontal recession. Vertical recession occurs when the inflow to the furrow has been turned off and water at the head of the furrow begins to drop while the water is still advancing at the lower end of the furrow, and ends when water drops to the soil surface at the upper end of the furrow. When water recedes along the furrow length, horizontal recession starts while the leading water is still advancing at the lower end of the furrow, and ends when water ceases to recede along the furrow length.

In most field observations, the recession length versus time is in linear relation, or it can be approximately represented by a linear relationship, according to Wu (1972), Coolidge et al. (1982), Walker and Humperys (1983) and Holzapfel (1984).

Fok (1964) suggested an approximation method to evaluate furrow water recession. He assumed that the averaged

depth of water absorbed into the furrow soil, when water is turned off at the upper end of a furrow, remains the same, while the horizontal water recession in a furrow is in process; and that the advance water follows its trend after the water is turned off. Then the maximum furrow length that can be covered by the flowing water can be evaluated by letting D_o of Equation (16) be equal to zero, and Equation (14) can be rewritten into Equation (21) with Equation (17) as input:

$$L_r = \frac{Q \, T \, n \, (1+n)}{W \, F \, k \, T^n} = \frac{Q \, n \, (1+n) \, T^{1-n}}{W \, F \, k} \qquad (21)$$

in which L_r = furrow water recession length; all other terms in Equation (21) have been defined previously. Since Q, n, W, F and k are known and used in the computation of furrow water advance, L_r can be calculated from T at the time when water is turned off at the upper end of the furrow. Since it was assumed that the water will follow the same advancing trend, the maximum recession time, e.g., the time needed by the soil to absorb all the water on soil surface, can be evaluated by substituting L_r into Equation (14) and solving for T_r the maximum recession time by trial and error method. Finally the point (L_r, T_r) can be plotted on a *L-T* graph as shown in Figure 3.

Following the procedures suggested by Criddle et al. (1956), in which the points (O, T) and (L_r, T_r) are connected by a straight line, thus an approximation of the furrow water recession curve is obtained, utilizing the example and data given by Criddle et al. (1956). Figure 3 shows in detail all the procedures necessary to determine a suitable length of a furrow. In this example, if the water was turned off at 230 minutes after it was turned on, it covered 305 m (1000 feet) in 348 minutes but a large amount of water was wasted. According to the infiltration study, it takes 40 minutes for the soil to absorb 10.16 cm (4 inches) of water to refill the root zone soil water reservoir. If the water was turned off at 98 minutes after it was turned on, the approximate recession curve with its intercept at 96 minutes on the time axis would have intersected the advance curve at 215 m (707 feet) with the time at 167 minutes. By comparing the approximate recession curve with the ideal recession curve (a curve with 40-minute lag above the advance curve), it is obvious that the length of run should have been selected at 168 m (550 feet), since the approximate recession curve intersects the ideal recession curve at that point. For actual design, this design is conservative, because the actual recession curve is always well above the approximate recession curve in this example. Therefore, the selection of 198 m (650 feet) for the length of run was appropriate. By converting the time intercepts between the advance curve and the recession curve at orderly locations into depth of water infiltrated into the soil at corresponding locations, the amount of water absorbed into the soil during the whole irrigation operation can be obtained. When this known amount of water is com-

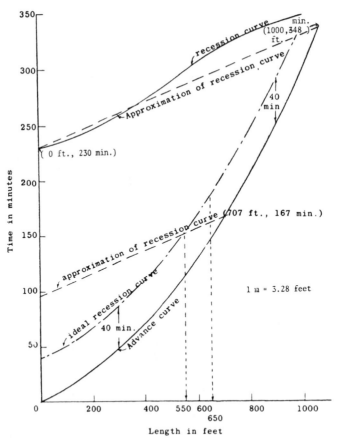

FIGURE 3. Determination of a suitable length of run by means of advance and recession curves [after Fok (1964) and Fok and Bishop (1969) and with permission from American Society of Agricultural Engineers for reproduction].

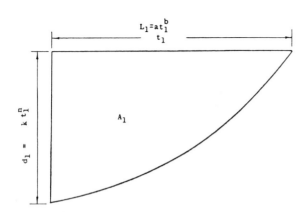

FIGURE 4. Profile of water distribution in soil at time t_1 [after Fok (1964) and Fok and Bishop (1965) and with permission from American Society of Agricultural Engineers for reproduction].

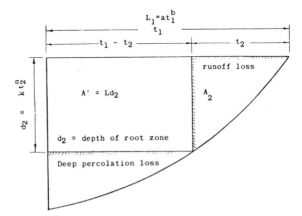

FIGURE 5. Profile of desired distribution of irrigation water, as related to runoff and deep percolation losses [after Fok (1964) and Fok and Bishop (1965) and with permission from American Society of Agricultural Engineers for reproduction].

pared to the total amount of water turned into the furrow and compared to the depth of water needed to refill the root zone soil reservoir, the application efficiency can be determined; likewise the distribution efficiency can also be evaluated because water infiltration depth at every point along the furrow can be evaluated. According to Fok (1964) and Fok and Bishop (1969), the application and distribution efficiencies can be expressed in terms of the infiltration exponent, n, the advance exponent, b, the correction factor, F, the time ratio, R (which is defined as the required infiltration time, t_2, divided by the advance time to cover the full furrow length with water, $t_1 - t_2$), and the factor G which is defined for the development of the distribution efficiency and will be shown later.

Expressing Application Efficiency in Terms of b, n, F and R

Figure 4 is a two-dimensional plot, which shows the water distribution profile area, A_1, in soil at a given time, t_1. In Figure 5, d_2 is the depth of water to be applied to refill the root zone reservoir; t_2 is the time needed for infiltration of d_2 into the soil. The term $t_1 - t_2$ is the time required for the water to reach the end of the furrow; A_2 is the area representing the runoff beyond the furrow length, L, at time t_1; and $A' = L\,d_2$ is the target area to be irrigated.

According to definition for the application efficiency, E_a, and Figure 4, E_a can be expressed in terms of A_1, A_2 and A' in Equation (22):

$$E_a = \left(\frac{A'}{A_1 - A_2}\right) 100\% \qquad (22)$$

Letting R = the ratio of infiltration time/the advance time = $t_2/(t_1 - t_2)$, A_2 and A' can be determined in terms of a, b, k, n and R by integration method, while $A_1 = D_a L$, in which D_a has been defined in Equation (17) and L has been defined in Equation (15). In other words, A_1 can be expressed in terms of a, b, k, n and F. When the above expressions of A_1, A_2 and A' are substituted into Equation (22), E_a can be expressed in Equation (23):

Equation (22) can be modified into Equation (24):

$$E_a = \frac{A'}{A_1} 100\% = \frac{(n+2)\,R^{n+1}}{F\,(R+1)^{b+n+1}} 100\% \qquad (24)$$

Equations (23) and (24) did not consider recession water on the soil surface. However, they are presented in the form of algebraic equations which may offer a very convenient way for furrow efficiency designs.

Expressing Distribution Efficiency in Terms of b, n, F, G, R and E_a

Following the same methodology in the development of the application efficiency, E_a, Fok (1964) and Fok and Bishop (1969) derived an expression for the distribution efficiency, E_d, in terms of b, n, F, G, R and E_a. In these terms only G has to been defined later, the rest having been defined previously.

According to Hansen (1960), the distribution efficiency, E_d, is defined as

$$E_d = (1 - s/D_d)\,100\% \qquad (25)$$

in which s = average numerical deviation of stored water depth versus the average stored water depth in the furrow after irrigation, and D_d = average stored water in the furrow after irrigation. According to Figures 4, 5, 6 and 7, $D_d = (A_1 - A_2)/L$ and $s = (A_1 - A_d)/L_d - (A_1 - A_2)/L$. Substituting these graphical relationships into Equation (25) and simplifying the resulting expression yields Equation (26):

$$E_d = \left[2 - \frac{\dfrac{A_1 - A_d}{L_d}}{\dfrac{A_1 - A_2}{L}}\right] 100\% \qquad (26)$$

Again because A_1, A_2, A_d, L and L_d can be expressed respectively in terms of a, b, F, k, n and R by integration method, E_d can also be expressed in terms of these terms

$$E_a = \frac{\left(\dfrac{1}{R+1}\right)^b \left(\dfrac{R}{R+1}\right)^{n+1} 100\%}{\dfrac{F}{n+2} - b\left(\dfrac{R}{R+1}\right)^{n+2}\left[\dfrac{1}{n+2} - \dfrac{(b-1)}{n+3}\dfrac{R}{R+1} + \dfrac{(b-1)(b-2)}{2(n+4)}\left(\dfrac{R}{R+1}\right)^2 - \cdots\right]} \qquad (23)$$

Equation (23) considers deep percolation as the only loss and assumes the runoff water can be reused, which includes surface storage water. In other words, recession water is considered reuseable.

If the runoff water, A_2, is considered also a loss, then

because of Equation (26). In addition, a term G and the application efficiency E_a are also included in the same expression to produce the resulting equation. According to Fok (1964) and Fok and Bishop (1969), the G term is defined in

Equation (27):

$$G = \left\{ \frac{\dfrac{F}{n+2} - b\left(\dfrac{R}{R+1}\right)^{n+2}\left[\dfrac{1}{n+2} - \dfrac{b-1}{n+3}\left(\dfrac{R}{R+1}\right) + \dfrac{(b-1)(b-2)}{2(n+4)}\left(\dfrac{R}{R+1}\right)^2 - \cdots\right]}{\left(\dfrac{1}{R+1}\right)^b} \right\}^{\frac{1}{n+1}} \qquad (27)$$

With the G defined in Equation (27), the distribution efficiency, E_d, can be expressed in Equation (28):

$$E_d = \left\{ 2 - \frac{E_a\left[\dfrac{F}{n+2} - bG^{n+2}\displaystyle\sum_{i=0}^{i}(-1)^i\frac{(b-1)!\,G^i}{(b-1-i)!\,i!\,(n+2+i)}\right]}{(1-G)^b\left(\dfrac{R}{R+1}\right)^{n+1}} \right\}100\% \qquad (28)$$

According to Equation (28), it can be seen that application efficiency, E_a, is part of the terms involved in the distribution efficiency. Furthermore, even if the application efficiency is 100%, the distribution efficiency may not attain 100%; however, $E_a = 100\%$ is not the prerequisite condition for distribution efficiency to attain a high level, as shown by Hansen (1960). With the application and distribution efficiencies developed in algebraic equations as Equations (23) and (28), irrigation engineers and planners can use these equations to design efficient furrow irrigation systems. This is possible because the advance of irrigation water in the furrows is considered in these equations. Perhaps it should be pointed out that researchers should put in more effort to help farmers, i.e., the irrigation water users,

to improve irrigation efficiency. The design and construction of an irrigation system can be made to attain very high irrigation efficiency; however, if operation and management of the same system were not matching with the designed specifications, the irrigation efficiency would not be high. As stated before, furrow irrigation systems need more labor during irrigation to regulate the flowing water. In other words, water may be wasted when there is a shortage of irrigation labor. In order to alleviate problems related to low irrigation efficiency in furrow systems, training of irrigation farmers and providing automated irrigation devices to control irrigation water application into the furrows in conjunction of the surge flow method would be a solution. The surge

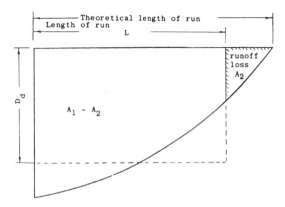

FIGURE 6. Average depth of water stored after irrigation: $D_d = (A_1 - A_2)/L$ [after Fok (1964) and Fok and Bishop (1965) and with permission from American Society of Agricultural Engineers for reproduction].

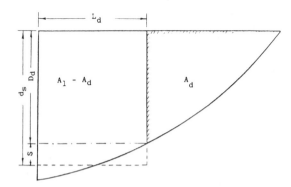

FIGURE 7. Average depth of water that deviates from the average depth of water stored after irrigation: $S = [(A_1 - A_d)/L_d] - D_d$ [after Fok (1964) and Fok and Bishop (1965) and with permission from American Society of Agricultural Engineers for reproduction].

flow irrigation method is relatively new; interested readers can refer to Stringham and Keller (1979) and Bishop et al. (1981) for details. Readers are reminded that the analyses presented in this chapter do not consider the aspects related to water quality. If irrigation water has high salinity, then the salinity balance in the irrigated farm land has to be considered. To maintain good balance of salinity in farm lands, additional quantities of irrigation water are applied and therefore drainage systems should be designed to carry excessive water from the root zone of the irrigated farm land. Perhaps treatment and reuse of drainage waters should also be included in the irrigated farm land planning and design.

DESIGN OF PROPER LENGTH OF FURROWS

In general the length of a furrow should be laid as long as the terrain permits. Especially for mechanized farming, fewer turn-arounds of the farming implements can save labor and productive land area. However, from the soil and water conservation standpoint, the length of a furrow should be properly designed, just reasoning from the simple fact that irrigation water in a furrow has to take time to advance from the head of the furrow to the lower end of the furrow. Therefore the longer the furrow, the longer the travel time (advance time) that is required. Obviously, in a long furrow the upper end is over irrigated while the lower end is under irrigated. This chapter presents the principles of the furrow irrigation systems by discussing each design element in the text. Now these design procedures can be summarized in the following:

1. Prepare a topographic map for the existing farm land.
2. Conduct soil survey to collect representative data of the root-zone soil profile including infiltration testing data following the method for furrow infiltration measurement described in this chapter. With these soil survey data—the root-zone soil water holding capacity (field capacity) minus allowable low soil water content before irrigation—the water depth required for irrigation can be determined.
3. Determine required infiltration time t_2 from the required irrigation water depth by means of field data evaluated infiltration equation—Equations (1) and (12).
4. Determine the maximum allowable non-erosive furrow flow rate with data from the topographic map of the existing farm land for the slope selected for the furrows and soil survey data with reference to Table 1.
5. Determine the shape and spacing of the furrows according to the kind of crops to be planted. From this information the number of furrows to be irrigated at one setting can be selected if the incoming irrigation water flow rate is known.
6. With the n value evaluated from infiltration data according to Equation (1), the b value can be evaluated from

Equations (19) and (20). With known values of n and b, the F factor can be estimated from Equation (18). With the input of D_s and D_a from Equations (16) and (17), the water advance length in the furrow versus time can be calculated. Using these calculated water advance length versus time data, the calculated b value can be determined by graphical method with reference to Equation (15). Now a check with the estimated and calculated b values can be made. If the difference between these two b values is more than 10%, then a new estimated b value should be made by following the procedures described within this step, and by trial-and-error method. If the error between estimated and calculated b values is less than 10%, then the average of them is the b value selected for uses in the design thereafter.

7. Once the values of b, n, and F are determined, the advance and recession of irrigation water can be evaluated for the furrows by following the methodology described in this chapter. The designer is now in the position to determine the time ratio R, which has been defined as the time needed for infiltration of the require water depth, d_2, to meet the irrigation requirement in proportion to the water to advance to the end of the furrow, $t_1 - t_2$. According to Criddle et al. (1956), if $R = t_2/(t_1 - t_2)$ is equal to or larger than 4, the application efficiency will be higher than 75% in most of the infiltration conditions. Designers are advised to use this $R = 4$ value as a guideline value and to use the application efficiency formula provided in this chapter as Equation (23) to calculate the application efficiency, E_a; or select the value of E_a and using trial-and-error method to determine the R value. Once the R value is determined, the designer can determine the irrigation time, t_1, from the definition of R, because the infiltration time, t_2, has been determined previously.
8. Once the irrigation time, t_1, is determined, the proper length of the furrows can be determined from Equation (14) because $T = t_1$, and all the other terms on the right-hand side of Equation (14) have been evaluated or can be estimated.

    Designers are reminded that the distribution efficiency is a function of the application efficiency according to Equation (28). In other words, if the field conditions of the furrow is the same for its full length, the functional relationship between application and distribution efficiencies should remain the same and can be examined by an inspection of Equation (28). This induction agrees with those elaborations made by Hansen (1960), because the derivation of Equation (28) assumes the field conditions as constant and uniform.
9. Once the proper length of the furrows has been determined, the plan for the preparation of the existing farm land can be made, the layouts of the irrigation ditches and drainage systems can be designed and constructed. Irrigation operational plan can also be designed.

POTENTIAL IMPROVEMENTS IN FURROW IRRIGATION

According to the discussions made in this chapter, the following potential improvements in furrow irrigation are suggested:

1. Research efforts for the operation and maintenance of the furrow irrigation systems stressing soil and water conservation
2. Researches of the maximum allowable non-erosive furrow discharge
3. Furrow surge flow irrigation method in conjunction with automation in irrigation water control may be considered a high technological study topic for future furrow irrigation development.
4. Methods to reduce flow resistance and infiltration along the furrow

CONCLUSION

According to the principles and design methodologies provided in this chapter, the following conclusions can be made:

1. The design of furrow irrigation systems is based on the principles of soil and water conservation. The proper length of the furrow irrigation systems can be designed according to the 9-step methodology provided in this chapter.
2. Furrow water advance equation and recession flow have been derived in this chapter based on the principle of conservation of mass. With the aid from the empirical advance equation, the derived furrow water advance equation can be checked against the empirical water advance equation by trial-and-error method to arrive at an acceptable value for design purpose.
3. Furrow infiltration equation, which is composed from segments of one- and two-dimensional infiltration, has been presented in this chapter. The resulting furrow infiltration equation has been reported to have the similar form as the Lewis-Kostiakov empirical infiltration equation.
4. The irrigation application efficiency and distribution efficiency have been shown to be in terms of parameters from the infiltration equation, the water advance equation, the irrigation time ratio and water recession.
5. The furrow water flow has been defined as a very complex hydrodynamic problem. This chapter is purposely tailored with all engineering methodologies for problem solutions. Aside from the mathematical approach, graphical, empirical, trial-and-error approximation and engineering selection approaches were employed such that the design equations can be presented in algebraic equations.

NOTATION

A = area in a furrow covered by irrigation water
A_d = wetted soil profile area beyond the average depth of stored water D_d
A_0 = furrow flow cross sectional area at the furrow inlet
A_1 = water distribution area in soil at a given time, t_1
A_2 = water runoff from a furrow which can be computed as wetted soil profile area beyond the infiltration water depth d_2 which can be computed from infiltration equation time, t_2
$A' = Ld_2$, in which L = length of the furrow d_2 = depth of irrigation water that can refill soil water to the root zone
a = empirical constant of the water advance empirical equation
B = segment of furrow length
b = empirical exponent of the advance equation
D = water depth in furrow irrigation
D_a = averaged infiltrated water depth in a furrow at time T
D_o = normal depth of furrow flow at the furrow inlet
D_s = averaged flow depth along the furrow length at time T
d = infiltration depth at time t
d_1 = infiltration depth at time t_1
d_2 = infiltration depth at time t_2
E_a = application efficiency
E_d = distribution efficiency
e = the base of natural logarithms
F = the Kiefer correction factor
G = a factor defined in Equation (27) to simplify the expression of the distribution efficiency
Σh = $h_c + h_o$ = total head
h_c = capillary head at wetting front
h_o = constant water depth on soil surface during infiltration
I = two-dimensional infiltration water volume into a segment of a furrow
K = hydraulic conductivity in the transmission zone during infiltration
k = empirical constant of the infiltration equation, Equation (1)
L = furrow length; furrow water advance length
L_d = furrow length from inlet to the point where the average infiltration depth is located
L_r = furrow water recession length
L_1 = water advance length at time t_1
$\ln$ = symbol of natural logarithms
$\log_e$ = natural logarithms
m = empirical exponent of the depth–area relationship of the furrow cross section evaluated from Manning Formula
n = exponent of the empirical infiltration equation, Equation (1)

Q = inflow rate to the furrow; $q = Q/W$

R = the time ratio of infiltration opportunity time (required infiltration time) to water advance time $= t_2/(t_1 - t_2)$

S = slope of the furrow

s = averaged numerical deviation of stored water depth versus the average water depth in the furrow after irrigation

T = irrigation time

T_r = recession time

t = infiltration time

t_1 = a given irrigation time, including the advance and infiltration times; also t_1, t_2, t_3 and t_4 are the indexes of infiltration time zones.

t_2 = required infiltration time to permit soil to absorb the required irrigation water depth, d_2

u = empirical constant of the depth–area relation of furrow section

W = width of the furrow water flow

x, y = axes of horizontal and vertical downward infiltration flows

$\Delta\theta = \theta_o + \theta_n$ = final soil water content after infiltration minus initial soil water content before infiltration.

α, β = empirical constants of the non-erosive furrow discharge and slope

REFERENCES

Agricultural Engineers Yearbook of Standards, Thirtieth Edition, American Society of Agricultural Engineers, St. Joseph, MI. USA. 30:526–529 (1983).

Bishop, A. A., W. R. Walker, N. L. Allen, and G. J. Poole, "Furrow Advance Rates Under Surge Flow Systems," *Jour. of the Irrigation and Drainage Div.,* ASCE, Vol. 107, No. IR3, Proc. Paper 16502, Sept., pp.257–264 (1981).

Buchta, H. G., D. E. Broberg, and Liggett. "Flat Channel Terraces," *Transactions of the ASAE.* 9(5): 571–573 (1966).

Chen, C. L., and V. E. Hansen, "Theory and Characteristics of Overland Flow," *Transactions of the ASAE.* 9(1): 20–24 (1966).

Christiansen, J. E., A. A. Bishop, F. W. Kiefer, and Y. S. Fok, "Evaluation of Intake Rate Constants as Related to Advance of Water in Surface Irrigation," *Transactions of the ASAE.* 9(5): 671–674 (1966).

Clemmens, A. J., "Infiltration Equations for Border Irrigation," *Advances in Infiltration.* Amer. Society of Agri. Engineers. 266–274 (1983).

Clothier, B., D. Scotter, and E. Harper, "Three-Dimensional Infiltration and Trickle Irrigation," *Transactions of the ASAE.* 28(2): 497–501 (1985).

Coolidge, P. S., W. R. Walker, and A. A. Bishop, "Advance and Runoff-Surge Flow Furrow Irrigation," *Jour. of the Irrigation and Drainage Division*, ASCE, Vol. 108, No. IR1, Proc. Paper 16930, Mar. pp. 35–42 (1982).

Criddle, W. D., S. Davis, C. H. Pair, and D. G. Shockley, "Methods for Evaluating Irrigation Systems," *Agricultural Handbook No. 82,* Soil Conservation Service, USDA, Washington, D. C. (April 1956).

Dominy, P. F. and L. D. Worley, "Design and Construction Techniques for Parallel Terrace," *Transactions of the ASAE.* 9(5): 580–582 (1966).

Fangmeier, D. D. and M. K. Ramsey, "Intake Characteristics of Irrigation Furrows," *Transactions of the ASAE.* 21(4): 696–700, 705 (1978).

Fok, Y.-S., "Analysis of Overland Flow on a Porous Bed with Application to the Design of Surface Irrigation Systems," Dissertation presented to Utah State University, Logan, Utah in partial fulfillment of the Requirements for the Degree of Doctor of Philosophy. No. 64-13735 *University Microfilm Inc.* 313 N. 1st Street, Ann Arbor, MI. 48104, USA (1964).

Fok, Y.-S., "Infiltration Equation in Exponential Forms," *Jour. of the Irrigation and Drainage Div.* ASCE, Vol. 93, No. IR4, Proc. Paper No. 5686, Dec., 125–135 (1967).

Fok, Y.-S., Discussion of "Surface Irrigation Hydraulics-Kinematics," *Jour. of the Irrigation and Drainage Div.* ASCE, Vol. 95, No. IR4, Proc. Paper 6930, Dec., 623–624 (1969).

Fok, Y.-S., "A Study of Two-Dimensional Infiltration," *Transactions of the ASAE,* Vol. 13, No. 5:676–677, 681, (Sept.-Oct. 1970).

Fok, Y.-S. and A. A. Bishop, "Analysis of Water Advance in Surface Irrigation," *Jour. of the Irrigation and Drainage Div.* ASCE. Vol. 91 No. IR1, Proc. Paper 4259, March. 99–116 (1965).

Fok, Y.-S. and A. A. Bishop, "Expressing Irrigation Efficiency in Terms of Application Time, Intake and Water Advance Constants," *Transactions of the ASAE,* Vol. 12, No. 4:438–442 (1969).

Fok, Y.-S. and S.-H. Chiang, "Three Dimensional Infiltration Equations," Abstract in *Advances in Infiltration,* American Society of Agricultural Engineers, Chicago, Illinois, 12–13 December. p. 375 (1983).

Fok, Y.-S. and S.-H. Chiang, "2-D Infiltration Equations for Furrow Irrigation," *J. Irrig. Drain. Eng., Am. Soc. Civ. Eng.* 110(2): 208–217 (1984).

Fok, Y.-S. and V. E. Hansen, "One-Dimensional Infiltration into Homogeneous Soil," *Jour. of the Irrigation and Drainage Div.* ASCE. Vol. 92, No. IR3, Proc. Paper 4912, Sept. 35–47 (1966).

Fok, Y.-S., S.-O. Chung, and C. C. K. Liu, "Two-Dimensional Exponential Infiltration Equations," *Jour. of the Irrigation and Drainage Div.* ASCE. Vol. 108, No. IR4, Proc. Paper 17565, Dec. 231–241 (1982).

Fry, J. W., H. G. Buchta and D. R. Vallicott, "Efficiencies and Benefits of Contour Bench Leveling for Irrigation," *Transactions of the ASAE.* 9(5): 574–575 (1966).

Green, W. H. and G. A. Ampt, "Studies on Soil Physics. Part I—The Flow of Air and Water Through Soils," *Jour. of Agricultural Science,* Vol. 4: 1–24 (1911).

Hamad, S. N. and G. E. Stringham, "Maximum Nonerosive Furrow Irrigation Stream Size," *Jour. of the Irrigation and Drainage Div.* ASCE. Vol. 104, No. IR3, Proc. Paper 14021, Sept. 275–281 (1978).

Hansen, V. E., "The Importance of Hydraulics of Surface Irrigation," *Jour. of the Irrigation and Drainage Div.* ASCE. Vol. 84, No. IR3, Proc. Paper 1788, (Sept. 1958).

Hansen, V. E., "New Concepts in Irrigation Efficiency," *Transactions of the ASAE.* 3(1): 55–57, 61, 64 (1960).

Haise, H. R., W. W. Donnan, J. T. Phelan, L. F. Lawhon, and D. G. Shockley, "The Use of Cylinder Infiltrometers to Determine the Intake Characteristics of Irrigated Soils," ARS 41-7, Agri. Res. Ser. and Soil Cons. Ser. U. S. Dept. of Agriculture, Washington D.C. (May 1956).

Harris, W., "Trend and Construction Techniques for Land Grading in Arkansas," *Transactions of the ASAE.* 9(5): 578–579, 582 (1966).

Hart, W. E., D. L. Bassett, and T. Strelkoff, "Surface Irrigation Hydraulics-Kinematics," *Jour. of the Irrigation and Drainage Div.* ASCE. Vol. 94, No. IR4, Proc. Paper 6284, Dec. 419–440 (1968).

Holzapfel, E. A., M. A. Marino, and J. Chavez-Morales, "Border Irrigation Model Selection," *Transactions of the ASAE.* 27(6): 1811–1816 (1984).

Houston, C. E., "Trends and Costs of Land Grading for Irrigation in California," *Transactions of the ASAE.* 9(5): 565–567, 570 (1966).

Israelsen, O. W., *Irrigation Principles and Practices* 2nd Edition, John Wiley and Sons, New York. pp. 18–19, 68–69, and 218–219 (1950).

Jacobson, P., "New Developments in Land-Terrace Systems," *Transactions of the ASAE.* 9(5): 576–577 (1966).

Jensen, M. E., "Evaluating Irrigation Efficiency," *Jour. of the Irrigation and Drainage Div.* ASCE. Vol. 93, No. IR1, Proc. Paper 5145, March, 83–98 (1967).

Kostiakov, A. N., "On the Dynamics of the Coefficients of Water Percolation in Soils," *Trans. 6th Commission, International Soc. Soil Sci.* Russian Part A: 15–21 (1932).

Lewis, M. R., "The Rate of Infiltration of Water in Irrigation-Practice," *Transactions Amer. Geophys. Union,* 18: 361–368 (1937).

Mickelson, R. H. "Level Pan Construction for Diverting and Spreading Runoff," *Transactions of the ASAE.* 9(5): 568–570 (1966).

Musick, J. T., Sletten, W. H. and Dusek, D. A., "Evaluation of Graded Furrow Irrigation with Length of Run on a Clay Loam Soil," *Transactions of the ASAE.* 15(6): 1075–1080, 1084 (1983).

Philip, J. R., "The Theory of Infiltration: 4.Sorptivity and Algebraic Infiltration Equations," *Soil Sci.* 84: 257–264 (1957).

Philip, J. R., "Absorption and Infiltration in Two- and Three-Dimensional Systems," *Proc. UNESCO Netherlands Symp. Water Unsaturated Zone,* Wageningen, The Netherlands. 1: 503–516 (1966).

Singh, P. and Chauhan, H. S., "Shape Factors in Irrigation Water Advance Equation," *Jour. of the Irrigation and Drainage Div.* ASCE. Vol. 98, No. IR3, Proc. Paper 9212, Sept. 443–458 (1972).

Stringham, G. E. and J. Keller, "Surge Flow for Automatic Irrigation," *Proc. Specialty Conference, Irrig. Drain. Div., Am. Soc. Civ. Eng.,* Albuquerque, N. M., pp. 132–142 (1979).

Su, H. H., "Hydraulics of Unsteady, Open Channel Flow over a Porous Bed." M. S. Thesis, Utah State University, Logan, Utah (1961).

Tinney, E. R. and D. L. Bassett, "Terminal Shape of a Shallow Liquid Front," *Jour. of the Irrigation and Drainage Div.* ASCE. Vol. 87, No. HY5. Proc. Paper 2934, Sept. 117–133 (1971).

Toksoz, S., D. Kirkham, and E. R. Baumann, "Two-Dimensional Infiltration and Wetting Fronts," *Jour. of the Irrigation and Drainage Div.* ASCE. Vol. 91, No. IR3, Proc. Paper 4477, Sept. 65–79 (1965).

Walker, W. R. and A. S. Humpherys, "Kinematic-Wave Furrow Irrigation Model," *Jour. of Irrigation and Drainage Eng.* ASCE. Vol. 109. No. 4. Dec. Proc. Paper 18460. 377–392 (1983).

Walker, W. R. and L. S. Willardson, "Infiltration Measurements for Simulating Furrow Irrigation," *Advances in Infiltration.* Amer. Soc. Agri. Engineers. Chicago, Illinois. 241–248 (1983).

Wu, I.-P., "Recession Flow in Surface Irrigation," *Jour. of the Irrigation and Drainage Div.* ASCE. Vol. 98. No. IR1, Proc. Paper 8764, March, 77–90 (1972).

CHAPTER 18

Border Irrigation: Dimensionless Runoff Curves and Reuse System Design

Delmar D. Fangmeier* **and Muluneh Yitayew***

INTRODUCTION

This chapter contains dimensionless runoff curves for irrigation borders and presents optional design procedures for reuse systems. The curves are based on runoff volume defined as a percentage of the total water applied at the head of the border.

In border irrigation, runoff at the end of the field is a major cause of inefficiency. Water can be saved by installing a properly designed reuse or runoff recovery system at a small additional cost. Such a system can improve application efficiencies from the present average of 55 percent to over 70 percent with improved uniformity. Energy savings may also be attained by controlling runoff [28].

The most important variable in the design of reuse systems is the quantity of runoff. Thus the design engineer must be able to predict the runoff quantity from a given system, preferably without actually measuring it. Several investigators measured runoff directly from the field and reported figures ranging from 10 percent to 35 percent runoff [10,14,20]. Few concentrated their effort on developing their own equations to predict runoff using the soil and hydraulic characteristics of the irrigation system [2,11,12,19,22,23].

Field measurements of runoff would be very difficult and expensive because of the large number of variables involved. However, the necessary data can be obtained through the use of mathematical models. A zero-inertia mathematical model developed by Strelkoff and Katopodes [18] was used to identify pertinent open channel hydraulic variables affecting runoff characteristics in sloping free outflowing borders and to develop predictive dimensionless runoff curves for calculating runoff from border irrigation

[18,25,26]. A design procedure was also developed based on the dimensionless runoff curves [25,27].

EQUATIONS OF BORDER IRRIGATION FLOW

Accurate determination of runoff quantity is part of the general problem of hydraulics of border irrigation. The basic hydrodynamic equations describing the flow of water in border irrigation are the two partial differential equations of first order usually called the De Saint-Venant equations [6,15,24]. These equations are basically continuity and momentum equations in a general form. The continuity equation applied to a border strip of unit width is given as:

$$\frac{\partial q}{\partial x} + \frac{\partial y}{\partial x} + I_x = 0 \qquad (1)$$

and the momentum equation:

$$\frac{V}{g}\frac{\partial v}{\partial x} + \frac{1}{g}\frac{\partial v}{\partial t} + \frac{\partial y}{\partial x} = S_o - S_f + \frac{V}{gA}I_v \qquad (2)$$

in which x and t = distance and time, respectively, q = flow rate, A = cross-sectional area of flow, y = flow depth, g = gravitational acceleration constant, v = velocity of individual particles, V = average velocity ($V = Q/A$), S_o = bottom bed slope, S_f = friction slope, I_x is infiltration rate ($I_x = I_v/B$) in which B is the width of the channel and I_v is the volumetric rate of infiltration.

The last term of the right side of Equation (2) is the result of net acceleration stemming from removal of zero velocity components of the surface stream at the bed of infiltration. If the Kostiakov-Lewis power law function is used, the cumulative infiltration is given by:

$$z = k\,t^a \qquad (3)$$

---

*Agricultural Engineering Department, University of Arizona, Tucson, AZ

where k and a are constants, and t is the opportunity time, i.e., the time the water has been in contact with the soil. The rate I_x is evaluated by taking the derivative of Equation (3) with respect to time:

$$I_x = \frac{\partial z}{\partial t} = a\,k\,t^{a-1} \qquad (4)$$

The Manning resistance formula, which is widely used in open channel hydraulics, can be used for S_f in Equation (2). Thus S_f is given by:

$$S_f = \frac{q^2\,n^2}{C_u^2\,y^{10/3}} \qquad (5)$$

where q = the flow rate per unit width, n = the Manning roughness coefficient, y = the depth of flow and C_u = a constant dependent on the units used.

Equations (1) and (2), which are quasilinear hyperbolic partial differential equations of first order, don't permit a closed form analytical solution because of their nonlinearity and nonhomogenity unless many simplifications are introduced. In irrigation channels the diversity in slope, roughness, boundary conditions, and complexity of lateral flow further complicates the analytical expressions that approximate these conditions to an extent that it is practically impossible to integrate the equations.

However, advances in numerical methods and increased capability of computers have made solutions of Equations (1) and (2) possible. Several works [1,7,24] have used this capability to solve both equations in their entirety using mathematical models. When all the terms in the equations are included, the models are complex and subject to limitations of computational instability.

Strelkoff and Katopodes [17] developed a mathematical model known hereafter as a zero-inertia model by assuming the water velocities in border irrigation are generally low and the acceleration terms in Equation (2) can be neglected. They demonstrated that for Froude numbers at normal depth and discharge below 0.3, the forces on the surface streams are essentially balanced. Accordingly, Equation (2) takes the form:

$$\frac{dy}{dx} = S_o - S_f \qquad (6)$$

This assumption leads to a pair of parabolic equations which makes the solution computationally less complicated and faster than hyperbolic equations. Solutions of the model have shown good agreement with field data [3].

The finite difference form of Equations (1) and (6) in dimensionless form can be written as [16]

$$V^*\,(\bar{q}_{rD}^* - \bar{q}_{rU}^*)\delta t^* = [(\tilde{y}^* + K^*\tilde{z}^*)\delta x^*]_{t^*} \\ - [(\tilde{y} + K^*\tilde{z}^*)\delta x^*]_{t^*+\delta t^*} \qquad (7)$$

and

$$y_U^* - y_D^* = (S_o^* - S_f^*)\,\delta x^* \qquad (8)$$

subject to conditions

$$q_o^* = q_{in}^* \qquad\qquad 0 < t^* < t_{co}^* \qquad (9)$$

$$q_o^* = 0 \qquad\qquad t^* > t_{co}^* \qquad (10)$$

$$q^* = 0 = y^* \qquad\qquad x^* = x_A^* \qquad (11)$$

in which

$$V^* = \frac{QT}{XY} \qquad (12)$$

$$K^* = Z/Y = \frac{kT^u}{Y} \qquad (13)$$

$$S_o^* = S_o X/Y \qquad (14)$$

$$q_{in}^* = q_{in}/Q \qquad (15)$$

$$S_f^* = S_f X/Y \qquad (16)$$

$$t_{co}^* = t_{co}/T \qquad (17)$$

$$L^* = L/X \qquad (18)$$

$$x^* = x/X \qquad (19)$$

$$z^* = z/Z \qquad (20)$$

and t_{co} is the time of cutoff of inflow. The reference variables, which are denoted by upper case letters, will be defined later.

The friction slope is computed using the reference depth and discharge in the Manning formula as:

$$S_f = \frac{Q^2 n^2}{C_u^2 Y^{10/3}} \qquad (21)$$

Simplification of Equations (7) and (8) is possible if the reference variables are properly selected. Katopodes and Strelkoff [7] after choosing first the characteristic discharge equal to the inflow discharge per unit width at the upstream end of the field, defined a characteristic depth equal to the normal depth for the characteristic discharge using the Manning equation:

$$Y = Y_n = \left(\frac{q_{in}n}{C_u\sqrt{S_o}}\right)^{3/5} \qquad (22)$$

Then the characteristic values T and X are related through the normal velocity $V_n = q/Y_n$, and by $X = V_nT$. As the result of this selection the dimensionless parameters in the governing equations are found to be K^*, a, and S_o^*.

Strelkoff and Clemmens [16] working with the same governing equations showed for sloping borders that if Y was set equal to normal depth and S_o^*, S_f^*, $\mathcal{V}^*$, and Q_{in}^* were all set to unity and the T is defined such that $T = Y_n/V_nS_o$ and correspondingly $K^* = kT^a/Y$ and $X = Y/S_o$, the dimensionless solutions would be governed by K^*, a, and t_{co}^*.

As one can observe, runoff from a given irrigation is a function of many variables expressed mathematically as:

$$R = f(q_{in}, S_o, n, t_{co}, k, a, L) \qquad (23)$$

where R = runoff expressed as percentage of the total applied water at the head of the border.

From dimensional analysis principles five dimensionless terms are needed. The choices are K^*, a, L^*, q_{in}^*, $|\mathbf{F}$, S_o^*, t_{co}^*, $\mathcal{V}^*$, and S_f^*. Setting the Froude number $|\mathbf{F} = 0$ by assumption of zero inertia, only four need to be selected from K^*, L^*, Q_{in}^*, S_o^*, t_{co}^*, $\mathcal{V}^*$, and S_f^*; the rest are arbitrarily set to constants. For this study q_{in}^*, $\mathcal{V}^*$, S_f^* were set to unity. This leaves K^*, L^*, S_o^*, t_{co}^* as possible choices.

Clemmens and Strelkoff [4] indicated that within the important ranges of Kostiakov's constant a, the ranges of other physical parameter values were found to be $2 < q_{in} < 25$ $(l/s - m)$; $0.02 < n < 0.25 \mathrm{m}^{1/6}$; $1 < K < 12 (\mathrm{cm/hr}^a)$; $0.0001 < S_o < 0.01$; and $100 < L < 500 (\mathrm{m})$. If the limit on t_{co} is $0.5 < t_{co} < 5 (\mathrm{hr})$ then the practical range for K^* is $0.1 < K^* < 10$ and for t_{co}^* is $0.1 < t_{co}^* < 100$. This was

TABLE 1. Adjustment of k for Different a Values with Same Depth of Infiltration and Same Time. Required Depth of Infiltration, z_r = 100 mm; Length of Border, L = 198 m; Manning Roughness, n = 0.15; Border Slope = 0.001; Assumed Efficiency, E_a = 70%.

Intake Family	k mm/hr^a	a	Runoff %
	68.99	.400	21.1
	62.56	.500	20.6
1.0	56.90	.600	20.0
	51.20	.706	20.2
	46.84	.800	18.8
	90.55	.400	13.6
	87.44	.500	13.8
2.0	85.50	.600	12.7
	82.40	.728	12.8
	80.72	.800	12.6

based on the assumption that all extremes will not occur at the same time.

This choice of dimensionless variables allows setting Y to normal depth, S_o^*, S_f^*, $\mathcal{V}^*$ and q_{in}^* to unity with K^*, t_{co}^*, a, and L^* as free parameters. The physical variables of the border irrigation are then related as follows:

$$Q = q_{in} \qquad (24)$$

$$Y = y_n = \left[\frac{q_{in}n}{C_u\sqrt{S_o}} \right]^{3/5} \qquad (25)$$

$$X = Y/S_o \qquad (26)$$

$$T = YX/Q \qquad (27)$$

$$K^* = kT^a/Y \qquad (28)$$

$$L^* = L/X \qquad (29)$$

$$t_{co}^* = t_{co}/T \qquad (30)$$

where the variables have been defined earlier.

Considering possible values in actual field conditions of the physical variables, the following ranges of the dimensionless variables used as an input were considered; $0.1 < K^* < 10$, $0.1 < t_{co}^* < 100$, and $0.1 < L^* < 100$ for a single value of a. The SCS [21] description of the intake families defines the values of a approximately between 0.6 and 0.8 with an average value of 0.7. This value was used to develop the dimensionless runoff curves such that adjustment for any other value can be made by simultaneously changing k to give the same required infiltration depth in the same time. Table 1 shows how an a value different from 0.7 can be adjusted using the required depth of infiltration and the intake opportunity time. As an example, a soil with an a value of 0.706 has been changed to different a values by changing the k value as shown in Table 1. There was no significant difference in the runoff percentage as the result of the adjustment. This is a great savings both in computer time and money, in that the result of the study made for an a value of 0.7 can be used for other values of a provided the same required infiltration depth is obtained in the same time.

Dimensionless Runoff Curves

Using the four governing free parameters, i.e., a, K^*, t_{co}^*, L^*, a generalized set of runoff curves was developed [26] for free outflowing borders. The only requirement for using these parameters is that the relationships established between the dimensional and dimensionless variables with the reference variables be maintained using Equations (24–30).

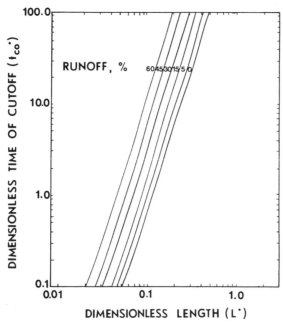

FIGURE 1. Runoff percentage (a = 0.7, K* = 10.0).

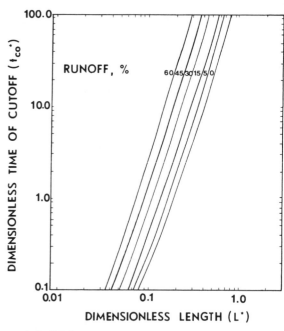

FIGURE 3. Runoff percentage (a = 0.7, K* = 6.0).

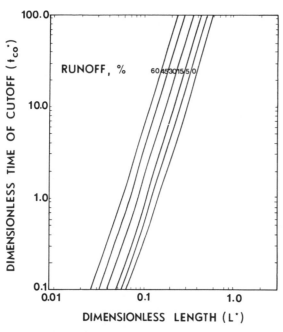

FIGURE 2. Runoff percentage (a = 0.7, K* = 8.0).

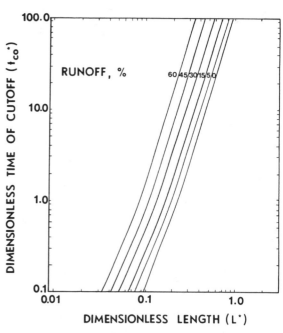

FIGURE 4. Runoff percentage (a = 0.7, K* = 5.0).

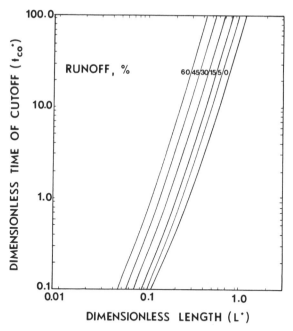

FIGURE 5. Runoff percentage (a = 0.7, K* = 4.0).

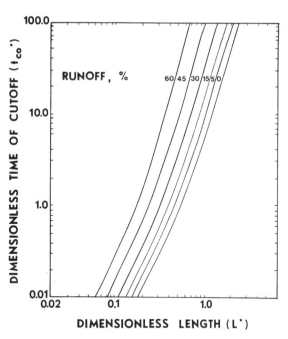

FIGURE 7. Runoff percentage (a = 0.7, K* = 2.0).

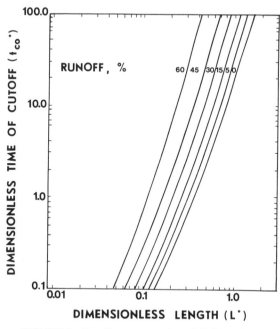

FIGURE 6. Runoff percentage (a = 0.7, K* = 3.0).

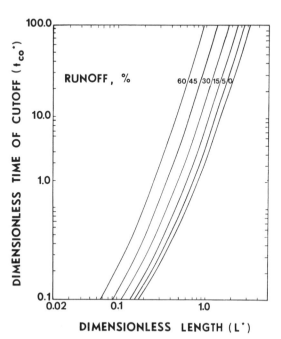

FIGURE 8. Runoff percentage (a = 0.7, K* = 1.50).

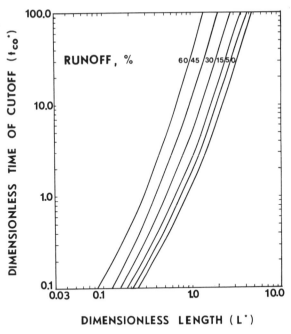

FIGURE 9. Runoff percentage (a = 0.7, K* = 1.00).

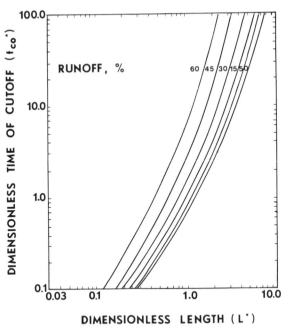

FIGURE 11. Runoff percentage (a = 0.7, K* = 0.60).

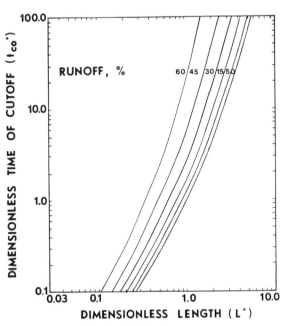

FIGURE 10. Runoff percentage (a = 0.7, K* = 0.80).

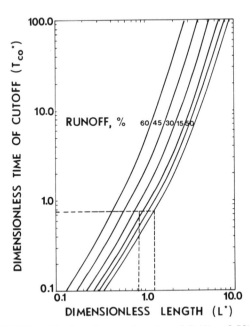

FIGURE 12. Runoff percentage (a = 0.7, K* = 0.50).

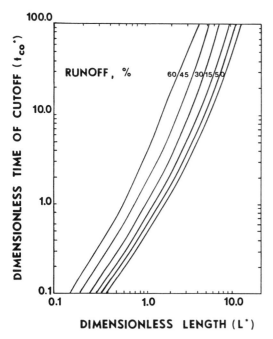

FIGURE 13. Runoff percentage (a = 0.7, K* = 0.40).

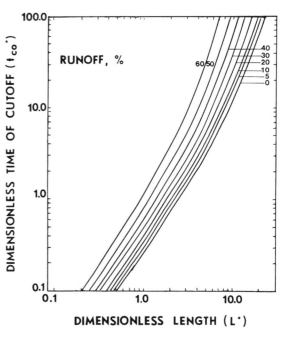

FIGURE 15. Runoff percentage (a = 0.7, K* = 0.20).

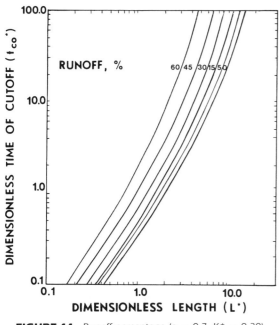

FIGURE 14. Runoff percentage (a = 0.7, K* = 0.30).

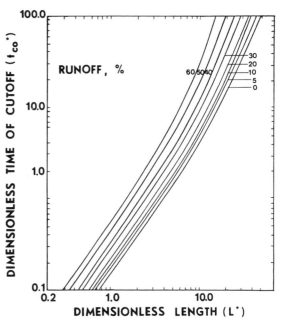

FIGURE 16. Runoff percentage (a = 0.7, K* = 0.10).

TABLE 2. Example of the Use of Dimensionless Runoff Curves in Sloping Borders.

Given: q_{in} = 2.995 l/s – m; L = 91.4 m; n = 0.25; S_o = 0.001; a = 0.7; k = 5.12 cm/hr³; t_{co} = 47 min;

Required:
  Runoff percentage
  Maximum advance distance

Solution:
  Reference and dimensionless variables are:

$$Y = \left(\frac{q_{in}n}{C_u\sqrt{S_0}} \right)^{3/5} = \left(\frac{0.002995 \times 0.25}{0.001} \right)^{3/5} = 0.1058 \text{m}$$

$$X = Y/S_o = \frac{0.1058}{0.001} = 105.8 \text{ m}$$

$$T = \frac{XY}{q_{in}} = \frac{105.8 \times 0.1058}{0.002995} = 3739.5 \text{ sec} = 62.32 \text{ min} = 1.04 \text{ hrs}$$

$$Z = kT^a \rightarrow K^* = \frac{kT^a}{Y} = \frac{(0.0512)(1.04)^{0.7}}{0.1058} = 0.497 = 0.5$$

$$t_{co}^* = \frac{t_{co}}{T} = \frac{47}{62.32} = 0.754$$

$$L^* = \frac{L}{X} = \frac{91.4}{105.8} = 0.864$$

Using K^* = 0.5, L^* = 0.864 and t_{co}^* = 0.754 from Figure 12, R = 22%. This matches the computer results for the same t_{co}^*, L^*, K^* and a values. The maximum advance distance is obtained for the same t_{co} value using the zero runoff line and indicates L^* = 1.24 from which L is computed to be 131 meters (i.e., 1.24 × 105.8).

The procedure followed in developing the curves is as follows. The runoff percentage R was first obtained by making several hundred computer runs for a single a value, different K^* values ranging from 0.1 to 10.0 and both L^* and t_{co}^* varying from 0.1 to 100. The percentage values obtained are then plotted against L^* for the different values of t_{co}^* and one value of K^* and a. This plot was then transformed graphically to a plot of t_{co}^* versus L^* by connecting points of equal runoff percentage but still for the same K^* and a values. Graphical transformation has the advantage over transformation through the use of computer runs in that the latter method uses a trial and error approach which requires many computer runs to get the right percentage values.

Only one value of a(= 0.7) was used in this study. Figures 1–16 are the dimensionless runoff curves developed for the range of values discussed earlier for K^*, L^*, t_{co}^*. An example, to show how to use the dimensionless curves to get the runoff percentage for a given hypothetical set of data, is given in Table 2.

Factors Affecting Runoff

As indicated by Equation (23), runoff from a sloping border is a function of inflow rate (q), time of cutoff (t_{co}), length of run (L), hydraulic drag or surface resistance expressed by the Manning roughness factor (n), border slope (S_o), and soil infiltration characteristics as represented by k and a in the Kostiakov-Lewis power law function. Runoff in general increases with increased q, t_{co}, and S_o and decreases with increased L, k, a, and n. Several hundred computer simulations were made to study the extent to which these variables affect runoff. The SCS curves [21] were used to determine the values of the various parameters needed for input to the model for the examples which follow.

Figure 17 shows an example of the effect of slope on runoff. There is an increase in runoff, as expected, with slope. It is interesting to note that once a certain steepness is reached, further slope increase does not seem to increase runoff. Also for low intake families, the change in runoff per unit change in slope is smaller than for higher intake families at high runoff percentage.

An example of the effect of roughness is shown in Figure 18 in which, as could be foreseen, runoff decreases with an increase in roughness. Again, the rate of decrease in runoff per unit increase in roughness is gradual for low intake families at high runoff percentage. This means that the values of both roughness and slope have to be determined carefully when they are to be used as independent variables with soils having a high intake rate.

The infiltration constants are the most difficult input parameters to determine in the field, but are very crucial in the design of any irrigation system. Model runs were made to observe the effects of an incorrect choice of intake family for a given soil by selecting the correct inputs for a given set of conditions. Then only the input values of k and a were changed to the next higher or lower intake family which simulates the effect of selecting the wrong intake family. The results in Table 3 indicate that if the actual intake family is higher than the family selected, the runoff is substantially decreased, while if the actual intake family is lower, runoff is substantially increased. Also of interest is the change in E_a (application efficiency). While the changes in runoff are relatively large, the application efficiency changes are relatively small. Therefore, whatever is gained by reducing the runoff is lost to deep percolation. This is also reflected by the change in uniformity. Low uniformity values associated with low runoff in all cases indicate that allowing higher runoff will provide an increased uniformity; the runoff water can then be reused to increase efficiency.

No significant change in runoff was observed when the slopes of the infiltration functions were changed while keeping the same cumulative depth of infiltration. That is, changing k and a in $z = kt^a$ for the same z and t values. This result indicates that the time to infiltrate a given depth, not the shape of the infiltration function, is more critical in determining the runoff volume for a given soil. This is demonstrated by Table 1 earlier.

Length of Run Consideration

MAXIMUM ADVANCE DISTANCE

An important design aid was obtained with respect to length of run of border irrigation and maximum dimensionless advance distance. This distance is defined as the distance from the upstream end of a given border to the location at which the water front ceases to advance. This distance increases with increase in flow rate, application time and bed slope, and decreases with increase in infiltration rate, and bed and vegetative drag. In dimensionless terms it can be viewed as a function of a, K^*, and t_{co}^*.

The importance of this distance is that for design purposes the length L^* should be less than the maximum advance distance x_{max}^*. Otherwise the stream will halt before reaching

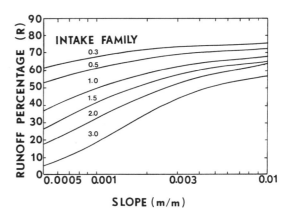

FIGURE 17. Example of effect of slope on runoff. Required depth of infiltration $z_r = 100$ mm; length of border = 198 m; Manning roughness n = 0.15; assumed efficiency $E_a = 70$ percent.

the downstream end of the field. For x_{max}^* greater than L^* there will be runoff in free outlet borders.

At the maximum advance distance, no runoff will exist. Thus, the zero runoff curve represents the maximum advance distance for a given dimensionless time of cutoff and the graphs from Figures 1 to 16 can be used to get either the runoff percentage R or the maximum advance length. To obtain the maximum advance distance, the user needs to extend the horizontal line for a given t_{co}^* to the zero runoff curve. The dimensionless distance that corresponds to the

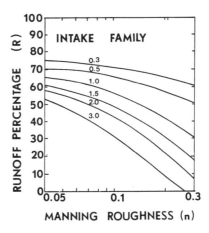

FIGURE 18. Example of effect of roughness on runoff. Required depth of infiltration $z_r = 100$ mm; length of border = 198 m; slope $S_o = 0.001$; assumed efficiency $E_a = 70$ percent.

TABLE 3. Effect of Selecting the Incorrect Intake Family. Required Depth of Infiltration z, = 100 mm; Length of Border, L = 198 m; Manning Roughness, n = 0.15; Slope, S₀ = 0.001; Assumed Efficiency E₀ = 70 Percent.

Intake Family	k mm/hr[a]	a	Runoff %	Uniformity Christiansens	Storage Eff., %	Application Eff., %
0.3	12.55	.595	61.7	.94	53.7	38.3
	22.66	.650	28.1	.87	98.6	69.7
	31.80	.684	3.4	.51	98.5	69.0
0.5	22.66	.650	46.8	.95	74.9	53.2
	31.80	.684	26.7	.85	99.1	69.6
	51.21	.706	0.0	.05	94.6	66.2
1.0	31.80	.684	46.9	.91	75.0	53.0
	51.21	.706	22.2	.83	99.5	70.6
	67.26	.718	7.2	.64	99.5	69.7
1.5	51.21	.706	33.7	.90	94.2	66.3
	67.26	.718	19.1	.80	99.8	70.7
	82.40	.726	9.2	.67	99.7	69.8
2.0	67.26	.718	26.0	.86	99.5	70.5
	82.40	.726	15.6	.76	99.8	71.1
	109.70	.735	2.5	.52	98.8	69.9
3.0	82.40	.726	27.4	.88	99.3	69.9
	109.70	.735	14.9	.80	100.0	70.4
	134.60	.740	6.2	.67	99.9	70.3

point on the zero runoff curve describes the maximum distance. The method was compared with the results of Shatanawi [13] and Clemmens and Strelkoff [4] and gives exactly the same results. It has the advantage of presenting both the runoff and the maximum distance in one set of curves rather than having two separate sets.

REUSE SYSTEMS

The primary purpose of a reuse system is to collect irrigation runoff water and control it for further use. The system requires a sump or reservoir to collect and store the water, pump facilities for pumping it back and a pipeline or canal for conveying water back to the irrigation system.

Reuse systems may be classified as reservoir or cycling sump systems depending on whether or not they accumulate and store runoff. They also are classified according to the method of handling runoff as return flow systems or sequence systems. The return flow system delivers runoff to a field at a higher elevation than the collecting point while the sequence system delivers water to a field at a lower elevation than the collection point.

System Design

Design of reuse systems involves the determination of various components and depends on the mode of operation

selected. Procedures to develop equations that go with different operational requirements for border irrigation systems are in the following:

1. Continuous Pumping with Variable Pumping Rate—The law of conservation of mass applied to runoff continuously pumped from borders having the same slope, length, width, retardance characteristics, and soil infiltration characteristics to other borders of the same characteristics yields a design equation for sizing sumps as:

$$\mathcal{V} = q_{in} t_{co} R + (N - 1)(\bar{Q}_R T_R - Q_p T_p) \qquad (31)$$

in which $\mathcal{V}$ = the volume of sump; q_{in} = constant inflow rate to the border; t_{co} = the cutoff time; R = the runoff percentage defined as the ratio of the amount of runoff to the amount of water applied to the border; $\bar{Q}_R$ = the average runoff rate, i.e., $\bar{Q}_R = q_{in}t_{co}R/T_R$, T_R = time of runoff from start to the end of runoff, Q_p = the rate of pumping from the sump; T_p = time of pumping from start of irrigation to time runoff ceases; and N = number of borders irrigated. First, if N equals one, the sump volume is equal to $Q_{in}t_{co}R$, which is the total runoff from one border. This runoff is not pumped back but stored for later use. Second, the minimum sump size occurs if $\bar{Q}_R T_R$ and $Q_p T_p$ are equal. Only a buffer storage is needed if the total runoff for that particular irrigation is equal to the volume reused. This system provides the simplest mode of operation.

The rate Q_p can be estimated by:

$$Q_p = \left(\frac{1}{N-1} - p \right) q_{in} \frac{t_{co}}{T_R} \qquad (32)$$

in which p represents the runoff fraction lost in the reuse system. Equation (32) is developed by assuming runoff from $N-1$ borders (if accumulated in a reservoir) would be enough to irrigate one border with the required depth in a time of pumping equal to the time of runoff. This can be adjusted by changing time from T_R to another time t depending on the need of the individual operator.

2. Steady Continuous Pumping—The size of reservoir for a reuse system with runoff pumped continuously at constant rate to another border for a time $T_p = Q_{in}t_{co}R/Qp$ can be determined by:

$$\Psi = q_{in}t_{co}R - \frac{q_{in}^2 R^2 t_{co}^2}{Q_p T_R} \qquad (33)$$

where the variables are as defined earlier.

If loss of runoff, p, is assumed due to the reuse system inefficiency, the volume of the sump will be given by:

$$\Psi = (1-p) \, q_{in}t_{co}R \left(1 - \frac{q_{in}t_{co}R}{Q_p T_R} \right) \qquad (34)$$

For N borders irrigated at the same time, the right side of Equations (33) or (34) should be multiplied by N to determine total volume.

3. Cycling Sump System—Larson [8] related inflow, pump capacity, sump volume and cycle time on a pumped drainage system:

$$C\Psi = 60 \, Q_R/Q_p \, (Q_p - Q_R) \qquad (35)$$

where Ψ is sump storage, C = number of cycles per hour, Q_R = inflow rate to the sump, and Qp = the pumping rate from the sump. Differentiating Equation (35) with respect to Q_R gives a maximum storage when Q_R is $\frac{1}{2} Q_p$. The result of substituting in Equation (35) is:

$$\Psi = 15 \, Q_p/C \qquad (36)$$

Larson and Manbeck [9] reported that manufacturers of single-phase motors recommend a maximum of 15 cycles per hour without decreasing pump efficiency. When this is substituted into Equation (14) the result is:

$$\Psi = Q_p \qquad (37)$$

Functional analysis of the system tells us that minimum sump volume would be obtained if $Q_p = Q_R$ maximum. Using this condition and assuming 15 cycles per hour, the minimum sump size can be obtained. For other than 15 cycles per hour, Equation (36) should be used.

Knowing the depth of usable storage (H) in the sump, the inside diameter of a circular sump (D) in centimeters can be obtained by:

$$D = 27.6 \sqrt{Q_p/H} \qquad (38)$$

in which Q_p is in liters/sec and H in meters. Equation (38) generalizes a table of minimum sump diameters by Davis [5].

4. Complete Control of Storage—If the entire runoff from irrigated borders is to be contained without pumping from the sump, the volume of storage needed is given by:

$$\Psi = Nq_{in}t_{co}R \qquad (39)$$

This system offers flexibility of use of the runoff water. It can be used as a separate supply source with a pump of any size. Its main drawback is that it requires a large sump, which requires more land and increases evaporation loss.

TIME OF RUNOFF

In Equations (31) to (34) the time of runoff, T_R, defined as the time from the beginning to the end of runoff, is the base of the runoff hydrograph for any one irrigation. The results from the zero inertia model showed that the intake opportunity time at the end of the field and the time base of the runoff hydrograph were almost identical. Thus T_R can be approximated by the intake opportunity time at the end of the field without any serious effect on the design of the system. The intake opportunity time, t, is defined by the Kostiakov-Lewis function Equation (3).

Shatanawi's [13] mathematical expression for the ultimate profile of infiltrated depth for free outflowing borders was used to determine T_R as related to the infiltrated depth at the end of the border by the infiltration function. The ultimate infiltrated depth at the end of the field is given by the expression

$$z_u(L) = Z' \, \bar{Z}* \, Y \qquad (40)$$

where $\bar{Z}*$ is given by t_{co}^*/X_{max}^* and is the dimensionless average depth infiltrated over the maximum distance, which resulted from the volume balance

$$Z' = a_o \, (1 - L^*/X_{max}^*)^{a/2} + a_1 \, (1 - L^*/X_{max}^*) \qquad (41)$$

where Z' is the dimensionless depth infiltrated at the lower end of the field, and a_o and a_1 are constants related to the Kostiakov-Lewis constant a by

$$a_1 = \frac{4 + 2a - 4a_o}{a + 2} \qquad (42)$$

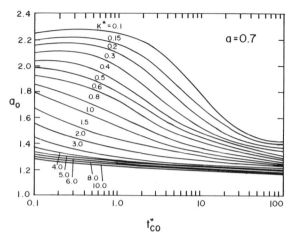

FIGURE 19. Shape factor variation (a = 0.7), from Shatanawi (1980).

The value of a_o to be used in conjunction with the runoff curves developed is given in Figure 19 as a function of the dimensionless border irrigation variables.

In Equation (41), L^* = the dimensionless field length and X^*_{max} is the dimensionless maximum advance distance defined by the zero runoff curves from Figures 1–16. Knowing z at the end of the field, the time of runoff is given by

$$T_R = \left(\frac{z_u(L)}{k}\right)^{1/a} \quad (43)$$

DESIGN DATA

In reuse systems, engineers are concerned with the sump, pump capacity, and conveyance mechanism based on border irrigation parameters. Other variables like the choice of the individual farmer or institutional factors affect the design. With due consideration to these factors only the border irrigation variables are discussed here.

The physical variables needed for the design of reuse systems are mainly the slope, S_o; length of run, L; type of crop cover; retardance coefficient, n; the system capacity which is described by the inflow rate, time of cutoff or application time for the given required depth of water to the border and the required level of application efficiency; and soil infiltration characteristics. These variables determine the percentage of runoff, R, through the dimensionless runoff curves, (Figures 1–16). Once the runoff, flow rate to the border, and time of application are defined, the equations developed in this section can be used to determine the sump volume for different pumping rates.

The pump size and pipeline system are determined from the pumping rate and the total pumping head. The total head for a reuse system can be calculated by

$$H = H_s + S_o L + h_f + h_m \quad (44)$$

in which H = total head (m); S_o = border slope (dimensionless); L = length of the border (m); H_s = suction lift (m); h_f = friction loss (m) and h_m = minor losses (m).

DESIGN PROCEDURE

The basic physical variables of the irrigation system must be known. These are the length, slope, type of crop to be grown, and the soil infiltration characteristics. The crop water requirement throughout the growing season can be calculated using available techniques and from it an irrigation schedule can be determined. The system capacity equation then can be applied in the form

$$qt_{co} = L \frac{z_r}{E_a} \quad (45)$$

or, in terms of the volume of water to be infiltrated, can be related by

$$q_{in}t_{co} = L\,d \quad (46)$$

z_r can be calculated from the Kostiakov-Lewis function; t can be approximated by the infiltration time t_r required at the upper end of the border to infiltrate z_r. Adjustment can be made to include the time lag at the upper end by using the Soil Conservation Service [21] experimental results. For most practical purposes as an initial estimate, t_{co} can be equated to t_r.

The design procedure is outlined as follows:

STEP ONE—Determine the crop water requirement as a function of time.

STEP TWO—Schedule the irrigation to satisfy the water requirement for given period of time and know the depth to be applied for each irrigation.

STEP THREE—Use the system capacity equation or design charts to determine the inflow rate.

STEP FOUR—Use crop, slope, and soil to determine n, K, and a.

STEP FIVE—Calculate the reference variables Y, X, T, and using these, calculate the dimensionless variables L^*, K^*, and t^*_{co}.

STEP SIX—Use dimensionless runoff curves to determine the runoff percentage for given t^*_{co}, K^*, L^*, and a values.

STEP SEVEN—Use the same figures as above for the given t^*_{co}, K^*, and a, to determine the X^*_{max} from the zero runoff line in the dimensionless curve.

STEP EIGHT—Use t^*_{co}, K^*, and a, to determine the shape function a_o from Figure 19.

STEP NINE—Calculate the infiltrated depth at the downstream end $z_u(L)$ using Equation (40).

STEP TEN – Determine the time of runoff by Equation (43).

STEP ELEVEN – Calculate the volume of sump required for the reuse system by Equations (31) to (34).

STEP TWELVE – (a) If a cycling sump is adaptable, use Equation (36) to determine the volume of sump required as a function of the pumping rate. (b) Determine the diameter of circular sump for different on and off float levels by Equation (38).

STEP THIRTEEN – If complete control of storage is required, calculate the sump volume using Equation (39).

STEP FOURTEEN – If continuous pumping is to be used, select a pump for the flow rate and total head in the system from Equation (44).

STEP FIFTEEN – Calculate the energy required for the system selected and use it for making management decisions.

Example: Design a reuse system for a field 198 m (650 ft) long with a slope $S_o = 0.001$. The cumulative infiltration $z = 5.12\ t^{0.7}$. The crop to be grown is alfalfa. The Manning roughness for fully grown alfalfa cover is estimated to be 0.15 $m^{1/6}$. The crop is to be grown in Tucson, Arizona. Border width of 12.2 m (40 ft) is an average for most of the farms in the area.

Solution:

STEP ONE – (a) The crop water requirement was calculated (see Table 4). (b) The moisture to be added for each irrigation was calculated based on 65% depletion (see Table 5). Assumed soil condition is given in Table 6.

STEP TWO – The maximum depth of water to be applied at any given period was 152 mm (6.1 inches). An estimate of the flow rate was obtained from the system capacity equation or design charts SCS [21]. The time of cutoff was calculated from (SCS) design charts as 260 minutes. Equation (45) was used with an entering efficiency of 65% to determine the flow rate q as 3.03 $l/s - m$ (0.033 cfs/ft).

STEP THREE – The reference variables Y, X, and T and the dimensionless variables L^*, K^*, and t_{co}^* were calculated with q_{in}, S_o, n, k, and a as follows:

$$Y = \left(\frac{q_{in}\, n}{C_u \sqrt{S_o}} \right)^{3/5} = \left(\frac{0.00303 \times 0.15}{\sqrt{0.001}} \right)^{3/5}$$

$$= 0.0784 \text{ m } (.257 \text{ ft})$$

$$X = Y/S_o = \frac{0.0784}{0.001} = 78.4 \text{ m } (257 \text{ ft})$$

$$T = \frac{XY}{q_{in}} = \frac{0.0784 \times 78.4}{0.00303} = 33.8 \text{ min } (2030 \text{ sec})$$

TABLE 4. Consumptive Use of Alfalfa for Tucson Area.

Month	Semi-monthly Consumptive Use (mm)	Cumulative Consumptive Use (mm)
February	46	46
	46	92
March	70	162
	82	244
April	90	334
	102	436
May	117	553
	136	689
June	136	825
	139	964
July	135	1099
	116	1215
August	113	1328
	127	1455
September	109	1564
	81	1645
October	63	1708
	47	1755
November	42	1797

TABLE 5. Assumed Soil Conditions and Available Moisture.

Soil Depth (m)	Soil .Texture	Available Moisture (mm)
0 – 0.3 (0 – 1 ft)	Loam	48.3
0.3 – 0.61 (1 – 2 ft)	Loam	48.3
0.61 – 0.91 (2 – 3 ft)	Fine Sandy Loam	40.6
0.91 – 1.22 (3 – 4 ft)	Fine Sandy Loam	40.6
1.22 – 1.52 (4 – 5 ft)	Fine Sandy Loam	40.6
1.52 – 1.83 (5 – 6 ft)	Loamy Sand	20.3
Total Available Moisture in the Entire Root Zone		239

$$K^* = \frac{kT^a}{Y} \frac{.0512 \left(\frac{33.8}{60}\right)^{0.7}}{0.0784} = 0.437$$

$$t_{co}^* = \frac{t_{co}}{T} = \frac{260}{33.84} = 7.68$$

$$L^* = \frac{L}{X} = \frac{198}{78.4} = 2.53$$

STEP FOUR—Using nondimensional curves (Figures 12 and 13), the runoff percentage was determined by interpolating for $K^* = 0.4$ and $K^* = 0.5$. An R value of 32% was obtained. Note $65 + 32 = 97$ gives 3% deep percolation loss.

STEP FIVE—Using the same Figures 12 and 13, the maximum dimensionless advance distance X_{max}^* was obtained by matching the zero runoff lines with the t_{co}^* value of 7.68 for both K^* values. An X_{max}^* value of 4.50 was obtained.

STEP SIX—The shape factor, a_o was found to be 1.51 from Figure 19 for $t_{co}^* = 7.68$, $K^* = 0.437$, and $a = 0.7$.

STEP SEVEN—The infiltrated depth at the downstream end (minimum for the given length) was determined as follows:

$$Z'(L) = a_o(\tau)^{a/2} + a_1 \tau$$

where

$$\tau = 1 - L'$$

$$L' = \frac{L^*}{X_{max}^*}$$

$$a_1 = \frac{4 + 2a - 4a_o}{2 + a}$$

Thus for $a_o = 1.51$

$$a_1 = \frac{4 + 2(0.7) - 4(1.51)}{2 + 0.7} = -0.237$$

$$L' = \frac{2.53}{4.5} = 0.562$$

$$\tau = 1 - 0.562 = 0.438$$

$$Z'(L) = 1.51 (.438)^{0.35} - 0.237 (.438) = 1.03$$

$$z_u(L) = \bar{Z}^* \times Z'(L) \times Y$$

where

$$\bar{Z}^* = \frac{t_{co}^*}{X_{max}^*} = \frac{7.68}{4.50} = 1.71$$

Therefore $z_u(L) = 1.71 \times 1.03 \times 0.0784 = 0.138$ m.

STEP EIGHT—The time of runoff was calculated by using Equation (43)

$$T_R = \left(\frac{z_u(L)}{k}\right)^{1/a}$$

$$= \left(\frac{0.138}{.0512}\right)^{1/0.7} = 4.11 \text{ hr.} = 247 \text{ min.}$$

STEP NINE—From known R, T_R, t_{co}, Q_{in}, ($Q_{in} = q_{in} \times B$), a relationship between sump volume, pumpage rate from the sump Q_p for different time of pumping T_p was developed using Equation (31).

Thus,

$$\bar{Q}_R = \frac{Q_{in}t_{co}R}{T_R} = \frac{0.00303 \times 12.2 \times 260 \times 0.32}{247}$$

$$= 0.0124 \text{ m}^3/\text{sec}$$

$$V = 0.0369 \times 260 \times 60 \times .32 + (N-1)(184 - Q_pT_p)$$

$$V = 184 + (N-1)(184 - Q_pT_p)$$

For two borders contributing to the sump

$$V = 184 + (184 - Q_pT_p)$$

Table 7 gives V as a function of Q_p and T_p. It should be noted that for pumpage rates of 0.05, 0.025, 0.017, and 0.012 m³/sec there was no additional storage required than the buffer storage, 184 m³, for time of pumping of 0.25 T_R, 0.5 T_R, 0.75 T_R, and T_R, respectively.

STEP TEN—If a cycling sump were to be used, Equation (36) would be applied with Equation (38) to obtain the results given in Table 8. Table 8 gives only the minimum value using 15 cycles per hour, but, as pointed out by Davis [5], there are other restrictions such as: (a) inside diameter at least five times the inside diameter of the pump column; (b) clearance between the sump floor and the strainer at least one-half the inside diameter of the pump column; (c) lowest water level able to provide submergence over the pump strainer; and (d) any local design requirements.

STEP ELEVEN—For complete storage of the total runoff, the reservoir size was calculated by: $V = 2(0.0369)$ $260 \times 60 \times .32 = 368$ m³.

STEP TWELVE—For the Q_p value given in Table 7, the

TABLE 6. Moisture to be Added per Irrigation.

Number of Irrigations	Active Root Depth (m)	Available Moisture (mm)	Moisture to be Added (mm)
1st	0 – 0.61 (0 – 2 ft)	96.5	62.7
2nd	0 – 0.91 (0 – 3 ft)	137.2	89.2
3rd	0 – 1.22 (0 – 4 ft)	177.8	115.6
4th	0 – 1.52 (0 – 5 ft)	218.4	142.0
5th	0 – 1.83 (0 – 6 ft)	233.7	151.9
6th – end	0 – 1.83 (0 – 6 ft)	233.7	151.9

TABLE 7. Volume of Sump for Different Pumping Rates and Times.

Pumping Rate (liters/sec)	Time of Pumping (T_p)			
	0.25 T_R	0.5 T_R	0.75 T_R	T_R
	Volume of Sump (V) (m^3)			
3.15	357	345	333	322
6.31	345	322	298	275
9.41	333	298	263	228
12.4	322	277	231	184
16.6	306	245	184	—
18.9	298	228	—	—
22.1	287	205	—	—
24.9	276	184	—	—
28.2	264	—	—	—
31.5	252	—	—	—
34.7	240	—	—	—
37.8	228	—	—	—
41.0	216	—	—	—
44.2	205	—	—	—
49.9	184	—	—	–

TABLE 8. Minimum Size of Sump for Reuse System (C = 15 Cycles per Hour).

Inflow to the Sump (liters/sec)	Depth in Storage Between On and Off Levels (m)					
	0.5	1.0	1.5	2.0	2.5	3.0
	Inside Diameter of Circular Sump (cm)					
1.0	39	28	23	20	17	16
2.5	62	44	36	31	28	25
5.0	87	62	50	44	39	36
10.0	124	87	71	62	55	50
15.0	151	107	87	76	68	62
20.0	175	124	101	87	78	71
25.0	195	138	113	98	87	80
30.0	214	151	124	107	96	87
35.0	231	164	134	116	103	94
40.0	247	175	143	124	111	101
45.0	262	185	151	131	117	107

next step is to calculate the total head required by Equation (44) as a function of the diameter by:

$$H = 2 + (0.001)(198) + (10.3)(10^{10})(0.011)^2(198)(12/4)^2/D^{16/3}$$

$$H = 2.20 + 3.79 \times 10^{11}/D^{16/3}$$

in which H = head in m and D = pipe diameter in mm.

STEP THIRTEEN—The energy required for the given discharge and head is then calculated using

$$\text{Energy } (E) = 9.8 \times 10^{-3} Q_p \cdot H \cdot T_R \qquad (47)$$

The energy required (assuming an 80% efficient pump) is 1.4 KWH with the diameter not making a significant difference.

ACKNOWLEDGMENT

Published as Paper No. 4105 of the Arizona Agricultural Experiment Station. Sponsored in part by the U.S. Water Conservation Laboratory, Phoenix, Arizona.

NOTATION

The following symbols are used in this paper:

a = unitless power in infiltration function
a_1 = constant
a_o = constant
A = cross-sectional area of flow
B = border width
C_u = units constant in Manning equation
d = depth of application
D = inside diameter of a circular sump
E_a = application efficiency
g = gravitational acceleration
H = head
I_x = infiltration rate (I_v/B)
I_v = volumetric rate of infiltration
k = constant in power law function
K^* = dimensionless relative infiltration
L = border length
L^* = dimensionless border length
n = Manning roughness coefficient
N = number of borders
p = percentage of runoff lost in the system
q = flow rate per unit width
q_o^* = dimensionless flow rate per unit width at the start of irrigation
q_{in}^* = dimensionless flow rate per unit width at upstream end of border
q_{rD}^* = dimensionless unit discharge crossing downstream boundary of flow element

q_{rU}^* = dimensionless unit discharge crossing upstream end of flow element
Q = characteristic unit discharge
Q_p = pumping rate from sump
R = runoff expressed as a percentage of water applied
S_o = field slope
S_f = friction slope
S_o^* = dimensionless relative field slope
S_f^* = dimensionless relative friction slope
t = time
t_{co} = cut off time
t_r = time to infiltrate z_r
T = characteristic time
T_n = time required to infiltrate a required depth
T_p = time of pumping rate from sump
T_R = time of runoff from start to end of runoff
t^* = dimensionless time
t_{co}^* = dimensionless time of cut off
V = sump volume
V^* = dimensionless relative volume
v = instantaneous velocity
V = average velocity of flow
V_n = normal velocity for characteristic discharge
x = distance
x_A^* = dimensionless advance distance
X = characteristic distance
x^* = dimensionless distance
y = depth of flow
Y = characteristic depth
Y_n = normal depth
Y^* = dimensionless depth average over distance
z = infiltrated depth
z_r = required depth of infiltration
z^* = dimensionless depth of infiltration
z_u = average ultimate infiltrated depth

REFERENCES

1. Bassett, D. L. and D. W. Fitzsimmons, "Simulating Overland Flow in Border Irrigation," *Transactions,* ASAE, Vol. 15, No. 5, pp. 992-995 (1976).

2. Bondurant, J. A., "Design of Recirculating Irrigation Systems," *Transactions,* ASAE, Vol. 12, No. 2, pp. 195-201 (1969).

3. Clemmens, A. J., "Verification of Zero-Inertia Model for Border Irrigation," *Transactions,* ASAE, Vol. 22, No. 6, pp. 1306-1309 (1979).

4. Clemmens, A. J. and T. Strelkoff, "Dimensionless Advance for Level Border Irrigation," *Journal of the Irrigation and Drainage Division,* ASCE, Vol. 105, No. IR3, pp. 259-273 (1979).

5. Davis, J. R., "Design of Irrigation Tailwater Systems," *Transactions,* ASAE, Vol. 7, No. 3, pp. 336-338 (1964).

6. Henderson, F. M., *Open Channel Flow,* McMillan, New York, NY (1966).

7. Katopodes, N. D. and T. Strelkoff, "Hydrodynamics of Border Irrigation—Computer Model," *Journal of the Irrigation and Drainage Division,* ASCE, Vol. 103, No. IR3, pp. 309–323 (1977).

8. Larson, C. L., "Planning Pump Drainage Outlets," *Agricultural Engineering,* Vol. 37, pp. 38–40 (1956).

9. Larson, C. L. and D. M. Manbeck, "Factors in Drainage Pumping Efficiency," *Agricultural Engineering,* Vol. 42, pp. 296–297, 305 (1961).

10. Marsh, A. W., F. M. Tileston, and J. W. Wolfe, "Improving Irrigation in Eastern Oregon," *Agricultural Experiment Station Bulletin No. 558,* Oregon State University, Corvallis, OR (1956).

11. Ohmes, F. E. and H. L. Manges, "Estimating Runoff From Furrow Irrigation," Presented at the June 1972, ASAE Summer Meeting, Paper 78-2008 (1972).

12. Pope, L. D. and A. D. Barefoot, "Reuse of Surface Runoff from Furrow Irrigation," *Transactions,* ASAE, Vol. 16, No. 6, pp. 1088–1091 (1973).

13. Shatanawi, M. R., "Analysis and Design of Irrigation in Sloping Borders," unpublished doctoral dissertation, University of California, Davis (1980).

14. Shockley, D. G., "Evaluating Furrow and Corrugation Irrigation," *Journal of the Irrigation and Drainage Division,* ASCE, Vol. 85, No. IR4, pp. 65–72 (1959).

15. Strelkoff, T., "One-Dimensional Equations of Open Channel Flow," *Journal of the Hydraulics Division,* ASCE, Vol. 130, No. HY3, pp. 357–378 (1969).

16. Strelkoff, T. and A. J. Clemmens, "Dimensionless Stream Advance in Sloping Borders," *Journal of the Irrigation and Drainage Division,* ASCE, Vol. 107, No. IR4, pp. 361–381 (1981).

17. Strelkoff, T. and N. D. Katopodes, "Border Irrigation Hydraulics with Zero-Inertia," *Journal of the Irrigation and Drainage Division,* ASCE, Vol. 103, No. IR3, pp. 325–342 (1977).

18. Strelkoff, T. and N. D. Katopodes, "End Depth Under Zero-Inertia Conditions," *Journal of the Hydraulics Division,* ASCE, Vol. 103, No. HY7, pp. 699–701 (1977a).

19. Stringham, G. E. and S. N. Hamad, "Design of Irrigation Runoff Recovery System," *Journal of the Irrigation and Drainage Division,* ASCE, Vol. 101, No. IR3, pp. 209–219 (1975).

20. Tyler, C. L., G. L. Corey, and L. B. Swarner, "Evaluating Water Use on a New Irrigation Project," Idaho Agricultural Experiment Station Research Bulletin No. 62 (1964).

21. U.S. Soil Conservation Service, "Border Irrigation," Chapter 4, Section 15, *SCS National Engineering Handbook,* USDA (1974).

22. Wilke, D. C., "Theoretical Irrigation Tailwater Volumes," *Journal of the Irrigation and Drainage Division,* ASCE, Vol. 99, No. IR3, pp. 415–420 (1973).

23. Willardson, L. S. and A. A. Bishop, "Analysis of Surface Irrigation Application Efficiency," *Journal of the Irrigation and Drainage Division,* ASCE, Vol. 93, No. IR2, pp. 21–36 (1967).

24. Yevjevich, V. and A. H. Barnes, "Flood Routing Through Storm Drains. Part I. Solution of Problems of Unsteady Free Surface Flow in Storm Drains," Hydrology Paper No. 43, Colorado State University, Fort Collins, CO (1970).

25. Yitayew, M., "Reuse System Design for Border Irrigation," unpublished doctoral dissertation, University of Arizona, Tucson, AZ, p. 122 (1982).

26. Yitayew, M. and D. D. Fangmeier, "Dimensionless Runoff Curves for Irrigation Borders," *Journal of Irrigation and Drainage Engineering,* ASCE, Vol. 110, No. 2, pp. 179–191 (1984).

27. Yitayew, M. and D. D. Fangmeier, "Reuse System Design for Border Irrigation," *Journal of Irrigation and Drainage Engineering,* ASCE, Vol. 111, No. 2, pp. 160–174 (1985).

28. Yitayew, M., M. Flug, and D. D. Fangmeier, Tailwater Systems Reduce Energy Use in Border Irrigation, Paper No. 81-2074, Presented at the June 1981 ASAE Summer Meeting, Orlando, FL (1981).

CHAPTER 19

Land and Water Use Planning in Alkali Soils Under Reclamation

N. K. TYAGI*

ABSTRACT

Evolution of alkalinity, which is generally associated with irrigation and water development, severely affects the productivity of agricultural land around the world. Amelioration of these degraded lands is possible through proper land and water management. Basic facts of nature and properties of alkali soils, water management practices and reclamation technology are presented. Various components of the irrigated system are synthesized to develop a methodology for optimal land and water use. The procedure involves: formulation of a deterministic linear programming model for allocation of water resources; development of appropriate techniques to estimate the system parameters which are of transient nature and generation of system responses at discrete stages of soil reclamation. The dynamic nature of the agricultural system and the non-linearly of the crop water production functions have been incorporated into the model by determining the input for small increment of time and by the introduction of multiple activities for each crop, respectively. The procedure so developed has been applied to a representative alkali land in Indo-Gangetic plain. Optimal cropping pattern over a period of 20 years at 5 discrete stages representing different stages of reclamation, have been determined. Use of parametric programming, in determing the impact of change in management strategies on the objective function, is explained. A sensitivity analysis on the cost coefficients of the objective function is also discussed.

INTRODUCTION

Land and water are the two basic resources that sustain life on earth and their optimal use is necessary for the survival of the ever increasing human population. The current statistics show that the availability of these resources is limited and is shrinking every day. According to the esti-

mates prepared by the United Nations (Szabolcs, 1976) a big encroachment on agricultural land is made by waterlogging, salinity and alkalinity. Alkali soils belong to that category of soils that have low fertility (Photoplate 1 and Figure 1) and as such are major constraints in achieving the food production targets. Realizing the fact that the geographical area cannot be increased, there is worldwide demographic pressure to bring the uncultivated but potentially productive alkali land under the plough. Concerted efforts have been made during the past three decades to develop technically sound and economically viable reclamation technologies throughout the world. As a result of these developments information on different aspects of alkali land reclamation has been generated. The awareness of the high production potentials of these soils being rather new, there is an urgent need to evolve appropriate land and water use planning procedures for making best use of the available resources.

DIMENSIONS OF THE PROBLEM

As a decision making process land and water use planning is a problem of multiple dimensions and includes different aspects of agriculture, engineering and economics. A schematic representation of the inter-acting variables is shown in Figure 2. The agricultural part includes the determination of soil physico-chemical properties, rate of reclamation, consumptive use pattern and crop-water production functions. The variables representing engineering part comprise of irrigation system efficiencies, temporal and spatial variability in water availability and several other elements of hydrologic cycle. The economic considerations include the benefit cost functions of differential water use, cost of reclamation and direct and indirect benefits of reclamation etc. These interacting variables complicate the decision making process.

*Head, Division of Agricultural Engineering, Central Soil Salinity Research Institute, Karnal, India

PHOTO PLATE 1. A vast expanse of alkali land in Hissar Distt. of Haryana (India).

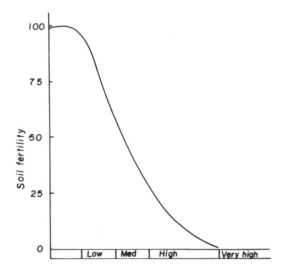

FIGURE 1. Effect of alkalinity on soil fertility.

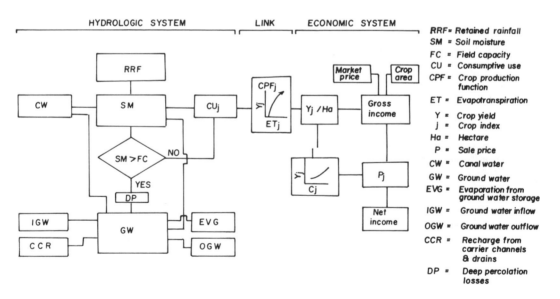

FIGURE 2. Hydrologic-economic flow system showing production function as a link between the two systems.

SCOPE OF THE PAPER

In order to be able to appreciate the scientific intricacies of land and water use planning basic understanding of the various facets of alkali land reclamation as well as of the factors involved in the decision making process is required. This chapter deals with the extent and distribution, nomenclature and classification, formation, nature and properties, reclamation technology and hydrologic and agricultural responses of land reclamation. The planning procedures are briefly discussed, and this is followed by development of a mathematical model for a typical alkali problem area. Finally the decision model has been validated and discussed with reference to an alkali area in the Indo-Gangetic plains in India.

EXTENT AND DISTRIBUTION

Soil alkalinity is of common occurrence in arid and semi-arid regions wherever irrigated agriculture has been practised. Reports indicate (Brinkman, 1980) that out of the total area of 985.7 million hectares occupied by problematic soils, nearly 322.9 million hectares is accounted for by saline and sodic soils (Table 1). This constitutes nearly 5 percent of the total geographical area of the earth and about 23 percent of the presently cultivated land in the world. No separate statistics regarding extent of saline and alkali area has been compiled so far. The worldwide distribution of alkali soils has been mapped by Food and Agriculture Organization (FAO) which shows that alkali soils are fairly well distributed throughout the globe (Figure 3).

NOMENCLATURE AND CLASSIFICATION

The word alkali is of Arabic origin meaning Ash-like, and was used for soils which looked like calcined wood ash. There are a number of terms used in USSR, USA, and the European countries. Solonetz and solod are the Russian terms applied to alkali and degraded alkali soils. In classifying the soils Russian Scientists have combined the principles of pedology, geochemistry, and plant physiology. In USA Hilgard introduced the term black alkali for alkali soils containing soluble calcium carbonates. In Hungary the term "SZIK" is applied to alkali soils. Recently the term alkali has been replaced by sodic to emphasize the major chemical content i.e. sodium, on the soil complex.

There is no single universally accepted classification for salt affected soils of which alkali soils form a part. However, the classification adopted by the United States Salinity Laboratory Staff (Richards, et al., 1968) is widely used for defining the different types of salt affected soils. According to this classification the salt affected soils are grouped into three classes, viz. (1) saline, (2) saline alkali, and (3) non-saline alkali. This grouping is done primarily on the basis of electrical conductivity (ECe), exchangeable sodium percentage (ESP) and pH of the saturation extract (Table 2). These terms are defined as follows:

Electrical Conductivity of Saturation Extract (ECe)—The reciprocal of electrical resistance of the saturation extract of soil which is one centimeter long and has a cross-sectional area of 1 cm^2 is defined as electrical conductivity of saturation extract. It is generally expressed in mmhos (Millimhos) per centimeter (cm) at 25°C. Recent trend is to use term siemens/m.

Exchangeable Sodium Percentage (ESP)—It is defined as the degree of saturation of the soil exchangeable complex with sodium. It can be computed by the following relationship.

$$ESP = \frac{\text{Exchangeable sodium (me/100 g)}}{\text{Cation-Exchange-Capacity (me/100 g)}} \times 100 \quad (1)$$

pH of Saturation Extract (pH$_s$)—The negative logarithm of the hydrogen ion activity of soil paste is defined as pH$_s$ and is used to distinguish the relative degree of acidity or

TABLE 1. Worldwide Distribution of a Number of Problem Soils (million hectares).

Type of Soils	Vertisols	Peat Soils	Acid Sulphate Soils	Planosols	Saline and Sodic Soils
Africa	105.0	12.2	3.7	15.9	69.5
Near and Middle East	5.7	0	0	0	53.1
Asia and Far East	57.8	23.5	6.7	2.7	19.5
Latin America	26.9	7.4	2.1	67.2	59.4
Australia	48.0	4.1	0	49.3	84.7
N. America	10.0	117.8	0.1	12.3	16.0
Europe	5.4	75.0	0	4.0	20.7
World total	258.8	240.0	12.6	151.4	322.9

Brinkman (1980).

FIGURE 3. Worldwide distribution of sodic soils (adapted from FAO/UNESCO soil map of the world 1971–1979).

alkalinity of soil. Recently, Soil Science Society of America (1974) have modified the definition and classification of salt affected soils. These soils are now defined on the basis of electrical conductivity of saturation extract (ECe), sodium adsorption ratio (SAR) and pHs. According to this classification, the ECe limits have been lowered from 4 to 2 mmhos/cm and the ESP has been replaced by SAR with value remaining unchanged at 15. The SAR is computed by the following formula.

$$SAR = \sqrt{\frac{Na^+}{\frac{Ca^{++} + Mg^{++}}{2}}} \quad (2)$$

where the ionic composition are expressed in milli-equivalents per litre (me/l).

TABLE 2. Classification of Salt Affected Soils.

Class	Limiting Values		
	ECe, (ds/m)	ESP	pHs
Saline soil	>4	<15	<8.5
Saline alkali soil	>4	>15	>8.5
Non-saline alkali	<4	>15	8.5–10.0

FORMATION OF ALKALI SOILS

The formation of alkali soils is influenced by a number of geo-chemical processes related to weathering. The chemical weathering of sodium and potassium aluminosilicate minerals is accompanied by formation of alkali solutions. The formation of carbonates and bicarbonates of alkali is particularly intensive in the case of rocks and minerals of recent volcanic origin in undrained basins. In dry-season when evaporation is high, the soil solution is concentrated resulting in increased sodium adsorption ratio (SAR). This leads to adsorption of sodium on the soil adsorbing complex up to 50–70 percent of the cation exchange capacity and raising of soil pH to 9–11. The replaced calcium precipitates at high pH. This process repeated over a period of time leads to the formation of alkali soils (Szabolcs, 1976 and Bhargava, et al., 1980). The chemical exchange reactions involved in the formation of alkali soils are given in (Gedroit, 1912 and Kelley, 1951).

As reported by Du-Khovny (1976) Kovda stated that the combination of salinity types of water along with those of soil has a great bearing on the formation of alkali soils. Slightly alkaline water characterized by the prevalence of salts of sodium over those of calcium, its total concentration equalling 0.7—1.5 grammes/litre, can cause alkalinization after several years of use. This is accompanied by aggregation and accumulation of sodium and magnesium and free

TABLE 3. Effect of Water Quality on Soil in Different Soil and Water Quality Combinations.

Type of Water	Type of Soil	Long Term Results as Related to Soil Fertility
Normal salinity i.e. 0.2–0.4 g/l, Ca prevails	a) Normal or acid	Favourable, steady, fertility
	b) Alkali	Favourable, gradual de-alkalinization
Alkaline salinity 0.6–2.0 g/l with high content of Na_2CO_3 and $Na\ HCO_3$	a) Normal	Unfavourable, alkalization
	b) Alkali	Permanent, unfavourable
	c) Acid	Neutralization
	d) Gypsiferous	Favourable, no alkalinization
Acid, containing H_2CO_3, H_2SO_4 etc.	a) Normal	Slowly increasing acid
	b) Alkali	Favourable reclaiming effect
	c) Acid	Rapidly increasing acidity

carbonates and bicarbonates of sodium. Such phenomena are common in the Nile Delta of Egypt, USA, India, and China. Based on Kovda's data, the types of the water effect on soils, depending on their different combination characteristics, are given in Table 3.

In summary, the process of alkalinization takes place due to:

1. Evaporation of ground water low in total salts but relatively high in carbonates (CO_3) and bicarbonates (HCO_3)
2. Waterlogging, causing anaerobic conditions leading to sulphate reduction and denitrification
3. Replacement of calcium ions by sodium ions in the process of removing soluble sodium salts in the absence of soluble calcium ions

NATURE AND PROPERTIES

The presence of excess exchangeable sodium has influence on the physico-chemical properties of these soils. A number of investigations have been made to increase our understanding of the phenomenon of alkalinization and its effect on nature and properties of these soils (Sigmond, 1938; Richards, 1950; Govindrajan and Murty, 1969; Szabolcs, 1977; Abrol and Acharya, 1975).

Chemical Properties

As a rule, in these soils, considerably high concentration of sodium salts capable of alkaline hydrolysis mainly of sodium carbonate, are found. When exchangeable sodium is accompanied by soluble salts, the electrical conductivity is more than 4 mmhos/cm. Depending upon the local conditions, the maximum concentration of salts may occur at the soil surface or in deeper layers. It has been found that in many cases, there is a close correlation between chemical composition of the salts in the soil profile and that of the ground water indicating the influence of mineralized groundwater on soil formation. Climatic factors influence

the development of soil profile to a great extent. The chemical composition of typical soil profile of saline-alkali soil of Indo-Gangetic plains, from Karnal (India), is given in Table 4. In general, excess soluble salts are present chiefly in upper 0–30 cm. Sodium carbonate and bicarbonate form an appreciable part of the soluble salts. The soils have a high pH (up to 10.2 in 1:2 soil water suspension). Excess sodium carbonate in the soil causes the precipitation of calcium in the soil solution and this increases the exchangeable sodium percentage (ESP) to a high level. The main chemical characteristics of these soils are summarized in Table 5.

Physical Characteristics

The presence of excess exchangeable sodium imparts these soils poor physical properties resulting in compaction of top layer, destruction of soil structure and extremely low transmission characteristics. The affected soil properties that have influence on soil-water behaviour are: infiltration, hydraulic conductivity, soil moisture retention and storage.

Infiltration—The process of water entry into the soil through the soil surface is called infiltration. To begin with the infiltration rate is high but it tends to decrease monotonically and approach a constant rate which is termed as basic intake rate. Due to the dispersed soil structure, the infiltration rate of alkali soil is very low as compared to normal soil (Figure 4). Besides the final intake rate, the cumulative infiltration depth also assumes importance because it limits the depth of irrigation per application. Higher depths of application are likely to cause damage to standing crops due to prolonged water inundation on the soil surface.

HYDRAULIC CONDUCTIVITY

The moisture supply to plants for meeting the evapotranspiration needs of the crops is through movement of water from the sub-soil layers under unsaturated phase. The rate of movement is a function of the unsaturated hydraulic conductivity of the soil which decreases very rapidly with increasing soil moisture suction (Abrol and Acharya, 1975). Results show (Figure 5) that at the same suction, hydraulic

TABLE 4. Physio-Chemical Characteristics of Profile Samples.

Depth (cm)	Mechanical Composition (%)				Exchangeable Cations (me/100 g)				ESP
	Clay	Silt	Sand	CaCO$_3$ (%)	Na	Ca	Mg	K	
0–10	11.8	18.5	67.4	1.6	5.1	Tr.	0.1	0.1	96.2
10–48	16.7	25.3	57.2	2.2	6.9	Tr.	0.3	0.1	89.6
48–88	23.5	29.2	46.1	2.9	9.5	0.2	0.1	0.1	91.3
88–110	28.2	31.5	28.6	9.4	10.2	1.4	1.2	0.5	80.9
110–175	17.6	23.4	43.6	14.1	5.2	0.6	1.5	0.7	66.6

Depth (cm)	pH (1:2)	ECe (mmhos/cm)	Composition of the Saturation Extract (me/l)							
			Na$^+$	Ca^{++}	Mg^{++}	K$^+$	CO$_3^{--}$	HCO$_3^-$	Cl$^-$	SO$_3^{--}$
0–10	10.5	19.82	235.0	0.6	0.2	0.2	143.2	88.4	8.4	4.3
10–48	10.5	10.92	128.0	0.7	0.3	0.1	92.4	30.0	1.6	3.7
48–88	10.1	3.72	44.5	0.4	0.3	0.1	21.2	17.4	0.6	1.5
88–110	9.4	1.42	15.5	0.4	0.3	0.1	3.6	9.2	0.9	0.6
110–175	9.4	0.92	10.3	0.4	0.2	0.2	2.8	5.8	0.4	0.6

conductivity of the normal soil is much higher as compared to alkali soil and this leads to reduced supply of moisture to the plants from lower soil layers in the latter case.

SOIL-MOISTURE RELATIONSHIP

The water availability to the plants is a function of the soil-moisture-suction relationship. The amount of water retained in a soil in equilibrium with a particular suction depends on the size and the relative pore size distribution. For a given texture, the structure of the soil greatly influences the soil-moisture suction. Investigations have shown (Abrol and Acharya, 1975) that in soils with high degree of alkalinity, the moisture retained at low suctions is relatively lower as compared to the soils with low level of alkalinity (Table 6). This situation is reversed at higher soil suction values. It implies that the available moisture (which is taken as the difference of moisture at 0.1 bar to 15 bar) would decrease with increase in alkalinity.

SOIL MOISTURE STORAGE

The soil aggregates are rendered structurally unstable due to the presence of exchangeable sodium. Due to this breakdown of aggregates and consequent dispersion of soil particles in the presence of water, a surface seal, which greatly impedes the water intake by the soil, is formed. As a result, the moisture extent in the sodic soil is considerably less as compared to normal soils (Figure 6). It seems that the total water storage is effectively reduced due to restricted entry from the surface layers. Since plants take water from the soil in the root zone, the availability of water to plants is considerably reduced.

RECLAMATION TECHNOLOGY

Reclamation of alkali soil essentially requires replacement of excess sodium from the exchange complex with calcium and subsequent leaching of the product of exchange from the crop root-zone. However, the land cannot be put to productive use by chemical reclamation alone. It requires a complete package of technology to realize the full production potential of these degraded lands. Yadav (1977) and Tyagi (1980) have listed the following components of reclamation technology.

1. Application of chemical amendments to reduce the exchangeable sodium percentage in the crop root-zone
2. Precision land levelling and field bunding for uniform leaching of exchange products as well as for efficient water application.

TABLE 5. Characteristics of Alkali Soils.

pH	>8.3 somewhere along the profile, or high ESP > 15 in horizon B
Chemistry of solution	Dominated mainly by HCO$_3$, CO$_3$ or both
Effect of electrolytes on soil particle	Dispersion
Main adverse or toxic effects on plants	Alkalinity of soil solution
First aim of reclamation	Lowering or neutralizing the high pH through chemical amelioration

Szabolcs (1980).

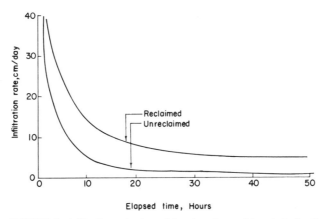

FIGURE 4. Infiltration rates in reclaimed and unreclaimed alkali soils.

3. Adopting irrigation methods that permit high frequency irrigation
4. Providing adequate drainage
5. Following suitable crop rotation and related agronomic practices

Common Amendments and Their Quantities

Materials that directly supply the soluble calcium for replacement of exchangeable sodium are called amendments. The choice of amendment and the quantity required for reclamation depends on the physico-chemical properties of the soil, the amount of exchangeable sodium to be replaced, the desired rate of improvement, the quantity and quality of water available for leaching and the cost of the amendment. Among the various chemical amendments in use, gypsum ($CaSO_4 \, 2 \, H_2O$) is by far the most popular because of its low cost and easy availability. The other common amendments are phosphogypsum, lime stone ($CaCO_3$) calcium chloride ($CaCl_2 \cdot 2 \, H_2O$). These chemicals directly provide soluble calcium. There are other amendments like sulphur (S), lime sulphur (9% Ca + 24% S) aluminum sulphate ($Al_2 \, SO4_3$ 18 H_2O) and ferric sulphate ($Fe_2 \, (SO_4)_3$ $9H_2O$) which react in calcareous soils to produce calcium. Because of its low solubility gypsum is slow acting as compared to sulphuric acid and calcium chloride (Overstreet, et al., 1951). Because of its high solubility calcium chloride produces a leaching solution of high electrolyte concentration and is advantageous in cases where quick reclamation is desired (Doering and Willis, 1975), but it is costly. Phosphogypsum also has considerably higher rate of dissolution than gypsum, which increases its effectiveness (Keren and Shainberg, 1981). The ability to increase soil permeability for ensuring sufficient leaching in a practical period of time is the most important consideration in selection of amendments. The common amendments are given in Table 7.

Prather, et al. (1978) have reported the advantages gained by using the different amendments in combination (Figure

7). It is seen that application of gypsum (75%) in combination with calcium chloride (25%) reclaimed the soil to a greater depth as compared to application of gypsum alone. The quantity of amendment needed to reclaim an alkali soil is determined as a product of gypsum requirement (the equivalent amount of exchangeable sodium to be replaced in the soil) which is multiplied by a factor (1.2−1.3) to compensate for the inefficiencies. Rhoades (1982) has shown that the quantity of amendment can be computed by the following equation.

$$\text{Kg gypsum/ha} = (8.5) \, dB \, E_c \, (R_{Na_i} - R_{Na_f}) \qquad (3)$$

where

d is depth of soil to be reclaimed, m
B is the bulk density of soil, mg/m³
E_c is cation exchange capacity mmol$_c$ per Kg of soil
R_{Na_i} and R_{Na_f} are initial and final sodium adsorption ratios

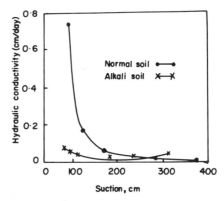

FIGURE 5. Hydraulic conductivity (cm/day) of alkali and normal soil as a function of soil water suction (Abrol and Acharya 1975).

TABLE 6. Moisture Retained (Percent by Weight) in a Sodic Soil Amended with Varying Quantities of Gypsum.

Gypsum Applied (t/ha)	Suction (bars)							
	15	10	5	0.75	0.5	0.33	0.1	0.001
0	11.77	12.13	13.92	18.88	21.45	23.07	28.50	34.06
7.5	11.04	11.75	13.45	17.17	20.85	22.85	28.88	36.23
15.0	10.26	10.00	12.21	15.51	21.54	23.92	30.67	37.39
22.5	9.04	9.14	10.64	14.05	19.17	22.39	30.92	40.64
30.0	9.36	9.46	11.18	14.51	19.74	22.24	31.11	39.03

Based on pH value of soil in 1:2 soil water suspension Abrol, et al. (1973) have developed a graphical relationship to determine the gypsum requirements of light, medium and heavy alkali soils (Figure 8). The quantities of gypsum computed by this method are, however, approximate.

Water Requirement for Leaching

Leaching of replaced sodium, which is essential for reclamation of alkali soils, requires considerable quantities of water. The limited data on leaching requirement (Quirk and Schofield 1955; Chaudhary and Warkentin, 1968; and Rhoades, 1974) show that it would require about 1.0 m/ha of water to dissolve 7.3 tonnes of agricultural grade gypsum

having a fineness such that 85 percent of it will pass through a 100-mesh sieve. Recently Hira, et al. (1981) have developed the following equation for determining gypsum dissolution and water requirements for reclaiming alkali soils.

$$Z^{-1/3} = 1 + KA_{Na_i} \sqrt{\frac{2}{3 \, \varrho \, D_o m}} \, I \qquad (4)$$

where

Z is $(m_o - m)/m$ with m being the amount of gypsum (in gram equivalent) dissolved

m_o is the initial amount of gypsum applied per unit surface area of soil

A_{Na_i} is initial amount of exchangeable sodium per unit of surface area of soil

ϱ is density of soil in Mg/m³

D_o is the initial diameter (cm) of the gypsum particle

I is the depth of irrigation water in cm, and

K is an empirical constant with dimensions of cm⁻²

It was found that it would require 22, 26, and 36 cm of water to dissolve 99 percent of the gypsum of particle size

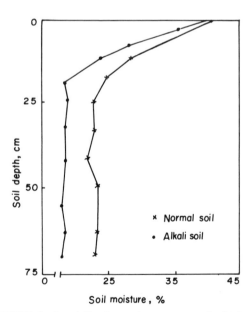

FIGURE 6. Post infiltration moisture content distribution in soil profiles of normal and alkali soils (Abrol and Acharya, 1975).

TABLE 7. Gypsum Equivalent of Different Amendments.

Amendment	Amount Equivalent to Gypsum
Gypsum (Ca SO₄ · 2H₂O)	1.00
Sulphur (S)	0.19
Sulphuric Acid (H₂SO₄)	0.57
Lime Sulphur (24% S)	0.77
Calcium Carbonate (CaCO₃)	0.58
Calcium Chloride Dihydrate (CaCl₂ · 2H₂O)	0.85
Ferrous Sulphate (FeSO₄ · 7H₂O)	1.61
Aluminium Sulphate (Al₂ (SO₄)₃, 18 H₂O)	1.29
Iron Pyrite (FeS₂ · 30% S)	0.63

0.26 mm, 0.26 to 0.5 mm, and 0.5 to 2.00 mm, respectively from a depth of 0–30 cm.

Precision Land Levelling

Land levelling is the precise operation that modified the land surface to a planned grade to achieve more efficient irrigation (ASAE, 1976 b). What is required is not to make the land level but to grade the surface to a uniform slope. The degree of levelling is generally expressed in terms of levelling index (Agarwal and Goel, 1981 and Tyagi, 1984) or the topography index (Khepar, et al., 1982). The levelling index is defined as the average deviation in elevation from the desired elevations required to achieve a planned grade. Mathematically, it is expressed as:

$$LI = \sum_{i=1}^{N} \frac{DLi - ALi}{N} \qquad (5)$$

where

LI is the levelling index representing a range of deviations from the desired elevations along the field length, cm
DLi is the desired elevation at grid point i, cm
ALi is the achieved elevation at grid point i, cm
N is the number of grid points
Σ is summation of the numerical values of the differences irrespective of their sign

It can be readily seen from Equation (5) that the maximum uniformity will be achieved at $LI = 0$ and any increase in the value of LI would mean deterioration in levelling quality.

Precision land levelling is beneficial to all types of soils but the benefits in alkali soils are more because the harmful effects of non-uniform irrigation are higher in these soils. Precision land levelling benefits the crop by permitting application of smaller irrigation depth at more frequent intervals. Investigations were made at CSSRI, Karnal to evaluate the effect of precision levelling on application depth, irrigation efficiencies and crop yield (Tyagi, 1984).

SYSTEM APPLICATION DEPTH AND IRRIGATION FREQUENCY

In surface irrigation, which is the commonly used method in alkali soils, depths of water to be applied to ensure a reasonable sufficiency is not a management controlled variable but, as shown in Figure 9, is greatly influenced by topographic uniformity (Tyagi and Narayana, 1983). This may be called the system application depth. The relationship between system application depth, and levelling index for a typical alkali soil of the Indo-Gangetic plain is shown in Figure 10. It is seen that the average system application depth increases with increase in the levelling in-

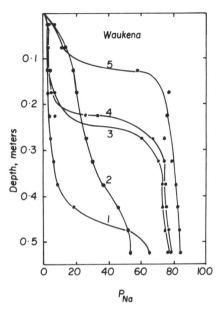

FIGURE 7. Relations between sodium adsorption ratio of Waukena soil versus soil depth after reclamation by use of sulphuric acid (1), calcium chloride solely (2), 25% sulfuric acid plus 75% gypsum (3), 25% calcium chloride plus 75% gypsum (4), and gypsum solely (5) (after Prather et al., 1978).

dex. Since for a constant seasonal irrigation depth, the depth per application and the irrigation frequency have inverse linear relationship the frequency of irrigation decreases with increase in system application depth.

It may be mentioned that besides levelling index, the frequency of irrigation also depends upon scheduling criterion.

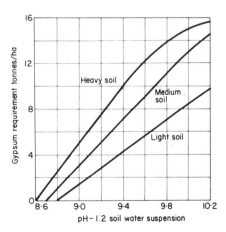

FIGURE 8. Nomogram for calculating gypsum requirement in alkali soil (Abrol et al., 1973).

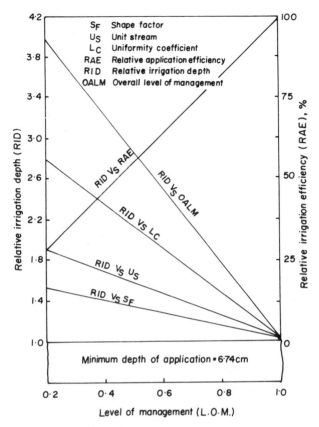

FIGURE 9. Effect of level of water management on irrigation efficiency.

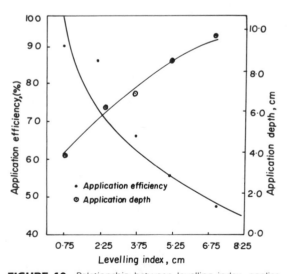

FIGURE 10. Relationship between levelling index, application efficiency and depth of irrigation.

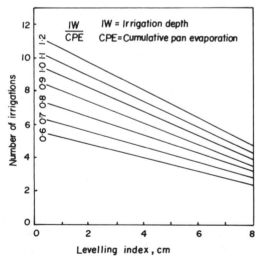

FIGURE 11. Relationship between levelling index, IW/CPE ratio and number of irrigations.

Among the various scheduling criteria, the one based on evaporation from open pan is the simplest to adopt (Prihar, et al., 1975 and Tyagi, et al., 1984). This procedure involves application of fixed depth *(IW)* after a predecided cumulative evaporation from USDA Open Pan *(CPE)*. Considering *IW/CPE* ratio as one parameter, graphical relationship such as the one shown in Figure 11, can be developed between levelling index and total number of seasonal irrigations. This relationship is useful in determining the attainable frequency of irrigation for different crops at varying levels of topographic uniformity.

TABLE 8. Grain Yield of Wheat at Different Levelling Indexes.

Levelling Index (cm)	Grain Yield (kg/ha)			
	1980–81	1981–82	1982–83	Average
0.0–1.5	3325	3128	3026	3226
1.5–3.0	3058	3079	2866	3001
3.0–4.5	2783	3832	2517	2711
4.5–6.0	2379	2359	2066	2268
6.0–7.5	2435	2246	1813	2169

LEVELLING UNIFORMITY AND APPLICATION EFFICIENCY

Application efficiency is defined as the ratio of depth of water stored in the root zone and the depth of water applied to the field. Mathematically,

$$AE = \frac{WS}{WD} \times 100 \qquad (6)$$

where

AE is the application efficiency, %
WS is depth of water stored in the root zone, cm
WD is the depth of water applied to the field, cm

High application efficiencies at the field level are generally associated with low application depth (Israelsen and Hansen, 1962). Since low application depth in surface water application methods are possible only in precisely levelled fields, only good quality levelling will ensure efficient application. Tyagi (1984) has shown that even in surface irrigation methods, application efficiencies as high as 80–85 percent (Figure 10) which are comparable to sprinkler and drip irrigation, are possible with levelling index in the range of 0–1.5 cm.

IRRIGATION FREQUENCY AND CROP YIELD

An irrigation cycle has two distinct phases which include (1) infiltration followed by (2) water extraction by the crop. In general infiltration phase is very brief as compared to extraction phase. However, in alkali soils, an infiltration dominated cycle which would keep the salt and moisture flux downward, is more preferable. Precision land levelling helps in attaining this objective by permitting low water applications at increased frequency, which has beneficial effect on crop yield (Gaul, et al., 1973). Just as the differential application of chemical amendments and fertilizers affects crop yields so does the uniformity of levelling. The relationship between crop yield and levelling uniformity can be represented by a polynomial of 2nd and 3rd degree. For the grain yield of wheat (Table 8) on a typical alkali soil of

Indo-Gangetic plain in India, Tyagi (1984) fitted the following second degree polynomial.

$$Y = 3297.4 - 96.7\,(L\,I) - 25.3\,(L\,I)^2$$

$$R^2 = 0.94 \qquad (7)$$

where

Y is grain yield in Kg/ha
R is correlation coefficient

This indicates that levelling quality affects the crop yield to a considerable extent with highest yield of 3128 Kg/ha occurring in plots with *L I* in the range of 0–1.5 cm as compared to a yield of only 2246 kg/ha in plots with *L I* in 6.0–7.5 cm range.

IRRIGATION METHOD

Basically water can be applied in four ways: (1) surface methods which include borders, furrows and check basin, (2) sub-surface application, (3) overhead applications with sprinklers and perforated pipes, and (4) drip or trickle system. Selection of irrigation method has to be made to suit the soil and crop requirements. In alkali soils where uniform leaching and downward movement of salt and water flux are the primary considerations, the use of furrows and sub-surface methods is precluded. The cropping pattern during the initial years being limited to rice during summer and wheat, berseem and barley during winter, the use of sprinkler has also not been found beneficial (Pandey and Singh, 1976). The ideal conditions for rice-wheat rotation are summarized in Table 9.

Admitting that border irrigation layout is desirable in alkali soils, the decision regarding border grades becomes important. The question whether land should be levelled to zero slope or should have some slope in the direction of irrigation flow, assumes importance. Tyagi (1983) studied the relative performance of level and graded borders in alkali soils and found that graded borders were efficient by as much as 30 percent (Figure 12).

TABLE 9. Ideal Land and Irrigation Management Practices for Rice-Wheat Rotation in Alkali Soils.

Practice	Rice	Wheat
Land development	1. Level land for uniform ponding	Land should be such that no water stagnation takes place.
	2. Long narrow strips commonly known as borders for soils with low infiltration rates.	Long narrow borders commonly known as borders for soils with low infiltration rates.
Irrigation practice	Surface flooding maintaining a minimum depth of 5 cm in the field.	For a given seasonal irrigation depth, light and frequent irrigations are preferred.

TRANSFORMATION OF GRADED BORDERS
INTO BASINS

Rice is usually the first crop to be taken in alkali soils. Meeting the ideal water management requirements of rice as indicated in Table 9 is possible only in level plot. In order to meet this ideal land requirement of the rice-wheat rotation and to minimize land forming costs, graded borders could be transformed into small basins by putting small earthen dykes provided with pipes connecting the compartments as shown in Figure 13. This procedure is essentially based on the principle of reducing slope length. The mathematical formulae establishing the relationship between the geometry of the border and the depth of standing water are as follows (Narayana and Tyagi, 1981).

Depth of water at the head end *(Hei)*

$$H_e = D - \frac{LS}{2N} \qquad (8)$$

where

D is the average depth of ponding
L is the length of border
S is the bed slope and
N is the number of compartments into which the border is divided

Depth of water at the tail end *(T_e)*

$$T_e = D + \frac{LS}{2N} \qquad (9)$$

The maximum deviation d_{max} of ponding depth either at the head end or tail end from the average depth "D" is

$$d_{max} = \frac{LS}{2N} \qquad (10)$$

UNIFORMITY OF PONDING

The uniformity of ponding is expressed in terms of uniformity coefficient *(UC)* which is computed as follows:

$$UC = \left(1 - \frac{LS}{2N} D\right) \qquad (11)$$

It is apparent from Equation (11) that for a given depth of ponding "D" and the border length "L" the uniformity of ponding can be increased by increasing the number of compartments. But the gain in uniformity becomes very small beyond four compartments (Figure 14).

The effectiveness of the graded borders transformed into check basin by adopting the procedure discussed above was investigated by Pandey, et al. (1977). It was shown that division of 93 m long borders into 1, 2, and 3 compartment by putting earthen dykes at 46.5 m in case of two and 31 m and

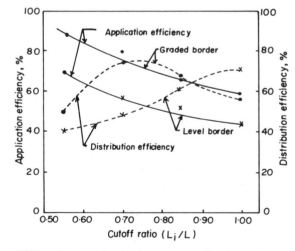

FIGURE 12. Relationship between cutoff ratio, application and distribution efficiency in graded and level borders.

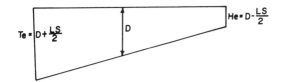

The sloping border with no compartment

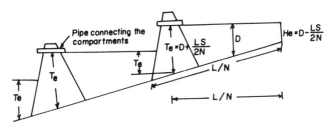

The sloping border with 'N' compartments

L   Border length

S   Slope
D   Average depth of ponding
H   Height of pipe line above ground level
N   Number of compartments

$$He = D - \frac{LS}{2N}$$

$$Te = D + \frac{LS}{2N}$$

FIGURE 13. Geometry of compartmentalized sloping border.

62 m in case of three compartments with pipe outlets at designed heights of $(D + LS/2N)$ increased the average grain yield of rice by 685 kg/ha. This increase was made possible due to increased uniformity of ponding (Table 10).

Drainage in Alkali Soils

The basic requirement of effective drainage implies that the surface congestion of water should be removed and the ground water table is maintained within permissible limits to provide favourable environment for crop growth. According to Luthin (1966) there are three basic methods of removing excess drainage water: (1) surface drainage, (2) subsurface or horizontal drainage, and (3) vertical drainage or drainage through wells. The applicability of one or several methods for providing adequate relief depends upon many factors, which are location specific.

SURFACE DRAINAGE

Raadsma and Schulze (1974) have listed the following situations where surface drainage will be useful:

1. Deep surface soils of low infiltration rate which is the case with alkali soils

2. High intensity daily rainfall (greater than 100 mm) occurring frequently at least once in five years

3. Flat topography

Surface drainage is accomplished through (1) land grading and smoothening to improve micro relief, (2) provision of

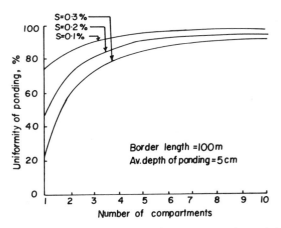

FIGURE 14. Effect of number of compartments in graded borders on uniformity of ponding.

TABLE 10. Uniformity of Ponding and Rice Yield with Different Border Compartments.*

Treatment	Uniformity Coefficient	Rice Yield kg/ha 1975	Rice Yield kg/ha 1976
Single compartment	42	6180	6320
Two compartments	71	6480	6540
Three compartments	82	7100	6770
C.D. (Critical difference) at 5%.		6.0	4.3

*L = 93 m, S = 0.001, D = 8 cm.

field and collector drain linked to a disposal outlet, and (3) facilities for storage of excess rainwater during heavy storms and its reuse during dry spells of the rainy season.

The most important step in arriving at a suitable design of surface drainage system is the drainage requirement of crops. The design discharge for any area is so chosen that an acceptable level of water is maintained in the crop land for a period not longer than the permissible number of days occurring with a frequency of 5–10 years. In general surface drainage is designed for a storm of 2 days duration and 5 year frequency. However, for high value crops with low tolerance to excess water, the magnitude of protection may be kept more. The value of runoff for each storm from catchment needing surface drainage is computed by Equation (12).

$$Q = P - (E + I) \, T - D \qquad (12)$$

where

Q is the runoff, cm
E is the daily evaporation during the storm, cm
I is the daily infiltration during the storm, cm
P is the storm precipitation, cm
D is the permissible depth of ponding in the catchment, cm
T is the duration of storm, days

The actual number of days "N" required to drain out this excess rain are computed by Equation (13).

$$N = \frac{Q}{(DD + E + I)} \qquad (13)$$

where DD is the design discharge, cm/day.

Determination of the maximum consecutive rainfall of different duration involves frequency analysis of rainfall data. Generally consecutive day rainfall of 2–3 days duration as shown in Table 11 is adequate for this purpose. Once, the drainage coefficient is determined, the design of the surface drainage system reduces to design of open channel which can be done by using Manning's or Chezy's formulae.

THREE TIER SURFACE DRAINAGE SYSTEM

In order to take care of the surface water stagnation during heavy storms and the drought conditions during dry spells of the rainy season Narayana (1980) has proposed a three tier surface drainage system for alkali soils. This system has application in cases where rice is grown during rainy season. The important features of this system are:

1. Collection of part of the rainfall in the cropland till such a time and extent that will not be harmful to the crops (Photo plate 2)
2. Collection of excess water from the cropland into the farm ponds in low lying areas. The water so stored is utilized for irrigation during the dry spells (Photo plate 3).
3. Draining the excess water from the farm pond into a regional drainage system

The effect of storing water up to different depths of rain water in alkali soil planted with rice was studied by Gupta and Pandey (1979). The yields and irrigation water applied at different levels of drainage during 1976 are shown in Figure 15. The slope of the curve for irrigation water requirements indicates a steep increase in irrigation water demand with increase in degree of drainage. The fairly flat slope of the crop yield curve shows that there is only a small yield reduction with reduced drainage. These results have significance because even in alkali soils with poor drainage characteristics excess rain water up to 15 cm values could be stored within field for use by the crop without significant loss in crop yields. The remaining excess water can be profitably stored in dugout farm ponds (Photo plate 3).

TABLE 11. Maximum Storm Rainfall and Dry Spells of Different Return Period.

Duration of Event	Return Period, in Years 1.01	2.33	5	10	25	100
Maximum 1-day rainfall in cms	4.1	12.0	15.2	18.3	22.1	28.2
Maximum 2-day rainfall in cms	5.1	15.5	20.1	23.8	28.5	35.5
Maximum 3-day rainfall in cms	6.1	17.1	21.9	25.8	30.7	38.1
Maximum 4-day rainfall in cms	6.7	17.9	22.8	26.8	31.8	39.4
Maximum dry spell in monsoon, in days	15	28	34	39	45	54

PHOTO PLATE 2. Storage of rainwater in rice fields.

SUB-SURFACE DRAINAGE IN ALKALI SOILS

In areas where the high ground water table is a problem throughout the year, provision of sub-surface drainage is basic to reclamation of alkali lands. The requisite drainage is provided either through deep open ditches or covered tile drains. The design of the system is based on the hydraulic conductivity, drainage coefficient and depth to impermeable layer.

Alkali soils behave like clay or clay loam soils and due to the poor hydraulic conductivity values the spacing between the two drains as per USDA specifications (Table 12) seldom exceeds 20 m. Since sub-surface drainage is expensive, lower spacing many times acts as a major constraint in

reclamation of the alkali soils. As shown by Narayana, et al. (1977), in situations where high water tables prevail only during a short period of time (1–2 months) provision of sub-surface drainage may not be advisable.

VERTICAL DRAINAGE OR DRAINAGE WELLS

In areas underlain by good water quality aquifers as is the case with alkali soils of the Indo-Gangetic plains in India, vertical drainage has shown greater promise in controlling the table. Vertical drainage is effected through shallow cavity or filter tubewells which draw water from the first aquifer layer. Geological and hydrological conditions are important in assessing the potentialities of vertical drainage.

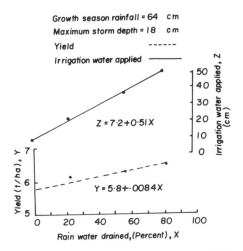

Growth season rainfall = 64 cm
Maximum storm depth = 18 cm
Yield   - - - - - -
Irrigation water applied   ⎯⎯⎯⎯

$Z = 7.2 + 0.51X$

$Y = 5.8 + .0084X$

FIGURE 15. Irrigation requirement and rice crop yield in relation to different levels of drainage (Gupta and Pandey, 1979).

PHOTO PLATE 3. Storage of rainwater in farm pond.

TABLE 12. Average Depth and Spacing of Tile Drains (U.S.D.A. S.C.S. specification).

| Soil Texture | Hydraulic Conductivity | | Spacing (m) | Depth (m) |
	Class	Rate m/day		
Clay	Very slow	0.03	9–15	0.9–1.0
Clay loam	slow	0.03–0.1	12–21	0.9–1.0
Average loam	Moderately slow	0.1 –0.5	18–30	1.0–1.2
Sandy loam	Moderate	0.5 –1.5	30–36	1.2–1.4
	Moderately rapid	1.5 –3.0	30–61	1.2–1.5
Peat and muck	Rapid	3.0–6.0	30–91	1.2–1.5
Irrigated soils	Variable	—	45–183	1.5–2.5

Presence of good aquifers at a depth of 15–20 m and favourable quality of ground water are the necessary conditions for success of vertical drainage. This type of drainage has found extensive application in California (USA), India, Pakistan, and USSR. Compared with horizontal drainage vertical drainage has the advantages of (1) low initial cost, (2) sharper change in drainage condition, and (3) reuse of drainage water. The operation cost is, however, high while the life span is shorter.

Crops in Alkali Soil

It is important to select proper crops and varieties during reclamation because growing tolerant crops during initial years can ensure reasonable economic returns. Important crops according to their level of tolerance to alkalinity are listed in Table 13.

The effect of different alkalinity levels on crop yields has been investigated by several workers (Pearson, 1960; and Bernstein, 1958; Singh, et al., 1979; and Abrol and Bhumbla, 1979). As a general rule, crops that can withstand excess moisture conditions are relatively more tolerant to alkalinity. Relating crop yield to alkalinity Abrol and Bhumbla (1979) have shown that rice could be successfully grown with only 10 percent reduction in yield even at ESP level of 55 whereas in case of wheat the corresponding relative yield would be obtained only if ESP were below 15. The high tolerance of rice to alkalinity is primarily due to its ability to withstand standing water throughout the growing season. Low permeability of alkali soil is of advantage in rice because it reduces the percolation, although in most of the cases it is sufficient to leach salts resulting from exchange of sodium. Pearl millet and ragi during summer season and wheat, berseem, sugarbeet, barley and raya (Brassica juncea) during winter season are the other important crops that can be successfully grown during initial years of reclamation. In crop selection, besides the tolerance to alkalinity, it is the relative cost benefit ratio that influences choice of crops.

TABLE 13. Relative Tolerance of Some Crops and Grasses to Exchangeable Sodium.

Tolerant	Semi-tolerant	Sensitive
Karnal grass (Diplachne fusca)	Wheat (Triticium vulgare)	Cowpeas (Vigna sinensis)
Rhodes grass (Chloris gayana)	Barley (Hordeium vulgare)	Gram (Cicer arietinum)
Para-grass (Brachiaria mutica)	Oats (Avena sativa)	Groundnut (Arachis hypogaea)
Bermuda grass (Cynodon dactylon)	Raya (Brassica juncea)	Lentil (Lens esculenta)
Rice (Oryza sativa)	Senji (Melilotus parviflora)	Mash (Phaseolus mungo)
Sugarbeet (Beta vulgaris)	Berseem (Trifolium alexandrinum)	Mung (Phaseolus aureus)
	Sugarcane (Sachharum officinarum)	Peas (Pisum sativum)
	Bajra (Pennisetum typhoides)	Maize (Zea mays)
	Cotton (Gossypium hirsutum)	Cotton at germination (Gossypium hirsutum)

Crop yields are seriously affected if the ESP is more than about 55, 35 and 10 in respect of tolerant, semi-tolerant and sensitive crops, respectively. Tolerance in each column decreases from top to bottom. The grasses listed are highly tolerant.

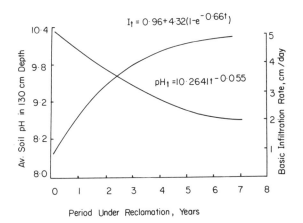

FIGURE 16. *Soil pH and basic infiltration rate during the span of reclamation.*

HYDROLOGIC RESPONSES OF RECLAMATION

Reclamation of alkali soil is a gradual process of improvement in the physico-chemical properties and the soil remains in a transient stage over a long period of time. Replacement of exchangeable sodium with more favourable calcium reduces the soil pH and increases the infiltration rate. These changes for typical alkali soils in the Indo-Gangetic plain are shown in Figure 16. Initially the reduc-

tion in pH or increase in infiltration rate is rapid but the rate of change gradually slows down.

Improvement in physico-chemical properties introduces changes in hydrologic behaviour of the alkali affected area which is reflected in the reduced runoff and increased retention of rain water. Typical runoff hydrographs recorded at Central Soil Salinity Research Institute, Karnal show (Figure 17a–c) that reclamation measures would decrease the peak and volume of runoff for storms of given intensity. The runoff hydrographs produced by the same storm in comparable alkali and normal watersheds show (Figure 17d) that peak and volume of runoff with normal soil were 0.43 and 0.25 times the corresponding values from alkali watershed. The reduction in surface runoff is primarily due to increased retention of rainwater in the reclaimed land. Field bunds around cultivated land and the standing crop which is mostly rice, provides ideal condition for increased retention. Investigations made by Singh, et al. (1979) have shown that for a rainfall of 80 mm/day the retention storage increased from 44 mm/day to 55 mm/day with increase of cultivated area from 67 to 70 percent (Figure 18).

Besides surface runoff and retention storage that undergo changes with reclamation, increase in ground water recharge is another favourable effect of reclamation. This is because it generates additional water supply that can be exploited through ground water development and used for irrigation. A field study conducted at Karnal has shown that reclamation could induce 30–40 percent of the seasonal rainfall to percolate and join the ground water storage (Table

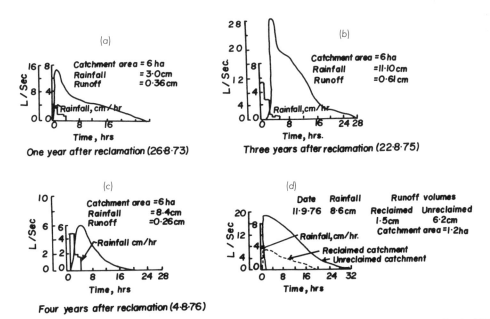

FIGURE 17. *Gauged runoff hydrographs from alkali and reclaimed soil watersheds (Narayana and Singh, 1976).*

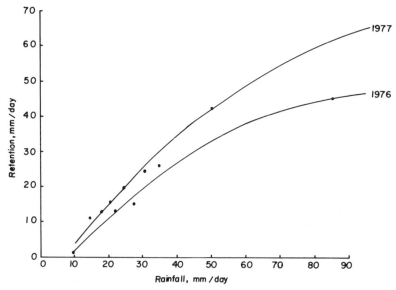

FIGURE 18. Changes in the catchment retention storage due to progressive reclamation (Singh et al., 1979).

14). It concludes that whereas reclamation of alkali lands creates the demand for the additional water supplies to meet the crop water requirements, it also helps in generating the water resources by way of reduced surface runoff, increased surface retention and higher ground water recharge. Singh

and Tyagi (1980) developed a mathematical model for estimating the rainwater available in reclaimed alkali soils and tested it in representative watershed at Karnal (India). The study showed that under the agro-climatic conditions prevailing at Karnal as much as 30 percent of the irrigation

TABLE 14. Water Balance Analysis of a Typical Alkali Catchment, in Karnal (June–September).

Particulars	Unit	1972	1973	1974	1975	1976	1977
Precipitation	cm	69.5	58.9	38.4	60.4	86.7	58.9
Canal seepage	cm	4.0	4.0	4.0	4.0	4.0	4.0
Total output (I)	cm	73.5	62.9	42.4	64.4	90.7	62.9
Runoff	cm	12.1	9.4	0.5	1.2	10.0	5.5
Evapotranspiration	cm	34.2	37.2	37.9	35.8	42.6	25.2
Addition to soil moisture storage	cm	9.0	8.0	6.8	8.8	10.9	7.5
Total output (0)	cm	55.3	54.6	45.2	45.8	63.5	38.2
Water surplus for local recharge $S = (I - 0)$	cm	18.2	8.3	− 2.8	18.6	27.2	23.7
	%	26	14	0	31	31	40
Computed input to ground water (G)	cm	29.1	18.7	17.1	23.9	25.5	22.8
	%	42	32	45	40	29	39
Probable recharge from outside the area (G-S)	cm	10.9	10.4	19.9	5.3	− 1.7	− 0.9
	%	16	18	52	8	0	0
Runoff (R)	%	17	16	1	2	12	9

% These are values expressed as percentage of rainfall.

PHOTO PLATE 4. Bumper rice crop in alkali soil under reclamation. (See barren alkali land by the side of rice field.)

water needs of rice-wheat rotation could be met from rainwater alone which otherwise goes waste to create drainage problem in the low lying area.

AGRICULTURAL RESPONSE OF RECLAMATION

To begin with, production potential of alkali soils is negligibly small. But as reclamation progresses, it brings in changes in the soil-water system by reduction of pH levels (Figure 16). This leads to gradual increase in the production levels of different crops as well as changes in irrigation requirements.

Changes in Yield Potential

Tolerant crops like rice (Photo plate 4) reach the production potential of normal soils in a comparatively shorter span of time as compared to semi-tolerant crops like wheat or sensitive crops like raya (Brassica Juncea L.) Safflower linseed, and groundnut. In a long-term study initiated in 1970 at C.S.S.R.I. Karnal (Abrol and Bhumbla, 1979) to determine the appropriate crops that can be introduced in alkali soils, it was found that rice attained the potential yield of about 7000 kg/ha even at ESP 30 whereas wheat could grow well only at ESP below 10 (Figure 19). Regular monitoring of soil pH (Figure 20) showed that if land were kept under continuous cropping and recommended reclamation technology were adopted, then it would take about 4-5 years to attain the pH of 8.8 and the corresponding ESP level of 30

(in 0-30 cm) and 8-10 years for ESP level of 10. Crops like safflower, linseed and raya also grow well only if the ESP levels in upper 50 cm depth are within 5-10 (Singh, et al., 1979, 1981).

Based on the rate of change in soils properties and the response of crops to varying pH levels, Tyagi (1980) computed the estimated yield of rice, berseem and wheat which are the important crops during the initial years of reclamation. The

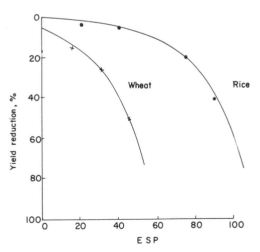

FIGURE 19. Effect of soil ESP on relative yield of wheat and rice (Abrol and Bhumbla, 1979).

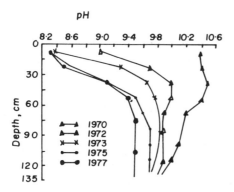

FIGURE 20. Reduction in pH in gypsum treated soil at different periods under reclamation.

yield estimates at periods of 1, 5, and 10 under reclamation are given in Table 15.

Changes in Irrigation Requirements

Among the factors that influence irrigation requirements, profile water use and application efficiencies undergo continuous changes with improvement in soils. In crops like wheat and berseem where irrigation is scheduled on the basis of moisture deficit in the crop root zone, there is no significant increase in the irrigation requirement with time. This is because in alkali soils at initial stage of reclamation, more frequent irrigation is practised leading to higher evaporative losses from the soil surface. But as the alkalinity decreases, water that was lost in the form of evaporation, is now lost as deep percolation and ultimately joins the ground water storage. Though irrigation requirements remain the same before and after irrigation in these crops, but the later case has the advantage in that, water is not completely lost to the system. However, the situation in case of rice is different because initially as well as after reclamation, the rice is grown under ponded conditions. Therefore, there is no

change in evaporation losses but the percolation losses increase with time resulting in increased irrigation requirements. Tyagi (1980) estimated the irrigation requirements of rice at different stage of reclamation (Figure 21) and predicted that for the agro-climatic conditions at C.S.S.R.I. Karnal, the irrigation requirement may increase from 75 cm in the first year to about 120 cm in the 20th year. The increased irrigation requirements have to be provided for in the proposed water use plans.

PLANNING PROCEDURES

The objective of planning is to be able to propose and evaluate the various alternatives to arrive at optimal decisions. The criterion of optimality is generally economic. Since the alternatives to be evaluated in a real world problem are many, resort is generally made to the principles of operations research or systems analysis. The application of system analysis involves the following steps.

1. Statement of objectives
2. Precise definition of the system under investigation
3. Definition of control parameters
4. Development of model which can represent the actual system as closely as possible
5. Validation of the model
6. Sensitivity analysis

There has been phenomenal growth in the techniques of systems analysis during the last five decades, but its application to water use planning and management of agricultural systems started only in the sixties. Since then it has been used, among others, to find answer to the following types of problems.

1. Optimal timing and amount of water allocations to individual crops
2. Optimal distribution of water among different crops on a farm

TABLE 15. Crop Yield and Income from Rice, Berseem and Wheat in Alkali Soils at Different Stages of Reclamation.

	Period Under Reclamation, Years						
	1		5		10		
Crop	Yield Tonnes/ ha	Income $/ha	Yield Tonnes/ ha	Income $/ha	Yield Tonnes/ ha	Income $/ha	Remarks
Rice	4.90	506.9	6.03	625.0	—	—	The yield becomes constant after
Berseem	65.00	792.7	75.00	914.6	88.0	1073.2	5 years in case of rice, 7 years in
Wheat	2.30	315.5	4.53	621.3	4.93	676.3	case of berseem, and 10 years in case of wheat.

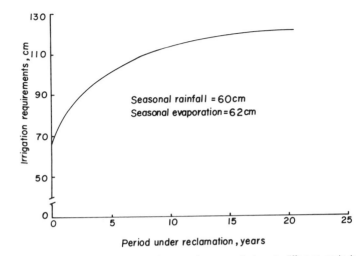

FIGURE 21. Estimated irrigation requirement of rice at different periods under reclamation.

3. Operation and management of surface water reservoirs or run of the river schemes

4. Operation and management of ground water reservoirs

5. Integrated operation of reservoirs and aquifers

Mathematical Modelling

Optimization of these problems has been accomplished by mathematical models which employ linear or dynamic programming techniques or simulation approaches or a combination of these methods. Detailed description of these programming techniques is beyond the scope of this chapter. The general principles of the optimizing techniques are given in Hall and Dracup (1971) and Buras (1972). Smith (1973), Yakowitz (1982), and Garelick (1983) have presented a good review on application of these techniques to practical problems.

Important Considerations in Water Use Planning

The following points need careful consideration and incorporation in the system model while seeking solution of practical field problems.

1. Crop production functions with dated inputs particularly with respect to water and fertilizer use. It is well known that dynamic programming is the best tool to optimize water use in individual crops. But very often, the non-availability of production function with dated inputs is a major constraint in its application.

2. Water use from the moisture stored in the soil profile at the end of a crop season. In areas with deficit water supply soil moisture stored in the recharged soil profile at the end of season contributes significantly towards the water requirements of crops. Due weightage should be given to this readily available moisture while computing the irrigation requirements.

3. Effective rainfall: Rainfall is the major source of direct water supply to the crops. However, the amount of effective rainfall differs from crop to crop as well as with the stage of crop growth. Precisely estimated values of effective rainfall improve the applicability of the model outputs.

4. Staggered planting and sowing of crops: This reduces the peak water requirement to a large extent and can minimize the design capacity of the irrigation system. This also facilitates better deployment of farm labour.

5. Transitional production function and water balance: The crop-water production function and the water balance of the project area containing alkali soils are time dependent. As the reclamation progresses, the crop-water-production function goes on changing and so does the water balance of the area.

In the water use planning model that follows these steps have been duly incorporated.

STATEMENT OF THE PROBLEM

The system under study may be assumed to consist of a command area containing both normal and alkali soils. In the normal soils, there is no restriction for growing crops like rice, wheat, berseem, sugarcane, and maize, etc. In the alkali soils in the initial years of reclamation, only crops like rice, wheat, and berseem can be grown.

Alkali soils have a high exchangeable sodium percentage at the start of the reclamation. As a result, they have poor physical properties such as low infiltration rates, adverse soil moisture retention and transmission characteristics, and a high surface runoff. During the reclamation process, an increase in infiltration rate and ground water recharge, and a reduction in surface runoff are brought about.

Rainfall meets part of the water requirements of the crops of the area. The remaining need is met from canal water supply and shallow tubewells. The water is delivered to the command area through a conventional canal system, and some seepage losses will, therefore, occur. Because the canal distributes water from a diversion structure of the river with no storage facilities the farmer can only use the water when it becomes available. If the geological conditions of the ground water aquifer are favourable, deep tubewells could be installed and ground water could be exported from the area.

The water resources utilization model for such an area must be capable of taking into account the transient nature of the various properties previously described. Linear programming, which has been used by a number of research workers (Lakshminarayana and Rajgopalan (1977), Matanga and Marino (1979), and Smith (1973) in such situations, may be used to arrive at optimal decisions. The solution of the problem involves the following steps:

1. Formation of a conceptual model for land and water use
2. Estimation of system parameters for the model
3. Generation of model responses for a given set of inputs to arrive at optimal cropping and water use plans

MATHEMATICAL MODEL

A model consisting of an objective function and a set of constraints is developed for the conjunctive use of surface and ground waters in the area for a period of one year

divided into 24 sub-periods of 15 days each. The objective is to allocate optimally the area for different crops to be grown in both normal and reclaimed soils, and simultaneously synchronize the water releases from different sources (canals and tubewells) so that the crop demands during different growth stages are met satisfactorily. The inputs to the model are considered to be of a deterministic nature. A schematic representation of the model is given in Figure 22.

OBJECTIVE FUNCTION

The objective function considers the income from crop production and water export and the cost of water from canals and shallow tubewell. The production cost for different crops and the water export are included only indirectly through their sale prices.

$$
\begin{array}{cc}
\text{Crop income} & \text{Water income}
\end{array}
$$

$$
Z_t = \max \sum_{j=1}^{m} P_{jt} A_{jt} + \sum_{i=1}^{n} P^{DT} DT_{it}
$$

$$
\begin{array}{cc}
\text{Canal water cost} & \begin{array}{c}\text{Shallow tube-}\\\text{well water cost}\end{array}
\end{array}
$$

$$
- \sum_{i=1}^{n} C^{CW} CW_{it} - \sum_{i=1}^{n} C^{ST} ST_{it} \qquad (14)
$$

in which, Z_t = maximized value of the objective function; A_{jt} = area under crop, ha; P_j = income from unit area under crop, \$/ha; DT_{it} = volume of water from deep tubewells (water export), $10^3 m^3$; P^{DT} = sale price of tubewell water, \$/$10^3 m^3$; ST_{it} = volume of ground water pumped by

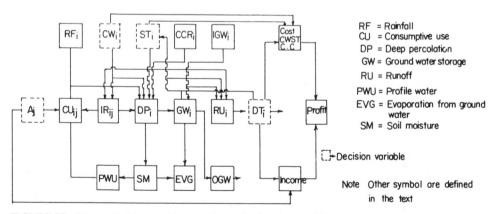

FIGURE 22. Schematic diagram of the model showing interaction of hydrologic, agricultural and economic parameters.

shallow tubewells, $10^3 m^3$; C^{ST} = cost of shallow, tubewell water, $/10^3 m^3$; CW_{it} = volume of canal water released, $10^3 m^3$; and C^{CW} = cost of canal water, $/10^3 m^3 \cdot i$ = time index for 15 days period having values 1, 2 . . . n. The first period (i = 1) is June 1–15, and the last period (i = 24) is May 16–31; j = crop index; t = year index.

CONSTRAINTS

The optimal policies are determined with the following constraint equations.

The irrigation requirements of all crops must be fully met at all stages of growth.

$$\sum_{i=1}^{n} (CEC\ CW_{it} + CET\ ST_{it}) = \sum_{i=1}^{n}$$

$$\sum_{j=1}^{m} \frac{NIR_{ijt}\ A_{ijt}}{AE_{jt}} \tag{15}$$

in which A_{ijt} = area under crops; CEC = conveyance efficiency of canal system; CET = conveyance efficiency of shallow tubewell system; NIR_{ijt} = net irrigation requirement; and AE_{jt} = field water application efficiency of crops.

The water release at any point and at any time cannot exceed the capacity of the irrigation:

$$CW_{it} \leq CAPCW \tag{16}$$

$$\frac{RPS\ ST_{it}}{CET} \leq CAPST \tag{17}$$

$$\frac{RPD\ DT_{it}}{CED} \leq CAPDT \tag{18}$$

in which CAP ($CW,\ ST,\ DT$) = installed capacity of the three system; canal water (CW) shallow tubewells (ST) and deep wells (DT); RPS = ratio of peak to average demand for shallow tubewell; RPD = ratio of peak to average demand for deep tubewell water; CET = efficiency of shallow tubewell system; and CED = efficiency of deep tubewell system.

The canal supply during any period cannot exceed the expected flow during that period:

$$CW_i \leq E\ (CW_i) \tag{19}$$

in which $E\ (CW_i)$ = expected canal flow.

The amount of ground water pumped cannot exceed the mining allowance:

$$\sum_{i=1}^{n} (ST_{it} + DT_{it} + OGW_{it} + EGW_{it})$$

$$- \sum_{i=1}^{n} \sum_{j=1}^{m} CPR_{ijt} - \sum_{i=1}^{n} \tag{20}$$

$$\times\ (CWR_{it} + STR_{it} + CCR_{it} + IGW_{it}) \leq MGW_t$$

in which IGW = ground water inflows; OGW = ground water outflow; EGW = evaporation from ground water in case of high water table; CPR = recharge from crop 1 and; STR = recharge from shallow tubewell conveyance system; CWR = recharge from canal conveyance system; CCR = recharge from carrier channel system and drains; and MGW = mining allowance.

A certain minimum quantity of water has to be exported for augmenting the canal supplies for use in other areas.

$$\sum_{i=1}^{n} DT_{it} \geq DT_{min} \tag{21}$$

in which DT_{min} = minimum quantity of water to be pumped by deep tubewell for export.

The cropped area under each category (normal and alkali) cannot exceed the available area for crop production.

$$\sum_{i=1}^{n} \sum_{j=1}^{m} L_{ijt}\ A_{ijt} \leq NL \tag{22}$$

$$\sum_{i=1}^{n} \sum_{j=m+1}^{m} L_{ijt}\ A_{ijt} \leq AL \tag{23}$$

in which L_{ij} = land use coefficient indicating the occupancy of land during a given period. L_{ij} = 1 if land is occupied otherwise it is zero; NL = area under normal soil, ha; AL = area under alkali soils, ha.

Minimum production of certain essential cereal crops has to be ensured by putting a lower limit on the area under these crops. Similarly, the area under certain high value crops has to be restricted to reflect the market demand.

The value of decision variables contained in the objective function cannot be less than zero.

$$A_j \geq 0;\ CW \geq 0;\ ST_i \geq 0;\ DT_i \geq 0 \tag{24}$$

MODEL INTERPRETATION

The linear programming model formulation in Equations (14–24) deals with a number of crop activities, for normal and alkali soils, irrigation water release by canal and shallow tubewells and the water export activities. The time dependency of the agricultural system is accounted for by adopting a shorter time interval of 15 days and entering the discrete values of all the inputs for each of the time increments during the growth cycle. In order to incorporate a non-linear crop water production function the curve for yield-water use is split up into discrete levels of water use with clearly defined maximum and minimum levels (Pomredo, 1978). In this way, each crop at different levels of water use, or when sown on different dates, is considered as

a separate activity; thus, in fact, a multiple activity for each crop is possible.

Hydrologic relationships, crop water requirements and crop yields in the alkali area undergo changes from year to year. This transient nature of the input variables and the response functions are incorporated into the system model by entering it with predetermined inputs based on time dependent functions for the various parameters at various stages of reclamation.

PHYSICAL SYSTEM

The physical system selected for testing and verification of the preceding model is an area commanded by the Jundla

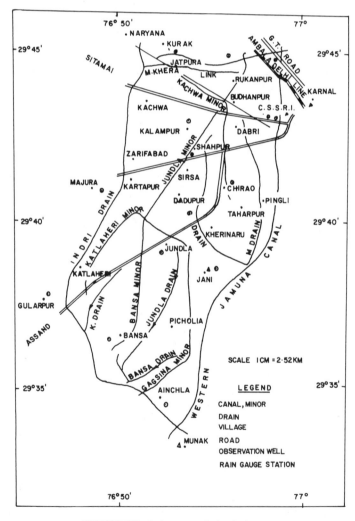

FIGURE 23. Index map of physical system.

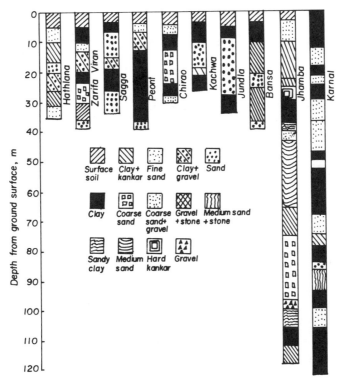

FIGURE 24. Logs of wells in Jundla command area.

canal system, a distributory of the Western Jamuna Canal system (WJC) in Haryana, India. This is a compact block of 21,500 ha (with cropable area of 16,000 ha) with canals and drains to separate its boundaries from the adjoining areas (Figure 23).

The climate of the area is sub-tropical and semi-arid with an average annual rainfall of 72 cm, while the annual pan evaporation value is 189.5 cm.

Geologically, the area forms a part of the Indo-Gangetic alluvial plain, with alluvial deposits varying from 100–2,000 m in depth (Tyagi 1980). The sub-soil formation makes developing cavity tubewells at 15–20 m depth possible.

The soil is broadly classified into normal and alkali categories. Normal soils have a pH below 8.5 and a comparatively high infiltration rate (5.0 cm/day). These soils have light texture with good internal drainage and occupy nearly 70% of the cropable land. All important crops can be grown in these soils.

The alkali soils account for the remaining 30% of the cropable 4,800 ha land. The alkali soils are usually interspersed with normal soils. These soils are now being reclaimed by adopting a reclamation technology.

The area receives irrigation water from the River Jamuna

through the Western Jamuna Canal. The canal water supplies are generally not adequate and meet only 30–35% of the irrigation need. The sub soil strata (Figure 24) is such that it is possible to install shallow tubewells. A large number of shallow cavity tubewells (15–25 m deep) have, therefore, been installed in the region. Besides shallow tubewells, some deep tubewells (75–150 m deep) have been installed by the government. Water from deep tubewells is used mostly for export to areas outside the study area. The increased use of ground water caused by the demands for reclaiming alkali soils has resulted in a decline of the ground water table (Figure 25) in the study area.

Estimation of Parameters

The three main components of the system consist of parameters related to: (1) water balance; (2) crop production and water use; and (3) costs and benefits associated with different activities.

The parameters relating to the water balance analysis (Equation 20) are CW, ST, DT, CWR, STR, and CPR. These parameters, incidentally, are functions of decision variables (Equation 14) and are actually computed within the program itself, once the recharge coefficients are known (Tables 16

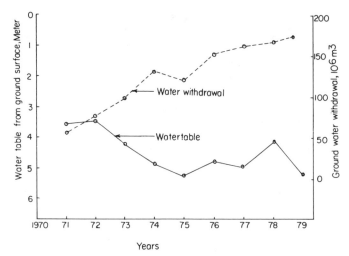

FIGURE 25. Water table fluctuations and ground water withdrawal in the project area.

and 17). However, the recharge from carrier channels and drains *(CCR)* in the area is assumed to be exogeneously given to be $66 \times 10^6 m^3$ per yr. The estimated values of *IGW*, *CGW*, and *EGW* were found to be small and, hence, neglected. Based on data from the past five years (1974–78), the value of DT_{min} (minimum quantity of water export) was taken as $35 \times 10^6 m^3/yr$.

The crop water yield relationship developed by Tyagi (1980) was included in the program. The irrigation practice which gave maximum yield was designated as high irrigation and practices at 80 and 60% of this high irrigation value in the case of rice, and 75 and 50% in other crops, were called medium and low irrigations. The water-crop production function for rice and wheat, which are the principal crops for the agroclimatic conditions prevailing in the study area, are shown in Figure 26.

Since the water requirement and yield potential of alkali soils under reclamation change with time, they must be known for different time periods. Rice, berseem, and wheat, the three recommended crops for the area, attain the potential yields (the yields on normal soils of the region) only after approximately 5, 7, and 10 years of reclamation, respectively (Tyagi, 1980). For determining the optimal policies at 1, 5, and 10 years after reclamation, the crop yield as a function of time is given in Table 14.

In case of crops like wheat and berseem, irrigation is generally scheduled on the basis of soil moisture deficit in the rootzone, and the changes in irrigation requirements with time are negligibly small (Tyagi, 1980). But in the case of rice, the situation is different. The evaporation remains constant, but the deep percolation losses increase continuously with the progress in reclamation; hence the irrigation requirements also increase (Figure 21).

The economic parameters contained in the objective function included: (1) Income from agricultural crops; (2) income from water; (3) cost of canal water; and (4) cost of shallow tubewell water. The sale prices of crops are fixed by the government every year, and those fixed for 1978–79 were used in the present study (Table 18). The maximum and minimum limits on area under different crops are given in Table 19. The sale price of water export (p^{DT}) is hypothetical, and it has been taken as $18.29/10^3 m^3$. It may be mentioned that the cost of water pumping from deep tubewells and its conveyance is $12.20/10^3 m^3$. The cost of canal water was taken as $1.22/10^3 m^3$ which is the actual rate charged by the government. The cost of shallow tubewell water was taken as $9.76/10^3 m^3$.

Using the values of the parameters, the conjunctive use model was run for computations at five discrete stages of

TABLE 16. Recharge Coefficients Used in Model.

System Component	Recharge Coefficient	Remarks
Canal conveyance and distribution system	0.80	The recharge coefficient is the fraction of water loss at various stages of irrigation system that joins the ground water.
Shallow tubewell conveyance and distribution system	0.80	
Carrier channels and drains	0.80	
Crop land		
a. Normal soils	0.50	
b. Alkali soils	0.25	

TABLE 17. Efficiencies of Different Irrigation System Components in Study Area.

System Component	Efficiency	Remarks
Canal system	0.72	
Shallow tubewells system	0.93	1. Efficiency under items 1–3 refers to conveyance efficiency while under item 4 it refers to field water application efficiency.
Carrier channels and drains	0.70	
Crop land		
1. Rice in normal soil	0.35	2. Field water application efficiency of crop not specifically mentioned in this table is taken as 0.60.
2. Rice in alkali soil	0.50	
3. Berseem in normal soil	0.50	
4. Berseem in alkali soil	0.58	3. *These crops are grown in normal soils only.
5. Wheat in normal soil	0.58	
6. Barley	0.60	
7. Gram*	0.60	
8. Potatoes*	0.60	
9. Sugarcane*	0.60	

reclamation, namely after 1, 5, 10, 15, and 20th year. The computed results were assessed for: (1) selecting the best cropping pattern; (2) utilizing the irrigation system capacity in the best possible manner; and (3) maintaining a favourable ground water balance. An analysis has been done to study the changes of the optimal solution found at the initial state of reclamation because of changes in the parameters concerning water resources management and the objective function cost coefficients.

The computed and existing cropping pattern at the initial stage of reclamation are presented in Figure 27. The present area under rice is 69%, but the optimal plan gives 58%. The area under wheat is brought down from 75% to 46%. The area gained in the optimal plan is used by sugarcane (20%) and potatoes (12.5%). In the optimal crop plan, this imbalance in different cropped areas seems to have been rectified. The area under these crops, particularly sugarcane

and potatoes was very small. In this analysis it is also noticed that the available quantity of water cannot meet the total irrigation requirements, which means that only low levels of irrigation is possible. As a result, the benefits realized through optimization of the cropping pattern are lower than the full potential.

The water diversion requirements from canals and ground water storage, at the initial stage of reclamation are shown in Figure 28. Of the total $165 \times 10^6 m^3$ water available in the project area 28.4 percent from canal and the remaining 71.6 percent is obtained from ground water storage. At zero mining allowance, the maximum utilizable ground water potential is $118.5 \times 10^6 m^3$. The peak diversion requirements for canals are 4.3 m³/sec and 18.07 m³/sec for tubewells. The peak demand falls in the first fortnight of July during the rice transplanting period.

During the reclamation, the soil improvement created a

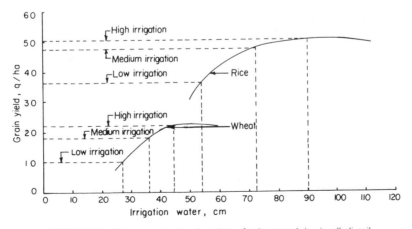

FIGURE 26. Water production function of wheat and rice in alkali soil.

TABLE 18. Sale Price of Different Crops.

Crop 1	Sale Price per Metric Ton, in Dollars 2	Crop 3	Sale Price per Metric Ton, in Dollars 4
Rice	103.66	Berseem (fodder)	12.20
Maize	121.95	Wheat	137.20
Jowar (fodder)	12.20	Barley	103.66
Sugarcane	15.24	Gram	231.71
Potatoes	30.49		

change in crops and cropping pattern. The following are the predicted changes in the optimal cropping pattern over a period of 20 years.

1. Progressive increase in the area under wheat, both in the normal as well as reclaimed alkali soil, raises the area cropped from 7,332 ha to 8,800 ha (20% increase) in the 15th year after reclamation.
2. The total area under rice remains constant during the entire period of reclamation (20 yr), but the level of irrigation shifts from low to medium.
3. The area of berseem in reclaimed soils decreases continuously and is simultaneously taken over by wheat. Higher productivity of wheat at the medium level of irrigation is the principal reason for this shift.
4. The area under gram (a crop generally grown only in normal soil) increases considerably in the final years of reclamation.

INCOME

The continuous improvement of the soil improves the crop yields. As a result, the annual income increases from an initial value of 13.23×10^6 to 15.6×10^6 in the 10th yr of reclamation (Figure 29). Any further improvement of the soil does not increase the crop yields but the irrigation re-

quirements continue to increase. This puts considerable strain on the limited water resources of the area and simultaneously increases the irrigation costs. The total income has, therefore, decreased after the 10th year. This means that suitable steps such as soil puddling, weed control, etc., which are not recommended in the initial years of reclamation, must be taken well in advance in order to maintain favourable soil properties after the 10th year.

IRRIGATION SYSTEM CAPACITY

The initial peak demand for shallow tubewell water during a 15 day period is $18 \times 10^6 m^3$, and it occurs during July which is the rice transplanting period. This peak demand can be met from 2,090 shallow tubewells of 10 L/s discharge. The requirement increases to 2,300 tubewells in the 15th yr, but there is no further increase at subsequent stages. This additional demand for water is met from increased ground water recharge (Figure 30). It should be mentioned that the ground water recharge is simulated to increase by as much as 17.5%. The variable DT, representing the requirement of deep tubewell water, appears in the solution at the minimum prescribed level of $35 \times 10^6 m^3/yr$, with a peak value of $6 \times 10^6 m^3/15$ day period (which is the existing installed capacity).

PARAMETRIC ANALYSIS

The parametric analysis has been applied to study the behaviour of the optimal plan with respect to changes in water resources management such as the level of canal water supply, quantity of water export, the mining allowance, and the change in cost or income coefficients.

The levels of water resources management parameters at which optimal solutions were obtained are given in Table 20. These values represent the possible ranges within which these parameters may vary. In the initial stage of reclamation, the total water available for irrigation is $165 \times 10^6 m^3/yr$. This value is denoted by IWO. The water availabil-

TABLE 19. Minimum and Maximum Limits of Area Under Certain Crops.

Crop (1)	Limit (2)	Area Hectares (3)
Rice in alkali soils	≥	4500
Wheat in alkali soils	≥	3000
Sugarcane	≤	3200
Potatoes	≤	2000
Berseem	≤	1500

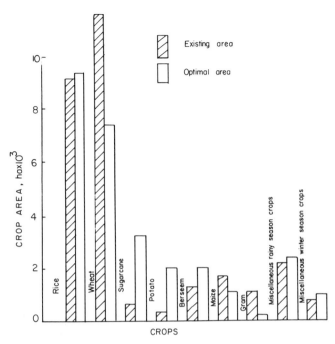

FIGURE 27. Existing and optimal area under different crops at the initial stage of reclamation.

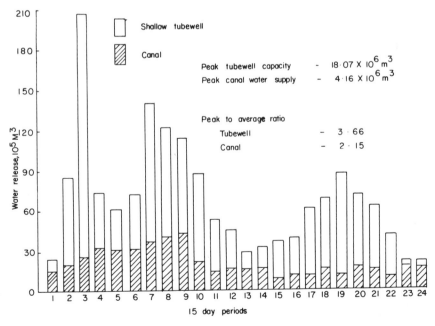

FIGURE 28. Irrigation water diversion requirements from canals and shallow tubewells, at 15 day interval.

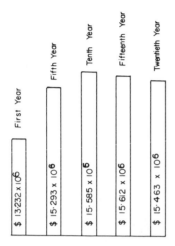

FIGURE 29. Income of project area at different stages of reclamation.

TABLE 20. Values of Water Resources Management Variables for Parametric Analysis.

Serial Number (1)	Parameter (2)	Value (3)
1.	Canal water	60, 80, 120 and 140 percent of expected level.
2.	Water export, 10^6 m³/year	0, 10, 25, 40 and 50
3.	Mining allowance, 10^6 m³/year	0, 10, 20 and 25

ity at each of the changed levels is called by *IWK*. The ratio *IWK/IWO* is taken as the independent parameter that changes for studying the response of the optimal solution. The results of the parametric analysis are discussed below.

An increase in the *IWK/IWO* ratio, irrespective of the source, predicts an increasing income. One of the factors responsible for this increase in income is the increase in area under rice and wheat because the level of irrigation could be raised to high level. For the values of *IWK/IWO* ranging from 1.00 to 1.18, the increase in income is only marginal, but for values from 1.18 to 1.25 the annual income increased at a high rate from 13.23×10^6 to 15.8×10^6/yr.

As expected, for values of *IWK/IWO* greater than one, the design capacity of the irrigation system increases. This in-

crease in design capacity of canal system (CAPCW) occurs only if the increase in the *IWK/IWO* ratio is due to changes in the canal supply *(CW)* level. But the capacity requirement for shallow tubewells increases with a positive change in any of the other water resource management parameters namely *CW*, *DT*, and *MGW*. For example, the number of shallow tubewells increases from 2,091 to 2,972 when the *IWK/IWO* ratio is increased from 1.0 to 1.3 by reducing water export to the zero level.

The fluctuations in water table that may occur in the area when the *IWK/IWO* ratio is changed was also investigated by changing mining allowance *(MGW)* from zero to 10, 20, and 25×10^6 m³/yr (Figure 31). The results show that if mining were allowed for a period of five year at the three rates under investigation, the water table would fall by about 1.8, 3.60, and 4.50 m, respectively, from the existing level of 4.65 m. This will make single stage centrifugal pumping inoperative due to higher suction lift and would require changing to multi-stage pumps.

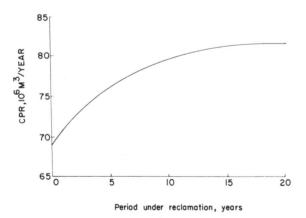

FIGURE 30. Recharge from crop land (CPR) at different periods under reclamation.

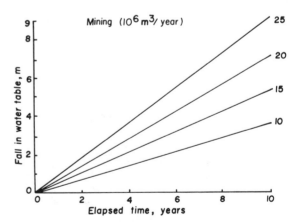

FIGURE 31. Fall in water table at different levels of mining (10^6 m³/year).

COST COEFFICIENTS OF THE OBJECTIVE FUNCTION

Only three coefficients, (namely costs of canal (C^{CW}) and shallow tubewell waters (C^{ST}) and the sale price of exported water (P^{DT}), were subjected to parametric analysis. A reduction in the cost of canal water (C^{CW}) from $1.22/10^3 m^3$ or an increase to $1.83/10^3 m^3$ does not alter the level of decision variables. Similarly, a reduction in cost of shallow tubewell water (C^{ST}) from $7.32/10^3 m^3$ to $3.66/10^3 m^3$ and an increase to $10.97/10^3 m^3$ produces no change in the optimal plan.

It indicates that the optimal plan is not very sensitive to changes in irrigation water charges as the marginal value of the irrigation water is quite high.

The parametric analysis with respect to the sale price of exported water (P^{DT}) indicates that it would be uneconomical to export water unless the sale price were higher than $89.76/10^3 m^3$, which is equivalent to its marginal value. Even at this price, the water export activity enters the solution at its minimum prescribed level of $35 \times 10^6 m^3/yr$. The continuation of the water export program can, therefore, be justified only as a social measure and not otherwise. It may be added that these optimal decisions are applicable to the area for which the study has been made.

CONCLUSIONS

Agriculturally, as well as hydrologically, alkali soils under reclamation continue to be in a transient stage over a long period of time. Therefore, policies that may be optimal at a given point of time will need review as the degree of soil reclamation improves. The methodology developed in this paper can be easily applied to determine appropriate land and water use policies for the period under transition. It is advantageous to initiate alkali land reclamation, even if it has to be done by reducing water allocation to areas with normal land or by shifting to less water requiring crops. The level of irrigation can be raised at subsequent stages when the water availability increases due to higher ground water recharge from cropland.

REFERENCES

1. Abrol, I. P. and C. L. Acharya, "Soil Water Behaviour and Irrigation Frequencies in Soils with Physical Constraints," *Proc. Second World Congress.* Water Resources Association, New Delhi, Vol. 1, pp. 335–342 (1975).
2. Abrol, I. P. and D. R. Bhumbla, "Crop Response to Differential Gypsum Applications in Highly Sodic Soil and the Tolerance of Several Crops to Exchangeable Sodium Under Field Conditions," *Soil Science, Vol. 127,* No. 2, pp. 79–85 (1979).
3. Abrol, I. P., K. S. Dargan, and D. R. Bhumbla, "Reclaiming Alkali Soils," CSSRI, Karnal Bulletin No. 2, p. 58 (1973).
4. Agarwal, M. C. and A. C. Goel, "Effect of Field Levelling Quality on Irrigation Efficiency and Crop Yield, *Agricultural Water Management, Vol. 4,* No. 3, pp. 457–467 (1981).
5. American Society of Agricultural Engineers Surface Drainage Committee (SW-233). "Design and Construction of Surface Drainage Systems on Peats in Humid Areas," *Engineering Practice E P-302, ASAE,* p. 8 (1976b).
6. Bernstein, L., "Crop Growth and Salinity," in *Drainage for Agriculture* (Ed. J. Van Schilfgaarde), American Society of Agronomy Monograph No. 17, Madison, Wisconsin, pp. 39–54 (1974).
7. Bhargava, G. P., R. C. Sharma, and D. K. Pal, "A Case Study of Distribution and Formation of Salt Affected Soils in Haryana State," *International Symposium on Salt Affected Soils,* CSSRI, Karnal (India), pp. 83–91 (1980).
8. Bresler, E., B. L. McNeal, and D. L. Carter, *Saline and Sodic Soils,* Springer Verlag Publishing Co., New York, p. 236.
9. Brinkman, R., "Saline and Sodic Soils," in *Land Reclamation and Water Management,* International Institute for Land Reclamation and Improvement, Wageningen, The Netherlands, Publication No. 27, pp. 62–72 (1980).
10. Buras, N., *Scientific Allocation of Water Resources,* American Elseweir Publishing Co., Inc., p. 208 (1972).
11. Chaudhary, G. H. and B. P. Warkentin, "Studies on Exchange of Sodium from Soils by Leaching with Calcium Sulphate," *Soil Science, Vol. 105,* pp. 190–197 (1968).
12. Davidson, J. L. and J. P. Quirk, "The Influence of Dissolved Gypsum on Pasture Establishment on Irrigated Sodic Clays," *Australian Journal of Agricultural Resources, Vol. 12,* No. 2, pp. 100–110 (1961).
13. Doering, E. J. and W. O. Willis, "Chemical Reclamation for Sodic Strip Mined Spoils," *USDA, ARSNC-20,* p. 8 (1975).
14. Dukhovny, V. K., "Saline and Alkali Soils, Their Use, Improvement and Related Problems," *State of Art: Irrigation, Drainage and Flood Control* (Ed. K. K. Framji), I.C.I.D. New Delhi, Vol. 1, pp. 392–458 (1976).
15. Gaul, B. L., I. P. Abrol, and K. S. Dargan, "Frequent and Light Irrigation: Best for Saline Sodic Soils," *Indian Farming, Vol. 24,* No. 2, pp. 14–15 (1973).
16. Gedroit, K. K., "Colloidal Chemistry as Related to Soil Science, I, Formation of Sodium Carbonate in Soils, Alkali and Saline Soils," *Journal of Experimental Agronomy, Vol. 13,* pp. 362–412 (1912).
17. Gorelick, S. M., "A Review of Distributed Parameter Ground Water Management Modelling Method," *Water Resources Research, Vol. 19,* No. 2, pp. 305–319 (1983).
18. Govindrajan, S. V. and R. S. Murty, "Physico-Chemical Properties of Alluvial Soils Containing High Sodium Carbonate in Indo-Gangetic Basin," *Agrochem. Talajt, Vol. 23,* No. 1, pp. 243–245 (1969).
19. Gupta, S. K. and R. N. Pandey, "Crop and Water Yields as Affected by Rainwater Storage in Rice-Fields—A Field Evaluation," *Field Crop Research, Vol. 2,* pp. 365–371 (1979).
20. Hall, W. A. and J. A. Dracup, *Water Resources Systems Engineering,* McGraw Hill Book Co., Inc., New York, p. 372 (1971).

21. Hira, G. S., M. S. Bajwa, and N. T. Singh, "Prediction of Water Requirements for Gypsum Dissolution in Sodic Soils," *Soil Science, Vol. 137*, pp. 353–358 (1981).

22. Israelsen, O. W. and V. E. Hansen, *Irrigation Principles and Practices,* John Wiley and Sons, Inc., New York, p. 437 (1962).

23. Kelley, W. P., *Alkali Soils: Their Formation, Properties and Reclamation,* Rainhold Publishing Corporation, New York (1951).

24. Keren, R. and I. Shainberg, "Effect of Dissolution Rate on the Efficiency of Industrial and Mined Gypsum in Improving Infiltration Rate of Sodic Soil," *Soil Science Society of America Journal, Vol. 45*, pp. 103–107 (1981).

25. Khepar, S. D., M. C. Chaturvedi, and B. K. Sinha, "Effect of Precision Land Levelling on Increase the Crop Yield and Related Economic Decisions, *Journal of Agricultural Engineering, Vol. 19*, No. 4, pp. 23–27 (1982).

26. Lakshminarayana, V. and S. P. Rajgopalan, "Optimal Cropping Pattern for the Basin in India,"*Journal of the Irrigation and Drainage Division, ASCE, Vol. 103*, No. IR2, pp. 53–70 (1977).

27. Luthin, J. N., *Drainage Engineering,* Wiley Eastern Pvt. Ltd., New Delhi, p. 250 (1966).

28. Matanga, G. B. and M. A. Marino, "Irrigation Planning: Cropping Pattern," *Water Resources Research, Vol. 15*, No. 3, pp. 672–678 (1979).

29. Narayana, V. V. D., R. N. Pandey, and S. K. Gupta, "Drainage of Alkali Soils," *Journal of Indian Association of Hydrologists, Vol. 1*, No. 2, pp. 21–28 (1977).

30. Narayana, V. V. D., "Reclaiming Alkali Soils—Engineering Aspects," *CSSRI, Karnal Bulletin No. 6*, p. 62 (1980).

31. Narayana, V. V. D. and N. K. Tyagi, "Irrigation Practices for Reclaimed Alkali Soils," *Journal of Irrigation and Drainage Engineering, ASCE Vol. 107*, No. 4, pp. 400–406 (1981).

32. Overstreet, R., J. C. Martin, and H. M. King, "Gypsum, Sulphur and Sulphuric Acid for Reclaiming an Alkali Soil of Fresno Series," *Hilgardia, Vol. 21*, pp. 113–120 (1951).

33. Pandey, R. N. and O. P. Singh, "Evaluation of Irrigation Methods Under Salt Affected Soils," *CSSRI, Karnal Annual Report*, p. 95 (1976).

34. Pandey, R. N., V. V. D. Narayana, N. K. Tyagi, and S. K. Gupta, "Management of Sloping Borders for Rice Cultivation in Alkali Soils," *Indian Journal of Agricultural Sciences, Vol. 47*, pp. 613–616 (1977).

35. Pearson, G. A. and L. Bernstein, "Influence of Exchangeable Sodium on Yield and Chemical Composition of Plants II, Wheat, Barley Oats, Rice, Tall Fescue and Tall Wheat Grass," *Soil Science, Vol. 96*, pp. 254–261 (1958).

36. Pearson, G. A., "Tolerance of Crops to Exchangeable Sodium," *USDA Information Bulletin 216*, p. 4 (1960).

37. Pomredo, C., "Economic Analysis of Irrigation Production Function: An Application to Linear Programming," *Water Resources Bulletin, Vol. 14*, No. 1, pp. 23–34 (1978).

38. Prather, R. J., J. O. Goertzen, J. D. Rhoades, and H. Frenkel, "Efficient Amendment Use in Sodic Soil Reclamation," *Soil Science Society of America Journal, Vol. 42*, pp. 782–786 (1979).

39. Prihar, S. S., B. S. Sandhu, and N. T. Singh, "A Critical Appraisal of Research on Irrigation Scheduling to Crops," *Proc. Second World Congress on Water Resources for Human Needs*, Water Resources Association, New Delhi, Vol. 1, pp. 343–357 (1975).

40. Quirk, J. P. and R. K. Schofield, "The Effect of Electrolyte Concentration on Soil Permeability," *Journal of Soil Science, Vol. 6*, pp. 163–178 (1955).

41. Raadsma, S. and F. E. Schulze, "Surface Field Drainage Systems—Drainage Principles and Applications," *Design and Management of Drainage Systems, Publication 16, Vol. 4*, p. 476 (1974).

42. Rhoades, J. D., "Drainage for Salinity Control," in *Drainage for Agriculture* (Ed., J. V. Schilfgaarde), American Society of Agronomy Monograph (4) pp. 433–461 (1974).

43. Rhoades, J. D., "Reclamation and Management of Salt Affected Soils After Drainage," *Proceedings First Annual Western Provincial Conference, Rationalization of Water and Soil Research and Management—*Soil Salinity, Alberta, Canada, pp. 123–198 (1982).

44. Richards, D. A., "Chemical and Physical Characteristics of Saline and Alkali Soils of Western United States," *Transactions of International Congress of Soil Science, Vol. 1*, pp. 378–383 (1950).

45. Richards, L. A. (ed.), "Diagnosis and Improvement of Saline and Alkali Soils, *USDA Handbook No. 60*, p. 160 (1968).

46. Signmond, A. A. J. De., *The Principles of Soil Science*, Illus., London, p. 362 (1938).

47. Singh, O. P., V. V. D. Narayana, and A. K. Tiwari, "Analysis of Hydrologic Changes Taking Place in Alkali Soil Watershed During Reclamation," *Canadian Journal of Agricultural Engineering, Vol. 21*, pp. 147–150 (1979).

48. Singh, O. P. and N. K. Tyagi, "Estimating Rainwater Availability in Reclaimed Alkali Soils," *International Symposium on Salt Affected Soils*, CSSRI, Karnal, pp. 549–554 (1980).

49. Singh, S. B., R. Chhabra, and I. P. Abrol, "Effect of Exchangeable Sodium on Yield and Chemical Composition of Raya (Brassica Juncea)," *Agronomy Journal, Vol. 71*, pp. 767–770 (1981).

50. Singh, S. B., R. Chhabra, and I. P. Abrol, "Effect of Exchangeable Sodium on the Yield, Chemical Composition and Oil Content of Safflower and Linseed," *Indian Journal of Agricultural Sciences, Vol. 51*, No. 12, pp. 885–889 (1981).

51. Smith, D. V., "Systems Analysis and Irrigation Planning," *Journal of Irrigation and Drainage Division, ASCE, Vol. 99*, No. IR 2, pp. 89–110 (1973).

52. Soil Science Society of America, "Glossary of Soil Terms," *Soil Science Am.*, 677, S. Segol Road, Madison, U.S.A. (1974).

53. Szabolcs, I., "Present and Potential Salt Affected Soils: An Introduction," *Prognosis of Salinity and Alkalinity, FAO Soils Bulletin 31*, pp. 9–13 (1976).

54. Szabolcs, I., "Salinity and Alkalinity of Soils, Extension,

Classification and Main Properties of Salt Affected Soils," *Proceedings Indo-Hungarian Seminar on Management of Salt Affected Soils,* CSSRI, Karnal (India), pp. 20–30 (1977).

55. Szabolcs, I., "Saline and Alkali Soils: Commonalities and Differences," *International Symposium on Salt Affected Soils,* CSSRI, Karnal, pp. 1–6 (1980).

56. Tyagi, N. K., "Crop Planning and Water Resources Management in Salt Affected Soils: A Systems Approach," Thesis submitted to the J. N. Technological University in Hyderabad in 1980, in Partial Fulfillment of the Requirements of the Degree of Doctor of Philosophy, p. 236.

57. Tyagi, N. K. and V. V. D. Narayana, "Evaluation of Some On-Farm Water Management Constraints in Surface Irrigation System," *I.C.I.D. Bulletin Vol. 32,* No. 8, pp. 11–16 (1983).

58. Tyagi, N. K., "Use Graded Borders in Alkali Soils," *Indian Farming, Vol. 33,* No. 8, pp. 14–16 (1983).

59. Tyagi, N. K. and V. V. D. Narayana, "Planning for Alkali Land Reclamation Under Rainfall Uncertainty," *Ecological Modelling, Vol. 20,* pp. 243–258 (1984).

60. Tyagi, N. K. and V. V. D. Narayana, "Water Use Planning for Alkali Soils Under Reclamation," *Journal of Irrigation and Drainage Engineering, ASCE, Vol. 110,* No. 2, pp. 192–207 (1984).

61. Tyagi, N. K., "Effect of Land Surface Uniformity on Some Economic Parameters of Irrigation in Sodic Soil Under Reclamation," *Irrigation Science, Vol. 5,* No. 3, pp. 151–160 (1984).

62. Tyagi, N. K., R. K. Gupta, and O. P. Singh, "A Probabilistic Approach to Irrigation Scheduling," *Journal of Institution of Engineers (India), Vol. 64,* Part A G1–2, pp. 116–120 (1984).

63. Yadav, J. S. P., "Reclamation of Salt Affected Soils," *Proceedings Indo-Hungarian Seminar on Management of Salt Affected Soils,* CSSRI, Karnal, pp. 134–139 (1977).

64. Yakowitz, S., "Dynamic Programming Applications in Water Resources," *Water Resources, Research, Vol. 18,* No. 4, pp. 673–696 (1982).

SECTION THREE
Environmental

CHAPTER 20

Industrial Waste

G. E. Ho*

INTRODUCTION

Industrial wastes are generated because raw materials used contain impurities, industrial processes are not one hundred percent efficient and the processes themselves may produce unwanted by-products or rejects. The quantities of wastes generated are therefore generally related to the amounts of products although within a particular industry there can be a wide variation of waste produced per unit product depending on the technology employed, plant operational practices and conservation measures adopted.

Industrial wastes can be classified into solid, liquid or gaseous. This classification is convenient since it is related to the land, water (including groundwater) and atmospheric environments into which the major innocuous portion of the wastes is to be discharged. The problems associated with industrial wastes are due to substances contained in the wastes which may be toxic to organisms or generally detrimental to ecosystems supporting life on earth.

Pollutants

Table 1 lists pollutants and their industrial sources. The list of pollutants in the table is by no means comprehensive. If a constituent of an industrial waste is suspected to be a pollutant, a check on the properties of the constituent is recommended [30]. Some waste products have also been known to be toxic even though the exact constituents causing the toxicity have not been fully identified.

Major pollutants of gaseous industrial wastes are particulates, sulfur oxides, nitrogen oxides, hydrocarbons and carbon monoxide, while of liquid industrial wastes are sus-

pended solids, oil and grease, organic matter requiring oxygen when decomposed by microorganisms in the environment, phosphorus and nitrogen, and acids and alkalis. Some industrial wastes consist of or contain substances which are dangerous to life, and a special term, hazardous wastes, has been given. These substances can present danger to human health because of their properties of flammability, corrosivity, reactivity, toxicity, or carcinogenecity.

Hazardous wastes

The U.S. Environmental Protection Agency has made use of the following criteria in determining hazardous waste [15]:

1. It is not excluded in listings which identify domestic, point source discharges to surface waters, overburden waste from mines and sewage sludge from publicly owned treatment works.
2. It exhibits any of the following characteristics: ignitibility (a flash point of less than 60°C); corrosivity (pH less than or equal to 2 or greater than or equal to 12.5, or which corrodes steel under specified conditions at a rate greater than 6.35 mm per year); reactivity (normally unstable and readily undergoes violent change, or when mixed with water reacts violently, or forms explosive mixtures, or generates toxic gases); toxicity (leachate produced at a pH of 5 contains a pollutant which exceeds the Federal Drinking Water Standard by 100-fold).
3. It contains any toxic constituents listed in Table 2, unless the waste is not capable of posing a substantial present or potential hazard to human health or the environment even when improperly treated, stored, transported, disposed or otherwise managed.
4. It is listed in prescribed lists, e.g., Tables 3 and 4. These lists are updated periodically and the most recent Federal Register and its indices should be consulted.

*School of Environmental and Life Sciences, Murdoch University, Murdoch, Australia

TABLE 1. Pollutants and Their Sources.

Chemical		Examples	Industrial source
Acids	L	Hydrochloric acid	Pickling
		Nitric acid	Chemical reagent
		Sulphuric acid	Byproducts, petrochemicals
		Acetic acid	Petrochemicals
Aldehydes	L	Acetaldehyde	Petrochemicals
Alkalis	L	Sodium hydroxide	Electroplating
		Lime	Beverage
			Photography
			Vegetable and fruit processing
Amonia, ammonium	G,L		Coal distillation
			Nitric acid
			Urea and ammonium nitrate works
			Food processing
Aniline and related compounds	L		Dyestuffs
Aromatic hydrocarbons	G,L	Benzene	Coal tar
		Toluene	Combustion
			Petrochemicals
			Pesticides
			Herbicides
Aromatic amines	G,L	4-aminodiphenyl	Dyes
		benzidine	Rubber
		2-naphthylamine	Coal gas
Arsenic	G,L,S	Arsine	Pigment and dye
		Arsenous acid and salts	Pesticide and herbicide
			Metallurgical processing of other metals
			Glass and ceramic
			Tanneries
Asbestos	S	Chrysotile	Equipment and building
		Amosite	Insulation
		Crocidolite	Fillers in various industries
			Motor vehicle assembly
			Fabric manufacture
			Polymers, plastics
Benzene	G,L		Glues
			Varnishes
Carbon monoxide	G		Coke ovens
			Incomplete combustion
			Smelting
			Metal extraction and refining
Chlorinated hydrocarbons	G,L,S	Trichlorethylene	Degreasing
		1,1,1-trichloroethane	Drycleaning
		DDT	Solvents
		BHC	Pesticides
		Aldrin	Wood treatment
		Dieldrin	
Chlorine and chlorides	G,L,S		Chlorinated hydrocarbons
			Chloralkali
			Paper and pulp
			Petrochemicals
			Metal extraction and refining
Chromium and compounds	L,S	Chromic acid	Anodising
		Sodium dichromate	Cement
			Dyes
			Electroplating
			Paint
			Tanneries

(continued)

TABLE 1 (continued).

Chemical		Examples	Industrial source
Cobalt and compounds	L,S	Cobalt oxide	Catalysts
			Fibres
			Paint
			Paper and pulp
			Pickling
Copper and compounds	L,S	Copper sulphate	Electroplating
		copper pyrophosphate	Electrical and electronics
		Cuprammonium compounds	Etching
			Pesticides
Cyanate	L		Coal distillation
			Oxidation of cyanide
Cyanides	L,S	Sodium cyanide	Heat treatment of metal
		Copper cyanide	Photographic
			Coal distillation
			Electroplating
			Synthetic fibre
Fluorides	G,L,S	Hydrogen fluoride	Bricks
		Calcium fluoride	Fertiliser
			Aluminum
Hydrocarbons, general			Coal distillation
			Petrochemicals
			Refineries
Ionizing radiation	G,L,S	Uranium	Nuclear power cycle industry
			Nuclear weapons industry
Iron and compounds	L,S	Iron oxide	Aluminum
		Ferrous chloride	Electroplating
			Pickling
			Pigments
			Electronics
			Titanium dioxide
Lead and compounds	L,S	Lead oxide	Batteries
		Tetraethyl lead (TEL)	Printing
			Combustion
			Explosives and pyrotechnics
			Pesticides
			Paint
			Refineries
			Petrochemicals
Manganese and compounds			Catalyst
			Batteries
			Glass
			Paint
			Pyrotechnics
Meat wastes	L,S		Meat processing and preparation
			Abattoirs
			Dairies
			Tanneries
Mercaptans	G		Refineries
			Coke ovens
Mercury	L,S	Methyl mercury	Herbicides
		Mercurous chloride	Bacterial activity on inorganic mercury
			Pesticides
			Electrical and electronic
			Pesticides
			Explosives
			Batteries

(continued)

561

TABLE 1 (continued).

Chemical	Examples		Industrial source
			Photographic
			Scientific instruments
			Chloralkali
			Paints
			Pharmaceuticals
			Paper and pulp
			Catalysts
			Cement
			Combustion of coal and oil
Methanol	L		Resins
			Paper
Nickel	L		Automobile
			Metal finishing
Nitrates	L,S	Potassium nitrate	Metals heat treatment
			Water treatment
Nitrogen oxides	G	Nitrogen dioxide	Combustion processes
			Electricity generation
Oil and soluble oil	L		Engineering
			Refineries
			Petrochemicals
Paraquat	L,S		Herbicide
Pesticides	L,S	Chlorinated hydrocarbons (q.v.)	Pesticides
(includes acaricides,		Carbamates (q.v.)	
avicides,		Organophosphorus	
batericides, insecticides		compounds, (q.v.)	
molluskicides, nematocides,			
piscicides, rodenticides)			
Pharmaceuticals	L,S	Asprin	Pharmaceutical industry
		Penicillin	
Phenol and related compounds	L,S	Phenol	Photographic
		Cresol	Coal distillation
			Dyestuffs
			Petrochemicals
			Pesticides
			Refineries
			Explosives
			Plastics
Phosphorus and compounds	L,S	Phosphoric acid	Detergents
			Fertilizers
			Corrosion protection
			Matches
			Boiler blowdown
			Metal finishing
			Food processing
Phthalates	L,S	Dibutyl phthalate	Plasticiser (polymers)
Polychlorinated biphenyls (PCB)	L,S		Pesticides
			Plasticiser in paint and polymers
			Adhesives
			Lubricants and hydraulic fluids
Silicates	L,G		Cement
			Metal extraction and refining
Sulphur oxides	G	Sulphur dioxide	Coal distillation
		Sulphur trioxide	Combustion of coal and heavy
			fuel oil
			Electricity generation
Tar	L		Refineries

(continued)

TABLE 1 (continued).

Chemical	Examples	Industrial source
		Coal distillation
Thiocyanate	L	Coal distillation
Tin and compounds	G,L,S	Tinplating
Titanium and compounds	L,S	Titanium dioxide	Paper
		Paint
		Astronautics
Vanadium and compounds	S	Vanadium pentoxide	Catalysts
		Oil ash
Vinyl chloride	G		Polyvinyl
		Polymer
Wax	S		Paper
		Refineries
		Fruit preserving
		Textiles
Zinc and compounds	G,L,S		Synthetic fibres
		Galvanising
		Electroplating
		Paper and pulp

Key: G—Pollutant occurs in gaseous waste S—Pollutant occurs in solid waste
  L—Pollutant occurs in liquid waste Adapted from Reference [2]; References [8,10].

TABLE 2. Hazardous Constituents.

Acetaldehyde
Acetonitrile
3-(alpha-Acetonylbenzyl)-4-hydroxy-
  coumarin and salts
2-Acetylaminofluorene
Acetyl chloride
1-Acetyl-2-thiourea
Acrolein
Acrylamide
Acrylonitrile
Aflatoxins
Aldrin
Allyl alcohol
Aluminum phosphide
4-Aminobiphenyl
6-Amino-1,1a,2,8,8a,8b-hexahydro-8-
  (hydroxymethyl)-8a-methoxy-5-methyl-
  carbamate azirino (2',3':3,4) pyrrolo
  (1,2-a)indole-4,7-dione (ester) (Mitomy-
  cin C)
5-(Aminomethyl)-3-isoxazolol
4-Aminopyridine
Amitrole
Antimony and compounds, N.O.S.[1]
Aramite
Arsenic and compounds, N.O.S.
Arsenic acid
Arsenic pentoxide
Arsenic trioxide
Auramine

Azaserine
Barium and compounds, N.O.S.
Barium cyanide
Benz(c)acridine
Benz(a)anthracene
Benzene
Benzene, 2 amino-1-methyl (o-toluidine)
Benzene, 2 amino-1-methyl (p-toluidine)
Benzenearsonic acid
Benzenethiol
Benzidine
Benzo(a)anthracene
Benzo(b)fluoranthene
Benzo(j)fluoranthene
Benzo(a)pyrene
Benzoquino and isomers
Benzotrichloride
Benzyl chloride
Beryllium and compounds, N.O.S.
Bis(2-chloroethoxy)methane
Bis(2-chloroethyl) ether
N,N-Bis(2-chloroethyl)-2-naphthylamine
Bis(2-chloroisopropyl) ether
Bis(chloromethyl)ether
Bis(2-ethylhexyl) phthalate
Bromoacetone
Bromomethane
4-Bromophenyl phenyl ether
Brucine
2-Butanone peroxide

Butyl benzyl phthalate
2-sec-Butyl-4,6-dinitrophenol (DNBP)
Cadmium and compounds, N.O.S.
Calcium chromate
Calcium cyanide
Carbon disulfide
Chlorambucil
Chlordane (alpha and gamma isomers)
Chlorinated benzenes, N.O.S.
Chlorinated ethane, N.O.S.
Chlorinated fluorocarbons
Chlorinated naphthalene, N.O.S.
Chlorinated phenol, N.O.S.
Chloroacetaldehyde
Chloroalkyl ethers
p-Chloroaniline
Chlorobenzene
Chlorobenzilate
1-(p-Chlorobenzoyl)-5-methoxy-2-
  methylindole-3-acetic acid
p-Chloro-m-cresol
1-Chloro-2,3-epoxybutane
2-Chloroethyl vinyl ether
Chloroform
Chloromethane
Chloromethyl methyl ether
2-Chloronaphthalene
2-Chloro-1,3-butadiene (chloropene)
2-Chlorophenol
1-(o-Chlorophenyl)thiourea

(continued)

563

TABLE 2 (continued).

3-Chloropropene (allyl chloride)
3-Chloropropionitrile
alpha-Chlorotoluene
Chlorotoluene, N.O.S.
Chromium and compounds, N.O.S.
Chrysene
Citrus red No. 2
Coal Tars
Copper cyanide
Creosote
Cresol
Cresylic acid
Crotonaldehyde
Cyanides (soluble salts and complexes)
   N.O.S.
Cyanogen
Cyanogen bromide
Cyanogen chloride
Cycasin
2-Cyclohexyl-4,6-dinitrophenol
Cyclophosphamide
Daunomycin
DDD
DDE
DDT
Diallate
Dibenz(a.h)acridine
Dibenz(a.j)acridine
Dibenz(a.h)anthracene(Dibenzo(a.h)
   anthracene)
7H-Dibenzo(c.g)carbazole
Dibenzo(a.e)pyrene
Dibenzo(a.h)pyrene
Dibenzo(a.j)pyrene
1,2-Dibromo-3-chloropropane
1,2-Dibromoethane
Dibromomethane
Di-n-butyl phthalate
Dichlorobenzene, N.O.S.
3,3'-Dichlorobenzidine
1,1-Dichloroethane
1,2-Dichloroethane
trans-1,2-Dichloroethene
Dichloroethylene, N.O.S.
1,1-Dichloroethylene
Dichloromethane
2,4-Dichlorophenol
2,6-Dichlorophenol
2,4-Dichlorophenoxyacetic acid (2,4-D)
Dichloropropane
Dichlorophenylarsine
1,2-Dichloropropane
Dichloropropanol, N.O.S.
Dichloropropene, N.O.S.
1,3-Dichloropropene

Dieldrin
Diepoxybutane
Diethylarsine
0,0-Diethyl-S-(2-ethylthio)ethyl ester of
   phosphorothioic acid
1,2-Diethylhydrazine
0,0-Diethyl-S-methylester phosphoro-
   dithioic acid
0,0-Diethylphosphoric acid. 0-p-
   nitrophenyl ester
Diethyl phthalate
0,0-Diethyl-0-(2-pyrazinyl)
   phosphorothioate
Diethylstilbestrol
Dihydrosafrole
3,4-Dihydroxy-alpha-(methylamino)-
   methyl benzyl alcohol
Di-isopropylfluorophosphate (DFP)
Dimethoate
3,3'-Dimethoxybenzidine
p-Dimethylaminoazobenzene
7,12-Dimethylbenz(a)anthracene
3,3'-Dimethylbenzidine
Dimethylcarbamoyl chloride
1,1-Dimethylhydrazine
1,2-Dimethylhydrazine
3,3-Dimethyl-1-(methylthio)-2-butanone-
   0-((methylamino) carbonyl)oxime
Dimethylnitrosoamine
alpha,alpha-Dimethylphenethylamine
2,4-Dimethylphenol
Dimethyl phthalate
Dimethyl sulfate
Dinitrobenzene, N.O.S.
4,6-Dinitro-o-cresol and salts
2,4-Dinitrophenol
2,4-Dinitrotoluene
2,6-Dinitrotoluene
Di-n-octyl phthalate
1,4-Dioxane
1,2-Diphenylhydrazine
Di-n-propylnitrosamine
Disulfoton
2,4-Dithiobiuret
Endosulfan
Endrin and metabolites
Epichlorohydrin
Ethylcarbamate (urethan)
Ethyl cyanide
Ethylene diamine
Ethylenebisdithiocarbamate (EBDC)
Ethyleneimine
Ethylene oxide
Ethylenethiourea
Ethyl methanesulfonate

Fluoranthene
Fluorine
2-Fluoroacetamide
Fluoroacetic acid, sodium salt
Formaldehyde
Glycidylaldehyde
Halomethane, N.O.S.
Heptachlor
Heptachlor epoxide (alpha, beta, and
   gamma isomers)
Hexachlorobenzene
Hexachlorobutadiene
Hexachlorocyclohexane (all isomers)
Hexachlorocyclopentadiene
Hexachlorodibenzo-p-dioxins
Hexachlorodibenzofurans
Hexachloroethane
1,2,3,4,10,10-Hexachloro-1,4,4a,5,8,8a-
   hexahydro-1,4:5,8-endo,endo-
   dimethanonaphthalene
Hexachlorophene
Hexachloropropene
Hexaethyl tetraphosphate
Hydrazine
Hydrocyanic acid
Hydrofluoric acid
Hydrogen sulfide
Indeno(1,2,3-c,d)pyrene
Iodomethane
Iron dextran
Isocyanic acid, methyl ester
Isosafrole
Kepone
Lasiocarpine
Lead and compounds, N.O.S.
Lead acetate
Lead phosphate
Lead subacetate
Maleic anhydride
Malononitrile
Melphalan
Mercury and compounds, N.O.S.
Methapyrilene
Methomyl
Methoxychlor
2-Methylaziridine
3-Methylcholanthrene
4,4'-Methylene-bis-(2-chloroaniline)
Methyl hydrazine
2-Methyllactonitrile
Methyl methacrylate
Methyl methanesulfonate
2-Methyl-2-(methylthio)propionaldehyde-
   o-(methylcarbonyl) oxime
N-Methyl-N'-nitro-N-nitrosoguanidine

(continued)

TABLE 2 (continued).

Methyl parathion	Pentachlorodibenzofurans	2,3,4,6-Tetrachlorophenol
Methylthiouracil	Pentachloroethane	Tetraethyldithiopyrophosphate
Mustard gas	Pentachloronitrobenzene (PCNB)	Tetraethyl lead
Naphthalene	Pentachlorophenol	Tetraethylpyrophosphate
1,4-Naphthoquinone	Phenacetin	Thallium and compounds, N.O.S.
1-Naphthylamine	Phenol	Thallic oxide
2-Naphthylamine	Phenyl dichloroarsine	Thallium (I) acetate
1-Naphthyl-2-thiourea	Phenylmercury acetate	Thallium (I) carbonate
Nickel and compounds, N.O.S.	N-Phenylthiourea	Thallium (I) chloride
Nickel carbonyl	Phosgene	Thallium (I) nitrate
Nickel cyanide	Phosphine	Thallium selenite
Nicotine and salts	Phosphorothioic acid, O,O-dimethyl	Thallium (I) sulfate
Nitric oxide	ester, O-ester with N,N-dimethyl ben-	Thioacetamide
p-Nitroaniline	zene sulfonamide	Thiosemicarbazide
Nitrobenzene	Phthalic acid esters, N.O.S.	Thiourea
Nitrogen dioxide	Phthalic anhydride	Thiuram
Nitrogen mustard and hydrochloride salt	2-Picoline	Toluene
Nitrogen mustard N-oxide and	Polychlorinated biphenyl, N.O.S.	Toluene diamine, N.O.S.
hydrochloride salt	Potassium cyanide	2,4 Toluene diamine
Nitrogen peroxide	Potassium silver cyanide	2,6 Toluene diamine
Nitrogen tetroxide	Pronamide	3,4 Toluene diamine
Nitroglycerine	1,2-Propanediol	o-Toluidine hydrochloride
4-Nitrophenol	1,3-Propane sultone	Tolylene diisocyanate
4-Nitroquinoline-1-oxide	Propionitrile	Toxaphene
Nitrosamine, N.O.S.	Propylthiouracil	Tribromomethane
N-Nitrosodi-N-butylamine	2-Propyn-1-ol	1,2,4-Trichlorobenzene
N-Nitrosodiethanolamine	Pyridine	1,1,1-Trichloroethane
N-Nitrosodiethylamine	Reserpine	1,1,2-Trichloroethane
N-Nitrosodimethylamine	Resorcinol	Trichloroethene (Trichloroethylene)
N-Nitrosodiphenylamine	Saccharin	Trichloromethanethiol
N-Nitrosodi-N-propylamine	Safrole	2,4,5-Trichlorophenol
N-Nitroso-N-ethylurea	Selenious acid	2,4,6-Trichlorophenol
N-Nitrosomethylethylamine	Selenium and compounds, N.O.S.	2,4,5-Trichlorophenoxyacetic acid (2,4,5-
N-Nitroso-N-methylurea	Selenium sulfide	T)
N-Nitroso-N-methylurethane	Selenourea	2,4,5-Trichlorophenoxypropionic acid
N-Nitrosomethylvinylamine	Silver and compounds, N.O.S.	(2,4,5-TP) (Silvex)
N-Nitrosomorpholine	Silver cyanide	Trichloropropane, N.O.S.
N-Nitrosonornicotine	Sodium cyanide	1,2,3-Trichloropropane
N-Nitrosopiperidine	Streptozotocin	0,0,0-Triethyl phosphorothioate
N-Nitrosopyrrolidine	Strontium sulfide	Trinitrobenzene
N-Nitrososarcosine	Strychnine and salts	Tris(1-azridinyl)phosphine sulfide
5-Nitro-o-toluidine	1,2,4,5-Tetrachlorobenzene	Tris(2,3-dibromopropyl) phosphate
Octamethylpyrophosphoramide	2,3,7,8-Tetrachlorodibenzo-p-dioxin	Trypan blue
Oleyl alcohol condensed with 2 moles	(TCDD)	Uracil mustard
ethylene oxide	Tetrachlorodibenzo-p-dioxins	Vanadic acid, ammonium salt
Osmium tetroxide	Tetrachlorodibenzofurans	Vanadium pentoxide (dust)
7-Oxabicyclo(2.2.1)heptane-2,3-	Tetrachloroethane, N.O.S.	Vinyl chloride
dicarboxylic acid	1,1,1,2-Tetrachloroethane	Vinylidene chloride
Parathion	1,1,2,2-Tetrachloroethane	Zinc cyanide
Pentachlorobenzene	Tetrachloroethene (Tetrachloroethylene)	Zinc phosphide
Pentachlorodibenzo-p-dioxins	Tetrachloromethane	

[1]The abbreviation N.O.S. signifies those members of the general class "not otherwise specified" by name in this listing.

Industry and EPA hazardous waste No.	Hazardous waste	Hazard code
Generic:		
F001	The following spent halogenated solvents used in degreasing: tetra-chloroethylene, trichloroethylene, methylene chloride, 1,1,1-trichloroethane, carbon tetrachloride, and chlorinated fluorocarbons; and sludges from the recovery of these solvents in degreasing operations.	(T)
F002	The following spent halogenated solvents: tetrachloroethylene, methylene chloride, trichloroethylene, 1,1,1-trichloroethane, chlorobenzene, 1,1,2-trichloro-1,2,2-trifuforoethane, ortho-dicholorobenzene, and trichlorofluoromethane; and the still bottoms from the recovery of these solvents.	(T)
F003	The following spent non-halogenated solvents: xylene, acetone, ethyl acetate, ethyl benzene, ethyl ether, methyl isobutyl ketone, n-butyl alcohol, cyclohexanone, and methanol; and the still bottoms from the recovery of these solvents.	(I)
F004	The following spent non-halogenated solvents: cresols and cresytic acid, and nitrobenzene; and the stiff bottoms from the recovery of these solvents.	(T)
F005	The following spent non-halogenated solvents: toluene, methyl ethyl ketone, carbon disulfide, isobutanol, and pyridine; and the still bottoms from the recovery of these solvents.	(I,T)
F006	Wastewater treatment sludges from electroplating operations except from the following processes: (1) sulfuric acid anodizing of aluminum; (2) tin plating on carbon steel; (3) zinc plating (segregated basis) on carbon steel; (4) aluminum or zinc-aluminum plating on carbon steel; (5) cleaning/stripping associated with tin, zinc and aluminum plating on carbon steel; and (6) chemical etching and milling of aluminum.	(T)
F019	Wastewater treatment sludges from the chemical conversion coating of aluminum.	(T)
F007	Spent cyanide plating bath solutions from electroplating operations.	(R,T)
F008	Plating bath sludges from the bottom of plating baths from electroplating operations where cyanides are used in the process.	(R,T)
F009	Spent stripping and cleaning bath solutions from electroplating operations where cyanides are used in the process.	(R,T)
F010	Quenching bath sludge from oil baths from metal heat treating operations where cyanides are used in the process.	(R,T)
F011	Spent cyanide solutions from salt bath pot cleaning from metal heat treating operations	(R,T)
F012	Quenching wastewater treatment sludges from metal heat treating operations where cyanides are used in the process (except for precious metals heat treating quenching wastewater treatment sludges).	(T)
F014	Cyanidation wastewater treatment tailing pond sediment from mineral metals recovery operations.	(T)
F015	Spent cyanide bath solutions from mineral metals recovery operations.	(R,T)
F020	Wastes (except wastewater and spent carbon from hydrogen chloride purification) from the production or manufacturing use (as a reactant, chemical intermediate, or component in a formulating process) of tri-, tetra-, or pentachlorophenol, or of intermediates used to produce their derivatives. (This listing does not include wastes from the production of Hexachlorophene from highly purified 2,4,5-trichlorophenol.	(H)
F021	Wastes (except wastewater and spent carbon from hydrogen chloride purification) from the manufacturing use (as a reactant, chemical intermediate, or component in a formulating process) of tetra-, penta-, or hexachlorobenzenes under alkaline conditions.	(H)

(continued)

Industry and EPA hazardous waste No.	Hazardous waste	Hazard code
Generic:		
F022	Wastes (except wastewater and spent carbon from hydrogen chloride purification) from the production of materials on equipment previously used for the production or manufacturing use (as a reactant, chemical intermediate or component in a formulating process) of materials listed under F020 and F021.	(H)
F023	Discarded unused formulations containing tri-, tetra-, or pentachlorophenol or discarded unused formulations containing compounds derived from these chlorophenols.	(H)
F024	Wastes, including but not limited to, distillation residues, heavy ends, tars, and reactor cleanout wastes from the production of chlorinated aliphatic hydrocarbons, having carbon content from one to five, utilizing free radical catalyzed processes. (This listing does not include light ends, spent filters and filter aids, spent dessicants, wastewater, wastewater treatment sludges, spent catalysts, and wastes listed in §261.32.).	(T)
F025	Light ends, spent filters and filter aids, and spent dessicant wastes from the production of chlorinated aliphatic hydrocarbons, having carbon content from one to five, utilizing free radical catalyzed processes.	(T)

Ignitable Waste . . . . (I)	EP Toxic Waste . . . . (E)
Corrosive Waste . . . . (C)	Acute Hazardous Waste . . . . (H)
Reactive Waste . . . . (R)	Toxic Waste . . . . (T)

TABLE 3b. Hazardous Wastes from Specific Sources.

Industry and EPA hazardous waste No.	Hazardous waste	Hazard code
Wood Preservation:		
K001	Bottom sediment sludge from the treatment of wastewaters from wood preserving processes that use creosote and/or pentachlorophenol.	(T)
Inorganic Pigments:		
K002	Wastewater treatment sludge from the production of chrome yellow and orange pigments.	(T)
K003	Wastewater treatment sludge from the production of molybdate orange pigments.	(T)
K004	Wastewater treatment sludge from the production of zinc yellow pigments.	(T)
K005	Wastewater treatment sludge from the production of chrome green pigments.	(T)
K006	Wastewater treatment sludge from the production of chrome oxide green pigments (anhydrous and hydrated).	(T)
K007	Wastewater treatment sludge from the production of iron blue pigments.	(T)
K008	Oven residue from the production of chrome oxide green pigments	(T)
Organic Chemicals:		
K009	Distillation bottoms from the production of acetaldehyde from ethylene	(T)
K010	Distillation side cuts from the production of acetaldehyde from ethylene	(T)
K011	Bottom stream from the wastewater stripper in the production of acrylonitrile.	(R,T)
K013	Bottom stream from the acetonitrile column in the production of acrylonitrile.	(R,T)

(continued)

Industry and EPA hazardous waste No.	Hazardous waste	Hazard code
Organic Chemicals:		
K014	Bottoms from the acetonitrile purification column in the production of acrylonitrile.	(T)
K015	Still bottoms from the distillation of benzyl chloride.	(T)
K016	Heavy ends or distillation residues from the production of carbon tetrachloride.	(T)
K017	Heavy ends (still bottoms) from the purification column in the production of epichlorohydrin.	(T)
K018	Heavy ends from the fractionation column in ethyl chloride production.	(T)
K019	Heavy ends from the distillation of ethylene dichloride in ethylene dichloride production.	(T)
K020	Heavy ends from the distillation of vinyl chloride in vinyl chloride monomer production.	(T)
K021	Aqueous spent antimony catalyst waste from fluoromethanes production.	(T)
K022	Distillation bottom tars from the production of phenol/acetone from cumene.	(T)
K023	Distillation light ends from the production of phthalic anhydride from naphthalene.	(T)
K024	Distillation bottoms from the production of phthalic anhydride from naphthalene.	(T)
K093	Distillation light ends from the production of phthalic anhydride from orthoxylene.	(T)
K094	Distillation bottoms from the production of phthalic anhydride from orthoxylene.	(T)
K025	Distillation bottoms from the production of nitrobenzene by the nitration of benzene.	(T)
K026	Stripping still tails from the production of methyl ethyl pyridines.	(T)
K027	Centrifuge and distillation residues from toluene diisocyanate production.	(R,T)
K028	Spent catalyst from the hydrochlorinator reactor in the production of 1,1,1-trichloroethane.	(T)
K029	Waste from the product steam stripper in the production of 1,1,1-trichloroethane.	(T)
K095	Distillation bottoms from the production of 1,1,1-trichloroethane.	(T)
K096	Heavy ends from the ends column from the production of 1,1,1-trichloroethane.	(T)
K030	Column bottoms or heavy ends from the combined production of trichloroethylene and perchloroethylene.	(T)
K111	Product wastewaters from the production of dinitrotoluene via nitration of toluene.	(C,T)
K112	Reaction by-product water from the drying column in the production of toluenediamine via hydrogenation of dinitrotoluene.	(T)
K113	Light ends from the purification of toluenediamine in the production of toluenediamine via hydrogenation of dinitrotoluene.	(T)
K114	Vicinals from the purification of toluenediamine in the production of toluenediamine via hydrogenation of dinitrotoluene.	(T)
K115	Heavy ends from the purification of toluenediamine in the production of toluenediamine via hydrogenation of dinitrotoluene.	(T)
K116	Organic condensate from the solvent recovery column in the production of toluene diisocyanate via phosgenation of toluenediamine.	(T)

(continued)

Industry and EPA hazardous waste No.	Hazardous waste	Hazard code
Pesticides:		
K031	By-product salts generated in the production of MSMA and cacodylic acid.	(T)
K032	Wastewater treatment sludge from the production of chlordane.	(T)
K033	Wastewater and scrub water from the chlorination of cyclopentadiene in the production of chlordane.	(T)
K034	Filter solids from the filtration of hexachlorocyclopentadiene in the production of chlordane.	(T)
K097	Vacuum stripper discharge from the chlordene chlorinator in the production of chlordane.	(T)
K035	Wastewater treatment sludges generated in the production of creosote.	(T)
K036	Still bottoms from toluene reclamation distillation in the production of disulfoton.	(T)
K037	Wastewater treatment sludges from the production of disulfoton.	(T)
K038	Wastewater from the washing and stripping of phorate production.	(T)
K039	Filter cake from the filtration of diethylphosphorodithioic acid in the production of phorate.	(T)
K040	Wastewater treatment sludge from the production of phorate.	(T)
K041	Wastewater treatment sludge from the production of toxaphene.	(T)
K098	Untreated process wastewater from the production of toxaphene.	(T)
K042	Heavy ends or distillation residues from the distillation of tetra-chlorobenzene in the production of 2,4,5-T.	(T)
K043	2,6-Dichlorophenol waste from the production of 2,4-D.	(T)
K099	Untreated wastewater from the production of 2,4-D.	(T)
Explosives:		
K044	Wastewater treatment sludges from the manufacturing and processing of explosives.	(R)
K045	Spent carbon from the treatment of wastewater containing explosives.	(R)
K046	Wastewater treatment sludges from the manufacturing, formulation and loading of lead-based initiating compounds.	(T)
K047	Pink/red water from TNT operations.	(R)
Petroleum Refining:		
K048	Dissolved air flotation (DAF) float from the petroleum refining industry.	(T)
K049	Slop oil emulsion solids from the petroleum refining industry.	(T)
K050	Heat exchanger bundle cleaning sludge from the petroleum refining industry.	(T)
K051	API separator sludge from the petroleum refining industry.	(T)
K052	Tank bottoms (leaded) from the petroleum refining industry.	(T)
Iron and Steel:		
K060	Ammonia still lime sludge from coking operations.	(T)
K061	Emission control dust/sludge from the primary production of steel in electric furnaces.	(T)
K062	Spent pickle liquor from steel finishing operations.	(C,T)
Primary Copper:		
K064	Acid plant blowdown slurry/sludge resulting from the thickening of blowdown slurry from primary copper production.	(T)
Primary Lead:		
K065	Surface impoundment solids contained in and dredged from surface impoundments at primary lead smelting facilities.	(T)

(continued)

Industry and EPA hazardous waste No.	Hazardous waste	Hazard code
Primary Zinc:		
K066	Sludge from treatment of process wastewater and/or acid plant blow-down from primary zinc production.	(T)
K067	Electrolytic anode slimes/sludges from primary zinc production.	(T)
K068	Cadmium plant leachate residue (iron oxide) from primary zinc production.	(T)
Secondary Lead:		
K069	Emission control dust/sludge from secondary lead smelting.	(T)
K100	Waste leaching solution from acid leaching of emission control dust/sludge from secondary lead smelting.	(T)

Ignitable Waste . . . . (I) EP Toxic Waste . . . . (E)
Corrosive Waste . . . . (C) Acute Hazardous Waste . . . . (H)
Reactive Waste . . . . (R) Toxic Waste . . . . (T)

TABLE 4a. Discarded Commercial Chemical Products, Off-Specification Species, Container Residues, and Spill Residues Thereof.

Hazardous Waste No.	Substance[1]	Hazardous Waste No.	Substance[1]	Hazardous Waste No.	Substance[1]
	AAF see U005	U015	Azaserine	U034	Chloral
U001	Acetaldehyde	U016	Benz(c)acridine	U035	Chlorambucil
U002	Acetone (I)	U017	Benzal chloride	U036	Chlordane
U003	Acetonitrile (I,T)	U018	Benz(a)anthracene	U037	Chlorobenzene
U004	Acetophenone	U019	Benzene	U038	Chlorobenzilate
U005	2-Acetylaminoflourene	U020	Benzenesulfonyl chloride (C,R)	U039	p-Chloro-m-cresol
U006	Acetyl chloride (C,T)			U040	Chlorodibromomethane
U007	Acrylamide	U021	Benzidine	U041	1-Chloro-2,3-epoxypropane
	Acetylene tetrachloride see U209		1,2-Benzisothiazolin-3-one, 1,1-dioxide see U202		CHLOROETHENE NU see U226
	Acetylene trichloride see U228		Benzo(a)anthracene see U018	U042	Chloroethyl vinyl ether
U008	Acrylic acid (I)	U022	Benzo(a)pyrene	U043	Chloroethene
U009	Acrylonitrile	U023	Benzotrichloride (C,R,T)	U044	Chloroform (I,T)
	AEROTHENE TT see U226	U024	Bis(2-chloroethoxy) methane	U045	Chloromethane (I,T)
	3-Amino-5-(p-acetamido-phenyl)-1H-1,2,4-triazole, hydrate see U011	U025	Bis(2-chloroethyl) ether	U046	Chloromethyl methyl ether
		U026	N,N-Bis(2-chloroethyl)-2-naphthylamine	U047	2-Chloronaphthalene
U010	6-Amino-1,1a,2,8,8a,8b-hexahydro-8-(hydroxymethyl) 8-methoxy-5-methylcarbamate azinno(2',3':3,4) pyrrolo(1,2-a) indole-4, 7-dione (ester)	U027	Bis(2-chloroisopropyl) ether	U048	2-Chlorophenol
		U028	Bis(2-ethylhexyl) phthalate	U049	4-Chloro-o-toluidine hydrochloride
		U029	Bromomethane	U050	Chrysene
		U030	4-Bromophenyl phenyl ether		C1 23060 see U073
		U031	n-Butyl alcohol (I)	U051	Cresote
		U032	Calcium chromate	U052	Cresols
			Carbolic acid see U188	U053	Crotonaldehyde
U011	Amitrole		Carbon tetrachloride see U211	U054	Cresylic acid
U012	Aniline (I)			U055	Cumene
U013	Asbestos				Cyanomethane see U003
U014	Auramine	U033	Carbonyl fluoride	U056	Cyclohexane (I)
				U057	Cyclohexanone (I)
				U058	Cyclophosphamide

(continued)

Hazardous Waste No.	Substance[1]	Hazardous Waste No.	Substance[1]	Hazardous Waste No.	Substance[1]
U059	Daunomycin	U099	1,2-Dimethylhydrazine	U141	Isosafrole
U060	DDD	U100	Dimethylnitrosoamine	U142	Kepone
U061	DDT	U101	2,4-Dimethylphenol	U143	Lasiocarpine
U062	Diallate	U102	Dimethyl phthalate	U144	Lead acetate
U063	Dibenz(a,h)anthracene	U103	Dimethyl sulfate	U145	Lead phosphate
	Dibenzo(a,h)anthracene see U063	U104	2,4-Dinitrophenol	U146	Lead subacetate
U064	Dibenzo(a,i)pyrene	U105	2,4-Dinitrotoluene	U147	Maleic anhydride
U065	Dibromochloromethane	U106	2,6-Dinitrotoluene	U148	Maleic hydrazide
U066	1,2-Dibromo-3-chloro-propane	U107	Di-n-octyl phthalate	U149	Malononitrile
		U108	1,4-Dioxane		MEK Peroxide see U160
		U109	1,2-Diphenylhydrazine	U150	Melphalan
U067	1,2-Dibromoethane	U110	Dipropylamine (I)	U151	Mercury
U068	Dibromomethane	U111	Di-n-propylnitrosamine	U152	Methacrylonitrile
U069	Di-n-butyl phthalate		EBDC see U114	U153	Methanethiol
U070	1,2-Dichlorobenzene		1,4-Epoxybutane see U213	U154	Methanol
U071	1,3-Dichlorobenzene	U112	Ethyl acetate (I)	U155	Methapyrilene
U072	1,4-Dichlorobenzene	U113	Ethyl acrylate (I)		Methyl alcohol see U154
U073	3,3'-Dichlorobenzidine	U114	Ethylenebisdithiocarbamate	U156	Methyl chlorocarbonate
U074	1,4-Dichloro-2-butene	U115	Ethylene oxide (I,T)		Methyl chloroform see U226
	3,3'-Dichloro-4,4'-diamino-biphenyl see U073	U116	Ethylene thiourea		
U075	Dichlorodifluoromethane	U117	Ethyl ether (I,T)	U157	3-Methylcholanthrene
U076	1,1-Dichloroethane	U118	Ethylmethacrylate		Methyl chloroformate see U156
U077	1,2-Dichloroethane	U119	Ethyl methanesulfonate		
U078	1,1-Dichloroethane		Ethylnitrile see U003	U158	4,4'-Methylene-bis-(2-chloroaniline)
U079	1,2-trans-dichloroethylene		Firemaster-T23P see U235		
U080	Dichloromethane	U120	Fluoranthene	U159	Methyl ethyl ketone (MEK) (I,T)
	Dichloromethylbenzene see U017	U121	Fluorotrichloromethane		
		U122	Formaldehyde	U160	Methyl ethyl ketone perox-ide (R)
U081	2,4-Dichlorophenol	U123	Formic acid (C,T)		Methyl iodide see U138
U082	2,6-Dichlorophenol	U124	Furan (I)	U161	Methyl isobutyl ketone
U083	1,2-Dichloropropane	U125	Furtural (I)	U162	Methyl methacrylate (R,T)
U084	1,3-Dichloropropene	U126	Glycidylaldehyde	U163	N-Methyl-N'-nitro-N-nitro-soguanidine
U085	Diepoxybutane (I,T)	U127	Hexachlorobenzene		
U086	1,2-Diethylhydrazine	U128	Hexachlorobutadiene	U164	Methylthiouracil
U087	0,0-Diethyl-S-methyl ester of phosphorodithioic acid	U129	Hexachlorocyclohexane		Mitomycin C see U010
		U130	Hexahclorocyclopenta-diene	U165	Naphthalene
		U131	Hexachloroethane	U166	1,4-Naphthoquinone
U088	Diethyl phthalate	U132	Hexachlorophene	U167	1-Naphthylamine
U089	Diethylstilbestrol	U133	Hydrazine (R,T)	U168	2-Naphthylamine
U090	Dihydrosafrole	U134	Hydrofluoric acid (C,T)	U169	Nitrobenzene (I,T)
U091	3,3'-Dimethoxybenzidine	U135	Hydrogen sulfide		Nitrobenzol see U169
U092	Dimethylamine (I)		Hydroxybenzene, see U188	U170	4-Nitrophenol
U093	p-Dimethylaminoazo-benzene	U136	Hydroxydimethyl arsine oxide	U171	2-Nitropropane (I)
				U172	N-Nitrosodi-n-butylamine
U094	7,12-Dimethylbenz(a) an-thracene		4,4'(Imidocarbonyl)bis(N,N-dimethyl)aniline see U014	U173	N-Nitrosodiethanolamine
				U174	N-Nitrosodiethylamine
U095	3,3'-Dimethylbenzidine			U175	N-Nitrosodi-n-propylamine
U096	alpha,alpha-Dimethyl-benzylhydroperoxide (R)	U137	Indeno(1,2,3-cd)pyrene	U176	N-Nitroso-n-ethylurea
U097	Dimethylcarbamoyl chloride	U138	Iodomethane	U177	N-Nitroso-n-methylurea
		U139	Iron Dextran	U178	N-Nitroso-n-methyl-urethane
U098	1,1-Dimethylhydrazine	U140	Isobutyl alcohol	U179	N-Nitrosopiperidine

(continued)

Hazardous Waste No.	Substance[1]	Hazardous Waste No.	Substance[1]	Hazardous Waste No.	Substance[1]
U180	N-Nitrosopyrrolidine	U206	Streptozotocin	U229	Trichlorofluoromethane
U181	5-Nitro-o-toluidine		2,4,5-T see U232	U234	Trinitrobenzene (R,T)
U182	Paraldehyde	U207	1,2,4,5-Tetrachlorobenzene	U235	Tris(2,3-dibromopropyl)
	PCNB see U185	U208	1,1,1,2-Tetrachloroethane		phosphate
U183	Pentachlorobenzene	U209	1,1,2,2-Tetrachloroethane	U236	Trypan blue
U184	Pentachloroethane	U210	Tetrachloroethene	U237	Uracil mustard
U185	Pentachloronitrobenzene		Tetrachloroethylene see	U238	Urethane
U186	1,3-Pentadiene (I)		U210		Vinyl chloride see U043
	Perc see U210	U211	Tetrachloromethane		Vinylidene chloride see
	Perchlorethylene see U210	U213	Tetrahydrofuran (I)		U078
U187	Phenacetin	U214	Thallium (I) acetate	U239	Xylene
U188	Phenol	U215	Thallium (I) carbonate	U248	3-(alpha-Acetonylbenzyl)
U189	Phosphorous sulfide (R)	U216	Thallium (I) chloride		-4-hydroxycoumarin and
U190	Phthalic anhydride	U217	Thallium (I) nitrate		salts, when present at
U191	2-Picoline	U218	Thioacetamide		concentrations of 0.3%
U192	Pronamide	U219	Thiourea		or less.
U193	1,3-Propane sultone	U220	Toluene	U248	Warfarin, when present at
U194	n-Propylamine (I)	U221	Toluenediamine		concentrations of 0.3%
U196	Pyridine	U222	o-Toluidine hydrochloride		or less.
U197	Quinones	U223	Toluene diisocyanate	U249	Zinc phosphide, when
U200	Reserpine	U224	Toxaphene		present at concentra-
U201	Resorcinol		2,4,5-TP see U233		tions of 10% or less.
U202	Saccharin	U225	Tribromomethane	U328	2-Amino-1-methylbenzene
U203	Safrole	U226	1,1,1-Trichloroethane	U353	4-Amino-1-methylbenzene
U204	Selenious acid	U227	1,1,2-Trichloroethane	U328	o-Toluidine
U205	Selenium sulfide (R,T)	U228	Trichloroethene	U353	p-Toluidine
	Silvex see U233		Trichloroethylene see U228		

[1]The Agency included those trade names of which it was aware; an omission of a trade name does not imply that it is not hazardous. The material is hazardous if it is listed under its generic name.

TABLE 4b. Discarded Commercial Chemical Products, Off-Specification Species, Container Residues, and Spill Residues Thereof.

Hazardous Waste No.	Substance[1]	Hazardous Waste No.	Substance[1]	Hazardous Waste No.	Substance[1]
	(Acetato)phenylmercury see P092	P003	Acrolein	P008	4-Aminopyridine
			Agarin see P007		Ammonium metavanadate
	Acetone cyanohydrin see P069		Agrosan GN 5 see P092		see P119
			Aldicarb see P069	P009	Ammonium picrate (R)
P001	3-(alpha-Acetonylbenzyl)		Aldifen see P048		ANTIMUCIN WDR see
	-4-hydroxycoumann and	P004	Aldrin		P092
	salts, when present at		Algimycin see P092		ANTURAT see P073
	concentrations greater	P005	Allyl alcohol		AQUATHOL see P088
	than 0.3%.	P006	Aluminum phosphide (R)		ARETIT see P020
P001	Warfarin, when present at		ALVIT see P037	P010	Arsenic acid
	concentrations greater		Aminoethylene see P054	P011	Arsenic pentoxide
	than 0.3%	P007	5-Aminomethyl)-3-isoxazo-	P012	Arsenic trioxide
P002	1-Acetyl-2-thiourea		lol		Athrombin see P001

(continued)

Hazardous Waste No.	Substance[1]	Hazardous Waste No.	Substance[1]	Hazardous Waste No.	Substance[1]
	AVITROL see P008	P035	2,4-Dichlorophenoxyacetic acid (2,4-D)	P050	Endosulfan
	Aziridene see P054			P051	Endrin
	AZOFOS see P061	P036	Dichlorophenylarsine		Epinephrine see P042
	Azophos see P061		Dicyanogen see P031	P052	Ethylcyanide
	BANTU see P072	P037	Dieldrin	P053	Ethylenediamine
P013	Barium cyanide		DIELDREX see P037	P054	Ethyleneimine
	BASENITE see P020	P038	Diethylarsine		FASCO FASCRAT POWDER see P001
	BCME see P016	P039	0,0-Diethyl-S-(2-(ethylthio) ethyl)ester of phosphoro-thioic acid		FEMMA see P091
P014	Benzenethiol			P055	Ferric cyanide
	Benzoepin see P050			P056	Fluonne
P015	Beryllium dust	P040	0,0-Diethyl-0-(2-pyrazinyl) phosphorathioate ·	P057	2-Fluoroacetamide
P016	Bis(chloromethyl) ether			P058	Fluoroacetic acid, sodium salt
	BLADAN-M see P071	P041	0,0-Diethyl phosphoric acid 0-p-nitrophenyl ester		FOLODOL-80 see P071
P017	Bromoacetone				FOLODOL M see P071
P018	Brucine	P042	3,4-Dihydroxy-alpha-(methylamino)-methyl benzyl alcohol		FOSFERNO M 50 see P071
P019	2-Butanone peroxide				FRATOL see P058
	BUFEN see P092				Fulminate of mercury see P065
	Butaphene see P020	P043	Di-isopropylfluorophos-phate		
P020	2-sec-Butyl-4,5-dinitro-phenol		DIMETATE see P044		FUNGITOX OR see P092
P021	Calcium cyanide		1,4:5,8-Dimethanonaph-thalene, 1,2,3,4,10,10, hexachloro-1,4,4a,5,8,8a -hexahydro endo. endo see P060		FUSSOF see P057
	CALDON see P020				GALLOTOX see P092
P022	Carbon disulfide				GEARPHOS see P071
	CERESAN see P092				GERUTOX see P020
	CERESAN UNIVERSAL see P092	P044	Dimethoate	P059	Heptachlor
	CHEMOX GENERAL see P020	P045	3,3-Dimethyl-1-(methylthio) -2-butanone-O-[(methyl-amino)carbonyl] oxime	P060	1,2,3,4,10,10-Hexachloro-1,4,4a,5,8,8a,-hexahydro 1,4,5:5,8-endo, endo-dimethanonaphthalene
	CHEMOX P E see P020				
	CHEM-TOL see P090	P046	alpha,alpha-Dimethyl-phenethylamine		1,4,5,6,7,7-Hexachloro-cyclic-5-norbornene-2,3-dimethanol sulfite see P050
P023	Chloroacetaldehyde				
P024	p-Chloroaniline		Dinitrocyclohexylphenol see P034		
P025	1-(p-Chlorobenzoyl)-5-methoxy-2-methylindole-3-acetic acid	P047	4,6-Dintro-o-cresol and salts	P061	Hexachloropropene
				P062	Hexaethyl tetraphosphate
P026	1-(o-Chlorophenyl)thiourea	P048	2,4-Dinitrophenol		HOSTAQUICK see P092
P027	3-Chloropropionitrile		DINOSEB see P020		HOSTAQUIK see P092
P028	alpha-Chlorotoluene		DINOSEBE see P020		Hydrazomethane see P068
P029	Copper cyanide		Disulfoton see P039	P063	Hydrocyanic acid
	CRETOX see P108	P049	2,4-Dithiobiuret		ILLOXOL see P037
	Coumadin see P001		DNBP see P020		INDOCI see P025
	Coumafen see P001		DOLCO MOUSE CEREAL see P108		Indomethacin see P025
P030	Cyanides		DOW GENERAL see P020		INSECTOPHENE see P050
P031	Cyanogen		DOW GENERAL WEED KILLER see P020		Isodnin see P060
P032	Cyanogen bromide			P064	Isocyanic acid, methyl ester
P033	Cyanogen chloride		DOW SELECTIVE WEED KILLER see P020		KILOSEB see P020
	Cyclodan see P050		DOWICIDE G see P090		KOP-THIODAN see P050
P034	2-Cyclohexyl-4,6-dinitro-phenol		DYANACIDE see P092		KWIK-KIL see P108
			EASTERN STATES DUOCIDE see P001		KWIKSAN see P092
	D-CON see P001				KUMADER see P001
	DETHMOR see P001				KYPFARIN see P001
	DETHNEL see P001		ELGETOL see P020		LEYTOSAN see P092
	DFP see P043				(continued)

573

Hazardous Waste No.	Substance[1]	Hazardous Waste No.	Substance[1]	Hazardous Waste No.	Substance[1]
	LIQUIPHENE see P092		OMPACIDE see P085	P102	2-Propyn-1-ol
	MALIK see P050		OMPAX see P085		PROTHROMADIN see
	MAREVAN see P001	P087	Osmium tetroxide		P001
	MAR-FRIN see P001	P088	7-Oxabicyclo(2,2,1)		QUICKSAM see P092
	MARTIN D MAR-FRIN		heptane-2,3-dicarboxylic		QUINTOX see P037
	see P001		acid		RAT AND MICE BAIT see
	MAVERAN see P001		PANIVARFIN see P001		P001
	MEGATOX see P005		PANORAM D-31 see P037		RAT-A-WAY see P001
P065	Mercury fulminate		PANTHERINE see P007		RAT-B-GON see P001
	MERSOLITE see P092		PANWARFIN see P001		RAT-O-CIDE #2 see P001
	METACID 50 see P071	P089	Parathion		RAT-GUARD see P001
	METAFOS see P071		PCP see P090		RAT-KILL see P001
	METAPHOR see P071		PENNCAP-M see P071		RAT-MIX see P001
	METAPHOS see P071		PENOXYL CARBON N see		RATS-NO-MORE se P001
	METASOL 30 see P092		P048		RAT-OLA see P001
P066	Methomyl	P090	Pentachlorophenol		RATOREX see P001
P067	2-Methylaziridine		Pentachlorophenate see		RATTUNAL see P001
	METHYL-E 605 see P071		P090		RAT-TROL see P001
P068	Methyl hydrazine		PENTA-KILL see P090		RO-DETH see P001
	Methyl isocyanate see		PENTASOL see P090		RO-DEX see P108
	P064		PENWAR see P090		ROSEX see P001
P069	2-Methyllactonitrile		PERMICIDE see P090		ROUGH & READY MOUSE
P070	2-Methyl 2-(methylthio)		PERMAGUARD see P090		MIX see P001
	propionaldehyde-o-		PERMATOX see P090		SANASEED see P108
	(methylcarbonyl) oxime		PERMITE see P090		SANTOBRITE see P090
	METHYL NIRON see P042		PERTOX see P090		SANTOPHEN see P090
P071	Methyl parathion		PESTOX III see P085		SANTOPHEN 20 see P090
	METRON see P071		PHENMAD see P092		SCHRADAN see P085
	MOLE DEATH see P108		PHENOTAN see P020	P103	Selenourea
	MOUSE-NOTS see P108	P091	Phenyl dichloroarsine	P104	Silver Cyanide
	MOUSE-RID see P108		Phenyl mercaptan see		SMITE see P105
	MOUSE-TOX see P108		P014		SPARIC see P020
	MUSCIMOL see P007	P092	Phenylmercury acetate		SPOR-KIL see P092
P072	1-Naphthyl-2-thiourea	P093	N-Phenylthiourea		SPRAY-TROL BRAND
P073	Nickel carbonyl		PHILIPS 1861 see P008		RODEN-TROL see P001
P074	Nickel cyanide		PHIX see P092		SPURGE see P020
P075	Nicotine and salts	P094	Phorate	P105	Sodium azide
P076	Nitric oxide	P095	Phosgene		Sodium coumadin see
P077	p-Nitroaniline	P096	Phosphine		P001
P078	Nitrogen dioxide	P097	Phosphorothioic acid,	P106	Sodium cyanide
P079	Nitrogen peroxide		O,O-dimethyl ester,		Sodium fluoroacetate see
P080	Nitrogen tetroxide		O-ester with N,N-di-		P056
P081	Nitroglycerine (R)		methyl benzene sulfona-		SODIUM WARFARIN see
P082	N-Nitrosodimethylamine		mide		P001
P083	N-Nitrosodiphenylamine		Phosphorothioic acid O,O-		SOLFARIN see P001
P084	N-Nitrosomethylvinylamine		dimethyl-O-(p-nitro-		SOLFOBLACK BB see
	NYLMERATE see P092		phenyl) ester see P071		P048
	OCTALOX see P037		PIED PIPER MOUSE SEED		SOLFOBLACK SB see
P085	Octamethylpyrophosphor-		see P108		P048
	amide	P098	Potassium cyanide	P107	Strontium sulfide
	OCTAN see P092	P099	Potassium silver cyanide	P108	Strychnine and salts
P086	Oleyl alcohol condensed		PREMERGE see P020		SUBTEX see P020
	with 2 moles ethylene	P100	1,2-Propanediol		SYSTAM see P085
	oxide		Propargyl alcohol see P102		TAG FUNGICIDE see P092
	OMPA see P085	P101	Propionitrile		TEKWAISA see P071

(continued)

TABLE 4b (continued).

Hazardous Waste No.	Substance[1]	Hazardous Waste No.	Substance[1]	Hazardous Waste No.	Substance[1]
	TEMIC see P070		THIFOR see P092	P119	Vanadic acid, ammonium salt
	TEMIK see P070		THIMUL see P092		
	TERM-I-TROL see P090		THIODAN see P050	P120	Vanadium pentoxide
P109	Tetraethyldithiopyrophos-phate		THIOFOR see P050		VOFATOX see P071
			THIOMUL see P050		WANADU see P120
P110	Tetraethyl lead		THIONEX see P050		WARCOUMIN see P001
P111	Tetraethylpyrophosphate		THIOPHENIT see P071		WARFARIN SODIUM see P001
P112	Tetranitromethane	P116	Thiosemicarbazide		
	Tetraphosphoric acid, hexaethyl ester see P062		Thiosulfan tionel see P050		WARFICIDE see P001
		P117	Thiuram		WOFOTOX see P072
	TETROSULFUR BLACK PB see P048		THOMPSON'S WOOD FIX see P090		YANOCK see P057
					YASOKNOCK see P058
	TETROSULPHUR PBR see P048		TIOVEL see P050		ZIARNIK see P092
		P118	Trichloromethanethiol	P121	Zinc cyanide
P113	Thallic oxide		TWIN LIGHT RAT AWAY see P001	P122	Zinc phosphide, when present at concentra-
	Thallium peroxide see P113				
P114	Thallium selenite		USAF RH-8 see P069		tions greater than 10%
P115	Thallium (I) sulfate		USAF EK-4890 see P002		ZOOCOUMARIN see P001

[1]The Agency included those trade names of which it was aware; an omission of a trade name does not imply that the omitted material is not hazardous. The material is hazardous if it is listed under its generic name.

Legislation

All industrialized countries now have legislation regulating the discharge of gaseous, liquid and solid wastes, including hazardous wastes. In the United States the major federal government legislation covering industrial waste includes the Resource Conservation and Recovery Act, Water Pollution Control Act, Clean Air Act, Clean Water Act and Toxic Substances Control Act. These are implemented through regulations drawn up by the U.S. Environmental Protection Agency. The legislation and regulations play an important role in the management of industrial waste, since they stipulate limits on concentration or load of pollutants that can be disposed to the environment, and the requirements for transportation, storage, treatment and ultimate disposal, particularly of hazardous wastes.

Toxicology

An understanding of the effect of pollutants on man and the ecosystem is essential to appreciate how standards for pollutants are set by regulatory authorities. Standards are usually determined based on scientific criteria developed for the pollutants, setting out the maximum concentration at which no harmful effect occurs. A high degree of uncertainty still exists on the precise effect of many chemicals, and criteria and standards are therefore set on the safe side. The latter may impose an unreasonable burden on the waste producer in treating the waste, but which has to be balanced against the risk of a later claim for damages when the effect is fully known. The toxicology of many pollutants on man (clinical toxicology) and animals [30] is well established though more work is being directed at the effect of pollutants on ecosystems [3,11].

INDUSTRIAL WASTE MANAGEMENT

The proper handling and safe disposal or re-use of industrial wastes employing the most economical means is the aim of sound industrial waste management. To achieve this objective it is necessary to characterize the waste, consider the best option from available treatment or reuse alternatives, and concurrently decide on the best ultimate disposal option.

The characteristics of industrial wastes vary from industry to industry and even within one industry there are differences due to the process and technology used, the age of the plant, and management practices. In developing the principles of industrial waste management in this chapter, a systems approach is adopted, where principles and methods that are generally applicable or common to most industrial wastes are separately described followed by how these can be integrated to solve a particular industry's waste problem. The only differentiation made for the wastes is that into gaseous, liquid and solid, since this facilitates groupings of monitoring instruments, treatment equipment and ultimate disposal options for the waste streams. References to particular industries should be consulted when this is desired (e.g., [12,18,19,20,22,24,26,31]).

Industrial Waste: Problem or Resource

Industrial wastes may be regarded as a problem when viewed from the point of meeting the regulatory requirements for disposal. This view point is the result of attention having been given primarily to the products by plant operators. Products are considered to bring income, whereas wastes incur costs. Optimization of plant operating conditions has generally been directed at maximizing production, and the resultant waste is then disposed at the minimum cost.

Industrial wastes should, however, be considered as a resource. Substances in the waste stream such as unrecovered products, process chemical and process and cooling waters may be able to be recycled after some treatment, and the treatment cost may be less than the cost of disposal when the value of the recovered products and cost of acquiring new chemicals or water are considered. Uses for unwanted by-products and inert residues should also be investigated (e.g., slags for road and building materials [6,20]).

Plant or process modifications may also reduce the production of waste and simultaneously increase the generation of products. There is a growing awareness by engineers of the need to consider both products and wastes in plant-optimization (cost and yield) and with the ideal being the development of the non-waste and non-polluting technology. There are case examples where this ideal can be achieved economically [29].

The first step in managing an industrial waste is to characterize it and to clearly define the problem.

Characterization of Industrial Wastes

An industrial waste survey is required to establish the quantities and characteristics of all waste streams from a factory or plant [13,19,28]. The information is essential in defining the problem, in determining the possibility of waste segregation, recycling and in the design and sizing of treatment and disposal systems.

To carry out an industrial waste survey, the engineer must become familiar with the production processes, the raw materials, additives, products, by-products and wastes. All sources of wastes must be identified, and this task is facilitated by having a diagram of the physical plant layout, process slowsheet and plant sewer (drainage) map, and the assistance of plant operating personnel.

For gaseous and liquid waste streams, quantities are determined from measurements of flow rate and temperature,

and operating pressure for gases. Variation in flow rate due to changes in operating schedules, particularly for batch processes, should be noted. The length of the survey and the frequency of measurements are governed by the degree of the variation. Gas flow in a duct can be measured using a Pilot tube, or as with liquid flow in a full pipe by placing a restriction in the flow (Orifice, Venturi) and measuring the pressure drop across the restriction. Liquid flow in open channels can be determined using a weir or a flume [33].

The characteristics of the wastes to be determined and the analyses to be run are governed by consideration of the pollutants to be removed from the waste streams; and their presence, in turn, can be deduced from the characteristics of the raw materials, the chemicals used and the nature of the process.

The common pollutants emanating from a particular industry can also be checked against literature data (Tables 1–4, [12,18,19,20,22,24,26,31]). Standard methods have been established for the analysis of pollutants in gaseous and liquid wastes [32,33,34,35]. The physical characteristics of the gas or liquid (density, viscosity) need also be determined for sizing pipes and treatment equipment.

For solid wastes, quantities are best expressed in weight, though weight can be estimated from volume and bulk density. The characteristics of solid wastes to be determined are again dictated by the pollutants to be controlled and the method of treatment to be employed or considered (compressibility for baling, heat value, moisture and ash contents for incineration, leachate production and characteristics for landfilling).

At the completion of the waste survey, a flow and materials diagram should be prepared to ascertain that the data obtained are consistent: flows from contributing sources must be equal to the total flow measured. And for many waste components (e.g., water used, inert residues and specific chemicals) the data can be checked against input to the plant and output via products and wastes. A statistical analysis is required for parameters which fluctuate widely during the sampling programme [13].

The results of the waste survey would usually suggest ways of reducing the volume and strength of wastes by improved housekeeping, in-plant water re-use, and plant modification.

For a plant under design, flow rates and characteristics of products and waste streams can be ascertained from the plant flowsheet and accompanying design calculations. These calculations are based on material balances around units and the overall plant and augmented by laboratory bench scale and pilot plant experimentation. Although, in the past, attention has been primarily given to ensuring that product specifications are met in the design of industrial plants, there is now a need to examine the flows and characteristics of the wastes and how to best meet discharge standards even at the design stage [27].

Waste Reduction Practices

Prior to considering treatment options, the reduction of waste volume and/or concentration of pollutants in the waste stream should be investigated. The results of the waste survey and the inspection undertaken during the survey usually point to ways and means by which waste reduction can be implemented. An indication of the potential of reduction that can be achieved may be obtained by comparing the results of the survey with published values for the amount of waste per unit of production. Factors such as plant age, process sequence and raw materials should, however, be taken into account.

Steps that can be taken to reduce waste fall into three categories:

1. Improved housekeeping (e.g., regular plant maintenance, repair of leaks, ensuring floor traps are in place and materials trapped are collected and disposed properly)
2. In-plant modification (e.g., installing automatic cut-off hoses, changing to high pressure, low volume sprays for cleaning, re-circulation of cooling and process of water)
3. Process modification (change of operating conditions to reduce by-products, substitution of chemicals, complete process route change)

A fair degree of in-plant recycling of solid wastes is already taking place in industry [2,21] and Waste Reduction practices commonly used for waste water are summarized in Table 5 [22].

The degree of change feasible is usually governed by the cost of implementing the change and the benefit from reducing the treatment cost of the remaining waste.

Segregation, Equalization and Neutralization

Segregation, equalization and neutralization of waste streams should be considered after ways of reducing wastes have been explored.

SEGREGATION

Segregation refers to keeping the waste streams separate, so that waste streams which contain hazardous pollutants or a high concentration of a pollutant are kept separate from other waste streams which can be disposed of safely without further treatment. The size of the treatment plant, which is proportional to the volume throughput, is kept small with a corresponding reduction in capital and operating costs.

In the design and construction of new plants provision should be made for segregation of all waste streams, particularly liquid streams; in the event that equipment/technology is available for separate treatment/re-use, expensive modification of plant drainage system is then not required.

TABLE 5. Waste Reduction Practices for Industrial Wastewater.

Practice	Description
Volume Reduction	
Classification of wastes	Separation of process wastewater, cooling water, sanitary wastewater, and rainfall run-off, so that the volume of water requiring intensive treatment is reduced.
Conservation of water	Changing from an open to a close system for process and cooling waters, good house-keeping and preventative maintenance: dry disposal of waste rather than using water for flushing, regulate water pressure, install meters in each department to make operators cost and quantity conscious.
Changing production to decrease waste	Improved process control, improved design, use of different or better quality raw materials.
Reusing industrial and municipal effluents	Adopting practices used in regions where water is scarce.
Strength reduction	
Process changes	Use a substitute processing chemical that is less polluting (e.g. enzymes for lime and sulfides in the leather industry, cellulosic sizing agents for starch in the textile industry, H_3PO_4 for H_2SO_4 in pickling).
Equipment modifications	Improve design to reduce carry over of solids to drain for example.
Segregation of wastes	See classification of waste in Volume Reduction above.
By-product recovery	Recovery of chemicals, solvents and other components for recycling or re-use for other purposes, e.g. blood from meat packing, chromium from metal plating, cellulose fibre from paper making, paint solvent from gaseous waste.

From Reference [22].

EQUALIZATION

Equalization refers to the provision of storage for solid waste or a holding tank or basin for liquid waste to even out fluctuation in flow or concentration so that subsequent treatment system does not have to be designed for the maximum flow or concentration. The larger the holding (equalization) container the smaller the amplitude of the fluctuation coming from it, and a balance needs to be reached between the saving in the cost of subsequent treatment and the cost of providing the equalization.

When a small stream is to be evenly mixed with a large stream it is generally more economical to hold only the small stream and add it in proportion to the flow of the large stream; this technique is commonly called proportioning.

NEUTRALIZATION

With liquid waste, acidic waste can generally be neutralized with alkaline waste rather than treating each separately with neutralizing agents. Wastes from nearby industry should be considered for the neutralizing agent.

INDUSTRIAL WASTE TREATMENT OPTIONS

Treatment options for wastes which cannot be reduced by practices outlined in the previous section are usually specific to a particular industry and even to a particular plant depending on the process used, availability of land, neighbouring industry, publicly operated treatment works and local factors. The treatment options, however, usually consist of units which are common to options used in other industry. These units can be classified into physical, chemical or biological operations or processes, and are described below.

The most common ways the units are combined for the treatment of solid wastes, liquid wastes and gaseous wastes are illustrated in Figures 1, 2, and 3. The relationship between the treatment trains is obvious from the common units that can be employed and from the conversion of a solid waste problem into a gaseous waste problem when hazardous solid waste is incinerated for example.

Solid Waste Treatment

Industrial solid wastes which do not contain hazardous constituents can be disposed of in the same manner as municipal solid wastes. Landfilling, without treatment, is the most common method of disposal for these wastes, since it is the most economical, and managed properly it is environmentally safe (see section on Ultimate Disposal). The potential of mechanical resource recovery from the wastes is similar to that of municipal solid wastes [9].

The available solid treatment processes are described in Table 6. Compaction and size reduction are processes aimed at reducing the volume of the waste when the waste is to be landfilled. Size reduction is also used as a preparatory step prior to composting and mechanical resource recovery. Composting and anaerobic digestion are processes relying on microorganisms to decompose organic substances in the waste; the first in the presence of oxygen while the latter in the absence of oxygen. The decomposition of the organics would prevent anaerobic conditions to be formed in a landfill with consequent production of acidic leachate and methane gas, both of which may not be desirable environmentally coming from a landfill. The products of composting and anaerobic digestion can also be applied (spread) on land as a method of ultimate disposal or of enriching the soil with organic carbon and nutrients (see section on Ultimate Disposal).

Composting and anaerobic digestion are suited to indus-

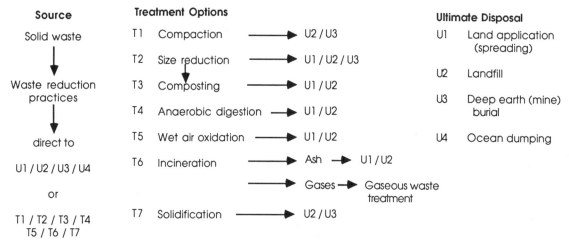

FIGURE 1. Solid waste treatment.

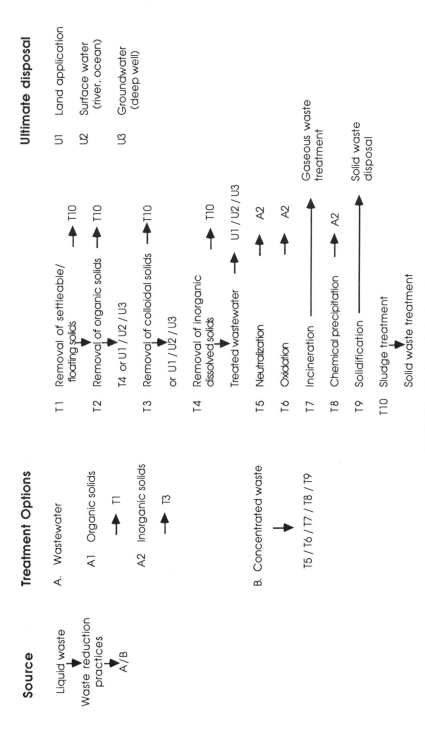

FIGURE 2. Liquid waste treatment.

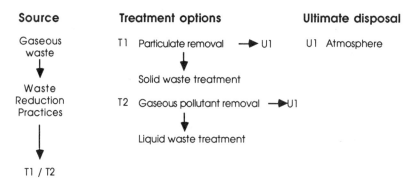

FIGURE 3. Gaseous waste treatment.

trial solid wastes which consist primarily of organic materials, including sludges from the treatment of industrial wastewaters which are organic in nature. Relying on microorganisms limits the processes to nontoxic organics, although many toxic organics could be decomposed and rendered harmless when present in not too high concentrations.

Wet air oxidation is a milder chemical oxidation process than incineration to satisfy the oxygen demand described above, while incineration is a severe oxidation process converting carbonaceous materials into carbon dioxide and water vapour suitable for both toxic and non-toxic containing organic solid wastes and sludges. In incineration, the operating temperature must be high enough for the particular solid waste components to achieve complete combustion; gaseous waste treatment is usually required to remove particulates and acids which may be generated. Of the solid waste treatment options (Table 6, Figure 1) incineration is the option most commonly used particularly for hazardous wastes, although solidification is increasingly employed.

Liquid waste treatment

Liquid wastes of industry may be classified into wastes that are aqueous, and concentrated wastes; the former consist of water which has been contaminated with pollutants, whereas the latter consist primarily of pollutants and are likely to be hazardous.

Aqueous wastes which contain pollutants in low enough concentrations may be disposed without treatment, and this may be the result of waste segregration practice as outlined in the waste reduction practices section.

Aqueous wastes which contain organics from biological sources (e.g., fruit and meat processing) can be treated using the same processes as for the treatment of sewage. The sequence of treatment usually starts with the removal of settleable or floating solids (employing one of several screening, sedimentation and flotation methods), followed by the removal of dissolved organic solids (employing one of several biological processes: activated sludge, trickling filtration, anaerobic digestion) accompanied by the treatment of

the sludge produced. Treatment can also be carried out by ponding or lagooning with a lower operating cost, when a large land area is available. Treatment and dewatering of the sludges produced are required prior to disposal as solid waste. Treatment is necessary to stabilize the organics so that its oxygen demand and potential for producing odour and acids are eliminated; processes which can be used are aeration, anaerobic digestion and wet air oxidation. Dewatering is carried by thickening, followed by filtration, centrifugation or sand bed drying. With lagooning or ponding the first lagoon in the series will eventually be filled and has to be cleaned of the sludge, or a new lagoon constructed.

Industrial organics (e.g., paper mill wastewater) may not contain balanced nutrients for microorganisms and the addition of nitrogen and phosphorus may be necessary or combined treatment with sewage may be desirable. The presence of substances toxic to microorganisms may limit the utility of biological processing, although waste segregation may overcome this problem (see section on Waste Reduction Practices).

Aqueous wastes containing primarily inorganic solids can be treated using the same processes as for the treatment of water for municipal water supply. The treatment consists of removing colloidal solids using coagulation or flocculation followed by sedimentation and filtration. Thickening of the sludge produced may be required prior to its disposal.

The removal of dissolved inorganic and refractory organics can be achieved if necessary using processes such as activated carbon adsorption, ion-exchange, reverse osmosis, electrodialysis, chemical precipitation, chlorination and other techniques which may be suited to the particular pollutants.

The removal of settleable and floating solids normally utilizes physical operations, the removal of colloidal and dissolved organic solids by biological processes and the removal of colloidal and dissolved inorganic solids by chemical or physico-chemical processes (Table 7). Dissolved nitrogen and phosphorus can, however, be removed by biological processes [14].

The treatment of concentrated liquid waste depends on whether the pollutants are inorganic (e.g., spent electroplating bath solutions) or organic (e.g., paint solvent) in nature. For the former acid-base neutralization, oxidation-reduction and chemical precipitation are commonly used to remove the pollutants from the wastewater, so that further treatment can be carried out using methods for dilute aqueous wastes. For concentrated organic liquid waste chemical processes, such as incineration and solidification, used in the treatment of hazardous solid wastes need be employed (see section on Solid Waste Treatment).

Gaseous Waste Treatment

Particulates in gaseous waste streams can be removed by one of several methods shown in Table 8. Mechanical collectors rely on the density difference between the particulates and the carrier gas, while fabric filters strain the particulates on the filter cake formed on the fabric filter. Wet scrubbing transfers the particulates to the scrubbing liquid, and electrostatic precipitation electrically charge the particulates and collecting them on electrodes. The choice of the equipment depends on the properties of the particulates

TABLE 6. Solid Waste Treatment Processes.

Treatment	Objective and Description	Treatment	Objective and Description
Physical operations		Solidification	The aim of the process to fix the waste in a solid matrix with low permeability, leachability and of sufficient compressive strength and durability prior to ultimate disposal. The waste is mixed with cement-like mixture and allowed to set.
Compaction	The aim of the operation is to reduce the volume of the solid waste prior to landfilling and hence reduce landfill space and cover material requirement. It is usually carried out using hydraulic compression to produce bales.		
		Biological processes	
Size reduction	The aim of the operation is to reduce the volume of bulky solid waste. It is also used as a preparation prior to composting and resource recovery. It is usually carried out using a hammer or other shredding machinery.	Composting	The aim of composting is to satisfy the oxygen demand of the biodegradable organic substances in the solid waste by aerating the waste and allowing microorganisms to convert the waste into compost (biological oxidation). Near neutral pH, carbon to nitrogen ratio of about 30 and a temperature of about 37°C are optimum process conditions. Composting can be carried out in open windrows, or in a tumbler for the initial rapid oxidation prior to windrowing. The composting process is exothermic and with insulation (or self insulation by the compost) maintains its temperature.
Chemical processes			
Incineration	The aim of the process is to convert all carbonaceous organic substances to gaseous carbon dioxide and water vapour. Additional fuel is required if the combustion is not self-sustaining due to high moisture content for example. Temperature required to achieve complete combustion depends on the nature of the organic substances; a second combustion chamber may also be required. Different incineration designs are available, but temperature, turbulence and time of combustion must be adequate for all waste components.		
		Anaerobic digestion	The aim of anaerobic digestion is, as in the case of composting, the decomposition of organic substances by microorganisms. Optimum conditions are as for composting, though for oxygen exclusion the process has to be carried in a closed vessel. Recent advances have shown that moisture content needs only be similar to composting, rather than as a slurry [17]. Methane produced can be used to heat up the digestor to maintain optimum temperature, with excess for other uses.
Wet air oxidation	The aim of the process is to oxidise organic substances chemically using air at a high temperature and pressure in the presence of moisture. It is carried out in a pressure vessel with compressed air; adequate retention time is essential.		

References: [7,19,20,25].

(size, shape, density, wettability, friability, stickiness, electrical conductivity) and the degree of removal desired.

The removal of gaseous pollutants depends on the nature of the pollutants and their concentration. Condensation is suitable for vapours with a high dew point; absorption is suited to substances which dissolve in water and hence concentrated in the liquid phase for further treatment, while activated carbon adsorption concentrates the pollutants on the surfaces of the carbon. Absorption may be enhanced by adding a chemical that will react with the pollutant and hence fixing it in the water phase.

Destruction of many organic pollutants can be carried out by oxidation using chlorine or other oxidizing agents. Combustion is a severe form of oxidation, and if the concentration of the combustible organic substance is high enough, the combustion can be self-sustaining (direct flame incineration). In most cases the gaseous waste stream needs to be contacted with the flame of an auxilliary fuel (thermal incineration). Temperature, time of contact and turbulence in the mixing zone should be sufficiently high to achieve complete combustion. The temperature of incineration can be lowered by the use of a catalyst (Catalytic incineration).

Specific physical and chemical methods have also been developed to treat pollutants in very small concentrations causing odour nuisance and for the treatment of sulfur dioxide from the combustion of fossil fuels [4,5].

TABLE 7. Liquid Waste Treatment Options.

Treatment	Objective and Description
A. For Aqueous Liquid waste (pollutants primarily organic in nature)	
1. Physical operations	
Screening	The removal of large solids at the start of a treatment sequence. Screening is carried out using coarse bars or racks which can be mechanically cleaned.
Sedimentation	The removal of settleable solids by reducing the flow of the wastewater in a sedimentation tank. Discrete fast settling particles are usually removed first, followed by the slower settling flocculent particles. The insertion of inclined tubes or plates in a sedimentation tank can increase the efficiency of flocculent particle removal.
Flotation	The removal of low density solids by introducing fine air bubbles to float the solids which are then skimmed. Air can be introduced through small perforations or more usually by dissolving air in the wastewater (or a recycle stream) under pressure and releasing the pressure in the flotation tank.
2. Biological processes	
Activated sludge process	The purpose of the process is to convert organics to microbial cells, which can be then separated out. The process is carried out in an aerated tank, where wastewater is mixed with microorganisms; the mixture is then separated in a sedimentation tank, where the microorganisms contained in flocs of solids (termed activated sludge) are sedimented from the clarified waste-water and part returned to the aeration tank.
Trickling filtration	The purpose of the process is the same as for the activated sludge process. The difference is that the microorganisms are immobilized on the surfaces of rocks or corrugated plastic sheets. The sloughed of microorgansisms are separated from the clarified wastewater in a sedimentation tank without any need for the return of the sludge.
Anaerobic digestion	The above biological processes can also be carried out in the absence of oxygen, with less sludge production and with the production of methane gas. The gas can be used for maintaining optimum digestion temperature (37°C), with any excess for other heating purposes.
3. Chemical processes	
Carbon-adsorption	Carbon adsorption is used to remove dissolved organic substances which are non-biodegradable. The clarified wastewater is commonly passed through a bed of granulated activated carbon, and the operation is carried out until the adsorption capacity of the bed is nearly exhausted and the breakthrough of the organics occurs. The activated carbon can usually be regenerated by heating to a high temperature (above 600°C).
Ion exchange	Ion-exchange is used to remove cations (e.g. Ni^{2+}) or anions (eg. chloride) by exchanging them with cations or anions respectivey attached to ion-exchange resin beads. The operation is usually carried out using beds of ion-exchanger, similar to the operation of granulated activated carbon (see above). Since ion-exchange is reversible, ion-exchanger can be regenerated using a brine solution, acid or alkali. The regenerant may require waste treatment prior to disposal.

(continued)

TABLE 7 (continued).

Treatment	Objective and Description
Reverse Osmosis	Reverse osmosis removes dissolved substances by forcing water through a semi-permeable membrane (e.g. cellulose acetate) that does not allow the dissolved substances to pass through. The semi-permeable membrane is generally supported on a tubular frame with pressure applied to the wastewater. A concentrated stream containing over 90% of the dissolved substances is produced, which may require treatment prior to disposal.
Electrodialysis	Electrodialysis removes cations and anions by attracting them from the wastewater using an anode and cathode respectively. The operation is carried out in a cell separated alternately by cation and anion resin membranes; the ions migrating towards the electrodes form in alternate compartments a concentrated solution, which may require treatment prior to disposal.
Chemical precipitation	Many pollutants (e.g. heavy metals) can be precipitated out of solution by the addition of a suitable reagent (e.g. lime). The precipitate formed can be removed by methods used for the removal of colloidal inorganic solids (see B below).
Chlorination	Chlorination is used to remove pathogenic pollutants (bacteria and viruses), but it also removes dissolved pollutants such as iron and manganese by oxidation and precipitation.

B. For aqueous liquid waste (pollutants primarily inorganic in nature)

1. Physical operations

Filtration	The removal of coagulated or flocculated particles by mechanical straining. The wastewater is passed through a filter medium under pressure (gravity, pumping or vacuum). The filter medium can be a sand bed or fabric filter.
Flocculation	Colloidal particles are flocculated (coagulated) by the addition of suitable cogulatingt agents (e.g. lime, alum) and/or polyelectrolytes (long chain polymers with active groups which can join to the colloidal particles).
Sedimentation	See A.1 above

2. Chemical processes — See A.3 above

C. Sludge dewatering and treatment prior to disposal as solid waste

1. Physical operations

Centrifugation	Sludge dewatering by applying centrifugal force on the sludge throwing the solids onto the wall of the centrifuge. Several types of centrifuge are available with different ways of discharging the collected solids.
Filtration	See B.1 above.
Thickening	See A.1 above under sedimentation.

2. Biological Processes

Aeration	To satisfy the oxygen demand of the organics similar to the Activated Sludge process but normally without sludge recycle.
Anaerobic digestion	See A.2 above.

3. Chemical processes

Wet air oxidation	See Solid Waste Disposal Options
Incineration	See Solid Waste Disposal Options

D. Concentrated Liquid Waste

1. Physical operation

Evaporation	Removal of water by applying heat (e.g. solar radiation) so produce solids, which may require further treatment as solid waste.

2. Chemical Processes

Chemical Precipitation	See A.3 above
Oxidation/Reduction	Specific pollutants can be converted to non-hazardous forms by oxidation (e.g. cyanide) or by chemical reduction (e.g. chromate) and followed by chemical precipitation.
Incineration	See Solid Waste Disposal Options
Solidification	See Solid Waste Disposal Options

References: [2,12,13,19,21,22].

TABLE 8. Gaseous Waste Treatment Options.

Treatment	Objective and Description
A. Particulate Removal	
1. Mechanical collectors	
Gravity settling chamber	Particulates are removed by slowing the flow of the gas in a chamber and allowing the particulates to settle into a hopper. It is effective as a precleaner for large particulates.
Cyclones	The gaseous waste is introduced tangentially into a cylindrical chamber and a centrifugal force is created throwing the particulates to the wall of the cylinder for collection at the base of the cylinder via a cone to a hopper. The cleaned gas spirals up the middle of the cylinder. Effective for particulates down to 3–5 microns.
2. Fabric filters	Particulates are collected continuously on fabric filters, which are periodically shaken to loosen the particles for collection into hoppers. Effective for particulates to less than 1 micron due to the straining action of the filter cake. Fabric filters cannot be operated at high temperatures due to the decomposition of the fabric, or for moist gas due to fabric clogging.
3. Wet scrubbers	Particulates are transferred from the gaseous waste to water by contacting the two in contractors such as packed bed scrubbers, plate scrubbers and venturi scrubbers. Droplets need to be removed using baffles or mechanical collecters (see A.1 above). Bled scrubbing water usually requires treatment prior to disposal.
4. Electrostatic precipitators	Particulates are negatively charged using high DC voltage applied on wires, and the particulates are collected on earthed plates. The latter are regularly rapped to loosen the collected particles for collection in hoppers.
B. Gaseous pollutant removal	
1. Physical operations	
Condensation	Cooling of the gaseous waste condenses a pollutant (e.g. solvent), which can be recycled or treated more economically as a liquid waste.
Absorption	A water soluble pollutant is dissolved by contacting the gas with the water in a wet scrubber (see A.3 above). The srubbing water may then be economically treated.
2. Chemical processes	
Adsorption	See Activated Carbon Adsorption in Liquid Waste Treatment Option.
Absoprtion & chemical reaction	The process of absorption (see B.1 above) may be enhanced by adding a chemical reagent that reacts with the pollutants in the water phase.
Chlorination	Chlorination and oxidation can be used to oxidise compounds which cause odour.
Incineration	The aim of incineration is to oxidise organic pollutants to carbon dioxide and water vapour. The incineration can be carried out without additional fuel with concentrated gaseous waste. In thermal incineration the gaseous waste is passed through the combustion zone of an auxilliary fuel. The temperature required for complete destruction may be lowered by the use of a catalyst bed placed in the combustion zone. (See also Incineration in Solid Waste Treatment).

References: [2,4,5,19,28].

TABLE 9. Convention on the Prevention of Marine Pollution by Dumping of Wastes and Other Matter.

ANNEX I

Not to be dumped

1. Organohalogen compounds.
2. Mercury and mercury compounds.
3. Cadmium and cadmium compounds.
4. Persistent plastics and other persistent synthetic materials, for example, netting and ropes, which may float or may remain in suspension in the sea in such a manner as to interfere materially with fishing, navigation or other legitimate uses of the sea.
5. Crude oil, fuel oil, heavy diesel oil, and lubricating oils, hydraulic fluids, and any mixtures containing any of these, taken on board for the purpose of dumping.
6. High-level radioactive wastes or other high-level radioactive matter, defined on public health, biological or other grounds, by the competent international body in this field, at present the International Atomic Energy Agency, as unsuitable for dumping at sea.
7. Materials in whatever form (e.g. solids, liquids, semi-liquids, gases or in a living state) produced for biological and chemical warfare.
8. The preceding paragraphs of this Annex do not apply to substances which are rapidly rendered harmless by physical, chemical or biological processes in the sea provided they do not:

    (i) make edible marine organisms unpalatable, or
    (ii) endanger human health or that of domestic animals.

    The consultative procedures for under Article XIV should be followed by a Party if there is doubt about the harmlessness of the substance.
9. This Annex does not apply to wastes or other materials (e.g. sewage sludges and dredged spoils) containing the matters referred to in paragraphs 1–5 above as trace contaminants. Such wastes shall be subject to the provisions of Annexes II and III as appropriate.

ANNEX II

The following substances and materials requiring special care are listed for the purposes of Article VI (1) (a).

A. Wastes containing significant amounts of the matters listed below:

arsenic
lead
copper   } and their compounds
zinc
organosilicon compounds
cyanides
fluorides
pesticides and their by-products not covered in Annex I.

B. In the issue of permits for the dumping of large quantities of acids and alkalis, consideration shall be given to the possible presence in such wastes of the substances listed in paragraph A and to the following additional substances:

beryllium
chromium } and their compounds
nickel
vanadium

C. Containers, scrap metal and other bulky wastes liable to sink to the sea bottom which may present a serious obstacle to fishing or navigation.
D. Radioactive wastes or other radioactive matter not included in Annex I. In the issue permits for the dumping of this matter, the Contracting Parties should take full account of the recommendations of the competent international body in this field, at present the International Atomic Energy Agency.

(continued)

TABLE 9 (continued).

ANNEX III

Provisions to be considered in establishing criteria governing the issue of permits for the dumping of matter at sea, taking into account Article IV (2), include:

A—Characteristics and composition of the matter

1. Total amount and average composition of matter dumped (e.g. per year).
2. Form, e.g. solid, sludge, liquid or gaseous.
3. Properties: physical (e.g. solubility and density), chemical and biochemical (e.g. oxygen demand, nutrients) and biological (e.g. presence of viruses, bacteria, yeasts, parasites).
4. Toxicity.
5. Persistence: physical, chemical and biological.
6. Accumulation and biotransformation in biological materials or sediments.
7. Susceptibility to physical, chemical and biochemical changes and interaction in the aquatic environment with other dissolved organic and inorganic materials.
8. Probability of production taints or other changes reducing marketability of resources (fish, shellfish, etc.)

B—Characteristics of dumping site and method of deposit

1. Location (e.g. coordinates of the dumping area, depth and distance from the coast), location in relation to other areas (e.g. amenity areas, spawning, nursery and fishing areas and exploitable resources).
2. Rate of disposal per specific period (e.g. quantity per day, per week, per month).
3. Methods of packaging and containment, if any.
4. Initial dilution achieved by proposed method of release.
5. Dispersal characteristics (e.g. effects of currents, tides and wind on horizontal transport and vertical mixing).
6. Water characteristics (e.g. temperature, pH, salinity, stratification, oxygen indices of pollution—dissolved oxygen (DO), chemical oxygen demand (COD), biochemical oxygen demand (BOD)—nitrogen present in organic and mineral form including ammonia, suspended matter, other nutrients and productivity).
7. Bottom characteristics (e.g. topography, geochemical and geological characteristics and biological productivity).
8. Existence and effects of other dumpings which have been made in the dumping area (e.g. heavy metal background reading and organic carbon content).
9. In issuing a permit for dumping, Contracting Parties should consider whether an adequate scientific basis exists for assessing the consequences of such dumping, as outlined in this Annex, taking into account seasonal variations.

C—General considerations and conditions

1. Possible effects on amenities (e.g. presence of floating or stranded material, turbidity, objectionable odour, discolouration and foaming).
2. Possible effects on marine life, fish and shell fish culture, fish stocks and fisheries, seaweed harvesting and culture.
3. Possible effects on other uses of the sea (e.g. impairment of water quality for industrial use, underwater corrosion of structures, interference with ship operations from floating materials, interference with fishing or navigation through deposit of waste or solid objects on the sea floor and protection of areas of special importance for scientific or conservation purposes).
4. The practical availability of alternative land-based methods of treatment, disposal or elimination, or of treatment to render the matter less harmful for dumping at sea.

ULTIMATE DISPOSAL

Although industry has the option of disposing of untreated or partially treated waste via a contractor or into a publicly owned treatment plant, the ultimate destination of the waste is one of the following: the atmosphere, surface water or land surface, by dispersion, dilution or spreading; or by storage in a shallow or deep confined land space.

Atmospheric Dispersion

Discharge through stacks is commonly regulated to meet ambient quality standards at ground level. Dispersion from a stack is governed by the design of the stack (height, diameter, exit velocity, temperature of the gas), meteorological conditions (wind speed and direction, intensity of turbulence, vertical temperature gradient) and local topographical features (ridges, water bodies). The stack then has to be designed to meet ambient ground level quality standards with local factors taken into account, including the presence of nearby stacks [4,5].

Ocean Disposal

The disposal of wastewater through an ocean outfall is regulated by the need to protect marine ecosystems (including fishing grounds) and to ensure that the diluted wastewater does not impact upon recreation and other near shore human activities.

The dispersion of wastewater from an outfall is governed by the depth of outfall, exit speed of the wastewater, density difference between the wastewater and the sea water and the geometry of the outfall. For wastewaters that are less dense than seawater the wastewater is diluted as it rises. Buoyancy force plays a major role and the degree of dilution at the surface is determined primarily by the density difference and the depth of the outfall [16].

Dumping of waste at sea is regulated by the London Convention, which has been ratified by many countries. Its provisions are listed in Table 9. The dumping of toxic, persistent and bio-accumulated pollutants is specifically prohibited.

Disposal into Estuaries, Rivers, Lakes and Wetlands

The assimilative capacities of these environments are very limited and stringent requirements must normally be met for disposal into those systems, particularly when the water does not flow to any appreciable extent such as at wetlands.

Land Spreading

Land spreading of treated wastewater and sludges, particularly those of biological origins, is practiced when large land areas are available. Physical, chemical and biological changes and decomposition take place after land application, converting the waste materials to normal soil organic constituents. Application rate should be determined from the rate of waste decomposition and whether the area is used for agricultural purposes [23,24].

Landfill, Deep Mine Burial and Deep Well Injection

Landfilling and deep mine burial of solid and liquid waste, and deep well injection of liquid waste are primarily storage operations, although physical, chemical and biological decomposition takes place during storage.

Landfilling is the most commonly used method for the disposal of solid waste since it is the most economical method. For organics that are biodegradable, aerobic decomposition takes place until the oxygen originally entrapped with the waste is depleted. Anaerobic decomposition then takes over and may last for many years. The leachate generated under aerobic conditions is generally neutral in pH, but under anaerobic conditions it is usually acidic since organic acids are produced during anaerobic decomposition. The acidic leachate can mobilise heavy metals disposed in the landfill. The realization that leachate may contaminate groundwater beneath the landfill has led to stringent requirements for lining the fill with clay or plastic liners. The deposited waste normally requires daily cover to prevent fire, litter, dust and odour and to reduce access to possible disease vectors. Once the landfill is completed a final cover is required and landscaping may be necessary to minimize rainfall run-off percolating into the landfill. During operation diversion of rainfall run-off may also be necessary. Leachate production can also be minimized by compaction of the waste and rapid filling of a smaller area rather than slower filling of a larger land area. Leachate that is produced during landfilling should be treated in the same way as for Industrial Liquid Waste. Clayey soils have varying capacities to attenuate pollutants by ion exchange, adsorption and chemical reaction, although acidic leachate can mobilise metals already adsorbed.

Methane gas is generated during anaerobic decomposition in a landfill and ventilation should be considered in the design and operation of a landfill. Volatilization of hazardous substances should be prevented and hazardous wastes to be landfilled should preferably be solidified [25].

The use of deep mine for the disposal of solid waste should be viewed in the same way as disposal by landfilling with the prevention of leachate from contaminating potable groundwater in mind, and should consider the stability of the geological structure of the mine area.

Deepwell injection into saline groundwater has been used for industrial wastewaters. The possible reactions between the wastewater, the existing groundwater and the aquifer material should be investigated to prevent precipitate forming which can clog the aquifer. Protection of potential potable groundwater aquifers should be ensured by cemented casings to protect the potable groundwater, and injection of the waste with a specified maximum pressure.

Monitoring is an essential part of the disposal of wastes to ensure that pollutants do not reach levels in the environment that can endanger human health and the integrity of ecosystems supporting life, and has been made mandatory in many countries as part of the granting of permits to operate disposal facilities.

REFERENCES

1. Barton, A. F. M., *Resource Recovery and Recycling,* John Wiley & Sons, New York (1979).
2. Bridgwater, A. V. and C. J. Mumford, *Waste Recycling and Pollution Control Handbook,* George Godwin, London (1979).
3. Butler, G. C. (ed.), *Principles of Ecotoxicology,* John Wiley & Sons, Chichester (1978).
4. Cheremisinoff, P. N. and R. A. Young, (eds.), *Industrial Odor Technology Assessment,* Ann Arbor Science, Ann Arbor, Michigan (1975).
5. Cheremisinoff, P. N. and R. A. Young, (eds.), *Air Pollution Control and Design Handbook,* Marcel Dekker, New York (1977).
6. Clifton, J. R., P. W. Brown, and G. Frohnsdorff, "Uses of Waste Materials and By-products, in Construction," *Resource Recovery and Recycling,* Vol. 5, pp. 139–160, 217–228 (1980).
7. Conway, R. A. and R. D. Ross, *Handbook of Industrial Waste Disposal,* van Nostrand Reinhold Co., New York (1980).
8. Department of the Environment, *Special Wastes: A Technical Memorandum Providing Guidance on Their Definition,* Waste Management Paper no. 23, HMSO, London (1981).
9. Diaz, L. F., G. M. Savage, and C. G. Golueke, *Resource Recovery from Municipal Solid Wastes* (2 volumes), CRC Press, Boca Raton, Florida (1982).
10. Doll, R. and R. Peto, *The Causes of Cancer,* Oxford University Press, Oxford (1981).
11. Duffus, J. H., *Environmental Toxicology,* Edward Arnold, London (1980).
12. Dyer, J. C. and N. A. Mignone, *Handbook of Industrial Residues,* Noyes Publications, Park Ridge, New Jersey (1983).
13. Eckenfelder, W. W., Jr., *Principles of Water Quality Management,* CBI Publishing, Boston, Massachusetts (1980).
14. Ekama, G. A. et. al., *Theory, Design and Operations of Nutrient Removal Activated Sludge Processes,* Water Research Commission, Pretoria, South Africa (1984).
15. Federal Register, *Hazardous Waste and Consolidated Permit Regulations,* US EPA, vol. 45, no. 98, Book 2 (May 19, 1980); also Federal Register, Vol. 45, No. 220 (November 12, 1980); Federal Register, Vol. 48, No. 65 (April 4, 1983); Federal Register, Vol. 49, No. 29 (February 10, 1984); Federal Register, Vol. 49, No. 90 (May 8, 1984); Federal Register, Vol. 49, No. 92 (May 10, 1984).
16. Fischer, H. B., E. J. List, R. C. Y. Koh, J. Imberger, and N. H. Brooks, *Mixing in Inland and Coastal Waters,* Academic Press, New York (1979).
17. Jewell, W. J., R. J. Cummings, S. Dell'Orto, K. J. Fanfoni, S. J. Fast, E. J. Gottung, D. A. Jackson, and R. M. Kabrick, *Dry Fermentation of Agricultural Residues,* Department of Agricultural Engineering, Cornell University, Ithaca, New York (1982).
18. Koziorowski, B. and J. Kucharski, *Industrial Waste Disposal,* Pergamon Press, Oxford (1972).
19. Liptak, B. G. (ed.), *Environmental Engineers' Handbook,* Chilton Book Co., Radnor, Pennsylvania (1974).
20. Mantell, C. L., *Solid Wastes: Origin, Collection, Processing and Disposal,* John Wiley and Sons, New York (1975).
21. Metcalf & Eddy, Inc. (Revised by Tschobanoglous, G.), *Wastewater Engineering: Treatment, Disposal and Reuse,* McGraw Hill Co., New York (1979).
22. Nemerow, N. L., *Industrial Water Pollution: Origins, Characteristics and Treatment,* Addison-Wesley Publishing Co., Reading, Massachusetts (1978).
23. Overcash, M. R. and D. Pal, *Design of Land Treatment Systems for Industrial Wastes—Theory and Practice,* Ann Arbor Science, Ann Arbor, Michigan (1981).
24. Parr, J. F., P. B. Marsh, and J. M. Kla, *Land Treatment of Hazardous Wastes,* Noyes Data Corporation, Park Ridge, New Jersey (1983).
25. Pojasek, R. B., *Toxic and Hazardous Waste Disposal* (6 volumes), Ann Arbor Science, Ann Arbor, Michigan (1980).
26. Powers, P. W., *How to Dispose of Toxic Substances and Industrial Wastes,* Noyes Data Corporation, Park Ridge, New Jersey (1976).
27. Ritcey, G. M., "Solvent Extraction—Projections to the Future," *Separation Science and Technology,* vol. 18, pp. 1617–1646 (1983).
28. Ross, R. D. (ed.), *Air Pollution and Industry,* van Nostrand Reinhold, New York (1972).
29. Royston, M. G., *Pollution Prevention Pays,* Pergamon Press, Oxford (1979).
30. Sax, N. I., *Dangerous Properties of Industrial Materials,* 6th edition, van Nostrand Reinhold Co., New York (1984).
31. Sittig, M., *How to Remove Pollutants and Toxic Materials from Air and Water,* Noyes Data Corporation, Park Ridge, New Jersey (1977).
32. *Standard Methods for the Examination of Water and Wastewater,* 16th edition, American Public Health Association (1984).
33. US Environmental Protection Agency, *Handbook for Monitoring Industrial Wastewater,* Technology Transfer, US EPA (August, 1973).
34. US Environmental Protection Agency, *Industrial Guide for Air Pollution Control,* Technology Transfer, EPA 625/6-78-004 (June 1978).
35. US Environmental Protection Agency, *Handbook Continuous Air Pollution Source Monitoring Systems,* Technology Transfer, EPA 625/6-79-005, Cincinnati, Ohio (June, 1979).

CHAPTER 21

Hazardous Waste

AARON A. JENNINGS*

INTRODUCTION

Hazardous Waste is a new discipline of scientific, technological, and legal expertise. It probably has its origins in the environmentalism of the 1960s, but did not receive the early attention paid to air and water pollution. It didn't evolve as an organized issue until early in the 1970s, and wasn't institutionalized until 1976. However, when the issue did arrive, it did so with an unprecedented outpouring of public concern.

Almost without fail, the term "hazardous waste" immediately conjures vivid negative images. Nearly everyone has heard of the Love Canal and can visualize the abandoned homes and protesting homeowners. The allegations of increased cancer and miscarriage rates in the local residents shocked the nation. Subsequent studies have cast serious doubt on the scientific merit of these early claims, but the image has lingered. Dramatic views of leaking drums stacked up around abandoned buildings, or news coverage of burning warehouses easily come to mind. This author will always remember the shock of discovering drums of cyanide waste stacked next to drums of concentrated sulfuric acid in a dilapidated shed in urban Chester, Pennsylvania. Names like Valley of the Drums (KY), Elizabeth (NJ), Times Beach (MO), Price's Pit (NJ), Rocky Mountain Arsenal (CO), The Wade Property (PA), Stringfellow Acid Pits (CA), Seymour (IN), Swartz Creek (MI), SILRESIM (MA), Iberville Parish (LA), and a host of others will haunt the issue of chemical waste management for years to come.

These dramatic images all come from a very brief era in the history of hazardous waste management. Industry traditionally disposed of chemical wastes (other than as air dis-

charges or wastewaters) by the avenues of conventional solid waste disposal. Most often these were landfilled or disposed of in surface impoundments. However, as environmental regulation intensified, industries became increasingly concerned about their disposal practices and began to seek alternatives. For a relatively brief time this led to the infamous uncontrolled hazardous waste disposal site.

Uncontrolled disposal sites came in many versions and operated under several guises ("waste recoverers" or "drum recyclers") or as surreptitious "midnight" operations. In general, they were short term projects doomed to eventual failure. Often they ended in conflagrations ignited by incompatible waste reactions. They existed because industries were willing to pay a lot of money to anyone able to relieve them of dangerous chemical wastes. Site operators could do this because it was generally held that title for the waste changed hands when the industry contracted for its diposal. Disposers operated for a brief time, often amassing tens of thousands of drums of anonymous chemical wastes. When eventually shut down, operators would simply declare bankruptcy, leaving behind their only asset, a huge inventory of hazardous waste.

Although we still have many of these sites to clean up, the era of the uncontrolled hazardous waste site should be over. Federal regulations and criminal penalties have put a stop to the practice for all but the most wildly irresponsible of industries. We have now entered an era where we are attempting to implement environmentally sound practices. The technologies and experience required to accomplish this are rapidly emerging.

This is an exciting time for hazardous waste professionals. We are developing the technologies and management strategies that will govern the fate of hazardous waste disposal for the next twenty years. Industries and communities are investing millions of dollars to improve their disposal efforts. More capital expenditures will be made in the 1980s than at any time in the future. Undoubtedly, future engineers will

*Department of Civil Engineering, University of Toledo, Toledo, OH

fine-tune our designs, but we are building the system they will work on.

Civil engineers have a critical role to play in this effort. The whole nation has been sensitized to the problems of hazardous waste. This is the driving force behind the federal legislation and industrial expenditures. Without this we could not solve the problem. Unfortunately, the public is extremely fearful of chemical hazards. In many states we have not been able to license new disposal facilities because of the public outcry that accompanies any mention of hazardous waste. Obviously, the industries of these states continue to generate and dispose of their wastes. These are either transported out of state, or disposed of at less desirable facilities allowed to operate because of the lack of alternatives. Therefore, public concern can be both a call for action and a barrier to action.

The solution of this type of public policy dilemma requires a careful balance between technical and legal expertise, and the realities of high visibility engineering. Great care must be taken to successfully address public concerns about the human and environmental dangers. This is a role particularly well suited to the civil engineering profession. As a result, many existing firms have added hazardous waste expertise, and numerous specialty firms have appeared. With this new opportunity comes new responsibility. In addition to engineering principles, the modern hazardous waste practitioner must be well versed in the chemistry and toxicology of chemical wastes, the principles of groundwater contamination (the most common avenue of environmental degradation), hazardous waste legislation and regulation, and the analysis and management of risk.

HAZARDOUS WASTE GENERATION

Although we have learned a great deal about hazardous waste generation in the United States over the last 15 years, it remains very difficult to characterize our national generation profile. There are several reasons for this. First, prior to RCRA regulations, there was no consistent definition of what should constitute a "hazardous" waste. This was determined at the state level with no consistency from state to state. Even now, wastes are classified on a case-by-case basis. Under RCRA the cumulative "definition" continues to evolve as wastes are listed and delisted. Secondly, pre-RCRA disposal practices were relatively effective in shielding the identity of generators. Title was allowed to change hands when wastes were disposed of. Therefore, it is very difficult to identify the generation source (type of industry, process of origin, etc.) in historical industrial waste data. Furthermore, most of the (sparse) state residual management programs were careful to shield the corporate identity of disposers. In addition, a large portion of industrial wastes were disposed of "on-site" (i.e., at locations under the ownership of the generator) with no serious effort at recordkeep-

ing. Finally, although some data are available on quantity, data on waste composition are extremely rare.

These current inadequacies of information will change under RCRA. However, this change is still to come. The Environmental Protection Agency and state environmental authorities are still struggling to implement RCRA's "cradle-to-grave" regulations. The manifest system will ultimately yield wonderful information on waste generation, composition, and disposal practices. Unfortunately, this will not occur until we have the solution well in hand. We are currently making the decisions that will establish the management facilities required to implement RCRA. This cannot be done effectively without some understanding of the generation source they are intended to serve.

The Historical Hazardous Waste Data Base

There are two broadly applicable sources of fundamental information on (potentially hazardous) industrial solid waste. The first is a series of assessment studies conducted on select industrial groups within the Manufacturing Division (Division D) of the Standard Industrial Classification (SIC) code. The second is a series of hazardous waste generation evaluations prepared by (or for) state waste management authorities. The industrial assessment documents integrate across state boundaries, but are incomplete in their scope. This second set of studies integrates across the total industrial sector, but not beyond state boundaries. Both must be used with caution because of the difficulties previously discussed.

EPA ASSESSMENT DOCUMENTS

In 1972, the EPA Office of Solid Waste Management Programs (OSWMP) commissioned to a series of studies under the mandate of The Resource Recovery Act of 1970 (P.L. 91-512). These were intended to develop an overview of the industrial hazardous waste problem. Studies in this series included:

a. "A Study of Hazardous Waste Materials, Hazardous Effects and Disposal Methods," Booz-Allen Applied Research, Inc., July, 1973.

b. "Recommended Methods of Reduction, Neutralization, Recovery or Disposal of Hazardous Waste," TRW Systems Group, August, 1973.

c. "Public Attitudes Towards Hazardous Waste Disposal Facilities," Human Resources Research Organization, June, 1973.

d. "Alternative to the Management of Hazardous Wastes at National Disposal Sites," Arthur D. Little, Inc., May, 1973.

e. "Program for the Management of Hazardous Wastes," Battelle Pacific Northwest Laboratories, July, 1973.

The last of these included estimates of the hazardous residual generation potential for each of the major industrial sectors of the SIC.

As a follow-up to this original survey, OSWMP sponsored a series of "Industrial Assessments" designed to yield additional detail on the industries judged to have the highest waste generation potential. Using mail surveys, telephone contacts, and plant visits of a statistical sampling of industries, waste generation projections were prepared for 1977 and 1983. In accomplishing this, a great deal of information was developed on waste generation factors, responsible unit operations, treatment and disposal practices, disposal costs, etc. This supplemental information has proven to be more valuable than the original waste projections. A list of the SIC codes covered by the assessment documents is provided in Table 1.

Limitations of this data base include: incomplete coverage of the manufacturing division, inconsistent definition of hazardous waste, inconsistent reporting of waste quantities, sparse information on waste composition, etc. Nevertheless, the industrial assessments have served as the basis for prediction of our national hazardous waste generation rate (ref. Booz-Allen & Hamilton, Inc. et al., 1980), and to support the analysis of our national hazardous waste disposal capacity.

STATE WASTE GENERATION DATA

The only substantial alternative to the assessment document series appears to be information generated by individual states. During the period of 1972 to 1974, several states conducted studies of their hazardous waste management problems. Notable among these are studies published by California, Idaho, Washington, and Oregon. Borrowing procedures and formats from these, EPA developed a State Implementation Guide (Porter, 1975) for conducting state surveys, and strongly recommended that each state develop a "local" information base. The response was not unanimous, but many states did compile generation and/or disposal information.

Unfortunately, these documents are more difficult to acquire. They are rarely included by computerized abstract services, and few can be purchased through the National Technical Information Service (NTIS). They must be gleaned from state sources using a judicious combination of patience and persistence.

In a study of "Municipal Disposal Methods For Industrially-Generated Hazardous Waste" (VanNoordwyk et al., 1979) thirty-one states were able to provide hazardous waste information. Unfortunately, the authors warned that: "These data were . . . inconsistent, both within individual reports and between reports from different states, extremely sketchy and incomplete, and reported in a non-uniform fashion." In part, this conclusion was due to the author's desire for information on specific disposal practices (as opposed to generation and composition), but in general, it is correct. Fortunately, the availability and quality of state generation information is improving rapidly.

In a study conducted two years later (Jennings, 1982), eight additional states were able to supply information, twenty-four states had made substantial updates of their previous surveys, and six were able to supply data directly from "manifest" tracking systems. Jennings provides citations to over 60 state hazardous waste generation documents. The location and type of these have been summarized in Figure 1. Based on these, the author was able to estimate the quantity, composition, spacial distribution, and industrial origin of our national hazardous waste generation profile. Current

TABLE 1. SIC Codes Covered by Industrial Assessment Documents.

SIC Code	Title	NTIS No.*	Date
22	Textile Mill Products	PB-258 953	1976
281	Industrial Inorganic Chemicals	PB-244 832	1975
282	Plastic Materials	PB-282 071	1978
283	Drugs	PB-258 800	1976
285	Paints, Varnishes, Lacquers	PB-251 669	1975
286,2879,2892	Industrial Organic Chemicals	PB-251 307	1975
2911	Petroleum Refining	PB-259 097	1976
2992	Lubricating Oils and Grease	PB-272 267	1977
30	Rubber and Plastic Products	PB-282 072	1978
3111	Leather Tanning and Finishing	PB-261 018	1976
331,332,339	Blast Furnaces, Steel Works	PB-276 171	1977
333,334	Primary/Secondary Smelting	PB-276 170	1977
3471	Electroplating, Plating	PB-264 349	1976
355,357	Special Industrial Machinery	PB-265 981	1977
367	Electrical Components	PB-265 532	1977
3691,3692	Storage/Primary Batteries	PB-241 204	1975

*National Technical Information Service.

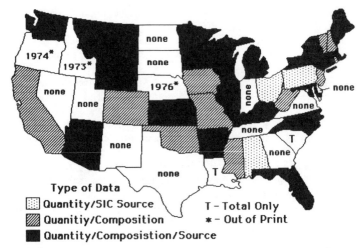

FIGURE 1. Availability of state hazardous waste generation information (source: Jennings, 1982).

(post-RCRA) state summaries should provide data of sufficient quality to support detailed state or local waste management planning.

National Hazardous Waste Generation

Commonly quoted estimates of our national hazardous waste generation rate range from 10 million (wet) metric tons to something over 350 million tons per year. Predictions near the top of this range generally include wastes that are regulated by several states, but not classified as hazardous under RCRA. These include huge volume wastes such as flyash, crop residues, animal manures, mining overburdens, and household wastes. The most reliable estimates of the generation of wastes currently classified as hazardous under RCRA (c. 1985) place the number in the range of 45 to 60 million metric tons per year.

GEOGRAPHICAL DISTRIBUTION OF WASTE GENERATION

In 1980, Booz-Allen and Hamilton, Inc. published estimates of the U.S. hazardous waste generation rate broken down by EPA region. This study was based on the distribution of industrial activity, and the "industrial assessment"

TABLE 2. Estimated Hazardous Waste Generation by EPA Region.

EPA Region	States	Estimate Based on Industrial Activity*	Estimate Based on State Data**
I	CT, MA, ME, NH, RI, VT	1,104,000	655,000
II	NY, NJ	3,113,000	1,728,000
III	MD, PA, WV, (DE, VA)†	4,354,000	8,114,000
IV	AL, FL, KY, MS, NC, SC, (GA, TN)	10,353,000	9,526,000
V	IL, MI, MN, OH, WI, (IN)	6,428,000	9,176,000
VI	AR, LA, OK, (NM, TX)	10,536,000	9,332,000
VII	IA, KS, MO, (NE)	1,201,000	1,340,000
VIII	CO, MT, WY, (ND, SD, UT)	318,000	1,577,000
IX	AZ, CA, (HI, NV)	2,838,000	1,551,000
X	ID, OR, WA, (AK)	955,000	1,112,000
Totals		41,235,000	44,111,000

*Booz-Allen & Hamilton, Inc. et al., (1980).
**Jennings, (1983).
†( )—states for which generation estimates were extrapolated.

FIGURE 2. Top twenty states in estimated hazardous waste generation (source: Epstein et al., 1982).

document series using updated waste generation factors. Results of this have been summarized in Table 2. Jennings (1983) published a similar study based on information reported by representative states. These estimates have also been included in Table 2. Although substantial regional differences can be seen, the overall agreement is quite good.

Several researchers have also published estimates of the ranked order of states. There is nearly universal agreement on the top ten generators (see Figure 2). It is believed that the top ten states produce in excess of 60 percent of the hazardous waste. The top 20 states (also see Figure 2) account for over 75 percent of the wastes produced.

INDUSTRIAL SOURCES OF HAZARDOUS WASTE

A great deal has also been published on the industrial sources of hazardous wastes. It is commonly believed that the majority of these are produced by Division D (manufac-

turing) industries of the Standard Industrial Classification code (SIC codes 2000−3999). These have been organized into the twenty major classifications listed in Table 3. This table also presents estimates of the magnitude of generation by each two-digit code group.

It is universally held that the chemical industry (SIC 28) produces the largest fraction of hazardous waste. Early estimates placed this as high as 70 to 80 percent of the total. More recent data have reduced this estimate to about 50 percent. This appears to be due to the underestimation of the generation of other industries (an artifact of the incomplete industrial assessment series) rather than an overestimation of SIC 28 generation.

It is understandable that the chemical industry generates a large fraction of the total. It is far less clear why the iron and steel industry (SIC 33, 34) should rival this production (ref. Table 3). The wastes of SIC 33 and 34 are relatively crude, unsophisticated products (bulk acids and bases, metallic

TABLE 3. Hazardous Waste Production of Division D Industries (Source: Jennings, 1982).

SIC Code	Title (abb.)	Rank	% of Total	SIC Code	Title (abb.)	Rank	% of Total
20	Food	8	3	30	Rubber	>10	1
21	Tobacco	>10	<1	31	Leather	>10	<1
22	Textiles	>10	<1	32	Stone	9	2
23	Apparel	>10	<1	33	Primary Metals	2	22
24	Lumber	>10	1	34	Fabricated Metals	3	6
25	Furniture	>10	<1	35	Machinery	4	5
26	Paper	5	4	36	Electrical	10	1
27	Printing	>10	<1	37	Transportation	6	3
28	Chemicals	1	48	38	Instruments	>10	<1
29	Petroleum	7	3	39	Misc. Manufacturing	>10	<1

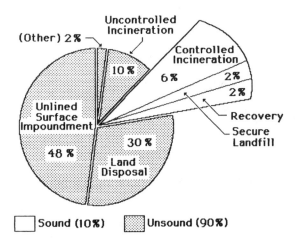

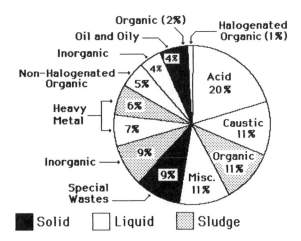

FIGURE 3. Distribution of historical waste disposal practices (source: EPA, 1980).

FIGURE 4. Approximate composition of industrial hazardous waste (source: Jennings, 1982).

sludges, etc.). These indicate an aging industry that has not yet implemented modern environmentally-sensitive technologies.

Historically, as much as 95% of this industrial waste was disposed of on-site at unregulated facilities. The vast majority went into unlined surface impoundments or landfills. The EPA has estimated that only 10% of the wastes generated prior to 1980 were disposed of by practices that would be considered sound by today's standards. The distribution of historical waste disposal practices is illustrated in Figure 3.

HAZARDOUS WASTE COMPOSITION

The concept of hazardous waste evolved to regulate dangerous industrial materials being disposed of by way of tra-

TABLE 4. Estimated National Hazardous Waste Composition Profile (Source: Jennings, 1982).

Class	% of Total	Composition of Class	% of Class
Solids	15	Organic	12
		Inorganic	28
		Special Solids	60
Liquids	60	Halogenated Organics	1
		Non-Halogenated Organics	9
		Acids	35
		Caustic Solutions	19
		Heavy Metal Solutions	12
		Oils and Oily Wastes	8
		Misc. Liquids	16
Sludges	25	Organic	42
		Inorganic	33
		Heavy Metal	25

ditional solid waste disposal routes. The concept does not apply to industrial wastewaters. Clearly these may be "hazardous" in the sense that they are dangerous to human health or the environment. However, they are regulated elsewhere and are not included as RCRA hazardous wastes. Nevertheless, recent data have indicated that as much as 60 percent of all RCRA hazardous wastes are liquids. True solids represent the smallest fraction (15%) with the remainder (25%) composed of semi-solids and sludges.

Using a statistical sample of actual waste disposal activities reported by 30 states (accounting for 13.9 million tons of waste), Jennings (1982) produced the waste composition profile presented in Table 4. Although the composition categories are relatively crude, this was found to be the most sophisticated classification scheme that could be evaluated with the available data. It has also been suggested that the categories of this scheme are sufficiently detailed to identify the basic disposal alternatives for each class of waste.

An overall analysis of these data is presented in Figure 4. Note that the wates of highest concern (i.e., halogenated organics) actually make up a very small fraction of the total. In fact, the largest category (acids) appears to be one that is relatively easy to treat and/or dispose of without long-term risk to the environment. It should also be clear from Figure 4 that no single facility could be expected to function well in universal service. Wise management requires a mixture of treatment and disposal facilities, each accepting the waste fraction for which it is ideally suited.

HAZARDOUS WASTE LEGISLATION

In the hazardous waste area, the flagship piece of legislation is The Resource Conservation and Recovery act of 1976

(P.L. 94-580), commonly referred to as RCRA. This superseded The Solid Waste Disposal Act of 1965 (P.L. 89-272), and amended The Resource Recovery Acts of 1970 and 1973 (P.L. 91-512 and P.L. 93-14). The fundamental structure of hazardous waste management as it is evolving today is dictated by RCRA and its attendant regulations. These have accumulated into an extremely large and complex set of documents. Volumes have been written simply to act as guides to the legislation or to answer specific questions that have commonly arisen about regulations. Anyone intending to work in this field will have to invest a substantial amount of time to become familiar with RCRA, and to stay up to date on its evolving regulations.

The Comprehensive Environmental Response, Compensation, and Liability Act of 1980 (CERCLA, P.L. 96-510) is another piece of legislation of particular interest to civil engineers. This act, often referred to simply as "Superfund," provides the authority and financial arrangements under which the federal government may act to clean up chemical spills or existing hazardous waste disposal sites. CERCLA is also responsible for introducing the concept of risk assessment (albeit in a modest version designed to prioritize "Superfund" sites) into hazardous waste management. Failure to directly address the definition, assessment, analysis, and management of hazardous waste risks has been one of the major criticisms of RCRA. Engineers should look for the necessity to address issues of risk as a growing part of modern hazardous waste management.

The most important features of RCRA and CERCLA will be discussed in the following sections. Obviously there are many other pieces of federal legislation that bear directly or indirectly on the hazardous waste problem. These include:

a. The Clean Air Act of 1970 (CAA, P.L. 89-272)
b. The Occupational Safety and Health Act (OSHA, P.L. 91-596)
c. The Federal Water Pollution Control Act of 1972 (FWPCA, P.L. 92-500)
d. The Marine Protection, Research, and Sanctuaries Act of 1972 (MPRSA, P.L. 92-532)
e. The Safe Drinking Water Act of 1974 (SDWA, P.L. 93-523)
f. The Federal Insecticide, Fungicide, and Rodenticide Act of 1975 (FIFRA, P.L. 94-140)
g. The Toxic Substances Control Act of 1976 (TSCA, P.L. 94-469)
h. Surface Mining Control and Reclamation Act of 1977 (SMCRA, P.L. 95-87)
i. The Clean Air Act of 1977 (CAA, P.L. 95-95)
j. The Clean Water Act of 1977 (CWA, P.L. 95-217)
k. The Hazardous Materials Transportation Act (HMTA, P.L. 95-403)
l. The Solid Wate Disposal Act Amendments of 1980 (P.L. 96-482)

The Resource Conservation and Recovery Act

The Resource Conservation and Recovery Act (P.L. 94-580) became law on October 21, 1976. This was intended to close the door on the uncontrolled land disposal of solid and hazardous wastes. This door had been left ajar by The Clean Air Act of 1970 (P.L. 89-272) and The Water Pollution Control Act Amendments of 1972 (P.L. 92-500). These were relatively effective in controlling pollutants being disposed of via air and surface water sinks. Unfortunately, the result was that many pollutants were simply diverted (in raw form, or as underflows from air or water treatment facilities) to land disposal alternatives. As this occurred, it slowly became obvious that we were creating another environmental problem of national magnitude. Uncontrolled chemical disposal sites were posing immediate, acute hazards to the local human and natural environments. It was also recognized that the potential for groundwater contamination represented a very serious, long term health and environmental risk. Therefore, RCRA was enacted with the intention of establishing sound engineering practices in the disposal of all solid wastes, and complete "cradle-to-grave" regulation of hazardous waste.

In the following section the major hazardous waste provisions of RCRA will be outlined with emphasis placed on its hazardous waste provisions. The reader is referred to the full text of the bill for additional detail. The reader is also cautioned that this is a complex piece of legislation that has been followed by the promulgation of numerous detailed regulations. Applications of RCRA to specific hazardous waste situations may well require the assistance of legal counsel in addition to specialized engineering expertise.

SUBTITLE A—GENERAL PROVISIONS (SECTION 1001–1008)
Subtitle A provides the general provisions of the law including a statement of the general objectives of the legislation and operative definitions of terms.

> The objectives of this Act are to promote the protection of health and the environment and to conserve valuable materials and energy resources by . . . (4) regulating the treatment, storage, transportation, and disposal of hazardous wastes which have adverse effects on health and the environment. . . .
>
> The term "hazardous waste" means a solid waste, or combination of solid wastes, which because of its quantity, concentration, or physical, chemical, or infectious characteristics may—(A) cause, or significantly contribute to an increase in mortality or an increase in serious irreversible, or incapacitating reversible, illness: or (B) pose a substantial present or potential hazard to human health or the environment when improperly treated, stored, transported, or disposed of, or otherwise managed.
>
> The term "hazardous waste management" means the systematic control of the collection, source separation, storage, transportation, processing, treatment, recovery, and disposal of hazardous wastes.

SUBTITLE B—OFFICE OF SOLID WASTE: AUTHORITIES OF THE ADMINISTRATOR (SECTION 2001–2006)

Subtitle B established the Office of Solid Waste within the Environmental Protection Agency to be headed by a Deputy Assistant Administrator. RCRA requires that EPA play the major federal role in promulgation and administering RCRA regulations.

SUBTITLE C—HAZARDOUS WASTE MANAGEMENT (SECTION 3001–3011)

Subtitle C contains RCRA's major hazardous waste regulatory provisions.

a. Section 3001. Identification and listing of hazardous waste.
b. Section 3002. Standards applicable to generators of hazardous waste.
c. Section 3003. Standards applicable to transporters of hazardous waste.
d. Section 3004. Standards applicable to owners and operators of hazardous waste treatment, storage, and disposal facilities.
e. Section 3005. Permits for treatment, storage, or disposal.
f. Section 3006. Authorized state hazardous waste programs.
g. Section 3007. Inspections.
h. Section 3008. Federal enforcement.
i. Section 3009. Retention of state authority.
j. Section 3010. Effective date.
k. Section 3011. Authorization and assistance to states.

Clearly the identification of the specific wastes to be covered, and the standards to be applied to management of these wastes embodies the heart of our hazardous waste control program. The critical specifics cannot be found in RCRA. Each of sections 3001 to 3005 begins with a statement similar to "No later than eighteen months after the date of enactment of this Act, the Administrator shall, after notice and opportunity for public hearing, and after consultation with appropriate Federal and State Agencies, develop and promulgate criteria for. . . ." It was recognized that this would be an immensely complex undertaking for the Administrator of EPA. It is unlikely that anyone really believed that this would be accomplished in merely eighteen months. In any case, it was not. The EPA struggled for several years to promulgate the mandated regulations. These will be addressed in the section on RCRA Regulation.

In general, section 3001 required EPA to develop a workable criteria for determining which wastes should be considered to be hazardous, and to develop a list of specific current wastes that should be regulated. Sections 3002–3005 established the manifest system designed to track individual waste lots from their point of origin to their ultimate disposal sink through a series of facilities specifically permitted to handle these wastes.

Section 3006 provided authorization for states to accept responsibility for the administration and enforcement of RCRA.

Section 3007 gave EPA and state officials reasonably strong rights of access to hazardous waste sites for the purposes of inspecting and/or copying records, or to inspect and/or sample wastes. This section also provided that this information (with some exceptions) must be made available to the public.

Section 3008 provided for civil and criminal penalties to aid in regulation enforcement. Violators were made liable for civil penalties of up to $25,000 for each day of continued noncompliance. In addition, anyone who knowingly transports waste to an unpermitted disposal facility, disposes of a hazardous waste without being permitted to do so, or makes any false statement or representation in the required documentation is, upon conviction, subject to a fine of up to $25,000 for each day of violation and/or imprisonment of up to one year. These penalties double after a person's first conviction. This authorization of federal criminal penalties was a major step forward for hazardous waste disposal enforcement. Prior to this, states had been forced to act against irresponsible waste disposers using nuisance charges such as "public endangerment."

SUBTITLE D—STATE OR REGIONAL SOLID WASTE PLANS (SECTION 4001–4009)

Subtitle D provides for the development of general state plans for the disposal of solid (nonhazardous) waste which are environmentally sound and which maximize the utilization of valuable resources. This subtitle also authorized EPA to promulgate criteria for determining which (existing) land disposal facilities may be classified as sanitary landfills and which must be classified as "open dumps" (to be upgraded or closed).

SUBTITLE E—DUTIES OF THE SECRETARY OF COMMERCE IN RESOURCE AND RECOVERY (SECTION 5001–5004)

Subtitle E requires the Secretary of Commerce to promote greater commercialization of proven resource recovery technologies by providing accurate specifications for recovered materials, stimulation of markets for recovered materials, and providing a forum for the exchange of technical and economic data on resource recovery.

SUBTITLE F—FEDERAL RESPONSIBILITIES (SECTION 6001–6004)

Subtitle F required all federal agencies to comply with the provision of RCRA unless specifically exempted by the President of the United States.

SUBTITLE G—MISCELLANEOUS PROVISIONS (SECTION 7001–7009)

Subtitle G provided protection for employees who might otherwise have been dismissed because of efforts to implement RCRA provisions at their place of employment.

Subtitle G also provided for "Citizen Suits." Under RCRA, any person may commence a civil action on his or her own behalf against any person (including the United States or governmental agency) alleged to be in violation of any permit, standard, regulation, condition, requirement, or order effective under RCRA. In addition, action may be taken against the Administrator of EPA for alleged failure to perform any nondiscretionary act or duty mandated under RCRA. Failure to meet the regulation promulgation deadlines mandated under Subtitle C gave rise to several "Citizen Suits."

SUBTITLE H—RESEARCH, DEVELOPMENT, DEMONSTRATION, AND INFORMATION (SECTION 8001–8007)

Subtitle H was designed to encourage research in a number of specific topic areas related to solid waste disposal, resource conservation and recovery, and hazardous waste management.

RCRA Regulations

The following section offers a brief summary of RCRA Subtitle C regulations. These are offered here for the purpose of defining the basic intent, and mechanisms of the "cradle-to-grave" regulatory effort. The reader is warned that a much more thorough analysis must be made of the current regulations before hazardous waste management decisions are made. This may be accomplished by obtaining the current annual edition of the *Code of Federal Regulations,* and by subscribing to the *Federal Register.*

IDENTIFICATION AND LISTING OF HAZARDOUS WASTE (40 CFR PART 261)

As required by RCRA Subtitle C, Sec. 3001, EPA published hazardous waste identification criteria and a listing of hazardous wastes in the *Federal Register* of May 19, 1980 (*Fed Reg., 45* (98), 33084–33133, 1980).

Subpart C, Section 261.20 provides that any waste is hazardous if it exhibits any of the characteristics of hazardous waste. These are defined as:

a. Characteristic of Ignitability (Section 261.21)—liquids with a flashpoint below 60 C., solids capable of vigorous and persistent ignition at standard temperatures, and ignitable compressed gas.
b. Characteristic of Corrosivity (Section 261.22)—aqueous solution with a pH below 2.0 or above 12.5, or a liquid capable of corroding steel at a rate of greater than 6.35 mm/yr at 55 C.
c. Characteristic of reactivity (Section 261.23)—any material that is normally unstable and readily undergoes violent change without detonating, reacts violently with water, forms explosive mixtures with water, generates toxic gases, vapors, or fumes when mixed with water, is a cyanide or sulfide bearing waste that can generate toxic

fumes, or is capable of detonation or explosive reaction when subjected to a strong initiating source or is heated.
d. Characteristic of EP Toxicity (Section 261.24)—based on an Extraction Procedure (EP) leaching test to identify wastes likely to leach hazardous concentrations of constituents into groundwater. The initial list of constituents for the EP toxicity were as follows:

Arsenic	5.0 mg/l	Barium	100.0 mg/l
Cadmium	1.0 mg/l	Chromium	5.0 mg/l
Lead	5.0 mg/l	Mercury	0.2 mg/l
Selenium	1.0 mg/l	Silver	5.0 mg/l
Endrin	0.02 mg/l	Lindane	0.4 mg/l
Methoxychlor	10.0 mg/l	Toxaphene	0.5 mg/l
2,4-D	10.0 mg/l	2,4,5-TP	1.0 mg/l

It is interesting to note that neither quantity nor degree of hazard are used in the determination of a hazardous waste. The EPA's justification for omitting a hazard ranking/rating was the lack of an existing, nonsubjective system that could be consistently administered. However, regulations acknowledge that some hazardous wastes are "acutely" hazardous and establish different exclusion limits for these wastes.

Exclusion limits set the amount of waste that may be generated before RCRA regulations must be applied. The original exclusion limit was set at 1000 kilograms per month with the intention of eventually reducing this to 100 kg/mo. It was believed that the 1000 kg/mo limit would force the regulation of in excess of 99% of the total waste being generated annually. Small generators are still required to dispose of their wastes at approved hazardous waste disposal sites. However, they are exempted from the manifest waste tracking system, and the annual reporting requirements of major generators.

Part 261-Subpart D of the regulations also provides a listing of specific wastes that must be treated as hazardous (i.e., "listed" wastes). Wastes are "listed" either because they exhibit one of the characteristics identified in Subpart C (Ignitability, Corrosivity, Reactivity, EP toxicity), or:

a. have been found to be fatal to humans in low doses
b. have an oral LD 50 (rat) toxicity of less than 50 mg/kg
c. have an inhalation LC 50 (rat) toxicity of less than 2 mg/kg
d. have a dermal LD 50 (rabbit) toxicity of less than 200 mg/kg
e. are capable of causing or contributing to an increase in serious irreversible, or incapacitating reversible, illness
f. contain any of a rather substantial list of hazardous constituents

The original hazardous waste "list" cited 16 wastes from nonspecific (generic) sources, 69 process wastes from sources such as the manufacture of inorganic pigments, organic chemicals, pesticides, explosives, petroleum refining, leather tanning, wood preservation, iron and steel, and copper, lead, and zinc refining. In addition 361 commercial products were listed which, when discarded, are defined to be hazardous wastes. On July 16, 1980, EPA added another 18 process wastes to the list. EPA's intention is to inventory and characterize all of the solid wastes produced in significant quantities by American industry and "list" hazardous wastes as they are identified. EPA anticipates that the list of "listed" wastes will increase several fold.

STANDARDS APPLICABLE TO GENERATORS OF HAZARDOUS WASTE (40 CFR PART 262)

As required by RCRA Subtitle C, Section 3002, EPA established standards for generators of hazardous waste in 1980 (*Fed. Reg.,* 45 (98), 33140 – 33149). In general, all generators are required to determine if their wastes are hazardous, and to ensure that their hazardous wastes are disposed of at permitted facilities.

Generators are responsible for the preparation of a manifest to accompany each lot of hazardous waste to be transported to an off-site treatment, storage, or disposal site. This must state the source and destination of the waste as well as its quantity and composition. Copies of the manifest must be signed at each stage of the disposal train. A copy of the manifest must also be returned to the generator from the designated disposal facility within 35 days of the date the waste was accepted by the initial transporter. Failing this, the generator must (within 10 days) submit an Exception Report to the EPA Regional Administrator explaining what efforts have been made to locate the waste shipment, and the results of those efforts.

Generators are also responsible for the proper packaging, labeling, and placecarding of hazardous wastes offered for disposal. Wastes may be accumulated on-site for up to 90 days (in proper containers). Accumulation times in excess of 90 days require the site to be permitted as a storage facility. All generators of hazardous waste are also required to submit annual reports summarizing their hazardous waste activities to the EPA.

STANDARDS APPLICABLE TO TRANSPORTERS OF HAZARDOUS WASTE (40 CFR PART 263)

As required by RCRA Subtitle C, Section 3003, EPA established standards for transporters of hazardous waste in 1980 (*Fed. Reg.,* 45 (98), 33150 – 33152). These standards focus on proper record keeping to ensure that the movement of hazardous waste shipments is properly documented.

In general, a transporter may not accept a hazardous waste unless it is accompanied by a proper manifest. All transporters accepting a waste shipment must return a signed copy of the manifest to the generator before leaving the site. The entire quantity of waste must then be delivered to the designated facility listed on the manifest. The transporter must obtain a signature from the operator of this facility, and retain one copy of the signed manifest for his permanent records. The remaining copies of the manifest must accompany the waste shipment.

A transporter is responsible for the cleanup of any hazardous waste discharged during transportation. Transporters are cautioned that they are also subject to the regulations of the Department of Transportation (DOT) governing the transportation of hazardous materials. These regulations concern labeling, marking, placecarding, using proper containers, and reporting discharges, and are codified in Title 49, *Code of Federal Regulations*, Subchapter C.

STANDARDS APPLICABLE TO OWNERS AND OPERATORS OF HAZARDOUS WASTE TREATMENT, STORAGE, AND DISPOSAL FACILITIES (40 CFR 264, 265)

As required by RCRA Subtitle C, Section 3004 and 3005, EPA established standards for the operation of hazardous waste management facilities in 1980 (*Fed. Reg.,* 45 (98), 33154 – 33258). In general, these provide that permits will not be issued for facilities (other than on an interim basis) that are not located, designed, constructed, and operated in accordance with EPA regulations. Provisions generally applicable to all facilities are discussed below. For specific waste processing technologies the reader is directed to the detailed interim or final regulations.

Prior to accepting wastes from a generator, the owner or operator of a hazardous waste facility must notify the generator (in writing) that he has the appropriate permits for the waste being generated. Upon receiving a waste, the operator must sign and date the manifest, immediately give a copy to the transporter, retain a copy for his permanent records, and within 30 days, send a copy of the manifest to the generator. Significant manifest discrepancies (in quantity or composition) must be reconciled with the generator and/or transported or reported to the Regional Administrator within 15 days.

All hazardous waste management facilities must be designed, constructed, maintained, and operated to minimize the possibility of fire, explosion, or other unplanned event that might allow the uncontrolled release of hazardous waste. Facilities must also be equipped with alarm systems designed to detect unintended releases, and a resonable supply of equipment for abating releases should they occur. A contingency plan must be prepared detailing actions to be taken by facility personnel in response to fires, explosions, or any unplanned (sudden or otherwise) release of hazardous materials. At all times there must be at least one qualified employee on site (or on call) with the responsibility of coordinating emergency response measures.

Hazardous waste facilities must be inspected for malfunctions and deteriorations which could lead to the release of wastes. Inspections must be often enough to identify problems in time to correct them before they harm human health or the environment.

Substantial record keeping efforts detailing the history of the site and its operation are required. These include a Facility Annual Report that must be submitted to the EPA.

EPA CONSOLIDATED PERMIT REGULATIONS (40 CFR PARTS 122, 123, 124 AND 125)

In an effort to consolidate EPA's environmental permit programs, consolidated regulations were announced in 1980 (*Fed. Reg.*, *45* (98), 33290–33588). These included permit regulations for the following permit programs.

a. The Hazardous Waste Management Program (HWM) under subtitle C of the Solid Waste Disposal Act as amended by the Resource Conservation and Recovery Act of 1976 (P.L. 94-580, as amended by P.L. 95-609)
b. The Underground Injection Control Program (UIC) under part C of the Safe Drinking Water Act (P.L. 93-523, as amended by P.L. 95-190)
c. The National Pollution Discharge Elimination System Program (NPDES) under sections 318, 402, and 405 of the Clean Water Act (P.L. 92-500 as amended by P.L. 95-217 and P.L. 95-576)
d. The Dredge or Fill Program (404) of the Clean Water Act
e. The Prevention of Significant Deterioration Program (PSD) under section 165 of the Clean Air Act (P.L. 88-206 as amended)

Under this consolidated program, anyone required to have a permit must complete, sign, and submit the proper forms to the Director (regional administrator of the appropriate EPA office or state director of approved state program). Specific procedures for application, review, issuance, duration, administration, transfer, modification, and termination of permits may be found in the *Federal Register*.

The Comprehensive Environmental Response, Compensation, and Liability Act of 1980 (CERCLA, P.L. 96–510)

The Comprehensive Environmental Response, Compensation, and Liability Act of 1980 (CERCLA or "Superfund") allows the federal government to respond directly (with remedial actions) whenever there is a release or threat of release of hazardous substances presenting an imminent and substantial danger to public health or the environment. CERCLA also allows the government to recover the costs of emergency responses or remedial actions on a strict liability basis.

Section 102 or CERCLA requires that the release of reportable quantities of hazardous wastes (1 lb. unless otherwise regulated) must be reported to the National Response Center (NRC). Failure to report such a release may result in a fine of $10,000 and/or imprisonment for up to one year. Section 103 required that the location of any unpermitted facility used for storing, treating, or disposing of hazardous wastes be reported to EPA. As the result of this reporting requirement, EPA is now aware of over 16,000 such sites.

THE NATIONAL CONTINGENCY PLAN

Section 105 of CERCLA required the EPA to revise and republish the National Contingency Plan (NCP) originally developed under The Clean Water Act (P.L. 95-217). The revised NCP was to include a new plan for responding to hazardous substance releases. This plan defines the extent to which remedial actions will be taken, the procedures to be used for guaranteeing that response actions are cost-effective, and the criteria to be used in establishing the National Priorities List (NPL). Short-term emergency actions costing up to 1 million dollars are authorized. Strict liability is imposed in the recovery of these costs. Long-term, "remedial" actions, however, are limited to sites on the National Priorities List.

THE NATIONAL PRIORITIES LIST

The National Priorities List was assembled by EPA from the total list of known or suspected hazardous waste disposal sites based on their relative degree of risk to public health or the environment. To accomplish this the EPA adopted the MITRE Hazard Ranking system (HRS).

$$\text{Score} = S_M + S_{FE} + S_{DC}$$

where

S_M = composite score for the potential to do human or environmental damage by migration via air, surface water, or groundwater
S_{FE} = score for the potential explosion/fire hazard
S_{DC} = score for the direct contact hazard

Each of these hazard evaluations is divided into a series of specific considerations, and a detailed scoring system is used to compute a cumulative factor score.

CERCLA required that the NPL contain at least 400 of the highest priority facilities with the top 100 of these ranked in order of priority. EPA published a list of the 115 highest priority sites in October 1981. Of these, 24 were rated as being potentially worse than Love Canal. The NPL of 418 sites was published in December of 1982. As of October 1984, 786 sites have been included on (or proposed for) the NPL. A map of these is presented as Figure 5. By law, at least one site was supposed to be listed for each site. However, even considering this, Figure 5 illustrates that the problem of uncontrolled hazardous waste sites is truly of national scope.

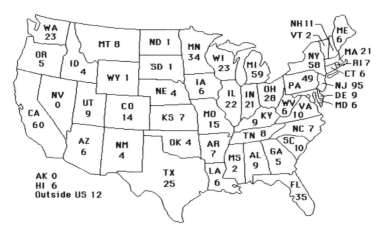

FIGURE 5. Summary of the 786 hazardous waste sites proposed for the National Priorities List (NPL) as of 1984 (source: Boraiko, 1985).

THE HAZARDOUS SUBSTANCES RESPONSE TRUST FUND

Section 221 of CERCLA established the Hazardous Substance Response Trust Fund. This trust (also referred to as "Superfund") is intended to pay for emergency response actions taken by the federal government. Superfund was financed by an initial appropriation of $220 million plus $1.38 billion to be recovered from oil and chemical taxes. In addition, Superfund is to receive any funds developed from cost-recovery reimbursements, penalties, or punitive damages authorized under CERCLA.

THE POST CLOSURE LIABILITY TRUST FUND

Section 107 of CERCLA established the Post Closure Liability Trust Fund. This fund is financed by a tax assessed on each ton of hazardous waste received by a permitted RCRA facility (if the waste will remain at the facility after it is closed). Therefore, the tax predominantly affects storage type facilities such as most land disposal technologies. The initial tax rate was established as $2.13 per (dry weight) ton.

Five years after closure of a RCRA facility, liability may be transferred from the owner to the fund if the facility complied with RCRA regulations specifying post-closure performance. After this transfer, the fund assumes all of the liability of the owner including the costs of monitoring, care, and maintenance.

The Toxic Substances Control Act of 1976 (TSCA, P.L. 94-469)

The Toxic Substances Control Act was enacted by Congress in 1976 in response to the growing fear that the manufacture, distribution, and disposal of dangerous indus-

trial chemicals was seriously under-regulated. Consider that tens of millions of pounds of PCBs were manufactured and widely distributed before it was realized how toxic and persistent they would be when released to the environment. It has been estimated that over 3 million chemical compounds are currently recognized (many of which are produced by American industry) and that approximately 3000 products are added each year. The intent of TSCA was to improve our ability to anticipate which of these may lead to unacceptable human and environmental risk, and to act (to control their manufacture, distribution, and disposal) before it is too late to undo the damage they cause.

Under TSCA, the Environmental Protection Agency was given the responsibility of identifying all chemicals that pose unacceptable risks to human health and the environment and to initiate actions to reduce all unacceptable risks. The EPA was also given the responsibility of reviewing new chemical products prior to their manufacture. This was designed to prevent the introduction of those that represent unacceptable risks.

Section 8(b) of TSCA requires that EPA identify, compile, keep current, and publish a list of chemical substances which are manufactured, imported, or processed for commercial purposes in the United States. EPA must be notified at least 90 days before the manufacture of new chemicals (i.e., substances that do not yet appear on the Master Inventory File) and prohibit, limit, or otherwise control their manufacture if they are found to pose an unacceptable risk.

Often, the standards for the treatment and disposal of toxic substances under TSCA are more stringent than those promulgated under RCRA. As an example, under TSCA, PCB incinerators must attain 99.9999% destruction efficiency (the "six nines" rule) while under RCRA only 99.99% reduction (i.e., "four nines") is generally required.

Hazardous Materials Warning Labels

DOMESTIC LABELING

Cargo Aircraft only   Magnetized Material   Package Orientation Markings

Handling Labels

General Guidelines on Use of Labels

- Labels illustrated above are normally for *domestic shipments*. However, some air carriers *may* require the use of International Civil Aviation Organization (ICAO) labels.

- Domestic Warning Labels *may* display UN Class Number, Division Number (and Compatibility Group for Explosives only.) Sec. 172.407(g).

- Any person who offers a hazardous material for transportation MUST label the package, if required. [Sec. 172.400(a)].

- Label(s), when required, must be printed on or affixed to the surface of the package near the proper shipping name. [Sec. 172.406(a)].

- When two or more different labels are required, display them next to each other. [Sec. 172.406(c)].

- Labels may be affixed to packages (even when not required by regulations) provided each label represents a hazard of the material in the package. [Sec. 172-401].

- The Hazardous Materials Tables, Sec. 172.101 and 172.102, identify the proper label(s) for the hazardous materials listed.

UN Class Numbers

Hazardous materials class numbers associated with the hazard classes.

Class 1—Explosives
Class 2—Gases (Compressed, Liquefied or dissolved under pressure)
Class 3—Flammable liquids
Class 4—Flammable solids or Substances
Class 5—Oxidizing Substances
Class 6—Poisonous and infectious Substances
Class 7—Radioactive Substances
Class 8—Corrosives
Class 9—Miscellaneous dangerous Substances

INTERNATIONAL LABELING

Substance liable to   Poisonous Substance   Poisonous Substance   Infectious Substance
Spontaneous Combustion

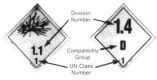

EXAMPLES OF INTERNATIONAL LABELS

- These are examples of International Labels not presently used for domestic shipments.

- Most of the domestic labels (illustrated above) *may* be used Internationally.

- Text, when used Internationally *may* be in the language of the country of origin.

- Text is *mandatory* on Radioactive Material, St. Andrews Cross and Infectious Substance labels.

EXAMPLES OF EXPLOSIVE LABELS

- The NUMERICAL DESIGNATION represents the CLASS or DIVISION.

- ALPHABETICAL DESIGNATION represents the COMPATIBILITY GROUP (for Explosives Only)

- DIVISION NUMBERS and COMPATIBILITY GROUP combinations can result in over 30 different "Explosives" labels (see IMDG Code/ICAO).

For complete details, refer to one or more of the following:

- Code of Federal Regulations, Title 49, Transportation. Parts 100-199. [All Modes]

- International Civil Aviation Organization (ICAO) Technical Instructions for the Safe Transport of Dangerous Goods by air. [Air]

- International Maritime Organization (IMO) Dangerous Goods Code. [Water]

- Canadian Transport Commission (CTC) Regulations. [Rail]

U.S. Department of Transportation

Research and Special Programs Administration

Materials Transportation Bureau
Washington, D.C. 20590

CHART 8
JANUARY 1985

FIGURE 6. DOT hazardous materials warning labels.

Hazardous Materials Warning Placards

DOMESTIC PLACARDING

Illustration numbers in each square ([1] through [18]) refer to TABLES 1 and 2 below.

1 EXPLOSIVES A — 1	2 EXPLOSIVES B — 1	3 BLASTING AGENTS — 1	4 POISON GAS — 2	5 FLAMMABLE GAS — 2	6 NON-FLAMMABLE GAS — 2	7 CHLORINE — 2
8 OXYGEN — 2	9 FLAMMABLE — 3	10 COMBUSTIBLE — 3	11 FLAMMABLE SOLID	12 FLAMMABLE SOLID W	13 OXIDIZER — 5	14 ORGANIC PEROXIDE — 5
15 POISON — 6	16 RADIOACTIVE — 7	17 CORROSIVE — 8	18 DANGEROUS			

HIGHWAY
- For "HIGHWAY ROUTE CONTROLLED QUANTITY" OF RADIOACTIVE MATERIALS [Sec. 172.507].
- For use of the words "GASOLINE" and "FUEL OIL" on placards [Sec. 172.542(c) or 172.544(c)].

RAIL
- For use of EXPLOSIVE A. POISON GAS AND POISON GAS-EMPTY placards. [Sec. 172.510(a)].

TABLE 1

HAZARD CLASSES	*NO.
Class A explosives	1
Class B explosives	2
Poison A	4
Flammable solid (DANGEROUS WHEN WET label only)	12
Radioactive material (YELLOW III label)	16
Radioactive material:	
Uranium hexafluoride fissile (containing more than 1.0% U^{235})	16 & 17
Uranium hexafluoride, low-specific activity (containing 1.0% or less U^{235})	16 & 17

NOTE: For details on the use of Tables 1 and 2, see Sec. 172.504 (See footnotes at bottom of tables.)

Guidelines

- Placard motor vehicles, freight containers, and rail cars containing any quantity of hazardous materials listed in TABLE 1.
- Placard motor vehicles and freight containers containing 1,000 pounds or more gross weight of hazardous materials classes listed in TABLE 2.
- Placard freight containers 640 cubic feet or more containing any quantity of hazardous material classes listed in TABLES 1 and/or 2 when offered for transportation by air or water. Under 640 cubic feet, see Sec. 172.512(b).
- Placard rail cars containing any quantity of hazardous materials classes listed in TABLE 2 except when less than 1,000 pounds gross weight of hazardous materials are transported in Trailers or Containers on Flat Car Service.

TABLE 2

HAZARD CLASSES	*NO.
Class C explosives	18
Blasting agent	3
Nonflammable gas	6
Nonflammable gas (Chlorine)	7
Nonflammable gas (Fluorine)	15
Nonflammable gas (Oxygen, cryogenic liquid)	8
Flammable gas	5
Combustible liquid	10
Flammable liquid	9
Flammable solid	11
Oxidizer	13
Organic peroxide	14
Poison B	15
Corrosive material	17
Irritating material	18

INTERNATIONAL PLACARDING

- Most international placards are identical (color and pictorial symbols) to the Domestic placards illustrated above.
- International placards are enlarged ICAO or IMO labels (See International Labeling—Otherside).
- Placard MUST correspond to hazard class of material.

- Placard ANY QUANTITY of hazardous materials when loaded in FREIGHT CONTAINERS, PORTABLE TANKS, RAIL CARS and HIGHWAY VEHICLES.
- International placards may be used in addition to DOT placards for international shipments.

When required, Subsidiary Risk placards must be displayed in the same manner as Primary Risk placards. Class numbers are not shown on Subsidiary Risk placards.

- COMPATIBILITY GROUP DESIGNATORS must be displayed on EXPLOSIVES PLACARDS.
- UN CLASS NUMBERS and DIVISION NUMBERS MUST be displayed on hazard class placards when required.

UN and NA Identification Numbers

- The four digit UN or NA numbers must be displayed on all hazardous materials packages.
- UN (United Nations) or NA (North American) numbers are found in the Hazardous Materials Tables, Sec. 172.101 and the Optional Hazardous Materials Tables, Sec. 172.102 (CFR, Title 49, Parts 100-199).
- UN numbers are displayed in the same manner for both Domestic and International shipments.
- NA numbers are used only in the USA and Canada.

When hazardous materials are transported in Tank Cars, Cargo Tanks and Portable Tanks, UN or NA numbers must be displayed on:

PLACARDS OR ORANGE PANELS

1090 and FLAMMABLE 3

Appropriate Placard must be used.

EUROPEAN NUMBERING SYSTEM—

Top Number—Hazard Index (Identification of Danger, 2 or 3 figures) Example: 33 = highly inflammable liquid.

Bottom Number—UN Number of substance Example: 1088 ACETAL

For more complete details on identification Numbers see Sec. 172.300 through 172.338.

FIGURE 7. DOT hazardous materials warning placards.

The Hazardous Materials Transportation Act (HMTA, P.L. 95-403)

The Hazardous Materials Transportation Act authorized the Department of Transportation (DOT) to establish regulations for the classification of hazardous materials, shipping container requirements, labeling and placard standards, handling procedure for various transportation vehicles, and requirements for the reporting of accidents for all interstate shipments of hazardous waste. Current DOT warning labels and placards for hazardous materials are illustrated in Figures 6 and 7.

Both the DOT and the American Society for Testing and Materials (ASTM STP 825) have published guides to emergency response actions to be taken for transportation accidents involving placarded materials. However, the reader is cautioned that placard classifications are insufficiently detailed to allow for the determination of proper emergency actions. All placarded containers must be treated with the utmost caution until their specific contents have been determined and proper emergency response actions have been implemented.

HAZARDOUS WASTE MANAGEMENT TECHNOLOGIES

There are two basic strategies for the disposal of hazardous waste. One may attempt to detoxify the material and release only non-hazardous residuals to the environment. This may be accomplished by treatment or thermal destruction. Alternatively, waste may be isolated from the environment by some form of permanent secure storage. Most land disposal technologies are designed to accomplish this. Dilution into the environment is nearly always unacceptable because many hazardous wastes can be dangerous at very low ambient concentrations.

Many of the technologies in these classes are well established procedures that have been upgraded to handle hazardous wastes. New technologies are also rapidly emerging that are designed to handle specific hazardous compounds. In general, there is no one technology that functions well for all wastes. Sound hazardous waste management requires a mixture of facility types, each accepting only those wastes for which it is ideally suited.

In the following sections, the basic disposal strategies will be outlined, and individual processes will be briefly discussed. The intent of this is to present the breadth of technologies available. The reader is also reminded that volume reduction or waste exchange programs can be powerful alternatives to disposal. There will also be briefly discussed.

ᐤ Land Disposal

Land disposal technologies are designed to attain permanent secure storage of hazardous wastes. The basic concept is that if wastes can be completely isolated from the environment, they can do no damage. Isolation is normally accomplished with a combination of natural barriers (low permeability soil, impermeable rock) and man-made materials (synthetic liners, steel containers, concrete, etc.). Distance may also be used as a barrier. Wastes are often disposed of hundreds of meters below ground or at remote surface locations. Since the wastes are not detoxified, they remain eternally dangerous. There always exists the danger that they will escape because of a facility failure or natural disaster and exert their environmental impact.

Groundwater contamination is normally the failure mode of highest concern. It is always possible that wastes may slowly escape into the subsurface and contaminate economic aquifers. Great care must be taken to isolate wastes from the hydrologic cycle and to detect and correct leaks. Nevertheless, it is unwise to use these facilities for disposal of any wastes that would be mobile in groundwater systems.

Historically, land disposal has dominated industrial waste disposal practices. This remains true even though the legislative intent of RCRA was to encourage resource recovery and innovative treatment. However, the role of land disposal is evolving. CERCLA has provided new economic incentives to detoxify rather than store wastes. In the future, engineers will have to be much more cautious about the kinds of hazardous wastes they inter at land disposal facilities.

SECURE LANDFILL

Secure hazardous waste landfills have evolved from the sanitary landfill and pre-RCRA chemical waste landfill. Essentially, they represent a greater intensity of engineering intended to guarantee the containment of interred wastes. It has been estimated that there are 300 secure landfills in use today, disposing of as much as ten million tons of hazardous waste per year. Many of these are commercially available.

Landfills have been used extensively for the disposal of nearly all types of chemical waste. Unfortunately, many early chemical landfills are now superfund sites because of their inability to permanently contain liquid wastes. In reaction to this, RCRA regulations first proposed would have prohibited the landfill disposal of all bulk liquids, semi-solids, and sludges. Ignitable or reactive wastes would also have been prohibited. The final regulations softened this stance, but bulk liquids or wastes containing free liquids may not be placed in a landfill unless the installation has a liner that is resistant to the liquid and a functioning leachate collection system, or the waste is stabilized (e.g., with an absorbent solid) so that free liquids are no longer present.

The components of a hazardous waste landfill design are as follows.

Site Selection

The importance (and difficulty) of site selection cannot be over-emphasized. The facility must be capable of preventing water run-on, infiltration, and groundwater contamination, and promoting surface run-off. An ideal site would be in an isolated location free of surface water bodies and underlain by a deep impermeable soil formation (i.e., clay) sitting atop unfractured bedrock in an area where annual evaporation exceeds annual precipitation. One should anticipate considerable public opposition to establishing a secure landfill at even an ideal site. Opposition will increase greatly as less ideal locations are proposed.

Leachate Containment

All hazardous waste landfills require some form of leachate containment system. This may be the underlying impermeable formation, but the integrity of natural formations is very difficult to guarantee. Most often an artificial barrier system (compacted clay layer, synthetic liner, etc.) must be installed on a carefully prepared subgrade. Liner materials must be tested for compatibility with the wastes to be disposed of. Unanticipated chemical reactions can greatly increase liner permeabilities. Great care must also be taken to guarantee their integrity during installation and as wastes are placed. Early liner experiences have confirmed that they can and do fail. Many designs now employ the double liner concept (see Figure 8).

Leachate Collection and Treatment

An underdrain system should be provided to collect any leachates that form in the landfill (see Figure 8). Even if all liquids are prohibited, water can enter the facility as wastes are placed, or because of cover failures. Leachate collection is essential to prevent large hydraulic gradients from building up within the landfill. This reduces the probability that the liner system will be breached and reduces the volume of any leakage that occurs. Absence of this system also makes corrective action much more difficult and expensive.

Landfill Cover

The landfill cover (temporary and permanent) is designed to minimize infiltration of surface water into the fill (see Figure 8). Covers are normally constructed of layers of synthetic membranes, impermeable soil, and topsoil. Vegetation must also be established to stabilize the final grade and prevent erosion. Great care must be taken to maintain the integrity of covers. They are installed on a far less stable foundation (landfilled chemical wastes) than bottom liners. They can fail by cracking or collapse due to subsidence of the filled materials, by surface erosion, excessive desiccation, animal borrowing, or root penetration. Without maintenance they are sure to leak.

Groundwater Monitoring

It is essential (and normally required) that a groundwater monitoring system be designed to provide early detection of underground leaks. This generally involves wells both up-gradient and "downstream" in the direction of the regional groundwater flow. Operation requires sampling and analysis

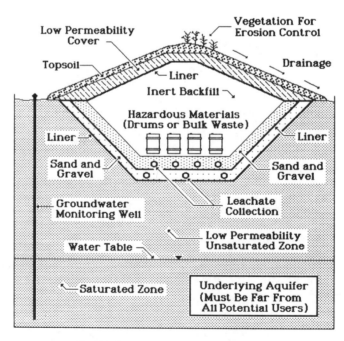

FIGURE 8. *Secure hazardous waste landfill.*

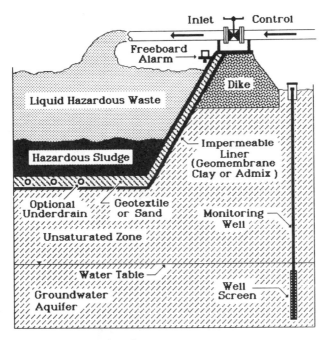

FIGURE 9. Surface impoundment.

at regular intervals to confirm proper containment or detect escaping chemicals.

Proper Operation

It is also essential that the facility be operated successfully. Incompatible wastes must be identified and isolated from one another in distinct disposal cells. The location, dimension, and contents of each cell must be recorded.

SURFACE IMPOUNDMENT

Surface impoundments are storage volumes, such as ponds, pits, basins, or lagoons, used to contain liquid or semi-solid hazardous wastes. They may be shallow excavations or above ground volumes contained by dikes and levees. A typical installation is illustrated in Figure 9. Normally impoundments remain open to the atmosphere until permanent closure. They are used for holding, compositing, and equalizing hazardous wastes before treatment, for dewatering wastes, and as a permanent disposal facility. In arid climates they can be effective in dewatering large volumes of sludge at minimal cost.

Surface impoundments have been used extensively to dispose of nearly any type of liquid waste, and have become notorious for their environmental impacts. Unlined ponds act as infiltration galleries that allow or actually force wastes to escape into the subsurface. Impoundments can also overflow due to uncontrolled discharge or extreme hydrologic conditions. Levees may fail releasing the entire contents of the impoundment. Finally, surface impoundments

are a constant source of odors due to uncontrolled atmospheric discharges. Most of these problems can be overcome with sound engineering, but the type of waste, the location, and operation mode must be considered very carefully to guarantee success.

Hazardous waste impoundments must be designed to provide a secure barrier against infiltration. Liners must be tested for compatibility. Great care must be taken to ensure proper liner installation. In addition, the facility must be designed and operated to prevent overtopping due to overfilling, wind and wave action, rainfall, equipment malfunctions, or human error. A groundwater monitoring system must also be provided to detect leaks.

Only non-volatile and non-malodorous liquids and semi-solids may be disposed of in surface impoundments. Great care must be exercised to ensure that wastes added to the impoundment are compatible with its existing contents. Failure to accomplish this can easily lead to the production of toxic gases posing an immediate threat to the local population. Dewatered solids must be periodically removed or interred in place when the impoundment is full.

DEEP WELL INJECTION

As its name implies, deep well injection is a procedure by which hazardous wastes are disposed of by injecting them into deep geological formations. This is illustrated in Figure 10. The key to a successful operation is the intregity of the geological formation and security of the injection well system. Both must be capable to complete waste contain-

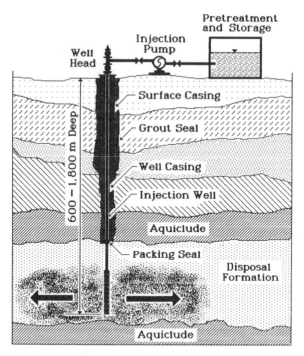

FIGURE 10. Deep well injection.

ment. Under appropriate conditions nearly any formation may be suitable, but sedimentary rock formations (sandstone, limestone, dolomite) are most attractive. Acceptable formations are generally found in the central United States.

Deep well injection has been used extensively in Texas, Louisiana, Illinois, Michigan, and Oklahoma. There are approximately 40,000 brine injection wells in operation and 500 industrial waste disposal facilities. Wells are generally 2,000 to 6,000 ft. (610 – 1,800 m) deep, but depths in excess of 10,000 ft (3,000 m) have been used. Injection rates vary from 75 to 1,000 (5 – 760 l/sec.) gallons per minute. Although initial costs can be high, operation costs are generally very low compared to other disposal technologies.

To be acceptable, a geological formation must be sufficiently deep, have sufficient permeability to accept wastes, adequate capacity, and be isolated by thick impermeable barriers (aquicludes). The reservoir material and packing seals must also be compatible with injected wastes. Unanticipated reactions such as the formation of metal hydroxides, carbonates, or sulfates can plug the domain and force premature facility closure.

A wide range of liquid wastes may be disposed of by deep well injection. This has been used extensively since the 1930s by the pretroleum industry to dispose of oil field brines. Acid and alkali solutions (both dilute and concentrated), heavy metal solutions, high BOD organic solutions, and waste solvents have also been disposed of. Generally,

the only process outputs (that must be routed to some other form of disposal) result from injection solution pretreatments to remove solids.

LAND TREATMENT

Land treatment (also referred to as landfarming or soil incorporation) is a relatively controversial treatment/disposal process involving the land application of hazardous materials. Wastes are incorporated into the first foot of soil by spreading them on or injecting them immediately below the surface. This is often followed by tilling to maximize soil contact and aeration. Time is then provided for soil microorganisms to decompose the biodegradable fraction of the waste. Essential nutrients may also be added to accelerate biological activity. Nondegradable wastes are (hopefully) incorporated into the soil matrix and permanently immobilized.

Land treatment has been used extensively for non-hazardous wastes such as animal manure, crop residues, municipal wastewaters and sludges. The method has also been employed extensively by the petroleum industry for the disposal of oily refinery wastes. In a 1981 study of the 197 hazardous waste land treatment systems that could be identified, over half were associated with the petroleum industry (SIC 29). The next largest fractions were treatment wastes from the chemical industry (SIC 28 – 15%), electrical, gas, and sanitary services (SIC 49 – 7%) and the steel fabrication industry (SIC 34 − 7%). Facilities were as large as 670 acres with a median of about 15 acres (Brown et al., 1983). It has been estimated that as much as half of all oily petroleum wastes are disposed of by landfarming at a cost of $15 − 20 per ton.

RCRA requires that no waste be disposed of by this procedure unless it will be rendered non-hazardous (or much less hazardous) by biological degradation or chemical reactions in the soil. Volatile wastes must be avoided because of their potential for air emissions. Ignitable and reactive wastes are also generally prohibited. In addition wastes with high inorganic content, particularly heavy metals, must be avoided.

The essential elements of a successful land treatment design are as follows.

Site Selection and Preparation

Land disposal requires a thick topsoil layer, preferably of poor agricultural value in a semi-arid climate (pH > 6.5, slope < 5%) well above the groundwater table. Careful run-on and run-off control is required.

Waste Analysis and Application

Strict front-end quality control is required to ensure that inappropriate wastes are not applied to the site. Application rates must be adjusted to maintain aerobic soil conditions at all times. Rates and schedules must also be carefully adjusted to suit weather conditions. Excessive application rates can permanently decrease soil permeability and treatment capacity.

Monitoring

In addition to operational monitoring of soil conditions, both air and groundwater monitoring must be provided. The groundwater monitoring plan must include sampling of the soils and pore water in the unsaturated zone below the application site in addition to conventional saturated zone monitoring.

Land disposal sites may discharge hazardous wastes to the environment as air emissions, surface water runoff, or via groundwater. The possibility of the release of wind-blown particulates and of major groundwater discharges are of particular concern.

WASTE PILES

RCRA provides for a class of land disposal practices defined as hazardous waste piles. These may be either disposal piles or treatment/storage piles. A disposal pile must be managed as a secure landfill (refer to the first section).

Hazardous waste storage piles are considered to be relatively small (less than 3 m high) temporary facilities. They are frequently used to accumulate wastes before shipment to a treatment or disposal facility and generally are composed of a single dry material. Under current RCRA regulations, wind dispersal, leachate generation, and run-off must be controlled. Generally this requires that the pile be placed on an impermeable base and be protected from rain and wind by a rigid enclosure or flexible cover. Waste analysis is also required to guarantee compatibility of pile components.

DEEP SECURE STORAGE

Hazardous wastes may be isolated from the environment by storing them in secure underground vaults mined from deep, impermeable geologic formations. These may be abandoned mines or may be carved from virgin formations specifically to confine hazardous waste (see Figure 11).

In Europe, deep secure storage has been used successfully for the last 15 years. The Herfa-Neurode facility in the Federal Republic of Germany, established in an abandoned portion of an 80 year old, 700 m deep salt mine, has been used to dispose of 270,000 tons of hazardous waste. The facility is currently handling approximately 40,000 tons of waste per year.

This form of disposal has several distinct advantages over most land technologies. Normally, wastes are sealed into steel drums by the generator and then carefully placed in disposal galleries where they may be periodically inspected. Compared to conventional landfilling, this provides a great improvement in container integrity and life. Also, because wastes remain accessible, they may be recovered for reuse or detoxification as new technologies become available. Herfa-Neurode has actually returned approximately 1,000 tons of waste to its producer for further use.

Because wastes must be handled safely in underground "warehouse" conditions, materials that could evolve explosive, toxic, or otherwise hazardous gases must be excluded. In addition, all liquid wastes must be solidified or confined by secure encapsulation. Reactive wastes must be carefully separated and must not be reactive with the native vault material. Although no wastes should escape this form of confinement, it is possible that volatile components could be released by the system's ventilation system.

DEEP SALT BED DISPOSAL

It has been proposed that deep salt bed formation could be used very successfully for the disposal of liquid and semi-solid hazardous wastes (Funderburk, 1984). Caverns would be solution mined into deep, confined salt formations at depths below those used for commercial salt mining. These would be filled with hazardous waste, solidified in place, and resealed. This process is illustrated in Figure 12.

Salt formations have been used for many years to store crude oil and natural gas. These formations have a very low permeability, are normally isolated from economic groundwaters, have compressible strength comparable to concrete, and are capable of withstanding high geologic and man-induced pressures. The Tatus dome in Mississippi proved capable of containing the explosion of a 5 kiloton nuclear device. The Department of Energy is developing designs for high-level radioactive waste repositories in salt formations.

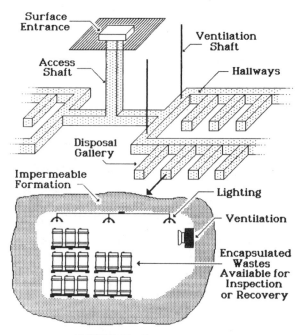

FIGURE 11. Deep secure storage.

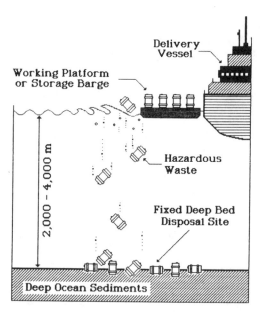

FIGURE 13. Ocean floor disposal (state-of-the-art).

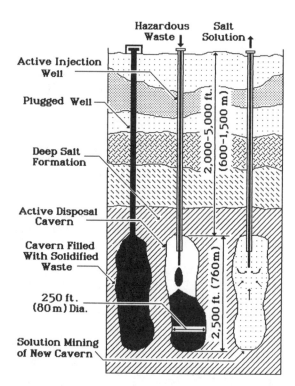

FIGURE 12. Deep salt bed disposal (source: Funderburk, 1984).

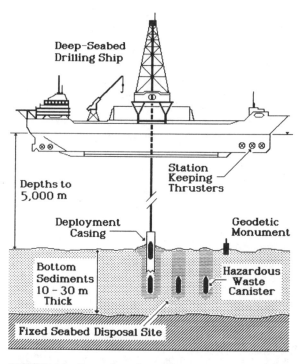

FIGURE 14. Ocean floor disposal (proposed seabed implantation).

Candidate salt formations for hazardous waste disposal are located along the Texas–Louisiana Gulf Coast, and in the Salina Basin underlying portions of Michigan, Ohio, Pennsylvania, and New York (areas with high hazardous waste generation rates). In addition to user proximity, it has been suggested that salt bed disposal would be inexpensive, less consumptive of surface area, and more palatable to the public than surface or near-surface alternatives.

OCEAN FLOOR DISPOSAL

Ocean floor disposal involves land disposal designed to use the ocean floor as the ultimate disposal sink. The intent is to make use of the impermeable properties of ocean floor sediments to help confine wastes and to add the nearly infinite dilution potential of the ocean as an additional barrier between man and his hazardous wastes.

In the past, bulk and drummed wastes were simply dumped into the ocean over predetermined sites. This is illustrated in Figure 13. Huge quantities of municipal and industrial sludges have been disposed of by this method. In addition, extremely dangerous materials such as nerve gas, explosives, and radioactive wastes have been disposed of by several nations (including the United States). Because of the fear of irreversible environmental damage, several international agreements have been signed on ocean floor disposal. Unfortunately, the process continues to be used by several European countries, Japan, and the United States.

In the Marine Protection, Research, and Sanctuaries Act of 1972 (P.L. 92-532), Congress declared its intention to prevent ocean disposal of any material that could adversely affect human health or the marine environment. However, proposals for ocean disposal sites persist. Figure 14 illustrates a method that has been suggested for the seabed implantation of non-destructible, highly hazardous wastes. The United States is also considering plans to resume the ocean disposal of low-level radioactive wastes. This is considered to be the prime alternative to mined repositories.

Thermal Destruction

Thermal destruction technologies attempt to destroy hazardous wastes by converting (oxidizing) them into non-hazardous products such as stack gases and inert ash. They may be designed for nearly any organic compound and are particularly well-suited for wastes that are classified as hazardous due to flammability. Often these systems operate at a net energy gain that helps to off-set their rather substantial capital and operation costs. With organics of lower thermal value or high water content, it becomes more difficult and expensive to accomplish complete destruction.

Occassionally non-combustible hazardous wastes are also "incinerated." Although this may seem illogical, it can be a very effective means of volume reduction (i.e., the removal of water and all volatile compounds) and often yields an inert or less reactive ash that may be more conveniently dis-

posed of. Care must be taken to ensure that the addition of non-combustible materials does not inhibit the incineration of hazardous organics.

All thermal destruction processes discharge "underflows" to some ultimate disposal sink. Most often these are non-hazardous stack gases discharged to the atmosphere and inert bottom and fly ash routed to land disposal. It is possible, however, to form dangerous stack gases such as free chlorine or hydrogen chloride. Quality control systems must be carefully designed to prevent escape of air pollutants. In addition, ashes and stack gas residues may still be classified as hazardous wastes and must be disposed of at another RCRA facility.

When combustion efficiency and stack gas control are guaranteed, thermal destruction is a very attractive (but expensive) option. Its application is increasing because of the ability to destroy waste. Industry may shed its liability for hazardous waste once it has been successfully incinerated. Given that "superfund" site clean-up operations often cost in excess of $10,000,000, this is a very powerful argument for thermal destruction.

LIQUID INJECTION INCINCERATION

Liquid injection incinerators are the most common hazardous waste combustion units. There are estimated to be 220 units in service in the United States. In general they may be used for any combustible liquid with a viscosity sufficient to allow atomization. Figure 15 presents a schematic flow diagram of a typical installation.

Prior to combustion, liquid wastes are converted to an aerosol to aid in the speed and efficiency of combustion. Atomization is normally accomplished by high pressure nozzle injection into a primary ignition chamber. Viscous liquids must be preheated to attain a viscosity sufficient to allow atomization (<750 SSU). Liquid injection incinera-

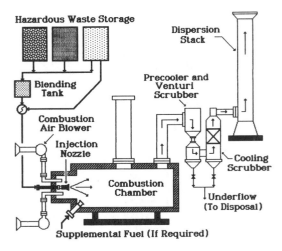

FIGURE 15. *Liquid injection incineration.*

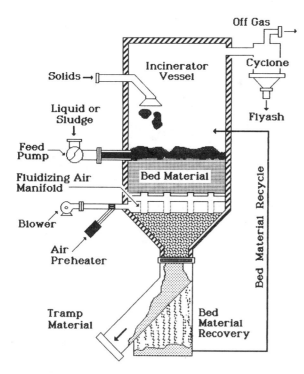

FIGURE 16. Fluidized bed incineration.

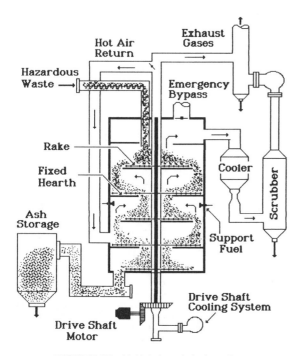

FIGURE 17. Multiple hearth incineration.

tors are generally designed to operate between 1,500 and 3,000 F. with residence times of 0.5 to 2 sec.

Incinerator efficiency is quite sensitive to waste composition and injection rate. Installations are normally designed with storage and blending tanks to help accomplish uniformity of the feed stream. In general, liquid injection incinerators have low maintenance requirements (few moving parts) and are reasonably cost effective.

ROTARY KILN INCINERATION

There are approximately 40 hazardous waste rotary kiln incinerators in the United States. Rotary kilns may be used for nearly any form of combustible hazardous material (solid, liquid, or gas) including intact drums of waste. Often this is accomplished in conjunction with the incineration of non-hazardous wastes such as municipal refuse or wastewater sludges. Special applications have included the incineration of explosives and chemical warfare agents.

The combustion chamber of a rotary kiln incinerator is a long, inclined cylindrical shell lined with firebrick or other refractory material. The kiln rotates around its central axis at 1 to 5 rpm. Rotation provides mixing of uncombusted materials, particularly when solids or sludges are being incinerated. Wastes are introduced through a charging chute and carried to the combustion chamber by traveling grates. These also help to heat and dewater wastes before they reach the combustion chamber. Liquid and gaseous wastes may be injected directly into the kiln. Combustion temperatures normally range from 1,500 to 3,000 F. Often combustion must be sustained by the addition of supplemental fuel. Residence times may be from seconds to hours. Offgases often pass into a secondary, fixed combustion chamber to ensure complete oxidation before discharge.

This is a very versatile (but expensive) form of incineration since it can accept nearly any form of waste. Rotary kilns are particularly attractive when size or volume precludes other incinerator types. Because inputs may be so diverse, it is difficult to guarantee high incineration efficiency for any one component. The process is generally unacceptable for highly toxic wastes such as halogenated hydrocarbons.

FLUIDIZED BED INCINERATION

Fluidized bed incinerators are versatile devices capable of incinerating solids, liquids, or gases. The unique feature of this technology is the presence of an inert granular medium (such as sand) which serves as a heat source/sink of tremendous magnitude within the combustion chamber (see Figure 16). This medium is fluidized by air introduced into the bottom of the incinerator through a plenum. Wastes are injected slightly above this air inlet. When hot, the medium rapidly transfers heat to the injected wastes. In turn, combustion of the waste reheats the medium. Turbulence within the reactor helps to promote efficient combustion, to break down solids, and to maintain a uniform bed temperature. The heat sink also helps to sustain combustion as waste composition varies. Ash is carried out of the reactor with the fluidizing air and must be separated for disposal.

Incineration temperatures are limited by the softening point of the granular medium. For sand this is approximately 2000 F. Normally, bed temperatures are maintained in the range of 1400 to 1600 F. The design and operation of the air delivery system is also critical to proper incineration. Air velocities in the neighborhood of 5 to 8 ft/sec (1.5−2.5 m/sec) must be maintained to keep the bed fluidized.

Fluidized bed incinerators have no moving parts in the high temperature zone. They do not, however, attain the temperatures required to incinerate all hazardous wastes. In addition, wastes containing relatively large inorganic fractions (e.g., sodium chloride) can lead to operating problems such as fusing of bed materials.

MULTIPLE HEARTH INCINERATION

The multiple hearth incinerator is a vertical, cylindrical chamber enclosing a series of fixed hearths (see Figure 17). They range in size from 6 to 25 feet (2−8 m) in diameter and from 12 to 75 feet (4−23 m) high. Wastes enter the incinerator at the top and are mixed by plow arms attached to a rotating central shaft. These arms gradually rake the waste to openings where it may drop to the next hearth. In the upper hearths, wastes are dried and mixed. Combustion occurs at intermediate levels. Often supplemental fuel is injected to help sustain combustion. The bottom hearths help to cool the ash and heat the incoming combustion air.

Multiple hearth incineration has been used extensively for biological wastewater treatment sludges. The process is ideal for sludge because of the dewatering that occurs in the top hearths. It may also be used successfully for solid wastes. Liquid or gaseous wastes are easily accommodated by injecting them into combustion chamber below the drying hearths.

Care must be taken to avoid solids or sludges that may yield fusible ashes. In addition, "cold" spots often form within the furnace allowing for incomplete combustion. Gases may also short circuit and escape without sufficient residence time. Multiple hearth incinerators are also relatively expensive and have high maintenance requirements.

MULTIPLE CHAMBER INCINERATION

Multiple chamber incinerators are normally designed to accomplish two-stage destruction of solid wastes. The first chamber accomplishes drying, ignition, and combustion of solids. The second chamber, normally provided with a secondary air supply, is designed for complete combustion of off-gases. Chambers may be in-line, or in the retort ("u" shape) configuration. Retort designs offer structural economy, but are limited in capacity (50−750 lb/hr.). The in-line design is better suited to high-capacity operation.

Multiple chamber incinerators are most appropriate for the incineration of batch-fed solids. Applications have included the incineration of plastics, PVC waste, acrylic and phenolic resins, and rubber wastes. With care they may also be used for the co-incineration of flammable liquids. They are normally inappropriate for the incineration of semi-solids or sludges.

OCEAN INCINERATION

Ocean incineration refers to the practice of incinerating hazardous wastes aboard ships (at designated locations on the high seas) or on oil-drilling platforms. Presently, there are several incinerator ships commercially available to incinerate liquid hydrocarbons (including halogenated hydrocarbons). Generally, these employ liquid injection incineration, but nearly any thermal destruction technology could be mounted aboard ship. It has been estimated that the U.S. hazardous waste generation rate could support as many as 50 offshore incinerator ships. Proposed designs include multiple incineration units (liquid injection, rotary kiln, etc.) to provide a broader range of service.

Incineration at sea is more economical because of the less stringent emission controls. As an example, ship incinerators are not required to scrub hydrogen chloride from the stack gases. This allows operators to maximize combustion at higher application rates. It was also believed that there would be less public resistance to ocean incineration than to land-based units. This has not proven to be the case.

Initial test burns aboard Vulcanus I demonstrated that it is very difficult to accomplish stack monitoring at sea (a task that is difficult enough at a shore-based unit). Destruction efficiencies in excess of 99.99% may be accomplished, but are not always maintained. Genuine concerns have been expressed about EPA's ability to regulate incineration on the high seas. There is also fear of large chemical spills due to navigation accidents or storms. In addition, this technology encourages the transportation of very dangerous wastes over long distances. As an example, dioxin contaminated organics were hauled from Niagra Falls, New York to Mobile, Alabama to conduct a "test-burn" of Vulcanus I in 1981.

EMERGING THERMAL DESTRUCTION TECHNOLOGIES

Recent incineration research has focused on improving thermal destruction technology. In particular, efforts have concentrated on more secure and efficient methods of destroying highly hazardous wastes such as halogenated hydrocarbons. Generally these are designed to operate at very high efficiency (99.9999 + %) on small amounts of well characterized materials. They are more likely to be employed as private, dedicated facilities than as commercial operations. Emerging technologies include the following.

Molten Salt Incineration

This incinerator is designed to hold a body of molten salt at high temperature (800−1000 F.). Waste and combustion air are fed into the molten bed. The heat of combustion sus-

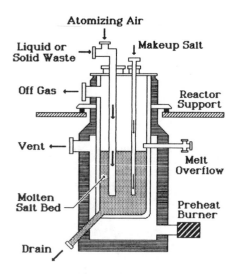

FIGURE 18. Molten salt incineration.

Additional processes such as catalytic incineration, pyrolysis, calcination, infrared furnace oxidation, ultraviolet (UV) light irradiation, etc., have also been proposed and applied to specific hazardous wastes.

Treatment

Treatment may involve any physical process or chemical (including biochemical) reaction used to accomplish some "treatment" objective. These may either be used to convert hazardous waste to a more convenient form (e.g., dewatering to accomplish volume reduction) before the waste is sent to ultimate disposal, may selectively attack hazardous components to render the waste non-hazardous (e.g., heavy metal extraction), or may purify the waste for reuse. Nearly any unit operation of chemical engineering can be adapted for use in hazardous waste treatment.

Obviously, this is a very broad class of technologies. Anything but the most superficial of treatments is beyond the scope of this work. The reader is also cautioned that these technologies nearly always rely on some unique physical or chemical property of the target hazardous waste. These are also relatively high technology processes that require a great deal of expert supervision. Few serve well in general service on arbitrary waste mixtures. Once waste streams have been composited, sophisticated treatment techniques must usually be eliminated because of their propensity to be fouled by unacceptable inputs.

Treatment processes are most effective on-site at strategic locations within the industrial process train. However, they nearly always yield a hazardous underflow that must be routed elsewhere for ultimate disposal. Civil engineers are most often involved with the off-site management of this underflow.

tains the molten salt body, and acidic by-products are automatically stripped from the off-gases by salt reactions. The salt body also acts as a heat source/sink helping to ignite wastes and sustain highly efficient combustion. This technology is illustrated in Figure 18. Salt must be wasted from the reactor as the inorganic ash content builds. This is easily solidified for land disposal. The technology has been successfully tested at pilot scale and is commercially available. Efficiencies in excess of 99.99998% have been achieved for the incineration of highly toxic materials such as chemical warfare agents.

Wet Air Oxidation

Wet air oxidation operates on the principle that combustible substances may be oxidized in the presence of a pressurized liquid (e.g., water) at temperatures in the range of 250 to 700 F. This offers the possibility of treating wastes with low thermal values without degradation of the combustion process. It also offers the possibility of eliminating atmospheric discharges since contaminants remain in aqueous solution.

Plasma Arc Reactor

A plasma arc reactor uses an electrical discharge to create an ionized gas plasma (argon, helium, hydrogen, oxygen, etc.) at ultrahigh temperature (50,000 to 100,000 F.). Organic compounds introduced into this plasma are instantly destroyed by high-energy free electrons that break molecular bonds. The procedure avoids the formation of potentially hazardous by-products. Initial tests demonstrated 99.9999999% (nine nines) destruction efficiency of pure transformer oil. Commercially available units are in development.

ORGANIC LIQUID TREATMENT

Organic liquid wastes may be treated by numerous physical and/or chemical processes. Often these are applied to concentrated aqueous solutions or non-aqueous mixtures for the purposes of material recovery. They may also be applied to dilute solutions to remove wastes that are difficult to treat biologically. Common processes include:

a. Adsorption
b. Air Stripping
c. Centrifugation
d. Dialysis
e. Electrophoresis
f. Evaporation
g. Freeze-Crystallization

h. Hydrolysis
i. Liquid/Liquid Extraction
j. Oxidation-Reduction
k. Ozonation
l. Reverse Osmosis
m. Ultrafiltration

Because the details of these processes are so diverse, no attempt will be made here to describe their individual operations. All are well-known industrial unit operations. Each can be used very effectively to treat a specific class of organic compounds. All are normally designed to handle a very specific and well characterized waste stream. Few serve well as general treatment technologies.

INORGANIC LIQUID TREATMENT

The treatment of inorganic solutions (such as metal solutions) may be accomplished by a number of conventional industrial operations. These are normally applied to accomplish volume reduction or material recovery. Common processes include:

a. Electrodialysis
b. Electrolytic Recovery
c. Evaporation
d. Ion-Exchange
e. Precipitation
f. Reverse Osmosis

Because the details of these are so diverse, no attempt will be made here to describe their individual operations. All are well known industrial processes and can be used on a wide variety of metal solutions. It is worth noting that, although material recovery is very desirable (and is encouraged under RCRA), it generally proves to be expensive. Only "valuable" or very dangerous metals (Ag, Au, Cr, Hg, Pb, Zn, etc.) are commonly recovered. The remainder are most often simply concentrated into an inorganic sludge and then routed to ultimate disposal.

❧ BIOLOGICAL TREATMENT

Biological treatment includes any process designed to capitalize upon the ability of organisms (generally microorganisms) to consume substrate. This material is either synthesized into new cell matter or used as an energy source for respiration. In either case, dissolved substrates are removed from solution and either destroyed or converted into colloidal material. Biological treatment is most often applied to organics in relatively dilute aqueous solutions, or to sludges of biological origin. Treatment may be accomplished in a wide variety of dispersed growth or fixed film reactor configurations. Since these are well known from their municipal and industrial wastewater applications, all process design details will be omitted here.

When biological processes are applied to hazardous wastes, great care must be taken to guarantee that wastes will not be toxic to the biological population. It has been found that nearly any organic compound may be degraded biologically if the organism population is sufficiently acclimated to the target substrate. Often this requires subjecting the population to carefully controlled pressures that select for naturally occurring organism mutations. Genetic engineering has also been proposed to manufacture "super bugs" designed to metabolize specific hazardous wastes. However, these very specialized populations can easily be destroyed by toxic wastes to which they have not been acclimated. In addition, it has often been discovered that even specialized populations will abandon their target substrate when provided with a more easily metabolized food source. Stringent front-end quality control is required for successful operation.

It should also be noted that "biological treatment" does not necessarily destroy hazardous waste. Wastes may reside (unaltered) within cells or be converted into hazardous metabolites. Aerobic processes may also strip volatile waste components into the atmosphere and release aerosols of hazardous mixed liquor. Large volumes of hazardous sludge may also be produced.

ACID–BASE NEUTRALIZATION

Bulk acids and base solutions are two of the most common and largest volume hazardous wastes produced. They may be disposed of by a number of processes, but it is also possible to eliminate them. Neutralization makes use of the fact that, when mixed, acids and bases can react to form a harmless salt and water. This process is illustrated in Figure 19.

Acid–base neutralization can be a very attractive technology, especially for an industry that produces both types of waste. This is commonly applied to sulfuric, hydrochloric, caustic soda, and ammonium hydroxide wastes. The necessity of purchasing the neutralizing agent makes this slightly less desirable. Acceptable wastes may be obtained from waste exchanges for this purpose. Neutralization is often accomplished by violent exothermic reactions. The composition of the neutralizing agent and the rate of reaction must be carefully controlled.

OIL/SOLVENT RECOVERY

Oils and industrial solvents may be recovered by redistillation. Materials commonly recovered include automotive oils, industrial lubrication oils, degreasing solvents, organic plating wastes, and paint sludges. It is more difficult to recover viscous wastes such as sludge or tars that require extensive pretreatment. There are currently about 150 commercial recovery facilities in operation in the United States and numerous private (on-site) facilities.

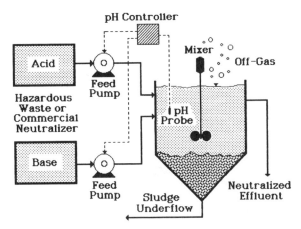

FIGURE 19. Acid-base neutralization.

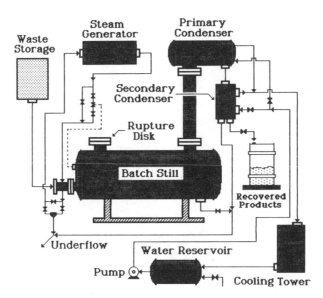

FIGURE 20. Organic recovery by redistillation.

Redistillation is used to segregate, separate, and purify liquid organic products. This requires heating the liquid (often under pressure) and then condensing the resulting vapors to recover its components. Figure 20 presents a schematic flow diagram of a typical process. Pretreatment is often required to screen out debris and free water. Centrifugation may also be used to remove entrained solids. Redistillation may be carried out in intermittent batches or as a continuous feed process. Following condensation, polishing is often required to produce commercial grade products.

Resource recovery is a very attractive option. Past federal regulations actually discouraged reuse by applying prohibitively stringent standards to recovered materials. Under RCRA this situation is changing. Many industries have now discovered that it is cost effective to recover concentrated organics, and that it can occasionally be done at a profit.

SOLIDIFICATION/FIXTURE

Solidification/Fixation processes are designed to convert hazardous wastes into as stable or insoluble a form as possible. This reduces the probability of the material escaping into the environment following ultimate disposal. Solidification involves the conversion of liquids, sludges, or solid debris into a solid, structurally sound material. Fixation generally involves the chemical binding of undesirable constituents making them less mobile in the environment. The basic techniques include the following.

Vitrification

Waste is incorporated into molten silica glass and then molded into blocks for ultimate disposal.

Thermoplastic Techniques

Diced, dried wastes are mixed with a molten organic material (paraffin wax, polyethylene, etc.) and placed in a container where the mixture may solidify.

Thermoset Techniques

A catalyst is dispersed into a mixture of monomer and waste sludge, causing it to harden. Wastes are trapped in the polymer crosslinking and immobilized. Thermoset materials include urea-formaldehyde and polyesters. Drying may be required prior to disposal.

Encapsulation

Waste is mixed with a binder, molded into solid blocks, and coated with an impermeable sheath. Polybutadiene is a common binder. Polyethylene is often used as the coating material. This process is used to stabilize very soluble inorganics.

Cementation

In this widely used process, Portland cement or lime-based mortars are employed to immobilize wastes. At the high pH of cement, many heavy metals react to form insoluble hydroxides or carbonates that are incorporated into the cement matrix. Other wastes are trapped in the matrix interstices and immobilized as the cement or mortar hardens. However, many organic wastes hinder setting and curing of the mixture.

VOLUME REDUCTION

Although not a classic treatment method, volume reduction can be very effective in reducing the total amount of hazardous waste to be treated or disposed of. This helps to prolong the life of storage facilities and can greatly reduce the cost of transportation and treatment. Volume reduction may be accomplished by any (or all) of the following.

Source Separation

The simplest, most obvious, and most cost-effective method of reducing hazardous waste volume is to segregate individual waste products. This prevents the contamination of large volume, non-hazardous wastes with chemicals that render them hazardous. It also allows wastes to be treated effectively by specialized processes rather than by inefficient crude alternatives. Although this may require the generator to invest in new equipment, it almost always results in a net financial gain. It also simplifies RCRA testing and reporting requirements and reduces the risks of disastrous incompatibility reactions.

Process Modification

Industrial processes can also be modified to produce less hazardous waste. Because past (uncontrolled) disposal practices were inexpensive, manufacturing processes were rarely optimized for minimum waste production. This is rapidly changing under RCRA regulations. Hazardous waste disposal is now very expensive. Engineers are discovering that by altering raw materials, operating conditions, and reactor configurations, waste generation rates (and disposal costs) can be greatly reduced.

Product Substitution

It can also be cost-effective to replace waste-intensive products with an alternative product line. Although this may seem drastic, the wisdom and long-term economic viability of any hazardous waste-intensive product should be seriously questioned.

TABLE 5. Industrial Wastes Exchanges, Alphabetical by State
(Source: Moore, 1982; Lindgren, 1983; Fawcett, 1984).

California Waste Exchange, 2151 Berkeley Way, Berkeley, CA 94704
World Association for Solid Waste Transfer and Exchange, 152 Utah Ave. "F", South San Francisco, CA 94080
Zero Waste Systems, Inc., 2928 Poplar St., Oakland, CA 94608
Colorado Waste Exchange, 1390 Logan, Denver CO 80203
World Association for Safe Transfer and Exchange, 130 Freight St., Waterbury, CT 06702
Florida Waste Information Exchange, P.O. Box 5497, Tallahassee, FL 32301
Georgia Waste Exchange, 181 Washington St. SW, Atlanta, GA 30303
Environmental Clearinghouse Organization—ECHO, 3426 Maple Lane, Hazel Crest, IL 60429
American Chemical Exchange, 4849 Golf Road, Skokie, IL 60076
Industrial Material Exchange Service, 2200 Curchill Rd., Springfield, IL 62706
Waste Materials Clearinghouse, 1220 Waterway Blvd. Indianapolis, IN 46202
Iowa Industrial Waste Information Exchange, 201 Bldg. E, Iowa State Univ., Ames, IA 50011
Louisville Area Waste Exchange, 300 West Liberty St., Louisville, KY 40202
The Exchange, 63 Rutland St., Boston, MA 02118
American Materials Exchange Network, 19489 Lahser Rd., Detroit, MI 48219
Great Lakes Regional Waste Exchange, 3250 Townsend NE, Grand Rapids, MI 49505
Minnesota Asso. of Commerce and Industry Waste Exchange Service, 480 Cedar St., St. Paul, MN 55101
Midwest Industrial Waste Exchange, 10 Broadway, St. Louis, MO 63102
Mecklenberg County Waste Exchange, 1501, I-85 N, Charlotte, NC 28216
Resource Conservation and Recovery Agency, P.O. Box 286, Stratham, NH 03885
Industrial Waste Information Exchange, 5 Commerce St., Newark, NJ 07102
Northeast Industrial Waste Exchange, 700 East Water St., Syracuse, NY 13210
The American Alliance of Resource Recovery Interests, Inc., 111 Wash. Ave., Albany, NY 12210
Atlantic Coast Exchange, 1905 Chapel Hill Rd., Durham, NC 27707
ORE Corp.—Ohio Resource Exchange, 2415 Woodmere Dr., Cleveland, OH 44106
Industrial Waste Information Exchange, 1646 W. Lane Ave., Columbus, OH 43221
Oregon Industrial Waste Information Exchange, 333 SW 5th, Suite 618, Portland, OR 97204
Pennsylvania Waste Information Exchange, 222 N. 3rd St., Harrisburg, PA 17101
National Waste Exchange, P.O. Box 190, Silver Springs, PA 17575
Tennessee Waste Swap, 708 Fidelity Federal Bldg., Nashville, TN 37219
Chemical Recycling Information Program, 1100 Milam Bldg., Houston, TX 77002
Inter-Mountain Waste Exchange, P.O. Box 1825, Salt Lake City, UT 84110
Information Center for Waste Exchange, 2112 3rd Ave., Suite 303, Seattle, WA 98121

Resource Recovery

Material that would otherwise become hazardous waste may be recovered and reused. Obviously this is not a new practice. Industries do not intentionally discard valuable materials. However, the economics of hazarous waste disposal are helping to redefine "value." A material has value if it can be reused at a cost less than its disposal cost.

The potential financial and environmental benefits of volume reduction cannot be underemphasized. As an example, the 3M Corporation has estimated that it saved $20 million over a 4 year period with its program to reduce hazardous waste generation.

WASTE EXCHANGE

It is also quite possible that wastes that cannot be reused within one industry may be reused by another. As an example, consider cleaning solvents used in the semiconductor industry. These must be of very high purity. Given the size and value of the product, it is generally uneconomical for the industry to treat and reuse these solvents. However, what is a "contaminated" solvent with respect to semiconductor manufacture may be perfectly acceptable for many other processes. Therefore, if the solvent can be sent on for use elsewhere, it need not become a hazardous waste.

Waste exchanges are information clearinghouses designed to accomplish this. Industries may list the properties and quantities of their wastes in the hope that someone else will want to use them. Industries may also use exchanges to shop for very inexpensive "raw" materials. Exchange organizations normally handle information only. Details of the actual material transfer must be arranged privately. A partial list of waste exchanges is presented in Table 5.

Not all wastes are "exchangeable." Transportation costs, quality control, and fear of liability tend to limit activity. The wastes most often exchanged include acids, catalysts, solvents, oils, and sludges with high (recoverable) heavy metal contents.

HAZARDOUS WASTE MANAGEMENT PLANNING

Knowledge of the technologies and regulation for hazardous waste generation, storage, transportation, treatment, and ultimate disposal is essential for sound hazardous waste management. However, this knowledge will not necessarily yield a wise, long-term management strategy. To date, RCRA has been less successful than originally hoped at diverting wastes from land disposal technologies. In part, this has been a problem of transition and implementation. It has been very difficult to permit any hazardous waste facility because of the high degree of public apprehension about safety. By default, this has favored sites already in the hazardous waste business.

It is also true that the implementation of RCRA facilities will not necessarily result in a wise management strategy.

This is because the regulations establish minimum standards for the design and operation of individual facilities. A higher level of performance may be necessary. There is absolutely no guarantee that an arbitrary mixture of facilities will result in a long-term solution to the problem of disposal. There is also no guarantee that this will lead to an equitable distribution of costs, human and environmental risk, or community responsibility. A higher level of management planning is required to accomplish this.

It is very important to keep in mind that hazardous waste is not a single material, but an incredibly diverse collection of chemical wastes. These have great variations in their physical and chemical properties, and represent vast differences in the type of hazards they present. No single facility could ever be successful at disposing of the whole inventory. A collection of facilities is required, each managing the type of waste for which it is best suited. These facilities must operate at reasonable cost, or wastes will be driven into less desirable disposal avenues.

Clearly this type of system requires both community and industrial cooperation. In the past, industries have spent minimal amounts on their waste disposal. Under RCRA they will spend a great deal more. Furthermore, given the penalties for noncompliance or poor compliance, industries are often willing to set aside cost as the major decision variable. Many are very concerned with minimizing the negative public relations impacts of hazardous waste activities. There is also strong motivation to minimize long-term legal liabilities in lieu of short-term cost savings.

Communities should also have strong motivation to cooperate in waste management planning. Hazardous waste disposal involves risks. These vary from the short-term acute risk borne by the site workers to the long-term chronic dangers imposed on the whole community. Cooperative

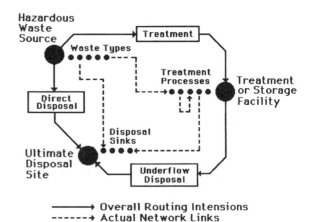

FIGURE 21. The hazardous waste management planning network concept.

management planning allows the community to exert positive control over the types, distributions, and intensities of these risks.

It has recently been demonstrated that hazardous waste management planning may be treated as a mass transfer optimization similar to a solid waste disposal network. This network analogy is illustrated in Figure 21. It is important to note that there are major differences between this type of disposal network and a conventional solid waste system. Possibly the most important distinction is that source nodes must be treated as producing a whole inventory of distinct chemical wastes. The network mathematics must be capable of distinguishing among these. Without distinct waste classifications, one cannot capitalize upon the unique attributes of the more sophisticated waste technologies. Processing nodes also take on much greater importance and complexity. These are no longer simple transfer stations, but major treatment facilities. They greatly reduce the amount of waste, and pass on underflows of distinctly different physical and chemical composition (solids, sludges, spent resins, etc.). Often, multiple processes are co-located, and on-site, process-to-process transfers must be accounted for. Ultimate sink nodes may also offer more than one disposal avenue. As an example, hazardous waste incinerators may be co-located with secure landfills. Finally, the network link structure is generally sparse. Regulations and physical limitations of planning goals generally eliminate many of the possible links. Generally it is best to remove these rather than to constrain them away with severe penalty functions. A complete mathematical model of the hazardous waste disposal network problem may be found in Jennings and Sholar (1984).

Hazardous waste planning models may be solved with several existing optimization programs including EPA's Waste Resources Allocation Program (WRAP). This is a reasonably well documented and supported code (Mitre, 1977 a,b) and has been used in several regional solid waste disposal studies (see Hasit and Warner, 1981). Jennings and Sholar (1984) have also demonstrated that all the added complexity of the hazardous waste problem may be handled (without reprogramming) by a judicious use of existing program features. However, the reader is cautioned that the WRAP is not infallible, and alternative simulation codes are available.

It is very convenient to study the problem of hazardous waste disposal by this network analysis approach. Recent works by Pierce and Davidson (1982), Jennings (1983), and Jennings and Sholar (1984) have illustrated that the estimated costs of technologies may be used to identify desirable sets of economically viable facilities. Cost functions for hazardous waste technologies may be found in A. D. Little, Inc. (1979), EPA (1980), and GCA Corp. (1980). Obviously these must be carefully updated for current use. Studies have also illustrated that risks may be used (in place of costs) as network penalty functions to examine the relative risks of proposed management plans and the trade-offs between cost and risks.

Risk analysis, particularly for management planning, is a new concept in hazardous waste management. This has tremendous possibilities for resolving our current public relations difficulties. Hazardous wastes expose the public to risk. They know this and are very concerned about it. Successful management plans must acknowledge this reality. A thorough and public analysis of the risks (including the current risk state and the dangers of inaction) is one of the most powerful vehicles for gaining public confidence and support.

REFERENCES

1. Arthur D. Little, Inc., "A Plan for Development of Hazardous Waste Management Facilities in the New England Region," prepared for the New England Regional Commission (Sept., 1979).
2. Booz-Allen & Hamilton, Inc., Putnam, Hayes & Bartlett, Inc., "Hazardous Waste Generation and Commercial Hazardous Waste Management Capacity," EPA SW-894 (Dec., 1980).
3. Boraiko, A. A., "Storing Up Trouble . . . Hazardous Waste," *National Geographic, 167* (3), 318−351 (March, 1985).
4. Brown, K. W., G. B. Evans, Jr., and B. D. Frentrup (ed.), *Hazardous Waste Land Treatment*, Butterworth Publishers (1983).
5. Environmental Protection Agency, "Everybody's Problem: Hazardous Waste," EPA SW-826 (1980).
6. Environmental Protection Agency, "Treatability Manual Volume IV, Cost Estimating," EPA-600/8-80-042d (July, 1980).
7. Epstein, S. S., L. O. Brown, and C. Pope, *Hazardous Waste in America*, Sierra Club Books (1982)
8. Fawcett, H. H., *Hazardous and Toxic Materials Safe Handling and Disposal*, John Wiley & Sons (1984).
9. Funderburk, R., "Disposal in Salt: The Fifth Alternative," *Pollution Engineering, XVI* (7), 41−43 (July, 1984).
10. GCA Corp., "Industrial Waste Management Alternatives Assessment for the State of Illinois," prepared for the Illinois Environmental Protection Agency (Nov. 1980).
11. Hasit, Y. and D. B. Warner, "Regional Solid Waste Planning With WRAP," *Journal of the Environmental Engineering Division, ASCE, 107* (EE3), 511−525 (June 1981).
12. Jennings, A. A., "Analysis of the National Industrial Residual Flow Problem," Final Report, University of Notre Dame, Notre Dame, IN (Jan., 1982).
13. Jennings, A. A., "Profiling Hazardous Waste Generation for Management Planning," *Journal of Hazardous Materials, 8,* 69−83 (1983).
14. Jennings, A. A. and R. L. Sholar, "Hazardous Waste Disposal Network Analysis," *Journal of Environmental Engineering, 110* (2), 325−342 (1984).
15. Lindgren, G. F., *Guide to Managing Industrial Hazardous Waste*, Butterworth Publishers (1983).

16. Mitre Corp., "WRAP: A Model for Regional Solid Waste Management Planning; Programmer's Manual," EPA/530/SW-573 (Feb., 1977).

17. Mitre Corp., "WRAP: A Model for Regional Solid Waste Management Planning; User's Guide," EPA/530/SW-574 (Feb., 1977).

18. Moore, G. F., "Industrial Waste Exchanges," *Pollution Engineering,* p. 33 (Jan., 1982).

19. Pierce, J. J. and G. M. Davidson, "Linear Programming in Hazardous Waste Management," *Journal of the Environmental Engineering Division, ASCE, 108* (EE5), 1014–1026 (Oct., 1982).

20. Porter, C. H., "State Program Implementation Guide: Hazardous Waste Surveys," EPA/530/SW-160 (July, 1975).

21. VanNoordwyk, H., L. Schalit, W. Wyss, and H. Atkins, "Quantification of Municipal Disposal Methods for Industrially Generated Hazardous Wastes," EPA-600/2-79-135 (Aug., 1979).

DESIGN BIBLIOGRAPHY

In a general reference work of this type it is difficult to present detailed design information. It is also true that many hazardous waste facilities are unique designs carefully tailored for specific sites. Most existing designs can not be treated as acceptable standards of practice.

Over the last few years, several reference texts have been published concentrating on specific aspects of hazardous waste design. A selection of these is presented here. This bibliography is by no means exhaustive, and no attempt has been made to cite the immense volume of work available in technical journals. The references cited are simply volumes that the author has recently reviewed and that are commonly available. They are suggested as supplemental reading on specific technical aspects of hazardous waste management.

1. Bennett, G. F., F. S. Feates, and I. Wilder, (ed.), *Hazardous Materials Spills Handbook,* McGraw-Hill Book Co. (1982).

2. Brown, K. W., G. B. Evans, and B. D. Frentrup, (ed.), *Hazardous Waste Land Treatment,* Butterwork Publishers (1983).

3. Cavaseno, V. (ed.), *Industrial Wastewater and Solid Waste Engineering,* McGraw-Hill Publications Co. (1980).

4. Cheremisinoff, N. P., P. N. Cheremisinoff, F. Ellerbusch, and A. J. Perna, *Industrial and Hazardous Wastes Impoundment,* Ann Arbor Science Publishers, Inc. (1979).

5. Cope, C. B., W. H. Fuller, and S. L. Willetts, *The Scientific Management of Hazardous Wastes,* Cambridge University Press (1983).

6. Epstein, S. S., L. O. Brown, and C. Pope, *Hazardous Wastes in America,* Sierra Club Books (1982).

7. Hackman, E. E., *Toxic Organic Chemicals, Destruction and Waste Treatment,* Noyes Data Corp. (1978).

8. Exner, J. H. (ed.), *Detoxification of Hazardous Waste,* Ann Arbor Science (1982).

9. Fung, R. (ed.), *Protective Barriers for Containment of Toxic Materials,* Noyes Data Corporation (1980).

10. Fawcett, H. H., *Hazardous and Toxic Materials, Safe Handling and Disposal,* John Wiley and Sons (1984).

11. Highland, J. H. (ed.), *Hazardous Waste Disposal, Assessing the Problem,* Ann Arbor Science (1982).

12. Jackson, L. P., A. R. Rohlik, and R. A. Conway, (ed.), *Hazardous and Industrial Waste Management and Testing: Third Symposium,* ASTM STP 851 (Dec., 1984). Also see *Second Symposium* (STP 805 1983) and *First Conference* (STP 760, 1981).

13. Kiang, Y-H. and A. A. Metry (ed.), *Hazardous Waste Processing Technology,* Ann Arbor Science (1982).

14. LaGrega, M. D. and L. K. Hendrian, (ed.), *Toxic and Hazardous Waste, Proceedings of the Fifteenth Mid-Atlantic Industrial Waste Conference,* Ann Arbor Science (1983).

15. Lehman, J. P. (ed.), *Hazardous Waste Disposal,* Plenum Press (1983).

16. Levine, S. P. and W. F. Martin (ed.), *Protecting Personnel at Hazardous Waste Sites,* Butterworth Publishers (1985).

17. Lindgren, G. F., *Guide to Managing Industrial Hazardous Waste,* Butterworth Publishers (1983).

18. Metry, A. A. (ed.), *The Handbook of Hazardous Waste Management,* Technomic Publishing Co., Inc. (1980).

19. Pierce, J. J. and P. A. Vesilind (ed.), *Hazardous Waste Management,* Ann Arbor Science (1981).

20. Pojasek, R. B. (ed.), *Toxic and Hazardous Waste Disposal, Volume 3: Impact of Legislation and Implementation of Disposal Management Practices,* Ann Arbor Science (1980).

21. Pojasek, R. B. (ed.), *Toxic and Hazardous Waste Disposal, Volume 4: New and Promising Ultimate Disposal Options,* Ann Arbor Science (1980).

22. Robinson, J. S. (ed.), *Hazardous Chemical Spill Cleanup,* Noyes Data Corporation (1979).

23. Sittig, M., *Handbook of Toxic and Hazardous Chemicals,* Noyes Data Corporation (1981).

24. Sittig, M., *Incineration of Industrial Hazardous Wastes and Sludges,* Noyes Data Corporation (1979).

25. Sittig, M., *Land Disposal of Hazardous Wastes and Sludges,* Noyes Data Corp. (1979).

26. Sweeney, T. L., H. G. Bhatt, R. M. Sykes, and D. J. Sproul (ed.), *Hazardous Waste Management for the 80's,* Ann Arbor Science (1982).

27. Weiss, G. (ed.), *Hazardous Chemical Data Book,* Noyes Data Corporation (1980).

CHAPTER 22

Modeling Organic Pollution of Streams

G. B. McBride*, J. C. Rutherford*, and R. D. Pridmore*

INTRODUCTION

Streams are commonly used for the conveyance and assimilation of wastes discharged from municipal sewers, urban drains, farms, and industries. Amongst the industries producing effluents containing substantial amounts of wastes are the food and textile industries, breweries, abattoirs, dairies, tanneries, and pulp and paper manufacturers.

Organic pollution results when large amounts of carbonaceous compounds, such as proteins, carbohydrates, and fats, are discharged into, and become mixed with, the stream water. Micro-organisms in the water and on the stream bed degrade these compounds and, in doing so, use up oxygen by their respiration. If the rate of respiration is greater than the rate at which oxygen is replenished by aquatic plant photosynthesis and atmospheric reaeration, then the oxygen content of the water will be lowered. In some situations, the dissolved oxygen may be depleted to such an extent as to kill fish or cause the migration of desirable forms of organisms from the polluted zone. Anoxic conditions may also develop, leading to the proliferation of anaerobic bacteria. These organisms can cause the stream water to become smelly, foul tasting, and, in some cases, toxic (usually through the production of ammonia and/or hydrogen sulphide).

Oxygen deficits, however, are not the only effect of organic pollution. The large number of micro-organisms which grow in response to the increased carbon supply can also be aesthetically displeasing as well as a nuisance to both water users and stream-dwelling organisms. For example, unsightly assemblages of "sewage fungus" may develop

(Figure 1). In addition to consuming dissolved oxygen, these growths can block water intake filters, interfere with recreational activities, smother fish eggs and invertebrates living on the stream bed, and produce unpleasant tastes and odours during their decomposition. Organic wastes also frequently contain substantial amounts of suspended solids and essential plant nutrients, like nitrogen and phosphorus. Suspended solids generally increase the turbidity of the water and reduce the amount of light available for aquatic plant photosynthesis. On settling out, these particles can destroy fish spawning grounds and infill the habitats of bottom-dwelling invertebrates. The effects of nitrogen and phosphorus are most noticeable after the particulate matter has been decomposed and the water regains its clarity. Under these conditions, the availability of light increases and the presence of nitrogen and phosphorus supports a higher level of primary production. Too much production, however, may again lead to an impairment of water use as proliferations of aquatic plants (i.e., algae, macrophytes) can cause many of the same problems already noted for microbial growths, particularly oxygen depletion. Organic wastes originating from man and/or domestic animals may also be rich in pathogenic organisms, such as *Salmonella, Shigella,* and *Leptospira,* and *Escherichia coli.* These organisms can cause a variety of disorders in man including acute gastro-enteritis, diarrhoea, kidney infections, and typhoid.

Because organic pollution can have such a dramatic effect on freshwater ecosystems and their water uses, it is important that one understands and approves of what will happen to a waterway before wastes are discharged into it. This chapter demonstrates how mathematical models can be used to describe the transport and dispersion of an organic pollutant down the course of a waterway and predict its subsequent effect on stream dissolved oxygen. Some comments are also given on different oxygen-related measures of organic pollution.

*Water Quality Centre, Ministry of Works and Development, Hamilton, New Zealand

619

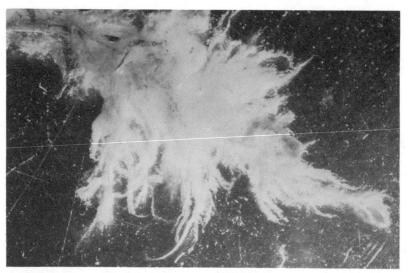

FIGURE 1. Nuisance growth of a microbial assemblage known as "sewage fungus." Sewage fungus consists of a mixture of bacteria, fungi, protozoans, and algae.

MODELING OF STREAM POLLUTANT TRANSPORT

Water resource management requires some means of predicting the effects of man's activities on pollution of streams, i.e., to predict whether stream standards will be breached. This is particularly so when the management agency is faced with an application either to change an existing waste discharge or to introduce a new discharge. Discharges may be from pipes (point sources) or from diffusers (line sources). Diffuse sources, from land runoff, can also be modeled [80] although the greatest application has been to point and line sources. The means is commonly a mathematical model that translates the *input* (the source) into the *output* (the resulting concentrations in the stream). Physical models [27] are of limited application for stream pollution. Models perform this translation by applying given values of *parameters* (or *coefficients*) to mathematical equations which are then solved.

There are three main types of mathematical models. *Empirical models* [10] are statistical summaries of existing stream pollution data and may be accurate for predicting output similar to that used to develop the empirical equations. But models are often needed to predict conditions somewhat dissimilar to existing data (e.g., for predicting the effects of a new point source) and empirical models may not then be trustworthy. *Stochastic models* take explicit account of the randomness or uncertainty in measurements of the various stream processes affecting waste concentration. They are data-intensive, since they deal with probability distributions of variables. *Deterministic models* deal with mean values of variables, and attempt to account for all the major stream processes affecting pollution. As such they offer the twin promises of not requiring excessive data (which are expensive to obtain), and being potentially trustworthy for predictions for stream conditions dissimilar to those already existing. For these reasons deterministic models are the most widely used. Fuller details of stream pollution modeling approaches are given in the recent book by Orlob [64].

This chapter concentrates on deterministic models.

Model Building and the Role of Data

We strongly advocate a two stage approach to modeling: *first,* making a preliminary "desk study" of a problem, with field studies and perhaps more complex modeling studies being undertaken *second* if necessary.

The initial "desk study" is conducted using the simplest possible deterministic model, together with whatever data are at hand supplemented where necessary with model coefficients taken from the literature. It is often appropriate to choose "worst case" conditions of low flow, high temperature and maximum waste discharge and to make predictions using a range of coefficient values. Such predictions may indicate that the added waste will not breach water quality standards or not have a deleterious effect on desired water uses. In this case, the discharge can be permitted and further modeling work is unnecessary. On the other hand, the predictions may indicate a potential problem (e.g., a large dissolved oxygen deficit), suggesting that further field work and perhaps more complex modeling should be undertaken.

The desk studies are also valuable for indicating which processes need to be modeled, which model should be used, and where field studies should be conducted.

At the second stage, special data collection studies are required. Routine monitoring data while highlighting pollution problems are rarely ideal for modeling [35] because they involve sampling infrequently over a sparse spatial grid over long periods of time. Sampling at several points below each major discharge intensively (e.g., at hourly intervals) over short periods (e.g., one day) is required for successful modeling.

If a decision is made to develop a more complex model, that model ought to be committed to paper and preliminary predictions made. That will facilitate the design of the data collection program. The interaction between the design and operation of data collection program(s), and model building, is vital to a successful pollution modeling study. If modeling is commenced only at the completion of the data collection programs, the chances of failure are high.

Reducing the Complexity of the Problem

The first question that should be considered in a stream modeling study is the appropriate dimensionality of the model. That is, do we need to model for lateral (direction z) or vertical (direction y) changes of pollutant concentration, or is it sufficient to model changes in the downstream direction only (direction x)? This is determined by how quickly the discharged pollutant mixes in the stream, compared with how quickly undesirable effects of the pollutant appear. In modeling the effects of pollutants on stream dissolved oxygen, it is generally satisfactory to use a one-dimensional model (direction x), because of the time taken for oxygen depletion to occur. Transverse mixing may be important in wide rivers. Other pollutants, e.g., bacteria, may have their maximum effect immediately so that near field mixing (lateral and vertical mixing) must be considered.

MIXING PROCESSES

When a slug of tracer is discharged into a stream, the resulting tracer patch is carried away from the outfall by the current, and as it does so it spreads out. Two fundamental processes cause this spreading: differential advection and diffusion. Differential advection arises because the distribution of downstream velocity over the stream cross-section is not uniform; surface water at the channel centre travels downstream faster than water near the bed or banks. Hence, the patch spreads further downstream in the channel centre, becoming horse-shoe shaped. Diffusion blurs the edge of this patch, spreading it in all directions. In laminar flow, this spreading is caused only by random molecular motions and is called molecular diffusion. The net diffusive transfer of tracer from a region of high concentration (the patch) to a region of lower concentration (uncontaminated water) proceeds at a rate proportional to the concentration gradient between the two regions. This is "Fick's Law," which in one dimension can be written mathematically as

$$R = -D \frac{dC}{dn} \qquad (1)$$

where R = transfer rate per unit area in the n direction, normal to the patch boundary, and D = molecular diffusion coefficient, a constant.

In turbulent stream flow, fluctuating eddies cause a considerable amount of transient differential advection on small spatial scales. This increases the opportunities for molecular diffusion, increasing the rate of spreading markedly. This combination of turbulent eddies and molecular diffusion we term "turbulent mixing" [25].

In some situations the rate of turbulent mixing can also be approximated by Fick's Law. However, the value of D may be several orders of magnitude larger than for molecular diffusion and is highly variable. We call D the "turbulent mixing coefficient" [25], though it is sometimes called the "turbulent diffusion coefficient" [19]. Its variability arises partly because the size and intensity of turbulent eddies vary considerably with position in the stream channel, with changes in flow or location, and from one channel to another. For example the rate of mixing can be expected to be smaller very close to the stream bed (where velocity and intensity of turbulence are small) than at mid-depth. Also as the size of the tracer patch increases, the velocity gradients change and larger eddies become involved in mixing. Thus very close to an outfall the rate of mixing can be expected to be smaller than it is further downstream.

On larger spatial scales, e.g., after averaging over the stream cross-section, the combined effects of differential advection and tubulent mixing are termed "dispersion."

In the most general problem, with mixing occurring in each of the three co-ordinate directions (x, y, z), the governing equations are comparatively complex. Secondary circulations (because of non-prismatic channels, and bends) may also give rise to lateral and vertical advection terms. In many practical problems, however, the analysis can be simplified by neglecting terms which are small. This can be done in three cases: if any of the velocities is small, if any of the concentration gradients is small, or if the nature of the discharge means that any of the concentration gradients is small. In particular, if the discharge is steady the effect of longitudinal dispersion is small [37].

In studying streams we can make the following simplifications: vertical and lateral average velocities are small; many streams are wide but shallow, and a tracer quickly becomes well-mixed vertically so that vertical mixing can be ignored. Table 1 summarises the characteristics of various problems in stream dispersion, and Figure 2 illustrates three types of stream dispersion problems. Several different analytical solutions are available for examining these problems, e.g., integral transforms [12,44], Green's functions [98], and linear superposition [99]. The approach used by Rutherford [77], based on the method of images, is outlined here.

Vertical Mixing

In channels with no secondary circulations, the principal mechanism causing vertical mixing is turbulence generated

TABLE 1. Summary of Important Mixing Problems in Streams and Terms Required to Study Them.

Type of source		Type of solution		Terms required[1]		Number of dimensions
time	space [1]	2	3	advection	mixing	
instantaneous	point source	near-field	$0 < x < [(U/2)d^2]/D_y$	x[4]	x,y,z	3
		mid-field	$[(U/2)d^2]/D_y < x < [(U/2)b^2]/D_z$	x	x,z	2
		far-field	$[(U/2)b^2]/D_z < x < \infty$	x	x	1
steady	point source	near-field	$0 < x < [(U/2)d^2]/D_y$	x[4]	y,z	3[6]
		mid-field	$[(U/2)d^2]/D_y < x < [(U/2)b^2]/D_z$	x	z	2[6]
		far-field	$[(U/2)b^2]/D_z < x < \infty$	−[5]	−[5]	0[5]
instantaneous	z-line source	near-field	$0 < x < [(U/2)d^2]/D_y$	x	x,y	2
		far-field	$[(U/2)d^2]/D_y < x < \infty$	x	x	1
steady	z-line source	near-field	$0 < x < [(U/2)d^2]/D_y$	x	y	2[6]
		far-field	$[(U/2)d^2]/D_y < x < \infty$	−[5]	−[5]	0[5]
instantaneous	y-line source	near-field	$0 < x < [(U/2)b^2]/D_z$	x	x,z	2
		far-field	$[(U/2)b^2]/D_z < x < \infty$	x	x	1
steady	y-line source	near-field	$0 < x < [(U/2)b^2]/D_z$	x	z	2[6]
		far-field	$[(U/2)b^2]/D_z < x < \infty$	−[5]	−[5]	0[5]

Notes:
[1] Co-ordinate directions are shown in Fig. 2.
[2] Near-field = very close to the outfall, mid-field = moderately, and far-field = some considerable distance away.
[3] D_x, D_y, D_z are turbulent mixing coefficients and U = mean cross section velocity.
[4] Secondary circulations may cause small y, z advections.
[5] Concentration is constant (fully mixed).
[6] The dimensionality can be reduced by one if the coordinate system used travels downstream at mean velocity.

by velocity shear at the bed. Elder [18] showed that the mixing coefficient varies parabolically with depth, and depends on both depth and shear velocity. Thus

$$D_y(y) = \frac{y}{d}\left(1 - \frac{y}{d}\right)\varkappa du^*$$ (2)

where D_y = vertical mixing coefficient, d = depth of flow, $\varkappa$ = von Kármán's constant ($\cong 0.4$), u^* = shear velocity = $\sqrt{gdS}$ where S = channel slope, and g = gravitational acceleration. The applicability of this equation has been confirmed in laboratory and field studies. For many practical problems, the depth average (denoted by just D_y) is used [24]; hence

$$D_y = 0.067du^*$$ (3)

Vertical secondary circulations can be expected to increase the rate of vertical mixing in natural channels. Few data are available to quantify their effect, but it appears that

$$0.067 < D_y/du^* < 0.33$$ (4)

VERTICAL MIXING DOWNSTREAM FROM A STEADY TRANSVERSE LINE SOURCE

A common problem is to predict tracer concentrations (denoted by C) downstream from a z-line source on the bed, such as a perforated pipe that extends across the entire channel width. Transverse and longitudinal concentration gradients are negligible. Thus, the problem includes only x-advection and y-mixing (see Table 1). The velocity and mixing coefficient are taken to be uniform over the depth as a rough approximation to turbulent flow.

Figure 3 shows lines of equal concentration downstream from the source. Variables are expressed in non-dimensional form so that many combinations of parameters appear on the same graph:

$$C^* = C/\bar{C} = CUbd/q$$ (5)

$$y^* = y/d$$ (6)

$$x^* = xD_y/Ud^2$$ (7)

where C^*, y^*, and x^* = non-dimensional concentration, vertical displacement, and downstream displacement, respectively; $\bar{C}$ = fully mixed concentration; U = mean cross-section velocity, d = stream depth (the mean depth should be used here if the channel is irregular); b = stream width; and q = tracer mass inflow rate. The bed and water surface are located at $y^* = 0$ and $y^* = 1$, but the problem is symmetrical in y since the flow velocity is assumed uniform, with depth. If the source is on the bed of a rough channel we suggest using 10% of the mean values for U and D_y to account for low velocities and mixing coefficients near the bed and for the effect of "dead zones."

Clearly,

$$0 \leq y^* \leq 1 \qquad (8)$$

and, a long way below the source,

$$C^* = 1 \qquad (9)$$

The regions to the left of the $C^* = 0.01$ contour do not contain any tracer, while in the region to the right of the $C^* = 1.01$ and 0.99 contours, the tracer is fully mixed. The figure shows that complete vertical mixing is attained within a distance

$$x_m \cong 0.4 \ Ud^2/D_y \qquad (10)$$

downstream from an outfall on the bed or at the surface.

This is similar to the result obtained by Sayre [79]. The figure also indicates the length and width of tracer plume in which concentrations exceed a specified level (see worked examples). Nomographs for other source depths are given by Rutherford [77].

VERTICAL MIXING DOWNSTREAM FROM A STEADY POINT SOURCE

A more complex problem is to predict concentrations downstream from a steady point source such as a single port outlet. Clearly, both transverse and vertical mixing are important close to the outfall. Thus the problem now includes both y- and z-mixing (see Table 1).

Three-dimensional cigar shaped surfaces of equal concentration are encountered below a point source. Contours of equal concentration on a vertical (x-y) plane which passes through the source are shown on Figure 4 for a typical combination of parameters. The figure also indicates the length and width of plumes in which concentrations exceed a specified level.

Clearly, higher concentrations are found immediately below a point source than below a line source of the same total output. However, tracer becomes vertically well mixed at much the same distance below point and line sources, and Equation (10) still applies. Thereafter, transverse mixing

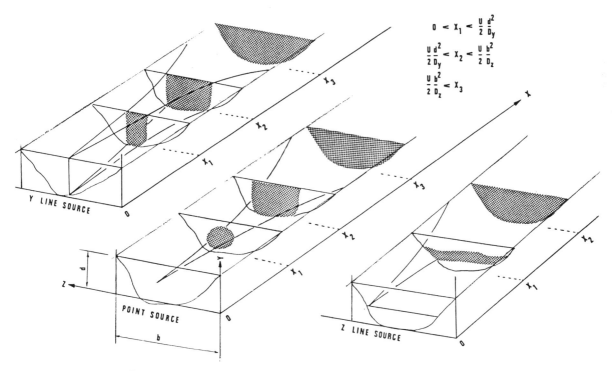

$$0 < X_1 < \frac{U}{2} \frac{d^2}{D_y}$$

$$\frac{U}{2} \frac{d^2}{D_y} < X_2 < \frac{U}{2} \frac{b^2}{D_z}$$

$$\frac{U}{2} \frac{b^2}{D_z} < X_3$$

FIGURE 2. Sketch of three types of stream mixing problems. From Rutherford [77].

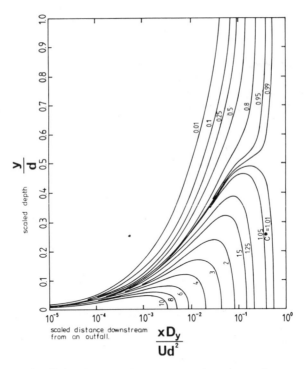

FIGURE 3. Concentration contours downstream from a steady transverse line source located on the stream bed. From Rutherford [77].

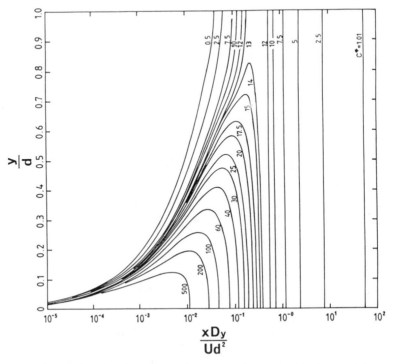

FIGURE 4. Concentration contours downstream from a steady point source in mid-channel on the stream bed: $b/d = 50$; $D_z/D_y = 3$. From Rutherford [77].

dominates and eventually mixes tracer throughout the stream channel.

If the source is located at either bank, theory indicates that concentrations are exactly twice those shown in Figure 4 with the exception of the $C^* = 1.01$ contour, which moves slightly to the right. For a source at the bed or bank, we again recommend using 10% of the mean values for U and D_y. For a multi-point source of width less than b, concentration contours can be determined for each point source separately using Figure 4. These are then added to produce the concentration contours for the multi-point source.

Nomographs for other source depths are given by Rutherford [77].

WORKED EXAMPLES

Consider a channel with $d = 1$ m; $b = 10$ m; $S = 10^{-4}$; $U = 1$ m $\cdot$ s^{-1}. Thus, the shear velocity $u^* = (9.81 \times 1 \times 10^{-4})^{1/2} = 0.0313$ m $\cdot$ s^{-1}, and D_y ranges from $0.067du^* \cong 20$ cm$^2 \cdot$ s^{-1} to $0.33du^* \cong 100$ cm$^2 \cdot$ s^{-1}.

Example 1

Taking $D_y = 20$ cm$^2 \cdot$ s^{-1}, determine the distance downstream from a surface outfall required for complete mixing for a point source and a transverse line source.

For both cases we can use Equation (10) to give $x_m \cong 0.4Ud^2/D_y = 200$ m.

Example 2

For an outfall at the surface with massflow 20 g $\cdot$ s^{-1} determine: (i) total length, x_s, and (ii) maximum spread, y_s, of the plume in which concentrations exceed 10 g $\cdot$ m^{-3}. Consider two cases: (a) a transverse line source and (b) a point source. Take $D_y = 20$ cm$^2 \cdot$ s^{-1}

Fully mixed concentration $\bar{C} = q/Ubd = 20/1 \times 10 \times 1 = 2$ g $\cdot$ m^{-3}. Then, from Equation (5) $C^* = 5$

(a) Line source

(i) From Figure 3, the $C^* = 5$ contour extends a maximum distance of $x_sD_y/Ud^2 \cong 0.015$ downstream. Thus $x_s \cong 7.5$ m

(ii) From Figure 3, the maximum spread of the $C^* = 5$ contour is $y_s/d \cong 0.10$. Thus $y_s \cong 0.1$ m

(b) Point source

(i) From Figure 4, $x_sD_y/Ud^2 \cong 2.15$, thus $x_s \cong 1,100$ m

(ii) $y_s/d = 1$, thus $y_s = 1$ m

Transverse Mixing

Laboratory studies on the transverse mixing coefficient D_z indicate that because larger eddies can develop in the transverse direction than in the vertical direction, D_z is greater than D_y by a factor of between 2 and 3:

$$0.08 < D_z/du^* < 0.18 \quad \text{average } 0.15 \quad (11)$$

D_z decreases with depth in much the same manner as mean velocity [24].

In straight natural channels the rate of transverse mixing is higher than in Equation (11) because the thalweg tends to meander and hence induce secondary currents. It appears that in such channels [24]

$$D_z/du^* \cong 0.24 \quad (12)$$

Bends and changes in channel cross-section result in stronger transverse circulations, which increase transverse mixing rates [24]

$$0.24 < D_z/du^* < 1.6 \quad (13)$$

Laboratory data in sinuous channels fit [24]

$$D_z \cong 0.25 \frac{U^2d^3}{x^5r^2u^*} \quad (14)$$

where r = radius of curvature.

In both laboratory and natural channels, D_z/du^* increases with the aspect ratio b/d [24]. This indicates that transverse mixing is affected by secondary currents whose importance depends on channel width. In streams the transverse mixing coefficient may, therefore, be more closely related to width and shear velocity than to depth and shear velocity [47].

EFFECTS OF NON-NEUTRALLY BUOYANT EFFLUENTS

When a non-neutrally buoyant effluent is discharged into a channel it either sinks or rises and, in so doing, induces transverse secondary circulations. These secondary circulations increase the rate of transverse mixing in the immediate vicinity of the outfall. Gradually, however, mixing spreads the effluent uniformly over the depth, the driving force of the secondary circulation diminishes, and the transverse mixing coefficient approaches that for a neutrally buoyant effluent.

This process has been quantified in laboratory channels [68] and it appears that the increase in spreading rate is greater for a rising than for a sinking effluent. Transverse mixing coefficients 2-4 times that for a neutrally buoyant effluent were observed within a distance $0 < x < 10$-$20b$ of the outfall.

TRANSVERSE MIXING DOWNSTREAM FROM A STEADY POINT SOURCE

In most streams, tracer quickly becomes vertically well mixed, so transverse mixing includes only x-advection and z-mixing (see Table 1). The velocity and mixing coefficient are assumed uniform across the channel, as an approximation to turbulent flow.

Figure 5 shows concentrations in the horizontal plane (x-z) downstream from a steady vertical line source on the stream bank. This approximates a point source everywhere

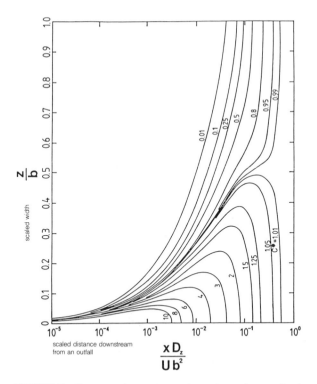

scaled width

$\frac{z}{b}$

scaled distance downstream
from an outfall

$$\frac{x\,D_z}{U\,b^2}$$

FIGURE 5. Concentration contours downstream from a steady vertical line source located on the stream bank. From Rutherford [77].

except very close to the source. The figure uses the non-dimensional variable C^* defined in Equation (5) and the dimensionless lateral and longitudinal distances given by

$$z^* = z/b \qquad (15)$$

$$x^* = xD_z/Ub^2 \qquad (16)$$

Again, we recommend taking 10% of the mean values for U and D_z for a bankside source.

Figures 3 and 5 are identical except for slightly different non-dimensional variables. By analogy with Equation (10) effluent becomes mixed across the whole channel within a distance

$$x_m \cong 0.4\ Ub^2/D_z \qquad (17)$$

Figure 5 can also be used to determine the length and width of a plume in which concentrations are above a specified level.

Nomographs for other source locations are given by Rutherford [77].

WORKED EXAMPLES

Consider the same channel as earlier, viz depth = 1 m, width = 10 m, mean velocity = 1 m · s⁻¹, shear velocity = 3.13 cm · s⁻¹.

Example 1

Select transverse mixing coefficients assuming the channel is (1) straight but rough, (2) a fairly straight natural stream channel, (3) sinuous with radius of curvature 100 m.
(1) From Equation (11) $D_z = 0.15du^* \cong 50$ cm² · s⁻¹
(2) From Equation (12) $D_z = 0.24\ du^* \cong 75$ cm² · s⁻¹
(3) From Equation (13) $D_z = 0.24 - 1.6du^* \cong 75$–$500$ cm² · s⁻¹

Also from Equation (14) $D_z = 780$ cm².s⁻¹

Example 2

Estimate how far downstream from an outfall at either bank a tracer becomes mixed to the other bank. Take $D_z = 200$ cm² · s⁻¹.

To account for reduced mixing near the bank take $D_z = 20$ cm² · s⁻¹. Then from Equation (17), $x_m \cong 0.4\ Ub^2/D_z = 20$ km.

Longitudinal Dispersion

Longitudinal dispersion is the spreading of tracer along the axis of flow. It is usually important only for instantaneous sources (e.g., a pollutant spillage) or perhaps for time-varying discharges (e.g., cyclic variations in waste treatment plant discharges). It results in the attenuation of peak concentrations, as illustrated in Figure 6. Mixing in the lateral direction is much more important than in the vertical direction in driving this process. This explains why early formulae for prediction of the longitudinal dispersion coefficient in infinitely wide channels [18] are much too low for application to streams; they were based on vertical mixing only.

Some way downstream from the source, a simple one-dimensional Fickian model can be used

$$\frac{\partial C}{\partial t} + U \frac{\partial C}{\partial x} = D_x \frac{\partial^2 C}{\partial x^2} \qquad (18)$$

where D_x is now the longitudinal dispersion coefficient, t is time, and C is now the *cross-section average tracer concentration*. It is important to note that use of this average does *not* imply that tracer is mixed uniformly in the stream cross-section. When a tracer patch reaches a cross-section, it appears first in mid-channel, but it is absent from water near the banks. Later, the mid-channel is clear, but some tracer is in the water near the banks. Later still, this tracer disappears, being mixed into the main channel and advected downstream. Nonetheless, it has been found that use of mid-channel concentration only does not induce too much error, but also reduces substantially the field work in longitudinal dispersion experiments.

Equation (18), first derived for pipe flow by Taylor [85,86], is remarkable because it has only one spatial dimension (x) yet it describes the effects of lateral and vertical mixing, as well as differential longitudinal advection. Its derivation is described lucidly in Fischer, *et al.* [25]. It does not apply within the "advective zone". The length of this zone, from Fischer [23], is

$$L \cong kb^2 U/Dz \qquad (19)$$

where k = constant with values given in Table 2. Within this zone, special modeling techniques can be used [25,58].

A wide range of values of D_x has been measured in rivers and laboratory channels (Table 3). Part of this variation may be attributable to measurements being made in the "advective zone" where D_x is not constant, but much of the varia-

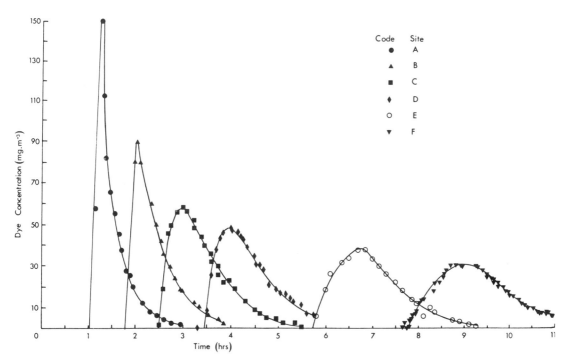

FIGURE 6. Longitudinal dispersion of cross-section mean dye concentrations in the Waikato River, New Zealand, resulting from an instantaneous source. From Rutherford, et al. [78].

TABLE 2. Value of k in Equation (19) for Various Types of Channels, $k = LD_z/b^2U$.

Uniform smooth[1]	Non-uniform smooth[1]	Non-uniform rough[2]
$0.1^3 - 0.25^4$	3.5^5	28^5

Notes:
[1] 10% "dead zones." "Dead zones" are zones where water is nearly stagnant, e.g., under rocks, in holes, behind obstacles.
[2] 38% "dead zones."
[3] From Fischer [23].
[4] From Chatwin [9].
[5] Extrapolated from Valentine [91].

TABLE 3. Summary of Reported Longitudinal Dispersion Coefficients.

Reference	Channel	Depth[1] d cm	Width b m	Shear velocity u* cm·s⁻¹	Discharge Q m³·s⁻¹	Dispersion[2] coefficient Dₓ m²·s⁻¹	Dₓ/du*
25	Chicago Ship Canal	807	48.8	1.91	106	3.0	20
25	River Derwent	25	—	14.0	—	4.6	130
57	Monocacy River	32	35.1	4.4	2.4	4.7	335
		45	36.6	5.1	5.2	13.9	605
		88	47.6	7.2	18.4	37.2	590
25	Green-Duwamish Rv	110	20	4.9	—	6.5–8.5	120–160
57	Comite River	26	12.5	4.4	1.0	7.0	610
		41	15.9	5.6	2.4	13.9	600
25	Clinch River, Va.	58	36	4.9	4.4	8.1	285
25	Clinch River, Tenn.	85	47	6.7	12.8	14	245
		210	53	10.7	92.4	47	210
		210	60	10.4	118.4	54	245
57	Antietam Creek	39	15.9	6.2	2.0	9.3	390
		52	19.8	7.1	4.4	16.3	440
		71	24.4	8.3	8.9	25.6	435
57	Elkhorn River	30	33	4.6	4.3	9.3	665
		42	51	4.7	9.9	20.9	1065
25	Powell River	85	34	5.5	4.3	9.5	200
25	Copper Ck, Va., (below gauge)	49	16	8.0	2.1	20	500
		49	16	8.0	2.0	9.5	245
		85	18	10.0	9.2	21	250
25	Copper Ck, Va. (above gauge)	40	19	11.6	1.2	9.9	220
25	Coachella Canal, Ca.	156	24	4.3	26.6	9.6	140
57	Bayou Anacoco	42	19.8	4.5	2.4	13.9	740
		94	25.9	6.8	8.2	32.5	510
		92	36.6	6.7	13.5	39.5	630
57	Muddy Creek	81	13.4	8.1	4.0	13.9	210
		120	19.5	9.9	10.6	32.5	275
57	John Day	56	25	14.0	14.2	13.9	175
		246	34	18.1	69	65	145
25	Sacramento River	400	—	5.1	—	15	75
25	South Platte River	46	—	6.9	—	16.2	510

(continued)

TABLE 3 (continued).

Reference	Channel	Depth[1] d cm	Width b m	Shear velocity u* cm · s^{-1}	Discharge Q m^3 · s^{-1}	Dispersion[2] coefficient D$_x$ m^2 · s^{-1}	D$_x$/du*
57	Amite River	81	36.6	7.0	8.6	23.2	410
		80	42.4	6.9	14.2	30.2	545
76	Manawatu River	100	25	10.0	26	26–58	260–580
57	White River	55	67	4.4	12.7	30.2	1250
57	Chattahoochee Rv.	113	65.6	7.6	29.0	32.5	380
78	Waikato River	200–300	70–130	5.4–5.8	160	33–70	240–510
	average	250	100	5.5	160	50	360
57	Nooksack River	76	64	27.0	32.6	34.8	170
		293	86	53.1	303	153	100
57	Sabine River, Texas	98	35	4.2	7.4	39.4	960
57	Sabine River	204	104	5.4	119	316	2850
		475	128	8.4	389	670	1680
57	Wind/Bighorn Rv.	98	67	11.2	58	41.8	380
		216	68.6	16.6	231	163	455
57	Susquehanna River	135	203	6.5	106	92.9	1060
57	Yadkin	233	70	10.0	71	111	480
		385	71.7	12.9	213	260	525
59	Mississippi	—	—	—	10,310	232	—
		—	—	—	22,260	700	900
25	Missouri	270	200	7.4	837	1500	7500
57	Missouri	233	183	6.6	380	465	3155
		356	201	8.4	913	837	2810
		311	197	7.8	935	892	3670

[1]Depth calculated from McQuivey and Keefer data [57] using d = Q/bU, where U is obtained from stream discharge measurement (their Ū).
[2]Dispersion ratios may not exactly agree with table entries D$_x$, d, u*, because of round-off errors.

tion reflects real differences between channels resulting from differences in channel geometry, turbulent mixing rates and "dead zones."

LONGITUDINAL MIXING BELOW AN INSTANTANEOUS POINT SOURCE

For an instantaneous point injection into a uniform channel, concentrations can be predicted [24] from

$$C = \frac{W}{A\sqrt{4\pi D_x t}} \exp\left[-\frac{(x - Ut)^2}{4D_x t}\right] \quad (20)$$

where A = channel cross-sectional area, and W = mass discharged at $x = 0$, $t = 0$. Strictly this solution is valid only when $x > 5L$, because it neglects the effects of the advective zone. However, it provides an approximate description of concentration even moderately close to the outfall, although D_x will vary with distance from the outfall (normally decreasing) and the shape of the profiles may be considerably less symmetrical than predicted.

Figure 7, from Rutherford [77], shows the times at which various concentrations occur at specified locations. The non-dimensional variables used are

$$C^* = \frac{CA}{W} \frac{D_x}{U} \quad (21)$$

$$x^* = \frac{xU}{D_x} \quad (22)$$

$$t^* = \frac{tU^2}{D_x} \quad (23)$$

Figure 7(a) indicates how much earlier than the mean travel time, x/U, concentrations first reach C^* while Figure 7(b) indicates how much later than the mean travel time concentrations drop below C^*. By comparing the lengths of vertical lines drawn through a particular value of x^* to a given C^* contour in Figures 7(a) and 7(b) it can be seen that concentration versus time profiles are markedly asymmetrical.

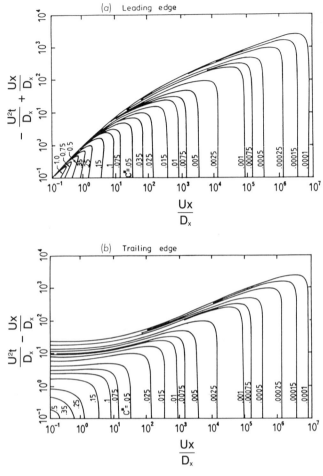

FIGURE 7. Concentration contours downstream from an instantaneous point discharge. From Rutherford [77].

The degree of asymmetry decreases as x^* increases but never entirely vanishes.

The peak concentration at a specified site, x^*, can be deduced from the largest value of C^* on a vertical line passing through x^*.

LONGITUDINAL MIXING BELOW A TIME-VARYING POINT SOURCE

For preliminary analysis of a problem in longitudinal dispersion, knowledge of the fully-mixed concentration and the behaviour of an instantaneous point discharge is often sufficient. For more detailed analysis, however, knowledge of how concentrations change with time below an unsteady point source may be required. These can be predicted by superposing solutions derived from Equation (20). This is straightforward, if somewhat tedious, to do manually.

WORKED EXAMPLES

Consider the same channel as before. No channel in Table 3 matches exactly the depth, width and shear velocity of the channel in this problem, but the Comite River and Muddy Creek are similar. Thus, D_x could be expected to fall in the range 7–14 $m^2 \cdot s^{-1}$. Assume an average value of 10 $m^2 \cdot s^{-1}$.

Example 1

Estimate the length of the advective zone. Take $D_z = 75$ $cm^2 s^{-1}$.

Assuming the channel is non-uniform with about 10% dead zones, from Equation (19) and Table 2

$$L = 3.5b^2U/D_z = 47 \text{ km}$$

Assuming the channel is uniform and smooth

$$L = 0.1 - 0.25 \, b^2 U/D_z = 1.3 - 3.3 \text{ km}$$

Example 2

Given $W = 100$ g, calculate the distance below the outfall where the peak concentration drops below (a) $0.1 \text{ g} \cdot \text{m}^{-3}$ and (b) $10 \text{ mg} \cdot \text{m}^{-3}$.

(a) From Equation (21)

$$C^* = \frac{CA}{W} \frac{D_x}{U} = 0.10$$

From Figure 7(b), the $C^* = 0.10$ contour reaches $x^* \cong 9$. Thus, from Equation (22), $x = 90$ m. Note: This is well within the advective zone calculated above and may be inaccurate.

(b) $C^* = 0.01$, $x^* \cong 900$, thus $x = 9$ km.

Example 3

Given $W = 100$ g and $x = 10$ km, calculate when the concentration:

(a) first reaches $5 \text{ mg} \cdot \text{m}^{-3}$ and (b) drops below $5 \text{ mg} \cdot \text{m}^{-3}$.

From Equation (21), $C^* = 0.005$, and from Equation (22), $x^* = 1,000$.

(a) From Figure 7(a), $x^* - t^* \cong 60$

Thus $t^* \cong 940$, so $t \cong 9,400$ s $= 2.6$ hours.

(b) From Figure 7(b), $t^* - x^* \cong 60$

Thus $t^* \cong 1,060$, so $t \cong 10,600$ s $= 2.94$ hours.

Extension to Non-Conservative Tracers

The preceding mixing calculations can be extended to tracers, such as coliforms, undergoing first-order decay (i.e., rate of decay being a constant (k) times the tracer concentration where k is the decay coefficient, measured in units of inverse time). The calculated tracer concentrations are merely multiplied by $\exp\{-kt\}$. Di Toro [14] has given a procedure for calculating the concentration of other tracers involved in sequential first order reactions with the first tracer.

Complex Modeling

The procedures in the previous sections make use of analytical solutions to the advection-dispersion equation, obtained from the principle of conservation of mass using an Eulerian frame of reference fixed to the stream bank. This equation is

$$\frac{\partial C}{\partial t} + u \frac{\partial C}{\partial x} + v \frac{\partial C}{\partial y} + w \frac{\partial C}{\partial z} = \frac{\partial}{\partial x}\left(D_x \frac{\partial C}{\partial x} \right)$$
$$+ \frac{\partial}{\partial y}\left(D_y \frac{\partial C}{\partial y} \right) + \frac{\partial}{\partial z}\left(D_z \frac{\partial C}{\partial z} \right) + S \tag{24}$$

with appropriate initial and boundary conditions. S is the source $(+)$ or sink $(-)$ of tracer mass per unit stream volume per unit time and u, v, w are the local velocities in the x, y and z directions. These solutions are accurate only for uniform stream flow (coefficients all constant) and for C being conservative $(S = 0)$ or subject to first-order processes only $(S = \pm kC)$. They are useful for making the initial simple calculations prior to developing more complex, hopefully more accurate, models and associated data collection programs. In some cases, they may also be useful for this subsequent modeling. Alternatively, e.g., because of nonuniformity of stream flow, numerical models may have to be used. Before discussing such models, some understanding of the model building process is necessary.

MODEL BUILDING

Before a model can be used to make trustworthy predictions, three separate phases should be completed: identification, calibration, and confirmation.

In *identification* (also called "construction"), we seek to identify the structure of the model appropriate to the problem to be solved. It is important to note that there is always a range of models applicable to any situation, and it is doubtful that any one of these is optimal. It can therefore be highly desirable to have several models available and to compare their predictions.

The model structure depends on three items: (1) the objectives of the modeling organisation (e.g., to set effluent standards on point sources, given a stream standard that is not to be breached); (b) the important in-stream processes operating (e.g., reaeration, benthic respiration, aquatic plant metabolism); (c) the desired level of complexity (the optimal complexity depends on the data available, which in turn determines how well the model parameters can be estimated). Note that a simple model may give a fairly good prediction, and a more complex model may refine this only slightly.

Calibration aims to determine the values of the parameters of a model. The value of some of these parameters may be able to be inferred from laboratory or field investigations (e.g., dispersion coefficient), while others may be estimated using standard empirical formulae (e.g., the reaeration coefficient, for a stream dissolved oxygen study). The model is then run with measured input data and the values of the remaining unknown parameters (e.g., deoxygenation coefficient) are adjusted until a good fit between observed and predicted stream dissolved oxygen values is obtained. Care must be exercised to ensure that the set of calibrated parameter values is unique, i.e., that there is not another plausible set of values that would give equally good agreement with the observed data. For that reason, algorithms for automatic fitting of parameters must be used with care. Clearly, complex models containing numerous parameters require more careful testing to ensure uniqueness than simpler models with fewer parameters. It is desirable to carry out a sensitivity analysis to investigate the uniqueness of the optimal set of parameters and the importance of each parameter and input.

In model *confirmation* (also variously called "verification," "validation," or "corroboration") parameter values are "frozen" and used with another set of measured stream data to see how well the observed and predicted values agree. These data should pertain to stream conditions different from those used in model calibration. It must be stressed that the field data used for model confirmation must be a different set from that used for model calibration. If agreement is poor, it will be necessary to revise the structure of the model and go through the calibration and confirmation steps again. If agreement is good, one can begin to consider the model as trustworthy, particularly for conditions within the range of the data used for calibration/confirmation.

Prediction makes use of a confirmed model. The inputs corresponding to the environmental conditions requiring simulation are specified and the model is used to obtain predictions. Note that even a carefully constructed model may not make accurate predictions for conditions significantly different from those under which it was developed. In general, it is not possible to invert the problem, i.e., given the parameters and output (e.g., as specified by a stream standard) then determine the maximum permissible input to achieve that standard (e.g., [70]). Rather, the allowable input is found by trial and error predictions. One exception to this is noted later, where an inverse stream dissolved oxygen problem is solved directly using a nomograph.

NUMERICAL MODELS

In numerical models, approximate numerical solutions to Equation (24) are obtained for situations when nonuniform stream flow, nonlinear sources and complex boundary geometries mean that analytical solutions cannot be found. A large number of such models has been developed, and these are listed (though their performance has not been quantified) by Basta and Bower [5].

While some numerical stream pollution models are used for near-field mixing problems [32,48], the majority are far-field models which solve the one dimensional equation

$$\frac{\partial AC}{\partial t} + \frac{\partial QC}{\partial x} = \frac{\partial}{\partial x}\left(AD_x\frac{\partial C}{\partial x}\right) + AS \qquad (25)$$

which may be obtained by integrating Equation (24) over the stream cross-section area A (e.g., [36]). The term AS represents the source or sink of tracer per unit channel length per

unit time. This equation is particularly appropriate to the modeling of stream DO or pH, because in these cases the effect of a point source is apparent only in the far-field region.

Most commonly the modeling is done for periods of steady flow (A and Q independent of t), both because critical pollution concentrations occur in steady low-flow periods, and because the data collection effort required to build a model for unsteady flow is enormous and is seldom undertaken. The unsteady stream temperature modeling by Keefer and Jobson [42] is an exception.

The transient term in Equation [25] is appropriate only when sources change strength with time (e.g., photosynthetic production of oxygen varies throughout the day; point sources may change rapidly). If the concentration profiles are steady then Equation (25) reduces to

$$\frac{dQC}{dx} = \frac{d}{dx}\left(AD_x\frac{dC}{dx}\right) + AS \qquad (26)$$

which may be solved rather simply by standard numerical methods (e.g., Runge-Kutta integration).

If Equation (25) is to be used, then obtaining accurate solutions may not be simple. The numerical solutions are obtained from a difference equation, usually obtained from the finite difference or finite element methods, that approximates Equation (25) on a computational grid. For example, for a grid with constant distance and time steps Δx and Δt, and for steady uniform flow (Q and A constant), Equation (25) could be approximated by

$$\frac{C_j^n - C_j^{n-1}}{\Delta t} + U\frac{C_{j+1}^n - C_{j-1}^n}{2\Delta x}$$

$$= D_x\frac{C_{j+1}^n - 2C_j^n + C_{j-1}^n}{\Delta x^2} + S_j^n \qquad (27)$$

where $C_j^n = C(j\Delta x, n\Delta t)$ and $U(= Q/A)$ is the mean cross-section velocity. This equation is known as the fully implicit scheme [72] and, once appropriate initial and boundary conditions are specified, can be solved for successive time levels (n) by the well-known tridiagonal (or "Thomas") algorithm [69]. It is implicit because the space derivative terms are all evaluated at the forward time level, n. Implicit schemes are not subject to any stability restraints, whereas explicit schemes (space derivatives evaluated at the backward time level, $n-1$) are usually unstable for cases where the Courant number α ($= U\Delta t/\Delta x$) > 1.

Considerable care is necessary in applying numerical schemes to Equation (25); a number of problems can arise. Most of these have been investigated assuming a conservative tracer [28], and they *may* be ameliorated, or obscured, when modeling nonconservative tracers. Nevertheless, we recommend paying careful attention to the following potential problems before using a numerical model.

Numerical Dispersion

If there are steep longitudinal concentration fronts in a stream, such as occur downstream of a rapidly varying point source, schemes may exhibit "numerical dispersion." Here the difference equation actually simulates an advection-dispersion equation with a longitudinal dispersion coefficient of $D_x + D_{num}$, where D_{num} is the numerical dispersion coefficient. This arises from the approximations made to the advection term on the left-hand-side of Equation (25). For the fully implicit scheme, Equation (27) $D_{num} = U^2 \Delta t/2$ [72].

As an example, McQuivey and Keefer [59], using accurate methods, estimated the longitudinal dispersion coefficient of the Mississippi River below Baton Rouge to be $D_x = 7,500$ ft²/s. Sparr [82] applied a numerical stream pollution model to the same data and inferred a value $D_x = 300$ ft²/s. The difference, which seems to have gone undetected, is probably attributable to numerical dispersion (i.e., for Sparr's case, $D_{num} = 7,200$ ft²/s). The model used by Sparr (QUALI) has a very high numerical dispersion coefficient (for uniform flow the scheme corresponds to the "backward-in-time, backward-in-space" description of Ponce, et al. [66] with $D_{num} = U\Delta x(1 + \alpha)/2$. Further evidence of the effects of D_{num} in the QUALII scheme (which involves the same scheme as QUALI) is given by Grenney, et al. [29].

Clearly one should beware of numerical dispersion. It can be calculated, and allowed for, as shown by Ponce, et al. [66] or Roache [72].

False Oscillations

In schemes where $D_{num} = 0$, e.g., the widely used scheme of Stone and Brian [83], as used for example by Lau and Krishnappen [48] and Miller and Jennings [60], steep concentration fronts can still be distorted. This is attributable to higher than second-order truncation errors in the numerical scheme. More significantly, one may then find substantial false oscillations upstream of the front. These become more severe as α increases, and as the dispersion number σ ($= D_x\Delta t/\Delta x^2$) decreases. For the Stone–Brian scheme, experiments suggest keeping $\alpha < 2$ and $\sigma > 0.1$. Other than this, methods for devising Eulerian schemes that minimise both numerical dispersion and false oscillations are somewhat complex [28] and are relatively untested in practical applications. To model problems with steep fronts we recommend that, where practicable, Lagrangian schemes (discussed later) be used.

Incorporation of Point Sources

Some further oscillations can be induced upstream of point sources [54], especially if central differences are used to approximate the advection term $U\partial C/\partial x$, such as in Equation (27). Surprisingly, this matter has received rather little attention in the literature. The QUALI and QUALII models use upstream differences for advection and hence, while

suffering from numerical dispersion, do not invoke oscillations at point sources. Local upstream differencing should be used if these oscillations are problematical in central difference Eulerian schemes; in so doing one must be careful to maintain mass conservation. This is shown in the following section.

Mass Conservation

The stream pollution equation may be written in two forms. The "conservative form" is that given as Equation (25); its derivation follows directly from the principle of conservation of mass of tracer. The "nonconservative form" is obtained by operating on Equation (25) with the conservation of water equation

$$\frac{\partial A}{\partial t} + \frac{\partial Q}{\partial x} = q \qquad (28)$$

where q is the rate of water inflow per unit channel length, from point or distributed sources. Subtracting Equation (28), multiplied by C, from Equation (25) and dividing by A gives the nonconservative form

$$\frac{\partial C}{\partial t} + U \frac{\partial C}{\partial x} = \frac{1}{A} \frac{\partial}{\partial x}\left(AD_x \frac{\partial C}{\partial x} \right) + S - \frac{qC}{A} \quad (29)$$

Difference schemes derived from Equation (25) will conserve mass, but schemes derived from Equation (29) may not. This depends on where the C value in the dilution term (the last term of Equation (29)) is evaluated. To see this, consider the pure advection ($D_x = 0$) of a conservative tracer in steady nonuniform flow. At node j, we admit a point source with a tracer massflow of m units. The fully implicit scheme representation of the conservative form, Equation (25), is then

$$A_j \frac{C_j^n - C_j^{n-1}}{\Delta t} + \frac{Q_{j+1} C_{j+1}^n - Q_{j-1} C_{j-1}^n}{2\Delta x} = \frac{m}{\Delta x} \quad (30)$$

The point source term on the right-hand-side follows from taking the source to be uniformly distributed over a cell surrounding node j, from $(j - \frac{1}{2})\Delta x$ to $(j + \frac{1}{2})\Delta x$. Equation (30) can be written as

$$A_j\Delta x(C_j^n - C_j^{n-1}) = \Delta t(F_{j-1/2}^n - F_{j+1/2}^n) + m\Delta t \quad (31)$$

where the flux term is given by

$$F_{j-1/2}^n = \frac{1}{2}(Q_{j-1}C_{j-1}^n + Q_j C_j^n) \qquad (32)$$

Equation (31) represents a mass balance statement for the cell; i.e., net mass stored in the cell during Δt = net mass advected into the cell during Δt + mass contributed by the point source. It therefore conserves mass. Note however that

Equation (32) is a poor estimator of flux; the flux at $(j - \frac{1}{2})\Delta x$ will be much closer to that at $(j - 1)\Delta x$ than to that at $j\Delta x$, where the point source has been incorporated. This poor flux estimate gives rise to false oscillations upstream of point sources. Local upstream differencing simply replaces Equation (32) by $F_{j-1/2}^n = Q_{j-1}C_{j-1}^n$ for the flux through the cell wall upstream of the point source. This flux estimate also must be used in the equation for the cell upstream of the source; if not, a spurious flux $[= \frac{1}{2}(Q_jC_j^n - Q_{j-1}C_{j-1}^n)]$ will be generated, destroying the mass balance.

The fully implicit version of the nonconservative form, Equation (29), is

$$\frac{C_j^n - C_j^{n-1}}{\Delta t} + U_j \frac{C_{j+1}^n - C_{j-1}^n}{2\Delta x} = \left(\frac{m}{\Delta x} - qC_\ell^n\right)\bigg/ A_j \quad (33)$$

where the concentration in the dilution term is evaluated at position $\ell\Delta x$, as yet unspecified. By noting that $Q_j - Q_{j-1} = q\Delta x$, we can write Equation (33) as

$$A_j\Delta x(C_j^n - C_j^{n-1}) = \Delta t(F_{j-1/2}^n - F_{j+1/2}^n)$$
$$+ m\Delta t + q\Delta x\Delta t(C_{j-1/2}^n - C_\ell^n) \quad (34)$$

where $C_{j-1/2}^n = \frac{1}{2}(C_{j-1}^n + C_j^n)$ and the flux is now estimated by $F_{j-1/2}^n = Q_{j-1}C_{j-1/2}^n$.

Equation (34) is the same form as Equation (31) except for the last term on the right-hand-side. This is another spurious flux; it will destroy the mass balance, unless set to zero. This is easily achieved by specifying $\ell = j - \frac{1}{2}$, i.e., $C_\ell^n = \frac{1}{2}(C_{j-1}^n + C_j^n)$, a choice that is not at all obvious from Equation (29). A similar analysis for upstream differencing of advective terms shows that in such schemes we should take $\ell = j - 1$.

It is not clear how large a mass falsification error can be introduced in practical applications of schemes derived from the nonconservative form, especially as the method of approximating the dilution term is seldom stated in the literature. We therefore recommend that, wherever practicable, schemes based on the conservative form should be used, and that some simple mass balance check be run for a single point source-conservative tracer test, to check on whether the scheme does conserve mass.

Lagrangian Schemes

These schemes are based on a difference equation written for a single segment of water as it moves downstream at velocity U. In contrast, more common Eulerian schemes are written for different segments of water moving past a fixed point on the stream bank. The advantage is that the resulting Lagrangian difference equation does not contain the troublesome advection fluxes that give rise to most of the numerical dispersion/false oscillation problems noted earlier. Lagrangian schemes are therefore ideal for handling steep concentration fronts. Solutions are generally very accurate.

Jobson [40] has devised an explicit Lagrangian scheme and computer code that offers excellent accuracy and versatility. The stability limit imposes a minimum time step of $\Delta t > 2D_x/U^2$. Successful field application has been reported [41].

McBride and Rutherford [56] have presented a simpler explicit or implicit Lagrangian scheme for nonuniform stream flow (called LAMBDA) and its successful field application. This scheme requires the user merely to specify the time step; it then calculates the grid distance steps, and the computations proceed in a straightforward manner. The implicit form avoids any stability restraints and also guarantees that no false oscillations will occur at point sources. The explicit form removes the need for any downstream boundary condition, which is generally not known in advance, but it too is subject to the restraint that $\Delta t > 2D_x/U^2$. It also facilitates incorporating nonlinear forms for sources and sinks; the difference equations remain linear. A computer code can be quickly assembled to produce accurate solutions using the LAMBDA scheme, and we recommend it as the starting point if a new stream pollution model is to be built. A brief discussion of this scheme for stream dissolved oxygen modeling is given at the end of this chapter.

MEASURES OF ORGANIC POLLUTION

Dissolved Oxygen

In most streams, dissolved oxygen (DO) is well distributed throughout the water column and often penetrates some way into the substratum. Typically the concentration of DO in natural surface water is less than $10 \text{ g} \cdot \text{m}^{-3}$ [2]. The oxygen dissolved in surface water is predominantly derived from the atmosphere and from the photosynthetic reactions of algae and higher aquatic plants. The total amount of oxygen that can be dissolved in water is dependent upon the water temperature, atmospheric pressure (which decreases with altitude), and salinity. The solubility of atmospheric oxygen in fresh water ranges from approximately 15 $\text{g} \cdot \text{m}^{-3}$ at $0°C$ to $8 \text{ g} \cdot \text{m}^{-3}$ at $25°C$ at sea level. Large reductions in DO concentration are usually caused by the microbial degradation of organic material and/or aquatic plant respiration. The oxidation of some inorganic compounds, such as ammonium, may also cause DO depletion in some situations.

Waters highly saturated with DO are acceptable for all uses except some industrial applications, where the presence of DO causes corrosion. If DO is very low, then anaerobic organisms may proliferate and produce unpleasant metabolites, such as sulphides, amines, and mercaptans. Anoxic conditions also lead to the release of inorganic substances, such as iron and manganese, which are known to produce unpleasant tastes and cause staining. A high DO concentration enhances the potability of water by precipitating many of these undesirable substances [8].

The importance of DO to aquatic organisms is well documented [33,52] and most guidelines for preserving natural aquatic ecosystems include a DO standard [20,21,67,90]. Oxygen requirements for aquatic life have been the subject of numerous investigations and several excellent reviews are available [1,13,93]. Although most of the work has concentrated on the oxygen requirements of fish, it appears that the requirements of fish and their forage animals are compatible [1]. Thus, the oxygen requirements of fish food organisms need not be evaluated separately when estimating the DO levels necessary for the protection of fish or fisheries.

The oxygen requirements of fish and other aquatic organisms vary considerably among species, with different life stages (eggs, larvae, and adults), and with the different life processes, e.g., feeding, growth, reproduction [1,51,61]. Environmental factors, such as pH, temperature and carbon dioxide, also influence DO requirements. Despite these sources of variation, a minimum DO concentration of 5 $g \cdot m^{-3}$ appears to ensure adequate protection for most sensitive life forms, provided that other environmental factors are favourable (1). Most minimum DO guidelines for the preservation of aquatic life fall within the range 4 to 6 $g \cdot m^{-3}$ [20,21,67,90].

Difficulties arise in formulating DO criteria for aquatic ecosystems because of the widely different patterns of DO fluctuations that can exist in streams. The changes due to added organic wastes must be assessed in relation to natural diurnal and seasonal variations associated with temperature, flow, and photosynthetic activity. The duration of low DO concentrations is also important; e.g., a low DO value occurring for one or two hours each day may not have the same effect as the same value held constant over a period of several days or weeks.

Three slightly different approaches to the formulation of DO standards are briefly described below. These represent the most recent level of informed opinion.

DOUDOROFF & SHUMWAY [15,89].

These authors recognise the need for differing levels of protection in different situations and relate their levels to existing seasonal DO minima in a given waterway. By providing several different levels of protection, they allow the decision maker a choice which is essentially a socio-economic one rather than a biological one. The criteria they suggest are summarised below.

Level	Protection
A	Maximum protection, unimpared productivity.
B_1	High level of protection for major spawning grounds of salmonids.
B	High level of protection but some risk of damage.
C	Moderate protection, some reduction in production expected.

Level	Protection
D	Low level of protection for unimportant fisheries. Permits persistance of tolerant species. Elimination of salmonids likely.

The DO levels pertaining to these classifications are presented graphically in Figure 8. The following example serves to illustrate this approach. Suppose we want to estimate the acceptable summer minimum DO concentration for the protection of a major salmonid spawning ground in a given stream (level B_1). Say the estimated natural summer DO minimum for the stream is 8.5 $g \cdot m^{-3}$; interpolation on Figure 8 gives an acceptable minimum summer DO value of approximately 8 $g \cdot m^{-3}$. Thus, if we decide to add organic wastes to the stream and still maintain the spawning ground, then efforts should be made to keep summer DO concentrations above 8 $g \cdot m^{-3}$.

DAVIS [13]

This author emphasises the importance of the oxygen pressure required to drive the dissolved gas across fish gills, as well as the amount available. He allows for three levels of protection:

Level	Protection
A	One standard deviation above the mean incipient oxygen response level for the group. There is little depression of oxygen from saturation, and a high degree of protection is assured for important fish stocks.
B	Based on the group mean, this level is where the average member of the community starts to exhibit distress. Some degree of risk to part of the community if the oxygen minimum is prolonged beyond a few hours.
C	One standard deviation below the group mean. A large proportion of the community affected by low oxygen. Deleterious effects may be severe if the minima are prolonged beyond a very few hours. To be applied only if fish populations are dispensable.

Table 4 summarises the above recommendations.

Davis notes that the application of his criteria on a nationwide basis is questionable. A more satisfactory approach would be to take regional variation into account.

ALABASTER & LLOYD [1]

These authors stress that it is inappropriate to put forward DO criteria based on a single minimum value never to be violated. Instead they adopt a probabilistic approach. They suggest that for resident populations of moderately tolerant

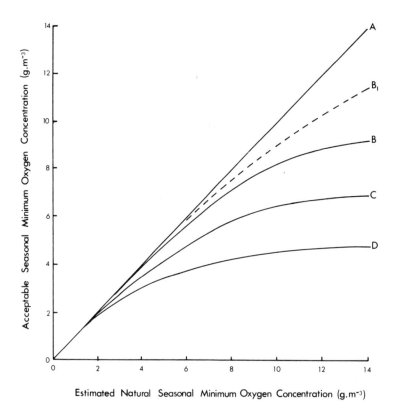

FIGURE 8. Proposed dissolved oxygen criteria for the protection of freshwater fisheries: curves relate "acceptable" seasonal dissolved oxygen minima, or minimum levels that are required for different specified levels of protection of fisheries, to estimated natural seasonal minima. Curves designated by A, B_1, B, C, and D correspond to the levels of protection described in text. From Warren, et al. [93].

freshwater species, such as roach, the annual 50- and 5-percentile dissolved oxygen values should be greater than $5 \text{ g} \cdot \text{m}^{-3}$ and $2 \text{ g} \cdot \text{m}^{-3}$, respectively, and for salmonids the percentiles should be $9 \text{ g} \cdot \text{m}^{-3}$ and $5 \text{ g} \cdot \text{m}^{-3}$, respectively. These suggestions are intended only for general guidance.

Oxygen Demand Tests

BIOCHEMICAL OXYGEN DEMAND

Biochemical oxygen demand (BOD) test is a measure of the amount of oxygen consumed by aerobic microorganisms in breaking down organic matter in a water sample. The more carbon broken down, the higher the BOD value. BOD is therefore not a pollutant, but an indirect measure of the microbial food source. BOD thus indicates the potential of a waterbody for deoxygenation. If there is a significant amount of ammonium in the sample, a nitrogenous oxygen demand may also be exerted as the ammonium is converted to nitrate by nitrifying bacteria.

BOD is usually reported as the amount of oxygen consumed in the dark over a specified incubation period and at a specified temperature. A 5-day time period at an incubation temperature of $20 \pm 1°C$ is the standard, and reported as BOD_5. In stream oxygen studies the BOD_5 has to be converted to the ultimate BOD. Several dilutions of the test sample are usually required before the BOD_5 test is performed. If this is not done, then all of the oxygen in the test sample may be consumed before the organic matter is completely degraded. The aim of the BOD_5 test is to make the carbon source, not the DO supply, the limiting factor. It is most important to use consistent methods in measuring BOD; e.g., test results will differ between stirred and unstirred samples or between the standard test and one using respirometers. It should also be noted that samples containing significant quantities of algae may give rise to false BOD results. Further details of the test and discussions of its limitations can be found in APHA [2], Gaudy [26], and Wilhm and Dorris [96].

Whereas DO relates directly to the survival of aquatic organisms, BOD relates more to aesthetic effects and the type of aquatic community which will be present. High BOD_5 values (i.e., >5 g $\cdot$ m^{-3}) are usually associated with high levels of suspended solids and/or sufficient carbon resources to promote nuisance microbial growths (such as sewage fungus). Values less than 3 g $\cdot$ m^{-3} are usually suitable for cold water fisheries, drinking water, irrigation and water contact recreation [21,45,49,67,73,81]. It is important to note, however, that BOD_5 values less than 3 g $\cdot$ m^{-3} can cause sewage fungus outbreaks if the carbon source consists largely of readily assimilable dissolved organic material, such as sucrose [17].

CHEMICAL OXYGEN DEMAND

Chemical oxygen demand (COD) is a measure of the amount of oxygen required to chemically oxidise the organic matter in a water sample. It is also referred to as dichromate oxygen demand, as a boiling mixture of potassium dichromate and sulphuric acid is used to oxidise the organic matter. The amount of oxidisable organic matter, measured as oxygen equivalents, is proportional to the potassium dichro-

mate consumed. An older method, of historical importance only, used potassium permanganate instead of dichromate.

The advantages of the COD test over the BOD_5 test are that it has greater precision as a test and may be completed in about 2 hours. The COD test, however, fails to include some simple straight chain alcohols and acids that are included in BOD_5 measurements, while including other organic compounds (e.g., cellulose) that are not so easily broken down microbially. Measurements of COD are almost always greater than corresponding measurements of BOD_5. Table 5 gives some typical values.

For a given effluent, a reasonable relationship between BOD_5 and COD may be established. However, the correlation can seldom be applied directly to another waste (even if the waste is of the same type) since any change in the fraction of the organic matter which is biodegradable will affect the correlation to the BOD_5 result. A comparison of COD and BOD_5 measurements is useful in that it gives some idea of the stability of the organic matter.

Total Organic Carbon

Total organic carbon (TOC) is composed of both dissolved and particulate organic carbon. It is often calculated

TABLE 4. Minimum Oxygen Criteria for Ecological Groups of Fish, with Three Levels of Protection. At the Lower Temperatures the Criteria are Derived from the Oxygen Partial Pressures Essential for Maintaining the Necessary Gradient Between Water and Blood for Proper Gas Exchange. At the Higher Temperatures the Criteria are Derived from the Oxygen Concentrations Necessary to Meet the Demands of Fish Respiration. From Davis [13].

| Group | Protection level | CRITERIA [expressed as percentage saturation at seasonal temperature maxima (°C)] | | | | | |
		0	5	10	15	20	25
Freshwater mixed	A	69	70	70	71	79	87
fish population	B	54	54	54	54	57	63
including salmonids	C	38	38	38	38	39	39
Freshwater mixed fish	A	60	60	60	60	60	66
populations with	B	47	47	47	47	47	48
no salmonids	C	35	35	35	35	35	36
Freshwater salmonid	A	76	76	76	76	85	93
population (including	B	57	57	57	59	65	72
steelhead)	C	38	38	38	42	46	51
Salmonid larvae and	A	98	98	98	98	100	100
mature eggs of	B	76	76	76	79	87	95
salmonids	C	54	54	57	64	71	78
Marine nonanadromous	A	88	88	95	100	100	100
species[1]	B	69	69	74	82	90	98
	C	50	51	51	55	60	65
Anadromous marine	A	100	100	100	100	100	100
species including	B	79	79	79	79	87	94
salmonids[1]	C	57	57	57	57	57	58

[1]Percentage saturation calculations based on salinity of 28°/₀₀.

TABLE 5. Comparative Strength of Effluents.[1]

Type of waste	Main pollutants	BOD_5 $(g \cdot m^{-3})$	COD $(g \cdot m^{-3})$
Abattoir	Suspended solids, protein	2600	4150
Beet sugar	Suspended solids, carbohydrate	850	1150
Cannery (meat)	Suspended solids, fat, protein	8000	17,940
Distillery	Suspended solids, carbohydrate, protein	7000	10,000
Domestic sewage[2]	Suspended solids, oil-grease, carbohydrate, protein	350	300
Pulp mill	Suspended solids, carbohydrate, lignin, sulphate	25,000	76,000
Tannery	Suspended solids, proteins, sulfide	2300	5100

[1]Adapted from Blakebrough [7].
[2]It is unusual for domestic sewage to have $COD < BOD_5$. In other cases the BOD_5/COD ratio may be 0.9 [16].

as the difference between total carbon, as measured on a carbon analyser, and total inorganic carbon.

Because the bulk of organic matter in water is comprised of humic substances and partly degraded plant and animal materials, which are often resistant to microbial degradation, measurements of stream water TOC do not usually relate well with BOD. Numerous workers have also found poor correlations between TOC and both BOD and COD for given effluent types [4].

Other Measures

A number of other measures may be used to assess aspects of stream pollution, including those for water appearance, nutrients (NH_4, NO_2, NO_3, PO_4) and biological indicators. These are covered elsewhere [2,39] and are not addressed in detail here.

MODELING STREAM DISSOLVED OXYGEN

Deterministic models of stream DO must account for the important processes operating, as summarised in Table 6. Tributary inflows of unpolluted water may provide significant DO and also dilute the BOD of polluted stream water. Polluted tributaries and waste inflows increase stream BOD and/or decrease stream DO. Reaeration is a physical process that occurs whenever stream water is depleted in oxygen; oxygen is transferred, by diffusion, from the atmosphere into the stream water. Aquatic BOD exertion refers to the oxygen consumed by planktonic organisms engaged in breaking down complex organic material to simple compounds. Oxygen consumption arises principally from the exertion of "carbonaceous" BOD in which organic carbon material is broken down, ultimately to carbon dioxide. In some situations a further "nitrogenous" oxygen demand is exerted by nitrogen compounds such as ammonia when they are oxidised to nitrate. Benthic BOD exertion refers to the action of organisms resident on the stream bed and banks. Benthic BOD supply from mud resuspension may be important in slow-flowing regions. Aquatic plants consume oxygen continuously in respiration but in the presence of sunlight also produce oxygen by photosynthesis. The result is a net production of oxygen during daylight, but a net consumption at night; consequently in some streams DO levels vary throughout the day being highest in late afternoon and lowest in early morning. Dispersion is important when inflows vary rapidly with time, for example, during a slug discharge of BOD.

Streeter-Phelps Model

This model was first developed in 1925 by Streeter and Phelps [84] following work carried out on the Ohio River, USA, and has been used many times since. It is a far-field model, which is justified because the maximum effect of a waste discharge on stream DO appears a considerable distance downstream from the inflow. In its simplest form the model predicts a "BOD decay curve" and a "DO sag curve" downstream from an inflow at the initial point. These are typified on Figure 9, which shows that the stream BOD continually decreases downstream from the initial point (where $t = 0$) and the stream DO sag reaches its maximum at time-of-travel t^* — the so-called "critical point" — where the stream DO is at a minimum. The extent of the sag at the critical point is usually of most concern since that is where environmental stress is greatest.

The Streeter-Phelps model assumes that the oxygen balance of any segment of water moving down a stream channel

TABLE 6. Important Stream Oxygen Transfer Process.

Always important	Sometimes important
Advection	Aquatic plant metabolism
Inflows	Benthic BOD supply from resuspension of mud
Rearation	Nitrogenous BOD exertion
BOD exertion	Dispersion

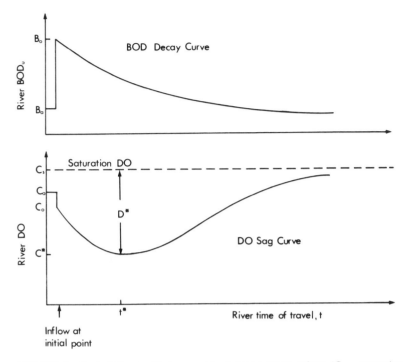

FIGURE 9. Typical Streeter-Phelps model solutions: single inflow (C_s = saturation DO; C^*, D^* = critical DO and DOD; B_a, C_a, B_0, C_0 = stream BOD_u and DO just above and just below the inflow). From McBride [53].

is the result of two major competing processes: removal of DO by exertion of BOD, and addition of DO by reaeration. The rate of exertion of BOD and the consequent decrease of DO is assumed to be proportional to the ultimate BOD concentration (BOD_u), the constant of proportionality being the deoxygenation coefficient, k_1. It is important to note that the rate of removal of DO is equal to the rate of exertion of the ultimate BOD, BOD_u, and it is necessary to estimate BOD_u from say BOD_5 measurements using a conversion factor (see later). The rate of reaeration is assumed to be proportional to the DO deficit ($DOD = C_s - C$), the constant of proportionality being the reaeration coefficient, k_2. Thus the basic Streeter-Phelps model has two important coefficients, k_1, and k_2. An increase in the value of k_1 corresponds to a decrease in the critical DO (C^* on Figure 9) and also an increase in the time-of-travel to the critical point (t^* on Figure 9). An increase in the value of k_2 corresponds to an increase in the critical DO and also a decrease in the time-of-travel to the critical point.

The model equations are

$$\frac{dL}{dt} = -k_1 L \tag{35}$$

$$\frac{dD}{dt} = \alpha k_1 L - k_2 D \tag{36}$$

with initial conditions

$$\left.\begin{array}{l} L = L_0 \\ D = D_0 \end{array}\right\} \text{at } t = 0$$

where

L = stream BOD_5, g · m^{-3}
D = stream DOD, g · m^{-3}
t = stream time-of-travel from the initial point, days
k_1 = stream deoxygenation coefficient (base e), days^{-1}
α = ratio of BOD_u:BOD_5, dimensionless
k_2 = stream reaeration coefficient (base e), days^{-1}

The values of t and α are defined by

$$t = \frac{x}{86.4\, U} \tag{37}$$

$$\alpha = \frac{1}{[1 - e^{-5k_L}]} \tag{38}$$

where

x = stream distance from the initial point, km

U = stream velocity, m · s^{-1}

k_L = BOD decay coefficient (base e) evaluated for the standard laboratory BOD test, days^{-1}

These equations can also be used for streams with nonuniform flow, or with multiple inflows. To do this the stream is subdivided into several reaches, the boundaries being located at each inflow and/or where a change in U, d, or T occurs. The model is then applied to each reach in turn starting at the most upstream point. All inflows are assumed to be constant (inflow rate and massflow do not vary with time). Even when this is not the case the model may still be useful provided time averages of inflow and stream data are used. Special modifications are required if the model predicts $D > C_s$, i.e., anoxic conditions [31]. The analytical solutions to Equations 35 and 36 are included in Table 7.

DATA REQUIREMENTS

The Streeter-Phelps model requires the following inputs and parameters to be specified by the user. The inputs may

TABLE 7. BOD-DO Model Equations and Solutions (From [30,64]).

Type	Differential equations	Analytical solutions
Streeter-Phelps (SP)	$BOD_5 : \dfrac{dL}{dt} = -k_1 L$	$L = L_o e^{-k_1 t}$
	$DOD : \dfrac{dD}{dt} = \alpha k_1 L - k_2 D$	$D = D_o e^{-k_2 t} + \dfrac{\alpha k_1 L_o}{k_2 - k_1}[e^{-k_1 t} - e^{-k_2 t}]$
SP + sedimentation	$BOD_5 : \dfrac{dL}{dt} = -k_r L : k_r = k_1 + k_3$	$L = L_o e^{-k_r t}$
	$DOD : \dfrac{dD}{dt} = \alpha k_1 L - k_2 D$	$D = D_o e^{-k_2 t} + \dfrac{\alpha k_1 L_o}{k_2 - k_r}[e^{-k_r t} - e^{-k_2 t}]$
SP + sedimentation + runoff + benthic demand + longitudinal dispersion	$BOD_5 : D_x \dfrac{d^2 L}{dx^2} - U\dfrac{dL}{dx} - k_r L + L_a = 0$	$L = L_o e^{mx} + \dfrac{L_a}{k_r}[1 - e^{mx}]$
	$DOD : D_x \dfrac{d^2 D}{dx^2} - U\dfrac{dD}{dx} + \alpha k_1 L - k_2 D + D_B = 0$	$D = D_o e^{rx} + \dfrac{\alpha k_1 [L_o - L_a/k_r]}{k_2 - k_r}[e^{mx} - e^{rx}]$
	where $k_r = k_1 + k_3$	$+ \left[\dfrac{D_B}{k_2} + \dfrac{\alpha k_1 L_a}{k_2 k_r}\right][1 - e^{rx}]$
		where $m = \{U - [U^2 + 4k_r D_x]^{0.5}\}/2D_x$
		$r = \{U - [U^2 + 4k_2 D_x]^{0.5}\}/2D_x$
SP + nitrogenous BOD	$CBOD_5 : \dfrac{dL}{dt} = -k_1 L$	$L = L_o e^{-k_1 t}$
	$NBOD : \dfrac{dN}{dt} = -k_N N$	$N = N_o e^{-k_N t}$
	$DOD : \dfrac{dD}{dt} = \alpha k_1 L + k_N N - k_2 D$	$D = D_o{}^{-k_2 t} + \dfrac{\alpha k_1 L_o}{k_2 - k_1}[e^{-k_1 t} - e^{-k_2 t}]$
		$+ \dfrac{k_N N_o}{k_2 - k_N}[e^{-k_N t} - e^{-k_2 t}]$

Definitions: $L = BOD_5(M/L^3)$; $D = DOD(M/L^3)$; $N = NBOD(M/L^3)$; subscript o means initial condition; $\alpha = BOD_u/BOD_5$; t = time of travel (T); k_1 = deoxygenation coefficient (T^{-1}, base e); k_2 = reaeration coefficient (T^{-1}, base e); k_N = nitrogenous deoxygenation coefficient (T^{-1}, base e); D_x = longitudinal dispersion coefficient (L^2/T); L_a = rate of addition of BOD_5 from runoff and/or bottom deposits ($M/L^3 T$); D_B = benthic oxygen demand ($M/L^3 T$).

be directly measured from field surveys and laboratory work. Estimation procedures for the model parameters (coefficients) are given later.

Inputs

Stream temperature (T); saturation DO (C_s–see Table 8); upstream rate of flow, BOD_5 and DO (Q_a, L_a, C_a); inflow rate of flow, BOD_5 and DO (Q_i, L_i, C_i); laboratory BOD decay coefficient (k_L, base e); cross-section mean velocity (U); and mean stream depth (d).

Parameters

Deoxygenation coefficient (k_1, base e); reaeration coefficient (k_2, base e).

Two points of caution must be made about the model coefficients k_1 and k_2. First, in the literature the coefficients may be quoted to base e or to base 10; sometimes the base is not identified. It is imperative that the correct base is identified; otherwise, gross errors will ensue. To illustrate, if k_2 is the reaeration coefficient to base e and K_2 is the coefficient to base 10, then $k_2 \cong 2.3\ K_2$. There is some considerable confusion in the literature on the base of coefficients (made worse by the lack of a uniform notation). For example, Eckenfelder [16] quotes three formulae for k_2; if one checks the original papers cited it is clear that the first formula is to base e, whereas the second and third are to base 10. Also, Fair, et al. [22] use a value of k_2 that may be shown to be to base 10 in a model that requires the coefficient to base e. This chapter deals exclusively with coefficients to the base e.

Second, the k_L and k_1 coefficients are distinct and should never be confused. The first describes BOD exertion in the BOD test bottle and is used to estimate BOD_u from BOD_5 data, while the second describes BOD exertion in the stream. Early literature on DO modelling [65] tended to use k_L as the stream deoxygenation coefficient, k_1, with some success. But this can be explained by the fact that early work was done on large rivers, such as the Ohio, which are dominated by aquatic respiration and in which one might expect k_1 and k_L to be similar. For smaller streams k_1 normally exceeds k_L because organisms on the bed and banks of the stream have a greater opportunity for "contact" with the organic matter in the stream water [97].

CONVERSION OF BOD_5 to BOD_u

The precise definition of the ultimate BOD is still a matter of some conjecture, but the following procedure is recommended. For a given k_L the ratio of BOD_u to BOD_5, α, required by the Streeter-Phelps model, can be read from Figure 10 [53]; then $B = \alpha L$ ($B = BOD_u$, $L = BOD_5$). A variety of methods for estimation of k_L have been developed and these are described in various texts [16,22,62,65,92]. These methods all employ a series of tests requiring the BOD of similar samples over different time intervals. Typical values are $k_L = 0.23$ day^{-1} for treated sewage,

$k_L = 0.35$ day^{-1} for primary treated Kraft pulp mill wastes, and $k_L = 0.5$–0.6 day^{-1} for primary treated meatworks wastes [55].

DEOXYGENATION COEFFICIENT, k_1

This coefficient may be expected to vary from stream to stream. As a first approximation it may be estimated from the formula presented by Wright and McDonnell [97]. In metric form this is

$$k_1 = 1.8Q^{-0.49}\ (\text{day}^{-1}) \qquad (39)$$

where Q = stream rate of flow in m$^3 \cdot$ s^{-1}. However it must be borne in mind that this equation is approximate and should only be used in preliminary "desk study" modeling. The actual value of k_1 will need to be calibrated from field data. For example, a value of $k_1 = 1$ day^{-1} was inferred by accurate modeling of DO in a stream flow at ~25 m$^3 \cdot$ s^{-1} [56], whereas Equation (39) would predict only a value of $k_1 = 0.37$ day^{-1}. Krenkel and Novotny [43] give an alternative procedure for calculating an absorption coefficient (B), from which, in our terminology, $k_1 = k_L + B$.

REAERATION COEFFICIENT, k_2

For desk studies this coefficient can be estimated satisfactorily using an empirical equation. Such equations have been derived from laboratory and field data and express k_2^{20} (at 20°C) as a function of stream velocity, U (m $\cdot$ s^{-1}), and depth, d (m). The equations listed here for k_2 all refer to 20°C, and values at other temperatures in the range 10–30°C may be calculated using

$$k_2 = 1.024^{T-20}k_2^{20} \qquad (40)$$

Because equations for k_2 are generally derived empirically, over limited ranges of flow conditions, care should be taken to ensure that the stream conditions match those used to derive the coefficient. The following equations may be used to calculate k_2^{20} (day^{-1}), in conjunction with Table 9, taken from Wilcock [94].

$$k_2^{20} = 3.87\ \sqrt{U}/d^{1.5} \quad [63] \qquad (41)$$

$$k_2^{20} = 5.01\ U^{0.969}/d^{1.673} \quad [11] \qquad (42)$$

$$k_2^{20} = 4.75\ U/d^{1.5} \quad [38] \qquad (43)$$

$$k_2^{20} = 5.13\ U/d^{1.33} \quad [46] \qquad (44)$$

Note that Equations (41–44) are valid only for U in units of metres per second and d in units of metres. Also, the constant 3.87 in Equation (41) is derived from O'Connor and Dobbins' equation $k_2 = \sqrt{UD_L}/d^{1.5}$, where D_L is the molecular diffusivity of oxygen in water. The best estimate of D_L at 20°C is $D_L = 2.01 \times 10^{-5}$ cm$^2 \cdot$ s^{-1} [34]. Recent

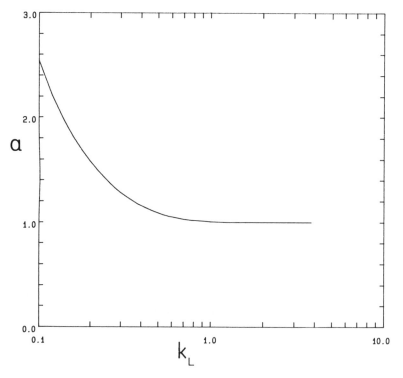

FIGURE 10. Graph of $BOD_u{:}BOD_5$ (α) versus k_L. From McBride [53].

work with gas tracers [95] has indicated that such equations may underestimate k_2 by up to 40%. For the time being, however, we recommend using Equations (41–44).

INCORPORATION OF INFLOWS

Each inflow is assumed to be immediately fully mixed with the stream water. The initial rate of flow, BOD_u and DO are then calculated from (using the notation on Figure 9)

$$Q_0 = Q_a + Q_i \tag{45}$$

$$B_0 = \frac{Q_a B_a + Q_i B_i}{Q_0} \tag{46}$$

$$C_0 = \frac{Q_a C_a + Q_i C_i}{Q_0} \tag{47}$$

CALCULATION OF DO SAG CURVE

Evaluation of the Streeter-Phelps analytical solution (Table 7) can be easily generated by programmable calculators [55], or by using available computer programs, especially the Texas Water Development Board's DOSAG [87]

and its metric upgrade, called DOSAGM, reported by Armstrong [3].

WORKED EXAMPLES

Example 1

Take a stream with temperature $T = 22°C$, so that the saturation DO is, from Table 8, $C_s = 8.74$ g $\cdot$ m^{-3}. The upstream rate of flow, BOD_5 and DO are $Q_a = 4.9$ m$^3 \cdot$ s^{-1}, $L_a = 1.3$ g $\cdot$ m^{-3}, and $C_a = 8.6$ g $\cdot$ m^{-3}, respectively. The inflow data are $Q_i = 0.1$ m$^3 \cdot$ s^{-1}, $L_i = 210$ g $\cdot$ m^{-3}, and $C_i = 3.6$ g $\cdot$ m^{-3}. For the stream water, $k_L = 0.4$ day^{-1}; so from Figure 10, $\alpha = 1.16$ and the upstream BOD_u is $B_a \cong 1.5$ g $\cdot$ m^{-3}. For the waste, take $k_L = 0.5$ day^{-1} so that from Figure 10, $\alpha = 1.09$, and the inflow BOD_u is $B_i = 1.09 \times 210 \cong 230$ g $\cdot$ m^{-3}. Equations (45–47) then give initial values of $Q_o = 5$ m$^3 \cdot$ s^{-1}, $B_o \cong 6$ g $\cdot$ m^{-3}, $C_0 = 8.5$ g $\cdot$ m^{-3}.

Assume that the stream deoxygenation coefficient has been calibrated at $k_1 \cong 2$ day^{-1}. The river velocity and depth are $U = 0.4$ m.s^{-1} and $d = 1.5$ m, for which Table 9 indicates that Equation (41) should be used to estimate the reaeration coefficient. Therefore, $k_2 = 1.024^2 \times 3.87 \times \sqrt{0.4}/1.5^{1.5} \cong 1.40$ day^{-1}.

Using the Streeter-Phelps solutions (Table 7), we obtain the following results (using a programmable calculator program [55]).

Stream Distance (km)	Stream BOD$_5$, L (g · m^{-3})	Stream DO, C (g · m^{-3})
0	5.17	8.50
5	3.87	7.19
10	2.90	6.45
15	2.17	6.11
19	1.72	6.03
20	1.63	6.02
21	1.53	6.03
25	1.22	6.09
30	0.91	6.26
40	0.51	6.71
50	0.29	7.18
60	0.16	7.58
70	0.09	7.90

Example 2

In this example take the same stream as above but also with (1) a meatworks waste inflow at 10 km where $Q_i = 0.15$ m^3 · s^{-1}, $L_i = 100$ g · m^{-3}, and $C_i = 0$ g · m^{-3}; (2) a change in stream depth from $d = 1.5$ m to $d = 1.8$ m, at 25

TABLE 8. Saturation DO Versus Temperature (From Benson and Krause [6]).

Temperature, T (°C)	Saturation DO (C$_s$) (g · m^{-3})
10	11.29
11	11.03
12	10.78
13	10.54
14	10.31
15	10.09
16	9.87
17	9.67
18	9.47
19	9.28
20	9.09
21	8.92
22	8.74
23	8.58
24	8.42
25	8.26
26	8.11
27	7.97
28	7.83
29	7.69
30	7.56

TABLE 9. k$_2$ Equation Numbers Appropriate for Different Stream Conditions.

Stream velocity U (m·s^{-1})	Stream depth, d(m)		
	0.2–0.5	0.5–1.0	>1.0
0.1–0.5	41	41	41
0.5–2.0	44	43	42

km; (3) a tributary inflow at 30 km where $Q_i = 1$ m^3 · s^{-1}, $L_i = 1$ g · m^{-3}, and $C_i = 8.7$ g · m^{-3}. The temperature of the tributary water is 22°C.

Before proceeding, we must calculate the BOD$_u$ for these new inflows. For the meatworks waste, we take a value of $k_L = 0.5$ day^{-1}; from Figure 10, $\alpha = 1.09$ and $B_i = 1.09 \times 100 = 109$ g · m^{-3}. For the tributary, we take the same k_L value as for the stream (0.4 day^{-1}) and from Figure 10, $\alpha = 1.16$, so that $B_i \cong 1.2$ g · m^{-3}. We must also compute the change in reaeration coefficient. Table 9 still indicates that Equation (41) should be used, so that the reaeration coefficient downstream from 25 km is

$$k_2 = 1.024^2 \times 3.87 \times \sqrt{0.4}/1.8^{1.5} \cong 1.06 \text{ day}^{-1}$$

Using the Streeter-Phelps solutions we obtain the following results.

Stream Distance (km)	Stream BOD$_5$, L (g · m^{-3})	Stream DO, C (g · m^{-3})
0	5.17	8.50
5	3.87	7.19
10	2.90	6.45
10*	5.55	6.27
15	4.16	5.26
20	3.11	4.81
24	2.47	4.71
25**	2.33	4.71
26	2.20	4.68
28	1.96	4.65
29	1.85	4.65
30	1.75	4.66
30*	1.63	5.31
35	1.22	5.36
40	0.91	5.51
50	0.51	5.97
60	0.29	6.48
70	0.16	6.95

*Refers to values calculated for full mixing of the inflow with the stream water
**Reaeration coefficient changed at 25 km to 1.06 day^{-1}

These results show that the effect of the meatworks inflow at 10 km is to about double the stream BOD$_5$ and slightly

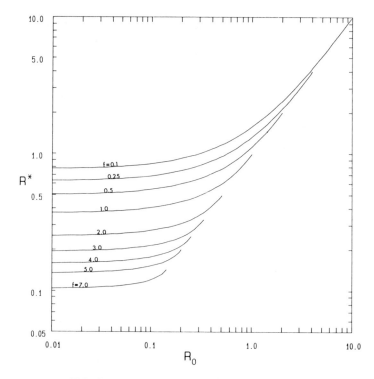

FIGURE 11. Deficit nomograph. From McBride [53].

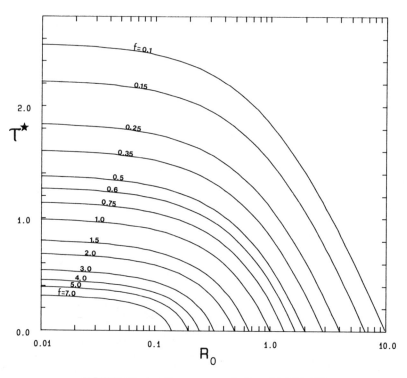

FIGURE 12. Location nomograph. From McBride [53].

reduce the stream DO at the point of inflow. Comparison with the single inflow case shows that the critical point is now located 25 km from the initial station and the DO has been reduced by about 1.3 g · m⁻³. The critical point occurs just upstream of the point where the stream depth increases to 1.8 m. The effect of this depth change is to create another critical point 5 km downstream. At the tributary inflow a substantial addition of DO is obtained by mixing and DO recovery is apparent.

NOMOGRAPHS FOR CALCULATION OF LOCATION AND MAGNITUDE OF CRITICAL OXYGEN SAG

In some cases it may be sufficient to calculate the location and magnitude of the critical DO sag caused by a particular waste inflow located at the initial point. This may be particularly so when an emergency waste discharge occurs or is contemplated. Simple nomographs may be used to perform these calculations rapidly [53], so long as two criteria are met: data required by the model (e.g., U, d, and T) are constant; and inflows downstream from the initial point can be neglected.

The procedure requires the values of α, k_1, k_2, D_0 ($= C_s - C_0$), L_0, and U. The steps are:

(a) Calculate the dimensionless "self-purification constant", f, from

$$f = k_2/k_1 \qquad (48)$$

(b) Calculate the initial "deficit-load ratio," R_o, from

$$R_0 = \frac{D_0}{\alpha L_0} \qquad (49)$$

(c) Using f and R_0 read the nomographs on Figures 11 and 12 to obtain values of R^* and τ^*.

(d) Calculate the critical DO from

$$C^* = C_s - R^* B_0 \qquad (50)$$

(e) Calculate the location of the critical point from

$$x^* = 86.4 \, \frac{U\tau^*}{k_1} \qquad (51)$$

where U is m · s⁻¹ and k_1 is in day⁻¹.

Example

Consider the single inflow case studied earlier, so that $C_s = 8.74°C$, $B_0 = 6$ g · m⁻³, $C_o = 8.5$ g · m⁻³, $k_1 = 2$ day⁻¹, $U = 0.4$ m · s⁻¹, and $k_2 = 1.40$ day⁻¹.

From the above procedure, $f = 1.40/2 = 0.7$ and $R_0 = (8.74 - 8.5)/6 = 0.04$. From Figures 11 and 12,

$R^* \cong 0.45$ and $\tau^* \cong 1.17$. Using Equations (50) and (51)

$$C^* = 8.74 - 0.45 \times 6 = 6.04 \text{ g} \cdot \text{m}^{-3}$$

$$x^* = 86.4 \times 0.4 \times 1.17/2 \cong 20.2 \text{ km}$$

These calculations agree with the calculations made in example 1 of the "Calculation of DO Sag Curve" section.

NOMOGRAPH FOR CALCULATION OF ASSIMILATIVE CAPACITY

The Streeter-Phelps stream DO model may be used to calculate the maximum allowable waste discharge in order to keep the downstream DO above a specified minimum level (C^*). This is the inverse problem noted earlier. For streams with multiple waste inflows, these maximum allowable discharges may be found by running the model with numbers of combinations of waste loads from each inflow. For a stream with a single waste inflow, located at the initial point, a simple nomograph approach is available [53], so long as data (U, d, T) are constant and downstream inflows can be neglected.

The procedure requires values of α, k_1, k_2, D_0 ($= C_s - C_0$), and D^* ($= C_s - C^*$). The steps are:

(a) Set the initial stream DO to the upstream DO, i.e., $C_0 = C_a$.

(b) Calculate the dimensionless "self-purification constant," f, from Equation (48).

(c) Calculate the dimensionless initial DO deficit, d_0, from

$$d_0 = \frac{D_0}{D^*} \qquad (52)$$

(d) Using f and d_o read the value of b_o from Figure 13.

(e) Calculate the allowable initial stream BOD_u from

$$B_0 = b_0 D^* \qquad (53)$$

(f) Calculate the allowable initial stream BOD_5 from

$$L_0 = B_0/\alpha \qquad (54)$$

Example

Consider again the single inflow case studied earlier. The minimum allowable stream DO is specified as 6 g · m⁻³. The given data are $C_a = 8.6$ g · m⁻³, $\alpha = 1.16$, $k_1 = 2$ day⁻¹ and $k_2 = 1.40$ day⁻¹. From step (a), we take $C_o = 8.6$ g · m⁻³.

From the procedure, then $f = 0.7$ and $d_o = (8.74 - 8.6)/(8.74 - 6) \cong 0.05$.

From Figure 13, $b_0 \cong 2.2$. Using Equations (53) and (54), $B_0 = 2.2 \times (8.74 - 6) \cong 6.0$ g · m⁻³ and $L_0 = 6.0/1.16 \cong 5.2$ g · m⁻³.

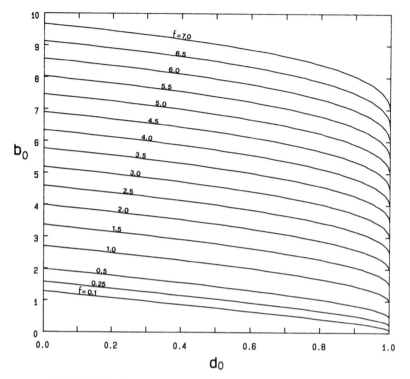

FIGURE 13. Assimilative capacity nomograph. From McBride [53].

Note that the predicted initial B_0 coincides with that specified in the earlier sag curve calculation example. This is as expected since the minimum DO calculated for that case (6.02 g · m⁻³ at 20 km) is very close to the specified minimum for this example ($C^* = 6$ g · m.⁻³). It should also be noted that this procedure, in step (a), assumes that $C_0 = C_a$. This assumption is necessary because C_0 cannot be specified in advance. However, for all but very large anoxic discharges it is a reasonable assumption.

FIELD DATA REQUIREMENTS FOR CALIBRATION AND CONFIRMATION

For model calibration it is necessary to undertake the following field work:

(a) Select a period of steady stream flow, usually a low flow when DO depletion is most noticeable.
(b) Where possible, select a period when inflows are steady in flow and composition. If this is not feasible (for example, because the inflow shows a marked diurnal variation), then it is necessary to make measurements at regular intervals over a long period (for diurnally varying inflows over at least 24 hours) and average the results obtained.
(c) Measure stream temperature over the reach being

modeled. It is desirable to make measurements at several locations and times and to average the results.
(d) Measure the average DO and BOD₅ at the upstream end of the reach.
(e) Measure the average flow, DO and BOD₅ of the inflows.
(f) Measure the average stream flow, mean depth and mean velocity.
(g) Measure the average DO and BOD₅ concentrations at a number of sites, at least four, downstream from the major inflows. These sites should be selected on the basis of preliminary model calculations so as to describe the DO sag and recovery as completely as practicable. Surface samples should be taken from as close to the thalweg as possible, using bridges, boats or even wading (on small streams). Do not sample within 20 × stream width of a major inflow.

There are no hard and fast rules about model calibration but the authors have found that the following procedure works fairly well:

(a) Using measured values of U, d, and T, estimate the reaeration coefficient, k_2.
(b) Predict stream DO concentrations using a range of k_1 values together with the k_2 value estimated in step (a). Select the value of k_1 which gives the best overall fit be-

tween observed and predicted DO concentrations. Goodness of fit can usually be gauged adequately by inspection of a graph of observed and predicted DO.

(c) Predict stream BOD_5 concentrations using the k_1 value estimated in step (b), and compare these with observed stream BOD_5 concentrations. If the observed and predicted BOD_5 concentrations match fairly closely then calibration has been achieved.

It should be noted that a failure to succeed in step (c) may be because stream BOD_5 concentrations do not always reflect accurately concentrations of oxidisable organic matter, especially where waste concentrations are low and large algal populations exist (say, greater than 10^4 cells.cm^{-3}). Consequently, it may not always be possible to achieve a good fit between observed and predicted stream BOD_5 concentrations. The user must exercise judgement in attempting to match observed and predicted BOD. Stream BOD_5 concentrations can normally be measured most accurately where concentrations are high (i.e., close to the outfall), but very close to an inflow (say, less than $20 \times$ stream width) mixing may be incomplete and a single sample may give a poor estimate of average concentrations.

Having undertaken calibration as described above, it is informative to make predictions of BOD and DO for a range of values of k_1 and k_2. In this manner, it is possible to gauge the sensitivity of predictions to uncertainties in the values of these model coefficients so that their likely range of values can be assessed.

In order to test whether or not the calibrated model can adequately predict DO depletion, attempts must be made to confirm the model. This necessitates conducting the same field work as for calibration under a different set of stream flow and/or inflow conditions. Model coefficients are then "frozen" at their calibrated values and predictions made of stream BOD and DO. These are then compared with observed stream BOD and DO concentrations, and, if the fit is considered acceptable, the model is said to be confirmed. If the fit is unacceptable, then it is advantageous to undertake a sensitivity analysis, which may indicate that a satisfactory match can be obtained using model coefficients somewhere within their likely range derived during calibration.

Failing this, it may be necessary to take account of other factors influencing stream DO and use a more complex modified Streeter-Phelps model.

Modifications to Streeter-Phelps Model

More complex modeling involves modifying the basic Streeter-Phelps model. A number of modifications have been made over the years and a good review of these (including differential equations and analytical solutions) has been given by Gromiec [30]. Some are given on Table 7. These modifications include the following.

SEDIMENTATION

Downstream of a poorly treated waste discharge, we may expect that some material may settle out, thus reducing stream BOD. This is modeled by adding a sedimentation coefficient, k_3, to k_1 in the BOD equation [Equation (35)], resulting in the removal coefficient $k_r = k_1 + k_3$. It is *not* added o k_1 in the DO equation [Equation (36)] because this BOD is assumed to be lost to the bed but not exerted. Its value may lie in the range -0.36 to 0.36 per day [88]; negative values describe resuspension of sediment.

BENTHIC OXYGEN DEMAND

This concerns the consumption of oxygen by benthic bacteria (on sediments, or in slimes). It is usually expressed on a streambed areal basis. Typically the magnitude of this demand is $0-10$ g $\cdot$ m^{-2} $\cdot$ d^{-1} (100). For fast flowing small streams with heavy infestations of sewage fungus, work with benthic chambers (C. W. Hickey,[1] *pers. comm.*) shows that the figure can be as high as 70 g $\cdot$ m^{-2} $\cdot$ d^{-1}. To obtain it as a per unit volume figure, the areal figure is divided by the depth.

In the basic Streeter-Phelps model both aquatic and benthic oxygen demand are modeled as a first-order process dependent on stream BOD concentration. Benthic oxygen demand can be modeled separately as a zero-order reaction; the prescribed per unit volume figure (D_B) is added to the DOD equation (see Table 7). The value of D_B at the time of a particular survey may be estimated from measured BOD and DO profiles [55] but before predictions can be made for different flows or BOD inputs it must be estimated, often a difficult task. It is probably more reliable, at least in "desk studies," to model benthic oxygen demand as a first order process, i.e., to use the basic Streeter-Phelps model.

LONGITUDINAL DISPERSION

Longitudinal dispersion is important for steady inflows only in the rare case that $U^2/kD_x \leq 100$, where $k = k_r$ or k_2. It is important for rapidly varying inflows [56].

NITROGENOUS OXYGEN DEMAND

For significant discharges of ammonia, oxygen may be consumed by nitrifying bacteria that convert the NH_4 into NO_2 and NO_3. In this process every gram of NH_4 converted to NO_3 consumes 4.33 grams of oxygen. In deep sluggish streams these nitrifiers tend to be planktonic, and then do not greatly influence the stream DO (A. B. Cooper,[1] *pers. comm.*), because their generation time is so long. However, benthic nitrifiers are prone to occur on the beds of swift shallow streams, and there exert a significant oxygen demand. From data presented by Lopez-Bernal, *et al.* [50] it can reach 15 g $\cdot$ m^{-2} $\cdot$ d^{-1}. Nitrogenous oxygen demand can be modeled as a first- or a zero-order reaction. First-

[1]Water Quality Centre, Ministry of Works and Development, Hamilton, New Zealand.

order is probably more reliable for prediction beyond the range of data used for model calibration/confirmation. First-order models have been reported by Miller and Jennings [60]; a simple model is given in Table 7.

PHOTOSYNTHESIS/RESPIRATION

Diurnal DO fluctuations caused by the photosynthesis of plants in the presence of sunlight, and their continuous respiration, can be incorporated into models [75]. This may include the effects of macrophytes, phytoplankton, and periphytes. The magnitude of these fluctuations can be used to infer the strength of a net photosynthesis term, as described elsewhere [2]. The photosynthesis term can be included by adding appropriate Fourier series terms to Equation (36) [30]. Calibration procedures for models including photosynthesis are given elsewhere [55].

NONLINEAR OXYGEN DEMAND KINETICS

The first-order descriptions of consumption of oxygen demand given in Table 7 may sometimes be inadequate to model the form of oxygen uptake, particularly if bacteria grow significantly during uptake. In this case, it may be necessary to construct ecological models, in which the oxygen uptake is modeled by the "Monod model" [71]. Ignoring advection, dispersion, sources, and sinks,

$$\frac{dS}{dt} = -\frac{\mu BS}{\beta + S} \tag{55}$$

$$\frac{dB}{dt} = -\gamma \frac{dS}{dt} \tag{56}$$

$$\frac{dC}{dt} = \delta \frac{dS}{dt} \tag{57}$$

where S is substrate concentration (proportional to BOD), B is bacteria concentration, β is a half saturation constant, C is DO, μ is the maximum specific growth rate of bacteria (day^{-1}), γ is the yield of bacteria produced per unit substrate consumed, and δ is oxygen consumed per unit of substrate consumed.

Monod stream models (including advection, dispersion, sources, and sinks) do not have analytical solutions; numerical solutions must be obtained [74].

MODEL CALIBRATION

The basic Streeter-Phelps model's simplicity means that its assumptions are transparent, the model is robust, and unique calibrations can usually be obtained with a modest amount of field work. As modifications are made extra parameters are included that must also be calibrated. As a result, the model's assumptions become less clear, and non-unique calibrations become possible.

We cannot offer a general procedure for calibrating

modified oxygen models over and above that already given for the basic Streeter-Phelps model, except to say that extra field work should be carried out to assess the magnitude for each extra parameter (e.g., use of benthic chambers to measure benthic oxygen demand), and that sensitivity tests *must* be employed to check the response of the model to changes in all of the important parameters.

NUMERICAL MODELS

All these modifications to the Streeter-Phelps equations have been incorporated into the QUALI model [88] and its subsequent variants [29]. These include an ability to model DO, BOD, temperature, nutrients, algae, and conservative materials. These models have been described excellently elsewhere [30,64]. Again, in using these models, one must beware of the presence of numerical dispersion.[2] Also, the facility to gain impressive-looking computer solutions should not be taken to mean that all problems of stream DO modeling have been solved. Careful calibration is still necessary.

As an alternative to using an established numerical model, an investigator may wish to construct his/her own model. This can be done rather simply with the accurate Lagrangian LAMBDA scheme of McBride and Rutherford [56]. For example, for uniform flow, the explicit version of this scheme downstream from an inflow for the Streeter-Phelps model with longitudinal dispersion is

$$C_j^n = \sigma C_j^{n-1} + (1 - 2\sigma - k\Delta t)C_{j-1}^{n-1} + \sigma C_{j-2}^{n-1} + g \tag{58}$$

where $\sigma = D_x \Delta t / \Delta x^2$ is a dispersion number. If, in Equation (58), $C \equiv$ BOD, then $k = k_1$, $g = 0$; if $C \equiv$ DOD then $k = k_2$ and $g = \alpha k_1 L$ ($L =$ BOD$_s$). This equation may be solved immediately to give the unknown C_j^n, requiring neither the tridiagonal algorithm nor a downstream boundary condition. Other modified Streeter-Phelps models can also be simply incorporated, including nonlinear Monod forms for oxygen uptake. The grid distance increments, Δx, are obtained by first specifying Δt, and then calculating $\Delta x = U\Delta t$. For stability, it is required that $\Delta t \geq 2D_x/U^2$. If this cannot be met, then an implicit form of the LAMBDA scheme must be used, requiring the use of the tridiagonal algorithm [69] and prescription of a downstream boundary condition. The extension to nonuniform flow is relatively straightforward [56].

SUMMARY

Stream pollutant modeling is potentially useful for water resource management. We recommend that a two-stage process be adopted: *first,* a simple "desk study" to determine the

---

[2]For uniform flow, the numerical dispersion coefficient is $U\Delta x$ $[1 + (U\Delta t/\Delta x)]/2$.

magnitude of the problem; *second,* a more detailed study, involving special data collection programs and possibly using more complex models, if that is deemed necessary. Modeling studies must proceed concurrently with the design and operation of data collection programs, to ensure that adequate model calibration/confirmation is obtained efficiently.

Considerable care is necessary in selecting or building models. Simple models are often the most satisfactory because they do not demand excessive data collection effort, they are easy to calibrate and test, they usually give unique solutions, and it is easy to describe the assumptions upon which they are based. If more complex models are to be used, in the hope that they will be more accurate and trustworthy for prediction, one must be aware of the danger of obtaining a non-unique calibration. If numerical models are to be used, the potential for numerical dispersion, false oscillation and mass conservation errors must first be investigated. We recommend consideration of Lagrangian schemes for problems with steep concentration gradients.

Recommendations are given for use of basic and modified Streeter-Phelps models of stream dissolved oxygen.

ACKNOWLEDGEMENTS

We thank our colleagues A. B. Cooper, C. W. Hickey, and R. J. Wilcock for useful information, and Mary Stokes for typing the manuscript. Permission to publish has been given by the Commissioner of Works, New Zealand.

REFERENCES

1. Alabaster, J. S., and R. Lloyd, *Water Quality Criteria for Freshwater Fish*, Butterworths, London, pp. 127–142 (1980).

2. American Public Health Association, *Standard Methods for the Examination of Water and Wastewater,* 15th ed., American Public Health Association, Washington, D.C. (1980).

3. Armstrong, N. E., "Development and Documentation of Mathematical Model for the Paraiba River Basin Study, Vol. 2—DOSAGM: Simulation of Water Quality in Streams and Estuaries," *Technical Report CRWR-145*, Centre for Research in Water Resources, The University of Texas at Austin, Texas (1977).

4. Aziz, J. A. and T. H. Y. Tebbutt, "Significance of COD, BOD, and TOC Correlations in Kinetic Models of Biological Oxidation," *Water Research*, Vol. 14, pp. 319–324 (1980).

5. Basta, D. J. and B. T. Bower, *Analyzing Natural Systems: Analysis for Regional Residuals—Environmental Quality Management,* Resources for the Future, Inc., Washington, D.C. (1982).

6. Benson, B. B., and D. Krause, "The Concentration and Isotopic Fractionation of Oxygen Dissolved in Freshwater and Seawater in Equilibrium with the Atmosphere," *Limnology and Oceanography,* Vol. 29, No. 3, pp. 620–632 (1984).

7. Blakebrough, N., *Biochemical and Biological Engineering Science,* Academic Press, London, pp. 318–319 (1967).

8. Bruvold, W. H. and R. M. Pangborn, "Dissolved Oxygen and the Taste of Water," *Journal of the American Water Works Association,* Vol. 62, pp. 721–722 (1970).

9. Chatwin, P. C., "The Cumulants of the Distribution of a Solution Dispersing in Solvent Flowing through a Tube," *Journal of Fluid Mechanics,* Vol. 51, No. 1, pp. 63–67 (1972).

10. Churchill, M. A. and R. A. Buckingham, "Statistical Method for Analysis of Stream Purification Capacity," *Sewage and Industrial Wastes,* Vol. 28, No. 4, pp. 517–537 (1956).

11. Churchill, M. A., H. L. Elmore, and R. A. Buckingham, "The Prediction of Stream Reaeration Rates," *Journal of the Sanitary Engineering Division,* ASCE, Vol. 88, No. SA4, pp. 1–46 (Aug. 1962).

12. Cleary, R. W., and D. D. Adrian, "New Analytical Solutions for Dye Diffusion Equations," *Journal of the Environmental Engineering Division,* ASCE, Vol. 99, No. EE3, pp. 213–227 (June 1973).

13. Davis, J. C., "Minimal Dissolved Oxygen Requirements of Aquatic Life with Emphasis on Canadian Species: A Review," *Journal of the Fisheries Research Board of Canada,* Vol. 32, pp. 2295–2332 (1975).

14. Di Toro, D. M., "Recurrence Relations for First Order Sequential Reactions in Natural Waters," *Water Resources Research,* Vol. 8, No. 1, pp. 50–57 (1972).

15. Doudoroff, P. and D. L. Shumway, "Dissolved Oxygen Requirements of Freshwater Fishes," *FAO Fisheries Technical Paper 86,* pp. 1–291 (1970).

16. Eckenfelder, W. W., Jr., *Water Quality Engineering for Practising Engineers,* Barnes and Noble, New York, N.Y. (1970).

17. Edelmann, W. and K. Wuhrmann, "Energy Balance of Running Water Systems," *Verhandlungen der Internationale Vereinigung fur Theoretische und Angewandte Limnologie,* Vol. 20, pp. 1800–1805 (1978).

18. Elder, J. W., "The Dispersion of Marked Fluid in Turbulent Shear Flow," *Journal of Fluid Mechanics,* Vol. 5, pp. 544–560 (1959).

19. Elhadi, N., A. Harrington, I. Hill, Y. L. Lau, and B. G. Krishnappan, "River Mixing—A State-of-the-Art Report," *Canadian Journal of Civil Engineering,* Vol. 11, pp. 585–609 (1984).

20. Environment Protection Authority of Victoria, *Manual of Recommended Water Quality Criteria,* Environment Protection Authority, East Melbourne, Australia (1983).

21. European Economic Community, "Council Directive of 18 July 1978 on the Quality of Fresh Waters Needing Protection or Improvement in Order to Support Fish Life", *Official Journal of the European Communities, No. L222/1–No. L222/10* (1978).

22. Fair, G. M., J. C. Geyer, and D. A. Okun, *Water and Wastewater Engineering Vol. 2—Water Purification and*

Wastewater Treatment and Disposal, John Wiley and Sons, New York, N.Y. (1968).

23. Fischer, H. B., "The Mechanics of Dispersion in Natural Streams," *Journal of the Hydraulics Division,* ASCE, Vol. 93, No. HY6, pp. 187–215 (Nov. 1967).

24. Fischer, H. B., "Longitudinal Dispersion and Turbulent Mixing in Open-Channel Flow," *Annual Review of Fluid Mechanics,* Vol. 8, pp. 59–78 (1973).

25. Fischer, H. B., J. Imberger, E. J. List, R. C. Y. Koh, and N. H. Brooks, eds., *Mixing in Inland and Coastal Waters,* Academic Press, New York, N.Y. (1979).

26. Gaudy, A. F., Jr., "Biochemical Oxygen Demand," *Water Pollution Microbiology,* R. Mitchell, ed., Wiley-Interscience, New York (1972).

27. Gaudy, A. F., Jr., "Prediction of Assimilation Capacity in Receiving Streams," *Water and Sewage Works,* pp. 62–65 (May 1975); pp. 78–79 (June 1975).

28. Gray, W. G., and G. F. Pinder, "An Analysis of the Numerical Solution of the Transport Equation," *Water Resources Research,* Vol., 12 No. 3, pp. 547–555 (June 1976).

29. Grenney, W. J., M. C. Teuscher, and L. S. Dixon, "Characteristics of the Solution Algorithms for the QUALII River Model," *Journal of the Water Pollution Control Federation,* Vol. 50, No. 1, pp. 151–157 (Jan. 1978).

30. Gromiec, M. J., "Biochemical Oxygen Demand – Dissolved Oxygen: River Models," *Application of Ecological Modelling in Environmental Management, Part A,* S. E. Jorgensen, ed., Elsevier Scientific Publishing Co., Amsterdam, pp. 131–225 (1983).

31. Gundelach, J. M. and J. E. Castillo, "Natural Stream Purification under Anaerobic Conditions," *Journal of the Water Pollution Control Federation,* Vol. 48, No. 7, pp. 1753–1758 (July 1976).

32. Harden, T. O. and H. T. Shen, "Numerical Simulation of Mixing in Natural Rivers," *Journal of the Hydraulics Division,* ASCE, Vol. 105, No. HY4, pp. 393–408 (Apr. 1979).

33. Hawkes, H. A., "Biological Surveillance of Rivers," *Journal of Water Pollution Control,* Vol. 81, pp. 329–342 (1982).

34. Himmelblau, D. M., "Diffusion of Dissolved Gases in Liquids," *Chemical Reviews,* Vol. 64, pp. 527–550 (1964).

35. Hines, W. G., D. A. Rickert, and S. W. McKenzie, "Hydrological Analysis and River-Quality Data Programs," *Journal of the Water Pollution Control Federation,* Vol. 49, No. 9, pp. 2031–2041 (Sep. 1977).

36. Holley, E. R. and D. R. F. Harleman, "Dispersion of Pollutants in Estuary Type Flows," *Hydrodynamics Report No. 74,* Massachusetts Institute of Technology, Cambridge, Mass. (1965).

37. Holly, F. M., Jr., "Two-dimensional Mass Dispersion in Rivers," *Hydrology Paper No. 78,* Colorado State University, Fort Collins, Colorado.

38. Isaacs, W. P., and A. F. Gaudy, Jr., "Atmospheric Oxidation in a Simulated Stream," *Journal of the Sanitary Engineering Division,* ASCE, Vol. 94, No. SA2, pp. 319–344 (Apr. 1968).

39. James, A. and L. Evison, *Biological Indicators of Water Quality,* Wiley, Chichester, England (1979).

40. Jobson, H. E., "Temperature and Solute-Transport Simulation in Streamflow Using a Lagrangian Reference Frame," *Water Resources Investigations 81-2,* United States Geological Survey, NSTL Station, Miss. (1981).

41. Jobson, H. E. and R. E. Rathbun, "Use of the Routing Procedure to Study Dye and Gas Transport in the West Fork Trinity River, Texas," *Water-Supply Paper 2252,* United States Geological Survey, Alexandria, Va. (1984).

42. Keefer, T. N. and H. E. Jobson, "River Transport Modeling for Unsteady Flows," *Journal of the Hydraulics Division,* ASCE, Vol. 104, No. HY5, pp. 635–647 (May 1978).

43. Krenkel, P.A. and V. Novotny, "River Water Quality Model Construction," *Modeling of Rivers,* H. W. Shen, ed., Wiley, New York, pp. 17-1–17-22 (1979).

44. Kuo, E. Y. T., "Analytical Solution for 3-D Diffusion Model," *Journal of the Environmental Engineering Division,* ASCE, Vol. 102, No. EE4, pp. 805–820 (Aug. 1976).

45. L' Assemblée Federale de la Confederation Suisse, "Ordinance for Waste Water Discharge," *814.225.21,* pp. 1–25 (8 December, 1975).

46. Langbein, W. B. and W. H. Durum, "The Aeration of Streams," *Circular No. 542,* United States Geological Survey, Washington, D.C. (1967).

47. Lau, Y. L., and B. G. Krishnappan, "Transverse Dispersion in Rectangular Channels," *Journal of the Hydraulics Division,* ASCE, Vol. 103, No. HY10, pp. 1173–1189 (Oct. 1977).

48. Lau, Y. L. and B. G. Krishnappan, "Modeling Transverse Mixing in Natural Streams," *Journal of the Hydraulics Division,* ASCE, Vol. 107, No. HY2, pp. 209–226 (Feb. 1981).

49. Lester, W. F., "River Quality Objectives," *Journal of the Institute of Water Engineers and Scientists,* Vol. 33, pp. 429–450 (1979).

50. Lopez-Bernal, F. F., P. A. Krenkel, and R. J. Ruane, "Nitrification in Free-Flowing Streams," *Progress in Water Technology,* Vol. 9, pp. 821–832 (1977).

51. Macan, T. T., *Freshwater Ecology,* John Wiley & Sons, New York, N.Y. (1974).

52. Mason, C. F., *Biology of Freshwater Pollution,* Longman, London (1981).

53. McBride, G. B., "Nomographs for Rapid Solutions for the Streeter-Phelps Equations," *Journal of the Water Pollution Control Federation,* Vol. 54, No. 4, pp. 378–384 (Apr. 1982).

54. McBride, G. B., "Incorporation of Point Sources in Numerical Transport Schemes," *Water Resources Research,* Vol. 21, No. 11, pp. 1791–1795 (Nov. 1985).

55. McBride, G. B. and J. C. Rutherford, "Handbook on Estimating Dissolved Oxygen Depletion in Polluted Rivers," *Water and Soil Miscellaneous Publication No. 51,* Ministry of Works and Development, Wellington, New Zealand (1983).

56. McBride, G. B. and J. C. Rutherford, "Accurate Modeling of River Pollutant Transport," *Journal of Environmental Engineering,* Vol. 110, No. 4, pp. 808–827 (Aug. 1984).

57. McQuivey, R. S. and T. N. Keefer, "Simple Method for Predicting Dispersion in Streams," *Journal of the Environmental Engineering Division, ASCE,* Vol. 100, No. EE4, pp. 997–1011 (Aug. 1974).

58. McQuivey, R. S. and T. N. Keefer, "Convective Model of Longitudinal Dispersion," *Journal of the Hydraulics Division, ASCE,* Vol. 102, No. HY10, pp. 1409–1424 (Oct. 1976).

59. McQuivey, R. S. and Keefer, T. N., "Dispersion-Mississippi River Below Baton Rouge, La.," *Journal of the Hydraulics Division, ASCE,* Vol. 102, No. HY10, pp. 1425–1437 (Oct. 1976).

60. Miller, J. E. and M. E. Jennings, "Modeling Nitrogen, Oxygen, Chattahoochee River, Ga.," *Journal of the Environmental Engineering Division, ASCE,* Vol. 105, No. EE4, pp. 641–653 (Aug. 1979).

61. Nebeker, A. V., "Effect of Low Oxygen Concentration on Survival and Emergence of Aquatic Insects," *Transactions of the American Fisheries Society,* Vol. 101, pp. 675–679 (1972).

62. Nemerow, N. L., *Scientific Stream Pollution Analysis,* McGraw-Hill, New York, N.Y. (1974).

63. O'Connor, D. J. and W. E. Dobbins, "Mechanisms of Reaeration in Natural Streams," *Transactions of the American Society of Civil Engineers,* Vol. 123, pp. 641–684 (1968).

64. Orlob, G. T., ed., *Mathematical Modelling of Water Quality—Streams, Lakes and Reservoirs,* Wiley, New York, N.Y. (1983).

65. Phelps, E. B., *Stream Sanitation,* John Wiley and Sons, New York, N.Y. (1944).

66. Ponce, V. M., Y. H. Chen, and D. B. Simons, "Unconditional Stability in Convection Computations," *Journal of the Hydraulics Division, ASCE,* Vol. 105, No. HY9, pp. 1079–1086 (Sept. 1979).

67. Price, D. R. H. and M. J. Pearson, "The Derivation of Quality Conditions for Effluents Discharged to Freshwaters," *Journal of Water Pollution Control,* Vol. 78, pp. 118–138 (1979).

68. Prych, E. A., "Effects of Density Differences on Lateral Mixing in Open Channels," *Report No. KH-R-21,* California Institute of Technology, Pasadena, California (1970).

69. Remson, I., G. M. Hornberger, and F. J. Molz, *Numerical Methods in Subsurface Hydrology,* Wiley-Interscience, New York, N.Y., pp. 168–170 (1971).

70. Rickert, D. A., "Use of Dissolved Oxygen Modeling Results in the Management of River Quality," *Journal of the Water Pollution Control Federation,* Vol. 56, No. 1, pp. 95–101 (1984).

71. Rinaldi, S., R. Soncini-Sessa, H. Stehfest, and H. Tamura, *Modeling and Control of River Quality,* McGraw-Hill, New York, N.Y., pp. 80–90 (1979).

72. Roache, P. J., *Computational Fluid Dynamics,* 3rd ed., Hermosa Publishers, Albuquerque, N.M. (1976).

73. Royal Commission on Sewage Disposal, "8th report", Vol. II, Appendix Pt II, *Cmd. 6973,* HMSO, London (1913).

74. Rutherford, J. C. and M. J. O'Sullivan, "Simulation of Water Quality in Tarawera River," *Journal of the Environmental En-*

gineering Division, ASCE, Vol. 100, No. EE2, pp. 369–390 (Apr. 1974).

75. Rutherford, J. C., "Modeling Effects of Aquatic Plants in Rivers," *Journal of the Environmental Engineering Division, ASCE,* Vol. 103, No. EE4, pp. 575–591 (1977).

76. Rutherford, J. C., "Investigations of Mechanisms Affecting BOD Concentrations in the Manawatu River near Palmerston North," *Internal Report No. 79/24,* Water Quality Centre, Ministry of Works and Development, Hamilton, New Zealand (1979).

77. Rutherford, J. C., "Handbook on Mixing in Rivers," *Water and Soil Miscellaneous Publication No. 26,* Ministry of Works and Development, Wellington, New Zealand (1981).

78. Rutherford, J. C., M. E. U. Taylor, and J. D. Davies, "Waikato River Pollutant Flushing Rates," *Journal of the Environmental Engineering Division, ASCE,* Vol. 106, No. EE6, pp. 1131–1150 (Dec. 1980).

79. Sayre, W. W., "Natural Mixing Processes in Rivers," *Environmental Impact on Rivers (River Mechanics III),* H. W. Shen, ed., H. W. Shen, Fort Collins, pp. 6-1–6-37 (1973).

80. Schaller, F. W. and W. Bailey, eds., *Agricultural Management and Water Quality,* Iowa State University Press, Ames, Iowa (1983).

81. Sladecek, V., "Continental Systems for the Assessment of River Water Quality," *Biological Indicators of Water Quality,* A. James and L. Evison, eds., Wiley-Interscience, Chichester, England (1977).

82. Sparr, T. M., "A Verification of the QUAL-I Water Quality Model for the Lower Mississippi River," *Water Resources Bulletin,* Vol. 15, No. 3, pp. 853–860 (June 1979).

83. Stone, H. L. and P. L. T. Brian, "Numerical Solution of Convective Transport Problems," *Journal of the American Institute of Chemical Engineers,* Vol. 9, No. 5, pp. 681–688 (Sept. 1963).

84. Streeter, H. W. and E. B. Phelps, "A Study of the Pollution and Natural Purification of the Ohio River," *Public Health Bulletin 146,* United States Public Health Service, Washington, D.C. (Feb. 1925).

85. Taylor, G. I., "Dispersion of Soluble Matter in Solvent Flowing Slowly through a Tube," *Proceedings of the London Mathematical Society, Series A,* Vol. 219, pp. 186–203 (1953).

86. Taylor, G. I., "The Dispersion of Matter in Turbulent Flow Through a Pipe," *Proceedings of the Royal Society of London, Series A,* Vol. 223, pp. 446–468 (1954).

87. Texas Water Development Board, "DOSAG-I Simulation of Water Quality in Streams and Canals," Program Documentation and Users Manual, prepared by Systems Engineering Division, Texas Water Development Board, Austin, TX (1970).

88. Texas Water Development Board, "Simulation of Water Quality in Streams and Canals. Theory and Description of the QUALI Modelling System," *Report 128,* Texas Water Development Board, Austin, TX (1971).

89. United States Environmental Protection Agency, "Water Qual-

ity Criteria 1972," *EPA-R3-73-033,* United States Environmental Protection Agency, Washington, D.C. (1973).

90. United States Environmental Protection Agency, "Quality Criteria for Water," *EPA-440/9-76-023,* United States Environmental Protection Agency, Washington, D.C. (1976).

91. Valentine, E. M., "Effects of Boundary Channel Roughness on Longitudinal Dispersion," thesis presented to the University of Canterbury, New Zealand, in 1978, in partial fulfillment of the requirements for the degree of Doctor of Philosophy.

92. Velz, C. J., *Applied Stream Sanitation,* Wiley-Interscience, New York, N.Y. (1970).

93. Warren, C. E., P. Doudoroff, and D. L. Shumway, "Development of Dissolved Oxygen Criteria for Freshwater Fish," *EPA-R3-73-019,* Environmental Protection Agency, Washington, D.C. (1973).

94. Wilcock, R. J., "Simple Predictive Equations for Calculating Stream Reaeration Rate Coefficients," *New Zealand Journal of Science,* Vol. 25, pp. 53–56 (1982).

95. Wilcock, R. J., "Methyl Chloride as a Gas-Tracer for Measuring Stream Reaeration Coefficients—II, Stream Studies," *Water Research,* Vol. 18, No. 1, pp. 53–57 (1984).

96. Wilhm, J. S. and T. C. Dorris, "Biological Parameters for Water Quality Criteria," *Bioscience,* Vol. 18, pp. 477–480 (1968).

97. Wright, R. M. and A. J. McDonnell, "In-Stream Deoxygenation Rate Prediction," *Journal of the Environmental Engineering Division,* ASCE, Vol. 105, No. EE2, pp. 323–335 (Apr. 1979).

98. Yeh, G. T. and Y. J. Tsai, "Analytical Three-Dimensional Modelling of Effluent Discharges," *Water Resources Research,* Vol. 12, No. 3, pp. 533–540 (June 1976).

99. Yotsukura, N. and F. A. Kilpatrick, "Tracer Simulation of Soluble Waste Concentration," *Journal of the Environmental Engineering Division,* ASCE, Vol. 99, No. EE4, pp. 499–515 (Aug. 1973).

100. Zison, S. W., W. B. Mills, D. Deimer, and C. W. Chen, "Rates, Constants, and Kinetics Formulations in Surface Water Quality Modeling," *EPA-600/3-78-105,* United States Environmental Protection Agency, Athens, Ga. (1978).

CHAPTER 23

Determination of Hazardous Waste Concentration as a Result of Accidental Discharge

PETER SEREICO*

Spills of hazardous chemicals have become a nationwide concern because of the possible risk to health and public safety caused by the accidental discharge of toxic materials into the environment. This concern will increase as the amount and kinds of chemical toxics being transported increases. With the occurrence of every accidental discharge episode there results an accompanying emotional outburst most likely to be completely out of proportion to the potential risks incurred.

Barges sinking in rivers, railroad tankcars accidentally discharging toxicants, and pesticides entering the water supplies are events which have occurred and are most likely to repeat. The concern is what will be the concentration of the discharged material "x" miles downstream away from the point of discharge and would that concentration pose an environmental hazard to the population involved.

Toxicants can be classified into two categories, those with a specific gravity greater than one and which sink, and those with a specific gravity less than one and which float on the water surface. Many toxicants are organic compounds which are only slightly soluble in water (less than 1%) or soluble (greater than 1%). In any event, at the point of discharge the solubility limits will be exceeded and no doubt greatly exceed existing water quality standards for that chemical. As the chemical spreads due to natural diffusion or convection it will gradually mix to include the body of water cross section and expose all aquatic biota within the water column and water bed to its effects. As the distance from the spill increases the toxicant concentration will decrease due to mixing plus chemical and physical processes taking place within the water environment.

Mathematical models will be presented that attempt to predict a concentration away from the point of discharge dependent on water conditions and the chemical and physical characteristics of the discharged pollutant. Water body conditions include reaeration, dispersion, and transport velocity. The chemical and physical properties include volatilization, biooxidation, photolysis, hydrolysis, and bioconcentration.

After a substance is discharged into a body of water events take place on the material which alters its concentration. Distinction must be made if the material is conservative or non-conservative. Conservative materials undergo no change in concentration due to chemical or biological reactions, or physical processes. These include salt solutions, suspended solids, etc. Non-conservative materials are capable of concentration changes in time. Organic materials are able to oxidize, evaporate, degrade, undergo photolysis, hydrolyze, and undergo biological decomposition. Radioactive materials also undergo constant decomposition. As these concentration changes are not instantaneous it is important to determine the temporal rates of change of these materials in their water environment as a means to predict their concentrations with time. A measure of the decomposition rate is given by k, the first order rate constant for a particular material.

The manner in which the material is discharged is equally important. Material can be discharged in one huge mass assuming that no time elapses during discharge. This condition is called instantaneous loading. The other situation is if the loading is continuous, around the clock, which is called continuous loading. For ease of discussion and calculations most continuous loading situations are assumed to occur under constant conditions.

The receiving body of water can be quiescent, as in a pond, pool, or reservoir, or fast moving, as in a stream or river. Discharge can also be into an estuary which is subject to cyclic tidal actions which cause materials to ebb and flow in a periodic manner.

*Division of Life Sciences, University of New England, Biddeford, ME

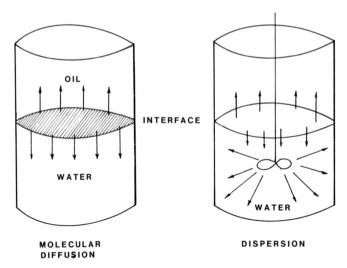

FIGURE 1. Diffusion and dispersion in an oil-water mixture.

Dispersion can be of two types, molecular or eddy. As an example consider an oil layer placed on a water surface making an interface between the two liquids. Natural molecular tendencies cause molecular penetration of molecules into the two phases and a measure of this movement can be described by a molecular diffusion term, D_x. If, on the other hand, an agitator is turned on there would be greater dispersal, both natural and mechanical, causing greater penetrations and higher concentrations of each material into the different phases. A measure of this movement is the dispersal coefficient, D_L.

Five possible cases can be presented based on the conditions described. These include:

CASE I
diffusion
no convection
instantaneous discharge
conservative,
  non-conservative

CASE II
diffusion
convection
instantaneous discharge
conservative,
  non-conservative

CASE III
diffusion
convection
continuous discharge
conservative

CASE IV
no diffusion
convection
continuous
non-conservative

CASE V
diffusion
convection
continuous
non-conservative

In CASES I–IV to change the condition from conservative to non-conservative the concentration expression is multiplied by the exponential raised to a power containing the decomposition rate constant, k, i.e., e^{-kt}.

The expressions for the different cases are the following:

CASE I
$$C = \frac{W}{\sqrt{4\pi D_x t}}\, e^{-\frac{x^2}{4D_x t}} \tag{1}$$

CASE II
$$C = \frac{W}{A\sqrt{4\pi D_x t}}\, e^{-\frac{(x - ut)^2}{4D_x t}} \tag{2}$$

CASE III $\quad C = W/UA$ downstream

$$\tag{3}$$

$$C = W/UA(e^{-\frac{ux}{D_x}})\ \text{upstream}$$

CASE IV $\quad C = \dfrac{W}{Au}\, e^{-kx/u} \tag{4}$

CASE V $\quad C = \dfrac{W}{Au\sqrt{M}}\, e^{\frac{u}{2D_x}}(1 \pm M)x - M$, downstream

$$\tag{5}$$

$$M = \sqrt{1 + 4\,kD_x/u^2} \qquad\qquad +M,\ \text{upstream}$$

where

C = concentration
W = loading, lb/day
t = time, days

D_x = diffusion coefficient, ft²/day
u = stream velocity, ft/sec.
x = distance, feet
A = cross-sectional area, ft²
k = decomposition constant, 1/day

Therefore, the fate of a material discharged into a body of water depends on its nature, how it is discharged, its dispersion potential, and the nature of the water it is being discharged into.

The focus of this chapter is to quantify the various decay mechanisms; biodegradation, hydrolysis, photolysis, evaporation, and oxidation by determining a rate constant describing that mechanism and modifying the expressions given in cases I to V to include these terms. Once each specific rate constant is determined they can be combined using the principle of superposition; i.e., the overall k is the sum of the individual k's.

$$k_{overall} = k_{ox} + k_e + k_h + k_p \ldots \quad (6)$$

where

k_{ox} = specific rate constant, oxidation
k_e = specific rate constant, evaporation
k_h = specific rate constant, hydrolysis
k_p = specific rate constant, photolysis

where all the rate constants are expressed in 1/day units.

The nature of each rate constant will be investigated. Thus, if a compound is inadvertently discharged and the k values can be determined it is possible to predict a concentration of a material as a function of time.

PHOTOLYSIS

Photolysis is the reaction of an organic compound with radiant energy sufficient to initiate a chemical reaction. Organic compounds absorb radiation in the ultraviolet regions to produce compounds with smaller molecular weight fractions or induce molecular changes. The rate at which a pollutant photolyzes depends on 1) intensity of solar radiation, 2) light absorption properties of the material, 3) photoreactivity of the material, and 4) depth of water. Also needed are values of solar radiation and the wavelengths of light responsible for photolytic changes.

Figure 2 shows the mean daily annual solar radiation expressed in langleys. Monthly charts are also available from the U.S. Department of Commerce.

Photochemical reactions are initiated by absorbing light but not all the energy absorbed is capable of producing a chemical reaction. Part of the absorbed energy may be lost to other mechanisms and the portion remaining, available for photolysis, is the quantum yield, $\varnothing_d$. Quantum yield is defined as:

$$\varnothing_d = \frac{\text{amount of chemical reacted}}{\text{amount of light absorbed}} \quad (7)$$

Quantum yield data is shown in Table 1.

When an organic pollutant is discharged into a body of water it could form two layers and rest on the surface, it could sink and cover the bottom, or if it is miscible, it could form a homogeneous mixture with no change in concentration as a function of depth. The body of water could also contain suspended solids and dissolved materials. As radiation penetrates deeper into water it can be absorbed and scattered by these contaminents and by the water itself. A term called the light attenuation coefficient, K, is a measure of these energy losses and is defined as:

$$K = D[a_w + (a_c \cdot C_c) + (a_o \cdot C_o) (a_{ss} \cdot C_{ss})] \quad (8)$$

where

D = radiation factor
a_w = absorptivity of water
a_c = absorptivity of chlorophyll-a
C_c = concentration of chlorophyll-a
a_o = absorptivity of organic material
C_o = concentration of organic material
a_{ss} = absorptivity of suspended solids
C_{ss} = concentration of suspended solids

If the body of water contains chlorophyll and suspended solids and dissolved organics with known concentrations the absorptivity can be determined as a function of wavelength in the range of 300 to 500 nm. Once K is determined it can be used to calculate the direct photolysis rate constant, k_d. An expression for the direct photolysis rate constant can be derived as:

$$k_d = 2.3(j)(\varnothing_d)(D) \int_{\lambda_o}^{\lambda} e(W) \frac{1 - e^{-Kz}}{Kz} d\lambda \quad (9)$$

where

j = conversion factor, 1.43×10^{-16} mole · cm³ · sec/ liter · day
$\varnothing_d$ = quantum yield
D = radiation factor, 1.6
e = molar extinction coefficient, liter/mol · cm
W = photon irradiance at the surface, photons/cm² · sec · nm
z = depth of water
K = diffuse light attenuation coefficient for water, 1/meter

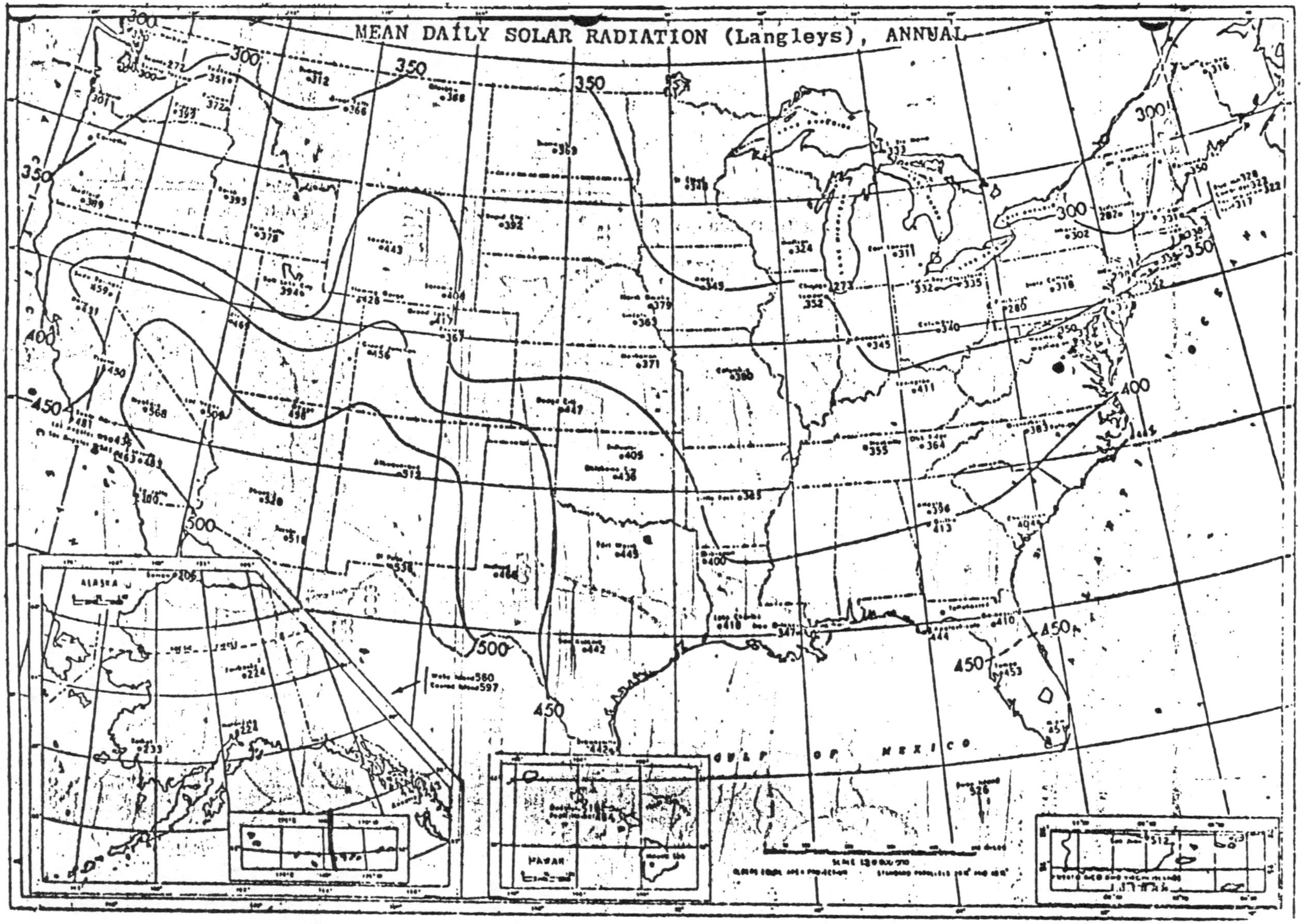

FIGURE 2. Solar radiation in the United States.

TABLE 1. Near Surface Direct Photolysis Rate Constants.

Compound	k_{do} 1/day	I_o langley/day	λ nm	$\emptyset_d$
Polycyclic Aromatics				
Napthalene	0.23	2100	310	0.015
1-Methylnapthalene	0.78	2100	312	0.018
2-Methylnapthalene	0.31	2100	320	0.005
Phenanthrene	2.0	2100	323	0.010
Anthracene	22.0	2100	360	0.003
9-Methylanthracene	130.0	2100	380	0.008
9,10-Dimethylanthracene	48.0	2100	400	0.004
Pyrene	24.0	2100	330	0.002
Fluoranthrene	0.79	2100	—	0.0001
Chrysene	3.8	2100	320	0.003
Napthecene	490.0	2100	440	0.013
Benzo(a)pyrene	31.0	2100	380	0.003
Benzo(a)anthracene	28.0	2100	340	0.001
Carbamate Pesticides				
Carbaryl	0.32	2100	313	0.006
Propham	0.003	740	—	
Chlorpropham	0.006	740	—	
Phthlate Esters				
dimethyl ester	0.005	600	—	0.031
diethyl ester	0.005	600	—	
di-n-butyl ester	0.005	600	—	
di-n-octyl ester	0.005	600	—	
di-(2-ethylhexyl) ester	0.005	600	—	
2, 4-D Esters				
butoxyethyl ester	.05	420	—	0.056
methyl ester	.03	420	—	0.031
Hexachlorocyclopentadiene	94.0	540	—	
Pentachlorophenol	0.46	600	—	
3,3'-dichlorobenzidine	670.0	2000	318	
N-nitrosoatrazine	300.0	1800	—	0.30
Trifluralin	30.0	1800	—	0.002
DMDE	17.0	2200	—	0.30

TABLE 2. Water Surface Irradiation Values.

Wavelength nm	Light Attenuation Coefficients 1/mg • meter				Photon Spectral Radiance 10^{14} photons/cm²sec • nm	
λ	a_w	a_c	a_o	a_{ss}	$W(\lambda)$	$W(\lambda)d\lambda$
300	0.141	69	6.25	0.35	.003	.03
310	0.105	67	5.41	0.35	.039	.39
320	0.0844	63	4.68	0.35	.113	1.13
330	0.0678	61	4.05	0.35	.181	1.81
340	0.0561	58	3.50	0.35	.211	2.11
350	0.0463	55	3.03	0.35	.226	2.26
360	0.0379	55	2.62	0.35	.241	2.41
370	0.0300	51	2.26	0.35	.268	2.68
380	0.0220	46	1.96	0.35	.294	2.94
390	0.0191	42	1.69	0.35	.366	3.66
400	0.0171	41	1.47	0.35	.526	5.26
410	0.0162	39	1.27	0.35	.692	6.92
420	0.0153	38	1.10	0.35	.712	7.12
430	0.0144	35	0.95	0.35	.688	6.88
440	0.0145	32	0.821	0.35	.814	8.14
450	0.0145	31	0.710	0.35	.917	9.17
460	0.0156	28	0.614	0.35	.927	9.27
470	0.0156	26	0.531	0.35	.959	9.59
480	0.0176	24	0.460	0.35	.983	9.83
490	0.0196	22	0.398	0.35	.930	9.30
500	0.0257	20	0.344	0.35	.949	9.49
510	0.0357	18	0.297	0.35	.962	9.62
520	0.0477	16	0.257	0.35	1.00	10.0
550	0.0638	10	0.167	0.35	1.04	10.4
600	0.244	6	0.081	0.35	1.07	10.7
650	0.349	8	—	0.35	1.08	10.8
700	0.650	3	—	0.35	1.07	10.7
750	2.47	2	—	0.35	1.03	10.3
800	2.07	0	—	0.35	0.98	9.9

Many times experimental data for direct photolysis is reported as a near surface rate constant independent of the properties of the water it is measured in. This value, k_{do}, can be modified to k_d by the following expression:

$$k_d = k_{do}(I/I_o)(D/D_o) \cdot 1 - e \frac{-K(\lambda^*)z}{K(\lambda^*)z} \quad (10)$$

where:

I = total solar irradiation, langleys/day
I_o = total solar radiation under which k_{do} was measured, langleys/day
λ^* = wavelength of maximum light absorption
D_o = radiation factor for water, 1.2

If the molar extinction coefficient, e, and the quantum yield are known it is possible to evaluate the integral in (9)

by the following modification:

$$k_d = 2.3 \, (j) \, (\varnothing_d) \, (D)\Sigma e(W') \cdot 1 - \frac{e^{-Kz}}{Kz} \quad (11)$$

where W' = photon irradiance, from Table 2.

To compute the values of the integral a table can be constructed with the following headings:

$$\varnothing, \, D, \, a_w, \, a_c, \, C_c, \, a_o, \, C_o, \, a_{ss}, \, C_{ss}, \, K, \, e, \, W' \, Kz, \, eW'/Kz$$

An example later will illustrate this procedure.

HYDROLYSIS

Certain organic compounds can react with water to alter the original compound. These reactions are called hydroly-

TABLE 3. Hydrolysis Rate Parameters.

Compounds	Hydrolysis Rate Values, 1/day			Environmental Rate Constants, 1/d
	k_a	k_n	k_b	k_H
Pesticides				
Endosulfon	—	—	3.3×10^5	3.5×10^{-2}
Heptachlor				0.7
Carbaryl			4.3×10^5	4.3×10^{-2}
Propham			0.66	6.6×10^{-8}
Chlorpropham			1.7	1.7×10^{-7}
2, 4-D	1.7		2.6×10^6	0.26
Parathion	130.0	.0036	2.4×10^3	3.9×10^{-3}
Captan		1.6	4.9×10^7	5.6
Atrazine	3.4	6.6		6.6
Methoxychlor	—	0.0026	31	2.6×10^{-3}
Halogenated HC's				
Chloromethane		0.0021	0.53	2.1×10^{-3}
Bromomethane		0.0035	12	3.5×10^{-2}
Chloroethane		0.0180		1.8×10^{-2}
Dichloromethane			1.8×10^3	2.8×10^{-6}
Trichloromethane			6.0	6.0×10^{-7}
Tribromeethane			28	2.6×10^{-6}
Halogenated Ethers				
Bis(chloromethyl) ether		1600		1.6×10^3
2-chloroethyl vinyl ether	380			3.8×10^{-5}
Phthalate Esters				
Dimethyl ester			6.0×10^3	6.0×10^{-4}
Diethyl ester			1.9×10^3	1.9×10^{-4}
Di-n-butyl ester			9.1×10^2	9.1×10^{-5}

sis reactions and usually the products are less complex and have smaller molecular weights. The general form of a hydrolysis reaction is:

$$RY + H_2O \rightarrow ROH + HY \qquad (12)$$

Examples of hydrolysis reactions include:

$$\text{starch } + \text{ water } \rightarrow \text{ glucose}$$

$$\text{fats } + \text{ water } \rightarrow \text{ stearic acid}$$

$$\text{p-nitro phenyl acetate } + H_2O \rightarrow \text{ p-nitrophenol } + \text{ acetic acid}$$

The rate of hydrolysis is temperature dependent and also is a function of pH. Hydrolysis products may be more or less light sensitive and are usually biodegradable. The

hydrolysis rate is assumed to be a first order reaction and can be expressed as:

$$\frac{dC}{dt} = k_h C \qquad (13)$$

k_h = first order hydrolysis constant, 1/day

But, k_h can be defined as:

$$k_h = k_n + k_a[H^+] + k_b[OH^-] \qquad (14)$$

where

k_n = neutral hydrolysis rate constant, 1/day
k_a = acid medium hydrolysis rate constant, 1/day
k_b = base medium hydrolysis rate constant, 1/day
$[H^+]$ = hydrogen ion concentration, mol/liter
$[OH^-]$ = hydroxide ion concentration, mol/liter

Values for k_n, k_a, k_b are shown in Table 3.

VOLATILIZATION

Volatilization is the ability of a substance to transfer from the dissolved state to the gaseous state. Molecules can acquire sufficient kinetic energy from ambient environmental conditions to break the bonds holding them to solvent molecules. The rate of volatilization is assumed to be first order and is computed in terms of the Henry's Law constant.

Henry's Law relates the partial pressure of a gaseous substance in the dissolved phase and in the atmosphere at equilibrium. It can be expressed as:

$$K_H' = C_a/C_w \qquad (15)$$

where

K_H' = Henry's Law constant, unitless
C_a = equilibrium concentration of pollutant in air, mol/m³
C_w = equilibrium concentration of pollutant in water, mols/m³

In terms of the partial pressure, P, it can be expressed as:

$$K_H = P/C_w \qquad (16)$$

where K_H = Henry's Law constant, atm · m³/mole.

If the values of the partial pressure and saturation concentration are known and using the Ideal Gas Relationship, K_H can be expressed as:

$$K_H' = 16(P/C_s)(M/T) \qquad (17)$$

where

C_s = saturation concentration, mol/m³
M = Molecular weight, gm/mol
T = temperature, degrees Kelvin

Once a value for the Henry's Law constant is known the rate of volatilization can be determined by this expression:

$$dC/dt = -k_v'(C - P/K_H) \tag{18}$$

where

k_v = volatilization rate, 1/day
C = concentration of pollutant

If the partial pressure of the pollutant in the atmosphere is low Equation (18) reduces to:

$$dC/dt = -k_v C \tag{19}$$

Finally, using the Ideal Gas Relationship K_H' can be expressed as a function of K_H as:

$$K_H' = 41.6\ K_H \tag{20}$$

The rate of volatilization can be determined by considering that the volatilization taking place from the liquid media and penetrating two distinct films at the water surface. This is shown in the figure and the process is described as the Two Film Penetration theory. The pollutant must pass thru a liquid film and then through a gas film before entering the atmosphere. The rates of penetration through the phases are

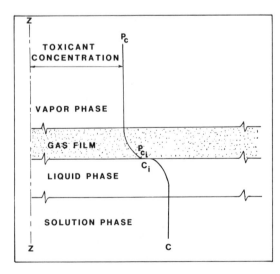

FIGURE 3. Two-film method diagram.

k_g and k_1. It has been shown that the rate of volatilization is equal to:

$$1/k_v = 1/k_1 + 1/K_H' k_g \tag{21}$$

If K_H' is large the second term on the right of expression (21) becomes small and $k_v = k_1$. This indicates that the rate of volatilization is limited by the rate passing through the liquid phase. If K_H' is small then $k_v = K_H'$ indicating that the rate of volatilization is limited passing through the gaseous phase. The limits set on Henry's Law constant are the following:

• liquid phase resistance occurs when $K_H' > 10^{-3}$
  atm · m³/mol

• gas phase resistance occurs when $K_H < 10^{-5}$
  atm · m³/mol

Another method to predict k_v involves knowing the diffusivity of a vapor through air and water to determine k_g and k_1 values. These relationships are:

$$k_g = D_g/v_g(0.0008)W \tag{22}$$

$$k_1 = D_1/v_1(0.0008)(W) \tag{23}$$

where

D_g = 1.9 $M^{-2/3}$
v_g = diffusivity of air, 0.15 cm²/sec
W = wind speed, meter/sec
D_1 = 22 × 10^{-5} $M^{-2/3}$
v_1 = diffusivity through water, 0.01 cm²/sec

If diffusivities are not known they can be approximated by using:

$$D_g/D_1 = (18/m)^{1/2} \tag{24}$$

Because many organic compounds are toxic to microorganisms biodegradation would not be an important decay mechanism. If these rates are low they may be masked by other processes taking place.

If enzymes are present in the water they may be capable of transforming the discharged organics into other molecular forms. While not really being a decay mechanism it does alter the original concentration and provides materials which may be more or less toxic than the original organic compound. However, this depends on the specific starting compound.

Depending on the compound discharged not all the described mechanisms may be taking place. Compounds may hydrolyze but may not be volatile, etc. Also, values for one specific process may be smaller than the others and would not be significant in the final value for the decay rate.

Also, if a reaction rate is small and the other processes are small it can be significant in the evolution of the final reaction rate constant. These are thoughts to keep in mind when attempting to determine the final rate constant to use in mathematical models presented.

EXAMPLE

On September 9th, 1980, a barge containing 10,000 lbs. of napthalene accidently discharged its cargo into the Illinois River at Peoria. Predict the concentration of nathalene at these locations:

Havana, Ill.	40 miles away
Beardstown, Ill.	70 miles away
Alton, Ill.	160 miles away

The following data pertain to napthalene and to the river:

Napthalene		River	
molecular weight	128	depth	3 meters
formula	$C_{10}H_8$	suspended solids	10 mg/l
vapor pressure, @ 25°C	1.5×10^{-4}atm		
solubility	32 mg/l	organics	5 mg/l
k_{do}	0.23/day	chlorophyll	1 mg/l
I_o	2100 L	area	100 ft²
λ	310 nm	D_x	100 ft²/hr
$\varnothing_d$	0.015	Wind Speed	4m/sec
k_b	0.14/day		

Napthalene is a bicyclic aromatic hydrocarbon which absorbs solar radiation at wavelengths above 300 nm. It is nonpolar white solid and not expected to undergo appreciable hydrolysis. Although toxic to microorganisms in the beginning after a sufficient lag time organisms are able to degrade the compound. The processes expected to take place are photolysis, vaporization, and biodegradation.

SOLUTION

From Table 2 the values of a_w, a_c, a_o, and a_{ss} are determined and using Equation (8) give:

$$K = 1.6[0.105 + 67(1) + 5.41(5) + .35(10)] = 162$$

Equation (10), from Figure 2 at Peoria where $I = 350$ langleys, gives

$$k_d = 0.23 \ (350/2100)(1.6/1.2)(1/162 \cdot 3) = 1 \times 10^{-4}/\text{day}$$

As a confirmation using the integral method, from Table 2 determine W and W' given λ.

λ	ϵ	$W' \times 10^{14}$
300	918	0.03
310	356	0.39
320	101	1.13
330	11	1.81

λ	D	a_w	+	a_cC_c	+	a_oC_o	+	$a_{ss}C_{ss}$	K	ϵ	W'	KZ	W'/KZ
300	1.6	.141		69		31.25		7	172	918	.03	512	.05
310	1.6	.105		67		27.05		7	162	356	.39	485	.29
320	1.6	.084		63		23.4		7	150	101	1.13	450	.76
330	1.6	.008		61		20.25		7	141	11	1.81	424	.05
													1.15 $\times 10^{14}$

$$k_d = 2.3 \ (1.43 \times 10^{-16})(350/540)(0.015)(1.6)(10^{-14})$$

$$= 6 \times 10^{-5}/\text{day}$$

With these low volatilization rates it can be assumed that photolysis may not be an important decay mechanism.

VOLATILIZATION

Napthalene is a compound capable of sublimation, i.e., it is able to pass from a solid to the vapor phase. Therefore, volatilization will be an important mechanism.
Equation (17):

$$K_H' = \frac{(16)(1.5 \times 10^{-4})(760)(128)}{(33)(290)} = 0.037$$

To determine k_v both k_g and k_l will be determined:
Equations (22) and (23):

$$k_g = \frac{(1.9)(128)^{-.67}(0.0008)(4)(86400)}{0.15} = 136 \text{ m/day}$$

$$k_l = \frac{(22 \times 10^{-5})(128)^{-.67}(0.0008)(4)(86400)}{0.01} = 0.235 \text{ m/d}$$

Equation (21):

$$1/k_v = 1/0.235 + 1/(136)(0.037)$$

$$k_v = 4.53 \text{ m} = 1.5/\text{day}$$

Therefore, $k_{overall}$ from Equation (6) equals:

$$k_{overall} = 0.14 + 1.5 + \delta + 1 \times 10^{-4}$$

$$k_{overall} = 1.64/\text{day}$$

The sinking of the barge can be considered an instantaneous load and the conditions given would indicate a case II situation: diffusion, convection, instantaneous loading, and a non-conservative material. Equation would apply:

$$C = \frac{10,000 \text{ lb/day}}{100 \text{ ft}^2 \ 4} (100 \text{ ft}^2/\text{hr})(1.67 \text{ day})e^{-}$$

$$\frac{(40 \ m - ut)^2}{4(100)(1.67)} e^{-k_o(1.67)}$$

For the maximum concentration to occur at the 40 mile point the first exponential term must approach zero. To do this the term $(x - ut)^2$ is zero at the location specified and the expression becomes:

$$C = \frac{10,000 \text{ lb/day}}{100 \ 4 \ (100)(24)(1.67)} e^{-(1.64)(1.67)}$$

$C = 462$ mg/l at Havana, Illinois

Location	Concentration
Havana	462 mg/l
Beardstown	60 mg/l
Alton	7 mg/l

CHAPTER 24

Physical Treatment Processes in Water and Wastewater Treatment

J. A. McCorquodale* and J. K. Bewtra*

INTRODUCTION

The conventional water and wastewater treatment practices include the removal of impurities which may be present in suspended, colloidal or dissolved forms. The common physical methods used in the removal or separation of these impurities are sedimentation, filtration, air stripping and carbon adsorption. The practice of flow equalization before or during treatment is becoming popular, particularly in wastewater treatment.

The success of any physical treatment process depends upon the nature of substances to be removed. These substances, organic or inorganic in nature, may exist in liquid as colloidal, suspended or dissolved impurities. Their characteristics may range from inert substances like grit or turbidity to complex toxic trace organics and heavy metals. Both the flow rates and the concentrations of substances reaching a treatment plant may vary considerably from hour to hour. Typical plots of hourly variations in flow rate, suspended solids concentration and the mass of suspended solids reaching West Windsor Pollution Control Plant are shown in Figures 1, 2 and 3 [36]. Similar plots for Delhi Wastewater Treatment Plant are shown in Figures 4 and 5 [13].

The sedimentation process, with or without chemical coagulation and flocculation can separate the inert colloidal and suspended solids, phosphates, organic matter and certain toxic substances from the influent. The filtration process is employed for the removal of finely divided impurities carried over from the clarification process. Air stripping has been used advantageously for the removal of dissolved gases such as hydrogen sulfide and ammonia. Most recently, carbon adsorption process has been applied for the removal of trace organic impurities from water.

The most common and effective unit process of sedimentation, along with flow equalization unit operation, is discussed in detail in this article. The new design approaches based on most recent research and scientific developments have been incorporated.

PROCESS DESCRIPTION

Flow Equalization

Both domestic and industrial wastewaters show considerable diurnal variations and it is useful to significantly dampen these variations in inflow to relieve hydraulic overload on both the biological and physical-chemical units. Flow equalization smooths out the variations in the influent characteristics.

Flow equalization basins are basically flow-through or side-line holding tanks and their capacity is determined by plotting inflow and outflow mass curves. These tanks are generally located after preliminary treatment and should be designed as completely-mixed basins, using either diffused air or mechanical surface aerators, to prevent settling of suspended impurities. If decomposable organic matter is present in the wastewater, aeration will prevent septicity. The preaeration can also reduce the biochemical oxygen demand on subsequent treatment units.

The flow equalization basins can also be used to neutralize the acidity or alkalinity in incoming wastewater. The neutralizing chemicals are added to the inflow wastewater stream before entering into the flow equalization basins and the retention period in these basins provides sufficient time for reaction. Any precipitates produced during neutralization are separated in subsequent sedimentation basins.

*Department of Civil Engineering, University of Windsor, Windsor, Ontario, Canada

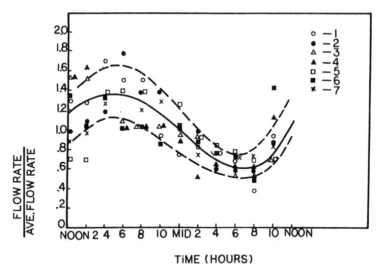

FIGURE 1. Hourly variations in flow rate—Windsor [36].

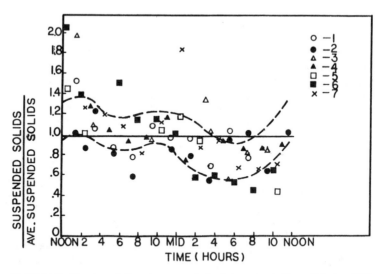

FIGURE 2. Hourly variations in suspended solids concentration—Windsor [36].

664

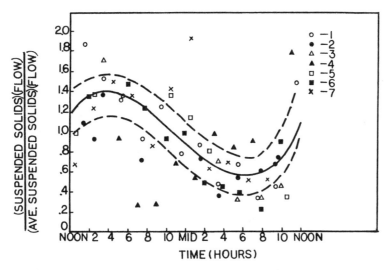

FIGURE 3. Hourly variations in suspended solids mass—Windsor [36].

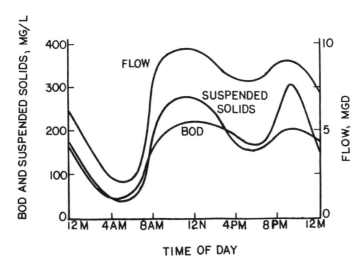

FIGURE 4. Hourly variations in flow rate and suspended solids concentration—Delhi [13].

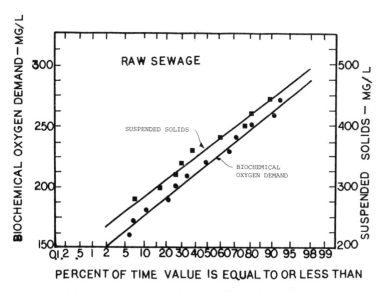

FIGURE 5. Frequency distribution of SS and BOD—Delhi [13].

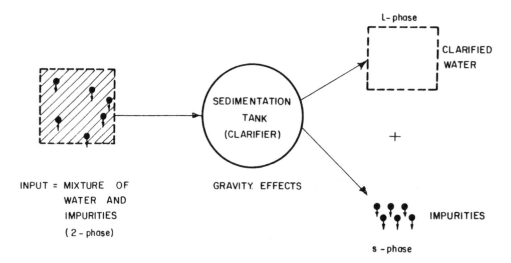

FIGURE 6. Clarifiers as solid-liquid separators [65].

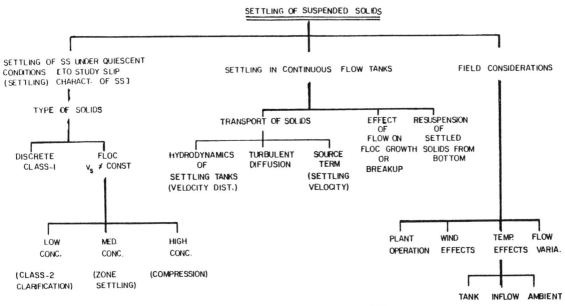

FIGURE 7. Settling of suspended solids [65].

Sedimentation

In sedimentation tanks, the mixture of water and impurities enters from the inlet zone and, as it proceeds towards the effluent zone, the suspended solids settle downward to the sludge zone due to the gravitational effect, as shown in Figure 6 [65]. Most clarifiers operate on a continuous flow basis. The factors affecting the removal process can be divided into three groups as shown on Figure 7 [65]: (i) the characteristics of the suspended solids and the transporting liquid; (ii) the hydrodynamic parameters in continuous flow operation; and (iii) the field conditions.

SETTLING CHARACTERISTICS OF SUSPENDED SOLIDS

The settlement of particles from a suspension depends on the suspension concentration and the flocculating properties of the particles. Four types of sedimentation can be identified: (i) Type-I sedimentation; (ii) Type-II sedimentation; (iii) Zone settling; and (iv) Compression as illustrated on Figure 8 [65]. Settling can also be classified on the basis of solids concentration in the suspension: (i) free settling when particles settle individually at different rates in a dilute suspension; (ii) hindered settling when the settling of particles is influenced by the presence of other particles; and (iii) compression settling when the rate of settling is reduced due to the physical contact between particles. The four types of sedimentation are described below:

Type-I Sedimentation

Type-I sedimentation is concerned with the settling of non-flocculating, discrete particles in a low concentration suspension. The settling of these particles is unhindered by the presence of other particles and every particle retains its individual characteristics. When a discrete particle is released in a quiescent fluid, it briefly accelerates until its fluid drag approaches its effective weight. After that, the particle settles out at a constant velocity called the fall or terminal or settling velocity which is a function of fluid properties (density and viscosity), and particle characteristics (size, shape, density, orientation). Khattab [77] has presented a detailed review of the various factors involved in discrete settling. The sedimentation of discrete spherical particles may be described by Newton's law, from which the terminal settling is found to be:

$$V_s = \left[\frac{4}{3}\left(\frac{g}{C_D}\right)\frac{\rho_s - \rho_L}{\rho_L} \cdot d\right]^{1/2} \quad (1)$$

where ρ_s is the particle density; ρ_L is the liquid density; g is the acceleration due to gravity; d is the diameter of spherical particle; C_D is the drag coefficient; and V_s is the terminal velocity.

The drag coefficient, C_D, is related to Reynolds Number. If the Reynolds Number is small, the settling velocity is given by Stoke's Law:

$$V_s = \frac{g}{18\mu}(\rho_s - \rho)d^2 \quad (2)$$

where μ is the absolute viscosity of the fluid.

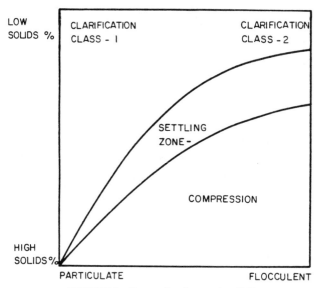

FIGURE 8. Types of sedimentation [65].

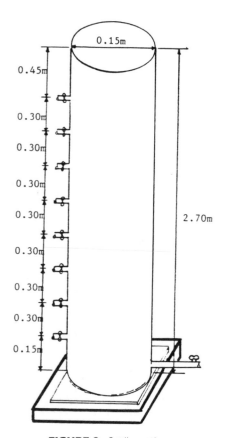

FIGURE 9. Settling column.

Although sewage particles are not spherical, the effect of irregular shape is not pronounced in low settling velocity. Most sedimentation devices are designed to remove small particles which settle slowly.

A typical suspension of particles may have a wide range of particle sizes and a corresponding range of settling velocities. A frequency distribution of settling velocities [19] can be determined by two different methods: (i) by the use of sieve analysis and hydrometer tests combined with the equilibrium equation that relates settling velocity with particle size [22,99]; or (ii) by a settling column test. Figure 9 shows a settling column while Figure 10 [65,6] shows the typical results of a settling column test.

Type-II Sedimentation

Often suspended solids in domestic and industrial wastewaters and coagulated particles in water treatment do not act as discrete particles. Generally, these solids are comprised of particles of different sizes and surface characteristics. Under quiescent conditions, large particles having high settling velocities overtake and coalesce with smaller, lighter particles to form still larger particles. The resulting velocities of the newly formed particles exceed the settling velocities of the original particles. This process is called flocculation.

The extent of flocculation in a settling tank depends on physical factors such as the opportunity for contact, which varies with the range of particle sizes, the concentration of particles, the depth of the tank, and the velocity gradients in the tank whereby particles in regions of higher velocity overtake and coalesce with particles in regions of lower

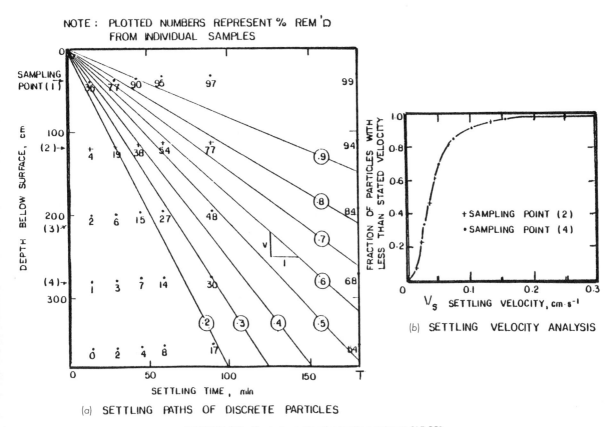

FIGURE 10. Typical results of a settling column [65,22].

velocity [65]. Flocculation also depends on chemical co-agulants and the zeta potential on particles. The floc shear resistance decreases with increase in the floc size and eventually is exceeded by the applied shear due to the local velocity gradients; thus the particle ceases to grow. The dynamics of flocs in presence of fluid shear has been reported by various writers [17,20,22,24].

Chemical Coagulation and Flocculation: The use of chemical treatment appeared early in the development of water and wastewater treatment technology. Aluminum sulfate, lime, and ferrous sulfate have been successful in producing an effluent of better quality than that obtained by plain sedimentation. The effluent is generally fairly clear, with only very fine suspended or colloidal solids and has practically all of the dissolved solids remaining.

The settling velocities of finely divided and colloidal particles in water and wastewaters are so small that removing them in a settling tank under ordinary conditions is impossible unless very long detention periods are provided. Therefore, it has been necessary to devise means to coagulate these very small particles into larger ones which will have higher settling velocities. The aggregation of dispersed par-

ticles in wastewater is induced by addition of chemical co-agulants to decrease the effects of the stabilizing factors, viz, hydration and zeta potential, and by agitation of the medium to encourage collisions between particles. In the past, the cost of the coagulants and the difficulty of disposing of bigger amounts of the sludge produced by this process discouraged its application in wastewater treatment. The recent revival of chemical treatment can be attributed to (i) the decrease in cost of chemicals; (ii) better understanding of floc formation and the factors affecting it; (iii) the development of methods of sludge filtration and processing which overcome, in part, the difficulty of greater sludge bulk; and (iv) the establishment of the relationship between eutrophication in streams and nutrients, particularly phosphorus, nitrogen oxides and organic matter. This has led to the need for final effluent wastes free of such pollutants regardless of the cost of additional treatment.

The effectiveness of chemical coagulation in the removal of biochemical oxygen demand, BOD, suspended solids, SS, and total phosphorus, P, from the influent to West Windsor Water Pollution Control Plant is shown in Figure 11 [78]. The samples were subjected to plain column settling, ferric chloride treatment with and without polymer, and the

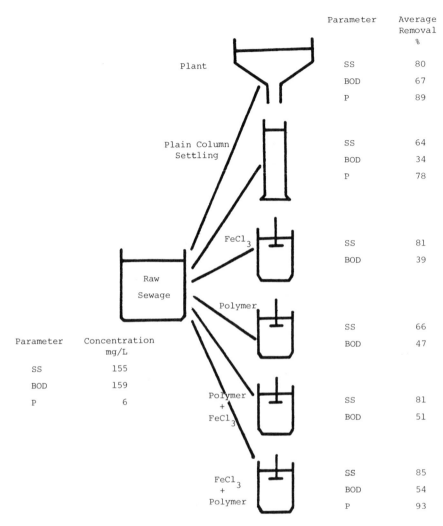

FIGURE 11. Effectiveness of chemical coagulation in removal of BOD, SS, and P [78].

results are compared with the performance of actual plant provided with chemical coagulation.

Since there is no adequate mathematical relationship to determine the effect of flocculation on the rate of change of settling velocity, settling column tests have been used to study the settling characteristics of flocculant suspensions. The suspension is placed in a column to a depth equal to the effective depth of the proposed tank and is allowed to settle under quiescent conditions. Samples are drawn off at various selected time intervals and different depths and analyzed for suspended solids concentrations. Concentrations are plotted with time and depth and the iso-removal lines are constructed as shown in Figure 12 [44]. The tests results are usually analyzed, as follows, to obtain the total removal of particles, for a given detention or residence time, t_* [42].

As shown in Figure 12, R_* of the suspended solids are completely removed. Particles in the range $(R_c - R_*)$ are removed proportional to V_s/V_o where V_o is the overflow rate, d_*/t_*, and V_s is d_1/t_* or alternatively in proportion to the average depth settled, d_1, to the total depth, d_*. Each subsequent range is computed in a similar manner, and the total removal, R_{TOTAL}, is given by

$$R_{TOTAL} = R_* + (R_c - R_*)\frac{d_1}{d_*}$$

$$+ (R_d - R_c)\frac{d_2}{d_*} + \cdots$$

(3)

The calculated total removal can be expected to occur in an ideal settling basin where settling takes place in exactly the same manner as in a quiescent settling column. The use of the settling column test in the manner discussed previously does not provide information on the settling velocities of the various classes of particle sizes and how they change both temporally and spatially. Obviously, such information is greatly needed if the settling of flocculated particles is to be understood.

Zone and Compression Settling

The settling characteristics exhibited by concentrated suspensions or sludges differ from those of dilute suspensions. Zone and compression settling are normally observed in clarifiers with activated sludge or flocculated chemical suspensions with solid concentrations exceeding 500 mg/L [128]. When solids concentration is high, particles settle in close proximity, their velocity fields interfere and the settling is hindered. In this case, the settling velocity of the suspension as a whole is less than the settling velocity of the single particles. If the suspension is non-flocculant and has a wide range of size distribution, the bulk settling rate will be between the fall velocities of the fastest and the slowest particles.

Zone and compression settling of a given sludge sample can be studied by considering its settling behaviour in a one-litre graduated cylinder settling column. Initially, the solids concentration has to be uniform throughout the column as shown in Figure 13 [65,117]. If the concentration is high enough, the sludge starts to settle out with a distinct interface (interface-1) between the mass of the settling sludge and the clarified liquid above. The zone below the clarified liquid is called the interfacial zone and the solids concentration in this zone is uniform and equal to the initial concentration.

The sludge particles in this zone settle as a blanket, maintaining the same relative position with respect to each other. Their hindered settling rate is constant and dependent on the initial sludge concentration as well as the unhindered settling rate of the particles. The zone settling velocity can be obtained by observing the rate of subsidence of interface-1 with time, as shown in Figure 13. As the settling proceeds, the solids on the bottom of the column build up at a constant rate. Subsequently, a transition zone is observed in which the settling velocity decreases due to the increase in solids concentration. The concentration of solids in the zone settling layer remains constant until interface-1 meets with the rising layers of compressed solids. At this time, t_2 in Figure

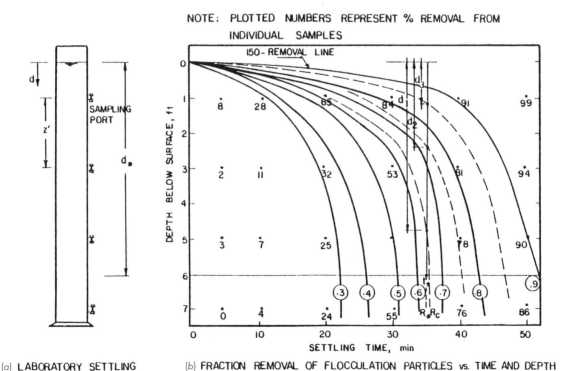

(a) LABORATORY SETTLING COLUMN

(b) FRACTION REMOVAL OF FLOCCULATION PARTICLES vs. TIME AND DEPTH

FIGURE 12. Typical settling column test for flocculated particles [44].

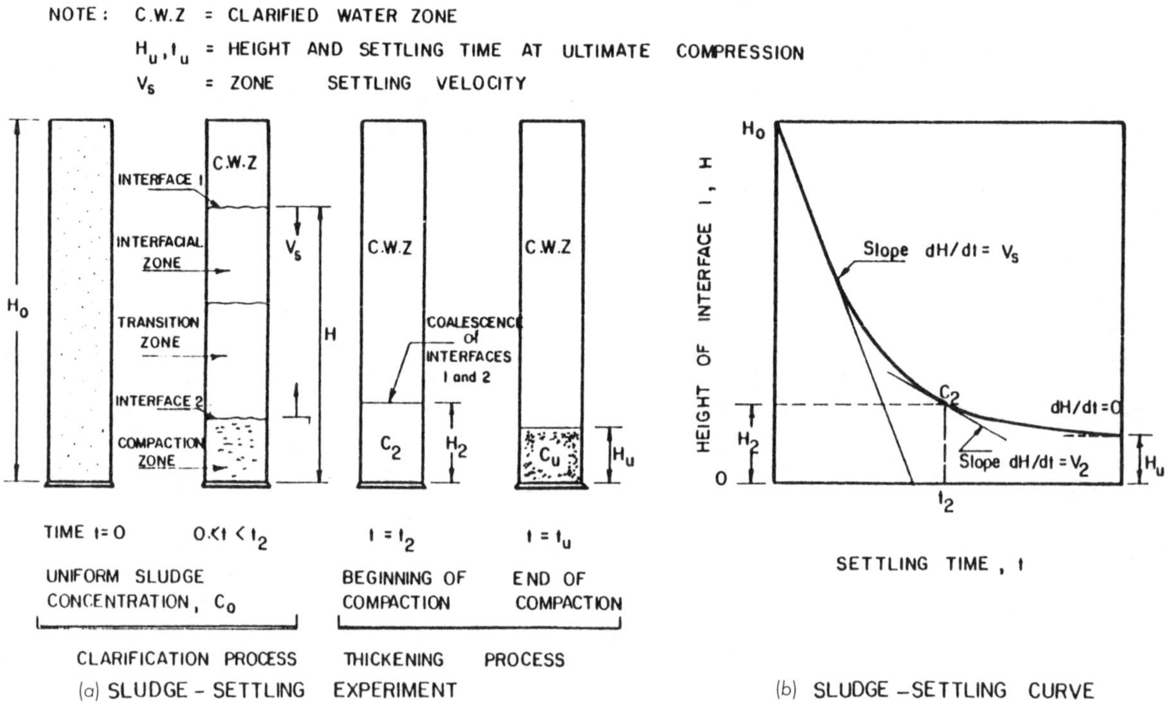

NOTE : C.W.Z = CLARIFIED WATER ZONE

H_u, t_u = HEIGHT AND SETTLING TIME AT ULTIMATE COMPRESSION

V_s = ZONE SETTLING VELOCITY

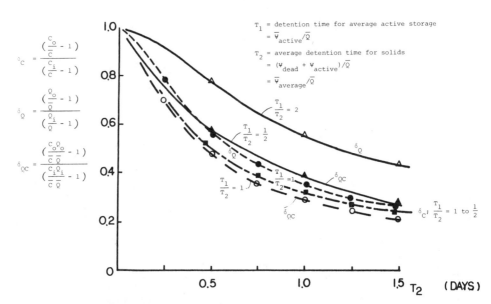

FIGURE 13. Zone and compression settling [65,117].

$$\delta_C = \frac{(\frac{C_o}{C} - 1)}{(\frac{C_i}{C} - 1)}$$

$$\delta_Q = \frac{(\frac{Q_o}{Q} - 1)}{(\frac{Q_i}{Q} - 1)}$$

$$\delta_{QC} = \frac{(\frac{C_o Q_o}{C Q} - 1)}{(\frac{C_i Q_i}{C Q} - 1)}$$

T_1 = detention time for average active storage
= $\overline{\Psi}_{active}/\overline{Q}$

T_2 = average detention time for solids
= $(\Psi_{dead} + \Psi_{active})/\overline{Q}$
= $\overline{\Psi}_{average}/\overline{Q}$

FIGURE 14. Performance of completely-mixed equalization basin.

13, the transition zone disappears and the column displays two zones: (a) clarified liquid zone; and (b) compression zone. The compression zone exhibits a uniform concentration called the critical concentration, C_2. With time, compression takes place and sludge begins to thicken, eventually reaching an ultimate concentration, C_u.

In summary, clarifiers handling suspensions of high concentrations experience zone settling and compression. The settling velocity in this case, depends on the unhindered settling rates of the solids involved as well as the sludge concentration. Larsen [80,81,82] used the following relationship to estimate the settling velocity of activated sludge solids:

$$V_S = V_{So}e^{-\alpha C} \tag{4}$$

where V_S = suspension settling rate; V_{So} = terminal settling velocity at infinite dilution; C = mass concentration of solid particles; and α = a constant depending on sludge properties and is determined using sludge volume index.

HYDRODYNAMICS

It has been stated earlier that particle characteristics and hydrodynamics of flow through basins are the two most important parameters in the performance behaviour of equalization tanks and sedimentation tanks. The influence of particle characteristics has already been discussed in process description and the significance of hydrodynamics of flow is presented in this section.

Hydraulics of Equalization Basins

The classical reservoir routing equations can be used to compute the flow equalization capability of a basin. The governing equations are:

$$\frac{d\Psi}{dt} = Q_i - Q_o \tag{5}$$

$$Q_o = f_1(\Psi) \text{ for weir control} \tag{6}$$

$$Q_o = f_2(t) \text{ for pump control} \tag{7}$$

where Ψ = volume of the basin; Q_i = inflow rate; and Q_o = outflow rate. For a completely mixed basin, the mass balance equation is:

$$\frac{dC}{dt} = \frac{Q_i}{\Psi}(C_i - C) \tag{8}$$

where C is the uniform concentration in the basin; C_i = influent concentration. In a completely mixed basin $C = C_o$ = effluent concentration.

Equations (5) and (8) can be readily solved by available integration subroutines such as the predictor-corrector or the Runge-Kutta methods. Figure 14 shows the results for a simple completely mixed rectangular basin provided with a weir control. Equations (5) to (8) were solved for sinusoidal diurnal variations in the inflow, Q_i, and the concentration, C_i. The following definitions are used to generalize the results:

$$T_1 = \overline{\Psi}_a/\overline{Q} \tag{9}$$

$\overline{\Psi}_a$ = average value of Ψ above the weir crest

$$= fcn \text{ (average depth over the weir)} \tag{10}$$

$\overline{Q}$ = average inflow

$$T_2 = \overline{\Psi}/\overline{Q} \tag{11}$$

$\overline{\Psi}$ = average total basin volume available for mixing

$$\delta C_* = \frac{(C_o/\overline{C} - 1)_{max}}{(C_i/\overline{C} - 1)_{max}} \tag{12}$$

$$\delta Q_* = \frac{(Q_o/\overline{Q} - 1)_{max}}{(Q_i/\overline{Q} - 1)_{max}} \tag{13}$$

$$\delta CQ_* = \frac{\left(\frac{C_oQ_o}{\overline{C}\,\overline{Q}} - 1\right)_{max}}{\left(\frac{C_iQ_i}{\overline{C}\,\overline{Q}} - 1\right)_{max}} \tag{14}$$

$\overline{C}$ = average concentration

Figure 14 can be used to estimate the value of T_2 in order to achieve a given reduction in variation in the concentration, flow or mass loading rate.

In the extreme case of no mixing in the equalization basin (plug flow), the fluctuations in concentration will not change significantly but the flow fluctuation in the outflow will be given by Figure 14. In most cases, this condition will result in higher peak mass loading on the treatment plant. The exception is when the tank is sized to make the peak concentration at the outflow match the minimum outflow.

Hydraulic Performance of Rectangular and Circular Tanks

The most common primary clarifiers are the centre-fed circular and the parallel flow rectangular systems as shown in Figure 15 [65]. There are many versions of these depending on inlet and outlet arrangements and sludge removal methods. Most circular tanks have a centre feed-well which

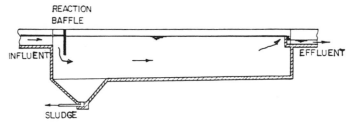

(a) **HORIZONTAL FLOW RECTANGULAR CLARIFIER**

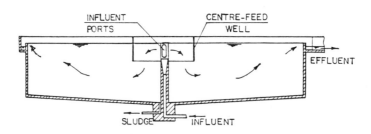

(b) **HORIZONTAL FLOW CIRCULAR CLARIFIER**

FIGURE 15. Typical clarifiers [65].

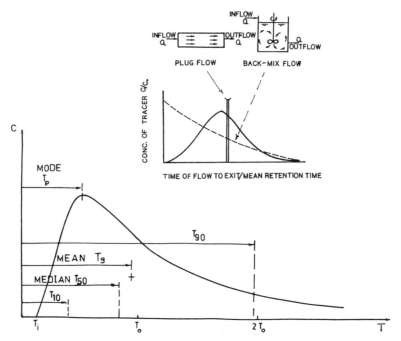

FIGURE 16. Time concentration curves showing hydraulic indicators [91].

dissipates some of the kinetic energy and deflects the flow downwards. A weak jet travels with some mixing towards and then over the effluent weirs provided around the periphery of the tank.

Some encouraging results were obtained by White [145], Dague and Baumann [34] by using peripheral-feed circular tanks which tend to have a more stable hydraulic performance than the centre-feed tanks. In some cases a spiral component is forced on the inflow to improve flow stability of circular tanks. Upflow tanks have also been used. In addition, tube settlers which maintain laminar flow have been installed in both circular and rectangular systems. High removal rates are possible but maintenance is a serious problem [33].

METHODS OF EVALUATING HYDRAULIC PERFORMANCE

The hydraulic behaviour of settling tanks can be studied by direct measurements in the prototype, by physical models and by mathematical models.

Prototype Studies

A common method of studying the hydraulic behaviour of tanks is to inject a known amount of tracer, such as salt, radioactive or fluorescent dye, into the influent and measure the variations in tracer concentration, with time, in the effluent. The resulting curve is called the flow through curve (FTC).

Figure 16 shows a typical flow-through curve. The important parameters are: c = concentration; C = normalized concentration = c/c_o; c_o = injected mass/tank volume = mass/V; t = time; T_g = mean flow through time; T_i = initial trace time ;T_p = modal time; T_{50} = median time; T_o = theoretical detention time = Q/V; T_{10} = time for 10% of the mass to exit; T_{90} = time for 90% of the mass to exit; Q = discharge of water; V = tank volume.

Hydraulic characteristics of a settling tank cannot be evaluated without reference to the resulting removal efficiency of the tank. There is no universal agreement amongst researchers on the relationship between the tank hydraulics (flow pattern and flow-through parameters) and the removal efficiency. Tanks of completely different designs and FTC's often show very similar removal efficiencies and there appears to be no simple explanation for this. If there are high velocities and thus high velocity gradients and turbulence, flocculation might be improved, but this could be offset by re-entrainment of sludge and turbulent mixing. Some researchers, e.g. Villemonte, Rohlich and Wallace [141], emphasize plug flow and the central tendency of the flow-through characteristics while others, e.g. Price and Clements [114], stress the time ratio concept which reflects the uniformity of the planwise distribution of the flow throughout the tank. White [144] comments, ''It seems highly probable that attempts to reduce lateral variations and to bring

the centroid of the curve (flow-through curve) near to the nominal detention time are worthwhile for nondiscrete as well as for discrete particles but the value of reducing the spread of the curve is unknown.'' One reason for this problem is that it is very difficult to obtain reliable flow-through curves under working conditions. White further states that, ''flow-through curves are of little value unless there is virtually complete detection of the tracer material at the outlet.'' Most available fluorescent dyes are subject to decay and adsorption which makes it difficult to interpret data from long duration tests.

In evaluating the hydraulic performance of tanks and modifications thereto, the following trends are considered desirable:

(a) uniform radial advance of the dye front throughout the tank (an indication of a good time ratio, i.e. near unity)
(b) high values of T_i/T_o, T_{50}/T_o and the mode
(c) low Morril Index, T_{90}/T_{10}
(d) high plug flow and low dead space
(e) acceptable velocities near the floor of the tank

Uniformity in the radial advance of the dye front can be evaluated by the time ratio suggested by Clements [28,29,30], i.e.

$$TR = \frac{\text{Average velocity}}{\text{Effective horizontal velocity}} \quad (15)$$

The time ratio is sensitive to poor distribution of flow in plan while being independent of the vertical velocity profile. Using Hazen's theory, he argues that the settling performance of a tank does not depend on the vertical distribution of the flow but is strongly dependent on the horizontal distribution. This is a useful argument but is not entirely justified since vertical velocity distribution can influence flocculation, turbulent dispersion and re-entrainment of sediment. The time ratio may be more useful for discrete particle sedimentation than flocculated sedimentation.

The values of T_i and T_{10} are used to indicate the severity of short circuiting; the lower the values of T_i/T_o and T_{10}/T_o, the worse is the short circuiting. The mode, T_p, is an indicator of the amount of plug flow in the tank. The T_{50} or median time is the probable flowing-through time and is a measure of the hydraulic efficiency of the tank. For ideal plug flow, T_{50}/T_o is unity. The parameter, T_{50}, has been used instead of the centroidal time, T_g, since T_g is more sensitive to incomplete dye recovery and the precise measurement of the long tail of the time concentration curve. Normally T_g will be between T_{50} and T_o.

The Morril Index, T_{90}/T_{10}, indicates the level of mixing or dispersion in the tank. A value of 1 shows no mixing (plug flow) and a value of 21.85 indicates complete mixing. Rebhun and Argaman [118] suggest a method for separating the flow in a settling tank into:

(a) the fraction of plug flow, p
(b) the fraction of mixed flow, $1 - p$
(c) the fraction of dead space, m

The accumulative flow-through curve is represented by

$$F(t/T_o) = \frac{\int_o^t cdt}{\int_o^\infty cdt} = 1 - e^{-a(t/T_o - \theta)} \qquad (16)$$

in which $a = 1/[(1 - p)(1 - m)]$ and $\theta = p(1 - m)$. A semi-log graphical analysis is used to solve for p and m.

According to Camp [23] bottom velocities of about 3 ft/min. (0.9 m/min) are common in practice but he states that much higher velocities, 12 to 18 ft/min. (3.7 to 5.5 m/min), can be used in some tanks. Heinke [59] recommends an upper limit of 8 ft/min. (2.4 m/min) for light chemical flocs.

Physical or Hydraulic Modelling

Several researchers, e.g. Camp [22,23], Dague and Baumann [34], Price *et al.* [114,115], Clements *et al.* [28,29,30], Thompson [138] and Bergman [11], have used scale models to study the hydraulics (flow patterns and flow-through characteristics) of both rectangular and circular tanks; however, there is no agreement on the scaling laws that should be used. Price and Clements have made numerous model and prototype studies using geometrically similar models operated at similar Reynolds Numbers to ensure similar turbulent conditions in both model and prototype. They report good model-prototype agreement. Thompson, who studied density currents, indicated that a geometrically similar model operated at the prototype densimetric Froude Number was sufficient to achieve similitude. Others, for example Camp, Dague and Baumann, insist that model and prototype must have the same Froude Number regardless of whether or not the model has similar turbulent conditions to the prototype. Bergman simulated the sedimentation process with models having the same surface loading (overflow rates) as the prototype when both used the same influent sediment. He showed that this criterion was intermediate between complete Froude and Reynolds scaling.

An important requirement for the modelling of deposition rates is kinematic similarity between the trajectories of the particles in the model and prototype, i.e.

$$(V_s)_R = V_R \qquad (17)$$

where V_s is the settling velocity; V is the fluid velocity and the subscript R means the ratio of the prototype to the model.

Delichatsios and Probstein [35] have presented some theoretical aspects of scaling the coagulation process including the effect of microscale and macroscale turbulence.

For complete similitude, a model must be geometrically similar to the prototype and have the same Froude number, Reynolds number, Weber number and Mach number. The Euler number is used as the dependent variable, e.g. to represent pressure drops or losses in the model. Since surface tension and compressibility are generally not important in settling tank models, both the Weber and Mach numbers can be omitted. If there are two phases or two components in the system each phase or component should satisfy the laws of similitude. Ideally, geometric similarity as well as the Froude and Reynolds laws should be used; however, in practice this is not possible, so a compromise must be found. In most free surface hydraulic models, with Froude Numbers of the order of 1, the Froude Law is used as the operating law but the model scale is set to ensure similar turbulent conditions in the model and prototype. In the case of settling tanks where the Froude Numbers are very low (10^{-5}) and the Reynolds Numbers are usually in the turbulent range, the normal modelling procedure, cited above, would yield large and costly models. Therefore, the rigid similitude requirements must be relaxed and the compromise suggested by Price and Clements [114], supported by field studies, seems to be reasonable, provided that density currents are not of interest. If density currents are important, the densimetric Froude number also could be reproduced by exaggerating density differences in the model relative to the prototype. This leads to the criterion for Reynolds number for the model,

$$R_{N_m} = \frac{D_m V_m}{\nu_m} \geq 2500 \qquad (18)$$

where D_m = depth in the model, V_m = velocity in the model, and ν_m = kinematic viscosity in the model.

In circular tanks, the Reynolds number is high in the centre and decreases towards the periphery. Thus in practice it may be necessary to use the minimum allowable Reynolds number at the periphery in order to avoid excessive exaggeration of the model Froude number.

In addition, for kinematic similarity, the ratio of the speed of a point on the scraper arm in a circular tank to the local fluid velocity should be the same in the model and the prototype [McCorquodale, 91], i.e.

$$\frac{\omega_m r_m}{(Q/A)_m} = \frac{\omega_p r_p}{(Q/A)_p} \qquad (19a)$$

or

$$\frac{Q_p}{Q_m} = \frac{\omega_p L_R^3}{\omega_m} \qquad (19b)$$

and

$$V_R/L_R = T_R^{-1} = \omega_R \qquad (19c)$$

where ω is the angular velocity of the arm; the subscript R denotes the ratio of prototype to model.

In the case of density currents, the densimetric Froude number should be the same in the model and the prototype. Using the definition given by Keulegan [74],

$$F' = \frac{V}{\sqrt{g'D}} \tag{20}$$

where $g' = g\Delta\rho/\rho$; ρ is a reference fluid density and $\Delta\rho$ is the difference in density between the density current and the main body of the fluid. Thus the prototype-model relationship is

$$\frac{V_p}{V_m} = \sqrt{\frac{\Delta\rho_p \cdot \rho_m \cdot D_p}{\Delta\rho_m \cdot \rho_p \cdot D_m}} \tag{21a}$$

$$\cong L_R^{1/2} \left(\frac{\Delta\rho_p}{\Delta\rho_m}\right)^{1/2} \tag{21b}$$

or

$$(\Delta\rho)_R = V_R^2/L_R \tag{21c}$$

since $\rho_p \cong \rho_m$. From Equation (21c) it is apparent that in a Froude Model, ($V_R^2 = L_R$), the density difference ratio is unity but in a Reynolds Model, ($V_R \cong L_R^{-1}$), this ratio will be very small and an exaggeration in the model density difference will be required. Temperature differences can be used to produce density currents. To represent a prototype temperature difference of about 0.1°C, a model difference of approximately 8°C is required for $L_R \cong 20$ and V_R satisfying the minimum turbulence requirements.

Hydrodynamic Simulation

In order to overcome some of the similitude problems associated with physical modelling of settling tanks, increasing attention is being given to mathematical modelling. The governing differential equations for the hydrodynamics have been presented by many researchers {Rodi [123], Schamber and Larock [125,126], Imam [65,66,67], Abdel-Gawad [1,2,3,4], Larsen [82] and Devantier and Larock [37]}. The constitutive equations for nearly horizontal ($U \gg V$) two dimensional turbulent flow in primary clarifiers with low concentration of suspended solids are:

Continuity: $\dfrac{\partial U}{\partial x} + \dfrac{\partial V}{\partial y} = 0$ (22)

Horizontal Momentum:
$$\frac{\partial U}{\partial t} + U\frac{\partial U}{\partial x} + V\frac{\partial U}{\partial x}$$
$$+ \frac{1}{\rho}\frac{\partial p}{\partial x} + \frac{\partial \overline{u^2}}{\partial x} + \frac{\partial \overline{uv}}{\partial y} \tag{23}$$
$$+ v\frac{\partial^2 U}{\partial y^2} = 0$$

Vertical Momentum:
$$\frac{\partial V}{\partial t} + U\frac{\partial V}{\partial x} + V\frac{\partial V}{\partial x}$$
$$+ \frac{1}{\rho}\frac{\partial p}{\partial y} + \frac{\partial \overline{uv}}{\partial x} \tag{24}$$
$$+ \frac{\partial \overline{v^2}}{\partial y} + g^* = 0$$

Temperature:
$$\frac{\partial T}{\partial t} + U\frac{\partial T}{\partial x} + V\frac{\partial T}{\partial y}$$
$$+ \frac{\partial \overline{uT'}}{\partial x} + \frac{\partial \overline{vT'}}{\partial y} = 0 \tag{25}$$

Concentration:
$$\frac{\partial C}{\partial t} + U\frac{\partial C}{\partial x} + V\frac{\partial C}{\partial y}$$
$$+ \frac{\partial \overline{uC'}}{\partial x} + \frac{\partial \overline{vC'}}{\partial y} \tag{26}$$
$$+ \frac{\partial}{\partial x}(-V_s C) = 0$$

State Equation: $\rho = \rho_o[1 - \beta_* T$
$$+ (S_s - 1)C_v] \tag{27}$$

in which U, V are, respectively, the horizontal and vertical mean local velocities; u, v are the turbulent components of the velocities U and V; p = local pressure; v = kinematic viscosity; ρ = local density; ρ_o = reference density of fluid; T = temperature relative to a reference temperature; T' = turbulent fluctuation in temperature; $g^* = \rho g/\rho_o$ = body force per unit mass of fluid; C = concentration; C' = fluctuation in concentration; C_v = concentration based on volume; β_* = thermal expansion coefficient; V_s = settling velocity of particles.

Before Equations (22–27) can be solved, a set of closure equations must be obtained to describe the turbulent correlations. A common approach is to neglect the viscous shear and represent the turbulent shear by introducing an eddy viscosity v_t such that,

$$-\overline{uv} = v_t\left(\frac{\partial U}{\partial y} + \frac{\partial V}{\partial x}\right) \tag{28}$$

To complete the closure scheme, Launder and Spalding [85,86] developed the $k - \varepsilon$ turbulence model in which

$$v_t = c_\mu \frac{k^2}{\varepsilon} \tag{29}$$

$$k = \overline{u_i u_i u}/2$$

$$\tag{30}$$

= turbulent kinetic energy

ε = dissipation rate of turbulent kinetic energy

c_μ = constant

For two dimensional flow the transport equations for k and ε are:

$$\frac{\partial k}{\partial t} + U \frac{\partial k}{\partial x} + W \frac{\partial k}{\partial y} - \frac{c_\mu}{\sigma_k}$$

$$\times \left[\frac{\partial}{\partial x} \left(\frac{k^2}{\varepsilon} \frac{\partial k}{\partial x} \right) + \frac{\partial}{\partial z} \left(\frac{k^2}{\varepsilon} \frac{\partial k}{\partial y} \right) \right] \tag{31}$$

$$- P_r + \varepsilon = 0$$

$$\frac{\partial \varepsilon}{\partial t} + U \frac{\partial \varepsilon}{\partial x} + W \frac{\partial \varepsilon}{\partial y} - \frac{c_\mu}{\sigma_\varepsilon}$$

$$\times \left[\frac{\partial}{\partial x} \left(\frac{k^2}{\varepsilon} \frac{\partial \varepsilon}{\partial x} \right) + \frac{\partial}{\partial z} \left(\frac{k^2}{\varepsilon} \frac{\partial \varepsilon}{\partial y} \right) \right] \tag{32}$$

$$- c_{\varepsilon_1} \frac{\varepsilon}{k} P_r + c_{\varepsilon_2} \frac{\varepsilon^2}{k} = 0$$

in which P_r = production rate and the constants are c_μ = 0.09, σ_k = 1.0, σ_ε = 1.3, c_{ε_1} = 1.45, c_{ε_2} = 1.90 and $\sigma_k = \sigma_\varepsilon = 1.0$ [84,85,123]. The turbulent temperature and concentration correlations are often treated by introducing a turbulent dispersion coefficient which is assumed to be proportional to v_t, e.g.

$$v_c = v_t/\sigma_c \tag{33}$$

where σ_c is the Prandtl-Schmidt number which is generally in the range of 0.5 to 1.0.

At this time a complete model incorporating Equations (22–33) has not been presented although some progress has been made. Stamou and Rodi [133] presented a review of research on modelling of sedimentation tanks. Table 1 which has been adapted from their report shows the chronological development of the subject.

Early settling basin design was based on experience along with simple methods of analysis. Hazen [57] introduced the concept of overflow rate using a plug flow hydraulic model. Dobbins [39] proposed a transport model using the same plug flow hydraulic model but accounting for the effects of wall generated turbulence on sedimentation. Camp [22] applied Dobbins' theory to the design of grit chambers and primary clarifiers.

Although the Camp-Dobbins approach was an improvement on the Hazen model, it failed to account for many of the hydraulic characteristics of a real settling tank {Abdel-Gawad et al. [1,2,3,4], Imam et al. [65,66,67], Larsen [80,81,82] and Schamber et al. [79,125,126]}. These are: (i) recirculation in plan; (ii) recirculation in profile; (iii) free stream turbulence in the influent; (iv) turbulent mixing of the influent with the flow in the tank; (v) possible laminar flow regions; (vi) turbulent boundary layers; (vii) separation or re-attachment of flow at the boundaries; (viii) stratification due to zone settling; (ix) density currents due to relatively warm or cool influent; (x) currents and turbulence related to wind shear; (xi) flow disturbances caused by the sludge removal system; (xii) effects of influent or effluent baffles; (xiii) the effects of plan and profile shape on the mean and turbulent motions of both fluid and sediment; (xiv) the magnitude and distribution of the dispersion coefficient within the tank; (xv) re-entrainment; (xvi) the distribution of the Prandtl-Schmidt number; and (xvii) the effect of sludge on the rheology of the fluid near the bed. Another important factor in determining the removal efficiency is the settling velocity distribution of the suspended solids in the influent.

There is no existing model, physical or numerical, that can simulate all of these effects. The best that can be done at the present is to simulate the major effects on suspended solids removal. The work of Larsen [82], Schamber and Larock [79,125,126], Imam, McCorquodale, et al. [66,67] have identified the following as major factors affecting solids removal: (i) the flow pattern including any recirculation and density currents; (ii) the turbulent mixing; (iii) resuspension; and (iv) the quiescent settling characteristics of the particle.

Simplified Model: As an example of a simple model, a brief description of a model developed by Abdel-Gawad and McCorquodale [1,2,3,4] is presented here. This model is computationally efficient and can be used to compute the removal efficiency of circular or rectangular clarifiers with inlet reaction baffles. The proposed model is restricted to the neutral density case. It is also assumed that equations of motion of the liquid phase can be decoupled from those of the solid phase, which is generally acceptable in primary clarifiers where the sediment concentrations are low (typically 200 mg/L). In this study, the Strip Integral Method (SIM) was used to solve the governing differential equations. The model consists of hydrodynamic submodel and a transport submodel. Figure 17 shows the assumed 2-D flow domain.

Formulation of the Hydrodynamic Submodel: Applying the boundary layer assumptions to the predominantly horizontal flow, it can be shown that the horizontal momentum equation can be reduced to

$$U \frac{\partial U}{\partial r} + V \frac{\partial U}{\partial y} = -g \left(\frac{\partial h}{\partial r} \right) + \frac{1}{\rho} \left(\frac{\partial \tau}{\partial y} \right) \tag{34}$$

TABLE 1. Theoretical Studies of Settling Tanks (after Stamou and Rodi [133]).

No.	Investigator	Type of Model
1	Hazen (1904)	Theory of continuous sedimentation. Theoretical analysis for settling of uniform discrete particles.
2	Camp (1945)	Extension of Hazen's model for quiescent settling of particles having a settling velocity distribution without turbulence or resuspension.
3	Dobbins (1961)	Development of a model for the transient concentration profile during settling, including turbulence.
4	Weidner (1967)	Model for velocity distribution in horizontal flow, rectangular tanks from an extension of basic hydraulic equations (flow between 2 parallel plates, etc.).
5	Vashel and Sak (1968)	Regression analysis for the prediction of removal efficiency based on influent and effluent data.
6	Smith (1969)	Regression analysis for the prediction of the solids removal based on influent and effluent data.
7	Takamatsu et al. (1974)	Dispersion model based on the hydrodynamics of fluid flow in a settling tank with no stratification. Scouring was taken into account.
8	Tebbut and Christoulas (1975)	Regression analysis for the removal efficiency. Influent SS-concentration and hydraulic loading were included.
9	Larsen and Gotthardson (1976)	Solution of hydrodynamic equation for rectangular tanks with activated sludge using DuFort-Frankel scheme.
10	Shiba and Inoue (1975)	Extension-simplification of Takamatsu et al's model. Use of empirical relationships to relate scouring to local fluid velocities.
11	Boadway (1978)	Non-linear regression analysis for settling rates for unhindered-discrete particle settling without flocculation.
12	Cordoba-Molina et al. (1978)	Extension of Dobbins' model.
13	Alarie et al. (1980)	Simulation model for the performance of circular primary clarifier.
14	Schamber and Larock (1981)	Numerical model for the prediction of velocity field with the finite element solution of 5 coupled non-linear partial differential equations ($\kappa - \varepsilon$ turbulence model).
16	Imam et al. (1983)	Finite difference model to simulate flow and concentration field. Use of constant eddy viscosity. Includes reaction baffles.
17	Abdel-Gawad and McCorquodale (1984)	Numerical model for rectangular basin. Use of Strip Integral Method.
18	Rechteler and Schrimpf (1984)	Numerical model for the dimensioning of horizontal flow tanks.

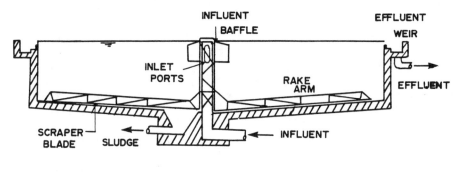

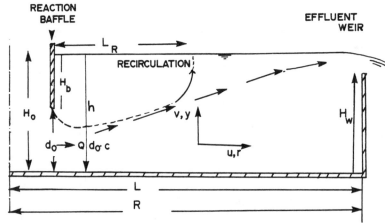

FIGURE 17. Actual and idealized two-dimensional solution domain (after Abdel-Gawad [1]).

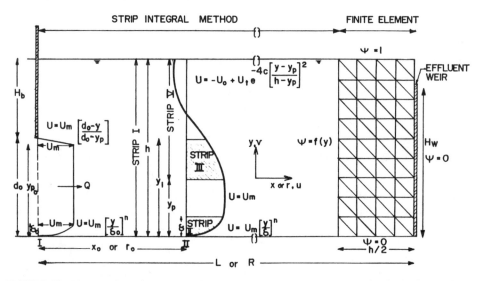

FIGURE 18. Velocity shape functions and strips for the strip integral method (after Abdel-Gawad [1]).

where τ = total shear; h = depth of flow; r = the horizontal coordinate. The pressure distribution is assumed to be hydrostatic. The continuity equation is

$$\frac{\partial}{\partial r}(r^m U) + \frac{\partial}{\partial y}(r^m V) = 0 \qquad (35)$$

in which radial tanks are represented by $m = 1$ and rectangular tanks by $m = 0$.

The strip integral method {Narayanan [106,107]; Moses [103]; McCorquodale and Khalifa [94]} is used to reduce Equations (34) and (35) to a set of ordinary differential equations that describes the longitudinal variation of selected characteristics of the velocity profiles. This method requires shape functions that have arbitrary functions of r that can be used to represent the flow pattern. Based on the observations of many researchers (see Stamou and Rodi [133]), the flow is divided into the strips shown in Figure 18 with the following velocity shape functions:

$$U = U_m \left(\frac{y}{\delta}\right)^n \qquad 0 < y \le \delta \qquad (36)$$

$$U = U_m \qquad \delta \le y \le y_p \qquad (37)$$

$$U = -U_0 + (U_0 + U_m)e^{-4c[(y - y_p/h - y_p)]^2}$$
$$\qquad (38)$$
$$y_p \le y \le h$$

in which U_m = maximum horizontal velocity; $U \to U_0$ as $y \to \infty$; δ = the inner layer thickness; y_p = distance from the bed to the top of the potential core; n = an exponent that varies with x and ranges from $\frac{1}{2}$ for laminar conditions to $\frac{1}{7}$ for turbulent conditions; c is a coefficient which may be a function of x, and h is the depth of flow. When the potential core ends, zone 2 will end and the mean velocity distribution will be represented by the $1/n$ power for the inner layer and the Gaussian distribution for the outer layer; with δ replacing y_p in Equation (38). Thus in the potential core region there are U_0, δ, y_p and h unknown functions of x; and in the fully developed region there are U_0, δ, h and U_m unknown functions of x. Four integral equations are required to solve for these unknowns. The integral continuity equation is:

$$\int_0^h \frac{\partial V}{\partial y} \, dy = -\int_0^h \frac{\partial U}{\partial x} \, dy = U_s \frac{dh}{dx} \qquad (39)$$

in which $U_s(dh/dx)$ is the kinematic condition at the free surface. Three other equations are obtained by integrating Equation (35) over the strips $0 \to \delta$; $y_p \to y$, and $y_p \to h$, e.g.:

$$\int_{y_1}^{y_2} U \frac{\partial U}{\partial x} \, dy + \int_{y_1}^{y_2} V \frac{\partial V}{\partial y} \, dy$$
$$\qquad (40)$$
$$= -g \int_{y_1}^{y_2} \frac{\partial h}{\partial x} \, dy + \frac{1}{\rho} \int_{y_1}^{y_2} \frac{\partial \tau}{\partial y} \, dy$$

in which y_1 and y_2 are the limits of the strips as shown in Figure 18. The bed shear stress, τ_0, at the strip limit $y = 0$ was defined using the boundary layer assumptions of Schlichting [127]. For the shear at the free mixing zone, τ_t, a modified Prandtl mixing concept [1] was used to define the shear at the limit $y = y_*$:

$$\tau_t = \rho \ell_m^2 \left[\frac{8cU_t}{(h - y_p)\sqrt{8c}} \right]^2 \qquad (41)$$

and

$$\ell_m = 0.048(h - y_p) + c_0 c_j (x - x_{00}) \qquad (42)$$

where x_{00} is the virtual origin; $U_t = U_0 + U_m$; $c_0 = 0.019$ from the free jet theory [116] and $c_j = 0.33$ from calibration of the model.

Substituting the velocities from Equations (36–38) and values of the shear at the limits of the strips into the integral equations, a set of four first-order ordinary differential equations can be obtained:

$$A_i \frac{dh}{dx} + B_i \frac{d\delta}{dx} + C_i \frac{dU_0}{dx} + D_i \frac{dy_p}{dx} = E_i$$
$$\qquad (43)$$
$$(i = 1,2,3,4)$$

The coefficients A_i, B_i, C_i, D_i and E_i are given elsewhere [1,2].

Inlet conditions are used together with macroscopic mass, momentum and energy balances in order to establish starting values for h, δ, U_∞ and y_p. A fifth-order Runge-Kutta method with an allowable step size $\Delta x/d_0 = 0.1$ is used to solve Equation (43), with a suitable inlet condition. The SIM is a forward marching scheme. To account for the curvilinear flow near the effluent weir, a standard finite element scheme is used to solve the nearly irrotational flow in the withdrawal zone. The finite element scheme is started when the forward marching scheme is at a distance of one-half water depth upstream from the effluent weir. This distance was chosen based on experimental observations and the work reported by Moursi [104] and Imam and McCorquodale [66].

Figure 19 shows comparisons of measured and predicted flow patterns in a radial flow tank. Figure 20 shows the predicted accumulated depth averaged eddy viscosities in a circular tank with different relative baffle heights (H_b/H_o).

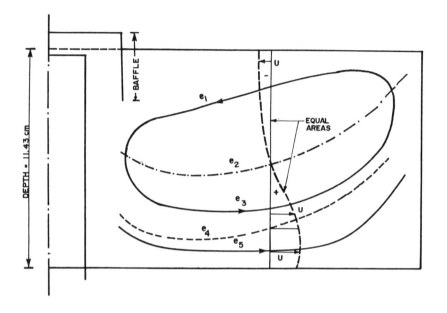

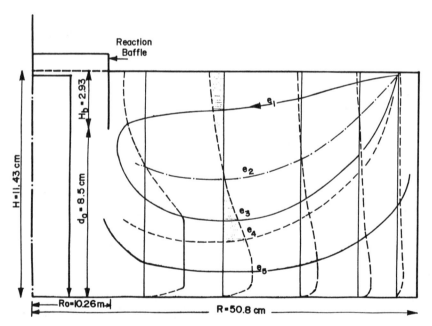

FIGURE 19. Predicted and measured flow patterns in a radial flow tank (measured data from Clements and Khattab [29,77,1,3]).

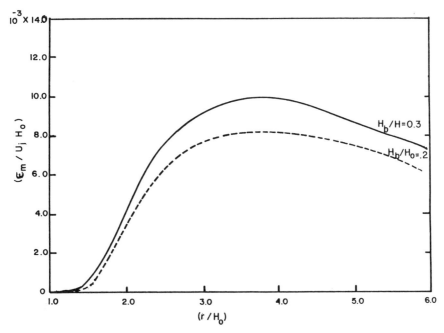

FIGURE 20. Accumulated average eddy viscosity in a radial flow tank (after Abdel-Gawad [1]).

The Transport Submodel: Assuming (a) steady state conditions; (b) discrete settling; (c) the effect of the longitudinal dispersion term can be neglected in relation to the vertical dispersion term; (d) a dilute suspension with $\varepsilon_{sy} \approx \varepsilon_m$; then Equation (26) can be reduced to

$$U \frac{\partial C}{\partial x} + V \frac{\partial C}{\partial y} = \frac{\partial}{\partial y} \left(\varepsilon_m \frac{\partial C}{\partial y} \right) + V_s \frac{\partial C}{\partial y} \qquad (44)$$

Based on experimental observations of Larsen [82], Kerssens et al. [73] and Heinke et al. [58,59,60,61,136], the following shape functions for $C_{(y)}$ are assumed (Figure 21):

$$C_{(y)} = C_{y_p} + (C_a - C_{y_p}) \left[\frac{(\delta - y)a}{y(\delta - a)} \right]^z \qquad (45)$$

$$a \leq y \leq \delta$$

$$C_{(y)} = C_{y_p} \qquad \delta \leq y \leq y_p \qquad (46)$$

$$C_{(y)} = C_{y_p} e^{-D_*(y - y_p)^2} \qquad y_p \leq y \leq h \qquad (47)$$

in which $z = V_s / \kappa U_*$; $U_* = \sqrt{\tau_0 / \rho}$; τ_0 = bed shear; κ = von Karman constant; $D_* = D_*(x)$; a = bed layer thickness which is taken equal to the thickness of the laminar sublayer [127], i.e. $a = 5\nu / U_*$; C_a is the concentration at $y = a$; and C_{y_p} is the concentration at $y = y_p$. Equations

(45–47) give the concentration C at any level provided that C_a, C_{y_p} and D_* are defined. The concentration C_a can be related to C_{y_p} by satisfying the boundary conditions at $y = a$. The other two parameters can be determined through the SIM by integrating Equation (44) twice, i.e. over the two strips shown in Figure 21. Thus the integral transport equations can be written as:

$$\int_a^h U \frac{\partial C}{\partial x} dy + \int_a^h V \frac{\partial C}{\partial y} dy$$
$$= \int_a^h \frac{\partial}{\partial y} \left(\varepsilon_m \frac{\partial C}{\partial y} \right) dy + \int_a^h V_s \frac{\partial C}{\partial y} dy \qquad (48)$$

$$\int_{y_p}^{y*} U \frac{\partial C}{\partial x} dy + \int_{y_p}^{y*} V \frac{\partial C}{\partial y} dy$$
$$= \int_{y_p}^{y*} \frac{\partial}{\partial y} \left(\varepsilon_m \frac{\partial C}{\partial y} \right) dy + \int_{y_p}^{y*} V_s \frac{\partial C}{\partial y} dy \qquad (49)$$

Substituting the concentration $C_{(y)}$ from Equations (45–47), the values of U_m, U_0, h, y_p, δ, y_* and their derivatives from the hydrodynamic submodel, a set of two ordinary differential equations are obtained in the form:

$$F_i \frac{dD_*}{dx} + \overline{G}_i \frac{dC_{y_p}}{dx} + M_i \frac{dC_a}{dx} = \overline{N}_i \qquad (i = 1,2) \qquad (50)$$

684   ENVIRONMENTAL

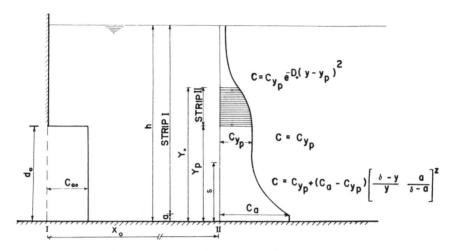

FIGURE 21. Concentration shape functions and strips (after Abdel-Gawad [1,5]).

where F_i, $\overline{G}_i$, M_i and $\overline{N}_i$ are functions of C_a, C_{y_p} and D_* and the hydrodynamic parameters.

The necessary third equation is derived from the condition at the bed. The boundary condition proposed by Takamatsu et al. [135] is used to simulate both deposition and scour of solids on the clarifier bottom, viz.:

$$\varepsilon_s \frac{\partial C}{\partial y} + K_r V_s C = 0 \qquad \text{at } y = a \qquad (51)$$

where K_r is a scour parameter which depends on the degree of scouring. If $K_r > 1$, there is a tendency of scouring; when $K_r = 1$, there is a balance between deposition and scouring. If $0 < K_r < 1$, there is a tendency for deposition, i.e., more deposition than scouring and the limit $K_r = 0$ signifies deposition only. In a settling basin, the rate of deposition exceeds that of scouring in normal operation; therefore, the range of K_r is usually $0 < K_r < 1$. In this study the value of K_r was fixed to be 0.1. Substituting Equation (45) into Equation (51), a relationship between C_a and C_{y_p} is established in the form:

$$C_a = C_{y_p} \cdot F_* \qquad (52)$$

Differentiating Equation (52) with respect to x and substituting in Equation (50), a closed set of equations is obtained

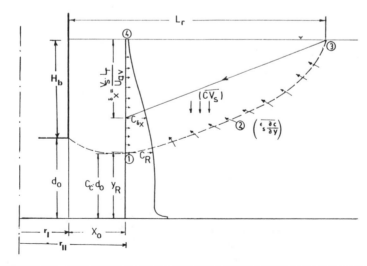

FIGURE 22. Initial conditions for the transport submodel (after Agdel-Gawad [1,5]).

in the form:

$$F_i \frac{dD*}{dx} + G_i \frac{dC_{y_p}}{dx} = N_i \qquad (i = 1,2) \qquad (53)$$

in which F_i, G_i, N_i and F_* are functions of C_a, C_{y_p}, D_* and the hydrodynamic parameters. Equation (53) can then be integrated with appropriate starting values for the unknowns.

Inlet Conditions for the Transport Submodel: To solve the system of Equation (53), starting values for D_*, C_{y_p}, and C_a are required. Also the backward movement of particles in the recirculating zone should be taken into account. An approximate mass balance for the recirculating zone is considered to estimate the rate of feedback of the sediment particles. A particle path is assumed, using the slip velocity, V_s, in the recirculation zone. The location of a recirculated particle at the inlet section is determined from the kinematic similarity of the velocity and the length, e.g., $\delta_x/V_s = L_r/U_{av}$, where δ_x is the slip distance by a particle with a settling velocity V_s and U_{av} is the average horizontal velocity in the recirculation zone, Figure 22. From the hydrodynamic submodel, the length of the recirculation zone, $L_r \simeq 5.5H_b$ where H_b is the baffle height. Considering the mass flux into and out of the recirculating zone, the following equation can be written:

$$\left(\frac{\partial C}{\partial t} \, \Psi \right) = L_r \left(-\varepsilon_s \frac{\partial C}{\partial y} \right) - \overline{CV_s} L_r + \int_{y_R}^{h} CUdy \qquad (54)$$

where Ψ is the volume of the recirculating zone (shown as 1–2–3–4–1 in Figure 22); $\overline{\varepsilon_s(\partial c/\partial y)}$ and $\overline{CV_s}$ are, respectively, the average of dispersion and settling from the zone and y_R is the depth to the recirculation zone at Section 1. For steady conditions, Equation (54) yields an implicit expression for D_*. The Newton-Raphson method was used to solve this equation with an appropriate starting value of D_*.

The inlet value of C_{y_p} can be obtained by considering the mass balance between Sections 1 and 2 (Figure 22). The equilibrium equation is:

$$\int_0^{d_0} C_0 Udy = \int_0^h CUdy$$
$$- \frac{(C_a + C_0)}{2} V_s x_0 - \varepsilon_s \frac{\partial C}{\partial y} \frac{x_0}{2} \qquad (55)$$

where C_0 is the uniform inlet concentration and x_0 is the distance to the first section ($= d_0$). The inlet value of C_a at the first section is related to C_{y_p} through Equation (52). These values of D_*, C_{y_p} and C_a are used as the starting condition

to solve Equation (53) by a fifth order Runge-Kutta method with an allowable step size $\Delta x/d_0 = 0.1$. The solution is executed as far as the withdrawal zone; it is assumed that whatever sediments reach this zone will leave the sedimentation basin. The flow in this zone is dominated by withdrawal by the weir and the vertical velocity is significant compared to the settling velocity of the remaining light solids. In addition, there may be a boundary layer separation in this zone which tends to inhibit settlement.

TYPICAL HYDRAULIC BEHAVIOUR OF RECTANGULAR AND CIRCULAR SETTLING TANKS

General

Several investigators have reported hydraulic efficiencies for various types of rectangular and circular tanks; however, most of the data have been obtained from models and thus may be subject to scale effects. The lack of prototype data stems from the difficulty of accounting for 100% of the injected dye or tracer and the interference of some dyes or tracers with the flow patterns as a result of density effects. Another problem is that there is no commonly accepted parameter for representing hydraulic efficiency. Some authors present the initial trace time while others use the mode, median or mean flowing-through times.

Tables 2 and 3 present a collection of hydraulic parameters from a number of sources for rectangular and circular tanks, respectively. Various tank arrangements and density conditions are included. These tables show that rectangular tanks generally have better hydraulic efficiencies than centre feed circular tanks. Most circular tanks have greater mixing than rectangular tanks but the probable flowing-through times are comparable for both type of tanks.

There is some evidence that peripheral feed tanks are hydraulically superior to centre feed systems, Bergman [11]. Bergman also reports that a warm influent in a circular tank causes severe short circuiting and lower removal efficiencies than a cool influent.

Physical model studies and a limited amount of field data have revealed the main hydraulic features of the major types of clarifiers under isothermal conditions. Figure 23 shows some observed velocity distributions in clarifiers that are rectangular in plan. In some clarifiers a reaction baffle is used to prevent short circuiting due to the high velocity influent jets. These baffles tend to deflect the flow towards and along the bed with a recirculating flow developing above this bottom current. Velocity data related to rectangular tanks have been collected by Larsen [82], Bretscher et al. [18], Tay et al. [136], Ditsios [38], Imam [65] and Moursi (El-Sebakhy) [104].

Flows in centre feed circular clarifiers under isothermal conditions are illustrated in Figure 24. Velocity data related to this type of clarifier have been presented by Anderson [7], Ponn [111], Heinke et al. [58,59,136], McCorquodale

TABLE 2. Hydraulic Efficiency of Rectangular Tanks.

Type of Tank	$\dfrac{T_i}{T_o}$	$\dfrac{T_{10}}{T_o}$	$\dfrac{T_{50}}{T_o}$	$\dfrac{T_{90}}{T_{10}}$	$\dfrac{T_g}{T_o}$	$\dfrac{T_p}{T_o}$	T.R.	p	$1-p$	m	Reference
Ideal Plug Flow Tank	1	1	1	1	1	1	1	1	0	0	Theoretical
Ideal Mixing Chamber	0	.11	.83	21.85	—	0	—	0	1	1	Theoretical
Rectangular, Long, N.D., Proto.	.52	.65	.90	2.3	~1	.82	—	.68	.32	.04	Camp
Long, B.D., Proto.	.12	.22	.44	5.0	—	.22	—	.25	.75	.46	Camp
Short, N.D., Proto.	.30	.70	.92	1.7	~1	.82	—	.76	.24	.05	Camp
Deep, N.D., Model	.24	.38	.78	4.7	—	.44	—	.33	.67	.04	University of Windsor
Weir Inflow N.D., Model-Proto.	.08	0.18	.60	11.0	—	0.24	—	.09	.91	.15	Villemonte et al.
Reaction Jets at Inflow, N.D. Model-Proto.	.22	.35	.63	3.9	—	0.50	—	.35	.65	.24	Villemonte et al.
No Inlet Baffle N.D., Model	.07	.15	.64	12.3	.83	0.10	—	.07	.93	.14	Rebhun et al.
Inlet baffle N.D., Model	.30	.48	.76	3.9	.96	.55	—	.31	.69	.01	Rebhun et al.
European, Long, N.D., Prototype	.22	.48	.89	4.0	—	.65	—	.32	.68	−0.7	Kalbskopf
Sarnia, N.D., Prototype	.33	.42	.65	3.7	—	.50	—	.40	.60	.17	Heinke et al.
Drowned Inlet Weir, N.D., Proto.	.06	0.1	0.16	2	—	.14	—	.35	.65	.80	Hamlin
British, Model	—	—	—	—	—	—	.3 to .75	No Baffling With Baffling			Clement, Price
British, Proto.	—	—	—	—	—	.30	.3 to .90	No Baffling With Baffling			Clement, Price
British Model No Wind	—	—	—	—	—	—	.71	—	—	—	Clement, Price
British Model Crosswind	—	—	—	—	—	—	.41	—	—	—	Clement, Price
Round-the-End N.D.	.74	.88	.99	1.3	~1	.95	—	.80	.20	.04	Fair, Geyer and Okun
Typical Range	.08 to .40	—	.44 to .90	2 to 8	—	—	—	—	—	—	

Note: N.D. ≡ Neutral Density; B.D. ≡ Buoyant Density; H.D. ≡ Heavy Density.

TABLE 3. Hydraulic Efficiency of Circular Tanks.

Type of Tank	$\dfrac{T_i}{T_o}$	$\dfrac{T_{10}}{T_o}$	$\dfrac{T_{50}}{T_{10}}$	$\dfrac{T_{90}}{T_{10}}$	$\dfrac{T_g}{T_o}$	$\dfrac{T_p}{T_o}$	T.R.	p	$1-p$	m	Reference
Ideal Plug Flow Tank											See Table 2
Ideal Mixing Chamber											See Table 2
Circular, Centre Feed, Proto. N.D.	.14	.37	.83	5.4	~1	0.50	—	.29	.71	−.10	Camp
Centre Feed, Proto. N.D. With Swirl	.24	—	.9*	—	—	—	—	—	—	—	Camp *Estimated
Centre Feed, Small Feedwell Model	.03	.12	.65	16	—	.13	—	.04	.96	+.16	Dague *et al.*
Burlington, 'N.D.' Deep, Small Proto.	.08	.19	.56	9.4	—	.26	—	.15	.85	.20	Heinke
Windsor, Prototype	.05	.09	.32	10.	—	.16					Heinke
Windsor, Prototype	.10	.17	.40	10.	—	.18					Heinke
Windsor, Prototype	.10	—	—	—	.4	.26	—	—	—	—	Heinke
Deep Centre Feed, H.D., Model	.15	.22	.48	4.8	—	—	—	.32	.68	.43	Bergman
Deep Centre Feed B.D., Model	.04	.08	.28	11.7	—	—	—	.07	.93	.60	Bergman
Windsor Primary Rotating Arm Strong Wind Low Flow H.D. or B.D.	.06	.25	0.60	8.0	—	.23	0.8	0.19	0.81	.06	McCorquodale
Centre Feed Model with Rotating Arm Design Flow N.D.	0.12 to 0.15	.23 to 0.45	0.84 to 0.83	9.0 to 4.4	—	.17 to 0.45	0.8	0.15 to 0.34	0.85 to 0.66	0 −.05	McCorquodale
Centre Feed Model with Rotating Arm Design Flow H.D.	0.14	.28	.76	6.8	—	.25	—	0.17 Mult. Peaks	0.83	.04	McCorquodale
Centre Feed Model with Rotating Arm Design Flow B.D.	0.06	0.16	0.60	12.	—	0.14	—	0.07	0.93	.13	McCorquodale
Centre Feed Model No Rotation Design Flow B.D.	0.50	0.12	0.45	18.	—	.13	—	.06	0.94	.14	McCorquodale

(continued)

TABLE 3 (continued).

Type of Tank	$\dfrac{T_i}{T_o}$	$\dfrac{T_{10}}{T_o}$	$\dfrac{T_{50}}{T_{10}}$	$\dfrac{T_{90}}{T_{10}}$	$\dfrac{T_g}{T_o}$	$\dfrac{T_p}{T_o}$	T.R.	p	$1-p$	m	Reference
Deep Peripheral Feed, B.D., Model	0.10	.16	.56	6.2	—	—	—	.24	.76	.40	Bergman
Peripheral Feed N.D., Prototype	—	—	.54 to .93	—	—	.4 to .5	—	—	—	—	Dague et al.
Peripheral Feed N.D., Model	.05 to .10	.28	.60 to .70	6.0	—	.3 to .4	—	.25	.75	+.11	Dague et al.
Typical Range for Centre Feed Tanks	.05 to .16		.4 to .8	5 to 12							

Note: N.D. ≡ Neutral Density; B.D. ≡ Buoyant Density; H.D. ≡ Heavy Density.

[91], Abdel-Gawad [1] and Groche [55]. Most of the studies indicated that a bottom current and surface recirculation occurs in the central part of the tank with counter eddy at the bed near the effluent weir.

The velocities in peripherally fed circular clarifiers were measured by Bretscher et al. [18]. Typical results for isothermal conditions are shown in Figure 25. A well defined bottom flow and a surface roller were observed.

Observations of velocity distributions in tanks subject to density currents are scarce. Density currents may result from thermal effects, concentration effects and the release of gas bubbles. Model studies by the authors have used hot and cold water to simulate buoyant and sinking density currents. In real tanks the diurnal variations in flow, temperature and concentration result in unsteady density currents; the density currents in the model study were also unsteady. Figures 26 and 27 illustrate the observed model density currents for buoyant and sinking conditions, respectively. The existence of bottom and surface density currents were confirmed by observations in the West Windsor primary clarifiers (McCorquodale [92]). It was found that prototype temperature differences of the order of 0.1°C were sufficient to induce strong thermal density currents.

Wind Effects

Wind can influence flow in a settling tank in a number of ways, such as:

(a) set-up and set-down, i.e., increasing the water level on one side of the tank and decreasing it on the other side
(b) interference with flow over the weirs
(c) causing waves which in turn result in increased dispersion at the top of the tank
(d) causing surging in the tank, i.e., a seiche effect
(e) inducing surface currents

Wind set-up can be estimated by a number of semi-theoretical equations, for example [75],

$$\Delta H = \frac{\frac{1}{2}kU^2 x}{gh} \qquad (56)$$

where $k = 3.3 \times 10^{-6}$ for lakes; h = effective depth; U = wind speed; x = fetch.

On the windward side of the tank the wind impinges on the water flowing over the weirs forcing the nappe towards the weir plate and visibly changing the pattern of the flow leaving the weir. The impingement pressure head of the air can be estimated by

$$\Delta H_{imp} \cong \frac{\gamma_a V_a^2}{\gamma_w g} \qquad (57)$$

where γ_a = specific weight of the air; V_a = velocity of air. For a 20 mph (32 km/h) wind on a circular tank, there would be about 6% decrease in flow on the windward side with a similar increase on the downwind side.

Both wind induced surface waves and surging have been observed in models and prototypes. The waves are mostly of short wave length (a few inches). The orbital size of a wave decreases exponentially with depth becoming almost insignificant at a depth of ½ the wave length. Thus the mixing effect in both the model and prototype would be restricted to the top few inches of the tank. This means that the wind-water models may be subjected to a disproportionately greater amount of wind induced mixing. Surging can be caused by wind gusts especially if these occur with average periods similar to one of the natural resonance periods for the tanks. Excessive surging would cause a fluctuation in the over-flow velocity at the weirs with the possibility of a slight increase in solids over-flow.

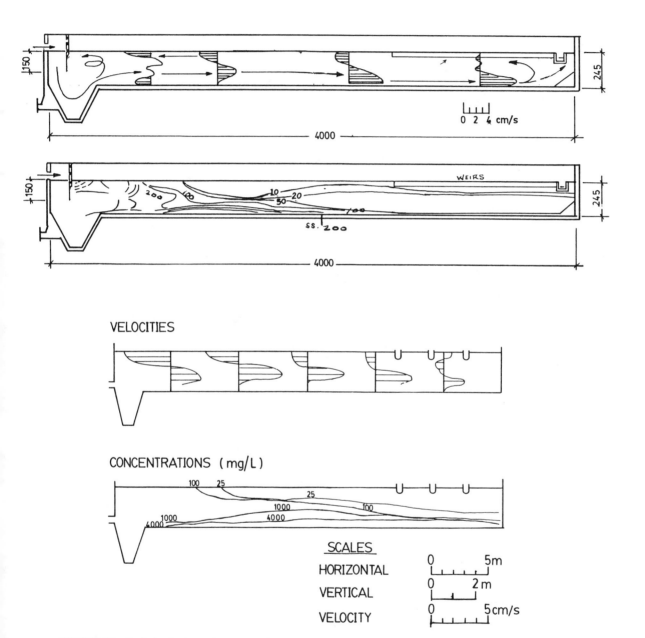

VELOCITIES

CONCENTRATIONS (mg/L)

SCALES

HORIZONTAL   0   5m

VERTICAL   0   2 m

VELOCITY   0   5 cm/s

FIGURE 23. Typical velocity and concentration distributions in rectangular clarifiers (after Stamou and Rodi [133]).

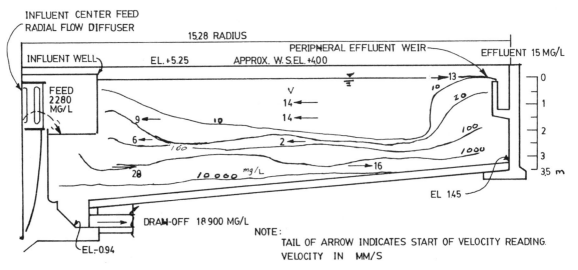

FIGURE 24. Typical velocity and concentration distribution in centre-fed circular clarifier (after Stamou and Rodi [133]).

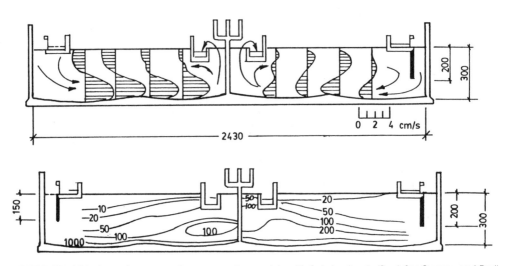

FIGURE 25. Velocity and concentration patterns in a peripherally fed circular clarifier (after Stamou and Rodi [133]).

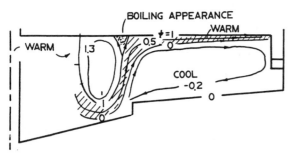

FIGURE 26. Buoyant density current in a centre-fed circular clarifier [91].

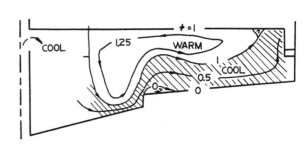

FIGURE 27. Heavy density current in a centre-fed circular clarifier [91].

The effects of wind induced currents have been mentioned by a number of researchers (see Camp [19,22] and Price [114,115]). Price made wind studies on rectangular tanks and found that horizontal uniformity (time ratio) of the flow was adversely affected by cross-winds but only slightly decreased by co-current or counter-current winds.

Keulegan [75] has estimated that the wind induced water surface current is approximately

$$U_s = V_a/30 \qquad (58)$$

where V_a is a representative over-water wind velocity. Thus a 20 mph wind could cause a 1 ft/sec surface current; this current would be confined to a very thin film (a fraction of an inch thick) but a significant shear force would be transmitted to the flow and could drastically change the flow pattern. Larsen [82] reports that wind is a major energy source in exposed tanks.

The importance of the wind shear can be appreciated if it is compared with a typical bed shear value. From the Darcy equation, the bed shear is

$$\tau_o = \frac{\gamma_w f V_w^2}{8g} \qquad (59)$$

and the wind shear

$$\tau_a = \frac{\gamma_a f V_a^2}{8g} \qquad (60)$$

Now $V_a \simeq 0(30)$; $V_w \simeq 0(.02)$; therefore, $\tau_a/\tau_w \simeq 0(10^3)$. Thus wind can be one of the most dominant forces affecting flow patterns in an exposed tank. Fencing and wind-breaks using trees should prove to be beneficial for highly exposed tanks.

Density Currents

When a fluid with a different density than the main fluid body is introduced it may flow in a defined stratum either above or below the main body of the fluid depending on whether its density is less or more than the rest of the fluid. The flow in this stratum is called a density current. The differences in density can be very small (a few parts per million) and can be caused by temperature differences, entrained air or gas bubbles, differences in concentration or density of solids, differences due to concentration of dissolved matter, e.g. salts, and differences in dispersed 'oily' wastes.

The hydraulic efficiency can be improved if these density currents are dispersed into the main body of the fluid; however, this does not necessarily mean that the removal efficiency would be improved. In general, the higher the densimetric Froude number and the higher the Reynolds number the more likely that the density current will be dispersed. Keulegan [74] suggests that in a rectangular basin instability

of a thin density layer would occur if

$$(F')^{3/2}R' > 400 \qquad (61)$$

and

$$(F'_\delta)^{3/2}R'_\delta > 77,000 \qquad (62)$$

where $F' = V'/\sqrt{g'y'}$; $R' = V'y'/v$; $F'_\delta = \Delta V'/\sqrt{g'\delta}$; $R'_\delta = \Delta V'\delta/v$; $y' = $ initial thickness of the density layer; $\delta = $ thickness of the mixing zone; $V' = $ velocity of the density current; $\Delta V' = $ change in velocity across the mixing zone.

Assuming that the radial density current behaves in a manner similar to a longitudinal current, it would appear that a prototype, with a temperature difference of 0.1°C and the design flow, would be unstable according to Equation (61) but stable according to Equation (62), whereas at double the design flow both criteria would indicate instability. Thus the effect of density should decrease with increasing flow.

PROCESS DESIGN

Equalization Basins

The purpose of flow equalization is to smooth the flow and mass loads to a treatment facility. The load smoothing generally improves the efficiency of the clarification and other flow related processes at wastewater treatment plants. The impact on receiving waters may also be lessened by flow equalization.

The flow and mass loading to a wastewater treatment are both stochastic processes (MacInnes, Middleton and Adamowski [89]). This should be considered when evaluating the probable performance of a particular basin design. The influent to an equalization basin may be variable in either or both the hydraulic load and the mass load. Ideally the effluent should have a constant hydraulic and mass loading. This may not be practical for systems with high influent variability, e.g. systems with high infiltration.

The inflow to equalization basins may be by gravity or through pumps. Similarly effluent may be removed by pumps or by gravity. If the outflow is pumped to the treatment facility, the pumping system must be designed so as not to exceed the plant capacity. According to the EPA [139], this can be achieved in one of the following ways:

(a) use of a variable flow division box with possible return of pumped flow to the wet well
(b) use of throttling valve to regulate the outflow
(c) use of variable speed pumps
(d) use of constant speed pumps designed to deliver flows between the average inflow and 1.5 times the average inflow

Another alternative is to pump the inflow and use gravity control for the outflow.

The flow equalization basins are generally aerated in order to prevent deposition of suspended solids. These basins improve the flocculation process thus improving the primary settling efficiency. However, if the outflow is pumped the floc may be sheared.

Sedimentation Basins

The performance of a sedimentation basin in removal of settleable particles depends upon two important parameters, the particle characteristics and the basin hydrodynamics. Until recently, an idealistic but very simple approach has been taken in designing these basins, with a result that most of the designs are carried out on an historical background.

PARTICLE CHARACTERISTICS

The settling velocity, V_s, of a discrete particle of diameter d and density ϱ_s in an infinite size vessel containing quiescent fluid of density ϱ and viscosity μ is given by Equation (2). However, in actual sedimentation basins, there are several other factors that influence the settling velocity of particles. These include the irregularity in shape and size, hinderance due to presence of other particles, influence of flocculating nature of particles, effect of turbulence in fluid moving through the basin and the change in viscosity both due to the presence of impurities and due to temperature change. One or more of these factors are changing at all times in actual basins, and since the impact of these factors on the settling velocity of particles is not well understood, it has become a common practice to either ignore them or use a safety factor in design.

IDEAL SEDIMENTATION BASIN BEHAVIOUR

Most of the sedimentation basins are designed based on surface loading rate originally proposed by Camp [19,21,22,23]. He considered a rectangular section, horizontal flow ideal settling basin and made the following assumptions:

1. Flow is steady.
2. Settling in sedimentation zone is ideal as for a discrete particle.
3. Concentration of suspended particles is same at all depths in the inlet zone.
4. Once the particle hits the sludge zone, it stays there.
5. Flow through time is equal to detention time.

Camp showed that under these conditions, all the particles with settling velocity, V_o, equal to or greater than the surface loading rate, Q/A, will be removed whereas the fractional removal of particles with settling velocity V_s less than V_o will be equal to V_s/V_o. Thus, in a suspension consisting of graded particles of various settling velocities, the total removal fraction by weight of all the particles is given by

$(1 - p_o) + (1/V_o) \int_o^{p_o} V_s \, dp$ where p is the fraction of particles by weight having a settling velocity equal to or less than V_s or V_o in the case of p_o. Ideal circular tanks with horizontal flow are expected to show a similar performance. However, in vertical flow tanks, the fractional removal becomes zero.

Based on Camp's theoretical analysis, sufficient surface area, A, must be provided so that the overflow rate, Q/A, is equivalent to the settling velocity of the smallest discrete particle to be completely removed. The depth or volume of a clarifier theoretically has no effect on clarification unless the particles to be removed are flocculent. For flocculent particles, the volume should provide sufficient detention time to achieve the desired degree of particle growth. It should, however, be recognized that sedimentation basins are poor flocculators.

In case the sedimentation basins are to act as thickening units also, e.g. secondary settling tanks, then the rate of application of solids per unit area should not exceed the rate at which they can be transported to the bottom of the tank. This area depends on the settling characteristics of the solids and the rate at which thickened sludge is removed from the tank. The volume for thickening is controlled by the need for containing sludge accumulated during periods of peak loading and by the need for uniformly removing thickened solids from the bottom of the tank.

The effect of variation in particle characteristics on their removal efficiency as determined by a column settling test is illustrated in Figure 10. It is obvious that the percentage removal rate in suspended solids increases with a decrease in surface loading rate. The effect of initial solids concentration on removal efficiency is shown in Figure 28 [88]. In addition, Figure 11 shows that under ideal settling conditions in a settling jar, up to an additional 5% in flocculated suspended solids can be achieved. The difference in 100% and the actual performance measured by percent removal of suspended solids is due to the combined effects of particle characteristics and basin hydrodynamics.

Camp had divided the ideal settling basin into inlet zone, outlet zone, settling zone and sludge zone. Each of these zones has a pronounced effect on the hydrodynamics of flow through the basin, resulting in turbulence, short circuiting, and density currents and consequently lower removal efficiency. In addition, fluctuations in flow rates can vary the surface loading rate. Some of the salient design factors are listed below:

1. Most of the sedimentation tanks are continuous flow type. These may be either rectangular tanks with longitudinal flow or circular tanks with radial flow. Flow in either case is considered as horizontal flow. In circular tanks, peripheral feed is better hydraulically, but central feed is more common.
2. Several units are provided in parallel for better operational control. This also reduces fluctuations in flow rate

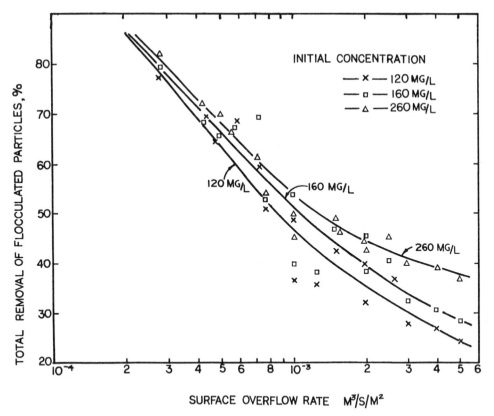

FIGURE 28. Effect of initial concentration on flocculation [88].

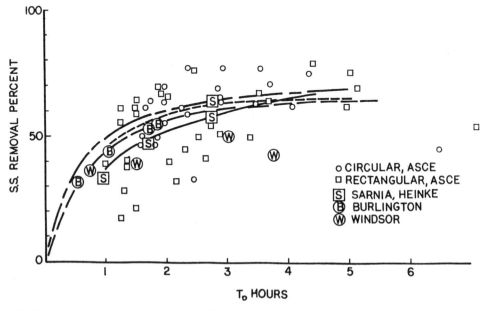

FIGURE 29. Comparison of suspended solids removal in rectangular and circular primary clarifiers [91].

TABLE 4. Simulation of Removal by Sarnia Rectangular Tank.

Fraction of solids (%)	Settling velocity (V_s) mm/s	$\dfrac{V_s}{V_o}$	Removal efficiency %			Measured (60)
			ideal	Camp	model	
20	2.25	3.26	100	100	100	
5	1.5	2.17	100	100	98	
5	1.1	1.6	100	99	90	N.A.
15	0.9	1.3	100	95	77	
15	0.56	0.81	81	73	52	
40	0.00	0.0	0.0	0.0	0.0	
Removal of settleable material			96	92	81	78
Total removal efficiency			57.5	55	48.5	47

Input parameters: L = 32.7 m, H_0 = 2.7 m, d_0 = 1.35 m, q = 0.28 L/s/m, V_0 = 60 m³/m²/day.

and thereby the surface loading rate in each unit. A flow equalization tank before distribution chamber will help in reducing the input fluctuations.

3. Ideally incoming flow and particles should be distributed uniformly between different basins and also within each basin. Inlet system should be designed such that it causes minimum turbulence and short circuiting while it is economical and practical.

4. The rate of flow over the outlet weirs should be as low as possible in order to prevent nonuniform flow caused by wind, obstruction to flow by stranded solids and by variation in weir elevation. Multiple outlet troughs with triangular V-notches are commonly used.

5. Some of the recent changes in designs are the use of upflow contact units and the tube settlers. In upflow contact units, mixing, flocculation and clarification are carried out in the same basin. The flow is in vertical upward direction where water or wastewater has to pass through a blanket of suspended floc and slurry. The performance of such units with influent of a constant

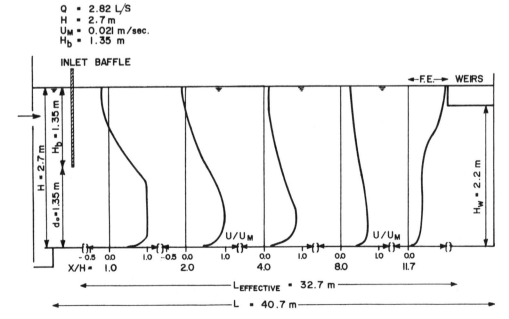

FIGURE 30. Predicted velocity profile in a rectangular clarifier [1,4].

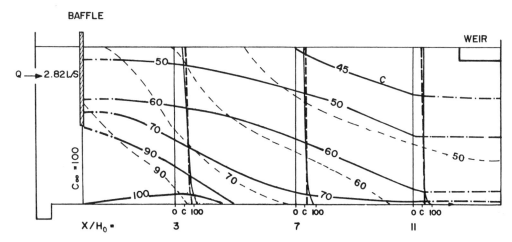

FIGURE 31. Predicted and measured concentration distributions in a rectangular clarifier [1,4].

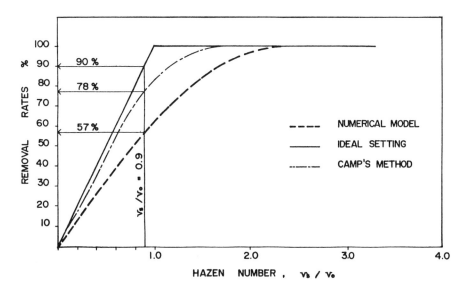

FIGURE 32. Comparison of suspended solids removal models [1].

quality is generally reported to be much better; however, these units are very sensitive to sudden changes in influent quality.

6. The larger of the surface areas, required for clarification and for thickening, and the sum of the two volumes, required for clarification of flocculent particles and for thickening, must be provided. The relationship between volume and surface area will determine the effective depth for the sedimentation basin.

TYPICAL REMOVAL EFFICIENCIES OF CIRCULAR AND RECTANGULAR TANKS

Figure 29 shows the removal efficiencies for several circular and rectangular tanks [8,9,58,59,60,61,136]. It is evident that removal efficiencies of either rectangular or circular tanks can vary considerably and that statistically one cannot say that the rectangular tanks are superior to the circular tanks. The ASCE Manual 36 [8] states that, "Plants showing abnormally high or low removal efficiencies of suspended solids usually can attribute these results to characteristics of the suspension encountered, a marked difference in the ability of the particles to flocculate, and/or the percentage of flocculant material in suspension."

Some researchers claim to have found a significant improvement in removal efficiencies as a result of improved hydraulic efficiency (Villemonte *et al.*). Bergman has shown that buoyant density currents reduce hydraulic and removal efficiencies in circular tanks. On the other hand, White and Allos [145] claim that tank shape and inlet conditions are secondary to the influent sewage concentration and characteristics.

It is particularly interesting to note from Figure 29 that the removal efficiency of circular tanks is very similar to rectangular tanks, although Tables 2 and 3 indicate the hydraulic efficiencies of the circular tanks are generally lower than rectangular tanks.

APPLICATION OF THE NUMERICAL MODEL

The SIM described in the hydrodynamic section was used to determine the velocity field, the concentration distribution and the removal efficiency of the rectangular tank in the City of Sarnia, Ontario. The settling velocity curve, and the measured removal efficiency were reported by Heinke *et al.* [58–61]. An equivalent condition to the inlet of the tank was used in the simulation, e.g. no flow over the baffle was allowed and the equivalent inlet restriction was taken as $H_b/H_0 = 0.5$ where H_b is the height of the baffle and H_0 is the initial water depth. The tank is 40.7 m long, 9.0 m wide and 2.7 m average depth. The effective settling zone from the inlet baffle to the effluent weir is 32.7 m. The overflow rate or surface loading (V_0) is 60 m^3/m^2/day which corresponds to a flowrate per unit width of 0.280 L/s/m. The particles in suspension were discrete and non-uniform. For modelling purposes, the suspension was created in classes.

The settling velocity curve was deduced from the S-curve given by Heinke *et al.* and is shown in Table 4.

Figure 30 shows the predicted velocity profiles at selected sections along the tank length. The recirculation zone extends to $L_r = 5.5H_b$ which again agrees with the finite difference results of Imam and McCorquodale [66]. The extent of the potential core and the growth of the boundary layer are well represented. The concentration contours, for the composite of all the particle size classes, are shown in Figure 31. The predicted and measured concentration profiles at selected sections are also shown in this figure. The distribution indicates the effect of strong mixing as the jet expands. There is a reasonable agreement between the predicted concentration distribution and those measured by Heinke *et al.* [60].

The discrepancy between the measured and the predicted concentration contours at the upper layer of the tank is due to the fact that in the real Sarnia tanks, there is small flow over the influent baffles as they extend from mid-depth to 0.4 m below the free surface. However, in the numerical model, it was assumed that the baffles extend all the way to the free surface. Therefore, the model predicts less sediment concentration at the upper layer than the measured values. This does not significantly affect the predicted final removal, since in the final portion of the tank the isoconcentration lines are in good agreement.

Table 4 also compares removal rates for each class of solids as predicted by (i) the proposed model, (ii) ideal settling theory (no turbulence), (iii) Camp's graph [6] using his formula for wall generated turbulence with a parabolic velocity distribution. The overall removal rates, as predicted by the SIM model (48.5%) and as measured by Heinke *et al.* (47%), are in very good agreement. Figure 32 shows a comparison of different predictive models. For high settling velocities, i.e. at high Hazen numbers ($V_s/V_0 > 2.0$), all the models predict similar removal rates. When the Hazen number was close to unity, the proposed numerical model predicted a lower removal rate than the other models. A comparison between the removal rates predicted by the model and the ideal settling shows the combined effects of both turbulence and recirculation. For the case of $V_s/V_0 = 0.9$, the removal rate as predicted by Camp's model was 77%, while the removal predicted by the proposed model was 57%. The difference in the removal rates may be attributed to the effect of the free mixing and the recirculating flow which are neglected in Camp's model but accounted for by the SIM.

ACKNOWLEDGEMENTS

The authors wish to acknowledge the research contributions of their former students Drs. E. Imam, A. Khalifa and S. M. Abdel-Gawad, that have been incorporated in this article.

REFERENCES AND BIBLIOGRAPHY

1. Abdel-Gawad, S. M., "Simulation of Settling Tanks by the Strip Integral Method," Ph.D. thesis, University of Windsor, Ontario, Canada, 1983.
2. Abdel-Gawad, S. M. and McCorquodale, J. A., "Strip Integral Method Applied to Settling Tanks," *Proc. ASCE,* Vol. 110, HY1, January, 1984.
3. Abdel-Gawad, S. M. and McCorquodale, J. A., "Hydrodynamics of Circular Primary Clarifiers," *Canadian Journal of Civil Engineering,* Vol. 11, No. 2, 1984.
4. Abdel-Gawad, S. M. and McCorquodale, J. A., "Numerical Simulation of Rectangular Settling Tanks," *Journal of Hydraulic Research,* Vol. 23, No. 2, 1985.
5. Abdel-Gawad, S. M. and McCorquodale, J. A., "Simulation of Particle Concentration Distribution in Primary Clarifiers," *Canadian Journal of Civil Engineering,* Vol. 12, No. 3, 1985.
6. Alarie, R. L., McBean, E. A. and Farquhar, G. J., "Simulation Modeling of Primary Clarifiers," *Journal of the Environmental Engineering Division, ASCE,* No. EE2, Vol. 106, Proc. Paper 15312, April, 1980, pp. 293–309.
7. Anderson, N. E., "Design of Final Settling Tanks for Activated Sludge," *Sew. Works J.,* Vol. 17, 50, 1945.
8. American Society of Civil Engineers, Sewage Treatment Plant Design, *ASCE Manual* No. 36, 1959.
9. American Society of Civil Engineers, "Sedimentation Engineering," *ASCE Manuals and Reports on Engineering Practice,* No. 54, reprinted 1977.
10. Ayden, M. and Sumer, B. M., "Solid Particle Settlement in Open-channel Flow," *International Association of Hydraulic Research Journal* (IAHR), Subject A-a (A13), pp. 95–104.
11. Bergman, B. S., "An Improved Circular Sedimentation Tank Design," *Journal and Proceedings, Inst. of Sewage Purification* (South African Branch), Part 1, 1958.
12. Bechteler, W. and Schrimpf, W., "Improved Numerical Model for Sedimentation," *Proc. ASCE,* Vol. 110, HY3, March, 1984.
13. Bewtra, J. K., "Characteristics of Sewage at Delhi," *Journal Water Pollution Control Federation,* Vol. 41, February, 1969, pp. 208–221.
14. Billmeier, E., "Verbesserte Bemessungsvorschlage fur horizontal durchstromte Nachklarbecken von Belebungsanlagen," *Berichte aus Wassergutewirtschaft und Gesundheitsingenieurwesen,* Institut fur Bauingenieurwesen, Technische Universitat Munchen, Nr. 21, 1978.
15. Bird, R. B., Steward, W. E. and Lightfoot, E. N., *Transport Phenomena,* John Wiley & Sons, Inc., 1960.
16. Boadway, J. D., "Determination of Sediment Settling Rates by Non-linear Regression," *Journal of WPCF,* Vol. 50, No. 4, April, 1978.
17. Boadway, J. D., "Dynamics of Growth and Breakage of Alum Floc in Presence of Fluid Shear," *J. of the Envir. Eng. Division,* Vol. 104, No. EE5, October, 1978, pp. 901–915.
18. Bretscher, U., Hager, W. und Hager, W. H., "Untersuchun-
gen uber die Stromungs- und Feststoff-Verteilungen in Nachklarbecken," *gwf-wasser/abwasser,* 125, 1984, H.2.
19. Camp, T. R., "A Study of the Rational Design of Settling Tanks," *Sewage Works Journal,* Vol. 8, 1936.
20. Camp, T. R. and Stein, P. C., "Velocity Gradients and Internal Work in Fluid Motion," *Journal Boston Society of Civil Engineers,* Vol. 30, No. 4, 1943.
21. Camp, T. R., "Discussion of Effect of Turbulence on Sedimentation by W. E. Dobbins," *ASCE Transactions,* Vol. 109, 1944, pp. 660–667.
22. Camp, T. R., "Sedimentation and the Design of Settling Tanks," *ASCE Transactions,* Vol. 111, 1946.
23. Camp, T. R., "Studies of Sedimentation Basin Design," *Sewage and Industrial Wastes,* Vol. 25, 1953.
24. Camp, T. R., "Flocculation and Flocculation Basins," *ASCE Transactions,* Vol. 120, Paper No. 2722, September, 1955, pp. 1–16.
25. Cordoba-Molina, J. F., Hudgins, R. R. and Silveston, P. L., "Settling in Continuous Sedimentation Tanks," *Proc. ASCE,* Vol. 104, EE6, December, 1978.
26. Cheong, H. F. and Shen, H. W., "Stochastic Characteristics of Sediment Motions," *Journal of Hydraulic Division, ASCE,* No. HY7, Paper No. 12269, July, 1976, pp. 1035–1049.
27. Clark, J. W., Viessman, W. and Hammer, M. J., *Water Supply and Pollution Control,* 3rd Edition, IEP-A Dun-Donnelly Publisher, New York, 1977.
28. Clements, M. S., "Velocity Variation in Rectangular Sedimentation Tanks," *Proc. Inst. Civil Engrs.,* Vol. 34, June 1966.
29. Clements, M. S. and Khattab, A. F. M., "Research into Time Ratio in Radial Flow Sedimentation Tanks," *Proc. Inst. Civil Engrs.,* 1968, pp. 471–494.
30. Clements, M. S. and Price, G. A., "A Two-Float Technique for Examination of Flow Characteristics of Sedimentation Tanks," *Journal of Munic. Engrs.,* Vol. 99, February, 1972.
31. Cozens, D. and Palanacki, M., "Improving the Efficiency of a Settling Tank by Means of a Dye Test," *B.A.Sc. Thesis,* Department of Civil Engineering, University of Windsor, 1973.
32. Crosby, R. M. and Bender, J. H., "Hydraulic Considerations that Affect Secondary Clarifier Performance," *U.S. EPA Technology Transfer,* March, 1980.
33. Culp, G., Hansen, S. and Richardson, G., "High Rate Sedimentation in Water Treatment Works," *Journal American Water Works Assoc.,* Vol. 60, June, 1968.
34. Dague, R. R. and Baumann, E. R., "Hydraulics of Settling Tanks Determined by Models," *Proceedings of the Iowa Water Pollution Control Association Annual Meeting,* 1961.
35. Delichatsois, M. A. and Probstein, R. F., "Scaling Laws for Coagulation and Sedimentation," *Journal WPCF,* Vol. 47, No. 5, May 1975.
36. DeRose, S. A. and Versage, P. M., Jr., "The Diurnal Variation of Certain Raw Sewage Characteristics," *B.A.Sc. thesis,* University of Windsor, Windsor, Ontario, 1971.
37. Devantier, B. A. and Larock, B. E., "Sediment Transport

in Stratified Turbulent Flow," *Proc. ASCE,* Vol. 109, HY12, December, 1983.

38. Ditsios, M., "Untersuchungen uber die erforderliche Tiefe von horizontal durchstromten rechteckigen Nachklarbecken von Belebungsanlagen," *Institut fur Siedlungs- und Industriewasserwirtschaft, Grundwasserhydraulik, Schutz- und Landwirtschaftlicher Wasserbau der Technischen Universitat Graz, Graz,* 1982.

39. Dobbins, W. E., "Effect of Turbulence on Sedimentation," *ASCE Transactions,* Vol. 109, No. 2218, 1944, pp. 629–656.

40. Dobbins, W. E., "Solid-Liquid Separation, Advances in Sewage Treatment Design," *Proceedings of a Symposium held at Manhattan College,* New York, May, 1961, pp. 1–23.

41. Eckenfelder, W. W. and O'Connor, D. J., *Biological Waste Treatment,* Pergamon Press, New York, 1961.

42. Eckenfelder, W. W., *Industrial Water Pollution Control,* McGraw-Hill Book Co., New York, 1966.

43. Eckenfelder, W. W. and Ford, D. L., *Water Pollution Control,* Jenkins Book Publishing Co., New York, 1970.

44. Eckenfelder, W. W., *Principles of Water Quality Management,* CBI Publishing Co., Boston, 1980.

45. El.Baroudi, H. M., "Characterization of Settling Tanks by Eddy Diffusion," *J. of Sanitary Eng., ASCE,* Vol. 95, No. SA3, June, 1969, pp. 327–344.

46. Fair, G. M., Geyer, J. C. and Okun, D. A., *Water and Waste Water Engineering,* Vol. 2, John Wiley and Sons, New York, 1968.

47. Fair, G. M., Geyer, J. C. and Okun, D. A., *Elements of Water Supply and Wastewater Disposal,* 2nd Edition, John Wiley & Sons, Inc., New York, 1971.

48. Fischerstrom, C., "Sedimentation in Rectangular Basins," *Proc. ASCE,* Paper No. 687, 1955.

49. Fitch, E. B., "Sedimentation Process Fundamentals," in *Biological Treatment of Sewage and Industrial Wastes,* Ed. J. McCabe and W. W. Eckenfelder, Jr., Reinhold Publishing Corp., 1958.

50. Geiger, G., "Stromungsvorgange im langsdurchflossenen Absetzbecken," *Berichte der Abwassertechnischen vereinigungen e.V.,* Heft 1, Stuttgarter Tagung 49.

51. Geinopolis, A. and Katz, W. J., "United States Practices in Sedimentation of Sewage and Solid Wastes," in *Water Quality Improvement by Physical and Chemical Processes,* University of Texas Press, Austin, Texas, 1970.

52. Gibson, M. M. and Launder, B. E., "On the Calculation of Horizontal Turbulent Free Shear Flows Under Gravitational Influence," *J. of Heat Transfer, Transactions of the ASME,* No. 76, HTS, February, 1976, pp. 81–87.

53. Girgidov, A. D., "Model for Motion of Solid Particles Suspended in Turbulent Flow," *IAHR*—(A12), Vol. 1, 1977, pp. 87–94.

54. Girling, R. M., "Experiences with High Rate Tube Settlers Applied to an Existing Final Clarifier, Winnipeg, Canada," *Intl., Assn. Water Poll. Res. Meeting,* Vienna, Austria, 1976.

55. Groche, D., "Die Messung von FlieBvorgangen in angefuhrten Bauwerken der Abwasserreinigung mit Hilfe von kunstlichen radioaktiven Isotopen und ihre Auswertung sowie Ruckschlusse auf die Konstruktionselemente," *Lehrstuhl fur Siedlungswasserwirtschaft,* Munchen, 1964.

56. Hamlin, M. J., Discussion of "Hydraulic and Removal Efficiencies in Sedimentation Basins," *Proceedings of the Third International Conference on Water Pollution Research,* Munich, 1966.

57. Hazen, A., "On Sedimentation," *ASCE Transactions,* 53, Paper No. 980, 1904, pp. 45–88 (with discussions).

58. Heinke, G. W., "Design and Performance Criteria for Settling Tanks for the Removal of Physical-Chemical Flocs," *Summary Report,* University of Toronto, March, 1974.

59. Heinke, G. W., Qazi, M. A. and Tay, A., "Design and Performance Criteria for Settling Tanks for Removal of Physical-Chemical Flocs," *Research Report,* University of Toronto, March, 1975.

60. Heinke, G. W., Qazi, M. A. and Tay, A., "Design and Performance Criteria for Settling Tanks for the Removal of Physical-Chemical Flocs," Research Program for the Abatement of Municipal Pollution Under Provisions of the Canada-Ontario Agreement on Great Lakes Water Quality, Vol. II, *Research Report* No. 56, 1977.

61. Heinke, G. W., Tay, A. and Qazi, M. A., "Effects of Chemical Addition on the Performance of Settling Tanks," *Journal of WPCF,* Vol. 52, No. 12, December, 1980.

62. Hinze, J. O., *Turbulence,* McGraw-Hill Book Company, Inc., New York, 1959.

63. Hinze, J. O., *Turbulence,* Revised Edition, McGraw-Hill Book Co., Inc., New York, 1975.

64. Hirsch, A. A., "Basin Tracer Curves Interpreted by Basin Analytics," *J. of Sanitary Eng., ASCE,* No. SA6, Vol. 95, Proc. Paper 6951, December, 1969, pp. 1031–1050.

65. Imam, E. H., "Numerical Modelling of Rectangular Clarifiers," thesis presented to the University of Windsor, at Windsor, Ontario, in 1981, in partial fulfillment of the requirements for the degree of Doctor of Philosophy.

66. Imam, E. and McCorquodale, J. A., "Simulation of Flow in Rectangular Clarifiers," *Proc. ASCE,* Vol. 109, EE3, June, 1983.

67. Imam, E., McCorquodale, J. A. and Bewtra, J. K., "Numerical Modelling of Sedimentation Tanks," *Proc. ASCE,* Vol. 109, HY12, December, 1983.

68. Ingersoll, A. C., McKee, J. E. and Brooks, N. H., "Fundamental Concepts of Rectangular Settling Tanks," *ASCE Transactions,* Paper No. 3837, 1956.

69. Ippen, A. T., *Estuary and Coastline Hydraulics,* Chapter 5, McGraw-Hill Book Co., New York, 1960. Also *Shore Protection Manual,* Coastal Engineering Research Center, U.S. Army, 1973.

70. Jobson, H. E. and Sayre, W. W., "Vertical Transfer in Open Channel Flow," *Journal of the Hydraulic Division, ASCE,* No. HY3, Vol. 96, Paper No. 7148, March 1970, pp. 703–724.

71. Jobson, H. E. and Sayre, W. W., "Predicting Concentration Profiles in Open Channels," *Journal of the Hydraulic Division, ASCE,* Vol. 96, No. HY10, Paper No. 17618, October, 1970, pp. 1983–1995.

72. Kalbskopf, K. H., "European Practices in Sedimentation," in *Water Quality Improvement by Physical and Chemical Processes,* University of Texas Press, Austin, Texas, 1970.

73. Kerssens, P. J. M., Van Rijn, L. C. and Van Wijngaarden, N. J., "Model for Non-Steady Suspended Sediment Transport, IAHR, *Modelling of Sediment Transport,* Subject A-a, Project Engineers Delft Hydraulics Lab., Vol. 1, No. A15, pp. 113–120.

74. Keulegan, G. H., "Wave Motion," in *Engineering Hydraulics,* Ed. H. Rouse, John Wiley and Sons, New York, 1950.

75. Keulegan, G. H., "Wind Tides in Small Closed Channels," *U.S. Bureau of Standards Journal of Research,* Vol. 46, May, 1951.

76. Khalifa, A., "Theoretical and Experimental Study of the Radial Hydraulic Jump," *Ph.D. thesis,* University of Windsor, Windsor, Ontario, 1980.

77. Khattab, A. F., "Flow Patterns in Circular Sedimentation Tanks," *Ph.D. thesis,* University of Sheffield, Sheffield, England, 1967.

78. Lai, K. M., Law, H. C. and Ngan, Y. H., "The Study of Coagulation and Sedimentation Processes for the West Windsor Pollution Control Plant," *B.A.Sc. thesis,* University of Windsor, Windsor, Ontario, 1979.

79. Larock, B. E. and Schamber, D. R., "Finite Element Computation of Turbulent Flows," *Third International Conference on Finite Element in Water Resources,* 1980, pp. 4.31–4.47.

80. Larsen, P., "Research on Settling Basin Hydraulics," *Tenth Anniversary Papers on Research in Progress,* Dept. of Water Res. Engg., Lund Ins. of Tech., Bull. Serie A, No. 55, Lund, Sweden, July, 1976, pp. 137–149.

81. Larsen, P. and Gotthardsson, S., "Om Sedimenteringsbassangers Hydraulik," Inst. for Teknisk Vattsresurslara, *Bulletin Serie* A NR 51, 1976.

82. Larsen, P., "On the Hydraulics of Rectangular Settling Basins—Experimental and Theoretical Studies," *Report No. 1001,* Department of Water Resources Engineering, Lund Institute of Technology, Lund, Sweden, 1977.

83. Larsen, P., "Apparat for maturing av sma densitetsskillnader hos vatskor," *Vatten* No. 2, 1974.

84. Launder, B. E. and Spalding, D. B., "Lectures in Mathematical Models of Turbulence," Academic Press, London and New York, 1972.

85. Launder, B. E. and Spalding, D. B., "The Numerical Computation of Turbulent Flows," *Computer Methods in Applied Mechanics and Engineering,* 1974, pp. 269–289.

86. Launder, B. E. and Spalding, D. B., *Lectures in Mathematical Models of Turbulence,* Third Printing, Academic Press, London, 1979.

87. Lee, C. R., Fan, L. T. and Takamatsu, T., "Optimization of Multistage Secondary Clarifier," *J.W.P.C.F.,* Vol. 48, No. 11, November, 1976, pp. 2578–2589.

88. Low, K. H., "Column Test on Flocculated Waste Water in the West Windsor Pollution Control Plant," *B.A.Sc. thesis,* Department of Civil Engineering, University of Windsor, Windsor, Ontario, 1984.

89. MacInnes, C. D., Middleton, A. C. and Adamowski, K., "Stochastic Design of Flow Equalization Basins," *Journal of the Environmental Engineering Division, ASCE,* Vol. 104, No. EE6, December, 1978, p. 1275.

90. Mazurczyk, A. L. et al., "Methodology for Assessing Clarifier Operation as Demonstrated at Pruskow, Poland," *Paper presented at the 53rd Annual Conf. of the WPCF,* Las Vegas, November, 1980.

91. McCorquodale, J. A., "Hydraulic Study of the Circular Settling Tanks at the West Windsor Pollution Control Plant," *Report submitted to Lafontaine, Cowie, Buratto and Associates, Limited,* Consulting Engineers, Windsor, Ontario, 1976.

92. McCorquodale, J. A., "Study of Temperature Distribution in the West Windsor Pollution Control Plant Primary Clarifiers," for *Lafontaine, Cowie, Buratto and Associates,* Windsor, Ontario, 1977.

93. McCorquodale, J. A., Abdel-Gawad, S. M. and Imam, E. H., "Modeling Sedimentation Basins," *International Conference on Computation Methods and Experimental Measurements,* Washington, D.C., July, 1982.

94. McCorquodale, J. A. and Khalifa, A., "Internal Flow in Hydraulic Jump," *Proceedings, Journal of the Hydraulics Division, ASCE,* Vol. 109, No. 5, May, 1983, pp. 684–701.

95. McLaughlin, R. T., "Settling Properties of Suspension," *ASCE Transactions,* No. 3275, 1961, pp. 1734–1786 (with discussion).

96. Mehaute, B., *An Introduction to Hydrodynamics and Water Waves,* Springer-Verlag, New York, Heidelberg, Berlin, 1976.

97. Mendis, J. B. and Benedek, A., "Tube Settlers in Secondary Clarification of Domestic Wastewaters," *Journal of WPCF,* Vol. 52, No. 7, July, 1980.

98. Menez, J. P. and Patterson, J. W., "Development and Validation of a Waste Strength Equalization Basin Design Technique," *Proc. 38th Industrial Waste Conference,* Purdue University, Ann Arbor Science Publishers, Woburn, Mass., 38, 919, 1984.

99. Metcalf and Eddy, *Wastewater Engineering—Collection, Treatment, Disposal,* McGraw-Hill Book Co., Inc., New York, 1972.

100. Miller, D. G., "Sedimentation. A Review of Published Work," *Water and Water Engineering,* February, 1964.

101. Mohart, J. L., "Model Dye Tracer Studies of Currents in Final Sedimentation Basins," *Special Problem Report,* Univ. of Kansas, Lawrence, November, 1978.

102. Morrill, A. B., "Sedimentation Basin Research and Design," *J.A.W.W.A.,* Vol. 24, No. 9, 1932, pp. 1442–1463.

103. Moses, Hal L., "A Strip-Integral Method for Predicting the Behavior of the Turbulent Boundary Layer," Proceedings,

Computation of Turbulent Boundary Layers, 1968, *AFOSR-IFP-Stanford Conference,* pp. 76–82.

104. Moursi, A. M. (El-Sebakhy), "Experimental Study of Flow in Rectangular Settling Tanks," *M.A.Sc. thesis,* Department of Civil Engineering, University of Windsor, Windsor, Ontario, 1982.

105. Munch, W. L. and Fitzpatrick, J. A., "Performance of Circular Final Clarifiers at an Activated Sludge Plant," *Journal of WPCF,* Vol. 50, No. 2, February, 1978.

106. Narayanan, R., "Theoretical Analysis of Flow Past Leaf Gate," *Journal of the Hydraulics Division, ASCE,* No. HY6, June, 1972, pp. 992–1011.

107. Narayanan, R., "Wall Jet Analogy to Hydraulic Jump," *Journal of the Hydraulics Division, ASCE,* Vol. 101, No. HY3, Proc. Paper 11172, March 1975, pp. 347–359.

108. O'Connor, D. J. and Eckenfelder, W. W., "Evaluation of Laboratory Settling Data for Process Design," in *Biological Treatment of Sewage and Industrial Wastes,* Ed. McCabe, B. J., 1958.

109. Orton, J. W., "Discussion on Sedimentation Basin Research and Design, by A. B. Morrill," *J.A.W.W.A.,* Vol. 24, No. 9, pp. 1442–1463.

110. Ouano, E. A. R., "Developing a Methodology for Design of Equalization Basins," *Water Engineering and Management,* November, 1977, p. 48.

111. Ponn, J., "Geschwindigkeitsverteilungen in radial durchstromten Nachklarbecken—Verwendung einer neu entwickelten Thermosonde," *Institut fur Siedlungs- und Industriewasserwirtschaft, Grundwasserhydraulik, Schutz- und Landwirtschaftlichen Wasserbau, Technische Universitat Graz, Graz,* Juni, 1975.

112. Parker, D. S. *et al.,* "Physical Conditioning of Activated Sludge Flow," *Journal of WPCF,* Vol. 43, 1971.

113. Patry, G. G., "Analysis and Design of Equalization Facilities," *Proc. of Environmental Engineering, ASCE,* July, 1980, p. 183.

114. Price, G. A. and Clements, M. S., "Some Lessons from Model and Full-Scale Tests in Rectangular Sedimentation Tanks," Paper No. 1, *Journal of Water Pollution Control* (London), Vol. 73, 1974.

115. Price, G. A., "Extension of the Time-Ratio Theory of Sedimentation Tanks," *Proc. Inst. of Civil Engrs.,* Vol. 57, Part 2, June, 1974.

116. Rajaratnam, N., *Turbulent Jets,* Elsevier Scientific Co., New York, 1976.

117. Ramalho, R. S., *Introduction to Wastewater Treatment Processes,* Academic Press, Inc., New York, 1977.

118. Rebhun, M. and Argaman, Y., "Evaluation of Hydraulic Efficiency of Sedimentation Basins," *Proc. ASCE,* Vol. 91, SA. 5, 1965.

119. Resch, H., "Untersuchungen an vertikal durchstromten Nachklarbecken von Belebungsanlagen," Berichte aus Wassergutewirtschaft und Gesundheitsingenieurwesen, *Institut fur Bauingenieurwesen, Technische Universitat Munchen,* Nr. 29, 1981.

120. Reynolds, A. J., *Turbulent Flows in Engineering,* John Wiley & Sons, New York, 1974.

121. Roach, P. J., *Computational Fluid Dynamics,* Hermosa Publishers, Albequerque, 1976.

122. Robinson, J. H., Jr., "A Study of Density Currents in Final Sedimentation Basins," *M.S. thesis,* University of Kansas, Lawrence, 1974.

123. Rodi, W., "Turbulence Models and Their Application in Hydraulics—A State-of-the-Art Review," *International Association of Hydraulics Research,* June, 1980.

124. Sarikaya, H. Z., "Numerical Model for Discrete Settling," *J. Hydraulic Division, ASCE,* Vol. 103, No. HY8, Proc. Paper 13150, August, 1977, pp. 865–877.

125. Schamber, D. R. and Larock, B. E., "A Finite Element Model of Turbulent Flow in Primary Sedimentation Basins," *Proceedings, Finite Element in Water Resources,* FE2, London, England, July, 1978, pp. 3.3–3.21.

126. Schamber, D. R. and Larock, B. E., "Numerical Analysis of Flow in Sedimentation Basins," *Journal of the Hydraulics Division, ASCE,* Vol. 107, No. HY5, May, 1981, pp. 575–591.

127. Schlichting, H., *Boundary-Layer Theory,* McGraw-Hill Series in Mechanical Engineering, 1968.

128. Schroeder, E. D., *Water and Wastewater Treatment,* McGraw-Hill Series in Water Resources and Environmental Engineering, New York, 1977.

129. Shiba, S. and Inoue, J., "Dynamic Response of Settling Basins," Journal of the Env. Eng. Division, ASCE, Vol. 101, No. EE5, Proc. Paper 11609, October, 1975, pp. 741–757.

130. Shiba, S., Inoue, J. and Ueda, Y., "Dynamic Estimation of Settling Basin Performance," *Journal of the Env. Eng. Division, ASCE,* Vol. 105, No. EE6, December, 1979, pp. 1113–1129.

131. Smart, P. L. and Laidlaw, J. M. S., "An Evaluation of Some Fluorescent Dyes for Water Tracing," *Water Resources Research,* Vol. 13, No. 1, February, 1977, pp. 15–32.

132. Smith, R., "Preliminary Design of Wastewater Treatment Systems," *Journal of the Sanitary Eng. Div., ASCE,* Vol. 95, No. SA1, February, 1969, pp. 117–145.

133. Stamou, A. I. and Rodi, W., "Review of Experimental Studies on Sedimentation Tanks," *StB* 210/E/2, Universitat Karlsruhe, August, 1984, p. 68.

134. Stukenberg, J. R., Rodman, L. C. and Tarslee, J. E., "Activated Sludge Clarifier Design Improvement," *Journal of WPCF,* Vol. 55, No. 4, April, 1983.

135. Takamatsu, T., Naito, M., Shiba, S. and Ueda, Y., "Effect of Deposit Resuspension of Settling Basins," *J. of the Env. Eng. Division, ASCE,* Vol. 100, No. EE4, August, 1974, pp. 883–903.

136. Tay, A. J. H. and Heinke, W. G., "Velocity and Suspended Solids Distribution in Settling Tanks," *Journal of the Water Pollution Control Federation,* 55, pp. 261–269.

137. Tebbutt, T. H. Y. and Christoulas, D. G., "Performance

Relationships for Primary Sedimentation," *Water Research,* Vol. 9, Pergamon Press, 1975, pp. 347–356.

138. Thomson, D. M., "Scaling Laws for Continuous Flow Sedimentation in Rectangular Tanks," *Proc. Inst. of Civil Engrs.,* Vol. 43, July, 1969.

139. U.S. Environmental Protection Agency, "Process Design Manual for Upgrading Existing Wastewater Treatment Plants," October, 1974, pp. 3–1 to 3–21.

140. U.S. Environmental Protection Agency, U.S. Government, *Process Design Manual for Suspended Solids Removal,* EPA 625/1-75-003a, January, 1975.

141. Villemonte, J. R., Rohlich, G. A. and Wallace, A. T., "Hydraulic and Removal Efficiencies in Sedimentation Basins," *Proceedings of the Third International Conference on Water Pollution Research,* Munich, 1966.

142. Voshel, D. and Sak, J. G., "Effect of Primary Effluent Suspended Solids and BOD on Activated Sludge Production," *Journal of WPCF,* Vol. 40, No. 5, May, 1968.

143. Weidner, J., "Zufluß, Durchfluß and Absetzwirkung zweckmäßig gestalteter Rechteckbecken," Kommissionsverlag R. Oldenbourg, Munchen, 1967.

144. White, J. B., *The Design of Sewers and Sewage Treatment Works,* Edward Arnold Press, London, 1970.

145. White, J. B. and Allos, M. R., "Experiments on Wastewater Sedimentation," *Journal of Water Pollution Control Fed.,* Vol. 48, July, 1976.

146. Zononi, A. E. and Blomquist, M. W., "Column Settling for Flocculant Suspensions," *J. of Env. Eng.,* Vol. 101, No. EE3, *Paper No.* 11356, June, 1975, pp. 309–318.

CHAPTER 25

Granular Activated Carbon in Water Treatment

CHANDUBHAI N. PATEL,* SU LING CHENG,* AND KASHI BANERJEE*

INTRODUCTION

The use of Granular Activated Carbon (GAC) for treatment of potable water supplies offers the water plant a simple, effective and economical way to remove taste, odor, color and other undesirable organics including toxic substances on a continuous basis without capital investment costs. Activated carbon removes organic contaminants from water by the process of adsorption or the attraction and accumulation of one substance on the surface of another. In general, high surface area and pore structure are the prime considerations in adsorption of organics from water; whereas, the chemical nature of the carbon surface is of relatively minor significance. Granular activated carbons typically have surface areas of 500–1,400 square meters per gram.

Much of the surface area available for adsorption in granular carbon particles is found in the pores within the granular carbon particles created during the activation process. The major contribution to surface area is located in pores of molecular dimensions. A molecule will not readily penetrate into a pore smaller than a certain critical diameter and will be excluded from pores smaller than this diameter. The most tenacious adsorption takes place when the pores are barely large enough to admit the adsorbing molecules. The smaller the pores with respect to the molecules, the greater the forces of attraction.

Activated carbons can be made from a variety of carbonaceous materials including wood, coal, peat, lignin, nut shells, bagasse (sugar cane pulp), sawdust, lignite, bone and petroleum residues. Activated carbon made from wood is normally termed charcoal and was the first type of granular carbon to find its way into municipal water treatment. Granular carbons made from coal are hard and dense, and are well-suited to water treatment because the carbon wets rapidly and does not float, but conforms to a neatly packed bed with acceptable pressure drop characteristics. Also, the carbon is quite dense, and thereby makes available more adsorption capacity in any given filter volume.

For most activated carbons, however, some properties arising from the nature of the starting material are masked by choice of production technique. These production techniques can form granular activated carbons by crushing or pressing. Crushed activated carbon is prepared by activating a lump material, which is then crushed and classified to desired particle size. Pressed activated carbon is formed prior to activation. The physical characteristics of hardness, a very important factor, are directly related to the nature of the starting material for crushed activated carbons.

MANUFACTURE AND CHARACTERISTICS OF ACTIVATED CARBON

The Activation Process

Activated carbon is manufactured by a process consisting of raw material dehydration and carbonization followed by activation. The starting material is dehydrated and carbonized by slowly heating in the absence of air, sometimes using a dehydrating agent such as zinc chloride or phosphoric acid. Excess water, including structural water, must be driven from the organic material. Carbonization converts this organic material to primary carbon, which is a mixture of ash (inert inorganics), tars, amorphous carbon, and crystalline carbon (elementary graphitic crystallites). Non-carbon elements (H_2 and O_2) are removed in gaseous form and the elementary carbon atoms are grouped into ox-

*Department of Civil and Environmental Engineering, New Jersey Institute of Technology, Newark, NJ

idized crystallographic formations. During carbonization, some decomposition products or tars will be deposited in the pores, but will be removed in the activation steps [1,2].

The essential steps in carbonization are as follows:

1. Dry the raw material at temperatures up to 170°C.
2. Heat the dried material above 170°C causing degradation with evolution of CO, CO_2 and acetic acid.
3. Exothermal decomposition of the material at temperatures of 270–280°C with formation of considerable amounts of tar, methanol and other by-products
4. Complete the carbonization process at a temperature of 400–600°C with a yield of approximately 80 percent primary carbon.

Activation is essentially a two-phase process requiring burn-off of amorphous decomposition products (tars), with an activating agent such as steam or carbon dioxide (steam is most widely used). Steam, at temperatures of 750–950°C, burns off the decomposition products exposing pore opening for subsequent enlargement. All pores are not plugged with amorphous carbon, therefore some pores are exposed to the activating agent for larger periods of time. Exposure to the activating agent results in the widening of existing pores, and development of the macroporous structure.

The Nature of Adsorption

Adsorption is the phenomenon whereby molecules adhere to a surface with which they come into contact, due to forces of attraction at the surface. The fact that activated carbon has an extremely large surface area per unit weight makes it an extremely efficient adsorptive material. The forces of attraction between the carbon and the adsorbed molecules are greater the closer the molecules are in size to the pores. The best adsorption takes place when the pores are just large enough to admit the molecules.

Crushing of carbon particles to produce smaller particles enhances the rate of adsorption by exposing more entrances to the carbon pores. However, it is evident that any restriction of pore openings or buildup of ash or other materials within the pore openings due to the presence of suspended or colloidal materials and their accumulation on or in the carbon particles might have an adverse effect upon the adsorptive capacity or service life of the carbon. These hazards can be minimized by applying water which has been pretreated to the highest practical clarity to the carbon.

Carbon Properties Relating to Adsorption

Because adsorption is a surface phenomenon, the ability of activated carbon to adsorb large quantities of organic molecules from solution stems from its highly porous structure, which provides a large surface area. Carbon has been activated with a surface area yield of some 2,500 m²/gram, but 1,000 m²/gram is more typical. Total surface area is normally measured by adsorption of nitrogen gas by the Brunauer-Emmett-Teller (BET) method. The distribution of this area into pores of different diameters is measured by determining the amount of nitrogen desorbed at intermediate pressures.

Carbon Particle Size

Activated carbons are classified according to their form: for example, granular or powdered; and according to their use: for example, water or wastewater purification, sugar decolorizing, and liquid or gas phase solvent extraction. Granular carbons are those which are larger than approximately U.S. Sieve Series No. 50, while powdered carbons are those which are smaller. The two most popular sizes of granular carbon are nominally 8 × 30 mesh and 12 × 40 mesh. The finer material has a higher rate of adsorption, but also has a higher head loss per unit depth of bed.

By changing the activating conditions, both the size and number of pores may be controlled to provide the macro and micropore distribution that is most suited for a particular use. The macro-pores are in excess of 100 Å and provide a ready acess to the micropores (100 Å) which are of molecular dimensions. These small micropores provide the major surface area for adsorption. In fact, in water grad carbons more than 70 percent of the available surface area occurs in pores having a radius of less than 50 Å. In good quality activated carbons the external surface area is negligible compared to that in the internal pores, and a particle size reduction would provide no appreciable increase in surface area. Therefore, the adsorption capacity of a given carbon is independent of particle size, exept for very large or insoluble molecules, while the rate of adsorption is particle size dependent.

Properties of Granular Carbon

Activated carbon contains chemically bonded elements, such as oxygen and hydrogen which can be derived from the starting material and remain as a result of imperfect carbonization, or they can become chemically bonded to the surface during activation. In addition it contains ash which is not an organic part of the product. The physical properties which are important in performance include resistance to breakage, particle size and degree of dustiness. For measuring resistance to breakage there are empirical tests such as Abrasion Number and hardness number. Particle size is determined by screen analysis. Density is simply weight per unit vol. of carbon. Typical specifications of GAC for water treatment are:

Total surface area	850–1,500 m²/g
Bulk density	26 lb/ft³
Particle density, wetted in water	1.3–1.4 g/cc
Effective size	0.8–0.9 mm
Uniformity coefficient	1.9 or less
Mean particle diameter	1.5–1.7 mm

Iodine number	Min 850
Abrasion number	Min 70
Ash	Max 8%
Moisture	Max 2%

MAJOR DESIGN CONSIDERATIONS

Three of the major considerations in the application of activated carbon for potable water treatment are:

1. The form of activated carbon
2. The carbon capacity
3. The rate of adsorption

The capacity and rate of adsorption are influenced by:

1. The nature and concentration of activated carbon
2. The nature and concentration of impurities present
3. The nature of the solvent (water)
4. The environmental and operating conditions

In treating large quantities of potable water, there is little control that can be exercised over the nature of the solvent, or the concentration of the impurities. In addition, environmental control has severe limitations due to costs and the requirements of the treated water. Therefore, the desired performance of the adsorption process is normally achieved by controlling the nature and concentration of the activated carbon, the method of operation, the allowable concentration of impurities in the final product and, to a minor extent, the nature of the impurities.

The first decision that must be made is the form of activated carbon to be employed. Powdered activated carbons are normally considered to be smaller than 150 mesh, while granular carbons are larger. After selecting the form of activated carbons, the capacity, rate of adsorption and method of application may be determined. Powdered activated carbons are normally more economical for minor or normal taste and odor problems, while granular carbon is desirable for more severe conditions or where high degrees of removal are required. When in doubt, both powdered and granular systems should be evaluated to ascertain which system is more cost-effective in achieving the desired results.

The equilibrium adsorption capacity may be determined from simple isotherm tests. Here, various dosages of activated carbon are added to different containers of the water to be treated, and the amount of organics removed calculated by the expression:

$$X = (C_o - C_E)V \qquad (1)$$

where

$X =$ the weight of organics adsorbed (mg)
$C_o =$ the initial organic concentration (mg/l)

$C_E =$ the equilibrium concentration achieved (mg/l)
$V =$ the test volume treated [1]

The specific adsorption capacity, X/M, is then calculated by dividing both sides of Equation (1) by the weight of activated carbon added to produce:

$$X/M = \frac{(C_o - C_E)\,V}{M} \qquad (2)$$

where $M =$ weight of activated carbon added (mg). Since the weight of carbon added per unit volume is the dosage, M/V, Equation (2) may also be expressed as:

$$X/M = \frac{C_o - C_E}{D} \qquad (3)$$

where $D =$ the carbon dosage (mg/l).

The carbon concentration which will adsorb a given quantity of adsorbate is related through an adsorption isotherm such as that shown in Figure 1. It should be noted that the amount of solute removed per unit weight of carbon increases with solute concentration initially, but approaches a limiting value at higher adsorbate concentrations.

Although there are several mathematical expressions that relate the amount of solute adsorbed per unit weight of adsorbent, the Langmuir and Freundlich isotherms are usually employed. The Langmuir equation for isothermal adsorption may be developed from a consideration of the thermodynamics of adsorption and is based on the assumptions that:

1. The maximum adsorption corresponds to a complete monolayer of solute on the adsorbent surface.
2. The energy of adsorption is constant at all adsorption sites.
3. No translational migration of solute occurs at the adsorbent site.

The Langmuir equation may be written as:

$$q = \frac{X}{M} = \frac{kKC_E}{1 + KC_E} \qquad (4)$$

where

$q =$ constant related to the energy of adsorption
$q =$ adsorbate required to produce a complete monolayer

The basic equation may be rearranged to produce two linear forms of the equation which are useful in evaluating the adsorption constants. The two linear forms are:

$$\frac{C_E M}{X} = \frac{1}{kK} + \frac{C_E}{k} \qquad (5)$$

$$\frac{1}{q} = \frac{M}{X} = \frac{1}{k} + \frac{1}{kKC_E} \qquad (6)$$

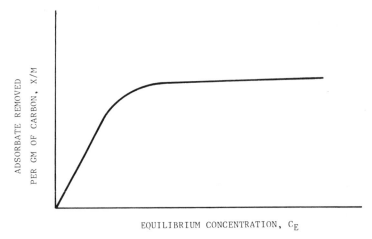

FIGURE 1. Typical adsorption isotherm.

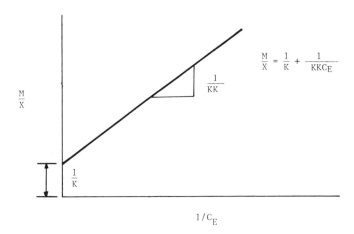

$$\frac{M}{X} = \frac{1}{K} + \frac{1}{KKC_E}$$

FIGURE 2. Evaluating the Langmuir constants.

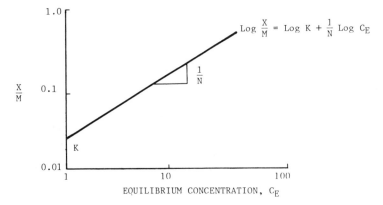

$$\text{Log } \frac{X}{M} = \text{Log } K + \frac{1}{N} \text{ Log } C_E$$

FIGURE 3. Evaluating the Freundlich constants.

Equation (6) is used in Figure 2 to demonstrate the evaluation of the constants k and K, but either equation may be employed. The choice is a matter of personal preference and the form of the data to be evaluated.

The Freundlich equation for isotherm adsorption was developed from empirical relations, but Adamson shows that it may be developed from the Langmuir relations by assuming a heterogeneous adsorption surface. The Freundlich equation may be written:

$$X/M = KC_E^{1/N} \qquad (7)$$

where k and N are constants.

Since

$$X/M = \frac{(C_O - C_E)V}{M}$$

and M/V is a dosage term the Freundlich equation may also be written

$$q = \frac{C_O - C_E}{D} = KC_E^{1/N} \qquad (8)$$

A linear form of the Freundlich equation may also be obtained by taking the logarithm of the Equation (7) which becomes,

$$\log \frac{X}{M} = \log K + \frac{1}{N} \log C_E \qquad (9)$$

This linear equation may then be used to evaluate the system constants K and $1/n$ as shown in Figure 3.

From the Figure 3, the slope of the line produced is equal to $1/n$ and the intercept equal to $\log K$ at a concentration $c = 1$, since $\log c = 0$.

Developing isotherms on a single water sample is helpful in comparing carbons, but it is inadequate for economic calculations or system design. In addition, the extraction of data from isotherms depends upon the type of process to be installed. For example, the carbon loading for a single stage powdered carbon system must be read at the desired effluent concentration, C_E, while granular columns can normally be designed to operate until the entire column is exhausted at the influent concentration, C_O. Countercurrent powdered carbon systems achieve carbon loading between these two extremes, and the actual loading depends upon the number of stages and slope of the adsorption isotherm. However, with the large flows encountered in potable water plants countercurrent powdered carbon staging would be difficult.

To achieve meaningful cost and design data, multiple isotherm tests should be conducted over a period of time to develop a statistical loading or carbon consumption rate. The latter is more useful and may be determined for each test by the equation:

$$R_{cc} = \frac{Q\,(C_O - C_E)\,8.34}{(X/M)} \qquad (10)$$

where

R_{cc} = carbon consumption rate (lb/day)
Q = flow rate (MGD)

The carbon consumption rates observed over a period of time may then be converted to a statistical plot. This is accomplished by arranging the carbon consumption rates obtained in descending order and calculating the percent of the

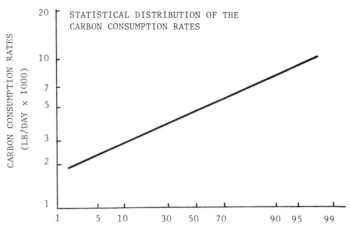

FIGURE 4. Percent of the exhaustion rates equal to or greater than the stated value.

time a given use rate is equalled or exceeded by the expression:

$$p = \frac{k \times 100}{n + 1} \qquad (11)$$

where

p = percent of the exhaustion rate
k = the rank of the sample
n = total number of sample

The data can then be presented on a statistical plot such as that shown in Figure 4. This is an informative plot in itself but when combined with rate studies forms a useful basis for design.

Rate studies on powdered carbon systems are normally conducted in batch studies while columns are used for granular carbon systems. In potable water systems, the time spent in mixing, flocculation and settling is so much longer than that required to reach equilibrium (5–45 min.) that rate studies are not required. However, rate studies for granular carbon systems form the basis for design and are, therefore, mandatory.

GRANULAR CARBON SYSTEM DESIGN AND SELECTION

Granular activated carbon may be installed in conventional sand filters or post-filtration columns. Carbon capacity and rate of adsorption must be determined for both installations, but the use of carbon as a filter media must also be considered when carbon is installed in the filter. Although the filtration properties of granular activated carbons are virtually the same as a comparable size sand or anthracite, many states still require the retention of some sand. This reduces the operating life and adsorption efficiency of the carbon due to reduced bed depths, but does provide the advantages of a reverse graded filter. Multi-media, or reverse graded filters employ a bed which decreases in particle size from top to bottom and in operation is somewhat analogous to a series of sieves. The large particles of floc are removed in the upper portion of the filter, and the smaller fraction passes through. As the material passes downward through the filter, it eventually encounters a restriction or is removed by gravity, inertia, or attractive forces. The mixed media bed, therefore, provides removal in depth as opposed to surface removal of normal filters. As a result, mixed beds are usually more efficient and subject to longer runs between washings. To provide a reverse graded filter, the various layers of media must be both larger and lighter than the material immediately below [3,4].

The major factor in designing a multi-media filter is backwash characteristics. If the carbon is too small, it may be washed out before appreciable sand expansion occurs; and a cumulative head loss could result. On the other extreme, if the carbon is too much larger than the sand, bed expansion would be retarded; and poor cleaning of the carbon would occur. For best results, the sand and carbon should have approximately the same expansion at a given flow rate [5].

Because of the inherent variability of waters, the determination of the carbon depth required to provide a given service life for organics removal must be determined for each individual installation. This is accomplished by using a granular carbon pilot filter. By measuring the influent and effluent concentration at various depths with time, breakthrough curves such as those shown in Figure 5 may be constructed.

The time of operation achieved before reaching an allowable breakthrough concentration CB through each column and the corresponding carbon depth may be determined from the various breakthrough curves. The operating time attained before reaching an unacceptable organic breakthrough concentration, CB, may then be plotted versus carbon bed depth as shown in Figure 6. It should be noted that at some critical bed depth, DC, zero operating time is attained before reaching an unacceptable taste or odor, CB. This dictates the minimum depth of carbon, and the actual depth to be employed may be determined by summarizing the test data in a plot similar to Figure 5. In water treatment plants the critical bed depth is measured in inches for taste and odor removal, but in feet for organics removal.

It should also be noted that the critical bed depth, DC, increases with flow rate and carbon particle size. If these two parameters are likely to be varied, they should be studied during the pilot study. Isotherm data should also be taken during the pilot tests to determine if the raw water conditions were typical, or more or less severe than normal conditions.

With breakthrough curves and isotherms available the selection and design of a granular carbon system may be completed. An economic analysis will dictate the selection of adsorbers based on a consideration of capital and operating costs.

In multi-stage adsorption systems the carbon may normally be exhausted at the influent concentration by designing each adsorber to retain at least one critical bed depth, DC. The lead column can then be operated until the effluent and influent concentrations are equal, and the second column would still be producing an effluent concentration equal to or less than the specified breakthrough concentration, CB. The adsorption columns can then be sized for various superficial velocities and the costs calculated to select the most economical system.

When the critical bed depths are several feet or less, single-stage contactors may be economically employed. Although the carbon cannot be exhausted at the influent concentration, CO, the increased operating costs may be offset by reduced fixed costs. Since the system will not be oper-

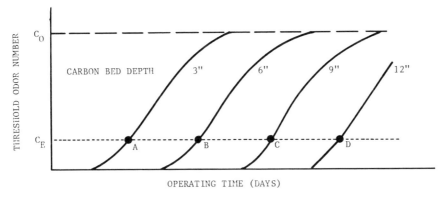

FIGURE 5. Granular carbon breakthrough curve.

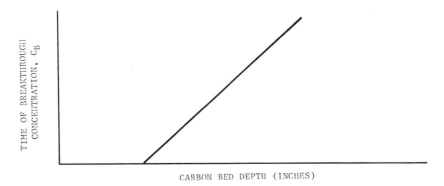

FIGURE 6. Carbon service time as a function of bed depth.

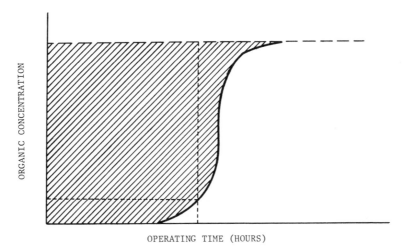

FIGURE 7. Evaluating the actual carbon loading.

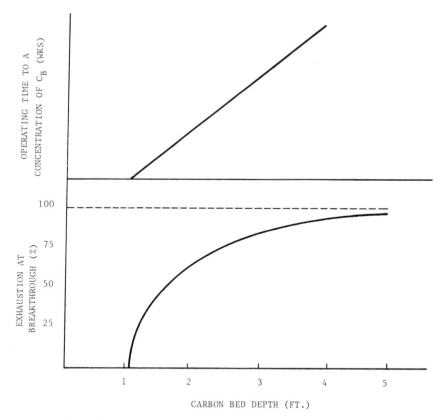

FIGURE 8. Evaluating the relative exhaustion as a function of bed depth.

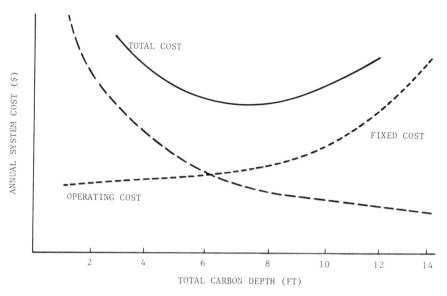

FIGURE 9. Cost analysis as a function of bed depth for single stage contactors.

ated to total exhaustion, the actual carbon loading must be calculated for several bed depths when the effluent concentration at that point reaches the allowable breakthrough concentration, CB. This is accomplished by integrating the area under the breakthrough curve as shown in Figure 7.

The actual carbon loading, X/M actual or the ratio of the actual carbon loading are obtained at C_O may then be plotted as shown in Figure 8.

It should be noted that immediate breakthrough occurs for all carbon depths less than the critical depth, so a zero loading results. As the actual bed depth increases above the critical bed depth, DC, the carbon loading increases sharply and then asymptotically approaches 100%. This demonstrates that greater bed depths result in better carbon utility and reduces operating costs. Carbon adsorbers only slightly greater than critical bed depth will give poor performance and result in abnormally high operating costs. This is a potential problem with replacing the sand with carbon in existing filters. The most cost-effective depth for a specific superficial velocity may be determined by plotting the fixed, operating and total annual costs as a function of bed depth as shown in Figure 9.

GRANULAR ACTIVATED CARBON

General

During the past few years, intensive interest in the problem of organic contaminants in water supplies has been shown. The problem is not new because the industry has been confronted with raw waters that contain odor and color derived from organics, both requiring removal. In a recently released report to Congress on the preliminary assessment of suspected carcinogens in drinking water, the Environmental Protection Agency (EPA) has tabulated 253 compounds that have been found in finished waters. While there is no evidence that a large number of these present any significant problem in water supplies, they likely represent only a small fraction of the total number of compounds present. These compounds originate from industrial and municipal discharges, agricultural runoff and other nonpoint sources decaying vegetation and reaction of water-treatment chemicals such as chlorine, with aqueous organic matter. Using MADAM (Michigan Adsorption Design and Application Model), Weber and Pirbazar [6] found that GAC can be used to treat trace quantities of specific toxic or carcinogenic compounds present in U.S. surface water or ground water.

In potable water applications, granular activated carbon can be used to remove taste and odor as well as trace organics and chloroform. In the case of taste and odor removal, the carbon may last for as long as three years before needing replacement. However, for trace organics or chloroform removal, carbon may need to be replaced quarterly

depending on the concentration of contaminants. In either case, it is common to remove the sand from the existing gravity filters and replace it with 0.5–1.0 M (20–40 in) of granular activated carbon. The resulting contact time is generally sufficient to achieve the desired removal efficiencies. The carbon can then serve as both a filter and an adsorber. Granular activated carbon appeared to be the most effective process in order to remove Trihalomethane present in the drinking water [7]. Pirbazar and Weber Jr. [8] studied adsorption of paradi-chlorobenzene and similar compounds by granular activated carbon which showed that GAC possesses fairly good adsorption capacity.

The principal advantages of GAC, namely ease of process control, more efficient use of the carbon, and the ability to regenerate the carbon for reuse, are specially significant if organics must be removed on a regular basis. To avoid the possible disadvantages of granular activated carbon filters and to use this process in a worthwhile manner, certain conditions have to be observed. Sontheimer summarized the most important rules which should be obeyed as follows:

1. Granular activated carbon filters should be included in the general treatment process scheme in such as way that an optimum efficiency will be achieved with this process. This would mean, for example,
   (a) Removal of all suspended, collodial and dissolved substances through flocculation and filtration before the activated carbon, so that these cannot cause any reduction in the absorption capacity.
   (b) Avoiding oxidation treatment with chlorine, leading to free chlorine and thus to organic chlorine compounds.
   (c) Transfer of non-biodegradable to biodegradable substances through ozone oxidation if biological activated carbon (BAC) is used.
2. Activated carbon used should be tested only under practical conditions. This means similar concentrations of the organics and also test conditions which give information for the adsorption competition between different compounds. Additional substances may be added to the water which are important for the purpose of the carbon in the type of water to be treated.
3. Activated carbon filters can be controlled in different ways. Controls should permit information on the behaviour of the filter at peak concentrations. Most worthwhile for a continuous control may be the measurement of the UV absorbance.

Evaluating GAC Adsorption Efficiency

Activated carbon has the ability to reduce the level of total organic matter in water as well as the levels of specific trace organics. Such removals are of particular importance if the carbon is being used to reduce the level of haloform precursors prior to chlorination. The humic substances, generally the major fraction of total organic matter, have been demonstrated to be precursors of the haloforms. The specific characteristic of the organics present will influence significantly the extent of removal. Weber Jr. and Jodellah [9] showed that the removal of Humic Acid by GAC can be improved after Alum coagulation.

An extensive evaluation of ten different commercial carbons to determine their ability to remove total organic matter was carried out in the laboratory. Absorption of ultraviolet light at 240 nm was used to indicate the amount of organic matter present. Typical isotherms using River Rhine (Germany) water that had been filtered through the river bank (this removes many biodegradable organics, and some dilution of river water with groundwater takes place), ozonated, and filtered are shown in Figure 10.

The conclusions that could be drawn from the results were,

1. The relative positions and slopes of the isotherms were dependent upon the point on the lower Rhine where the sample was taken, and upon the time of the year and rate of flow of the river.
2. The phenol number and the Brunauer-Emmett-Teller (BET) surface area do not provide a good indication of carbon effectiveness for total organic removal for carbons prepared from different raw materials or by different activation processes. For example, carbon 3 had a

surface area of 800 m²/g while the area for carbon 5 was 1050 m²/g. Also, the phenol numbers for carbons 10 and 2 were 5.0 and 2.7 respectively. If the same raw materials and same activation process are used, however, such as was the case for carbons 1 and 7, 5 and 8, and 4 and 9, better adsorption properties are associated with the higher values of these parameters. It is also likely, however, that the more extensively activated carbons will cost more.

3. Carbon 10, which in the past had been proven to be very effective for odor removal, performed the worst for total organic removal.

It was also concluded that the better carbons based on the isotherm evaluation generally removed more material in the beds as well. All of the carbons tested in beds seemed equally effective for odor removal, but that saturation of the carbon's capacity for total organics was reached much before its capacity for odorous compounds was reached. Some of the odors were removed in the upper part of the bed by biological activity.

In other experiments that were carried out at the Lawrence Experiment Station, Lawrence, MA, four types of GAC from three different manufacturers were used. GAC Types 1, 2 and 4 were of bituminous base with an approximate mesh size of 12 × 40. GAC Type 3 was of lignite base with a 10 × 30 mesh size.

The four different types of GAC tested are plotted as a function of time in Figure 11. From these data an initial adsorption rate constant (k) was calculated in terms of mg/g-hour (see Table 1).

As shown in Table GAC Type 1 exhibited the greatest adsorption rate constant followed closely by Type 2, while Type 4 showed the poorest adsorption among the virgin carbons tested. The rate constant for exhausted Type 4 GAC was found to be 6 percent of virgin type 4. Based on estimates of instrumental accuracy and inherent system errors, differences in the adsorption rate constant of less than 5 percent were not considered significant.

Because of density differences among the many GAC types and manufacturers "cost per unit volume of GAC" should be calculated which results in a Cost-Efficiency Number (CE):

$$CE = \frac{\text{relative adsorption rate constant}}{\text{cost per unit volume}}$$

The grade of GAC which has the highest CE number represents the best compromise between cost and adsorption efficiency. Alternatively, an abrasion number could be incorporated into the formula instead of the per cent loss figure to obtain:

$$CE = \frac{(\text{relative } k) \, (\text{abrasion number})}{\text{cost per unit volume}}$$

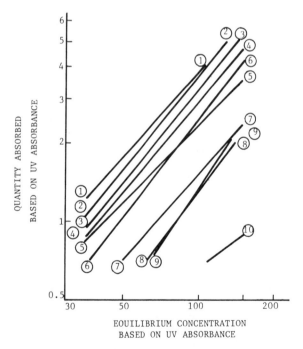

FIGURE 10. Quantity adsorbed vs. equilibrium concentration.

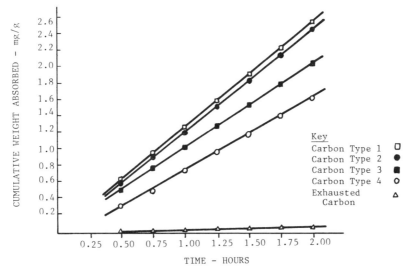

FIGURE 11. Cumulate weight absorbed vs. time.

In cases where various GAC types have similar densities, cost per pound may be substituted for cost per unit volume.

Pretreatment Effects on the Adsorption Process

CLARIFICATION

Good particulate removal is especially important for a number of reasons. The humic substances that cause color may be particulate in nature. Thus, removal of color of this type also should reduce the concentration of trihalomethanes that form when the water is chlorinated. Also trace organics, such as pesticides, may be associated with silt trace organics, such as pesticides, may be associated with silt and particulate fractions of humic substances. Thus simply removing particulates may result in good removals of organics. Greve and Wit found that nearly all of the 0.2–0.6 mg/l of endosulfan present in River Rhine water was removed by clarification (coagulation, flocculation, and filtration). Good removal of humic substances (pH range 4 to 6) and TOC was also achieved.

OZONATION

Study report reductions of dissolved organic carbon (DOC), COD and UV absorbance of 0–25 percent, 0–20 percent, and 11–38 percent respectively when Rhine River bank filtrate is ozonated. When both ozonation and clarification were used, removals were significantly better, indicating that ozonation was improving the removal of organics by clarification. Breakpoint chlorination and clarification resulted in a 15 percent removal of TOC from River Rhine water. The total COD removal was approximately 50 percent. Research at Bremen has shown that an activated carbon which removed 1 mg/l of TOC before ozonation removed 3 mg/l after preozonation.

BANK FILTRATION

In the Federal Republic of Germany, a common pretreatment procedure is to filter River Rhine water through the river bank prior to application of treatment chemicals. Sontheimer and Maier report removals on the order of 65–75 percent based on DOC or UV absorbance.

TABLE 1. Adsorption Efficiency Studies—Merrimack River.

Source	GAC type	$k = \dfrac{mg}{g\text{-}hr.}$	Relative $k = \dfrac{kx}{kL^*}$
Merrimack River	1	1.28	1.51
at Lawrence,	2	1.23	1.45
Mass.	3	1.01	1.19
CCE conc. 1.25	4	0.85	1.00
mg/l	Exhausted 4	0.05	0.06

*kL = lowest adsorption rate constant found for virgin carbon.

CHLORINATION

Prechlorination may render some substances more insoluble and, thus, enhance their adsorption. For example, as phenol becomes substituted with chlorine, the solubility of the neutral species decreases, and the chlorine is adsorbed more strongly on carbon. However, it is evident that adsorption increases with decrease in pH.

SOFTENING AND MANGANESE REMOVAL

Lime softening of water can have a detrimental effect on adsorption by carbon if the water is not stabilized properly before adsorption. Stabilization is necessary to avoid deposition of precipitate on the carbon and the accompanying loss of efficiency. At Piqua, Ohio, lime softening prior to carbon filtration resulted in a calcium carbonate encrustation that cemented the filter media together. Hundreds of gallons of HCl were required, but the problem was solved by replacing carbon by 1 m layer of filter media.

Organics – Carbon Adsorbable (OCA) Method

The presence of organic compounds in drinking water that produce odor, taste, or toxicity problems is of concern to the consumer and thus to the water utility. The carbon adsorption method (CAM) introduced in 1951–1952, has been modified from a high-flow to a low-flow method and a miniaturized modification has recently been adopted. Each modification has improved the efficiency of the method and has decreased the volume of water sampled, the amount of carbon required for adsorption of organics and the time required for analysis. Due to the simplicity and low cost it is favored for monitoring organic content of water [10–12].

As part of a general program concerned with assessing the quality of Canadian drinking water, the OCA method has been evaluated by monitoring Ottawa tap water for one year. Ottawa tap water was sampled and analyzed periodically during a one-year period. Carbon chloroform extract (CCE) residue weights, uncorrected for blank values, were obtained by means of the OCA method and are listed in Table 2.

The method stipulates correction of the sample residue weight by the dry carbon blank value in order to calculate CCE-M for a water sample. It has been shown by the authors that correction by a pre-wetted and dried carbon blank residue weight would be more appropriate. The difference between the two types of values has been reported to be due to the extraction of larger quantities of elemental sulfur from pre-wetted carbon than from non-wetted carbon. This finding was confirmed by the author's results (Table 3).

Although typical sample residue weights (Table 2) changed only slightly during the year, there were occasional abnormally high values. These anomalous values were not due to flow changes during sampling or other sampling irregularities. Often only one value in a triplicate set was abnormal; therefore, changes in water quality could not be the cause. Although wide variation in percentage of sulfur in extracts has been reported, the results of this study (Table 3) indicate that the amount of sulfur percent was essentially constant (18 mg) for residue weights ranging from 58.2 to 100.1 mg. Since the anomalous residue weights could not be accounted for by minor variations in the workup procedure, each of the main steps in the procedure was evaluated (Table 4) for reproducibility.

The method essentially consists of passing 60 L of water through 70 g of carbon; drying the carbon under controlled conditions; extracting the carbon with chloroform in a Soxhlet extractor; and concentrating the extract by distillation followed by compressed air-drying.

Application of a wet carbon blank value to the data in Table 2 gave CCE-M values ranging from 0.35 mg/l to 1.07 mg/l for Ottawa tap water during the monitoring period. No correlation could be found between variation in CCE-M and changes in temperature, turbidity, total alkalinity or conductivity of raw or treated water.

CONCLUSIONS

Although the OCA method is relatively simple and inexpensive, it has inherent drawbacks both in analysis, time and accuracy of results. Some lack of precision due to variations in sampling, drying, and extraction procedures is acceptable, but the gross corrections required for sulfur content, as well as considerations of the observed variation in sulfur mobilization with water quality, make the method less acceptable for routine monitoring. Because the OCA method detects less than 20 percent of the total organics, gives no estimation of volatile organics and is sensitive only to gross water quality changes, it cannot be recommended for general use. It remains useful in special situations or for collection of residues when identification of individual compounds is required.

Fluorometry

Determination of microorganics in drinking waters is difficult because in some cases a minimum detectable limit of 1 ug/l is required for specific compounds. Several methods of analysis have been employed to detect and identify the trace organics. These techniques include gas chromatography-mass spectrometry, infrared spectrometry, electromechanical methods, ultraviolet-visible absorption spectrometry and fluorometry.

The development of a simple and inexpensive monitoring test to use as a control parameter for the operation of GAC beds has been cited by EPA as a major research need. Although such a simplified process control procedure is not envisioned to identify specifically all organics in water basins and treated waters, nonetheless, a realistic approach to the problem should be attempted.

TABLE 2. Ottawa Tap Water, 1975–1976: Carbon Chloroform Extract Residue Weights.

Date 1975–1976	Carbon* Lot Number	Residue Weight** - mg				
		Sample 1	Sample 2	Sample 3	Mean	σ
8/22	1	52.07	51.69	53.54	52.43	0.98
9/2	1	54.98	46.86	50.80	50.88	4.06
9/29	1	57.99	60.79	58.62	59.13	1.46
10/6	1	66.28	62.86	62.77	63.97	2.00
10/14	2	64.34	64.84	63.78	64.32	0.53
11/3	2	68.83	71.22	73.09	71.05	2.14
11/17	2	67.05	62.05	70.30	66.47	4.16
11/24	2	68.05	65.75	68.35	67.38	1.42
12/1	2	67.65	70.10	—	68.88	1.73
12/8	2	69.90	72.10	—	71.00	1.56
12/15	2	66.65	62.00	67.50	65.38	2.96
12/22	2	64.90	63.45	66.20	64.85	1.38
12/29	2	118.35	69.45	86.70	91.50	24.80
1/5	3	65.45	97.35	88.70	83.83	16.50
1/19	3	71.45	77.85	—	74.65	4.53
2/9	4	65.65	80.85	65.99	70.83	8.68
2/16	4	71.65	—	—	71.65	—
2/23	4	58.20	105.40	67.65	77.08	24.97
3/22	4	62.65	61.55	71.10	65.10	5.23
4/12	5	58.20	60.90	59.10	59.40	1.37
4/20	5	95.00	97.50	100.15	97.55	2.58
5/25	5	54.79	57.54	—	56.17	1.94
6/7	5	60.00	59.30	45.45	54.92	8.21
6/21	5	62.69	63.36	70.72	65.59	4.46
7/5	5	70.46	65.61	69.34	68.47	2.54
7/19	5	61.92	73.29	70.08	68.43	5.86
8/3	6	67.23	67.96	71.10	68.76	2.06
8/16	6	66.42	67.83	68.93	67.73	1.26

*Filtrasorb 200, Calgon Corp., Pittsburgh, PA.
**CCE-m-uncorrected for blank.

TABLE 3. Sulfur Content of Thimble Extracts, Wet and Dry Blank CCE Residues, and Potable Water CCE Residues.

Sample		Residue Weight—mg.				Sulfur Content—mg.			
Type	No.	Min.	Max.	Mean	σ	Min.	Max.	Mean	σ
T*	6	5.07	60.49	15.93	21.91	0	0	—	—
T + dry carbon[1]	8	2.83	13.62	7.64	4.44	2.3	3.8	2.94	0.50
T + wet carbon[2]	6	19.35	33.69	24.68	5.14	11.8	17.5	15.98	2.20
T + monitor carbon[3]	9	58.20	100.15	74.91	17.76	17.1	19.8	18.11	0.83

*Cellulose thimble extract residues
[1]Dry carbon blank (70 g) residues
[2]Wet carbon blank (70 g) residues
[3]Monitor sample residues

TABLE 4. Evaluation of OCA Method Steps.

Step	No. of Samples	Residue weight—mg.			
		Min.	Max.	Mean	σ
CHCl₃ blank*	3	1.15	1.70	1.35	0.30
Air drying	12	61.85	62.80	62.39	0.29
Distillation & air drying	12	55.50	56.80	56.42	0.37
Mixed dry exposed carbon	12	57.90	86.35	67.18	8.32
Mixed wet exposed carbon	9	52.00	124.90	70.59	22.36
	12	57.20	89.80	65.73	8.10

*Chloroform (300 ml) carried through Soxhlet (no thimble), distillation and drying steps.

Ideally a process control monitor for operation of GAC beds should possess the following characteristics.

1. Response measured by the instrument should be derived from some of the organic contaminants already identified in drinking water.
2. Detection limit in the microgram-per-liter range so that preconcentration is not needed.
3. Amenability to continuous operation.
4. Relative inexpensiveness.
5. Simplicity of operation and maintenance.

Fluorometry may meet the criteria for a simplified monitoring procedure for evaluation of the adequacy of activated carbon for removal of organics in drinking water. Fluorescence appears to be an intrinsic property of drinking waters caused by contamination from traces of organic fluorophors. Fluorometric methods can detect concentrations of organics in water as low as one part in ten billion without preconcentration.

Water fluorescence intensity IWF is a control parameter for the operation of activated carbon filters. IWF measurements are related to source water origin, and pH, quenching, a carbon adsorption isotherm, carbon measured by CCE and CAE, volatility of the trace organics, and dissolved vs. suspended organics. In addition, IWF values are correlated with total organic carbon (TOC) measurements in the evaluation of the efficiency of two carbon home filter units in removing organics from finished waters.

Experiments were carried out to determine water fluorescence intensity. Figure 12 shows the fluorescence spectra of finished New Orleans municipal water sample from a tap, and a bottled water that is treated with GAC prior to marketing.

As shown in Figure 12 the bottled water sample showed a very weak fluorescence, due to treatment of activated carbon. Table 5 summarizes IWF measurements from several different source waters. The maximum wavelength of fluorescence and the shape of the fluorescence spectra of tap waters from New Orleans and Destrehan, LA were similar. This is not surprising in view of the fact that both finished waters originated from the same source supply, raw surface water from the Mississippi River. Similarly, finished tap water from the municipal supply of Slidell, LA and a private source from the same area showed similar fluorescence spectra and the same maximum wavelength of fluorescence. An investigation revealed that both samples originated from an underground source.

Table 6 shows the effect of pH on IWF measurements of source tap waters. At natural pHs the finished tap waters exhibited a large IWF than at acidic or basic conditions. Apparently, the nature of the organic fluorescent impurities in the water sample is such that either strong protonation or deprotonation of fluorophors reduces the overall fluorescence.

QUENCHING

As noted in Table 5, treatment of the various source waters with activated carbon resulted in an apparent decrease in IWF values relative to untreated waters. The decrease of the IWF values is caused by either removal of fluorescent organic impurities from the water samples by activated carbon or quenching of the fluorescence by impurities contained in the carbon. Therefore, a quenching study was proposed to elucidate the mechanism for the reduction of IWF values in source waters treated with activated carbon.

The Stern-Volmer quenching equation was used to determine if quenching was responsible for the decreased IWF after treatment with carbon. The Stern-Volmer equation, expressed in terms of IWF, can be given by

$$IWF^p/IWF^q = 1 + kq^T(Q)$$

where

IWF^p = the water fluorescence intensity without the presence of quencher

IWF^q = the water fluorescence intensity in the presence of a quencher

kq^T = constants

Q = concentrations of quencher

The equation predicts that if the concentration of Q increases, the ratio IWF^p/IWF^q becomes larger than 1 if Q is a true quencher. On the other hand, if Q is not a quenching agent then IWF^p/IWF^q remains constant at 1.0 as Q is increased.

CCE AND CAE

It seemed desirable to utilize IWF measurements in studying the recovery of organics by the CCE and CAE techniques. In this technique, a portable sample is filled with 70 g of GAC, 14–40 mesh, and connected to a water tap; water is allowed to pass through at a rate of about 20 ml/mm providing a contact time of about 4.5 min, for a 48-hour period. The total volume passing through the activated carbon is measured, and the carbon is then dried. The carbon is first extracted with chloroform, followed by ethanol. Evaporation of the chloroform leaves a residue known as CCE; evaporation of the alcohol leaves a residue known as CAE. For demonstration, 57.6 l of water from the EPA pilot plant at Lake Tahoe were passed through the miniaturized GAC sampler. The residue from the chloroform and ehtanol extracts was redissolved in a liter of warm fluorescent free water and then each solution was diluted to a final volume of 57.6 l. The resultant IWF intensities were measured at $\lambda em = 410$ nm and compared with the controlled IWF value Lake Tahoe water, which bypassed the miniaturized carbon adsorption method (CAM) filter. The data are summarized in Table 7. From this data, the CCE and CAE techniques gave an apparent recovery of 90.8% of the fluorescent adsorbed to the carbon.

CONCLUSION

IWF values were higher at natural pH of source waters than at other pHs. The IWF values from a municipal source

utilizing a surface water were caused by dissolved, nonvolatile fluorophors contained in the water supply. This work should lend further support to the potential application of IWF measurements as a continuous screening technique for evaluation of adequacy of activated carbon for removal of organics in drinking water. Based on the observed variation of the IWF with various basic parameters there appears to be justification for further research into this rapid, simple, and very sensitive technique.

Chlorination

Recently there has been great interest in the study of organic compounds in drinking water—interest that stems largely from the results of a study of New Orleans drinking water, and the publicity that followed. About the same time, two studies called attention to the presence in finished drinking water of some trihalomethanes (mostly chloroform) which were not found in the respective raw waters at the locations of study. Both reports concluded that the trihalomethanes were formed during the chlorination step of the water treatment process. The study also showed, there were six halogenated compounds in raw and finished water. Those six included four trihalomethanes (chloroform, bromodichloromethane, dibromochloromethane, bromoform) suspected of being formed during chlorination, plus carbon tetrachloride and 1,2-dichloroethane, known contaminants at New Orleans, but not necessarily formed on chlorination. The goal was then to develop general conclusions applicable to rational modification of water treatment processes if removal of trihalomethanes was finally deemed important for public health reasons.

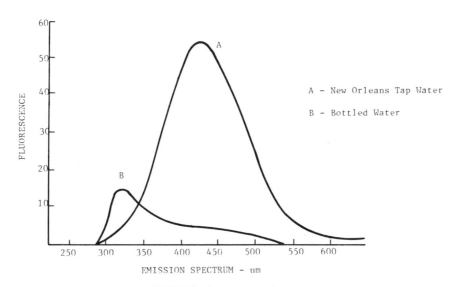

FIGURE 12. Source water fluorescence.

TABLE 5. Fluorescence Maximum of Various Source Waters.

Source	Comments	λem nm	IWF Units	IWF Normalized
Quartz cell			No fluorescence	
Fluorescent free water			No fluorescence	
Bottled water	Ground water from La. treated with GAC	315	3	0.05
Dallas, Texas		435	10.5	0.19
Lake Tahoe, California		420	18.5	0.34
Finished New Orleans municipal water	obtained from tap in GSRI lab.	430	55	1.00
Finished Destrehan, La. municipal water	located 25 mi. up-river from New Orleans	430	74	1.35
Slidell, La. private drinking water	located 25 mi. north of New Orleans	440	86	1.56
Finished municipal water		440	160	2.91

TABLE 6. Effect of pH on IWF Measurements of Municipal Tap Waters.

Source	pH	λem nm	IWF Units	IWF Normalized
New Orleans	1	425	31.7	0.83
	natural	430	38.0	1.00
	13	425	34.0	0.90
Destrehan, La.	1	425	44.3	0.68
	natural	430	65.5	1.00
	13	425	51.0	0.77
Slidell, La.	1	435	28.7	0.32
	natural	440	88.0	1.00
	13	435	54.2	0.61
Slidell, La. (private source)	1	435	26.7	0.56
	natural	440	47.5	1.00
	13	435	31.0	0.65

TABLE 7. Recovery of Fluorescent Organic Impurities from Activated Carbon by CCE and CAE.

Sample	IWF	IWF Recovered per cent
Control water	19.5	
CAE diluted to 57.6 ℓ	14.7	75.4
CCE diluted to 57.6 ℓ	3.0	15.4
Total		90.8

PRECURSOR AT pH 7

Trihalomethanes must result from a reaction or series of reactions of chlorine with a precursor material. Because control of trihalomethane production by precursor removal or control of precursor reaction rate was considered the best approach, some knowledge of precursor identity was required. However, identity of precursor varied from complex humic materials to simple methyl ketones or simple compounds with acetyl moiety.

Monochloramine is becoming a popular alternative to chlorine for the disinfection of water supplies because it does not react with the humic substances to form trihalomethanes. However, chloramines have recently been found to cause hemolytic anemia in patients undergoing kidney dialysis. Hence, it was found necessary to remove monochloramine prior to supply for domestic use. Recent study showed that granular activated carbon could be a good adsorbent to treat water containing monochloramine [13].

An experiment was performed to determine whether GAC adsorption had any effect on precursor concentration. In this work, samples of water taken from the pilot plant were chlorinated at a dose of 8 mg/l to satisfy chlorine demand and maintain a free residual in the distribution system. In this experiment, not only were settled and activated carbon filtered water samples chlorinated to determine the effect of the carbon, but dual media filtered and raw water samples were also chlorinated at the same concentration for comparison. All four samples were buffered at pH 7. The results in Figure 13 show that when the result of chlorination of fresh GAC-filtered water was compared with the result of chlorinating the settled water, removal of precursor was indicated. The effectiveness of GAC filtration, however, was shown

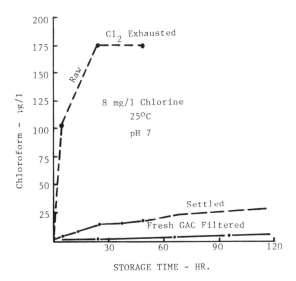

FIGURE 13. Effect of treatment on chloroform production.

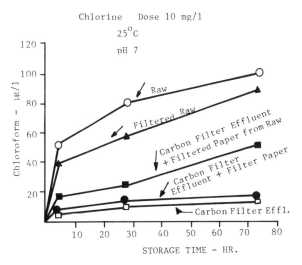

FIGURE 14. Effects of simple filtration on trihalomethane production.

later to be relatively shortlived—a matter of only a few weeks under conditions of pilot plant operation.

Comparison of the reaction rate curves for raw and filtered raw water is shown in Figure 14 which illustrates a reduction of the rate of trihalomethane production caused by removal of particulates. The rate curve for GAC filter effluent with resuspended filter paper and particulates from the raw water indicates that essentially all of the difference between the raw and filtered raw water rate curves can be accounted for by the substances trapped on the resuspended filter paper. The curves for GAC filter effluent and GAC filter effluent plus filter paper are simply the appropriate controls and are nearly identical.

Because humic substances are more likely to be found in natural waters as small particulates or sorbed on clay particles than are soluble simple methyl ketones, a direct test of Rook's hypothesis was attempted using commercially available humic acid, both suspended at pH 7 and dissolved at high pH, which was later adjusted to pH 7. At concentrations of humic acid representing a nonvolatile total organic carbon (NVTOC) concentration similar to that found for Ohio River water (approximately 3 mg/l of NVTOC) the rate curve for formation of trihalomethanes was observed to be very similar to that seen for chlorination of the natural water (Figure 15).

EFFECT OF pH ON REACTION RATE AND PRECURSOR IDENTITY

Because the rate determining step of the classical haloform reaction is enolization of a ketone, the rate of trihalomethane formation is pH dependent. For example, the reaction of acetone with hypochlorite to form chloroform

proceeds at a faster rate at pH 11.5 than at pH 6.5. Experimentally, a sample of settled water was buffered at pH 6.5 and another at pH 11.5; both were chlorinated at an initial concentration of 10 mg/l. The results show that the rate of formation of chloroform increases with an increase in pH (Figure 16). This could be explained simply by an increase in the humic acid reaction rate, as would be expected by

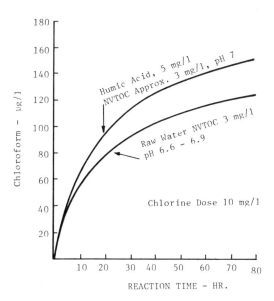

FIGURE 15. Comparison of humic acid, raw water reaction roles.

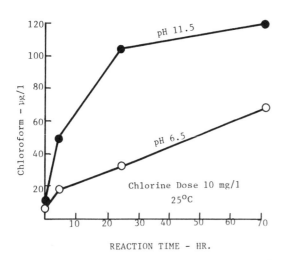

FIGURE 16. Effect of pH on chloroform production, settled water.

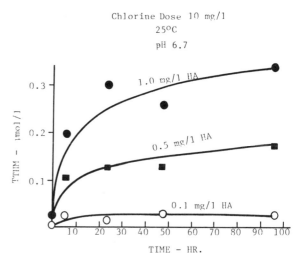

FIGURE 17. Effect of humic acid conc. on trihalomethane production.

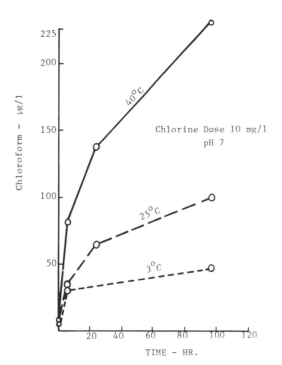

FIGURE 18. Chloroform production.

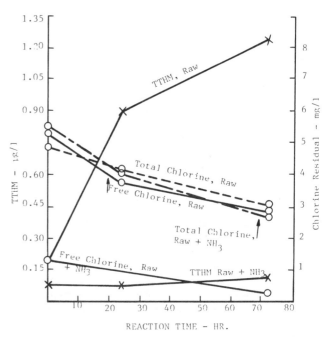

FIGURE 19. Free vs. combined chlorine and TTHM production with and without NH$_3$ addition pH7.

classical mechanism. Another possibility, however, is that other compounds in the source water (settled), such as acetone that do not react readily at pH 6.5, become significant contributors to the overall reaction rate at pH 11.5.

Figure 17 illustrates the reaction rate curves for formation of total Trihalomethanes (TTHM) from three concentrations of humic acid (0.1, 0.5, 1.0 mg/l) spiked in GAC-filtered water in presence of excess chlorine (10 mg/l with less than 10% change during the course of the experiment). An apparent first order rate dependence on initial humic acid concentration is demonstrated; that is, at any given time between any two curves, the ratios of concentration of TTHM produced are equal to the respective ratios of initial humic acid concentrations.

EFFECT OF TEMPERATURE

The effect of temperature on the rate of reaction of precursors present in Ohio River water was investigated to assess the potential effect of wide seasonal temperature variations in raw and treated waters. As shown in Figure 18 chloroform concentrations increases with increase in temperature.

DISINFECTANT

Figure 18 illustrates the result of an attempt to form trihalomethanes with chlorine added in the presence of added ammonia. Chlorine was added at 5.5 mg/l to raw water and to raw water spiked with 20 mg/l NH_4Cl (ammonia nitrogen, 5.2 mg/l). The results of the measurements for trihalomethane production and free and combined (mostly NH_2Cl) chlorine residuals in Figure 19 show that when combined chlorination was practiced, trihalomethane production was minimized. Therefore, during chlorination of water where the ammonia breakpoint is not achieved, trihalomethane production may not be a problem.

CONCLUSION

The precursor to trihalomethane production during the chlorination process in drinking water treatment is probably a complex mixture of humic substances and simple low-molecular-weight compounds containing the acetyl moiety. The relative importance and contribution to trihalomethane production of each of the specific precursor compounds are pH dependent. The point of chlorination in the treatment process, being a significant factor in trihalomethane production, probably represents the most important variable to be considered for change in attempts to reduce ultimate trihalomethane concentrations in finished drinking water.

GAC for Different Types of Organics

BIOLOGICALLY-DERIVED ODOR

A monitoring study conducted by the EPA of six GAC Field installations since 1968 has shown that odor break-through, probably of biological origin, occurs much later than breakthrough of general organic compounds. Biological odor problems are generally intermittent, and the compounds causing the odor are generally present at low concentrations.

Odors at Piqua, Ohio, that sometimes required dosages of over 100 mg/l of powdered activated carbon (PAC) for control were found to be controlled effectively by the replacement of 51 cm (20 m) of sand in 81 cm (32 m) deep rapid sand filters with GAC. After 33 weeks of service the beds still were producing odorless water. The experiment, however, showed that the capacity for CCE was essentially exhausted after ten weeks of operation.

Laboratory research was conducted using geosmin and 2-methylisoborneol (MIB), two causative agents of earthy-musty odor and of biological origin. These compounds were adsorbed strongly on GAC, even in the presence of natural organic matter, although the presence of natural organic matter did reduce the adsorption capacity somewhat. Batch adsorption tests were conducted with geosmin in a manner similar to MIB. The results are shown in Figure 20 (for MIB) and Figure 21 (for geosmin). The data show that at an equilibrium geosmin concentration of 0.1 ug/l the capacity of carbon is 0.54 mg/g in deionized, distilled water, twice the capacity shown in Figure 20 for MIB at the same equilibrium concentration.

ODORS OF INDUSTRIAL ORIGIN

Nitro, WV, obtained its water supply from the Kanawha River, which is heavily polluted with various organic industrial wastes. A series of small scale experiments and a full scale test were conducted to evaluate GAC filters for purifying this supply. GAC beds of 8 × 30 US Standard mesh were able to produce an odor-free water for as long as 26 days, depending on the contact time and the threshold odor number (TON) of the water applied to the bed. Empty bed contact times varied from 3.8 min to 15 min. Additionally, a finer carbon, 20 × 50 mesh, could produce an odor-free water for twice as long as the 8 × 30 mesh, and the degree of activation of the carbon did not appear to have a noticeable effect on the efficiency of odor removal.

SULFIDE ODOR

In order to control a sulfide odor problem, the Goleta Water District in California installed four pressure GAC adsorbers following diatomaceous earth filters at the Patterson plant. The cylindrical pressure tanks were loaded with 2 m (8 feet) of carbon resting on stainless steel screens and were operated at a rate of 3×10^{-3} m/s (5 gpm/sq ft). These beds were in service for two years before reactivation was necessary. The problem of sulfide odor was attributed to hydrogen polysulfides in the water, which are not removed by aeration, and GAC was only partially effective in removing them. In general, influent TONs varied from 6 to 1000, while effluent TONs ranged from odor-free to 35.

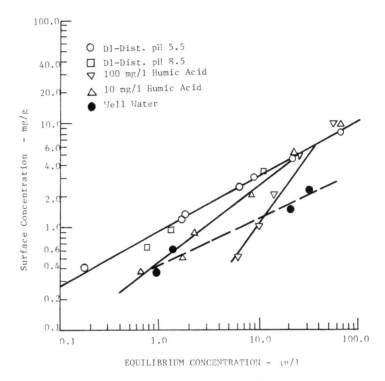

FIGURE 20. Adsorption of MIB.

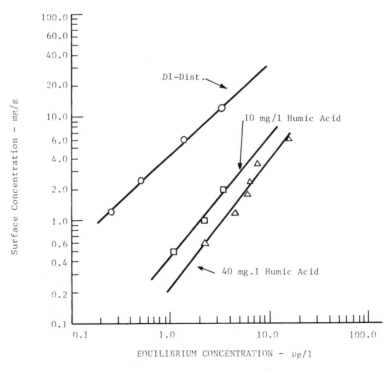

FIGURE 21. Adsorption of GEOSMIN.

PESTICIDES

GAC beds were evaluated for the removal of selected pesticides spiked into river water. One carbon column was exhausted for removal of CCE and COD. Spiked, sand-filtered river water was applied to the exhausted carbon column, and the effluent from it was applied to a second, fresh column. With concentrations of dieldrin as high as 4.3 ug/l, the effluent from the first column was reduced to 0.3 ug/l. Further reduction in the second column reached concentrations as low as 0.05 ug/l and often below the detection limit of 0.01 ug/l. This happened in spite of sudden increases in the COD of the influent river water. The CCE and COD of the effluent from the first column indicated little removal of other organics as expected. Experiments have shown that the chemicals like lindane, methoxychlor, endrin, toxaphene, and two chloropheroxys, 2,4,-D and 2,4,5-TP can be removed in excess of 99% by using GAC. The conclusion was made that carbon beds are advantageous in removing pesticides because of the preferential adsorption.

HYDROCARBONS

A coal base GAC column 71 cm (2 in) deep was operated at a flow rate of 1×10^{-3} m/s (2 gpm/sq ft) with Cincinnati, Ohio tap water spiked with 50 ug/l of naphthalene. After seven months of operation, the non-volatile total organic carbon (NVTOC) front for 50 percent removal had penetrated the first 51 cm (20 in) of the bed, while the 50 percent removal point for napthalene was only about 5 cm (2 in) down the column, indicating that this compound is adsorbed very strongly. Studies have shown that river-bank filtration followed by activated carbon treatment was very effective in that the concentration of polynuclear aromatic hydrocarbons in water was reduced by about 99 percent.

HALOFORMS

The EPA currently is carrying out research to determine the effectiveness of GAC for removing haloforms [14]. In one experiment, two 71 cm (28 in) deep glass columns were operated in parallel, one with a coal-base GAC and the other with a lignite-base GAC, to observe removal of chloroform, bromodichloromethane, dibromochloromethane, and bromoform from Cincinatti, Ohio tap water. The columns were operated at a flow rate of 1.4×10 m/s (2 gpm/sq ft). After four weeks the ability of the coal-base GAC appeared to be somewhat reduced, to remove chloroform while the lignite-base GAC appeared to be better. A similar result was found for the bromodichloromethane. The concentration of dibromochloromethane in the influent varied widely, but in all cases it was removed. Also, no bromoform was observed in the influent water throughout the experiment.

TASTE AND ODOR

The Plains Region of the American Water Works Corp. (AWWC) system operates 18 plants in seven states. Of these, 11 showed an increasing problem with taste and odor begin-

ning in the later 1960s. Granular activated carbon has worked successfully to end taste and odor complaints in these 11 plants [15].

The installation of GAC at the Granite City, IL plant was the first installation in the Plains Region. Complete conversion of the sand filters to granular carbon was completed in February 1971. Daily threshold tests performed on both raw and treated water revealed the ability of the carbon to consistently produce a plant effluent with a threshold odor value of less than one, regardless of fluctuations in the river supply. In addition to providing for the removal of taste and odor causing constituents, granular carbon has also served the secondary role of filtration. The 12×40 standard sieve carbon (Calgon Corp., Filtrasorb 400) has an effective size of 0.45–0.55 mm. As a result, a high degree of suspended solids removal was achieved while keeping the filter head loss to a minimum.

Clarence A. Blanck [16] has reported that the overall cost reduction in these 11 plants for taste and odor control was 34%. This varied from plant to plant, ranging from 11% in Granite City to 70% at Terre Haute (see Table 8).

Removal of Heavy Metals

CADMIUM (Cd)

Biologically, cadmium is considered a non-essential, non-beneficial element of high toxic potential. In addition to its association with cardiovascular disease, particularly hypertension, cadmium poisoning from contaminated food and beverages has been reported. Cadmium is not expected to pose a serious problem for the water utility industry. Only minute traces have been found in natural waters, generally in areas where lead, copper, and zinc are mined and processed. 0.01 mg/l is mandatory according to USPHS.

For the removal of cadmium, the Drinking Water Research Div. (DWRD) pilot plants tests used virgin and exhausted GAC filter media in parallel with conventional dual-media filters. Except for high pH runs, when cadmium removals were also achieved with the dual media filter, cadmium removals for virgin and exhausted GAC filters were always significantly higher than those of the dual media filter. Cadmium removals by the exhausted GAC filters was equal to or greater than that achieved by the virgin GAC filters. GAC will remove moderate amounts of cadmium (30–50 percent) and therefore may be useful for surface water treatment.

LEAD (Pb)

Lead is well known for its toxicity, and no beneficial effects have been found on human or animal development. Although acute lead poisoning is somewhat rare, chronic lead toxicity is common because lead accumulates in bone and tissue and people are exposed to lead in food, air and water, including that used in cooking and beverages. Another source is inhaled tobacco smoke. World Health Or-

TABLE 8. AWWC Plains Region Plant Operating Data, 1974.

Plant	Capacity (mgd)	Raw water source	Carbon depth. (in.)	Filtration rate gpm/ft^2	Raw water turbidity (ftu)	Taste & odor	Finished turbidity	Water Taste & odor	Cost ($/mg)
Ashtabula, Ohio	4	Lake Erie	24	2.0	22	1	0.1	<1	4.54
Davenport, Iowa	20	Mississippi River	18	2.0	124	4	0.23	<1	3.23
E. St. Louis, Ill.	28	Mississippi River	18	2.0	147	10	0.40	<1	2.31
Granite City, Ill.	10	Mississippi River	18	2.0	97	5	0.15	<1	3.77
Kokomo, Indiana	10	Wildcat Creek	20	2.0	54	4	0.56	<1	3.90
Lexington, Ky.	16	Private reservoir	12	2.0	15	4	0.57	<1	2.08
Marion, Ohio	6	Wells & Scioto River	18	2.0	38	1	0.11	<1	4.25
Muncie, Indiana	5	White River	24	2.0	44	2	0.50	<1	6.17
Peoria, Illinois	10	Illinois River	42	3.0	107	26	0.16	<1	5.01
Richmond, Ind.	8	Private reservoir	24	2.0	11	3	0.26	1.5	2.86
Terre Haute, Ind.	5	Wells & Wabash River	30	3.0	91	30	0.16	1.3	3.47

ganization (WHO) recommends 5 ug of lead per kg per day as a safe total daily intake.

Although GAC may be effective for lead removal, the DWRD pilot-plant studies showed such high removals by conventional coagulation treatment that no advantages are seen for lead removal by substituting GAC for dual-media filtration for surface water treatment. No other information was found on removal of lead from water by GAC.

SILVER (Ag)

The principal effect of silver on the body is cosmetic. Silver causes an unsightly permanent blue-grey color of the skin, eyes, and mucous membranes. The amount of silver required to produce this condition (argyria) is unknown, but the amount of siver from injected Ag-arsphenamine, a substance which produces argyria, is precisely known to be any amount greater than 1 g of silver, 8 g Ag-arsphenamine in an adult. According to EPA (1975–76 study) maximum contaminant level (MCL) for silver is 0.05 mg/l (primary).

The OWRD studies show that removals of silver by the GAC filters were at least equivalent to those by dual-media filters. Silver can be removed in excess of 81 percent by GAC from water.

ARSENIC (As)

The EPA National Interim Primary Drinking Water Regulations (NIPDWR) established MCL of 0.05 mg/l for arsenic. Arsenic is believed to be a potential carcinogen, with inorganic arsenic compounds linked to the develop-

ment of cancer of the skin, respiratory system, and the gastrointestinal tract. The National Academy of Science (NAS) report on the NIPDWR concluded that "the absence of positive results from controlled animal studies makes it impossible to estimate quantitatively a risk of cancer from intake of arsenic in any form of concentration."

Experiments on removal of arsenic from water by GAC has shown that both virgin and exhausted GAC has nearly the same efficiency. Arsenic removal of 98 percent was reported.

SELENIUM (Se)

EPA established a primary MCL of 0.01 mg/l in 1975–76. Selenium is widely believed to have symptoms similar to arsenic poisoning and has been associated with increased dental caries. The Water Supply Research Division (WSRD) pilot plant results have shown that GAC does not adsorb selenium.

GAC System Costs

When it first proposed the regulation, EPA asserted that the anticipated capital expenditures nationally would be in the range of $291–685 million. It estimated annual revenue increases of $57–145 million. Recently, EPA substantially revised those estimates. EPA now estimates capital expenditures nationally of $616–831 million and annual revenue requirements of $124–169 million, premised upon the assumption that only 61 systems will have to install GAC facilities.

EPA's estimates are still far short of reality. Many costs are not included in EPA's "revised" cost estimates, some of which will be substantial. EPA has not attempted to determine the substantial costs of obtaining a variance or other governmental permits for on-site regeneration of GAC, which will often require siting of large furnaces in park-like settings. EPA also ignores the cost of the requirement that water systems repalce the existing filter media with GAC during the interim period prior to the installation of the GAC facilities.

The preferred course would be to estimate costs using data obtained from the systems which EPA asserts will be affected. However, EPA has not identified the 61 systems which it anticipates will be required by the proposed regulation to install GAC facilities. In fact, EPA has failed to define the criteria that would enable one to obtain a variance from the GAC treatment requirement.

Advantages and Disadvantages Using GAC

ADVANTAGES

Granular carbon offers the potable water plant many advantages as follows:

* The reserve adsorption capacity of granular carbon eliminates the need to adjust carbon dosage to compensate for sudden changes in the level of organic contaminants
* Beds or columns of granular carbon are in constant operation year-round and prevent undetected odors, tastes, color and toxic organics from reaching the distribution system.
* Improved water quality—better taste, odor, safety and appearance
* Increased plant capacity in terms of gallons per minute due to dual media filtration effects.
* Less backwashing required than with sand alone.
* Improved filtration of suspended solids due to dualmedia filtration.
* Disposal, storage and dust problems associated with powdered carbons can be eliminated.
* Labor costs for handling granular carbon are usually less than for powdered carbon.
* Granular carbon helps automate small and moderate size plants, as only minimum attention and adjustment are necessary.

DISADVANTAGES

Granular carbon has several disadvantages as follows:

* GAC treatment facility is, overall, an expensive process.
* When GAC is used for removing organics, frequent regeneration is required, thus boosting the cost.
* Precursor leakage occurs when GAC is used.
* Eighteen inorganic elements, including lead, mercury

and cadmium have been found in a composite sample of GAC.
* Consecutive regeneration of the carbon could lead to formation of polycyclic aromatic hydrocarbons (PAHs) and other toxic materials.
* For GAC running tests properly must be the problem, because, as disclosed by many presentations, laboratory results are not scaling up.
* Use of the material as supplied can also be a problem because of the difficulty in obtaining a representative sample from the manufacturer.
* GAC performs poorly at removing the likes of chloroform.
* In the U.S. at present time, GAC generally serves the dual purpose of filtration and adsorption. Often the granular material in a rapid filter bed is simply partially or entirely replaced with GAC. These gravity beds have several inherent disadvantages, however, in that (1) frequent backwashing may upset the classical "front of organic penetration" and lead to premature breakthrough, (2) removal of GAC for reactivation is difficult, (3) use of GAC in pressure beds could permit counter-current operation and insure greater efficiencies (4) adsorption efficiency is best if the carbon granules are small, but headloss buildup is too rapid if the media used for suspended solid removal is too small, and (5) contact time is limited and may be less than optimal.

Conclusion

GAC has the ability to remove many organic compounds from water supplies very efficiently in well designed and operated beds. Use of GAC for reduction of TOC should be proceded by good clarification because many organics are removed by this step and the load on the carbon process thereby is reudced. Good operation is essential to avoid losses of carbon through backwashing and deposition of materials on the carbon granules. The development and improvement of monitoring procedures that can be used in different locations to achieve different objectives is especially important. Biological activity in the bed may lead to prolonged lifetimes of the carbon beds and to removal of nonadsorbable but degradable compounds that otherwise might cause problems in the distribution system. Care must be taken, however, to avoid high levels of bacteria in the bed effluent.

BIOLOGICAL ACTIVATED CARBON

The EPA's proposed regulations for control of organic chemicals in drinking water consists of two parts.

1. A MCL for total trihalomethanes (TTHM) in drinking water of 0.10 mg/l (l00 ppb). Initially, this would be ap-

plicable only to community water systems serving a population of greater than 75,000 people. Community water systems serving populations between 10,000 and 75,000 would not have to comply with the MCL, but would be required to monitor THM levels. Community water systems serving fewer than 10,000 persons are not affected.

2. A treatment technique requirement, calling for use of granular activated carbon for the control of synthetic organic chemicals.

The presence in drinking water of chloroform and other trihalomethanes and synthetic organic chemicals may have an adverse effect on people's health. The basic idea behind EPA's sweeping new proposals is to reduce human exposure to suspected cancer-causing chemicals by requiring major changes in the treatment of given drinking water. Treatment for control of organics, other than those causing taste and odor, is not practiced in the U.S., although some organics are removed by conventional treatment. The National Academy of Sciences, in its recent report to the EPA, said 309 volatile organic compounds plus 55 pesticides have been identified in drinking water. Yet these represent no more than 10% by weight of the total organic materials in drinking water. Today the number of identified organics is well over 400.

There are only four approaches to reducing trihalomethanes in drinking water. (1) The chlorine added must be reduced. (2) The time THM precursors and chlorine are in contact must be reduced. (3) Concentration of precursors must be reduced. (4) A disinfectant other than chlorine must be used. The last approach is not being seriously considered in the U.S. at this time.

Accordingly, a sound approach is to remove or reduce organics before applying chlorine. This is an art many European cities have been practicing for years. European water treatment plants are using ozone (a powerful oxidizing agent and disinfectant than chlorine), chlorine dioxide (another powerful oxidant and disinfectant) and activated carbon. The concept of granular activated carbon columns, treating water void of chlorinated organics and subjected to preozonation without requiring regeneration is termed as Biological Activated Carbon (BAC). Most experience to date with ozonation has been in France, Germany, Switzerland, Russia, Holland, Belgium, Austria, other western European countries, Japan and Canada. Over 1000 drinking water plants in Europe use ozone for one or more purposes. Today, more than 460 water supply systems in France are using ozone for a variety of purposes. In Canada, 20 water plants use ozone. In the U.S. there are currently two municipal plants using ozone—Whiting, Indiana (taste and odors) and Strasburg, Pennsylvania (disinfection).

U.S. and European water-treatment philosophies differ markedly. Europeans are brought up with the understanding that when there are any tastes in water (especially chlorine tastes) the water is contaminated. Americans are brought up with the understanding that when chlorine cannot be tasted, the water may be contaminated. Therein lies the major reason for the two basic approaches to treating water supplies. Americans deliberately produce chlorinous taste in the water to assure bacteriological (but not necessarily chemical) safety; Europeans reduce the chlorine demand of water (insuring chemical safety) so that the amount of residual chlorine or chlorine dioxide finally used to maintain bacteriological safety in the distribution systems will be so small as to be tasteless [17–19].

In Germany, because of the historically good experiences using groundwater, efforts are made to use treatment procedures that include passage of water through the ground. This practice of drawing water from wells located 50 to 250 meters from the bank of the river is known as river sand bank filtration. Groundwater is extracted in vertical and horizontal wells from large 10 to 30 meter water-bearing diluvial gravel and sand sediments. In the Duesseldorf area, the ratio of bank filtrate water to groundwater in the water to be treated is about 3 to 2, varying with the water level of the river and changes in the water table. The effect of natural treatment by ground passage between the Rhine River and the wells results in a 65–75% reduction in the amount of dissolved organic carbon (DOC).

Because of increasing levels of pollution and the demands for higher quality drinking water, however, use of ozone as an oxidant and disinfectant has grown rapidly in Europe. Ozone is the second most powerful oxidant. In many cases, it is the only available oxidizing agent capable of attacking certain organic contaminants. The combination of ozonation followed by GAC has been suggested as a means to improve the cost effectiveness of the GAC process for organic removal [20].

OZONE USES

Ozone is capable of performing many more tasks than disinfection. In Europe, it is always regarded as a water treatment technique and seldom simply as a disinfectant. The uses of ozone in Europe water treatment can be divided into five major categories:

1. Iron and manganese removal.
2. Oxidation of organics including color bodies, taste and odor bodies, algae, suspended solids, and dissolved solids.
3. Microflocculation.
4. Bacterial disinfection.
5. Viral inactivation.

In conjunction with BAC columns, ozone removes ammonia biologically, simultaneously reducing dissolved organics and greatly extending the operating life of the activated carbon. Ozone can be used safely as a terminal disinfection step only if:

1. Residence time in the distribution system is short.
2. The temperature of the finished water is relatively low.
3. The quality of the finished water is high, preferably having a dissolved organic carbon concentration of less than 0.1 mg/l.
4. The ammonia content of the finished water is very low.

POINTS OF OZONE APPLICATION

Ozone is applied at different points in the treatment process, depending on the purpose. For iron and manganese oxidation and for flocculation, ozone is applied early in the process. When ozone is used for organics oxidation, it is usually applied near the middle of the process, following a filtration step for removal of coarse materials. When ozone is used for viral inactivation or bacterial disinfection, it is applied near the end of the process.

EFFECTIVENESS OF OZONE

Gomella has reported on the "fast and full inactivation of enterovirsues along with a perfect bacterial effect." Nice's (France) distribution network has been fed ozonized water exclusively, with no chlorine residual, for more than 70 years without incident.

Schalekamp cites these advantages of using ozone, depending on the dosage level: inactivates viruses within seconds; 100% effective kill of coliform bacteria; removes up to 65% color; reduces the amount of organics based on humic acid by 30–50%; reduces 30–90% of certain organic halogen compounds; improves taste and reduces odors; causes flocculation.

Kuehn, Sontheimer and Kurz note the increased biodegradability of humic material after ozonization, which furthers natural biological processes. Sontheimer has shown at the Dohne plant in Muelheim that an initial break-point chlorination step (using up to 30–50 mg/l of chlorine) can be eliminated through use of pre-ozonation (1 mg/l) followed by flocculation and filtration, then postozonation (2 mg/l) for oxidation, followed by BAC, then 0.2–0.3 mg/l of chlorine for residual. Thus, the effectiveness of ozone for various treatment purposes has been clearly established through research and over 70 years of successful operation. Howard M. Neukrug et al. [21] showed through their experiment that the cost of preozonation does not sufficiently offset by lower GAC operating costs when removal of volatile halogenated organics in the controlling criterion.

COST OF OZONE

The most often cited reason for U.S. water supply plants not using ozone is the high capital cost of an ozonation system. Large ozone systems cost $600–800/lb of ozone generation capacity/day (from air) in Europe. Intermediate size systems cost up to $1000/lb/of daily ozone production.

These costs do not include costs for housing and installation. When ozone costs are compared with those of chlorine and chlorine dioxide on the basis of oxidation potentials, states Schalekamp, ozone costs 3.5 times more than chlorine and 1.5 times more than chlorine dioxide.

SAFETY OF OZONATION SYSTEMS

In the 71 years of use of ozone in treating European drinking water, there has never been a single recorded instance of an accident resulting in death of an operator due to accidental exposure to ozone. Because ozone is the second most powerful oxidizing agent readily available to mankind, however, care must be and is taken to avoid unnecessary exposure to this gas. As soon as the electrical power for the ozone generator is turned off, production of ozone ceases. Modern European water treatment plants include a sensor that monitors ozone concentrations in the ozoneur room atmospheres. If the allowable limit of 0.1 mg/l of ozone is exceeded, an alarm rings and power to the ozone generators can be stopped. Besides this safety feature, other plants maintain air in their ozonator rooms under slightly negative pressure.

Rouen-La-Chapelle, France

The La-Chapelle plant at Rouen, France, is the world's longest operating biological activated carbon water treatment facility specifically designed for removing ammonia and synthetic organics. The 13 mgd plant has been operating for more than 26 months and has produced excellent water—with no need as yet to regenerate carbon beds [22].

Rouen water-treatment plant key steps are: preozonation, sand filtration, carbon adsorption, postozonation, and a very low dose of chlorine immediately before water enters the distribution system.

PREOZONATION

Ozone (1% in air) is injected into raw water by a turbine in one of two contacting chambers (each 5 m high), providing a dosage of 0.7 mg/l. The ozonated water then flows through a second chamber, insuring a total contact time of 3 minutes. The ozone comes from two sources: the off-gases from the post-ozonation contactor, and the ozone generator.

Preozonation prepares raw water for double-filtration. Specifically it:

- Satisfies the oxygen demand of the water. Organic pollutants are oxidized and their concentrations partially reduced.
- Reoxygenates the water, partly by decomposition of the ozone, mostly by the oxygen contained in the air added with the ozone.
- Breaks down many of the complex, biorefractory organic molecules present in the raw water (i.e. these molecules

are partially oxygenated by the ozone) making them biodegradable.

- Reduces the concentration of various micropollutants, by oxidation and flocculation.
- Converts soluble manganese and soluble iron to insoluble oxides, so they can be removed by the downstream sand filter. The sand filter retains all precipitated manganese, iron and flocculated organics. This is important, for it protects the downstream granular activated carbon: the active sites on the carbon will not become blocked. If that were to happen, the capacity of GAC to adsorb dissolved organics would be reduced.

SAND FILTRATION

The primary function of the sand filters is to remove precipitated manganese and iron, plus some flocculated organics rendered insoluble by ozone oxidation. A secondary function is initial nitrification of the ammonia. Empty-bed contact time through the sand beds is 12 minutes.

Sand filtration is accomplished by six filters with total surface of 454.8 m² and a filtration velocity of 4.84 m/hr. These filters are charged with 100 cm of a special Dutch quartz sand, 8 to 12 mesh (1 mm) supported by porous false bottoms (strainers).

To backwash the sand filters, forward water flow is stopped and air is introduced (65 m³/m² of filter) on the underside of the filter for 2 to 3 minutes. As soon as air emerges from the top layers of sand, water is also introduced at 25 m/hr. The concurrent flow of air and water continues for 15 minutes. Then, air flow is stopped and water flow alone is continued for an additional 5 minutes. This technique of using air, then air + water, then water alone for filter backwashing prevents bubble problems and the formation of mud balls.

GRANULAR ACTIVATED CARBON TREATMENT

The granular activated carbon is only 75 cm deep and is coconut carbon, with 1000 m² of surface area per gram.

When ozonized water containing ammonia, dissolved oxygen, dissolved organics, and residual ozone reaches the GAC, several reactions occur. Empty bed contact time is 9 minutes. First, the dissolved ozone reacts immediately with carbon and is converted to oxygen. At the same time, some organics are adsorbed onto the surface and into the pores of the activated carbon. Those organics not adsorbed would be expected to pass through. However, because the carbon contains bacterial growth, these non-adsorbed organics (and ammonia) are attacked by the bacteria. The organics adsorbed by the GAC are also attached by bacteria. As these adsorbed organics are converted to carbon dioxide and water, the carbon sites again become available to adsorb more sorbable materials. Thus, the bacteria effectively regenerate the granular activated carbon.

Every three months, tests of the activated carbon are made to see what organics are being adsorbed onto the carbon surfaces. Table 9 shows average plant performance at the Rouen-La-Chapelle.

POSTOZONATION

Postozonation is accomplished by porous diffusers placed in two of the three contact chambers (each 4.5 m high) in each treatment train. The third contact chamber is merely a holding tank, where ozonized water is degassed and off gases collected for recycle into the preozonation stage. The purposes of postozonation are: bacterial disinfection; viral inactivation; and further reduction in concentration of dissolved organics.

Average ozone dosage at this point is 1.4 mg/l. 67% of this post-ozone dosage is applied in the first diffuser contact chamber, the balance in the second chamber.

STORAGE AND DISTRIBUTION OF TREATED WATER

Storage of treated water is provided by two underground reservoirs, each having a capacity of 1250 m³, and one pumping tank of 250 m³. When treated water is sent to the distribution system, a dosage of 0.4 to 0.5 mg/l of chlorine

TABLE 9. Average Plant Performance at the Rouen-La-Chapelle Biological Activated Carbon Water Treatment Plant.

	Raw Water	Pre-ozonized	Filtered (sand and carbon	Post-ozonized	Percentage elimination
Turbidity (JTU)	4	—	—	2	—
NH₄⁺, mg/ℓ	1.80	1.80	0.40	0.26	86
Mn, mg/ℓ	0.15	0.07	0.04	0.02	87
Detergents (mg/ℓ DBS)	0.12	0.09	0.06	0.03	75
Phenols, µg/ℓ	6.50	4.0	1.50	0	100
SEC, µg/ℓ	590	470	250	150	75
Substance extractable with cyclohexane (µg/ℓ)	1335	740	535	410	69

Average NH₃ content of raw water: 0.3 mg/ℓ in 1968; 2.6 mg/ℓ in 1975.

is added. There are two reasons for not using high concentrations of final disinfectants. First, if the water has been treated properly nearly all of the dissolved organic compounds and other reactants will have been removed and the disinfectant dosages needed to provide residuals are very low. Second, the Germans believe water-treatment processes should be designed to remove bacteria prior to disinfection. The belief is that since chlorine is harmful to bacteria, it may likewise be harmful to humans.

COSTS

Cost for building the La-Chapelle plant was $4 million. The cost of treating water at the plant is $.46/1000 gallons. But the cost of water to the consumer averages $2.42/1000 gallons because of added sewer charges and taxes.

Advantages of BAC

In European pilot studies and in drinking water treatment plants it has been shown by many workers that preozonation followed by activated carbon adsorption results in:

1. More effective removal (up to 200%) of dissolved organics from solution by the BAC
2. Increased capacity of the carbon to remove organics (by a factor of about 10)
3. Increased operating life of the carbon columns before having to be regenerated (up to 3 years), especially if the GAC can be kept free of halogenated organics
4. Biological conversion of ammonia in the GAC columns
5. Use of less ozone for removing a given amount of organics than using ozonation alone (BAC is cost-effective over ozonation in removing dissolved organic carbon)
6. Filtrates from BAC columns in drinking water plants can be treated with small quantities (0.1–0.5 mg/l) of chlorine or chlorine dioxide, which produces drinking water of acceptable bacterial quality and provides a residual disinfection for distribution systems

Disadvantages of BAC

The main disadvantage of using BAC is the highly expensive process. Another disadvantage is that in "removing" ammonia, nitrates are formed. Thus, with drinking waters containing 10 mg/l of ammonia-nitrogen, 10 mg/l of nitrate nitrogen can be anticipated in the treated waters. An additional "disadvantage"—the utility not only has to install GAC, still considered a new technology in the U.S., it must also install ozonation, another "new" technology.

BAC: What Significance for the U.S. Water Supply Industry?

Incorporation of BAC into U.S. drinking water treatment systems holds out several promises.
- Eliminating the need for breakpoint chlorination to remove ammonia, thus eliminating formation of large quantities of THMS and other halogenated organics, and saving on chemical costs
- Lowering total organic carbon (TOC) in product water
- For those systems that may be required by EPA to install GAC for removal of organics, BAC may be more cost-effective than GAC alone.

Conclusion

The use of ozone, chlorine dioxide, and ozone plus activated carbon are established and proven technologies in Europe. For polluted raw waters, ozone or ozone coupled with activated carbon appears to be the best techniques available for reduction of organics concentrations in drinking water. This statement can be reinforced by assertively stating that a process consisting of screening → flocculation → sedimentation → rapid sand filtration → ozonation → granular activated carbon → chlorine or chlorine dioxide would ensure a safe drinking water, given proper design, operation and maintenance.

POWDERED ACTIVATED CARBON

The powdered carbons are those which are smaller than U.S. Sieve Series No. 50. The better grades of powdered carbons are made from lignin and lignite.

Powdered activated carbon (PAC) is usually added to the water with automatic chemical feeders. These feeders meter in the carbon at a predetermined rate at various points in the system. The point of carbon application varies with the treatment process and the desired results.

To deal with the problem of odor, PAC most commonly has been selected. A recent survey conducted by two committees of the American Water Works Association (AWWA) has shown that approximately 25 percent of the 645 U.S. utilities that were surveyed use PAC. While the use is predominantly for odor control, removals of non-odorous compounds also may take place. The extensive use of PAC probably developed because PAC can readily be applied as needed to remove odors that appear periodically through the year. Its dosage is readily controlled using the threshold-odor test. Whether PAC has significant advantages when the objectives do not include or are broader than odor removal on an occasional basis, however, is questionable. For example, when trace compounds or natural organic matter that has been implicated as the precursor of haloforms must be removed on a continuous basis, the use of PAC is questionable. The mixing achieved in the flocculator and sedimentation basin is not sufficient to permit the adsorption capacity of the PAC to be achieved. Flash mixing of a coagulant and PAC can lead to incorporation of carbon particles into the floc thereby decreasing the rate of adsorption of organics.

Also, even if equilibrium were achievable, it is possible to adsorb more matter using a column than in the concurrent flow process used for PAC.

Reduction in Trihalomethanes

Powdered activated carbon feed facilities were installed in 1977 at all six major East Bay Municipal Utility District (EBMUD) filter plants for the control of anticipated taste and odors from the delta water added to the system during the drought. During prechlorination PAC has proved relatively ineffective in the net reduction of THMS and total organic carbon (TOC).

During post-chlorination a PAC dosage of approximately 10 mg/l produced a reduction in terminal THM formation potential of approximately 30 percent. The reliability of post-chlorination at the direct filtration plants, however, will depend upon the effectiveness of the disinfection of the raw water in the aqueducts. Treatment with PAC at the levels necessary to produce significant reductions in the THMs would be more costly than disinfection with chlorine dioxide, ammonia chlorination, or ozonation and would not be as efficient in the reduction of THMs.

A test of PAC's effectiveness in removing THMs and TOC was conducted at the Walnut Creek filter plant. A summary of the results is presented in Table 10.

PAC dosages from 5 to 30 mg/l were added and the plant's raw and filtered waters analyzed for THM and TOC. Raw water TTHMs ranged from 25.3 to 84.5 ug/l depending upon the amount of exposure to chlorine the water received in the Mokelumne aqueducts. During the test pre-chlorination was used and, as might be expected, increased chlorine dosages were required as the PAC dosage increased. During this test there was some concern that without the sedimentation step the PAC would pass the filters, increasing the particulate matter in the treated water. There was virtually no particulate degradation of the treated water with PAC dosages up to 30 mg/l at a filtration rate of 1.2 mm/s (1.8 gpm/sq ft).

The PAC probably served as nuclei for flocculation and, for this reason, did not pass through to the treated water. On the other hand, rapid floc growth on the PAC particles may have effectively shielded the carbon from dissolved organics in the raw water, limiting its effectiveness in the reduction of THMs and TOC.

Removal of Pesticides

The six chemicals—endrin, lindane, methoxychlor, and toxaphene which are chlorinated hydrocarbons and two chlorophenoxys, 2,4,-D and 2,4,5-TP (silvex)—can be grouped under the general term "pesticides." It has been found that PAC can remove all these chemicals effectively from drinking water.

Most of the information on reducing various concentrations of endrin has been gathered through laboratory studies and pilot-scale water treatment plant experiments. Robeck, Dostal, Cohen and Kreissl has shown that by using PAC concentrations of 5, 10 and 20 mg/l endrin removal of 85, 92, and 94 percent can be achieved respectively. Robeck also concluded that lindane removal of 30, 55, and 80 percent was achieved by using PAC concentrations of 5, 10, and 20 mg/l respectively. Adsorption with activated carbon was more encouraging to remove toxaphene. An initial concentration of 0.1 mg/l toxaphene was reduced to 0.007 mg/l by 5 mg/l PAC, so the authors concluded that "no common treatment other than that with activated carbon will remove toxaphene." Treatment information is not available on methoxychlor in 1977. It is very likely, however, that PAC would remove this contaminant effectively. Whitehouse conducted laboratory studies to determine the effect of pH and types of PAC in removing 2,4-D from solution; however, the concentrations of adsorbants and adsorbates (100 mg/l level) are thought to be too atypical in water treatment to warrant further details on the results. By using PAC concentrations of 5, 10, and 20 mg/l, 2,4,5-Tester removal of 80, 80 and 95 percent were achieved respectively.

The following assumptions are made in estimating the

TABLE 10. Powdered Activated Carbon Test: Direct Filtration.

Carbon Dosage (mg/ℓ)	Raw Chlorine Residual (mg/ℓ)	Prechlorination Dosage (mg/ℓ)	Total Available Chlorine (mg/ℓ)	Particle counts 2.5 m (no./ml)	Raw Water TTHM (μg/ℓ)	Filtered Water TTHM (μg/ℓ)	Raw Water TOC (mg/ℓ)	Filtered Water TOC (mg/ℓ)
0	0.30	0.60	0.90	16	84.5	85.1	2.38	1.69
5	0.31	1.08	1.39	40	73.4	81.0	2.42	1.60
10	0.28	1.20	1.48	5	75.9	74.3	2.20	1.53
20	0.03	2.88	2.91	10	25.3	46.2	2.02	1.36
30	0.06	2.76	2.82	11	37.7	41.0	2.18	1.12

cost of adding PAC for controlling pesticides in drinking water. (1) A filtration plant already exists. (2) PAC is already being fed for odor control. (3) Sludge-handling facilities and disposal are adequate for the additional imposed loadings. (4) Analytical costs for pesticide analysis are excluded. (5) PAC costs 30 cents per pound. (6) A PAC dose of 5–80 mg/l would allow for a contamination level from 2.5 to 50 times the MCL.

Removal of Heavy Metals

CADMIUM (Cd)

Powdered activated carbon has some ability to remove cadmium from water. Thiem and O'Connor studied various types of PAC for removing cadmium from a synthetic drinking water. The best removals were 29 and 26 percent with 100 mg/l of PAC 1 and PAC 2, respectively, with a solution containing 0.05 mg/l of cadmium at pH 7. Removals increased with increasing pH. At pH 9 and the same carbon dose, cadmium removal with PAC 2 doubled, from 26 to 53 percent and removal by PAC 1 rose to 43 percent. A third PAC material was the least effective of the three. At pH 7 and 9, removals were 7 percent and 27 percent respectively, with a 100 mg/l dose. Because large amounts of PAC removed only small additional amount of cadmium, the use of PAC is neither practical nor economical for cadmium removal.

LEAD (Pb)

Netzer and Norman studied the removal of lead at varying pH with large doses (10,000 mg/l) of PAC. In the pH range 3–5, where pH has no effect on lead solubility, lead reductions from 100 to 10 mg/l to less than 0.1 mg/l were achieved with PAC alone. At pH 5–10, the combination of pH adjustment and PAC continued to produce very high removals (greater than 99 percent), but in this range pH adjustment was responsible for 8–84 percent of the reduction with removal percentage increasing with increase in pH.

In a Drinking Water Research Division (DWRD) conducted jar test to evaluate the removal of 0.15 mg/l of lead from river water (45 tu) at pH 7.3–7.4 with PAC, varying doses of 10 to 200 mg/l of PAC were introduced into six jars and the water stirred rapidly for 1 hour. Lead measurement on centrifuged samples showed removals increasing from 89 percent with 10 mg/l of PAC to 98 percent with 200 mg/l of PAC.

SILVER (Ag)

A DWRD jar test on Ohio river water spiked with 0.15 mg/l silver showed low removals by PAC. Removal increased with increasing doses of PAC, but the best results were only 66 percent with 100 mg/l PAC. Smith and Sigworth, however, report laboratory studies on silver nitrate solution showing that activated carbon could adsorb 9 percent of its own weight at pH 2.1 and 12.5 percent at pH 5.4. No data were presented on removal at high pH, but the authors suggested good potential for removal.

ARSENIC (As)

Jar-test studies by Logsdon showed that PAC is not effective for the removal of either arsenic III or V. Less than 3 percent removals were achieved on water containing 0.5 mg/l of arsenic V with doses of 100 mg/l PAC. Very poor removals (less than 8 percent) were obtained with PAC doses up to 300 mg/l on water containing 0.5 mg/l of arsenic III. Studies of the effect of pH in the 6–8.2 range indicated that pH did not affect removal of arsenic III. Based upon the results of these limited studies, PAC cannot be considered useful for arsenic removal.

SELENIUM (Se)

The Water Supply Research Division (WSRD) conducted jar-test studies to determine the removal of selenium IV and VI from water by PAC, which showed that less than 4 percent of either selenium species could be removed with PAC doses up to 100 mg/l on well water spiked with selenium concentration of 0.03 and 0.1 mg/l. Tests done at several pH values showed that pH had no effect on removals.

In jar tests done by Thiem and O'Connor with three different types of PAC on the removal of 0.05 mg/l of selenium IV at pH 7, 8 and 9, the results were almost identical to the results of the WSRD, with 4 percent or less removal achieved with doses up to 100 mg/l.

MERCURY (Hg)

The results of the studies to determine the removal of mercury by adsorption on to PAC are seen in Figure 22.

From these curves, it can be seen that less than 30 percent of the 10 ug/l solution of mercury was removed at a carbon dosage of 10 mg/l at pH 7. This would seem to indicate that the activated carbon that is currently being added to water supplies in the range of 5 mg/l to control taste and odors may be accomplishing little in the way of mercury removal. Only when 100 mg/l of carbon was applied to the test solution did the residual concentration approach the established drinking water standard. Removals also appear to be quite sensitive to pH. Roughly twice as much mercury was removed at pH 7 as at pH 9. The addition of chelating agents such as EDTA or tannic acid prior to contact with PAC measurably improved the removals [23].

Advantages of PAC

The major advantages of PAC are that it can effectively remove odor, that it is cheaper than granular carbon, and that extensive plant-construction costs are not required in

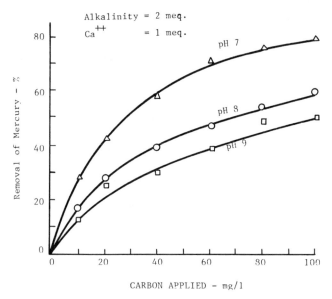

FIGURE 22. Influence of pH on the removal of Mercury II.

order to use it. PAC simply can be fed into existing treatment basins to provide the necessary contact for removal of the undersired contaminants. PAC can also remove pesticides and some of the heavy metals.

Disadvantages of PAC

The percentage increase in the cost of PAC has been greater than those for GAC. Since attention now is being given to removal of organics on a continuous basis, the fact that GAC can be used in columns or beds—permitting higher adsorptive capacity to be achieved and easier process control than is the case for PAC. PAC, of course, does increase the sludge handling problems and does not, as presently practiced, make use of the reactivation potential of GAC.

**TRANSPORT AND REGENERATION OF
ACTIVATED CARBON**

Transport

The objectives in any carbon transfer operation are to complete the transfer in the least time, at the lowest cost, with the least disruption of plant operation, and—most importantly—with the least attrition to the carbon or damage to the transfer system.

One of the most important steps toward achieving this goal is the use of abrasion-resistant carbon. When carbon

beds must be changed once a month, even when the best technology is used in the transfer-reactivation system, the largest single cost element is carbon makeup for attrition losses. A second physical property fundamental to reducing operating costs is the capability of the spent carbon for reactivation. If a good working capacity is not restored upon reactivation, the react frequency will be high, which raises losses and thus operating costs.

The primary use of air or pneumatic transport of carbon is in bulk-handling of makeup carbon. Once carbon is introduced into the adsorption-regeneration system, it is usually transported hydraulically in slurry form. Handling characteristics have been experimentally studied by using water slurries of 12 × 40 mesh granular carbon in a 2 inch pipeline. The data indicate that a maximum of 3 lb of carbon per gal of water could be transported hydraulically but that it is better to use 1 gal of water for moving each lb of carbon. The velocity necessary to prevent settling of carbon is a function of pipe diameter, granule size, and liquid and particle density. The minimum linear velocity to prevent carbon settling was found to be 3.9 ft/sec. It is recommended that a linear velocity between 3.5 to 5.0 ft/sec be used. Velocities of over 10 ft/sec are objectionable due to carbon abrasion and pipe erosion. Carbon delivery rates are a function of pipe diameter, slurry concentration, and linear velocity. Data are shown in Figure 23.

Pilot plant tests indicate that after an initial higher rate, the rate of attrition for activated carbon in moving water slurries is approximately constant for any given velocity, reaching an approximate value of 0.12 percent fines, generated per exhaustion-regeneration cycle. This deterioration of the carbon with cyclic operation has been reported to be independent of the velocity of slurry (3.5–5 ft/sec). Loss of carbon by attrition in hydraulic handling apparently is not related to the type of pump (diaphragm or centrifugal) used.

Carbon slurries can be transported by using water or air pressure (blowcase), centrifugal pumps, eductors or diaphragm pumps. The choice of motive power is a combination of owner preference, turndown capabilities, economics, and differential head requirements.

ON-SITE REACTIVATION

At the present time no potable water facility in the U.S. has a full-scale, on-site reactivation process for spent granular carbon. On-site reactivation involves several transfer steps and the special carbon transfer equipment outline here.

1. At the end of a service run, spent carbon in a filter is transferred as a slurry to a storage tank that holds a minimum of 1.5 times the carbon volume contained in a single filter.
2. As soon as the filter bed has been emptied, a fresh

charge of reactivated carbon from a second storage tank is transferred as a slurry into the empty filter, backwashed to remove fines, and places in service.
3. The spent carbon in the storage tank is transferred as a slurry to the dewatering screw, where the carbon is drained and metered to the react furnace.
4. From the react furnace the carbon falls into a quench tank and is transferred, again as a slurry, to the reactivated carbon storage tank, where it accumulates until another filter is ready to be taken off-line for carbon reactivation.

OFF-SITE REACTIVATION

The transfer steps and equipment for off-site reactivation are:

1. Spent carbon is emptied out of the filter into the storage tank.
2. Fresh carbon is delivered directly to the filter by a bulk hopper truck from the supplier contracted to reactivate the carbon.
3. The spent carbon in the storage tank is transferred to the bulk hopper truck to be hauled away for reactivation.

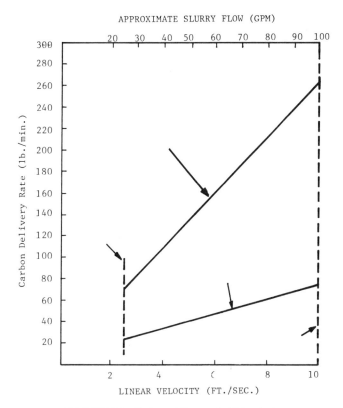

FIGURE 23. Carbon delivery rate (2 in pipe).

Regeneration

There are four general methods for regenerating granular carbons: solvent wash, acid or caustic wash, steam reactivation, and thermal regeneration. The fourth method, that of thermal regeneration, is univerally used because of a simple, easily controlled, reliable, practical, and economical means of regenerating carbon.

A typical carbon regeneration system is shown in Figure 24.

The thermal regeneration process itself involves three steps.

1. Drying,
2. Baking (pyrolysis of adsorbates), and
3. Activating (oxidation of the residue from the adsorbate).

The total regeneration process requires about 30 minutes. The first 15 minutes is a drying period during which the water retained in the carbon pores is evaporated, a 5 minutes period during which the adsorbed material is pyrolyzed and the volatile portions thereof are driven off, and a 10 minute period during which the adsorbed material is oxidized and the granular carbon reactivated.

The theoretical required furnace capacity can be determined simply by multiplying the carbon dosage (in lbs of carbon per million gallons) by the daily flow rate in millions of gallons per day. This will determine the lbs of carbon per day that must be regenerated. This computation does not allow for furnace downtime or other contingencies that must be considered when actually specifying the furnace size to

be installed. Multiple hearth furnaces used for regenerating carbon should provide a hearth area of about one square foot per 40 lbs of carbon to be regenerated per day. The regeneration furnace should be designed so that the hearth temperatures, furnace feed rate, rabble arm speed, and steam addition can be controlled.

As carbon is removed from service for regeneration, the spent carbon is usually hydraulically transported to a drain bin. The drained carbon is dried during the first step in a furnace which heats the carbon to less than 212°F. During baking, the temperature increases from 212°F to 1,500°F, by which time adsorbed organics are thoroughly carbonized. This is accomplished by evolution of gases and by the formation of a carbon residue in the micropores of the activated carbon. The objective of this activating step is to oxidize the carbon residue with minimum resultant damage to the basic pore structure, thereby effecting maximum restoration of the original properties of the carbon. The activating gas temperature during this step is about 1,700°F, while the carbon temperatures range from 1,500°F to 1,650°F. Fuel gas supplemented by varying amounts of additional steam and limitation of oxygen produces the desired atmosphere. The most important phase of the regeneration process is that of activation, with the critical parameters being carbon temperature, duration of activation, and steam or carbon dioxide concentration in the activating gas mixture. Since most installations use direct-fired multiple hearth furnaces for regeneration, the combustion of natural gas with air provides the required heat, while carbon dioxide, oxygen and steam as part of the products of combustion, are the activat-

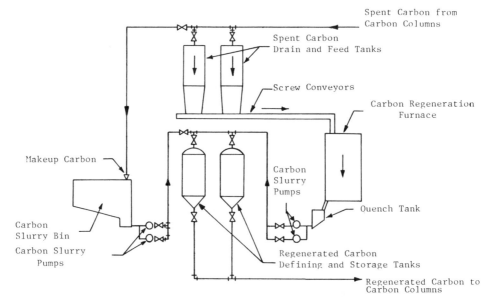

FIGURE 24. Carbon regeneration system.

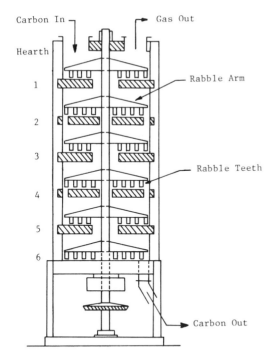

Carbon In — Gas Out

Hearth

1

2

3

4

5

6

Rabble Arm

Rabble Teeth

Carbon Out

FIGURE 25. Cross-sectional view of multiple hearth furnace.

ing agents. Extra steam at approximately one pound per pound of carbon regenerated is supplied. This requires auxiliary steam generating equipment.

The temperatures on the various hearths are approximately as follows: No. 1, 800°; No. 2, 1,000°; No. 3, 1,300°; No. 4, 1,680°; No. 5, 1,600°; and No. 6, 1,680°F. It was found that the addition of steam on hearths 4 and 6 gave a more uniform distribution of temperatures throughout the furnace. The effect of steam is to reduce the apparent density and increase the Iodine Number of the regenerated carbon. About 1 lb of steam per lb of dry carbon was used. Figure 25 shows cross-sectional view of multiple hearth furnace.

The regenerated carbon is discharged from the bottom of the furnace into a quench tank. The water jets must be placed to keep the carbon moving in the quench tank to prevent carbon build-up and plugging of the tank's discharge lines. Partially dewatered carbon from drain bins can be fed into the furnace by a screw conveyor.

The total fuel requirement for the regeneration system is estimated to be 4,250 Btu per pound of carbon regenerated, exclusive of fuel used in the afterburner. This figure is based on the following:

1. 3,000 Btu/lb carbon is required for furnace heat.
2. 1,250 Btu will generate one pound of steam which is sufficient for one pound of carbon under regeneration.

MISCELLANEOUS

The EPA-Proposed GAC Treatment Requirement: Panacea or Pandora's Box?

There are 390 community water systems in the U.S. serving populations greater than 75,000 persons. These systems provide water for a total population of 100 million people—52% of the total population served by community water systems. Of these 390 public water systems, 60 purchased their water from other systems, and another 18 do not presently add chlorine. Of the remaining 312 systems, estimates EPA, only 86 have TTHM levels above 0.10 mg/l (100 ppb) [11,24,25].

These trihalomethanes may be hazardous to health. That's what activated carbon "is all about." On February 9, 1978, the U.S. EPA published a proposed amendment to the National Interim Primary Drinking Water Regulations. The first part of the proposed regulation sets an MCL of 0.1 mg/l for TTHM in drinking water. The second part of the proposed regulation prescribes the use of GAC treatment for control of synthetic organic chemicals in drinking water. These proposed regulations are initially applicable only to systems serving more than 75,000 persons, but all systems serving more than 10,000 persons are required to monitor for THMs.

The Environmental Defense Fund's Robert Harris told: "I'm convinced there is a relationship between drinking water quality and cancer." Harris as well as numbers of other environmental engineers believe there is a safe, inexpensive, and highly effective means of removing most organics from water: passing the water in a treatment plant through a bed of granular activated carbon. "The costs for installing carbon treatment," argues Harris, "are so ridiculously low that I can't believe the values of the American public have eroded so much that they wouldn't be willing to spend $10 to $15 per family per year to protect themselves."

On the other hand, Philadelphia's hard-driving water commissioner, Carmen Guarino, believes EPA's cost estimates of $4 to $9 per year for carbon treatment for a typical household are too low. According to his rough estimates, adding carbon beds to the city's water plants and regenerating the carbon frequently enough to remove synthetic organics would cost $13.3 million/yr. This would boost the water bill for a typical household, Guarino says, from the present $45/yr to $71/yr—over a 50% increase. He argues, "Money can be better spent on more pressing urban problems."

Robert McGarry, a Washington Suburban Sanitary Commission general manager says, "the EPA interim requirement to replace the sand in the filter with GAC is not a good move. It would be difficult to remove carbon from such a sand bed for regeneration—and costly."

The existing GAC facilities in the U.S. do not and could not satisfy the EPA-proposed GAC criteria. Nor do or could the European GAC systems. Many of the European GAC systems and virtually all of those in West Germany, use river bank filtration before treatment or ground passage after treatment to effect removal of organics. These data suggest that river bank filtration or ground passage, not carbon treatment, is responsible for much of the original removal in European waterworks. European GAC practice is not concerned with design criteria similar to those proposed by EPA.

Macroreticular resins are more effective than GAC in removing synthetic organic chemicals such as the volatile halogenated organics from drinking water; it is noteworthy that they cannot be substituted in lieu of the GAC treatment as proposed. The resins are specifically designed to remove small compounds, such as the volatile halogenated organics, but do not remove TOC.

An increasing number of scientists are questioning whether heavy metals or polycyclic aromatic hydrocarbons (PAHs) contained in the virgin or, more particularly, the regenerated GAC, may be released from the GAC into the drinking water. Joshua Lederberg, a Nobel Laureate and noted authority in the health protection field, warned the National Drinking Water Advisory Council in 1976: "I have to tell you, you could make a horror story of the chemistry of what's in charcoal. . . . If you were to take charcoal and disaggregate it and hydrogenate it, you would end up with a lot of horror substances that we would really seriously like to avoid." He declared further, "It is quite possible that we could be generating pollution of a kind and scope far worse than we are solving with this particular panacea."

It has been reported to EPA that eighteen inorganic elements, including lead, mercury, and cadmium have been found in a composite sample of GAC. Indeed, the elution of such materials from carbon has been reported as a problem in the CCE method. Most importantly, there is concern that consecutive regeneration of the carbon could lead to formation of PAHs and other toxic materials.

GAC's large surface area also may provide a site for reaction, among others, between chlorine and precursors in the effluent so as to catalyze reactions that lead to the formation of harmful substances. John T. Cookson reports that activated carbon behaves as a catalyst and forms compounds not originally present in the water stream. Cookson points out that this catalysis by GAC may result in the formation of compounds that pose more insidious health risks than their precursors.

GAC provides an ideal location for bacterial growth that will subsequently be discharged into the finished drinking water.

There also may be problems with the wastewater generated by the use of GAC facilities. It is estimated that more than 50,000 gallons of wastewater will be generated for each million gallons of water treated through GAC facilities. Thus for a 100 mgd system an additional five million gallons of wastewater per day will result from the use of GAC facil-

ities. Although some of this wastewater may be reclaimed and recycled, at least half will need to be discharged. Not only will this create additional problems and expenses for the water systems, it will also create additional burdens on wastewater or sewage treatment plants if the wastewater from the GAC process is discharged into the municipal sewer systems.

A discussion of American Chemical Society (ACS) meeting has reported: In spite of many studies, a complete understanding of GAC process . . . is not at hand . . . many of the studies still view the application of GAC as purely an adsorption process while in fact adsorption is but one of several removal mechanisms (including) . . . filtration . . . and biological degradation. Furthermore, these mechanisms . . . interact synergistically. One conclusion stemming from the presentations was that GAC definitely is not always the answer to THM removal, GAC may not be an answer in many situations, and GAC may actually cause more harm than good in some situations. Steam sterilization was said to improve the adsorability of activated carbon—as much as 50%, according to results of one of the presentations. In particular, steam reportedly not only increases the adsorbability of the activated carbon, it reduces leakage of THM from the carbon.

Dr. Geldreich, Chief of the Microbiological Treatment Branch at EPA's Municipal Environmental Research Laboratory in Cincinnati, concluded: We don't want to close our minds to the fact that there may be a problem even though we don't see it at present. I'm not trying to be an alarmist. I'm not trying to build a dynasty. I'm just trying to introduce a word of caution.

In light of the serious health concerns that may be associated with the use of GAC, additional research and testing are absolutely necessary before plunging this nation's major waterworks into investing millions of dollars in GAC systems. GAC treated water may well pose a risk of harm to humans greater than the assumed risk it is supposed to reduce.

EPA vs. AWWA

As an alternative to EPA proposals for reducing organics in drinking water, AWWA proposes:

- Expanded and accelerated health-effects research on trihalomethanes and synthetic organics. A scientific basis for regulation must be established before—not after—rules are proposed.
- Establishment of a 100 ppb level of total trihalomethanes as a goal—not a requirement—for all public water supply systems. Efforts to achieve that goal, suggests AWWA, should be limited to modifications of existing treatment plant processes, e.g. relocating the point of chlorine application. In addition, all public water systems—not just

those with more than 75,000 customers – should be urged to meet this goal.

- Elimination of EPA's proposed requirement of GAC as a treatment technique. In its place, provide at least four-plant-size research projects underwritten by the government to gather financial and operating, as well as scientific, data that do not now exist.
- Adoption of EPA's proposed monitoring program for TTHM, except that public notification should not be required. Establishment of a monitoring program for synthetic organics is a good idea. Many treatment plants have used chloramines for decades with great success. Chloramines are an excellent means for reducing TTHM formation. It would be a serious msitake, says AWWA, to limit their use now. In fact, AWWA recommends chloramine experimentation and use be encouraged as a means of preventing TTHM formation – provided satisfactory disinfection is achieved.

Briefly, here are AWWA's reasons for recommending the GAC treatment technique requirement be dropped in favor of further experimentation.

- The need for such a technique has not been demonstrated.
- Design data for synthetic organics removal using GAC are lacking.
- Reliable cost information on plant construction and operation is woefully lacking.
- Data on efficient regeneration system design are lacking.
- Research on alternative adsorption, disinfection and other treatment techniques need to be done.

Market of Activated Carbon

The major producers of activated carbon are Westvaco, ICI Americas, Calgon, Husky Industries, Barnebey-Cheney, Union Carbide and Witco. Total capacity for activated carbon is about 290 million lbs per year but the industry is producing only about 190 million lb per year. Much of the excess is in the granular area. The production was up by 15% in 1977 than in 1976. Thomas A. McConomy, director of Calgon's adsorption systems division says, "If EPA doesn't do anything, the water and watewater markets will grow at a rate of about 15% per year. Activated carbon producers estimate that in the next few years, the percentage of activated carbon production used for potable water treatment could increase from about 20% today to about 35 to 40%. Besides water and wastewater treatment which makes up about a third of the activated carbon market, carbon is used in air pollution control, catalyst applications, sugar decolorizing, solvent recovery, purification of chemicals and gases, dry cleaning, rubber reclamation, auto evaporative control systems, and cigarette filters. It is water and wastewater markets that offer the greatest potential for activated carbon.

REFERENCES

1. Cheremisinoff, P. N. and F. Ellerbusch, eds. *Carbon Adsorption Handbook.* Ann Arbor Sciences (1978).
2. "Process Design Manual for Carbon Adsorption," U.S. Environmental Protection Agency (Oct. 1973).
3. Baer, D., "Taste an Odor Removal Enhanced by Switch to Granular Carbon," *Public Works* (Nov. 1978).
4. Janecek, K. *Carbon Transfer Techniques.* American Water Association, p. 581 (Oct. 1978).
5. Kornegay, B. H., "Control of Organic Chemical Contaminants in Drinking Water," Westvaco Corporation – Nuclear Activated Carbons (Feb. 1979).
6. Weber, Jr., W. J. and Massoud Pirbazari, "Adsorption of Toxic and Carcinogenic Compounds from Water," *Journal of American Water Works Association,* p. 203 (April 1982).
7. Glaze, W. and J. Wallace, "Control of Trihalomethane Precursors in Drinking Water: Granular Activated Carbon With and Without Preozonation," *Journal of American Water Works Association,* p. 68 (Feb. 1984).
8. Pirbazar, M. and W. J. Weber, Jr., "Adsorption of p-dichlorobenzene from Water," *Journal of American Water Works Association,* p. 82 (Feb. 1984).
9. Weber, N. J. and A. M. Jodellah, "Removing Humic Substances by Chemical Treatment and Adsorption," *Journal of American Water Works Assoc.,* p. 132 (Apr. 1985).
10. Otson, R., P. D. Bothwell and D. T. Williams, "An Evaluation of the Organics – Carbon Adsorbable Method for Use with Potable Water," *Journal of American Water Works Association,* p. 42 (Jan. 1979).
11. Siemak, R. C., R. R. Trussell, A. R. Trussell and M. D. Umphres, "How to Reduce Trihalomethanes in Drinking Water," *Civil Engineering – ASCE* (Feb. 1979).
12. Carns, K. and K. Stinson, "Controlling Organics: The East Bay Municipal Utility District Experience," *Journal of American Water Works Association,* p. 637 (Nov. 1978).
13. Kondorita, D., J. Snoeyink and V. L. Snoeyink, "Monochloramine Removal from Water by Activated Carbon," *Journal of American Water Works Association,* p. 62 (Jan. 1985).
14. "Water Industry Fighting EPA's Drinking Water Regulations," *Civil Engineering – ASCE* (July 1978).
15. Blanck, C. A., "End Taste and Odor Complaints with Granulated Activated Carbon," *American City and County* (Oct. 1977).
16. Blank, H., "Special Features of Adsorption Equilibria of the System Benzene," *Journal of Physical Chemistry,* 260 (3), p. 395 (1979).
17. Miller, W. and R. Rice, "European Water Treatment Practices – and What We Can Learn from Them," *Civil Engineering – ASCE* (Nov. 1977).
18. Miller, W. and R. Rice, "European Water Treatment Practices – Their Experience with Ozone," *Civil Engineering – ASCE* (Jan. 1978).

19. Miller, W. and R. Rice, "European Water Treatment Practices— The Promise of Biological Activated Carbon," *Civil Engineering— ASCE* (Feb. 1978).

20. Maloney, S., I. H. Suffet, K. Bancroft and H. M. Neukrug, "Ozone—GAC Following Conventional vs. Drinking Water Treatment," *Journal of American Water Works Association*, p. 66 (Aug. 1985).

21. Neukrug, H., M. G. Smith, S. N. Maloney and I. H. Suffet, "Biological Activated Carbon—at What Cost?" *Journal of American Water Works Association*, p. 158 (Apr. 1984).

22. Rice, R., C. Gomella and W. Miller, "Rouen, France, Water Treatment Plant: Good Organics and Ammonia Removal with No Need to Regenerate Carbon Beds," *Civil Engineering—ASCE* (May 1978).

23. Thiem, L., D. Badorek and J. T. O'Connor, "Removal of Mercury from Drinking Water Using Activated Carbon," *Journal of American Water Works Association*, p. 452 (Aug. 1976).

24. "The Activated Carbon Dilemma," *American Chemical Society Meeting Report, Water and Sewage Works* (Dec. 1978).

25. Pendygraft, G. W., "The EPA-Proposed Granular Activated Carbon Treatment Requirement: Panacea or Pandora's Box?" *Journal of American Water Works Association*, p. 52 (Feb. 1979).

26. Stopka, K., "Ozone-Activated Carbon can Remove Organics," *Water and Sewage Works* (May 1978).

27. McGeary, J. J. and V. L. Snoeyink, "Granular Activated Carbon in Water Treatment," *Journal of American Water Works Association* (Aug. 1977).

28. Sylvia, A. E., D. A. Bancroft and J. D. Miller, "Analytical Note—A Method for Evaluating Granular Activated Carbon Adsorption Efficiency," *Journal of American Water Works Association*, p. 99 (Feb. 1977).

29. Stevens, A. A., C. J. Slocum, D. R. Seeger and G. G. Robeck, Chlorination of Organics in Drinking Water," *Journal of American Water Works Association* (Nov. 1976).

30. Montalvo, J. G. and C. G. Lee, "Analytical Notes—Removal of Organics from Water: Evaluating Activated Carbon," *Journal of American Water Works Association*, p. 211 (Apr. 1976).

CHAPTER 26

Trace Contaminants in Streams

JAMES S. KUWABARA* AND PAUL HELLIKER**

INTRODUCTION

Water pollution by trace contaminants can result from a variety of sources, including industrial discharges, sewage treatment plant discharges, agricultural or urban runoff, sedimentation from logging or construction and so on. The effects of such pollution are just as diverse, ranging from changes in the appearance or hydrology of water bodies to fish kills and other drastic effects on aquatic biota (Forstner and Wittman, 1979; U.S. Environmental Protection Agency, 1980). An assessment of these effects has been determined in a number of ways, using gross indicators of contamination such as biochemical or chemical oxygen demand (BOD or COD), biological indicators such as species diversity or productivity, analyses of the quantity of macronutrients present, such as ammonium or phosphate, or measurements of the concentrations of specific chemicals present in water bodies. Other chapters in this handbook addressed the topics of macronutrients in lakes ("Lake Eutrophication") and gross measures of pollution in streams, with associated predictive transport models ("Stream Pollution"). This chapter focuses on trace contaminants in streams and is composed of two fairly distinct sections. The first section presents background on the development of current guidelines and water quality criteria in the United States and then briefly discusses methods from U.S. Environmental Protection Agency (EPA) publications to monitor and determine the concentration of trace contaminants. The second section discusses methods to model the transport and fate of trace contaminants with emphasis on a process interactive approach. Formulation of the process in-

teractive model begins with the conservative physical submodel for advection and dispersion. Additional physical processes such as evaporation, streambed influx/efflux and volatilization are then discussed. The review of a class of chemical reactive terms then addresses sorption, complexation/dissociation, precipitation/dissolution, photolysis and redox processes. Finally, the biological reaction section examines modeling of uptake/release, toxicity and adaptation, and community structure/species interactions. An appendix at the end of the chapter defines model parameters and describes how values have been determined in recent studies. Further elaboration of the topics discussed in this chapter is available in the literature cited.

DEVELOPMENT OF GUIDELINES AND CRITERIA

Efforts to control water pollution extend at least as far back as the times of the Roman Empire. Major aqueducts were constructed to supply Rome with high quality drinking water and waste waters were carried away through an extensive sewer system. The Middle Ages saw a decline in sanitation, with concomitant epidemics of diseases such as cholera, typhoid fever, and dysentery. It was not until the early years of the nineteenth century that efforts were again expended to provide safe drinking water with the employment of filters in Scotland and England for municipal supplies.

In 1849, Sir John Snow published his report on the Broad Street well in London, linking an outbreak of cholera to poor water quality. Snow's study followed an earlier report by Edwin Chadwick, which laid out a number of axioms, the most germane to this discussion being the cause and effect relationship between "insanitation, defective drainage, inadequate water supply, and overcrowded housing" and "disease, high mortality rates, and low expectation of life"

*U.S. Geological Survey, Menlo Park, CA
**U.S. Environmental Protection Agency, San Francisco, CA

(Flinn, 1965). With rapid advances in analytical chemistry and biology during the past century, transport and environmental effects of a wide spectrum of aquatic contaminants, including many that exist at trace concentrations, may be closely studied. Evolution of such studies in the United States is indicative of this rapid advancement.

Water Pollution Control in the United States

The first major compilation of contaminants of fresh water was written by Ellis (1937), in which he reviewed the existing literature and listed lethal concentrations found by various researchers for 114 substances. In 1952, the state of

TABLE 1. 1976 EPA Maximum Contaminant Levels for Drinking Water.

Constituent	Concentration
Arsenic	50 ug/l
Barium	1000
Cadmium	10
Chromium	50
Lead	50
Mercury	2
Nitrate (as N)	10,000
Selenium	50
Silver	50
Endrin	0.2
Lindane	4.0
Methoxychlor	100
Toxaphene	5.0
2,4-D	100
2,4,5-TP Silvex	10
TTHM	100
Turbidity	1 TU
Microbiological	1/100 ml
Rad-226 & Rad-228	5 pCi/l
Gross Alpha	15 pCi/l

1984 Recommended Maximum Contaminant Levels for Volatile Organic Compounds

Constituent	Concentration
Tetrachloroethylene	85 ug/l
Trichloroethylene	257
1,1,1 Trichloroethane	1000
Carbon Tetrachloride	25
1,2 Dichloroethane	260
Vinyl Chloride	60
Benzene	25
1,1 Dichloroethylene	350
p-Dichlorobenzene	3.75

California published a seminal document (1,369 references) reviewing all water quality criteria promulgated to that date by state and interstate agencies, along with the application of the criteria to protect the safe use of natural waters.

McKee and Wolf (1963) expanded this effort with a massive summarization of the world's literature on water quality criteria. This extensive document (3,827 water quality references) formed the basis for tabulation of the effects on aquatic biota, in ascending order of severity, of a wide range of contaminants and water quality measures. Specific values were reported as either "damaging" or "not harmful" to fish, during the indicated time and conditions of exposure. The results were not always consistent, in that values reported as harmful often overlapped those described as having no effect on fish. Such inconsistencies are not unusual, however, given the varying test organisms and techniques used by the authors, the physiological state of the organisms, the varying constituents of dilution waters and the varying temperatures of the tests (Rai, et al., 1981). Despite these inconsistencies, McKee and Wolf's work provided a significant step toward a comprehensive understanding of the concentration ranges within which water contaminants begin to have adverse effects on receiving waters.

In 1965, Congress amended the Federal Water Pollution Control Act of 1948 to include requirements that States adopt:

- water quality criteria applicable to interstate waters
- a plan for the implementation and enforcement of the water quality criteria adopted

These criteria and plans adopted by the States would, upon approval of the federal government (the Federal Water Pollution Control Administration), become the legally enforceable water quality standards. Subsequent to these amendments, the FWPCA published a compilation of recommended water quality criteria, chosen to protect five general categories of uses: recreation and aesthetics; public water supplies; fish, aquatic life, and wildlife; agricultural uses and industry. Commonly known as the "Green Book," *Water Quality Criteria* (FWPCA, 1968), prepared by the National Technical Advisory Committee to the FWPCA, represented the first nationally recommended criteria for water quality. Included in that book are both numerical limits for contaminants, as well as narrative descriptions of physical and chemical characteristics which must be maintained in the hydrologic regime to protect the uses listed above. Where a specific recommendation for a criterion for a particular water pollutant was infeasible due to little or conflicting information, authors of the Green Book suggested that designated application factors be used, based upon data generated by 96-hour bioassays using sensitive aquatic species and receiving water as the diluent.

TABLE 2. 1980 EPA Water Quality Criteria.

Compound	Freshwater		Saltwater		Human Consumption
	Acute	Chronic	Acute	Chronic	
Acenapthene	1700	520	970	710	Org.
Acrolein	68	21	55	—	320
Acrylonitrile	7550	2600	—	—	0.058[a]
Aldrin	3.0	—	1.3	—	0.074 ng/l[a]
Antimony	9000	1600	—	—	146
Arsenic	360	190	69	36	2.2 ng/l[a]
Asbestos	—	—	—	—	30,000 fib/l[a]
Benzene	5300	—	5100	700	0.66[a]
Benzidine	2500	—	—	—	0.12 ng/l[a]
Beryllium	130	5.3	—	—	6.8 ng/l[a]
Cadmium	[c]	[c]	43	9.3	10
Carbon Tetrachloride	35,200	—	50,000	—	0.40[a]
Chlordane	2.4	0.0043	0.09	0.0040	0.46 ng/l[a]
Chlor. Benzenes	250	50	160	129	0.72 ng/l[a]
Chlor. Ethanes	980[b]	540[b]	940[b]	281[b]	0.17[ab]
Chlor. Napthalenes	1600	—	7.5	—	—
Chlor. Phenols	30[a]	—	440[a]	—	1.2[ab]
Chloroalkyl Ethers	238,000	—	—	—	0.0038 ng/l[ab]
Chloroform	28,900	1240	—	—	0.19[a]
Chlorophenol	4380	2000	—	—	Org.
Chromium (VI)	16	11	1100	50	50
Copper	[c]	[c]	2.9	2.9	Org.
Cyanide	22	5.2	1.0	1.0	200
DDT	1.1	0.001	0.13	0.001	0.024 ng/l[a]
DDE	1050	—	14	—	0.024 ng/l[a]
TDE	0.6	—	3.6	—	0.024 ng/l[a]
Dichlorobenzenes	1120	763	1970	—	400
Dichlorobenzidines	—	—	—	—	0.0103[a]
Dichloroethylenes	11,600	—	224,000	—	0.033[a]
2,4-Dichlorophenol	2020	365	—	—	3090
Dichloropropanes	23,000	5700	10,300	3040	—
Dichloropropenes	6060	244	790	—	87
Dieldrin	2.5	0.0019	0.71	0.0019	0.071
2,4-Dimethylphenol	2120	—	—	—	Org.
Dinitrotoluene	330	230	590	370	0.11[a]
1,2-Diphenylhydrazine	270	—	—	—	42 ng/l[a]
Endosulfan	0.22	0.056	0.034	0.0087	74
Endrin	0.18	0.0023	0.037	0.0023	1
Ethylbenzene	32,000	—	430	—	1400
Fluoranthene	3980	—	40	16	42
Haloethers	360	122	—	—	—
Halomethanes	11,000	—	12,000	6400	0.19
Heptachlor	0.52	0.0038	0.053	0.0036	0.28 ng/l[a]
Hexachlorobutadiene	90	9.3	32	—	0.45[a]
Hexachlorocyclohexane:					
Lindane	2.0	0.080	0.16	—	9.2 ng/l[ab]
BHC	100	—	0.34	—	9.2 ng/l[ab]
Hexachlorocyclopentadiene	7.0	5.2	7.0	—	206
Isophorone	117,000	—	12,900	—	5200
Lead	[c]	[c]	140	5.6	50
Mercury	2.4	0.012	2.1	0.025	0.144
Napthalene	2300	620	2350	—	—
Nickel	[c]	[c]	75	8.3	13.4

(continued)

TABLE 2 (continued).

Compound	Freshwater		Saltwater		Human Consumption
	Acute	Chronic	Acute	Chronic	
Nitrobenzene	27,000	—	6680	—	19,800
Nitrophenols	230	150	4850	—	13.4[b]
Nitrosamines	5850	—	3,300,000	—	0.8 ng/l[ab]
Pentachlorophenol	20	13	13	7.9	1010
Phenol	10,200	2560	5800	—	3500
Phthalate Esters	940	3	2944	3.4	15,000[b]
PCB	2.0	0.014	10	0.030	0.79 ng/l[a]
PAH	—	—	300	—	2.8 ng/l[a]
Selenium	260	35	410	54	10
Silver	[c]	0.12	2.3	—	50
2,3,7,8-TCDD (1984)	0.01	0.00001	—	—	0.014 pg/l[a]
Tetrachloroethylene	5280	840	10,200	450	0.8[a]
Thallium	1400	40	2130	—	13
Toluene	17,500	—	6300	5000	14,300
Toxaphene	0.73	0.0002	0.21	0.0002	0.71 ng/l[a]
Trichloroethylene	45,000	21,900	2000	—	2.7
Vinyl Chloride	—	—	—	—	2.0
Zinc	[c]	[c]	95	86	Org.

All units are ug/l unless otherwise specified.

[a]Compound is carcinogenic, so that the recommended criterion is zero, with the specified value representing 1/1,000,000 risk.

[b]Isomers are of various toxicities, value is most toxic compound.

[c]Criterion is based on hardness of water.

Org.: Organoleptic data: not information on toxicity of compound to humans.

—: Inadequate data is available to support a criterion.

Clean Water Act

The Federal Water Pollution Control Act Amendments of 1972 (P.L. 92-500) added another requirement to the provisions of the 1965 Amendments mentioned above, namely, that States must adopt water quality criteria applicable to intrastate waters, which would become legal standards upon federal approval. Responsibility for developing national criteria had been shifted in 1970 to the Environmental Protection Agency, into which the FWPCA had been incorporated. The EPA contracted in 1972 with the National Academy of Sciences and the National Academy of Engineering to update and expand the 1968 Green Book. *Water Quality Criteria 1972,* also referred to as the "Blue Book," was published in essentially the same format as that of the Green Book, with criteria for contaminants grouped under six headings, each associated with a different use of water resources: recreation and aesthetics, public water supply, freshwater aquatic life and wildlife, marine aquatic life and wildlife, agricultural uses and industrial supply.

EPA again updated water quality criteria in 1976, in a document entitled *Quality Criteria for Water* (the "Red Book"). Criteria based on avoiding toxic effects to aquatic organisms or on providing safe drinking water for humans were em-phasized in the Red Book, because these values were usually most stringent. Most of the data concerning aquatic biota, upon which these 1976 criteria were based, were derived from acute toxicity studies in which the toxic concentration of a contaminant was determined as that which caused 50% mortality of test organisms after a certain period of exposure (e.g., 96 hour LC50's). Information concerning chronic or sublethal effects had substantially increased by 1976, however, and the EPA included these results, where available, into the Red Book values. The Red Book also included the Maximum Contaminant Levels for 20 pollutants specified in the National Interim Primary Drinking Water Regulations, established by the EPA pursuant to the Safe Drinking Water Act of 1974 (Table 1).

Congress amended the Clean Water Act in 1977, adding provisions emphasizing the control and elimination of toxic pollutants in the Nation's waters. The Natural Resources Defense Council sued EPA for lack of action in developing criteria for these contaminants, as per the intent of Congress. In fulfillment of the consent decree negotiated to settle the NRDC suit, EPA published in November of 1980 water quality criteria for 64 of the 65 classes of toxic pollutants. Criteria for the 65th category, 2,3,7,8 tetrachloro-dibenzo-p-dioxin (commonly known as "dioxin") were published in a subsequent document, dated February, 1984. The 1980 and 1984 documents give short summaries of the criteria recommended to protect freshwater and marine aquatic life and human health (see Table 2 for a complete listing of these criteria). Full discussion of both the data and the rationale used to develop these criteria are contained in EPA Criteria Development Documents, available for each of the 65 classes of priority pollutants.

Scientific data concerning the chronic toxicity to aquatic biota of a number of water contaminants improved considerably between 1976 and 1980. As a result, criteria for the 65 priority pollutant classes are specified in terms of maximum levels and 24-hour average values, to protect against acute and chronic toxicity, respectively. Human health data is presented for both carcinogenic and non-carcinogenic contaminants, based on the exposure pathway assumptions of a 70 kg adult consuming 2 liters of water and an average of 6.5 grams of aquatic organisms per day. For the non-carcinogens, "no observable adverse effect levels" are determined from a survey of the toxicological literature. The recommended concentration of carcinogens in water is zero, since there is no "safe" threshold level, below which such substances are non-carcinogenic. Since some waters in the U.S. already contain these contaminants, EPA estimated concentrations corresponding to incremental lifetime cancer risks of 1 in 10^5, 1 in 10^6, and one in 10^7. These are shown in Table 2 at the 1 in 1 million level.

In sum, the Red Book and the 1980/84 documents contain the current EPA-recommended national ambient water quality criteria. These criteria form the basis from which legally enforceable state-adopted standards are developed. Few states, however, have adopted standards for all the criteria listed in the EPA documents. Even where standards exist, they often vary from state to state and within states themselves, due to differing hydrologic conditions and background levels. The maximum contaminant levels listed in Table 1 are nationally mandated standards, with which water used for public water supply must comply. Both drinking water standards and ambient water quality criteria are constantly reevaluated by the EPA, as scientific research expands the range of information relating to the effects of water contamination on human health and aquatic biota.

MONITORING WATER QUANTITY AND QUALITY

Monitoring trace contaminants in streams is important in two respects. First, streams act as a source of these contaminants to large bodies of water with longer residence times. Secondly, the stream itself supports a biological community that may be drastically affected by the presence of trace contaminants. Therefore, an understanding of both the quantity and quality of stream water is of obvious significance. On a practical level, water quantity and quality are monitored for a variety of purposes, including establishment of baseline

data and determination of long-term trends, assessment of compliance with criteria and standards, and verification of stream flow and contaminant transport models. A number of techniques are available to determine the velocity and flowrate of streams and the chemical composition of their waters. Some of the more commonly used techniques will be described here. It should be stressed that the monitoring methods discussion below represents a brief summary of a very active and quickly developing area of research. A comprehensive discussion of all the widely acknowledged methods is beyond the scope of this chapter. The reader is urged to consult source documents before implementing a particular method.

Water Quantity

Throughout the United States, the U.S. Geological Survey (USGS) has been measuring stream and river flow since 1888. The USGS currently operates over 16,000 gaging stations, distributed throughout the country on waters ranging from mere creeks to the Mississippi River. These measurements are reported annually in USGS publications (e.g., Water Resources Data Reports by State).

Most streamflow measurements performed by the USGS are made with current meters of two basic types (vertical-axis and horizontal-axis types). The latter, commonly used in Europe, have the advantage of being less likely to disturb flow around the meter or to become entangled with debris. Vertical-axis flow meters are more durable, however, and are accurate over a wider range of velocities (particularly low velocities) than the horizontal-axis type. Two makes of the vertical-axis current meter, the Price type AA[1] and the Price pygmy meter, are used extensively by the USGS. A four-vane vertical-axis meter was developed by the USGS (Buchanan and Somers, 1966) for use in iced-over streams.

All these meters are rated for the Survey by the National Bureau of Standards. Tables of stream velocities are provided for field use, relating the revolutions of the meter (measured by clicks heard in headphones) over certain time intervals, to specified velocities. These ratings are normally given for velocities between 0.25 and 8 feet/second, with equations relating time and revolutions to velocities outside that range.

The other component of discharge measurement is a determination of the cross-sectional area of the stream. Depth and position in the vertical are measured by rigid rods, by sounding weights suspended from cables or by acoustical sounders. Rods are normally used in streams which can be waded, while the other two types of equipment are for larger or less accessible streams. Width of the stream is measured either with previously marked cableways or bridges or with metallic tape lines or tag lines. These methods are all described in Buchanan and Somers (1966).

In order to characterize the velocity of each cross-sectional element, a velocity profile composed of current meter measurements taken at a number of different depths, would ideally be performed. Such extensive measurements are made occasionally to verify the following simplified procedures, but normally they are not necessary. Based upon studies of numerous actual observations and upon mathematical theory, the mean velocity of a particular vertical segment of streamflow can be accurately gaged by the average of the values taken at 0.2 and 0.8 of the depth below the surface. For streams which are shallower than 2.5 feet, a reliable measure of the velocity of a vertical segment can be taken at 0.6 of the depth below the surface. The 0.6-depth method is also useful for quick measurements of velocity when the stage in a stream is changing rapidly. Buchanan and Somers (1966) also discuss techniques to measure or estimate discharge in streams when either the cross-sectional area or the velocity profile are difficult or impossible to measure, such as during flood flows or rapidly changing stages (see also Dalrymple and Benson, 1967; Benson and Dalrymple, 1967; USGS, 1980).

Sampling Water Quality

Just as the measurement of the quantity of water flowing in a stream consists of compositing measurements at discrete points into a unique value representing the whole flow, so does the measurement of water quality rest on the assumption that sampling a small part of a particular water body accurately characterizes the much larger whole. Consequently, site selection and sampling technique are crucial elements in the process of determining water quality. There are two basic ways to approach this problem of heterogeneity in water quality: either take numerous grab samples and analyze each, or take composite samples which consist of mixtures of discrete samples taken at different places or at different times. Composite samples are generally more cost-effective, since the costs of laboratory analyses are generally more expensive than the cost of sampling. Sampling sites should be selected for easy access, with well-mixed hydrologic conditions, to minimize the number of grab samples necessary for a composite sample. Consideration should also be given to the availability of supplemental or historical data at a site. For studies in which more rigorous sampling is required, statistical methods for optimizing site selection are given in the *Handbook for Sampling and Sample Preservation of Water and Wastewater* (EPA, 1982).

The stream constituents of concern in this chapter generally occur at such low concentrations that the possibility of unintentional contamination of the sample is higher than

with other water quality parameters. To minimize this potential problem, sampling equipment should be constructed of materials inert to the solutes being analyzed (e.g., Teflon for trace metal analysis), and they should be easy to clean and to operate at a variety of depths. Clean technique should be routinely used in all phases of preparation, sampling, and analysis (Bruland, et al., 1979; Kuwabara and North, 1980). For example, all glassware and plasticware used in sampling and analysis of trace substances should be acid washed after cleaning of coarse material with detergent (Goerlitz and Brown, 1972; Fitzwater, et al., 1982). High purity water (> 10 Mohm resistance) should obviously be used for all cleaning and reagent preparation. Five basic types of surface water samplers are available and are described below: cylindrical samplers (open at both ends), bottle samplers (open at one end), US-series integrating samplers, bag samplers, and pumps.

Cylindrical samplers generally consist of a cylinder with stoppers at each end that can be closed by either a messenger weight or an electric solenoid mechanism. These samplers are well suited for sampling deep water, but can also be very susceptible to contamination by surface slicks. These samplers are usually constructed of brass, polyvinyl chloride (PVC), or acrylic plastic. Brass samplers may contaminate samples with metals, while PVC or acrylic plastic samplers may introduce phthalate esters into samples, so that the former should be used for samples to be analyzed for organics and the latter for samples to be analyzed for metals (Table 3).

Bottle samplers, with spring-loaded lids or other mechanisms for underwater opening, are popular and useful for shallow waters. The common material from which bottle samplers are made is borosilicate glass, a material that may be inappropriate for trace metal analysis due to metal adsorption to vessel walls. The drawbacks of bottle samplers are that they require ballast, which can contaminate samples or stir up sediment if not lowered carefully, and they are useful only in shallow water.

A more sophisticated sampler, the US-series integrating sampler, is most often used for suspended solids samples, but is useful for trace contaminants, as well. The integrating sampler has a filling rate which is proportional to stream velocity, so it is the best choice for making flow-weighted samples in rivers and streams.

The two other samplers, the bag sampler and the pump, are normally used for oceanographic work and sampling wastewater streams, respectively. Further description of these samplers is available in the EPA Handbook (1982), referred to above.

In rivers and streams, discharge-weighted samples are the most accurate. The most effective procedure for taking discharge-weighted samples is to use the US-series integrating samplers with either of the following two depth integrating methods: the equal-discharge increment (EDI) method or the equal-width increment (EWI) method. Quot-

TABLE 3. Container Type, Sample Volume and Preservation (EPA, 1981).

Sample Parameter	Size and Type of Container	# of Samples	Preservation Technique	Holding Times
Cyanide	1 liter polyethylene or glass bottle	1	NaOH (1:10 dilution) to pH > 12; refrigerate to 4°C	14 days
Metals	1 liter Teflon or polyethylene bottle	1	Nitric acid (1:1 dilution) to pH < 2* Refrigerate to 4°C	6 months (except mercury**)
Asbestos	1 liter plastic bottle	1	None required; recommended refrigeration or storage in dark	None specified
Volatile Organics	40 ml or 125 ml glass screw top vial	2	Refrigerate to 4°C	7 days†
Extractable	1 gallon amber flint glass bottle	1	Refrigerate to 4°C Extraction within 48 hours‡	
Total Phenolics	1 liter glass bottle	1	H2SO4 to pH < 2; refrigerate to 4°C	28 days

*Where HNO3 cannot be used because of shipping restrictions, the sample may be initially preserved by refrigerating and should be immediately shipped to the laboratory. Upon receipt in the laboratory, the sample must be acidified to pH < 2 with HNO3. At the time of analysis, the sample container should be thoroughly rinsed with 1:1 HNO3 and the washings added to the sample (volume correction may be required). According to FAA regulations, other potential corrosives including H3PO4 and NaOH are allowed on all commercial passenger airlines at least up to 1 quart. However, some packaging specifications may be required depending on the liquid and the particular airline.

**13 days holding time for mercury sample held in plastic container.

†VOA samples not analyzed within 7 days of collection must be prepared for duplicate analysis. The first duplicate should be refrigerated and analyzed for the alkyl compounds. The second duplicate should be preserved with 1:1 HCl to a pH < 2 and analyzed for the aromatic compounds. Maximum holding time for duplicate analysis is 14 days.

‡If extractable organics samples are not extracted within 48 hours of collection and residual chlorine is present, they must be preserved by adding 35 mg of sodium thiosulfate per 1 ppm of free chlorine per liter of sample. All samples must be extracted within 7 days and completely analyzed within 40 days.

ing from the *National Handbook of Recommended Methods for Water Data Aquisition* (USGS, 1980):

In the EDI method, the cross-sectional area is divided laterally into a series of subsections, each of which conveys the same water discharge. Depth integration is then carried out at the vertical in each subsection where half of the subsection discharge is on one side and half is on the other side. In each individual subsection, a vertical transit rate is used that will provide a sample volume for the vertical which is equal to the sample volumes for every other volume. . . . Generally, if more than five verticals (more than five subsections) are sampled, an accurate mean discharge-weighted concentration will be obtained.

In the EWI method, depth integration is performed at a series of verticals in the flow section that are equally spaced across the transect to obtain a series of subsamples. Unlike the EDI method, however, the vertical transit rate used at each vertical is exactly the same as that used at every other vertical, and the subsamples are composited even though they are of different volumes. This procedure provides a transect sample whose concentration is discharge weighted both vertically and laterally and whose volume is proportional to the water discharge in the sampled zone. An advantage of the EWI method is that a knowledge of the lateral distribution of discharge is not required.

The primary disadvantage of the EDI method is that the lateral distribution of water discharge must be known or measured each time prior to sampling. With the EWI method, on the other hand, (1) it is sometimes difficult to maintain the same vertical transit rated at all verticals, (2) more verticals must be sampled for a given accuracy than with the EDI method, and (3) wherever the flow is not essentially perpendicular to the transect, the width increment between sampling verticals must be adjusted by dividing it by the sine of the angle between the flow lines and the transect. Generally, 10 to 20 verticals will provide an accurate mean discharge-weighted concentration by the EWI method.

As mentioned above, the selection of the appropriate container is essential for assuring accurate and representative results. Table 3 gives the recommended sample container, preservation technique and holding times for the different categories of priority pollutants. When dividing samples into containers for different analyses, samples should be well-mixed either by agitating the sample or by stirring with a Teflon rod. A churn splitter and cone splitter developed by the USGS are probably among the best devices for dividing samples, however, their plastic construction may make them unsuitable for priority pollutant sampling, since phthalate esters may leach into solution. With respect to volatile organics samples, care should be taken to fill sample vials

TABLE 4. Target Species for Warm Water, Cold Water and the Great Lakes and Other Cold Water Lentic Systems (EPA, 1981).

Predators	Bottom Feeders
Warm Water	
Largemouth bass** (Micropterus salmoides)	White sucker** (Catostomus commersoni)
Smallmouth bass** (M. solomieui)	Carp** (Cyprinus carpio)
Yellow perch** (Perca flavescens)	Stoneroller* (Campostoma anomalum)
Bluegill* (Lepomus macrochirus)	Spotted sucker* (Minytrema melanops)
Channel catfish* (Ictalurus punctatus)	Silver redhorse* (Moxostoma anisurum)
Chain pickerel* (Esox niger)	Freshwater drum* (Aplodinotus grunniens)
Cold Water	
Rainbow trout** (Salmo gairdneri)	White sucker** (Catostomus commersoni)
Brown trout** (S. trutta)	Largescale sucker** (C. macrochheilus)
Brook trout** (Salvelinus fontinalis)	Carp* (Cyprinus carpio)
Squawfish* (Ptychocheilus oregonensis)	Silver redhorse* (Moxostoma anisurum)
Round whitefish* (Prosopium cylindraceum)	Northern (shorthead) redhorse* (M. macrolepidotum)
Mountain whitefish* (P. williamsoni)	Freshwater drum* (Aplodinotus grunniens)
Great Lakes and Other Cold Water Lentic Systems	
Lake trout** (Salvelinus namaycush)	White sucker** (Catostomus commersoni)
Yellow perch** (Perca flavescens)	Carp** (Cyprinus carpio)
Walleye* (Stizostedian vitreum)	Silver redhorse* (Moxostoma anisurum)
Sauger* (S. canadense)	Northern (shorthead)red horse* (M. macrolepidotum)
Northern pike* (G sox lucius)	Freshwater drum* (Aplodinotus grunniens)
Round whitefish* (Prosopium cylindraceum)	
Lake whitefish* (Coregonus clupeaformis)	
Rainbow smelt* (Osmerus mordax)	

*Good target species.
**Preferred target species.

completely, and to create as little turbulence as possible during filling.

All samples should be clearly labelled with the location, date, and time of sampling, the analysis requested and the required preservation treatment, as well as an identification of the sampling personnel. Duplicate samples should be sent for quality control checks. All samples should be refrigerated to 4 degrees centigrade or otherwise preserved (e.g., acidification with nitric acid to pH < 2 for trace metals) during transport to the laboratory, and should be analyzed as quickly as possible.

Bed Sediment Sampling

Unlike water samples, which are only indicators of instantaneous conditions, bed sediment can indicate long-term trends in the concentration of some of the priority pollutants. Most of the contaminants discussed in this chapter partition more readily onto sediment than in water, particularly onto sediment of small grain size and high organic carbon content. Iron and manganese content of sediment can also affect the sorption of trace contaminants onto sediment particles (EPA, 1981).

Sampling sites at which clays and silts predominate should be chosen. These are normally found where stream velocity drops from a faster to a slower current, such as downstream of islands or obstructions or on the insides of river bends. Sites of easy access are preferable, with shallow waters being best, due to the fact that corers are the recommended collection tool. In order to accurately characterize a site, a number of samples should be collected, generally along a cross section.

Methods for sampling and storage of stream sediment have been reviewed by Forstner and Wittman (1979). Bed sediment samplers generally fall into three categories: coring devices, mechanical grabs and scoops or drag buckets. Since scoops and drag buckets cause the greatest amount of disturbance of bed sediment, particularly the fine overlying material, they are not recommended for sampling priority pollutants. Plastic devices should be avoided, due to potential contamination of the sample, and metal devices should be either high quality stainless steel or should be used in a way to minimize metal interference. Glass or Teflon are the preferred materials for sediment sampler construction, and since the use of these materials is restricted to coring devices, they are the recommended tools for sediment sampling.

Corers of approximately 5 cm (2 in) in diameter and 13 cm (5 in) in length are best for sampling silt or clay. Cores are taken by pushing the corer into the sediment, then extracting the corer and capping the core tube. Gravity and piston corers are other examples of this type of instrument. The former is designed for deeper bodies of water and utilizes the momentum derived from free fall to penetrate the sediment. Piston corers are similar to hand corers except they include a piston which rests on the sediment surface as the corer is pushed in, creating a vacuum which helps retain the sample. The piston also allows rapid extraction of the core.

Various types of mechanical grabs are available, each with its particular features. All are similar, though, in that they are devices with jaws which are forced shut by weights, lever arms, springs, or cords. Mechanical grab samples must be subsampled with glass or Teflon containers, both to avoid contamination by the sampler parts and to provide a container for transporting the sample to the laboratory. Mechanical grabs are normally used in oceanographic work or for sampling performed in lakes and reservoirs. A more complete discussion of the types available and their various features is given in EPA, *Sampling Protocols,* 1981.

Core samples should be handled as little as possible to reduce the risk of contamination. Transport of bulk samples to the laboratory is recommended. Samples should be labelled and refrigerated during transport, preferably using "blue ice" or dry ice.

Biological Samples

Due to the tendency of many of the priority pollutants to bioconcentrate, fish tissue can often be a useful indicator of the presence of trace contaminants in streams or rivers. Sampling fish tissue is complicated, however, by differences is size, age, lipid content, uptake, and clearing rates and position in the food web. Due to these factors, certain species have been suggested as "target species" to be used for trace contaminant analysis, since they represent fish which are broadly distributed, non-migratory, easy to identify and capture, and pollution tolerant. They are listed in Table 4.

Fish can be collected either actively or passively, as described below. Active methods include electrofishing, seining, trawling, angling, and chemical poisoning. Passive methods include netting (gill, Fyke, trammel, or hoop) and using D-traps. Active methods are generally more efficient than passive methods for covering a large number of sites, from each of which a small amount of fish is required. Active methods are especially useful in shallow water such as streams, coastal areas, or shorelines.

Electrofishing consists of applying an AC, DC, or pulsed DC current to water to stun fish. This method is the most efficient but is also the least selective (with the exception of poisoning). Electrofishers can be either boat mounted or carried on the back. Seining is done with a wall-like net held upright by floats on the surface and lead weights on the bottom. One end of the net is held steady, while the other is brought around to form a circle, which is then held steady while the trapped fish are brought to shore. Mesh sizes are generally selected to just catch the smallest fish desired. Seines are most effective in shallow waters, although a regular bottom free of snags is required. Trawling is done with a specialized form of seine netting which is shaped like a conical bag, with the mouth of the net held open by a

wooden beam or by otter boards. Trawls also require regular bottoms.

Angling is the most selective form of fish collection, using the hook and line method. Either single lines from individual rods or drop lines from floating mainlines are generally used. Finally, poisoning is the most severe method for sampling fish populations, and should be avoided for priority pollutant analyses due to physiological changes induced in the fish.

As for passive collection systems, gill nets and trammel nets are both methods which entail snaring the fish in the net and holding them until collected. These two methods are similar to seine nets, with the net held vertically by floats and weights. Both types eventually kill the fish, so that only live or recently dead fish from the nets should be collected.

Hoop nets consist of a series of conical nets spaced evenly within a large cylindrical net which is closed at one end. Fish swim through one or more of the cones into the larger space at the base of the next cone and are caught without a ready means of escape. Fyke nets are a modification of the hoop net, with side wings attached to the outer cone, to make it more efficient.

D-traps are constructed of wire, wood slats and cotton or synthetic mesh, in the shape of a *D*. These traps consist of two compartments, the "chamber," which fish enter through a funnel, and the "parlor," into which the fish pass from the chamber through a second funnel. D-traps are normally placed on or just above the river, lake or estuary bottom, and are effective at catching slow moving fish and crustaceans which inhabit that area.

Generally, a minimum of 300 grams of fish should be collected from each site. These samples should not be collected during spawning season, due to low lipid content and altered physiology of fish during that season. Autumn is usually recommended for sampling, except for certain salmonids and other fish which spawn during the fall. After stunning the fish with a sharp blow, it should be handled with clean forceps, to reduce the possibility of contamination by the sampler's hands, and should be packaged in Teflon bags or cleaned aluminum foil. Plastic packaging should not be used, due to the possibility of contamination. If aluminum foil is used, it should be previously cleaned with acetone, then with pesticide grade hexane and allowed to dry in a contaminant-free area. Sample packages should be labelled, then frozen with dry ice, and sent to the analytical laboratory as soon as possible. These techniques are more fully discussed in *Sampling Protocols* (EPA, 1981), *Handbook for Sampling and Sample Preservation of Water and Wastewater* (EPA, 1982a) and Shack, et al. (1973).

CHEMICAL ANALYTICAL TECHNIQUES

Upon receipt at the analytical laboratory, samples should be preserved in the appropriate manner for respective analyses (see Table 3) and should be kept refrigerated until actual performance of the analyses. The two standard manuals commonly used for physical and chemical analyses are *Standard Methods for the Examination of Waters and Wastewaters* (APHA, 1985) and *Methods for Chemical Analysis of Water and Wastes* (EPA, 1979a). Two other EPA manuals are also available for analyzing organic contaminants and pesticides: *Methods for Organic Chemical Analysis of Municipal and Industrial Wastewater* (EPA, 1982b) and *Manual of Analytical Methods for the Analysis of Pesticide Residues in Human and Environmental Samples* (EPA, 1979b). It is strongly advised that source documents from these reviews be referred to before directly employing any analytical procedure.

Analysis for trace metals commonly is performed using an atomic absorption spectrophotometer (AAS). This technique essentially consists of injecting microliter volumes of sample into a flame or graphite furnace of known temperature and then recording the absorbence of emitted light at given wavelengths with a spectrophotometer. The wavelengths and intensity of light emitted by the flame/aerosol or furnace atomization are then compared to patterns produced by standardized solutions. Dissolved trace metal concentrations may be concentrated by various extraction procedures to bring sample concentrations above detection limits and to minimize chemical matrix effects (Bruland et al., 1979; Bloom and Crecelius, 1985). AAS may also be used to analyze trace metal concentrations in fish tissue and in sediments, as well. Fish tissue is first prepared by homogenizing it with a blender, and then digesting (heating) it in acid. The reconstituted sample is then analyzed by AAS. Sediment is often handled similarly, with acid digestion preceding analysis. A number of other sediment extraction methods have been developed to estimate biological availability of particle-bound solutes (Luoma and Jenne, 1976; Tessier et al., 1984). Other methods for trace inorganic analyses, such as anodic stripping voltammetry and ion specific electrode analysis, may be useful in quantifying chemical speciation of certain solutes. Limitations to these methods have been reviewed by Luoma (1983).

Analyses for organic chemicals usually involves an initial sample fractionation between particulate and what is operationally termed "soluable" phases (efficiency of fractionation is highly dependent on procedure). The "soluable" phase is then further separated into molecular weight or molecular size fractions. A multitude of techniques may be used to physically and chemically characterize each fraction. The reader is referred to a comprehensive review of these techniques by Hunter and Rickert (1972) that serves as a valuable starting point for organic contaminant studies. Some of the most frequently used analytical techniques employ a gas chromatograph/mass spectrometer. Organic contaminants may be solvent-extracted (often with methylene chloride) from the water, sediment or homogenized biological tissue. In many of the tests, the organics are further partitioned into another solvent (e.g., hexane). Microliter to milliter

volumes of this mixture are then injected into the heated inlet of the chromatograph. The vaporized sample is carried through the packed column by an inert gas. Different compounds in the sample are retarded by the column at different rates, so that they reach the detector at different times.

The detector can be either a thermal conductivity type or a flame ionization type. As the gas passes the filament of the first type, the detector measures the thermal conductivity of the gas, which depends on the composition of the gas. The flame ionization type mixes the carrier gas with air and hydrogen and burns it, producing ions which alter the current output of the detector. This type of detector is more sensitive and, therefore, more commonly used. Both methods measure the mass of a particular component passing the detector at a particular time. This characteristic curve (chromatogram), consisting of peaks at different intervals, is compared to the curve from a standard sample, in order to determine the identity and concentration of the contaminant.

The priority pollutants which partition readily into the gaseous phase, such as purgeable aromatics, purgeable halocarbons and acrolein and acrylonitrile, are first extracted from samples with an inert gas, which is then injected into the gas chromatograph. As described above, the resulting concentrations of the different compounds in the gas are then analyzed by the detector and compared to standardized sample results.

Monitoring techniques like those mentioned above provide useful information about the status of a particular site at a particular instant in time. It is clear, however, that an overall understanding and quantitative description of the fate of a trace contaminant in a stream environment demands the consideration and accurate description of interactive processes that control transport of that solute. The following is a review of recent attempts to that end.

MODELING TRANSPORT OF TRACE CONTAMINANTS

Engineers involved in formulating models to describe transport of trace contaminants in streams are faced with a number of problems:

1. A model must be selected whose structure reflects those processes (as submodels) that control transport of the contaminant(s) in question,
2. Parameters for each submodel must be estimated, and
3. The resulting transport model must be tested to clarify limitations to its applicability.

Concepts in model selection will be discussed in the following sections using models that have been developed for a wide range of stream environments. Types of submodels will also be discussed along with a description of how submodel parameters can be estimated. For ease in referencing and in lieu of a lengthy discussion and description of the estimation of each parameter, Appendix I gives an alphabetic listing of all presented parameters with their definitions and with references for their determinations. For a more detailed discussion of the theory of parameter estimation, refer to Rinaldi et al. (1979). The third topic of model testing is accomplished by monitoring studies or perturbation experiments using techniques described earlier in this chapter and will not be discussed further.

Model Selection

The structure of a transport model for trace contaminants may vary with time, location and contaminant, but is generally of the form:

$$\hat{u}(\hat{z},t) \rightarrow \boxed{F = \sum_i \sum_j f_{ij}} \rightarrow \hat{y}(\hat{z},t) \qquad (1)$$

where $u(z,t)$ is an input (state) vector at position z and time t. F is a matrix of model terms f_{ij}'s that quantify mechanisms for each process j within class i, and y is the model's "yield" vector (e.g., trace contaminant concentration or organism occurrence with respect to position and time). The transport model may be visualized as a matrix of f_{ij} terms as depicted in Table 5. Complexities in model selection arise from the fact that the matrix members represent physical, chemical and biological processes that are highly interdependent (Figure 1; Kuwabara, et al., 1984; Schindler, et al., 1985). Based on available information, a transport model is selected (or often gradualy built) that considers all relevant processes. The model is basically limited by the processes that can be quantified. The term "relevant processes," therefore, is dependent on the modeler's choice of acceptable accuracy. Unfortunately, relevant processes are seldom known before model building begins. Two approaches are typically employed in this situation. If an interdisciplinary knowledge of potentially important processes exists or is assessible, processes may be screened to focus down on

essential submodels. This approach may not be desirable or even feasible for engineering application. The second approach assumes the simplest conceivable model and then expands until the model may be successfully tested. The following sections will review various types of submodels so that the reader may relate the application of these submodels to his or her problem. Because of the large number of interdisciplinary processes involved in contaminant transport, there is an inherent risk that a discussion of this type may appear disjoint. In an attempt to present this discussion in some logical fashion, and to emphasize the interdependence of processes, submodels will be presented by class (physical, chemical, and biological) in the order they appear in Table 5. The reader should note that the time and length scales for these processes may vary greatly with the solute and stream reach in question. An understanding of these time and length scales can be very useful in identifying potentially important transport processes. Methods to determine the need to incorporate kinetic terms, as opposed to models that assume local equilibrium, have been examined recently (Jaffe and Ferrara, 1983; Valocchi, 1985). Although these process-scaling studies dealt with solute transport in porous media, the processes considered in these models maybe pertinent in stream systems as well.

Physical Models

Engineers probably view physical models with considerable familiarity due to standard curriculum requirements in fluid mechanics. The standard one-dimensional advection-dispersion model is often used in stream modeling and is shown below:

$$\frac{\partial(Ay)}{\partial t} + \frac{\partial}{\partial x}(Avy) - \frac{\partial}{\partial x}\left[\frac{(AD\partial y)}{\partial x}\right] = 0 \qquad (2)$$

where y is again the "yield" or tracer concentration, A the cross sectional channel area, v the velocity, the longitudinal dispersion coefficient and x the longitudinal position. Under

TABLE 5. Matrix of processes that affect trace contaminant stream transport.*

Term j	Physical	Chemical	Biological
		Model Class i	
1	Advection/Dispersion	Adsorption/Desorption	Uptake/Release
2	Evaporation	Complexation/Dissociation	Toxicity/Adaptation
3	Streambed Influx/Efflux	Precipitation/Dissolution	Metabolism/Storage
4	Atmospheric Input/Volatilization	Photolysis	Community Structure/Species Interaction
5	Thermal Variation	Oxidation/Reduction	

*The matrix members are quantitatively described as model terms (f_{ij}'s in Equation (1)). The processes presented here are by no means an exhaustive listing.

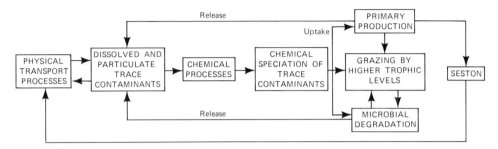

FIGURE 1. Interdependence of physical, chemical and biological processes controlling stream transport of trace contaminants.

a uniform channel assumption, Equation (2) simplifies to:

$$\frac{\partial y}{\partial t} = \frac{-Q}{A}\left(\frac{\partial y}{\partial x}\right) + D\left(\frac{\partial^2 y}{\partial x^2}\right) \qquad (3)$$

Discharge Q and area A are typically measured (Buchanan and Somers, 1969) while D is calculated by empirical formulae. For a uniform channel, D may be estimated by:

$$D = \frac{(0.11 \ vw^2)}{u \cdot d} \qquad (4)$$

in which w = average stream width, u^* = shear velocity, and d = average channel depth (Fisher et al., 1979).

The one-dimensional conservative model above is limited in its applicability because no sources or sinks for y are considered and there is an assumption that the solute is well mixed vertically and transversely (i.e., this model fails on at least two counts close to a point source). Fisher et al. (1979) discussed methods for determining the plume width and transverse mixing coefficient of a solute or thermal plume. An additional limitation to the one-dimensional conservative model is the typical existence of zones of little or no flow along river banks (called "transient," "dead," "storage," or "stagnant" zones). These zones are handled by solving two simultaneous equations of the form:

$$\frac{\partial y}{\partial t} = \frac{-Q}{A}\frac{\partial y}{\partial x} + \frac{1}{A}\frac{\partial}{\partial x}\left(\frac{AD\partial y}{\partial x}\right) + S - \frac{\alpha}{A}(y - y_s) \qquad (5)$$

(Stream Channel)

$$\frac{\partial y_s}{\partial t} = \frac{\alpha}{A_s}(y - y_s) + S_s \qquad (6)$$

(Storage Channel)

where

S = volume, surface (atmospheric) and lateral

(ground water or overland) stream channel solute sources (generally measured quantities)

S_s = storage zone sources (measured or fitted parameter)

Q,A,D,y = discharge, cross sectional area, dispersion coefficient and tracer concentration in the stream channel as previously defined

α = first order exchange coefficient for the storage zone (usually a calculated or fitted parameter)

A_s = storage zone cross sectional area (measured or fitted parameter)

y_s = trace concentration in the storage zone

Solutions for these equations are numerically determined by finite difference or finite element methods (Isaacson and Keller, 1966). A comprehensive review of such simulations was presented by Onishi et al. (1982).

Volatilization can be an important transport process for some inorganic solutes and for many trace organics. Keith and Telliard (1979) pointed out that 31 of the 129 USEPA priority pollutants are purgeable organics. A two-film model (Lewis and Whitman, 1924) for volatilization is commonly used that assumes well mixed water and atmospheric phases separated by water and air films through which mass transfer occurs by molecular diffusion (Figure 2). The relative resistance to transport in the water and air films is dependent on the Henry's Law constant (H) for the organic solute (Liss and Slate, 1974; Smith et al., 1980). For example, most of the mass transport resistance by volatilization is in the liquid phase for organic compounds with $H > 0.1$ kPa m³/g·mole. Compounds with $H > 0.5$ kPa m³/mole are termed "high volatility compounds" (e.g., benzene, chloroform and trichloroethylene). Flux of these volatile compounds may be modeled by the equation:

$$N = k[y - (P/H)] \qquad (7)$$

where N is the mass volatilization flux, k is the liquid phase mass transfer coefficient, and P is the atmospheric partial pressure of the solute (Mackay et al., 1979). There are quasi-volatile compounds (e.g., ethylene dibromide) where the total transport resistance is dependent on both liquid and gas film contributions. This condition has been examined by Rathbun and Tai (1986).

Evaporation concentrates solute mass per unit stream water volume and also represents a loss of latent heat to the atmosphere. Evaporation E is generally modeled as a linear function of wind speed w:

$$E = 4.33 \times 10^{-5} w (e_o - e_a) \qquad (8)$$

where e_o = saturation vapor pressure at the water surface temperature (mm Hg) and e_a = vapor pressure of air (mm Hg) (Marciano and Harbeck, 1954). A modeling discussion of the thermal effects of evaporation is presented by Jacquet (1983).

Sedimentation and resuspension are physical processes that are closely coupled with chemical processes of precipitation and adsorption later to be discussed. Computer program SERATRA (Onishi and Wise, 1982) considers sedimentation and streambed scour rates for sand, silt, and clay by solving mass conservation equations for each size class using a finite element method. Note that particle size distribution and hence adsorbing surface area may also change with time (Chapman et al., 1982) under conditions of high pH variability due to coagulation (e.g., along acid mine drainage streams where pH may fluctuate around the isoelectric point of colloidal oxyhydroxides). Wolf (1971) suggested a more simple first order sedimentation submodel:

$$\frac{\partial s}{\partial t} = \frac{-b[s - (s_o V^4)]}{C + v^4} \qquad (9)$$

where

s = suspended solids concentration

s_o = initial suspended solids concentration

v = stream velocity

b,c = fitted parameters

Rinaldi et al. (1979) suggested that Equation (9) may be applied with different b and c values to consider sediment resuspension.

Chemical Reactive Terms

Chemical reactive terms are a logical extension to the physical transport models discussed previously, because of the high reactivity of most trace contaminants. It is clear from recent studies (Sunda and Guillard, 1976; Kuwabara, 1982) that chemical reactions involving a trace contaminant can significantly affect the chemical form (speciation) of that contaminant and hence its availability to biota. Various types of reactions may occur including: adsorption and desorption, complexation and dissociation, precipitation and dissolution, oxidation and reduction, and photolysis. Submodels that consider such chemical reactions will be discussed in the following section.

Solution interactions with stream sediment (adsorption/desorption and precipitation/dissolution reactions) can markedly affect transport of trace contaminants. Onishi et al. (1982) cited studies that demonstrate a drastic reduction of dissolved radionuclides due to adsorption to streambed and suspended sediment (up to 90% over a 10 mile reach). Factors affecting solute adsorption and desorption include:

1. Sediment particle size—smaller particles generally have a larger reactive surface to volume ratio and are more easily transported downstream
2. Particle concentration and surface composition—reac-

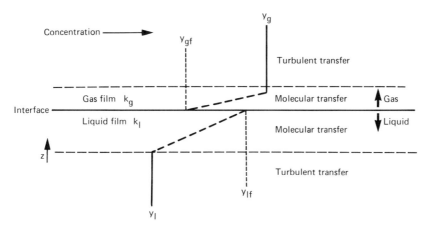

FIGURE 2. Two layer model of the gas-liquid interface. Total resistance to volatilization (l/k) is the sum of the series resistance by the gas film (l/k_g) and the liquid film (l/k_l) (from Liss and Slater, 1974).

tivity of clays, oxides and organically-coated minerals varies considerably (Wauchope and McDowell, 1984)

3. Stream water pH—adsorption site density and chemical speciation of solutes is affected

4. Presence of dissolved organics—competes with surfaces for solute complexation and forms organic films over mineral surfaces

5. Solute and its chemical speciation—reactivity of the chemical forms of metals and ligands as they exist in natural water samples may markedly affect adsorption. Solutes may also compete for sediment adsorption sites (Benjamin and Leckie, 1980), and

6. Rates of reaction or of diffusive mass transfer to the site of reaction.

A number of models have been developed that consider solute adsorption and desorption (Onishi et al., 1982). Kuwabara et al. (1984) modeled copper adsorption and desorption to streambed sediment with a simple first order equation:

$$\frac{\partial y}{\partial t} = \text{Physical terms} + \varrho_s R_s \qquad (10)$$

$$\text{(Main channel)}$$

$$\frac{\partial y_s}{\partial t} = -R_s = \lambda_s (y_s - K_{ds} y) \qquad (11)$$

$$\text{(Streambed)}$$

where

R_s = kinetic sediment sorption term

y_s = tracer concentration associated with the sediment

K_{ds} = partitioning coefficient between solution and sediment phases

λ_s = first order rate coefficient (different empirically-determined values for adsorption and desorption may be used)

ϱ_s = mass of accessible sediment per unit volume stream water

The sediment partitioning coefficient K_{ds} was assumed constant in this case and determined from solution and sediment analyses from samples collected before solute injection (i.e., at assumed thermodynamic equilibrium). In general, however, K_{ds} varies with the extent of adsorption site coverage. Bencala (1984) defined distribution coefficients based on a Freudlich Isotherm assumption for strontium, postassium and lithium:

$$K_{ds} = K_{ds}' \bigg/ \left(\frac{y}{\eta_s}\right)^2 \qquad (12)$$

where

y' = tracer concentration in a reference sample

K_{ds}' = distribution coefficient at y', and

n = fitted dimensionless parameter

A summary of adsorption information derived in laboratory experiments for various metals and ligands is presented by Onishi et al. (1982). It is often difficult to extrapolate adsorption and desorption results from one study site to the next, because of the variability of sediment composition and solute speciation. Sediment is usually a composite of many minerals that must be characterized before independent adsorption data may be applied. An additional difficulty is that results from laboratory studies may underestimate the biological availability of particle-bound solutes. A number of extraction procedures have been developed and revised to estimate availability of adsorbed toxicants and nutrients (Luoma and Jenne, 1976; Tessier et al., 1984).

Complexation and dissociation affect solute availability to biota. Zinc, for example, is actively transported into a biological cell. Metabolic energy is expended to take up Zn against a concentration gradient using a specific carrier protein (i.e., a specific biochemical ligand). Uncomplexed Zn^{2+} is therefore most biologically available (Anderson et al., 1978). Silver and mercury, on the other hand, are thought to be passively transported through the cell wall and membrane by diffusion and osmotic pressure. The uncharged chloro-complexes $AgCl°$ and $HgCl_2°$ are therefore thought to be biologically available forms (Engel et al., 1981; Gutknecht, 1981). Complexation involves association of metals and ligands and this association is modeled by stability constants (K) determined for each reaction. Chelation and hydrolysis are examples of complexation reactions involving multidentate ligands and hydroxide ions, respectively. Acid-base reactions may involve a proton H^+ as the metal or a hydroxide ion as the ligand. An example is given below for the case of i atoms and j molecules of a metal and ligand, respectively:

$$i(\text{Metal}) + j(\text{Ligand}) \rightarrow \text{Complex} \qquad (13)$$

$$\text{(Association reaction)}$$

$$K = [\text{Complex}]/[\text{Metal}]^i[\text{Ligand}]^j \qquad (14)$$

where brackets denote concentrations of each reactant. A discussion of acid-base reactions specific to trace organic contaminants has been presented by Mills et al. (1982). Stability constants for complexation reactions involving trace metals as well as trace organics have been compiled by Martell and Smith (1974).

The dissociation reaction, given by Equation (13) when the direction is reversed, and is quantified by stability constant K^{-1}. More than one type of metal and ligand may be involved in a complexation reaction. It is quickly evident that a long list of complexation reactions may occur in natural waters. To solve a system of complexation reactions (i.e., to determine concentrations of reactants for all complexes) computer programs have been created (Westall et al., 1978; Felmy et al., 1983). A flow chart is presented to describe the general structure of such programs (Figure 3). There are limitations to the use of such computations. First, the stability constants are dependent on temperature and ionic strength. Subroutines within chemical speciation models have been devised to make corrections for K before chemical speciation is computed (Krupka and Jenne, 1982). Even with such corrections, the user is advised to carefully review corrected stability constants used in computations, because there are limits to the applicability of these correcting subroutines. For example, the extended Debye-Huckel equation often used for ionic strength (I) corrections is useful for $I < 10^{-1}$ (Stumm and Morgan, 1970). Second, the use of stability constants in these programs implies the assumption of thermodynamic equilibrium between all reactants. Therefore, application of these chemical submodels with physical submodels demands coupling at a number of analytical nodes along the study reach. The node spacing between chemical speciation calculations is dependent on the kinetics of other simultaneous reactions within the study reach (i.e., closer node spacing to increase resolution of kinetic effects). Chapman et al. (1982) coupled a dispersive transport model for non-uniform channels with the chemical equilibrium program MINEQL (Westall et al., 1978). The resulting transport model called RIVEQL was applied to a mine drainage stream in New South Wales to describe metal transport.

Precipitation and dissolution reactions are closely coupled with adsorption/desorption reactions, because both involve interactions between solid and solution phases. In addition, precipitation and dissolution directly affect concentration of adsorption sites per unit volume suspension. Solubility reactions maybe modeled by equilibrium constants in much the same way as complexation/dissociation reactions:

$$\text{Solid} \rightarrow i(\text{Metal}) + j(\text{Ligand}) \quad (15)$$

(Dissolution reaction)

$$K_{so} = [\text{Metal}]^i[\text{Ligand}]^j \quad (16)$$

When the product of the metal and ligand activities exceed K_{so}, the system is said to be supersaturated with respect to the reactants and the solid should precipitate although the kinetics of such reactions may be significantly slower than complexation reactions. In these cases the local equilibrium assumption is inappropriate and precipitation/dissolution kinetic terms should be incorporated. If K_{so} is not exceeded, the submodel predicts solid dissolution until the solubility product is reestablished. The number of possible solubility expressions for a natural water sample may necessitate the use of the chemical speciation programs previously discussed. The limitation on such computations

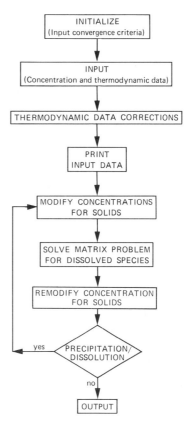

FIGURE 3. Program flow chart for the chemical speciation computation (adapted from Westall et al., 1978).

mentioned for complexation reactions apply here also. A more detailed discussion of precipitation and dissolution reactions is given by Stumm and Morgan (1970).

Photolytic reactions can significantly transform the structure of many trace organics by conversion of light to chemical energy. The transformation has been simply modeled by a first order reaction (Wolfe et al., 1976):

$$\frac{-\partial y}{\partial t} = ky \quad (17)$$

where k = photolysis constant, a function of the molecular light extinction coefficient, intensity of incident light, efficiency of light energy conversion. For detailed discussion of photochemical reactions the reader is directed to Balzani and Carassiti (1970).

Oxidation and reduction (redox) reactions are analogous to acid-base reaction except electron rather than proton transfer is involved. Unlike acid-base reactions, redox reactions occur in coupled reactions because free electrons do not exist in the way that H^+ ions do. An example is given below for a simple redox couple involving the transfer of i electrons:

$$(\text{Oxidant})^{+j} + i(\text{Electrons}) \rightarrow (\text{Reduced Product})^{-i}$$
$$(\text{Reduction with equilibrium constant } K_1)$$
$$(18)$$

$$(\text{Reductant})^{+k} \rightarrow (\text{Oxidized Product})^{+k+i} + i(\text{Electrons})$$
$$(\text{Oxidation with equilibrium constant } K_2)$$
$$(19)$$

$$K_1 K_2 = \frac{[\text{Reduced Product}][\text{Oxidized Product}]}{[\text{Reductant}][\text{Oxidant}]} \quad (20)$$

The similarity between redox reactions and complexation reactions is evident upon comparison of Equation (19) with Equation (14) above. This similarity facilitates the use of chemical speciation programs previously discussed. The importance of such reactions is unique in nature in that they often involve biological mediation. Transport model TOX-IWASP (Ambrose et al., 1983) uses a second order rate equation to define contaminant redox reactions:

$$\frac{\partial y_2}{\partial t} = k_{or} \, y_1 y_2 \quad (21)$$
$$(\text{Oxidation of } y_2)$$

where

k_{or} = second order redox rate constant
y_1 = oxidant concentration
y_2 = reductant concentration

The fact that the water column in most streams is oxygenated, might lead one to assume that elements only occur in oxidized states (e.g., $NO_3^- - N$). It is clear from the ubiquitous presence of dissolved humic materials that this assumption is inappropriate. Autotrophs convert light to chemical energy thereby producing reduced states as organic molecules from oxidized inorganic nutrients. Conversely, heterotrophs through respiration use organic molecules to generate energy for growth and metabolism. As cells die, settle out and decay, microbial degradation mediates the regeneration of oxidized nutrients. The biological availability and distribution of selenium (Se), for example, is strongly regulated by biological processes. Certain microbes are able to catalyze the oxidation of inorganic Se and utilize the energy released from that reaction. Other microbes mediate the reduction of oxidized, inorganic Se to organoselenides or elemental Se°. This reduction reaction is particularly important in water quality management because it represents an immobilization reaction that decreases the biologically available and potentially toxic selenite (SeO_3^{2-}) and selenate (SeO_4^{2-}) concentrations and generates reduced forms that are less available to

higher plants and animals. A thorough discussion of microbial mediation in Se cycling was presented by Doran (1982). Microbial transformation processes for trace organic contaminants have been discussed by Mills et al. (1982).

Biological mediation in redox reactions is to some degree present in all stream environments. One might therefore expect that thermodynamic equilibrium with respect to redox reactions [Equation (19)] may depend heavily on rates of biological processes. This interaction between chemical and biological processes provides an appropriate lead into the final class of biological reactive terms.

Biological Reactive Terms

Biological processes, represented by the third and final class of modeling terms to be discussed here, have long been considered in nutrient (e.g., oxygen, carbon, nitrogen and phosphorus) cycling models (Zison et al., 1978). However, effects of biological processes on transport of trace contaminants have only recently been closely examined. The irony of such transport models is that the yield vector $y(z,t)$ often defines toxicant concentrations in space and time, while water quality management decisions are ultimately based on biological parameters of growth and community structure. For example, toxicant transport model TOXIWASP (Ambrose et al., 1983) combines biological transport effects with chemical processes of hydrolysis, photolysis and redox reactions) into an overall first order rate equation:

$$\frac{\partial y}{\partial t} = k_b y \quad (\text{Combined nonhydrologic transport}) \quad (22)$$

where

$$k_b = \sum_{j=i}^{i} k_j$$

is the overall first order rate constant for combined nonconservative processes l through i. The following discussion will emphasize modeling terms that consider biological and ecological processes.

Kuwabara, et al. (1984) used a first order rate equation to describe biological uptake and release of copper during a 24 hr stream injection study in the following way:

$$\frac{\partial y}{\partial t} = \text{Physical Terms} + \text{Chemical Terms} + \varrho_b R_b \quad (22)$$
$$(\text{Main channel})$$

$$\frac{\partial y_b}{\partial t} = -R_b = \lambda_b (y_b - K_{db} y) \quad (23)$$
$$(\text{Biological Uptake and Release})$$

where

R_b = kinetic biological term
y_b = tracer concentration in biota
K_{db} = partitioning coefficient for the contaminant between solution and biota
λ_b = first order rate coefficient (different values for uptake and release may be used)
ϱ_b = mass of biota per unit volume stream water

The uptake and release constants λ_b's, the standing crop of biological material ϱ_b and the partitioning coefficient K_{db} were assumed constant over the injection monitoring period. This assumption is, in general, inappropriate over the long term (i.e., over many generations of growth). Each parameter is affected by the interactive processes presented in Figure 1. A discussion of how these biological parameters are affected by environmental conditions is presented in terms of biological uptake and release, community structure and species interactions, toxicity and adaptation, and metabolism and storage models.

Uptake and Release

Biological uptake rate may be controlled by a long list of abiotic and biotic parameters, but the following are usually at the top of that list: temperature, light, limiting nutrient concentration, and contaminant concentration.

Temperature, for example, significantly affects all biological processes (Heiman and Knight, 1975; Terry, 1983). Methods for modeling temperature effects on biological rates have been reviewed by Jorgensen (1983) and are summarized in Table 6. Most formulae require an *apriori* knowledge of the biological process or at least an experimental determination of an optimal temperature. The effect of light on photosynthesis imposes an indirect effect on trace contaminant uptake. The methods for modeling light effects on photosynthesis and algal growth are highly diverse and the reader is referred to a comprehensive review by Jorgensen (1983) on this topic.

Tracer uptake as a function of nutrient (substrate) concentration is conventionally modeled by Michaelis-Menten Kinetics (Cloern, 1978):

$$\lambda_b = \lambda_{max} \frac{[\text{Substrate}]}{[\text{Half Saturation Const.}] + [\text{Substrate}]} \quad (22)$$
(Monod equation)

where the maximum uptake rate λ_{max} and the half saturation constant are experimentally determined. Although this expression is very useful in describing uptake of limiting nutrients (e.g., nitrogen or phosphorus uptake), it is generally not applicable for uptake of toxic substances, because chemical speciation (as opposed to total concentration) of the substrate plays a more prominent role in determining the

TABLE 6. Modeling of temperature effects on biological rates. Note that in many cases an optimum temperature for the process (T_{opt}) must be experimentally determined (from Jorgensen, 1983).

(1) Chen and Orlob (1975)

$$K(T) = K_{20} \cdot KOT^{T-20}$$

(2) Lassiter and Kearns (1975)

$$K(T) = K_{opt} \exp[a(T - T_{opt})] \left(\frac{T_{max} - T}{T_{max} - T_{opt}} \right)^{a(T_{max} - T_{opt})}$$

(3) Lehman et al. (1975)

$$K(T) = K_{opt} \exp[-2.3(T - T_{opt})^2/(T_{max} - T_{opt})^2] \quad T > T_{opt}$$

$$K(T) = K_{opt} \exp[-2.3(T_{opt} - T)^2/(T_{opt} - T_{min})^2] \quad T > T_{opt}$$

(4) Jorgensen (1976)

$$K(T) = K_{opt} \exp\left(-2.3 \left| \frac{T - T_{opt}}{15} \right| \right)$$

$$K(T) = K_{opt} \exp(KT)$$

(5) Lamanna and Malette (1965)

$$K(T) = K_{opt} \left(\frac{T}{T_{opt}} \right)^n \exp\left[1 - \left(\frac{T}{T_{opt}} \right)^n \right] 0 < T < T_{opt}$$

$$K(T) = K_{opt} \left[1 - \left(\frac{T - T_{opt}}{T_{max} - T_{opt}} \right)^n \right] T_{opt} < T < T_{max}$$

(6) Park et al (1979)

$$PHOTO(T) = \exp[K(K_1 T^2 - (K_2 T^{K_3} - 1)]$$

$$K = -\ln PHOTO(0°C)$$

$$K_1 = K_2(T_{max})^{K_3-2}$$

$$K_2 = \frac{1 + (\ln PHOTO(T_{opt}))/K}{(T_{max})^{K_3-2}(T_{opt})^{K_1} - (T_{opt})^{K_3}}$$

$$\frac{2^{1/(K_3-2)}}{K_3} = \frac{T_{opt}}{T_{max}}$$

(7) Straskraba (1976)

$$T_{opt} = T + 28 \exp(-0.115T)$$

$$PHOTO(T) = PHOTO(T_{opt}) \cdot \exp[-K(T_{opt} - T)^2]$$

availability of trace contaminants. Secondly, the concept of saturation kinetics is by definition inappropriate here (i.e., contaminant concentrations imply overabundance not deficiency). Finally, λ_{max} and the half saturation constant in Equation (22) are themselves a function of other parameters that fluctuate in stream environments (Zison, et al., 1978).

A number of authors have determined that rates of trace contaminant accumulation by algae are rapid (i.e., $\lambda_b \geq 10^{-4}s^{-1}$). Fisher (1984) found that metal uptake by marine phytoplankton reached an equilibrium after 12 hr. of exposure. Button and Hostetter (1977) observed rapid Cu^{2+} sorption by both a diatom and chlorophyte within the first 5 minutes and neglible sorption after 2 hr. Bates, et al. (1982) observed no increase in adsorbed Zn by the chlorophyte *Chlamydomonas* after 3 hr. Kuwabara, et al. (1984) determined a λ_b value of $10^{-4}s^{-1}$ for copper uptake by periphyton in a mountain stream. Release rates for contaminant from biota have not been extensively studied. Similar λ_b values ($10^{-5}s^{-1}$) were estimated for Cu release from algae (Button and Hostetter, 1977; Kuwabara, et al., 1984); an order of magnitude slower than Cu uptake rates for periphyton. By analogy, slower desorption rates from inorganic particles relative to adsorption rates have been observed for anions (Hingston, et al., 1974; Hachiya, et al., 1979; Lijklema, 1980).

Trace metal accumulation by marine algal species was examined by Fisher (1984) and described by the Freundlich isotherm:

$$\log y_b = a\log(y) + b \quad \text{(Algal accumulation)} \quad (23)$$

where a and b are empirically-determined constants for a given contaminant under fixed light and temperature conditions. Note that this equation may be rearranged to yield:

$$\frac{(y_b)_{eq}/}{y} = K_{db} = K'y^{a-1} \quad (24)$$
$$\text{(Algal partitioning coefficient)}$$

where $K' = $ fitted constant and K_{db} is an estimate for the partitioning coefficient in the biological submodel [Equation (23)]. Values for K_{db} can be greatly affected by solute interactions. For example, algal studies have shown that elevated Zn^{2+} activities can increase phosphorus K_{db} (Bates, et al., 1982; Kuwabara, 1985). Kuwabara (1985) observed a strong linear dependence of accumulated P per cell with increasing Zn^{2+} in activity, presumably due to an inability to completely utilize intracellular P. For organic contaminants Neeley, et al. (1974) suggested that K_{db} could be determined as a function of the n-octanol/water partition coefficient (P_{ow}):

$$\log K_{db} = a_b\log P_{ow} + b_b \quad (27)$$

where a_b and b_b are fitted parameters that vary between species (Veith, et al., 1979; Veith, et al., 1980). Compiled K_{db} data (Zaroogian, 1985) illustrate the wide variability in bioconcentration of various organic chemicals.

For organisms in higher trophic levels, toxic substances may be taken up from solution as discussed above or through particulate (food) injestion. The modeling of such uptake is basically in the following form:

$$\frac{dy_b}{dt} = \sum_{j=i}^{k} (P_j + S_j - E_j)\varrho_j \text{ (Biological uptake)} \quad (25)$$

where

P_j = tracer uptake by particle injestion per unit mass of food group or organism functional group j

S_j = tracer uptake by sorption processes

E_j = tracer excretion rate

ϱ_j = mass of food group j per unit volume stream water

It is clear from Equation (25) that biological uptake may involve changes in community structure and species interactions and will consequently be discussed in more detail in a later section of this chapter.

Toxicity and Adaptation

Organisms respond to (i.e., metabolize and accumulate) trace contaminants in various ways and thereby affect biological submodel parameters such as uptake rate λ_b, standing crop ϱ_b and partitioning coefficient K_{db}. The difficulty in quantifying such effects is that they are often contaminant and species specific (McKim, 1978; Spehar, et al., 1978; Leland and Kuwabara, 1984). For example, elevated free Cu^{2+} ion concentrations adversely affect growth rate by diatoms (Gavis, et al., 1981), while chlorophytes are affected by an extended lag phase, i.e., temporary inhibition of ϱ_b (Kellner, 1955; Steeman Nielsen and Wium-Anderson, 1970). Zinc is another example of a metal that induces a variety of toxicological effects that consequently affect transport parameters. To complicate matters further, elevated concentrations of certain trace contaminants (e.g., lead and benzene) may simply result in contaminant accumulation in one trophic level [i.e., no demonstrable effect on biological transport parameters (Weber, et al., 1978; Sakanari, et al., 1984)], while inhibition of growth and development in higher trophic levels may significantly decrease ϱ_b (Hrs-Brenko, 1977). This diversity in toxicological effects ultimately exhibits itself as changes in the community structure of stream fauna (Weatherley, et al., 1980). A method to account for community changes in the biological submodel is presented in the following section.

In addition to toxicity mechanisms, aquatic organisms have evolved a variety of mechanisms to accommodate to elevated contaminant concentrations. Certain algal species produce organic substances that reduce biologically available Cu if released extracellularly in sufficient amounts (McKnight and Morel, 1979; Lumsden and Florence, 1983). The effect of this adaptation mechanism on the transport model is to alter the chemical speciation of the contaminant (i.e., production of an organic ligand as input to the chemical submodel discusseed previously). If chemical speciation models are not integrated into the transport model, an alternative, but less accurate method of modeling the effects of this adaptive mechanism would be to gradually decrease K_{db} based on total contaminant concentration. Cell wall exclusion and intracellular detoxification have also been exhibited (Silverberg, et al., 1976; Stokes, et al., 1977; Sicko-Goad, 1982; Reed and Moffat, 1983; Bates, et al., 1983). Animals (e.g., mollusks and crustaceans) produce organic complexing agents (e.g., metallothioneins) to inhibit metabolism of intracellularly-accumulated metals (Overnell and Coombs, 1978; Roesijadi, 1980; Fujita, et al., 1982; McCormick, 1984). Work by Cain and Luoma (1985) indicate that environmental history of clams results in significant differences in metal accumulation and release rates. An example is given below for estimation of Cu content in the filter-feeding clam *Macoma balthica* relative to sediment-bound Cu assuming the following model:

$$y_{b(k)} = \gamma_o + \sum_{i=0}^{q} \theta_i y_{s(k-i)}$$

where $y_{b(k)}$ and $y_{s(k)}$ are Cu content in clam tissue and sediment, respectively, and γ_o and the θ_i's are fitted "quasi" moving average parameters. A process of this type is not strictly a moving average process, because it does not fulfill the stationarity criterion. Sediment-bound tracer, y_s is seldom treated as a random variable in solute transport models, but rather as a parameter exhibiting distinct trends (Hirsch, et al., 1982). The model and analysis of variance is given (Table 7) for the second order ($q = 2$) model using clam content data (ug/size-standardized clam) (Cain and Luoma, 1985) and HCl-leached, sediment-bound Cu (Thomson, personal communication). Data represent analytical concentrations determined at monthly intervals in the Winter of 1978 and Spring of 1979 at a site in South San Francisco Bay, California near a municipal sewage treatment plant. The fact that θ_o and θ_2 are statistically significant for $p > 0.99$ (Table 7) suggest that Cu content in these clams (y_b) was affected by short term changes (~months) in sediment-bound Cu. The model does a reasonable job of describing y_b accounting for 91% of the total variation, but Cain and Luoma (1985) demonstrated an even stronger dependence

TABLE 7. Modeled Cu Content in Tissues of a Filter-Feeding Clam.*

Variance due to	df	SS	MS	F ratio
Regression				
θ_0	1	529846	529846	325
θ_2	1	31991	31991	20
Residual	32	52090	1628	
Total	34	613927		

*(y_b in ug Cu/size-standardized clam) at monthly sampling interval "i" relative to Cu adsorbed onto surficial sediment (y_s in ug/g) at interval "i" and two intervals prior "i-2". Symbol theta denotes a fitted coefficient. The resident population of clams at a site in South San Francisco Bay was sampled over a seven month period (n = 35). The example represents a simple approach to test effects of environmental history on tracer content in biota. Model: $y_{b(i)} = -581 + 52.5\ y_{s(i)} - 9\ y_{s(i-2)}$, $r^2 = 0.91$, $F_{(1,32,0.99)} = 7.50$.

on environmental history. They observed a significant difference between Cu (and Ag) accumulation in transplanted and resident clams. Their calculations suggested that transplanted clams from a pristine environment would require a year or longer to attain metal burdens similar to resident clams at the contaminated site. Their estimate and the modeling example described above strongly suggest that for modeling purposes the rate coefficient for tracer accumulation by injestion [within the P_j term of Equation (25)] is considerably smaller than that cited previously for λ_b by sorption [within the S_j term of Equation (25)]. They also suggest that in dynamic aquatic systems a K_{db} value would be very difficult to determine in the field. It is clear that in addition to contaminant concentration $y(t)$, submodel parameters at time t may also be affected by the time sequence of contaminant concentrations $y(t, t - T, t - 2T, \ldots)$ due to adaptation mechanisms. Time series analysis including "moving average" models are discussed in detail by Box and Jenkins (1976) and subroutines to compute moving average parameters in stochastic models (IMSL, 1982) may be accessed from transport models. Application of spectral analysis for biological processes has been reviewed by Platt and Denman (1975).

Community Structure and Species Interactions

Solute fluxes due to biological processes obviously involve the interactions between trophic levels within the stream community. It is therefore clear that each term in Equation (25) regulating tracer concentration for a given food group (injestion, sorption and excretion) affects and is affected by these very biological processes in other food groups. At discrete time intervals, a transport model may access a subroutine to iterate on the injestion, sorption and excretion rates for each food group based upon the values computed for the other groups. For example, solute uptake by particle injestion for food group $j(P_j)$ may be determined by:

$$P_j = \left(\sum_{i=1}^{k_j} \epsilon_{ji} y_{bi} w_{ji} \right) Z_j \qquad \text{(particle injestion)} \qquad (28)$$

k_j = number of particulate (food) groups preyed upon by group j

ϵ_{ji} = efficiency of solute uptake by injestion of particulate type i

y_{bi} = solute concentration of particulate i (note that this may also include injestion of inorganic particles with concentration y_{si})

w_{ji} = weight percentage for injestion of particulate i by group j

Z_j = total mass injestion rate per unit mass of group j

Net uptake by food group j through sorption-excretion processes may be modeled in a manner analogous to Equation (23):

$$(S_j - E_j) = \lambda_j(y - K_{dbj} y_{bj}) \qquad \text{(sorption-excretion)} \qquad (29)$$

where K_{dbj} and y_{bj} are the partitioning coefficient and solute concentration for group j, respectively. The rate coefficient λ_j takes on the value λ_{fj} for $y > K_{dbj} y_{bj}$ (sorption) and λ_{rj} for $y < K_{dbj} y_{bj}$ (excretion).

The time sequence for mass concentration of food group j may be computed as follows:

$$\varrho_j(t) = [\epsilon_f Z_j - \epsilon_s V_j]\varrho_j(t - \Delta t)\Delta t - M_j(t) \qquad (30)$$
$$\text{(mass concentration)}$$

where ϵ_f and ϵ_s are efficiencies for injested and respired mass conversions (conversion ratios), V_j is the mass respiration rate, Z_j is the mass injestion rate, delta t is the selected time increment, and M_j is the mortality rate [generally modeled as a power function of $\varrho_j(t)$] (Namkung, et al., 1983). The system of equations for all food groups [Equation (25)] is then used to determine total solute flux by biological processes.

The above approach, although ostensibly straight forward to apply, is deceptively problematic. Although consideration of the effects of physical parameters (e.g., light, temperature) and chemical processes on biological parameters have been previously addressed, there is also a large number of biological processes that contribute to the complexity of accurately determining biological contaminant flux. For example, the life history of an organism may shift it from one food group to another (Allan, 1982; Ross and Wallace, 1983). Changes in community structure due to chemical or physical processes may in turn affect the weight percentages for injestion (the w_{ji}'s) due to preferential predation (Chapman and Demory, 1963; Siegfried and Knight, 1976a,b,; McIntyre and Colby, 1978). For example, sublethal Cu and Cd concentrations decrease swimming velocities of the copepod *Eurythemora affinis* and thus reduces its ability to avoid predation (Sullivan, et al., 1983). Use of a mass concentration parameter (j) also implies a homogeneity in organism distribution or particle characteristics. In truth, organisms often exist in patches in association with other organisms and flow regimes (Reice, 1981). Strategies for consideration of this pattern have been presented by Gauch (1973) and Field, et al. (1982).

CONCLUDING REMARKS

Monitoring trace contaminant in streams involves a myriad of sampling and analytical techniques, some of which have been described here. References cited in the initial sections of this chapter should provide a good starting point for studying contaminants that have not been discussed. Strategies for such studies should take into account processes that may potentially regulate the fate of contaminants within the study reach.

Although a large number of trace contaminant transport processes have been discussed in this section, this was by no means intended to represent an exhaustive treatment of all possible terms in trace contaminant transport models. Rather this discussion was intended to present some of the very different types of submodels that are presently used and are available for incorporation in existing transport models. Secondly, it is hoped that some of the limitations on the applicability of present models and needs for further submodel development especially for non-hydrologic processes have been clarified. Finally, the authors wish to impress the importance of applying or at least becoming familiar with process interactive models as they may offer the only means to generate accurate predictions for the fate of trace contaminants in streams.

ACKNOWLEDGEMENTS

Editorial comment and suggestions by Drs. K. E. Bencala and J. E. Cloern, and by J. M. Bahr, J. L. Carter, S. V. Fend, K. Richard-Haggard, M. A. Sylvester, and P. C. Woods are gratefully appreciated. Assistance in manuscript preparation and graphics by J. S. Andrews and D. R. Jones is also acknowledged.

REFERENCES

1. Allan, D. J., "Feeding Habits and Prey Consumption of Three Setipalpian Stoneflies (Plecoptera) in a Mountain Stream," *Ecology, Vol. 63*, pp. 26–34 (1982).
2. Ambrose, R. B., S. I. Hill, and L. E. Mulkey, "User's Manual for the Chemical Transport and Fate Model TOXIWASP Version 1," USEPA 600/3-83-005 (1983).
3. American Public Health Association (APHA), *Standard Methods for the Examination of Water and Wastewater*, 16th Edition, Washington, D.C. (1985).
4. Anderson, D. M., F. M. M. Morel, and R. R. L. Guillard, "Growth Limitation of a Coastal Diatom by Low Zinc Ion Activity," *Nature* (London.), *Vol. 276*, pp. 70–71 (1978).
5. ASTM, "Sampling Atmospheres for Analysis of Gases and Vapors," in *Annual Book of ASTM Standards, Part 26*, American Society for Testing and Materials, Philadelphia, pp. 469–490.
6. Balzani, V. and V. Carasseti, *Photochemistry of Coordination Compounds*, Academic Press, London, 432 pp. (1970).
7. Bates, S. S., A. Tessier, P. G. Campbell, and J. Buffle, "Zinc Adsorption and Transport by *Chlamydomonas variabilis* and *Scenedesmus subspicatus* (Chlorophyceae) Grown in Semicontinuous Culture," *J. Phycol., Vol. 18*, pp. 521–529 (1982).
8. Bencala, K. E., "Interactions of Solutes and Streambed Sediment 2. A Dynamic Analysis of Coupled Hydrologic and Chemical Processes that Determine Solute Transport," *Water Resour. Res., Vol. 20*, pp. 1804–1814 (1984).
9. Benjamin, M. M. and J. O. Leckie, "Adsorption of Metals of Oxide Surfaces: Effect of the Concentrations of Adsorbate and Competing Metals," in R. A. Baker (ed.), *Contaminants and Sediments, Vol. 2*, Ann Arbor Science, pp. 305–322 (1980).
10. Benson, M. A. and T. Dalrymple, "General Field and Office Procedures for Indirect Measurements," *Techniques of Water Resources Investigations of the U.S. Geological Survey*, Book 3, Chapter A1.
11. Bloom, N. S. and E. A. Crecelius, "Determination of Subnanogram per Liter Concentrations of Silver in Seawater by Cobalt APDC Coprecipitation and Zeeman Graphite Furnace Atomic Absorption Spectrometry," *Anal. Chim. Acta* (1985).
12. Boiling, R. H., et al., "Toward a Model of Detritus Processing in a Woodland Stream," *Ecology, Vol. 56*, pp. 141–151 (1975).
13. Borg, H., A. Edin, K. Holm, and E. Skold, "Determination of Metals in Fish Livers by Flameless Atomic Absorption Spectroscopy," *Water Res., Vol. 15*, pp. 1291–1295 (1981).
14. Box, G. E. P. and G. M. Jenkins, *Time Series Analysis: Forecasting and Control*, Holden-Day, 575 pp. (1976).
15. Brown, B. E., "Uptake of Copper and Lead by a Metal-Tolerant Isopod *Asellus meridianus* Rac.," *Freshwater Biology, Vol. 7*, pp. 235–244 (1977).
16. Bruland, K. W., R. P. Franks, G. A. Knauer, and J. H. Martin, "Sampling and Analytical Methods for the Determination of Copper, Cadmium, Zinc and Nickel at the Nanogram per Liter Level in Seawater," *Anal. Chim. Acta, Vol. 50*, pp. 233–245 (1979).
17. Bryan, G. W. and L. G. Hummerstone, "Adaptation of the Polychaete *Nerias diversicolor* to Estuarine Sediments Containing High Concentrations of Heavy Metals," *J. Mar. Biol. Ass. U.K., v. 51*, pp. 845–883 (1971).
18. Buchanan, T. J. and W. P. Somers, "Discharge Measurements at Gaging Stations," *Techniques of Water Resources Investigations of the U.S. Geological Survey*, Chapter A8, 65 pp. (1966).
19. Button, K. S. and H. P. Hostetter, "Copper Sorption and Release by *Cyclotella meneghiniana* (Bacillariophyceae) and *Chlamydomonas reinhardii* (Chlorophyceae)," *J. Phycol., Vol. 17*, pp. 198–202 (1977).
20. Cain, D. J., S. N. Luoma, "Copper and Silver Accumulation in Transplanted and Resident Clams *(Macoma balthica)* in South San Francisco Bay," *Mar. Environ. Res., Vol. 15*, pp. 115–135 (1985).
21. Chapman, B. M., R. O. James, R. F. Jung, and H. G. Washington, "Modeling the Transport of Reacting Chemical Contaminants in Natural Streams," *Aust. J. Mar. Freshw. Res., Vol. 33*, pp. 617–628 (1982).
22. Chapman, D. W. and R. L. Demory, "Seasonal Changes in the Food Ingested by Aquatic Insect Larvae and Nymphs in Two Oregon Streams," *Ecology, Vol. 44*, pp. 140–146 (1963).
23. Coffman, W. P., K. W. Cummins, and J. C. Wuycheck, "Energy Flow in a Woodland Stream Ecosystem: I. Tissue Support Trophic Structure of the Autumnal Community," *Arch. Hydrobiol., Vol. 68* (2), pp. 232–276 (1971).
24. Cummins, K. W., R. C. Peterson, F. O. Howard, J. C. Wuycheck, and V. I. Holt, "The Utilization of Leaf Litter by Stream Detritivores," *Ecology, Vol. 54*, pp. 336–345 (1973).
25. Dalrymple, T. and M. A. Benson, "Measurement of Peak Discharge by the Slope-Area Method," *Techniques of Water Resources Investigations of the U.S. Geological Survey*, Book 3, Chapter A2 (1967).
26. Dean, J. M., "The Accumulation of ^{65}Zn and Other Radionuclides by Tubificid Worms," *Hydrobiologia, Vol. 45*, (1), pp. 33–38 (1974).
27. Doran, J. W., "Microorganisms and the Biological Cycling of Selenium," in K. C. Marshall (ed.), *Advances in Microbial Ecology, Vol. 6*, Plenum Publ., pp. 1–32 (1982).
28. Dorsey, N. E., *Properties of Ordinary Water Substance*, Hafner Publ., 673 pp. (1968).
29. Dressing, S. A., R. P. Maas, and C. M. Weiss, "Effect of Chemical Speciation on the Accumulation of Cadmium by the Caddisfly, *Hydropsyche* sp.," *Bull. Environm. Contam. Toxicol., Vol. 28*, pp. 172–180 (1982).
30. Ellis, M. M., "Detection and Measurement of Stream Pollution," *Bulletin of the U.S. Bureau of Fisheries*, 48:365–437 (1937).
31. Elwood, J. W., S. G. Hildebrand, and J. J. Beauchamp, "Contribution of Gut Contents to the Concentration and Body Bur-

den of Elements in *Tipula* ssp. from a Spring-fed Stream," *J. Fish. Res. Board Can., Vol. 33* (1930–1938).

32. Felmy, A. R., D. C. Girvin, and E. A. Jenne, "MINTEQ—A Computer Program for Calculating Aqueous Geochemical Equilibrium," U.S. EPA, Athens, GA (1983).

33. Feltz, H. R., "Significance of Bottom Material Data in Evaluating Water Quality," *Contaminants and Sediments, Volume I,* R. A. Baker (ed.), Ann Arbor Science Publications, Ann Arbor, Michigan, pp. 271–287 (1980).

34. Field, J. G., K. R. Clarke, and R. M. Warwick, "A Practical Strategy for Analysing Multispecies Distribution Patterns," *Mar. Ecol., Vol. 8,* pp. 37–52 (1982).

35. Fisher, H. B., et al., *Mixing in Inland and Coastal Waters,* Academic Press, N.Y. (1979).

36. Fisher, N. S., M. Bohe, and J. Teyssie, "Accumulation and Toxicity of Cd, Zn, Ag, and Hg in Four Marine Phytoplankters," *Mar. Ecol. Prog. Ser., Vol. 18,* pp. 201–213 (1984).

37. Fitzwater, S. E., G. A. Knauer, and J. H. Martin, "Metal Contamination and Its Effect on Primary Production Measurements," *Limnol. Oceanogr., Vol. 27,* pp. 544–551 (1982).

38. Flinn, D. W. (ed.), "Report on the Sanitary Condition of the Laboring Population of Great Britain, 1842," by Edwin Chadwick, Edinburgh at the University Press (1965).

39. Forstner, U. and G. T. W. Wittman, *Metal Pollution in the Aquatic Environment,* Springer-Verlag, Berlin, 486 pp. (1979).

40. Fujita, M., E. Takabatake, K. Tsuchiya, "Effects of Maternal Exposure to Zinc and Cadmium on Metallothionein in Fetal Rat Liver," *Arch. Environ. Contam. Toxicol., Vol. 11,* pp. 645–649 (1982).

41. Gauch, H. G., "The Relationship Between Sample Similarity and Ecological Distance," *Ecology, Vol. 54,* pp. 618–622 (1973).

42. Goerlitz, D. F. and E. Brown, "Methods for the Analysis of Organic Substances in Water," *Techniques of Water Resources Investigations of the U.S. Geological Survey,* Book 5, Chapter A3 (1972).

43. Gutknecht, J., "Inorganic Mercury (Hg^{2+}) Transport through Lipid Bilayer Membranes, *J. Membr. Biol., Vol. 61,* pp. 61–66 (1981).

44. Guy, H. P. and V. W. Norman, "Field Methods for Measurements of Fluvial Sediment," *Techniques of Water Resources Investigations of the U.S. Geological Survey,* Chap. C2, 59 pp. (1970).

45. Hachiya, K., M. Ashida, et al., "Study of the Kinetics of Adsorption-Desorption of Pb^{2+} on γ-Al_2O_3 Surface by Means of Relaxation Techniques," *J. Phys. Chem., Vol. 83,* pp. 1866–1971 (1979).

46. Heiman, D. R. and A. W. Knight, "The Influence of Temperature on the Bioenergetics of the Carnivorous Stonefly Nymph, *Acroneuria californica* Banks (Plecoptera: Perlidae)," *Ecology, Vol. 56,* pp. 105–116 (1975).

47. Hingston, F. J., A. M. Posner, and J. P. Quirk, "Anion Adsorption by Goetite. II. Desorption of Anions from Hydrous Oxide Surfaces," *J. Soil Sci., Vol. 25,* pp. 16–26 (1974).

48. Hirsch, R. M., J. R. Slack, and R. A. Smith, "Techniques in Trend Analysis for Monthly Water Quality Data," *Water Resour. Res., Vol. 18,* pp. 107–121 (1982).

49. Hrs-Brenko, M., C. Claus, and S. Bubic, "Synergistic Effects of Lead, Salinity and Temperature on Embryonic Development of the Mussel *Mytilus galloprovincialis," Mar. Biol., Vol. 44,* pp. 109–115 (1977).

50. Hunter, J. V. and D. A. Rickert, "Organic Analytical Chemistry in Aqueous Systems," in L. L. Ciaccio (ed.), *Water and Water Pollution Handbook, Vol. 3,* Marcel Dekker, Inc., New York, pp. 1021–1164.

51. IMSL, *IMSL Library Reference Manual, Edition 9, Vol. 1–4,* IMSL, Inc., Houston, TX (1982).

52. Isaacson, E. and H. B. Keller, *Analysis of Numerical Methods,* Wiley and Sons, N. Y. (1966).

53. Jacquet, J., "Simulation of the Thermal Regime in Rivers," in G. T. Orlob (ed.), *Mathematical Modeling of Water Quality: Streams, Lakes and Reservoirs,* John Wiley and Sons, Chichester, pp. 150–175 (1983).

54. Jaffe, P. R. and R. A. Ferrara, "Desorption Kinetics in Modeling of Toxic Chemicals," *J. Environ. Eng., Vol. 109,* pp. 859–867 (1983).

55. Jorgensen, S. E., "Modeling the Ecological Processes," in G. T. Orlob (ed.), *Mathematical Modeling of Water Quality: Streams, Lakes and Reservoirs,* John Wiley and Sons, Chichester, pp. 116–149 (1983).

56. Keith, L. H. and W. A. Telliard, "Priority Pollutants I—A Perspective View," *Environ. Sci. Technol., Vol. 13,* pp. 416–423 (1979).

57. Krause, G. and K. Ohm, "A Method to Measure Suspended Load Transports in Estuaries," *Estuar. Coast. Shelf Sci., Vol. 19,* pp. 611–618 (1984).

58. Krupka, K. M. and E. A. Jenne, "WATEQ3 Geochemical Model: Thermodynamic Data for Several Additional Solids." PNL-4276. Pacific Northwest Laboratory, Richland, WA (1982).

59. Kuwabara, J. S., "Micronutrient and Kelp Cultures: Evidence for Cobalt and Manganese Deficiency in Southern California Deep Seawater," *Science, Vol. 216,* pp. 1219–1221 (1982).

60. Kuwabara, J. S., H. V. Leland, and K. E. Bencala, "Copper Transport along a Sierra Nevada Stream," *J. Environ. Eng., Vol. 110,* pp. 646–655 (1984).

61. Kuwabara, J. S. and W. J. North, "Culturing Microscopic Stages of *Macrocystis pyrifera* (Phaeophyta) in Aquil, a Chemically Defined Medium," *J. Phycol., Vol. 16,* pp. 546–549 (1980).

62. Leland, H. V. and J. S. Kuwabara, "Trace Metals," *Fundamentals of Aquatic Toxicology,* G. M. Rand and S. R. Petrocelli (ed.), Hemisphere Publ., Washington, pp. 374–415.

63. Lewis, W. K. and W. G. Whitman, "Principals of Gas Absorption," *Ind. Eng. Chem., Vol. 16,* pp. 1215–1220 (1924).

64. Lijklema, L., "Interaction of Orthophosphate with Iron (III) and Aluminum Hydroxides," *Environ. Sci. Technol., Vol. 14,* pp. 537–541 (1980).

65. Liss, P. S. and P. G. Slater, "Flux of Gases Across the Air-Sea Interface," *Nature* (London), *Vol. 247,* pp. 181–184 (1974).

66. Lumsden, B. R. and T. M. Florence, "A New Algal Assay Procedure for the Determination of the Toxicity of Copper Species in Seawater," *Environ. Technol. Letters, Vol. 4*, pp. 271–276 (1983).

67. Luoma, S. N., "Bioavailability of Trace Element to Aquatic Organisms," *Sci. Total Environ., Vol. 28*, pp. 1–22 (1983).

68. Luoma, S. N. and J. A. Davis, "Requirements for Modeling Trace Metal Partitioning in Oxidized Estuarine Sediments," *Mar. Chem., Vol. 12*, pp. 159–181 (1983).

69. Luoma, S. N. and E. A. Jenne, "Estimating Bioavailability of Sediment-Bound Trace Metals with Chemical Extractants," in D. D. Hemphill (ed.), *Trace Substances in Environmental Health*, Univ. Missouri, pp. 343–351 (1976).

70. Mackay, D., W. Y. Shiu, and R. P. Sutherland, "Determination of Air-Water Henry's Law Constants for Hydrophobic Pollutants," *Environ. Sci. Technol., Vol. 13*, pp. 333–337 (1979).

71. Marciano, T. T. and G. E. Harbeck, Jr., "Mass Transfer Studies, Water Loss Investigations, Lake Hefner Studies," Technical Rpt., U.S. Geological Survey Professional Paper 269 (1954).

72. Martell, A. E. and R. M. Smith, *Critical Stability Constants, Vol. 1*, Plenum Press, N.Y. (1974).

73. McAuliffe, J. R., "Resource Depression by a Stream Herbivore: Effects on Distributions and Abundances of Other Grazers," *OIKOS, Vol. 42*, pp. 327–333.

74. McCormick, C. C., "Induction and Accumulation of Metallothionein in Liver and Pancreas of Chicks Given Oral Zinc: A Tissue Comparison," *J. Nutr.*, Vol. 114, pp. 191–203 (1984).

75. McIntire, C. D. and J. A. Colby, "A Hierarchical Model of Lotic Ecosystems," *Ecol. Monogr., Vol. 48*, pp. 167–190 (1978).

76. McIntire, C. D. and H. K. Phinney, "Laboratory Studies of Periphyton Production and Community Metabolism in Lotic Environments," *Ecol. Monogr., Vol. 35*, pp. 237–258 (1965).

77. McKee, J. E. and H. H. Wolf, "Water Quality Criteria," State Water Quality Control Board, Sacramento, California (1963).

78. McKim, J. M., R. L. Anderson, D. A. Benoit, R. L. Spehar, and G. N. Stokes, "Effects of Pollution on Freshwater Fish," *J. Water Pollut. Control Fed., Vol. 47*, pp. 1711–1820 (1975).

79. McKnight, D. M. and F. M. M. Morel, "Release of Weak and Strong Copper-Complexing Agent by Algae," *Limnol. Oceanogr., Vol. 24*, pp. 823–837 (1979).

80. Mill, T., W. R. Mabey, P. C. Bomberger, T. W. Chou, D. G. Hendry, and J. H. Smith, "Laboratory Protocols for Evaluation the Fate of Organic Chemicals in Air and Water," USEPA 600/3-82-0220 (1982).

81. Mills, W. B., J. D. Dean, D. B. Porcella, S. A. Gherini, R. J. M. Hudson, W. E. Frick, G. L. Rubb, and G. L. Bowie, "Water Quality Assessment: A Screening Procedure for Toxic and Conventional Pollutants, Parts 1 and 2," U.S. Environmental Protection Agency EPA 600/6-82-004a,b (1982).

82. Namkung, E., R. G. Stratton, and B. E. Rittman, "Predicting Removal of Trace Organic Compounds by Biofilms," *J. Water Pollut. Contr. Fed., Vol. 55*, pp. 1366–1372 (1983).

83. National Interim Primary Drinking Water Regulations, 40 Federal Register 59566 (December 24, 1975).

84. Neeley, W. G., D. R. Branson, and G. E. Blau, "The Use of the Partition Coefficient to Measure the Bioconcentration Potential of Organic Chemicals in Fish," *Environ. Sci. Technol., Vol. 8*, pp. 1113–1115 (1974).

85. Nehring, R. B., "Aquatic Insects as Biological Monitors of Heavy Metal Pollution," *Bull. Environ. Contam. Toxicol., Vol. 15*, pp. 147–154 (1976).

86. Onishi, Y. and S. E. Wise, "User's Manual for the Instream Sediment-Contaminant Transport Model SERATRA," USEPA 600/3-82-055, Athens, GA (1982).

87. Onishi, Y., R. J. Serne, E. M. Arnold, C. E. Cowan, and F. L. Thompson, "Critical Review: Radionuclide Transport, Sediment Transport, and Water Quality Mathematical Modeling; and Radionuclide Adsorption/Desorption Mechanisms," Rep. NUREG/CR-1322, PNL-2901, Pacific Northwest Lab., Richland, WA (1982).

88. Overnell, J. and T. L. Coombs, "Cadmium in the Marine Environment: The Importance of Metallothionein," in *Actualites de Biochimie Marine*, pp. 207–218 (1978).

89. Platt, T. and K. L. Denman, *Spectal Analysis in Ecology*, pp. 189–210 (1975).

90. Rai, L. C., J. P. Gaur, and H. D. Kumar, "Phycology and Heavy Metal Pollution," *Biol. Rev., Vol. 56*, pp. 99–151 (1981).

91. Rathbun, R. E. and D. Y. Tai, "Technique for Determining the Volatilization Coefficients of Priority Pollutants in Streams," *Water Res., Vol. 15*, pp. 243–250 (1981).

92. Rathbun, R. E. and D. Y. Tai, "Gas-Film Coefficients for the Volatilization of EDB from Water," *Environ. Sci. Technol.*, Vol. 20, pp. 949–952 (1986).

93. Reice, S., "Interspecific Associations in a Woodland Stream," *Can. J. Fish. Aquat. Sci.*, Vol. 38, pp. 1271–1280 (1981).

94. Rinaldi, S., R. Soncini-Sessa, H. Stehfest, and H. Tamura, *Modeling and Control of River Quality*, McGraw-Hill Inc., London, 380 pp. (1979).

95. Roesijadi, G., "The Significance of Low Molecular Weight, Metallothionein-Like Proteins in Marine Invertebrates: Current Status," *Mar. Environ. Res., Vol. 4*, pp. 167–179 (1980).

96. Ross, D. H. and J. B. Wallace, "Longitudinal Patterns of Production, Food Consumption and Seston utilization by Net-Spinning Caddisflies *(Trichoptera)* in a Southern Appalachian Stream (USA), *Holarctic Ecol., Vol. 6*, pp. 270–284 (1983).

97. Sakanari, J. A., M. Moser, C. A. Reilly, and T. P. Yoshino, "Effects of Sublethal Concentrations of Zinc and Benzene on Striped Bass, *Morone saxatilis* (Walbaum), infected with larval *Anisakis* nematodes, *J. Fish. Biol.*

98. Schindler, D. W., K. H. Mills, D. F. Malley, D. L. Findley, J. A. Shearer, I. J. Davies, M. A. Turner, G. A. Lindsey, and D. R. Cruikshank, "Long-Term Ecosystem Stress: The Effects of Years of Experimental Acidification on a Small Lake," *Science, Vol. 228*, pp. 1395–1401 (1985).

99. Siegfried, C. A. and A. W. Knight, "Tropic Relations of *Acroneuria (Calineuria) californica* (Plecoptera: Perlidae) in

a Sierra Foothill Stream," *Environ. Entomol.*, Vol. 5, pp. 575–581 (1976a).

100. Siegfried, C. A. and A. W. Knight, "Prey Selection by a Setipalpian Stonefly Nymph, *Acroneuria (calineuria) californica* Banks (plecoptera: Perlidae)," *Ecology*, Vol. 57, pp. 603–608 (1976b).

101. Shack, K. V., R. C. Averett, P. E. Greeson, and R. G. Lipscomb, "Methods for Collection and Analysis of Aquatic Biological and Microbiological Samples," *Techniques of Water Resources Investigations of the U.S. Geological Survey*, Book 5, Chapter A3 (1973).

102. Short, R. A., S. P. Canton, and J. V. Ward, "Detrital Processing and Associated Macroinvertebrates in a Colorado Mountain Stream," *Ecology*, Vol. 61 (4), pp. 727–732 (1980).

103. Smith, J. H., D. C. Bomberger, and D. L. Hayes, "Prediction of the Volatilization Rates of High-Volatility Chemicals from Natural Water Bodies," *Environ. Sci. Technol.*, Vol. 14, pp. 1332–1337 (1980).

104. Smock, L. A., "Relationships between Metal Concentrations and Organism Size in Aquatic Insects," *Freshw. Biol.*, Vol. 13, pp. 313–321 (1983a).

105. Smock, L. A., "The Influence of Feeding Habits on Whole-Body Metal Concentrations in Aquatic Insects," *Freshwater Biology*, Vol. 13, pp. 301–311 (1983b).

106. Spehar, R. L., R. L. Anderson, and J. T. Fiandt, "Toxicity and Bioaccumulation of Cadmium and Lead in Aquatic Invertebrates," *Environ. Pollut.*, Vol. 15, pp. 195–207 (1978).

107. State Water Pollution Control Board, *Water Quality Criteria*, Sacramento, California (1952).

108. Stumm, W. and J. J. Morgan, *Aquatic Chemistry*, Wiley Interscience, N.Y. (1970).

109. Sumner, W. T., C. D. McIntire, "Grazer—Pheriphyton Interactions in Laboratory Streams," *Arch. Hydrolbiol.*, Vol. 93, pp. 135–157 (1982).

110. Sullivan, B. K., E. Buskey, D. C. Miller, and P. J. Ritacco, "Effects of Copper and Cadmium on Growth, Swimming and Predator Avoidance in *Eurythemora affinis* (Copepoda)," *Mar. Biol.*, Vol. 77, pp. 299–306 (1983).

111. Sunda, W. G. and R. R. L. Guillard, "The Relationship between Cupric Ion and the Toxicity of Copper to Phytoplankton," *J. Mar. Res.*, Vol. 34, pp. 511–529 (1976).

112. Terry, K. L., "Temperature Dependence of Ammonium and Phosphate Uptake, and their Interactions, in the Marine Diatom *Phaeodactylum tricornutum* Bohlin," *Mar. Biol. Letters*, Vol. 4, pp. 398–320 (1983).

113. Tessie, A. P. G. C. Campbell, J. C. Auclair, and M. Bisson, "Relationship between Partitioning of Trace Metals in Sediment and Their Accumulation in the Tissues of the Freshwater Mollusc *Elliptio complanata* in a Mining Area," *Can. J. Fish. Aquat. Sci.*, Vol. 10, pp. 1463–1472 (1984).

114. U.S. Department of Interior, *Water Quality Criteria, A Report of the National Technical Advisory Committee to the Secretary of the Interior*, U.S. Government Printing Office, Washington, D.C. (1968).

115. U.S. Environmental Protection Agency, *Water Quality Criteria 1972, A Report to the Committee on Water Quality Criteria*, U.S. Government Printing Office (1973).

116. U.S. Environmental Protection Agency, *Quality Criteria for Water*, EPA-440/9-76-023, U.S. Government Printing Office (1976).

117. U.S. Environmental Protection Agency, *Methods for the Chemical Analysis of Water and Wastes*, Environmental Monitoring and Research Laboratory, Cincinnati, Ohio (1979a).

118. U.S. Environmental Protection Agency, *Manual of Analytical Methods for the Analysis of Pesticide Residues in Human and Environmental Samples*, Health Effects Research Laboratory, Research Triangle Park, North Carolina (1979b).

119. U.S. Environmental Protection Agency, *Water Quality Criteria Documents; Availability*, 40 Federal Register 70313, et. seq. (1980).

120. U.S. Environmental Protection Agency, *Sampling Protocols for Collecting Surface Water, Bed Sediment, Bivalves and Fish for Priority Pollutant Analysis* (1981).

121. U.S. Environmental Protection Agency, *Handbook for Sampling and Sample Preservation of Water and Wastewater*, EPA-600/4-82-029, Environmental Monitoring and Research Laboratory, Cincinnati, Ohio (1982a).

122. U.S. Environmental Protection Agency, *Methods for Organic Chemical Analysis of Municipal and Industrial Wastewater*, Environmental Monitoring and Research Laboratory, Cincinnati, Ohio (1982b).

123. U.S. Geological Survey, *National Handbook of Recommended Methods for Water Data Acquisition*, Office of Water Data Co-ordination, Reston, Virginia (1980).

124. Valocchi, A. J., "Validity of the Local Equilibrium Assumption for Modeling Sorbing Solute Transport through Homogeneous Soils," *Water Resourc. Res.*, Vol. 21, pp. 808–820 (1985).

125. Veith, G. D., D. L. Defoe, and B. V. Bergstedt, "Measuring and Estimating the Bioconcentration Factor of Chemicals in Fish," *J. Fish. Res. Board Can.*, Vol. 36, pp. 1040–1048 (1979).

126. Veith, G. D., K. J. Macek, S. R. Petrocelli, and J. Carroll, "An Evaluation of Using Partition Coefficients and Water Solubility to Estimate Bioconcentration Factors for Organic Chemicals in Fish," in J. G. Eaton, P. R. Parrish, and A. C. Hendricks (eds.), *Aquatic Toxicology*, ASTM/STP707, Philadelphia, pp. 116–129 (1980).

127. Wauchope, R. D. and L. L. McDowell, "Adsorption of Phosphate, Arsenate, Methanearsonate, and Cacodylate by Lake and Stream Sediments: Comparisons with Soils," *J. Environ. Qual.*, Vol. 13, pp. 499–504 (1984).

128. Wolf, P., "Incorporation of Latest Findings into Oxygen Budget Calculations for Rivers," *GWF-Wasser/Abwasser*, Vol. 112, pp. 200–203, 250–254 (1971).

129. Wolfe, N. L., R. G. Zepp, R. C. Baughmann, and J. A. Gordon, "Chemical and Photochemical Transformation of Selected Pesticides in Aquatic Systems," USEPA 600/3-76-067, Corvallis, OR (1976).

130. Zaroogian, G. E., J. F. Heltshe, and M. Johnson, "Estimation of Bioconcentration in Marine Species using Structural-Activity Models," *Environ. Toxicol. Chem.*, *Vol. 4*, pp. 3–12 (1985).

131. Zison, S. W., W. B. Mills, D. Deimer, and C. W. Chen, "Rates, Constants and Kinetics Formulations in Surface Water Quality Modeling, U.S. Environmental Protection Agency EPA 600/3-78-105 (1978).

APPENDIX

Notation and modeling parameters presented in this chapter along with references for their determination. Dimensional units are given in parentheses in terms of length (l), mass (m), time (t), temperature (T) and pressure (P) units.

Parameter Symbol	Description (units)	Method for Estimation
PHYSICAL SUBMODELS		
A	Stream channel cross sectional area (l^2)	Transverse depth indications (Buchanan and Somers, 1969)
A_s	Storage zone cross sectional area (l^2)	Usually a fitted parameter after conducting conservative tracer studies (Onishi, et al., 1981; Bencala, et al., 1984)
D	Dispersion coefficient (l^2/t)	Empirical determination (Fisher, et al., 1979)
E	Evaporation rate (l/t)	Calculated by Equation (8)
e_o	Saturation vapor pressure (P)	Tabulated as a function of temperature (Dorsey, 1968)
e_a	Air vapor pressure (P)	Thermocouple psychrometer (Marciano and Harbeck, 1954)
H	Henry's Law constant (Pl^3/m)	Isothermal gas flow stripping (Mackay, et al., 1979)
k	Liquid phase mass transfer coefficient (l/t)	Modified tracer technique (Rathbun and Tai, 1981)
N	Mass volatilization flux ($m/l^2 t$)	Calculated by Equation (7)
P	Atmospheric partial pressure (P)	Mass spectrometry (ASTM, 1977)
Q	Stream discharge, Volume rate of flow (l^3/t)	Current meter, weir, Parshall flume (Buchanan and Somers, 1969)
s	Suspended solid concentration (m/l^3)	Gravimetric (Guy and Norman, 1970), optical measurements (Krause and Ohm, 1984)
S	Volume, surface and lateral solute source (l^3/t)	Outfall discharge measurement, atmospheric traps and Fickian diffusion model, conservative trace study (Fisher, et al., 1979; Bencala, et al., 1984)
t	Time (t)	
T	Temperature (T)	
v	Average velocity (l/t)	Current meters (Buchanan and Somers, 1969)
x	Longitudinal distance (l)	
y	Tracer or solute concentration (m/l^3)	Tracer dependent (see monitoring methods section of this chapter).
w	Wind velocity (l/t)	Anemometer

(continued)

Parameter Symbol	Description (units)	Method for Estimation
CHEMICAL SUBMODELS		
k	Photolysis constant $(1/t)$	Estimated by laboratory experiments (Wolfe, et al., 1976)
k_{or}	Second order redox constant (l^3/mt)	Laboratory estimates (Mill, et al., 1982)
K	Stability constant (units are reaction dependent)	K values for various reactions have been compiled by Martell and Smith (1974)
K_d	Partitioning coefficient (units vary but often dimensionless)	Measured solute distribution between phase (Kuwabara, et al., 1984; Bencala, et al., 1984; Luoma and Jenne, 1976; Tessier, et al., 1984)
λ_s	First order rate coefficient for sediment sorption $(1/t)$	Generally computed from laboratory sorption studies on sediment size fraction (Kuwabara, et al., 1984; Bencala, et al., 1984)
ϱ_s	Mass of accessible sediment per unit volume of stream water (m/l^3)	Fitted or calculated parameter (Kuwabara, et al., 1984; Bencala, et al., 1984)
R_s	Kinetic sediment sorption term (m/l^3t)	Computed by first order rate Equation (9)
y_s	Solute concentration associated with sediment $(m/m$ or $m/l^3)$	Sediment extracts and digests (Luoma and Jenne, 1976; Luoma and Davis, 1983; Tessier, et al., 1984)
BIOLOGICAL SUBMODELS		
E_j	Solute excretion rate by organism j $(m/t{\cdot}m)$	Net sorption computed by Equation (29)
K_{db}	Partitioning coefficient (units vary but often dimensionless)	Measured solute distribution between phases at equilibrium (Nehring, 1976; Smock, 1983a; Kuwabara, et al., 1984; Bencala, et al., 1984)
M_j	Mortality rate for organism j (m/t)	Empirically determined function of organism concentration (Namkung, et al., 1983); Measured values (Cummins, et al., 1973)
P_j	Solute uptake through particle injestion by organism j $(m/t{\cdot}m)$	Laboratory measurement or gut analysis (Elwood, et al., 1976; Brown, 1977; Sullivan, et al., 1983)
P_{ow}	n-Octanol/water partitioning coefficient dimensionless)	Tabulated values (Neeley, et al., 1974)
R_b	Kinetic biological uptake (m/l^3t)	Computed by rate Equation (23)
S_j	Solute sorption rate by organism j $(m/t{\cdot}m)$	Net sorption computed by Equation (29); Measured effects of chemical speciation (Dressing, et al., 1982); Measured uptake (Dean, 1974; Smock, 1983b)
V_j	Mass respiration rate by organism j $(m/t{\cdot}m)$	Generally modeled as a function of temperature (McIntire and Colby, 1978)
w_{ji}	Weight percentage for particle i injestion by organism j	Measured gut content diversity (Coffman, et al., 1971; Siegfried and Knight, 1976a,b; Ross and Wallace, 1983; Smock, 1983a)
y_b	Solute concentration associated with biota $(m/m$ or $m/l^3)$	Tissue digestion (Bryan and Hummerstone, 1971; Borg, et al., 1981)
Z_j	Total particle injestion rate per unit mass of organism j $(m/t{\cdot}m)$	Measured gut content (Coffman, et al., 1971; Allan, 1982); Leaf pack mass reduction (Boling, et al., 1975; Short, et al., 1980)

(continued)

Parameter Symbol	Description (units)	Method for Estimation
BIOLOGICAL SUBMODELS		
ϵ	Efficiency of mass conversion by biota (dimensionless)	Production/food consumption estimates (Cummins, et al., 1973; Heiman and Knight, 1975; Ross and Wallace, 1983)
λ_b	First order rate coefficient for biological uptake or $(1/t)$	Measured by laboratory studies on plankton or periphyton samples (Kuwabara, et al., 1984)
ϱ_b	Mass of accessible biota per unit volume of stream water (m/l^3)	Gravimetric, chlorophyll-a or ATP measurements (McIntire and Phinney, 1965; Sumner and McIntire, 1982; McAuliffe, 1984, Kuwabara, et al., 1984)
γ_o	Estimate for "quasi" moving average constant (m/l^3)	Fitted parameter (IMSL, 1982)
θ_i	Estimate for ith order moving average parameter (typically dimensionless)	Fitted parameter (IMSL, 1982)

CHAPTER 27

Use of Wetlands for Wastewater Treatment

PETER SEREICO* AND CAROL LARNEO**

INTRODUCTION

Background

The use of wetlands to dispose of wastewater is not a new concept. Wetlands typically have dense vegetation, uncertain ground, mosquitoes, and other features that historically have been considered undesirable by the public. Because of this perception and their occupation of low areas in the landscape, wetlands historically have been a logical wastewater disposal point. There are a number of sites where wastewater has been discharged since the 1920s. Where surface waters are remote, discharge to wetlands is, in many cases, the only means of disposing of a community's waste.

The purposeful use of wetlands in the wastewater treatment and disposal process began in the early 1960s, given impetus by strong evidence that wetland systems can degrade and eliminate various waterbourne pollutants and by major legislative steps to control point source pollution—the Federal Water Pollution Control Act Amendments of 1972 (P.L. 92-500) and the Clean Water Act Amendments of 1977 (P.L. 95-217). The Clean Water Act encourages the use of "innovative and alternative" technologies for:

- reclamation and reuse of wastewater and wastewater constituents
- recovery and conservation of energy
- reduction in costs compared to existing conventional technologies

Under the Construction Grants Program for wastewater

*Division of Life Sciences, University of New England, Biddeford, ME
**Ecol-Sciences, Rahway, NJ

treatment facilities, the United States Environmental Protection Agency (EPA) defined "alternative" technologies as proven methods which provide for reclamation and reuse of wastewater, recycling of wastewater constituents or recovery of energy. "Innovative" technologies are developed methods which offer an advancement in the state-of-the-art but have not been fully proven. A number of innovative and alternative technologies have been investigated, including:

- improved on-site sewage disposal methods to eliminate the need for centralized sewerage systems (septic tank-soil absorption systems, mound systems, aerobic tank-soil absorption systems, split systems, cluster systems)
- wastewater reclamation via land application and reuse
- wetlands and natural biological treatment systems

The interest in wetlands for wastewater treatment can be attributed to four basic factors

1. Public demand for more stringent wastewater effluent standards, including removal of nutrients and trace contaminants, as well as organics and suspended solids
2. Rapidly escalating costs of construction and operation and maintenance of conventional wastewater treatment facilities, compared to the low cost, low energy consuming wetlands treatment process
3. Recognition of the natural treatment functions of wetlands, particularly as nutrient sinks and buffer zones
4. Emerging or renewed appreciation of aesthetic, wildlife and other environmental benefits associated with the preservation and enhancement of wetlands

Definition of Wetlands

Wetlands occur in a wide range of physical settings at the interface of terrestrial and aquatic ecosystems. Due to the diversity of wetland types and the fact that the interface is not sharply defined, but rather a gradient, there is no single, indisputable definition for the term "wetland." Wetlands is a relatively new term which includes areas formerly referred to as marshes, swamps and bogs. Generally, wetlands are lands where saturation with water is the dominant factor determining soil development and biota; the single feature that most wetlands share is soil or substrate that is at least periodically saturated with or covered by water. Recent definitions are made by Executive Order 11990, Protection of Wetlands (May 25, 1977) and the United States Fish and Wildlife Service (USFWS).

Executive Order 11990 defines wetlands as "those areas that are inundated by surface or ground water with a frequency sufficient to support and under normal circumstances does or would support a prevalence of vegetative or aquatic life that requires saturated or seasonally saturated soil conditions for growth and reproduction. Wetlands generally include swamps, marshes, bogs, and similar areas such as sloughs, potholes, wet meadows, river overflows, mud flats, and natural ponds."

Currently the most widely used definition of wetlands is that of the U.S. Fish and Wildlife Service. Under this definition, an area may be designated as a wetland if it meets one of the following criteria:

- At least periodically, the land supports predominately hydrophytes (plants which can withstand periodic oxygen deficiency due to excessive water content of the soil).
- The substrate is predominately undrained hydric soil (soils becoming anaerobic periodically due to extended inundation).
- The substrate is non-soil and is saturated with water or covered by shallow water at some time during the growing season of each.

In recent years, the feasibility of wastewater treatment and discharge via wetland systems has been shown to have merit. The success of a wetland system depends upon site specific characteristics and the design engineer's understanding of the system. In addition to technical factors, such as treatment efficiency and appropriate design parameters, both environmental and regulatory factors must be considered in developing an effective wetland treatment system.

Greatest attention to date has been given to the use of wetlands for tertiary treatment of municipal wastewater, necessary to meet stringent nutrient removal requirements. In a few instances, raw sewage and primary effluent have been applied to wetlands for treatment purposes. Recently, there has been a great deal of interest in utilizing wetlands as secondary treatment systems. The U.S. EPA, in order to encourage wetlands use for wastewater treatment, published a technical report on the effects of wastewater treatment on wetlands [1] and is currently initiating research at various wetlands sites which receive wastewater in order to gain a better understanding of how such systems function.

MECHANISMS FOR POLLUTANT REMOVAL

Introduction

Municipal wastewater is characterized by relatively high levels of suspended solids, BOD, and nutrients. Typical ranges for constituents of raw wastewater are summarized in Table 1. While it is unlikely that wetlands systems will be implemented to treat raw wastewater due to health factors, odor and putrescibility, it is necessary to first know raw wastewater quality in order to choose appropriate treatment process (including wetlands systems) and to predict effluent quality. Prior to wetlands discharge, primary treatment, in which the bulk of settleable solids and a portion of the organic matter are removed by sedimentation, is necessary and secondary treatment, in which organic matter is significantly reduced by biological oxidation, is common.

TABLE 1. Composition of Raw Municipal Wastewater [2].

Constituent	Raw Wastewater Range, mg/l	
	Metcalf & Eddy, Inc.	U.S. EPA
Solids, total	350–1,200	700–1,000
Dissolved solids, total	250–850	400–700
Mineral	145–525	250–450
Volatile	105–325	150–250
Suspended solids, total	100–350	180–300
Mineral	30–75	40–70
Volatile	70–275	140–230
Settleable solids, (ml/liter)	5–20	—
Settleable solids, total	—	150–180
Mineral	—	40–50
Volatile	—	110–130
Biochemical oxygen demand, (BOD$_5$) 20°C	100–300	160–280
Ultimate carbonaceous	—	240–420
Ultimate nitrogenous	—	80–140
Total oxygen demand (TOD)	—	400–500
Chemical oxygen demand (COD)	250–1,000	550–700
Total organic carbon (TOC)	100–300	200–250
Nitrogen, total as N	20–85	40–50
Organic	8–35	15–20
Ammonia	12–50	25–30
Nitrites	0	0
Nitrates	0	0
Phosphorus, total as P	6–20	10–15
Organic	2–5	3–4
Inorganic	4–15	7–11
Chlorides added	30–100	50–60
Alkalinity added, as CaCO$_3$	50–200	100–125
Grease	50–150	90–110

TABLE 2. Composition of Treated Municipal Wastewaters [3].

Constituent	Primary effluents, mg/l			Secondary effluents, mg/l		
	Average	Range	Range	Average	Range	Range
BOD₅	136	70–200	68–243	25	15–45	4.4–241
COD	330	165–500	—	55	25–80	—
Suspended solids	80	40–120	27–100	15	10–30	19–129
Nitrogen, total as N	35	—	16–45	30	—	23–48
Ammonia, as N	—	—	9–27	—	—	5–29
Phosphorus, total as P	7.5	—	1.2–9.8	5.0	—	1.2–9.1
Total coliforms, MPN/100 ml	2×10^6	5×10^6–5×10^8	—	1×10^3	1×10^2–1×10^4	—

Characteristics of primary and secondary effluent are summarized in Table 2.

Wetland systems have properties which seem to provide the ability to degrade and eliminate wastewater contaminants, hence making them attractive for wastewater treatment. Such properties include:

- high plant productivity and nutrient requirements
- high decomposition activity
- large absorptive area in the substrate
- low oxygen content of the substrate

Recent studies [4–7] have indicated that wetlands are able to effectively remove wastewater contaminants including BOD, suspended solids, nutrients, trace organics, and heavy metals. Removal mechanisms in wetlands for these contaminants have been identified from field observation, laboratory and pilot scale experimentation and are summarized in Table 3. The principal removal mechanisms involved in wetland treatment of wastewater can be classified as physical, chemical, or biological. Table 3 summarizes the primary, secondary and incidental mechanisms responsible for the removal of specific contaminants. It should be noted that within a wetlands treatment system these mechanisms are operative in the water column, within the underlying substrate, and at the water-substrate interface. From the literature, it is apparent that processes associated with pollutant removal are well known, although relatively few researchers have attempted to quantify removal rates and treatment efficiencies.

The following discussion focuses on the constituents of concern in wastewater and presents a summary of removal mechanisms for each within the wetland treatment system.

BOD Reduction

Biodegradable organics, composed primarily of proteins, carbohydrates, and fats, are commonly measured in terms of BOD or COD. If discharged to the environment, biological stabilization of the organics can lead to depletion of natural oxygen resources and to the development of septic conditions.

Typical BOD removal efficiencies are reported as 70 to 96% for natural wetlands receiving secondary effluent and 50 to 90% for artificial wetlands receiving primary effluent. It must be noted that the type of BOD leaving the treatment plant and that leaving the wetland system differ. Materials comprising BOD in the plant effluent are the degradation products of human waste, while material exerting BOD in wetland effluent is partially composed of algae and other organisms at the base of the food chain.

It is assumed that bacterial metabolism is the primary factor in BOD/COD reduction within a wetland system. The large surface area of plant stems and litter form the substrate for bacterial populations. Settling removes BOD associated with solids from the surface waters; these solids often decay anaerobically upon settling. Algae within the wetland system provide high levels of dissolved oxygen (D.O.), which further enhances BOD removals. Based on a review of the data in the literature on natural and artificial wetlands and overland flow systems (which are similar in function to wetland systems), Tchobanoglous and Culp [6] concluded that the removal of BOD₅, TOC and COD in a wetland system can be described by first order reaction kinetics expressed as:

$$C_t = C_o e^{-kt} \tag{1}$$

where:

C_t = concentration remaining at time t, mg/l
C_o = concentration at time $t = 0$, mg/l
k = specific first order removal rate constant for given constituent at 20°C, 1/day
t = detention time in wetland, d

TABLE 3. Removal Mechanisms in Wetlands for the Contaminants in Wastewater [6].

Mechanism	Settleable Solids	Colloidal Solids	BOD	Nitrogen	Phosphorus	Heavy Metals	Refractory Organics	Bacteria and Virus	Description
Physical									
Sedimentation	P	S	I	I	I	I	I	I	Gravitational settling of solids (and constituent contaminants) in pond/marsh settings.
Filtration	S	S							Particulates filtered mechanically as water passes through substrate, root masses, or fish.
Adsorption		S							Interparticle attractive force (van der Waals force).
Chemical									
Precipitation					P	P			Formation of or co-precipitation with insoluble compounds.
Adsorption					P	P	S		Adsorption on substrate and plant surfaces.
Decomposition							P	P	Decomposition or alteration of less stable compounds by phenomena such as UV irradiation, oxidation, and reduction.
Biological									
Bacterial Metabolism		P	P	P			P		Removal of colloidal solids and soluble organics by suspended, benthic, and plant-supported bacteria. Bacterial nitrification/denitrification.
Plant Metabolism							S	S	Uptake and metabolism of organics by plants. Root excretions may be toxic to organisms of enteric origin.
Plant Absorption				S	S	S	S		Under proper conditions significant quantities of these contaminants will be taken up by plants.
Natural Die-Off								P	Natural decay of organisms in an unfavorable environment.

P = primary effect, S = secondary effect, I = incidental effect (effect occurring incidental to removal of another contaminant). The term metabolism includes both biosynthesis and catabolic reactions.

If it is assumed that 95% of the BOD_5 in primary wastewater is removed in 10 days, the value of k is on the order of 0.3 per day. Because of the areal extent of most wetlands, the value of k will depend largely on the temperature. From experience with other biological systems, the effect of temperature can probably be modelled using the following equation:

$$k_T = k_{20}\theta(T - 20) \qquad (2)$$

where:

k_T = removal rate constant at temperature, T, l/d
k_{20} = removal rate constant at 20°C, l/d
θ = temperature coefficient, 1.05 to 1.08
T = temperature of water in wetland, °C

Temperature, therefore, is a factor of great significance in the design of wetland treatment systems in temperate and cold regions which exhibit seasonal variation. The area of a typical wetland treatment system must be increased by a factor of two during the winter to achieve the same level of treatment.

The dissolved oxygen regime of the wetland should be considered with respect to BOD removal. Typically, D.O. levels in a wetland vary diurnally and seasonally, ranging from less than 0.1 mg/l to supersaturation. While many wetland components, including plant species and microfauna and flora, are adapted to low and periodically anoxic regimes, such regimes may interfere with the degradation and respiration processes in the sediments. The magnitude of the D.O. fluctuation would be expected to increase with the addition of wastewater.

Suspended Solids Removal

Under natural conditions a wetland receives suspended solids from two major sources—runoff and litterfall from component vegetation. Additionally, decomposition of the litter results in high levels of fine particulate organic matter. It has been demonstrated that natural suspended solids content of wetland waters is high, in the range of 100 to 300 mg/l, as compared to secondary effluent discharge standards, in the 10 to 30 mg/l range.

In a wetland system, the most important suspended solids removal mechanism is sedimentation. Three physical processes cause the movement of particles in water: Brownian motion, frictional resistance, and gravity. Brownian motion, the random vibratory movement of extremely small particles, has little effect in solids removal as there is no net movement toward the bottom.

Gravity moves particles with a specific gravity greater than one downward while turbulence tends to cause resuspension. The competition between settling and resuspension determines the rate of settling for any size/density class of suspended matter. Because wetlands systems are continuously subject to varying external forces such as wind, water depth, and the straining effects of vegetation it is impossible to predict a general settling rate for suspended solids. Generally, the larger the particle, the higher the settling velocity. To illustrate the settling velocity of suspended solids in wastewater which normally range in diameter from 0.005μ to 100μ, particles with a diameter less than 10μ have a terminal settling velocity less than 0.01 cm/s.

Under the stagnant or slowly moving hydrologic conditions which typify wetlands, materials are deposited. Long detention times allow for significant particle agglomeration and settling of finely divided solids. The nature of flow patterns through wetlands has been found to influence deposition of suspended matter and overall pollutant removal efficiency. Sheet flow and meandering channels lead to large effective flow areas and the best deposition rates. Resuspension, however, occurs in response to even slight changes in the flow regime which may be associated with storm events, seasonal runoff or regulated discharges (such as wastewater).

It should be noted that the wastewater discharged to the wetland is likely to have a lower solids content than the wetland itself, particularly if it is secondary effluent. Incoming wastewater solids will generally be quickly deposited and a long period would be expected before any of these solids arrive at the outflow point of the wetland. The subject of suspension, settling and transport of particles through the wetland system has not been well-studied.

Nutrient Removal

Municipal wastewater is characterized by relatively high levels of nutrients. Because nitrogen and phosphorus are the primary cause of eutrophication, effluent discharge requirements to lakes and streams are becoming increasingly stringent, necessitating tertiary wastewater treatment. The goal of such treatment is to remove nitrogen and phosphorus. Several recent studies have shown that natural wetlands are able to remove these nutrients effectively, resulting in improved water quality. Nutrient levels in incoming wastewater have been observed to decrease rapidly as the water flows from the discharge point (refer to Figure 1).

Nutrient capture processes are limited in retention capacity (particularly physical adsorption of phosphorus) and saturation occurs in the vicinity of the wastewater discharge point. Nutrients pass through saturated areas, resulting in a slowly moving dissolved nutrient front, beyond which the concentrations drop sharply.

Nutrient removal is a result of both physiochemical and biological processes and is accomplished by one or a combination of three general mechanisms: vascular plant uptake, microbial processes, and sediment reactions.

Because of the complex interrelationships between the components of a wetland system and the uniqueness of each wetland system, there is little information regarding nutrient

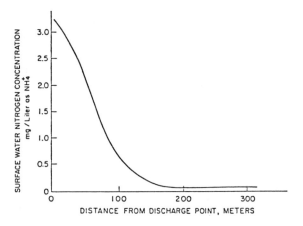

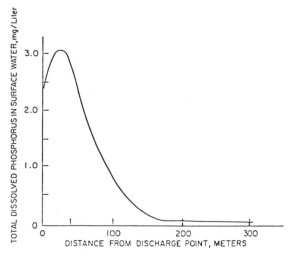

FIGURE 1. Typical removal of nitrogen and phosphorus from wastewater by wetland irrigation [5].

dynamics, removal kinetics, and the relative importance of individual components of the nutrient removal model. Nutrient removal within a wetland system can, however, be described in general terms.

VASCULAR PLANT UPTAKE

Nutrients, required for plant metabolism, growth and reproduction, are absorbed from the environment. Upland vegetation obtains nutrients from the soil through the root system and uptake capability is directly proportional to underground root volume. Wetland plants have morphological and physiological adaptations to their low oxygen waterlogged environments which enhance nutrient uptake. Wetland plants can absorb nutrients from either water or substrate via shoots, leaves, and water roots (finely

branched basal parts of the stem) as well as through true roots.

Nutrient uptake rates are related to plant growth rates and thus vary seasonally. Uptake of nutrients is probably affected by the same factors that influence other plant physiological processes, such as light, temperature, and dissolved oxygen. At low nutrient concentrations, an active form of nutrient uptake process occurs which requires an expenditure of energy to transport nutrients into the root system. Under such conditions uptake rates have been shown, for a wide taxonomic range, to follow classic Michaelis-Menten kinetics.

$$1/v = (k_m/v_m)(1/C_o) + 1/V_m \qquad (3)$$

where:

v = observed rate of nutrient uptake at external nutrient concentration C_o
V_m = maximum rate of uptake
K_m = Michaelis constant equal to the nutrient concentration at $v = 1/2V_m$
$v = 1/2V_m$

At higher nutrient concentrations, as would be the case in a wetland receiving wastewater, a passive diffusion process predominates. Nutrient uptake rates under such conditions have been observed to far exceed what would be predicted by Michaelis-Menten kinetics. Each plant has a critical nutrient concentration in the environment, below which growth is limited. If nutrients are available in excess of plant metabolic requirements, the plant continues to absorb nutrients in a phenomenon known as luxury uptake. In wetlands reviewing wastewater, luxury uptake would be expected for nitrogen and phosphorus and a shift in species composition may occur favoring species able to utilize excess amounts of these nutrients.

MICROBIAL PROCESSES

Bacterial metabolism is a significant mechanism for nitrogen removal from wastewater discharged to a wetland system. Bacterial populations are restricted to substrate surfaces, including litter, soils, and plant stems. Three microbial processes are of significance with respect to nitrogen removal from wastewater: ammonification, nitrification, and denitrification.

Microorganisms in the substrate and water column break down organic nitrogen to ammonia forms in the ammonification process. The reaction proceeds faster during the growing season and at high temperatures. Ammonia produced may be taken up by wetland plants directly, as discussed in the previous section, or converted in the nitrification process.

Nitrification occurs only in the presence of oxygen and would be limited to the uppermost surfaces of wetlands sediments where oxygen can readily diffuse to the reaction site. Under aerobic conditions ammonia is converted in a two step process to nitrate.

$$2NH_3 + 3O_2 \xrightarrow{\text{Nitrosomonas}} 2NO_2 + 2H^+ + 2H_2O \quad (A)$$

$$2NO_2 + O_2 \xrightarrow{\text{Nitrobacter}} 2NO_3 \quad (B)$$

Nitrates produced in excess of plant uptake potential do not bind to the soils but, rather, percolate into the groundwater. Minimal nitrification occurs where the concentration of dissolved oxygen is less than 1 to 2 mg/l.

Denitrification, which occurs only under anaerobic conditions, may contribute significantly to overall permanent nitrogen removal from the wetland system. In the process of denitrification anaerobic bacteria reduced nitrate-N to gaseous nitrogen which escapes from the system into the atmosphere:

$$NO_3\text{-}N \xrightarrow{\text{anaerobic bacteria}} N_2 + N_2O \quad (C)$$

This reaction occurs at the anaerobic substrate-water interface and within the sediments. Wetlands typically provide excellent conditions for denitrification due to their water-logged, stagnant anaerobic characteristics and abundant organic matter to provide energy for the bacteria. Denitrification rates have been found to decrease with lower temperatures, lower pH, and shallow depths at which mixing creates aerobic conditions.

SEDIMENT REACTIONS

The incorporation of nutrients into the substrate structure by adsorption and net soil building represents a mechanism for relatively permanent immobilization. Phosphorus is removed from the water column by chemical precipitation and transferred to the bottom sediments which act as a sink. Under aerobic conditions inorganic phosphorus in the water column forms complexes with various metal compounds, such as oxidized forms of iron, calcium, and manganese. In general, phosphorus precipitates above pH 8.

Wetlands soils are characterized by a thin, oxidized mud-water layer and the absence of molecular oxygen and reduced conditions in the deeper layers. The oxidized soil-water interface is of particular importance to nutrient cycling in wetlands. This layer absorbs and retains phosphate from the water column and the deeper anaerobic layers of the sediment. The anaerobic zone acts as a release for phosphate when the surrounding water is low in soluble phosphate, but will absorb phosphate from solutions high in this ion. This is caused mainly by the process of adsorption to ferrous and manganese colloidal gels.

The nutrient removal capacity and sink function of wetland soils and sediments is limited, gradually decreasing over time due to saturation of available absorption sites.

Thereafter, a steady state is attained so that the ultimate fate of nutrients in wetland soils would include remaining attached and incorporated in the sediment; leaching out; release by cation exchange, microbial action, or plant uptake; and loss by erosion.

Nutrients contained in plant tissues, unless harvested, are incorporated into the sediment, primarily by peat formation. Peat is a general soil type resulting from the gradual accumulation of slowly decaying plants and other organic matter. Peat formation is a slow process; rates are affected by hydrologic regime, nutrient levels, climate and basin morphometry. Various types of peat soils may be formed, depending on the plant species, climate, and types of decomposing microorganisms. Nutrient retention ability varies between different peat types and is a factor affecting the performance of a wetland receiving wastewater.

SUMMARY

Wetlands, because of great variation in species composition, substrate type, water chemistry, hydrologic conditions, and other factors, vary in nutrient removal capacity. Hence, the feasibility of using a wetland system for nutrient removal must be determined based on specific characteristics of a given wetland. As a general overview, wetlands conditions favoring nutrient removal are summarized.

- conditions favoring phosphorus removal
  - organic soil in poor nutrient regimes
  - vegetation limited by phosphorus supply or with the ability for luxury consumption
  - presence of iron and aluminum compounds
- conditions favoring nitrogen removal
  - reduced soil-water interface (denitrification)
  - vegetation limited by nitrogen or with the ability for luxury consumption
- conditions favoring nitrogen and phosphorus removal
  - low or no energy subsidy to the wetland by tides, wave action, streamflow, etc.

Metals Removal

Municipal wastewater may contain varying amounts of trace metals as summarized in Table 4. Of particular concern are mercury, lead, arsenic, cadmium, copper, and zinc due to their toxicity and tendency to accumulate in the food chain.

In wetlands systems physical and chemical reactions are the primary mechanisms for trace metal removal, with vegetative uptake secondary. Wetland sediments generally act as a sink for trace metals and many wetland plant species can absorb and bioaccumulate them. Little, however, is known concerning how metals are cycled within the wetland system.

Key physical-chemical mechanisms of trace metal removal include absorption and sedimentation and the forma-

TABLE 4. Trace Element Levels in Raw and Treated Municipal Effluents [3].

Element	Untreated Wastewater, mg/l	Primary Effluents, mg/l National	Secondary Effluents, mg/l National
Arsenic	0.003	0.002	0.005–0.01
Cadmium	0.004–0.14	0.004–0.028	0.0002–<0.02
Chromium	0.02–0.700	<0.001–0.30	<0.010–0.17
Copper	0.02–3.36	0.024–0.13	0.05–0.22
Iron	0.9–3.54	0.41–0.83	0.04–3.89
Lead	0.05–1.27	0.016–0.11	0.0005–<0.20
Manganese	0.11–0.14	0.032–0.16	0.021–0.38
Mercury	0.002–0.044	0.009–0.035	0.0005–0.0015
Nickel	0.002–0.105	0.063–0.20	<0.10–0.149
Silver	—	—	—
Zinc	0.030–8.31	0.015–0.75	0.047–0.35

tion of metal complexes and precipitates. Loss to the atmosphere through volatilization is significant for elemental mercury only and this process is controlled by temperature, wind speed, and agitation. Heavy metals readily adsorb to particulates in the water column and sediments. The adsorptive capacity is dependent upon water chemistry, the nature of the particles, ionic strength, pH, temperature, redox potential, and presence of competing cations and, thus, will vary considerably between wetland systems. The following examples illustrate how various parameters affect heavy metal absorption:

• Cadmium is better adsorbed to suspended solids in low ionic strength water than in high ionic strength water under aerobic conditions. Under anoxic conditions, cadmium exchanges with iron at the surface of ferrous sulfide substrates.
• Copper, unlike cadmium, is adsorbed more strongly in high ionic strength water than in low ionic strength water.
• Lead adsorbs to metal-organic complexes and to clay minerals.
• Mercury readily adsorbs to solids but the route is usually from oil rather than water.
• Zinc adsorbs to colloidal hydrous oxides in salt water environments.

Heavy metals may also be bound by chelation, in which a complex is formed with anionic molecules (ligands). The wetlands environment provides many sources of ligands, including chloride, sulfate, and bicarbonate which form soluble complexes (primarily in salt waters); products of organic decomposition; proteinaceous compounds excreted by microorganisms; and other components found in wastewater. Trace metals, including cadmium, copper, lead mercury, zinc, and silver form insoluble sulfide precipitates under reducing conditions which are common to wetlands.

Oxides and hydroxides of cadmium, chromium, iron, manganese, and zinc will form under aerobic conditions. Metal retention by particles and sediments is dependent upon a number of factors, including type of metal, pH, organic content, and loading rate. Within a wetland system low pH and low redox potentials tend to favor stabilization of trace metals while sulfide precipitation and complexing with organics tends to favor deposition.

Biochemical interactions are a secondary mechanism for heavy metal removal from the system. Heavy metals are absorbed by aquatic vegetation both from the sediment via roots and from the water via leaf surfaces. Certain trace metals are plant micronutrients and will promote growth in low concentrations but may be phytotoxic in higher concentrations.

Uptake rates generally vary seasonally and metal storage capacity varies between tissues of the same plant. Cattails, reeds, and grasses, for example, show maximum metals concentrations in above ground parts with no translocation of metals back to roots and rhizomes in the growing season. Metals would be released to the environment upon death and decomposition unless external harvesting was practiced. Floating aquatic vegetation has been noted for its ability to absorb and accumulate trace metals far beyond ambient water concentrations. Water hyacinth has been shown to accumulate 50 to 200 times the water concentration of arsenic, chromium, copper, lead and nickel; 700 times the concentration of mercury; and over 6000 times the concentration of iron.

Bacteria and Virus Reduction

Pathogens survive the wastewater treatment process although density generally does decrease as a result of treatment. Discharge of wastewater, particularly where minimum in-plant treatment has occurred, to the wetlands environment is of concern because of the potential for virus and bacterial transmission.

Treatment of wastewater will reduce, but not eliminate, viruses which are generally present in secondary effluent even after chlorination. The lack of standardized, reliable, and economical means to quantify viruses in water has resulted in the lack of virus research at wetlands sites. It appears, from several studies, that viruses tend to remain trapped within the wetland system due to absorption processes.

Bacterial pathogens are the most susceptible to the treatment process but are still present in effluent. Fecal and total coliform counts are the standard parameters tested and used as indicators of other bacteriologic pathogens. Based on a variety of coliform studies, which concluded that the total and fecal coliform content of a natural wetland is extremely variable and that when significant levels of fecal coliform are introduced into the wetland with wastewater, these levels are dramatically reduced with passage through the wetland.

Little information is available regarding the removal of pathogens in a wetland treatment system. Removal from the water column is primarily by adsorption onto suspended particles followed by sedimentation. Adsorption of viruses to suspended particles is a function of pH and dissolved materials; viruses can remain within the soil matrix, particularly when clay content is high, for extended periods. Desorption could occur in response to changing environmental conditions. Other mechanisms for virus and bacterial removal within a wetland system are exposure to ultraviolet radiation, attack by chemicals, and predation. Reduction rates and potential for disease transmission have not been established.

DESIGN CRITERIA

Introduction

From the discussions of pollutant removal mechanisms within a wetland system it is evident that standard generalized design for wetland treatment systems is neither available nor feasible. Major differences in climate, hydrology, water chemistry, substrate, and vegetation significantly affect wetland response to pollutants. Because of these differences and the limited use of wetland treatment, there are no standard process design criteria or procedures to date for either natural or artificial wetland treatment systems.

This section presents parameters which should be considered in the development of a wetland treatment system and where possible includes criteria developed from examples found in the literature. Topics of primary importance which should be considered during design are:

- use and treatment objectives
- system type and site selection

- design parameters
- operational requirements
- treatment efficiency

Use and Treatment Objectives

Establishment of objectives is a critical initial step in the design of any treatment system. The principal objective of the wetland system is treatment of applied wastewater; however, the rationale for selecting wetland treatment over conventional treatment and levels of treatment to be achieved must be addressed. Obviously, objectives will vary from project to project. Primary objectives can include:

- Wetlands treatment is an innovative and alternative technology and as such is a preferred strategy due to greater available funding under the Construction Grants Program.
- lower capital and operation and maintenance costs than conventional treatment
- Water conservation: seepage through wetland may recharge an underlying aquifer.
- Economic return from use of water and nutrients present in wastewater for marketable crops: harvested wetland vegetation may be used as animal feed, woody plants for pulp in the paper industry, or digested to produce biogas.
- preservation, and perhaps enhancement, of wetlands habitat

In terms of treatment objectives, the most common use of natural systems is for advanced (tertiary) treatment of wastewater following conventional secondary treatment. There is little information in the literature regarding use of natural wetlands for treating either raw sewage of primary effluent. Because of this, any proposal to use a natural wetland for less than secondary effluent should first be substantiated with a pilot study. Artificial wetland systems have been used with some degree of success to treat primary as well as secondary effluent. In addition to identifying the incoming wastewater quality, target pollutants should be identified so that the system can be designed to provide mechanisms to remove pollutants of concern. For municipal wastewater, target pollutants will be one or a combination of the following: oxygen demanding organic wastes, nitrogen, phosphorus, heavy metals. The various mechanisms within a wetland system associated with removal of the above have been described. Table 5 briefly summarizes the best situations for their removal.

System Type and Site Selection

The initial decision in development of a wetland treatment system is whether to select a natural or artificial wetland. Both wetland types have advantages and disadvantages with respect to creating a wastewater treatment system as summarized in Table 6. While the relative importance of factors

TABLE 5. Characteristics to Be Emphasized in the Design of Wetlands Systems for the Treatment of Various Pollutants.

BOD Reduction

- large surface area to support microorganism populations
- uniform distribution of BOD load
- adequate dissolved oxygen
- open surfaces in the wetland to increase oxygen transfer to water

Suspended Solids Removal

- meandering channels with slow moving water and large surface areas to enhance sedimentation
- thick vegetation to enhance filtering

Nitrogen Removal

- rapid plant growth
- plant harvesting
- deep areas to provide anaerobic conditions for denitrification

Phosphorus Removal

- shallow flow and seepage conditions
- rapid plant growth
- plant harvesting
- organic substrate

Heavy Metals Removal

- shallow flow and seepage conditions
- species chosen for uptake and bioaccumulation ability of specific metal
- species diversity to ensure good overall uptake
- organic substrate with low pH and low redox potential

listed in Table 6 is dependent on specific circumstances associated with a particular facilities planning effort, it is the intent of this section to provide an overview of wetland selection factors.

No guidance is available for the selection of one particular natural wetland over another. Local conditions and specific needs are the prime factors in selection. Geographical location and climate are the primary factors determining which wetland types will be available for consideration. Certain wetland types have limited ranges and could not be established successfully outside of them. In New Jersey, for example, neither peatlands nor cypress swamps are native wetland types and thus should not be considered as potential treatment systems.

Assuming that the natural wetland system will be utilized to provide advanced treatment of secondary effluent and further assuming that the secondary treatment facility is already in existence, the wetland chosen must be within a reasonable distance from the treatment plant. Sutherland [8] determined that capital costs were related primarily to the

distance between the treatment plant and wetland according to the relationship:

$$C = 129\,D + 288 \qquad (4)$$

where C equals cost in thousands of dollars and D equals distance in miles.

Figure 2 illustrates the break-even point for cost versus distance. The break-even distance represents the radius within which a suitable natural wetland must be located in order to make wetland treatment more economical than other options.

Site identification and preliminary evaluation can be accomplished using readily available information. Wetland sites can be identified using USGS topographic maps in con-

TABLE 6. Comparison of Natural and Artificial Wetlands.

NATURAL WETLANDS

Advantages

- immediate availability, no extended construction or vegetation establishment required
- no new land requirements

Disadvantages

- considered to be environmentally sensitive
- may connect to sensitive water bodies
- inadequate size to assimilate wasteload
- stringent requirements for incoming wasteload to protect wetland value or because wetland is permitted as a water of the state
- operation may change wetland ecosystem structure and balance
- inconvenient or distant location

ARTIFICIAL WETLANDS

Advantages

- flexibility in site location
- size can be designed for projected wasteload
- features to improve pollutant removals and facilitate maintenance can be incorporated into design (e.g., channels, shallow areas, levees)
- less stringent criteria for incoming wastewater
- supplements existing wetland areas; creates habitat
- not viewed as an alteration of a natural wetland ecosystem
- easier to control water level and flow rates

Disadvantages

- cost and availability of suitable sites
- construction costs for grading and planting site
- reduced performance during vegetation establishment period
- difficulties in establishing wetland vegetation

junction with National Wetlands Inventory Map overlays prepared by the U.S. Fish and Wildlife Service. Wetland types are identified on these overlays using the Cowardin System which has been described. Recent aerial photographs are also useful in identifying wetland areas, extent, and surrounding land use. Preliminary sites should be verified in the field as discrepancies in vegetation type and wetland size are common and land use changes may have occurred (e.g., the wetland has been filled in for development or drained for agriculture). Site selection should be based on the following:

- Vegetation type — Plants exhibiting good removal efficiencies for target pollutants and tolerance to increased nutrient levels and to an altered hydrologic regime would be preferred.
- Soil or substrate — These should provide mechanisms for target pollutant removal and retention.
- Implementability factors — Factors such as ownership, current use, downstream water use, and adjacent land use should be considered during the site selection stage.

ARTIFICIAL WETLANDS

An artificial wetland system may be selected because of its technical advantages (refer to Table 7) or for lack of an appropriate natural wetland within a reasonable distance. Artificial wetlands offer flexibility in siting, requiring only a vacant parcel of sufficient size. Suitable grading, substrate, and vegetation can be created during construction of the system. Vacant parcels can be identified using USGS topographic maps, land use maps from local or county planning boards, and local tax maps. As in the site selection process described for natural sites, all candidate sites should be field checked prior to selection and implementability factors such as ownership, zoning, and adjacent land use should be considered.

Four wetland types are currently considered for artificial wetlands: marshes, marsh/pond, ponds, and trenches. These are commonly planted with reeds, rushes and emergent wetlands vegetation. Indigenous species, as well as species most efficient in removal of target pollutants and tolerant to wastewater additions, should be selected.

Design Parameters

Wetland treatment system design is a relatively new technology, and apart from specific case-study information, little data is available on design of wetland wastewater treatment facilities. Design parameters developed for artificial wetlands systems from literature data by Tchobanoglous and Culp [6] are shown in Table 7. These and other parameters found elsewhere in the literature are discussed below.

HYDROLOGIC LOADING

The hydrologic loading rate is a key engineering consideration in the design of a wetland system, natural or ar-

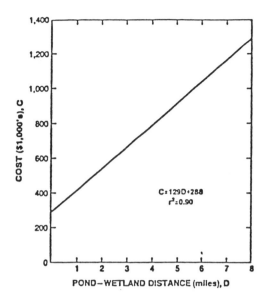

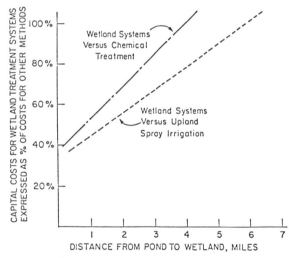

FIGURE 2. Relationship between cost and distance from treatment plant [8].

tificial, for wastewater treatment. In a natural system, appropriate hydrologic loadings are dependent on existing hydrologic regime: surface flow rates, soil infiltration rates, and depth. In order to properly assess the performance of a wetland, a water budget must be prepared which identifies and quantifies all inflows and outflows throughout the year. The annual water budget can be expressed as unsteady-state material balance of the following form:

$$Q_i - Q_o + A + P - E - I = M \qquad (5)$$

TABLE 7. Preliminary Design Parameters for Planning Artificial Wetland Wastewater Treatment Systems[a] [4].

Type of System	Flow Regime[b]	Detention Time, d Range	Typ.	Depth of Flow, ft (m) Range	Typ.	Loading Rate g/ft · d (cm/d) Range	Typ.
Trench (with reeds or rushes)	PF	6–15	10	1.0–1.5 (0.3–0.5)	1.3 (0.4)	0.8–2.0 (3.25–8.0)	1.0 (4.0)
Marsh (reeds rushes, others)	AF	8–20	10	0.5–2.0 (0.15–0.6)	0.75 0.25	0.2–2.0 (0.8–8.0)	0.6 (2.5)
Marsh-Pond							
1. Marsh	AF	4–12	6	0.5–2.0 (0.15–0.6)	0.75 (0.25)	0.3–3.8 (0.8–15.5)	1.0 (4.0)
2. Pond	AF	6–12	8	1.5–3.0 (0.5–1.0)	2.0 (0.6)	0.9 2.0 (4.2–18.0)	1.8 (7.5)
Lined Trench	PF	4–20 (hr.)	6 (hr.)	—	—	5–15 (20–60)	12 (50)

[a]Based on the application of primary or secondary effluent.
[b]PF = plug flow, AF = arbitrary flow.

where

Q_i = flow in, m³
Q_o = flow out, m³
A = wastewater additions, m³
P = precipitation, m³
E = evapotranspiration, m³
I = downward infiltration through the soil, m³
M = above ground water accumulation, m³

Surface water inflow, Q_i, may enter the wetland as streamflow, tidal flow, or runoff. Streamflow and tidal flow are easily measured. Runoff, the more important source of nutrients, cannot be measured with accuracy. However, methods are available for the prediction of runoff quantity and quality based on watershed area, precipitation, and land use. Surface water outflow, Q_o, may leave the wetland through drainage streams or as tidal volume, both easily measured.

Wastewater addition, A, is actually a component of the surface inflow. It is of significance because it can greatly alter both the water and the nutrient budget of a wetland. The contribution of wastewater to the water and nutrient budget can be estimated based on flow records and effluent monitoring records for an existing treatment plant or projected flows and effluent quality for a proposed facility.

Precipitation, P, is easily obtainable from meteorological records. A good source of data is from the National Oceanic and Atmospheric Administration (NOAA) which maintains a number of monitoring stations in every state.

Evapotranspiration, E, removes water from both land and water surfaces and can account for a substantial amount of water loss from wetlands, especially during the growing season. Loss due to evapotranspiration can range from 30 to 90 percent in a wetland and therefore can significantly affect water quality by increasing concentrations of pollutants. A number of predictive equations are available to estimate evapotranspiration based on solar radiation, wind, relative humidity, soil temperature, air temperature, and cover type. In addition, evapotranspiration rates are commonly given in the NOAA Climatological Data.

Infiltration/percolation, I, through the soil is highly dependent on soil type and structure. Infiltration can be estimated by examining soil cores or by using soil data from the Soil Conservation Service. It should be noted that this term can be either positive or negative. Commonly, surface water is lost by seepage or infiltration through the soil and contributes to groundwater recharge. However, a wetland site may be a groundwater discharge area in which the groundwater moves up toward the water table.

DEPTH

Surface water depth, h, can be related directly to the above groundwater accumulation, M, as long as M is positive. Where M is a negative term, no surface water exists and the substrate is no longer saturated.

$$h = (M/S\phi_s) \tag{6}$$

where

h = water depth, m
S = surface area, m²
ϕ_s = fraction of wetland volume available above ground for water storage, dimensionless

Typical water depths in a wetland may range from a few centimeters to about one meter. Spatial variations in depth within a specific area are common and primarily due to changes in the elevation of underlying soils. Changes in water depth associated with wastewater input will be limited by the tolerance of existing vegetation to flooding. Greater than usual depths may limit some wetland types for long term wastewater application because seeds will not germinate even though mature members of the same species survive. Additionally, greater than usual water depths limit diffusion of gases and metabolites. Many wetland communities are adapted to cyclic flooding—tidal wetlands on a daily cycle; riverine wetlands on a seasonal cycle. Matching wastewater application to natural flooding cycles may avoid stress created by maintaining consistent water depths.

RESIDENCE TIME
The period available for treatment within the wetland system is the residence time. The rate at which nutrients and other pollutants can be removed from the water column will dictate residence times that should be designed for. Wastewater can be introduced into the wetland treatment system by batch of continuous flow. In an enclosed area wastewater can be treated batchwise—introduced, held until desired effluent quality is achieved and released. Residence time for such a process can be easily defined. However, wetland systems are generally used as continuous flow systems, with residence time defined as:

$$\theta = \frac{\text{volume of surface water}}{\text{volumetric flow rate of surface water}} \quad (7)$$

The volumetric flow will vary from point to point in the wetland due to rain, runoff, evapotranspiration, and streamflow. If these changes are not extreme an average flow rate will suffice, otherwise the wetland must be divided into components. For a rectangular wetland with a discharge along one side,

$$\theta = \frac{\phi h w z}{Q} \quad (8)$$

where

ϕ = void fraction of wetland for water flow, experimentally estimated as $0.1-0.3$ for shallow meadows and 1.0 for deep, open water

h = surface water depth, m
w = width of flow path, m
z = distance from discharge point, m
Q = average surface water flow rate, m³/s

Some wetland types, such as bottomland hardwood forests, are adapted to infrequent flooding and tolerate standing water for short periods only. Patustrine wetlands, which are isolated from other water bodies, are more amenable to wastewater treatment than tidal, riverine, or lacustrine wetlands because palustrine residence times are naturally high due to hydraulic separation.

POLLUTANT LOADING
The wastewater loading rate to a wetland must consider application of individual pollutants as well as hydrologic loading. Loading rates for constituents of concern (e.g., BOD, nutrients, suspended solids, heavy metals) can be calculated from the liquid loading rate and the concentration of each particular constituent:

$$\text{Loading Rate} = QC_i \quad (9)$$

where

Q = average annual wastewater addition, m³/yr
C_i = average concentration of contaminant in wastewater, gm/m³

Permissible loading rates will depend on wetland type, vegetation, substrate, hydrologic factors, climate and season. During design the selected loading rate can be compared to those in Table 8 but should be evaluated, preferably in a pilot study, for the specific project.

SYSTEM CAPACITY
The addition of nutrients to a wetlands will cause a zone of increased vegetative growth (Figure 3). This zone will ex-

TABLE 8. Applied Pollutant Concentrations [4].

Pollutant	Treated Wastewaters, Range mg/l		
	Marshes	Hyacinth Ponds	Experimental
BOD₅	19–44	19	110–173
Suspended Solids	22–154	46	97–353
Total Kjeldahl Nitrogen—N	4–23	4	12–25
Ammonium Nitrogen—N	7–11	2	8
Total Phosphorus—P	0.4–11	15	4–7

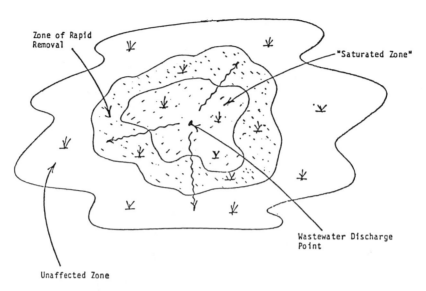

FIGURE 3. Schematic of the zone of affected soil and biomass.

pand with time until permanent capacity of the zone equals the loading rate or boundaries of the wetland are reached. The advance of this zone can be described mathematically as a mass balance equation for the expanding loaded zone:

addition rate = permanent removal rate + sorption rate + temporary binding rate + discharge rate

where

$$\text{addition rate} = QC_i$$

permanent removal rate = amount immobilized by soil building and development of woody material or lost to the atmosphere; mechanisms do not saturate and rate is proportional to area of zone

sorption rate = physical or chemical binding to soil substrate; found to reach equilibrium in short periods of time; may be zero for some constituents

temporary binding rate = storage in an expanding biomass compartment

The wetland area may be smaller than required for complete removal of constituents, particularly nutrients, to natural background levels. This may be found initially or after a period of years. In either case, sorption, biomass, and permanent removal mechanisms remain fixed at a level

insufficient to handle loadings. Additional required area can be determined using the rather complex mass transport and saturation models developed by Hammer [5].

DISCHARGE DURATION, FREQUENCY AND TIMING

In addition to hydrologic loading (the quantity of flow through the wetland), duration, frequency and timing of flow are important considerations. Treatment plant discharges can provide a relatively steady source of flow to the receiving wetland as compared to variable inputs from runoff and precipitation. The selection of a natural wetland or design of an artificial system should be made in light of flow regime.

Increased flow or flooding of the wetland should coincide with the growth requirements of the vegetation present. For example, permanently flooded wetlands such as cypress swamps, lacustrine wetlands, and artificial marsh/pond systems, are not dependent on wastewater for flow and can accommodate continuous or intermittent discharges. Nonpermanently flooded wetlands, on the other hand, are more sensitive to hydrologic change and may be disturbed by a continuous flow.

Small wastewater flows can be stored for seasonal application, either to take advantage of the growth phase of the vegetation or to alleviate poor treatment during certain times of year. In northern climates with freezing winters, removals of BOD and phosphorus substantially decrease during the cold season and in warm climates annual dormancy of vegetation produces similar effects. Physical removal processes (e.g., sedimentation, adsorption on substrate) are not significantly affected by temperature or season. Year-round operation can be designed for wetland systems relying on these processes. Biological removal processes can be substantially decreased in cold temperatures when bacterial metabolic rates drop thereby reducing BOD removal and plant species die back thereby reducing nutrient uptake. Where insufficient removals are found, seasonal storage of wastewater should be considered. Where seasonal dormancy is a problem, a diverse vegetative mix may minimize the effects.

SYSTEM CONFIGURATION

The manner in which wastewater flows are introduced and distributed through the wetland system is an important consideration. The objective is to maximize contact between wastewater and soils and vegetation and to avoid short-circuiting. This is generally accomplished by wide distribution and circuitous flow paths. In natural wetlands little can be done to change the flow regime without affecting the vegetative community. However, in artificial or reconstructed wetland systems, alterations in flow pattern through the system are possible. Wetland arrangements which enhance removals of nutrients and sediments are illustrated in Figure 4. For artificial wetlands, work at the Brookhaven National Laboratory indicates that water routed through a mix of wetland types (meadow-marsh-pond) provides greater nutrient removal than passage through a single wetland type. It has also been found that configuration was important to performance systems. Dye studies indicated that internal flow patterns in an open marsh resulted in short circuiting of flow, whereas marsh trenches were shown to have superior treatment efficiency. Based on these findings, a geometric configuration of a 20:1 length-to-width ratio was recommended.

Operational Requirements

The techniques used to manage the wetland treatment system will influence the design and operation of the system

Wetland with straight-through flow least desirable.

Island wetland meadows with increased trapping surface and shape intercepts flow and sediments.

Wetlands placed in meandering sequence. Flow forced to migrate around edges.

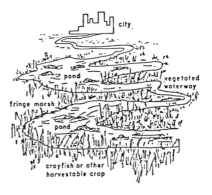

Wetland drainage channels with intermittent and terminal ponds.

FIGURE 4. *Alternative wetland arrangements designed for increased nutrient and sediment retention [4].*

and should be identified in the initial stages of design process. Management techniques are employed to improve performance, maintain environmental conditions, and utilize by-products of the treatment process. Common techniques include the following:

- Pretreatment—Type and degree of pretreatment depends on the target pollutants as well as public health considerations and odor control. If greater removals than provided by the wetland are required, some BOD and nutrient removal may be necessary before wetland discharge. Disinfection may be required to reduce pathogens and stabilization to alleviate odor problems. Low pretreatment levels have been implicated in clogging of the treatment system, difficulty in establishing vegetation in artificial systems, and poor treatment efficiency.
- Seasonal application—Wetland system removals may be related to season and temperatures; wastewater application must be timed accordingly and provisions made for necessary storage.
- Outfall regulation—This is necessary to control the residence time and depth within the system, factors which are critical to the control of nitrogen, seasonally released nutrients and toxic compounds.
- Flushing—Periodic flushing can be used to control the buildup of silt and the release of nutrients and other pollutants due to overloading.
- Varying the application point—This may achieve improved removals.

TABLE 9. Reported Removal Efficiency Ranges for the Constituents in Wastewater in Natural and Artificial Wetlands [6].

| | Removal Efficiency, % | | | |
| | Natural Wetlands | | Artificial Wetlands | |
Constituent	Primary	Secondary	Primary	Secondary
Total solids	40–75			
Dissolved solids	5–20			
Suspended solids	60–90			
BOD_5	70–96		50–90	
TOC	50–90			
COD	50–80		50–90	
Nitrogen (total as N)	40–90		30–98	
Phosphorus (total as P)	10–50		20–90	
Refractory organics				
Heavy metals[1]	20–100			
Pathogens				

[1]Removal efficiency varies with each metal.

- Dredging—Sediment accumulation is primarily related to litter fall from vegetation; the minor amount contributed by wastewater solids is not usually a problem. Channel dredging of cleaning would be an infrequent operation.
- Harvesting vegetation—This can be an important factor in maintaining removal capacity of wetland. Timing and extent of harvest is dependent on vegetation type.
- By-product utilization—Food and other valuable crops may be grown in wetlands enhanced by wastewater nutrients. Harvested biomass could be used as livestock feed, compost, soil amendment, or digested for biogas production. Where local markets are available, criteria favoring by-product utilization should be designed into the system.

Treatment Efficiency

Treatment efficiency and system reliability are key considerations in assessing the feasibility of a wetland treatment system. Currently, there are not sufficient long term data to enable generalized predictions on wetland system performance. Studies to date have been site specific and, as has been discussed previously, each wetland system is unique in its vegetation, substrate, flow regime, water chemistry and climatic conditions, making findings of one study difficult to apply to another system with any degree of accuracy. Nevertheless, there is sufficient data in the literature to indicate that wetland systems can accomplish removals of BOD, suspended solids, nutrients and other wastewater constituents which equal or exceed conventional treatment systems. Removal efficiency ranges for natural and artificial wetlands are shown in Table 9. It should be noted that the difference in concentration values measured at the inlet and outlet of the wetland may not accurately represent system performance. Infiltration, dilution of wastewater by rain and surface flow, or concentration due to evapotranspiration must be considered in the interpretation of system performance. The lack of data for some constituents, the relatively wide removal ranges exhibited for others, and the method used in determining efficiencies shown in Table 9 point to the necessity of establishing a pilot system prior to full scale design and operation. Furthermore, little is known concerning the long term ability of wetlands to remove nutrients; removal efficiency may decline greatly over long periods of time. Results of short term studies may be misleading.

An important design consideration is reliability, defined as consistent treatment without system failure. Reliability problems in wetlands systems are due primarily to environmental factors such as climatic conditions, changes in species composition, diseases affecting microorganisms or plants responsible for treatment. On the other hand, wetlands systems are less prone to processing upsets that occur in conventional treatment plants where a high degree of operator control is necessary.

FEASIBILITY OF WETLAND SYSTEM USE IN NEW JERSEY

When dealing with Innovative and Alternative wastewater treatment technologies, two basic problems must be addressed—development and implementation. The previous sections of this chapter have addressed development of the wetland treatment technology in general terms. After examination of the mechanisms present in a wetland system which contribute to wastewater treatment and data on removal rates and treatment efficiency, it can be concluded that wetland treatment systems offer a promising alternative to conventional advanced waste treatment (AWT) processes and may have applications for the treatment of less than secondary wastewaters. This section explores the possibility of and the development of a wetlands treatment system design to meet the needs of an existing New Jersey treatment plant currently experiencing problems.

Design Example

To illustrate the benefit of evaluating wetlands treatment as an alternative to conventional treatment and discharge, the following case study has been included.

In 1980 a facilities planning study was initiated to address the water pollution problems in Jefferson Township (Morris County), New Jersey. Currently septic systems serve the majority of the Township; however, one area is served by a municipal treatment plant, White Rock Lake Wastewater Treatment Plant (WWTP). The White Rock Lake WWTP is an extended aeration plant with a design flow of 129,500 gallons per day (gpd) and an existing average daily flow of 66,000 gpd. The WWTP discharges to a tributary to Mitt Pond which in turn is tributary to a series of downstream lakes along the Rockaway Riber. Because of downstream eutrophication problems, the WWTP was originally designed as a tertiary facility to provide phosphorus removal. Due to poor operation and maintenance practices, White Rock Lake WWTP provided marginal treatment, no phosphorus removal and was in consistent violation of its NPDES discharge permit. Influent and effluent characteristics compared to permit requirements are summarized below:

	Influent (mg/l)	Effluent (mg/l)	NPDES Permit Requirement (mg/l)
BOD_5	94	28	14.2[2]
SS	22.4	1	3.4[2]
Total P	15[1]	6.25	0.6[3]

[1]Estimated using average values in Table 2-1.
[2]The permit reads 30 mg/l (30 day average) for BOD_5 and SS or 85% removal, whichever is more stringent.
[3]NJDEP wasteload allocation at the time of study
(Source: Jefferson Township Health Department, 1982 [9]).

Under the above discharge conditions, at an average daily flow of 66,000 gpd, the White Rock Lake WWTP contributes the following loadings to its receiving waters:

BOD_5	15 lb/day
SS	7.7 lb/day
Total P	3.4 lb/day

The phosphorus loading has been implicated as the primary cause of massive seasonal duckweed blooms, dense rooted aquatic plants, and strong odor due to decaying plant material in Mitt Pond.

The Jefferson Township Facilities Plan (Elam and Popoff/ESEI [10]) identified the following alternatives for the White Rock Lake WWTP:

- Abandon the facility; divert flows to another treatment plant.
- Upgrade the facility using one of the following processes:
  —conventional activated sludge
  —extended aeration/oxidation ditch
  —rotating biological contactors

The selected alternative was the oxidation ditch process with continued discharge to the tributary to Mitt Pond. The cost of plant modifications alone is in excess of one million dollars, a cost which cannot be borne by the service area population. In addition, the proposed alternative does not address nutrient removal, known to be a factor critical to the downstream environment. For these reasons, the White Rock Lake WWTP seems to lend itself to a wetlands treatment alternative, which, as discussed in previous sections of this chapter, is attractive for two reasons: (1) lower cost and (2) effectiveness in nutrient removal. The following sections outline the development of a wetland treatment alternative for the White Rock Lake facility.

USE AND TREATMENT OBJECTIVES

The White Rock Lake WWTP is not accomplishing designed for secondary treatment levels with phosphorus removal; however, a substantial BOD_5 and suspended solids and some phosphorus removal is occurring. Rather than reconstruct the WWTP, at excessive costs and questionable increases in phosphorus removal efficiency (as no phosphorus removal process is included in the proposed alternative), a wetland treatment system is proposed into which sub-secondary, disinfected effluent produced by existing treatment units would be discharged. The primary objectives in utilizing a wetland treatment system for the White Rock Lake situation are:

- lower capital and operation and maintenance costs than the alternative proposed in the Facilities Plan
- to meet NPDES discharge permit requirements for BOD_5 and suspended solids
- maximize phosphorus removal to wasteloads allocation of 0.6 mg/l or below

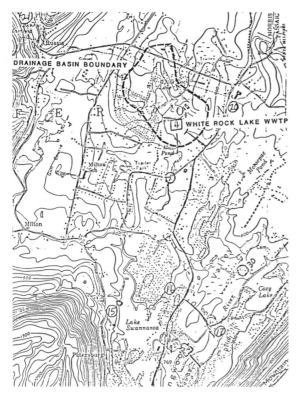

FIGURE 5. USGS topographic map, Franklin, NJ, Quadrangle, showing White Rock Lake WWTP site.

WETLAND SITE IDENTIFICATION

The existing White Rock Lake WWTP is located on the outskirts of a residential development and is surrounded by a considerable amount of vacant land (Figure 5).

It appears that approximately 20 acres of wetland are located immediately south and east of the WWTP. The wetland is classified as palustrine forested/scrub-shrub vegetation characterized by broad leaved deciduous vegetation (map symbol PFO/SS 1). Based on field observation, this designation was confirmed and the vegetation was identified as red maple (dominant species), alder, and vibernum, with cattails and reedgrass in the open previously disturbed area directly adjacent to the treatment plant site. The Soil Survey for Morris County (SCS, 1976) indicates that the substrate of this wetland is Adrian muck, a nearly level, poorly drained organic soil with a high water table at or near the surface year-round.

Because a natural wetland exists so close the WWTP, there is no reason to consider constructing an artificial wetland at this time. Should a pilot scale operation of the natural system indicate poor treatment or adverse effects upon the wetland, artificial wetlands and conventional treat-

ment units should be evaluated for cost and estimated treatment efficiency. The proximity of available vacant land would be advantageous for creation of an artificial system. At this point, however, it is assumed that natural wetland treatment is a viable alternative for White Rock Lake.

HYDROLOGIC LOADING

Using the water budget equation developed by Hammer and Kadlec [5] existing conditions (no wastewater input) and future conditions wastewater input at the rate of 66,000 gpd can be estimated.

From the topographic map it is apparent that the wetland may be hydraulically connected to the intermittent stream from White Rock Lake to Mitt Pond. No flow monitoring data are available for this intermittent tributary to the Rockaway River; however, the wetland is in the headwaters area, and assuming that any stream flow is a direct result of runoff events, flow calculations can be based on runoff within the drainage basin. The drainage basin for the wetland, shown as a dotted line in Figure 5, is approximately 186 acres (166 acres excluding the wetland).

Several simplifying assumptions can be made before applying the water budget equation to existing conditions:

- depth (M) = 0
  Because water table is at surface but there is no standing water, there is no depth (positive or negative); any seasonal flooding would not be accounted for in this assumption.
- inflow into wetland = outflow from wetland
- wastewater inflow (A) = 0

With these assumptions the water balance equation identified as Equation (5) becomes:

$$Q_i + P - E - I = Q_o$$

where

Q_i = inflow from runoff
  = ($P-E$) × basin area × runoff coefficient; basin area = 166 ac and average runoff coefficient for basins within Jefferson Township is 0.4 as specified in Township Ordinance 78-1
  = (3.8 − 1.9) ft/yr × 166 ac × 0.4 = 126.2 ac-ft/yr
  = 112,645 gpd

P = precipitation on wetland surface at an average rainfall of 46 in/yr
  = 3.8 ft/yr × 20 ac = 76 ac-ft/yr
  = 68,454 gpd

E = evapotranspiration from wetland which in this area of New Jersey is approximately 50% of precipitation; 23 in/yr
  = 1.9 ft/yr × 20 ac = 38.3 ac-ft/yr
  = 34,227 gpd

I = infiltration into wetland
  = $(P-E)$ × wetland area × $(1-\text{runoff coefficient})$
  average runoff coefficient for unimproved area with
  heavy soils and 2–7% slope is 0.2 (Chow, 1964)
  = 34,227 gpd × $(1 - 0.2)$
  = 27,382 gpd
Q_o = outflow from wetland; unknown

Substituting these values into the equation, existing outflow
can be determined:

$$Q_o = (112,645 + 68,454 - 34,227 - 27,382) \text{ gpd}$$
$$= 119,490 \text{ gpd}$$

DEPTH

Under future conditions where wastewater discharge
from the White Rock lake facility (A) is equal to 66,000 gpd
and assuming that outflow from the wetland will remain unchanged, the hydrologic budget can be used to predict
changes in water depth within the wetland.

$$M = Q_i - Q_o + A + P - E - I$$
$$= (112,645 - 119,490 + 66,000 + 68,454 - 34,227$$
$$- 27,382) \text{ gpd}$$
$$= 66,000 \text{ gpd over 20 ac} = 8824.2 \text{ cuft/d} - 871,200$$
$$\text{sqft}$$
$$= 0.01 \text{ ft/d} = 0.12 \text{ in/d} = 43.8 \text{ in/yr}$$

As demonstrated by the water budget, increasing the inflow to the wetland by adding a wastewater discharge
without a compensating increase in outflow will cause water
to accumulate within the wetland at a rate of 0.12 inches per
day or 43.8 inches per year. Increases in water depth will
significantly change the biological structure of the wetland.
The type and degree of injury caused by flooding differs
with species, soil type and flooding regime. Red maples, the
dominant species in the wetland, are generally considered
flood tolerant, able to withstand flooding for most of one
growing season. However, continuous flooding and increasing water depths which would result from the wastewater
discharge would cause the woody species which now characterize the site to give way to nonwoody emergent species
more tolerant to flooded conditions and, over the years as
the depth increases, the nonwoody emergents would give
way to floating vegetation.

To mitigate this impact, both the discharge volume and
the outflow volume can be regulated such that the change is
biologically insignificant and the discharge timing can be
optimized. In order to determine the critical depth of water
allowed, below which the red maples will not be significantly affected, several findings regarding tree response to
water level changes should be noted. While many species
can survive temporary soil saturation and partial inundation
(some even show increased growth rates), prolonged flood-

ing is usually injurious and there are few, if any, temperate
species that can survive an indefinite period of even partial
inundation. To survive indefinitely, even the most flood
tolerant species generally needs to be unflooded for at least
55–60 percent of the growing season.

Year-found root inundation can be tolerated in isolated
years. Periodic flooding is of little significance in the survival of bottomland trees such as red maples; flooding during the growing season is mainly responsible for death. Discharge timing, rather than a critical depth, appears to be the
most important factor in maintaining the character of the existing wooded wetland.

RESIDENCE TIME

As defined, residence time equals volume divided by flow
rate:

$$\theta = V/Q$$

The volume of accumulated water in the wetland is dependent upon the magnitude and duration of the wastewater discharge, assuming the flow rate out of the wetland (Q_o) remains unchanged from initial conditions (119,490 gpd).
Based on the volume delivered to the wetland in one day
(119,490 + 66,000) gpd, the residence time would be:

$$\theta = \frac{185,490 \text{ gal}}{119,490 \text{ gpd}}$$

$$= 1.5 \text{ d}$$

From Table 7 it appears that the 1.5 d residence time predicted above may not be adequate. Typical residence times
shown are on the order of 10 days. Then, using 10 days as
the desired residence time, the design volume can be predicted as follows:

$$V = \theta Q_o$$
$$= 10\text{d}(119,490 \text{ gpd})$$
$$= 1,194,900 \text{ gal}$$

By distributing this volume over the 20 acre wetland, the
desired depth can be determined:

$$M = \frac{(1.194,900 - 3.259 \times 10^5)\text{ac-ft}}{20 \text{ ac}}$$

$$= 0.18 \text{ ft or 2.2 in}$$

POLLUTANT LOADING

The pollutant loading to the wetland will be derived from
several sources, including the White Rock Lake WWTP discharge and runoff constituents. Input concentrations of
BOD_5, suspended solids, and total phosphorus in mg/l are

summarized below:

	WWTP (@ 66,000 gpd)	Runoff (@ 112,645 gpd)*
BOD$_5$	28	70
SS	14	3,400
Total P	6.25	1.3

*Source: USEPA, 1976.

From this data loading rate can be calculated using the equation:

$$\text{Loading} = \frac{Q_1C_1 + Q_2C_2}{\text{Total } Q}$$

BOD$_5$
$$\text{loading} = $$
$$\frac{(28 \text{ mg/l} \times 249{,}810 \text{ l}) + (70 \text{ mg/l} \times 426{,}361 \text{ l})}{702{,}080 \text{ l}}$$
$$= 52.2 \text{ mg/l}$$

SS
$$\text{loading} = $$
$$\frac{(14 \text{ mg/l} \times 249{,}810 \text{ l}) + (3400 \text{ mg/l} \times 426{,}361 \text{ l})}{702{,}080}$$
$$= 2070 \text{ mg/l}$$

Total P
$$\text{loading} = $$
$$\frac{(6.25 \text{ mg/l} \times 249{,}810 \text{ l}) + (1.3 \text{ mg/l} \times 426{,}361 \text{ l})}{702{,}080}$$
$$= 3.0 \text{ mg/l}$$

Comparison of these projected pollutant loadings with Tables 4 and 5 indicates that BOD$_5$ and total phosphorus loadings are within ranges reported in other studies, but the suspended solids loading is 5 to 90 times greater. Because runoff is contributing approximately 99 percent of the suspended solids loading intercepting runoff in a settling basin before discharge to the wetland should reduce the loading to a level within reported ranges.

DISCHARGE DURATION, TIMING, AND FREQUENCY

Since effluent discharge will affect the hydrology, it is desirable to determine whether the discharge should be managed to stabilize year-to-year cycles, to mimic such cycles, or to enhance the cycles. This determination is required for each hydrologic, biologic, and climatic season: snow melt, spring flood, summer moisture deficit, fall flood, snow and ice season; spring flora, spring growth, summer reproduction, and overwintering. The wooded character of the wetland chosen to receive the White Rock

Lake WWTP discharge may pose a problem with respect to wastewater treatment. Flooding (due to wastewater discharge in this case) during growing season may significantly affect tree health, yet maximum nutrient uptake occurs during the same season making this period the most desirable for wastewater discharge. This conflict must be resolved by testing the seasonality of treatment effectiveness by discharging effluent year round while monitoring several treatment parameters and wetland sensitivity (condition and health of component species). Results of this testing can be used to refine discharge duration, frequency and timing.

SYSTEM CAPACITY AND TREATMENT

The capacity of the wetland and its efficiency in treating the wastewater discharge from White Rock Lake WWTP are key factors in assessing whether or not the wetland treatment alternative is feasible. The following estimates of BOD$_5$, suspended solids and total phosphorus removals have been made for the White Rock Lake situation:

	Input to Wetland (mg/l) (WWTP + Runoff)	Removal Efficiency, %	Predicted Wetland Effluent (mg/l)
BOD$_5$	52.2	70–96	2.1–15.8
SS	21*	40–75	5.2–12.6
Total P	3.0	10–50	1.5–2.7

*Assumes most of SS in runoff is removed in settling basin.

The testing program prescribed in the previous section, wherein treatment parameters would be monitored during a year round study, will also be necessary to determine the treatment efficiency of the proposed system. As discussed previously, the necessity of a pilot system must be emphasized. The paucity of wetland data, lack of developed design criteria, and the uniqueness of each wetland system makes predictions on treatment performance, at best, guesses. A good monitoring program will confirm predictions and serve as a guide in refining design and operating parameters.

REFERENCES

1. USEPA, 1983, Technical Report, "The Effects of Wastewater Treatment Facilities on Wetlands in the Midwest," EPA 905/2/83-002.
2. Metcalf and Eddy, *Wastewater Engineering, Treatment, Disposal and Reuse*, 2nd ed., McGraw-Hill, NY (1979).
3. USEPA, 1977, "Process Design Manual—Wastewater Treatment Facilities for Sewered Communities," EPA 625/1-77-009.
4. Chan, E., T. Bursztynsky, N. Hantzsche and Y. Litwin, "The Use of Wetlands for Water Pollution Control," EPA 600/2/82-036.
5. Hammer, D. and R. Kadlec, "Wetland Utilization for Manage-

ment of Community Wastewater," Industrial Development Division, Institute of Science and Technology.

6. Tchobanlous, G. and G. Culp, "Wetland Systems for Wastewater Treatment: An Assessment," EPA 430/9/80-007.

7. Kadlec, R. and D. L. Tilton, "The Use of Freshwater Wetlands as a Tertiary Wastewater Treatment Alternative," CRC Critical Reviews in Environmental Control (1980).

8. Sutherland, J. C., "Wetland-Wastewater Economics, In Wetlands Treatment of Municipal Wastewater," *Proc. USFWS/ USEPA Workshop*, Univ. of Mass. (June 1982).

9. Jefferson Township Health Department, "Wastewater Quality Data for White Rock Lake, WWTP" (1982).

10. Elam and Popoff, "Wastewater Facilities Plan for Jefferson Township," Glen Rock, NJ (1984).

CHAPTER 28

Methane Recovery from Landfill: Practical Gas Processing—An Overview

NORMAN L. BENECKE*

An interesting article appeared in the *Wall Street Journal* on Feb. 9, 1987. New York—"This city has decided to prospect for energy in its mountains of garbage. And it hopes to mine a few royalties along the way. Plans call for tapping the decayed debris in landfills for methane gas—."

The captioned article was referring to the 3,000 acre Fresh Kills landfill in Staten Island, reputed to be one of the world's largest sanitary disposal sites. In a joint venture, the Getty Oil Company and the Brooklyn Union Gas Company planned and built a $20 million plant capable of producing 4×10^6 cubic feet of methane gas a day for 15 years. This quantity, the City of New York claims, will provide the needs of 10,000 homes in the borough of Richmond.

NATURAL GAS

Natural gas is primarily used as a fuel because of its very clean-burning characteristics. It does not dispense objectional pollutants into the atmosphere as products of combustion. It is also a major raw material for numerous chemical processes, notably as a feedstock for the production of ammonia, hydrogen and methanol, which are in turn the intermediate raw materials for a great many synthetic products.

The constituents of natural gas vary with geographic location. No one gas analysis can be considered typical. However, Table 1 gives the analysis of a sample Natural Gas taken from a gas field in Texas. This pipeline quality gas (consumer ready gas) illustrates two important parameters.

1. The principal component of the gas is methane. This simplest of hydrocarbons burns cleanly producing only

carbon dioxide and water vapor. Small fractions of other hydrocarbons are also listed in the analysis, Ethane through Hexane plus ($C_2 - C_6 +$). These components also contribute to the heating value of the gas, but, because the ratio of hydrogen atoms to carbon atoms in the molecular structure changes (becomes less) in the higher order paraffins, the possibility of carbon escaping as black smoke increases [1]. This can produce a particulate emission pollution problem in gas supplies containing a large fraction of these hydrocarbons.

2. The sour gas components in the analysis are very low. The term "sour gas" is sometimes called "acid gas" and is generally applied to compounds which form an acid if dissolved in water. In process technology, the gases of Carbon dioxide—CO_2, Hydrogen sulfide—H_2S, Carbon oxysulfide—COS, Nitrogen oxides—NO_x and Mercaptans—CH_3SH are sour gases. In this example, the troublesome sour gas component, Hydrogen sulfide, is 0.0 mole %.

BIOGAS

The natural process of anaerobic microbiological respiration of municipal wastes in landfills produces biogas, a mixture of methane and carbon dioxide and small amounts of other gases.

Respiration refers to the biochemical processes in which microorganisms oxidize organic matter and extract the stored energy needed for growth and reproduction. Respiration methods may be subdivided into two major groups, aerobic and anaerobic, depending upon the nature of the ultimate electron acceptor [2]. There are various pathways for the degradation of organic substrates, providing for the metabolism of sugars and the interconversions between other organic materials, but the convenient definition is the degradation of glucose. The breakdown of glucose via the

---

*Lummus Crest Inc., a subsidiary of Combustion Engineering Inc., Bloomfield, NJ

glycolytic pathway which yields 2 moles each of pyruvate, adenosine triphosphate (ATP) and nicotinamide adenine dinucleotide (NAD) per mole of glucose (See Appendix A).

The first major group reacts under aerobic conditions. The pyruvate is oxidized to CO_2 and H_2O via the tricarboxylic acid or Krebs cycle and the electron transport system. About 620 kcal per mole glucose are liberated, much of its as stored ATP.

The second major group, and the one of interest in biogas production, is under anaerobic conditions. Various pathways exist for pyruvate metabolism which reoxidize the reduced hydrogen carriers formed during glycloysis. The ultimate acceptor builds up as a waste product in the culture medium. The end products are 1) CO_2 and ethanol (alcohol production), 2) CO_2 and CH_4, 3) CO_2 and acetate, 4) CO_2 and 2,3-butylene glycol, and 5) other end products including ATP, acetone, butanol, succinate and lactate. The pathway that produces the end product depends on the microorganisms and the culture conditions.

The anaerobic biochemical respiration process is very inefficient in terms of energy liberation. The end products are reduced and possess high heats of combustion. It is this heat value of these end products for fuels which makes the anaerobic oxidation of organic substances attractive. Some examples are shown in Table 2.

The biogas generated at the Fresh Kills landfill has the analysis given in Table 3. This analysis will change as the site ages and matures, and may not be the exact assay at any

TABLE 1. Analysis of a Sample of Natural Gas.*

Component	Mole %
Methane	76.2
Ethane	6.4
Propane	3.8
n-Butane	1.3
Isobutane	0.8
n-Pentane	0.3
Isopentane	0.3
Cyclopentane	0.1
Hexane + hydrocarbons	0.3
Nitrogen	9.8
Oxygen	Trace
Argon	Trace
Hydrogen	0.0
Hydrogen sulfide	0.0
Carbon dioxide	0.2
Helium	0.45

*Prepared by Henry E. Duckham, Jr., The M. W. Kellogg Company, a Division of Pullman Incorporated, Houston—for the *Chemical and Process Technology Encyclopedia*, D. M. Considine, Editor-in-Chief, McGraw-Hill, New York, 1974.

TABLE 2. Gas Combustion Constants.*

Products	Heat of Combustion (BTU/cu ft)
CO_2	0
CH_4	913
C_2H_3OH	1450

*Data taken from *Chemical Engineers Handbook*, John J. Perry, Editor, 2nd Edition, McGraw-Hill, New York, 1941 (Table 29, page 2378).

point in time, but this analysis was the result of the best composite determinations made over a period of time. The estimated heat value of the design gas analysis is 450 BTU per cu ft.

LANDFILL METHANE TECHNOLOGY

The natural production of methane within the landfill requires the harvesting and treating of the gas. Existing methane-from-landfill recovery projects require five steps;

TABLE 3. Design Charge Feed Analysis.*

Component	Mole %
Methane	55.6410
Oxygen	0.0100
Nitrogen	0.3500
Ethane	0.0
Propane	0.0
Pentane	0.0001
Hexane	0.0004
Heptane	0.0105
Octane	0.0178
Nonane	0.0675
Decane	0.0322
Unedecane	0.0031
Benzene	0.0182
Toluene	0.0869
Xylene	0.1338
Methylene Chloride	0.0047
Carbonyl Sulfide	0.0077
Sulfur Dioxide	0.0035
Trichloroethylene	0.074
Perchloroethylene	0.0010
Hydrogen Sulfide	0.0020
Hexanol	0.400
Carbon Dioxide	43.2022
H_2O content: Saturated at 110°F & 14.7 psia	

*Prepared by George Sauer, Project Manager for Getty Synthetic Fuels, Inc., for this report, October 1984.

Collection, Inlet Compression, Pretreatment, Acid Gas Removal and Delivery Compression.

Collection

The collection process involves drilling wells at various surface locations to various depths within the landfill. The wells are fitted with laterals and header pipes to transport the gas from the wells to a treatment plant. It is important to draw the gas out of the landfill at a rate which does not pull air into the landfill [4]. Production of methane will decline if air is pulled in because methanogenic bacteria poisoned by oxygen.

At the Fresh Kills landfill, 100 wells drilled to a depth of 60 to 75 feet over a 400 acre portion of the site withdraw the gas. The gas is then transported via a piping system to the Feed Gas Compressor. The pressure at the compressor suction nozzle is 12.2 psia.

Inlet Compression

The main plant compressor creates the vacuum causing the gas in the landfill to flow to the wells and provides the pressure needed by the treatment process to make the gas suitable as a fuel.

At the Fresh Kills site, raw gas from the landfill is mixed with the overhead vapor from the Stripper column and sent, under slight vacuum, to the Feed Gas Receiver for liquid separation and removal (See Figure 1, Process Flow Scheme). The vapor from this drum passes to the Feed Gas Compressor where the gas is compressed by two stages to about 150 psia. Air coolers are used for interstage cooling. Condensate from the Compressor First Stage K.O. (knockout) Drum is sent to the Feed Gas Receiver. The liquid from the Feed Gas Receiver, together with the condensate from the Second Stage K.O. Drum and Feed Gas K.O. Drum is pumped back to the landfill. The vapor is compressed in the third stage of the Feed Gas Compressor to 480 psia.

Pretreatment

Gas from landfills contains between 45 percent and 60 percent methane and 30 percent 50 percent carbon dioxide plus small amounts of various contaminants. To make the gas suitable for CO_2 removal processes, the gas must be pretreated to remove the contaminants. Removal of contaminants without significant corrosion of the pretreatment equipment is a problem. Corrosion at the Palos Verdes site (Los Angeles County) was so severe that 1/4 inch thick carbon steel valves were down to wafer thickness in less than two weeks [5]. Gas from landfills contain 0.5 to 1 mole % trace contaminants, not readily identified because of their small concentrations. The pretreatment equipment can remove up to 50 contaminants which builds up and concen-

trates enough to hydrolyze and decompose to sulfurous and hydrochloric acids. Materials of Incoloy and Inconel alloys have been used, but it is an area where ongoing research is very active.

At the Fresh Kills site the pretreater uses the solvent of the acid gas wash. This unit operations is quite extensive and complex, but a brief description is as follows.

Hot gas from the compressor third stage discharge is cooled to about 85° by multiple heat exchangers. The cooled gas then passes through the Feed Gas K.O. (knockout) Drum where condensate is withdrawn and returned to the landfill. The vapor is then mixed with liquid solvent from the acid gas wash and chilled to 10°F. The condensate formed in the Phase Separator drum contains the heavy hydrocarbons and contaminants. In the Fresh Kills Plant, the condensate is sent to a solvent recovery system where the hydrocarbons, contaminants and water are separated from the solvent liquid and returned to the landfill while the solvent is filtered and reused. (Note: The Solvent Recovery System is a proprietary Selexol recovery process of Getty Synthetic Fuels Inc. and is considered confidential information. For a discussion of other regeneration and recovery processess, see Appendix B.)

Carbon Dioxide Removal

Removal of CO_2 is usually accomplished by wash processes, especially at the higher concentrations of the acid gas. Adsorption systems such as molecular sieves have been used for CO_2 removal. The original design at the Palos Verde site used three pairs of towers packed with molecular sieves. They operated on a pressure-swing adsorption principle—one pair on stream, one pair vented, and the third pair being regenerated [5].

A large number of acid gas removal processes are available. The original solvent wash process was developed at the U.S. Bureau of Mines by Benson, Field and coworkers in the 1950s [6]. This process is based on the use of hot potassium carbonate (HPC) solutions to scrub CO_2 and other acid gases from a gas stream. Innovations have been incorporated into the process, such as activating chemicals and mechanical vapor compression, to remain efficient and competitive with newly developed organic solvent washes.

Three of the more successful physical wash systems using organic solvents are the Rectisol process, the Selexol wash and the Sepasolv wash. In Rectisol washes, methanol is usually used as a solvent; however the Rectisol wash is not limited to methanol. In the Selexol wash and the Sepasolv wash, ethers of polyethyleneglycol are used as the solvent [7].

A comparison of different types of washes is difficult. There is interaction with process operations upstream and downstream of the wash cycle and any differences have to be included in the comparison. Utilities and chemical con-

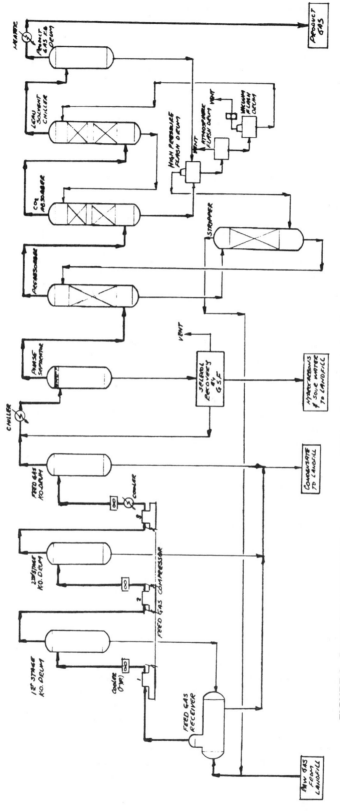

FIGURE 1. Process flow scheme methane gas recovery GSF Inc., Staten Island, New York City. Drawn by N. L. Benecke, Nov. 24, 1984.

sumptions must be considered. These include 1) steam consumption, 2) refrigeration requirements, 3) electric power consumption, and 4) solvent losses (solvent costs).

At the Fresh Kills site, acid gas removal is affected by physical wash with Selexol organic solvent. This chemical is specified and supplied by Allied Chemical Company.

Selexol is a copyright by the Norton Chemical Company (1983). The process is based on the capability of the dimethyl ether of polyethyleneglycol to physically absorb gases of CO_2, H_2S, COS and mercaptans from natural or synthesis gas streams. The Selexol solvent is manufactured under license.

The Selexol process uses little or no heat for solvent regeneration and is therefore more energy efficient (See Appendix B).

Sour gas compounds are removed from the feed gas in a counter-current absorption column. The process needs pressure; the higher the better [8]. At Fresh Kills, the feed gas is sent to the CO_2 Absorber at about 457 psia where it is counter currently contacted with chilled lean Selexol solvent. The overhead from this column passes through the Solvent Chiller Tower to the Product Gas K.O. Drum where any entrained liquid is removed and returned to the process. The vapor is heated by a final heat exchanger and delivered as product gas.

Delivery Compression

In order to deliver to a gas company pipeline, the methane gas must be compressed to 180 to 200 psia minimum. If the pressure from the gas cleanup system is insufficient, mechanical compressors will need to boost the pressure. However, with most physical wash systems, the gas pressure is sufficiently high at the inlet compression step.

The product gas from the Fresh Kills site is delivered to the pipeline at 400 psig minimum at a temperature of 100° maximum.

The Fresh Kills system processes 10 MM SCFD of raw, wet landfill gas and produces pipeline quality gas with the following characteristics:

Heating Value, BTU/SCF	960 min. HHV
CO_2, mole %	3.5 max.
Water Content, lbs/MM SCF	7.0 max.
H_2S Content, grains/100 SCF	0.25 max.
Oxygen, mole %	0.5 max.
Oxygen & Nitrogen, mole %	1.0 max.

Advantages

One of the great advantages of biogas from landfills is the availability of an existing distribution and consumption system. The most common arrangement is independent ownership of the operations. Gas is sold under contract to a local utility and the owner of the landfill, usually a municipality or county, receives a royalty. Sales to the utility is the market outlet for the gas, where the utility sells the gas as part of its system supply.

Producing gas from landfills would provide the United States with a renewable domestic energy source. Because gas is interchangeable with fuel oil for many purposes, each equivalent of biogas produced could replace a barrel of imported oil. The economic value of gas from landfills can then be compared to the cost of imported oil. As long as the cost of production and distribution is less than the cost of importing oil, the production of biogas from landfill projects is attractive.

In addition to the economic advantages listed above, additional environmental benefits can be achieved by proper siting, design and operation of a facility. However, even where development could be economical, some barrier policies may prevent full development. For example, lower gas prices from deregulation in a free market can reduce margin and slow development; competition from coal-gasification gas producers can also reduce margin; air quality standards may be a problem with State air quality controls because of NO_x emissions from landfills (with or without a biogas recovery system); and finally, landfill zoning may require variances to permit operation of a recovery system.

ACKNOWLEDGEMENT

Mr. George Sauer, Project Manager for Getty Synthetic Fuels, Inc., Fresh Kills Landfill Gas Project, provided the detail information and data during an interview on October 22, 1984.

REFERENCES

1. Reed, R. D., *Furnace Operations*, Gulf Publishing Company, p. 19 (1973).
2. Gaudy, A. F., Jr. and E. T. Gaudy, *Microbiology for Environmental Scientists and Engineers*, McGraw-Hill Book Company, Chapter 7 (1980).
3. Perry, J. J. (Ed.), *Chemical Engineers' Handbook*, 2nd Edition, McGraw-Hill Book Company (1941).
4. Emcon Associates, *Methane Generation & Recovery From Landfills*, for Consolidated Concrete Limited & Alberta Environment, Ann Arbor Science, p. 40 (1980).
5. *Chemical and Engineering News*, p. 25 (August 28, 1978).
6. Benson, H. E. and J. H. Field, "Method of Separating CO_2 and H_2S from Gas Mixtures," U.S. Patent 2,886,405 (May 1959).
7. *Chemical Engineering Progress*, p. 30 (October 1984).
8. "Selexol Solvent Process," bulletin R-01-09 (40C), Norton Co. Chemical Process Products.

APPENDIX A

Discussion of Metabolic Mechanisms

Metabolism is a term that includes all the diverse reactions by which a cell processes food materials to obtain energy and the compounds needed to produce new cell components. The degradative reaction sequences that yield energy and material for biosynthesis are collectively called catabolism. The biosynthetic reactions are known as anabolism.

Catabolism involves the conversion of a variety of substrates belonging to various classes of compounds into intermediate products that can enter into several catabolic pathways. These pathways degrade organic compounds to carbon dioxide under aerobic conditions.

Anabolism involves the use of various intermediates as starting points for a variety of biosynthetic pathways that

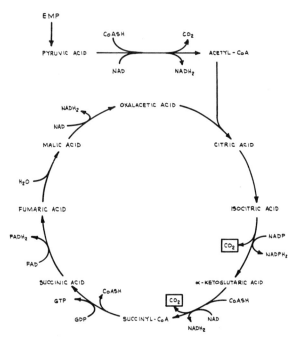

FIGURE A-2. The tricarboxylic acid (TCA) cycle for terminal oxidation to CO_2 in aerobic cells. From Gaudy, A. F. and Gaudy, E. T., *Microbiology for Environmental Scientists and Engineers*, McGraw-Hill, Inc., New York, 1980, Figure 9-3, page 396.

produce the many compounds required for the manufacture of new cells.

All living cells are composed of the same types of materials. These materials include proteins, lipids, carbohydrates, RNA (Ribonucleic acid) and DNA (Deoxyribonucleic acid). All living cells require the elements in approximately the same proportions and all heterotrophs, organisms that require carbon as source material, share some of the common anabolic and catabolic reactions.

Microorganisms are capable of degrading many compounds that are not metabolized by higher organisms. For example, hydrocarbons are metabolized by various species of bacteria and fungi, but only very slowly or not at all by animal tissue. Therefore, microorganisms display a greater capability in breaking down or preparing compounds for entry into the central pathway using the same pathway as do the higher organisms. It is this versatility that makes microorganisms useful in the biological treatment of many domestic and industrial wastes.

Three principal pathways are employed by most heterotrophic organisms to provide energy and the initial materials for biosynthesis of the pathway: 1) The Embden-Meyerhoff-Parnas (EMP) pathway; 2) the cyclic pathway for terminal aerobic oxidation or tricarboxylic acid (TCA) pathway, also

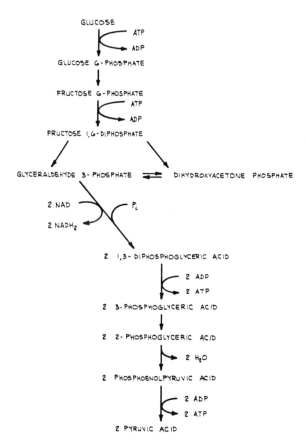

FIGURE A-1. Embden-Meyerhof-Parnas (EMP) pathway for catabolism of glucose. From Gaudy, A. F. and Gaudy, E. T., *Microbiology for Environmental Scientists and Engineers*, McGraw- Hill, Inc., New York, 1980, Figure 9-2, page 388.

known as the Krebs cycle; and 3) the hexose monophosphate (HMP) pathway.

EMP PATHWAY

The EMP (glycolytic) pathway is the major route for catabolism of carbohydrates and related compounds (See Figure A-1). In this pathway, a molecule of glucose is converted into two molecules of pyruvic acid through the reactions and the process is utilized equally well by anaerobic organisms. However, the further metabolism of pyruvic acid differs greatly in aerobic or anaerobic cells.

TCA CYCLE

The TCA cycle is a cyclic series of reactions that accomplish complete aerobic oxidation of the two-carbon acetic acid. This is fed into the cycle as acetyl-CoA in accordance with the following:

where:

NAD is Nicotinamide adenine dinucleotide
$NADH_2$ is Reduce nicotinamide dinucleotide

Here the pyruvate undergoes an oxidative decarboxylation catabolyzed by the pyruvate dehydrogenase. Other hydrocarbons also form acetyl-CoA. Therefore, the TCA cycle serves a vital function in recycling organic carbon. The overall cycle is shown in Figure A-2.

HMP PATHWAY

The hexose monophosphate pathway is rather complex, involving several splitting and condensation reactions (See Figure A-3). It is not necessary that the glucose entering the pathway proceed through the entire sequence of reactions. There are alternative routes utilizing various combinations of enzymes that can be used to form the biosynthetic processors and $NADPH_2$. The pathway can operate in a cyclic pattern producing glucose 6-phosphate, or the fructose 6-phosphate from which it is formed, which could enter the EMP pathway when $NADPH_2$ is not needed. Therefore, in cells that have both pathways, which is usually the case, the fraction of glucose that is metabolized through the HMP pathway is established by the momentary needs of the cell.

FORMATION OF METHANE

The two central catabolic pathways, EMP and HMP, lead to the formation of pyruvate which is the pivotal compound in catabolism. Two other pathways, the Entner-Douderoff (ED) and the heterolactic fermentation process lead to the formation of glyceraldehyde 3-phosphate. These pathways

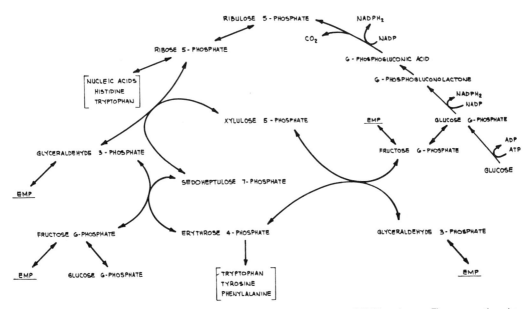

FIGURE A-3. The nonoxidative segment of the hexose monophosphate (HMP) pathway. The connections between the HMP and EMP (Embden-Meyerhof-Parnas) pathways and the reversibility of the reactions allow carbon to flow between the pathways or into the biosynthetic systems shown in brackets. From Gaudy, A. F. and Gaudy, E. T., *Microbiology for Environmental Scientists and Engineers*, McGraw-Hill, Inc., New York, 1980, composite of Figures 9-4 and 9-5, pages 400 and 401.

use the common reaction sequence of the lower EMP to form the pyruvate, although in the ED pathway, one molecule of pyruvate is formed directly from the hexose 2-keto-3-deoxy-6-phosphogluconate.

In anaerobic metabolism, pyruvate is decarboxylated without oxidation to form acetaldehyde and this is the hydrogen acceptor for the reoxidation of the $NADH_2$ formed in the pathway to pyruvate. The reactions are:

The reaction is balanced when glucose (or another hexose) is converted to two molecules each of CO_2 and ethanol. Acetaldehyde and ethanol are the major fermentation products.

Depending upon the strain of organisms and the environmental conditions of the culture, (for example—if a hydrogen acceptor is not needed and the organism can conserve the bond energy of acetylphosphate), many different products can be formed, including butanediol, acetic acid, formic acid, acetone and isopropanol, butyric acid and butanol, and succinic acid. Most of these product formations are active in a pathway and can be withdrawn from a medium.

Studies of methane formation have shown that the range of substrates used by methanogenic bacteria is very limited. The Methanobacterium MOH cannot metabolize ethanol but is able to use hydrogen as its energy source and electron donor and carbon dioxide as its carbon source and electron acceptor forming methane. Other bacteria are able to use formic acid (HCOOH) and some species of Methanosarcina are able to form methane from methanol or acetic acid. If hydrogen is available, CO_2 is reduced by:

$$CO_2 + 4H_2 \rightarrow CH_4 + 2H_2O$$

The removal of hydrogen by the methane bacteria pulls the reaction forming hydrogen. Therefore fatty acids and other compounds that are metabolized to acetyl-CoA in aerobic systems can be fermented to acetate and the hydrogen needed for the formation of methane. In the absence of the hydrogen using methanogens, the hydrogen would accumulate in sufficient concentrations to be toxic or to prevent further formation due to product accumulation.

References

McKinney, R. E., *Microbiology for Sanitary Engineers*, McGraw-Hill Book Company, Chapter 8 (1962).

Gaudy, A. F., Jr. and E. T. Gaudy, *Microbiology for Environmental Scientist and Engineers*, McGraw-Hill Book Company, Chapters 7, 9, and 11 (1980).

APPENDIX B

The Selexol Solvent Process

Using SELEXOL solvent as a physical absorbent offers several advantages over older processes based on chemical reaction between the gas and solvent. The SELEXOL process uses little or no heat for solvent regeneration and is inherently more energy efficient. It also provides selectivity between H_2S and CO_2 to produce a concentrated acid-gas stream. This results in a less expensive and more efficient plant.

ABSORPTION

Sour gas compounds are removed from the feed gas in a counter-current absorption column. Most SELEXOL process designs involve recycle of gases from a high pressure flash stage.

The process needs high pressure. The feed gas in most cases is compressed to 300–500 psia or more. Much of the energy consumed here is offset by reduced compression for the final product gas.

SOLVENT REGENERATION

Various processes can be used to regenerate the solvent. Selection depends on the application and required gas purity. Downstream operations can include one or more of the following:

1. High pressure flash. Less soluble components coabsorbed with the acid gases are flashed off. These gases are compressed and recycled to the absorber to minimize methane loss and improve process selectivity.
2. Low pressure flash. This removes most of the absorbed acid gas. In many cases, solvent becomes lean enough for recycle to the absorber without stripping. Several stages of low pressure flash can be cascaded and a modest vacuum used in the last stage to yield an even leaner solvent.
3. Stripping. When the product gas purity requires an especially lean solvent, this is achieved by a stripping column. The stripping gas is either air, inert gas, one of the flash gases, or steam.

ENERGY RECOVERY

The SELEXOL process becomes more efficient as temperatures in the absorber are lowered. The heat associated with absorbing the acid gas can be compensated by heat exchange with the self-refrigeration produced when these dissolved components are vaporized during flashing.

Reference

"Selexol Solvent Process," bulletin R-01-09 (40C), Norton Co., Chemical Process Products.

CHAPTER 29

The Flare as a Safe Pollution Abatement Device

NORMAN L. BENECKE* AND PAUL N. CHEREMISINOFF**

INTRODUCTION

The primary function of a flare is to safely dispose to atmosphere any relieved and unwanted obnoxious, toxic or flammable vapors or gases from a chemical process by combustion, thereby ensuring the safety of personnel together with the protection of the plant and equipment.

Relieved vapors and gases may be continuous or intermittent, frequently or rarely, and can range from controlled venting to an instantaneous voluminous dump, a major discharge caused by an emergency or total plant shutdown.

Important considerations must be given to the effect of the flare on the surrounding community. These considerations include health, safety, and aesthetics. Governmental regulations often dictate the design requirements of the flare, to control or limit the generation of pollutants including smoke, specific gas concentrations, heat, light, noise, and odor.

TYPES OF FLARES

There are three basic types of flares available for disposal of waste gases:

1. Elevated fires
2. Ground flares
3. Burning pits

Elevated flares are the most common type used. They are found installed in small chemical or petrochemical plants to large refineries or on off-shore oil drilling platforms. This type of flare produces light and smoke, although smokeless

combustion may be achieved by the injection of steam or air into the combustion zone to provide turbulence and air inspiration. In addition to the production of light and smoke, the flare may discharge uncombusted material or unwanted combustion products into the atmosphere.

Ground flares are open box structures built on a slab or pad on grade. The structure is fitted with a floor of piping manifolds. Air is drawn into the combustion zone by natural draft and the whole structure is lined with refractory insulation. This box is surrounded by a high wind break to shield against the emission of light at ground level. The main disadvantages of this type of flare are very high capital costs and the high probability of atmospheric pollution caused by the relatively low level of gas discharge. Therefore this type of flare is normally installed for aesthetic reasons, to handle the small minor continuous or intermittent discharges.

Burning pits are the cheapest form of a gas (and liquid) combustor. This type of flare consists of a depressed or below grade rectangular basin fitted with gas manifolding and ignitor system. Flames are propagated horizontally. Burning pits can handle extremely large flows of waste gases and, with proper nozzles, flammable liquids. Burning pits are installed in areas where atmospheric pollution is considered unimportant and where real estate is plentiful in fairly remote locations. Although burn pits were at one time used in the United States, especially in the south-western oil fields, I know of no active burn pit in the U.S. at this time. There are installations in the world where this type of flare is in use, such as GOSP (Gas-Oil Separation Plants) installations in the Middle East.

Since the Ground Flare and Burning Pit Flares[1] are used

---

*Lummus Crest Inc., a subsidiary of Combustion Engineering Inc., Bloomfield, NJ

**Department of Civil and Environmental Engineering, New Jersey Institute of Technology, Newark, NJ

[1]Authors' Note: The definition of burning pits does not include the concept of pit flares where low level multiple oxidizers are installed in a dry basin with staged burning for high gas exit velocity and air entrainment. These are smokeless devices spread over a large area, usually in remote regions. The flare pit may be fitted with an acoustic wall to reduce noise levels and visible light.

relatively infrequently compared to the Elevated Flare, we will confine the following discussions to the design, operating, and polluting aspects of the Elevated Flare.

Another combustion appliance which is increasing in popularity as a waste gas destruction device is the incinerator, or specifically, the vapor incinerator. This is a more sophisticated contrivance capable of controlling the three T's: Temperature, Turbulence, and Time (residence). Although modern incinerators can be designed for a large turn-down, they are not practical to accept high voluminous reliefs from vented discharges. Incinerators will not be considered in this discussion.

ELEVATED FLARE

The design of an elevated flare involves the selection of diameter and height appropriate to satisfying the requirements of the relieving gas volume and the radiant heat intensity produced by the gas. In the past decade, another essential criterion has been included in the flare design process: the effect of ground level concentration (GLC) of pollutants due to atmospheric dispersion of combusted and uncombusted discharges.

The selection of the flare stack diameter involves assessing many factors. Diameter will establish the velocity of the gases. High discharge velocities offer some advantages in stack design, create other problems which limit stack diameters. Some of the advantages and disadvantages of high exit velocity are listed as follows:

Advantages:

1. At higher velocities, the physical dimension of the stack diameter will be reduced, thereby affecting a direct capital equipment cost saving.
2. High exit velocities tend to promote complete combustion by mixing with air at the stack tip.
3. The higher the velocity, the higher the resultant plume rise. Therefore the lower the ground level concentration (GLC) downwind of the stack of possible emissions.
4. The greater the velocity, the higher the flame will be projected vertically. This will reduce possible flame damage to the tip of the flare and diminish the heat radiation intensity at ground level.

Disadvantages:

1. At higher velocities, the gas pressure drop in the system increases with associated increased back pressure on the relieving devices. This will increase the collection system sizing requirements affecting a direct increase in equipment costs.
2. High velocities tend to promote liquid droplet entrainment and could possibly lead to the formation of burning liquid carry over or "flaming rain."
3. High exit velocities could produce the phenomenon of "blow-off" from the tip. The fire lifts from the tip and ex-

tinguishes the flame. This could cause delayed reignition with a resultant explosion.
4. Higher exit speeds will increase noise levels.

A second criterion for establishing the geometry of the elevated flare is the radiant heat produced by combustion. Flaring situations are usually brought about by over-pressure of process vessels initiated by electrical or cooling water failure. This produces a sudden and rapid release of the gaseous contents of the process system which must be conducted through piping to the flare for safe disposal. The hugh fire produced requires a tall stack to satisfy the heat intensity considerations at ground level.

In calculating the stack height for the flaring of gases, the prime consideration is the welfare of personnel likely to be exposed to high levels of heat radiation. There is also the risk of damage to the equipment, although equipment can usually tolerate much higher heat levels without permanent damage.

The intensity of heat radiation at any distance from the epicenter of the flame is calculated by the inverse square, assuming equal radiation in a sphere from the center of the flame:

$$I = \frac{QE}{4\pi R^2}$$

where I is the intensity, BTU/sq.ft.hr. R is the distance from the point source of the heat to the point of the receiver in feet. Q is the heat release. For most hydrocarbons, this may be estimated at a calorific value of about 21,000 BTU/lb (see Table 1). E is the fraction of total heat appearing as radia-

TABLE 1. Emissivity and Heat Values for Some Common Gases and Vapors [5].

	Emissivity Value	Heating Value BTU/LB
Carbon Monoxide	0.075	4,350
Hydrogen	0.075	61,000
Hydrogen Sulfide	0.070	7,080
Ammonia*	0.070	—
Methane	0.100	23,900
Propane	0.110	21,570
Butane	0.120	21,230
Ethylene	0.125	21,450
Propylene	0.135	21,020
Maximum E is 0.135		

*Ammonia flaring is difficult because of the presence of bound nitrogen which has a quenching effect on the flare flame. Ammonia has a heating value of 365 BTU/ft_3 but it requires assist gas to increase its heat content and insure complete combustion. NOx production is also a problem with ammonia combustion. Assist gas quantities must be selected to discourage NOx formation and by initial fuel premixing and attempting to maintain low oxidation temperatures.

TABLE 2. Emissivity Tests—Flaregas Prototype FS-24.

Test No.	Propylene #/Hr.	Steam #/Hr.	ST/HC	E(Average)
1	45,600	17,900	0.39	0.10
2	36,200	16,700	0.46	0.078
3	25,500	12,500	0.49	0.074
4	16,800	7,150	0.42	0.106
5	10,300	3,270	0.32	0.115
6	10,300	—	—	0.135
7	10,300	15,000	1.45	0.067

Source: ICI Wilton, Flaregas Bulletin viii/73A, 1973.

tion. Theoretical values indicated levels of about 35%, however practical values are generally much lower due to the following:

1. Apparent emissivity decreases with distance from the flame epicenter. Some of the heat is absorbed by the gases of combustion and some of the heat is absorbed by the air layers between the flame and the point of measurements.

2. The injection of steam (or air) to produce turbulence and reduce smoking reduces the emissivity considerably. This is illustrated in Table 2. The values of E decrease as the steam/hydrocarbon ratio increases. (Note: The actual base ratio of steam to hydrocarbon depends on the gas composition or specifically on the H/C weight ratio.) Also, note run No. 6. Without steam, the apparent emissivity does not exceed 15% of the total heat release. This is in keeping with the other data [5].

A third criterion that must be considered in the design of a stack in a flaring system is the discharge of pollutants contained in toxic or noxious vapors released by the flare effluents.

ENVIRONMENTAL CONCERNS

Within the past several years, the U.S. Federal Government has promulgated stringent regulations for pollution control. These rules are the result of increasing public awareness and concerns about the environment. Today, industry, with the support and participation of the general public, must comply with the many different pollution control laws.

Two pieces of legislation for air pollution are important and influence the design and operation of process equipment. The two laws are the Clean Air Act (CAA) of 1970 and the Clean Air Act Amendments of 1977. The 1970 legislation, with later modifications, is the basis for the regulations currently in effect.

Six air pollutants were chosen as critical in the CAA: particulates, sulfur dioxide, carbon monoxide, nitrogen oxides, photochemical oxidants, and hydrocarbons. For each pollutant, the U.S. Environmental Protection Agency (EPA) was required to develop criteria documents including emission standards. These emission standards are known as the National Ambient Air Quality Standards (NAAQS).

Two levels of NAAQS were developed: primary and secondary. These were issued in 1972. (See Table 3.) The primary NAAQS standards protect the public health without regard to any factors, such as technical feasibility or costs. Secondary standards protect the public welfare, to avoid adverse effects on personal comfort and well-being.

The NAAQS standards are given as Ground Level Concentrations (GLC). Therefore, in order to predict the resultant GLC from a specific emitter, gas dispersion modeling must be used. These predictions can be confirmed by monitoring with instrumental analysis.

TYPES OF EMISSIONS

Since the primary purpose of the flare is to safely dispose of emergency reliefs, the air quality emission standards are

TABLE 3. National Ambient Air Quality Standards [8].

Pollutant	Primary	Secondary
PARTICULATE MATTER		
Annual geometric mean	75 $\mu g/m^3$	60 mg/m^2
Maximum, 24 hours (not to be exceeded more than once per year)	260 $\mu g/m^3$	150 $\mu g/m^3$
SULFUR OXIDES (as SO_2)		
Annual arithmetic mean	80 $\mu g/m^3$	—
Maximum, 24 hours (not to be exceeded more than once per year)	365 $\mu g/m^3$	
CARBON MONOXIDE		
Maximum, 8 hours (not to be exceeded more than once per year)	10 $\mu g/m^3$	10 $\mu g/m^3$
Maximum, 1 hour (not to be exceeded more than once per year)	40 $\mu g/m^3$	40 mg/m^3
NITROGEN DIOXIDE		
Annual Arithmetic mean	100 $\mu g/m^3$	100 $\mu g/m^3$
HYDROCARBONS		
Maximum, 3 hours (6 to 9 AM) (not to be exceeded more than once per year)	160 $\mu g/m^3$	160 $\mu g/m^3$
PHOTOCHEMICAL OXIDANTS		
Maximum, 1 hour (not to be exceeded more than once per year)	160 $\mu g/m^3$	160 $\mu g/m^3$

TABLE 4. Threshold Limits for Certain Toxic Substances, Gases, and Vapors.

Gas or Vapor	PPM	Gas or Vapor	PPM
Acetaldehyde	200	Chloroform	100
Acetic acid	10	Cresol (all isomers)	5
Acetic anhydride	5	Cyclohexane	400
Acetone	1,000	Cyclohexanol	100
Acrolein	0.5	Cyclohexanone	100
Acrylonitrile	20	Cyclohexene	400
Ammonia	100	Cyclopropane	400
Amyl acetate	200	Diacetone alcohol	50
Amyl alcohol	100	0-Dichlorobenzene	50
Aniline	5	1,1-Dichloroethane	100
Arsine	0.05	Diethylamine	25
Benzene	35	Diisobutyl ketone	50
Benzyl chloride	1	Dimethylaniline	5
Bromide	1	Dimethylsulfate	1
Butadiene	1,000	Diethylene dioxide	100
Butyl alcohol	100	Ethyl acetate	400
Butylamine	5	Ethyl alcohol (ethanol)	1,000
Carbon dioxide	5,000	Ethylamine	25
Carbon disulfide	20	Ethylbenzene	200
Carbon monoxide	100	Ethyl bromide	200
Carbon tetrachloride	25	Ethyl chloride	1,000
Chlorobenzene	75	Ethyl ether	400
Ethylene chlorohydrin	4	2-Methoxyethanol	25
Ethylenedriamine	10	Methyl chloride	100
Ethylene dibromide	25	Methylcyclohexane	500
Ethylene dichloride	100	Methylcyclohexanol	100
Ethylene oxide	100	Methylcyclohexanone	100
Fluorine	0.1	Methylene chloride	500
Formaldehyde	5	Methyl formate	100
Gasoline	500	Methyl amyl alcohol	25
Hydrazine	1	Methylene chloride	500
Hydrogen bromide	5	(dichloromethane)	
Hydrogen chloride	5	Naptha (coal tar)	200
Hydrogen cyanide	10	Naptha (petroleum)	500
Hydrogen fluoride	3	pNitroaniline	1
Hydrogen peroxide	1	Nickel carbonyl	0.001
90%		Nitrobenzene	1
Hydrogen selenide	0.05	Nitroethane	100
Hydrogen sulfide	20	Nitrogen dioxide	5
Isodine	0.1	Nitromethane	100
Isophorone	25	Nitrotoluene	5
Isopropylamine	5	Ozone	0.1
Mesityl oxide	50	Pentane	1,000
Methyl acetate	200	Propyl Ketone	200
Methyl acetylene	1,000	Phenol	5
Methyl alcohol	200	Phenylhydrazine	5
Methyl bromide	20	Phosgene (carbonyl	1
Phosphine	0.05	chloride)	
Phosphorus trichloride	0.5	Sulphur monochloride	1
Propyl acetate	200	Sulphur pentafluoride	0.025
Propyl alcohol	400	1,1,2,2-Tetrachloro-	5
Propyl ether	500	ethane	

(continued)

TABLE 4. (continued).

Gas or Vapor	PPM	Gas or Vapor	PPM
Propylene dichloride	75	Tetranitromethane	1
Pyridine	10	Toluene (toluol)	5
Quinone	0.1	o-Toluidine	200
Stibine	0.1	Trichloroethylene	200
Styrene	200	Trichloroethane	500
Sulphur dioxide	10	Turpentine	100
Sulphur hexafluoride	1,000	Vinyl chloride	500
		Xylene (xylol)	200

Source: Olishifski, J. B. and McElroy, F. E., *Fundamentals of Industrial Hygiene*, National Safety Council, Chicago, Illinois (1971).

rarely exceeded since such relief loads are infrequent. These abnormal conditions do not occur often enough with telling impact on air quality. However, two conditions must be considered in the design and application of flare systems.

1. If the flare system is designed to handle normal operating poops and vents on a continuous or semicontinuous basis, the emitted discharges must not violate the NAAQS standards.

Of the six pollutants cited by the NAAQS, usually one sulfur oxides, and possibly particulates, pose a problem. For those operating reliefs containing a sulfur rich gas, the flared discharge may exceed the sulfur oxide limits. Therefore, this gas stream cannot be flared, but either must be sent to a sulfur recovery unit or must be combusted in a gas incinerator with appropriate off-gas scrubbing to remove SO_2.

Particulate emission is usually the result of unburned carbon particles or soot emanating from the flame. This tendency to produce black smoke in flare burning of hydrocarbons is influenced by the Hydrogen to Carbon ratio (by weight), but it is not directly proportional [1]. For example, methane (H/C = 0.33) produces no smoke but propane (H/C = 0.222) will produce comparatively heavy smoke. As the H/C ratio by weight decreases the smoking intensity increases. Therefore, flares designed for smokeless operation are required for clean disposal of hydrocarbon streams above methane in the paraffin group and for all olefins and aromatics. These flares use steam, high pressure fuel gas, water spray, or an air blower to produce the "Coanda Effect" for eduction of combustion air. (The Coanda Effect is the effect by which a flow of fluid is attracted to a surface adjacent to the fluid stream due to the negative pressure created by the flow of fluid over the surface.) This produces turbulance and gas/air mixing with subsequent improvement to the combustion chemistry.

2. If the flare system is designed to handle toxic or noxious gases or vapors, the emitted discharges must not exceed the Threshold Limit Values at GLCs. A list of some toxic substances, with their respective TLVs, are given in Table 4.

These values, whether for normal operational discharges or emergency releases, must not be exceeded at ground level.

For the flaring of toxic material a somewhat different approach to the calculation of stack height has to be taken from that which considers only heat radiation. Although the gaseous material may be completely combusted to harmless products, allowance must be made for possible incomplete combustion in the flare flame. It must be assumed that some small fraction escapes combustion at the end of the flame, where the flame becomes too lean to burn. The amount of escaped gas must be sufficiently diluted with air so that the maximum concentration at ground level never exceeds the acceptable limit.

The emission of incompletely burned hydrocarbons from flare flames are not known accurately. The amount of material flared is uncertain and the combustion efficiency of the flare flame is difficult to measure.

PROGRAMS TO EVALUATE FLARES

Government and industry environmental officials are concerned with the effects of flared hydrocarbons on air quality. however, flares do not lend themselves to conventional testing techniques. Flare emissions testing problems are characterized by meandering irregular flames, high temperature radiant heat zones, and underfined dilution with ambient air of the flare emission plume. Because of these conditions, suitable sampling locations are difficult to specify.

Despite these problems, an experimental study, sponsored in part by the U.S. Environmental Protection Agency (EPA) and the Chemical Manufacturer's Association (CMA), was performed to determine the efficiencies of flare burners [2]. This work, reported in March 1983, included rigorous testing of flares under 34 different operating conditions during a three week period in 1982. The primary objectives of the study were to determine the combustion efficiency and hydrocarbon destruction efficiency for air and steam assisted flares.

The test methods used during the study employed a special 27 foot long sample probe suspended by a crane over the flare flame. Extractive flare sampling was used whereby the sample was analyzed by continuous emission monitors to determine concentrations of carbon dioxide (CO_2), carbon monoxide (CO), total hydrocarbons (THC), sulfur dioxide (SO_2), oxides of nitrogen (NO_x), and oxygen (O_2). Also, probe tip temperature, ambient air temperature, wind speed, and wind direction were measured. Particulate matter samples were collected during smoking flare tests.

Combustion efficiency was used as the primary measure of flare performance during this study. This term characterizes the percentage of flare emissions that are completely oxidized to carbon dioxide (CO_2). The combustion efficiency is defined as

$$\%CE = \frac{CO_2}{CO_2 + CO + THC} \times 100$$

where:

CO_2 = ppmv of carbon dioxide
CO = ppmv of carbon monoxide
THC = ppmv of total hydrocarbon (as methane)

The following conclusions and observations were compiled from this study:

- When flares are operated under conditions which are representative of industrial practices, excluding emergency releases, the combustion efficiencies in the flare plume are greater than 98%.
- Steam and air assisted flares are generally an efficient means of hydrocarbon control over the range of operating conditions evaluated.
- Varying flow rates of relief gas have no effect on steam-assisted flare combustion efficiencies over the flow ranges evaluated.
- Varying Btu content of relief gases has no observed effect on steam-assisted flare combustion efficiencies for relief gases above 300 Btu/SCF. A slight decline in combustion efficiency was noted for relief gases below 300 Btu/SCF.
- Excessive steam-to-relief gas ratios lower the combustion efficiencies.
- Flaring low Btu content gases at high exit velocities, causing flare liftoff, may result in lower combustion efficiencies for air-assisted flares.
- Smoking flares achieve high gaseous hydrocarbon combustion efficiencies.
- In many cases, where high combustion efficiencies were observed, the carbon monoxide and hydrocarbon concentrations observed in the flare plume were equal to or lower than those found in ambient air.
- Concentrations of NO_x emissions in the flare plume were observed to range from 0.5 to 8.16 ppm.
- The combustion efficiency data were insensitive to sampling probe height within the normal operating heights of the probe when the probe was not in the flame.
- Further development of a technique to use sulfur or another material as a tracer material to determine the flame dilution ratios is required.
- Flares burning relief gases with less than 450 Btu/SCF lower heating value did not smoke, even with zero steam assistance.
- Although hydrocarbon destruction efficiencies were not directly measured because of the lack of dilution ratio data, the total hydrocarbon destruction efficiencies would, by definition, be higher than the calculated gaseous combustion efficiencies.
- The meandering of the flame's position relative to the sampling probe with varying wind conditions affected the continuous measurements but had no apparent effect on the combustion efficiency values.

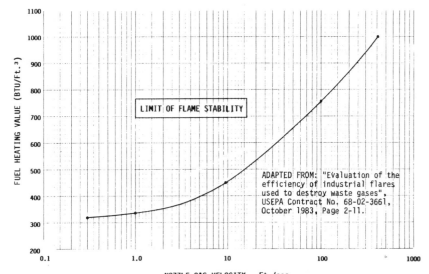

FIGURE 1.

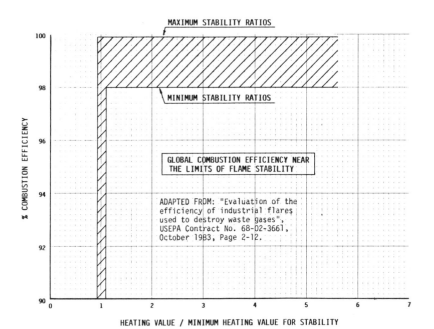

FIGURE 2.

In summary, this study provides a data base for defining the air quality impact of flaring operations.

Almost concurrent with the above cited study, another independent work was undertaken on the west coast sponsored by the EPA [3]. This study, first reported in 1984, was designed to evaluate the efficiency of industrial flares to destroy waste gases. In the absence of direct measurements of emissions on field installed operating flares, measurements were made on pilot scale and small commercial flare heads. The studies indicated that the combustion efficiency from flares could be very high. However, these studies also showed that the combustion efficiency of flare flames could be very low under some specific operating conditions. The combustion efficiency correlated with a dimensionless heating value of the fired gas. The combustion efficiency of a flare flame, as an envelope or global, was correlated with the minimum heating value required to stabilize a flame, shown in Figure 1. The combustion efficiencies were then correlated with the ratio of the heating of the actual heating value to minimum heating value required to maintain a flame at that velocity, shown in Figure 2. Low combustion efficiencies are predicted for ratios of the heating values which approach 1. Higher combustion efficiencies may occur for ratios of the heating value near the stability limit and, when disturbed (i.e. wind) can easily produce high emissions of unburned material.

The following conclusions were listed in this study:

- Flare flames usually destroy waste gases efficiently. However, the flames become less efficient for conditions of velocities and heating values of the gases near the stability limits of the flames.
- Flare flames can be maintained efficiently by altering the flame stability through use of aerodynamic stabilizers, pilot flames, or the addition of combustible gases.
- The combustion efficiency of flare flames is more dependent on the operating conditions of the flame than on the size of the head. Hence, combustion efficiencies measured on representative pilot scale flare heads can be applied to commercial flare leads with reasonable confidence.
- Flare flames are controlled by both buoyance and momentum forces. Hence, the structural features of flare flames can be correlated using a combination of Richardson Number,[2] flare head size, and properties of the gases flared.

The data included in this chapter indicates global combustion efficiencies in the range of 99 + % for most of the studied operating parameters. This data is listed for both prototype and commercial flares. This data is generally in agreement with the results of the CMA study.

[2]Authors' Note: Richardson Number = $R_i = gd/u^2$ where: g = acceleration of gravity; d = diameter; u = axial velocity.

CONCLUSIONS

The results of the studies, both the CMA work and EPA work, are limited to the conditions of the investigations. However, the correlations are expected to be valid for many other gases and flare head designs, especially those commercial designs which use external mixing.

Steam and air assisted flares are efficient devices to control non-emergency hydrocarbon emissions. When flares are properly designed, applied and operated, combustion efficiencies in excess of 98% can be achieved. These high destruction capabilities can ensure emission discharges well within the regulated limits currently required by the U.S. Environmental Protection Agency.

REFERENCES

1. Reed, R. D. *Furnace Operations*. Gulf Publishing Company (1973).
2. Chemical Manufacturers Association, "A Report on a Flare Efficiency Study," *Engineering-Science*, Austin, Texas (March 1983).
3. Industrial Environmental Research Laboratory, U.S. Environmental Protection Agency, "Evaluation of the Efficiency of Industrial Flares: Test Results," EPA-600/2-84-095, Research Triangle Park, NC 27711 (May 1984).
4. Brzustowski, T. A., S. R. Gallahole, M. P. Gupta, M. Koptein, and H. F. Sullivan, "Radiant Heating from Flares," ASME Heat Transfer Division, Paper 75-HT-4, Heat Transfer Conference, San Francisco, California (August 11–13, 1975).
5. Straitz, J. F., III and R. J. Altube, "Flares: Design and Operation," Second Seminatio de Utilitades, Sao Paulo, Brazil (November 1977).
6. "Guide for Pressure Relief and Depressurizing Systems," RP-521, American Petroleum Institute (1969).
7. Sutton, P., "Air Pollution in Petroleum Refining," *Chemical and Process Engineering* (April 1968).
8. *Manual on Disposal of Refinery Wastes*, API Publication 931 (1975).
9. Banerjee, K., N. P. Cheremisinoff, and P. N. Cheremisinoff, *Flare Gas Systems Pocket Handbook*, Gulf Publishing Co., Houston, TX (1985).

APPENDIX A

Design for Thermal Radiation

Thermal radiation calculations must be performed to determine radiant heat values in order to avoid dangerous exposure to personnel and equipment. The following procedures is a convenient way to find the intensity of radiant at different locations and to determine the required height of the flare stack.

Spherical radiation formula:

$$I = \frac{QE}{4\pi R^2} \qquad (1)$$

where I = radiation intensity at point x, BTU/hr ft²; Q = heat release = Flow × HV = flow lb/hr × heating value BTU/lb; E = emissivity factor; R = distance from flame center to point x.

The emissivity factor E is the fraction of total heat appearing as radiation. This value has been experimentally observed by Brzustowski, et al. and Straitz, III, et al., and others for various gases. A list of recommended emissivity values for the most frequently flared gases, without steam or air assist, as listed in Table 1.

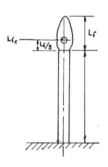

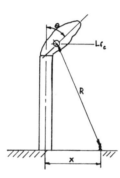

To calculate the intensity of radiation at different locations, it is necessary to determine the length of the flame and the angle of the tilt of the flame due to wind effects. A convenient formula to estimate flare length is:

$$L_f = 10D \sqrt{\frac{P}{55}} \qquad (2)$$

where L_f = length of flame, ft; D = tip diameter, inches; P = pressure drop at the tip, inches.

The center of the flame is assumed to be one-third the length of the flame from the flare tip. The angle of the flame is the vectorial addition of the wind velocity and the gas exit velocity at the flare tip

$$\Theta = \tan^{-1} \left(\frac{V \text{ wind}}{V \text{ gas}} \right) \qquad (3)$$

where:

V wind = wind velocity, ft/sec.

$$V \text{ gas} = 550 \sqrt{\frac{P}{55}}, \text{ ft/sec}$$

The coordinates of the flame center relative to the tip are:

$$X_c = L_f/3 \sin \Theta$$
$$X_c = L_f/3 \cos \Theta$$

Therefore the distance from the center of the flame to any point on the ground:

$$R = \sqrt{(X-X_c)^2 + (H + Y)^2} \qquad (4)$$

Inserting the value of R into Equation (1) will determine the radiation intensity at any location.

WORST CASE

The stack height required when considering the worst case and worst location under the flame occurs when the wind speed is 30 mph. Higher wind speeds will tend to lay the flame down lower (shorten Y_c), but wind cooling effects will reduce radiation levels. Any winds over 50 to 60 miles per hour will tend to shorten the flame.

The worst location under the flame occurs directly under the eye or center of the flame. The stack height selected for design of heat radiation must consider this worst location, vertically below the flame center.

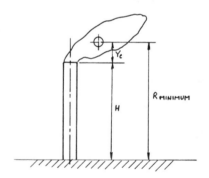

Radiation levels for design:

Personnel short time exposure: 2000 BTU/ft² (includes Solar radiation)

Equipment Protection: 3000 BTU/ft²

APPENDIX B

Rapid Determination of Approximate GLC

There is a considerable amount of literature available to determine plume rise and dispersion of furnace gases emitted from smoke stacks. There has been little work done to determine plume rise and dispersion from flare stacks.

Many authorities have suggested that the rules for chimney off-gases be applied to flare stacks, that the emission be considered as furnace gases, containing the same amount of

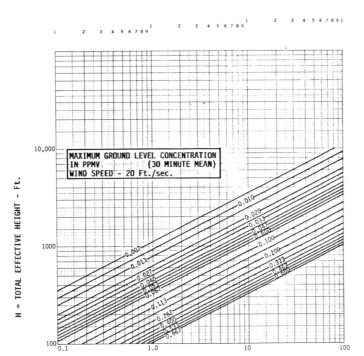

FIGURE 3. Source: Sutton, P., 1968, p. 104 (modified).

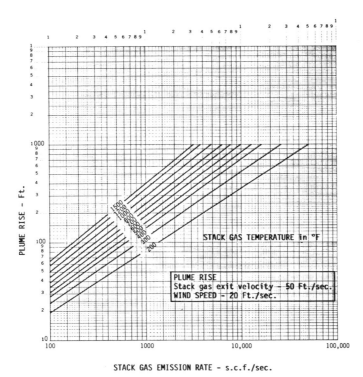

FIGURE 4. Source: Sutton, P., 1968, p. 104.

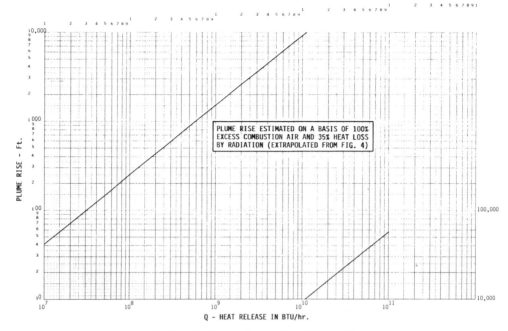

FIGURE 5. *Source: Sutton, P., 1968, p. 104.*

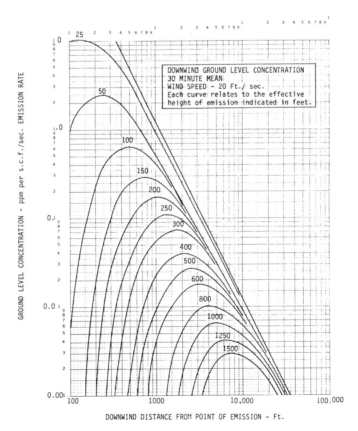

FIGURE 6.

heat. Since the combustion is not instantaneous and is not confined to a furnace box, but proceeds along the length of the flame envelope, the concept of a buoyant column of furnace gases can be visualized. The concept of this column of gases penetrating upward into the atmosphere should be modified to that of a plume extending some distance from the flare tip. How much the actual delayed combustion differs from the theoretical instantaneous combustion can be established by tests, such as infra-red photography or a similar technique.

Figures 3 and 4 are published curves for smoke stack calculations. These data may be adapted for use with flare stacks to give an approximation of ground level concentrations. Figure 4 indicates plume rise of hot gas column for various stack gas temperatures. For a given heat release the temperature of exhaust gases is closely related to the gas flow rate. By assuming 100% excess air and allowing 35% heat loss by radiation, one can arrive at a gas temperature of about 1500°F. Since plume rise is dependent upon total heat and not on actual temperature, the error involved is this determination is small. This concept permits the construction of Figure 5.

Figure 5 gives a conservative estimate of plume rise when compared with other models using a similar basis and is, therefore, reasonable. It should be noted that the plume rise does not consider higher atmospheric disturbances or influencing conditions which are important and may be controlling factors in a specific case.

The pollutants fall on the downwind side of the flare stack. A typical downwind gradient of GLC is shown in Figure 6. The point of maximum GLC is located at five times the total effective height of the plume. The wind speed selected is 20 feet/second which probably produces the greatest actual ground level concentration.

The above information is a simple graphical solution which can be used to determine pollution conditions quickly for practical applications. However, for final design work, mathematical models must be used with formulae which consider wind effects, changes in climatic conditions, day/night effects and the influence of topographical features.

INDEX

809